Modern Flight Dynamics

David K. Schmidt
Professor Emeritus
University of Colorado

McGraw-Hill
Connect Learn Succeed™

The McGraw-Hill Companies

MODERN FLIGHT DYNAMICS

Published by McGraw-Hill, a business unit of The McGraw-Hill Companies, Inc., 1221 Avenue of the Americas, New York, NY 10020.

This book is printed on acid-free paper.

1 2 3 4 5 6 7 8 9 0 DOC/DOC 1 0 9 8 7 6 5 4 3 2 1

ISBN 978-0-07-339811-2
MHID 0-07-339811-X

Vice President & Editor-in-Chief: *Marty Lange*
Vice President EDP/Central Publishing Services: *Kimberly Meriwether David*
Publisher: *Raghothaman Srinivasan*
Executive Editor: *Bill Stenquist*
Marketing Manager: *Curt Reynolds*
Development Editor: *Lorraine Buczek*
Project Manager: *Melissa M. Leick*
Design Coordinator: *Brenda A. Rolwes*
Cover Designer: *Studio Montage, St. Louis, Missouri*
Cover Images: © *Photo courtesy of NASA and Wrights Take Flight: Courtesy of Library of Congress/RF*
Buyer: *Kara Kudronowicz*
Media Project Manager: *Balaji Sundararaman*
Compositor: *Laserwords Private Limited*
Typeface: *10.5/12 Times Roman*
Printer: *RR Donnelley*

Library of Congress Cataloging-in-Publication Data

Schmidt, D. K. (David K.)
Modern flight dynamics/David K. Schmidt.—1st ed.
p. cm.
Includes bibliographical references.
ISBN-13: 978-0-07-339811-2 (alk. paper)
ISBN-10: 0-07-339811-X (alk. paper)
1. Aerodynamics. 2. Flight engineering. 3. Flight control. I. Title.
TL570.S298 2010
629.132'3—dc22 2010043961

www.mhhe.com

ABOUT THE AUTHOR

David Schmidt was born in Lafayette, Indiana, and attended Purdue University where he received the B. S. degree, cum laude, in Aeronautical Engineering. He later received the M. S. degree from the University of Southern California and the Ph.D from Purdue, both in aerospace engineering. Prior to his graduate studies he served on the technical staff of the Douglas, and then the McDonnell Douglas Missiles and Space Corporation. After first supporting the Apollo program in the development of the Saturn booster, he became Engineering Lead in a preliminary vehicle-design group of the Advanced Systems and Technology Division. Upon completion of his graduate education, he served on the technical staff of the Stanford Research Institute, focusing on research in systems analysis and optimization of air transportation systems.

Dr. Schmidt's academic career began when he joined the faculty of the School of Aeronautics and Astronautics at Purdue, where he served as professor of aeronautics and astronautics for 14 years. He then joined the faculty of Arizona State University, where he served as professor of mechanical and aerospace engineering for six years. He later moved to the University of Maryland at College Park, where he served as professor of aerospace engineering for an additional six years. Lastly, he was invited to join the faculty at the University of Colorado, Colorado Springs, where he helped establish the brand new Department of Mechanical and Aerospace Engineering. He retired from the University of Colorado in 2006, and was appointed Professor Emeritus. While at Arizona State Dr. Schmidt served as the founding director of the Aerospace Research Center in the College of Engineering, and while at the University of Maryland he served as the founding director of the Flight Dynamics and Control Laboratory in the Department of Aerospace Engineering. His teaching was recognized at several of these institutions through many prestigious teaching awards.

In addition to his earlier industrial experience, in 1978 Dr. Schmidt was invited to serve as a summer faculty fellow at the USAF Flight Dynamics Laboratory, Wright-Patterson AFB, and in 1984–85 he served as a visiting sabbatical professor at NASA's Langley Research Center.

He has been an invited member of several national review panels, including the National Academy of Engineering's (NAE) National Research Council (NRC) Review Panel for a Decadal Study of NASA Aeronautics Research, the NAE's NRC Committee on Advanced Supersonic Technology, and the NAE's NRC Committee on High-Speed Research. Furthermore, he has served as an invited member of the USAF Scientific Advisory Board's Science and Technology Panel on Vehicles and Power. In 1996 he served as the General Conference Chair for the Guidance, Navigation and Control Conference of the American

Institute of Aeronautics and Astronautics (AIAA). In 1991–93 he also chaired the AIAA National Technical Committee on Guidance, Navigation, and Control.

Dr. Schmidt is the author of over 200 research articles on flight dynamics, air-traffic control systems, and man-machine control systems, and he has been invited to lecture worldwide on his research. From 2001–2009 he was a member of the AIAA's Education Editorial Board, and from 1988–1991 he was associate editor of the AIAA's *Journal of Dynamics, Guidance, and Control.* He is listed in *Who's Who in America,* and is a member of Tau Beta Pi and Sigma Gamma Tau engineering honor societies. In 1997, Dr. Schmidt received AIAA's highest honor in the field of flight dynamics and control when he was awarded the national Mechanics and Control of Flight Award. He is a fellow of the AIAA.

Dedication

To my parents
and to Karalee, my wife and my best friend

BRIEF TABLE OF CONTENTS

CONTENTS

ACKNOWLEDGEMENTS

I am indebted to many people who have greatly contributed to this effort, and I will not be able to mention them all. But to all of you, I am genuinely grateful. I would first like to acknowledge my colleagues and students, both graduate and undergraduate, at Purdue University. My knowledge of the subject of flight dynamics was greatly enhanced during my tenure there. I learned a great deal from my students, and I was constantly inspired by their dedication to learning and to aerospace engineering. Next I would like to acknowledge all my colleagues in the flight dynamics community, by whom I am always being educated. Thanks also to Professor John D. Anderson, Jr. for his first encouraging me to undertake writing this book and for his continued support.

I would also like to thank the terrific staff of McGraw-Hill, especially my editors Bill Stenquist and Lorraine Buczek, for their professionalism and encouragement. Finally, I would like to acknowledge the priceless contributions of my colleagues who gave of their valuable time to review chapters of the manuscript and make many valuable suggestions. The list includes:

Professor Dominic Andrisani, Purdue University
Professor Jewel Barlow, University of Maryland
Professor Riccardo Bonazza, University of Wisconsin, Madison
Professor David Bridges, Mississippi State University
Professor Ron Hess, University of California, Davis
Professor Ki Dong Lee, University of Illinois at Urbana-Champaign
Professor Benjamin Liaw, City College of New York
Professor Cornel Sultan, Virginia Polytechnic Institute and State University
Professor John Valasek, Texas A&M University
Professor Tom Zeiler, University of Alabama
Professor Zvi Rusak, Rensselaer Polytechnic Institute
Professor Bruce Walker, University of Cincinnati

Finally, any errors remaining in the book are my responsibility alone. Instructors using the book who uncover errors or who have suggestions are asked to bring them to my attention at Schmidt.Flight.Dynamics@gmail.com. I would appreciate hearing from you.

On the Subject

For those who love flying machines, the study of *flight dynamics* is most exciting indeed. When introduced to the topic, at whatever age, students discover that flight dynamics is the essence of aeronautical engineering because it involves the study of the motion—the flight—of the vehicle. This motion defines the vehicle's performance, a topic of enormous significance to the ultimate success of the machine. Thus the excitement generated from the pursuit of this subject follows from both romantic as well as practical reasons.

Flight dynamics is also a study of the complete vehicle, rather than just a component of a vehicle. Hence it is fundamentally the study of a multicomponent system and its dynamics. This study is by necessity multidisciplinary, as depicted in Figure 1. When first introduced to the subject, students may see how some of the other disciplines they've been asked to master actually fit together. Thus it also provides an integrating function.

Although flight dynamics is multidisciplinary, it is still a basic aerospace science. The study is relatively new, compared to aerodynamics, for example. But it has become quite clear that the study of aeronautical or aerospace engineering is incomplete without the thorough treatment of flight dynamics. Many more traditional undergraduate and post-graduate aeronautical engineering curricula have been modified over the last several years to reflect this fact.

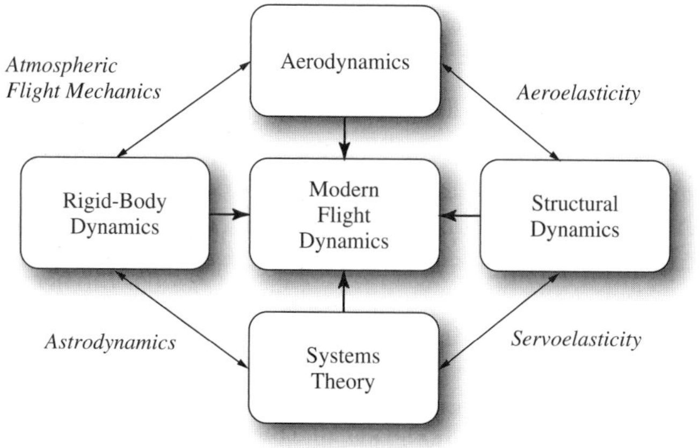

Figure 1 Components of modern flight dynamics.

The theories and methodologies developed in modern flight dynamics are critical in several aspects of flight-vehicle design and development. Notable among these are the determination of the vehicle's configuration, or geometry, the tailoring of its dynamics, especially its handling characteristics (i.e., its "controllability" on the part of a pilot or autopilot), and the design of the autopilot itself.

On the Wrights

The subject of flight dynamics goes back to the Wright brothers, who may be considered the fathers of airplane flight dynamics. When the Wrights developed their revolutionary aircraft, reciprocating engines had already been developed, and the aerodynamics (lift, drag) of airfoils was fairly well understood. (Aerodynamics had been studied by the French and Germans, for example, for quite a while.) The tremendous breakthrough made by the Wrights was how to control the vehicle in flight. In the words of Wilbur Wright[1] two years before their first powered flight:

> Men already know how to construct wings or aeroplanes, which when driven through the air at sufficient speed, will not only sustain the weight of the wings themselves, but also that of the engine, and of the engineer as well. Men also know how to build engines and screws of sufficient lightness and power to drive these planes at sustaining speed. . . . Inability to balance and steer still confronts students of the flying problem. . . . When this one feature has been worked out, the age of flying machines will have arrived, for all other difficulties are of minor importance.

Prior to the Wright's first flight, the only known ways to turn a vehicle were by using a rudder, as with ships, or by shifting the vehicle's center of mass by shifting the pilot's body weight, as Lilienthal tried to do with his gliders. But as we now know, if you only use a rudder to try to turn an airplane, it slips sideways. And shifting the pilot's body weight severely limits the size and maneuverability of the vehicle. The Wrights discovered that to efficiently turn an airplane, they had to bank the vehicle (like a bicycle!). In addition, the Wrights invented wing warping (the precursor of ailerons) to initiate and control the banking.

The Wrights also demonstrated that an unstable vehicle could be flown successfully (although with difficulty). By using a large canard, the Wright's pitch-control surface located forward of the pilot, they produced a vehicle that was neutrally stable, or slightly unstable in pitch. But the Wrights were not too concerned. Since they were bicycle mechanics, they were accustomed to dealing with unstable machines—machines that had to be stabilized by the human operating them! One wonders if the Wrights could have been as successful as they were, had they not been bicycle mechanics.

---

[1] McFarland, M. W., ed., *The Papers of Wilber and Orville Wright,* Vol. 1, McGraw-Hill, 1953.

The Subject Revisited

Many years after the Wrights' historic flights, researchers began to develop an analytical theory of *flight dynamics.* Like the more classical aerospace sciences of aerodynamics and structures, modern flight dynamics has its genesis in dynamics, or mechanics, with the focus on the dynamics of nonlinear, multidimensional systems. So it is applied physics. But, with reference to Figure 1, it also relies heavily on the tools that have been developed in the area of mathematical systems theory. Furthermore, the subject deals not only with describing the vehicle's dynamics, but also with ways to affect these dynamics in some desired way—to tailor them. This is accomplished either through the art of vehicle-configuration design, or through the introduction of feedback control systems.

However, this "new" aerospace science is neither a subset of aerodynamics, nor of feedback-control theory. Though modern flight dynamics draws from these topics, they do not define the discipline. The definition of the discipline has been the source of confusion sometimes. Historically, when a airplane's dynamics were determined almost entirely by the aerodynamic forces acting on it, aerodynamicists coined the phrase *aerodynamic stability and control.* This topic was one aspect of applied aerodynamics. The focus was almost exclusively on the estimation of the aerodynamic forces and moments. The rigorous treatment of the system's rigid-body dynamics was absent, and there was no mention of feedback control or structural deformation.

Conversely, the majority of aerospace vehicles developed since around 1970 include some sort of active feedback-control system for flight guidance and/or attitude stabilization and control. Furthermore, the vehicle's dynamics are greatly affected, if not completely dominated, by the actions of such feedback systems. That is, the vehicle geometry alone no longer determines the dynamics of the vehicle, or its stability. Therefore, the topic of modern flight dynamics has evolved in large part due to the introduction of active feedback systems into the vehicle design.

Just as the subject of modern flight dynamics is not only applied aerodynamics, neither is it just applied control theory. According to the definitions from experts in the field, feedback control theory deals little, if at all, with defining (or modeling) the dynamics of the system to be controlled. Furthermore, in the feedback-control design problem defined in texts dealing with that subject, the system to be controlled is taken as a given. It cannot be changed. Only feedback loops can be added around it to augment its dynamics. And frequently the modeling of the dynamics of the system is performed independently from the design of the feedback loops.

However, in the design of modern aerospace vehicles, the entire system is being designed, both the vehicle to be controlled as well as its feedback loops. And the use of feedback expands the design space the vehicle designer has to work with. So in modern flight dynamics, the modeling and the control design are inexorably intertwined. The subject cannot be decoupled in some artificial way into pure aerodynamics, pure dynamics, pure vibrations, and pure feedback control, for modern flight dynamics lies specifically in the intersection of these areas. To force any decoupling—if you will—throws the baby out with the bath water.

On This Book

All these considerations have significantly influenced the writing of this book, which attempts to do justice to the beauty and uniqueness of the field of modern flight dynamics, while paying tribute to the rich history of the subject. The book is a result of the author's 30-years of experience in teaching flight dynamics, plus his years of experience as a practitioner and researcher in the field.

Most of the material herein arises from undergraduate and graduate courses developed by the author at Purdue University over several years. Portions of the material have also been presented by the author in courses at Arizona State University, the University of Maryland, College Park, and finally the University of Colorado, Colorado Springs. In addition to its use as a textbook, we intend the book to be a useful reference for the practicing engineer and researcher.

Since the thorough treatment of modern flight dynamics requires the student to have mastered prerequisite material in several areas, the material presented herein is intended for fourth-year undergraduates and/or graduate students in aerospace or mechanical engineering. We assume that undergraduate students have completed coursework in rigid-body dynamics and aerodynamics. Portions of the prerequisite subjects key to the development of modern flight dynamics are reviewed in early chapters and in several other sections of this book, but these are only reviews. In addition, key material with which undergraduate students may not have sufficient familiarity is presented in *just-in-time tutorials* immediately prior to the relevant topic.

With regard to the treatment of feedback control in this text, the approach to feedback-systems design is that of "dynamics-based" synthesis, as opposed to "algorithmic-based" synthesis. That is, the feedback control laws are synthesized based on an intimate knowledge and understanding of the vehicle's dynamics—its physics. The feedback systems are designed to act in concert with, and to naturally exploit and enhance, those dynamics. Frequency-domain tools are relied upon heavily, and the approach will appear to be quite "conventional," in the terminology of control theorists.

There are key reasons for taking this conventional, dynamics-based approach. First, a general consensus exists among modern flight dynamicists that this approach has been demonstrated over the years to work very well, and it is used throughout the industry. In fact, this design approach has been the basis for much if not all of the design of every operational aircraft's flight-control system. Other synthesis approaches have been attempted in research projects, but if the project led to flight of an actual vehicle, a great deal of conventional, dynamics-based control concepts were invariably relied upon in the development of its flight-control systems. So understanding this philosophy is critical to students, if they want to master this subject. Second, though they may not have used them, almost all undergraduate engineering students have been introduced to frequency-domain tools in their curriculum. So these tools are not foreign to them.

Finally, the decision to use English Engineering units in this book is based on the fact that these units are used exclusively in industry today. The historical databases so important in design, and legacy software packages for analysis, were all developed in English units, so there would be a tremendous cost involved with any organizationwide unit conversion. Thus there is an understandable reluctance on the part of the industry to such a conversion. Consequently, it is to the students' advantage to be familiar with working with these units, as they will certainly use them throughout their careers.

To the Instructor

This book could serve as the text for a two-course sequence in modern flight dynamics, the first an undergraduate course, the second at the undergraduate or graduate level. But other course usage is quite possible. The flowcharts in Figure 2 present possibilities for grouping the subject matter into three courses.

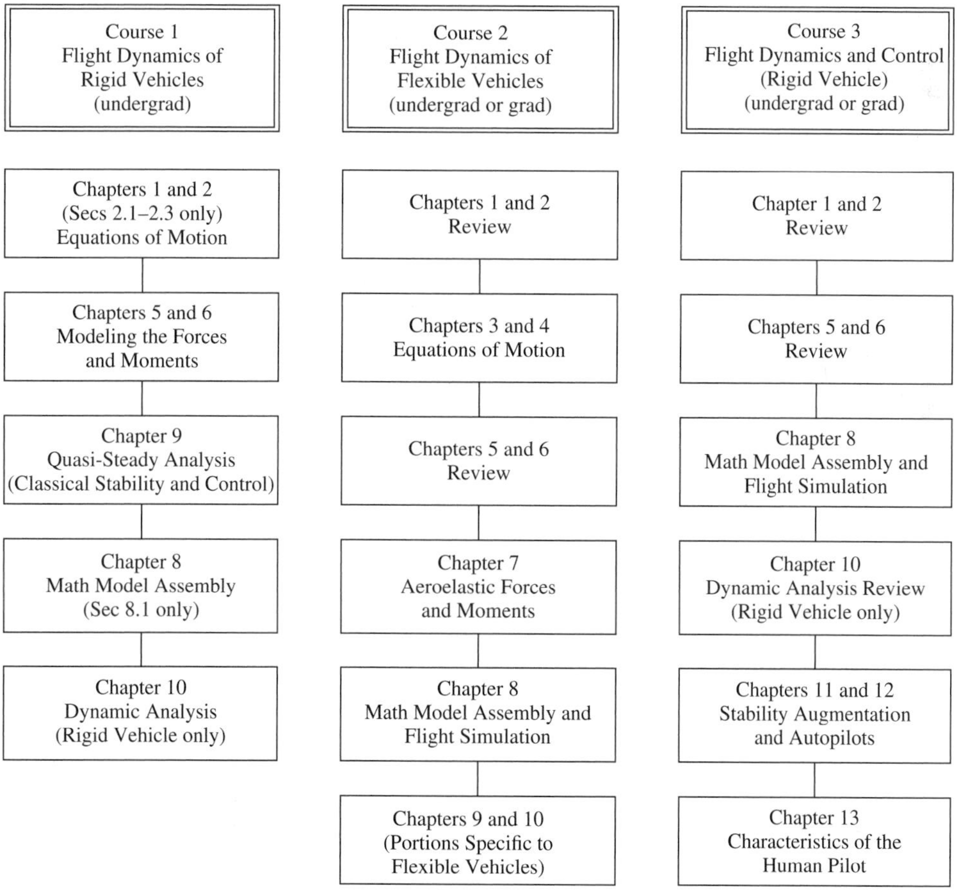

Figure 2 Possible course content.

In addition, a "Chapter Road Map" has been included at the beginning of each chapter. The primary purpose of this road map is to indicate to the student and the instructor which sections in that chapter should be covered if the course is a first course in flight dynamics.

The author has presented a course similar to Course 1 at the undergraduate level for many years. With regard to the order of the material in Course 1, the author is fully aware of a conundrum faced by many instructors teaching the subject to undergraduates. This issue has greatly influenced the writing of this book. The conundrum deals with the fact that frequently the students need to be introduced to "classical static stability and control" early in the semester, to support their senior design course. And yet, it has been the experience of the author having taught such a course for 30 years, that if these classical topics are presented first, the students frequently don't understand where the concepts stem from. That is, they don't have a context to help them see how flight dynamics fits any logical framework. Hence, the student needs to be aware of the concept of small-perturbation analysis and the analysis of the equilibrium conditions. This fact leads to the order of presentation used in this book.

That being said, however, instructors wishing to cover classical stability and control early in the semester can also use this book. Specifically, as shown in Figure 3, one could start with Chapter 1 (Section 1.1), and then move quickly to Chapter 9, classical stability and control, while using key material from Chapters 5 and 6 as necessary. Then Chapter 2 (Sections. 2.1–2.3), the remainder of Chapter 6, and Chapter 10 could be used to cover dynamic analysis.

Either Course 2 or Course 3 could follow Course 1. A fourth course would be like Course 3, but with the material on flexible vehicles substituted for flight simulation. Such a course would most likely be best offered at the graduate level to allow for more rapid movement through the material.

As noted in several places, this book was written with the assumption that the student would have already mastered basic dynamics and had some exposure to classical control theory prior to beginning this course sequence. A previous course in aerodynamics would be helpful, but could be taken concurrently.

Software Accompanying the Book

Several software modules, written in MATLAB® or Simulink®, have been developed to support sections of this book, such as some of the examples. These routines have been made available for educational purposes only—they are not supported, and are not sufficiently developed for actual design use in practice.

The first is a module called MATAERO, which is a routine for estimating the aerodynamic characteristics of two lifting surfaces in close proximity (e.g., wing and horizontal tail). A document is included that contains the theoretical background and a user's manual. Other routines include files that assemble the math

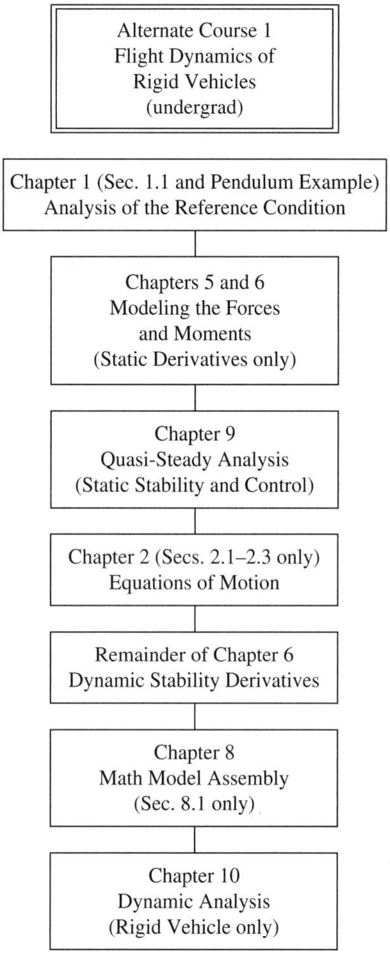

Figure 3 Reordering the sequence in Course 1.

models for several aircraft, using the data in Appendix B, while still others include the development of control laws discussed in several examples in the book. All these modules are identified in a file named "Read Me," and are available from the book's website at www.mhhe.com/schmidt.

MCGRAW-HILL DIGITAL OFFERINGS

McGraw-Hill Create™

Craft your teaching resources to match the way you teach! With McGraw-Hill Create™, www.mcgrawhillcreate.com, you can easily rearrange chapters, combine material from other content sources, and quickly upload content you have written like your course syllabus or teaching notes. Find the content you need in Create by searching through thousands of leading McGraw-Hill textbooks. Arrange your book to fit your teaching style. Create even allows you to personalize your book's appearance by selecting the cover and adding your name, school, and course information. Order a Create book and you'll receive a complimentary print review copy in 3–5 business days or a complimentary electronic review copy (eComp) via email in minutes. Go to www.mcgrawhillcreate.com today and register to experience how McGraw-Hill Create™ empowers you to teach your students your way.

McGraw-Hill Higher Education and Blackboard have teamed up

Blackboard, the Web-based course-management system, has partnered with McGraw-Hill to better allow students and faculty to use online materials and activities to complement face-to-face teaching. Blackboard features exciting social learning and teaching tools that foster more logical, visually impactful and active learning opportunities for students. You'll transform your closed-door classrooms into communities where students remain connected to their educational experience 24 hours a day.

This partnership allows you and your students access to McGraw-Hill's Create™ right from within your Blackboard course—all with one single sign-on. McGraw-Hill and Blackboard can now offer you easy access to industry leading technology and content, whether your campus hosts it, or we do. Be sure to ask your local McGraw-Hill representative for details.

Electronic Textbook Options

This text is offered through CourseSmart for both instructors and students. CourseSmart is an online resource where students can purchase the complete text online at almost half the cost of a traditional text. Purchasing the eTextbook allows students to take advantage of CourseSmart's web tools for learning, which include full text search, notes and highlighting, and email tools for sharing notes between classmates. To learn more about CourseSmart options, contact your sales representative or visit www.CourseSmart.com.

NOMENCLATURE*

Symbol	Definition (Typical Units)	Chapter Introduced
$\{\mathbf{1}\}$	Column of ones, rigid-body mode shape in vibration problem	3
a	Sonic velocity in air (fps)	5
ac	Aerodynamic center	5
$A = b^2/S$	Lifting-surface aspect ratio	5
$\mathbf{A}$	System matrix in state-variable description of system	8
b	Span of a lifting surface (ft)	5
$\mathbf{B}$	Control-distribution matrix in state-variable description of system	8
c	Airfoil section chord length (ft)	5
$\mathbf{C}$	Centripetal acceleration (vector) (ft/sec^2)	2, 9
$\mathbf{C}$	Constraint matrix in vibration problem	3
$\mathbf{C}$	Response matrix in state-variable description of system	8
$\bar{c}$	Mean aerodynamic chord length (ft)	5
$c_d = d/qc$	Aerodynamic 2-D section drag coefficient	5
c_p	Pressure coefficient	5
$C_D = D/q_\infty S$	Aerodynamic drag coefficient	5
C_{D_i}	Induced-drag coefficient	5
C_{D_P}	Parasite drag coefficient	5
$C_{D_\alpha} = \dfrac{\partial C_D}{\partial \alpha}$	Angle-of-attack drag effectiveness (/rad)	5, 6
$C_{D_{\dot{\alpha}}} = \dfrac{\partial C_D}{\partial \dot{\alpha}}$	Angle-of-attack-rate drag effectiveness (sec)	5, 6

* **Bold** indicates a vector or matrix quantity.

Symbol	Definition (Typical Units)	Chapter Introduced
$C_{D_{\delta_E}} = \dfrac{\partial C_D}{\partial \delta_E}$	Elevator drag effectiveness (/rad)	5, 6
$C_{D_{M_\infty}} = \dfrac{\partial C_D}{\partial M_\infty}$	Mach-number drag effectiveness	5, 6
$C_{D_q} = \dfrac{\partial C_D}{\partial q}$	Pitch-rate drag effectiveness (sec)	5, 6
$C_{D_u} = \dfrac{\partial C_D}{\partial u}$	Surge-velocity drag effectiveness (sec/ft)	5, 6
C_f	Aerodynamic friction coefficient	5
C_h	Hinge-moment coefficient for control surface	9
$C_{h_\alpha} = \dfrac{\partial C_h}{\partial \alpha}$	Angle-of-attack hinge-moment effectiveness (/rad)	9
$C_{h_\delta} = \dfrac{\partial C_h}{\partial \delta}$	Control-surface-deflection hinge-moment effectiveness (/rad)	9
$c_l = \dfrac{l}{qc}$	Aerodynamic 2-D section lift coefficient	5
$c_{l_\alpha} = \dfrac{\partial c_l}{\partial \alpha}$	Aerodynamic 2-D angle-of-attack lift effectiveness (/rad)	5
$c_{l_\delta} = \dfrac{\partial c_l}{\partial \delta_{\text{Flap}}}$	Aerodynamic 2-D flap lift effectiveness (/rad)	5
c_{l_0}	Aerodynamic 2-D section lift at zero angle of attack	5
$C_L = L/q_\infty S$	Aerodynamic lift coefficient	5
C_{L_0}	Aerodynamic lift coefficient at zero angle of attack	5
$C_{L_\alpha} = \dfrac{\partial C_L}{\partial \alpha}$	Angle-of-attack lift effectiveness (/rad)	5, 6
$C_{L_{\dot\alpha}} = \dfrac{\partial C_L}{\partial \dot\alpha}$	Angle-of-attack-rate lift effectiveness (sec)	5, 6
$C_{L_{\delta_E}} = \dfrac{\partial C_L}{\partial \delta_E}$	Elevator lift effectiveness (/rad)	5, 6
$C_{L_{M_\infty}} = \dfrac{\partial C_L}{\partial M_\infty}$	Mach-number lift effectiveness	5, 6
$C_{L_q} = \dfrac{\partial C_L}{\partial q}$	Pitch-rate lift effectiveness (sec)	5, 6

Symbol	Definition (Typical Units)	Chapter Introduced
$C_{L_u} = \dfrac{\partial C_L}{\partial u}$	Surge-velocity lift effectiveness (sec/ft)	5, 6
$C_{L_{\eta_i}} = \dfrac{\partial C_L}{\partial \eta_i}$	Modal-deflection lift effectiveness	7
$C_{L_{\dot{\eta}_i}} = \dfrac{\partial C_L}{\partial \dot{\eta}_i}$	Modal-rate lift effectiveness (sec)	7
$C_{L\text{roll}} = L_{\text{Roll}}/q_\infty Sb$	Aerodynamic rolling-moment coefficient	5
$C_{L_\beta} = \dfrac{\partial C_{L\text{roll}}}{\partial \beta}$	Sideslip roll effectiveness (/rad)	5, 6
$C_{L_{\delta_A}} = \dfrac{\partial C_{L\text{roll}}}{\partial \delta_A}$	Aileron roll effectiveness (/rad)	5, 6
$C_{L_{\delta_R}} = \dfrac{\partial C_{L\text{roll}}}{\partial \delta_R}$	Rudder roll effectiveness (/rad)	5, 6
$C_{L_p} = \dfrac{\partial C_{L\text{roll}}}{\partial p}$	Roll-rate roll effectiveness (sec)	5, 6
$C_{L_r} = \dfrac{\partial C_{L\text{roll}}}{\partial r}$	Yaw-rate roll effectiveness (sec)	5, 6
$C_{L\text{roll}_{\eta_i}} = \dfrac{\partial C_{L\text{roll}}}{\partial \eta_i}$	Modal-displacement roll effectiveness	7
$C_{L\text{roll}_{\dot{\eta}_i}} = \dfrac{\partial C_{L\text{roll}}}{\partial \dot{\eta}_i}$	Modal-rate roll effectiveness (sec)	7
c.m.	Center of mass	2
c.g.	Center of gravity	2
$c_m = m/qc^2$	2-D section pitching-moment coefficient	5
$c_{m_{\text{ac}}}$	2-D section pitching-moment coefficient about its *ac*	5
$c_{m_\alpha} = \dfrac{\partial c_m}{\partial \alpha}$	2-D angle-of-attack moment effectiveness (/rad)	5
$c_{m_\delta} = \dfrac{\partial c_m}{\partial \delta_{\text{Flap}}}$	2-D trailing-edge-flap moment effectiveness (/rad)	5
$C_M = M/q_\infty S\bar{c}$	Aerodynamic pitching-moment coefficient	5
$C_{M_{\text{ac}}}$	Aerodynamic pitching-moment coefficient about a surface's or body's *ac*	5, 6

Symbol	Definition (Typical Units)	Chapter Introduced
$C_{M_\alpha} = \dfrac{\partial C_M}{\partial \alpha}$	Angle-of-attack pitching-moment effectiveness (/rad)	5, 6
$C_{M_{\dot\alpha}} = \dfrac{\partial C_M}{\partial \dot\alpha}$	Angle-of-attack-rate pitching-moment effectiveness (sec)	5, 6
$C_{M_{\delta_E}} = \dfrac{\partial C_M}{\partial \delta_E}$	Elevator pitching-moment effectiveness (/rad)	5, 6
$C_{M_{M_\infty}} = \dfrac{\partial C_M}{\partial M_\infty}$	Mach-number pitching-moment effectiveness	5, 6
$C_{M_q} = \dfrac{\partial C_M}{\partial q}$	Pitch-rate pitching-moment effectiveness (sec)	5, 6
$C_{M_u} = \dfrac{\partial C_M}{\partial u}$	Surge-velocity pitching-moment effectiveness (sec/ft)	5, 6
$C_{M_{\eta_i}} = \dfrac{\partial C_M}{\partial \eta_i}$	Modal-deflection pitching-moment effectiveness	7
$C_{M_{\dot\eta_i}} = \dfrac{\partial C_M}{\partial \dot\eta_i}$	Modal-velocity pitching-moment effectiveness (sec)	7
$C_N = N/q_\infty Sb$	Aerodynamic yawing-moment coefficient	5
$C_{N_\beta} = \dfrac{\partial C_N}{\partial \beta}$	Sideslip yawing-moment effectiveness (/rad)	5, 6
$C_{N_{\delta_A}} = \dfrac{\partial C_N}{\partial \delta_A}$	Aileron yawing-moment effectiveness (/rad)	5, 6
$C_{N_{\delta_R}} = \dfrac{\partial C_N}{\partial \delta_R}$	Rudder yawing-moment effectiveness (/rad)	5, 6
$C_{N_p} = \dfrac{\partial C_N}{\partial p}$	Roll-rate yawing-moment effectiveness (sec)	5, 6
$C_{N_r} = \dfrac{\partial C_N}{\partial r}$	Yaw-rate yawing-moment effectiveness (sec)	5,6
$C_{N_{\eta_i}} = \dfrac{\partial C_N}{\partial \eta_i}$	Modal-displacement yawing-moment effectiveness	7
$C_{N_{\dot\eta_i}} = \dfrac{\partial C_N}{\partial \dot\eta_i}$	Modal velocity yawing-moment effectiveness (sec)	7
$C_{Q_i} = Q_i/q_\infty S\bar{c}$	Generalized-force aerodynamic coefficient	7

Symbol	Definition (Typical Units)	Chapter Introduced
$C_{Q_{i_\alpha}} = \dfrac{\partial C_{Q_i}}{\partial \alpha}$	Angle-of-attack generalized-force effectiveness (/rad)	7
$C_{Q_{i_\beta}} = \dfrac{\partial C_{Q_i}}{\partial \beta}$	Sideslip generalized-force effectiveness (/rad)	7
$C_{Q_{i_p}} = \dfrac{\partial C_{Q_i}}{\partial p}$	Roll-rate generalized-force effectiveness (sec)	7
$C_{Q_{i_q}} = \dfrac{\partial C_{Q_i}}{\partial q}$	Pitch-rate generalized-force effectiveness (sec)	7
$C_{Q_{i_r}} = \dfrac{\partial C_{Q_i}}{\partial r}$	Yaw-rate generalized-force effectiveness (sec)	7
$C_{Q_{i_{i_H}}} = \dfrac{\partial C_{Q_i}}{\partial i_H}$	Tail-incidence generalized-force effectiveness (/rad)	7
$C_{Q_{i_{\delta_E}}} = \dfrac{\partial C_{Q_i}}{\partial \delta_E}$	Elevator generalized-force effectiveness (/rad)	7
$C_{Q_{i_{\delta_A}}} = \dfrac{\partial C_{Q_i}}{\partial \delta_A}$	Aileron generalized-force effectiveness (/rad)	7
$C_{Q_{i_{\delta_R}}} = \dfrac{\partial C_{Q_i}}{\partial \delta_R}$	Rudder generalized-force effectiveness (/rad)	7
$C_{Q_{i_{\eta_j}}} = \dfrac{\partial C_{Q_i}}{\partial \eta_j}$	Modal-displacement generalized-force effectiveness	7
$C_{Q_{i_{\dot{\eta}_j}}} = \dfrac{\partial C_{Q_i}}{\partial \dot{\eta}_j}$	Modal-velocity generalized-force effectiveness (sec)	7
$C_S = \mathsf{S}/q_\infty S$	Aerodynamic side-force coefficient	5
$C_{S_\beta} = \dfrac{\partial C_S}{\partial \beta}$	Sideslip side-force effectiveness (/rad)	6
$C_{S_{\delta_A}} = \dfrac{\partial C_S}{\partial \delta_A}$	Aileron side-force effectiveness (/rad)	6
$C_{S_{\delta_R}} = \dfrac{\partial C_S}{\partial \delta_R}$	Rudder side-force effectiveness (/rad)	6
$C_{S_p} = \dfrac{\partial C_S}{\partial p}$	Roll-rate side-force effectiveness (sec)	6

Symbol	Definition (Typical Units)	Chapter Introduced
$C_{S_r} = \dfrac{\partial C_S}{\partial r}$	Yaw-rate side-force effectiveness (sec)	6
$C_{S_{\eta_i}} = \dfrac{\partial C_S}{\partial \eta_i}$	Modal-displacement side-force effectiveness	7
$C_{S_{\dot{\eta}_i}} = \dfrac{\partial C_S}{\partial \dot{\eta}_i}$	Modal-velocity side-force effectiveness (sec)	7
c_r	Wing root-chord length (ft)	5
c_t	Wing tip-chord length (ft)	5
C_X	Coefficient for aerodynamic force along the vehicle X axis	6
$C_{X_{\eta_i}} = \dfrac{\partial C_X}{\partial \eta_i}$	Modal-displacement axial-force effectiveness	7
$C_{X_{\dot{\eta}_i}} = \dfrac{\partial C_X}{\partial \dot{\eta}_i}$	Modal-velocity axial-force effectiveness (sec)	7
C_Y	Coefficient for aerodynamic force along the vehicle Y axis	6
$C_{Y_{\eta_i}} = \dfrac{\partial C_Y}{\partial \eta_i}$	Modal-displacement lateral-force effectiveness	7
$C_{Y_{\dot{\eta}_i}} = \dfrac{\partial C_Y}{\partial \dot{\eta}_i}$	Modal-velocity lateral-force effectiveness (sec)	7
C_Z	Coefficient for aerodynamic force along the vehicle Z axis	6
$C_{Z_{\eta_i}} = \dfrac{\partial C_Z}{\partial \eta_i}$	Modal-displacement vertical-force effectiveness	7
$C_{Z_{\dot{\eta}_i}} = \dfrac{\partial C_Z}{\partial \dot{\eta}_i}$	Modal-velocity vertical-force effectiveness (sec)	7
d	Aerodynamic 2-D section drag (lb/ft)	5
$\mathbf{d}_E$	Elastic-displacement vector of mass element or point on vehicle	4
D,d	Aerodynamic drag, *perturbation* (lb)	5
$\mathbf{D} = \mathbf{M}^{-1}\mathbf{K}$	Dynamic matrix in vibration problem (ft/sec^2)	3
$\mathbf{D}$	Control-distribution matrix for responses in state-variable system description	8

Symbol	Definition (Typical Units)	Chapter Introduced
df_{ext}	Infinitesimal external force acting on vehicle (lb)	2
δW	Virtual work in Lagrange's equation (ft-lb)	3, 4
$\delta \mathbf{p}$	Perturbation parameter vector in Taylor-series expansion of forces and moments	6
$\frac{d\mathbf{a}}{dt}\Big\|_1$	Time rate of change of vector **a** with respect to Frame 1	1
e	Oswald's drag efficiency factor, feedback error	5, 12
F, f	Force, *perturbation* (lb)	2
$\mathbf{F}$	Force vector (lb)	2, 4
$F_\bullet$	Force component along • axis (lb)	2, 4
$F_{A\bullet}, f_{A\bullet}$	Aerodynamic force component along • axis, *perturbation* (lb)	2, 4
$F_{P\bullet}, f_{P\bullet}$	Propulsive force component along • axis, *perturbation* (lb)	2, 4
F_S	Stick, control-wheel, or rudder pedal force, (lb)	9
$\mathbf{g}$	Earth's gravitational-acceleration vector, 32.174 $\mathbf{k}_E$ (ft/sec^2)	1
$g(s)$	Transfer function of a dynamic element	10
$g(j\omega)$	Frequency response or Bode representation of dynamic element	12
G	Control-surface gearing ratio	9
h, h	Altitude above the surface of the earth, *perturbation* (ft)	2
h_x, h_y, h_z	Three components of angular momentum due to rotating machinery (sl-ft^2/sec)	2
$\mathbf{H}_V$	Angular momentum of vehicle about its *c.m.* (sl-ft^2/sec)	2
$\mathbf{i}$	Unit vector defining direction of X axis	1
i_H	Incidence angle of horizontal tail relative to fuselage reference line	5, 6
i_W	Incidence angle of the wing relative to fuselage reference line	5, 6
j	Imaginary number $\sqrt{-1}$	1

Symbol	Definition (Typical Units)	Chapter Introduced
$[\mathbf{I}]$	Inertia matrix	2
I_{xx} etc.	Products of inertia, elements of $[\mathbf{I}]$ (sl-ft^2)	2
$\mathbf{j}$	Unit vector defining direction of Y axis	1
J	Moment of inertia of rotating machinery (sl-ft^2)	2
$k(s)$	Feedback-control compensation	12
$\mathbf{k}$	Unit vector defining direction of Z axis	1
K_i	Feedback-control gain on *i*'th response	11
$\mathbf{K}$	Stiffness matrix in vibration problem	3, 4
$\mathcal{K}$	Generalized stiffness matrix	3, 4
l	Aerodynamic 2-D section lift (lb/ft), characteristic length (ft)	5
L	Aerodynamic lift (lb)	5
L_A, L_{A_0}, l_A	Aerodynamic rolling moment, reference value, *perturbation* (ft-lb)	2
L_P, L_{P_0}, l_P	Propulsive rolling moment, reference value, *perturbation* (ft-lb)	2
L_{Roll}	Rolling moment on lifting surface or vehicle (ft-lb)	5, 6
$L_\beta = \dfrac{q_\infty Sb}{I_{xx}} C_{L_\beta}$	Rolling-moment-due-to-sideslip dimensional derivative (/sec^2)	8
$L_{\delta_A} = \dfrac{q_\infty Sb}{I_{xx}} C_{L_{\delta_A}}$	Rolling-moment-due-to-aileron dimensional derivative (/sec^2)	8
$L_{\delta_R} = \dfrac{q_\infty Sb}{I_{xx}} C_{L_{\delta_R}}$	Rolling-moment-due-to-rudder dimensional derivative (/sec^2)	8
$L_p = \dfrac{q_\infty Sb}{I_{xx}} C_{L_p}$	Rolling-moment-due-to-roll-rate dimensional derivative (/sec)	8
$L_r = \dfrac{q_\infty Sb}{I_{xx}} C_{L_r}$	Rolling-moment-due-to-yaw-rate dimensional derivative (/sec)	8

Symbol	Definition (Typical Units)	Chapter Introduced
$L'_\beta = \left(L_\beta + N_\beta I_{xz}/I_{xx}\right)/\left(1 - I_{xz}^2/\left(I_{xx}I_{zz}\right)\right)$	Rolling-moment-due-to-sideslip primed dimensional derivative (/sec^2)	8
$L'_{\delta_A} = \left(L_{\delta_A} + N_{\delta_A} I_{xz}/I_{xx}\right)/\left(1 - I_{xz}^2/\left(I_{xx}I_{zz}\right)\right)$	Rolling-moment-due-to-aileron primed dimensional derivative (/sec^2)	8
$L'_{\delta_R} = \left(L_{\delta_R} + N_{\delta_R} I_{xz}/I_{xx}\right)/\left(1 - I_{xz}^2/\left(I_{xx}I_{zz}\right)\right)$	Rolling-moment-due-to-rudder primed dimensional derivative (/sec^2)	8
$L'_p = \left(L_p + N_p I_{xz}/I_{xx}\right)/\left(1 - I_{xz}^2/\left(I_{xx}I_{zz}\right)\right)$	Rolling-moment-due-to-roll-rate primed dimensional derivative (/sec)	8
$L'_r = \left(L_r + N_r I_{xz}/I_{xx}\right)/\left(1 - I_{xz}^2/\left(I_{xx}I_{zz}\right)\right)$	Rolling-moment-due-to-yaw-rate primed dimensional derivative (/sec)	8
m	Mass (sl), aerodynamic 2-D section pitching moment (ft-lb/ft)	2, 5
$\dot{m}, \ddot{m}$	Rate of change of mass, and its rate, for variable-mass vehicle (sl/sec), (sl/sec^2)	2
$M_A, \mathrm{M}_{A_0}, m_A$	Aerodynamic pitching moment, reference value, *perturbation* (ft-lb)	2
$M_P, \mathrm{M}_{P_0}, m_P$	Propulsive pitching moment, reference value, *perturbation* (ft-lb)	2
M	Pitching moment on a lifting surface or body	5, 6
M_∞	Free-stream Mach number	5
$\mathbf{M}$	Moment vector (ft-lb)	2
$\mathbf{M}$	Mass matrix in vibration problem (sl)	3, 4
$\mathbf{M}$	Modal matrix of eigenvectors	10
$\mathcal{M}$	Generalized mass matrix in vibration problem (sl)	3, 4
MAC	Mean aerodynamic chord (ft)	5
$M_\alpha + M_{P_\alpha} = \dfrac{q_\infty S\bar{c}}{I_{yy}}\left(C_{M_\alpha} + C_{P_{M_\alpha}}\right)$	Pitching-moment-due-to-angle-of-attack dimensional derivative (/sec^2)	8
$M_{\dot{\alpha}} = \dfrac{q_\infty S\bar{c}}{I_{yy}} C_{M_{\dot{\alpha}}}$	Pitching-moment-due-to-angle-of-attack-rate dimensional derivative (/sec)	8

Symbol	Definition (Typical Units)	Chapter Introduced
$M_{\delta_E} = \dfrac{q_\infty S\bar{c}}{I_{yy}} C_{M_{\delta_E}}$	Pitching-moment-due-to-elevator dimensional derivative (/sec^2)	8
$M_{\delta T} = \dfrac{(d_T \cos\phi_T - x_T \sin\phi_T)}{I_{yy}}$	Pitching-moment-due-to-thrust dimensional derivative (rad/lb-sec^2)	8
$M_q = \dfrac{q_\infty S\bar{c}}{I_{yy}} C_{M_q}$	Pitching-moment-due-to-pitch-rate dimensional derivative (/sec)	8
$M_u + M_{P_u} = \dfrac{q_\infty S\bar{c}}{I_{yy}}\left(\left(C_{M_u} + \dfrac{2}{U_0} C_{M_0}\right) + \left(C_{P_{M_u}} + \dfrac{2}{U_0} C_{P_{M_0}}\right)\right)$	Pitching-moment-due-to-surge-velocity dimensional derivative (rad/ft-sec)	8
N_A, N_{A_0}, n_A	Aerodynamic yawing moment, reference value, *perturbation* (ft-lb)	2
N_P, N_{P_0}, n_P	Propulsive yawing moment, reference value, *perturbation* (ft-lb)	2
N, n	Yawing moment, 2-D aerodynamic yawing moment or *perturbation* (ft-lb)	5, 6
$N_\beta = \dfrac{q_\infty Sb}{I_{zz}} C_{N_\beta}$	Yawing-moment-due-to-sideslip dimensional derivative (/sec^2)	8
$N_{\delta_A} = \dfrac{q_\infty Sb}{I_{zz}} C_{N_{\delta_A}}$	Yawing-moment-due-to-aileron dimensional derivative (/sec^2)	8
$N_{\delta_R} = \dfrac{q_\infty Sb}{I_{zz}} C_{N_{\delta_R}}$	Yawing-moment-due-to-rudder dimensional derivative (/sec^2)	8
$N_p = \dfrac{q_\infty Sb}{I_{zz}} C_{N_p}$	Yawing-moment-due-to-roll-rate dimensional derivative (/sec)	8
$N_r = \dfrac{q_\infty Sb}{I_{zz}} C_{N_r}$	Yawing-moment-due-to-yaw-rate dimensional derivative (/sec)	8

Symbol	Definition (Typical Units)	Chapter Introduced
$N'_\beta = \left(N_\beta + L_\beta I_{xz}/I_{zz}\right)/\left(1 - I_{xz}^2/\left(I_{xx}I_{zz}\right)\right)$	Yawing-moment-due-to-sideslip primed dimensional derivative ($/\sec^2$)	8
$N'_{\delta_A} = \left(N_{\delta_A} + L_{\delta_A} I_{xz}/I_{zz}\right)/\left(1 - I_{xz}^2/\left(I_{xx}I_{zz}\right)\right)$	Yawing-moment-due-to-aileron primed dimensional derivative ($/\sec^2$)	8
$N'_{\delta_R} = \left(N_{\delta_R} + L_{\delta_R} I_{xz}/I_{zz}\right)/\left(1 - I_{xz}^2/\left(I_{xx}I_{zz}\right)\right)$	Yawing-moment-due-to-rudder primed dimensional derivative ($/\sec^2$)	8
$N'_p = \left(N_p + L_p I_{xz}/I_{zz}\right)/\left(1 - I_{xz}^2/\left(I_{xx}I_{zz}\right)\right)$	Yawing-moment-due-to-roll-rate primed dimensional derivative (/sec)	8
$N'_r = \left(N_r + L_r I_{xz}/I_{zz}\right)/\left(1 - I_{xz}^2/\left(I_{xx}I_{zz}\right)\right)$	Yawing-moment-due-to-yaw-rate primed dimensional derivative (/sec)	8
$\mathbf{p}$	Position (vector) of mass element relative to vehicle *c.m.*	2
$\mathbf{p}$	Parameter vector in Taylor-series expansion of forces or moments	6
$\mathbf{p}_{RB}$	Position of mass element relative to vehicle *c.m.*, for undeformed vehicle	4
$\mathbf{p}_V$	Inertial position (vector) of vehicle's *c.m.*	2
$\mathbf{p}'$	Inertial position (vector) of mass element	2
$\mathbf{p}_E$	Position of nozzle exit plane relative to vehicle *c.m.* for variable-mass vehicle	2
P, P_0, p	Roll rate about X_V axis, reference value, *perturbation* (rad/sec)	2
$\mathbf{P}(s)$	System matrix in polynomial-matrix system description	10
p_∞	Free-stream atmospheric pressure (psf)	5
Q, Q_0, q	Pitch rate about Y_V axis, reference value, *perturbation* (rad/sec)	2
$\mathbf{Q}(s)$	Control-distribution matrix in polynomial-matrix system description	10
Q_i	Generalized force in Lagrange's equation associated with *i*'th degree of freedom (lb)	3, 4

Symbol	Definition (Typical Units)	Chapter Introduced
$Q_{i_\alpha} = \dfrac{\partial Q_i}{\partial \alpha}$	Angle-of-attack generalized-force effectiveness (lb/rad)	7
$Q_{i_\beta} = \dfrac{\partial Q_i}{\partial \beta}$	Sideslip generalized-force effectiveness (lb/rad)	7
$Q_{i_p} = \dfrac{\partial Q_i}{\partial p}$	Roll-rate generalized-force effectiveness (lb-sec/rad)	7
$Q_{i_q} = \dfrac{\partial Q_i}{\partial q}$	Pitch-rate generalized-force effectiveness (lb-sec/rad)	7
$Q_{i_r} = \dfrac{\partial Q_i}{\partial r}$	Yaw-rate generalized-force effectiveness (lb-sec/rad)	7
$Q_{i_{i_H}} = \dfrac{\partial Q_i}{\partial i_H}$	Stabilizer-incidence generalized-force effectiveness (lb/rad)	7
$Q_{i_{\delta_E}} = \dfrac{\partial Q_i}{\partial \delta_E}$	Elevator generalized-force effectiveness (lb/rad)	7
$Q_{i_{\delta_A}} = \dfrac{\partial Q_i}{\partial \delta_A}$	Aileron generalized-force effectiveness (lb/rad)	7
$Q_{i_{\delta_r}} = \dfrac{\partial Q_i}{\partial \delta_R}$	Rudder generalized-force effectiveness (lb/rad)	7
$Q_{i_{\eta_j}} = \dfrac{\partial Q_i}{\partial \eta_j}$	Modal-displacement generalized-force effectiveness (lb)	7
$Q_{i_{\dot{\eta}_j}} = \dfrac{\partial Q_i}{\partial \dot{\eta}_j}$	Modal-velocity generalized-force effectiveness (lb-sec)	7
$q_\infty = \dfrac{1}{2}\rho_\infty V_\infty^2$	Dynamic pressure (psf)	5
$\mathbf{q}$	Vector of generalized coordinates in Lagrange's equation	3, 4
R, R_0, r	Yaw rate about Z_V axis, reference value, *perturbation* (rad/sec)	2
R_E	Radius of the earth (mi)	2
R_l	Reynolds number	5
S	Aerodynamic reference area, lifting-surface planform area (ft^2)	5
SM	Static margin	9

Symbol	Definition (Typical Units)	Chapter Introduced
S_{Wet}	Total, or wetted area, of lifting surface (ft^2)	5
S	Aerodynamic side force (lb)	6
s	Complex variable in Laplace transform (/sec)	10
T	Kinetic energy in Lagrange's equation (ft-lb)	3, 4
T	Thrust (lb), modal time constant ($= -1/\lambda$, sec)	2, 10
$\mathbf{T}_{\text{I–II}}$	Direction-cosine matrix relating unit vectors in Frame I to those in Frame II	1
U, U_0, u	Surge velocity component along X_V axis, reference value, *perturbation* (fps)	2
U	Potential (strain) energy (ft-lb)	3, 4
$\mathbf{U}$	Unit-conversion matrix	10
$\mathbf{u}$	Control input vector in state-variable or polynomial-matrix system descriptions	8, 10
V, V_0, v	Lateral velocity component along Y_V axis, reference value, *perturbation* (fps)	2
$\mathbf{V}_V$	Vehicle velocity vector, $\lvert\mathbf{V}_V\rvert = V_V = \sqrt{U^2 + V^2 + W^2}$ (fps)	2
W, W_0, w	Plunge velocity component along Z_V axis, reference value, *perturbation* (fps)	2
$w_E(y)$	Wing plunge velocity at span y due to elastic deformation (fps)	7
$\mathbf{x}$	State vector in state-variable system description	8
$x_{\text{ac}}(y)$	Chord-wise distance of section *ac* aft of leading edge at span location y (ft)	5
X_{ac}	Chord-wise location of wing *ac* (ft)	5
X_I	Inertial position (North)	2, 4
$X_{\text{LE}}(y)$	Chord-wise distance of leading edge aft of wing apex at span (y) (ft)	5

Symbol	Definition (Typical Units)	Chapter Introduced
$X_\alpha = \frac{q_\infty S}{m}(-C_{D_\alpha} + C_{L_0})$	Axial-force-due-to-angle-of-attack dimensional derivative (ft/rad-sec^2)	8
$X_{\dot{\alpha}} = \frac{q_\infty S}{m}(-C_{D_{\dot{\alpha}}})$	Axial-force-due-to-angle-of-attack-rate dimensional derivative (ft/rad-sec)	8
$X_{\delta_E} = \frac{q_\infty S}{m}(-C_{D_{\delta_E}})$	Axial-force-due-to-elevator dimensional derivative (ft/rad-sec^2)	8
$X_{\delta T} = \frac{\cos\phi_T}{m}$	Axial-force-due-to-thrust dimensional derivative (ft/lb-sec^2)	8
$X_q = \frac{q_\infty S}{m}(-C_{D_q})$	Axial-force-due-to-pitch-rate dimensional derivative (ft/rad-sec)	8
$X_u + X_{P_u} = \frac{q_\infty S}{m}\left(-\left(C_{D_u} + \frac{2}{U_0}C_{D_0}\right) + \left(C_{P_{X_u}} + \frac{2}{U_0}C_{P_{X_0}}\right)\right)$	Axial-force-due-to-surge-velocity dimensional derivative (/sec)	8
$\mathbf{y}$	Response vector in state-variable or polynomial-matrix system descriptions	8, 10
Y_I	Inertial position (East)	2, 4
Y_{MAC}	Span-wise distance from root-chord plane to MAC (ft)	5
$Y_C(j\omega)$	Controlled element in pilot-in-the-loop feedback system	13
$Y_P(j\omega)$	Pilot's describing function in pilot-in-the-loop feedback system	13
$Y_\beta = \frac{q_\infty S}{m}C_{S_\beta}$	Lateral-force-due-to-sideslip dimensional derivative (ft/rad-sec^2)	8
$Y_{\delta_A} = \frac{q_\infty S}{m}C_{S_{\delta_A}}$	Lateral-force-due-to-aileron dimensional derivative (ft/rad-sec^2)	8

Symbol	Definition (Typical Units)	Chapter Introduced
$Y_{\delta_R} = \frac{q_\infty S}{m} C_{S_{\delta_R}}$	Lateral-force-due-to-rudder dimensional derivative (ft/rad-sec^2)	8
$Y_p = \frac{q_\infty S}{m} C_{S_p}$	Lateral-force-due-to-roll-rate dimensional derivative (ft/rad-sec)	8
$Y_r = \frac{q_\infty S}{m} C_{S_r}$	Lateral-force-due-to-yaw-rate dimensional derivative (ft/rad-sec)	8
Z_{ac}	Vertical location of the aerodynamic center of the vertical tail (ft)	6
$z_{ac}(y)$	Vertical location of the section aerodynamic center at span y on the vertical tail (ft)	6
Z_I	Inertial position ($= -$h for flat earth)	2, 4
$Z_\alpha = \frac{q_\infty S}{m}(-C_{L_\alpha} + C_{D_0})$	Vertical-force-due-to-angle-of-attack dimensional derivative (ft/rad-sec^2)	8
$Z_{\dot{\alpha}} = \frac{q_\infty S}{m}(-C_{L_{\dot{\alpha}}})$	Vertical-force-due-to-angle-of-attack-rate dimensional derivative (ft/rad-sec)	8
$Z_{\delta_E} = \frac{q_\infty S}{m}(-C_{L_{\delta_E}})$	Vertical-force-due-to-elevator dimensional derivative (ft/rad-sec^2)	8
$Z_{\delta T} = -\frac{\sin\phi_T}{m}$	Vertical-force-due-to-thrust dimensional derivative (ft/lb-sec^2)	8
$Z_q = \frac{q_\infty S}{m}(-C_{L_q})$	Vertical-force-due-to-pitch-rate dimensional derivative (ft/rad-sec)	8
$Z_u + Z_{P_u} = \frac{q_\infty S}{m}\left(-\left(C_{L_u} + \frac{2}{U_0}C_{L_0}\right) + \left(C_{P_{Z_u}} + \frac{2}{U_0}C_{P_{Z_0}}\right)\right)$	Vertical-force-due-to-surge-velocity dimensional derivative (/sec)	8

Symbol	Definition (Typical Units)	Chapter Introduced
Greek		
α, α_0, α	Aerodynamic angle of attack, reference value, *perturbation* (deg, rad)	5, 6
$\alpha_\delta = C_{L_\delta}/C_{L_\alpha}$	Flap lift effectiveness	5
α_0	Angle of attack for zero lift (deg)	5
β, β_0, β	Aerodynamic sideslip angle, reference value, *perturbation* (deg, rad)	5, 6
β	Prandtl–Glauert compressibility correction factor	5
δ, δ_0, δ	Control-surface or flap deflection, reference value, *perturbation* (deg, rad)	5
$\delta\bullet$	Perturbation in parameter •	6
$\Delta\bullet$	Change in variable •	5
ε	Downwash angle (deg)	5
$\varepsilon(y)$	Wing twist angle at span location y (deg)	5
ϕ, Φ_0, ϕ	Euler angle (rotation about X_V), "bank angle," reference value, *perturbation* (rad)	2, 8
$\mathbf{\Phi}$	Modal matrix	3, 4, 10
$\mathbf{\Phi}(t-t_0)$	State-transition matrix	8
ϕ_T	Angle between thrust vector and X_F axis, (deg)	6
ϕ_W	Velocity- or wind-axis bank angle (deg)	2
γ, γ_0, γ	Flight-path angle, reference value, *perturbation* (deg, rad)	2
Γ	Dihedral angle (deg)	5
η, H_o, η	Generalized coordinate associated with elastic degree of freedom, reference, *perturbation*	3, 4
η_i	Span location of inboard edge of flap (ft)	5
η_o	Span location of outboard edge of flap (ft)	5
λ	Eigenvalue (/sec)	3, 4, 10

Symbol	Definition (Typical Units)	Chapter Introduced
Λ	Eigenvalue matrix	3, 4, 10
$\Lambda_\bullet$	Wing sweep angle at chord location • (e.g., leading edge, mid-chord) (deg)	5
λ_L	Latitude of local-horizontal-local-vertical coordinate frame (deg)	2
μ_L	Longitude of local-horizontal-local-vertical coordinate frame (deg)	2
$\boldsymbol{\mu}$	Left eigenvector	10
$\boldsymbol{\nu}$	Eigenvector, vibration mode shape, right eigenvector	3, 4, 10
ν'	Slope of vibration (displacement) mode shape (rad)	7, 10
θ, Θ_0, θ	Euler angle, (rotation about an intermediate Y axis), "pitch-attitude angle," reference value, *perturbation* (rad)	2, 8
$\theta_E(y)$	Wing twist at span *y* due to elastic deformation (deg)	7
Θ_{Ref}	Reference, or rigid-body angle of reference axis	3
ρ_V	Density of vehicle material (sl/ft^3)	2
ρ_∞	Density of air (sl/ft^3)	5
σ	Real part of a complex number (/sec)	10
τ_E	Effective time delay of the human pilot (sec)	13
τ_{NM}	Neuromotor time constant of the human pilot (sec)	13
τ_{IP}	Information-processing time constant of the human pilot (sec)	13
ω	Frequency (rad/sec)	3, 10
$\boldsymbol{\omega}_{\text{I,II}}$	Angular velocity (vector) of Frame I with respect to Frame II (rad/sec)	1
$\boldsymbol{\omega}_{\text{Earth}}$	Angular velocity of earth about its axis (2π rad/day)	2
ψ, Ψ_0, ψ	Euler angle (rotation about Z_E axis), "heading angle," reference value, *perturbation* (rad)	2, 8
$\dot{\Psi}_0$	Steady rate of turn (deg/sec)	9
ψ_W	Velocity- or wind-axes heading angle (deg)	2

Symbol	Definition (Typical Units)	Chapter Introduced
ζ	Modal damping	3, 10
Ξ	Constraint matrix in vibration problem	3
Ξ_i	Dimensional stability derivative associated with i'th elastic degree of freedom (/sec^2)	10
Subscript		
0	Reference condition	2
A	Aileron, aerodynamic	5
C	Canard	6
E	Earth-fixed axes	2
E	Elevator, elastic	5, 7
F	Fuselage-referenced axes	6
F	Flap	5
H	Horizontal tail	5
I	Inertial	2
P	Propulsive	6
R	Rudder	6
S	Stability axes	6
T	Thrust	6
V	Vertical tail, vehicle-fixed axes	2, 6
W	Wing, wind axes	2, 6
X	Component along X axis	2, 6
Y	Component along Y axis	2, 6
Z	Component along Z axis	2, 6
∞	Free-stream flow condition	5
Overbar		
$\bar{\bullet}$	Distance or length •, normalized by the wing's mean aerodynamic chord. For example, $\bar{x}_{cg} = \frac{x_{cg}}{\bar{c}_W}$	5, 6

CHAPTER 1

Introduction and Topical Review

Chapter Roadmap: *This chapter presents a review of topics that are especially important to the student undertaking a first course in flight dynamics. Sections 1.1 and 1.6 should be covered before addressing material in any of the other chapters, and Sections 1.2–1.5 should be covered prior to addressing the derivations of the equations of motion (Chapter 2).*

To begin our study of modern flight dynamics, it will be useful for the student to review some key concepts. These concepts include types of coordinate systems, kinematics and direction-cosine matrices, vector differentiation, Newton's laws, and, very importantly, small-perturbation theory for the study of nonlinear dynamic systems. Most likely the student is already familiar with the underlying principles upon which these topics are based, if not the explicit theories themselves.

Although review of these concepts may seem unnecessary to some, it has been found over the years that many students have critical misunderstandings regarding the theoretical bases underpinning these concepts. These misunderstandings make it difficult for the student to really master flight dynamics. Therefore, this review is intended to eliminate any such misunderstandings.

Perhaps even more important to the student is that the concepts in this chapter, especially the material on small-perturbation theory, provide a rigorous overarching framework for developing and presenting the subject of flight dynamics. Since flight dynamics involves the integration of several classical disciplines, and since the serious study of flight dynamics involves a myriad of details, establishing such a framework has been found to be very enlightening to the student.

1.1 SMALL PERTURBATION THEORY FOR NONLINEAR SYSTEMS

Throughout the study of flight dynamics, we fundamentally deal with the study of a nonlinear dynamic system—the dynamics of an atmospheric flight vehicle. It is both important to realize this fact, and to understand the overall approach taken in our study. That approach is based on small-perturbation theory (Ref. 1), which will be reviewed in this section, and applied to an illustrative example in Section 1.6.

Abstractly speaking, small-perturbation theory involves the investigation of a nonlinear system's dynamic behavior (e.g., stability) in the neighborhood of selected reference conditions, frequently taken to be its equilibrium conditions. Unlike linear systems, which have only one equilibrium condition, nonlinear systems frequently have multiple equilibrium conditions. Further, if the investigation is restricted to the motion of the nonlinear system "in the neighborhood of reference conditions," then some simplifying assumptions can be made.

There are five formal steps in the perturbation analysis:

1. Derive the (nonlinear) equations governing the dynamic behavior of the system.
2. Make a formal change of variables, expressing all the degrees of freedom governed by the equations of motion in terms of a reference condition, plus deviations from this reference condition. Express the governing equations found in Step 1 in terms of these new variables.
3. Under the small-perturbation assumption, extract two sets of equations from the results from Step 2. One set (the Reference Set) is expressed only in terms of the reference variables defined above, and one set (the Perturbation Set) is expressed in terms of the reference variables plus the perturbation variables. The Perturbation Set will be linear in the perturbation variables.
4. Using the Reference Set of equations, characterize the reference condition(s) of interest.
5. Using the Perturbation Set of equations, plus the reference conditions selected in Step 4, analyze the behavior of the system in the neighborhood of the reference conditions(s).

Under small-perturbation theory, in the neighborhood of a reference condition the nonlinear system behaves like the linear perturbation system. This all sounds rather abstract at this point, but the steps will become more clear when we apply the procedure to our example in Section 1.6.

Now, in order to proceed with Step 1, the derivation of the equations governing the system's dynamics, we first need to review some other basic concepts. Then we will return to the remaining steps in the procedure.

1.2 COORDINATE SYSTEMS

There are fundamentally three types of coordinate systems that must be considered in flight dynamics—inertial, vehicle fixed, and intermediate. Within each type, several variations may also be defined. But for now we will consider the definitions of these three types.

A coordinate system, or coordinate frame, is inertial if it is undergoing neither rectilinear acceleration nor rotation. Examples of a truly inertial frame referenced to some physical body, such as a star, may be difficult to find, but in treating different classes of flight-dynamics problems, certain frames are chosen, or assumed, to be inertial. In such cases, assumptions about some accelerations are made. Whether these assumptions are sufficiently valid depends on the problem addressed.

For example, consider a frequently used coordinate system that is fixed to a point on the surface of the earth (e.g., the end of a runway), with two of its three mutually orthogonal axes oriented true north and east, respectively. In the analysis of aircraft dynamics, this coordinate system is frequently assumed to be inertial. This is not strictly true, of course, since under such an assumption the rotation of the earth about its axis, as well as the orbital motion of the earth in the solar system, are being ignored. But if the accelerations associated with these motions are sufficiently small compared to those resulting from the forces acting on the vehicle, and if the time of flight being considered is small enough, such assumptions may be sufficiently valid.

But neither of the above assumptions is usually valid when considering orbiting spacecraft, for example. In this case, the ignored accelerations are not typically small compared to other accelerations experienced by the spacecraft, and the time of flight is usually much longer than that for an aircraft. Therefore, selection of a different coordinate system as inertial is typically necessary in the analysis of spacecraft. In the study of the dynamics of spacecraft operating in the neighborhood of the earth, the inertial frame is frequently defined as one with its origin at the geometric or mass center of the earth, and one axis either always directed at the sun, or at another star. In either case, the unit vector directed at the sun or the other star is then defined (or assumed) to be "fixed in inertial space."

Finally, even if such an inertial coordinate system were defined in a particular study, we would also frequently introduce another reference frame fixed on the earth's surface. It is on the earth's surface that airports, tracking stations, and/or launch sites are located. So we are usually interested in defining a vehicle's positions and velocities relative to this frame. If such a frame were not defined as inertial, it would be an intermediate frame, neither inertial nor vehicle fixed.

A vehicle-fixed frame is one that is fixed to the vehicle in some specified way and therefore moves with the vehicle. Several of these frames will be defined and used later in our study of flight dynamics.

To clarify these ideas further, consider Figure 1.1, in which the earth is shown schematically. Three coordinate systems are also shown. Frame S has its origin fixed at the earth center, with the X_S axis directed at a star of choice such as the sun. Frame E has its origin fixed at coordinates λ degrees latitude, and μ degrees longitude, on the earth's surface. It has X_E axis directed north, parallel to the earth meridian μ, and Y_E axis directed east. Finally, Frame V is fixed to the vehicle, with origin fixed to a selected point on the vehicle, and with X_V axis directed toward the front of the vehicle. So this latter vehicle-fixed frame both translates and rotates with the vehicle.

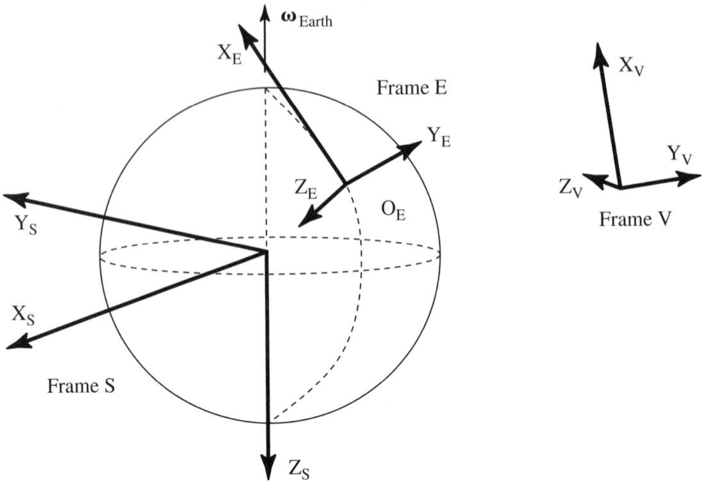

Figure 1.1 Three typical coordinate systems.

Now with these definitions in mind, note that Frame E and, in general, Frame V are both moving with respect to Frame S. Consider, for example, the velocity of Frame E with respect to Frame S. Assuming no deformation of the earth, this relative velocity, denoted $\boldsymbol{\omega}_{E,S}$, is simply related to the earth's rotation about its axis $\boldsymbol{\omega}_{\text{Earth}}$. Or

$$\boldsymbol{\omega}_{E,S} = -\boldsymbol{\omega}_{\text{Earth}}\, \mathbf{k}_S = -(2\pi \text{rad/day})\, \mathbf{k}_S \tag{1.1}$$

where $\mathbf{k}_S$ is the unit vector defining the direction of the Z_S axis.

A coordinate system, or frame, is vehicle-fixed simply if it moves with the vehicle in some well-defined way. For example, the origin of the frame maybe taken to be fixed at some point on a vehicle, such as the vehicle's instantaneous mass center. A vehicle-fixed frame also rotates with the vehicle in a specified way. We will see several examples of such frames in Chapters 2, 4, and 6.

An intermediate coordinate system, or frame, may be inertial, vehicle-fixed, or neither of these. Intermediate frames are frequently defined as part of an analysis or derivation to aid in that analysis. An example of an intermediate frame would be a coordinate system with origin at an aircraft's propeller hub, and rotating with the propeller.

1.3 VECTORS, COORDINATE TRANSFORMATIONS, AND DIRECTION-COSINE MATRICES

Students are no doubt familiar with vectors, and with several vector quantities such as forces, velocities, and accelerations. But frequently students are confused when required to deal rigorously with vector quantities and vector components. This section is intended to reduce or eliminate that confusion.

We will define a vector here as a directed line segment with specified magnitude or length. It could represent a physical quantity like a force or a velocity.

As a line in space, such a vector is independent of a coordinate system. One only needs to be concerned about a coordinate system when doing two things with this vector: (1) working with the components of the vector or (2) differentiating the vector with respect to time. In these two cases (components and differentiation), one must carefully identify the coordinate system of interest.

Now that several critical coordinate systems have been discussed, we will turn our attention to components of a vector, and the relationship between components of vectors expressed in different frames. Specifically, let **a** be any vector, and assume it is expressed in terms of its coordinates in some frame, say Frame S. Or, if the unit vectors defining Frame S are $\mathbf{i}_S$, $\mathbf{j}_S$, and $\mathbf{k}_S$, then

$$\mathbf{a} = a_x\,\mathbf{i}_S + a_y\,\mathbf{j}_S + a_z\,\mathbf{k}_S$$

where the coordinates of **a** in Frame S are a_x, a_y, and a_z. Now assume we wish to determine, in a straightforward and rigorous manner, the coordinates of **a** in some other frame. Note first that the vector **a** itself is not changing as we consider different frames—**a** is a line in space. We are simply finding components of **a** in the different frames. This is accomplished easily by means of *direction-cosine matrices,* which will be denoted as **T**.

NOTE TO STUDENT

Regarding the notation used in this book, **bold** lowercase letters will typically indicate vector quantities, while **BOLD** capital letters will typically indicate matrix quantities. Also, the first time an important word or phrase is introduced, it will appear in *italicized* letters.

The direction-cosine matrix is defined as the matrix that relates unit vectors defining one frame to unit vectors defining another frame. For example, consider two arbitrary frames, Frames 1 and 2 shown in Figure 1.2, with Frame 2 rotated an angle θ_3 about its Z_2 axis (into the page) relative to Frame 1.

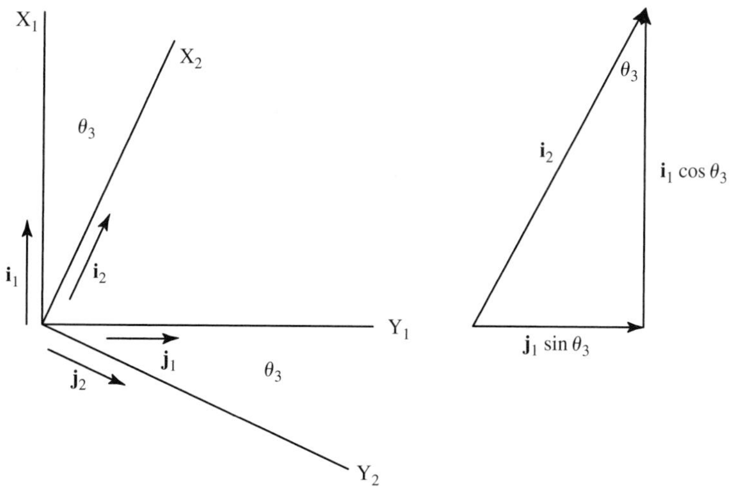

Figure 1.2 Frames 1 and 2, with relative angle θ_3.

Clearly, with all unit vectors having unity length one can see by inspection of the figure on the right in Figure 1.2 that the following relationship must hold between the two sets of unit vectors:

$$\mathbf{i}_2 = \cos\theta_3\,\mathbf{i}_1 + \sin\theta_3\,\mathbf{j}_1 \tag{1.2}$$

Through a similar analysis, one can show that

$$\mathbf{j}_2 = -\sin\theta_3\,\mathbf{i}_1 + \cos\theta_3\,\mathbf{j}_1 \tag{1.3}$$

And finally, it is obvious that

$$\mathbf{k}_2 = \mathbf{k}_1 \tag{1.4}$$

Writing the above three expressions in matrix notation yields the direction-cosine matrix $\mathbf{T}_{1\text{-}2}$ for this case, or

$$\begin{Bmatrix}\mathbf{i}_2\\ \mathbf{j}_2\\ \mathbf{k}_2\end{Bmatrix} = \begin{bmatrix}\cos\theta_3 & \sin\theta_3 & 0\\ -\sin\theta_3 & \cos\theta_3 & 0\\ 0 & 0 & 1\end{bmatrix}\begin{Bmatrix}\mathbf{i}_1\\ \mathbf{j}_1\\ \mathbf{k}_1\end{Bmatrix} \triangleq \mathbf{T}_{1\text{-}2}(\theta_3)\begin{Bmatrix}\mathbf{i}_1\\ \mathbf{j}_1\\ \mathbf{k}_1\end{Bmatrix} \tag{1.5}$$

Here the subscripts on $\mathbf{T}$ remind us of the two sets of unit vectors, or frames, related through $\mathbf{T}$. And the order of the subscripts is important, as we shall see below. So clearly the unit vectors of Frame 1 are related to the unit vectors of Frame 2 through $\mathbf{T}_{1\text{-}2}(\theta_3)$. Since the two frames are related through the single angle θ_3, $\mathbf{T}_{1\text{-}2}$ is only a function of that angle.

Note also that, in general, direction-cosine matrices have the property of *orthogonality,* or their inverse equals their transpose. Since the transpose operation is much simpler to perform than inversion, this is a convenient property for computation. Using the orthogonality property, from Equation (1.5) we obtain

$$\begin{Bmatrix}\mathbf{i}_1\\ \mathbf{j}_1\\ \mathbf{k}_1\end{Bmatrix} = \begin{bmatrix}\cos\theta_3 & -\sin\theta_3 & 0\\ \sin\theta_3 & \cos\theta_3 & 0\\ 0 & 0 & 1\end{bmatrix}\begin{Bmatrix}\mathbf{i}_2\\ \mathbf{j}_2\\ \mathbf{k}_2\end{Bmatrix} = \mathbf{T}^T_{1\text{-}2}(\theta_3)\begin{Bmatrix}\mathbf{i}_2\\ \mathbf{j}_2\\ \mathbf{k}_2\end{Bmatrix} \triangleq \mathbf{T}_{2\text{-}1}\begin{Bmatrix}\mathbf{i}_2\\ \mathbf{j}_2\\ \mathbf{k}_2\end{Bmatrix} \tag{1.6}$$

This relation can also be verified from the geometry in Figure 1.2. Note further that $\mathbf{T}_{2\text{-}1} = \mathbf{T}^T_{1\text{-}2}$, and that the order of the subscripts of $\mathbf{T}$ is now reversed. The student should now understand the meaning of these subscripts.

The above example considers two frames with relative orientation defined by an angle θ_3 about the Z axes of both frames. We now introduce a third frame, which is oriented relative to Frame 2 above an angle θ_2 about the Y_2 axis. (See Figure 1.3.) One can verify, using a graphical analysis similar to that above, that the unit vectors defining Frames 2 and 3 are related as follows:

$$\begin{Bmatrix}\mathbf{i}_3\\ \mathbf{j}_3\\ \mathbf{k}_3\end{Bmatrix} = \begin{bmatrix}\cos\theta_2 & 0 & -\sin\theta_2\\ 0 & 1 & 0\\ \sin\theta_2 & 0 & \cos\theta_2\end{bmatrix}\begin{Bmatrix}\mathbf{i}_2\\ \mathbf{j}_2\\ \mathbf{k}_2\end{Bmatrix} \triangleq T_{2\text{-}3}(\theta_2)\begin{Bmatrix}\mathbf{i}_2\\ \mathbf{j}_2\\ \mathbf{k}_2\end{Bmatrix} \tag{1.7}$$

Finally, let a fourth frame be defined such that it is oriented relative to Frame 3 through an angle θ_1 about the X_3 axis. Again, one can verify that the unit vectors of Frames 3 and 4 are related by

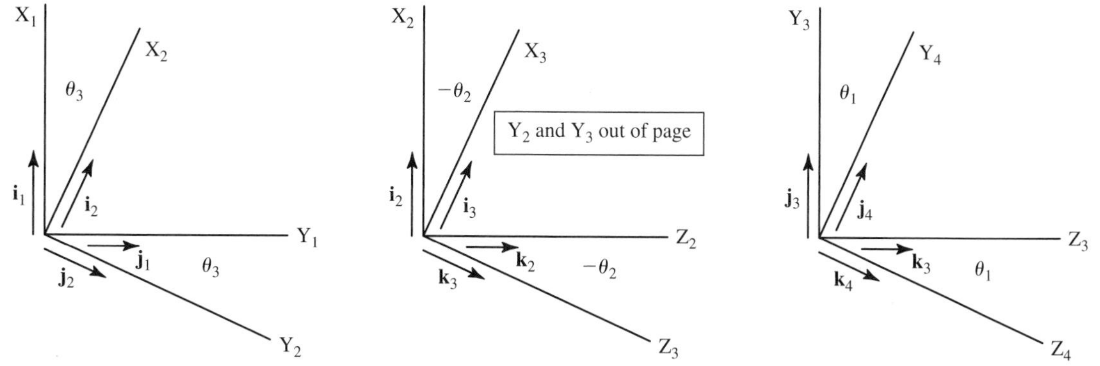

Figure 1.3 Relative orientation between Frames 1, 2, 3, and 4.

$$\begin{Bmatrix} \mathbf{i}_4 \\ \mathbf{j}_4 \\ \mathbf{k}_4 \end{Bmatrix} = \begin{bmatrix} 1 & 0 & 0 \\ 0 & \cos\theta_1 & \sin\theta_1 \\ 0 & -\sin\theta_1 & \cos\theta_1 \end{bmatrix} \begin{Bmatrix} \mathbf{i}_3 \\ \mathbf{j}_3 \\ \mathbf{k}_3 \end{Bmatrix} \triangleq \mathbf{T}_{3\text{-}4}(\theta_1) \begin{Bmatrix} \mathbf{i}_3 \\ \mathbf{j}_3 \\ \mathbf{k}_3 \end{Bmatrix} \tag{1.8}$$

The four frames and three angles (*Euler angles*) are all shown in Figure 1.3. The three angles arise from a "3-2-1 rotation," referring to the axes of rotation and the order of the rotations.

Investigating these direction-cosine matrices further, we see that they can relate the unit vectors between any pair of the four frames. For example, if we are interested in the relationship between the unit vectors of Frames 1 and 4, we have

$$\begin{Bmatrix} \mathbf{i}_4 \\ \mathbf{j}_4 \\ \mathbf{k}_4 \end{Bmatrix} = \mathbf{T}_{3\text{-}4}(\theta_1) \begin{Bmatrix} \mathbf{i}_3 \\ \mathbf{j}_3 \\ \mathbf{k}_3 \end{Bmatrix} = \mathbf{T}_{3\text{-}4}(\theta_1)\mathbf{T}_{2\text{-}3}(\theta_2) \begin{Bmatrix} \mathbf{i}_2 \\ \mathbf{j}_2 \\ \mathbf{k}_2 \end{Bmatrix}$$

$$= \mathbf{T}_{3\text{-}4}(\theta_1)\mathbf{T}_{2\text{-}3}(\theta_2)\mathbf{T}_{1\text{-}2}(\theta_3) \begin{Bmatrix} \mathbf{i}_1 \\ \mathbf{j}_1 \\ \mathbf{k}_1 \end{Bmatrix} \tag{1.9}$$

or

$$\begin{Bmatrix} \mathbf{i}_4 \\ \mathbf{j}_4 \\ \mathbf{k}_4 \end{Bmatrix} = \mathbf{T}_{1\text{-}4}(\theta_1, \theta_2, \theta_3) \begin{Bmatrix} \mathbf{i}_1 \\ \mathbf{j}_1 \\ \mathbf{k}_1 \end{Bmatrix}$$

where

$$\mathbf{T}_{1\text{-}4}(\theta_1, \theta_2, \theta_3) \triangleq \mathbf{T}_{3\text{-}4}(\theta_1)\mathbf{T}_{2\text{-}3}(\theta_2)\mathbf{T}_{1\text{-}2}(\theta_3) \tag{1.10}$$

$$= \begin{bmatrix} 1 & 0 & 0 \\ 0 & \cos\theta_1 & \sin\theta_1 \\ 0 & -\sin\theta_1 & \cos\theta_1 \end{bmatrix} \begin{bmatrix} \cos\theta_2 & 0 & -\sin\theta_2 \\ 0 & 1 & 0 \\ \sin\theta_2 & 0 & \cos\theta_2 \end{bmatrix} \begin{bmatrix} \cos\theta_3 & \sin\theta_3 & 0 \\ -\sin\theta_3 & \cos\theta_3 & 0 \\ 0 & 0 & 1 \end{bmatrix}$$

is the relevant direction-cosine matrix. Note also that this direction-cosine matrix is necessarily a function of all three angles. And note that all direction-cosine matrices possess the orthogonality property.

Two major points emerge from the above development. The first is that the orientation between any two coordinate systems, or frames, can be defined in terms of, at most, three angles. This fact is known as *Euler's theorem* (Ref. 2). In the case presented above, the Euler angles θ_1, θ_2, and θ_3, defined by rotations about the X_4 (or X_3), Y_3 (or Y_2), and Z_1 (or Z_2) axes, respectively, were used to define the relative orientation between Frame 1 and Frame 4. Note, further, that both the axis about which the angle is defined, and the order of the rotations are critical. In the example, the order of rotations was $\theta_3 \Rightarrow \theta_2 \Rightarrow \theta_1$ to get from Frame 1 to Frame 4. Conversely, to get from Frame 4 to Frame 1, the rotations required are $-\theta_1 \Rightarrow -\theta_2 \Rightarrow -\theta_3$ about the respective axes.

Other sets of Euler angles can be used as well, and in fact there are 13 possible combinations of such angles (Ref. 2). Different sets of angles lead, of course, to different direction-cosine matrices (DCMs), so care should be used when a DCM is provided. The set of Euler angles used to define the DCM must be known.

The second major point is that the DCMs $\mathbf{T}_{3\text{-}4}(\theta_1)$, $\mathbf{T}_{2\text{-}3}(\theta_2)$, and $\mathbf{T}_{1\text{-}2}(\theta_3)$ provide us with the means to obtain the components of a vector defined in any of the four frames, given the components of the same vector defined in any other of the frames. For example, we previously considered an arbitrary vector **a** expressed in terms of its components in some frame S, or

$$\mathbf{a} = a_{x_S}\mathbf{i}_S + a_{y_S}\mathbf{j}_S + a_{z_S}\mathbf{k}_S$$

Note that this expression can also be written in terms of the following vector (array) inner product.

$$\mathbf{a} = \begin{bmatrix} a_{x_S} & a_{y_S} & a_{z_S} \end{bmatrix} \begin{Bmatrix} \mathbf{i}_S \\ \mathbf{j}_S \\ \mathbf{k}_S \end{Bmatrix} \tag{1.11}$$

Now assume the orientation of the vehicle-fixed frame (Frame V in Figure 1.1) is defined in terms of the three angles θ_1, θ_2, and θ_3 defined above, relative to Frame S. (Or let the S frame correspond to Frame 4 above, and the V frame correspond to Frame 1.) Then the components of **a** in the vehicle-fixed frame V are determined as follows.

Applying Equation (1.9) yields

$$\mathbf{a} = \begin{bmatrix} a_{x_S} & a_{y_S} & a_{z_S} \end{bmatrix} \begin{Bmatrix} \mathbf{i}_S \\ \mathbf{j}_S \\ \mathbf{k}_S \end{Bmatrix} = \begin{bmatrix} a_{x_S} & a_{y_S} & a_{z_S} \end{bmatrix} \mathbf{T}_{1\text{-}4}(\theta_1, \theta_2, \theta_3) \begin{Bmatrix} \mathbf{i}_V \\ \mathbf{j}_V \\ \mathbf{k}_V \end{Bmatrix}$$

$$\triangleq \begin{bmatrix} a_{x_V} & a_{y_V} & a_{z_V} \end{bmatrix} \begin{Bmatrix} \mathbf{i}_V \\ \mathbf{j}_V \\ \mathbf{k}_V \end{Bmatrix} \tag{1.12}$$

and so the following must hold:

$$\begin{bmatrix} a_{x_V} & a_{y_V} & a_{z_V} \end{bmatrix} = \begin{bmatrix} a_{x_S} & a_{y_S} & a_{z_S} \end{bmatrix} \mathbf{T}_{1\text{-}4}(\theta_1, \theta_2, \theta_3)$$

Or by transposing both sides we have the relationship between the components of **a** in the two frames.

$$\begin{bmatrix} a_{x_V} \\ a_{y_V} \\ a_{z_V} \end{bmatrix} = \mathbf{T}^T_{1\text{-}4}(\theta_1, \theta_2, \theta_3) \begin{bmatrix} a_{x_S} \\ a_{y_S} \\ a_{z_S} \end{bmatrix} = \mathbf{T}_{4\text{-}1}(\theta_1, \theta_2, \theta_3) \begin{bmatrix} a_{x_S} \\ a_{y_S} \\ a_{z_S} \end{bmatrix} \tag{1.13}$$

NOTE TO STUDENT

Expressions such as that above are used frequently in texts on dynamics, but they are seldom, if ever, derived from the fundamental expressions developed here between the unit vectors.

In a special case of particular interest in flight dynamics, now let Frame 1 correspond to Frame E fixed on the earth's surface, and Frame 4 correspond to the vehicle-fixed Frame V. Now the three angles θ_1, θ_2, and θ_3 are by convention denoted as the following angles:

$$\phi = \theta_1; \text{ "vehicle bank angle"}$$
$$\theta = \theta_2; \text{ "vehicle pitch angle"}$$
$$\psi = \theta_3; \text{ "vehicle heading angle"}$$

In this case then

$$\mathbf{T}_{1\text{-}4}(\theta_1, \theta_2, \theta_3) = \mathbf{T}_{E\text{-}V}(\phi, \theta, \psi) \tag{1.14}$$

And from Equation (1.10), we have the <u>direction-cosine matrix relating the unit vectors defining the earth- and vehicle-fixed frames</u>:

$$\mathbf{T}_{E\text{-}V}(\phi, \theta, \psi) = \begin{bmatrix} 1 & 0 & 0 \\ 0 & \cos\phi & \sin\phi \\ 0 & -\sin\phi & \cos\phi \end{bmatrix} \begin{bmatrix} \cos\theta & 0 & -\sin\theta \\ 0 & 1 & 0 \\ \sin\theta & 0 & \cos\theta \end{bmatrix} \begin{bmatrix} \cos\psi & \sin\psi & 0 \\ -\sin\psi & \cos\psi & 0 \\ 0 & 0 & 1 \end{bmatrix}$$

$$= \begin{bmatrix} (\cos\theta\cos\psi) & (\cos\theta\sin\psi) & (-\sin\theta) \\ (\sin\phi\sin\theta\cos\psi - \cos\phi\sin\psi) & (\sin\phi\sin\theta\sin\psi + \cos\phi\cos\psi) & (\sin\phi\cos\theta) \\ (\cos\phi\sin\theta\cos\psi + \sin\phi\sin\psi) & (\cos\phi\sin\theta\sin\psi - \sin\phi\cos\psi) & (\cos\phi\cos\theta) \end{bmatrix} \tag{1.15}$$

EXAMPLE 1.1

Components of the Gravity Vector

As a particularly useful application of the above direction-cosine matrix, we will determine the components of the gravity vector in a vehicle-fixed frame, or Frame V. Assuming the earth to be spherical, and its center of mass to be at its geometric center, the vector of gravitational attraction is

$$\mathbf{g} = g\mathbf{k}_E \tag{1.16}$$

where g is the earth's gravitational constant at its surface. In general, this term is a function of distance from the earth, but for flight within the atmosphere the value used is typically taken to be simply the value at sea level, or 32.174 ft/sec^2.

To determine the components of this vector in Frame V, note that

$$\mathbf{g} = \begin{bmatrix} 0 & 0 & g \end{bmatrix} \begin{Bmatrix} \mathbf{i}_E \\ \mathbf{j}_E \\ \mathbf{k}_E \end{Bmatrix} = \begin{bmatrix} 0 & 0 & g \end{bmatrix} \mathbf{T}^T_{E\text{-}V}(\phi, \theta, \psi) \begin{Bmatrix} \mathbf{i}_V \\ \mathbf{j}_V \\ \mathbf{k}_V \end{Bmatrix} = \begin{bmatrix} g_{x_V} & g_{y_V} & g_{z_V} \end{bmatrix} \begin{Bmatrix} \mathbf{i}_V \\ \mathbf{j}_V \\ \mathbf{k}_V \end{Bmatrix} \tag{1.17}$$

Using the transpose of Equation (1.15) and carrying out the vector-matrix multiplication, we have

$$\mathbf{g} = \begin{bmatrix} -g\sin\theta & g\cos\theta\sin\phi & g\cos\theta\cos\phi \end{bmatrix} \begin{Bmatrix} \mathbf{i}_V \\ \mathbf{j}_V \\ \mathbf{k}_V \end{Bmatrix} \tag{1.18}$$

And so the components we seek in the vehicle-fixed frame are

$$\boxed{\begin{aligned} g_{x_V} &= -g\sin\theta \\ g_{y_V} &= g\cos\theta\sin\phi \\ g_{z_V} &= g\cos\theta\cos\phi \end{aligned}} \tag{1.19}$$

These expressions will be utilized in the development of the equations of motion in Chapter 2.

1.4 VECTOR DIFFERENTIATION

Next we review and discuss vector notation and the differentiation of vectors. As noted previously, a vector is a line in space defined in terms of two points in space. As an example, let one point, denoted O, be taken to be the origin of some reference Frame F, and let the other point be denoted as P, which has coordinates (x,y,z) in that frame. Then the vector from O to P is the position vector of P relative to O, which we will denote as $\mathbf{p}$. Of course, from the definition of a vector, it has both magnitude and direction.

When one wants to define the rate of change of any vector, one must first define a reference frame to use to determine change of direction. In the above discussion, **p** was the position vector of the point P with respect to the point O, the origin of Frame F. But one can also consider **p** changing over time with respect to a frame other than F, since again, **p** is just a line in space. For this reason, when we discuss the rate of change of a vector, we will also define the frame used to determine the rate of change of direction. And we will identify this frame using the following notation for the time-derivative of **p**.

$$\frac{d\mathbf{p}}{dt}\Big|_F$$

This expression denotes the rate of change of vector **p** with respect to Frame F.

As a simple example, consider an earth-fixed Frame E with origin O_E located at the base of the Washington Monument. Let the position of O_E with respect to the origin of Frame S (see Figure 1.1) located at the center of the earth be given as a position vector **w** (for Washington). Now recall that Frame S is sun-directed and therefore does not rotate with the earth as it spins on its axis, while Frame E does so rotate. Here the rate of change of **w** with respect to Frame E is zero, while the rate of change of **w** with respect to Frame S is nonzero. That is,

$$\frac{d\mathbf{w}}{dt}\Big|_E = \mathbf{0}$$

while

$$\frac{d\mathbf{w}}{dt}\Big|_S \neq \mathbf{0}$$

The relationship between the above two vector rates of change is, of course, the familiar "chain-rule" expression

$$\frac{d\mathbf{w}}{dt}\Big|_S = \frac{d\mathbf{w}}{dt}\Big|_E + \boldsymbol{\omega}_{E,S} \times \mathbf{w} \tag{1.20}$$

where the vector $\boldsymbol{\omega}_{E,S}$ is the angular velocity of Frame E with respect to Frame S. Of course, in the situation above, $\boldsymbol{\omega}_{E,S} = \boldsymbol{\omega}_{\text{Earth}}\,\mathbf{k}_S$.

It is important to note that the "chain-rule" expression is valid when evaluating the rates of change of any vector with respect to any two frames. The only thing that must be known is the relative angular velocity between the two frames. That is, in general when considering any arbitrary vector **a,** and any two Frames 1 and 2,

$$\frac{d\mathbf{a}}{dt}\Big|_1 = \frac{d\mathbf{a}}{dt}\Big|_2 + \boldsymbol{\omega}_{2,1} \times \mathbf{a} \tag{1.21}$$

Finally, note that $\boldsymbol{\omega}_{2,1}$ always equals $-\boldsymbol{\omega}_{1,2}$. Sometimes students get the impression that Equation (1.21) is only valid if one of the reference frames is inertial, or that this equation is somehow part of the definition of an inertial frame. Neither impression is correct, and this equation can be shown intuitively to be valid just through geometry.

EXAMPLE 1.2

Inertial Velocity of an Aircraft's Wing Tip

Consider an aircraft shown in Figure 1.4, with instantaneous translational velocity **V** and instantaneous rotational velocity $\boldsymbol{\omega}_{V,E}$. We wish to determine the velocity of one of its wing tips relative to the reference frame fixed to the earth's surface, or Frame E in Figure 1.1, and **V** and $\boldsymbol{\omega}_{V,E}$ are the velocities of Frame V relative to Frame E. The vehicle is assumed to be rigid.

The position of the origin of the vehicle-fixed Frame V with respect to the origin of Frame E is defined by the vector $\mathbf{R}_V$, and the position of the wing tip relative to the origin of Frame E is defined by the vector $\mathbf{R}_{\text{tip}}$. The position of the wing tip relative to the origin of the vehicle-fixed Frame V is defined by the vector $\mathbf{r}_{\text{tip}}$. Note that

$$\mathbf{R}_{\text{tip}} = \mathbf{R}_V + \mathbf{r}_{\text{tip}}$$

So the wing tip velocity we seek is

$$\mathbf{V}_{\text{tip}} = \frac{d\mathbf{R}_{\text{tip}}}{dt}\Big|_E$$

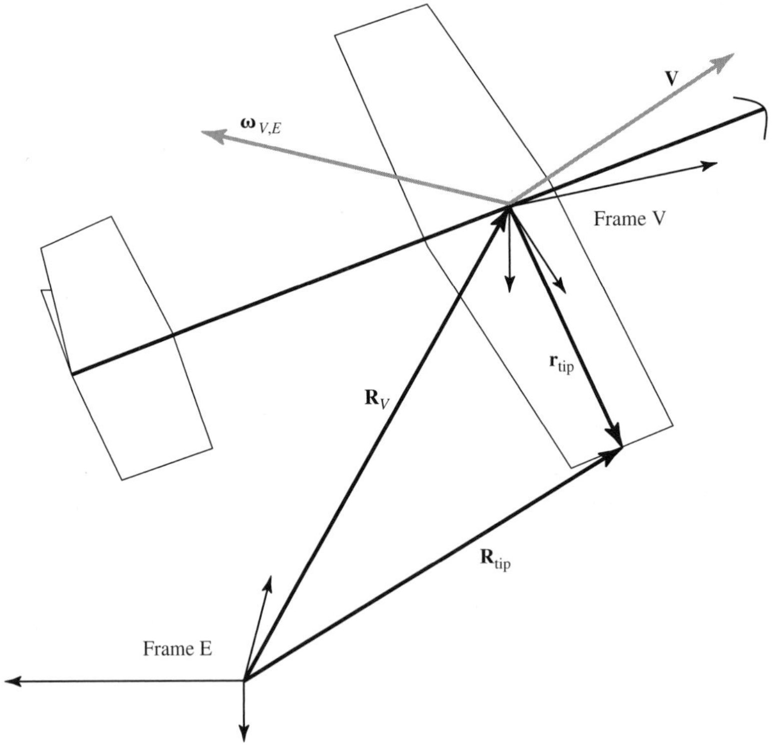

Figure 1.4 Vehicle, reference frames, and wing-tip position vectors.

This velocity can be expressed in terms of the vehicle's translational and rotational velocities as shown here.

$$\frac{d\mathbf{R}_{\text{tip}}}{dt}\Big|_E = \frac{d\mathbf{R}_V}{dt}\Big|_E + \frac{d\mathbf{r}_{\text{tip}}}{dt}\Big|_E$$

$$= \mathbf{V} + \left(\frac{d\mathbf{r}_{\text{tip}}}{dt}\Big|_V + \boldsymbol{\omega}_{V,E} \times \mathbf{r}_{\text{tip}}\right) = \mathbf{V} + \left(\boldsymbol{\omega}_{V,E} \times \mathbf{r}_{\text{tip}}\right) \text{ For rigid vehicle}$$

So the wing-tip velocity relative to Frame E is

$$\frac{d\mathbf{R}_{\text{tip}}}{dt}\Big|_E = \mathbf{V} + \left(\boldsymbol{\omega}_{V,E} \times \mathbf{r}_{\text{tip}}\right) \tag{1.22}$$

EXAMPLE 1.3

Velocity of an Aircraft's Propeller Tip

Consider an aircraft shown in Figure 1.5, with instantaneous translational velocity $\mathbf{V}$ and instantaneous rotational velocity of $\boldsymbol{\omega}_{V,E}$. (To simplify the figure, $\mathbf{V}$ and $\boldsymbol{\omega}_{V,E}$ are not shown here.) We wish to determine the velocity of the propeller tip relative to the

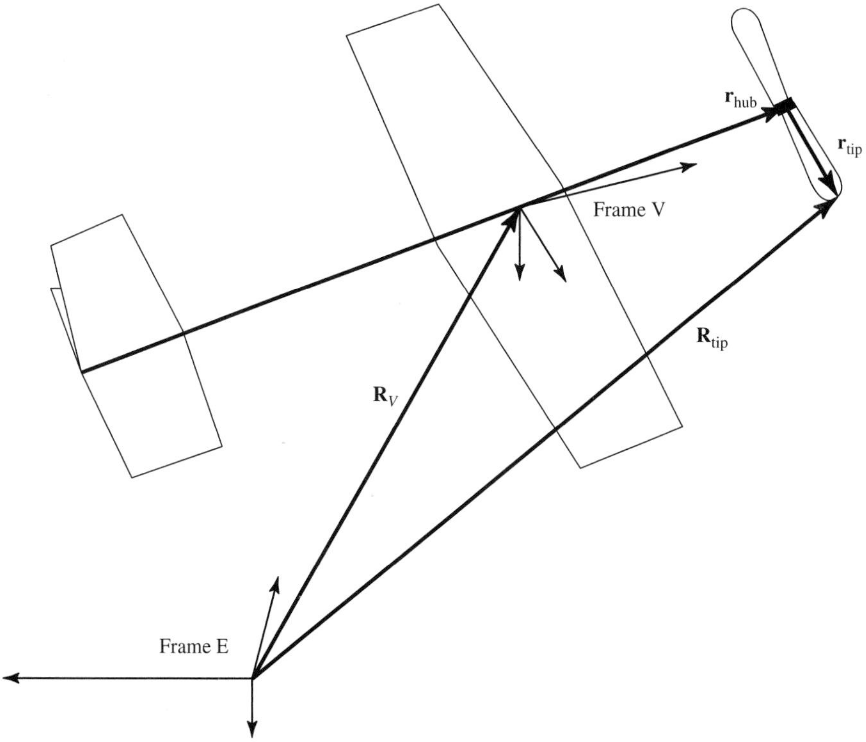

Figure 1.5 Propeller-tip position vectors.

reference frame fixed to the earth's surface, or Frame E, and $\mathbf{V}$ and $\boldsymbol{\omega}_{V,E}$ are defined relative to this frame. The vehicle and propeller are assumed rigid.

The position of the origin of the vehicle-fixed Frame V with respect to the origin of Frame E is again defined by the vector $\mathbf{R}_V$, and the position of the propeller tip relative to the origin of Frame E is defined by the vector $\mathbf{R}_{\text{tip}}$. The position of the propeller tip relative to the origin of the vehicle-fixed Frame V is defined by the vectors $\mathbf{r}_{\text{hub}}$ and $\mathbf{r}_{\text{tip}}$. The vector $\mathbf{r}_{\text{hub}}$ defines the position of the propeller hub relative to the origin of Frame V, and the vector $\mathbf{r}_{\text{tip}}$ defines the position of the propeller tip relative to the propeller hub and rotates with the propeller. So then

$$\mathbf{R}_{\text{tip}} = \mathbf{R}_V + \mathbf{r}_{\text{hub}} + \mathbf{r}_{\text{tip}}$$

and the propeller tip velocity we seek is

$$\mathbf{V}_{\text{tip}} = \frac{d\mathbf{R}_{\text{tip}}}{dt}\bigg|_E$$

Before we determine this velocity, it will be helpful to introduce a third intermediate reference frame, Frame P (not shown), with origin at the propeller hub, and rotating with the propeller such that $\mathbf{r}_{\text{tip}}$ is fixed in this frame. The angular velocity of Frame P with respect to Frame V, or the propeller speed, is denoted $\boldsymbol{\omega}_{P,V}$. The desired velocity can now be expressed in terms of the vehicle and propeller translational and rotational velocities as shown here.

$$\begin{aligned}
\frac{d\mathbf{R}_{\text{tip}}}{dt}\bigg|_E &= \frac{d\mathbf{R}_V}{dt}\bigg|_E + \frac{d\mathbf{r}_{\text{hub}}}{dt}\bigg|_E + \frac{d\mathbf{r}_{\text{tip}}}{dt}\bigg|_E \\
&= \mathbf{V} + \left(\frac{d\mathbf{r}_{\text{hub}}}{dt}\bigg|_V + \boldsymbol{\omega}_{V,E} \times \mathbf{r}_{\text{hub}}\right) + \left(\frac{d\mathbf{r}_{\text{tip}}}{dt}\bigg|_V + \boldsymbol{\omega}_{V,E} \times \mathbf{r}_{\text{tip}}\right) \\
&= \mathbf{V} + \left(\boldsymbol{\omega}_{V,E} \times \mathbf{r}_{\text{hub}}\right) + \left(\left(\frac{d\mathbf{r}_{\text{tip}}}{dt}\bigg|_P + \boldsymbol{\omega}_{P,V} \times \mathbf{r}_{\text{tip}}\right) + \boldsymbol{\omega}_{V,E} \times \mathbf{r}_{\text{tip}}\right) \\
&= \mathbf{V} + \left(\boldsymbol{\omega}_{V,E} \times \mathbf{r}_{\text{hub}}\right) + \left(\left(\boldsymbol{\omega}_{P,V} + \boldsymbol{\omega}_{V,E}\right) \times \mathbf{r}_{\text{tip}}\right)
\end{aligned}$$

So the propeller tip velocity relative to Frame E is

$$\frac{d\mathbf{R}_{\text{tip}}}{dt}\bigg|_E = \mathbf{V} + \left(\boldsymbol{\omega}_{V,E} + \mathbf{r}_{\text{hub}}\right) + \left(\boldsymbol{\omega}_{V,E} + \boldsymbol{\omega}_{P,V}\right) \times \mathbf{r}_{\text{tip}} \tag{1.23}$$

1.5 NEWTON'S SECOND LAW

Everyone recalls learning about Newton's second law, usually stated as

$$\mathrm{F} = \mathrm{ma}$$

or force equals mass times acceleration. But in flight dynamics we will need a more rigorous statement of this law to accomplish our objectives. First, the acceleration must be acceleration with respect to an inertial frame. And second, we will need this law stated in vector form.

Using our vector-derivative notation, define $\mathbf{p}$ to be the position vector of a point mass m (with the mass constant for now) relative to the origin of an inertial Frame I. And define $\mathbf{F}$ as the external-force vector acting on m. Then with the mass constant, we can more rigorously restate Newton's second law (for constant mass) as

$$\boxed{\mathbf{F} = m\frac{d}{dt}\bigg|_I\left(\frac{d\mathbf{p}}{dt}\bigg|_I\right)} \tag{1.24}$$

Note here that

$$\frac{d\mathbf{p}}{dt}\bigg|_I \triangleq \mathbf{V}_I \tag{1.25}$$

where $\mathbf{V}_I$ is the <u>inertial</u> velocity (a vector quantity) of m. And

$$\frac{d\mathbf{V}_I}{dt}\bigg|_I \tag{1.26}$$

is m's <u>inertial</u> acceleration. Finally, $m\mathbf{V}_I$ is, of course, the translational momentum of m.

Equation (1.24) is Newton's second law in translational form. We can derive Newton's second law in rotational form by crossing the position vector $\mathbf{p}$ into both sides of Equation (1.24). Or

$$\mathbf{p} \times \mathbf{F} = \mathbf{p} \times m\frac{d}{dt}\bigg|_I\left(\frac{d\mathbf{p}}{dt}\bigg|_I\right) = m\left(\mathbf{p} \times \frac{d}{dt}\bigg|_I\left(\frac{d\mathbf{p}}{dt}\bigg|_I\right)\right) \tag{1.27}$$

But note that

$$\frac{d}{dt}\bigg|_I\left(\mathbf{p} \times \frac{d\mathbf{p}}{dt}\bigg|_I\right) = \left(\frac{d\mathbf{p}}{dt}\bigg|_I \times \frac{d\mathbf{p}}{dt}\bigg|_I\right) + \left(\mathbf{p} \times \frac{d}{dt}\bigg|_I\left(\frac{d\mathbf{p}}{dt}\bigg|_I\right)\right)$$

and that

$$\frac{d\mathbf{p}}{dt}\bigg|_I \times \frac{d\mathbf{p}}{dt}\bigg|_I = \mathbf{0}$$

since any vector crossed into itself is zero. So we therefore can write Equation (1.27) as

$$\boxed{\mathbf{p} \times \mathbf{F} = m\frac{d}{dt}\bigg|_I\left(\mathbf{p} \times \frac{d\mathbf{p}}{dt}\bigg|_I\right)} \tag{1.28}$$

Now noting that $\mathbf{p} \times \mathbf{F} = \mathbf{M}$, the external moment (vector) taken about the origin of the inertial frame, and that $m\left(\mathbf{p} \times \frac{d\mathbf{p}}{dt}\Big|_I\right) = \mathbf{H}$, the angular momentum (vector) about the inertial origin, we have the desired result for <u>Newton's law in rotational form</u>.

$$\boxed{\mathbf{M} = \frac{d\mathbf{H}}{dt}\bigg|_I} \tag{1.29}$$

Or the rate of change of the angular momentum of m about the inertial origin is equal to the external moment acting on m taken about the inertial origin.

EXAMPLE 1.4

Equations of Motion of a Pendulum

To demonstrate all of the above concepts in the context of a comprehensive example, we will derive the equation of motion of a simple pendulum. The results of this derivation will be used in Section 1.6 in the remaining steps of small-perturbation analysis.

The schematic is shown in Figure 1.6, with an inertial reference frame, Frame I, and a pendulum-fixed frame, Frame M, that is translating and rotating with the pendulous mass m. The rod, of length l, is taken as rigid and massless. If the rod is pivoting in the absence of friction, the forces (vectors) acting on m are gravity, $m\mathbf{g}$, and $\mathbf{C}$, the internal force on m carried by the rod.

Newton's law in translational form will be employed, although the rotational form could be used as well. Letting $\mathbf{r}$ denote the position vector of the pendulous point mass m relative to the origin of the inertial Frame I we have

$$\frac{d}{dt}\Big|_I\left(m\frac{d\mathbf{r}}{dt}\Big|_I\right) = m\mathbf{g} + \mathbf{C}$$

Note that we have brought m inside the parenthesis to reveal the translational momentum, but since m is constant, m can be moved in front of the terms on the left-hand side of the equation, as in Equation (1.24). Also, from trigonometry we can express $\mathbf{r}$, $\mathbf{g}$, and $\mathbf{C}$ in terms of their components in the inertial frame as shown below.

$$\begin{aligned}
\mathbf{r} &= l\sin\theta\mathbf{i}_I + l\cos\theta\mathbf{k}_I = \begin{bmatrix} l\sin\theta & l\cos\theta \end{bmatrix}\begin{Bmatrix} \mathbf{i}_I \\ \mathbf{k}_I \end{Bmatrix} \\
\mathbf{g} &= g\mathbf{k}_I = \begin{bmatrix} 0 & g \end{bmatrix}\begin{Bmatrix} \mathbf{i}_I \\ \mathbf{k}_I \end{Bmatrix} \\
\mathbf{C} &= -C\sin\theta\mathbf{i}_I - C\cos\theta\mathbf{k}_I
\end{aligned} \tag{1.30}$$

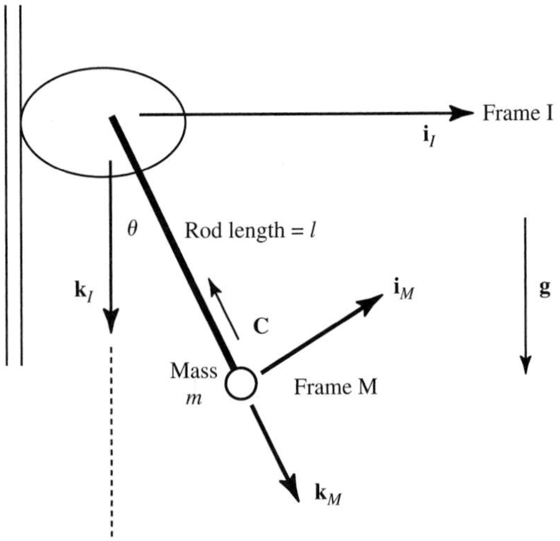

Figure 1.6 Simple pendulum.

Before proceeding further, the direction-cosine matrix relating Frames I and M will be identified. In this case, one can show that

$$\begin{Bmatrix} \mathbf{i}_M \\ \mathbf{k}_M \end{Bmatrix} = \begin{bmatrix} \cos\theta & -\sin\theta \\ \sin\theta & \cos\theta \end{bmatrix} \begin{Bmatrix} \mathbf{i}_I \\ \mathbf{k}_I \end{Bmatrix} = \mathbf{T}_{I\text{-}M}(\theta) \begin{Bmatrix} \mathbf{i}_I \\ \mathbf{k}_I \end{Bmatrix}$$

The inertial velocity vector may now be found easily using Equation (1.21), or

$$\frac{d\mathbf{r}}{dt}\Big|_I = \frac{d\mathbf{r}}{dt}\Big|_M + \boldsymbol{\omega}_{M,I} \times \mathbf{r} \tag{1.31}$$

where

$$\boldsymbol{\omega}_{M,I} = \dot{\theta}\mathbf{j}_I = \dot{\theta}\mathbf{j}_M \tag{1.32}$$

(Note that $\mathbf{j}_I$ and $\mathbf{j}_M$ are unit vectors directed out of the page.) Since the rod is rigid,

$$\frac{d\mathbf{r}}{dt}\Big|_M = \mathbf{0} \tag{1.33}$$

and therefore the inertial velocity vector is

$$\frac{d\mathbf{r}}{dt}\Big|_I = \boldsymbol{\omega}_{M,I} \times \mathbf{r} = l\dot{\theta}(\cos\theta\mathbf{i}_I - \sin\theta\mathbf{k}_I) \tag{1.34}$$

The inertial acceleration of m is

$$\frac{d}{dt}\Big|_I\left(\frac{d\mathbf{r}}{dt}\Big|_I\right) \triangleq \frac{d^2\mathbf{r}}{dt^2}\Big|_I = \frac{d\boldsymbol{\omega}_{M,I}}{dt}\Big|_I \times \mathbf{r} + \boldsymbol{\omega}_{M,I} \times (\boldsymbol{\omega}_{M,I} \times \mathbf{r}) \tag{1.35}$$

Noting that $\frac{d\boldsymbol{\omega}_{M,I}}{dt}\Big|_I = \ddot{\theta}\mathbf{j}_I$ yields

$$\frac{d^2\mathbf{r}}{dt^2}\Big|_I = l\ddot{\theta}(\cos\theta\mathbf{i}_I - \sin\theta\mathbf{k}_I) - l\dot{\theta}^2(\sin\theta\mathbf{i}_I + \cos\theta\mathbf{k}_I) \tag{1.36}$$

Now all the necessary vector quantities are available and expressed in terms of their components in Frame I, and they could be substituted directly into the vector force equation. It is instructive, however, to express these same vectors in terms of their components in Frame M, and then substitute them into the vector force equation. To accomplish this, the direction-cosine matrix is employed.

Addressing the gravitational vector first we have

$$\mathbf{g} = g\mathbf{k}_I = [0 \quad g]\begin{Bmatrix} \mathbf{i}_I \\ \mathbf{k}_I \end{Bmatrix} = [0 \quad g]\mathbf{T}^T_{I\text{-}M}(\theta)\begin{Bmatrix} \mathbf{i}_M \\ \mathbf{k}_M \end{Bmatrix} \tag{1.37}$$

So the vector $\mathbf{g}$, in terms of its components in Frame M, is

$$\mathbf{g} = [-g\sin\theta \quad g\cos\theta]\begin{Bmatrix} \mathbf{i}_M \\ \mathbf{k}_M \end{Bmatrix} = -g\sin\theta\mathbf{i}_M + g\cos\theta\mathbf{k}_M \tag{1.38}$$

Similarly, for the force vector $\mathbf{C}$ we have

$$\mathbf{C} = -C\sin\theta\mathbf{i}_I - C\cos\theta\mathbf{k}_I = [-C\sin\theta \quad -C\cos\theta]\begin{Bmatrix} \mathbf{i}_I \\ \mathbf{k}_I \end{Bmatrix}$$

$$= [-C\sin\theta \quad -C\cos\theta]\mathbf{T}^T_{I\text{-}M}(\theta)\begin{Bmatrix} \mathbf{i}_M \\ \mathbf{k}_M \end{Bmatrix} \tag{1.39}$$

or

$$\mathbf{C} = \begin{bmatrix} 0 & -C \end{bmatrix} \begin{Bmatrix} \mathbf{i}_M \\ \mathbf{k}_M \end{Bmatrix} = -C\mathbf{k}_M \tag{1.40}$$

And for the acceleration vector we have

$$\begin{aligned} \frac{d^2\mathbf{r}}{dt^2}\Big|_I &= l(\ddot{\theta}\cos\theta - \dot{\theta}^2\sin\theta)\mathbf{i}_I - l(\ddot{\theta}\sin\theta + \dot{\theta}^2\cos\theta)\mathbf{k}_I \\ &= \begin{bmatrix} l(\ddot{\theta}\cos\theta - \dot{\theta}^2\sin\theta) & -l(\ddot{\theta}\sin\theta + \dot{\theta}^2\cos\theta) \end{bmatrix} \begin{Bmatrix} \mathbf{i}_I \\ \mathbf{k}_I \end{Bmatrix} \\ &= \begin{bmatrix} l(\ddot{\theta}\cos\theta - \dot{\theta}^2\sin\theta) & -l(\ddot{\theta}\sin\theta + \dot{\theta}^2\cos\theta) \end{bmatrix} \mathbf{T}^T_{I\text{-}M}(\theta) \begin{Bmatrix} \mathbf{i}_M \\ \mathbf{k}_M \end{Bmatrix} \end{aligned} \tag{1.41}$$

so

$$\frac{d^2\mathbf{r}}{dt^2}\Big|_I = l\ddot{\theta}\mathbf{i}_M - l\dot{\theta}^2\mathbf{k}_M \tag{1.42}$$

Of course, this is recognizable as the familiar two components of the acceleration of a point moving along a circular path—the tangential and the centripital.

Now substituting this acceleration vector, along with **C** and **g**, into the vector force equation yields

$$\left(ml\ddot{\theta}\mathbf{i}_M - ml\dot{\theta}^2\mathbf{k}_M\right) = -C\mathbf{k}_M + \left(-mg\sin\theta\mathbf{i}_M + mg\cos\theta\mathbf{k}_M\right)$$

Equating $\mathbf{i}_M$ and $\mathbf{k}_M$ components yields two differential equations:

$$\ddot{\theta} + (g/l)\sin\theta = 0 \tag{1.43}$$

and

$$C = ml\dot{\theta}^2 - mg\cos\theta \tag{1.44}$$

The first of these two is the desired equation of motion. The second is an auxiliary equation, from which the magnitude of the force carried by the rod may be determined if desired.

1.6 SMALL PERTURBATION ANALYSIS REVISITED

As noted in Section 1.1, small-perturbation analysis involves the investigation of a nonlinear system's dynamic behavior (stability, etc.) in the neighborhood of a reference condition. There are five steps in perturbation analysis. These steps are repeated below.

1. Derive the (nonlinear) equations governing the behavior of the system.
2. Make a formal change of variables, expressing all the degrees of freedom governed by the equations of motion in terms of a reference condition, plus deviations from this reference condition. Express the governing equations found in Step 1 in terms of these new variables.
3. By making a small-perturbation assumption, extract two sets of equations from the results from Step 2. One set (Reference Set) is expressed only in terms of the reference state, and one set (Perturbation Set) is expressed in terms of the reference state plus the deviation from the reference.

4. Using the Reference Set of equations, characterize the reference condition(s) of interest.
5. Using the Perturbation Set and the reference conditions selected in Step 4, analyze the behavior of the system's perturbation dynamics in the neighborhood of the reference conditions(s).

Now consider the simple pendulum again, for which the equation of motion was given in Equation (1.43). (Step 1 has already been completed.) The single degree of freedom of the pendulum (its rotational displacement about the pivot) is governed by the nonlinear equation

$$\ddot{\theta} + (g/l)\sin\theta = 0$$

Following Step 2, make the change of variable by redefining θ as follows:

$$\theta = \Theta_0 + \theta_p \tag{1.45}$$

where Θ_0 is some reference angle, to be selected later, and θ_p represents some deviation from the reference angle.

Formally substituting the above change of variable into the governing equation yields

$$(\ddot{\Theta}_0 + \ddot{\theta}_p) + (g/l)\sin(\Theta_0 + \theta_p) = 0 \tag{1.46}$$

Using the trigonometric identity for the sine of the sum of two angles yields

$$(\ddot{\Theta}_0 + \ddot{\theta}_p) + (g/l)(\sin\Theta_0\cos\theta_p + \cos\Theta_0\sin\theta_p) = 0 \tag{1.47}$$

This completes Step 2.

In Step 3 we make the small-perturbation assumption, or we assume (and always remember later that we did this) that θ_p is sufficiently small such that $\cos\theta_p \cong 1$ and $\sin\theta_p \cong \theta_p$. Under this assumption, the governing equation now becomes

$$\begin{aligned}&(\ddot{\Theta}_0 + \ddot{\theta}_p) + (g/l)(\sin\Theta_0 + (\cos\Theta_0)\theta_p)\\&= (\ddot{\Theta}_0 + (g/l)\sin\Theta_0) + (\ddot{\theta}_p + (g/l)(\cos\Theta_0)\theta_p) = 0\end{aligned} \tag{1.48}$$

(If the equations of motion included higher-order terms in the perturbations, such as products of perturbation quantities, these higher-order terms would be assumed small, and neglected at this point.) If we require the reference variable, or the system's reference condition, to satisfy its original governing equation (and why shouldn't it?), then

$$\ddot{\Theta}_0 + (g/l)\sin\Theta_0 = 0 \tag{1.49}$$

Equation (1.49) is the Reference Set of equations (only one equation in this case). Note that as required, this equation only involves the reference variables.

Now, if the Reference Set is satisfied, then Equation (1.48) becomes

$$\ddot{\theta}_p + (g/l)(\cos\Theta_0)\theta_p = 0 \tag{1.50}$$

And Equation (1.50) is the Perturbation Set of equations called out in Step 3 above. This equation governs the deviations from the reference, or perturbation angle. Note also that the term $(g/l)\cos\Theta_0$ is a coefficient of θ_p, the variable governed by the equation, and that Equation (1.50) is linear in θ_p! Finally, it is important to remember as we use Equation (1.50) that it is only valid as long as the small-perturbation assumption used to derive it is satisfied.

Moving to Step 4, we use Equation (1.49), or the Reference Set, to analyze and select reference conditions of interest. If we are interested in system stability, we invariably want to consider *equilibrium* reference conditions of the system, which means the system is not undergoing accelerations. So if equilibrium of the reference condition implies no accelerations, then $\ddot{\Theta}_0 = 0$, and Equation (1.49) becomes simply $\sin\Theta_0 = 0$. This equation has an uncountable number of solutions, $\Theta_0 = n\pi$ with n an integer including zero. (Note that often students attempting to perform small-perturbation analysis quickly assume that $\Theta_0 = 0$, and forget or do not consider the other equilibrium conditions. Careful analysis of the Reference Set helps to assure that we find all possible equilibrium conditions.)

Now we can move to Step 5, and use the Perturbation Set to analyze the behavior of the system in the neighborhood of the two unique equilibrium reference conditions, $\Theta_0 = 0$ and π. First consider the condition $\Theta_0 = 0$. In this case, Equation (1.50) becomes

$$\ddot{\theta}_p + (g/l)\theta_p = 0 \tag{1.51}$$

This equation is second order, linear, with constant coefficients, and linear analysis techniques may be employed.

Recall from the theory of ordinary differential equations that the solution of Equation (1.51) depends on its characteristic roots. These two characteristic roots, or the roots of the characteristic polynomial $s^2 + (g/l) = 0$, are complex since g/l is positive, and are

$$s_{1,2} = \pm j\sqrt{g/l} \tag{1.52}$$

where $j = \sqrt{-1}$. The homogeneous solution to Equation (1.51) is

$$\theta(t) = C_1 e^{s_1 t} + C_2 e^{s_2 t} \tag{1.53}$$

with C_1 and C_2 the constants of integration, which depend on the system's initial conditions. The fact that the characteristic roots $s_{1,2}$ are complex with real parts equal to zero tells us that the behavior of the system (pendulum) in the neighborhood of the equilibrium condition $\Theta_0 = 0$ is characterized by undamped oscillations with a natural frequency of $\sqrt{g/l}$. But the system does not diverge from this equilibrium condition, and hence it is not exponentially unstable.

In contrast, consider the behavior of the system in the neighborhood of the other unique equilibrium condition, or $\Theta_0 = \pi$. Equation (1.50) governing the small deviations from this equilibrium condition becomes

$$\ddot{\theta}_p - (g/l)\theta_p = 0 \tag{1.54}$$

since $\cos\pi = -1$. The two characteristic roots of this equation, or roots of the characteristic polynomial $s^2 - (g/l) = 0$, are now real, and are

$$s_{1,2} = \pm\sqrt{g/l} \tag{1.55}$$

The solution of Equation (1.54) is again of the form

$$\theta_p = C_1 e^{s_1 t} + C_2 e^{s_2 t} \tag{1.56}$$

And, therefore, since one of the characteristic roots $s_{1,2}$ is real and positive, θ_p must diverge, or the system is unstable in the neighborhood of this equilibrium condition.

In summary, we have applied perturbation theory step by step to analyze the dynamic behavior of the pendulum in the neighborhood of its two unique equilibrium conditions. We found that the pendulum was unstable in the neighborhood of $\Theta_0 = \pi$, and oscillatory but not divergent in the neighborhood of $\Theta_0 = 0$. Note that perturbation analysis does not help us analyze the system's dynamics at all values of θ, only for small angular displacements from the reference conditions. For large displacements, we must resort to a full nonlinear system simulation. This is why nonlinear, real-time piloted simulations and flight tests of aircraft are required.

In using small-perturbation theory for the study of atmospheric flight dynamics, the steps described above require a great deal of analysis, and it is possible to lose sight of which step one is addressing. That is one reason the pendulum example is useful: It is easier to keep track of progress. But as a preview, the relationships between the steps of perturbation analysis and the flight-dynamics topics are listed below.

1. Derivations of the nonlinear equations of motion (Step 1) are presented in Chapters 2 and 4.
2. Development of the Reference Set and Perturbation Set of equations (Steps 2 and 3) is presented in Chapters 2 and 4, and in Chapters 6, 7, and 8.
3. Analysis of the Reference Set of equations and characterization of the reference conditions (Step 4) is presented in the discussions of quasi-steady flight in Chapter 9.
4. Analysis of the perturbation dynamics using the Perturbation Set of equations (Step 5) is the topic of discussion in Chapters 10-13.
5. Finally, linear and nonlinear simulation are discussed in Chapter 8.

1.7 Summary

Several key concepts were reviewed in this chapter, including types of coordinate systems, kinematics and direction-cosine matrices, vector differentiation, Newton's laws, and, very importantly, small-perturbation theory for the study of nonlinear dynamic systems. Also presented was a review of vector differential calculus, along with the introduction of notation found helpful in vector differentiation. Especially important to the student is that the concepts in this chapter, particularly the material on small-perturbation theory, provide a rigorous overarching framework for developing and presenting the subject of flight dynamics. This framework will be applied throughout the book.

1.8 Problems

1.1 Consider the vehicle discussed in Example 1.2. Assume that the vehicle's translational and rotational velocities $\mathbf{V}$ and $\boldsymbol{\omega}_{V,E}$ are constant. Show that the acceleration of the wing tip with respect to Frame E is $\mathbf{a}_{\text{tip}} = \boldsymbol{\omega}_{V,E} \times (\boldsymbol{\omega}_{V,E} \times \mathbf{r}_{\text{tip}})$.

1.2 Consider a tilt-rotor vehicle shown in Figure 1.7. The engines and rotors on the ends of the wings are being rotated relative to the wings at the rate $\boldsymbol{\omega}_{\text{tilt}}$. Using the vehicle and propeller velocities discussed in Example 1.3, and appropriately defined position vectors, show that the velocity of the tip of a propeller may be expressed as

$$\mathbf{V}_{\text{tip}} = \mathbf{V} + \left(\boldsymbol{\omega}_{V,E} \times \mathbf{r}_{\text{engine}}\right) + \left(\boldsymbol{\omega}_{V,E} + \boldsymbol{\omega}_{\text{tilt}}\right) \times \mathbf{r}_{\text{hub}} + \left(\boldsymbol{\omega}_{V,E} + \boldsymbol{\omega}_{\text{tilt}} + \boldsymbol{\omega}_{\text{prop}}\right) \times \mathbf{r}_{\text{tip}}$$

where $\boldsymbol{\omega}_{\text{prop}}$ is the propeller speed, or rate of rotation about its hub.

1.3 Consider a manned simulator consisting of a sphere on the end of an arm that rotates at a velocity $\omega(t)$ relative to the ground. Inside the sphere is the seated subject of the experiment, and under the seat is a three-axis accelerometer measuring the acceleration environment experienced by the subject, or the acceleration of the seat with respect to a ground-based coordinate frame (assumed inertial). Sketch the simulator showing the relevant coordinate frame(s), vectors, velocities, and accelerations, and determine the components of the inertial acceleration vector experienced by the subject.

1.4 A point mass, with constant mass m and instantaneous position vector $\mathbf{p}$ relative to the origin of an inertial Frame I, is acted upon by an external force $\mathbf{f}$ as well

Figure 1.7 NASA XV-15 tilt-rotor vehicle. *(Photograph courtesy of NASA Dryden Flight Research Center.)*

as gravity **g**. Assume that another coordinate Frame M has its origin fixed to *m*, and has instantaneous orientation relative to the inertial frame given by the Euler angles ϕ, θ, and ψ as defined in this chapter. Let the inertial velocity of *m* be denoted as **v**, a vector with components in Frame M given as

$$\frac{d\mathbf{p}}{dt}\Big|_I = \mathbf{v} = \begin{bmatrix} v_{x_M} & v_{y_M} & v_{z_M} \end{bmatrix} \begin{Bmatrix} \mathbf{i}_M \\ \mathbf{j}_M \\ \mathbf{k}_M \end{Bmatrix}$$

Likewise, assume **f** has components in Frame M given as f_{x_M}, f_{y_M}, and f_{z_M}. Also assume that Frame M is rotating with respect to Frame I at the rate $\boldsymbol{\omega}_{M,I}$, which has components in Frame M given as ω_{x_M}, ω_{y_M}, and ω_{z_M}. Using Newton's second law, derive the three scalar differential equations governing v_{x_M}, v_{y_M}, and v_{z_M}.

1.5 Consider the pendulum discussed in Example 1.4. Now add a torsional spring at the pivot point such that a moment $\mathbf{M}_S$ is applied to the rod about the pivot, where

$$\mathbf{M}_S = K_T\left(\frac{\pi}{2} - \theta\right)\mathbf{j}_I$$

with K_T the torsional spring constant.

a. Show that the nonlinear equation of motion is now

$$ml^2\ddot{\theta} + K_T\left(\theta - \frac{\pi}{2}\right) + mgl\sin\theta = 0$$

Hint: You can use whatever method you like, but one way is to simply sum moments about the pivot and include the "inertial force" F = ma acting on the mass (with the acceleration appropriately defined).

b. Using small-perturbation theory, derive the Reference and Perturbation Sets of equations associated with the pendulum's equation of motion.

c. Find all equilibrium conditions for this pendulum.

d. Assess the stability of each of the equilibrium conditions.

1.6 Consider an inverted pendulum on a cart, acted on by a force $F(t)$, as shown in Figure 1.8.

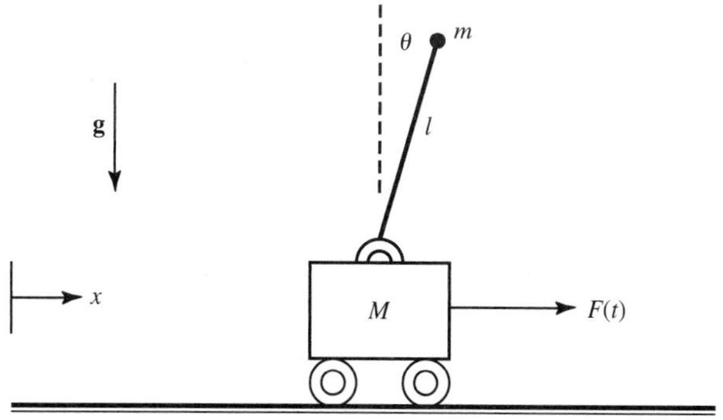

Figure 1.8 Inverted pendulum on rolling cart.

a. Show that ignoring friction, the two equations of motion governing this system are

$$l\ddot{\theta} + \ddot{x}\cos\theta - g\sin\theta = 0$$

$$(M + m)\ddot{x} + ml\left(\ddot{\theta}\cos\theta - \dot{\theta}^2\sin\theta\right) = F(t)$$

b. Determine if this system has an equilibrium condition for which the reference force $F_0 = C$ (a constant other than zero).

References

1. Vidyasagar, M.: *Nonlinear Systems Analysis,* Prentice-Hall, Upper Saddle River, NJ, 1978, Section 5.4.

2. Wertz, James, ed.: *Spacecraft Attitude Determination and Control,* Reidel Publishing, Dordrecht, the Netherlands, 1986, Chapter 12.

CHAPTER 2

Equations of Motion of the Rigid Vehicle

Chapter Roadmap: *For a first course in flight dynamics, Sections 2.1—2.3 are critical, not only for the development of the equations of motion (for a flat, nonrotating earth), but also for introducing important notation and terminology. The remaining sections are typically treated in more advanced courses. However, it is useful for the student in a first course to be aware of the existence of the more advanced material.*

We will now present in detail the derivation of the differential equations of motion for a rigid vehicle, initially making the assumption of a flat, nonrotating earth. While these "flat-earth" equations of motion are well known and available from many texts, it is important to understand the details of the derivation. The student will then better appreciate the assumptions made, and will be better prepared to address the more unfamiliar cases discussed later in this chapter, such as the effects of a rotating, spherical earth. The student will also be prepared to modify the equations of motion as needed to model future novel vehicles that may be encountered.

We recommend that the student review the material in Chapter 1 prior to beginning this chapter, as we will be applying all the techniques discussed in Chapter 1, including various coordinate frames, direction-cosine matrices, vector differentiation, and Newton's laws in vector form, to rigorously derive the equations of motion. If the student is not sufficiently familiar with these techniques, this chapter may be quite difficult to follow.

Finally, the modeling of the (aerodynamic and propulsive) forces acting on the vehicle, for incorporation into the equations derived in this chapter, will be presented later in Chapter 6.

2.1 VECTOR EQUATIONS OF MOTION—FLAT EARTH

Under the flat, nonrotating-earth assumption, a coordinate frame fixed to the earth's surface (and therefore rotating with the earth) is taken to be inertial. Thus, the effect of this rotation is being ignored. In addition, gravity

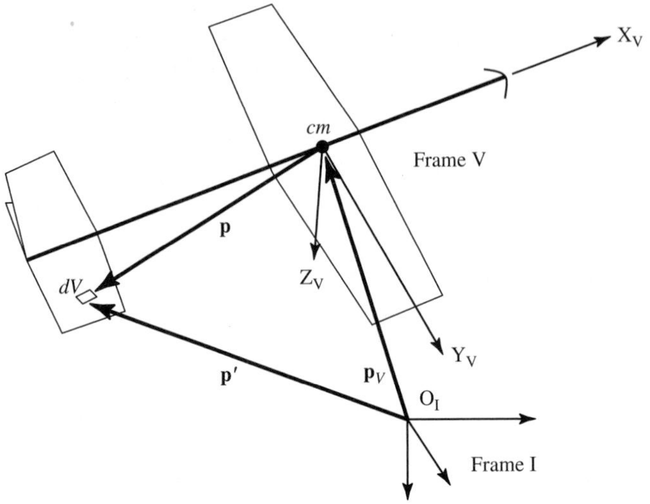

Figure 2.1 Inertial and vehicle-fixed frames and position vectors.

is taken to be constant, and thus independent of altitude or distance from the earth.

Consider a vehicle as shown in Figure 2.1. Let the inertial Frame I be fixed to the earth's surface, and let Frame V, for vehicle, be some vehicle-fixed frame defined in more detail later. Also consider an infinitesimal volume of the vehicle's material, dV, having position $\mathbf{p}'$ with respect to O_I, the origin of Frame I. If the density of the material at this location in the vehicle is denoted ρ_V, and the mass $dm = \rho_V dV$, Newton's second law may be applied to this mass.

Let the forces acting on the mass dm consist of gravitational plus some other external force $d\mathbf{f}_{ext}$, which is only mechanical in nature and does not include any field effects such as electromagnetic forces. Therefore, for the mass dm we have the following two expressions for Newton's second law (from Equations (1.24) and (1.28)). The equations are expressed in terms of the translational and rotational momentum of the mass dm, respectively.

$$\frac{d}{dt}\Big|_I \left(\rho_V \frac{d\mathbf{p}'}{dt}\Big|_I dV \right) = \rho_V \mathbf{g} dV + d\mathbf{f}_{ext}$$

$$\frac{d}{dt}\Big|_I \left(\mathbf{p}' \times \rho_V \frac{d\mathbf{p}'}{dt}\Big|_I dV \right) = \mathbf{p}' \times \rho_V \mathbf{g} dV + \mathbf{p}' \times d\mathbf{f}_{ext}$$

By integrating the above expressions over all the infinitesimal volumes dV in the vehicle, one obtains

$$\int_{\mathrm{Vol}} \frac{d}{dt}\Big|_I \left(\rho_V \frac{d\mathbf{p}'}{dt}\Big|_I dV \right) = \int_{\mathrm{Vol}} \rho_V \mathbf{g} dV + \int_{\mathrm{Vol}} d\mathbf{f}_{ext}$$

and

$$\int_{\text{Vol}} \frac{d}{dt}\Big|_I \left(\mathbf{p}' \times \rho_V \frac{d\mathbf{p}'}{dt}\Big|_I dV\right) = \int_{\text{Vol}} \mathbf{p}' \times \rho_V \mathbf{g} dV + \int_{\text{Vol}} \mathbf{p}' \times d\mathbf{f}_{\text{ext}}$$

All internal forces in the vehicle material will now be assumed in equilibrium, so the only net forces on the vehicle must act on the surface of the vehicle. As a result, the last terms on the right-hand sides of the above two expressions become integrals over the vehicle surface. This now yields the two fundamental equations of motion.

$$\int_{\text{Vol}} \frac{d}{dt}\Big|_I \left(\rho_V \frac{d\mathbf{p}'}{dt}\Big|_I dV\right) = \int_{\text{Vol}} \rho_V \mathbf{g} dV + \int_{\text{Surface}} d\mathbf{f}_{\text{ext}} \tag{2.1}$$

$$\int_{\text{Vol}} \frac{d}{dt}\Big|_I \left(\mathbf{p}' \times \rho_V \frac{d\mathbf{p}'}{dt}\Big|_I dV\right) = \int_{\text{Vol}} \mathbf{p}' \times \rho_V \mathbf{g} dV + \int_{\text{Surface}} \mathbf{p}' \times d\mathbf{f}_{\text{ext}} \tag{2.2}$$

But expressed in terms of the position of mass particles relative to the origin of the inertial frame, these equations are not in the most useful form. We will now address this issue. From Figure 2.1 it can be seen that

$$\mathbf{p}' = \mathbf{p}_V + \mathbf{p} \tag{2.3}$$

where $\mathbf{p}$ is the position of dV relative to the origin of the vehicle-fixed Frame V, or O_V. Also, the instantaneous mass of the vehicle, m, is

$$\int_{\text{Vol}} \rho_V dV = m \tag{2.4}$$

Now <u>define</u> Frame V such that its origin O_V is at the instantaneous center of mass cm of the vehicle (which is also the center of gravity here since $\mathbf{g}$ is assumed constant), and the equations above will take on a more familiar form. With regard to the center of mass, write the first mass moment of the vehicle about the inertial origin as

$$\int_{\text{Vol}} \rho_V \mathbf{p}' dV = \int_{\text{Vol}} \rho_V \mathbf{p}_V dV + \int_{\text{Vol}} \rho_V \mathbf{p} dV = \mathbf{p}_V \int_{\text{Vol}} \rho_V dV + \int_{\text{Vol}} \rho_V \mathbf{p} dV$$

$$= m\mathbf{p}_V + \int_{\text{Vol}} \rho_V \mathbf{p} dV$$

Now <u>define</u> the center of mass to be that point in the vehicle about which the first mass moment is zero, or

$$\int_{\text{Vol}} \rho_V \mathbf{p} dV = \mathbf{0} \tag{2.5}$$

As a result, we now can express the inertial position of the center of mass as

$$\mathbf{p}_V = \frac{1}{m}\int\limits_{\mathrm{Vol}} \rho_V \mathbf{p}' dV$$

Looking now at the left side of Equation (2.1), we have

$$\int\limits_{\mathrm{Vol}} \frac{d}{dt}\Big|_I \left(\rho_V \frac{d\mathbf{p}'}{dt}\Big|_I dV\right) = \int\limits_{\mathrm{Vol}} \frac{d}{dt}\Big|_I \left(\rho_V \frac{d\mathbf{p}_V}{dt}\Big|_I dV\right) + \int\limits_{\mathrm{Vol}} \frac{d}{dt}\Big|_I \left(\rho_V \frac{d\mathbf{p}}{dt}\Big|_I dV\right) \quad (2.6)$$

but

$$\int\limits_{\mathrm{Vol}} \frac{d}{dt}\Big|_I \left(\rho_V \frac{d\mathbf{p}_V}{dt}\Big|_I dV\right) = \frac{d}{dt}\Big|_I \left(\frac{d\mathbf{p}_V}{dt}\Big|_I \int\limits_{\mathrm{Vol}} \rho_V dV\right) = \frac{d}{dt}\Big|_I \left(m \frac{d\mathbf{p}_V}{dt}\Big|_I\right)$$

This last term is recognized as the rate of change of the translational momentum of the vehicle, and it is expressed in terms of the position and velocity of the vehicle's center of mass. In obtaining this result, note that the velocity of the vehicle's center of mass (or gravity)

$$\frac{d\mathbf{p}_V}{dt}\Big|_I$$

may be brought outside the volume integral since the velocity of that point is not a function of the infinitesimal volumes dV over which the integration operation is being performed. That is, the position and velocity of the center of mass (gravity) is constant with respect to the operation of volume integration over the vehicle.

With regard to the second term on the right side of Equation (2.6), application of Equation (1.21), the vector-derivative "chain rule," yields

$$\frac{d\mathbf{p}}{dt}\Big|_I = \frac{d\mathbf{p}}{dt}\Big|_V + \boldsymbol{\omega}_{V,I} \times \mathbf{p} \quad (2.7)$$

Substituting this into the integral in question yields

$$\int\limits_{\mathrm{Vol}} \frac{d}{dt}\Big|_I \left(\rho_V \frac{d\mathbf{p}}{dt}\Big|_I dV\right) = \int\limits_{\mathrm{Vol}} \frac{d}{dt}\Big|_I \left(\rho_V \frac{d\mathbf{p}}{dt}\Big|_V dV\right) + \int\limits_{\mathrm{Vol}} \frac{d}{dt}\Big|_I \left(\rho_V (\boldsymbol{\omega}_{V,I} \times \mathbf{p}) dV\right) \quad (2.8)$$

The first integral on the right-hand side is zero since if the vehicle is rigid

$$\frac{d\mathbf{p}}{dt}\Big|_V = 0$$

for all mass particles $\rho_V dV$ of the vehicle.

Plus, the last integral in Equation (2.8) can be manipulated as follows.

$$\int_{Vol} \frac{d}{dt}\Big|_I \left(\rho_V\left(\boldsymbol{\omega}_{V,I} \times \mathbf{p}\right)dV\right) = \frac{d\boldsymbol{\omega}_{V,I}}{dt}\Big|_I \times \int_{Vol} \rho_V \mathbf{p} dV + \boldsymbol{\omega}_{V,I} \times \int_{Vol} \frac{d}{dt}\Big|_I \left(\rho_V \mathbf{p} dV\right)$$

$$= \frac{d\boldsymbol{\omega}_{V,I}}{dt}\Big|_I \times \int_{Vol} \rho_V \mathbf{p} dV + \boldsymbol{\omega}_{V,I} \times \left(\int_{Vol} \frac{d}{dt}\Big|_V \left(\rho_V \mathbf{p} dV\right) + \boldsymbol{\omega}_{V,I} \times \int_{Vol} \rho_V \mathbf{p} dV \right) = 0$$

This is due to two facts. First, recall Equation (2.5) and the fact that O_V was chosen to be coincident with the vehicle's center of mass (or gravity), about which the first mass moment is zero. Second, the expression

$$\int_{Vol} \frac{d}{dt}\Big|_V \left(\rho_V \mathbf{p} dV\right) = 0 \tag{2.9}$$

is zero if the densities of the volumes dV's, or ρ_V, are constant (and therefore the vehicle mass is constant), the vehicle is assumed rigid, and Frame V is a *vehicle-fixed frame* in that it does not change orientation with respect to the vehicle. This latter point follows directly from the definition of the vector

$$\frac{d\mathbf{p}}{dt}\Big|_V = 0$$

which is the rate of change of position of a mass element of the vehicle with respect to the vehicle-fixed frame. In other words, the position of the frame is invariant with respect to any point on the vehicle and vice versa. Finally, note that since Equation (2.8) equals zero, the second integral on the right side of Equation (2.6) vanishes.

This completes the discussion of Equation (2.1), or Newton's law in translational form, which may now be expressed as

$$\boxed{\frac{d}{dt}\Big|_I \left(m\frac{d\mathbf{p}_V}{dt}\Big|_I\right) \triangleq m\frac{d\mathbf{V}_V}{dt}\Big|_I = \int_{Vol} \rho_V \mathbf{g} dV + \int_{Surface} d\mathbf{f}_{ext}} \tag{2.10}$$

Note here that the vehicle's inertial velocity (vector) $\mathbf{V}_V$ has been introduced.

Now consider Equation (2.2), or Newton's law in rotational form. Note that the left side of the equation may be written as

$$\int_{Vol} \frac{d}{dt}\Big|_I \left(\mathbf{p}' \times \rho_V \frac{d\mathbf{p}'}{dt}\Big|_I dV\right) = \int_{Vol} \frac{d\mathbf{p}'}{dt}\Big|_I \times \rho_V \frac{d\mathbf{p}'}{dt}\Big|_I dV + \int_{Vol} \mathbf{p}' \times \frac{d}{dt}\Big|_I \left(\rho_V \frac{d\mathbf{p}'}{dt}\Big|_I dV\right)$$

(the first integral on the right side is struck through)

$$= \int_{Vol} \mathbf{p}' \times \frac{d}{dt}\Big|_I \left(\rho_V \frac{d\mathbf{p}'}{dt}\Big|_I dV\right)$$

since the cross product between two parallel vectors equals zero. Next, by invoking Equation (2.3), and by noting that the vector $\mathbf{p}_V$ is independent of the volume-integral operation, the left side of Equation (2.2) can be written as

$$\int_{Vol} \frac{d}{dt}\Big|_I \left(\mathbf{p}' \times \rho_V \frac{d\mathbf{p}'}{dt}\Big|_I dV\right) = \int_{Vol} \left(\mathbf{p}' \times \frac{d}{dt}\Big|_I \left(\rho_V \frac{d\mathbf{p}'}{dt}\Big|_I dV\right)\right)$$

$$= \int_{Vol} \left((\mathbf{p}_V + \mathbf{p}) \times \frac{d}{dt}\Big|_I \left(\rho_V \frac{d\mathbf{p}'}{dt}\Big|_I dV\right)\right)$$

$$= \boxed{\mathbf{p}_V \times \int_{Vol} \frac{d}{dt}\Big|_I \left(\rho_V \frac{d\mathbf{p}'}{dt}\Big|_I dV\right)} + \int_{Vol} \left(\mathbf{p} \times \frac{d}{dt}\Big|_I \left(\rho_V \frac{d\mathbf{p}'}{dt}\Big|_I dV\right)\right) \tag{2.11}$$

Further, the right side of Equation (2.2) can be written as

$$\int_{Vol} \mathbf{p}' \times \rho_V \mathbf{g} dV + \int_{Surface} \mathbf{p}' \times d\mathbf{f}_{ext} = \int_{Vol} (\mathbf{p}_V + \mathbf{p}) \times \rho_V \mathbf{g} dV + \int_{Surface} (\mathbf{p}_V + \mathbf{p}) \times d\mathbf{f}_{ext} \tag{2.12}$$

$$= \boxed{\mathbf{p}_V \times \left(\int_{Vol} \rho_V \mathbf{g} dV + \int_{Surface} d\mathbf{f}_{ext}\right)} + \int_{Vol} \mathbf{p} \times \rho_V \mathbf{g} dV + \int_{Surface} \mathbf{p} \times d\mathbf{f}_{ext}$$

Now the terms in the boxes in Equations (2.11) and (2.12) can be recognized as the vector $\mathbf{p}_V$ crossed into Equation (2.1), or $\mathbf{p}_V \times$ (Equation (2.1)). Since Equation (2.1) is always satisfied (it's Newton's law), the terms in the boxes will cancel when equating Equation (2.11) and (2.12). As a result, Equation (2.2) becomes

$$\int_{Vol} \left(\mathbf{p} \times \frac{d}{dt}\Big|_I \left(\rho_V \frac{d\mathbf{p}'}{dt}\Big|_I dV\right)\right) = \int_{Vol} \mathbf{p} \times \rho_V \mathbf{g} dV + \int_{Surface} \mathbf{p} \times d\mathbf{f}_{ext} \tag{2.13}$$

Now from the property of the center of mass (Equation (2.5)), and the fact that the gravity vector is taken as constant with respect to the operation of volume integration over the vehicle, we have

$$\int_{Vol} \mathbf{p} \times \rho_V \mathbf{g} dV = \int_{Vol} \rho_V \mathbf{p} dV \times \mathbf{g} = 0$$

Therefore, Equation (2.2) becomes

$$\int_{Vol} \left(\mathbf{p} \times \frac{d}{dt}\Big|_I \left(\rho_V \frac{d\mathbf{p}'}{dt}\Big|_I dV\right)\right) = \int_{Surface} \mathbf{p} \times d\mathbf{f}_{ext} \tag{2.14}$$

By again invoking Equation (2.3), we can eliminate $\mathbf{p}'$ from this equation, and the left side of Equation (2.14) can be written as

$$\int_{\text{Vol}} \left(\mathbf{p} \times \frac{d}{dt}\bigg|_I \left(\rho_V \frac{d\mathbf{p}'}{dt}\bigg|_I dV\right)\right) = \int_{\text{Vol}} \left(\mathbf{p} \times \frac{d}{dt}\bigg|_I \left(\rho_V \frac{d\mathbf{p}_V}{dt}\bigg|_I dV\right)\right) + \int_{\text{Vol}} \left(\mathbf{p} \times \frac{d}{dt}\bigg|_I \left(\rho_V \frac{d\mathbf{p}}{dt}\bigg|_I dV\right)\right)$$

But if the vehicle's material density ρ_V is not a function of time, and recalling Equation (2.5), we have

$$\int_{\text{Vol}} \left(\mathbf{p} \times \frac{d}{dt}\bigg|_I \left(\rho_V \frac{d\mathbf{p}_V}{dt}\bigg|_I dV\right)\right) = \int_{\text{Vol}} (\rho_V \mathbf{p} dV) \times \frac{d}{dt}\bigg|_I \left(\frac{d\mathbf{p}_V}{dt}\bigg|_I\right) = 0$$

So Equation (2.14) finally becomes

$$\boxed{\int_{\text{Vol}} \left(\mathbf{p} \times \frac{d}{dt}\bigg|_I \left(\rho_V \frac{d\mathbf{p}}{dt}\bigg|_I dV\right)\right) = \int_{\text{Surface}} \mathbf{p} \times d\mathbf{f}_{\text{ext}}} \tag{2.15}$$

Note that we have just converted the expression for Newton's law in rotation form, originally expressed in terms of moments and momentum about the origin of the <u>inertial</u> frame (Equation (2.2)), into an equation expressed in terms of moments and momentum about the origin of the <u>vehicle</u>-fixed frame (Equation (2.15)). This is critical. It leads to much simpler and more useful expressions for mass properties and for the moments acting on the vehicle.

We can further expand the left side of Equation (2.15) in the following manner. Invoking the vector-derivative chain rule, Equation (2.7), several times, we have

$$\int_{\text{Vol}} \left(\mathbf{p} \times \frac{d}{dt}\bigg|_I \left(\rho_V \frac{d\mathbf{p}}{dt}\bigg|_I dV\right)\right) = \int_{\text{Vol}} \left(\mathbf{p} \times \frac{d}{dt}\bigg|_I \left(\rho_V \left(\frac{d\mathbf{p}}{dt}\bigg|_V + \boldsymbol{\omega}_{V,I} \times \mathbf{p}\right) dV\right)\right)$$

$$= \int_{\text{Vol}} \mathbf{p} \times \left(\rho_V \left(\frac{d^2\mathbf{p}}{dt^2}\bigg|_V + 2\left(\boldsymbol{\omega}_{V,I} \times \frac{d\mathbf{p}}{dt}\bigg|_V\right) + \left(\frac{d\boldsymbol{\omega}_{V,I}}{dt}\bigg|_I \times \mathbf{p}\right) + \boldsymbol{\omega}_{V,I} \times (\boldsymbol{\omega}_{V,I} \times \mathbf{p})\right) dV\right)$$

By again noting that <u>for a rigid vehicle</u>,

$$\frac{d\mathbf{p}}{dt}\bigg|_V = 0$$

Equation (2.15) becomes the following equation for Newton's law in rotational form

$$\int_{\text{Vol}} \mathbf{p} \times \left(\rho_V \left(\left(\frac{d\boldsymbol{\omega}_{V,I}}{dt}\Big|_I \times \mathbf{p} \right) + \boldsymbol{\omega}_{V,I} \times (\boldsymbol{\omega}_{V,I} \times \mathbf{p}) \right) dV \right) = \int_{\text{Surface}} \mathbf{p} \times d\mathbf{f}_{\text{ext}} \tag{2.16}$$

Now we will look further at the right sides of Equations (2.10) and (2.16), which include the forces and moments acting on the vehicle. First we have

$$\int_{\text{Vol}} \rho_V \mathbf{g} dV = \mathbf{g} \int_{\text{Vol}} \rho_V dV = m\mathbf{g}$$

where again we note that the gravity vector is assumed constant with respect to the volume integration over the vehicle (or gravity is simply constant). Next, let

$$\int_{\text{Surface}} d\mathbf{f}_{\text{ext}} = \mathbf{F}_{\text{Aero}} + \mathbf{F}_{\text{Prop}}$$

which implies the forces acting on the vehicle arise from the integration of a pressure distribution $\mathbf{f}_{\text{ext}}$ over the vehicle that arises due to aerodynamic and propulsive effects. Likewise, let

$$\int_{\text{Surface}} \mathbf{p} \times d\mathbf{f}_{\text{ext}} = \mathbf{M}_{\text{Aero}} + \mathbf{M}_{\text{Prop}}$$

which is simply the expression for the moments, taken about the vehicle center of mass, due to the same external pressure distribution giving rise to the aerodynamic and propulsive forces.

Combining all these results yields the equations of motion given below for a rigid vehicle with a constant density distribution. The first, Equation (2.17), expressed in terms of the vehicle's translational momentum, governs the translational motion of the vehicle, while the second, Equation (2.18), expressed in terms of the vehicle's angular momentum about its center of mass, governs the rotational motion.

$$\frac{d}{dt}\Big|_I \left(m \frac{d\mathbf{p}_V}{dt}\Big|_I \right) = m \frac{d\mathbf{V}_V}{dt}\Big|_I = m\mathbf{g} + \mathbf{F}_{\text{Aero}} + \mathbf{F}_{\text{Prop}} \tag{2.17}$$

$$\int_{\text{Vol}} \mathbf{p} \times \left(\rho_V \left(\left(\frac{d\boldsymbol{\omega}_{V,I}}{dt}\Big|_I \times \mathbf{p} \right) + \boldsymbol{\omega}_{V,I} \times (\boldsymbol{\omega}_{V,I} \times \mathbf{p}) \right) dV \right) = \mathbf{M}_{\text{Aero}} + \mathbf{M}_{\text{Prop}} \tag{2.18}$$

Recall that the first of these two equations was derived using a coordinate frame with origin fixed at the vehicle's center of mass. The second of these two equations also required that the selected coordinate frame, Frame V, was vehicle-fixed, where "vehicle-fixed" implies each point on the vehicle remains at a fixed position relative to this frame. Or

$$\frac{d\mathbf{p}}{dt}\Big|_V = 0.$$

2.2 SCALAR EQUATIONS OF MOTION—FLAT EARTH

Six scalar equations of motion will now be developed from the above two vector relationships (Equations (2.17) and (2.18)). These equations of motion will not specifically include the dynamic effects of rotating machinery on the vehicle, such as propellers or turbojet engines. Nor will they include the effects of a variable-mass vehicle. Necessary modifications to the equations of motion to include these effects are presented in Sections 2.4 and 2.5. Nevertheless, the equations derived in this section are widely used because they are sufficiently accurate to model the dynamics of the majority of conventional aircraft in existence today. Exceptions may include the tilt-rotor vehicle mentioned in the problems in Chapter 1, novel micro air vehicles (MAVS), rockets and missiles, or future hypersonic vehicles similar to NASA's X-43 (to be discussed in Chapters 10 and 11.).

The following equations define the components of the necessary vectors in the vehicle-fixed Frame V. (See Figure 2.1.)

$$\begin{aligned}
&\frac{d\mathbf{p}_V}{dt}\Big|_I \triangleq \mathbf{V}_V = U\mathbf{i}_V + V\mathbf{j}_V + W\mathbf{k}_V\\
&\mathbf{p} = x\mathbf{i}_V + y\mathbf{j}_V + z\mathbf{k}_V\\
&\mathbf{g} = g_x\mathbf{i}_V + g_y\mathbf{j}_V + g_z\mathbf{k}_V\\
&\boldsymbol{\omega}_{V,I} = P\mathbf{i}_V + Q\mathbf{j}_V + R\mathbf{k}_V\\
&\mathbf{F}_{\text{Aero}} = F_{A_X}\mathbf{i}_V + F_{A_Y}\mathbf{j}_V + F_{A_Z}\mathbf{k}_V\\
&\mathbf{F}_{\text{Prop}} = F_{P_X}\mathbf{i}_V + F_{P_Y}\mathbf{j}_V + F_{P_Z}\mathbf{k}_V\\
&\mathbf{M}_{\text{Aero}} = L_A\mathbf{i}_V + M_A\mathbf{j}_V + N_A\mathbf{k}_V\\
&\mathbf{M}_{\text{Prop}} = L_P\mathbf{i}_V + M_P\mathbf{j}_V + N_P\mathbf{k}_V
\end{aligned}\tag{2.19}$$

NOTE TO STUDENT

Sometimes students are confused by the first of Equations (2.19). This equation defines the components of the vehicle's inertial-velocity vector (or the inertial velocity of its center of mass), which is defined as the rate of change of the position vector with respect to the inertial frame. But the components of the vector given here are its components in the vehicle-fixed frame. This is but one example of the need to carefully specify the frame used to define the rate of change of a vector and the frame used to define the components of a vector.

The angular-velocity vector $\boldsymbol{\omega}_{V,I}$ represents the angular velocity of Frame V with respect to inertial, and its components in the vehicle-fixed frame are defined above to be P, Q, and R. P is called the *roll rate,* Q the *pitch rate,* and R the *yaw rate* of the vehicle.

With mass constant, we have the term on the left side of Equation (2.17), or the vehicle's translational momentum, given by

$$\frac{d}{dt}\Big|_I\left(m\frac{d\mathbf{p}_V}{dt}\Big|_I\right) = m\left(\frac{d\mathbf{V}_V}{dt}\Big|_I\right) = m\left(\frac{d\mathbf{V}_V}{dt}\Big|_V + \boldsymbol{\omega}_{V,I} \times \mathbf{V}_V\right) \tag{2.20}$$

Substituting Equations (2.19) into Equation (2.17), invoking Equation (2.20), carrying out the vector operations, and equating $\mathbf{i}_V$, $\mathbf{j}_V$, and $\mathbf{k}_V$ components yields the following three scalar equations governing vehicle translation

$$\begin{aligned} m(\dot{U} + QW - VR) &= mg_x + F_{A_X} + F_{P_X} \\ m(\dot{V} + RU - PW) &= mg_y + F_{A_Y} + F_{P_Y} \\ m(\dot{W} + PV - QU) &= mg_z + F_{A_Z} + F_{P_Z} \end{aligned} \tag{2.21}$$

(Compare these expressions to your answers to Problem 1.3 in Chapter 1.) Finally, by recalling Equations (1.19) for the components of gravity in Frame V, the three scalar equations governing translational motion then become

$$\boxed{\begin{aligned} m(\dot{U} + QW - VR) &= -mg\sin\theta + F_{A_X} + F_{P_X} \\ m(\dot{V} + RU - PW) &= mg\cos\theta\sin\phi + F_{A_Y} + F_{P_Y} \\ m(\dot{W} + PV - QU) &= mg\cos\theta\cos\phi + F_{A_Z} + F_{P_Z} \end{aligned}} \tag{2.22}$$

Note that these equations are expressed in terms of the three Euler angles, ψ, θ, and ϕ—the 3-2-1 rotation angles defining the orientation of the vehicle-fixed Frame V with respect to the inertial Frame I. (See Section 1.3 for review.)

We now turn our attention to the development of the scalar equations governing the vehicle's rotation, starting with Equation (2.18) repeated here.

$$\int_{\text{Vol}} \mathbf{p} \times \left(\rho_V\left(\left(\frac{d\boldsymbol{\omega}_{V,I}}{dt}\Big|_I \times \mathbf{p}\right) + \boldsymbol{\omega}_{V,I} \times \left(\boldsymbol{\omega}_{V,I} \times \mathbf{p}\right)\right)dV\right) = \mathbf{M}_{\text{Aero}} + \mathbf{M}_{\text{Prop}}$$

First, note that

$$\frac{d\boldsymbol{\omega}_{V,I}}{dt}\Big|_I = \frac{d\boldsymbol{\omega}_{V,I}}{dt}\Big|_V + \boldsymbol{\omega}_{V,I} \times \boldsymbol{\omega}_{V,I} = \frac{d\boldsymbol{\omega}_{V,I}}{dt}\Big|_V$$

and that

$$\frac{d\boldsymbol{\omega}_{V,I}}{dt}\Big|_V = \dot{P}\mathbf{i}_V + \dot{Q}\mathbf{j}_V + \dot{R}\mathbf{k}_V.$$

Then, from the vector-triple-product identity[1] we can write

$$\mathbf{p} \times \left(\left(\frac{d\boldsymbol{\omega}_{V,I}}{dt}\Big|_V \times \mathbf{p} \right) + \left(\boldsymbol{\omega}_{V,I} \times \left(\boldsymbol{\omega}_{V,I} \times \mathbf{p} \right) \right) \right)$$

$$= \frac{d\boldsymbol{\omega}_{V,I}}{dt}\Big|_V (\mathbf{p}\cdot\mathbf{p}) - \mathbf{p}\left(\mathbf{p}\cdot \frac{d\boldsymbol{\omega}_{V,I}}{dt}\Big|_V\right) + \mathbf{p} \times \boldsymbol{\omega}_{V,I}(\boldsymbol{\omega}_{V,I}\cdot\mathbf{p}) - \cancel{\mathbf{p} \times \mathbf{p}(\boldsymbol{\omega}_{V,I}\cdot\boldsymbol{\omega}_{V,I})} \tag{2.23}$$

while noting that the last term on the right is zero since it is the cross product of two parallel vectors.

Using the defined components of the selected vectors we have

$$\begin{aligned} \mathbf{p}\cdot\mathbf{p} &= x^2 + y^2 + z^2 \\ \mathbf{p}\cdot\boldsymbol{\omega}_{V,I} &= Px + Qy + Rz \\ \mathbf{p}\cdot\frac{d\boldsymbol{\omega}_{V,I}}{dt}\Big|_V &= \dot{P}x + \dot{Q}y + \dot{R}z \end{aligned} \tag{2.24}$$

Now define the following moments and products of inertia to be

$I_{xx} = \int_{Vol} (y^2 + z^2)\rho_V dV$	$I_{xy} = I_{yx} = \int_{Vol} xy\rho_V dV$
$I_{yy} = \int_{Vol} (x^2 + z^2)\rho_V dV$	$I_{xz} = I_{zx} = \int_{Vol} xz\rho_V dV$
$I_{zz} = \int_{Vol} (x^2 + y^2)\rho_V dV$	$I_{yz} = I_{zy} = \int_{Vol} yz\rho_V dV$

(2.25)

and note that if the density distribution over the vehicle, ρ_V, is constant over time, the inertia terms in Equations (2.25) are constant. With these inertias defined, one can show (an exercise for the student) that the integral on the left-hand side of Equation (2.18) can be expressed as the vector quantity given in Equation (2.26) below.

---

[1] Vector-triple-product identity: $\mathbf{a} \times (\mathbf{b} \times \mathbf{c}) = \mathbf{b}\,(\mathbf{a}\cdot\mathbf{c}) - \mathbf{c}\,(\mathbf{a}\cdot\mathbf{b})$.

$$\int_{\text{Vol}} \mathbf{p} \times \left(\left(\frac{d\boldsymbol{\omega}_{V,I}}{dt}\bigg|_I \times \mathbf{p} \right) + \boldsymbol{\omega}_{V,I} \times (\boldsymbol{\omega}_{V,I} \times \mathbf{p}) \right) \rho_V dV$$

$$\begin{aligned} &= \left(I_{xx}\dot{P} - I_{xz}\left(\dot{R} + PQ\right) - I_{yz}\left(Q^2 - R^2\right) - I_{xy}\left(\dot{Q} - RP\right) + \left(I_{zz} - I_{yy}\right)RQ\right)\mathbf{i}_V \\ &+ \left(I_{yy}\dot{Q} + \left(I_{xx} - I_{zz}\right)PR - I_{xy}\left(\dot{P} + QR\right) - I_{yz}\left(\dot{R} - PQ\right) + I_{xz}\left(P^2 - R^2\right)\right)\mathbf{j}_V \\ &+ \left(I_{zz}\dot{R} - I_{xz}\left(\dot{P} - QR\right) - I_{xy}\left(P^2 - Q^2\right) - I_{yz}\left(\dot{Q} + RP\right) + \left(I_{yy} - I_{xx}\right)PQ\right)\mathbf{k}_V \end{aligned} \tag{2.26}$$

Finally, note that the two moment vectors on the right-hand side of Equation (2.18) have components in Frame V as given in Equations (2.19). Using these definitions, we can equate the $\mathbf{i}_V$, $\mathbf{j}_V$, and $\mathbf{k}_V$ components of Equation (2.18) and Equation (2.26) to obtain the desired <u>three scalar equations of rotational motion</u> given here.

$$\boxed{\begin{aligned} &I_{xx}\dot{P} - I_{xz}\left(\dot{R} + PQ\right) - I_{yz}\left(Q^2 - R^2\right) - I_{xy}\left(\dot{Q} - RP\right) + \left(I_{zz} - I_{yy}\right)RQ = L_A + L_P \\ &I_{yy}\dot{Q} + \left(I_{xx} - I_{zz}\right)PR - I_{xy}\left(\dot{P} + QR\right) - I_{yz}\left(\dot{R} - PQ\right) + I_{xz}\left(P^2 - R^2\right) = M_A + M_P \\ &I_{zz}\dot{R} - I_{xz}\left(\dot{P} - QR\right) - I_{xy}\left(P^2 - Q^2\right) - I_{yz}\left(\dot{Q} + RP\right) + \left(I_{yy} - I_{xx}\right)PQ = N_A + N_P \end{aligned}} \tag{2.27}$$

Note that if the XZ plane of the vehicle is a plane of symmetry then $I_{xy} = I_{yz} = 0$, and the above equations may be simplified considerably. A vehicle that does <u>not</u> possess such a plane of symmetry is NASA's oblique-wing vehicle shown in Figure 2.2. For this vehicle, with the wing in an oblique position, all the products of inertia are nonzero.

COMMENT

An alternate and sometimes more convenient approach can be taken to obtain Equations (2.27). We begin again with Equation (2.15) repeated here

$$\int_{\text{Vol}} \left(\mathbf{p} \times \frac{d}{dt}\bigg|_I \left(\rho_V \frac{d\mathbf{p}}{dt}\bigg|_I dV \right) \right) = \int_{\text{Surface}} \mathbf{p} \times d\mathbf{f}_{\text{ext}}$$

and note that the left-hand side can be written as

$$\begin{aligned} \int_{\text{Vol}} \left(\mathbf{p} \times \frac{d}{dt}\bigg|_I \left(\rho_V \frac{d\mathbf{p}}{dt}\bigg|_I dV \right) \right) &= \int_{\text{Vol}} \frac{d}{dt}\bigg|_I \left(\mathbf{p} \times \left(\rho_V \frac{d\mathbf{p}}{dt}\bigg|_I dV \right) \right) \\ = \int_{\text{Vol}} \frac{d}{dt}\bigg|_I (\mathbf{p} \times (\rho_V(\boldsymbol{\omega}_{V,I} \times \mathbf{p})dV)) &= -\int_{\text{Vol}} \frac{d}{dt}\bigg|_I (\mathbf{p} \times (\rho_V(\mathbf{p} \times \boldsymbol{\omega}_{V,I})dV)) \end{aligned} \tag{2.28}$$

Note also that the integrand in the above expression represents the inertial rate of change of the angular momentum of a mass element about the vehicle's center of mass.

Figure 2.2 Multiple exposure of NASA's oblique-wing research vehicle. *(Photograph courtesy of NASA.)*

We now introduce the *cross-product matrix.* By using the familiar determinant operation to evaluate the innermost cross product we have

$$\mathbf{p} \times \boldsymbol{\omega}_{V,I} = \begin{vmatrix} \mathbf{i}_V & \mathbf{j}_V & \mathbf{k}_V \\ x & y & z \\ P & Q & R \end{vmatrix} = (yR - zQ)\mathbf{i}_V + (zP - xR)\mathbf{j}_V + (xQ - yP)\mathbf{k}_V$$

However, we can also evaluate this cross product using a cross-product matrix as shown here

$$\mathbf{p} \times \boldsymbol{\omega}_{V,I} = \left[\begin{bmatrix} 0 & -z & y \\ z & 0 & -x \\ -y & x & 0 \end{bmatrix}\begin{bmatrix} P \\ Q \\ R \end{bmatrix}\right]^T \begin{Bmatrix} \mathbf{i}_V \\ \mathbf{j}_V \\ \mathbf{k}_V \end{Bmatrix} = \begin{bmatrix}(yR - zQ) & (zP - xR) & (xQ - yP)\end{bmatrix}\begin{Bmatrix} \mathbf{i}_V \\ \mathbf{j}_V \\ \mathbf{k}_V \end{Bmatrix}$$

where we have used the vector-array notation introduced in Section 1.3. Or

$$\mathbf{p} \times \boldsymbol{\omega}_{V,I} = \tilde{\mathbf{p}} \otimes \boldsymbol{\omega}_{V,I} \tag{2.29}$$

where

$$\tilde{\mathbf{p}} = \begin{bmatrix} 0 & -z & y \\ z & 0 & -x \\ -y & x & 0 \end{bmatrix}$$

is a cross-product matrix, and $\otimes$ is the *cross-product-matrix operation.* As required, the result of the cross product between two vectors is always a vector, and to carry out a cross product the two vectors must be expressed in terms of their components defined in the same frame.

One advantage of introducing the cross-product matrix and operation lies in the associative property of the operation. For example, we can see that

$$\mathbf{p}\times(\mathbf{p}\times\boldsymbol{\omega}_{V,I})=\tilde{\mathbf{p}}\otimes(\tilde{\mathbf{p}}\otimes\boldsymbol{\omega}_{V,I})=\left[\begin{bmatrix}0&-z&y\\z&0&-x\\-y&x&0\end{bmatrix}\begin{bmatrix}0&-z&y\\z&0&-x\\-y&x&0\end{bmatrix}\begin{bmatrix}P\\Q\\R\end{bmatrix}\right]^T\begin{Bmatrix}\mathbf{i}_V\\\mathbf{j}_V\\\mathbf{k}_V\end{Bmatrix}$$

$$=\tilde{\mathbf{p}}\tilde{\mathbf{p}}\otimes\boldsymbol{\omega}_{V,I}=\begin{bmatrix}-(y^2+z^2)&xy&xz\\xp&-(x^2+z^2)&yz\\xz&yz&-(x^2+y^2)\end{bmatrix}\otimes\boldsymbol{\omega}_{V,I}$$

or the triple cross product can be expressed in terms of the <u>product</u> of two cross-product matrices. And that product is also a cross-product matrix.

As a result, Equation (2.28) can be written as

$$\int_{\text{Vol}}\frac{d}{dt}\Big|_I\left(\mathbf{p}\times\left(\rho_V\frac{d\mathbf{p}}{dt}\Big|_I dV\right)\right)=-\int_{\text{Vol}}\frac{d}{dt}\Big|_I(\mathbf{p}\times(\rho_V(\mathbf{p}\times\boldsymbol{\omega}_{V,I})dV))$$

$$=-\int_{\text{Vol}}\frac{d}{dt}\Big|_I\left(\left(\rho_V\begin{bmatrix}-(y^2+z^2)&xy&xz\\xp&-(x^2+z^2)&yz\\xz&yz&-(x^2+y^2)\end{bmatrix}\otimes\boldsymbol{\omega}_{V,I}\,dV\right)\right)$$

$$=\frac{d}{dt}\Big|_I\left[\int_{\text{Vol}}\rho_V\begin{bmatrix}(y^2+z^2)&-xy&-xz\\-xp&(x^2+z^2)&-yz\\-xz&-yz&(x^2+y^2)\end{bmatrix}dV\right]\otimes\boldsymbol{\omega}_{V,I}$$

$$=\frac{d}{dt}\Big|_I[\mathbf{I}]\otimes\boldsymbol{\omega}_{V,I}=\frac{d}{dt}\Big|_I\mathbf{H}_V$$

where $[\mathbf{I}]$ is the inertia matrix, also a cross-product matrix, and $\mathbf{H}_V$ is the vector representation of the vehicle's angular momentum about its center of mass. Note that $[\mathbf{I}]$ is written as

$$[\mathbf{I}]=\begin{bmatrix}I_{xx}&-I_{xy}&-I_{xz}\\-I_{yx}&I_{yy}&-I_{yz}\\-I_{zx}&-I_{zy}&I_{zz}\end{bmatrix}\tag{2.30}$$

with the elements as given in Equations (2.25).

Consequently, Equations (2.27) may also be derived from the following expression for Newton's law (which is equivalent to Equation (2.15) if the applied moments arise only from aerodynamic and propulsive effects).

$$\boxed{\frac{d\mathbf{H}_V}{dt}\Big|_I=\mathbf{M}_{\text{Aero}}+\mathbf{M}_{\text{Prop}}}\tag{2.31}$$

EXAMPLE 2.1

MATLAB Calculation of Angular Momentum

Another advantage of using the cross-product matrix stems from the fact that it allows for vector cross products to be computed easily in software, such as MATLAB (Ref. 1). In this example, we will demonstrate such a computation by calculating the angular momentum of an aircraft undergoing a maneuver.

Let the inertias of an A-4D aircraft be

$$I_{xx} = 8090 \text{ sl-ft}^2 \qquad I_{yy} = 25{,}900 \text{ sl-ft}^2$$

$$I_{zz} = 29{,}200 \text{ sl-ft}^2 \qquad I_{xz} = 1300 \text{ sl-ft}^2$$

At some instant during a maneuver, the aircraft us undergoing the following angular rates

$$P(\text{roll rate}) = 0$$

$$Q(\text{pitch rate}) = 0.00685 \text{ rad/sec}$$

$$R(\text{yaw rate}) = 0.02950 \text{ rad/sec}$$

Determine the vehicle's angular momentum $\mathbf{H}_V$ at this instant.

■ Solution

For the data given, the aircraft's inertia matrix is

$$[\mathbf{I}] = \begin{bmatrix} 8090 & 0 & -1300 \\ 0 & 25{,}900 & 0 \\ -1300 & 0 & 29{,}200 \end{bmatrix}$$

and the angular-velocity vector is

$$\boldsymbol{\omega}_{V,I} = 0\,\mathbf{i}_V + 0.00685\,\mathbf{j}_V + 0.02950\,\mathbf{k}_V$$

We know that the angular momentum is

$$\mathbf{H}_V = [\mathbf{I}] \otimes \boldsymbol{\omega}_{V,I} = \left[[\mathbf{I}] \begin{bmatrix} P \\ Q \\ R \end{bmatrix} \right]^T \begin{Bmatrix} \mathbf{i}_V \\ \mathbf{j}_V \\ \mathbf{k}_V \end{Bmatrix}$$

The MATLAB calculations follow:

```
»I=[8090 0 -1300;
  0 25900 0;
  -1300 0 29200]

I =

   8090       0   -1300
      0   25900       0
  -1300       0   29200
```

```
»omega=[0;0.00685;0.02950]

omega =

        0
  6.8500e-03
  2.9500e-02

»H=I*omega

H =
 -3.8350e+01
  1.7741e+02
  8.6140e+02
```

Therefore, the vehicle's angular momentum vector is

$$\mathbf{H}_V = -38.35\,\mathbf{i}_V + 177.41\,\mathbf{j}_V + 861.40\,\mathbf{k}_V \text{ sl-ft}^2\text{/sec}$$

To summarize the development so far in this section, the six scalar equations of motion for the rigid vehicle are given in Equations (2.22) (translation *U, V,* and *W* and forces) and Equations (2.27) (rotation *P, Q,* and *R* and moments). These six equations govern the six degrees of freedom of the rigid vehicle.

The final sets of equations to be derived are referred to as the kinematic equations, since they do not model dynamic but kinematic relationships between the angular and translational rates. First, consider the angular rates. From the definition of the three 3-2-1 Euler angles ψ, θ, and ϕ in Section 1.3, it is clear that

$$\boldsymbol{\omega}_{V,I} = \dot{\phi}\mathbf{i}_V + \dot{\theta}\mathbf{j}_2 + \dot{\psi}\mathbf{k}_I \tag{2.32}$$

(Note the different subscripts on the unit vectors.) And we had previously defined the components of $\boldsymbol{\omega}_{V,I}$ in Frame V to be

$$\boldsymbol{\omega}_{V,I} = P\mathbf{i}_V + Q\mathbf{j}_V + R\mathbf{k}_V \tag{2.33}$$

One can show that one of the relationships between the unit vectors of Frame V and those of the intermediate Frame 2 introduced in the definition of the Euler angles is

$$\mathbf{j}_2 = \cos\phi\,\mathbf{j}_V - \sin\phi\,\mathbf{k}_V \tag{2.34}$$

Plus, from the direction-cosine matrix in Equation (1.15) we have

$$\mathbf{k}_I = -\sin\theta\,\mathbf{i}_V + \sin\phi\cos\theta\,\mathbf{j}_V + \cos\phi\cos\theta\,\mathbf{k}_V \tag{2.35}$$

Substituting the above two expressions into Equation (2.32), and equating $\mathbf{i}_V$, $\mathbf{j}_V$, and $\mathbf{k}_V$ components with those in Equation (2.33), yields the desired kinematic equations relating the angular rates.

$$\boxed{\begin{aligned} P &= \dot{\phi} - \dot{\psi}\sin\theta \\ Q &= \dot{\theta}\cos\phi + \dot{\psi}\cos\theta\sin\phi \\ R &= \dot{\psi}\cos\theta\cos\phi - \dot{\theta}\sin\phi \end{aligned}} \tag{2.36}$$

Or, inverting the above set of equations yields the alternate form of angular kinematic equations used in nonlinear simulations, for example. It is clear that the Euler-angle rates are governed by these expressions.

$$\boxed{\begin{aligned}\dot{\phi} &= P + Q\sin\phi\tan\theta + R\cos\phi\tan\theta \\ \dot{\theta} &= Q\cos\phi - R\sin\phi \\ \dot{\psi} &= (Q\sin\phi + R\cos\phi)\sec\theta\end{aligned}} \tag{2.37}$$

Now, to address the translational kinematics, recall that we had expressed the vehicle's inertial velocity vector in terms of its components in the vehicle-fixed frame as

$$\mathbf{V}_V = U\,\mathbf{i}_V + V\,\mathbf{j}_V + W\,\mathbf{k}_V = \begin{bmatrix} U & V & W \end{bmatrix}\begin{Bmatrix}\mathbf{i}_V \\ \mathbf{j}_V \\ \mathbf{k}_V\end{Bmatrix} \tag{2.38}$$

where we have again used the vector-array notation introduced in Section 1.3. And let us define the components of the same vector in the inertial frame to be

$$\mathbf{V}_V = \begin{bmatrix} \dot{X}_I & \dot{Y}_I & -\dot{\mathrm{h}} \end{bmatrix}\begin{Bmatrix}\mathbf{i}_I \\ \mathbf{j}_I \\ \mathbf{k}_I\end{Bmatrix} \tag{2.39}$$

where h is the vehicle's altitude above the earth's surface. Now recall the direction-cosine matrix Equation (1.15), repeated here

$$\mathbf{T}_{E\text{-}V}(\phi,\,\theta,\,\psi) = \begin{bmatrix} 1 & 0 & 0 \\ 0 & \cos\phi & \sin\phi \\ 0 & -\sin\phi & \cos\phi \end{bmatrix}\begin{bmatrix} \cos\theta & 0 & -\sin\theta \\ 0 & 1 & 0 \\ \sin\theta & 0 & \cos\theta \end{bmatrix}\begin{bmatrix} \cos\psi & \sin\psi & 0 \\ -\sin\psi & \cos\psi & 0 \\ 0 & 0 & 1 \end{bmatrix}$$

$$= \begin{bmatrix} (\cos\theta\cos\psi) & (\cos\theta\sin\psi) & (-\sin\theta) \\ (\sin\phi\sin\theta\cos\psi - \cos\phi\sin\psi) & (\sin\phi\sin\theta\sin\psi + \cos\phi\cos\psi) & (\sin\phi\cos\theta) \\ (\cos\phi\sin\theta\cos\psi + \sin\phi\sin\psi) & (\cos\phi\sin\theta\sin\psi - \sin\phi\cos\psi) & (\cos\phi\cos\theta) \end{bmatrix}$$

with

$$\begin{Bmatrix}\mathbf{i}_V \\ \mathbf{j}_V \\ \mathbf{k}_V\end{Bmatrix} = \mathbf{T}_{E\text{-}V}(\phi,\,\theta,\,\psi)\begin{Bmatrix}\mathbf{i}_E \\ \mathbf{j}_E \\ \mathbf{k}_E\end{Bmatrix} = \mathbf{T}_{I\text{-}V}(\phi,\,\theta,\,\psi)\begin{Bmatrix}\mathbf{i}_I \\ \mathbf{j}_I \\ \mathbf{k}_I\end{Bmatrix}$$

In the above relationships we are using the fact that in this analysis we have taken Frame E on the earth's surface to be inertial.

Substituting these expressions relating the unit vectors into Equation (2.38), equating to Equation (2.39), and transposing the result yields

$$\begin{bmatrix} \dot{X}_I \\ \dot{Y}_I \\ -\dot{h} \end{bmatrix} = \mathbf{T}_{I\text{-}V}^T(\phi, \theta, \psi) \begin{bmatrix} U \\ V \\ W \end{bmatrix}$$

$$= \begin{bmatrix} \cos\psi & -\sin\psi & 0 \\ \sin\psi & \cos\psi & 0 \\ 0 & 0 & 1 \end{bmatrix} \begin{bmatrix} \cos\theta & 0 & \sin\theta \\ 0 & 1 & 0 \\ -\sin\theta & 0 & \cos\theta \end{bmatrix} \begin{bmatrix} 1 & 0 & 0 \\ 0 & \cos\phi & -\sin\phi \\ 0 & \sin\phi & \cos\phi \end{bmatrix} \begin{bmatrix} U \\ V \\ W \end{bmatrix}$$

So the desired kinematic expressions relating the translational rates are given by

$$\boxed{\begin{aligned} \dot{X}_I &= U\cos\theta\cos\psi + V(\sin\phi\sin\theta\cos\psi - \cos\phi\sin\psi) \\ &\quad + W(\cos\phi\sin\theta\cos\psi + \sin\phi\sin\psi) \\ \dot{Y}_I &= U\cos\theta\sin\psi + V(\sin\phi\sin\theta\sin\psi + \cos\phi\cos\psi) \\ &\quad + W(\cos\phi\sin\theta\sin\psi - \sin\phi\cos\psi) \\ \dot{h} &= U\sin\theta - V\sin\phi\cos\theta - W\cos\phi\cos\theta \end{aligned}} \tag{2.40}$$

Summarizing the key results in this section, we have developed 12 nonlinear equations of motion governing the six degrees of freedom of the rigid vehicle.

1. Equations (2.22) govern the translational-velocity components U, V, and W, driven by the forces on the vehicle.
2. Equations (2.27) govern the rotational-velocity components P, Q, and R, driven by the moments acting on the vehicle.
3. Equations (2.36) (or (2.37)) relate P, Q, and R to $\dot{\phi}$, $\dot{\theta}$, and $\dot{\psi}$.
4. Equations (2.40) relate U, V, and W to $\dot{X}_I$, $\dot{Y}_I$, and $\dot{h}$.

In a full nonlinear simulation of the vehicle, for example, all 12 of these differential equations would be integrated simultaneously. But for much of the analysis of the vehicle's dynamics in this book, Equations (2.40) are frequently not used. This is due to the fact that (assuming gravity is independent of altitude) none of the other nine differential equations are functions of inertial position X_I, Y_I, or h.

2.3 REFERENCE AND PERTURBATION EQUATIONS—FLAT EARTH

We have just completed the derivation of 12 nonlinear equations of motion governing the vehicle's dynamics, which is also the first step in small-perturbation analysis (see Section 1.1). Here we will address the second step—a change of variables and the identification of the Reference and the Perturbation Sets of equations. We begin by defining a change of variables involving reference and perturbation quantities. Then we will substitute this change of variables into the Equations (2.22), governing translation, and then into Equations (2.27), governing rotation, and finally into Equations (2.36) (or (2.37)), the kinematic relationships between the angular rates.

For the change of variables, define the following:

$U = \mathrm{U}_0 + u$	$V = \mathrm{V}_0 + v$	$W = \mathrm{W}_0 + w$
$P = \mathrm{P}_0 + p$	$Q = \mathrm{Q}_0 + q$	$R = \mathrm{R}_0 + r$
$\theta = \Theta_0 + \theta$	$\phi = \Phi_0 + \phi$	$\psi = \Psi_0 + \psi$
$F_{A_X} = \mathrm{F}_{A_{X_0}} + f_{A_X}$	$F_{A_Y} = \mathrm{F}_{A_{Y_0}} + f_{A_Y}$	$F_{A_Z} = \mathrm{F}_{A_{Z_0}} + f_{A_Z}$
$F_{P_X} = \mathrm{F}_{P_{X_0}} + f_{P_X}$	$F_{P_Y} = \mathrm{F}_{P_{Y_0}} + f_{P_Y}$	$F_{P_Z} = \mathrm{F}_{P_{Z_0}} + f_{P_Z}$
$L_A = \mathrm{L}_{A_0} + l_A$	$M_A = \mathrm{M}_{A_0} + m_A$	$N_A = \mathrm{N}_{A_0} + n_A$
$L_P = \mathrm{L}_{P_0} + l_P$	$M_P = \mathrm{M}_{P_0} + m_P$	$N_P = \mathrm{N}_{P_0} + n_P$

(2.41)

Here, the quantities with the zero subscript are the reference variables, and the quantities denoted in lower-case italics are the perturbation quantities.

Under this change of variables, Equations (2.22) become

$$\begin{aligned}
&m\big((\dot{\mathrm{U}}_0 + \dot{u}) + (\mathrm{Q}_0\mathrm{W}_0 + \mathrm{Q}_0 w + \mathrm{W}_0 q + wq) - (\mathrm{V}_0\mathrm{R}_0 + \mathrm{V}_0 r + \mathrm{R}_0 v + rv)\big)\\
&= -mg\sin(\Theta_0 + \theta) + (\mathrm{F}_{A_{X_0}} + f_{A_X}) + (\mathrm{F}_{P_{X_0}} + f_{P_X})\\
&m\big((\dot{\mathrm{V}}_0 + \dot{v}) + (\mathrm{R}_0\mathrm{U}_0 + \mathrm{R}_0 u + \mathrm{U}_0 r + ru) - (\mathrm{P}_0\mathrm{W}_0 + \mathrm{P}_0 w + \mathrm{W}_0 p + pw)\big)\\
&= mg\cos(\Theta_0 + \theta)\sin(\Phi_0 + \phi) + (\mathrm{F}_{A_{Y_0}} + f_{A_Y}) + (\mathrm{F}_{P_{Y_0}} + f_{P_Y})\\
&m\big((\dot{\mathrm{W}}_0 + \dot{w}) + (\mathrm{P}_0\mathrm{V}_0 + \mathrm{P}_0 v + \mathrm{V}_0 p + pv) - (\mathrm{Q}_0\mathrm{U}_0 + \mathrm{Q}_0 u + \mathrm{U}_0 q + qu)\big)\\
&= mg\cos(\Theta_0 + \theta)\cos(\Phi_0 + \phi) + (\mathrm{F}_{A_{Z_0}} + f_{A_{Z_0}}) + (\mathrm{F}_{P_{Z_0}} + f_{P_Z})
\end{aligned} \tag{2.42}$$

Noting the following identities for the sine and cosine of the sum of two angles

$$\sin(\Theta_0 + \theta) = \sin\Theta_0\cos\theta + \cos\Theta_0\sin\theta$$
$$\cos(\Theta_0 + \theta) = \cos\Theta_0\cos\theta - \sin\Theta_0\sin\theta$$

(with similar expressions for ϕ), Equations (2.42) become

$$\begin{aligned}
&m\big((\dot{\mathrm{U}}_0 + \dot{u}) + (\mathrm{Q}_0\mathrm{W}_0 + \mathrm{Q}_0 w + \mathrm{W}_0 q + wq) - (\mathrm{V}_0\mathrm{R}_0 + \mathrm{V}_0 r + \mathrm{R}_0 v + rv)\big)\\
&= -mg(\sin\Theta_0\cos\theta + \cos\Theta_0\sin\theta) + (\mathrm{F}_{A_{X_0}} + f_{A_{X_0}}) + (\mathrm{F}_{P_{X_0}} + f_{P_X})
\end{aligned}$$

$$m\left((\dot{V}_0 + \dot{v}) + (R_0U_0 + R_0u + U_0r + ru) - (P_0W_0 + P_0w + W_0p + pw)\right)$$
$$= mg(\cos\Theta_0\cos\theta - \sin\Theta_0\sin\theta)(\sin\Phi_0\cos\phi + \cos\Phi_0\sin\phi) + (F_{A_{Y_0}} + f_{A_Y})$$
$$+ (F_{P_{Y_0}} + f_{P_Y})$$

$$m\left((\dot{W}_0 + \dot{w}) + (P_0V_0 + P_0v + V_0p + pv) - (Q_0U_0 + Q_0u + U_0q + qu)\right)$$
$$= mg(\cos\Theta_0\cos\theta - \sin\Theta_0\sin\theta)(\cos\Phi_0\cos\phi - \sin\Phi_0\sin\phi) + (F_{A_{Z_0}} + f_{A_Z})$$
$$+ (F_{P_{Z_0}} + f_{P_Z})$$

We now invoke the *small-perturbation assumption,* which implies that

products of perturbation quantities ~ 0

sin (perturbation angle) ~ perturbation angle

cos (perturbation angle) ~ 1.

Under these assumptions, Equations (2.42) finally become

$$m\left((\underline{\dot{U}_0} + \dot{u}) + (\underline{Q_0W_0} + Q_0w + W_0q) - (\underline{V_0R_0} + V_0r + R_0v)\right)$$
$$= -mg(\underline{\sin\Theta_0} + \cos\Theta_0\theta) + (\underline{F_{A_{X_0}}} + f_{A_X}) + (\underline{F_{P_{X_0}}} + f_{P_X})$$

$$m\left((\underline{\dot{V}_0} + \dot{v}) + (\underline{R_0U_0} + R_0u + U_0r) - (\underline{P_0W_0} + P_0w + W_0p)\right) \tag{2.43}$$
$$= mg(\underline{\cos\Theta_0} - \sin\Theta_0\theta)(\underline{\sin\Phi_0} + \cos\Phi_0\phi) + (\underline{F_{A_{Y_0}}} + f_{A_Y}) + (\underline{F_{P_{Y_0}}} + f_{P_Y})$$

$$m\left((\underline{\dot{W}_0} + \dot{w}) + (\underline{P_0V_0} + P_0v + V_0p) - (\underline{Q_0U_0} + Q_0u + U_0q)\right)$$
$$= mg(\underline{\cos\Theta_0} - \sin\Theta_0\theta)(\underline{\cos\Phi_0} - \sin\Phi_0\phi) + (\underline{F_{A_{Z_0}}} + f_{A_Z}) + (\underline{F_{P_{Z_0}}} + f_{P_Z})$$

The underlined terms are discussed further below.

We are now ready to extract the two sets of equations—the Reference Set and the Perturbation Set. Looking closely at the three equations above, first note that the underlined terms are not only functions of reference variables alone, but are in fact the original equations of motion given in Equations (2.22). This is not surprising since at the reference condition all perturbations are zero. Therefore, the following equations must be satisfied.

$$\boxed{\begin{aligned} m(\dot{U}_0 + Q_0W_0 - V_0R_0) &= -mg\sin\Theta_0 + F_{A_{X_0}} + F_{P_{X_0}} \\ m(\dot{V}_0 + R_0U_0 - P_0W_0) &= mg\cos\Theta_0\sin\Phi_0 + F_{A_{Y_0}} + F_{P_{Y_0}} \\ m(\dot{W}_0 + P_0V_0 - Q_0U_0) &= mg\cos\Theta_0\cos\Phi_0 + F_{A_{Z_0}} + F_{P_{Z_0}} \end{aligned}} \tag{2.44}$$

This constitutes the <u>Reference Set</u> of equations corresponding to Equations (2.22).

Referring again to Equations (2.43), we see that after extraction of Equations (2.44) from the three equations in question, and neglecting any terms involving products of perturbation quantities, we obtain the Perturbation Set of equations below.

$$\begin{aligned}
&m\left(\dot{u} + (\mathrm{Q}_0 w + \mathrm{W}_0 q) - (\mathrm{V}_0 r + \mathrm{R}_0 v)\right) = -mg\cos\Theta_0\theta + f_{A_X} + f_{P_X}\\
&m\left(\dot{v} + (\mathrm{R}_0 u + \mathrm{U}_0 r) - (\mathrm{P}_0 w + \mathrm{W}_0 p)\right) = mg\left(\cos\Theta_0\cos\Phi_0\phi\right.\\
&\qquad\qquad\left.- \sin\Theta_0\sin\Phi_0\theta\right) + f_{A_Y} + f_{P_Y}\\
&m\left(\dot{w} + (\mathrm{P}_0 v + \mathrm{V}_0 p) - (\mathrm{Q}_0 u + \mathrm{U}_0 q)\right) = -mg\left(\cos\Theta_0\sin\Phi_0\phi\right.\\
&\qquad\qquad\left.+ \sin\Theta_0\cos\Phi_0\theta\right) + f_{A_Z} + f_{P_Z}
\end{aligned} \tag{2.45}$$

These three equations constitute the Perturbation Set of equations corresponding to Equations (2.22), the equations governing translational motion. Note that the Perturbation Set is linear in the perturbation variables, and the coefficients of these perturbation variables are functions of the reference variables.

We now turn our attention to the equations governing rotational motion, or Equations (2.27). As with the equations governing translation, we first express the rotational equations in terms of the reference and perturbation variables via the change of variables, and then extract the Reference and Perturbation Sets of equations.

Substituting the change of variables (Equations (2.41)) into Equations (2.27) yields

$$\begin{aligned}
&I_{xx}(\dot{\mathrm{P}}_0 + \dot{p}) - I_{xz}\left((\dot{\mathrm{R}}_0 + \dot{r}) + (\mathrm{P}_0 + p)(\mathrm{Q}_0 + q)\right) + (I_{zz} - I_{yy})(\mathrm{Q}_0 + q)(\mathrm{R}_0 + r)\\
&= (\mathrm{L}_{A_0} + l_A) + (\mathrm{L}_{P_0} + l_P)\\
&I_{yy}(\dot{\mathrm{Q}}_0 + \dot{q}) + (I_{xx} - I_{zz})(\mathrm{P}_0 + p)(\mathrm{R}_0 + r) + I_{xz}\left((\mathrm{P}_0 + p)^2 - (\mathrm{R}_0 + r)^2\right)\\
&= (\mathrm{M}_{A_0} + m_A) + (\mathrm{M}_{P_0} + m_P)\\
&I_{zz}(\dot{\mathrm{R}}_0 + \dot{r}) - I_{xz}\left((\dot{\mathrm{P}}_0 + \dot{p}) - (\mathrm{Q}_0 + q)(\mathrm{R}_0 + r)\right) + (I_{yy} - I_{xx})(\mathrm{P}_0 + p)(\mathrm{Q}_0 + q)\\
&= (\mathrm{N}_{A_0} + n_A) + (\mathrm{N}_{P_0} + n_P)
\end{aligned} \tag{2.46}$$

(Note that here we have assumed that the XZ plane of the vehicle is a plane of symmetry, and hence $I_{xy} = I_{yz} = 0$. If that is not the case, the additional terms must be added to the above expressions.) We will now rearrange the above three equations. After neglecting terms involving products of perturbation quantities (the small-perturbation assumption), we obtain

$$\begin{aligned}
&I_{xx}(\underline{\dot{\mathrm{P}}_0} + \dot{p}) - I_{xz}\left((\underline{\dot{\mathrm{R}}_0} + \dot{r}) + (\underline{\mathrm{P}_0\mathrm{Q}_0} + \mathrm{Q}_0 p + \mathrm{P}_0 q)\right) + (I_{zz} - I_{yy})(\underline{\mathrm{Q}_0\mathrm{R}_0} + \mathrm{R}_0 q + \mathrm{Q}_0 r)\\
&= (\underline{\mathrm{L}_{A_0}} + l_A) + (\underline{\mathrm{L}_{P_0}} + l_P)\\
&I_{yy}(\underline{\dot{Q}_0} + \dot{q}) + (I_{xx} - I_{zz})(\underline{\mathrm{P}_0\mathrm{R}_0} + \mathrm{R}_0 p + \mathrm{P}_0 r) + I_{xz}\left((\underline{\mathrm{P}_0^2} + 2\mathrm{P}_0 p) - (\underline{\mathrm{R}_0^2} + 2\mathrm{R}_0 r)\right)\\
&= (\underline{\mathrm{M}_{A_0}} + m_A) + (\underline{\mathrm{M}_{P_0}} + m_P)
\end{aligned}$$

$$I_{zz}(\underline{\dot{R}_0}+\dot{r})-I_{xz}\left((\underline{\dot{P}_0}+\dot{p})-(\underline{Q_0R_0}+R_0q+Q_0r)\right)+(I_{yy}-I_{xx})(\underline{P_0Q_0}+Q_0p+P_0q)$$
$$=(\underline{N_{A_0}}+n_A)+(\underline{N_{P_0}}+n_P) \qquad (2.47)$$

Note that again we have underlined the terms in the above expressions that correspond to the original equations of motion (Equations (2.27)), and that these underlined terms are expressed in terms of reference variables only. At the reference condition, all perturbations are zero, and so we are left with

$$\boxed{\begin{aligned} &I_{xx}\dot{P}_0-I_{xz}(\dot{R}_0+P_0Q_0)+(I_{zz}-I_{yy})Q_0R_0=L_{A_0}+L_{P_0}\\ &I_{yy}\dot{Q}_0+(I_{xx}-I_{zz})P_0R_0+I_{xz}(P_0^2-R_0^2)=M_{A_0}+M_{P_0}\\ &I_{zz}\dot{R}_0-I_{xz}(\dot{P}_0-Q_0R_0)+(I_{yy}-I_{xx})P_0Q_0=N_{A_0}+N_{P_0}\end{aligned}} \qquad (2.48)$$

These constitute the <u>Reference Set</u> of equations corresponding to Equations (2.27).

Now, after extracting Equations (2.48) (the underlined terms) from Equations (2.47), we are left with

$$\boxed{\begin{aligned} &I_{xx}\dot{p}-I_{xz}\left(\dot{r}+(Q_0p+P_0q)\right)+(I_{zz}-I_{yy})(R_0q+Q_0r)=l_A+l_P\\ &I_{yy}\dot{q}+(I_{xx}-I_{zz})(R_0p+P_0r)+2I_{xz}(P_0p-R_0r)=m_A+m_P\\ &I_{zz}\dot{r}-I_{xz}\left(\dot{p}-(R_0q+Q_0r)\right)+(I_{yy}-I_{xx})(Q_0p+P_0q)=n_A+n_P\end{aligned}} \qquad (2.49)$$

These three equations constitute the <u>Perturbation Set</u> of equations corresponding to Equations (2.27), which govern the rotational motion of the vehicle.

And finally, we address the three rotational kinematic equations, Equations (2.36). Substituting the change of variables (Equations (2.41)) into these kinematic equations yields

$$\begin{aligned} &(P_0+p)=(\dot{\Phi}_0+\dot{\phi})-(\dot{\Psi}_0+\dot{\psi})\sin(\Theta_0+\theta)\\ &(Q_0+q)=(\dot{\Theta}_0+\dot{\theta})\cos(\Phi_0+\phi)+(\dot{\Psi}_0+\dot{\psi})\sin(\Phi_0+\phi)\cos(\Theta_0+\theta)\\ &(R_0+r)=-(\dot{\Theta}_0+\dot{\theta})\sin(\Phi_0+\phi)+(\dot{\Psi}_0+\dot{\psi})\cos(\Phi_0+\phi)\cos(\Theta_0+\theta)\end{aligned} \qquad (2.50)$$

Using the identities for the sums of two angles, and setting

$$\sin(\text{perturbation angle})\sim\text{perturbation angle}$$
$$\cos(\text{perturbation angle})\sim 1$$

(i.e, making the small-perturbation assumption) yields

$$\begin{aligned} &(P_0+p)=(\dot{\Phi}_0+\dot{\phi})-(\dot{\Psi}_0+\dot{\psi})(\sin\Theta_0+\cos\Theta_0\theta)\\ &(Q_0+q)=(\dot{\Theta}_0+\dot{\theta})(\cos\Phi_0-\sin\Phi_0\phi)\\ &\qquad+(\dot{\Psi}_0+\dot{\psi})(\sin\Phi_0+\cos\Phi_0\phi)(\cos\Theta_0-\sin\Theta_0\theta)\end{aligned}$$

$$(R_0 + r) = -(\dot{\Theta}_0 + \dot{\theta})(\sin\Phi_0 + \cos\Phi_0\,\phi) + (\dot{\Psi}_0 + \dot{\psi})(\cos\Phi_0 - \sin\Phi_0\,\phi)(\cos\Theta_0 - \sin\Theta_0\,\theta)$$

Rearranging yields

$$\begin{aligned}
(\underline{P_0} + p) &= (\underline{\dot{\Phi}_0} + \dot{\phi}) - \dot{\Psi}_0(\underline{\sin\Theta_0} + \cos\Theta_0\,\theta) - \sin\Theta_0\,\dot{\psi} \\
(\underline{Q_0} + q) &= \dot{\Theta}_0(\underline{\cos\Phi_0} - \sin\Phi_0\,\phi) + \cos\Phi_0\,\dot{\theta} \\
&\quad + \dot{\Psi}_0(\underline{\sin\Phi_0\cos\Theta_0} + \cos\Phi_0\cos\Theta_0\,\phi - \sin\Phi_0\sin\Theta_0\,\theta) + \sin\Phi_0\cos\Theta_0\,\dot{\psi} \\
(\underline{R_0} + r) &= -\dot{\Theta}_0(\underline{\sin\Phi_0} + \cos\Phi_0\,\phi) - \sin\Phi_0\,\dot{\theta} \\
&\quad + \dot{\Psi}_0(\underline{\cos\Phi_0\cos\Theta_0} - \sin\Phi_0\cos\Theta_0\,\phi - \cos\Phi_0\sin\Theta_0\,\theta) + \cos\Phi_0\cos\Theta_0\,\dot{\psi}
\end{aligned} \tag{2.51}$$

Collecting the underlined terms, and noting that these are the original nonlinear equations now governing the reference variables, we have the following Reference Set of equations

$$\boxed{\begin{aligned}
P_0 &= \dot{\Phi}_0 - \dot{\Psi}_0\sin\Theta_0 \\
Q_0 &= \dot{\Theta}_0\cos\Phi_0 + \dot{\Psi}_0\sin\Phi_0\cos\Theta_0 \\
R_0 &= -\dot{\Theta}_0\sin\Phi_0 + \dot{\Psi}_0\cos\Phi_0\cos\Theta_0
\end{aligned}} \tag{2.52}$$

Or, inverting the above and consistent with Equations (2.37), we also have

$$\boxed{\begin{aligned}
\dot{\Phi}_0 &= P_0 + Q_0\sin\Phi_0\tan\Theta_0 + R_0\cos\Phi_0\tan\Theta_0 \\
\dot{\Theta}_0 &= Q_0\cos\Phi_0 - R_0\sin\Phi_0 \\
\dot{\Psi}_0 &= (Q_0\sin\Phi_0 + R_0\cos\Phi_0)\sec\Theta_0
\end{aligned}} \tag{2.53}$$

Finally, extracting Equations (2.52) from Equations (2.51) leaves the following Perturbation Set of equations corresponding to Equations (2.36).

$$\boxed{\begin{aligned}
p &= \dot{\phi} - \dot{\Psi}_0\cos\Theta_0\,\theta - \sin\Theta_0\,\dot{\psi} \\
q &= \cos\Phi_0\,\dot{\theta} - \dot{\Theta}_0\sin\Phi_0\,\phi + \dot{\Psi}_0(\cos\Phi_0\cos\Theta_0\,\phi - \sin\Phi_0\sin\Theta_0\,\theta) \\
&\quad + \sin\Phi_0\cos\Theta_0\,\dot{\psi} \\
r &= \cos\Phi_0\cos\Theta_0\,\dot{\psi} - \dot{\Theta}_0\cos\Phi_0\,\phi - \sin\Phi_0\,\dot{\theta} \\
&\quad - \dot{\Psi}_0(\sin\Phi_0\cos\Theta_0\,\phi + \cos\Phi_0\sin\Theta_0\,\theta)
\end{aligned}} \tag{2.54}$$

If instead of starting with Equations (2.36) we began with the inverted set, or Equations (2.37), one can show that the resulting Perturbation Set of equations governing the perturbation kinematics is

Table 2.1 Reference Sets, Perturbation Sets, and Genesis Nonlinear Equations

Genesis Equations	Reference Set	Perturbation Set
Translational dynamics Equations (2.22)	Equations (2.44)	Equations (2.45)
Rotational dynamics Equations (2.27) (with $I_{xy} = I_{yz} = 0$)	Equations (2.48)	Equations (2.49)
Rotational kinematics Equations (2.36) or (2.37)	Equations (2.52) or (2.53)	Equations (2.54) or (2.55)

$$\begin{aligned}
\dot{\phi} &= p + \tan\Theta_0\big(\sin\Phi_0 q + \cos\Phi_0 r + (\mathrm{Q}_0\cos\Phi_0 - \mathrm{R}_0\sin\Phi_0)\phi\big)\\
&\quad + \big(\mathrm{Q}_0\sin\Phi_0 + \mathrm{R}_0\cos\Phi_0 + (\dot{\Phi}_0 - \mathrm{P}_0)\tan\Theta_0\big)\theta\\
\dot{\theta} &= \cos\Phi_0 q - \sin\Phi_0 r - (\mathrm{Q}_0\sin\Phi_0 + \mathrm{R}_0\cos\Phi_0)\phi\\
\dot{\psi} &= \dot{\Psi}_0\tan\Theta_0\theta + \big(\sin\Phi_0 q + \cos\Phi_0 r - (\mathrm{R}_0\sin\Phi_0 - \mathrm{Q}_0\cos\Phi_0)\phi\big)/\cos\Theta_0
\end{aligned} \tag{2.55}$$

In summary, using the small-perturbation assumption we have extracted the Reference and Perturbation Sets of equations associated with the original nonlinear equations of motion. The reference and perturbation equations and their genesis nonlinear equations are summarized in Table 2.1 above. As required, the Reference Sets are identical to the original nonlinear equations of motion, and involve only reference variables, while the Perturbation Sets are linear in the perturbation variables.

As noted at the end of Section 2.2, the kinematic equations relating translational velocities U, V, and W to the vehicle's inertial position are frequently of little interest in the perturbation analysis of the vehicle's dynamics (though there are exceptions to this note, such as with altitude). Hence the Reference and Perturbation Sets of equations associated with these kinematic equations are not derived here. This is left as an exercise for the student.

We will use the results in this section to perform the small-perturbation analysis of the vehicle in Chapters 9 and 10, after discussing the aerodynamic and propulsive forces in Chapters 5 and 6.

2.4 EFFECTS OF ROTATING MASSES

As a special case of frequent interest in flight dynamics, we will now develop the equations of motion of a rigid vehicle (i.e., no structural deformation) that possesses rotating machinery such as turbojet engines or propellers. For certain vehicles, such as the tilt-rotor vehicle mentioned in the problems in Chapter 1, the effects of these rotating masses can significantly influence the vehicle's attitude dynamics. We will develop the equations by considering an aircraft with

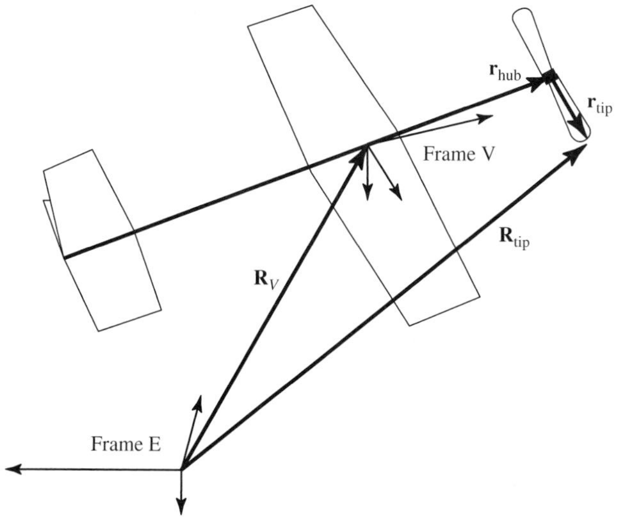

Figure 2.3 Propeller position vectors.

a propeller, but, as we will see, the results may be applied directly to aircraft with any machinery rotating about an axis of symmetry.

Consider the vehicle schematic shown in Figure 2.3, which is the same as Figure 1.5 used in Example 1.3. We will employ here the same techniques used in the example, such as expressing the position of a propeller mass particle $\rho_V dV$ in terms of the vectors $\mathbf{r}_{\text{hub}}$ and $\mathbf{r}_{\text{tip}}$, and breaking up the volume integrals over the vehicle into integrals over the rigid vehicle plus integrals over the propeller. However, since we will be interested in all positions of the propeller mass particles rather than just at the tip, we will use the position vector $\mathbf{r}_{dV}$ rather than $\mathbf{r}_{\text{tip}}$, where dV refers to an infinitesimal volume of the propeller's material. Note that $\mathbf{r}_{dV}$ defines the position of a rotating (propeller) mass particle relative to the hub of the propeller, or the center of mass (gravity) of the rotating mass (propeller). Finally, for convenience, an intermediate coordinate frame, Frame P, is introduced that is fixed to, and rotates with, the rotating mass (propeller) at an angular velocity relative to Frame V of $\boldsymbol{\omega}_{\text{Prop}}$.

We begin with Newton's law in translational form, or Equation (2.1), repeated here.

$$\int_{\text{Vol}} \frac{d}{dt}\Big|_I \left(\rho_V \frac{d\mathbf{p}'}{dt}\Big|_I dV \right) = \int_{\text{Vol}} \rho_V \mathbf{g} dV + \int_{\text{Surface}} d\mathbf{f}_{\text{ext}}$$

The integral expressions on the right-hand side of this equation remain unchanged due to propeller rotational dynamics, although this rotation may affect the pressure distribution $\mathbf{f}_{\text{ext}}$ over the vehicle. As always, one must properly track which aspects of the pressure distributions are associated with aerodynamic effects and which are associated with propulsive effects. But this has to do with evaluating the integral expression, not the form.

However, we must look closely at the left-hand side of Equation (2.1), in light of the rotating mass. Starting with Equation (2.6), repeated here, we had

$$\int_{\text{Vol}} \frac{d}{dt}\Big|_I \left(\rho_V \frac{d\mathbf{p}'}{dt}\Big|_I dV\right) = \int_{\text{Vol}} \frac{d}{dt}\Big|_I \left(\rho_V \frac{d\mathbf{p}_V}{dt}\Big|_I dV\right) + \int_{\text{Vol}} \frac{d}{dt}\Big|_I \left(\rho_V \frac{d\mathbf{p}}{dt}\Big|_I dV\right)$$

and we noted in Section 2.1 that the first integral on the right-hand side becomes

$$\int_{\text{Vol}} \frac{d}{dt}\Big|_I \left(\rho_V \frac{d\mathbf{p}_V}{dt}\Big|_I dV\right) = \frac{d}{dt}\Big|_I \left(\frac{d\mathbf{p}_V}{dt}\Big|_I \int_{\text{Vol}} \rho_V dV\right) = \frac{d}{dt}\Big|_I \left(m \frac{d\mathbf{p}_V}{dt}\Big|_I\right)$$

This result is not changed due to the rotating propeller mass. (We will, however, note that the vehicle's center of mass, which has inertial position $\mathbf{p}_V$, may be located by treating the propeller as a disk with the same mass as the propeller.)

Now recalling Equation (2.7), or

$$\frac{d\mathbf{p}}{dt}\Big|_I = \frac{d\mathbf{p}}{dt}\Big|_V + \boldsymbol{\omega}_{V,I} \times \mathbf{p}$$

we will look carefully at the last term in Equation (2.6), which can be written as in Equation (2.8):

$$\int_{\text{Vol}} \frac{d}{dt}\Big|_I \left(\rho_V \frac{d\mathbf{p}}{dt}\Big|_I dV\right) = \int_{\text{Vol}} \frac{d}{dt}\Big|_I \left(\rho_V \frac{d\mathbf{p}}{dt}\Big|_V dV\right) + \int_{\text{Vol}} \frac{d}{dt}\Big|_I \left(\rho_V (\boldsymbol{\omega}_{V,I} \times \mathbf{p}) dV\right)$$

We had previously shown that for the rigid vehicle Equation (2.8) equaled zero. We need to determine if this is still the case with the rotating mass.

Note first that Equation (2.7) is valid for defining the positions and inertial velocities of any of the mass elements of the vehicle. But for the mass elements making up the rotating propeller, for which $\mathbf{p} = \mathbf{r}_{\text{hub}} + \mathbf{r}_{dV}$, we can also write

$$\begin{aligned} \frac{d\mathbf{p}}{dt}\Big|_I &= \left(\frac{d\mathbf{r}_{\text{hub}}}{dt}\Big|_V + \frac{d\mathbf{r}_{dV}}{dt}\Big|_V\right) + \left(\boldsymbol{\omega}_{V,I} \times (\mathbf{r}_{\text{hub}} + \mathbf{r}_{dV})\right) \\ &= \left(\frac{d\mathbf{r}_{\text{hub}}}{dt}\Big|_V + \boldsymbol{\omega}_{V,I} \times \mathbf{r}_{\text{hub}}\right) + \left(\frac{d\mathbf{r}_{dV}}{dt}\Big|_P + (\boldsymbol{\omega}_{\text{Prop}} + \boldsymbol{\omega}_{V,I}) \times \mathbf{r}_{dV}\right) \end{aligned} \tag{2.56}$$

Now breaking the volume integral on the left-hand side of Equation (2.8) into integrals over the rigid vehicle plus integrals over the rotating mass (propeller), instead of the first term on the right-hand side of Equation (2.8) we now have

$$\begin{aligned} \int_{\text{Vol}} \frac{d}{dt}\Big|_I \left(\rho_V \frac{d\mathbf{p}}{dt}\Big|_V dV\right) \Rightarrow & \int_{\text{Rigid Vol}} \frac{d}{dt}\Big|_I \left(\rho_V \frac{d\mathbf{p}}{dt}\Big|_V dV\right) \\ & + \int_{\text{Prop Vol}} \frac{d}{dt}\Big|_I \left(\rho_V \frac{d\mathbf{r}_{\text{hub}}}{dt}\Big|_V dV\right) + \int_{\text{Prop Vol}} \frac{d}{dt}\Big|_I \left(\rho_V \frac{d\mathbf{r}_{dV}}{dt}\Big|_P dV\right) = 0 \end{aligned} \tag{2.57}$$

Each of these three integrals equals zero since for mass elements $\rho_V dV$ of the rigid vehicle or the rigid propeller, and for the hub position on a rigid vehicle,

$$\frac{d\mathbf{p}}{dt}\Big|_V = \frac{d\mathbf{r}_{\text{hub}}}{dt}\Big|_V = \frac{d\mathbf{r}_{dV}}{dt}\Big|_P = 0$$

And instead of the last term on the right-hand side of Equation (2.8) we have

$$\begin{aligned}\int_{\text{Vol}} \frac{d}{dt}\Big|_I \left(\rho_V(\boldsymbol{\omega}_{V,I}\times\mathbf{p})dV\right) \Rightarrow & \int_{\text{Rigid Vol}} \frac{d}{dt}\Big|_I \left(\rho_V(\boldsymbol{\omega}_{V,I}\times\mathbf{p})dV\right) \\ &+ \int_{\text{Prop Vol}} \frac{d}{dt}\Big|_I \left(\rho_V(\boldsymbol{\omega}_{V,I}\times\mathbf{r}_{\text{hub}})dV\right) + \int_{\text{Prop Vol}} \frac{d}{dt}\Big|_I \left(\rho_V\left((\boldsymbol{\omega}_{\text{Prop}}+\boldsymbol{\omega}_{V,I})\times\mathbf{r}_{dV}\right)dV\right)\end{aligned} \tag{2.58}$$

The first term on the right of Equation (2.58) can be expressed as

$$\begin{aligned}\int_{\text{Rigid Vol}} \frac{d}{dt}\Big|_I \left(\rho_V(\boldsymbol{\omega}_{V,I}\times\mathbf{p})dV\right) = & \frac{d\boldsymbol{\omega}_{V,I}}{dt}\Big|_I \times \int_{\text{Rigid Vol}} \rho_V\mathbf{p}dV \\ &+ \boldsymbol{\omega}_{V,I}\times\left(\int_{\text{Rigid Vol}} \frac{d}{dt}\Big|_V \left(\rho_V\mathbf{p}dV\right) + \boldsymbol{\omega}_{V,I}\times\int_{\text{Rigid Vol}} \rho_V\mathbf{p}dV\right)\end{aligned} \tag{2.59}$$

Now since the densities are constant, and since the above equation involves integration over just the rigid vehicle, then

$$\frac{d\mathbf{p}}{dt}\Big|_V = 0 \Rightarrow \int_{\text{Rigid Vol}} \frac{d}{dt}\Big|_V \left(\rho_V\mathbf{p}dV\right) = 0 \tag{2.60}$$

So we are left with the following:

$$\int_{\text{Rigid Vol}} \frac{d}{dt}\Big|_I \left(\rho_V(\boldsymbol{\omega}_{V,I}\times\mathbf{p})dV\right) = \left(\frac{d\boldsymbol{\omega}_{V,I}}{dt}\Big|_I + (\boldsymbol{\omega}_{V,I}\times\boldsymbol{\omega}_{V,I})\right)\times\int_{\text{Rigid Vol}} \rho_V\mathbf{p}dV \tag{2.61}$$

Likewise, the second term on the right-hand side of Equation (2.58) can be written

$$\begin{aligned}\int_{\text{Prop Vol}} \frac{d}{dt}\Big|_I \left(\rho_V(\boldsymbol{\omega}_{V,I}\times\mathbf{r}_{\text{hub}})dV\right) = & \frac{d\boldsymbol{\omega}_{V,I}}{dt}\Big|_I \times\mathbf{r}_{\text{hub}} \int_{\text{Prop Vol}} \rho_V dV \\ &+ \boldsymbol{\omega}_{V,I}\times\left(\frac{d}{dt}\Big|_V \mathbf{r}_{\text{hub}} \int_{\text{Prop Vol}} (\rho_V dV) + \boldsymbol{\omega}_{V,I}\times\mathbf{r}_{\text{hub}} \int_{\text{Prop Vol}} \rho_V dV\right)\end{aligned} \tag{2.62}$$

where, since $\mathbf{r}_{\text{hub}}$ is constant with respect to the volume-integration operation, it has been moved outside the integrals. Again, since the densities of the propeller mass elements are constant, and since for a rigid vehicle the position of the hub remains fixed with respect to Frame V, we have

$$\frac{d\mathbf{r}_{\text{hub}}}{dt}\Big|_V = 0 \Rightarrow \frac{d}{dt}\Big|_V \mathbf{r}_{\text{hub}} \int_{\text{Prop Vol}} (\rho_V dV) = 0 \tag{2.63}$$

And we are left with the following:

$$\int_{\text{Prop Vol}} \frac{d}{dt}\Big|_I \left(\rho_V(\boldsymbol{\omega}_{V,I} \times \mathbf{r}_{\text{hub}})dV\right) = \left(\frac{d\boldsymbol{\omega}_{V,I}}{dt}\Big|_I + (\boldsymbol{\omega}_{V,I} \times \boldsymbol{\omega}_{V,I})\right) \times \mathbf{r}_{\text{hub}} \int_{\text{Prop Vol}} \rho_V dV \tag{2.64}$$

Finally, the third term on the right-hand side of Equation (2.58) can be shown to equal

$$\int_{\text{Prop Vol}} \frac{d}{dt}\Big|_I \left(\rho_V\left((\boldsymbol{\omega}_{\text{Prop}} + \boldsymbol{\omega}_{V,I}) \times \mathbf{r}_{dV}\right)dV\right) = \frac{d(\boldsymbol{\omega}_{\text{Prop}} + \boldsymbol{\omega}_{V,I})}{dt}\Big|_I \times \int_{\text{Prop Vol}} \rho_V \mathbf{r}_{dV} dV$$
$$+ (\boldsymbol{\omega}_{\text{Prop}} + \boldsymbol{\omega}_{V,I}) \times \left(\int_{\text{Prop Vol}} \frac{d}{dt}\Big|_P (\rho_V \mathbf{r}_{dV} dV) + (\boldsymbol{\omega}_{\text{Prop}} + \boldsymbol{\omega}_{V,I}) \times \int_{\text{Prop Vol}} \rho_V \mathbf{r}_{dV} dV\right) \tag{2.65}$$

Since the densities of the propeller mass elements are constant, the propeller is rigid, and the first mass moment of the propeller about its hub (i.e., center of mass) is zero, we have

$$\frac{d\mathbf{r}_{dV}}{dt}\Big|_P = 0 \Rightarrow \frac{d}{dt}\Big|_P \int_{\text{Prop Vol}} (\rho_V \mathbf{r}_{dV} dV) = 0 \tag{2.66}$$

Therefore, we are left with the following result

$$\int_{\text{Prop Vol}} \frac{d}{dt}\Big|_I \left(\rho_V\left((\boldsymbol{\omega}_{\text{Prop}} + \boldsymbol{\omega}_{V,I}) \times \mathbf{r}_{dV}\right)dV\right)$$
$$= \left(\frac{d(\boldsymbol{\omega}_{\text{Prop}} + \boldsymbol{\omega}_{V,I})}{dt}\Big|_I + \left((\boldsymbol{\omega}_{\text{Prop}} + \boldsymbol{\omega}_{V,I}) \times (\boldsymbol{\omega}_{\text{Prop}} + \boldsymbol{\omega}_{V,I})\right)\right) \times \int_{\text{Prop Vol}} \rho_V \mathbf{r}_{dV} dV \tag{2.67}$$

Now we note that Equation (2.67) equals zero since the last integral term in the equation is the first mass moment of the propeller taken about its hub, the propeller's center of mass. Also we note finally that the <u>sum</u> of Equations (2.61) and (2.64) equals zero because the sum of the two integral terms in these equations equals zero, or

$$\int_{\text{Rigid Vol}} \rho_V \mathbf{p} dV + \mathbf{r}_{\text{hub}} \int_{\text{Prop Vol}} \rho_V dV = \int_{\text{Rigid Vol}} \rho_V \mathbf{p} dV + \mathbf{r}_{\text{hub}} m_{\text{Prop}} = 0 \tag{2.68}$$

This result follows from the fact that the <u>sum</u> of these two integral terms represents the first mass moment of the vehicle, including the mass of the rotating propeller, taken about the vehicle's center of mass.

Consequently, we have just proven that Equation (2.58) equals zero, and hence Equation (2.8) equals zero. Therefore, the presence of a mass rotating about its center of mass has no effect on Equation (2.1), which governs the translational motion of the vehicle. And furthermore, Equations (2.22), the three scalar equations governing translation, will also remain unchanged.

We will now show, however, that the rotating mass does affect the equations governing rotational motion. This should be expected since a rotating mass affects the total angular momentum of the vehicle. We begin with the basic equation of rotational motion, Equation (2.15), repeated here:

$$\int_{\text{Vol}} \left(\mathbf{p} \times \frac{d}{dt}\Big|_I \left(\rho_V \frac{d\mathbf{p}}{dt}\Big|_I dV\right)\right) = \int_{\text{Surface}} \mathbf{p} \times d\mathbf{f}_{\text{ext}}$$

As in the case of translational motion, the right-hand side of the above equation remains unchanged.

Now note that the left-hand side of the equation is an expression for the total angular momentum of the vehicle $\mathbf{H}_V$ about its center of mass (gravity). However, instead of using Equation (2.15) above, it will be convenient to use the alternate form of the equation of motion given in Equation (2.31), developed in the Comment in Section 2.2, and repeated here:

$$\frac{d\mathbf{H}_V}{dt}\Big|_I = \mathbf{M}_{\text{Aero}} + \mathbf{M}_{\text{Propulsion}}$$

The vehicle's angular momentum $\mathbf{H}_V$ now consists of the angular momentum of the rigid vehicle including the propeller mass (perhaps taken to be a disk), but without the mass rotating with respect to the vehicle, plus the additional momentum due to the mass rotating with respect to the vehicle. That is, let

$$\mathbf{H}_V = \mathbf{H}_{\text{Rigid}} + \mathbf{H}_{\text{Prop}} \tag{2.69}$$

Again, as in the Comment, using the cross-product-matrix notation we have the rigid-vehicle momentum given as

$$\mathbf{H}_{\text{Rigid}} = [\mathbf{I}] \otimes \boldsymbol{\omega}_{V,I} = \left[\begin{bmatrix} I_{xx} & -I_{xy} & -I_{xz} \\ -I_{yx} & I_{yy} & -I_{yz} \\ -I_{zx} & -I_{zy} & I_{zz} \end{bmatrix} \begin{bmatrix} P \\ Q \\ R \end{bmatrix}\right]^T \begin{Bmatrix} \mathbf{i}_V \\ \mathbf{j}_V \\ \mathbf{k}_V \end{Bmatrix} \tag{2.70}$$

And we have the additional angular momentum due to the mass's rotation relative to the vehicle (see Example 2.2 below) given as

$$\mathbf{H}_{\text{Prop}} = J_{\text{Prop}} \omega_{\text{Prop}} \mathbf{i}_{\text{Prop}} \tag{2.71}$$

where $\mathbf{i}_{\text{Prop}}$ is the unit vector directed along the axis of propeller rotation, J_{Prop} is the moment of inertia of the propeller about its axis of rotation, and ω_{Prop} is the magnitude of the angular velocity of the propeller with respect to the vehicle.

This additional momentum due to propeller rotation will have components in the vehicle-fixed frame, Frame V, defined as

$$\mathbf{H}_{\text{Prop}} = \begin{bmatrix} h_x & h_y & h_z \end{bmatrix} \begin{Bmatrix} \mathbf{i}_V \\ \mathbf{j}_V \\ \mathbf{k}_V \end{Bmatrix} \tag{2.72}$$

For example, if the unit vector $\mathbf{i}_{\text{Prop}}$ makes an angle τ up from the X_V axis of the vehicle (i.e., a positive τ rotation about the vehicle's Y_V axis), $\mathbf{H}_{\text{Prop}}$ could be expressed as

$$\mathbf{H}_{\text{Prop}} = J_{\text{Prop}}\omega_{\text{Prop}}\begin{bmatrix} \cos\tau & 0 & -\sin\tau \end{bmatrix} \begin{Bmatrix} \mathbf{i}_V \\ \mathbf{j}_V \\ \mathbf{k}_V \end{Bmatrix} = \begin{bmatrix} h_x & h_y & h_z \end{bmatrix} \begin{Bmatrix} \mathbf{i}_V \\ \mathbf{j}_V \\ \mathbf{k}_V \end{Bmatrix} \tag{2.73}$$

thus defining the momentum components h_x, h_y, and h_z.

Then, still using the cross-product-matrix notation we have the following for the rate of change of the vehicle's angular momentum

$$\begin{aligned}
\frac{d\mathbf{H}_V}{dt}\Big|_I &= \frac{d\mathbf{H}_V}{dt}\Big|_V + \boldsymbol{\omega}_{V,I} \times \mathbf{H}_V \\
&= \left(\frac{d\mathbf{H}_{\text{Rigid}}}{dt}\Big|_V + \frac{d\mathbf{H}_{\text{Prop}}}{dt}\Big|_V\right) + \boldsymbol{\omega}_{V,I} \times \left(\mathbf{H}_{\text{Rigid}} + \mathbf{H}_{\text{Prop}}\right) \\
&= \left[\left[\begin{bmatrix} I_{xx} & -I_{xy} & -I_{xz} \\ -I_{yx} & I_{yy} & -I_{yz} \\ -I_{zx} & -I_{zy} & I_{zz} \end{bmatrix}\begin{bmatrix} \dot{P} \\ \dot{Q} \\ \dot{R} \end{bmatrix} + \begin{bmatrix} \dot{h}_x \\ \dot{h}_y \\ \dot{h}_z \end{bmatrix}\right]^T\right. \\
&\quad \left. + \left[\begin{bmatrix} 0 & -R & Q \\ R & 0 & -P \\ -Q & P & 0 \end{bmatrix}\left[\begin{bmatrix} I_{xx} & -I_{xy} & -I_{xz} \\ -I_{yx} & I_{yy} & -I_{yz} \\ -I_{zx} & -I_{zy} & I_{zz} \end{bmatrix}\begin{bmatrix} P \\ Q \\ R \end{bmatrix} + \begin{bmatrix} h_x \\ h_y \\ h_z \end{bmatrix}\right]\right]^T\right] \begin{Bmatrix} \mathbf{i}_V \\ \mathbf{j}_V \\ \mathbf{k}_V \end{Bmatrix} \\
&= \left(I_{xx}\dot{P} - I_{xz}(\dot{R} + PQ) - I_{yz}(Q^2 - R^2) - I_{xy}(\dot{Q} - RP) + (I_{zz} - I_{yy})RQ\right. \\
&\quad \left. + \dot{h}_x - Rh_y + Qh_z\right)\mathbf{i}_V \\
&\quad + \left(I_{yy}\dot{Q} + (I_{xx} - I_{zz})PR - I_{xy}(\dot{P} + QR) - I_{yz}(\dot{R} - PQ) + I_{xz}(P^2 - R^2)\right. \\
&\quad \left. + \dot{h}_y + Rh_x - Ph_z\right)\mathbf{j}_V \\
&\quad + \left(I_{zz}\dot{R} - I_{xz}(\dot{P} - QR) - I_{xy}(P^2 - Q^2) - I_{yz}(\dot{Q} + RP) + (I_{yy} - I_{xx})PQ\right. \\
&\quad \left. + \dot{h}_z - Qh_x + Ph_y\right)\mathbf{k}_V
\end{aligned} \tag{2.74}$$

where, similar to Equation (2.72), $\dot{h}_x$, $\dot{h}_y$, and $\dot{h}_z$ are defined by the equation

$$\frac{d\mathbf{H}_{\text{Prop}}}{dt}\Big|_V = \begin{bmatrix} \dot{h}_x & \dot{h}_y & \dot{h}_z \end{bmatrix} \begin{Bmatrix} \mathbf{i}_V \\ \mathbf{j}_V \\ \mathbf{k}_V \end{Bmatrix}$$

These three components arise from changes in the rate of propeller rotation and/or changes in the orientation of $\mathbf{i}_{\text{Prop}}$ (such as with a tilt-rotor vehicle).

Finally, by equating the $\mathbf{i}_V$, $\mathbf{j}_V$, and $\mathbf{k}_V$ components in Equation (2.74) with like components of the moments on the right-hand-side of Equation (2.31), we obtain the three scalar equations of rotational motion below (which do not include the assumption that the XZ plane of the vehicle is a plane of symmetry).

$$\boxed{\begin{aligned} &I_{xx}\dot{P} - I_{xz}(\dot{R} + PQ) - I_{yz}(Q^2 - R^2) - I_{xy}(\dot{Q} - RP) + (I_{zz} - I_{yy})RQ \\ &+ \underline{\dot{h}_x - Rh_y + Qh_z} = L_A + L_P \\ &I_{yy}\dot{Q} + (I_{xx} - I_{zz})PR - I_{xy}(\dot{P} + QR) - I_{yz}(\dot{R} - PQ) + I_{xz}(P^2 - R^2) \\ &+ \underline{\dot{h}_y + Rh_x - Ph_z} = M_A + M_P \\ &I_{zz}\dot{R} - I_{xz}(\dot{P} - QR) - I_{xy}(P^2 - Q^2) - I_{yz}(\dot{Q} + RP) + (I_{yy} - I_{xx})PQ \\ &+ \underline{\dot{h}_z - Qh_x + Ph_y} = N_A + N_P \end{aligned}} \tag{2.75}$$

The underlined terms, representing additional gyroscopic torques about each axis, are the terms that arise due to the presence of the rotating mass. Note that typically the largest component of the angular momentum due to mass rotation is h_x. So a pitch rate Q induces a gyroscopic yaw torque, and a yaw rate R induces a gyroscopic pitch torque on the vehicle.

Since the additional terms in Equations (2.75) due to the rotating mass are the only changes to the equations of motion of the vehicle, it is relatively easy to derive the necessary changes to the Reference and Perturbation Sets of equations associated with these equations of motion. And Equations (2.48) and (2.49) would be the only reference and perturbation equations changed due to the rotating mass. The additional terms to be added to these Reference and Perturbation Sets of equations are left as an exercise to the student.

Thus we have derived the scalar equations of rotational motion, including the effects of a rotating mass (propeller). These equations would also be appropriate for a vehicle with a turbojet engine, for example. Using the techniques presented above, similar equations could be developed for vehicles with more than one rotating mass, although multi-engine aircraft frequently have engines or propellers rotating in opposite directions to cancel the gyroscopic torques.

EXAMPLE 2.2

Propeller Angular Momentum

Show that the additional angular momentum due to the propeller's rotation is

$$\mathbf{H}_{\text{Prop}} = J_{\text{Prop}}\omega_{\text{Prop}}\mathbf{i}_{\text{Prop}}$$

with J_{Prop} the moment of inertia of the propeller about its hub.

■ Solution

The angular momentum of a body about its center of mass is expressed as

$$\mathbf{H}(t) = \int_{\text{Vol}} \mathbf{p}(t) \times \rho_V \frac{d\mathbf{p}(t)}{dt}\bigg|_{\text{Ref}} dV$$

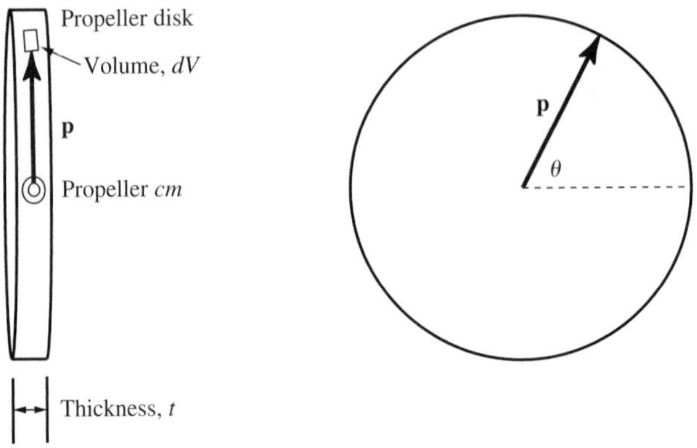

Figure 2.4 Propeller "disk" and mass-element position.

where $\mathbf{p}$ is the position of a mass element $\rho_V dV$ with respect to the center of mass, and the vector derivative is taken with respect to an appropriate reference frame. We will apply this expression to determine the angular momentum of the propeller about its center of mass, due to the propeller's rotation relative to the vehicle.

Let us idealize the propeller as a disk with the same diameter as the propeller, and with a radial mass distribution to match that of the propeller. (One might think of varying the density of the notional disk material radially to accomplish this matching.) This disk, with a mass-element position vector $\mathbf{p}$ is shown in Figure 2.4. Note that the thickness t is constant, and the mass-element volume at a radius p (the magnitude of $\mathbf{p}$) is $dV = tpdpd\theta$.

We can write this mass-element's velocity with respect to a vehicle-fixed frame as

$$\frac{d\mathbf{p}}{dt}\Big|_V = \frac{d\mathbf{p}}{dt}\Big|_P + \left(\boldsymbol{\omega}_{P,V} \times \mathbf{p}\right)$$

where for convenience we have introduced an additional frame fixed to, and rotating with, the propeller, Frame P. Assuming the propeller is rigid, the above expression reduces to

$$\frac{d\mathbf{p}}{dt}\Big|_V = \left(\boldsymbol{\omega}_{P,V} \times \mathbf{p}\right) = \omega_{\text{Prop}} p \, \mathbf{k}_P$$

In obtaining this result we have taken the vector $\mathbf{i}_P$ to be aligned with the axis of propeller rotation, and the radial vector $\mathbf{p}$ to be directed along the $\mathbf{j}_P$ direction.

With the vectors so defined, the cross product in the integrand above becomes

$$\mathbf{p} \times \frac{d\mathbf{p}}{dt}\Big|_{\text{Ref}} = \mathbf{p} \times \frac{d\mathbf{p}}{dt}\Big|_V = \omega_{\text{Prop}} p^2 \, \mathbf{i}_P$$

Consequently, the propeller's angular momentum, with respect to the propeller's center of mass, due to its rotation relative to the vehicle can now be expressed as

$$\mathbf{H}_{\text{Prop}} = \omega_{\text{Prop}}\left(2\pi t \int_0^R p^3 \rho_{\text{Disk}}(p)\, dp\right)\mathbf{i}_P = \omega_{\text{Prop}} J_{\text{Prop}} \, \mathbf{i}_P$$

where R is the radius of the propeller, $\rho_{\text{Disk}}(p)$ is the radial prop-material-density distribution discussed above, and J_{Prop} is thus the propeller's inertia about its hub—its center of mass.

EXAMPLE 2.3

Rotating Engine Mass

Consider an aircraft like the NASA KESTRAL research aircraft, as shown in Figure 2.5. This aircraft is capable of vertical takeoff and landing due to its large thrust-to-weight ratio and thrust-vectoring capability. Let its jet engine have an inertia about its axis of rotation, J_{Engine}, of 3000 slug-ft^2, and have an operating rotational velocity of 20,000 rpm. If the aircraft is experiencing a pitch rate of 5 deg/sec while hovering, what is the gyroscopic torque being exerted on the aircraft, and in what direction does it act?

■ Solution

Taking the engine's axis of rotation to be aligned with the X_V axis of the vehicle, the additional angular momentum due to the engine rotation is

$$\mathbf{H}_{\text{Engine}} = J_{\text{Engine}}\omega_{\text{Engine}}\mathbf{i}_V = h_x\,\mathbf{i}_V$$

Using the third of Equations (2.75), and assuming the engine is rotating at a constant speed, the torque being applied to the aircraft arising due to the engine rotation is

$$\begin{aligned} N_{\text{Gyro}} &= \omega_{\text{Engine}} J_{\text{Engine}} Q \\ &= 20{,}000\left(\frac{2\pi}{60}\right) \times 3000 \times 5\left(\frac{2\pi}{360}\right) \\ &= 5.48 \times 10^5 \text{ ft-lb} \end{aligned}$$

Figure 2.5 NASA KESTRAL VTOL aircraft. *(Photo courtesy of NASA.)*

Note that the gyroscopic term $-h_xQ$ in Equation (2.75) may be moved to the right-hand side of the equation, and thus acts like an applied torque. The above torque represents a positive torque applied about the vehicle's Z_V axis, tending to create a positive yaw rate (nose rotating to the right). If the engine is rotating in the opposite direction (negative X_V), the torque would be of opposite sign and would lead to a negative yaw. Note also that 5.48×10^5 ft-lb is an appreciable torque, and this engine-rotation effect tends to make such a vehicle difficult to operate in hover.

2.5 EFFECTS OF VARIABLE MASS

We will now develop the equations of motion for a rigid vehicle that is expelling mass, associated with the production of propulsive thrust of a missile, for example. By doing so we will be able to identify the necessary modifications to the equations of motion due to the change in mass. Variable-mass effects are not usually significant for air-breathing vehicles (vehicles using air as the working fluid for propulsion). However, variable-mass effects can be significant when modeling the dynamics of a missile, the mass of which includes both fuel and oxidizer to be expelled.

Consider the sketch shown in Figure 2.6, which shows the position vectors of the vehicle's instantaneous center of mass, $\mathbf{p}_V$, and those of an arbitrary mass element of the vehicle $\rho_V dV$, $\mathbf{p}$ and $\mathbf{p}'$. Also shown is a mass element of the vehicle Δm that has just been expelled from the vehicle by passing through the plane at the nozzle exit. This mass element has position vectors $\mathbf{p}_E$ and $\mathbf{p}_o$. Finally, note that $\mathbf{p}_V$, $\mathbf{p}'$, and $\mathbf{p}_o$ are inertial position vectors, while $\mathbf{p}$ and $\mathbf{p}_E$ are positions relative to the vehicle's instantaneous center of mass.

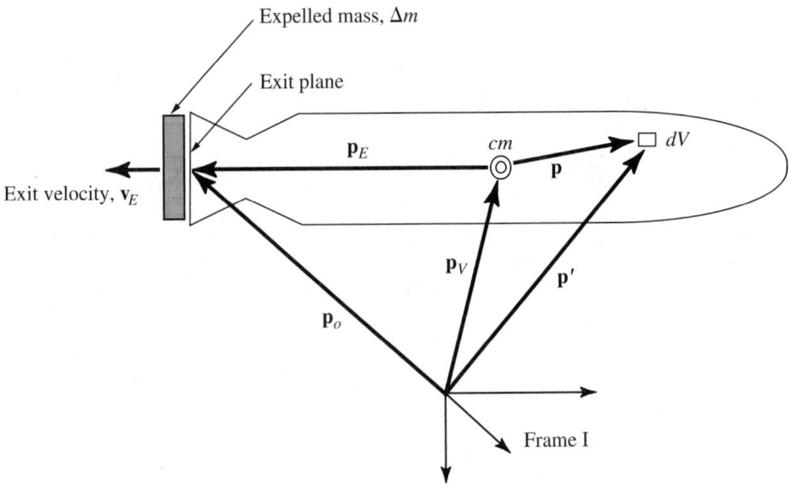

Figure 2.6 Schematic for a variable-mass system (missile).

Now at time t, the translational momentum $\mathbf{Q}(t)$ of the variable-mass system (vehicle) is

$$\mathbf{Q}(t) = \int_{\text{Vol}} \rho_V \frac{d\mathbf{p}'}{dt}\Big|_I \, dV \tag{2.76}$$

while at time $t + \Delta t$ the system's momentum is

$$\mathbf{Q}(t + \Delta t) = \int_{\text{Vol}} \left(\rho_V \frac{d\mathbf{p}'}{dt}\Big|_I + \Delta\left(\rho_V \frac{d\mathbf{p}'}{dt}\Big|_I\right)\right) dV + \Delta m\left(\frac{d\mathbf{p}_o}{dt}\Big|_I + \Delta\frac{d\mathbf{p}_o}{dt}\Big|_I\right) \tag{2.77}$$

The last term represents the change in the system's momentum due to the change in mass (with $\Delta m < 0$ if mass is being expelled). Taking the limit as Δt goes to zero yields the rate of change of momentum, or

$$\lim_{\Delta t \to 0} \frac{\Delta \mathbf{Q}}{\Delta t} = \frac{d\mathbf{Q}}{dt}\Big|_I = \underline{\int_{\text{Vol}} \frac{d}{dt}\Big|_I \left(\rho_V \frac{d\mathbf{p}'}{dt}\Big|_I\right) dV} + \dot{m}\frac{d\mathbf{p}_o}{dt}\Big|_I \tag{2.78}$$

since

$$\lim_{\Delta t \to 0} \frac{\Delta m\left(\Delta \frac{d\mathbf{p}_o}{dt}\Big|_I\right)}{\Delta t} \to 0$$

By comparing the underlined term in Equation (2.78) with Equation (2.1), Newton's law, we see that the total rate of change in translational momentum may now be written as

$$\frac{d\mathbf{Q}}{dt}\Big|_I = \int_{\text{Vol}} \rho_V \mathbf{g} dV + \int_{\text{Surface}} d\mathbf{f}_{\text{ext}} + \dot{m}\frac{d\mathbf{p}_o}{dt}\Big|_I \tag{2.79}$$

Likewise, from the definition of the vehicle's center of mass we have at time t

$$m\mathbf{p}_V = \int_{\text{Vol}} \rho_V \mathbf{p}' dV \tag{2.80}$$

and at time $t + \Delta t$ we have

$$m\mathbf{p}_V + \Delta(m\mathbf{p}_V) = \int_{\text{Vol}} \rho_V(\mathbf{p}' + \Delta\mathbf{p}')dV + \Delta m(\mathbf{p}_o + \Delta\mathbf{p}_o) \tag{2.81}$$

Again taking the limit as Δt goes to zero yields

$$\lim_{\Delta t \to 0} \frac{\Delta(m\mathbf{p}_V)}{\Delta t} = \frac{d(m\mathbf{p}_V)}{dt}\Big|_I = \int_{\text{Vol}} \rho_V \frac{d\mathbf{p}'}{dt}\Big|_I \, dV + \dot{m}\mathbf{p}_o = \mathbf{Q}(t) + \dot{m}\mathbf{p}_o \tag{2.82}$$

So now the momentum $\mathbf{Q}$ can be expressed in terms of the velocity of the vehicle's center of mass. Or, since $d(m\mathbf{p}_V)/dt = \dot{m}\mathbf{p}_V + md\mathbf{p}_V/dt$ and $\mathbf{p}_o = \mathbf{p}_V + \mathbf{p}_E$,

$$\mathbf{Q}(t) = m\frac{d\mathbf{p}_V}{dt}\Big|_I - \dot{m}\mathbf{p}_E$$

Differentiating the above yields

$$\frac{d\mathbf{Q}}{dt}\Big|_I = m\frac{d^2\mathbf{p}_V}{dt^2}\Big|_I + \dot{m}\left(\frac{d\mathbf{p}_V}{dt}\Big|_I - \frac{d\mathbf{p}_E}{dt}\Big|_I\right) - \ddot{m}\mathbf{p}_E \tag{2.83}$$

Also, note that even though the inertial position of the infinitesimal expelled mass dm is

$$\mathbf{p}_o = \mathbf{p}_V + \mathbf{p}_E \tag{2.84}$$

its inertial velocity is

$$\begin{aligned}\frac{d\mathbf{p}_o}{dt}\Big|_I &= \frac{d\mathbf{p}_V}{dt}\Big|_I + \frac{d\mathbf{p}_E}{dt}\Big|_I + \mathbf{v}_E \\ &= \mathbf{V}_V + (\boldsymbol{\omega}_{V,I} \times \mathbf{p}_E) + \mathbf{v}_E\end{aligned} \tag{2.85}$$

where $\mathbf{v}_E$ is the velocity of dm relative to the vehicle (the exit velocity), and $\mathbf{V}_V$ is the inertial velocity of the vehicle's center of mass.

Now, equating Equations (2.79) and (2.83), and using the above relationships between the vectors $\mathbf{p}_o$, $\mathbf{p}_V$, and $\mathbf{p}_E$ yields

$$\begin{aligned}&m\frac{d^2\mathbf{p}_V}{dt^2}\Big|_I + \dot{m}\left(\frac{d\mathbf{p}_V}{dt}\Big|_I - \frac{d\mathbf{p}_E}{dt}\Big|_I\right) - \ddot{m}\mathbf{p}_E \\ &= \int_{\text{Vol}} \rho_V \mathbf{g}\, dV + \int_{\text{Surface}} d\mathbf{f}_{\text{ext}} + \dot{m}\left(\frac{d(\mathbf{p}_V + \mathbf{p}_E)}{dt}\Big|_I + \mathbf{v}_E\right)\end{aligned} \tag{2.86}$$

We now take gravity to be constant, plus we note that the aerodynamic force due to the pressure distribution over the body of the vehicle is

$$\mathbf{F}_{\text{Aero}} = \int_{\text{Body Surface}} d\mathbf{f}_{\text{aero}} \tag{2.87}$$

We also <u>define</u> the propulsive thrust vector, positive acting forward on the vehicle, to be

$$\textbf{Thrust} = \dot{m}\mathbf{v}_E + \int_{\text{Exit Area}} d\mathbf{f}_{\text{exit}} \tag{2.88}$$

which includes the "pressure thrust," the integral of the pressure difference across the nozzle exit plane through which the mass is expended. (Here note that with propellant mass being expelled, $\dot{m}$ is negative, and that if the exit pressure at the nozzle is greater than atmospheric, $d\mathbf{f}_{\text{exit}}$ is negative—outwards pressure instead of pressure on the vehicle. So **Thrust** acts in the direction opposite to that of $\mathbf{v}_E$ and $d\mathbf{f}_{\text{exit}}$, or forward.)

Taking all the above into account, we have finally obtained the equation of motion governing vehicle translation, or

$$\boxed{\begin{aligned} m\frac{d\mathbf{V}_V}{dt}\Big|_I &= m\mathbf{g} + \textbf{Thrust} + \mathbf{F}_{\text{Aero}} + \underline{2\dot{m}\frac{d\mathbf{p}_E}{dt}\Big|_I + \ddot{m}\mathbf{p}_E} \\ &= m\mathbf{g} + \textbf{Thrust} + \mathbf{F}_{\text{Aero}} + \underline{2\dot{m}\left(\frac{d\mathbf{p}_E}{dt}\Big|_V + \boldsymbol{\omega}_{V,I} \times \mathbf{p}_E\right) + \ddot{m}\mathbf{p}_E} \end{aligned}} \tag{2.89}$$

Comparing Equation (2.89) to Equation (2.17), and noting that $\mathbf{F}_{\text{Prop}} = \textbf{Thrust}$ here, we see that the new terms arising due to the changing mass are those underlined in Equation (2.89). It should be noted that the movement of the exit plane with respect to the vehicle's center of mass, or $\frac{d\mathbf{p}_E}{dt}\Big|_V$, is not zero here, even if the vehicle is rigid. This is due to the fact that as mass is expelled, the location of the *cm* is not constant with respect to the vehicle, and the *cm* is the origin of the vehicle-fixed frame, Frame V.

In practice, the apparent forces represented by these underlined terms may be small when compared to the propulsive thrust and aerodynamic forces, since $\ddot{m}$ is typically small, and since $\mathbf{p}_E$ (the position of the nozzle exit plane relative to the vehicle center of mass) and its inertial rate of change ~($\boldsymbol{\omega}_{V,I} \times \mathbf{p}_E$) are typically small compared to propellant exit velocity. But the variable mass can have an appreciable effect on the vehicle's rotational dynamics, which we will address next.

Following the approach presented above, the system's (vehicle's) angular momentum $\mathbf{H}_I(t)$ about the inertial origin at time t is

$$\mathbf{H}_I(t) = \int_{\text{Vol}} \mathbf{p}' \times \rho_V \frac{d\mathbf{p}'}{dt}\Big|_I dV \tag{2.90}$$

And the angular momentum at time $t + \Delta t$ may be expressed as

$$\begin{aligned} \mathbf{H}_I(t + \Delta t) = &\int_{\text{Vol}} \left((\mathbf{p}' + \Delta\mathbf{p}') \times \rho_V\left(\frac{d\mathbf{p}'}{dt}\Big|_I + \Delta\frac{d\mathbf{p}'}{dt}\Big|_I\right)\right)dV \\ &+ (\mathbf{p}_o + \Delta\mathbf{p}_o) \times \Delta m\left(\frac{d\mathbf{p}_o}{dt}\Big|_I + \Delta\frac{d\mathbf{p}_o}{dt}\Big|_I\right) \end{aligned} \tag{2.91}$$

where the last term represents the momentum associated with the expelled mass element ($\Delta m < 0$). Again taking the limit as $\Delta t \to 0$, and noting that second- and higher-order Δ terms vanish, yields the rate of change of angular momentum, or

$$\lim_{\Delta t \to 0} \frac{\Delta \mathbf{H}_I}{\Delta t} = \frac{d\mathbf{H}_I}{dt}\Big|_I = \int_{\text{Vol}} \mathbf{p}' \times \rho_V \frac{d^2\mathbf{p}'}{dt^2}\Big|_I dV + \int_{\text{Vol}} \frac{d\mathbf{p}'}{dt}\Big|_I \times \rho_V \frac{d\mathbf{p}'}{dt}\Big|_I dV$$
$$+ \left(\mathbf{p}_o \times \dot{m}\frac{d\mathbf{p}_o}{dt}\Big|_I\right) \tag{2.92}$$
$$= \underline{\int_{\text{Vol}} \frac{d}{dt}\Big|_I \left(\mathbf{p}' \times \rho_V \frac{d\mathbf{p}'}{dt}\Big|_I\right) dV} + \left(\mathbf{p}_o \times \dot{m}\frac{d\mathbf{p}_o}{dt}\Big|_I\right)$$

By comparing the underlined term above with Equation (2.2), Newton's law, we can now write the rate of change of angular momentum as

$$\frac{d\mathbf{H}_I}{dt}\Big|_I = \int_{\text{Vol}} \mathbf{p}' \times \rho_V \mathbf{g} dV + \int_{\text{Surface}} \mathbf{p}' \times d\mathbf{f}_{\text{ext}} + \left(\mathbf{p}_o \times \dot{m}\frac{d\mathbf{p}_o}{dt}\Big|_I\right) \tag{2.93}$$

Then by recalling Equation (2.90), and noting that

$$\mathbf{p}' = \mathbf{p}_V + \mathbf{p}$$

we can write the vehicle's inertial angular momentum $\mathbf{H}_I$ in terms of its angular momentum about its center of mass $\mathbf{H}_V$, or

$$\mathbf{H}_I = \int_{\text{Vol}} (\mathbf{p}_V + \mathbf{p}) \times \rho_V \left(\frac{d\mathbf{p}_V}{dt}\Big|_I + \frac{d\mathbf{p}}{dt}\Big|_I\right) dV \tag{2.94}$$
$$= \int_{\text{Vol}} \mathbf{p}_V \times \rho_V \left(\frac{d\mathbf{p}_V}{dt}\Big|_I + \frac{d\mathbf{p}}{dt}\Big|_I\right) dV + \int_{\text{Vol}} \mathbf{p} \times \rho_V \frac{d\mathbf{p}_V}{dt}\Big|_I dV + \int_{\text{Vol}} \mathbf{p} \times \rho_V \frac{d\mathbf{p}}{dt}\Big|_I dV$$
$$= \mathbf{p}_V \times \int_{\text{Vol}} \rho_V \frac{d\mathbf{p}'}{dt}\Big|_I dV + \int_{\text{Vol}} \mathbf{p} \times \rho_V \frac{d\mathbf{p}}{dt}\Big|_I dV = \mathbf{p}_V \times \int_{\text{Vol}} \rho_V \frac{d\mathbf{p}'}{dt}\Big|_I dV + \mathbf{H}_V$$

In obtaining the above result, we have made use of the fact that

$$\int_{\text{Vol}} \mathbf{p} \times \rho_V \frac{d\mathbf{p}_V}{dt}\Big|_I dV = \left(\int_{\text{Vol}} \rho_V \mathbf{p} dV\right) \times \frac{d\mathbf{p}_V}{dt}\Big|_I = 0 \tag{2.95}$$

since the term in parentheses is the first mass moment of the vehicle about its instantaneous center of mass.

Differentiating Equation (2.94) with respect to the inertial frame yields

$$\frac{d\mathbf{H}_I}{dt}\Big|_I = \frac{d\mathbf{p}_V}{dt}\Big|_I \times \left(\int_{\text{Vol}} \rho_V \frac{d\mathbf{p}'}{dt}\Big|_I dV\right) + \mathbf{p}_V \times \frac{d}{dt}\Big|_I \left(\int_{\text{Vol}} \rho_V \frac{d\mathbf{p}'}{dt}\Big|_I dV\right) + \frac{d\mathbf{H}_V}{dt}\Big|_I$$
$$= \frac{d\mathbf{p}_V}{dt}\Big|_I \times \mathbf{Q} + \mathbf{p}_V \times \frac{d\mathbf{Q}}{dt}\Big|_I + \frac{d\mathbf{H}_V}{dt}\Big|_I \tag{2.96}$$

since we had defined the translational momentum to be

$$\mathbf{Q}(t) = \int_{\text{Vol}} \rho_V \frac{d\mathbf{p}'}{dt}\Big|_I dV$$

But we had also previously shown that the translational momentum could be written as

$$\mathbf{Q}(t) = m\frac{d\mathbf{p}_V}{dt}\Big|_I - \dot{m}\mathbf{p}_E$$

and that its rate of change could be expressed as

$$\frac{d\mathbf{Q}}{dt}\Big|_I = \int_{\text{Vol}} \rho_V \mathbf{g} dV + \int_{\text{Surface}} d\mathbf{f}_{\text{ext}} + \dot{m}\frac{d\mathbf{p}_o}{dt}\Big|_I$$

Substituting these latter two expressions into Equation (2.96) yields

$$\begin{aligned}
\frac{d\mathbf{H}_I}{dt}\Big|_I &= \frac{d\mathbf{p}_V}{dt}\Big|_I \times \left(m\frac{d\mathbf{p}_V}{dt}\Big|_I - \dot{m}\mathbf{p}_E\right) \\
&\quad + \mathbf{p}_V \times \left(\int_{\text{Vol}} \rho_V \mathbf{g} dV + \int_{\text{Surface}} d\mathbf{f}_{\text{ext}} + \dot{m}\frac{d\mathbf{p}_o}{dt}\Big|_I\right) + \frac{d\mathbf{H}_V}{dt}\Big|_I \qquad (2.97) \\
&= -\frac{d\mathbf{p}_V}{dt}\Big|_I \times \dot{m}\mathbf{p}_E + \mathbf{p}_V \times \left(\int_{\text{Vol}} \rho_V \mathbf{g} dV + \int_{\text{Surface}} d\mathbf{f}_{\text{ext}} + \dot{m}\frac{d\mathbf{p}_o}{dt}\Big|_I\right) + \frac{d\mathbf{H}_V}{dt}\Big|_I
\end{aligned}$$

Now, equating Equations (2.93) and (2.97) yields the equation governing the vehicle's angular momentum <u>about its center of mass</u>, $\mathbf{H}_V$, or

$$\begin{aligned}
&\frac{d\mathbf{H}_V}{dt}\Big|_I - \frac{d\mathbf{p}_V}{dt}\Big|_I \times \dot{m}\mathbf{p}_E + \mathbf{p}_V \times \left(\int_{\text{Vol}} \rho_V \mathbf{g} dV + \int_{\text{Surface}} d\mathbf{f}_{\text{ext}} + \dot{m}\frac{d\mathbf{p}_o}{dt}\Big|_I\right) \\
&= \int_{\text{Vol}} \mathbf{p}' \times \rho_V \mathbf{g} dV + \int_{\text{Surface}} \mathbf{p}' \times d\mathbf{f}_{\text{ext}} + \left(\mathbf{p}_o \times \dot{m}\frac{d\mathbf{p}_o}{dt}\Big|_I\right)
\end{aligned} \qquad (2.98)$$

But this expression can be considerably simplified by recalling that

$$\mathbf{p}' = \mathbf{p}_V + \mathbf{p}$$

$$\mathbf{p}_o = \mathbf{p}_V + \mathbf{p}_E$$

$$\frac{d\mathbf{p}_o}{dt}\Big|_I = \frac{d\mathbf{p}_V}{dt}\Big|_I + \frac{d\mathbf{p}_E}{dt}\Big|_I + \mathbf{v}_E$$

Substituting these three relations into Equation (2.98), noting that the first mass moment about the instantaneous center of mass is zero, and rearranging yields

$$\frac{d\mathbf{H}_V}{dt}\Big|_I = \int_{\text{Surface}} \mathbf{p} \times d\mathbf{f}_{\text{ext}} + \mathbf{p}_E \times \dot{m}\left(\frac{d\mathbf{p}_E}{dt}\Big|_I + \mathbf{v}_E\right) \tag{2.99}$$

Now just as we did previously regarding the forces, we need to do a little bookkeeping regarding the moments acting on the vehicle. We first break the integral of the pressure distribution $\mathbf{f}_{\text{ext}}$ into two parts, one due to the pressure over the body, or due to aerodynamic effects, and one due to the pressure differential across the exit plane through which the expended mass travels. Or

$$\int_{\text{Surface}} \mathbf{p} \times d\mathbf{f}_{\text{ext}} = \int_{\text{Body Surface}} \mathbf{p} \times d\mathbf{f}_{\text{aero}} + \mathbf{p}_E \times \int_{\text{Exit Area}} d\mathbf{f}_{\text{exit}}$$

And since we had defined **Thrust** to be

$$\textbf{Thrust} = \dot{m}\mathbf{v}_E + \int_{\text{Exit Area}} d\mathbf{f}_{\text{exit}}$$

we have the following expression for the rate of change of angular momentum.

$$\boxed{\begin{aligned}\frac{d\mathbf{H}_V}{dt}\Big|_I &= \mathbf{M}_{\text{Aero}} + (\mathbf{p}_E \times \textbf{Thrust}) + \left(\mathbf{p}_E \times \dot{m}\frac{d\mathbf{p}_E}{dt}\Big|_I\right)\\ &= \mathbf{M}_{\text{Aero}} + \mathbf{M}_{\text{Prop}} + \underline{\mathbf{p}_E \times \dot{m}\left(\frac{d\mathbf{p}_E}{dt}\Big|_V + \boldsymbol{\omega}_{V,I} \times \mathbf{p}_E\right)}\end{aligned}} \tag{2.100}$$

This constitutes the equation of motion governing the vehicle's angular momentum, including variable-mass effects. These effects are captured in the underlined terms, and the vector triple product is sometimes referred to as the "jet-damping" effect because it tends to retard the vehicle's angular motion, and is proportional to the vehicle's angular velocity. Note that this damping is proportional to the propulsive mass flow rate and the square of the distance from the instantaneous center of mass to the nozzle exit. Hence, the damping is more significant for long vehicles with large mass flow rates (e.g., missiles).

Finally, it should again be noted that the movement of the exit plane with respect to the vehicle's center of mass, or $\frac{d\mathbf{p}_E}{dt}\Big|_V$, is not zero here, even if the vehicle is rigid.

EXAMPLE 2.4

Variable-Mass Effects on a Rocket

Consider a rocket of length l, having an engine with a constant propellant-flow rate of $\dot{m}_{\text{Prop}}$. Further, let the position of the nozzle exit be defined by the vector $\mathbf{p}_E$, where $\mathbf{p}_E = -l/2\,\mathbf{i}_V$, and the vehicle's angular velocity $\boldsymbol{\omega}_{V,I}$ be defined as

$$\boldsymbol{\omega}_{V,I} = P\,\mathbf{i}_V + Q\,\mathbf{j}_V + R\,\mathbf{k}_V$$

Evaluate the extra terms in Equations (2.89) and (2.100) that arise due to variable-mass effects. (Ignore the movement of the *cm* with respect to the nozzle exit due to mass expulsion.)

■ Solution

First, regarding Equation (2.89), the variable-mass terms in the equation governing translational momentum are

$$2\dot{m}\left(\frac{d\mathbf{p}_E}{dt}\bigg|_V + \boldsymbol{\omega}_{V,I}\times\mathbf{p}_E\right) + \ddot{m}\mathbf{p}_E = -2\dot{m}_{\text{Prop}}\left(\boldsymbol{\omega}_{V,I}\times\mathbf{p}_E\right)$$

$$= -2\dot{m}_{\text{Prop}}\begin{vmatrix}\mathbf{i}_V & \mathbf{j}_V & \mathbf{k}_V\\ P & Q & R\\ -l/2 & 0 & 0\end{vmatrix} = \dot{m}_{\text{Prop}}l\left(R\,\mathbf{j}_V - Q\,\mathbf{k}_V\right) \qquad (2.101)$$

Hence, the right-hand sides of the second and third equations in Equations (2.22) must be modified by adding the terms $\dot{m}_{\text{Prop}}\, lR$ and $-\dot{m}_{\text{Prop}}\, lQ$, respectively. These apparent forces due to variable-mass effects are proportional to vehicle length, propellant flow rate, and pitch and yaw angular velocities.

Next, regarding Equation (2.100), the variable-mass terms are

$$\mathbf{p}_E\times\dot{m}\left(\frac{d\mathbf{p}_E}{dt}\bigg|_V + \boldsymbol{\omega}_{V,I}\times\mathbf{p}_E\right) = -\dot{m}_{\text{Prop}}\left(l/2\mathbf{i}_V\right)\times\left(\boldsymbol{\omega}_{V,I}\times l/2\mathbf{i}_V\right)$$

$$= -\dot{m}_{\text{Prop}}\begin{vmatrix}\mathbf{i}_V & \mathbf{j}_V & \mathbf{k}_V\\ l/2 & 0 & 0\\ 0 & Rl/2 & -Ql/2\end{vmatrix} = -\left(\dot{m}_{\text{Prop}}l^2/4\right)\left(Q\,\mathbf{j}_V + R\,\mathbf{k}_V\right) \qquad (2.102)$$

Hence, the right-hand sides of the second and third equations in Equations (2.27) must be modified by adding the terms $-\dot{m}_{\text{Prop}}\, l^2Q/4$ and $-\dot{m}_{\text{Prop}}\, l^2R/4$, respectively. These apparent pitch and yaw moments due to variable-mass effects are proportional to the square of vehicle length, propellant flow rate, and pitch and yaw angular velocities. A pitch rate induces a pitch-damping effect, while a yaw rate induces a yaw-damping effect.

The additional terms that appear in the equations of motion (Equations (2.89) and (2.100)) due to variable-mass effects will, or course, give rise to additional terms in the corresponding Reference Set and Perturbation Set of equations. One may follow an approach similar to that demonstrated in Example 2.4 and derive the additional terms explicitly for a given vehicle geometry. Then the necessary modifications to the Reference and Perturbation Sets of equations may be developed from these additional terms.

2.6 EFFECTS OF A SPHERICAL, ROTATING EARTH

In all of the previous sections, we have ignored the effects of an oblate rotating earth. We assumed that the ground-fixed reference frame was inertial, and that the earth had no curvature. As noted in Section 2.2, these assumptions are valid when modeling the dynamics of the majority of conventional aircraft. However, for vehicles that operate at high supersonic or hypersonic velocities (such as the SR-71 or vehicles similar to the X-43), and/or fly long distances (such as the Space Shuttle), these assumptions may not be sufficiently valid.

In this section, we will consider the effects of a rotating earth, and we will assume the earth is spherical, with a spherically uniform mass distribution. This spherical-earth assumption is sufficiently valid when dealing with vehicle dynamics, but it may not be so when addressing precision guidance, for example. For vehicle-guidance analysis the effects of the oblate shape of the earth may be significant.

To begin this discussion, we will first refer to Figure 2.7, which is similar to Figure 1.1. Shown are four coordinate frames. Frame V is a typical vehicle-fixed frame with its origin at the vehicle center of mass, and rotates with the vehicle. An earth-centered frame, Frame I, with its X_I axis always directed at some chosen star, is taken to be inertial. Frame E is fixed to a chosen point of reference on the surface of the earth, with X_E directed north and Z_E directed towards the earth's center. Finally, a new frame, Frame L, is introduced. The origin of Frame L is also at the vehicle center of mass, but this frame is always aligned such that its X_L axis is directed north, parallel to the local meridian, its Y_L axis is directed east, and its Z_L axis is always directed towards the center of the earth. Frame L is referred to as a *local-vertical–local-horizontal* frame, and it is a vehicle-carried rather than a vehicle-fixed frame since it does not rotate with the vehicle.

The position of Frame L, and therefore the vehicle's center of mass, may be defined in terms of its latitude (positive north), λ_L, its longitude (positive east),

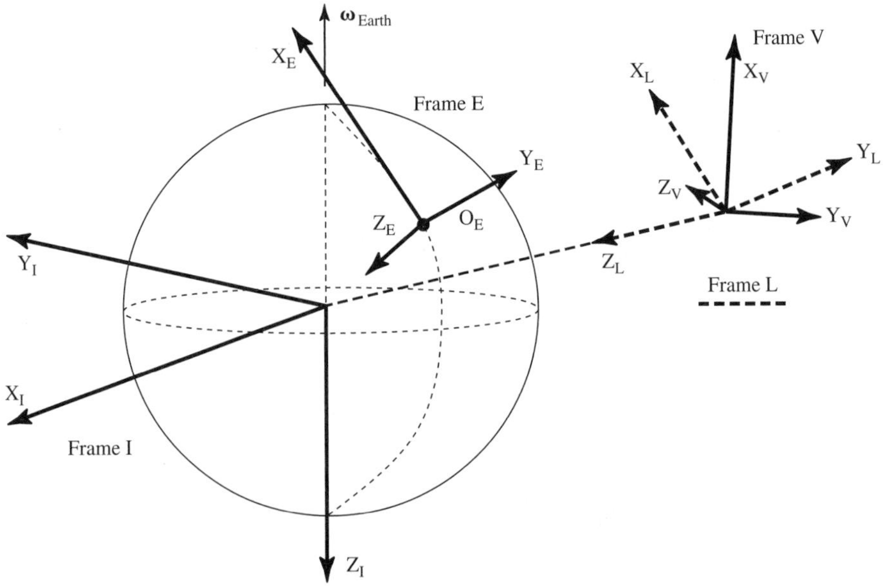

Figure 2.7 Frames I, E, L, and V.

μ_L (both are angles), and its altitude above the surface of the earth, h. The orientation of the vehicle, or Frame V, is defined relative to Frame L by the familiar 3-2-1 Euler angles ψ, θ, and ϕ discussed in Section 1.3.

We are interested here in developing the equations of motion of a constant-mass vehicle with no rotating parts, governing both the translational velocity relative to the origin of Frame E and the rotational velocity of the vehicle relative to Frame I. We will then compare these equations of motion to those derived in Section 2.2 to identify additional terms that arise due to the effects of the rotating, spherical earth. The aspect of this formulation that requires particular attention is keeping track of the various rotating coordinate frames.

The translational velocity of the vehicle relative to Frame E is of interest because that would be the velocity measured by a radar site, for example, or the vehicle's "ground speed." This velocity is also closely related to the velocity of the vehicle relative to the atmosphere, or its "airspeed." However, because Frame E is not inertial, the velocity of the vehicle relative to Frame E is not an inertial velocity. The rotational velocity of the vehicle, or Frame V, relative to Frame I, is an inertial velocity. This inertial angular velocity is of interest because it would be the angular velocity measure by an on-board inertial navigation system, for example.

With all the necessary reference frames defined, we can now define the various angular rates. From the geometry of the problem we can see that the earth's angular-velocity vector, $\boldsymbol{\omega}_{\text{Earth}}$, or the rate of rotation of Frame E relative to Frame I, is

$$\boldsymbol{\omega}_{\text{Earth}} = \boldsymbol{\omega}_{E,I} = -\omega_{\text{Earth}}\,\mathbf{k}_I = \omega_{\text{Earth}}\begin{bmatrix}\cos\lambda_L & 0 & -\sin\lambda_L\end{bmatrix}\begin{Bmatrix}\mathbf{i}_L\\ \mathbf{j}_L\\ \mathbf{k}_L\end{Bmatrix} \tag{2.103}$$

The Euler angles are defined analogous to those in the flat-earth case, in that they specify the orientation of Frame V with respect to Frame L. So we have the same direction-cosine matrix relating unit vectors of Frames L and V, or from Equation (1.15)

$$\begin{Bmatrix}\mathbf{i}_V\\ \mathbf{j}_V\\ \mathbf{k}_V\end{Bmatrix} = \mathbf{T}_{L\text{-}V}(\phi,\theta,\psi)\begin{Bmatrix}\mathbf{i}_L\\ \mathbf{j}_L\\ \mathbf{k}_L\end{Bmatrix}$$

$$\mathbf{T}_{L\text{-}V}(\phi,\theta,\psi) = \begin{bmatrix}1 & 0 & 0\\ 0 & \cos\phi & \sin\phi\\ 0 & -\sin\phi & \cos\phi\end{bmatrix}\begin{bmatrix}\cos\theta & 0 & -\sin\theta\\ 0 & 1 & 0\\ \sin\theta & 0 & \cos\theta\end{bmatrix}\begin{bmatrix}\cos\psi & \sin\psi & 0\\ -\sin\psi & \cos\psi & 0\\ 0 & 0 & 1\end{bmatrix}$$

$$= \begin{bmatrix}(\cos\theta\cos\psi) & (\cos\theta\sin\psi) & (-\sin\theta)\\ (\sin\phi\sin\theta\cos\psi - \cos\phi\sin\psi) & (\sin\phi\sin\theta\sin\psi + \cos\phi\cos\psi) & (\sin\phi\cos\theta)\\ (\cos\phi\sin\theta\cos\psi + \sin\phi\sin\psi) & (\cos\phi\sin\theta\sin\psi - \sin\phi\cos\psi) & (\cos\phi\cos\theta)\end{bmatrix} \tag{2.104}$$

So $\boldsymbol{\omega}_{E,I}$ can also be expressed as

$$\boldsymbol{\omega}_{E,I} \triangleq \begin{bmatrix} P_E & Q_E & R_E \end{bmatrix} \begin{Bmatrix} \mathbf{i}_V \\ \mathbf{j}_V \\ \mathbf{k}_V \end{Bmatrix} = \omega_{\text{Earth}} \begin{bmatrix} \cos\lambda_L & 0 & -\sin\lambda_L \end{bmatrix} \mathbf{T}^T_{L\text{-}V}(\phi, \theta, \psi) \begin{Bmatrix} \mathbf{i}_V \\ \mathbf{j}_V \\ \mathbf{k}_V \end{Bmatrix} \quad (2.105)$$

or

$$\boxed{\begin{bmatrix} P_E \\ Q_E \\ R_E \end{bmatrix} = \mathbf{T}_{L\text{-}V}(\phi, \theta, \psi) \begin{bmatrix} \omega_{\text{Earth}} \cos\lambda_L \\ 0 \\ -\omega_{\text{Earth}} \sin\lambda_L \end{bmatrix}} \quad (2.106)$$

and P_E, Q_E, and R_E are thus defined in terms of the Euler angles and the vehicle's latitude.

The relative angular velocities of the V, L, and I frames are related through the fact that

$$\boldsymbol{\omega}_{V,I} = \boldsymbol{\omega}_{V,L} + \boldsymbol{\omega}_{L,I} \quad (2.107)$$

and one can show that the rate of rotation of Frame L relative to Frame I is

$$\boldsymbol{\omega}_{L,I} = \begin{bmatrix} (\omega_{\text{Earth}} + \dot{\mu}_L)\cos\lambda_L & -\dot{\lambda}_L & -(\omega_{\text{Earth}} + \dot{\mu}_L)\sin\lambda_L \end{bmatrix} \begin{Bmatrix} \mathbf{i}_L \\ \mathbf{j}_L \\ \mathbf{k}_L \end{Bmatrix} \quad (2.108)$$

But if we also write

$$\boldsymbol{\omega}_{L,I} \triangleq \begin{bmatrix} P_L & Q_L & R_L \end{bmatrix} \begin{Bmatrix} \mathbf{i}_V \\ \mathbf{j}_V \\ \mathbf{k}_V \end{Bmatrix} \quad (2.109)$$

then from Equation (2.108) and the direction-cosine matrix in Equation (2.104) we have

$$\begin{aligned} \boldsymbol{\omega}_{L,I} &= \begin{bmatrix} P_L & Q_L & R_L \end{bmatrix} \begin{Bmatrix} \mathbf{i}_V \\ \mathbf{j}_V \\ \mathbf{k}_V \end{Bmatrix} \\ &= \begin{bmatrix} (\omega_{\text{Earth}} + \dot{\mu}_L)\cos\lambda_L & -\dot{\lambda}_L & -(\omega_{\text{Earth}} + \dot{\mu}_L)\sin\lambda_L \end{bmatrix} \mathbf{T}^T_{L\text{-}V}(\phi, \theta, \psi) \begin{Bmatrix} \mathbf{i}_V \\ \mathbf{j}_V \\ \mathbf{k}_V \end{Bmatrix} \end{aligned} \quad (2.110)$$

so then P_L, Q_L, and R_L are defined in terms of the Euler angles, latitude, and longitude as

$$\boxed{\begin{bmatrix} P_L \\ Q_L \\ R_L \end{bmatrix} = \mathbf{T}_{L\text{-}V}(\phi, \theta, \psi) \begin{bmatrix} (\omega_{\text{Earth}} + \dot{\mu}_L)\cos\lambda_L \\ -\dot{\lambda}_L \\ -(\omega_{\text{Earth}} + \dot{\mu}_L)\sin\lambda_L \end{bmatrix}} \quad (2.111)$$

We will also define the angular velocity of the vehicle, or Frame V, with respect to Frame L to be

$$\boldsymbol{\omega}_{V,L} \triangleq \begin{bmatrix} P_V & Q_V & R_V \end{bmatrix} \begin{Bmatrix} \mathbf{i}_V \\ \mathbf{j}_V \\ \mathbf{k}_V \end{Bmatrix} \tag{2.112}$$

And since we are using the same 3-2-1 Euler angles as defined in the flat-earth case, we have the same kinematic equations relating angular rates, or Equations (2.36) or (2.37). That is,

$$\boxed{\begin{aligned} P_V &= \dot{\phi} - \dot{\psi}\sin\theta \\ Q_V &= \dot{\theta}\cos\phi + \dot{\psi}\sin\phi\cos\theta \\ R_V &= \dot{\psi}\cos\phi\cos\theta - \dot{\theta}\sin\phi \end{aligned}} \tag{2.113}$$

or by inverting we have

$$\boxed{\begin{aligned} \dot{\phi} &= P_V + Q_V\sin\phi\tan\theta + R_V\cos\phi\tan\theta \\ \dot{\theta} &= Q_V\cos\phi - R_V\sin\phi \\ \dot{\psi} &= \left(Q_V\sin\phi + R_V\cos\phi\right)\sec\phi \end{aligned}} \tag{2.114}$$

These equations govern the Euler angles, but are now expressed in terms of P_V, Q_V, and R_V instead of the inertial rates P, Q, and R.

And, finally, writing Equation (2.107) as

$$\boldsymbol{\omega}_{V,I} = \boldsymbol{\omega}_{V,L} + \boldsymbol{\omega}_{L,I} \triangleq \begin{bmatrix} P & Q & R \end{bmatrix} \begin{Bmatrix} \mathbf{i}_V \\ \mathbf{j}_V \\ \mathbf{k}_V \end{Bmatrix} \tag{2.115}$$

we have thus defined P, Q, and R, the components of the vector representing the inertial rotational velocity of the vehicle, or Frame V, with respect to the inertial, Frame I. From Equation (2.115), and by noting Equation (2.112), we can now determine P_V, Q_V, and R_V, since $\boldsymbol{\omega}_{V,L} = \boldsymbol{\omega}_{V,I} - \boldsymbol{\omega}_{L,I}$. Furthermore, we see that the difference between the flat- versus the spherical-rotating-earth cases is simply the term $\boldsymbol{\omega}_{L,I}$. In the flat-earth case, this term is taken to be zero.

EXAMPLE 2.5

Effect of Earth's Rotation

Consider a helicopter hovering at a fixed position of 0 deg longitude, 45 deg north latitude. The vehicle's attitude remains level, and it maintains a constant heading of 90 deg (east). Find the vehicle's inertial angular velocity as well as its angular velocity relative to its Frame L, $\boldsymbol{\omega}_{V,L}$.

■ **Solution**

First, if the vehicle's attitude and heading remain constant, then $\dot{\phi} = \dot{\theta} = \dot{\psi} = 0$. And from Equation (2.113) we see that P_V, Q_V, and R_V must all be zero. So $\boldsymbol{\omega}_{V,L} = \mathbf{0}$. Second, from Equation (2.115), $\boldsymbol{\omega}_{V,I}$ must then equal $\boldsymbol{\omega}_{L,I}$. And from Equation (2.111) we see that the components of $\boldsymbol{\omega}_{L,I}$ are

$$\begin{bmatrix} P_L \\ Q_L \\ R_L \end{bmatrix} = \mathbf{T}_{L\text{-}V}(\phi, \theta, \psi) \begin{bmatrix} (\omega_{\text{Earth}} + \dot{\mu}_L)\cos\lambda_L \\ -\dot{\lambda}_L \\ -(\omega_{\text{Earth}} + \dot{\mu}_L)\sin\lambda_L \end{bmatrix}$$

Since $\phi = \theta = 0$, and $\psi = 90$ deg here, the direction-cosine matrix $\mathbf{T}_{L\text{-}V}$ is

$$\mathbf{T}_{L\text{-}V}(\phi, \theta, \psi)$$

$$= \begin{bmatrix} (\cos\theta\cos\psi) & (\cos\theta\sin\psi) & (-\sin\theta) \\ (\sin\phi\sin\theta\cos\psi - \cos\phi\sin\psi) & (\sin\phi\sin\theta\sin\psi + \cos\phi\cos\psi) & (\sin\phi\cos\theta) \\ (\cos\phi\sin\theta\cos\psi + \sin\phi\sin\psi) & (\cos\phi\sin\theta\sin\psi - \sin\phi\cos\psi) & (\cos\phi\cos\theta) \end{bmatrix}$$

$$= \begin{bmatrix} 0 & 1 & 0 \\ -1 & 0 & 0 \\ 0 & 0 & 1 \end{bmatrix}$$

Now since $\lambda_L = 45$ degrees, and when hovering in a fixed location $\dot{\lambda}_L = \dot{\mu}_L = 0$, we have

$$\begin{bmatrix} P_L \\ Q_L \\ R_L \end{bmatrix} = \begin{bmatrix} 0 & 1 & 0 \\ -1 & 0 & 0 \\ 0 & 0 & 1 \end{bmatrix} \begin{bmatrix} \omega_{\text{Earth}}\cos 45° \\ 0 \\ -\omega_{\text{Earth}}\sin 45° \end{bmatrix} = \begin{bmatrix} 0 \\ -\omega_{\text{Earth}}\cos 45° \\ -\omega_{\text{Earth}}\sin 45° \end{bmatrix} = \begin{bmatrix} 0 \\ -2\pi(0.707)\text{rad/day} \\ -2\pi(0.707)\text{rad/day} \end{bmatrix}$$

Therefore, the vehicle's inertial angular velocity is

$$\boldsymbol{\omega}_{V,I} = \begin{bmatrix} 0 \\ -2.9 \times 10^{-3} \\ -2.9 \times 10^{-3} \end{bmatrix} \text{deg/sec}$$

and we see that the vehicle is slowly pitching down and yawing left with respect to the inertial frame. In the flat-earth case, this angular velocity would have been zero. Although the rates are small, the vehicle's inertial pitch attitude and heading will both change -63 degrees in six hours.

Now, as shown in Figure 2.8, the vehicle's inertial position (vector) may be expressed as

$$\mathbf{p}_{V,I} = \mathbf{R}_E + \mathbf{p}_{V,E}$$

and the magnitude of $\mathbf{p}_{V,I}$ is

$$\text{p}_{V,I} = \text{R}_E + \text{h}$$

where $\mathbf{R}_E$ is the inertial position of the origin of Frame E, R_E is the earth's radius, and h (not shown in the figure) is the vehicle's altitude above the surface of the earth.

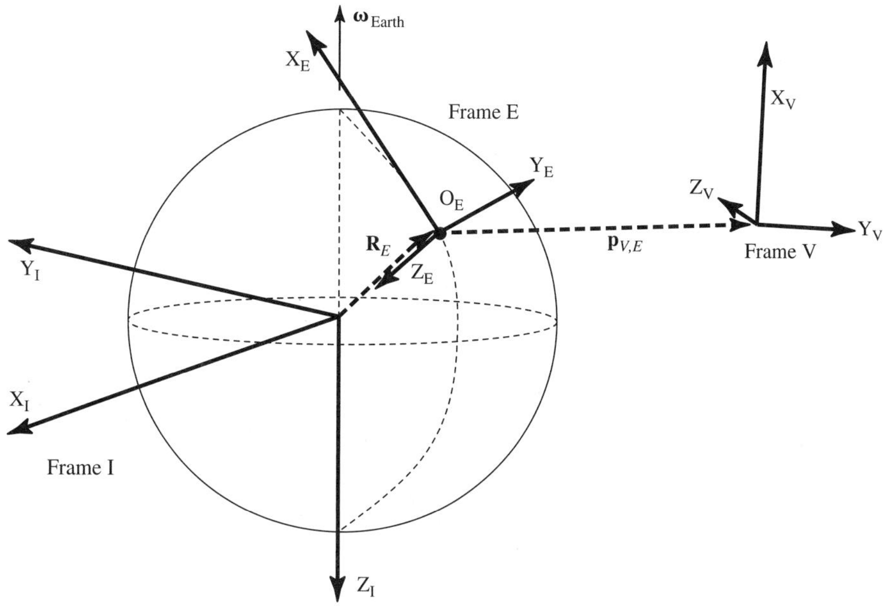

Figure 2.8 Position vectors $\mathbf{R}_E$ and $\mathbf{p}_{V,E}$.

(Note the difference between R_E and R_E here. R_E is the earth's radius, while R_E (given in Equation (2.106)) is the $\mathrm{Z_V}$ component of the angular velocity of Frame E.)

We can now define *U, V,* and *W* as the Frame-V components of the vehicle's translational velocity relative to Frame E, or

$$\frac{d\mathbf{p}_{V,E}}{dt}\Big|_E \triangleq \mathbf{V}_{V,E} = \begin{bmatrix} U & V & W \end{bmatrix} \begin{Bmatrix} \mathbf{i}_V \\ \mathbf{j}_V \\ \mathbf{k}_V \end{Bmatrix} \tag{2.116}$$

Note that *U, V,* and *W* are defined in a manner analogous to the flat-earth case, but they are no longer components of an inertial-velocity vector. The inertial translational velocity of the vehicle written in terms of $\mathbf{V}_{V,E}$ is

$$\begin{aligned} \frac{d\mathbf{p}_{V,I}}{dt}\Big|_I \triangleq \mathbf{V}_{V,I} = \frac{d\mathbf{R}_E}{dt}\Big|_I + \frac{d\mathbf{p}_{V,E}}{dt}\Big|_I &= \left(\boldsymbol{\omega}_{E,I} \times \mathbf{R}_E\right) + \left(\mathbf{V}_{V,E} + \left(\boldsymbol{\omega}_{E,I} \times \mathbf{p}_{V,E}\right)\right) \\ &= \mathbf{V}_{V,E} + \boldsymbol{\omega}_{E,I} \times \left(\mathbf{R}_E + \mathbf{p}_{V,E}\right) \end{aligned} \tag{2.117}$$

So in the flat-earth case, Frame E was taken to be inertial, and $\boldsymbol{\omega}_{E,I}$ would then be zero. Under this condition, Equation (2.117) reveals that $\mathbf{V}_{V,I} = \mathbf{V}_{V,E}$, as expected.

We may now develop the equations governing translational motion (*U, V,* and *W*) beginning with Equation (2.10), Newton's second law, repeated here.

$$\frac{d}{dt}\Big|_I \left(m \frac{d\mathbf{p}_{V,I}}{dt}\Big|_I\right) = m \frac{d\mathbf{V}_{V,I}}{dt}\Big|_I = \int_{\text{Vol}} \rho_V \mathbf{g}\, dV + \int_{\text{Surface}} d\mathbf{f}_{\text{ext}}$$

Using Equation (2.117) we have the inertial acceleration of the vehicle given as

$$
\begin{aligned}
\frac{d\mathbf{V}_{V,I}}{dt}\Big|_I &= \frac{d\mathbf{V}_{V,E}}{dt}\Big|_I + \boldsymbol{\omega}_{E,I}\times\left(\frac{d\mathbf{R}_E}{dt}\Big|_I + \frac{d\mathbf{p}_{V,E}}{dt}\Big|_I\right)\\
&= \frac{d\mathbf{V}_{V,E}}{dt}\Big|_V + \left(\boldsymbol{\omega}_{V,I}\times\mathbf{V}_{V,E}\right) + \boldsymbol{\omega}_{E,I}\times\left(\frac{d\mathbf{p}_{V,E}}{dt}\Big|_E + \left(\boldsymbol{\omega}_{E,I}\times\left(\mathbf{R}_E+\mathbf{p}_{V,E}\right)\right)\right)\\
&= \frac{d\mathbf{V}_{V,E}}{dt}\Big|_V + \left(\boldsymbol{\omega}_{V,I} + \underline{\boldsymbol{\omega}_{E,I}}\right)\times\mathbf{V}_{V,E} + \underline{\boldsymbol{\omega}_{E,I}\times\left(\boldsymbol{\omega}_{E,I}\times\left(\mathbf{R}_E+\mathbf{p}_{V,E}\right)\right)}
\end{aligned}
\tag{2.118}
$$

Again note the result when $\boldsymbol{\omega}_{E,I}$ is taken to be zero. The new terms arising due to the rotating earth are those underlined.

Noting that the vehicle's inertial position may also be written as

$$
\mathbf{p}_{V,I} = \mathbf{R}_E + \mathbf{p}_{V,E} = -\mathrm{p}_{V,I}\mathbf{k}_L = -\left(\mathrm{R}_E + \mathrm{h}\right)\mathbf{k}_L \tag{2.119}
$$

and using Equation (2.103), we have the centripetal acceleration vector $\mathbf{C}$ given as

$$
\mathbf{C} \triangleq \boldsymbol{\omega}_{E,I}\times\left(\boldsymbol{\omega}_{E,I}\times\mathbf{p}_{V,I}\right) = \omega^2_{\text{Earth}}\left(\mathrm{R}_E+\mathrm{h}\right)\cos\lambda_L\left[\sin\lambda_L \quad 0 \quad \cos\lambda_L\right]\begin{Bmatrix}\mathbf{i}_L\\ \mathbf{j}_L\\ \mathbf{k}_L\end{Bmatrix} \tag{2.120}
$$

So defining the components of $\mathbf{C}$ in Frame V as

$$
\mathbf{C} \triangleq \left[C_x \quad C_y \quad C_z\right]\begin{Bmatrix}\mathbf{i}_V\\ \mathbf{j}_V\\ \mathbf{k}_V\end{Bmatrix} \tag{2.121}
$$

we have these components given as

$$
\boxed{
\begin{aligned}
&\begin{bmatrix}C_x\\ C_y\\ C_z\end{bmatrix} = \mathbf{T}_{L\text{-}V}(\phi,\theta,\psi)\begin{bmatrix}\omega^2_{\text{Earth}}\left(\mathrm{R}_E+\mathrm{h}\right)\cos\lambda_L\sin\lambda_L\\ 0\\ \omega^2_{\text{Earth}}\left(\mathrm{R}_E+\mathrm{h}\right)\cos^2\lambda_L\end{bmatrix}\\
&C_x = \omega^2_{\text{Earth}}\left(\mathrm{R}_E+\mathrm{h}\right)\cos\lambda_L\left(\sin\lambda_L\cos\theta\cos\psi - \cos\lambda_L\sin\theta\right)\\
&C_y = \omega^2_{\text{Earth}}\left(\mathrm{R}_E+\mathrm{h}\right)\cos\lambda_L\left(\sin\lambda_L\left(\sin\phi\sin\theta\cos\psi - \cos\phi\sin\psi\right)\right.\\
&\qquad\qquad \left. + \cos\lambda_L\sin\phi\cos\theta\right)\\
&C_z = \omega^2_{\text{Earth}}\left(\mathrm{R}_E+\mathrm{h}\right)\cos\lambda_L\left(\sin\lambda_L\left(\cos\phi\sin\theta\cos\psi + \sin\phi\sin\psi\right)\right.\\
&\qquad\qquad \left. + \cos\lambda_L\cos\phi\cos\theta\right)
\end{aligned}
}
\tag{2.122}
$$

Note that the non-underlined terms on the right-hand side of Equation (2.118) are analogous to the right-hand side of Equation (2.20), and that Equation (2.20) led to Equations (2.22), the scalar translational equations of motion in the flat-earth case. Next, note that $\mathbf{g} = g\mathbf{k}_L$. And finally, noting Equations (2.118) and (2.122), we have from Equation (2.10) the following scalar equations of motion governing the translational velocity of the vehicle relative to Frame E.

$$\begin{aligned} m\left(\dot{U} + W\left(Q + \underline{Q_E}\right) - V\left(R + \underline{R_E}\right) + \underline{C_x}\right) &= -mg\sin\theta + F_{A_X} + F_{P_X} \\ m\left(\dot{V} + U\left(R + \underline{R_E}\right) - W\left(P + \underline{P_E}\right) + \underline{C_y}\right) &= mg\cos\theta\sin\phi + F_{A_Y} + F_{P_Y} \\ m\left(\dot{W} + V\left(P + \underline{P_E}\right) - U\left(Q + \underline{Q_E}\right) + \underline{C_z}\right) &= mg\cos\theta\cos\phi + F_{A_Z} + F_{P_Z} \end{aligned} \tag{2.123}$$

The differences between these equations and Equations (2.22) are the underlined terms.

EXAMPLE 2.6

Apparent Forces

Consider an aircraft in level flight at sea level, heading due east over the equator at 200 fps. Letting the vehicle mass equal 500 sl and its lateral velocity V equal zero, find the apparent force acting on the vehicle in the Z_V direction due to the effects of a rotating earth.

■ Solution

"The apparent force acting in the Z_V direction" refers to the additional terms in the $\dot{W}$ equation in Equations (2.123) that arise due to earth's rotation. These terms, when moved to the right-hand side of the equation, are

$$-m\left(VP_E - UQ_E + C_z\right)$$

which are zero for a flat nonrotating earth. In this problem, $V = 0$, and U will be taken to equal the ground speed of 200 fps. Also, note that $\phi = \theta = 0$, and $\psi = 90$ deg, so the direction-cosine matrix is the same as that in Example 2.5, or

$$\mathbf{T}_{L\text{-}V}(\phi, \theta, \psi) = \begin{bmatrix} 0 & 1 & 0 \\ -1 & 0 & 0 \\ 0 & 0 & 1 \end{bmatrix}$$

Using Equation (2.106), and with the latitude $\lambda_L = 0$, we have

$$\begin{bmatrix} P_E \\ Q_E \\ R_E \end{bmatrix} = \mathbf{T}_{L\text{-}V}(\phi, \theta, \psi)\begin{bmatrix} \omega_{\text{Earth}}\cos\lambda_L \\ 0 \\ -\omega_{\text{Earth}}\sin\lambda_L \end{bmatrix} = \begin{bmatrix} 0 & 1 & 0 \\ -1 & 0 & 0 \\ 0 & 0 & 1 \end{bmatrix}\begin{bmatrix} 2\pi\cos\lambda_L \\ 0 \\ 0 \end{bmatrix}$$

$$= \begin{bmatrix} 0 \\ -2\pi\text{rad/day} \\ 0 \end{bmatrix}$$

So $P_E = R_E = 0$, and $Q_E = -7.3 \times 10^{-5}$ rad/sec. This yields an apparent force of

$$mUQ_E = 7.3 \text{ lbs}$$

From Equation (2.122) we see that

$$\begin{bmatrix} C_x \\ C_y \\ C_z \end{bmatrix} = \mathbf{T}_{L\text{-}V}(\phi, \theta, \psi) \begin{bmatrix} \omega^2_{\text{Earth}}(\mathrm{R}_E + \mathrm{h})\cos\lambda_L \sin\lambda_L \\ 0 \\ \omega^2_{\text{Earth}}(\mathrm{R}_E + \mathrm{h})\cos^2\lambda_L \end{bmatrix}$$

$$= \begin{bmatrix} 0 & 1 & 0 \\ -1 & 0 & 0 \\ 0 & 0 & 1 \end{bmatrix} \begin{bmatrix} 0 \\ 0 \\ (2\pi/\text{day})^2\, 4000 \text{ mi} \end{bmatrix} = \begin{bmatrix} 0 \\ 0 \\ 0.11 \text{ ft/sec}^2 \end{bmatrix}$$

So the centripetal acceleration $C_z = 0.11$ ft/sec^2. This corresponds to an apparent force of

$$mC_z = 55 \text{ lbs}$$

So the total apparent force acting on the vehicle in the Z_V direction is

$$-m(VP_E - UQ_E + C_z) = 7.3 - 55 = -47.7 \text{ lbs}$$

For a 16,000 lb vehicle, this force corresponds to 0.3 percent of the weight of the vehicle. In the flat-earth case, this apparent force is assumed to be zero.

We will now address the kinematics relating *U, V,* and *W* to the rate of change of geoposition. From the geometry defined in Figures 2.7 and 2.8, it should be apparent that the instantaneous velocity of the vehicle, or the velocity of the origin of Frame V relative to Frame E on the earth's surface, could also be written in terms of the rates of change of longitude, latitude, and altitude, or

$$\mathbf{V}_{V,E} = \left[\dot{\lambda}_L(\mathrm{R}_E + \mathrm{h}) \quad \dot{\mu}_L(\mathrm{R}_E + \mathrm{h})\cos\lambda_L \quad -(\dot{\mathrm{R}}_E + \dot{\mathrm{h}})\right] \begin{Bmatrix} \mathbf{i}_L \\ \mathbf{j}_L \\ \mathbf{k}_L \end{Bmatrix} \tag{2.124}$$

Therefore, using the above expression, along with Equations (2.104) and (2.116), the kinematic relationships governing the <u>vehicle's position relative to the surface of the earth</u> are

$$\begin{bmatrix} \dot{\lambda}_L(\mathrm{R}_E + \mathrm{h}) \\ \dot{\mu}_L(\mathrm{R}_E + \mathrm{h})\cos\lambda_L \\ -\dot{\mathrm{h}} \end{bmatrix} = \mathbf{T}^T_{L\text{-}V}(\phi, \theta, \psi) \begin{bmatrix} U \\ V \\ W \end{bmatrix}$$

$$= \begin{bmatrix} (\cos\theta\cos\psi) & (\cos\theta\sin\psi) & (-\sin\theta) \\ (\sin\phi\sin\theta\cos\psi - \cos\phi\sin\psi) & (\sin\phi\sin\theta\sin\psi + \cos\phi\cos\psi) & (\sin\phi\cos\theta) \\ (\cos\phi\sin\theta\cos\psi + \sin\phi\sin\psi) & (\cos\phi\sin\theta\sin\psi - \sin\phi\cos\psi) & (\cos\phi\cos\theta) \end{bmatrix}^T \begin{bmatrix} U \\ V \\ W \end{bmatrix} \tag{2.125}$$

We now turn our attention to rotational motion. The equations of motion governing the vehicle's <u>inertial rotational</u> velocity are, of course, developed from Equation (2.18), Newton's law in rotational form, repeated here.

$$\int_{\text{Vol}} \mathbf{p} \times \left(\rho_V\left(\left(\frac{d\boldsymbol{\omega}_{V,I}}{dt}\Big|_I \times \mathbf{p}\right) + \boldsymbol{\omega}_{V,I} \times (\boldsymbol{\omega}_{V,I} \times \mathbf{p})\right)dV\right) = \mathbf{M}_{\text{Aero}} + \mathbf{M}_{\text{Prop}}$$

But since the same position vector **p** would be used here to define the position of the vehicle mass particles relative to the vehicle center of mass, and since we defined the vehicle's inertial angular velocity to be $\boldsymbol{\omega}_{V,I}$ as given in Equation (2.115), we conclude that the equations of motion governing the vehicle's rotation are unchanged from those for a flat, nonrotating earth. Or, (assuming the XZ plane is a plane of symmetry, and hence $I_{xy} = I_{xz} = 0$), we have from Equations (2.27).

$$\boxed{\begin{aligned} I_{xx}\dot{P} - I_{xz}(\dot{R} + PQ) + (I_{zz} - I_{yy})QR &= L_A + L_P \\ I_{yy}\dot{Q} + (I_{xx} - I_{zz})PR + I_{xz}(P^2 - R^2) &= M_A + M_P \\ I_{zz}\dot{R} - I_{xz}(\dot{P} - QR) + (I_{yy} - I_{xx})PQ &= N_A + N_P \end{aligned}}$$

To summarize the results obtained in this section,

1. Equations (2.27) govern the inertial rotational velocity, or P, Q, and R.
2. Equations (2.123) govern the earth-referenced translational velocity, or U, V, and W.
3. Equations (2.125) govern the rate of change of geoposition, or $\dot{\lambda}_L$, $\dot{\mu}_L$, and $\dot{h}$.
4. Equations (2.114) govern the Euler-angle rates, or $\dot{\phi}$, $\dot{\theta}$, and $\dot{\psi}$.
5. Knowing P, Q, and R along with P_L, Q_L, and R_L (given by Equations (2.111)), one can determine the angular rates P_V, Q_V, and R_V from Equation (2.115).

Thus we have a set of 15 equations—12 nonlinear differential equations of motion that in a numerical simulation would be integrated simultaneously, plus the three algebraic constraints relating P_V, Q_V, and R_V to the other angular rates and the Euler angles. So even though the rotational equations of motion (Equations (2.27)) are the same as in the flat-earth case, it is now more involved to determine the vehicle's attitude and position. For example, one needs to solve for P, Q, and R, P_L, Q_L, and R_L, and P_V, Q_V, and R_V before the attitude (ϕ, θ, and ψ) can be determined.

Finally, by appropriately defining reference and perturbation variables for the new variables introduced in this (spherical, rotating earth) case, one can use the same techniques presented in Section 2.3 to determine the Reference and Perturbation Sets of equations corresponding to this case. This derivation is straightforward, but tedious, aided by the fact that many of the terms in the equations of motion are of the same form as in the case of a flat earth. This derivation is left as an exercise to the student.

2.7 POINT-MASS PERFORMANCE EQUATIONS

In this section, we will derive another set of nonlinear equations governing the translational motion of a rigid vehicle. A flat, nonrotating earth is again assumed, as in Sections 2.1 and 2.2. Plus, the effects of variable mass are ignored. The main difference here is that we derive the equations governing the components of the velocity vector expressed in a special vehicle-carried, rather than a vehicle-fixed frame. This frame, Frame W shown in Figure 2.9, is referred to as the *wind axes,* for reasons that will soon become apparent. A second, "normal" vehicle-fixed frame, denoted Frame V, is also shown in the figure. The difference between these frames will be discussed below.

The equations to be derived here are particularly useful in trajectory or performance analysis and optimization. They govern the three translational degrees of freedom, but not the three rotational degrees of freedom. Therefore, these equations are also referred to as "point-mass" equations. We will not use these equations in our perturbation analysis later in this text, but since they are so common in the analysis of flight performance, it is important to include them in this chapter dealing with equations of motion. Doing so also helps to emphasize the methodologies employed throughout this chapter.

Considering Figure 2.9, Frame I is inertial (i.e., fixed to the surface of the earth), while Frame W moves with the vehicle in the following manner. The origin of the frame is fixed to the vehicle center of mass (gravity), as required in the derivation of the vector equation of motion. However, the orientation of Frame W is such that the X_W axis is always aligned with the vehicle's instantaneous velocity vector $\mathbf{V}_V$, or aligned into the relative "wind" being encountered by the vehicle. The Z_W axis is always in the XZ plane of the vehicle, and the Y_W axis is directed along the right wing. The angles α and β define the direction of the

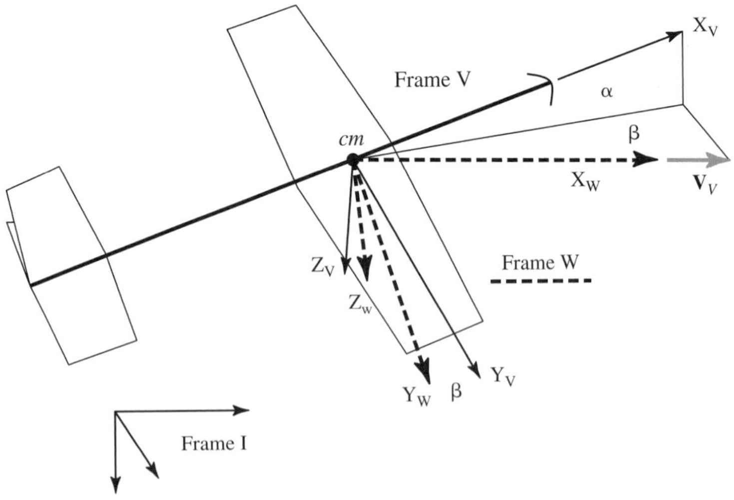

Figure 2.9 Vehicle-fixed and wind axes—angles α and β.

velocity vector $\mathbf{V}_V$ relative to the X_V axis, as well as the orientation of Frame V relative to Frame W. The angle α lies in the $X_V Z_V$ plane, while β lies in the $X_W Y_W$ plane.

The orientation of the wind axes relative to inertial is defined in terms of three Euler angles denoted ϕ_W, γ, and ψ_W. These Euler angles are analogous to, but differing from, the three Euler angles defined in Section 1.3. The three angles here are still defined as three ordered (3-2-1), successive rotations to get from Frame I to Frame W, instead of Frame V. So ψ_W is a rotation about the Z_I axis, γ a rotation about the intermediate Y_2 axis, and ϕ_W is a rotation about the X_W axis. (See Section 1.3 for further details about 3-2-1 Euler angles.) For the Euler angles defined in this fashion, ψ_W is called the "velocity-heading angle," γ the "flight-path angle," and ϕ_W the "velocity-bank angle." The unit vectors defining Frames I and W are related through the following direction-cosine matrix

$$\begin{Bmatrix} \mathbf{i}_W \\ \mathbf{j}_W \\ \mathbf{k}_W \end{Bmatrix} = \mathbf{T}_{I\text{-}W}(\phi_W, \gamma, \psi_W) \begin{Bmatrix} \mathbf{i}_I \\ \mathbf{j}_I \\ \mathbf{k}_I \end{Bmatrix} \tag{2.126}$$

with

$$\mathbf{T}_{I\text{-}W}(\phi_W, \gamma, \psi_W) = \begin{bmatrix} 1 & 0 & 0 \\ 0 & \cos\phi_W & \sin\phi_W \\ 0 & -\sin\phi_W & \cos\phi_W \end{bmatrix} \begin{bmatrix} \cos\gamma & 0 & -\sin\gamma \\ 0 & 1 & 0 \\ \sin\gamma & 0 & \cos\gamma \end{bmatrix} \begin{bmatrix} \cos\psi_W & \sin\psi_W & 0 \\ -\sin\psi_W & \cos\psi_W & 0 \\ 0 & 0 & 1 \end{bmatrix}$$

With the wind axes defined, we may now define the components of the necessary vectors in Equation (2.17), Newton's law governing translation, as follows:

$$\begin{aligned} \text{Vehicle inertial velocity vector, } \mathbf{V}_V &= V_V\,\mathbf{i}_W + 0\,\mathbf{j}_W + 0\,\mathbf{k}_W \\ \text{Aerodynamic force vector, } \mathbf{F}_{\text{Aero}} &= -D\,\mathbf{i}_W + \mathsf{S}\,\mathbf{j}_W - L\,\mathbf{k}_W \\ \text{Propulsive force vector, } \mathbf{F}_{\text{Prop}} &= F'_{P_X}\,\mathbf{i}_W + F'_{P_Y}\,\mathbf{j}_W + F'_{P_Z}\,\mathbf{k}_W \end{aligned} \tag{2.127}$$

(In the above equations, the primes on some of the components are used to distinguish them from similar (but different) vector components defined in Equations (2.19).) Note that by using the wind axes, the only component of velocity is along the X_W axis. Plus, the components of the aerodynamic force vector are easily expressed in terms of the familiar aerodynamic lift, L, drag, D, and side force, S, which will be discussed in more detail in Chapters 5 and 6. In fact, lift, drag, and side force are defined by the equation above. As we shall see in later chapters, expressions for the aerodynamic-force components are more complicated when using other vehicle-fixed frames. All these facts lead to the advantage of using the wind-axes equations in performance analysis.

With everything now defined, the wind-axes scalar equations of motion resulting from Equation (2.17) may now be developed. We start with this vector equation of motion considered in Section 2.1, repeated here for convenience.

$$\frac{d}{dt}\Big|_I\left(m\frac{d\mathbf{p}_V}{dt}\Big|_I\right) = m\frac{d\mathbf{V}_V}{dt}\Big|_I = m\mathbf{g} + \mathbf{F}_{\text{Aero}} + \mathbf{F}_{\text{Prop}}$$

Recalling the "chain rule" for vector differentiation we have

$$\frac{d\mathbf{V}_V}{dt}\Big|_I = \frac{d\mathbf{V}_V}{dt}\Big|_W + \boldsymbol{\omega}_{W,I} \times \mathbf{V}_V$$

Letting the angular-velocity components be defined by

$$\boldsymbol{\omega}_{W,I} \triangleq P_W\,\mathbf{i}_W + Q_W\,\mathbf{j}_W + R_W\,\mathbf{k}_W \tag{2.128}$$

and carrying out the cross product yields

$$\frac{d\mathbf{V}_V}{dt}\Big|_I = \dot{V}_V\mathbf{i}_W + \left(V_V R_W \mathbf{j}_W - V_V Q_W \mathbf{k}_W\right) \tag{2.129}$$

Substituting Equations (2.127) and (2.129) into the original vector equation (Equation (2.17)), and equating $\mathbf{i}_W$, $\mathbf{j}_W$, and $\mathbf{k}_W$ components, yields the three following scalar equations of motion:

$$\begin{aligned} m\dot{V}_V &= mg_{x_W} + F'_{P_X} - D \\ mV_V R_W &= mg_{y_W} + F'_{P_Y} + S \\ -mV_V Q_W &= mg_{z_W} + F'_{P_Z} - L \end{aligned} \tag{2.130}$$

We now need to address the angular rates. From the definitions of ϕ_W, γ, and ψ_W here, it is clear that

$$\boldsymbol{\omega}_{W,I} = \dot{\phi}_W\,\mathbf{i}_W + \dot{\gamma}\,\mathbf{j}_2 + \dot{\psi}_W\,\mathbf{k}_I \tag{2.131}$$

(Note the different subscripts on the unit vectors.) And one can show that one of the relationships relating the unit vectors of Frame W and those of the intermediate Frame 2 is

$$\mathbf{j}_2 = \cos\phi_W\,\mathbf{j}_W + \sin\phi_W\,\mathbf{k}_W \tag{2.132}$$

Plus, from the direction-cosine matrix in Equation (2.126), we have

$$\mathbf{k}_I = -\sin\gamma\,\mathbf{i}_W + \sin\phi_W\cos\gamma\,\mathbf{j}_W + \cos\phi_W\cos\gamma\,\mathbf{k}_W \tag{2.133}$$

Substituting the above two expressions into Equation (2.131), and equating $\mathbf{i}_W$, $\mathbf{j}_W$, and $\mathbf{k}_W$ components with those in Equation (2.128) yields the following three kinematic relations.

$$\begin{aligned} P_W &= \dot{\phi}_W - \dot{\psi}_W\sin\gamma \\ Q_W &= \dot{\gamma}\cos\phi_W + \dot{\psi}_W\sin\phi_W\cos\gamma \\ R_W &= \dot{\psi}_W\cos\phi_W\cos\gamma - \dot{\gamma}\sin\phi_W \end{aligned} \tag{2.134}$$

Finally, by recalling Equations (1.19) governing the components of gravity in a vehicle-fixed or vehicle-carried frame, and noting that Equations (1.19) are developed in terms of a set of 3-2-1 rotation angles analogous to ϕ_W, γ, and ψ_W used here, we have

$$
\begin{aligned}
g_{x_W} &= -g\sin\gamma \\
g_{y_W} &= g\sin\phi_W\cos\gamma \\
g_{z_W} &= g\cos\phi_W\cos\gamma
\end{aligned}
\tag{2.135}
$$

Substituting Equations (2.134) and (2.135) into Equations (2.130) yields the following scalar equations of motion, developed in the wind axes, governing vehicle translation.

$$
\boxed{
\begin{aligned}
&m\dot{V}_V = -mg\sin\gamma + F'_{P_X} - D \\
&mV_V\left(\dot{\psi}_W\cos\phi_W\cos\gamma - \dot{\gamma}\sin\phi_W\right) = mg\sin\phi_W\cos\gamma + F'_{P_Y} + \mathsf{S} \\
&mV_V\left(\dot{\gamma}\cos\phi_W + \dot{\psi}_W\sin\phi_W\cos\gamma\right) = -mg\cos\phi_W\cos\gamma - F'_{P_Z} + L
\end{aligned}
}
\tag{2.136}
$$

Note that the relative orientation between Frames V and W must be defined (in terms of α and β) to apply these equations. This relative orientation, of course, directly affects the aerodynamic forces acting on the vehicle. But equally important is the fact that this relative orientation determines the components of the propulsive force in the wind axes $\left(F'_{P_X},\, P_Y,\text{ and } P_Z\right)$.

Referring again to Figure 2.9, note that α and β define the relative orientation between Frames V and W. The vehicle-fixed axes, or Frame V, are fixed to the vehicle and rotate with it exclusively, while the wind axes rotates with the velocity vector $\mathbf{V}_V$ as well as the vehicle. In other words, one might think of Frame V as clamped to the vehicle structure, but Frame W only has its origin clamped to the vehicle center of mass. Finally, α is defined in terms of a rotation about the Y_V axis, while β is defined in terms of a rotation about the Z_W axis.

Now, for example, if the only propulsive force acting on the vehicle is thrust, here denoted as **T**, and if this force acts directly along the vehicle X_V axis, then the components of **T** in Frame W are

$$
\begin{aligned}
F'_{P_X} &= T\cos\alpha\cos\beta \\
F'_{P_Y} &= T\cos\alpha\sin(-\beta) \\
F'_{P_Z} &= -T\sin\alpha
\end{aligned}
$$

So as long as **T** acts directly along X_V, the equations governing translational motion become

$$
\boxed{
\begin{aligned}
&m\dot{V}_V = T\cos\alpha\cos\beta - D - mg\sin\gamma \\
&mV_V\left(\dot{\psi}_W\cos\phi_W\cos\gamma - \dot{\gamma}\sin\phi_W\right) = \mathsf{S} + T\cos\alpha\sin(-\beta) + mg\sin\phi_W\cos\gamma \\
&mV_V\left(\dot{\gamma}\cos\phi_W + \dot{\psi}_W\sin\phi_W\cos\gamma\right) = L + T\sin\alpha - mg\cos\phi_W\cos\gamma
\end{aligned}
}
\tag{2.137}
$$

Next, to determine the position of the vehicle we need to develop the three kinematic equations relating translational velocities, using the direction-cosine matrix, Equation (2.126). Let the inertial components of the vehicle's inertial-position vector be defined as

$$\mathbf{p}_V \triangleq X_I\,\mathbf{i}_I + Y_I\,\mathbf{j}_I - \mathrm{h}\,\mathbf{k}_I \tag{2.138}$$

where h is the altitude. We have previously defined the inertial velocity vector to be $\mathbf{V}_V$, so then

$$\frac{d\mathbf{p}_V}{dt}\Big|_I = \mathbf{V}_V = V_V\,\mathbf{i}_w = \dot{X}_I\,\mathbf{i}_I + \dot{Y}_I\,\mathbf{j}_I - \dot{\mathrm{h}}\,\mathbf{k}_I \tag{2.139}$$

But from Equation (2.126), the direction-cosine matrix, we have

$$\mathbf{i}_W = \cos\gamma\cos\psi_W\,\mathbf{i}_I + \cos\gamma\sin\psi_W\,\mathbf{j}_W - \sin\gamma\,\mathbf{k}_W \tag{2.140}$$

Substituting this into Equation (2.139), and equating the coefficients of $\mathbf{i}_I$, $\mathbf{j}_I$, and $\mathbf{k}_I$ yields the desired kinematic equations:

$$\boxed{\begin{aligned} \dot{X}_I &= V_V\cos\gamma\cos\psi_W \\ \dot{Y}_I &= V_V\cos\gamma\sin\psi_W \\ \dot{\mathrm{h}} &= V_V\sin\gamma \end{aligned}} \tag{2.141}$$

Consequently, the scalar equations of motion, developed in the wind axes, that govern the translation of a rigid vehicle have now been developed. These equations are given by Equations (2.136) (or (2.137)) and (2.141). The equations governing the attitude of the vehicle are frequently not used in trajectory analysis. In such cases, the thrust magnitude T and the angles α, β, and ϕ_W would be used as inputs, or forcing functions, for Equations (2.136) (or (2.137)). The angles of course determine the directions of action of the forces, as well as the magnitudes of the aerodynamic lift, drag, and side force. When the attitude dynamics are ignored, as described here, the assumption is made that the vehicle can instantaneously achieve the commanded attitude. In performance analyis, this assumption is often sufficiently valid.

If the attitude-equations are required, however, the equations used are those expressed in the vehicle's vehicle-fixed axes, as developed in Section 2.2, so the inertia terms (products of inertia) are constant. However, if the use of these attitude equations is necessary, one might just as well use the complete flat-earth equations (translation and rotation) developed in Section 2.2, because any advantage of using the wind-axes equations developed in this section would be essentially lost.

2.8 Summary

In this chapter we derived in detail several sets of nonlinear differential equations of motion for a rigid vehicle. The first set relied on the assumption of a flat, nonrotating earth, while another set resulted when we relaxed this assumption and considered a rotating, spherical earth. The "flat-earth" equations were also modified to yield two additional sets that capture the effects of rotating masses or variable mass.

To aid in small-perturbation analysis later, the nonlinear, flat-earth equations were also manipulated to yield the corresponding reference and perturbation equations. A change of variable was used in each case, which involved the introduction of reference and perturbation variables.

Finally, a fifth set of equations was also derived using the wind axes. This set governs only vehicle translation, and it was noted that this set is used frequently in aircraft performance analysis.

2.9 Problems

2.1 Using the defined components of the mass-element position vector and of the vehicle's angular-velocity vector, and the products of inertia given in Equations (2.25), verify Equation (2.26).

2.2 An aircraft in a steady, level, right turn is flying at a constant altitude and at a constant turning rate ω_{Turn}. If P is the vehicle's roll rate, Q the pitch rate, and R the yaw rate, and assuming a flat, nonrotating earth, show that the vehicle's roll, pitch, and yaw rates in such a steady, level turn will equal

$$P = -\omega_{\text{Turn}} \sin\theta$$

$$Q = \omega_{\text{Turn}} \sin\phi \cos\theta$$

$$R = \omega_{\text{Turn}} \cos\phi \cos\theta$$

where ϕ is the aircraft's bank, or roll-attitude (Euler) angle, and θ is the vehicle's pitch-attitude (Euler) angle. (Note that frequently pilots will claim that they are not undergoing any pitch rate while executing a steady, level turn, since they do not see the outside horizon moving up and down in their windscreen. But clearly a steady pitch rate is being experienced since Q is not zero. This is one reason why mathematical precision is required in flight-dynamic analysis.)

2.3 Prove the final result in the Comment in Section 2.2.

2.4 By defining appropriate reference and perturbation variables, derive the Reference and Perturbation Sets of equations corresponding to Equations (2.40).

2.5 Assuming a flat, nonrotating earth, what additional terms must be added to the Reference and Perturbations Sets of equations for a rigid vehicle to account for the presence of rotating machinery on the vehicle?

2.6 An aircraft flying over a point on the earth located at 45 degrees north latitude is told by air traffic control that its velocity is $\mathbf{V}_{V,E}$, which is known to differ from its true inertial velocity $\mathbf{V}_{V,I}$. Show that in the case of a spherical, rotating earth the theoretical difference $(\mathbf{V}_{V,I} - \mathbf{V}_{V,E})$ between these two velocities amounts to a difference of about 740 mph in an easterly direction.

2.7 Derive the complete Reference and Perturbation Sets of equations for all the equations of motion under the assumption of a spherical, rotating earth. (Note that this will require many algebraic manipulations.)

2.8 The wind axes used to develop the translational equations of motion developed in Section 2.7 is a "vehicle-carried" frame, and not a "vehicle-fixed" frame. Is this valid, or is a vehicle-fixed frame actually required for this derivation to be correct? Why?

2.9 If only flight in a vertical plane is to be considered in the performance analysis of an aircraft, and if the propulsive thrust acts only along the X_V axis, show that the equations governing the vehicle's trajectory may be expressed as

$$m\dot{V}' = T\cos\alpha\cos\beta - D - mg\sin\gamma$$

$$mV'\dot{\gamma}\cos\phi_W = L + T\sin\alpha - mg\cos\phi_W\cos\gamma$$

along with the following constraint that must be satisfied

$$\mathsf{S} + T\cos\alpha\sin\beta + (L + T\sin\alpha)\tan\phi_W = 0$$

Note that one way to satisfy this constraint is to require S, β, and ϕ_W all to be zero. Does this seem reasonable, and why?

References

1. MATLAB and Simulink, products of The Mathworks, Natick, MA.
2. Roskam, J.: *Airplane Flight Dynamics and Automatic Flight Controls,* Roskam Aviation and Engineering Corp., Lawrence, KS 1979.
3. Etkin, B.: *Dynamics of Atmospheric Flight,* Wiley, New York 1972.
4. Meriam, J. L.: *Dynamics,* Wiley, New York 1966.

CHAPTER 3

Structural Vibrations—A "Just-In-Time Tutorial"

Chapter Roadmap: *This chapter would typically not be covered in a first course in flight dynamics. The material would more likely be included in a second course.*

To understand the development of the equations of motion of elastic vehicles, as presented in Chapter 4, it is critical for students to be exposed to the theory of structural vibrations. Therefore, this chapter presents a "Just-In-Time Tutorial" on the topic. Even if students are familiar with the subject, they are encouraged to at least quickly review the tutorial to acquaint themselves with the notation and terminology used in this book, as well as the treatment of rigid-body degrees of freedom. In typical texts and courses on structural vibrations, rigid-body degrees of freedom receive little attention, although they are critical to the subject of flight dynamics.

In this topical review of lumped-mass vibrations we will discuss the important concepts of natural modes, vibration frequencies and mode shapes, and the important orthogonality property of the modes. We will also introduce the concepts of generalized coordinates and generalized forces, which will be applied in the derivations of the equations of motion of elastic vehicles discussed in Chapter 4.

3.1 LUMPED-MASS IDEALIZATIONS AND LAGRANGE'S EQUATION

In this discussion, general properties of vibrating systems can be stated after they are observed in two generic examples. In the absence of examples, the properties may seem quite abstract.

The first generic example is that of a cantilevered beam, similar to a wing, as shown in the left side of Figure 3.1. One form of the partial differential equation governing the transverse deformation Z of the beam is given as

$$\frac{\partial^2}{\partial x^2}\left(EI(x)\frac{\partial^2 Z(x,t)}{\partial x^2}\right) + m(x)\frac{\partial^2 Z(x,t)}{\partial t^2} = 0 \tag{3.1}$$

where E is the elastic modulus of the beam material, I is the area moment of inertia of the beam cross section about the its neutral axis, and $m(x)$ is the beam's mass distribution. Using the technique known as *separation of variables* to solve this partial differential equation, one can write the solution in the form

$$Z(x,t) = \sum_{i=1}^{\infty} \nu_i(x)\eta_i(t) \tag{3.2}$$

or an infinite sum of terms, each consisting of the product of a purely time-dependent function $\eta_i(t)$ and a purely space-dependent function $\nu_i(x)$. The functions $\eta_i(t)$ are referred to as *modal coordinates,* and the functions $\nu_i(x)$ are called *mode shapes* or *eigenfunctions.* Since the solution contains an infinite sum, the beam-vibration problem is referred to as an *infinite-dimensional* problem.

In practice, the infinite sum in Equation (3.2), or the solution to the vibration problem, is approximated by truncating it to a finite number of terms. This truncation is analogous to idealizing the continuous beam on the left in Figure 3.1 as a finite number of masses and springs, as shown on the right. Such an approximation is called a *lumped-mass* approximation, and for the remainder of our discussion of elastic vehicles, we will use lumped-mass approximations. We will further note that the solution to a vibration problem obtained from a finite-element structural analysis, for example, yields a lumped-mass approximation. So such approximations are always used when dealing with real, complex, flight-vehicle structures.

Referring to the right side of Figure 3.1, the beam has been idealized in terms of lumped masses m_i, and torsional springs with spring stiffnesses k_i. The lumped masses are connected by massless rigid rods of lengths l_i. The angular displacements of the springs and rods due to beam elastic deformation are denoted by the angles θ_i. In this example, the beam is idealized using two masses,

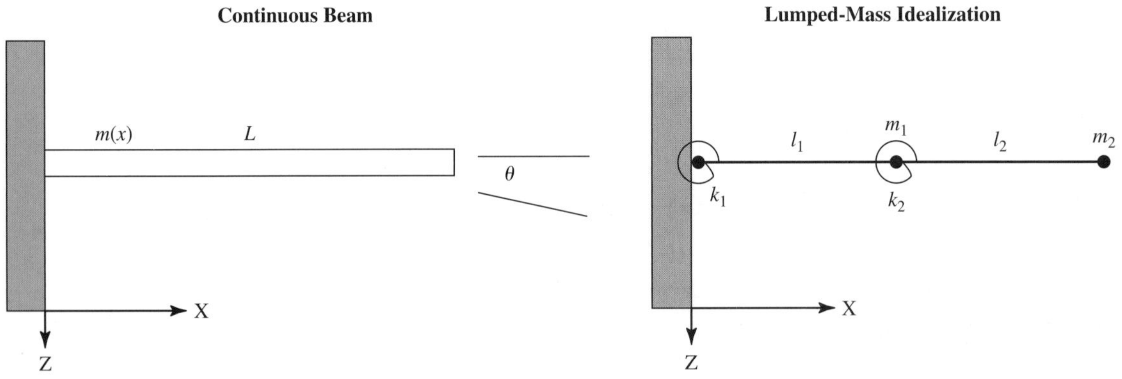

Figure 3.1 Cantilevered beam model with lumped-mass idealization.

two springs, and two rods. More masses could be used, which would, of course, improve the accuracy of the lumped-mass idealization, but that is not necessary here to demonstrate the key concepts.

We will now apply the Lagrangian method to derive this beam's equations of motion. This method will again be used in Chapter 4 to derive the equations of motion of the elastic vehicle. In the absence of external forces acting on the beam, Lagrange's equation states that

$$\boxed{\frac{d}{dt}\left(\frac{\partial \mathrm{T}}{\partial \dot{q}_i}\right) - \frac{\partial \mathrm{T}}{\partial q_i} + \frac{\partial \mathrm{U}}{\partial q_i} = 0} \tag{3.3}$$

where T is the kinetic energy of the system (beam), U the system's potential (or strain) energy, and q_i is the i'th generalized coordinate used to describe the system. These coordinates are called *generalized* because they can include both physical and nonphysical coordinates, as we will see later in Section 3.8. For now, possible coordinates used to describe the motion of this beam include relative angular displacements at the nodes $\theta_i(t)$ or transverse displacements of the masses $Z_i(t)$.

EXAMPLE 3.1

Beam Equations of Motion

Using the Lagrangian approach, find the equations of motion of the lumped-mass cantilever beam shown in Figure 3.1.

■ Solution

The potential or strain energy of the beam may be written as

$$\mathrm{U} = \frac{1}{2}\left(k_1\theta_1^2 + k_2\theta_2^2\right) = \frac{1}{2}\begin{Bmatrix}\theta_1\\ \theta_2\end{Bmatrix}^T \begin{bmatrix}k_1 & 0\\ 0 & k_2\end{bmatrix}\begin{Bmatrix}\theta_1\\ \theta_2\end{Bmatrix} \tag{3.4}$$

While the kinetic energy of the beam may be written as

$$\mathrm{T} = \frac{1}{2}\left(m_1\dot{Z}_1^2 + m_2\dot{Z}_2^2\right) = \frac{1}{2}\begin{Bmatrix}\dot{Z}_1\\ \dot{Z}_2\end{Bmatrix}^T \begin{bmatrix}m_1 & 0\\ 0 & m_2\end{bmatrix}\begin{Bmatrix}\dot{Z}_1\\ \dot{Z}_2\end{Bmatrix} \tag{3.5}$$

However, note that the vertical and angular displacements are not independent. In fact, assuming linear vibrations, which imply small-angle approximations are valid, we have the two sets of coordinates related by

$$\begin{Bmatrix}Z_1\\ Z_2\end{Bmatrix} = \begin{bmatrix}l_1 & 0\\ (l_1 + l_2) & l_2\end{bmatrix}\begin{Bmatrix}\theta_1\\ \theta_2\end{Bmatrix} \tag{3.6}$$

Using this constraint relation, we can rewrite the kinetic energy, Equation (3.5), in terms of angular displacements, or

$$\mathrm{T} = \frac{1}{2}\left(m_1 l_1^2\dot{\theta}_1^2 + m_2\left((l_1 + l_2)\dot{\theta}_1 + l_2\dot{\theta}_2\right)^2\right) \tag{3.7}$$

Using the angular displacements as the coordinates q_i to describe the system, we apply Lagrange's equation, Equation (3.3), to obtain the following two ordinary differential equations governing the system's (beam's) motion.

$$\begin{bmatrix} m_1 l_1^2 + m_2(l_1 + l_2)^2 & m_2 l_2(l_1 + l_2) \\ m_2 l_2(l_1 + l_2) & m_2 l_2^2 \end{bmatrix} \begin{Bmatrix} \ddot{\theta}_1 \\ \ddot{\theta}_2 \end{Bmatrix} + \begin{bmatrix} k_1 & 0 \\ 0 & k_2 \end{bmatrix} \begin{Bmatrix} \theta_1 \\ \theta_2 \end{Bmatrix} = \begin{Bmatrix} 0 \\ 0 \end{Bmatrix} \tag{3.8}$$

Or, in a more general form, we have the matrix-vector differential equation

$$[\mathbf{M}]\{\ddot{\mathbf{q}}\} + [\mathbf{K}]\{\mathbf{q}\} = \{\mathbf{0}\} \tag{3.9}$$

where **M** is referred to as the *mass matrix,* and **K** is referred to as the *stiffness matrix.* Both of these matrices will always be real, symmetric matrices. In the case of this cantilevered beam, both matrices will also always be positive definite, indicating that an inverse always exists and all eigenvalues are strictly positive.

Example 3.1 demonstrates that the equations of motion of the vibrating beam can be written in the form given in Equation (3.9). In fact, all lumped-mass vibration problems may be written in this generic form (although all lumped-mass vibration problems do not yield positive-definite stiffness matrices). Hence a vibration problem is completely described by selecting the system's coordinates q_i, knowing the initial conditions on those coordinates, and finding the mass and stiffness matrices.

3.2 MODAL ANALYSIS

We will now continue with our general development, and introduce the *modal analysis* of the general lumped-mass vibration problem, which will always have the form given in Equation (3.9). Note first that since the mass matrix **M** is positive definite, its inverse exists, and therefore

$$\{\ddot{\mathbf{q}}\} + [\mathbf{D}]\{\mathbf{q}\} = \{\mathbf{0}\} \tag{3.10}$$

where the *dynamic matrix* is $\mathbf{D} = \mathbf{M}^{-1}\mathbf{K}$. Using the fact that the solution will have the form given in Equation (3.2), we introduce the following linear relationship between the physical displacement coordinates (θ_i's in the beam example) and the modal coordinates.

$$\{\mathbf{q}\} = [\boldsymbol{\Phi}]\{\boldsymbol{\eta}\} \tag{3.11}$$

where the matrix $\boldsymbol{\Phi}$ and the modal coordinates $\eta_i(t)$ still need to be determined. Substituting this relationship into Equation (3.10) yields

$$[\boldsymbol{\Phi}]\{\ddot{\boldsymbol{\eta}}\} + [\mathbf{D}][\boldsymbol{\Phi}]\{\boldsymbol{\eta}\} = \{\mathbf{0}\} \tag{3.12}$$

or

$$\{\ddot{\boldsymbol{\eta}}\} + [\boldsymbol{\Phi}]^{-1}[\mathbf{D}][\boldsymbol{\Phi}]\{\boldsymbol{\eta}\} = \{\mathbf{0}\} \tag{3.13}$$

We will now select $\boldsymbol{\Phi}$ to diagonalize the dynamic matrix **D**. That is, we will choose $\boldsymbol{\Phi}$ such that

$$[\boldsymbol{\Phi}]^{-1}[\mathbf{D}][\boldsymbol{\Phi}] = [\boldsymbol{\Lambda}] = [diag(\lambda_i)],\ i = 1 \ldots n \tag{3.14}$$

where the λ_i's are the n eigenvalues of **D** for which

$$\det\left[\lambda_i[\mathbf{I}] - [\mathbf{D}]\right] = 0$$

and where in this chapter $[\mathbf{I}]$ is the identity matrix.

The desired $\mathbf{\Phi}$ is the *modal matrix* of **D**, consisting of the n eigenvectors of **D**. That is, if λ_i is an eigenvalue of **D**, then the eigenvector $\boldsymbol{\nu}_i$ corresponding to λ_i satisfies the following relation.

$$\left[\lambda_i[\mathbf{I}] - [\mathbf{D}]\right]\{\boldsymbol{\nu}_i\} = \{\mathbf{0}\} \tag{3.15}$$

Finally, the modal matrix $\mathbf{\Phi}$ is constructed from the eigenvectors of **D** as follows.

$$[\mathbf{\Phi}] = [\boldsymbol{\nu}_1|\boldsymbol{\nu}_2|\ldots|\boldsymbol{\nu}_n] \tag{3.16}$$

That is, the eigenvectors of **D** are the columns of the modal matrix $\mathbf{\Phi}$.

With the modal matrix $\mathbf{\Phi}$ selected as just described, Equation (3.13) becomes

$$\boxed{\{\ddot{\boldsymbol{\eta}}\} + [\mathbf{\Lambda}]\{\boldsymbol{\eta}\} = \{\mathbf{0}\}} \tag{3.17}$$

and these n differential equations are now uncoupled. Consequently, we have n independent ordinary differential equations of the form

$$\ddot{\eta}_i + \lambda_i\eta_i = 0 \tag{3.18}$$

each with a solution of the form

$$\eta_i(t) = A_i \cos\left(\sqrt{\lambda_i}t + \Gamma_i\right) \tag{3.19}$$

The constants A_i and Γ_i are found from the initial conditions. Therefore, we have n natural modes, each oscillating at its natural frequency $\sqrt{\lambda_i}$. Hence, the eigenvalues are frequently denoted as $\lambda_i = \omega_i^2$.

Armed with the system's eigenvalues and eigenvectors, we can now determine the system's response. Recall we had the following relationship between the system coordinates (angles θ_i, in our example) and the modal coordinates.

$$\boxed{\begin{aligned}\{\mathbf{q}(t)\} &= [\mathbf{\Phi}]\{\boldsymbol{\eta}(t)\} = [\boldsymbol{\nu}_1|\ldots|\boldsymbol{\nu}_n]\{\boldsymbol{\eta}(t)\} \\ &= \sum_{i=1}^{n}\{\boldsymbol{\nu}_i\}\eta_i(t)\end{aligned}} \tag{3.20}$$

and therefore we have

$$\begin{aligned} q_1(t) &= \nu_{1,1}\eta_1(t) + \ldots + \nu_{1,n}\eta_n(t) \\ &\vdots \\ q_n(t) &= \nu_{n,1}\eta_1(t) + \ldots + \nu_{n,n}\eta_n(t) \end{aligned} \tag{3.21}$$

From these expressions we can see that each i,j element of an eigenvector defines the amount each modal response $\eta_j(t)$ contributes to the system response q_i (e.g., angular displacement). These eigenvectors $\boldsymbol{\nu}_i$'s are also called *mode shapes.*

Elements of the eigenvectors, or mode shapes, correspond to a physical displacement (e.g., transverse displacement Z_i, or angular displacement θ_i) of a lumped-mass element. Hence, they may be plotted against the mass-element location along the beam x. (See Example 3.2 below.) Furthermore, since eigenvectors have arbitrary magnitude, they must be normalized somehow. For example, they may all be normalized to a unit length (as in Example 3.2), or to unity displacement of a selected element, or to a unity generalized mass (also discussed below).

Finally, it was stated that the $2n$ constants of integration A_i and Γ_i were found from the initial conditions (on the modal coordinates η_i). For example, if the initial conditions on the physical coordinates are given as

$$\begin{aligned} \{\mathbf{q}(t_1)\} &= \{\mathbf{C}_1\} \\ \{\dot{\mathbf{q}}(t_1)\} &= \{\mathbf{C}_2\} \end{aligned} \tag{3.22}$$

the following initial conditions are imposed on the modal coordinates

$$\begin{aligned} \{\boldsymbol{\eta}(t_1)\} &= [\boldsymbol{\Phi}]^{-1}\{\mathbf{q}(t_1)\} = [\boldsymbol{\Phi}]^{-1}\{\mathbf{C}_1\} \\ \{\dot{\boldsymbol{\eta}}(t_1)\} &= [\boldsymbol{\Phi}]^{-1}\{\dot{\mathbf{q}}(t_1)\} = [\boldsymbol{\Phi}]^{-1}\{\mathbf{C}_2\} \end{aligned} \tag{3.23}$$

MATLAB EXAMPLE 3.2

Modal Analysis of Cantilevered Beam Model

From a modal analysis, determine the vibration response of the cantilever beam considered in Example 3.1. Let m_1, m_2, l_1, l_2, k_1, and k_2 all equal unity.

■ Solution

We had previously determined that the equations of motion describing the lumped-mass system were

$$\begin{bmatrix} m_1 l_1^2 + m_2(l_1+l_2)^2 & m_2 l_2(l_1+l_2) \\ m_2 l_2(l_1+l_2) & m_2 l_2^2 \end{bmatrix} \begin{Bmatrix} \ddot{\theta}_1 \\ \ddot{\theta}_2 \end{Bmatrix} + \begin{bmatrix} k_1 & 0 \\ 0 & k_2 \end{bmatrix} \begin{Bmatrix} \theta_1 \\ \theta_2 \end{Bmatrix} = \begin{Bmatrix} 0 \\ 0 \end{Bmatrix} \tag{3.24}$$

or

$$\begin{bmatrix} m_{1,1} & m_{1,2} \\ m_{1,2} & m_{2,2} \end{bmatrix} \begin{Bmatrix} \ddot{q}_1 \\ \ddot{q}_2 \end{Bmatrix} + \begin{bmatrix} k_1 & 0 \\ 0 & k_2 \end{bmatrix} \begin{Bmatrix} q_1 \\ q_2 \end{Bmatrix} = \begin{Bmatrix} 0 \\ 0 \end{Bmatrix}$$

with obvious definitions for the elements of the mass matrix **M** and stiffness matrix **K**. Consequently, we have the dynamic matrix given by

$$\mathbf{D} = \mathbf{M}^{-1}\mathbf{K} = \frac{1}{\det \mathbf{M}} \begin{bmatrix} m_{2,2}k_1 & -m_{1,2}k_2 \\ -m_{1,2}k_1 & m_{1,1}k_2 \end{bmatrix} \tag{3.25}$$

Note that even though **M** and **K** are symmetric, **D** is not always symmetric. Substituting the given numerical values we have

$$\mathbf{M} = \begin{bmatrix} 5 & 2 \\ 2 & 1 \end{bmatrix} \qquad \mathbf{K} = \begin{bmatrix} 1 & 0 \\ 0 & 1 \end{bmatrix} \qquad \mathbf{D} = \begin{bmatrix} 1 & -2 \\ -2 & 5 \end{bmatrix}$$

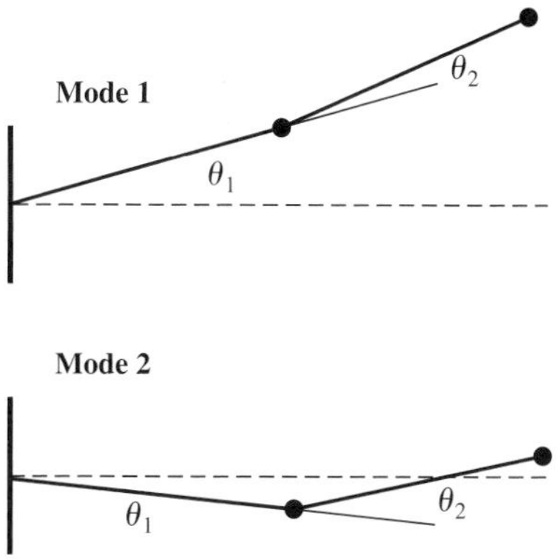

Figure 3.2 Mode shapes for Example 3.2.

Using MATLAB, we find that **D** has the following two sets of eigenvalues and eigenvectors.

$$\text{Mode 1: } \lambda_1 = 0.1716 \text{ sec}^{-2}, \ \boldsymbol{\nu}_1 = \begin{Bmatrix} 0.9239 \\ 0.3827 \end{Bmatrix} \text{ rad}$$

$$\text{Mode 2: } \lambda_2 = 5.8234 \text{ sec}^{-2}, \ \boldsymbol{\nu}_2 = \begin{Bmatrix} -0.3827 \\ 0.9239 \end{Bmatrix} \text{ rad}$$

Consequently, the vibration response is

$$\begin{Bmatrix} q_1(t) \\ q_2(t) \end{Bmatrix} = \begin{Bmatrix} \theta_1(t) \\ \theta_2(t) \end{Bmatrix} = \begin{Bmatrix} 0.9239 \\ 0.3827 \end{Bmatrix} A_1 \cos\left(\sqrt{0.1716}t + \Gamma_1\right) + \begin{Bmatrix} -0.3827 \\ 0.9239 \end{Bmatrix} A_2 \cos\left(\sqrt{5.8234}t + \Gamma_2\right) \text{ rad}$$

where the four constants of integration (A's and Γ's) may be determined from initial conditions. The two mode shapes (eigenvectors) are plotted in Figure 3.2 above.

3.3 ORTHOGONALITY OF THE VIBRATION MODES

We will now address another property of the vibration modes, that of orthogonality. From the definition of the eigenvalues and eigenvectors of the dynamic matrix **D**, we have

$$\left[\lambda_i[\mathbf{I}] - [\mathbf{D}]\right]\{\boldsymbol{\nu}_i\} = \{\mathbf{0}\} \tag{3.26}$$

so then

$$\lambda_i\{\boldsymbol{\nu}_i\} = [\mathbf{D}]\{\boldsymbol{\nu}_i\}$$

or

$$\lambda_i[\mathbf{M}]\{\boldsymbol{\nu}_i\} = [\mathbf{K}]\{\boldsymbol{\nu}_i\} \tag{3.27}$$

Now let $\boldsymbol{v}_i$ and $\boldsymbol{v}_j$ be eigenvectors of **D** associated with two eigenvalues λ_i and λ_j, respectively, with each pair satisfying Equation (3.27). Multiplying each equation of the form of Equation (3.27) by the transpose of the other eigenvector yields

$$\begin{aligned}\lambda_i\{\boldsymbol{v}_j\}^T[\mathbf{M}]\{\boldsymbol{v}_i\} &= \{\boldsymbol{v}_j\}^T[\mathbf{K}]\{\boldsymbol{v}_i\}\\ \lambda_j\{\boldsymbol{v}_i\}^T[\mathbf{M}]\{\boldsymbol{v}_j\} &= \{\boldsymbol{v}_i\}^T[\mathbf{K}]\{\boldsymbol{v}_j\}\end{aligned} \tag{3.28}$$

Noting that both **M** and **K** are symmetric, and by transposing one of the above equations and subtracting the result from the other equation, we have

$$(\lambda_i - \lambda_j)\{\boldsymbol{v}_j\}^T[\mathbf{M}]\{\boldsymbol{v}_i\} = 0 \tag{3.29}$$

Therefore, if λ_i differs from λ_j (i.e., the eigenvalues are unique), the following *orthogonality* property must hold.

$$\{\boldsymbol{v}_j\}^T[\mathbf{M}]\{\boldsymbol{v}_i\} = 0,\ i \neq j \tag{3.30}$$

Plus, if the same eigenvalue and eigenvector are used in Equation (3.29), then

$$\boxed{\{\boldsymbol{v}_i\}^T[\mathbf{M}]\{\boldsymbol{v}_i\} \triangleq \mathcal{M}_i} \tag{3.31}$$

where $\mathcal{M}_i$ is defined as the i'*th generalized mass.*

Through a similar analysis (left as an exercise to the student), one can also show that for unique eigenvalues the following orthogonality property also holds

$$\{\boldsymbol{v}_j\}^T[\mathbf{K}]\{\boldsymbol{v}_i\} = 0,\ i \neq j \tag{3.32}$$

and that

$$\boxed{\{\boldsymbol{v}_i\}^T[\mathbf{K}]\{\boldsymbol{v}_i\} = \mathcal{K}_i} \tag{3.33}$$

where $\mathcal{K}_i$ is defined as the i'*th generalized stiffness.*

Finally, recalling the definition of the modal matrix $\boldsymbol{\Phi}$, and using Equation (3.12), we can write

$$[\boldsymbol{\Phi}]^{-1}[\mathbf{M}][\boldsymbol{\Phi}]\{\ddot{\boldsymbol{\eta}}\} + [\boldsymbol{\Phi}]^{-1}[\mathbf{K}][\boldsymbol{\Phi}]\{\boldsymbol{\eta}\} = \{\mathbf{0}\} \tag{3.34}$$

Now since the eigenvectors (mode shapes) are orthogonal, using Equations 3.31 and (3.33) the above expression becomes

$$[\mathcal{M}]\{\ddot{\boldsymbol{\eta}}\} + [\mathcal{K}]\{\boldsymbol{\eta}\} = \{\mathbf{0}\} \tag{3.35}$$

where $[\mathcal{M}]$ is the (diagonal) *generalized-mass matrix* and $[\mathcal{K}]$ is the (diagonal) *generalized-stiffness matrix.* Continuing with Equation (3.35) we see that

$$\{\ddot{\boldsymbol{\eta}}\} + [\mathcal{M}]^{-1}[\mathcal{K}]\{\boldsymbol{\eta}\} = \{\mathbf{0}\}$$

and then, consistent with Equation (3.17), we have

$$\boxed{\{\ddot{\boldsymbol{\eta}}\} + [\boldsymbol{\Lambda}]\{\boldsymbol{\eta}\} = \{\mathbf{0}\}} \tag{3.36}$$

Or the eigenvalue λ_i equals $\mathcal{M}_i/\mathcal{K}_i$.

In summary, we have shown that the n-lumped-mass vibration problem may be expressed in terms of n orthogonal modes, each described by its modal coordinate, mode shape (eigenvector), and frequency (square root of its eigenvalue). Furthermore, for the restrained vibration problem considered so far, these modes are orthogonal with respect to both the mass and stiffness matrices (Equations (3.30) and (3.32)).

3.4 RIGID-BODY DEGREES OF FREEDOM

We will now turn our attention to vibration problems involving bodies that are unrestrained, or free to translate and/or rotate, unlike the cantilever beam considered previously. This whole-body motion is associated with *rigid-body degrees of freedom,* and these degrees of freedom are obviously critical to describing the motion of vehicles in flight.

Consider an unrestrained beam represented by three lumped masses, as shown in Figure 3.3. This beam model might represent a launch vehicle, for example. As in the case with the cantilevered beam, here we will only deal with transverse displacement Z. The displacements of the three masses will be denoted as Z_1, Z_2, and Z_3, while the bending displacement of the elastic beam is defined by the relative angle θ. With only transverse displacements considered, and three lumped masses, there are three degrees of freedom.

The kinetic energy of the beam is

$$\begin{aligned} \mathrm{T} &= \frac{1}{2}\left(m_1\dot{Z}_1^2 + m_2\dot{Z}_2^2 + m_3\dot{Z}_3^2\right) \\ &= \frac{1}{2}\{\dot{\mathbf{Z}}\}^T[\mathbf{M}]\{\dot{\mathbf{Z}}\} \end{aligned} \qquad (3.37)$$

where

$$\{\dot{\mathbf{Z}}\} = \begin{Bmatrix} \dot{Z}_1 \\ \dot{Z}_2 \\ \dot{Z}_3 \end{Bmatrix} \qquad (3.38)$$

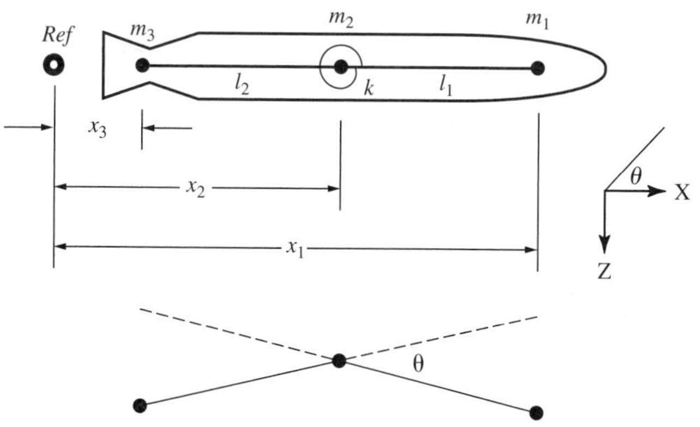

Figure 3.3 Unrestrained three-lumped-mass beam model.

The beam's potential, or strain energy, is

$$U = \frac{1}{2}k\theta^2 = \frac{1}{2}\theta^T k\theta \tag{3.39}$$

However, we can again express the deflection angle θ in terms of the transverse displacements as follows. From geometry, and assuming small angles, we see that

$$\theta = \frac{Z_1 - Z_2}{x_1 - x_2} - \frac{Z_2 - Z_3}{x_2 - x_3}$$

or

$$\begin{aligned} \theta &= \left[\frac{1}{x_1 - x_2} \quad \left(-\frac{1}{x_1 - x_2} - \frac{1}{x_2 - x_3} \right) \quad \frac{1}{x_2 - x_3} \right] \{\mathbf{Z}\} \\ &= [\mathbf{C}]\{\mathbf{Z}\} \end{aligned} \tag{3.40}$$

So **C** is a constraint matrix that relates the beam-displacement coordinates. As a result, the potential energy can also be expressed in terms of the transverse displacements, or

$$\begin{aligned} U &= \frac{1}{2}[\mathbf{Z}]^T[\mathbf{C}]^T k[\mathbf{C}][\mathbf{Z}] \\ &= \frac{1}{2}[\mathbf{Z}]^T[\mathbf{K}_c][\mathbf{Z}] \end{aligned} \tag{3.41}$$

with obvious definition of the *constrained stiffness matrix* $\mathbf{K}_c$.

With both the kinetic and the potential energies now defined in terms of transverse displacements **Z** and velocities $\dot{\mathbf{Z}}$, we can again apply Lagrange's equation, (Equation (3.3)), to obtain the equations of motion of the beam, expressed here in terms of the mass and constrained-stiffness matrices.

$$[\mathbf{M}]\{\ddot{\mathbf{Z}}\} + [\mathbf{K}_c]\{\mathbf{Z}\} = \mathbf{0}$$

Or

$$\begin{aligned} \{\ddot{\mathbf{Z}}\} + [\mathbf{M}]^{-1}[\mathbf{K}_c]\{\mathbf{Z}\} &= \mathbf{0} \\ \{\ddot{\mathbf{Z}}\} + [\mathbf{D}_c]\{\mathbf{Z}\} &= \mathbf{0} \end{aligned} \tag{3.42}$$

where now the dynamic matrix is denoted $\mathbf{D}_c$ to indicate that the constraint matrix has been employed. The eigenvalues and eigenvectors of $\mathbf{D}_c$ may now be determined.

If λ_i and $\boldsymbol{\nu}_i$ are an eigenvalue/eigenvector pair of $\mathbf{D}_c$, then from the definitions of eigenvalues and eigenvectors (as discussed in the case of the cantilever beam)

$$\lambda_i \boldsymbol{\nu}_i = \mathbf{D}_c \boldsymbol{\nu}_i$$

or

$$\lambda_i \mathbf{M} \boldsymbol{\nu}_i = \mathbf{K}_c \boldsymbol{\nu}_i$$

Also, for a second eigenvalue/eigenvector pair (pair j),

$$\lambda_j \mathbf{M}\boldsymbol{\nu}_j = \mathbf{K}_c \boldsymbol{\nu}_j$$

So, as with Equation (3.29), repeated here, and with $\mathbf{K}_c$ symmetric we have

$$(\lambda_i - \lambda_j)\{\boldsymbol{\nu}_j\}^T[\mathbf{M}]\{\boldsymbol{\nu}_i\} = 0$$

With the restrained cantilever beam, all the eigenvalues were distinct, so Equation (3.29) assured that all the mode shapes were mutually orthogonal with respect to the mass matrix. That is,

$$\{\boldsymbol{\nu}_j\}^T[\mathbf{M}]\{\boldsymbol{\nu}_i\} = 0, \, i \neq j \tag{3.43}$$

But with the unrestrained beam considered here, two of the eigenvalues of $\mathbf{D}_c$ will be zero, and hence equal, due to the existence of two rigid-body degrees of freedom—vertical translation and rotation of the entire beam. Thus, Equation (3.29) does <u>not</u> assure that the two eigenvectors associated with these two zero eigenvalues will be orthogonal with respect to **M**. However, we shall see that for other reasons, all the mode shapes are still mutually orthogonal with respect to the mass matrix. Since there are three degrees of freedom, the solution yields two rigid-body modes and a single vibration mode. The single nonzero eigenvalue, and associated eigenvector, corresponds to the beam's vibration mode for the three-mass beam.

MATLAB EXAMPLE 3.3

Preliminary Modal Analysis of the Unrestrained Beam

Consider the three-mass unrestrained beam as shown in Figure 3.3. Let the spring stiffness k be unity, the masses be

$$m_1 = m_2 = 1, \, m_3 = 2 \text{ mass units}$$

and the lengths be $l_1 = l_2 = 1$ length units. Find the frequency and mode shape of the beam's single vibration mode. Also note the characteristics of the additional two eigenvectors.

■ Solution

The mass matrix for this beam is

$$[\mathbf{M}] = \begin{bmatrix} 1 & 0 & 0 \\ 0 & 1 & 0 \\ 0 & 0 & 2 \end{bmatrix}$$

while the constrained stiffness matrix, using Equations (3.40) and (3.41), is

$$[\mathbf{K}_c] = \begin{bmatrix} 1 \\ -2 \\ 1 \end{bmatrix}[1][1 \quad -2 \quad 1] = \begin{bmatrix} 1 & -2 & 1 \\ -2 & 4 & -2 \\ 1 & -2 & 1 \end{bmatrix}$$

The constrained dynamic matrix is $\mathbf{D}_c = \mathbf{M}^{-1}\mathbf{K}_c$.

Using MATLAB we have

```
»M=[1 0 0;0 1 0;0 0 2]

M =

   1  0  0
   0  1  0
   0  0  2

»Kc=[1 -2 1; -2 4 -2;1 -2 1]

Kc =

   1  -2   1
  -2   4  -2
   1  -2   1

»Dc=M\Kc

Dc =

   1.0000e+00  -2.0000e+00   1.0000e+00
  -2.0000e+00   4.0000e+00  -2.0000e+00     Constrained dynamic matrix
   5.0000e-01  -1.0000e+00   5.0000e-01

»[V,lambda]=eig(Dc)

V =

  -9.0914e-01  -4.3644e-01  -2.2454e-01     Three eigenvectors (columns)
  -4.0406e-01   8.7287e-01   3.4366e-01     (Note that MATLAB normalized the
   1.0102e-01  -2.1822e-01   9.1186e-01     eigenvectors to unity length)

lambda =

   0           0            0
   0  5.5000e+00            0               Three eigenvalues on diagonal
   0           0   2.9620e-17               (two are zero)
```

So we have determined that there is one vibration mode with a frequency equal to $\sqrt{5.5}$ rad/sec, and it can be shown that the two eigenvectors associated with the rigid-body degrees of freedom (Columns 1 and 3 of the matrix V) are orthogonal (with respect to the mass matrix) to the vibration mode shape (Column 2 of the matrix V). However, it is significant to note that these two eigenvectors associated with the two zero eigenvalues are <u>not</u> mutually orthogonal with respect to the mass matrix. Specifically, from MATLAB:

```
»V(:,1)'*M*V(:,3)

ans =

  2.4950e-01
```

At this point structural dynamicists might stop their analysis, since they are frequently only interested in the vibration frequencies and mode shapes. But it is important to note that we seek a modal representation of the beam such that the beam's complete motion can be described in terms of mutually orthogonal or *normal* modes. In Example 3.3, even though the two eigenvectors of $\mathbf{D}_c$ associated with the rigid-body degrees of freedom were not originally mutually orthogonal, two new orthogonal eigenvectors can be found from the initial modal analysis of $\mathbf{D}_c$.

From linear algebra we know that if a matrix has repeated eigenvalues (both zero here), any linear combination of the eigenvectors associated with the repeated eigenvalues are also eigenvectors of the given matrix. This fact can be used to obtain three mutually orthogonal or normal modes for our unrestrained beam.

It should be clear that one of the rigid-body degrees of freedom is rigid-body translation, and the mode of motion associated with this degree of freedom would have the following mode shape

$$\boldsymbol{\nu}_{\text{translation}} = 1/\sqrt{3}\begin{Bmatrix} 1 \\ 1 \\ 1 \end{Bmatrix} \tag{3.44}$$

where this vector has been normalized to unity length, consistent with the MATLAB results. So we should be able to obtain $\boldsymbol{\nu}_{\text{translation}}$ by linearly combining the two eigenvectors associated with the two repeated eigenvalues. Finally, we should also be able to linearly combine the two eigenvectors associated with the repeated eigenvalues and obtain a third eigenvector of unity length that is orthogonal to $\boldsymbol{\nu}_{\text{translation}}$.

That is, if $\boldsymbol{\nu}_1$ and $\boldsymbol{\nu}_3$ are the initial eigenvectors associated with the two zero eigenvalues in Example 3.3, then we should be able to find coefficients c_1 and c_2 such that

$$\{c_1\boldsymbol{\nu}_1 + c_2\boldsymbol{\nu}_3\} = \boldsymbol{\nu}_{\text{translation}} \tag{3.45}$$

In addition, we seek a coefficient c_3 such that the following orthogonality condition is satisfied.

$$\{\boldsymbol{\nu}_1 + c_3\boldsymbol{\nu}_3\}^T[\mathbf{M}]\{\boldsymbol{\nu}_{\text{translation}}\} = 0 \tag{3.46}$$

And finally, we may find the coefficient c_4 to normalize the resulting eigenvector. That is, find the c_4 such that

$$c_4\{\boldsymbol{\nu}_1 + c_3\boldsymbol{\nu}_3\}^T\{\boldsymbol{\nu}_1 + c_3\boldsymbol{\nu}_3\} = 1 \tag{3.47}$$

The sought rigid-body mode shapes are then $\boldsymbol{\nu}_{\text{translation}}$ and

$$\boldsymbol{\nu}_{3\text{ new}} = c_4\{\boldsymbol{\nu}_1 + c_3\boldsymbol{\nu}_3\} \tag{3.48}$$

MATLAB EXAMPLE 3.4

Finding the Rigid-Body Mode Shapes

For the constrained dynamic matrix $\mathbf{D}_c$ found in MATLAB Example 3.3, find the two mutually orthogonal rigid-body mode shapes.

■ Solution

From MATLAB, we found that

```
»Dc=M\Kc

Dc =

  1.0000e+00  -2.0000e+00   1.0000e+00
 -2.0000e+00   4.0000e+00  -2.0000e+00      Constrained dynamic matrix
  5.0000e-01  -1.0000e+00   5.0000e-01
```

And we had determined that its eigenvalues and eigenvectors are

```
»[V,lambda]=eig(Dc)

V =

 -9.0914e-01  -4.3644e-01  -2.2454e-01
 -4.0406e-01   8.7287e-01   3.4366e-01    Initial set of eigenvectors (columns)
  1.0102e-01  -2.1822e-01   9.1186e-01

lambda =

  0          0            0
  0  5.5000e+00           0                Eigenvalues on the diagonal
  0          0   2.9620e-17
```

NOTE TO STUDENT

Be aware that different versions of MATLAB may yield different eigenvectors here. However, the indicated procedure will still produce the desired results.

Next, we find the linear combination of $\boldsymbol{v}_1$ (V(:,1))and $\boldsymbol{v}_3$ (V(:,3)) that equals $\boldsymbol{v}_{\text{translation}}$. In other words, we find c_1 and c_2 such that

$$c_1\boldsymbol{v}_1 + c_2\boldsymbol{v}_3 = 1/\sqrt{3}\begin{Bmatrix}1\\1\\1\end{Bmatrix} = \boldsymbol{v}_{\text{translation}}$$

The coefficients may be found by solving the two simultaneous equations

$$\begin{aligned} c_1\boldsymbol{v}_3^T\boldsymbol{v}_1 + c_2(1) &= \boldsymbol{v}_3^T\boldsymbol{v}_{\text{translation}} \\ c_1(1) + c_2\boldsymbol{v}_1^T\boldsymbol{v}_3 &= \boldsymbol{v}_1^T\boldsymbol{v}_{\text{translation}} \end{aligned} \tag{3.49}$$

From MATLAB we let

```
»A=[V(:,3)'*V(:,1) 1;1 V(:,1)'*V(:,3)]
A=

   1.5739e-01   1.0000e+00
   1.0000e+00   1.5739e-01

»B=[V(:,3)'*[1/√3 1/√3 1/√3]';V(:,1)'*[1/√3 1/√3 1/√3]']
B=

   5.9524e-01
  -6.9985e-01

»CC=A\B
CC =

  -8.1370e-01
   7.2330e-01
```

Coefficients c_1 and c_2 that yield $\boldsymbol{\nu}_{\text{translation}}$

```
»CC(1)*V(:,1)+CC(2)*V(:,3)
ans =

   5.7735e-01
   5.7735e-01
   5.7735e-01
```

Checking the result

Next, we find the desired third mode shape, orthogonal to $\boldsymbol{\nu}_{\text{translation}}$, from the linear combination of the two initial eigenvectors associated with the zero eigenvalues. Or we find c_3 such that

$$\{\boldsymbol{\nu}_1 + c_3\boldsymbol{\nu}_3\}^T[\mathbf{M}]\{\boldsymbol{\nu}_{\text{translation}}\} = 0$$

And finally, we normalize that result by finding c_4 such that

$$c_4\{\boldsymbol{\nu}_1 + c_3\boldsymbol{\nu}_3\}^T\{\boldsymbol{\nu}_1 + c_3\boldsymbol{\nu}_3\} = 1$$

From MATLAB

```
»CC3=-(V(:,1)'*M*[1/√3 1/√3 1/√3]')/(V(:,3)'*M*[1/√3 1/√3 1/√3]')
CC3=

   5.7193e-01
```

CC3 is c_3

```
»V3new=V(:,1)+.57193*V(:,3)
V3new=

  -1.0376e+00
  -2.0751e-01
   6.2253e-01
```

Desired second rigid-body mode shape

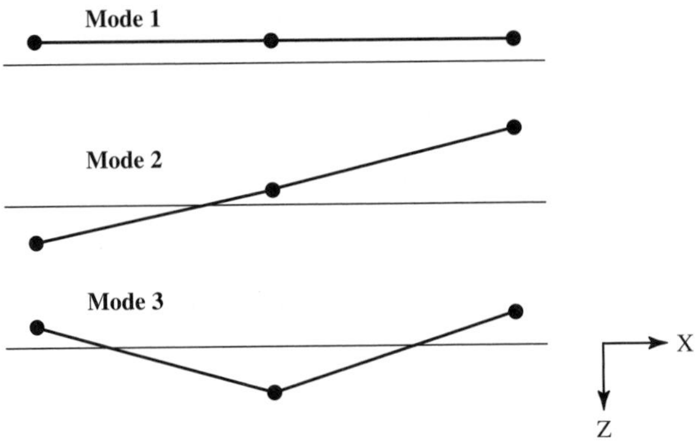

Figure 3.4 Mode shapes of the unrestrained, three-mass beam.

```
»CC4=1/(V3new'*V3new)
CC4 =
  6.6351e-01
```

Scale factor to normalize the second rigid-body mode shape to unity length

Finally, we check to make sure that the two new rigid-body mode shapes are both orthogonal to the vibration mode shape. From MATLAB we have

```
»[1/√3 1/√3 1/√3]*M*V(:,2)
ans =
  -7.5191e-17
»V3new'*M*V(:,2)
ans =
  5.9466e-17
```

The three mode shapes are plotted in Figure 3.4. The first represents rigid-body translation, the second, rigid-body rotation, and the third, the vibration mode.

3.5 REFERENCE AXES AND RELATIVE MOTION

In contrast to the presentation in Section 3.4, an alternate approach will now be developed, one that deals more directly with rigid-body degrees of freedom. Inspired by the results in Section 3.4, let the total or inertial transverse position of the masses be described in terms of their displacement relative to some reference line, or axis, plus the vertical and angular position of that reference axis.

As shown in Figure 3.5, the total displacements of the masses are now given by

$$Z_1 = Z_{\text{Ref}} + x_1\Theta_{\text{Ref}} + z_1$$

$$Z_2 = Z_{\text{Ref}} + x_2\Theta_{\text{Ref}} + z_2$$

$$Z_3 = Z_{\text{Ref}} - x_3\Theta_{\text{Ref}} + z_3$$

or

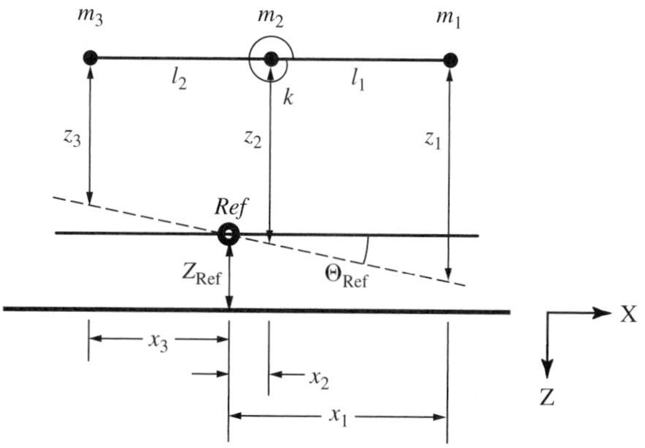

Figure 3.5 Unrestrained three-lumped-mass beam model with reference axis.

$$\boxed{\{\mathbf{Z}\} = \{\mathbf{1}\}Z_{\text{Ref}} + \{\mathbf{X}\}\Theta_{\text{Ref}} + \{\mathbf{z}\}} \tag{3.50}$$

where the vector **X** is defined by the positions of the masses along the beam relative to the reference point, as if the origin was shifted to that point. For the three-mass beam

$$\{\mathbf{X}\} = \{x_1 \quad x_2 \quad -x_3\}^T$$

and the location of the reference point will be defined further in the following discussion.

Again we will use Lagrange's equation to derive the equations of motion of the beam. The kinetic energy is still

$$\mathrm{T} = \frac{1}{2}\left(m_1\dot{Z}_1^2 + m_2\dot{Z}_2^2 + m_3\dot{Z}_3^2\right)$$

or

$$\mathrm{T} = \frac{1}{2}\{\dot{\mathbf{Z}}\}^T[\mathbf{M}]\{\dot{\mathbf{Z}}\} \tag{3.51}$$

And the strain energy is still

$$\mathrm{U} = \frac{1}{2}k\theta^2 = \frac{1}{2}\theta^T k\theta \tag{3.52}$$

Similar to before (see Equations (3.40)), the displacement θ is now written in terms of the transverse relative displacements **z** as follows. From Equations (3.40) and (3.50) (assuming small displacements and small angles) we can see that

$$\theta = \frac{Z_1 - Z_2}{x_1 - x_2} - \frac{Z_2 - Z_3}{x_2 + x_3}$$

$$\frac{Z_1 - Z_2}{x_1 - x_2} = \frac{z_1 - z_2}{x_1 - x_2} + \Theta_{\text{Ref}} \tag{3.53}$$

$$\frac{Z_2 - Z_3}{x_2 + x_3} = \frac{z_2 - z_3}{x_2 + x_3} + \Theta_{\text{Ref}}$$

Therefore, the bending displacement may be expressed in terms of <u>relative</u> displacements, or

$$\begin{aligned}\theta &= \begin{bmatrix} \dfrac{1}{x_1 - x_2} & \left(-\dfrac{1}{x_1 - x_2} - \dfrac{1}{x_2 + x_3}\right) & \dfrac{1}{x_2 + x_3} \end{bmatrix} \begin{Bmatrix} z_1 \\ z_2 \\ z_3 \end{Bmatrix} \\ &= [\mathbf{C}]\{\mathbf{z}\}\end{aligned} \tag{3.54}$$

and the strain energy may now be expressed as

$$\begin{aligned}U &= \frac{1}{2}\{\mathbf{z}\}^T[\mathbf{C}]^T[k][\mathbf{C}]\{\mathbf{z}\} \\ &= \frac{1}{2}\{\mathbf{z}\}^T[\mathbf{K}_c]\{\mathbf{z}\}\end{aligned} \tag{3.55}$$

where we again use the notation $\mathbf{K}_c$ to denote a constrained stiffness matrix. Note at this point that the kinetic energy T, Equation (3.51), is defined in terms of the total, or inertial velocities, while the strain energy U, Equation (3.55), is a function of only the relative displacements.

Two additional constraints among the system variables are now imposed. First, in the absence of external forces or moments acting on the beam, both translational and rotational momenta of the beam must be constant. Taking these two arbitrary constants to be zero, indicating that the beam as a whole is neither translating nor rotating, yields the following two constraints, with obvious definitions of the vectors and matrices.

$$\begin{aligned}\text{C1: }\underline{\text{Translational momentum}} = 0 &\Rightarrow m_1\dot{Z}_1 + m_2\dot{Z}_2 + m_3\dot{Z}_3 = 0 \\ &\Rightarrow \{\mathbf{1}\}^T[\mathbf{M}]\{\dot{\mathbf{Z}}\} = 0\end{aligned} \tag{3.56}$$

$$\begin{aligned}\text{C2: }\underline{\text{Angular momentum}} = 0 &\Rightarrow m_1x_1\dot{Z}_1 + m_2x_2\dot{Z}_2 - m_3x_3\dot{Z}_3 = 0 \\ &\Rightarrow \{\mathbf{X}\}^T[\mathbf{M}]\{\dot{\mathbf{Z}}\} = 0\end{aligned} \tag{3.57}$$

These two constraints imply that under free vibration, the motion must be such that the vector $\dot{\mathbf{Z}}$ is orthogonal (with respect to the mass matrix **M**) to the vectors $\{\mathbf{1}\}$ and $\{\mathbf{X}\}$, both defined through Equation (3.50). Let's <u>define</u> the $\dot{\mathbf{Z}}$ vector that is so constrained to be $\dot{\mathbf{Z}}_c$, and the so constrained relative velocity $\dot{\mathbf{z}}_c$.

The above two constraints, along with Equation (3.50), also suggest that the vectors $\{\mathbf{1}\}$ and $\{\mathbf{X}\}$ may be appropriate rigid-body mode shapes. If that is the case, then these two mode shapes must be mutually orthogonal with respect to the mass matrix. That is,

$$\begin{aligned}\text{C3: }\underline{\text{Orthogonality of rigid-body modes}} &\Rightarrow m_1x_1 + m_2x_2 - m_3x_3 = 0 \\ &\Rightarrow \{\mathbf{1}\}^T[\mathbf{M}]\{\mathbf{X}\} = 0\end{aligned} \tag{3.58}$$

But this is equivalent to stating that the reference point in Figure 3.5 is at the location of the beam's <u>center of mass</u>. Finally, the <u>total mass</u> of the beam M_{Tot} can be written as

$$M_{\text{Tot}} = \{\mathbf{1}\}^T [\mathbf{M}] \{\mathbf{1}\} \tag{3.59}$$

Rewriting Equation (3.50) in terms of the displacements constrained by C1–C3, we have

$$\{\mathbf{Z}_c\} = \{\mathbf{1}\} Z_{\text{Ref}} + \{\mathbf{X}\} \Theta_{\text{Ref}} + \{\mathbf{z}_c\} \tag{3.60}$$

from which we can see that if we invoke the above constraints, and if we can express the relative motion $\mathbf{z}_c$ in terms of mutually orthogonal or normal modal responses, we will have the desired solution to the vibration problem.

Differentiating Equation (3.60) with respect to time, and recalling Equation (3.56), yields

$$\{\mathbf{1}\}^T [\mathbf{M}] \{\{\mathbf{1}\} \dot{Z}_{\text{Ref}} + \{\mathbf{X}\} \dot{\Theta}_{\text{Ref}} + \{\dot{\mathbf{z}}_c\}\} = 0 \tag{3.61}$$

Noting Equations (3.58) (center-of-mass constraint) and (3.59) (total mass), the above expression becomes

$$\dot{Z}_{\text{Ref}} = -\frac{1}{M_{\text{Tot}}} \{\mathbf{1}\}^T [\mathbf{M}] \{\dot{\mathbf{z}}_c\} \tag{3.62}$$

Again differentiating Equation (3.60) with respect to time, and recalling Equation (3.57), yields

$$\{\mathbf{X}\}^T [\mathbf{M}] \{\{\mathbf{1}\} \dot{Z}_{\text{Ref}} + \{\mathbf{X}\} \dot{\Theta}_{\text{Ref}} + \{\dot{\mathbf{z}}_c\}\} = 0 \tag{3.63}$$

Noting Equation (3.58) (and that the transpose of Equation (3.58) is also zero), the above expression becomes

$$\dot{\Theta}_{\text{Ref}} = -\frac{1}{I} \{\mathbf{X}\}^T [\mathbf{M}] \{\dot{\mathbf{z}}_c\} \tag{3.64}$$

where

$$I = \{\mathbf{X}\}^T [\mathbf{M}] \{\mathbf{X}\} = \sum_{i=1}^{3} m_i x_i^2 \tag{3.65}$$

is the moment of inertia of the beam about its center of mass.

By applying the two constraints above (Equations (3.62) and (3.64)) we can finally write the (constrained) total velocities in terms of the (constrained) relative velocities, or

$$\begin{aligned} \{\dot{\mathbf{Z}}_c\} &= \left[[\mathbf{I}] - \frac{1}{M_{\text{Tot}}} \{\mathbf{1}\}\{\mathbf{1}\}^T [\mathbf{M}] - \frac{1}{I} \{\mathbf{X}\}\{\mathbf{X}\}^T [\mathbf{M}] \right] \{\dot{\mathbf{z}}_c\} \\ &\triangleq [\boldsymbol{\Xi}] \{\dot{\mathbf{z}}_c\} \end{aligned} \tag{3.66}$$

where $[\mathbf{I}]$ is a three-by-three identity matrix. We may now write the kinetic energy in terms of the constrained <u>relative</u> velocities. Or, from Equations (3.51) and (3.66), we have

$$\boxed{\begin{aligned} \mathrm{T} &= \frac{1}{2} \{\dot{\mathbf{z}}_c\}^T [\boldsymbol{\Xi}]^T [\mathbf{M}] [\boldsymbol{\Xi}] \{\dot{\mathbf{z}}_c\} \\ &= \frac{1}{2} \{\dot{\mathbf{z}}_c\}^T [\mathbf{M}_c] \{\dot{\mathbf{z}}_c\} \end{aligned}} \tag{3.67}$$

where we have introduced the matrix $\mathbf{M}_c$ to denote a *constrained mass matrix.* The potential energy, written in terms of the constrained relative displacements, is just

$$\boxed{U = \frac{1}{2}\{\mathbf{z}_c\}^T[\mathbf{K}_c]\{\mathbf{z}_c\}} \tag{3.68}$$

With Equations (3.67) and (3.68) for kinetic and potential energy, and using the constrained relative displacements $\mathbf{z}_c$ as the generalized coordinates $\mathbf{q}$, we may now apply Lagrange's equation, written here in vector form:

$$\frac{d}{dt}\left(\frac{\delta T}{\delta \dot{\mathbf{q}}}\right) - \frac{\delta T}{\delta \mathbf{q}} + \frac{\delta U}{\delta \mathbf{q}} = \{\mathbf{0}\}^T \tag{3.69}$$

to obtain the equations of motion for the unrestrained vibrating beam model. The resulting matrix equation of motion is[1]

$$[\mathbf{M}_c]\{\ddot{\mathbf{z}}_c\} + [\mathbf{K}_c]\{\mathbf{z}_c\} = \{\mathbf{0}\} \tag{3.70}$$

Unlike the situation discussed in Section 3.4, the constrained mass matrix above, or $\mathbf{M}_c$, is now singular. Hence the inverse of $\mathbf{M}_c$ does not exist, and a dynamic matrix $\mathbf{D}_c$ cannot be formed. And previously we had determined the vibration frequencies and mode shapes from the eigenvalues and eigenvectors of this dynamic matrix. In the situation being considered now, we will instead determine the vibration frequencies and mode shapes from the *generalized eigenvalues and eigenvectors* of the matrix pair consisting of $\mathbf{M}_c$ and $\mathbf{K}_c$. That is, we will find the n λ_i's and $\boldsymbol{\nu}_i$'s that satisfy the following relationship.

$$\lambda_i \mathbf{M}_c \boldsymbol{\nu}_i = \mathbf{K}_c \boldsymbol{\nu}_i$$

or

$$(\lambda_i \mathbf{M}_c - \mathbf{K}_c)\boldsymbol{\nu}_i = \mathbf{0}$$

This problem is also easily solved using the eig routine in MATLAB.

3.6 MODAL ANALYSIS OF THE GENERALIZED EIGENSOLUTION

A modal analysis of the generalized eigensolution corresponding to Equation (3.70) may now be performed. Each generalized eigenvalue/eigenvector pair satisfies the expression

$$\lambda_i \mathbf{M}_c \boldsymbol{\nu}_i = \mathbf{K}_c \boldsymbol{\nu}_i$$

So, as before, we have for any two i,j pairs of generalized eigenvalues and eigenvectors

$$(\lambda_i - \lambda_j)\{\boldsymbol{\nu}_i\}^T[\mathbf{M}_c]\{\boldsymbol{\nu}_j\} = 0 \tag{3.71}$$

[1] In obtaining the above result, we have made use of the fact that the derivative of the quadratic form $f = \mathbf{y}^T\mathbf{Q}\mathbf{x}$ with respect to the vector $\mathbf{x}$ is $df/d\mathbf{x} = \mathbf{y}^T\mathbf{Q}$.

This expression appears to be the same as discussed previously. However, when $i \neq j$ and the two eigenvalues are distinct (e.g., not both zero), this equation assures that the two associated eigenvectors are orthogonal with respect to the constrained mass matrix, not the mass matrix **M**, as required.

However, looking further at Equation (3.71) we see from the definition of $\mathbf{M}_c$ that we can write

$$(\lambda_i - \lambda_j)\{\boldsymbol{\nu}_i\}^T[\boldsymbol{\Xi}]^T[\mathbf{M}][\boldsymbol{\Xi}]\{\boldsymbol{\nu}_j\} = 0 \tag{3.72}$$

Therefore, when $i \neq j$, and the two eigenvalues are distinct (e.g., not both zero), the above expression states that the constrained, or transformed eigenvectors, given by

$$\{\boldsymbol{\nu}_{c_i}\} = [\boldsymbol{\Xi}]\{\boldsymbol{\nu}_i\} \tag{3.73}$$

are orthogonal with respect to the mass matrix **M**.

Note that for the unrestrained beam, two of the transformed generalized eigenvectors will be null, and these arise due to the two rigid-body modes. Therefore, for the three-mass beam only one mode of motion satisfies the three orthogonality constraints, the single vibration mode shape $\boldsymbol{\nu}_{\text{vib}_1}$. For an n-mass beam we would have the relative displacements given by

$$\boxed{\{\mathbf{z}_c(t)\} = \sum_{i=1}^{n-2}\{\boldsymbol{\nu}_{\text{vib}_i}\}A_i\cos\left(\omega_{\text{vib}_i}t + \Gamma_i\right)} \tag{3.74}$$

consisting of n–2 vibration modes.

All the vibration mode shapes obtained from Equation (3.73) will be mutually orthogonal with respect to the mass matrix. To prove that these vibration mode shapes are also orthogonal to the rigid-body mode shapes **1** and **X**, first express the relative motion $\mathbf{z}_c$ in terms of the original eigenvectors (before transformation). That is, let

$$\mathbf{z}_c(t) = \sum_{i=1}^{n}\boldsymbol{\nu}_i\eta_i(t) \tag{3.75}$$

Now recall that Equations (3.56) and (3.57) assure that the beam's constrained velocity $\dot{\mathbf{Z}}_c$ is orthogonal to the rigid-body mode shapes **1** and **X**, respectively. Plus, from Equation (3.66) we have the relation

$$\{\dot{\mathbf{Z}}_c\} = [\boldsymbol{\Xi}]\{\dot{\mathbf{z}}_c\}$$

Therefore, from Equation (3.56) we have

$$\{\mathbf{1}\}^T[\mathbf{M}]\{\dot{\mathbf{Z}}_c\} = \{\mathbf{1}\}^T[\mathbf{M}][\boldsymbol{\Xi}]\{\dot{\mathbf{z}}_c\} = \{\mathbf{1}\}^T[\mathbf{M}][\boldsymbol{\Xi}]\sum_{i=1}^{n}\{\boldsymbol{\nu}_i\}\dot{\eta}_i(t) = 0 \tag{3.76}$$

and from Equation (3.57)

$$\{\mathbf{X}\}^T[\mathbf{M}]\{\dot{\mathbf{Z}}_c\} = \{\mathbf{X}\}^T[\mathbf{M}][\boldsymbol{\Xi}]\{\dot{\mathbf{z}}_c\} = \{\mathbf{X}\}^T[\mathbf{M}][\boldsymbol{\Xi}]\sum_{i=1}^{n}\{\boldsymbol{\nu}_i\}\dot{\eta}_i(t) = 0 \tag{3.77}$$

For Equation (3.76) to be true requires that

$$\{\mathbf{1}\}^T[\mathbf{M}][\boldsymbol{\Xi}]\{\boldsymbol{\nu}_i\} = \{\mathbf{1}\}^T[\mathbf{M}]\{\boldsymbol{\nu}_{c_i}\} = 0 \text{ for all } i \tag{3.78}$$

or all the constrained eigenvectors are orthogonal to **1**, with respect to the mass matrix. Likewise, for Equation (3.77) to be true requires that

$$\{\mathbf{X}\}^T[\mathbf{M}][\boldsymbol{\Xi}]\{\boldsymbol{\nu}_i\} = \{\mathbf{X}\}^T[\mathbf{M}]\{\boldsymbol{\nu}_{c_i}\} = 0 \text{ for all } i \tag{3.79}$$

or all the constrained eigenvectors are orthogonal to **X** with respect to the mass matrix. Thus, these constrained eigenvectors must be the normal vibration mode shapes we seek. And all these eigenvectors (including **1** and **X**) are mutually orthogonal with respect to **M**.

MATLAB EXAMPLE 3.5

Rigid-Body Formulation Using the Reference Axis

For the unrestrained, three-mass beam model considered in Example 3.3, and shown in Figure 3.3, describe the transverse displacements of the beam explicitly in terms of a reference axis, and determine the mutually orthogonal, or normal rigid-body and vibration mode shapes and vibration frequency. Compare with the results from Example 3.4.

■ Solution

The total or inertial transverse displacements may be described as in Equation (3.50), or

$$\{\mathbf{Z}\} = \{\mathbf{1}\}Z_{\text{Ref}} + \{\mathbf{X}\}\Theta_{\text{Ref}} + \{\mathbf{z}\}$$

where the vector **X** consists of the positions of the three masses along the beam relative to the beam's center of mass. So first we must find the center of mass, using the property that the first mass moment about the center of mass is zero.

Letting the absolute position of the masses relative to some arbitrary reference be X_1, X_2, and X_3, and the absolute position of the center of mass be X_{Ref}, we have the first mass moment

$$\{\mathbf{1}\}^T[\mathbf{M}]\{\mathbf{X}\} = 0 = \{\mathbf{1}\}^T[\mathbf{M}]\{\mathbf{X}' - \{\mathbf{1}\}X_{\text{Ref}}\}$$

where

$$\{\mathbf{X}'\} = \begin{Bmatrix} X_1 \\ X_2 \\ X_3 \end{Bmatrix}$$

Therefore, we have

$$X_{\text{Ref}} = \frac{\{\mathbf{1}\}^T[\mathbf{M}]\{\mathbf{X}'\}}{\{\mathbf{1}\}^T[\mathbf{M}]\{\mathbf{1}\}} = \frac{\{\mathbf{1}\}^T[\mathbf{M}]\{\mathbf{X}'\}}{M_{\text{Tot}}}$$

For this beam, let the absolute positions be $X_1 = 2$, $X_2 = 1$, and $X_3 = 0$, or the chosen arbitrary reference is the left end of the beam. Plus, recall that the masses are $m_1 = m_2 = 1, m_3 = 2$, so $M_{\text{Tot}} = 4$.

From MATLAB we have

```
»Xprime=[2 1 0]
Xprime =
   2  1  0
»(one'*M*Xprime')/Mtot
ans =
   7.5000e-01
```

This is X_{Ref}, the location of the center of mass from the left end of the beam

So we see that the distances of the masses relative to the center of mass are $x_1 = 1.25$, $x_2 = 0.25$, and $x_3 = -0.75$. The vector **X** is then

$$\{\mathbf{X}\} = \begin{Bmatrix} 1.25 \\ 0.25 \\ -0.75 \end{Bmatrix}$$

Finally, recall that the rigid-body mode shapes are **{1}** and **{X}**.

Now we form the constrained mass and stiffness matrices. Equation (3.54) yields the constraint relating the beam's bending displacement to its relative transverse displacements. So here,

$$[\mathbf{C}] = [1 \quad -2 \quad 1]$$

and the constrained stiffness matrix is

$$[\mathbf{K}_c] = [\mathbf{C}]^T[k][\mathbf{C}]$$

with $k = 1$.

To obtain the constrained mass matrix, we use the beam's total mass, M_{Tot}, and its moment of inertia about its center of mass, I. From Equation (3.59) we found that $M_{\text{Tot}} = 4$, and from Equation (3.65) we know that

$$I = \{\mathbf{X}\}^T[\mathbf{M}]\{\mathbf{X}\}$$

From MATLAB we have

```
»C(1)=1/(x(1)-x(2));
»C(3)=1/(x(2)-x(3));
»C(2)=-C(1)-C(3)
C =
   1  -2  1
»Kc=C'*k*C
Kc=
   1  -2   1
  -2   4  -2
   1  -2   1
```

$k = 1$ here

Constrained stiffness matrix

```
»X=[1.25;.25;-.75];
»I=X'*M*X
I=
   2.7500e+00
```

Moment of inertia

From Equation (3.67) we have the constrained mass matrix

$$[\mathbf{M}_c] = [\boldsymbol{\Xi}]^T [\mathbf{M}][\boldsymbol{\Xi}]$$

and from Equation (3.66) we have the constraint matrix

$$[\boldsymbol{\Xi}] = \left[[\mathbf{I}] - \frac{1}{M_{\text{Tot}}}\{\mathbf{1}\}\{\mathbf{1}\}^T[\mathbf{M}] - \frac{1}{I}\{\mathbf{X}\}\{\mathbf{X}\}^T[\mathbf{M}] \right]$$

Turning to MATLAB, we have

```
»P=[eye(3)-(1/Mtot)*one*one'*M-(1/I)*X*X'*M]
P =
   1.8182e-01  -3.6364e-01   1.8182e-01
  -3.6364e-01   7.2727e-01  -3.6364e-01
   9.0909e-02  -1.8182e-01   9.0909e-02
```

Constraint matrix

```
»Mc=P'*M*P
Mc =
   1.8182e-01  -3.6364e-01   1.8182e-01
  -3.6364e-01   7.2727e-01  -3.6364e-01
   1.8182e-01  -3.6364e-01   1.8182e-01
```

Constrained mass matrix

```
»[V,lamda]=eig(Kc,Mc)
V=
   1   0.75166        0
   0   0.04689  0.44721
   0  -0.65788  0.89443
```

Generalized eigenvectors (columns)

```
lamda=
   5.5000e+00       0   0
        0      9.0407   0
        0           0   0
```

Generalized eigenvalues on the diagonal

So the vibration frequency is either $\sqrt{5.5}$ or $\sqrt{9.04}$ rad/sec, and the generalized eigenvectors are the three columns of the above matrix **V** (i.e., V). To determine the vibration frequency and mode shape, we must determine the transformed eigenvectors from the three columns of the matrix $\mathbf{V}_c$, where

$$[\mathbf{V}_c] = [\boldsymbol{\Xi}][\mathbf{V}]$$

Again, from MATLAB

```
»Vc=P*V
Vc =
   0.18182   9.8332e-17  -6.3119e-17
  -0.36364  -6.2648e-17  -2.8801e-17
   0.09091   4.9166e-17  -3.1560e-17
```

Constrained eigenvectors

With only the first column non-null, the single vibration mode shape is the first column of Vc, and the vibration frequency is $\sqrt{5.5}$ rad/sec. The rigid-body mode shapes were previously specified to be

$$\boldsymbol{\nu}_{\text{translation}}^T = \{1 \quad 1 \quad 1\}$$

$$\boldsymbol{\nu}_{\text{rotation}}^T = \{1.25 \quad 0.25 \quad -0.75\}$$

These three mode shapes, two rigid-body and one vibration, are equivalent to those plotted in Figure 3.4. The results obtained here agree with those from Example 3.4, subject to the fact that the mode shapes in the two examples differ by a constant factor. That is, $\boldsymbol{\nu}_{\text{Eg. 3.4}} = C\,\boldsymbol{\nu}_{\text{Eg. 3.5}}$, where C is a constant. (In this case, C is approximately -2.4.) And we know from linear algebra that a matrix's eigenvector scaled by a constant is also an eigenvector of that matrix.

3.7 MULTI-DIRECTIONAL MOTION

Prior to this point, we have only considered motion in one direction—Z. But, of course, general vibration may involve motion in all directions. As we shall see, the approach for dealing with multi-directional motion is basically the same as that for one-directional motion. But since each element of a mode shape can only correspond to one direction of motion, the size of the vectors and matrices expand considerably.

To make the discussion more specific, we will address bi-directional motion of a truss—or motion in two directions (X and Z). Consider a model with lumped-mass representation shown in Figure 3.6. This model might represent

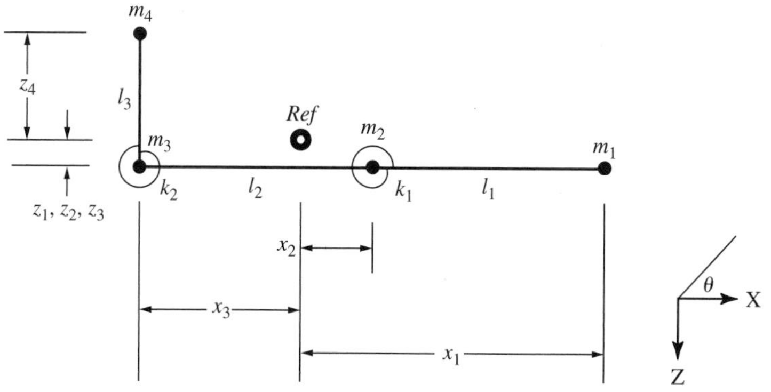

Figure 3.6 Bi-directional truss model.

the fuselage and vertical tail of an aircraft, for example. Each of the four masses has degrees of freedom in both the X and Z directions, and there are two idealized springs. The locations of the masses are again defined relative to a reference location, the center of mass of the truss.

The kinetic energy for this model is

$$\begin{aligned} T &= \frac{1}{2}\left(m_1\dot{X}_1^2 + m_2\dot{X}_2^2 + m_3\dot{X}_3^2 + m_4\dot{X}_4^2\right) + \frac{1}{2}\left(m_1\dot{Z}_1^2 + m_2\dot{Z}_2^2 + m_3\dot{Z}_3^2 + m_4\dot{Z}_4^2\right) \\ &= \frac{1}{2}\{\dot{\mathbf{X}}'\}^T[\mathbf{M}]\{\dot{\mathbf{X}}'\} + \frac{1}{2}\{\dot{\mathbf{Z}}'\}^T[\mathbf{M}]\{\dot{\mathbf{Z}}'\} = \frac{1}{2}\{\dot{\mathbf{Z}}'^T \quad \dot{\mathbf{X}}'^T\}\begin{bmatrix} \mathbf{M} & \mathbf{0} \\ \mathbf{0} & \mathbf{M} \end{bmatrix}\begin{Bmatrix} \dot{\mathbf{Z}}' \\ \dot{\mathbf{X}}' \end{Bmatrix} \end{aligned} \tag{3.80}$$

where the diagonal mass matrix **M** has been introduced. Note that X_i and Z_i are the absolute positions of the masses, while x_i and z_i (in Figure 3.6) are the positions of the masses relative to the indicated reference location. The potential (strain) energy of the truss is

$$U = \frac{1}{2}k_1\theta_1^2 + \frac{1}{2}k_2\theta_2^2 = \frac{1}{2}\{\theta_1 \quad \theta_2\}\begin{bmatrix} k_1 & 0 \\ 0 & k_2 \end{bmatrix}\begin{Bmatrix} \theta_1 \\ \theta_2 \end{Bmatrix} \tag{3.81}$$

where the two deflection angles are depicted in Figure 3.7.

But from geometry, we can develop the following geometric constraints among the mass elements (see Section 3.4).

$$\begin{aligned} \theta_1 + \frac{Z_2 - Z_3}{x_2 + x_3} &= \frac{Z_1 - Z_2}{x_1 - x_2} \\ \theta_2 + \frac{X_3 - X_4}{z_3 + z_4} &= \frac{Z_3 - Z_2}{x_2 + x_3} \\ X_1 = X_2 &= X_3 \text{ (equal vibration displacement)} \\ Z_3 &= Z_4 \text{ (equal viration displacement)} \end{aligned} \tag{3.82}$$

or

$$\begin{aligned} \theta_1 &= \left[\left(\frac{1}{x_1 - x_2}\right) \quad \left(\frac{-1}{x_1 - x_2} - \frac{1}{x_2 + x_3}\right) \quad \left(\frac{1}{x_2 + x_3}\right) \quad 0\right]\begin{Bmatrix} Z_1 \\ Z_2 \\ Z_3 \\ Z_4 \end{Bmatrix} \\ &= [\mathbf{C}_1]\{\mathbf{Z}'\} \end{aligned} \tag{3.83}$$

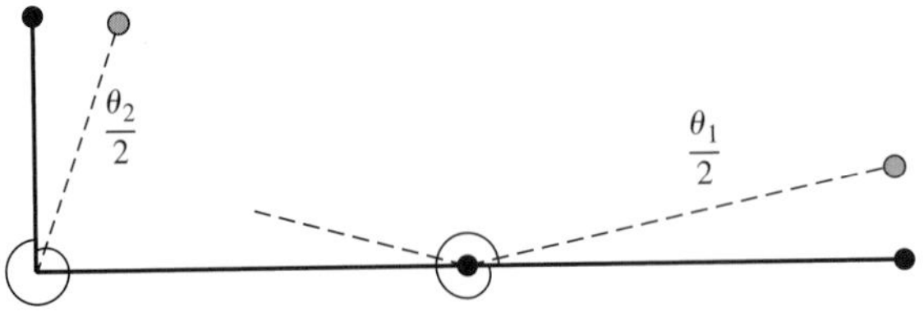

Figure 3.7 Two beam-displacement angles.

$$\theta_2 = \begin{bmatrix} 0 & \left(\dfrac{-1}{x_2+x_3}\right) & \left(\dfrac{1}{x_2+x_3}\right) & 0 \end{bmatrix} \begin{Bmatrix} Z_1 \\ Z_2 \\ Z_3 \\ Z_4 \end{Bmatrix} + \begin{bmatrix} 0 & 0 & \left(\dfrac{-1}{z_3+z_4}\right) & \left(\dfrac{1}{z_3+z_4}\right) \end{bmatrix} \begin{Bmatrix} X_1 \\ X_2 \\ X_3 \\ X_3 \end{Bmatrix}$$

$$= \begin{bmatrix} \mathbf{C}_2 & \mathbf{C}_3 \end{bmatrix} \begin{Bmatrix} \mathbf{Z}' \\ \mathbf{X}' \end{Bmatrix}$$

In terms of the constraints expressed in Equations (3.83), the potential energy becomes

$$\mathrm{U} = \frac{1}{2}\{\theta_1 \;\; \theta_2\} \begin{bmatrix} k_1 & 0 \\ 0 & k_2 \end{bmatrix} \begin{Bmatrix} \theta_1 \\ \theta_2 \end{Bmatrix} = \frac{1}{2}\{\mathbf{Z}'^T \;\; \mathbf{X}'^T\} \begin{bmatrix} \mathbf{C}_1 & \mathbf{0} \\ \mathbf{C}_2 & \mathbf{C}_3 \end{bmatrix}^T \begin{bmatrix} k_1 & 0 \\ 0 & k_2 \end{bmatrix} \begin{bmatrix} \mathbf{C}_1 & \mathbf{0} \\ \mathbf{C}_2 & \mathbf{C}_3 \end{bmatrix} \begin{Bmatrix} \mathbf{Z}' \\ \mathbf{X}' \end{Bmatrix}$$

$$= \frac{1}{2}\{\mathbf{Z}'^T \;\; \mathbf{X}'^T\}[\mathbf{K}_c] \begin{Bmatrix} \mathbf{Z}' \\ \mathbf{X}' \end{Bmatrix} \tag{3.84}$$

where a constrained stiffness matrix $\mathbf{K}_c$ is introduced.

The remaining two constraints in Equations (3.82) may be written as

$$\begin{Bmatrix} Z_1 \\ Z_2 \\ Z_3 \\ Z_4 \end{Bmatrix} = \begin{bmatrix} 1 & 0 & 0 \\ 0 & 1 & 0 \\ 0 & 0 & 1 \\ 0 & 0 & 1 \end{bmatrix} \begin{Bmatrix} Z_1 \\ Z_2 \\ Z_3 \end{Bmatrix} \qquad \begin{Bmatrix} X_1 \\ X_2 \\ X_3 \\ X_4 \end{Bmatrix} = \begin{bmatrix} 1 & 0 \\ 1 & 0 \\ 1 & 0 \\ 0 & 1 \end{bmatrix} \begin{Bmatrix} X_3 \\ X_4 \end{Bmatrix}$$

or

$$\begin{Bmatrix} \mathbf{Z}' \\ \mathbf{X}' \end{Bmatrix} = \begin{bmatrix} \mathbf{C}'_z & \mathbf{0} \\ \mathbf{0} & \mathbf{C}'_x \end{bmatrix} \begin{Bmatrix} \mathbf{Z}'' \\ \mathbf{X}'' \end{Bmatrix} \tag{3.85}$$

In terms of the constraints expressed in Equation (3.85), and the double mass matrix **MM** defined in Equation (3.80), the kinetic and potential energies now become

$$\mathrm{T} = \frac{1}{2}\{\dot{\mathbf{Z}}'^T \;\; \dot{\mathbf{X}}'^T\}[\mathbf{MM}] \begin{Bmatrix} \dot{\mathbf{Z}}' \\ \dot{\mathbf{X}}' \end{Bmatrix} = \frac{1}{2}\{\dot{\mathbf{Z}}''^T \;\; \dot{\mathbf{X}}''^T\} \begin{bmatrix} \mathbf{C}'_z & \mathbf{0} \\ \mathbf{0} & \mathbf{C}'_x \end{bmatrix}^T [\mathbf{MM}] \begin{bmatrix} \mathbf{C}'_z & \mathbf{0} \\ \mathbf{0} & \mathbf{C}'_x \end{bmatrix} \begin{Bmatrix} \dot{\mathbf{Z}}'' \\ \dot{\mathbf{X}}'' \end{Bmatrix}$$

$$= \frac{1}{2}\{\dot{\mathbf{Z}}''^T \;\; \dot{\mathbf{X}}''^T\}[\mathbf{MM}'] \begin{Bmatrix} \dot{\mathbf{Z}}'' \\ \dot{\mathbf{X}}'' \end{Bmatrix} \tag{3.86}$$

$$\mathrm{U} = \frac{1}{2}\{\mathbf{Z}'^T \;\; \mathbf{X}'^T\}[\mathbf{K}_c] \begin{Bmatrix} \mathbf{Z}' \\ \mathbf{X}' \end{Bmatrix} = \frac{1}{2}\{\mathbf{Z}''^T \;\; \mathbf{X}''^T\} \begin{bmatrix} \mathbf{C}'_z & \mathbf{0} \\ \mathbf{0} & \mathbf{C}'_x \end{bmatrix}^T [\mathbf{K}_c] \begin{bmatrix} \mathbf{C}'_z & \mathbf{0} \\ \mathbf{0} & \mathbf{C}'_x \end{bmatrix} \begin{Bmatrix} \mathbf{Z}'' \\ \mathbf{X}'' \end{Bmatrix}$$

$$= \frac{1}{2}\{\mathbf{Z}''^T \;\; \mathbf{X}''^T\}[\mathbf{K}'_c] \begin{Bmatrix} \mathbf{Z}'' \\ \mathbf{X}'' \end{Bmatrix} \tag{3.87}$$

Note that the constrained mass matrix **MM′** has been introduced, along with a modified constrained stiffness matrix $\mathbf{K}_c'$.

Application of Lagrange's equation yields the equations of motion governing the vibration problem. Or

$$\left[\mathbf{MM}'\right]\begin{Bmatrix}\ddot{\mathbf{Z}}''\\ \ddot{\mathbf{X}}''\end{Bmatrix} + \left[\mathbf{K}_c'\right]\begin{Bmatrix}\mathbf{Z}''\\ \mathbf{X}''\end{Bmatrix} = \{\mathbf{0}\} \tag{3.88}$$

So the system's dynamic matrix is

$$\mathbf{D}_c' = \left[\mathbf{MM}'\right]^{-1}\left[\mathbf{K}_c'\right] \tag{3.89}$$

MATLAB EXAMPLE 3.6

A Bi-Directional Vibration Problem

For the lumped-mass model shown in Figure 3.6, find the vibration modal frequencies and mode shapes, and show that these mode shapes are orthogonal (with respect to the mass matrix) to the system's rigid-body mode shapes. Let each mass equal one unit, each rod have one length unit, and each spring stiffness equal unity (torque unit per radian).

■ Solution

We will use the direct formulation, as in Example 3.3, to find the vibration frequencies and mode shapes. Then we will then determine the rigid-body mode shapes, and show that they are orthogonal to the vibration mode shapes.

Turning to MATLAB, the stiffness and mass matrices are

```
»K=eye(2)
K =
    1  0
    0  1

»M=eye(4)
M=
    1  0  0  0
    0  1  0  0
    0  0  1  0
    0  0  0  1

»MM=[M 0*eye(4);0*eye(4) M]
MM=
    1  0  0  0  0  0  0  0
    0  1  0  0  0  0  0  0
    0  0  1  0  0  0  0  0
    0  0  0  1  0  0  0  0
    0  0  0  0  1  0  0  0
    0  0  0  0  0  1  0  0
    0  0  0  0  0  0  1  0
    0  0  0  0  0  0  0  1
```

The geometric constraints, Equations (3.83), lead to the following matrices.

```
»C1=[1 -2 1 0]
C1 =
   1  -2   1   0

»C2=[0 1 -1 0]
C2 =
   0   1  -1   0

»C2=-C2
C2 =
   0  -1   1   0

»C3=[0 0 -1 1]
C3 =
   0   0  -1   1

»Czprime=[eye(3);0 0 1]                 Constraints from Equation (3.85)
Czprime =
   1   0   0
   0   1   0
   0   0   1
   0   0   1

»Cxprime=[1 0;1 0;eye(2)]
Cxprime =
   1   0
   1   0
   1   0
   0   1

»ones=[1 1 1 1]'
ones =
   1
   1
   1
   1

»C13=[C1 0*ones';C2 C3]                 Constraint matrix in Equation (3.84)
C13 =
   1  -2   1   0   0   0   0   0
   0  -1   1   0   0   0  -1   1
```

So the constrained stiffness matrix in Equation (3.84) is

```
»Kc=C13'*K*C13

Kc =

   1  -2   1   0   0   0   0   0
  -2   5  -3   0   0   0   1  -1
   1  -3   2   0   0   0  -1   1
   0   0   0   0   0   0   0   0
   0   0   0   0   0   0   0   0
   0   0   0   0   0   0   0   0
   0   1  -1   0   0   0   1  -1
   0  -1   1   0   0   0  -1   1
```

The additional geometric constraints yield

```
»zero42=[0*ones 0*ones]

zero42 =

   0   0
   0   0
   0   0
   0   0

»zero43=[zero42 0*ones]

zero43 =

   0   0   0
   0   0   0
   0   0   0
   0   0   0

»Cprime=[Czprime zero42;zero43 Cxprime]

Cprime =

   1   0   0   0   0
   0   1   0   0   0
   0   0   1   0   0
   0   0   1   0   0
   0   0   0   1   0
   0   0   0   1   0
   0   0   0   1   0
   0   0   0   0   1
```

So the constrained mass matrix in Equation (3.86) is

```
»MMPrime=Cprime'*MM*Cprime

MMPrime =

   1   0   0   0   0
   0   1   0   0   0
   0   0   2   0   0
   0   0   0   3   0
   0   0   0   0   1
```

And the modified constrained stiffness matrix in Equation (3.87) is

```
»Kcprime=Cprime'*Kc*Cprime
Kcprime =
   1  -2   1   0   0
  -2   5  -3   1  -1
   1  -3   2  -1   1
   0   1  -1   1  -1
   0  -1   1  -1   1
```

The constrained dynamic matrix in Equation (3.89) becomes

```
»Dcprime=MMPrime\Kcprime
Dcprime =
   1.0000e+00  -2.0000e+00   1.0000e+00            0            0
  -2.0000e+00   5.0000e+00  -3.0000e+00   1.0000e+00  -1.0000e+00
   5.0000e-01  -1.5000e+00   1.0000e+00  -5.0000e-01   5.0000e-01
            0   3.3333e-01  -3.3333e-01   3.3333e-01  -3.3333e-01
            0  -1.0000e+00   1.0000e+00  -1.0000e+00   1.0000e+00
```

And the modal analysis of the constrained dynamic matrix for the five-degree-of-freedom system yields the following eigenvectors (in V) and eigenvalues (in Lambda).

```
»[V,Lambda]=eig(Dcprime)
V =
   8.4444e-01   3.3710e-01  -4.8176e-01  -3.5265e-01  -3.3549e-01
   4.1139e-01  -8.7646e-01   1.6059e-01  -1.7180e-01   1.8961e-01
  -2.1652e-02   2.6968e-01   1.6059e-01   9.0424e-03   7.1470e-01
  -1.0826e-01  -6.7420e-02  -2.6764e-01  -5.5366e-01  -5.5913e-02
   3.2478e-01   2.0226e-01   8.0293e-01  -7.3451e-01  -5.8101e-01
Lambda =
   0           0            0             0            0
   0  7.0000e+00            0             0            0
   0           0   1.3333e+00             0            0
   0           0            0   -1.2326e-32            0
   0           0            0             0   1.2176e-16
```

Note that the two vibration modes have squared frequencies of 1.333 and 7.0 $(\text{rad/sec})^2$. The mode shapes given in terms of the absolute Z and X displacements of the four masses are

```
»Vfull=Cprime*V
Vfull =
   8.4444e-01   3.3710e-01  -4.8176e-01  -3.5265e-01  -3.3549e-01
   4.1139e-01  -8.7646e-01   1.6059e-01  -1.7180e-01   1.8961e-01
  -2.1652e-02   2.6968e-01   1.6059e-01   9.0424e-03   7.1470e-01
  -2.1652e-02   2.6968e-01   1.6059e-01   9.0424e-03   7.1470e-01
  -1.0826e-01  -6.7420e-02  -2.6764e-01  -5.5366e-01  -5.5913e-02
  -1.0826e-01  -6.7420e-02  -2.6764e-01  -5.5366e-01  -5.5913e-02
  -1.0826e-01  -6.7420e-02  -2.6764e-01  -5.5366e-01  -5.5913e-02
   3.2478e-01   2.0226e-01   8.0293e-01  -7.3451e-01  -5.8101e-01
```

The second and third columns of the above matrix Vfull are the vibration mode shapes.

It is important to recall here that in our formulation each of these mode shapes (eigenvectors) is of the form

$$\boldsymbol{\nu}_i = \begin{Bmatrix} \boldsymbol{\nu}_{Z_i} \\ \boldsymbol{\nu}_{X_i} \end{Bmatrix}$$

That is, the first four elements of the eigenvector correspond to transverse (Z) displacements of the four masses for the i'th mode, and the last four elements correspond to axial (X) displacements of the four masses for that mode.

To check the orthogonality between the rigid-body and vibration modes, we must first identify the rigid-body mode shapes. These mode shapes may be identified in this problem by determining the expressions for the translational and rotational momentum of the truss.

The translational momentum may be expressed as

$$\{\mathbf{1}\}^T[\mathbf{M}]\{\dot{\mathbf{X}}'\} + \{\mathbf{1}\}^T[\mathbf{M}]\{\dot{\mathbf{Z}}'\} = \{\mathbf{1}^T \quad \mathbf{1}^T\}\begin{bmatrix} \mathbf{M} & \mathbf{0} \\ \mathbf{0} & \mathbf{M} \end{bmatrix}\begin{Bmatrix} \dot{\mathbf{Z}}' \\ \dot{\mathbf{X}}' \end{Bmatrix}$$

where

$$\{\mathbf{1}\} = \begin{Bmatrix} 1 \\ 1 \\ 1 \\ 1 \end{Bmatrix}$$

So one of the rigid-body mode shapes may be taken as

$$\boldsymbol{\nu}_{\mathrm{RB}_1} = \begin{Bmatrix} \{\mathbf{1}\} \\ \{\mathbf{1}\} \end{Bmatrix}$$

The rotational momentum of the truss about its center of mass may be expressed as

$$-\{\mathbf{X}\}^T[\mathbf{M}]\{\dot{\mathbf{Z}}'\} + \{\mathbf{Z}\}^T[\mathbf{M}]\{\dot{\mathbf{X}}'\} = \{-\mathbf{X}^T \quad \mathbf{Z}^T\}\begin{bmatrix} \mathbf{M} & \mathbf{0} \\ \mathbf{0} & \mathbf{M} \end{bmatrix}\begin{Bmatrix} \dot{\mathbf{Z}}' \\ \dot{\mathbf{X}}' \end{Bmatrix} \tag{3.90}$$

where

$$\mathbf{X} = \begin{Bmatrix} x_1 \\ x_2 \\ -x_3 \\ -x_4 \end{Bmatrix} \qquad \mathbf{Z} = \begin{Bmatrix} z_1 \\ z_2 \\ z_3 \\ -z_4 \end{Bmatrix}$$

Referring to Figure 3.6, note that x_i and z_i correspond to the X and Z locations of the i'th mass, relative to the center of mass of the truss. From Equation (3.90) we can express the second rigid-body mode shape as

$$\boldsymbol{\nu}_{\mathrm{RB}_2} = \begin{Bmatrix} -\mathbf{X} \\ \mathbf{Z} \end{Bmatrix}$$

The negative sign in this mode shape arises due to the definition of positive angular momentum used in Equation (3.90) and the other sign conventions defined in Figure 3.6.

Turning again to MATLAB we have

```
»ones=[ones' ones']'          First rigid-body mode shape
ones =
    1
    1
    1
    1
    1
    1
    1
    1
»ones'*MM*Vfull(:,2)
ans =
   -2.4980e-16
»ones'*MM*Vfull(:,3)
ans =
   -2.2204e-16
```

So we see that the first rigid-body mode shape is orthogonal to both vibration mode shapes.

To consider the second rigid-body mode shape, we must first locate the center of mass.

```
»ones=[1 1 1 1]'
ones =
    1
    1
    1
    1
»Mtot=ones'*M*ones             Total mass of the truss
Mtot =
    4
»Mx=1*1+2*2              First mass moment of the truss about its right end (Z displacement)
Mx =
    5
»Xcm=Mx/Mtot
Xcm =
    1.2500e+00
```

```
»Mz=1*1          First mass moment of the truss about its right end (X displacement)
Mz =
   1
»Zcm=Mz/Mtot
Zcm =
   2.5000e-01
```

So the center of mass is 1.25 length units to the left of mass m_1, and 0.25 length units above masses $m_1 - m_3$. This means that the two vectors **X** and **Z** are as follows:

```
»X=[1.25 0.25 -0.75 -0.75]'
X =
   1.2500e+00
   2.5000e-01
  -7.5000e-01
  -7.5000e-01
»Z=[0.25 0.25 0.25 -0.75]'
Z =
   2.5000e-01
   2.5000e-01
   2.5000e-01
  -7.5000e-01
```

So the second rigid-body mode shape is

```
»XZ=[-X' Z']'
XZ =
  -1.2500e+00
  -2.5000e-01
   7.5000e-01
   7.5000e-01
   2.5000e-01
   2.5000e-01
   2.5000e-01
  -7.5000e-01
»XZ'*MM*Vfull(:,2)
ans =
   0                Orthogonal to first vibration mode shape
»XZ'*MM*Vfull(:,3)
ans =
  -6.6613e-16       Orthogonal to second vibration mode shape
```

Consequently, we see that both of the rigid-body mode shapes, defined in terms of the translational and rotational momentum, are orthogonal (with respect to the mass matrix) to the two vibration mode shapes.

NOTE TO STUDENT

In Example 3.6, as in other examples in this chapter, we have shown that the vibration modes are orthogonal to the rigid-body modes of the structure. Thus, the coordinate frames defined in the examples satisfy what are known as the *mean-axis constraints.* Mean axes and the mean-axis constraints are discussed further in Chapter 4 when dealing with the derivation of the equations of motion for elastic vehicles.

3.8 PREFERRED DERIVATION OF EQUATIONS OF MOTION

A key result of orthogonality among the modes is the fact that the physical responses of the unrestrained beam can be expressed in terms of the linear combination of mutually orthogonal modes—rigid-body and vibration. To use a concrete example, consider only transverse displacements of the straight beam model in Example 3.5. The beam's response can be expressed as

$$\begin{aligned}\{\mathbf{Z}(t)\} &= \{\mathbf{1}\}Z_{\text{Ref}}(t) + \{\mathbf{X}\}\Theta_{\text{Ref}}(t) + \{\mathbf{z}_{\perp}(t)\}\\ &= \{\mathbf{1}\}Z_{\text{Ref}}(t) + \{\mathbf{X}\}\Theta_{\text{Ref}}(t) + \sum_{i=1}^{n-2}\{\boldsymbol{\nu}_{\perp}\}_i\eta_i(t)\end{aligned} \tag{3.91}$$

where the notation $\boldsymbol{\nu}_{\perp}$ has been introduced to remind us that the vibration mode shapes are both mutually orthogonal (with respect to the mass matrix), as well as orthogonal to the appropriately defined rigid-body mode shapes **1** and **X**.

In addition, the rigid-body coordinates Z_{Ref} and Θ_{Ref} plus the vibration modal coordinates η_i form a set of perfectly legitimate generalized coordinates for describing the motion of the unrestrained beam model of the flight vehicle. Therefore, the potential and kinetic energies may be expressed in terms of these coordinates, and Lagrange's equation may be applied directly to obtain the equations of motion of the flexible beam (vehicle).

Rewriting the beam's responses, Equation (3.91), in terms of the modal matrix and the chosen generalized coordinates, we have

$$\begin{aligned}\{\mathbf{Z}(t)\} &= \left[\{\mathbf{1}\}\ \{\mathbf{X}\}\ \{\boldsymbol{\nu}_{\perp}\}_1\ \cdots\ \{\boldsymbol{\nu}_{\perp}\}_{n-2}\right]\begin{Bmatrix} Z_{\text{Ref}}\\ \Theta_{\text{Ref}}\\ \eta_1\\ \vdots\\ \eta_{n-2}\end{Bmatrix}\\ &= [\boldsymbol{\Phi}]\{\mathbf{q}\}\end{aligned} \tag{3.92}$$

In terms of these coordinates, the beam's kinetic energy is

$$\mathrm{T} = \frac{1}{2}\{\dot{\mathbf{Z}}\}^T[\mathbf{M}]\{\dot{\mathbf{Z}}\} = \frac{1}{2}\{\dot{\mathbf{q}}\}^T[\boldsymbol{\Phi}]^T[\mathbf{M}][\boldsymbol{\Phi}]\{\dot{\mathbf{q}}\} = \frac{1}{2}\{\dot{\mathbf{q}}\}^T[\mathcal{M}]\{\dot{\mathbf{q}}\} \tag{3.93}$$

where now the generalized mass matrix $[\mathcal{M}]$ includes the beam's total mass and inertia about the center of mass. Or

$$[\mathcal{M}] = \begin{bmatrix} M_{\text{Tot}} & 0 & \mathbf{0} \\ 0 & I & \mathbf{0} \\ \mathbf{0} & \mathbf{0} & [\mathcal{M}_{\text{Vib}}] \end{bmatrix}$$

Note that one useful form of the kinetic energy given in Equation (3.93) is

$$\boxed{\begin{aligned} \text{T} &= \left(\frac{1}{2}M_{\text{Tot}}\dot{Z}_{\text{Ref}}^2 + \frac{1}{2}I\dot{\Theta}_{\text{Ref}}^2\right) + \left(\frac{1}{2}\{\dot{\boldsymbol{\eta}}_{\text{Vib}}\}^T[\mathcal{M}_{\text{Vib}}]\{\dot{\boldsymbol{\eta}}_{\text{Vib}}\}\right) \\ &= (\textit{Rigid Body}) + (\textit{Elastic}) \end{aligned}} \tag{3.94}$$

where

$$\{\dot{\boldsymbol{\eta}}_{\text{Vib}}\} = \begin{Bmatrix} \dot{\eta}_1 \\ \vdots \\ \dot{\eta}_{n-2} \end{Bmatrix}$$

Likewise, we have for the potential or strain energy

$$\text{U} = \frac{1}{2}\{\mathbf{z}_\perp\}^T[\mathbf{K}_c]\{\mathbf{z}_\perp\} = \frac{1}{2}\{\boldsymbol{\eta}_{\text{Vib}}\}^T[\boldsymbol{\Phi}_{\text{Vib}}]^T[\mathbf{K}_c][\boldsymbol{\Phi}_{\text{Vib}}]\{\boldsymbol{\eta}_{\text{Vib}}\} = \frac{1}{2}\{\boldsymbol{\eta}_{\text{Vib}}\}^T[\mathcal{K}_{\text{Vib}}]\{\boldsymbol{\eta}_{\text{Vib}}\}$$

where the generalized stiffness matrix $[\mathcal{K}_{\text{Vib}}]$ has been introduced, and the vibration modal matrix is

$$[\boldsymbol{\Phi}_{\text{Vib}}] = \left[\{\boldsymbol{\nu}_\perp\}_1 \quad \cdots \quad \{\boldsymbol{\nu}_\perp\}_{n-2}\right]$$

consisting only of the vibration mode shapes (eigenvectors).

Finally, we may now apply Lagrange's equation directly to the above expressions for kinetic and potential energy and obtain the unrestrained beam's equations of motion.

$$\boxed{\begin{aligned} &M_{\text{Tot}}\ddot{Z}_{\text{Ref}} = 0 \\ &I\ddot{\Theta}_{\text{Ref}} = 0 \\ &[\mathcal{M}_{\text{Vib}}]\{\ddot{\boldsymbol{\eta}}_{\text{Vib}}\} + [\mathcal{K}_{\text{Vib}}]\{\boldsymbol{\eta}_{\text{Vib}}\} = \mathbf{0} \end{aligned}} \tag{3.95}$$

The first two equations are the familiar equations of motion for a rigid body in translation and rotation, respectively, plus the third corresponds to the vibrating structure. In addition, note that all these equations are decoupled—as a result of modal orthogonality. This formulation is fundamental to that used in Chapter 4 to develop the equations of motion for the flexible flight vehicle.

3.9 FORCED MOTION AND VIRTUAL WORK

Equations (3.95) don't look much like the equations of motion of a flight vehicle developed in Chapter 2 because these equations do not include the forces and moments acting on the vehicle. We will now introduce the concepts necessary for treating these external forces and moments.

Rather than begin directly with virtual-work concepts, it is instructive to first use Newtonian mechanics to consider forced motion of the two-mass cantilever beam model considered in Examples 3.1 and 3.2. Shown in Figure 3.8 we have the beam acted upon by two forces F_1 and F_2, applied at the locations of masses m_1 and m_2, respectively. From Newton's law, we obtain the following equations of motion governing the bending displacements θ_1 and θ_2.

$$[\mathbf{M}]\begin{Bmatrix}\ddot{\theta}_1\\ \ddot{\theta}_2\end{Bmatrix} + [\mathbf{K}]\begin{Bmatrix}\theta_1\\ \theta_2\end{Bmatrix} = \{\mathbf{F}\} \tag{3.96}$$

with

$$[\mathbf{M}] = \begin{bmatrix} m_1 l_1^2 + m_2(l_1 + l_2)^2 & m_2 l_2(l_1 + l_2) \\ m_2 l_2(l_1 + l_2) & m_2 l_2^2 \end{bmatrix}$$

$$[\mathbf{K}] = \begin{bmatrix} k_1 & 0 \\ 0 & k_2 \end{bmatrix}$$

$$\{\mathbf{F}\} = \begin{Bmatrix} F_1 l_1 + F_2(l_1 + l_2) \\ F_2 l_2 \end{Bmatrix}$$

Now assume a solution to the <u>unforced</u> vibration problem is available, yielding the eigenvalues and modal matrix $\mathbf{\Phi}$ of the dynamic matrix, consisting of the (orthogonal) vibration mode shapes. Writing the physical displacements in terms of the vibration modal coordinates η_i, we have

$$\{\boldsymbol{\theta}\} = [\mathbf{\Phi}]\{\boldsymbol{\eta}\}$$

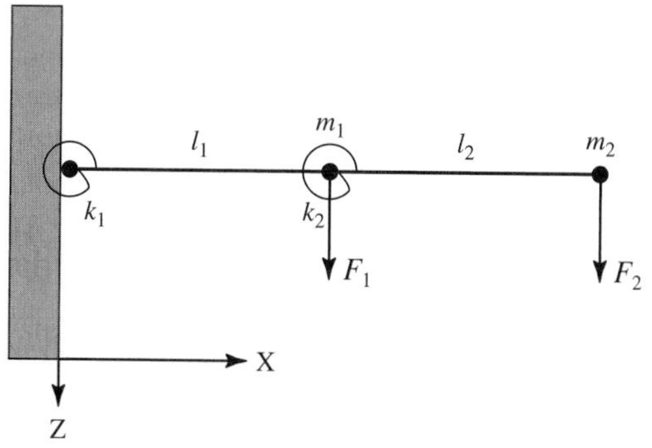

Figure 3.8 Forces acting on the cantilever beam model.

And now Equation (3.96) can be rewritten as

$$[\boldsymbol{\Phi}]^T[\mathbf{M}][\boldsymbol{\Phi}]\{\ddot{\boldsymbol{\eta}}\} + [\boldsymbol{\Phi}]^T[\mathbf{K}][\boldsymbol{\Phi}]\{\boldsymbol{\eta}\} = [\boldsymbol{\Phi}]^T\{\mathbf{F}\} \tag{3.97}$$

or, since $\boldsymbol{\Phi}$ is the modal matrix,

$$[\mathcal{M}]\{\ddot{\boldsymbol{\eta}}\} + [\mathcal{K}]\{\boldsymbol{\eta}\} = [\boldsymbol{\Phi}]^T\{\mathbf{F}\}$$

or

$$\{\ddot{\boldsymbol{\eta}}\} + [\boldsymbol{\Lambda}]\{\boldsymbol{\eta}\} = [\mathcal{M}]^{-1}[\boldsymbol{\Phi}]^T\{\mathbf{F}\} \tag{3.98}$$

The above expression is the vibration equation of motion including the external forces **F**.

Now consider an alternate derivation using Lagrange's equation and virtual work. With external forces applied, Lagrange's equation now states that

$$\frac{d}{dt}\left(\frac{\partial \mathrm{T}}{\partial \dot{q}_i}\right) - \frac{\partial \mathrm{T}}{\partial q_i} + \frac{\partial \mathrm{U}}{\partial q_i} = Q_i \tag{3.99}$$

with the *generalized forces* Q_i defined as

$$Q_i = \frac{\partial(\delta W)}{\partial(\delta q_i)}$$

Here, δW is the virtual work, defined below, and δq_i is a virtual displacement of the generalized coordinate q_i.

Virtual work is defined as

$$\delta W = \sum_{i=1}^{m} \mathbf{F}_i \cdot \delta \mathbf{d}_i \tag{3.100}$$

or the sum over all forces acting on the body of the vector dot product between each force $\mathbf{F}_i$ and the virtual physical displacement of the point of application of the force $\delta\mathbf{d}_i$. For example, for the two-mass cantilever beam, and using a unit vector **k** acting in the vertical direction, let

$$\mathbf{F}_1 = F_1\mathbf{k} \qquad \mathbf{F}_2 = F_2\mathbf{k}$$

$$\delta\mathbf{d}_1 = \delta Z_1\mathbf{k} \qquad \delta\mathbf{d}_2 = \delta Z_2\mathbf{k}$$

Consequently,

$$\delta W = F_1\delta Z_1 + F_2\delta Z_2$$

with

$$\delta Z_1 = l_1\delta\theta_1$$

$$\delta Z_2 = l_2\delta\theta_2 + (l_1 + l_2)\delta\theta_1 \tag{3.101}$$

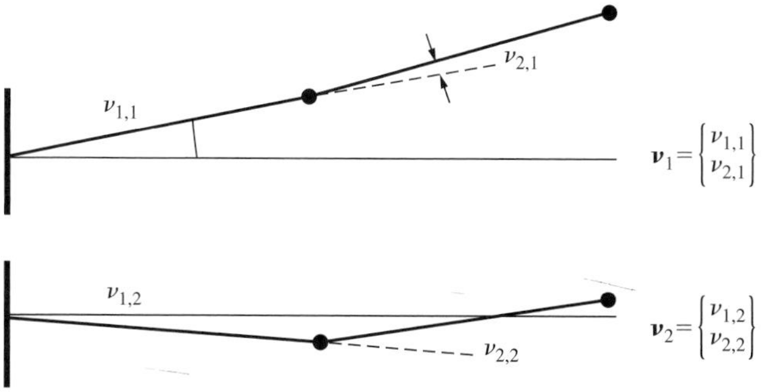

Figure 3.9 Mode shapes corresponding to angular displacements.

Now, assume the unforced or free vibration problem had been formulated and solved in terms of the angular displacements θ_1 and θ_2. Therefore, the elements of the free-vibration mode shapes would correspond to angular displacements (see Figure 3.9 above). The virtual displacements $\delta\theta_1$ and $\delta\theta_2$ in Equations (3.101) can be expressed in terms of these mode shapes and the two vibration modal coordinates η_i, or

$$\begin{aligned} \delta\theta_1 &= \nu_{1,1}\delta\eta_1 + \nu_{1,2}\delta\eta_2 \\ \delta\theta_2 &= \nu_{2,1}\delta\eta_1 + \nu_{2,2}\delta\eta_2 \end{aligned} \tag{3.102}$$

or

$$\begin{Bmatrix} \delta\theta_1 \\ \delta\theta_2 \end{Bmatrix} = [\boldsymbol{\Phi}] \begin{Bmatrix} \delta\eta_1 \\ \delta\eta_2 \end{Bmatrix}$$

Using the above result to rewrite the virtual work, we have

$$\begin{aligned} \delta W &= \begin{Bmatrix} \nu_{1,1}(F_1 l_1 + F_2(l_1 + l_2)) + \nu_{2,1} F_2 l_2 \\ \nu_{1,2}(F_1 l_1 + F_2(l_1 + l_2)) + \nu_{2,2} F_2 l_2 \end{Bmatrix}^T \begin{Bmatrix} \delta\eta_1 \\ \delta\eta_2 \end{Bmatrix} \\ &= \{\mathbf{F}\}^T [\boldsymbol{\Phi}] \begin{Bmatrix} \delta\eta_1 \\ \delta\eta_2 \end{Bmatrix} \end{aligned} \tag{3.103}$$

where $\{\mathbf{F}\}$ is the array of applied forces introduced in Equation (3.96), and should not be confused with $\mathbf{F}_i$ introduced in Equation (3.100), the general expression for virtual work.

Finally, we can express the kinetic and potential energies in terms of the free-vibration modal coordinates as follows:

$$\begin{aligned} T &= \frac{1}{2} \begin{Bmatrix} \dot{\theta}_1 \\ \dot{\theta}_2 \end{Bmatrix}^T [\mathbf{M}] \begin{Bmatrix} \dot{\theta}_1 \\ \dot{\theta}_2 \end{Bmatrix} = \frac{1}{2} \begin{Bmatrix} \dot{\eta}_1 \\ \dot{\eta}_2 \end{Bmatrix}^T [\boldsymbol{\Phi}]^T [\mathbf{M}][\boldsymbol{\Phi}] \begin{Bmatrix} \dot{\eta}_1 \\ \dot{\eta}_2 \end{Bmatrix} = \frac{1}{2} \{\dot{\boldsymbol{\eta}}\}^T [\mathcal{M}] \{\dot{\boldsymbol{\eta}}\} \\ U &= \frac{1}{2} \{\boldsymbol{\theta}\}^T [\mathbf{K}] \{\boldsymbol{\theta}\} = \frac{1}{2} \{\boldsymbol{\eta}\}^T [\mathcal{K}] \{\boldsymbol{\eta}\} \end{aligned} \tag{3.104}$$

Applying Lagrange's equation in Equation (3.99) (or Equation (3.115)), with the modal coordinates used as the generalized coordinates q_i, yields

$$\boxed{[\mathcal{M}]\{\ddot{\boldsymbol{\eta}}\} + [\mathcal{K}]\{\boldsymbol{\eta}\} = \{\mathbf{Q}\}} \tag{3.105}$$

where

$$\{\mathbf{Q}\} = [\boldsymbol{\Phi}]^T\{\mathbf{F}\}$$

Note that this result is identical to that in Equation (3.98), developed from Newtonian mechanics. Although it may seem like the method of virtual work introduces undue complications, it is the preferred systematic approach in studying the flight mechanics of flexible flight vehicles, which have much more complex geometry than the simple beam.

3.10 FORCED MOTION OF THE UNRESTRAINED BEAM MODEL

We are now finally ready for the derivation of the equations of motion of the unrestrained beam acted upon by external forces. (The analogy between this case and the case of a missile, for example, should be apparent.) Consider a three-mass beam acted upon by three forces, as depicted in Figure 3.10. We again assume that a solution to the unforced vibration problem and a modal analysis is available. If not, one must be performed before proceeding further.

We know that we can express the total, or inertial displacements of the beam in terms of the vertical position of the center of mass, $Z_{\text{Ref}} = Z_{\text{CM}}$, the angular position of the reference axis Θ_{Ref}, and the (constrained) relative displacements $\mathbf{z}_\perp$. Or, from Equation (3.91), we have

$$\begin{aligned}\{\mathbf{Z}(t)\} &= \{\mathbf{1}\}Z_{\text{CM}}(t) + \{\mathbf{X}\}\Theta_{\text{Ref}}(t) + \{\mathbf{z}_\perp(t)\} \\ &= \{\mathbf{1}\}Z_{\text{CM}}(t) + \{\mathbf{X}\}\Theta_{\text{Ref}}(t) + \sum_{i=1}^{n-2}\{\boldsymbol{\nu}_\perp\}_i\eta_i(t)\end{aligned} \tag{3.106}$$

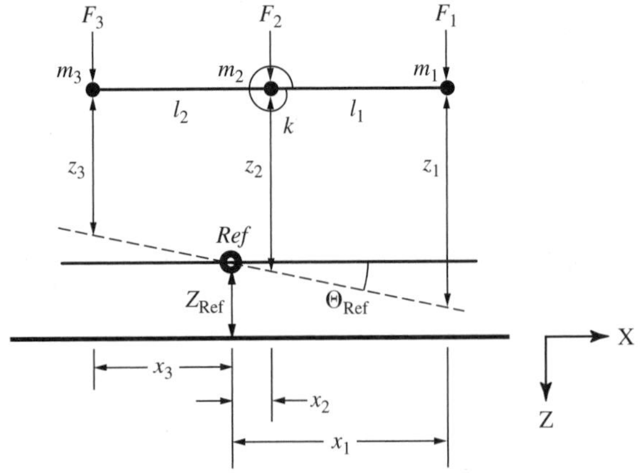

Figure 3.10 Unrestrained beam model with external forces.

The subscript $\perp$ on the relative displacements $\mathbf{z}$ and the vibration mode shapes $\boldsymbol{\nu}_i$ remind us that we have invoked the transformation given in Equation (3.73) in developing this modal representation for the relative displacements.

NOTE TO STUDENT

It is important to remember in the development to follow that the mode shapes and modal coordinates used in Equations (3.106) are those from the unforced vibration problem (no forces acting on the beam model). Students are sometimes confused about this fact. These unforced vibration frequencies and mode shapes are properties of the structure, independent of any forces acting on it.

Using Equation (3.100), the definition of virtual work, we can write the virtual work here as

$$\delta W = F_1 \delta Z_1 + F_2 \delta Z_2 + F_3 \delta Z_3 = \{\mathbf{F}\}^T \{\delta \mathbf{Z}\} \tag{3.107}$$

But the virtual inertial displacements δZ_i can be written in terms of the coordinates used in Equation (3.106). Or

$$\begin{aligned} \delta Z_1 &= \delta Z_{\text{CM}} + x_1 \delta \Theta_{\text{Ref}} + \delta z_{c_{\perp 1}} \\ \delta Z_2 &= \delta Z_{\text{CM}} + x_2 \delta \Theta_{\text{Ref}} + \delta z_{c_{\perp 2}} \\ \delta Z_3 &= \delta Z_{\text{CM}} - x_3 \delta \Theta_{\text{Ref}} + \delta z_{c_{\perp 3}} \end{aligned} \tag{3.108}$$

Furthermore, we can express the kinetic energy as

$$\mathrm{T} = \frac{1}{2}\left(M_{\text{Tot}} \dot{Z}_{\text{CM}}^2 + I \dot{\Theta}_{\text{Ref}}^2 + \{\dot{\mathbf{z}}_\perp\}^T [\mathbf{M}] \{\dot{\mathbf{z}}_\perp\}\right) \tag{3.109}$$

with $\mathbf{M}$ the beam's mass matrix. And we can express the potential (strain) energy as

$$\mathrm{U} = \frac{1}{2} \{\mathbf{z}_\perp\}^T [\mathbf{K}_c] \{\mathbf{z}_\perp\} \tag{3.110}$$

with the constrained stiffness matrix $\mathbf{K}_c$ given by Equation (3.55). So all the information necessary for the derivation is contained in Equations (3.107)–(3.110), which are expressed in terms of the rigid-body coordinates and relative displacements.

The relative displacements now may be expressed in terms of the (unforced) vibration modal coordinates, as indicated in Equation (3.106). Or

$$\{\mathbf{z}_\perp(t)\} = \sum_{i=1}^{n-2} \{\boldsymbol{\nu}_\perp\}_i \eta_i(t)$$

where n is the number of lumped masses in the beam's representation. We know there are $n - 2$ vibration modes, since there are two rigid-body degrees of freedom.

The total displacements may now be written as

$$\{\mathbf{Z}(t)\} = \left[\{\mathbf{1}\} \ \ \{\mathbf{X}\} \ \ \{\boldsymbol{\nu}_\perp\}_1 \ \ \cdots \ \ \{\boldsymbol{\nu}_\perp\}_{n-2}\right] \begin{Bmatrix} Z_{\text{CM}} \\ \Theta_{\text{Ref}} \\ \eta_1 \\ \vdots \\ \eta_{n-2} \end{Bmatrix}$$

$$= \left[\{\mathbf{1}\} \ \ \{\mathbf{X}\} \ \ [\boldsymbol{\Phi}_{\text{Vib}}]\right] \begin{Bmatrix} Z_{\text{CM}} \\ \Theta_{\text{Ref}} \\ \{\boldsymbol{\eta}\} \end{Bmatrix} \tag{3.111}$$

$$= [\boldsymbol{\Phi}]\{\mathbf{q}\}$$

with obvious definition of the modal matrix $\boldsymbol{\Phi}$. Note that the vibration modal matrix $\boldsymbol{\Phi}_{\text{Vib}}$ here is not square but an n-by-$(n-2)$ matrix, since the two null eigenvectors associated with the two zero eigenvalues have been removed. Finally, the vector of generalized coordinates $\mathbf{q}$ (in Lagrange's equation) has also been defined in the above equation.

For the virtual displacements we have

$$\{\delta\mathbf{Z}\} = \left[\{\mathbf{1}\} \ \ \{\mathbf{X}\} \ \ [\boldsymbol{\Phi}_{\text{Vib}}]\right] \begin{Bmatrix} \delta Z_{\text{CM}} \\ \delta\Theta_{\text{Ref}} \\ \{\delta\boldsymbol{\eta}\} \end{Bmatrix}$$

$$= [\boldsymbol{\Phi}]\{\delta\mathbf{q}\}$$

which may be substituted into Equation (3.107) to obtain the virtual work. Or

$$\boxed{\delta W = \{\mathbf{F}\}^T[\boldsymbol{\Phi}]\{\delta\mathbf{q}\}} \tag{3.112}$$

For the kinetic energy we now have

$$\mathrm{T} = \frac{1}{2}\left(M_{\text{Tot}}\dot{Z}_{\text{CM}}^2 + I\dot{\Theta}_{\text{Ref}}^2 + \{\dot{\boldsymbol{\eta}}\}^T[\boldsymbol{\Phi}_{\text{Vib}}]^T[\mathbf{M}][\boldsymbol{\Phi}_{\text{Vib}}]\{\dot{\boldsymbol{\eta}}\}\right)$$

$$= \frac{1}{2}\left(M_{\text{Tot}}\dot{Z}_{\text{CM}}^2 + I\dot{\Theta}_{\text{Ref}}^2 + \{\dot{\boldsymbol{\eta}}\}^T[\mathcal{M}_{\text{Vib}}]\{\dot{\boldsymbol{\eta}}\}\right) \tag{3.113}$$

$$= \frac{1}{2}\{\dot{\mathbf{q}}\}^T \begin{bmatrix} M_{\text{Tot}} & 0 & \mathbf{0} \\ 0 & I & \mathbf{0} \\ \mathbf{0} & \mathbf{0} & [\mathcal{M}_{\text{Vib}}] \end{bmatrix} \{\dot{\mathbf{q}}\} = \frac{1}{2}\{\dot{\mathbf{q}}\}^T[\mathcal{M}]\{\dot{\mathbf{q}}\}$$

with obvious definition for the generalized mass matrix $[\mathcal{M}]$, and with $[\mathcal{M}_{\text{Vib}}]$ the square generalized mass matrix of dimension $n-2$ associated with the vibration-modal representation. Finally, the potential energy may now be written as

$$\begin{aligned} U &= \frac{1}{2}\{\boldsymbol{\eta}\}^T[\boldsymbol{\Phi}_{\text{Vib}}]^T[\mathbf{K}_c][\boldsymbol{\Phi}_{\text{Vib}}]\{\boldsymbol{\eta}\} \\ &= \frac{1}{2}\{\boldsymbol{\eta}\}^T[\mathcal{K}]\{\boldsymbol{\eta}\} = \frac{1}{2}\{\mathbf{q}\}^T\begin{bmatrix} \mathbf{0} & \mathbf{0} \\ \mathbf{0} & [\mathcal{K}] \end{bmatrix}\{\mathbf{q}\} \end{aligned} \tag{3.114}$$

with obvious definition for the generalized stiffness matrix $[\mathcal{K}]$.

Recalling Lagrange's equation, written here in vector form, we have

$$\frac{d}{dx}\left(\frac{\partial T}{\partial \dot{\mathbf{q}}}\right) - \frac{\partial T}{\partial \mathbf{q}} + \frac{\partial U}{\partial \mathbf{q}} = \mathbf{Q}^T = \frac{\partial(\delta W)}{\partial(\delta \mathbf{q})} \tag{3.115}$$

Applying this equation to the expressions developed above, Equations (3.112)–(3.114), yields

$$\{\ddot{\mathbf{q}}\}^T[\mathcal{M}] + \{\mathbf{q}\}^T\begin{bmatrix} \mathbf{0} & \mathbf{0} \\ \mathbf{0} & [\mathcal{K}] \end{bmatrix} = \{\mathbf{F}\}^T[\boldsymbol{\Phi}]$$

Or transposing the above we have

$$\boxed{[\mathcal{M}]\{\ddot{\mathbf{q}}\} + \begin{bmatrix} \mathbf{0} & \mathbf{0} \\ \mathbf{0} & [\mathcal{K}] \end{bmatrix}\{\mathbf{q}\} = [\boldsymbol{\Phi}]^T\{\mathbf{F}\}} \tag{3.116}$$

From the definitions of the vectors and matrices in Equation (3.116), one finds that the equations of motion may be written as

$$\boxed{\begin{aligned} &M_{\text{Tot}}\ddot{Z}_{\text{CM}} = F_1 + F_2 + F_3 \\ &I\ddot{\Theta}_{\text{Ref}} = F_1x_1 + F_2x_2 - F_3x_3 \\ &\{\ddot{\boldsymbol{\eta}}\} + [\boldsymbol{\Lambda}_{\text{Vib}}]\{\boldsymbol{\eta}\} = [\mathcal{M}_{\text{Vib}}]^{-1}[\boldsymbol{\Phi}_{\text{Vib}}]^T\{\mathbf{F}\} \end{aligned}} \tag{3.117}$$

with $\boldsymbol{\Lambda}_{\text{Vib}} = [\mathcal{M}_{\text{Vib}}]^{-1}[\mathcal{K}]$. For the three-mass beam model considered in Example 3.5, the third equation in Equations (3.117) is simply

$$\ddot{\eta} + 5.5\eta = 1/\mathcal{M}_{\text{Vib}}[0.18 \quad -0.36 \quad 0.09]\begin{Bmatrix} F_1 \\ F_2 \\ F_3 \end{Bmatrix}$$

with

$$\mathcal{M}_{\text{Vib}} = \{\boldsymbol{\nu}_{\text{Vib}}\}^T[\mathbf{M}]\{\boldsymbol{\nu}_{\text{Vib}}\} = 0.178 \text{ mass units}$$

3.11 Summary

In this chapter we have introduced the basic concepts associated with structural vibrations, along with the derivation of the equations of motion of elastic bodies. Key concepts include the modal analysis of the vibration problem (including rigid-body modes), modal orthogonality, virtual work, degrees of freedom, and generalized coordinates. We showed, for example, that if the solution to the unforced vibration problem was available for an unrestrained elastic body (vehicle), the unforced vibration modes and mode shapes could be used to derive the equations of motion for that body (vehicle). Finally, we noted that due to orthogonality of the rigid-body and vibration modes, the reference frames used in all examples in this chapter satisfied the mean-axis constraints to be introduced in Chapter 4.

3.12 Problems

3.1 Given that the restrained unforced vibration problem may be described by the equation

$$[\mathbf{M}]\{\ddot{\mathbf{q}}\} + [\mathbf{K}]\{\mathbf{q}\} = \{\mathbf{0}\} \Rightarrow \quad \{\ddot{\mathbf{q}}\} + [\mathbf{D}]\{\mathbf{q}\} = \{\mathbf{0}\}$$

show that the eigenvectors of the dynamic matrix **D**, or the vibration mode shapes, are orthogonal with respect to the stiffness matrix **K** in addition to the mass matrix **M**.

3.2 Consider the lumped-mass system depicted in Figure 3.11, where the carts roll without friction. Letting the potential (strain) energy be

$$\mathrm{U} = \frac{1}{2}k\left(x_2 - x_1\right)^2$$

write the kinetic energy of the system. Noting that a rigid-body mode with mode shape

$$\boldsymbol{\nu}_{\mathrm{RB}} = \begin{Bmatrix} 1 \\ 1 \end{Bmatrix}$$

can be one of the system's modes, find the system's vibration frequency in terms of k, m_1, and m_2. Also, plot the vibration mode shape and express the general motion of the system in terms of its modes.

3.3 Consider a four-mass unrestrained straight beam, with a mass at each end, and the remaining two masses evenly spaced along the beam (e.g., $l = 1$ for each mass). Let the four masses be

$$m_1 = m_4 = 1, m_2 = m_3 = 2$$

and let the two spring constants for the springs co-located with masses 2 and 3 both equal unity. Considering only transverse vibrations of the beam, use the alternate formulation for dealing with rigid-body degrees of freedom discussed in Sections 3.5 and 3.6, and perform a modal analysis of the beam. Show that the

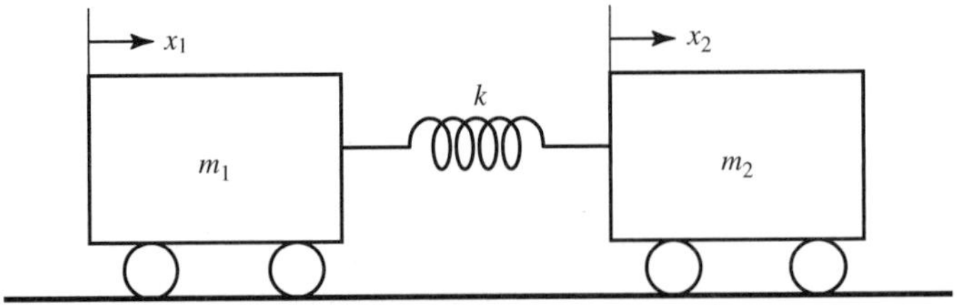

Figure 3.11 Two-cart elastic system.

vibration modes are mutually orthogonal (with respect to the mass matrix), and are also orthogonal to the rigid-body mode shapes **{1}** and **{X}**. Hence, find the system's normal modes.

3.4 For the beam considered in Problem 3.3, let four forces F_1–F_4 act vertically on the four masses, respectively. Derive the equations of motion of the elastic beam, governing rigid-body translation and rotation and the elastic degrees of freedom.

References

1. Meirovitch, L.: *Analytical Methods in Vibrations,* Macmillan., New York, 1967.

2. Greenwood, D. T.: *Principles of Dynamics,* 2nd ed., Prentice-Hall, Upper Saddle River, NJ, 1988.

4 CHAPTER

Equations of Motion for Elastic Vehicles

Chapter Roadmap: *As with Chapter 3, the material in this chapter would most likely not be covered in a first course in flight dynamics but in a more advanced course. In addition, much of this material would not be covered in a typical course on aeroelasticity, which seldom, if ever, treats the subject of rigid-body degrees of freedom.*

In this chapter we will extend the concepts presented in Chapters 1 and 2, in which only rigid vehicles were considered, and will apply the concepts developed in Chapter 3 to derive the equations of motion for flexible vehicles. A flat, nonrotating earth will be assumed. Students may want to review Chapter 3 before beginning this chapter, especially those less familiar with the subject of structural vibrations. Concepts such as modal orthogonality, vibration mode shapes and frequencies, the vibration eigenvalue problem, relative motion, and rigid-body degrees of freedom are all discussed in the previous chapter.[1]

In this book we are interested in the effects of elastic deformation on the flight dynamics of the vehicle. We are not interested in purely structural/aeroelastic phenomena such as flutter or divergence, for example. As a result, only the lower-frequency elastic modes will typically be of interest, and the higher-frequency modes will be truncated from the models we will develop.

Since all real vehicles are flexible, one must ultimately address the question of when a rigid-body assumption is appropriate. Frequently that question can only be answered after obtaining a model of the flexible vehicle. For now we can state that flexibility can be a significant issue in the flight dynamics of larger vehicles because the elastic-mode frequencies of such vehicles are relatively low.

---

[1] Much of the material in this chapter and in Chapter 7 is based in large part on research performed by Mr. Marty Waszak while he was a graduate research assistant at Purdue University.

Hence, those modes are more likely to couple with the rigid-body modes of the vehicle, and/or to interact with the flight-control systems. We will discuss this in more detail in Chapters 10–12.

Newtonian mechanics formed the basis for the derivations of the equations of motion of a rigid vehicle presented in Chapter 2. In contrast, Lagrangian mechanics and energy concepts were presented in Chapter 3 for dealing with elastic bodies, and the Lagrangian approach will again be used here. The basic theories behind Lagrangian mechanics will be stated without proof and the interested student is encouraged to refer to the many excellent texts that deal with the proofs of the fundamental equations.

As in Chapter 2, the modeling of (aerodynamic and propulsive) forces acting on the vehicle, for incorporation into the equations derived in this chapter, will be presented in Chapters 5, 6, and 7.

4.1 LAGRANGE'S EQUATION—KINETIC AND POTENTIAL ENERGIES

As noted in Chapter 3, Lagrange's equation states that

$$\frac{d}{dt}\left(\frac{\partial \mathrm{T}}{\partial \dot{\mathbf{q}}}\right) - \frac{\partial \mathrm{T}}{\partial \mathbf{q}} + \frac{\partial \mathrm{U}}{\partial \mathbf{q}} = \mathbf{Q}^T = \frac{\partial(\delta W)}{\partial(\delta \mathbf{q})} \tag{4.1}$$

where

T is the system's or body's total kinetic energy

U is the potential energy of the system, including the strain energy of the elastically deformed body

$\mathbf{q}$ is the vector of generalized coordinates used to describe the system, such as the vehicle's position, velocity, and elastic deformation

$\mathbf{Q}$ is the vector of generalized forces acting on the body

δW is the virtual work done by external forces acting on the body

$\delta\mathbf{q}$ is the vector of virtual displacements of the generalized coordinates

We begin this discussion as we did in Chapter 2, by considering a mass element of the vehicle, as depicted in Figure 2.1, repeated here as Figure 4.1. The inertial (Frame I) and vehicle-fixed (Frame V) frames are shown, with the precise definition of the vehicle-fixed frame to be considered later. Note that the inertial position of a mass element of the vehicle, $\mathbf{p}'$ may be expressed in terms of the inertial position of the origin of the vehicle-fixed frame, $\mathbf{p}_V$, plus the position of the mass element relative to the vehicle-fixed origin, $\mathbf{p}$.

The total kinetic energy may then be expressed as the volume integral over the energies of each mass element. The inertial velocity of a mass element is

$$\mathbf{V}_{\text{element}} = \frac{d\mathbf{p}'}{dt}\Big|_I \tag{4.2}$$

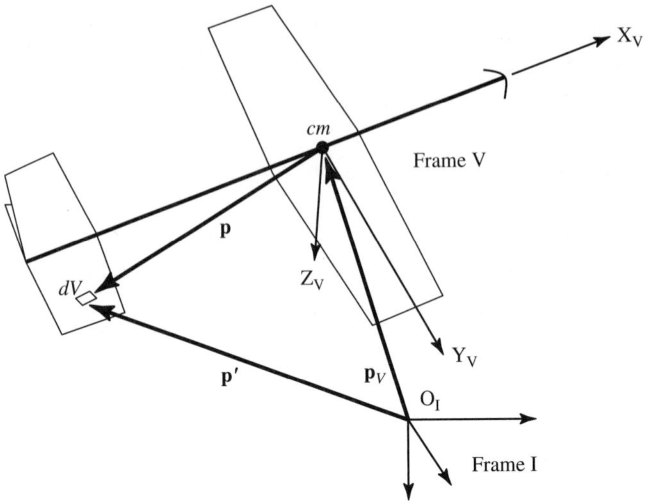

Figure 4.1 Inertial and vehicle-fixed frames and position vectors.

So the kinetic energy of the element is

$$T_{\text{element}} = \frac{1}{2}|\mathbf{V}_{\text{element}}|^2 \rho_V dV = \frac{1}{2}\frac{d\mathbf{p}'}{dt}\Big|_I \cdot \frac{d\mathbf{p}'}{dt}\Big|_I \rho_V dV \tag{4.3}$$

where, as in Chapter 2, ρ_V is the material density of the mass element with volume dV. Integrating over all the mass elements of the vehicle yields the vehicle's total kinetic energy

$$T = \frac{1}{2}\int_{\text{Vol}} \frac{d\mathbf{p}'}{dt}\Big|_I \cdot \frac{d\mathbf{p}'}{dt}\Big|_I \rho_V dV \tag{4.4}$$

Now note that

$$\mathbf{p}' = \mathbf{p}_V + \mathbf{p} \tag{4.5}$$

and that the inertial velocity of the origin of Frame V is

$$\mathbf{V}_V = \frac{d\mathbf{p}_V}{dt}\Big|_I \tag{4.6}$$

In addition,

$$\frac{d\mathbf{p}}{dt}\Big|_I = \frac{d\mathbf{p}}{dt}\Big|_V + \boldsymbol{\omega}_{V,I} \times \mathbf{p} \tag{4.7}$$

where $\boldsymbol{\omega}_{V,I}$ is the angular velocity of Frame V with respect to Frame I. So the kinetic energy now becomes

$$\begin{aligned} T = \frac{1}{2}\int_{\text{Vol}} \Bigg(& \frac{d\mathbf{p}_V}{dt}\Big|_I \cdot \frac{d\mathbf{p}_V}{dt}\Big|_I + 2\left(\frac{d\mathbf{p}_V}{dt}\Big|_I \cdot \frac{d\mathbf{p}}{dt}\Big|_V + \left(\frac{d\mathbf{p}_V}{dt}\Big|_I + \frac{d\mathbf{p}}{dt}\Big|_V\right) \cdot (\boldsymbol{\omega}_{V,I} \times \mathbf{p})\right) \\ & + \frac{d\mathbf{p}}{dt}\Big|_V \cdot \frac{d\mathbf{p}}{dt}\Big|_V + (\boldsymbol{\omega}_{V,I} \times \mathbf{p}) \cdot (\boldsymbol{\omega}_{V,I} \times \mathbf{p})\Bigg) \rho_V dV \end{aligned} \tag{4.8}$$

The potential energy of the vehicle includes gravitational potential energy U_g, plus the elastic strain energy of the vehicle's elastic structure U_e. Here we will write the gravitational potential energy of the vehicle as the integral of the potential energies of the mass elements, or

$$U_g = -\int_{Vol} (\mathbf{g}\cdot\mathbf{p}')\rho_V dV = -\int_{Vol} \big(\mathbf{g}\cdot(\mathbf{p}_V + \mathbf{p})\big)\rho_V dV \tag{4.9}$$

where $\mathbf{g}$ is the acceleration due to gravity.

The elastic strain energy is the energy stored in an elastic structure due to its deformation resulting from some applied force. The strain energy is the negative of the work done on the structure by the applied force, and work is force acting over a distance (or displacement). Now let the position $\mathbf{p}$ of a mass element of the vehicle with respect to the origin of Frame V be represented in terms of its undeformed, or rigid-body position $\mathbf{p}_{RB}$, plus the elastic displacement of that point on the structure $\mathbf{d}_E(x,y,z,t)$. Or, let

$$\mathbf{p} = \mathbf{p}_{RB} + \mathbf{d}_E(x,y,z,t) \tag{4.10}$$

where it is noted that $\mathbf{d}_E$ depends on the (x,y,z) (undeformed) location of the mass element of the structure in Frame V. Also note that

$$\frac{d\mathbf{p}}{dt}\Big|_V = \frac{d\mathbf{d}_E}{dt}\Big|_V \tag{4.11}$$

since the vector $\mathbf{p}_{RB}$ is invariant with respect to Frame V. Using D'Alembert's principle to express the force on a mass element in terms of the mass of the element and its acceleration, we have for the elastic strain energy

$$U_e = -\frac{1}{2}\int_{Vol}\left(\frac{d^2\mathbf{d}_E}{dt^2}\Big|_V\cdot\mathbf{d}_E\right)\rho_V dV \tag{4.12}$$

This expression is consistent with those in Refs. 1–3 at the end of this chapter.

4.2 VEHICLE-FIXED FRAME—THE MEAN AXES

We will now address the vehicle-fixed Frame V specifically. When dealing with rigid vehicles, as in Chapter 2, the selection of the vehicle-fixed frame is arbitrary, except that the origin of the frame must coincide with the vehicle's center of mass. But for flexible vehicles, more care must be taken. In Chapter 3 we discussed the existence of n mutually orthogonal modes, the normal modes, of an n degree-of-freedom unrestrained elastic body. For this unrestrained body with rigid-body degrees of freedom, we used the position of a reference axis plus the body's displacement relative to this axis to define the motion of the body. We will now use the same concepts when dealing with a flight vehicle.

In Refs. 1–2, it is noted that for an elastic body a coordinate frame called *mean axes* always exists such that the relative translational and angular momenta (about the center of mass) due to elastic deformation of the structure undergoing unforced vibrations are zero. We will refer to these properties as the *mean-axis constraints,* which imply that if the vehicle-fixed Frame V are mean axes, then

$$\int_{\text{Vol}} \frac{d\mathbf{p}}{dt}\Big|_V \rho_V dV = \int_{\text{Vol}} \mathbf{p} \times \frac{d\mathbf{p}}{dt}\Big|_V \rho_V dV = \mathbf{0} \tag{4.13}$$

Furthermore, the origin of the mean axes must coincide with the instantaneous center of mass of the vehicle. Thus the mean axes constitute a special "vehicle-fixed" frame in that the origin is fixed to the center of mass, not to any particular material point on the vehicle.

Recalling from Equation (4.10) that the position of each mass element relative to the origin of Frame V can we written as

$$\mathbf{p} = \mathbf{p}_{\text{RB}} + \mathbf{d}_E$$

and invoking Equation (4.11), we can also write Equations (4.13) as

$$\int_{\text{Vol}} \frac{d\mathbf{d}_E}{dt}\Big|_V \rho_V dV = \int_{\text{Vol}} \mathbf{p}_{\text{RB}} \times \frac{d\mathbf{d}_E}{dt}\Big|_V \rho_V dV + \int_{\text{Vol}} \mathbf{d}_E \times \frac{d\mathbf{d}_E}{dt}\Big|_V \rho_V dV = \mathbf{0} \tag{4.14}$$

Now if the elastic displacement $\mathbf{d}_E$ is sufficiently small such that only linear effects are considered, the third integral above may be neglected. The remaining two expressions in Equations (4.14) are called the *practical mean-axis constraints.*

Consequently, Equations (4.13) may be used in the theoretical development of the equations of motion, while Equations (4.14), the *practical mean-axis constraints,* may be used to confirm that the selected axes are mean-axes. Note that these two practical constraints are analogous to Equations (3.56) and (3.57), which are orthogonality conditions between the rigid-body and vibration modes. In practice, then, by satisfying modal orthogonality among all the modes, we assure the practical mean-axis constraints are satisfied.

To demonstrate this fact, let us assume that a free-vibration analysis of the structure has been performed consistent with the methods presented in Sections 3.4–3.6, yielding the *n* free-vibration mode shapes and frequencies including both the rigid-body and elastic modes. And recall that all these modes are mutually orthogonal. Consequently, the elastic displacement of the structure can be expressed in terms of a modal expansion using the *n* free-vibration modes, or

$$\mathbf{d}_E = \sum_{i=1}^{n} \boldsymbol{\nu}_i(x,y,z)\eta_i(t) \tag{4.15}$$

Here, $\boldsymbol{\nu}_i(x,y,z)$ is the vibration mode shape and $\eta_i(t)$ the generalized coordinate associated with the *i*'th vibration mode. In general, each mode shape $\boldsymbol{\nu}_i(x,y,z)$ is a vector with **i**, **j**, and **k** components defined in Frame V, with each component a function of the (x,y,z) location on the undeformed structure. Now, the first of Equations (4.14), can be written as

$$\int_{\text{Vol}} \frac{d\mathbf{d}_E}{dt}\Big|_V \rho_V dV = \int_{\text{Vol}} \sum_{i=1}^{n} \boldsymbol{\nu}_i(x,y,z)\dot{\eta}_i(t)\rho_V dV = \sum_{i=1}^{n} \dot{\eta}_i(t)\left(\int_{\text{Vol}} \boldsymbol{\nu}_i(x,y,z)\rho_V dV\right) = \mathbf{0} \tag{4.16}$$

But each integral term in parentheses above, analogous to Equation (3.56), will be zero since the vibration modes are orthogonal (with respect to the mass distribution) to the rigid-body translation mode(s). Hence, this mean-axis constraint will be satisfied. Likewise, the second integral in Equations (4.14) may be written as

$$\int_{\text{Vol}} \mathbf{p}_{\text{RB}} \times \frac{d\mathbf{d}_E}{dt}\Big|_V \rho_V dV = \sum_{i=1}^{n} \dot{\eta}_i(t)\left(\int_{\text{Vol}} \mathbf{p}_{\text{RB}} \times \boldsymbol{\nu}_i(x,y,z)\rho_V dV\right) = \mathbf{0} \tag{4.17}$$

Where each integral term in parentheses above, analogous to Equation (3.57), will also be zero since the vibration modes are orthogonal (with respect to the mass distribution) to the rigid-body rotation mode(s). Hence the second practical mean-axis constraint will be satisfied as well.

We will now apply the mean-axis or othogonality constraints expressed in Equation (4.13) to the kinetic and gravitational potential energies expressed in Equations (4.8) and (4.9), respectively. The expression for kinetic energy, Equation (4.8), can be considerably simplified since now

$$\int_{\text{Vol}} \frac{d\mathbf{p}_V}{dt}\Big|_I \cdot \frac{d\mathbf{p}}{dt}\Big|_V \rho_V dV = \frac{d\mathbf{p}_V}{dt}\Big|_I \cdot \int_{\text{Vol}} \frac{d\mathbf{p}}{dt}\Big|_V \rho_V dV = 0 \tag{4.18}$$

and

$$\int_{\text{Vol}} \frac{d\mathbf{p}}{dt}\Big|_V \cdot (\boldsymbol{\omega}_{V,I} \times \mathbf{p})\rho_V dV = \int_{\text{Vol}} \mathbf{p} \times \frac{d\mathbf{p}}{dt}\Big|_V \rho_V dV \cdot \boldsymbol{\omega}_{V,I} = 0 \tag{4.19}$$

Here, as in Chapter 2, some of the above terms are invariant with respect to the volume integration, and hence can be brought outside the integral. Now the expression for the vehicle's kinetic energy becomes

$$\begin{aligned} \mathrm{T} = \frac{1}{2}\int_{\text{Vol}} \Bigg(& \frac{d\mathbf{p}_V}{dt}\Big|_I \cdot \frac{d\mathbf{p}_V}{dt}\Big|_I + 2\left(\frac{d\mathbf{p}_V}{dt}\Big|_I \cdot (\boldsymbol{\omega}_{V,I} \times \mathbf{p})\right) + \frac{d\mathbf{p}}{dt}\Big|_V \cdot \frac{d\mathbf{p}}{dt}\Big|_V \\ & + (\boldsymbol{\omega}_{V,I} \times \mathbf{p}) \cdot (\boldsymbol{\omega}_{V,I} \times \mathbf{p})\Bigg)\rho_V dV \end{aligned} \tag{4.20}$$

The requirement that the origin of Frame V (mean axes) is located at the instantaneous center of mass requires that the first mass moment about the origin be zero. Or

$$\int_{\text{Vol}} \mathbf{p}\rho_V dV = 0 \tag{4.21}$$

(Note that this requirement can also be interpreted as another orthogonality constraint, analogous to Equation (3.58).) Looking at the second term in Equation (4.20), we have

$$\int_{\mathrm{Vol}} \frac{d\mathbf{p}_V}{dt}\Big|_I \cdot (\boldsymbol{\omega}_{V,I} \times \mathbf{p})\rho_V dV = \frac{d\mathbf{p}_V}{dt}\Big|_I \cdot \left(\boldsymbol{\omega}_{V,I} \times \int_{\mathrm{Vol}} \mathbf{p}\rho_V dV\right) = 0 \tag{4.22}$$

and the kinetic energy, Equation (4.20), further simplifies to

$$\mathrm{T} = \frac{1}{2}\int_{\mathrm{Vol}} \left(\frac{d\mathbf{p}_V}{dt}\Big|_I \cdot \frac{d\mathbf{p}_V}{dt}\Big|_I + \frac{d\mathbf{p}}{dt}\Big|_V \cdot \frac{d\mathbf{p}}{dt}\Big|_V + (\boldsymbol{\omega}_{V,I} \times \mathbf{p}) \cdot (\boldsymbol{\omega}_{V,I} \times \mathbf{p})\right)\rho_V dV \tag{4.23}$$

At this point, similar to the development in Chapter 2, we write

$$\int_{\mathrm{Vol}} \left(\frac{d\mathbf{p}_V}{dt}\Big|_I \cdot \frac{d\mathbf{p}_V}{dt}\Big|_I\right)\rho_V dV = \frac{d\mathbf{p}_V}{dt}\Big|_I \cdot \frac{d\mathbf{p}_V}{dt}\Big|_I \int_{\mathrm{Vol}} \rho_V dV = m\frac{d\mathbf{p}_V}{dt}\Big|_I \cdot \frac{d\mathbf{p}_V}{dt}\Big|_I \tag{4.24}$$

where m is the total mass of the vehicle. In addition, we write

$$\int_{\mathrm{Vol}} (\boldsymbol{\omega}_{V,I} \times \mathbf{p}) \cdot (\boldsymbol{\omega}_{V,I} \times \mathbf{p})\rho_V dV = \boldsymbol{\omega}_{V,I}^T[\mathbf{I}]\boldsymbol{\omega}_{V,I} \tag{4.25}$$

where $[\mathbf{I}]$ is the inertia matrix of the vehicle, introduced in Chapter 2. This matrix is, in general, time dependent due to the elastic deformation of the structure $\mathbf{d}_E$. But this time dependence is frequently ignored due to the assumption of small elastic deformations. Using Equations (4.24) and (4.25), the kinetic energy finally becomes

$$\mathrm{T} = \frac{1}{2}m\frac{d\mathbf{p}_V}{dt}\Big|_I \cdot \frac{d\mathbf{p}_V}{dt}\Big|_I + \frac{1}{2}\boldsymbol{\omega}_{V,I}^T[\mathbf{I}]\boldsymbol{\omega}_{V,I} + \frac{1}{2}\int_{\mathrm{Vol}} \frac{d\mathbf{p}}{dt}\Big|_V \cdot \frac{d\mathbf{p}}{dt}\Big|_V \rho_V dV \tag{4.26}$$

Finally, since the first mass moment about the origin of Frame V (the center of mass) is zero, we have from Equation (4.9) that the gravitational potential energy becomes

$$\mathrm{U}_g = -\int_{\mathrm{Vol}} \mathbf{g} \cdot (\mathbf{p}_V + \mathbf{p})\rho_V dV = -\mathbf{g} \cdot \mathbf{p}_V \int_{\mathrm{Vol}} \rho_V dV - \mathbf{g} \cdot \int_{\mathrm{Vol}} \mathbf{p}\rho_V dV = -\mathbf{g} \cdot \mathbf{p}_V m \tag{4.27}$$

4.3 MODAL EXPANSION USING FREE-VIBRATION MODES

The modal expansion for the elastic deformation, expressed as in Equation (4.15), is

$$\mathbf{d}_E = \sum_{i=1}^{n} \boldsymbol{\nu}_i(x,y,z)\eta_i(t)$$

This expression may now be used to rewrite the last term in the kinetic energy, Equation (4.26). Or

$$\begin{aligned}\int_{Vol} \frac{d\mathbf{p}}{dt}\Big|_V \cdot \frac{d\mathbf{p}}{dt}\Big|_V \rho_V dV &= \int_{Vol} \frac{d\mathbf{d}_E}{dt}\Big|_V \cdot \frac{d\mathbf{d}_E}{dt}\Big|_V \rho_V dV \\ &= \int_{Vol} \left(\sum_{i=1}^{n} \boldsymbol{\nu}_i \frac{d\eta_i}{dt} \cdot \sum_{i=1}^{n} \boldsymbol{\nu}_i \frac{d\eta_i}{dt} \right) \rho_V dV\end{aligned} \tag{4.28}$$

But this expression can be simplified due to the mutual orthogonality of the free-vibration modes, or

$$\begin{aligned}&\int_{Vol} \boldsymbol{\nu}_i \cdot \boldsymbol{\nu}_j \rho_V dV = 0, \quad i \neq j \\ &\int_{Vol} \boldsymbol{\nu}_i \cdot \boldsymbol{\nu}_i \rho_V dV = \mathcal{M}_i\end{aligned} \tag{4.29}$$

where $\mathcal{M}_i$ is the generalized mass of the i'th vibration mode. Therefore we now have

$$\int_{Vol} \frac{d\mathbf{p}}{dt}\Big|_V \cdot \frac{d\mathbf{p}}{dt}\Big|_V \rho_V dV = \int_{Vol} \left(\sum_{i=1}^{n} \boldsymbol{\nu}_i \cdot \boldsymbol{\nu}_i \left(\frac{d\eta_i}{dt} \right)^2 \right) \rho_V dV = \sum_{i=1}^{n} \mathcal{M}_i \dot{\eta}_i^2 \tag{4.30}$$

Using the above expression, the kinetic energy, Equation (4.26), becomes

$$\mathrm{T} = \frac{1}{2} m \frac{d\mathbf{p}_V}{dt}\Big|_I \cdot \frac{d\mathbf{p}_V}{dt}\Big|_I + \frac{1}{2} \boldsymbol{\omega}_{V,I}^T [\mathbf{I}] \boldsymbol{\omega}_{V,I} + \frac{1}{2} \sum_{i=1}^{n} \mathcal{M}_i \dot{\eta}_i^2 \tag{4.31}$$

Finally, the free-vibration modes may also be used to write the elastic strain energy in the structure. From Equation (4.12), repeated here, we have the strain energy given by

$$\mathrm{U}_e = -\frac{1}{2} \int_{Vol} \left(\frac{d^2\mathbf{d}_E}{dt^2}\Big|_V \cdot \mathbf{d}_E \right) \rho_V dV$$

But from Equation (4.15) we note that

$$\frac{d^2\mathbf{d}_E}{dt^2}\Big|_V = \sum_{i=1}^{n} \boldsymbol{\nu}_i(x,y,z) \ddot{\eta}_i(t)$$

and now the strain energy may be expressed as

$$\begin{aligned}\mathrm{U}_e &= -\frac{1}{2} \int_{Vol} \left(\frac{d^2\mathbf{d}_E}{dt^2}\Big|_V \cdot \mathbf{d}_E \right) \rho_V dV \\ &= -\frac{1}{2} \int_{Vol} \left(\sum_{i=1}^{n} \boldsymbol{\nu}_i(x,y,z) \ddot{\eta}_i(t) \cdot \sum_{i=1}^{n} \boldsymbol{\nu}_i(x,y,z) \eta_i(t) \right) \rho_V dV\end{aligned} \tag{4.32}$$

Again due to orthogonality of the vibration modes (Equations (4.29)), Equation (4.32) can be simplified. Invoking orthogonality, we now have for strain energy

$$\begin{aligned} U_e &= -\frac{1}{2}\int_{Vol}\left(\sum_{i=1}^{n}\boldsymbol{\nu}_i(x,y,z)\ddot{\eta}_i(t)\cdot\sum_{i=1}^{n}\boldsymbol{\nu}_i(x,y,z)\eta_i(t)\right)\rho_V dV \\ &= -\frac{1}{2}\int_{Vol}\left(\sum_{i=1}^{n}\boldsymbol{\nu}_i(x,y,z)\cdot\boldsymbol{\nu}_i(x,y,z)\ddot{\eta}_i(t)\eta_i(t)\right)\rho_V dV \end{aligned} \tag{4.33}$$

Recall that under free vibration, the time response of each generalized coordinate takes the form

$$\eta_i(t) = A_i\cos(\omega_i t + \Gamma_i)$$

where A_i and Γ_i are constants of integration. Therefore,

$$\ddot{\eta}_i(t) = -\omega_i^2 A_i\cos(\omega_i t + \Gamma_i) = -\omega_i^2\eta_i(t) \tag{4.34}$$

Consequently, if the structure were undergoing only free (unforced) vibrations, its elastic strain energy would be

$$U_e = -\frac{1}{2}\int_{Vol}\left(\sum_{i=1}^{n}\boldsymbol{\nu}_i(x,y,z)\cdot\boldsymbol{\nu}_i(x,y,z)\ddot{\eta}_i(t)\eta_i(t)\right)\rho_V dV = \frac{1}{2}\sum_{i=1}^{n}\omega_i^2\eta_i^2(t)\mathcal{M}_i \tag{4.35}$$

where the i'th generalized mass is given by

$$\mathcal{M}_i = \int_{Vol}\boldsymbol{\nu}_i(x,y,z)\cdot\boldsymbol{\nu}_i(x,y,z)\rho_V dV \tag{4.36}$$

For the same structural deformation, the strain energy of the structure undergoing free vibration is identical to the structure undergoing forced vibration. Hence, the strain energy of the deformed flexible vehicle must be that given in Equation (4.35).

4.4 SELECTION OF THE GENERALIZED COORDINATES

With the kinetic and potential energies expressed in terms of rigid-body and modal coordinates, one may apply Lagrange's equation to develop the vehicle's equations of motion. However, care must be taken in dealing with quantities defined in the inertial Frame I versus those defined in the vehicle-fixed Frame V. Let the inertial position of the origin of the vehicle-fixed Frame V be given as

$$\mathbf{p}_V = X_I\,\mathbf{i}_I + Y_I\,\mathbf{j}_I + Z_I\,\mathbf{k}_I \tag{4.37}$$

where $\mathbf{i}_I$, $\mathbf{j}_I$, and $\mathbf{k}_I$ are the unit vectors defining the inertial Frame I. (Note that for a flat earth $Z_I = -\mathrm{h}$, altitude) So the vehicle's inertial velocity vector is simply

$$\frac{d\mathbf{p}_V}{dt}\bigg|_I = \dot{X}_I\,\mathbf{i}_I + \dot{Y}_I\,\mathbf{j}_I + \dot{Z}_I\,\mathbf{k}_I \tag{4.38}$$

Also, consistent with the treatment of rigid vehicles in Chapter 2, we will take the 3-2-1 Euler angles ψ, θ, and ϕ to define the orientation of Frame V with respect to the inertial Frame I. In addition, as in Chapter 2, let the vector defining the angular velocity of Frame V with respect to the Frame I be

$$\boldsymbol{\omega}_{V,I} = P\mathbf{i}_V + Q\mathbf{j}_V + R\mathbf{k}_V \tag{4.39}$$

And so the Euler equations relating the angular rates are from Equations (2.36)

$$\begin{aligned} P &= \dot{\phi} - \dot{\psi}\sin\theta \\ Q &= \dot{\psi}\cos\theta\sin\phi + \dot{\theta}\cos\phi \\ R &= \dot{\psi}\cos\theta\cos\phi - \dot{\theta}\sin\phi \end{aligned}$$

or from Equation (2.37)

$$\begin{aligned} \dot{\phi} &= P + Q\sin\phi\tan\theta + R\cos\phi\tan\theta \\ \dot{\theta} &= Q\cos\phi - R\sin\phi \\ \dot{\psi} &= \left(Q\sin\phi + R\cos\phi\right)\sec\theta \end{aligned} \tag{4.40}$$

The generalized coordinates of the system (vehicle motion) may then be chosen to be

$$\mathbf{q} = \left\{X_I \quad Y_I \quad Z_I \quad \phi \quad \theta \quad \psi \quad \eta_i, i = 1, 2, \cdots\right\}^T \tag{4.41}$$

Using the vector components defined in Equations (4.38–4.39), we may write the kinetic energy, Equation (4.31), as

$$\boxed{\begin{aligned} \mathrm{T} &= \frac{1}{2}\{\dot{X}_I \quad \dot{Y}_I \quad \dot{Z}_I\}m\begin{Bmatrix}\dot{X}_I\\ \dot{Y}_I\\ \dot{Z}_I\end{Bmatrix} + \frac{1}{2}\{P \quad Q \quad R\}[\mathbf{I}]\begin{Bmatrix}P\\ Q\\ R\end{Bmatrix} + \frac{1}{2}\sum_{i=1}^{n}\mathcal{M}_i\dot{\eta}_i^2 \\ &= \left(\textit{inertial translation}\right) + \left(\textit{inertial rotation}\right) + \left(\textit{elastic displacement}\right) \end{aligned}} \tag{4.42}$$

Note that as a result of using the (mutually orthogonal) normal modes, the kinetic energy consists of just three decoupled terms.

For the gravitational potential energy we have simply

$$\boxed{\mathrm{U}_g = -m\mathbf{p}_V\cdot\mathbf{g} = -mgZ_I = mgh} \tag{4.43}$$

And for the strain energy, from Equation (4.35), we have

$$\boxed{\mathrm{U}_e = \frac{1}{2}\sum_{i=1}^{n}\omega_i^2\eta_i^2(t)\mathcal{M}_i} \tag{4.44}$$

4.5 EQUATIONS OF MOTION GOVERNING RIGID-BODY TRANSLATION

Using the Euler equations, Equations (4.40), one can clearly see that the kinetic energy, Equation (4.42), can be expressed in terms of the generalized coordinates defined in Equation (4.41). Lagrange's equation, Equation (4.1), may now be applied directly to obtain the equations of motion.

More specifically, for a constant-mass vehicle, consider for now only the inertial coordinates X_I, Y_I, and Z_I and kinetic energy. Applying Lagrange's equation we have

$$\begin{aligned}\frac{d}{dt}\left(\frac{\partial \mathrm{T}}{\partial \dot{X}_I}\right) &= \frac{d}{dt}(m\dot{X}_I) = m\ddot{X}_I \\ \frac{d}{dt}\left(\frac{\partial \mathrm{T}}{\partial \dot{Y}_I}\right) &= \frac{d}{dt}(m\dot{Y}_I) = m\ddot{Y}_I \\ \frac{d}{dt}\left(\frac{\partial \mathrm{T}}{\partial \dot{Z}_I}\right) &= \frac{d}{dt}(m\dot{Z}_I) = m\ddot{Z}_I\end{aligned} \tag{4.45}$$

Considering the gravitational potential energy and the same inertial coordinates X_I, Y_I, and Z_I we have

$$\frac{\partial \mathrm{U}_g}{\partial X_I} = \frac{\partial \mathrm{U}_g}{\partial Y_I} = 0, \quad \frac{\partial \mathrm{U}_g}{\partial Z_I} = -mg \tag{4.46}$$

Consequently, including the generalized forces, the three equations of motion associated with rigid-body translation are

$$\begin{aligned} m\ddot{X}_I &= Q_X \\ m\ddot{Y}_I &= Q_Y \\ m\ddot{Z}_I - mg &= Q_Z \end{aligned} \tag{4.47}$$

But the above equations, though valid, are certainly not in the most useful form for flight-dynamics analysis. To address this issue, rewrite Equations (4.47) using Equations (4.45–4.46) as follows:

$$m\frac{d}{dt}\{\dot{X}_I \quad \dot{Y}_I \quad \dot{Z}_I\} = \{Q_X \quad Q_Y \quad Q_Z\} + \{0 \quad 0 \quad mg\}$$

Now post-multipling the left and rights sides of the above with the set of unit vectors for the inertial Frame I yields the following vector equation of motion

$$m\frac{d}{dt}\Big|_I\{\dot{X}_I \quad \dot{Y}_I \quad \dot{Z}_I\}\begin{Bmatrix}\mathbf{i}_I\\ \mathbf{j}_I\\ \mathbf{k}_I\end{Bmatrix} = \{Q_X \quad Q_Y \quad Q_Z\}\begin{Bmatrix}\mathbf{i}_I\\ \mathbf{j}_I\\ \mathbf{k}_I\end{Bmatrix} + \{0 \quad 0 \quad mg\}\begin{Bmatrix}\mathbf{i}_I\\ \mathbf{j}_I\\ \mathbf{k}_I\end{Bmatrix} \tag{4.48}$$

Note that the derivative on the left-hand side is simply the time rate of change of the inertial-velocity vector with respect to the inertial frame (compare to Equations (4.45)). That is,

$$m\frac{d}{dt}\Big|_I\{\dot{X}_I \quad \dot{Y}_I \quad \dot{Z}_I\}\begin{Bmatrix}\mathbf{i}_I\\ \mathbf{j}_I\\ \mathbf{k}_I\end{Bmatrix} = m\{\ddot{X}_I \quad \ddot{Y}_I \quad \ddot{Z}_I\}\begin{Bmatrix}\mathbf{i}_I\\ \mathbf{j}_I\\ \mathbf{k}_I\end{Bmatrix} = m\frac{d}{dt}\Big|_I\left(\frac{d\mathbf{p}_V}{dt}\Big|_I\right) \tag{4.49}$$

So we now have the translational equation of motion in vector form given by

$$m\frac{d}{dt}\Big|_I\left(\frac{d\mathbf{p}_V}{dt}\Big|_I\right) = \{Q_X \quad Q_Y \quad Q_Z\}\begin{Bmatrix}\mathbf{i}_I\\ \mathbf{j}_I\\ \mathbf{k}_I\end{Bmatrix} + \{0 \quad 0 \quad mg\}\begin{Bmatrix}\mathbf{i}_I\\ \mathbf{j}_I\\ \mathbf{k}_I\end{Bmatrix} \tag{4.50}$$

To address the generalized forces on the right-hand side of Equation (4.50), we will first consider the generalized force vector comprising the first term on the right-hand side. Let the total resultant force $\mathbf{F}$ acting on the vehicle due to aerodynamic and propulsive effects be expressed as

$$\mathbf{F} = F_{X_V}\mathbf{i}_V + F_{Y_V}\mathbf{j}_V + F_{Z_V}\mathbf{k}_V \tag{4.51}$$

and note that the above force components are in the vehicle-fixed Frame V. Consequently, the virtual work associated with this force may be written

$$\delta W_{\mathbf{F}} = F_{X_V}\delta x + F_{Y_V}\delta y + F_{Z_V}\delta z = \{F_{X_V} \quad F_{Y_V} \quad F_{Z_V}\}\begin{Bmatrix}\delta x\\ \delta y\\ \delta z\end{Bmatrix} \tag{4.52}$$

where δx, δy, and δz are virtual displacements in the x, y, and z directions of Frame V, respectively. Recalling that the generalized force associated with the i'th generalized coordinate is defined as

$$Q_{q_i} = \frac{\partial(\delta W)}{\partial(\delta q_i)}$$

the generalized-force vector acting on the vehicle must be

$$\{Q_{X_V} \quad Q_{Y_V} \quad Q_{Z_V}\}\begin{Bmatrix}\mathbf{i}_V\\ \mathbf{j}_V\\ \mathbf{k}_V\end{Bmatrix} = \{F_{X_V} \quad F_{Y_V} \quad F_{Z_V}\}\begin{Bmatrix}\mathbf{i}_V\\ \mathbf{j}_V\\ \mathbf{k}_V\end{Bmatrix} \tag{4.53}$$

This expression may now be substituted for the generalized-force vector appearing as the first term on the right-hand side of Equation (4.50).

The above assertion may be verified through the following argument. The virtual displacements in the vehicle-fixed directions are related to virtual displacements in the inertial directions through the same direction-cosine matrix introduced in Chapter 1 that relates unit vectors in Frame V to unit vectors in Frame I. Specifically,

$$
\begin{Bmatrix} \delta x \\ \delta y \\ \delta z \end{Bmatrix} = \begin{bmatrix} 1 & 0 & 0 \\ 0 & \cos\phi & \sin\phi \\ 0 & -\sin\phi & \cos\phi \end{bmatrix} \begin{bmatrix} \cos\theta & 0 & -\sin\theta \\ 0 & 1 & 0 \\ \sin\theta & 0 & \cos\theta \end{bmatrix} \begin{bmatrix} \cos\psi & \sin\psi & 0 \\ -\sin\psi & \cos\psi & 0 \\ 0 & 0 & 1 \end{bmatrix} \begin{Bmatrix} \delta X_I \\ \delta Y_I \\ \delta Z_I \end{Bmatrix}
$$

$$
= \begin{bmatrix} (\cos\theta\cos\psi) & (\cos\theta\sin\psi) & (-\sin\theta) \\ (\sin\phi\sin\theta\cos\psi - \cos\phi\sin\psi) & (\sin\phi\sin\theta\sin\psi + \cos\phi\cos\psi) & (\sin\phi\cos\theta) \\ (\cos\phi\sin\theta\cos\psi + \sin\phi\sin\psi) & (\cos\phi\sin\theta\sin\psi - \sin\phi\cos\psi) & (\cos\phi\cos\theta) \end{bmatrix} \begin{Bmatrix} \delta X_I \\ \delta Y_I \\ \delta Z_I \end{Bmatrix} \tag{4.54}
$$

$$
= \mathbf{T}_{I\text{-}V}(\phi, \theta, \psi) \begin{Bmatrix} \delta X_I \\ \delta Y_I \\ \delta Z_I \end{Bmatrix}
$$

So substituting Equations (4.54) into the expression for virtual work due to the force **F**, or Equation (4.52), we can now express the virtual work in terms of virtual displacements in the inertial directions, or

$$
\delta W_{\mathbf{F}} = \{F_{X_V} \quad F_{Y_V} \quad F_{Z_V}\}[\mathbf{T}_{I\text{-}V}(\phi, \theta, \psi)] \begin{Bmatrix} \delta X_I \\ \delta Y_I \\ \delta Z_I \end{Bmatrix} \tag{4.55}
$$

Consequently, the generalized forces associated with virtual displacements in the <u>inertial</u> (X_I, Y_I, Z_I) directions are

$$
\begin{aligned}
Q_X &= F_{X_V}(\cos\theta\cos\psi) + F_{Y_V}(\sin\phi\sin\theta\cos\psi - \cos\phi\sin\psi) \\
&\quad + F_{Z_V}(\cos\phi\sin\theta\cos\psi + \sin\phi\sin\psi) \\
Q_Y &= F_{X_V}(\cos\theta\sin\psi) + F_{Y_V}(\sin\phi\sin\theta\sin\psi + \cos\phi\cos\psi) \\
&\quad + F_{Z_V}(\cos\phi\sin\theta\sin\psi - \sin\phi\cos\psi) \\
Q_Y &= F_{X_V}(-\sin\theta) + F_{Y_V}(\sin\phi\cos\theta) + F_{Z_V}(\cos\phi\cos\theta)
\end{aligned} \tag{4.56}
$$

Additionally, from Chapter 1, the unit vectors in Frame I are related to those in Frame V through the same direction-cosine matrix, or

$$
\begin{Bmatrix} \mathbf{i}_I \\ \mathbf{j}_I \\ \mathbf{k}_I \end{Bmatrix} = \mathbf{T}^T_{I\text{-}V}(\phi, \theta, \psi) \begin{Bmatrix} \mathbf{i}_V \\ \mathbf{j}_V \\ \mathbf{k}_V \end{Bmatrix} \tag{4.57}
$$

where $\mathbf{T}_{I\text{-}V}(\phi, \theta, \psi,)$ is defined in Equation (4.54). Consequently, the first term on the right-hand side of Equation (4.50) becomes

$$
\begin{aligned}
\{Q_X \quad Q_Y \quad Q_Z\} \begin{Bmatrix} \mathbf{i}_I \\ \mathbf{j}_I \\ \mathbf{k}_I \end{Bmatrix} &= \{F_{X_V} \quad F_{Y_V} \quad F_{Z_V}\}[\mathbf{T}_{I\text{-}V}(\phi, \theta, \psi)][\mathbf{T}^T_{I\text{-}V}(\phi, \theta, \psi)] \begin{Bmatrix} \mathbf{i}_V \\ \mathbf{j}_V \\ \mathbf{k}_V \end{Bmatrix} \\
&= \{F_{X_V} \quad F_{Y_V} \quad F_{Z_V}\} \begin{Bmatrix} \mathbf{i}_V \\ \mathbf{j}_V \\ \mathbf{k}_V \end{Bmatrix}
\end{aligned} \tag{4.58}
$$

as claimed. The product $\mathbf{T}_{I\text{-}V}\mathbf{T}_{I\text{-}V}^T$ equals the identity matrix since, as discussed in Chapter 1, $\mathbf{T}_{I\text{-}V}$ is an orthogonal matrix and its transpose equals its inverse.

Now the last term in Equation (4.50) may be written

$$\begin{aligned}\{0 \quad 0 \quad mg\}\begin{Bmatrix}\mathbf{i}_I\\ \mathbf{j}_I\\ \mathbf{k}_I\end{Bmatrix} &= \{0 \quad 0 \quad mg\}\mathbf{T}_{I\text{-}V}^T(\phi,\theta,\psi)\begin{Bmatrix}\mathbf{i}_V\\ \mathbf{j}_V\\ \mathbf{k}_V\end{Bmatrix}\\ &= \{-mg\sin\theta \quad mg\cos\theta\sin\phi \quad mg\cos\theta\cos\phi\}\begin{Bmatrix}\mathbf{i}_V\\ \mathbf{j}_V\\ \mathbf{k}_V\end{Bmatrix}\end{aligned} \tag{4.59}$$

and the vector equation of motion becomes

$$\begin{aligned}m\frac{d}{dt}\Big|_I\left(\frac{d\mathbf{p}_V}{dt}\Big|_I\right) &= \{F_{X_V} \quad F_{Y_V} \quad F_{Z_V}\}\begin{Bmatrix}\mathbf{i}_V\\ \mathbf{j}_V\\ \mathbf{k}_V\end{Bmatrix}\\ &\quad + \{-mg\sin\theta \quad mg\cos\theta\sin\phi \quad mg\cos\theta\cos\phi\}\begin{Bmatrix}\mathbf{i}_V\\ \mathbf{j}_V\\ \mathbf{k}_V\end{Bmatrix}\end{aligned} \tag{4.60}$$

But if we write

$$\frac{d\mathbf{p}_V}{dt}\Big|_I \triangleq \mathbf{V}_V = \{U \quad V \quad W\}\begin{Bmatrix}\mathbf{i}_V\\ \mathbf{j}_V\\ \mathbf{k}_V\end{Bmatrix} \tag{4.61}$$

and note that

$$\frac{d\mathbf{V}_V}{dt}\Big|_I = \frac{d\mathbf{V}_V}{dt}\Big|_V + \boldsymbol{\omega}_{V,I} \times \mathbf{V}_V$$

we have the following vector equation of motion from Equation (4.60).

$$\begin{aligned}&m\{(\dot{U} - VR + WQ) \quad (\dot{V} + UR - WP) \quad (\dot{W} - UQ + VP)\}\begin{Bmatrix}\mathbf{i}_V\\ \mathbf{j}_V\\ \mathbf{k}_V\end{Bmatrix}\\ &= \{(F_{X_V} - mg\sin\theta) \quad (F_{Y_V} + mg\cos\theta\sin\phi) \quad (F_{Z_V} + mg\cos\theta\cos\phi)\}\begin{Bmatrix}\mathbf{i}_V\\ \mathbf{j}_V\\ \mathbf{k}_V\end{Bmatrix}\end{aligned} \tag{4.62}$$

Finally, equating the $\mathbf{i}_V$, $\mathbf{j}_V$, and $\mathbf{k}_V$ components in the above vector equation yields the desired scalar equations of motion governing vehicle translation.

$$\begin{aligned}\dot{U} - VR + WQ &= -mg\sin\theta + F_{X_V}\\ \dot{V} + UR - WP &= mg\cos\theta\sin\phi + F_{Y_V}\\ \dot{W} - UQ + VP &= mg\cos\theta\cos\phi + F_{Z_V}\end{aligned} \tag{4.63}$$

Now the forces arise from two sources, aerodynamic and propulsive effects. Therefore, we can write

$$F_{X_V} = F_{A_X} + F_{P_X} \qquad F_{Y_V} = F_{A_Y} + F_{P_Y} \qquad F_{Z_V} = F_{A_Z} + F_{P_Z} \tag{4.64}$$

Substituting these three expressions into Equations (4.63), the equations of motion governing rigid-body translation (i.e., the translation of Frame V) are

$$\boxed{\begin{aligned} m(\dot{U} - VR + WQ) &= -mg\sin\theta + F_{A_X} + F_{P_X} \\ m(\dot{V} + UR - WP) &= mg\cos\theta\sin\phi + F_{A_Y} + F_{P_Y} \\ m(\dot{W} - UQ + VP) &= mg\cos\theta\cos\phi + F_{A_Z} + F_{P_Z} \end{aligned}} \tag{4.65}$$

These, of course, are identical to Equations (2.22) that govern the translation of the rigid vehicle. Any effects of elasticity on these equations will appear through the effects on the forces on the right-hand sides. These forces will be discussed in more detail in Chapters 6 and 7.

4.6 EQUATIONS OF MOTION GOVERNING RIGID-BODY ROTATION

We will now consider the next three generalized coordinates: the Euler angles ϕ, θ, and ψ. But we must again do so with care since these angles are defined as rotations about three axes in three different coordinate frames.

Recall from Chapter 2, Equations (2.32), that the angular velocity of Frame V with respect to the inertial Frame I may be written as

$$\boldsymbol{\omega}_{V,I} = \dot{\phi}\,\mathbf{i}_V + \dot{\theta}\,\mathbf{j}_2 + \dot{\psi}\mathbf{k}_I \tag{4.66}$$

which also defines the axes of rotations. The intermediate Frame 2 arises due to the ordered rotations of the Euler angles. Also, in Equation (4.39), repeated here, we defined the components of the angular-velocity vector expressed in Frame V to be

$$\boldsymbol{\omega}_{V,I} = P\mathbf{i}_V + Q\mathbf{j}_V + R\mathbf{k}_V$$

Finally recall from Equations (2.34) and (2.35), when developing the direction-cosine matrices the unit vectors are related by

$$\begin{aligned} \mathbf{j}_2 &= \cos\phi\,\mathbf{j}_V - \sin\phi\,\mathbf{k}_V \\ \mathbf{k}_I &= -\sin\theta\,\mathbf{i}_V + \sin\phi\cos\theta\,\mathbf{j}_V + \cos\phi\cos\theta\,\mathbf{k}_V \end{aligned}$$

From these facts we derived Equations (4.40), repeated here, that relate the rigid-body angular rates.

$$\begin{aligned} P &= \dot{\phi} - \dot{\psi}\sin\theta \\ Q &= \dot{\psi}\cos\theta\sin\phi + \dot{\theta}\cos\phi \\ R &= \dot{\psi}\cos\theta\cos\phi - \dot{\theta}\sin\phi \end{aligned}$$

It is convenient at this juncture to introduce the virtual angular displacements necessary for later expressing virtual work associated with torques and rotation. In terms of the Euler angles, we have virtual rotations $\delta\phi$, $\delta\theta$, and $\delta\psi$, which are rotations about axes in three different frames. Plus, we will now introduce three additional virtual rotations about the three axes of the vehicle-fixed Frame V. These rotations are denoted $\delta\phi_V$, $\delta\theta_V$ and $\delta\psi_V$. Just as we did above to relate the angular rates, we can again use Equations (2.34) and (2.35) to relate the two sets of virtual rotations, and we find that

$$\begin{aligned} \delta\phi_V &= \delta\phi - \delta\psi \sin\theta \\ \delta\theta_V &= \delta\psi \cos\theta \sin\phi + \delta\theta \cos\phi \\ \delta\psi_V &= \delta\psi \cos\theta \cos\phi - \delta\theta \sin\phi \end{aligned} \tag{4.67}$$

This is possible because just like angular rates, infinitesimal angular rotations may be expressed as vector quantities having direction defined with unit vectors. These three additional virtual-rotation angles will be used later in this section when discussing virtual work due to torques acting on the vehicle.

We will now apply Lagrange's equation to develop the equations of motion governing rotations of Frame V. Since the algebra involved can be onerous, it is convenient to use matrix and vector-array notation. Specifically, define

$$\begin{aligned} \{\mathbf{C}_\omega\} &= \{P \quad Q \quad R\}^T \\ \{\mathbf{q}_{\measuredangle}\} &= \{\phi \quad \theta \quad \psi\}^T \end{aligned} \tag{4.68}$$

where $\mathbf{C}_\omega$ consists of the coefficients of the vector $\boldsymbol{\omega}_{V,I}$ in Frame V, and $\mathbf{q}_{\measuredangle}$ includes the generalized coordinates consisting of the three Euler angles.

To apply Lagrange's equation, first note that

$$\frac{\partial \mathrm{T}}{\partial \mathbf{q}_{\measuredangle}} = \frac{\partial \mathrm{T}}{\partial \mathbf{C}_\omega} \frac{\partial \mathbf{C}_\omega}{\partial \mathbf{q}_{\measuredangle}} \tag{4.69}$$

where, from Equation (4.42), we see that[2]

$$\frac{\partial \mathrm{T}}{\partial \mathbf{C}_\omega} = \{\mathbf{C}_\omega\}^T [\mathbf{I}]$$

and from Equations (4.40) we have

$$\frac{\partial \mathbf{C}_\omega}{\partial \mathbf{q}_{\measuredangle}} = \begin{bmatrix} 0 & -\dot{\psi}\cos\theta & 0 \\ \dot{\psi}\cos\theta\cos\phi - \dot{\theta}\sin\phi & -\dot{\psi}\sin\theta\sin\phi & 0 \\ -\dot{\psi}\cos\theta\sin\phi - \dot{\theta}\cos\phi & -\dot{\psi}\sin\theta\cos\phi & 0 \end{bmatrix} = \begin{bmatrix} 0 & -\dot{\psi}\cos\theta & 0 \\ R & -\dot{\psi}\sin\theta\sin\phi & 0 \\ -Q & -\dot{\psi}\sin\theta\cos\phi & 0 \end{bmatrix}$$

---

[2] From linear algebra, given the quadratic form $S = \mathbf{x}^T\mathbf{M}\mathbf{x}$, with $\mathbf{x}$ an n vector and $\mathbf{M}$ an n-by-n matrix, then $dS/d\mathbf{x} = 2\mathbf{x}^T\mathbf{M}$.

The inertia matrix $\left[\mathbf{I}\right]$ is assumed constant, due to small structural deformations. Likewise, we can write

$$\frac{\partial \mathrm{T}}{\partial \dot{\mathbf{q}}_{\measuredangle}} = \frac{\partial \mathrm{T}}{\partial \mathbf{C}_{\omega}} \frac{\partial \mathbf{C}_{\omega}}{\partial \dot{\mathbf{q}}_{\measuredangle}} \tag{4.70}$$

with

$$\frac{\partial \mathbf{C}_{\omega}}{\partial \dot{\mathbf{q}}_{\measuredangle}} = \begin{bmatrix} 1 & 0 & -\sin\theta \\ 0 & \cos\phi & \cos\theta\sin\phi \\ 0 & -\sin\phi & \cos\theta\cos\phi \end{bmatrix}$$

We may now compute the terms called for in Lagrange's equation involving the kinetic energy, namely

$$\frac{d}{dt}\left(\frac{\partial \mathrm{T}}{\partial \dot{\mathbf{q}}_{\measuredangle}}\right) - \frac{\partial \mathrm{T}}{\partial \mathbf{q}_{\measuredangle}} \tag{4.71}$$

Carrying out the matrix multiplications, and with considerable algebra, one obtains

$$\begin{aligned}
&\frac{d}{dt}\left(\frac{\partial \mathrm{T}}{\partial \dot{\phi}}\right) - \frac{\partial \mathrm{T}}{\partial \phi} = C_1 \\
&\frac{d}{dt}\left(\frac{\partial \mathrm{T}}{\partial \dot{\theta}}\right) - \frac{\partial \mathrm{T}}{\partial \theta} = C_2\cos\phi - C_3\sin\theta \\
&\frac{d}{dt}\left(\frac{\partial \mathrm{T}}{\partial \dot{\psi}}\right) - \frac{\partial \mathrm{T}}{\partial \psi} = -C_1\sin\theta + C_2\cos\theta\sin\phi + C_3\cos\theta\cos\phi
\end{aligned} \tag{4.72}$$

where

$$\begin{aligned}
C_1 &= I_{xx}\dot{P} - \left(I_{yy} - I_{zz}\right)QR - I_{xy}\left(\dot{Q} - PR\right) - I_{yz}\left(Q^2 - R^2\right) - I_{xz}\left(\dot{R} + PQ\right) \\
C_2 &= I_{yy}\dot{Q} + \left(I_{xx} - I_{zz}\right)PR - I_{xy}\left(\dot{P} + QR\right) - I_{yz}\left(\dot{R} - PQ\right) + I_{xz}\left(P^2 - R^2\right) \\
C_3 &= I_{zz}\dot{R} + \left(I_{yy} - I_{xx}\right)PQ + I_{xy}\left(Q^2 - P^2\right) - I_{yz}\left(\dot{Q} + PR\right) - I_{xz}\left(\dot{P} - QR\right)
\end{aligned} \tag{4.73}$$

So, using Equations (4.71–4.73), we can determine the set of equations associated with the three Euler angles.

From Lagrange's equation, or

$$\frac{d}{dt}\left(\frac{\partial \mathrm{T}}{\partial \dot{\mathbf{q}}_{\measuredangle}}\right) - \frac{\partial \mathrm{T}}{\partial \mathbf{q}_{\measuredangle}} + \frac{\partial \mathrm{U}}{\partial \mathbf{q}_{\measuredangle}} = \mathbf{Q}_{\measuredangle}^{T}$$

the three equations of motion governing rotation become

$$\frac{d}{dt}\left(\frac{\partial \mathrm{T}}{\partial \dot{\phi}}\right) - \frac{\partial \mathrm{T}}{\partial \phi} + \cancel{\frac{\partial \mathrm{U}}{\partial \phi}} = C_1 = Q_{\phi}$$

$$\frac{d}{dt}\left(\frac{\partial T}{\partial \dot{\theta}}\right) - \frac{\partial T}{\partial \theta} + \cancel{\frac{\partial U}{\partial \theta}} = C_2 \cos\phi - C_3 \sin\theta = Q_\theta \tag{4.74}$$

$$\frac{d}{dt}\left(\frac{\partial T}{\partial \dot{\psi}}\right) - \frac{\partial T}{\partial \psi} + \cancel{\frac{\partial U}{\partial \psi}} = -C_1 \sin\theta + C_2 \cos\theta \sin\phi + C_3 \cos\theta \cos\phi = Q_\psi$$

(Note that the potential energy U is independent of the Euler angles.)

Now Equations (4.74) look nothing like the equations governing rigid-body rotations derived in Chapter 2. But Equations (4.74) can be rearranged and simplified. First, we must develop the expressions for the generalized forces on the right-hand sides of the equations.

Writing the resultant external moment on the vehicle in terms of its components about the three axes of the vehicle-fixed Frame V, let

$$\mathbf{M} = L\mathbf{i}_V + M\mathbf{j}_V + N\mathbf{k}_V \tag{4.75}$$

And hence the virtual work associated with this moment is

$$\delta W_{\mathbf{M}} = L\delta\phi_V + M\delta\theta_V + N\delta\psi_V \tag{4.76}$$

where again, virtual angular displacements about the unit vectors in Frame V are employed. Taking the approach used in Section 4.5, we substitute Equations (4.67) into the above, and noting that the generalized forces associated with the Euler angles are defined as

$$\mathbf{Q}_{\measuredangle}^T = \frac{\partial(\delta W_{\mathbf{M}})}{\partial(\mathbf{q}_{\measuredangle})}, \tag{4.77}$$

we obtain the following expressions for the generalized forces associated with the external moments acting on the vehicle

$$\begin{aligned} Q_\phi &= L \\ Q_\theta &= M\cos\phi - N\sin\phi \\ Q_\psi &= -L\sin\theta + M\cos\theta\sin\phi + N\cos\theta\cos\phi \end{aligned} \tag{4.78}$$

Therefore, from Equations (4.74) and (4.78), we have the rotational equations of motion

$$\begin{aligned} &C_1 = L \\ &C_2\cos\phi - C_3\sin\theta = M\cos\phi - N\sin\phi \\ &-C_1\sin\theta + C_2\cos\theta\sin\phi + C_3\cos\theta\cos\phi \\ &= -L\sin\theta + M\cos\theta\sin\phi + N\cos\theta\cos\phi \end{aligned} \tag{4.79}$$

with C_1, C_2, and C_3 given in Equations (4.73). By comparing the forms of the three expressions on the left-hand side of the above expressions with those on the right-hand side, and noting Equations (4.73), we see that another form of the equations of motion is

$$
\begin{aligned}
&I_{xx}\dot{P} - (I_{yy} - I_{zz})QR - I_{xy}(\dot{Q} - PR) - I_{yz}(Q^2 - R^2) - I_{xz}(\dot{R} + PQ) = L \\
&I_{yy}\dot{Q} + (I_{xx} - I_{zz})PR - I_{xy}(\dot{P} + QR) - I_{yz}(\dot{R} - PQ) + I_{xz}(P^2 - R^2) = M \\
&I_{zz}\dot{R} + (I_{yy} - I_{xx})PQ + I_{xy}(Q^2 - P^2) - I_{yz}(\dot{Q} + PR) - I_{xz}(\dot{P} - QR) = N
\end{aligned}
\tag{4.80}
$$

Now let the resultant moment on the vehicle arise only from aerodynamic and propulsive effects, or

$$\mathbf{M} = \mathbf{M}_{\text{Aero}} + \mathbf{M}_{\text{Prop}}$$

And as in Equations (2.19), define the components of $\mathbf{M}_{\text{Aero}}$ and $\mathbf{M}_{\text{Prop}}$ to be

$$
\begin{aligned}
\mathbf{M}_{\text{Aero}} &= L_A\mathbf{i}_V + M_A\mathbf{j}_V + N_A\mathbf{k}_V \\
\mathbf{M}_{\text{Prop}} &= L_P\mathbf{i}_V + M_P\mathbf{j}_V + N_P\mathbf{k}_V
\end{aligned}
\tag{4.81}
$$

Now the equations governing the rigid-body rotational motion of Frame V are

$$
\boxed{
\begin{aligned}
&I_{xx}\dot{P} - (I_{yy} - I_{zz})QR - I_{xy}(\dot{Q} - PR) - I_{yz}(Q^2 - R^2) - I_{xz}(\dot{R} + PQ) = L_A + L_P \\
&I_{yy}\dot{Q} + (I_{xx} - I_{zz})PR - I_{xy}(\dot{P} + QR) - I_{yz}(\dot{R} - PQ) + I_{xz}(P^2 - R^2) = M_A + M_P \\
&I_{zz}\dot{R} + (I_{yy} - I_{xx})PQ + I_{xy}(Q^2 - P^2) - I_{yz}(\dot{Q} + PR) - I_{xz}(\dot{P} - QR) = N_A + N_P
\end{aligned}
}
\tag{4.82}
$$

These equations, of course, are identical to Equations (2.27), developed in Chapter 2 for the rigid vehicle. And as with the equations governing translation, elastic effects will enter in the above equations through changes in the moments on the right-hand side. These moments will be discussed in more detail in Chapters 6 and 7.

4.7 EQUATIONS OF MOTION GOVERNING ELASTIC DEFORMATION

The equations of motion governing elastic deformation are the most straightforward to develop, due to the use of the vibration modes. Recall that the elastic deformation may be expressed in terms of the free-vibration mode shapes and modal coordinates. Or we have Equation (4.15), repeated here.

$$\mathbf{d}_E = \sum_{i=1}^{n} \boldsymbol{\nu}_i(x,y,z)\eta_i(t)$$

where $\boldsymbol{\nu}_i$ is the mode shape associated with the i'th free-vibration mode. Also recalling Equation (4.42), the kinetic energy of the elastic vehicle, and Equation (4.44), the elastic (potential) strain energy, both repeated here, we have

$$\mathrm{T} = \frac{1}{2}\{\dot{X}_I \quad \dot{Y}_I \quad \dot{Z}_I\}m\begin{Bmatrix}\dot{X}_I\\ \dot{Y}_I\\ \dot{Z}_I\end{Bmatrix} + \frac{1}{2}\{P \quad Q \quad R\}[\mathbf{I}]\begin{Bmatrix}P\\ Q\\ R\end{Bmatrix} + \frac{1}{2}\sum_{i=1}^{n}\mathcal{M}_i\dot{\eta}_i^2$$

and

$$\mathrm{U}_e = \frac{1}{2}\sum_{i=1}^{n}\omega_i^2\eta_i^2(t)\mathcal{M}_i$$

Using the modal coordinates, Lagrange's equation, given as

$$\frac{d}{dt}\left(\frac{\partial \mathrm{T}}{\partial \dot{\eta}_i}\right) - \frac{\partial \mathrm{T}}{\partial \eta_i} + \frac{\partial \mathrm{U}}{\partial \eta_i} = Q_i = \frac{\partial(\delta W)}{\partial(\delta \eta_i)} \tag{4.83}$$

may now be applied to obtain the n equations of motion governing these coordinates, which are

$$\boxed{\ddot{\eta}_i + \omega_i^2\eta_i = \frac{Q_i}{\mathcal{M}_i} \qquad i = 1\cdots n} \tag{4.84}$$

Here $\mathcal{M}_i$ is the generalized mass (see Equation (4.29)), and Q_i is the generalized force, each associated with the i'th free-vibration mode.

The generalized force indicated in Equation (4.84) is again derived from virtual work. Let the external pressure distribution acting on the surface of the vehicle's structure be denoted

$$\mathbf{P}(x,y,z)$$

where the local pressure $\mathbf{P}$ depends on the (x,y,z) location defined in Frame V, and $\mathbf{P}$ is a vector quantity having both magnitude and direction of action. Of course, integrating this pressure distribution over the surface of the vehicle leads to the resultant aerodynamic and propulsive forces and moments $\mathbf{F}_{\mathrm{Aero}}$, $\mathbf{F}_{\mathrm{Prop}}$, $\mathbf{M}_{\mathrm{Aero}}$, and $\mathbf{M}_{\mathrm{Prop}}$, with components given in Equations (4.64) and (4.81).

The local elastic virtual deformation of the structure may be expressed in terms of the modal expansion

$$\delta\mathbf{d}_E(x,y,z) = \sum_{i=1}^{n}\boldsymbol{\nu}_i(x,y,z)\delta\eta_i(t) \tag{4.85}$$

consistent with Equation (4.15). Therefore, the infinitesimal virtual work due to the pressure $\mathbf{P}$ acting at a point located at (x,y,z) on the structure is

$$d(\delta W_{\mathbf{P}}) = \mathbf{P}(x,y,z)\cdot\sum_{i=1}^{n}\boldsymbol{\nu}_i(x,y,z)\delta\eta_i(t)\,dS \tag{4.86}$$

where dS is an infinitesimal surface area. So the total virtual work performed by the pressure distribution becomes

$$\begin{aligned} \delta W_{\mathbf{P}} &= \int_{\text{Area}} \mathbf{P}(x,y,z) \cdot \sum_{i=1}^{n} \boldsymbol{\nu}_i(x,y,z)\delta\eta_i(t)\,dS \\ &= \sum_{i=1}^{n} \int_{\text{Area}} \mathbf{P}(x,y,z) \cdot \boldsymbol{\nu}_i(x,y,z)\,dS\,\delta\eta_i(t) \end{aligned} \tag{4.87}$$

Using Equation (4.87), the equations of motion, Equations (4.84), become

$$\ddot{\eta}_i + \omega_i^2 \eta_i = \frac{1}{\mathcal{M}_i} \int_{\text{Area}} \mathbf{P}(x,y,z) \cdot \boldsymbol{\nu}_i(x,y,z)\,dS \quad i = 1 \cdots n \tag{4.88}$$

These equations govern the deformation of the elastic vehicle. The discussion of the generalized forces on the right hand side will be discussed in more detail in Chapter 7.

EXAMPLE 4.1

Equations of Motion of a Flexible Missile

Consider a flexible missile operating outside the atmosphere with geometry shown in Figure 4.2. The only force acting on the vehicle is the propulsive thrust T, which can be vectored by deflecting the nozzle to generate moments about the Y axis of the vehicle. Let the missile length L, mass per unit length m, and thrust T all be normalized to unity, with the vehicle center of mass (gravity) at $x = L/2$. Also let the first vibration mode have a natural frequency of 1 rad/sec, and a mode shape given by

$$\nu_z(x) = \sin(\pi x/L) - 0.5$$

With this mode shape, the first generalized mass is 0.27 mass units. Including only this first vibration mode in the model, find the equations governing the motion of the vehicle in its XZ plane. Ignore the mass of the engine nozzle, and assume the initial bank angle and roll rate are zero.

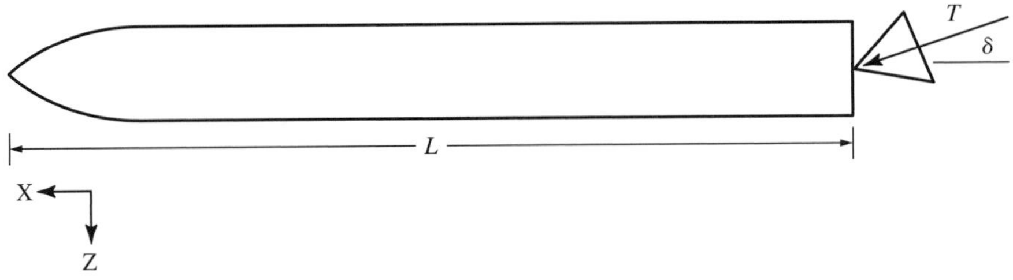

Figure 4.2 Missile schematic.

■ Solution

The pitch inertias of the vehicle about its Y and Z axes are

$$I_{yy} = I_{zz} = \int_0^L mx^2 dx = m\frac{L^3}{3} = \frac{1}{3}$$

while I_{xx} is small and all other products of inertia are zero. The forces acting on the vehicle in its XZ plane are

$$F_{P_X} = T\cos\delta = \cos\delta$$

$$F_{P_Z} = T\sin\delta = \sin\delta$$

Likewise, the relevant moments acting on the vehicle are

$$M_{P_X} = 0$$

$$M_{P_Y} = \frac{TL}{2}\sin\delta = \frac{1}{2}\sin\delta$$

Substituting for the inertias, moments, and zero roll rate in Equations (4.82), we have the rigid-body equations governing the angular rates

$$I_{xx}\dot{P} = 0$$

$$\frac{1}{3}\dot{Q} = \frac{1}{2}\sin\delta$$

From Equations (4.65) we also have the equations governing rigid-body translation given by

$$(\dot{U} + WQ) = -g\sin\theta + \cos\delta$$

$$(\dot{W} - UQ) = g\cos\theta + \sin\delta$$

And from Equations (4.40) with roll rate P and bank angle ϕ equal to zero, we have the equations governing the Euler angles

$$\dot{\phi} = 0$$

$$\dot{\theta} = Q$$

Finally, from Equation (4.84) we have

$$\ddot{\eta}_i + \omega_i^2\eta_i = \frac{Q_i}{\mathcal{M}_i}$$

where Q_i is the i'th generalized force given by

$$Q_i = \frac{\partial(\delta W)}{\partial(\delta\eta_i)}$$

With thrust the only force acting on the vehicle, the virtual work due to virtual transverse bending is

$$\delta W = (T\sin\delta)\delta z = (T\sin\delta)\nu_z(0)\delta\eta = (-0.5\sin\delta)\delta\eta$$

Consequently, the equation of motion governing the elastic degree of freedom is

$$\ddot{\eta} + \eta = \frac{-0.5\sin\delta}{0.27} = -1.85\sin\delta$$

The five differential equations above govern the motion of the elastic vehicle in its XZ plane, with two of these being trivial (which ones?).

4.8 MOTION OF A PARTICULAR POINT ON THE ELASTIC VEHICLE

We noted when discussing the mean axes that these axes are not fixed to a particular material point on the structure of the vehicle. Rather, the origin of the mean axes coincides with the vehicle's instantaneous center of mass, which, for a deforming structure, is not at a fixed point on the structure. This fact sometimes raises the question as how to locate or track a particular point on the vehicle's structure, such as the pilot's location or the location of a sensor such as an accelerometer. We shall see that a rather straightforward method exists to accomplish this task.

Let the location of the material point of interest on the structure be denoted by its inertial vector $\mathbf{p}'$, and consistent with the notation used in Figure 4.1, note that $\mathbf{p}' = \mathbf{p}_V + \mathbf{p}$. Recall that $\mathbf{p}_V$ is the inertial position of the origin of Frame V, which is at the instantaneous center of mass of the vehicle, and $\mathbf{p}$ is the position of the material point of interest on the structure relative to the origin of Frame V. Now, as introduced in Section 4.1, Equation (4.10), we can express the position of the point of interest as

$$\mathbf{p} = \mathbf{p}_{\mathrm{RB}} + \mathbf{d}_E(x,y,z,t)$$

where $\mathbf{p}_{\mathrm{RB}}$ is the location relative to the center of mass of that point of interest on the <u>undeformed vehicle</u>, and $\mathbf{d}_E$ represents the elastic displacement of the point of interest. Finally we know from the modal expansion for the structural deformation, Equation (4.15), that

$$\mathbf{d}_E(x,y,z,t) = \sum_{i=1}^{n} \mathbf{v}_i(x,y,z)\eta_i(t)$$

Therefore, the elastic deformation $\mathbf{d}_E$ of the point of interest is given in terms of the free-vibration mode shapes and modal coordinates.

The inertial position of the material point of interest may then be expressed as

$$\mathbf{p}' = \mathbf{p}_V + \mathbf{p}_{\mathrm{RB}} + \sum_{i=1}^{n} \mathbf{v}_i(x,y,z)\eta_i(t) \tag{4.89}$$

Note that each of the terms on the right-hand side of the above expression is either known or can be determined from the solution of the equations of motion derived in this chapter. First, by selecting the material point of interest in the structure, its location on the undeformed structure, $\mathbf{p}_{RB}$, is known. Since a solution to the unforced (free) vibration problem is assumed available throughout this chapter, the free-vibration mode shapes $\boldsymbol{\nu}_i$ are also known. This leaves $\mathbf{p}_V$ and $\eta_i(t)$ to be determined.

But $\mathbf{p}_V$ is the inertial position of the instantaneous center of mass, which is governed by the rigid-body translational equations, Equations (4.65), along with the kinematic equations arising from Equation (4.61). We now develop these kinematic equations. From Equation (4.61) we have

$$\frac{d\mathbf{p}_V}{dt}\bigg|_I \triangleq \mathbf{V}_V = \{U \quad V \quad W\}\begin{Bmatrix}\mathbf{i}_V\\ \mathbf{j}_V\\ \mathbf{k}_V\end{Bmatrix}$$

But since $\mathbf{V}_V$ may also be expressed as

$$\mathbf{V}_V = \{\dot{X}_I \quad \dot{Y}_I \quad \dot{Z}_I\}\begin{Bmatrix}\mathbf{i}_I\\ \mathbf{j}_I\\ \mathbf{k}_I\end{Bmatrix} \tag{4.90}$$

we have

$$\{\dot{X}_I \quad \dot{Y}_I \quad \dot{Z}_I\}\begin{Bmatrix}\mathbf{i}_I\\ \mathbf{j}_I\\ \mathbf{k}_I\end{Bmatrix} = \{U \quad V \quad W\}\begin{Bmatrix}\mathbf{i}_V\\ \mathbf{j}_V\\ \mathbf{k}_V\end{Bmatrix} \tag{4.91}$$

But from Equation (4.57) we know that the unit vectors in the two frames are related by the direction-cosine matrix $\mathbf{T}_{I\text{-}V}$, or

$$\begin{Bmatrix}\mathbf{i}_I\\ \mathbf{j}_I\\ \mathbf{k}_I\end{Bmatrix} = \mathbf{T}^T_{I\text{-}V}(\phi, \theta, \psi)\begin{Bmatrix}\mathbf{i}_V\\ \mathbf{j}_V\\ \mathbf{k}_V\end{Bmatrix}$$

where $\mathbf{T}_{I\text{-}V}$ is given in Equation (4.54). Therefore, the desired kinematic equations relating the translational-velocity components are

$$\{\dot{X}_I \quad \dot{Y}_I \quad \dot{Z}_I\}\begin{Bmatrix}\mathbf{i}_I\\ \mathbf{j}_I\\ \mathbf{k}_I\end{Bmatrix} = \{U \quad V \quad W\}\mathbf{T}_{I\text{-}V}(\phi, \theta, \psi)\begin{Bmatrix}\mathbf{i}_I\\ \mathbf{j}_I\\ \mathbf{k}_I\end{Bmatrix}$$

or

$$\begin{Bmatrix}\dot{X}_I\\ \dot{Y}_I\\ \dot{Z}_I\end{Bmatrix} = \mathbf{T}^T_{I\text{-}V}(\phi, \theta, \psi)\begin{Bmatrix}U\\ V\\ W\end{Bmatrix} \tag{4.92}$$

Thus, from the solutions of Equations (4.65) and (4.92), the position vector $\mathbf{p}_V$ may be determined. (Note that obtaining the solution of these two sets of equations involves the simultaneous solutions of Equations (4.65), (4.82) (governing rigid-body rotation), and Equations (4.40) (governing the three Euler angles). Obtaining such solutions is discussed in Chapter 8, which deals with simulation of the vehicle's dynamics.)

The final quantities in Equation (4.89) that must be determined are the modal coordinates $\eta_i(t)$. These, of course, are determined from the solutions of Equations (4.88). Therefore, the instantaneous position of any point on the vehicle, including the elastic deformation, may be determined from the solutions of the equations of motion derived in this chapter. And the instantaneous inertial velocity or acceleration of that location on the vehicle may likewise be determined.

EXAMPLE 4.2

Equations Governing Local Pitch Rate of a Flexible Missile

Consider the flexible missile addressed in Example 4.1. Assume a rate gyro is to be mounted at the nose of the vehicle to sense local pitch rate $\dot{\theta}_{\text{Sensed}} = \dot{\theta} + \dot{\theta}_{\text{Elastic}}$, where $\dot{\theta}$ is the rigid-body pitch rate and $\dot{\theta}_{\text{Elastic}}$ is the local pitch rate at the nose due to elastic deformation. Find the equation(s) that govern $\dot{\theta}_{\text{Sensed}}$.

■ Solution

First, from Example 4.1, the equations governing rigid-body pitch rate are

$$\frac{1}{3}\dot{Q} = \frac{1}{2}\sin\delta$$

$$\dot{\theta} = Q$$

Next, the local pitch deflection due to elastic deformation may be found from Equation (4.15), which, considering only z deflections and one vibration mode, reduces to

$$z(x) = \nu_z(x)\eta(t)$$

And recall that in Example 4.1, the mode shape was given as

$$\nu_z(x) = \sin(\pi x/L) - 0.5$$

Now note that the local elastic pitch deflection at location x along the vehicle may be written in terms of a local slope, as shown in Figure 4.3. Or

$$\theta_{\text{Elastic}}(x) = \frac{\partial z}{\partial x}(x) = \lim_{\delta x \to 0} \frac{\left(z(x+\delta x) - z(x)\right)}{\delta x} = \lim_{\delta x \to 0} \frac{\left(\nu_z(x+\delta x) - \nu_z(x)\right)}{\delta x}\eta(t) = \frac{\partial \nu_z(x)}{\partial x}\eta(t)$$

Therefore the local pitch rate due to the elastic deformation at the nose of the vehicle is

$$\dot{\theta}_{\text{Elastic}}(L) = \frac{\partial \nu_z(L)}{\partial x}\dot{\eta}(t) = \pi\cos\pi\,\dot{\eta}(t) = -\pi\dot{\eta}(t)$$

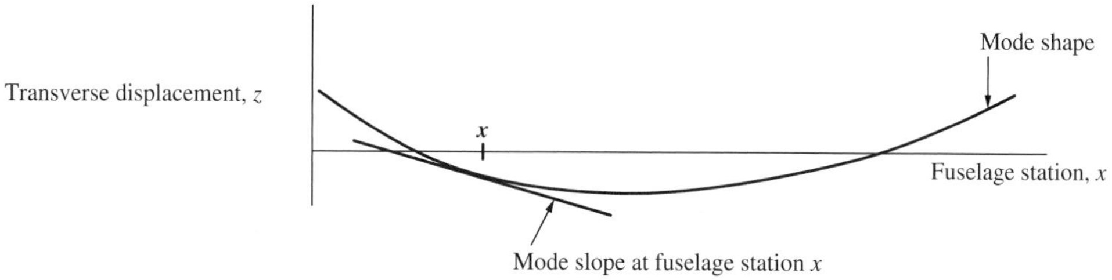

Figure 4.3 Local mode slope.

And from Example 4.1, the equation of motion governing the modal coordinate is

$$\ddot{\eta} + \eta = -1.85 \sin \delta$$

Consequently, the equations of motion governing the sensed pitch rate, given by

$$\dot{\theta}_{\text{Sensed}} = Q - \pi \dot{\eta}$$

are

$$\dot{Q} = \frac{3}{2} \sin \delta$$

$$\ddot{\eta} + \eta = -1.85 \sin \delta$$

4.9 REFERENCE AND PERTURBATION EQUATION SETS FOR PERTURBATION ANALYSIS

The complete set of nonlinear equations governing the motion of the elastic vehicle consists of Equations (4.40), (4.65), (4.82), and (4.88). If perturbation analysis is to be performed, as discussed in Chapter 1, both the Reference and the linearized Perturbation Sets of equations must be obtained. However, this is a relatively simple task since the equations governing both rigid-body translation and rotation (Equations (4.40), (4.65), and (4.82)) are identical to those for the rigid vehicle (Equations (2.22), (2.27), and (2.37)). This is one key advantage of using the mean-axis formulation of the equations of motion.

Since the nonlinear equations in Chapter 2 and in this chapter governing the rigid-body degrees of freedom are identical, the Reference and Perturbation Sets of equations are identical to those given in Chapter 2. Table 4.1 summarizes the sets of equations governing these rigid-body coordinates.

Table 4.1 Reference and Perturbation Equation Sets Governing Rigid-Body Coordinates

Nonlinear Equations	Reference Equation Sets	Perturbation Equation Sets
Equations (4.40), (4.65), (4.82)	Equations (2.44), (2.48), (2.52) or (2.53)	Equations (2.45), (2.49), (2.54) or (2.55)

To develop the Reference and Perturbation Sets of equations governing the deformation of the elastic vehicle, we begin with Equations (4.88). Introducing the change of variables for the modal coordinates and pressure distribution, let

$$\eta_i(t) = \mathrm{H}_{0_i}(t) + \eta_i(t) \text{ and } \mathbf{P}(x,y,z) = \mathbf{P}_0(x,y,z) + \mathbf{p}(x,y,z) \tag{4.93}$$

where the variables with zero subscript are the variables associated with the reference condition, and the other two variables without the zero subscript represent perturbations from the reference condition. Substituting this change of variables into Equations (4.88) yields

$$\begin{aligned}(\ddot{\mathrm{H}}_{0_i} + \ddot{\eta}_i) + \omega_i^2(\mathrm{H}_{0_i} + \eta_i) \\ = \frac{1}{\mathcal{M}_i}\int_{\text{Area}} \left(\mathbf{P}_0(x,y,z) + \mathbf{p}(x,y,z)\right) \cdot \boldsymbol{\nu}_i(x,y,z)dS, \quad i = 1\cdots n\end{aligned} \tag{4.94}$$

By inspection, we see that the Reference Set of equations is

$$\boxed{\ddot{\mathrm{H}}_{0_i} + \omega_i^2 \mathrm{H}_{0_i} = \frac{1}{\mathcal{M}_i}\int_{\text{Area}} \mathbf{P}_0(x,y,z)\cdot \boldsymbol{\nu}_i(x,y,z)dS, \quad i = 1\cdots n} \tag{4.95}$$

and the Perturbation Set is

$$\boxed{\ddot{\eta}_i + \omega_i^2 \eta_i = \frac{1}{\mathcal{M}_i}\int_{\text{Area}} \mathbf{p}(x,y,z)\cdot \boldsymbol{\nu}_i(x,y,z)dS, \quad i = 1\cdots n} \tag{4.96}$$

4.10 Summary

In this chapter we developed the equations governing the motion of flexible flight vehicles, including the equations governing the elastic deformation of that vehicle. Throughout the derivation, we assumed the existence of the solution to the unforced- (free-) vibration problem for the vehicle's structure. We used the modal coordinates arising in the free-vibration solution as generalized coordinates to describe the motion of the flexible vehicle. We introduced the concept of mean axes, and we demonstrated that by using the mean-axis derivation, the equations of motion are decoupled into those governing rigid-body degrees of freedom and those governing the elastic degrees of freedom. In addition, the equations of motion governing the rigid-body degrees of freedom are identical to those for a rigid vehicle, with elastic effects contributing only to the forces and moments.

4.11 Problems

4.1 Using the properties of modal orthogonality, along with the definitions of the inertia matrix, show that the kinetic energy of the elastic vehicle may be expressed as in Equation (4.42).

4.2 Derive Equations (4.67).

4.3 Perform the necessary algebra to derive Equations (4.72).

4.4 Assume an accelerometer is mounted at a particular location on a flexible aircraft, with the sensor location given in terms of its location on the undeformed vehicle structure. Let this position, defined relative to the center of mass, be denoted by the vector $\mathbf{p}_{\text{Accel}}$. Beginning with Equation (4.89), derive the equation(s) for the inertial acceleration at the location of the accelerometer.

References

1. Milne, R. D.: "Some Remarks on the Dynamics of Deformable Bodies," *AIAA Journal,* vol. 6, March 1968, p. 556.

2. Milne, R. D.: "Dynamics of the Deformable Airplane," Her Majesty's Stationary Office, Reports and Memoranda No. 3345, September 1962.

3. Waszak, M. R. and D. K. Schmidt: "Flight Dynamics of Aeroelastic Vehicles," *AIAA Journal of Aircraft,* vol. 25, no. 6, June 1988, pp. 563–571.

4. Dusto, A. R. et al.: "A Method for Predicting the Aeroelastic Characteristics of an Elastic Airplane, Vol. 1: FLEXSTAB Theoretical Description," NASA CR-114712, October 1974.

5. Etkin, B.: *Dynamics of Atmospheric Flight,* Wiley, New York, 1972.

6. Meirovitch, L.: *Analytical Methods in Vibrations,* Macmillan, New York, 1967.

5 CHAPTER

Basic Aerodynamics of Lifting Surfaces

Chapter Roadmap: *Students taking their first course in flight dynamics must be familiar with the material in this chapter, either from another course or by covering it in a flight dynamics course. The author has had success letting students read this chapter primarily on their own, with only a few suggestions from the instructor.*

At this juncture, and for the next three chapters, our attention turns to the modeling of the aerodynamic forces and moments acting on the vehicle. These forces and moments will then be incorporated into the vehicle's equations of motion, developed in earlier chapters, for further analysis.

There are three main approaches to the determination of a flight vehicle's aerodynamic characteristics—analytical, computational, and experimental. Analytical techniques range from sophisticated theoretical approaches to those based on simple empirical methods. Computational techniques include computer-based numerical methods to estimate pressure distributions. And experimental techniques include the use of wind tunnels and vehicle flight tests to measure or infer the forces, moments, and/or pressure distributions on the vehicle. The methods—analytical, computational, and experimental—are, in general, listed in the order of increasing costs (and time), and of increasing accuracy or "face validity."

Frequently, however, simple, empirically based analytical methods yield quite accurate results, and are quick and easy to use. As such, they are used often, especially in the preliminary or conceptual design of flight vehicles. Such methods aid in providing insight into the complex relationships between a vehicle's geometry and its aerodynamic characteristics. This chapter is based on such empirical or semi-empirical methods. A sample computational technique, including a MATLAB code that may be used for analysis of a wing and horizontal tail combination, is available from this book's website located at www.mhhe.com/schmidt.

The overall technique used in this book for estimating the aerodynamic characteristics of a flight vehicle is sometimes referred to as the *component build-up method*. That is, key characteristics of components of a wing or a vehicle are first estimated, and then these results are aggregated to estimate the characteristics of the complete wing or vehicle.

Basic concepts from aerodynamics and important notation used throughout this book are introduced in this chapter. This information is meant to supplement that from other course work, to provide additional perspective, and to offer some simple techniques for quickly estimating aerodynamic characteristics key to the study of flight dynamics. However, a student's mastery of the material presented here does not replace the need for additional course work in aerodynamics.

The methodologies included in this chapter will also not be sufficiently inclusive so as to allow the aerodynamic characterization of all flight vehicle geometries. Instead, we present techniques so that the student may preliminarily characterize the aerodynamics of common vehicle geometries, may better understand the general approach taken, and may gain qualitative insight into flight vehicle aerodynamic characteristics that are important in flight dynamics. More details on the methodology used may be obtained from Ref. 1 listed at the end of this chapter.[1]

5.1 SUBSONIC AIRFOIL SECTION CHARACTERISTICS

The basic component of a lifting surface is the airfoil cross section of that lifting surface. And wings, horizontal and vertical tails, canards (forward surfaces), and so on, are all lifting surfaces. A tremendous amount of work was performed by the National Advisory Committee on Aeronautics (NACA), the precursor of NASA, to characterize the aerodynamics of airfoils. This work led to a large catalog of airfoil aerodynamic characteristics, documented in terms of the two-dimensional airfoil cross section. The geometries of these cross sections were grouped into what became known as the *4-, 5-, and 6-digit NACA airfoil series.* Major aircraft companies like Boeing and Airbus have developed their own proprietary airfoils, but some general aviation aircraft companies still use the NACA airfoils. Regardless, the type of aerodynamic data associated with airfoil sections, discussed below, is all the same.

The important geometric properties of an arbitrary airfoil are depicted in Figure 5.1. The definitions of the parameters defining the geometry are given in Table 5.1. Of particular importance are chord length, c, thickness, t, mean line, and airfoil camber quantified in terms of the maximum distance between the mean line and chord, $y_{c_{max}}$. If the mean line lies on the chord line $\left(y_{c_{max}} = 0\right)$,

---

[1] In fact, most of the data and figures included in this chapter are reproduced from Ref. 1. Permission of the U.S. Air Force Flight Dynamics Laboratory to use this material is gratefully appreciated.

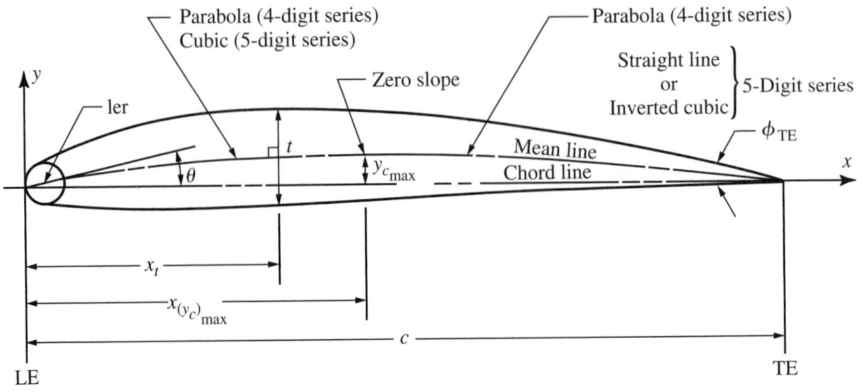

Figure 5.1 2-D airfoil cross-section geometry.

Table 5.1 Definitions of Airfoil Cross-Section Parameters.

Basic Symmetric Airfoil	Camber Mean Line
c = chord of airfoil section	$y_{c\max}$ = maximum ordinate of mean line
x = distance along chord measured from LE	$y_c(x)$ = shape of mean line
y = ordinate at some value of x (measured normal to and from the chord line for symmetric airfoils, measured normal to and from the mean line for cambered airfoils)	$x_{y_{c\max}}$ = position of maximum camber
	θ = slope of ler through LE equals the slope of the mean line at the LE
$y(x)$ = thickness distribution of airfoil	c_l = section lift coefficient
t = maximum thickness of airfoil	c_{l_i} = design section lift coefficient
x_t = position of maximum thickness	
ler = leading-edge radius	
ϕ_{TE} = trailing-edge angle (included angle between the tangents to the upper and lower surfaces at the trailing edge)	

the airfoil is symmetric about its chord. Note the origin of the x,y airfoil coordinate system is at the leading edge of the chord, and that x is positive aft.

Air passing over the airfoil creates a pressure distribution acting on the airfoil's upper and lower surfaces, as depicted in Figure 5.2. The net pressure between the upper and lower surfaces, when integrated over the chord length, gives rise to a force per unit span (into the page). This force F, shown in Figure 5.3, is usually decomposed into two components, lift, L, and drag, D. Lift, L, is defined as that force component normal to the free-stream velocity vector, $\mathbf{V}_\infty$, while drag, D, is that component acting along the direction of free-stream velocity. Also defined in

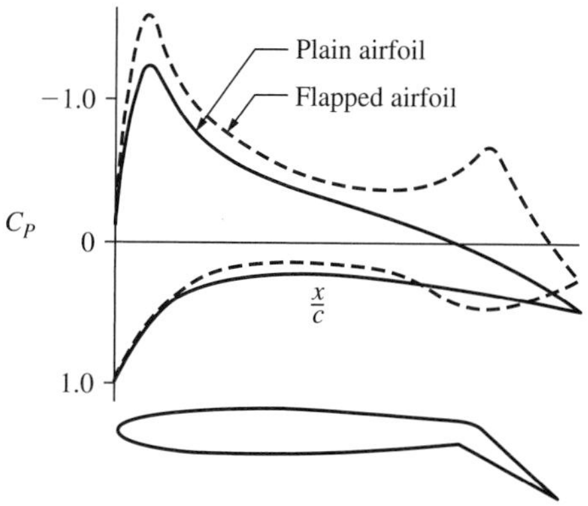

Figure 5.2 Pressure distributions over a plain and flapped airfoil.

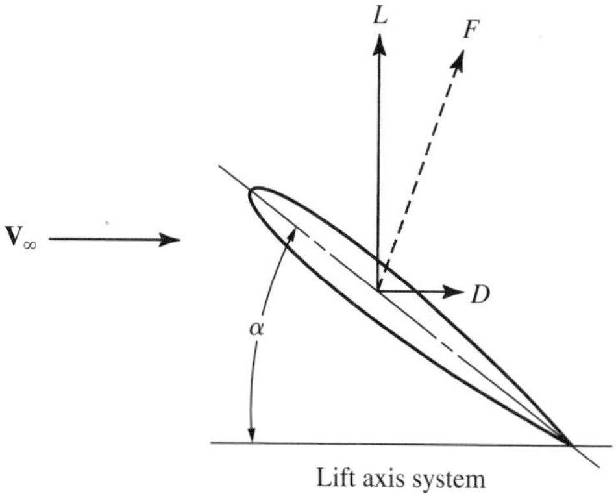

Figure 5.3 Lift and drag components of the resultant aerodynamic force.

this figure is the aerodynamic *angle of attack,* α, which is the angle between the airfoil chord line and the free-stream velocity, $\mathbf{V}_\infty$.

Aerodynamic forces and moments are conveniently expressed in terms of nondimensional coefficients. Use of such coefficients allows for easy scaling of forces between the full-scale vehicle and a smaller vehicle model of identical shape, for example. A force coefficient, C_F, is defined by the relation

$$F = C_F q_\infty S$$

where F is the force in question and q_∞ is the dynamic pressure of the free-stream air flow. In terms of the free-stream air density ρ_∞, this dynamic pressure is given as

$$q_\infty = \frac{1}{2}\rho_\infty V_\infty^2 \tag{5.1}$$

Here S is a reference area chosen in the nondimensionalization process. (Think of a force measured in a wind tunnel, and the force coefficient being obtained through nondimensionalization.) A moment coefficient, C_M, is likewise defined by the relation

$$M = C_M q_\infty S l$$

where M is the moment in question, and l is a reference length also chosen in the nondimensionalization process. And finally, a pressure coefficient, C_p, is defined by the relation

$$C_p = \frac{p - p_\infty}{q_\infty}$$

Critical aerodynamic characteristics of an airfoil include the lift, drag, and pitching moment generated, and the location of its *aerodynamic center.* As noted, the forces and moment arise from integrating the pressure distribution acting on the airfoil. The aerodynamic center, x_{ac}, is a specific point along the chord of the airfoil about which the aerodynamic pitching moment is invariant with angle of attack. In contrast, the *center of pressure,* or x_{cp}, is that location along the chord about which the aerodynamic pitching moment is zero. The lift, drag, and pitching moment are frequently presented graphically, plotted versus airfoil angle of attack. This important graphical data will be discussed below.

NOTE TO STUDENT

But first consider an important note on notation. When we are dealing with 2-D section aerodynamic characteristics, we will use lower-case letters for the forces, moments, and all coefficients. Upper case letters will be reserved for the characteristics of 3-D components such as wings or the entire vehicle.

5.1.1 Section Lift and Drag

Let us now consider the lift and drag characteristics of 2-D airfoil sections. The lift characteristics are plotted schematically in Figure 5.4, in which the section lift coefficient, c_l, is plotted versus section angle of attack, α. Note that over a rather wide range of angle of attack, up to the angle of attack denoted α^*, the section lift coefficient varies linearly with α, with a slope denoted by c_{l_α}. The lift coefficient achieves its maximum, $c_{l_{\max}}$, at an angle of attack denoted $\alpha_{c_{l_{\max}}}$. Finally, the angle of attack at which the lift coefficient is zero is denoted α_0. By knowing these five quantities for a particular airfoil section, the curve depicting the lift coefficient versus angle of attack may easily be sketched.

In flight dynamics, the most important of the parameters just cited is the slope of the lift curve, c_{l_α}, or *lift effectiveness,* which is defined as

$$\boxed{c_{l_\alpha} \triangleq \frac{\partial c_l}{\partial \alpha}} \tag{5.2}$$

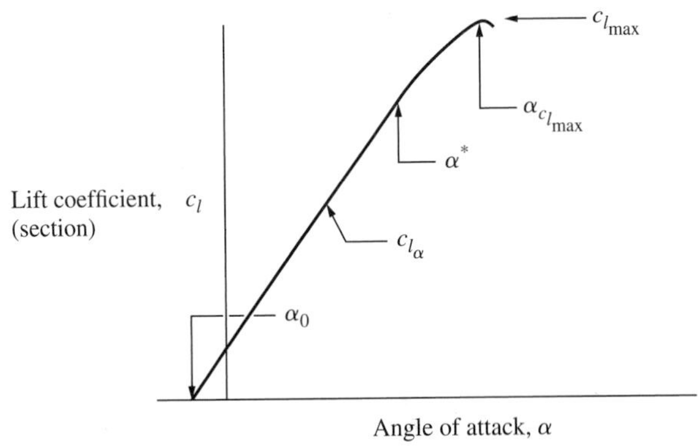

Figure 5.4 Section lift coefficient versus angle of attack.

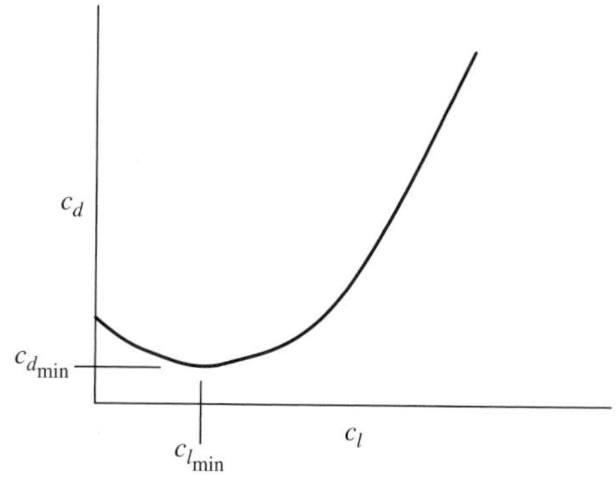

Figure 5.5 Section drag polar.

Under incompressible thin-airfoil theory, an airfoil's section lift effectiveness $c_{l_\alpha} = 2\pi$ /rad. For subsonic flow (Mach $= M_\infty < 1.0$) and including compressibility effects, the section lift effectiveness is given by

$$c_{l_\alpha}\big|_{\text{theory}} = \frac{2\pi}{\sqrt{1 - M_\infty^2}}, \; M_\infty < 1.0 \tag{5.3}$$

where the term $\sqrt{1 - M_\infty^2}$ is the *Prandl–Glauert subsonic compressibility factor.* Supersonically ($M_\infty > 1.0$), thin-airfoil theory also yields

$$c_{l_\alpha}\big|_{\text{theory}} = \frac{4}{\sqrt{M_\infty^2 - 1}}, \; M_\infty > 1.0 \tag{5.4}$$

where $\sqrt{M_\infty^2 - 1}$ is the *Prandl–Glauert supersonic compressibility factor.*

Finally, the section drag coefficient, c_d, is usually shown plotted versus angle of attack or lift coefficient. This plot is called the *drag polar* of the section, and it is somewhat parabolic in shape. A typical section drag polar is shown in Figure 5.5. In this figure the minimum drag coefficient is denoted $c_{d_{\min}}$, and the lift coefficient corresponding to the minimum drag is denoted $c_{l_{\min}}$.

5.1.2 Section Pitching Moment

The pitching moment created by the pressure distribution over the airfoil is another key characteristic. This moment, taken about a point along the chord line, is shown schematically in Figure 5.6, and may be expressed in terms of a section

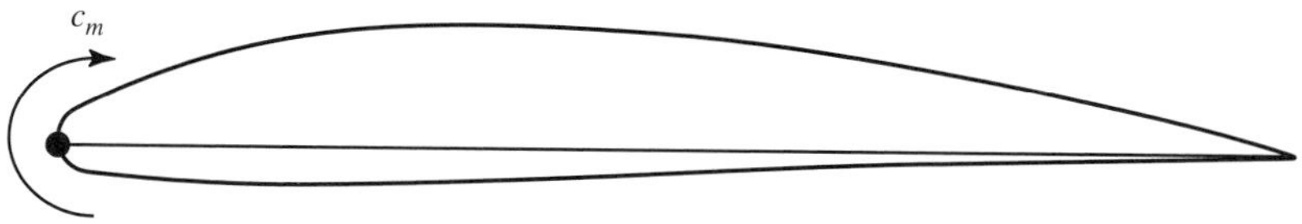

Figure 5.6 Section aerodynamic pitching moment taken about the leading edge.

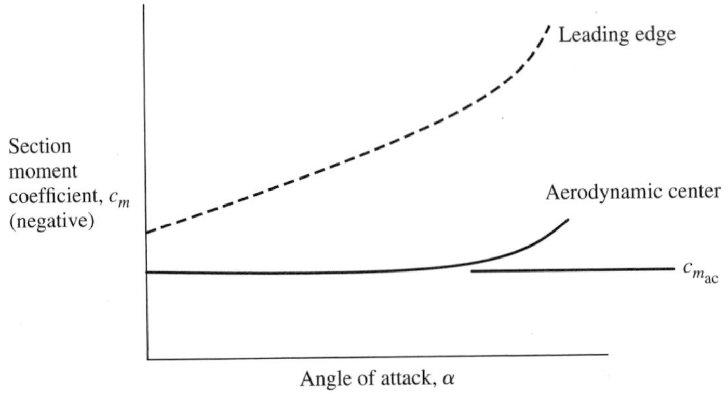

Figure 5.7 Section aerodynamic pitching-moment coefficient.

moment coefficient, c_m. This coefficient is shown plotted against angle of attack in Figure 5.7. Two curves are shown, one for the moment taken about the leading edge of the airfoil, and a second for the moment taken about the airfoil's aerodynamic center. Again, the aerodynamic center of an airfoil section is that location along the chord about which the moment is invariant with angle of attack, or

$$\boxed{\frac{\partial c_m}{\partial \alpha}\Big|_{x_{ac}} = 0} \tag{5.5}$$

Incompressible thin-airfoil theory indicates that a section's aerodynamic center lies at its quarter-chord point. The section moment coefficient about the section's aerodynamic center is denoted $c_{m_{ac}}$.

If the lift per unit span (into the page), l, and moment (per unit span), m_1, about a point on the chord x_1 are known, the moment (per unit span) at any other point along the chord x_2 may be determined. From Figure 5.8, note that

$$m_2 = m_1 + l(x_2 - x_1)$$

or

$$c_{m_2} q_\infty c c = c_{m_1} q_\infty c c + c_l q_\infty c (x_2 - x_1)$$

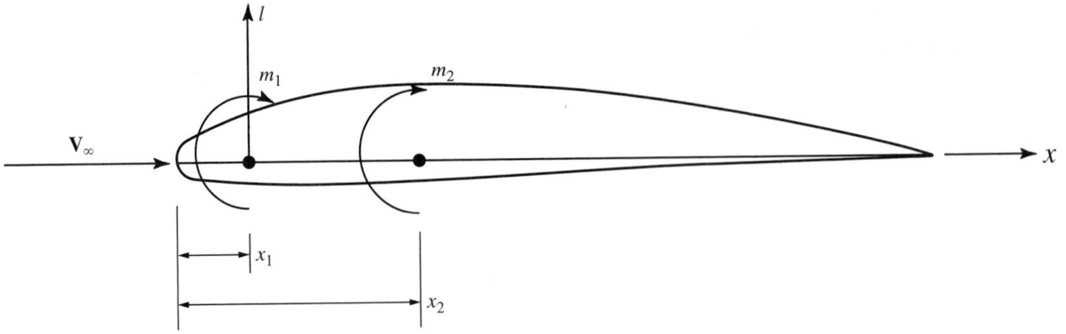

Figure 5.8 Section moment referenced at different locations along chord.

where we have defined the section reference area (per unit span into the page) as the chord length, and the section reference length as the chord length as well. (Note that 2-D section lift is always denoted as *l,* not to be confused with the characteristic length used to nondimensionalize moment coefficients. The context should make the usage clear.)

Therefore we have the moment coefficient about reference location x_2 given by

$$\boxed{c_{m_2} = c_{m_1} + c_l \frac{(x_2 - x_1)}{c} = c_{m_1} + c_l(\bar{x}_2 - \bar{x}_1)} \tag{5.6}$$

(Note here that we have denoted the x distance nondimensionalized by the section chord length as $\bar{x}$. Such a nondimensionalization of lengths will always be denoted with an overbar.) Consequently, the moment (coefficient) about the aerodynamic center $c_{m_{ac}}$, the center of pressure $\bar{x}_{cp}$, and the aerodynamic center $\bar{x}_{ac}$ are related by the expression

$$c_{m_{ac}} = -c_l(\bar{x}_{cp} - \bar{x}_{ac}) \tag{5.7}$$

which yields

$$\boxed{\frac{\partial c_{m_{ac}}}{\partial c_l} = -(\bar{x}_{cp} - \bar{x}_{ac})} \tag{5.8}$$

Another moment coefficient is also frequently used, the *moment (coefficient) at zero lift,* c_{m_0}. This moment is always taken about the aerodynamic center, and due to the definition of aerodynamic center we have

$$c_{m_0} = c_{m_{ac}}$$

Unless otherwise specified, pitching moments of airfoil sections will always be defined as taken about the section's aerodynamic center.

5.1.3 Section Data

A collection of 2-D airfoil aerodynamic characteristics for several NACA 4- and 5-digit and 6-series airfoils are summarized in Table 5.2. For symmetric airfoils, $\alpha_0 = c_{m_0} = 0$, while for cambered airfoils these parameters take on nonzero values. Note that the section lift coefficient is linear up to at least 8–10 deg, and that in most cases $c_{l_{max}} \approx 1.5$. It is especially significant to note that for all these airfoil cross sections, the lift-curve slope, or lift effectiveness, $c_{l_\alpha} \approx 0.1$ per deg (5.7 per rad), and the location of the aerodynamic center is approximately at the quarter chord (measured from the leading edge of the section). Finally, the thickness ratio, *t/c,* is indicated by the last two digits in the airfoil designation.

The data in Table 5.2 is for low subsonic Mach number. At higher subsonic speeds ($M_\infty > 0.7$–0.8) local Mach numbers near the surface of the airfoil approach Mach $= 1$. The free-stream velocity at which local Mach number just reaches unity is called the *critical Mach number* for that airfoil. Many effects of compressibility on an airfoil's aerodynamic characteristics are well captured by

Table 5.2 Experimental Low-Speed Airfoil Section Aerodynamic Characteristics

NACA Airfoil	α_0 (deg)	c_{m_0}	c_{l_α} (per deg)	ac	$\alpha_{c_{l\max}}$ (deg)	$c_{l\max}$	α^* (deg)
0006	0	0	0.108	0.250	9.0	0.92	9.0
0009	0	0	0.109	0.250	13.4	1.32	11.4
1408	0.8	−0.028	0.109	0.250	14.0	1.35	10.0
1410	−1.0	−0.020	0.108	0.247	14.2	1.50	11.0
1412	−1.1	−0.025	0.108	0.252	15.2	1.58	12.0
2412	−2.0	−0.047	0.105	0.247	16.8	1.68	9.5
2415	−2.0	−0.049	0.106	0.248	16.4	1.62	10.0
2418	−2.8	−0.050	0.103	0.241	14.0	1.47	10.0
2421	−1.8	−0.040	0.103	0.241	16.0	1.47	8.0
2424	−1.8	−0.040	0.098	0.281	16.0	1.29	8.4
4412	−3.8	−0.092	0.105	0.247	14.0	1.67	7.5
4415	−4.3	−0.098	0.105	0.246	15.0	1.64	8.0
4418	−8.8	−0.088	0.105	0.242	14.0	1.52	7.2
4421	−8.8	−0.085	0.103	0.236	16.0	1.47	6.0
4424	−8.8	−0.082	0.100	0.239	16.0	1.38	4.8
28012	−1.4	−0.014	0.107	0.247	18.0	1.79	12.0
28015	−1.0	−0.007	0.107	0.248	18.0	1.72	10.0
28018	−1.2	−0.005	0.104	0.242	14.0	1.60	11.8
28021	−1.2	0	0.103	0.238	15.0	1.50	10.3
28024	−0.8	0	0.097	0.231	15.0	1.40	9.7
63-006	0	0.005	0.112	0.253	10.0	.87	7.7
-009	0	0	0.111	0.252	11.0	1.15	10.7
63-206	−1.9	−0.037	0.112	0.254	10.5	1.06	6.0
-209	−1.4	−0.032	0.110	0.262	13.0	1.4	10.8
-210	−1.2	−0.035	0.113	0.261	14.5	1.56	9.6
63_1-012	0	0	0.116	0.265	14.0	1.45	12.8
-212	−2.0	−0.035	0.114	0.263	14.5	1.08	11.4
-412	−2.8	−0.075	0.117	0.271	15.0	1.77	9.6

Lift coefficients are based on chord.

Definitions:

c_{m_0} Section zero-lift pitching moment taken about the aerodynamic center
ac Section aerodynamic center chord-wise location, in fraction of chord ($=x_{ac}/c$)

the Prandtl–Glauert compressibility rule. For example, at subsonic speeds below the critical Mach number, consistent with Equation (5.3), the lift-curve slope can be assumed to obey the relationship

$$c_{l_\alpha}\big|_{M_\infty \neq 0} = \frac{c_{l_\alpha}\big|_{M_\infty = 0}}{\sqrt{1 - M_\infty^2}} \tag{5.9}$$

where $c_{l_\alpha}\big|_{M_\infty = 0}$ is the lift effectiveness given in Table 5.2. Above the critical Mach number, the slope of the lift curve reduces dramatically for conventional subsonic airfoils.

Another very important effect of flow velocity above the critical Mach number, as well as at supersonic velocities, is the shift in the location of the aerodynamic center of the airfoil. For subsonic flight we noted that the aerodynamic center was near the quarter chord of a well-designed airfoil. But for supersonic flight ($M_\infty > 1.5$), the aerodynamic center shifts aft near 40–45 percent of chord. As we shall see in Chapters 6, 9, and 10, this can cause large changes in the dynamics of an aircraft, and was the cause of the loss of aircraft in early flight tests before this phenomenon was understood.

To estimate other characteristics of airfoils in supersonic flow, it is noted that shock-expansion theory yields rather good results. In the hypersonic speed range, Newtonian theory has sometimes been used. The interested reader is referred to Ref. 2, which notes, for example, that Newtonian theory has been used in the preliminary design of hypersonic vehicles. The estimation of airfoil aerodynamic characteristics in the transonic speed range ($0.7 < M_\infty > 1.5$) is quite difficult, as the flow is very sensitive to small changes in shape. Consequently, empirical methods are limited, and experimental and/or computational techniques are recommended.

EXAMPLE 5.1

2-D Aerodynamic Characteristics of a Hypersonic Lifting Body

To compliment the discussion of subsonic airfoil section characteristics, this example addresses 2-D characteristics in hypersonic flow ($M_\infty \geq 5$). Consider a symmetric wedge-shaped hypersonic lifting body with cross-section geometry as shown in Figure 5.9. Using modified Newtonian impact theory, determine the 2-D lift, drag, and pitching moment coefficients at a Mach number $M_\infty = 5$, and plot the results versus angle of attack.

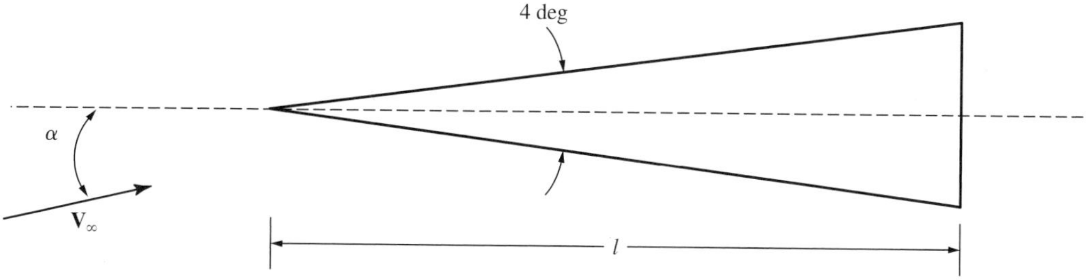

Figure 5.9 Hypersonic lifting body.

■ Solution

According to modified Newtonian impact theory (Ref. 2), the pressure (coefficient) on a body's surface as a function of the local flow-inclination angle is simply given by

$$c_p = c_{p,\max} \sin^2\theta \tag{5.10}$$

where θ is the local flow-inclination angle relative to the body surface, and at $M_\infty = 5$,

$$c_{p,\max} \triangleq \frac{2}{\gamma M_\infty^2}\left(\frac{p_{0,2}}{p_\infty} - 1\right) = \frac{2}{1.4(5^2)}(32.65 - 1) = 1.81 \tag{5.11}$$

(Here, $p_{0,2}/p_\infty$ is the stagnation pressure ratio across a normal shock wave at the given Mach number.)

So, for the lower and upper surfaces, the flow-inclination angles are given by

$$\theta_{\text{lower}} = 2 + \alpha \text{ deg}$$

$$\theta_{\text{upper}} = 2 - \alpha \text{ deg}$$

Now from their definitions, we can write the 2-D lift, drag, and moment coefficients as

$$c_l = c_{l_{\text{lower}}} + c_{l_{\text{upper}}} = c_{p_{\text{lower}}} \cos\theta_{\text{lower}} - c_{p_{\text{upper}}} \cos\theta_{\text{upper}}$$

$$c_d = c_{d_{\text{lower}}} + c_{d_{\text{upper}}} = c_{p_{\text{lower}}} \sin\theta_{\text{lower}} + c_{p_{\text{upper}}} \sin\theta_{\text{upper}}$$

$$c_m = c_{m_{\text{lower}}} + c_{m_{\text{upper}}} = -\left(c_{p_{\text{lower}}} - c_{p_{\text{upper}}}\right)\frac{1}{2\cos 2°}$$

where for $\theta_{\bullet} \leq 0$, $c_p = 0$. Note that the pitching moment is taken about the leading edge, and the characteristic length used is the body length.

The results are listed and plotted below. (The negative of the moment coefficient is plotted in Figure 5.10 to allow the use of the same scale.)

Angle of Attack, α, deg	Lower Surface	Upper Surface	Total
0	$c_p = 2.20$	$c_p = 2.20$	$c_l = 0$
	$c_l = 2.20$	$c_l = -2.20$	$c_d = 0.154$
	$c_d = 0.077$	$c_d = 0.077$	$c_m = 0$
	$c_m = -1.10$	$c_m = 1.10$	
1	$c_p = 4.96$	$c_p = 0.551$	$c_l = 4.40$
	$c_l = 4.95$	$c_l = -0.551$	$c_d = 0.269$
	$c_d = 0.259$	$c_d = 0.0096$	$c_m = -2.20$
	$c_m = -2.48$	$c_m = 0.276$	

(continued)

2	$c_p = 8.81$	$c_p = 0$	$c_l = 8.79$
	$c_l = 8.79$	$c_l = 0$	$c_d = 0.614$
	$c_d = 0.614$	$c_d = 0$	$c_m = -4.40$
	$c_m = -4.40$	$c_m = 0$	
3	$c_p = 13.7$	$c_p = 0$	$c_l = 13.7$
	$c_l = 13.7$	$c_l = 0$	$c_d = 1.20$
	$c_d = 1.20$	$c_d = 0$	$c_m = -6.87$
	$c_m = -6.87$	$c_m = 0$	
4	$c_p = 19.8$	$c_p = 0$	$c_l = 19.7$
	$c_l = 19.7$	$c_l = 0$	$c_d = 2.07$
	$c_d = 2.07$	$c_d = 0$	$c_m = -9.89$
	$c_m = -9.89$	$c_m = 0$	
5	$c_p = 26.9$	$c_p = 0$	$c_l = 26.7$
	$c_l = 26.7$	$c_l = 0$	$c_d = 3.28$
	$c_d = 3.28$	$c_d = 0$	$c_m = -13.4$
	$c_m = -13.4$	$c_m = 0$	

All data $\times 10^{-3}$.

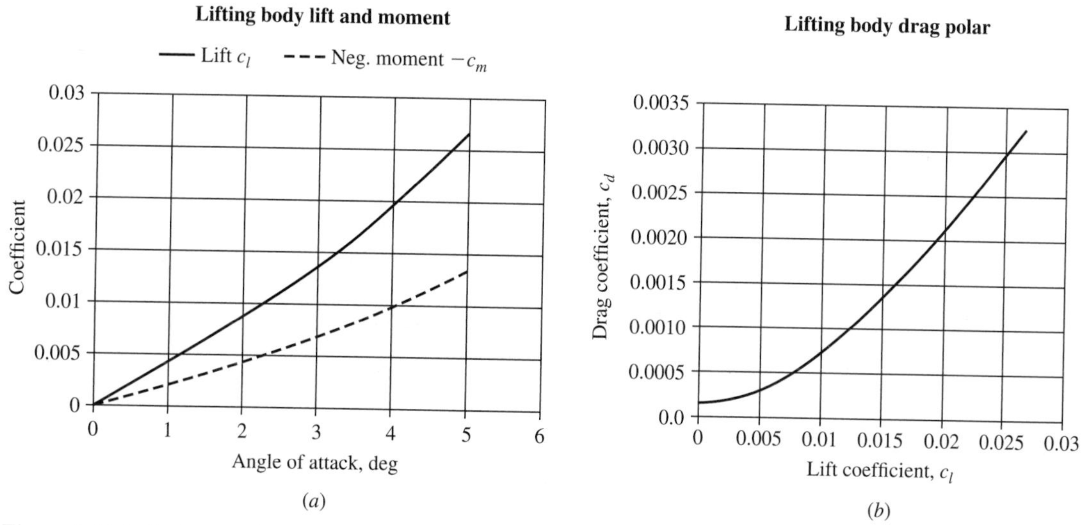

Figure 5.10 Lift, drag, and moment coefficients of hypersonic lifting body.

Note that for this case involving hypersonic flow, the lift coefficient is (approximately) proportional to the square of the sine of angle of attack, instead of linear with angle of attack—the case in subsonic flow.

5.2 EFFECTS OF FLAPS ON SUBSONIC AIRFOIL SECTION CHARACTERISTICS

Conventional aerodynamic high-lift devices, and control surfaces such as elevators and ailerons, are similar in that each functions by changing the effective camber and/or chord length of the airfoil. Although the design purpose of these two types of devices may be quite different, no distinction is made here as to the aerodynamic characteristics of these devices. Hence, in this section we will use the terms "flap" and "control surface" interchangeably.

Several examples of types of flaps are shown in the sketches in Figure 5.11. The most common is the plain flap. High-lift devices include both trailing-edge (e.g., flaps) and leading-edge (e.g., slots) devices. And both have similar effects on section aerodynamic characteristics. Here we will focus only on trailing-edge devices, and those interested in the leading-edge type are referred to Ref. 1.

The effect of a high-lift device on section lift is revealed in Figure 5.12. Basically, use of the high-lift device shifts the entire lift curve up, thus both increasing the lift at a given angle of attack, as well as increasing the maximum lift $c_{l_{\max}}$. The lift-curve slope c_{l_α} in the linear range is essentially unaffected by flap deflection.

The effect on the lift curve diminishes at higher flap-deflection angles, as shown in Figure 5.13. And at higher flap-deflection angles the increase in maximum lift, $c_{l_{\max}}$, and the upper limit on the linear range of lift behavior can actually reverse, due to stalling of the flow on the upper surface of the airfoil.

The aerodynamic effectiveness of a flap is quantified in terms of two related parameters. Knowing one, the other can be determined. First, the change in lift (coefficient) per unit flap deflection is defined as

$$\boxed{c_{l_\delta} \triangleq \frac{\partial c_l}{\partial \delta}\Big|_{\alpha=\text{const}}} \tag{5.12}$$

This parameter is depicted graphically in Figure 5.12. The second effectiveness parameter is the change of angle of attack per unit flap deflection, given by

$$\boxed{\alpha_\delta \triangleq \frac{\partial \alpha}{\partial \delta}\Big|_{c_l=\text{const}} = \frac{c_{l_\delta}\big|_{\alpha=\text{const}}}{c_{l_\alpha}\big|_{\delta=\text{const}}}} \tag{5.13}$$

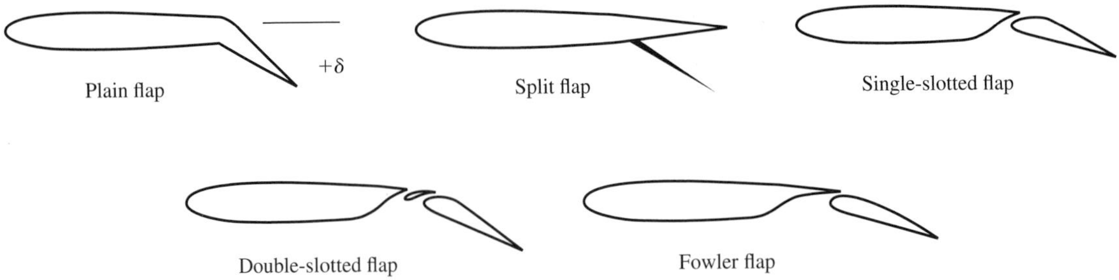

Figure 5.11 Examples of common types of flaps.

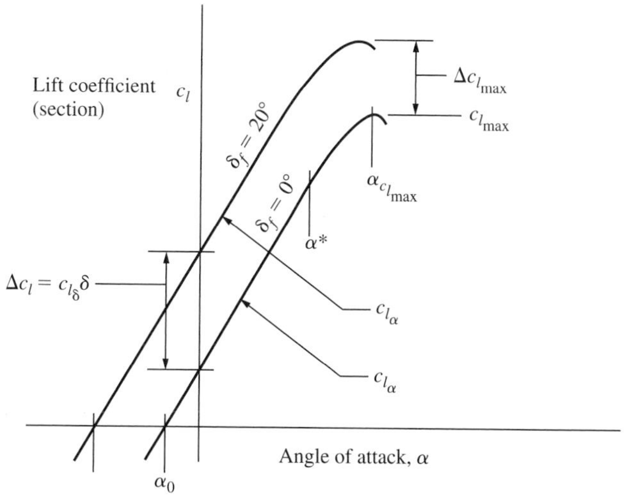

Figure 5.12 Effect of high-lift devices on section lift.

The sketch is typical of all types of trailing-edge flaps except split flaps.

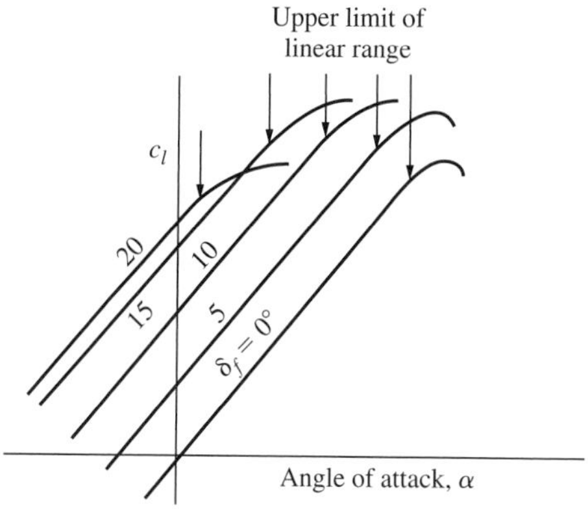

Figure 5.13 Effect of high-lift devices at higher deflections.

Again referring to Figure 5.12, for the case shown,

$$\alpha_\delta = \frac{\Delta\alpha_0(\text{deg})}{20(\text{deg})}$$

(Note that this parameter may actually be determined graphically at c_l's other than zero.)

For plain flaps in subsonic flow, the section flap effectiveness may be estimated from the following relation.

$$c_{l_\delta} = \frac{1}{\beta}\left(\frac{c_{l_\delta}}{c_{l_\delta}|_{\text{theory}}}\right)c_{l_\delta}|_{\text{theory}} \tag{5.14}$$

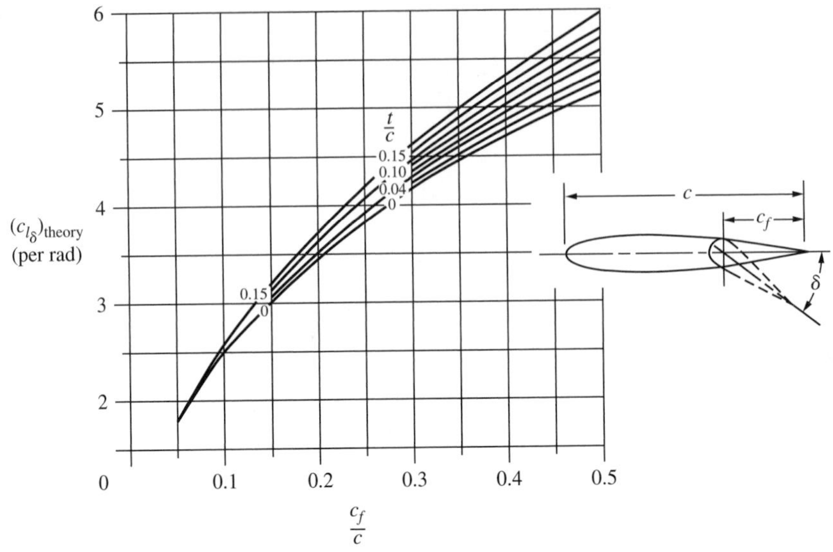

Figure 5.14 Theoretical section-lift effectiveness of plain flaps.

where

$\beta = \sqrt{1 - M_\infty^2}$ is the Prandtl–Glauert subsonic compressibility factor

$c_{l_\delta}|_{\text{theory}}$ is the effectiveness from thin-airfoil theory, taken from Figure 5.14 (Note: t/c is the thickness ratio of the airfoil section.)

$\left(\frac{c_{l_\delta}}{c_{l_\delta}|_{\text{theory}}}\right)$ is the ratio of experimental-to-theoretical flap effectiveness found from Figure 5.15 (Note: To use Figure 5.15, one must use the ratio of c_{l_α} taken from Table 5.2, for example, to the theoretical value of c_{l_α}, which is 2π/.)

Flap deflection also affects the pitching moment of the airfoil section, due to the change of pressure distribution depicted schematically in Figure 5.16. The resultant change in moment is leading edge down, or negative for a positive flap deflection. According to thin-airfoil theory, the change in moment due to deflection of a plain flap is given by

$$c_{m_\delta} = -2\sqrt{\frac{c_f}{c}\left(1 - \frac{c_f}{c}\right)^3} \text{ /rad} \tag{5.15}$$

where c_f is the chord of the flapped section, and c is the chord of the airfoil section including the undeflected flap. This expression yields reasonably accurate results for flap deflections less that about 25 deg. Although the magnitude of the moment is affected by flap deflection, the slope of the moment with angle of attack, c_{m_α}, is

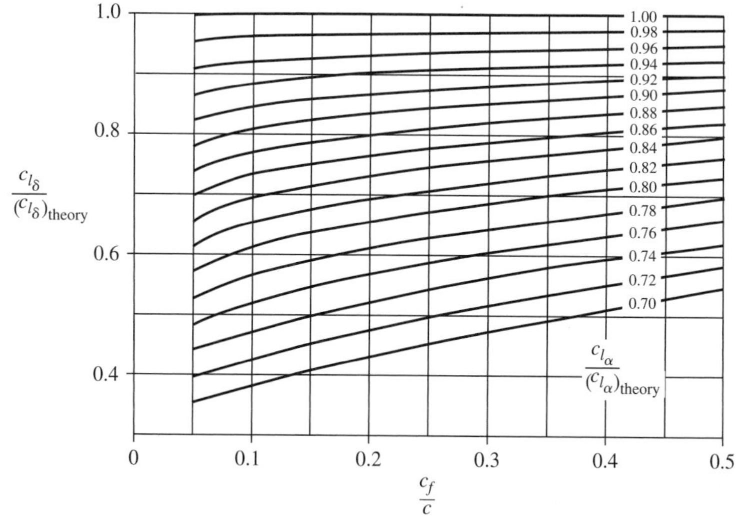

Figure 5.15 Empirical correction for plain-flap section-lift effectiveness.

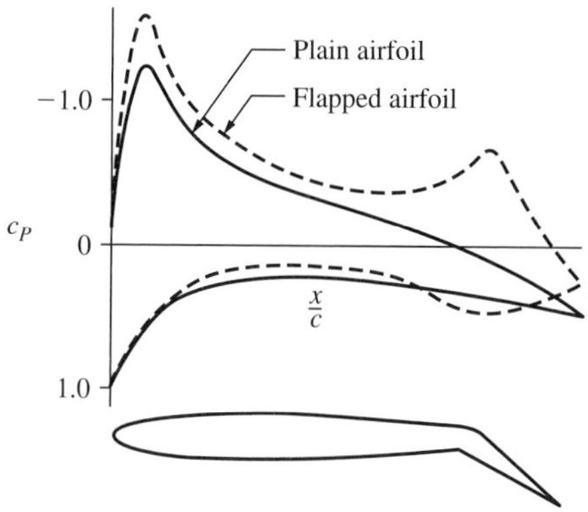

Figure 5.16 Change in section pressure distribution due to flap deflection.

not appreciably affected. Hence, the location of the section's aerodynamic center, x_{ac}, is not appreciably affected by flap deflection.

The final effect to be considered is the change on section drag due to flap deflection. This effect may be estimated for plain flaps using Figure 5.17. Drag will be discussed further in Section 5.3 where we deal with three-dimensional wings.

The above discussion of the effects of high-lift devices was restricted to the case of subsonic flow. Supersonically, again, shock-expansion theory tends to give reasonable results, while transonic flow presents difficulties. The interested reader is referred to Refs. 1–3, for more information.

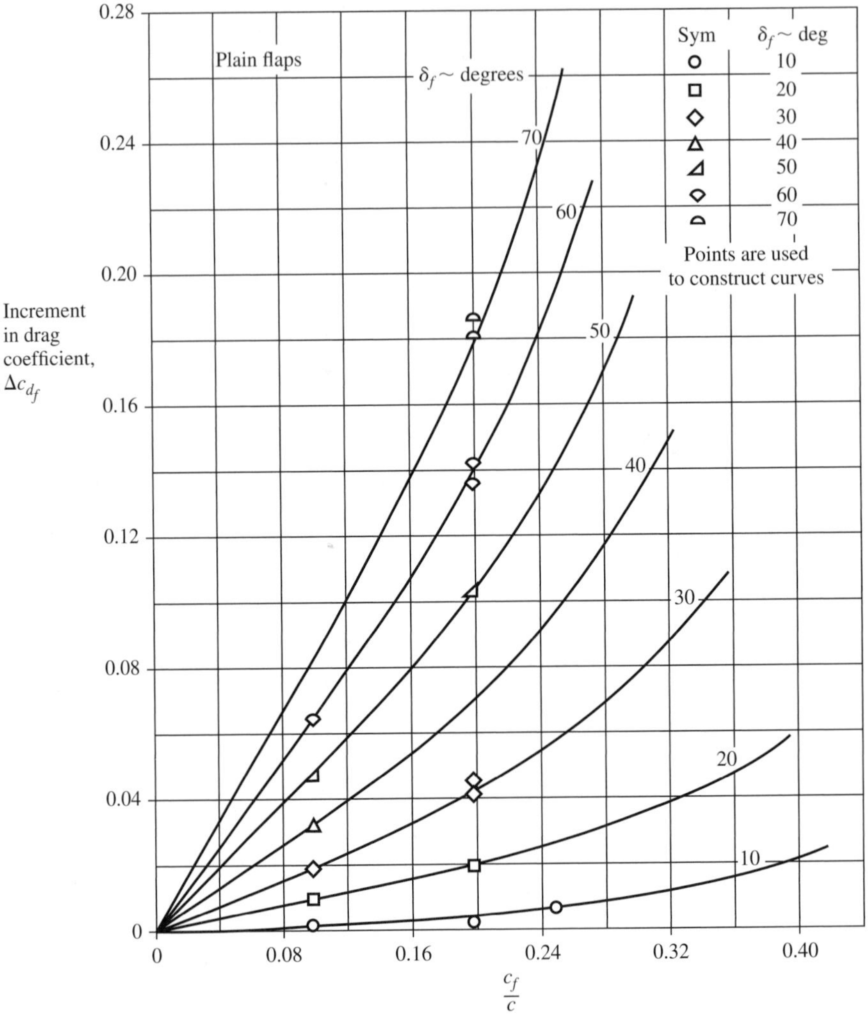

Figure 5.17 Section drag increment due to plain flap deflection.

EXAMPLE 5.2

Aerodynamic Characteristics of a Subsonic Airfoil with Flap

Consider an NACA 2412 2-D airfoil with a chord length of 2 ft. The airfoil has a trailing-edge flap with a flap chord of 0.6 ft. If the airfoil section is at an angle of attack of 3 deg, the flap is deflected 20 deg, and the free-stream velocity, V_∞, is 600 fps, estimate the section lift and moment coefficients, c_l and c_m, and the location of the airfoil's aerodynamic center, x_{ac}. (Assume sonic velocity a here is 1,100 fps.)

■ Solution

From Table 5.2 we have for the NACA 2412 2-D airfoil: $\alpha_0 = -2\,\text{deg}$, $c_{m_0} = -0.047$, $c_{l_\alpha} = 0.108\text{ /deg}$, $\bar{x}_{ac} = 0.247$, and $\alpha^* = 9.5\,\text{deg}$. The thickness ratio (indicated by the last two digits of the NACA number) is $\frac{t}{c} = 0.12$.

Therefore, noting that the angle of attack is within the linear range, <u>without</u> any flap deflection we have

$$c_l = c_{l_\alpha}(\alpha - \alpha_0) \text{ with } c_{l_\alpha} = \frac{c_{l_\alpha}|_{M_\infty=0}}{\beta}$$

Here

$$M_\infty = 600/1{,}100 = 0.55$$

So

$$\beta = \sqrt{1 - M_\infty^2} = \sqrt{1 - 0.55^2} = \sqrt{0.7} = 0.84$$

and

$$c_l = c_{l_\alpha}(\alpha - \alpha_0) = \frac{0.108}{0.84}(3 + 2) = 0.64$$

Plus without flap deflection

$$c_m = -0.047$$

From Equation (5.14), the flap lift effectiveness is

$$c_{l_\delta} = \frac{1}{\beta}\left(\frac{c_{l_\delta}}{c_{l_\delta}|_{\text{theory}}}\right)c_{l_\delta}|_{\text{theory}}$$

With a flap-to-section chord ratio of 0.3, and a thickness ratio of 0.12, Figure 5.14 indicates that

$$c_{l_\delta}|_{\text{theory}} = 4.5\text{ /rad} = 0.079\text{ /deg.}$$

For this airfoil,

$$\frac{c_{l_\alpha}}{c_{l_\alpha}|_{\text{theory}}} = \frac{0.108}{2\pi/57.3} = 0.985$$

So from Figure 5.15 we have

$$\frac{c_{l_\delta}}{c_{l_\delta}|_{\text{theory}}} = 0.98$$

Therefore,

$$c_{l_\delta} = \frac{1}{0.84}(0.98)0.079 = 0.092\text{ /deg}$$

Equation (5.15) indicates that the flap moment effectiveness is

$$c_{m_\delta} = -2\sqrt{\frac{c_f}{c}\left(1 - \frac{c_f}{c}\right)^3} = -2\sqrt{0.3(1 - 0.3)^3} = -0.66\text{ /rad} = -0.012\text{ /deg}$$

So at a 3 deg angle of attack and with a 20 deg flap deflection, the section lift and pitching-moment coefficients are, respectively,

$$c_l = c_l|_{\text{flap}=0} + c_{l_\delta}\delta = 0.64 + 0.092 \times 20 = 2.48$$

$$c_m = c_m|_{\text{flap}=0} + c_{m_\delta}\delta = -0.047 - 0.012 \times 20 = -0.287$$

Finally, for a chord length of 2 ft the aerodynamic center of the section is located at

$$x_{ac} = \bar{x}_{ac} \times c = 0.247 \times 2 = 0.494 \text{ ft aft of the leading edge}$$

5.3 WING PLANFORM CHARACTERISTICS

Although this section is titled "Wing Planform Characteristics," all the concepts to be discussed apply as well to any three-dimensional lifting surface such as horizontal and vertical tail surfaces and canards (surfaces forward of the wing). Also, though we will continue to refer to trailing-edge high-lift devices as "flaps" here, the techniques for dealing with high-lift devices also apply to trailing-edge control surfaces such as elevators, ailerons, and rudders.

As in the discussion of airfoil section characteristics in Section 5.2, we begin with the physical description of the wing planform. An arbitrary wing planform is shown looking down from the top in Figure 5.18, in which the x,y wing coordinates, with origin at the wing apex, are defined. Key geometric parameters are also defined in Table 5.3. Of special note are the wing planform or projected area S, span b, aspect ratio A, and chords at the wing tip c_t and wing root, or root chord, c_r. This root chord is the chord at the center of the wing in its plane of symmetry. Also key is the wing's *mean aerodynamic chord* (MAC) (which is actually a mean geometric chord), denoted $\bar{c}$. (By convention we use the overbar here, but the MAC is not nondimensional—it has the units of length.)

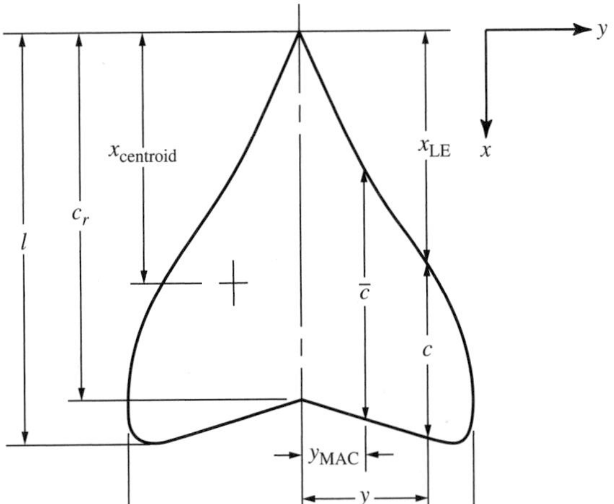

Figure 5.18 Arbitrary wing planform.

If the leading and trailing edges of the wing are straight, additional parameters may used to describe its shape. Such a wing is shown in Figure 5.19, with additional geometric parameters defined in Table 5.4. Of special note is the taper ratio $\lambda = c_t/c_r$ and the various sweep angles Λ. In addition (although not shown), wing twist angle $\varepsilon(y)$ is frequently used, with twist varying from wing root to tip. That is, with increasing span location the airfoil-section chord line is twisted about the span coordinate y.

Table 5.3 Geometric Properties of an Arbitrary Wing Planform

Variable	Definition
A	Aspect ratio $= b^2/S$
b	Wing span
$b/(2l)$	Wing-slenderness parameter
c	Chord (parallel to axis of symmetry) at any given span station y
$\bar{c}$	Mean aerodynamic chord (MAC) (or mean geometric chord) $\bar{c} = \frac{2}{S}\int_0^{b/2} c^2\,dy$
c_r	Root chord
l	Over-all length from wing apex to most aft point on trailing edge
p	Planform-shape parameter $= S/(bl)$
S	Wing area $= 2\int_0^{b/2} c\,dy$ Note: projected or planform area
x_{LE}	Chordwise location of leading edge at span station y
x_{centroid}	Chordwise location of centroid of area (chordwise distance from apex to $\bar{c}/2$) $x_{\mathrm{centroid}} = \frac{2}{S}\int_0^{b/2} c\left(x_{\mathrm{LE}} + \frac{c}{2}\right)dy$
y	General span station measured perpendicular to plane of symmetry
y_{MAC}	spanwise location of MAC (equivalent to spanwise location of centroid of area) $y_{\mathrm{MAC}} = \frac{2}{S}\int_0^{b/2} cy\,dy$

For straight tapered wings, one can show that the planform area is

$$S = \left(\frac{b}{2}\right)c_r(1 + \lambda) \tag{5.16}$$

and that the MAC length is given by

$$\bar{c} = \frac{2}{3}c_r\frac{1 + \lambda + \lambda^2}{1 + \lambda}$$

To characterize a 3-D wing's aerodynamic properties, we seek to find the following parameters.

Wing lift effectiveness (lift-curve slope), C_{L_α}
Wing zero-lift angle of attack, α_0, or the wing lift coefficient at $\alpha = 0$, C_{L_0}
Axial location of the wing's aerodynamic center, X_{ac}
Wing pitching moment about its aerodynamic center, $C_{M_{\mathrm{ac}}}$
Wing rolling moment about its centerline $C_{L_{\mathrm{Roll}}}$

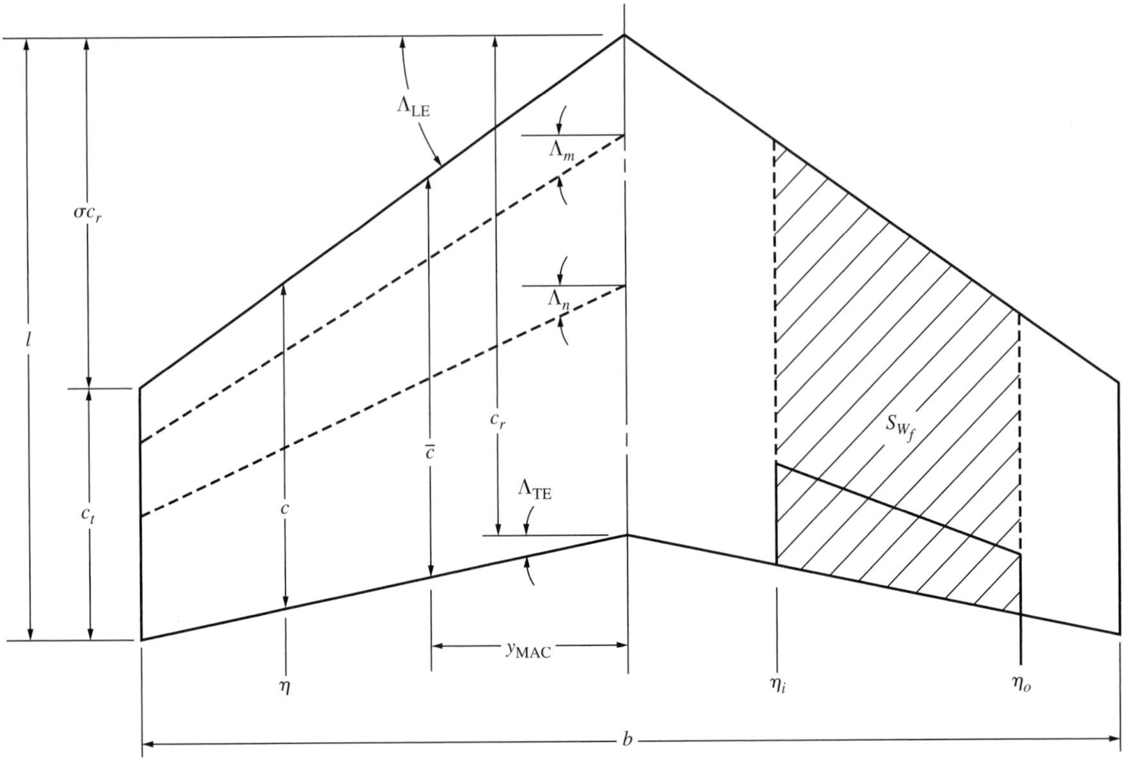

Figure 5.19 Planform of straight, tapered wing.

Table 5.4 Geometric Properties of a Straight, Tapered Wing

Variable	Definition
c_r	Root chord
c_t	Tip chord
m, n	Nondimensional chordwise stations (e.g., 50% chord)
S_{W_f}	Wing area affected by trailing-edge deflection
η	Nondimensional span-wise station
λ	Taper ratio $= c_t/c_r$
Λ_{LE}	Sweep angle of leading edge
$\Lambda_{m \text{ or } n}$	Sweep angle of line at position m or n along the chord (e.g., mid chord)
Λ_{TE}	Sweep angle of trailing edge
$X_{\text{LE}_{\text{MAC}}}$	Axial position of the wing's MAC (i.e., the position of its leading edge) where $X_{\text{LE}_{\text{MAC}}} = \dfrac{2}{S}\displaystyle\int_0^{b/2} x_{\text{LE}}(y)c(y)dy$ (5.17)

One way to determine these parameters is to use the aerodynamic characteristics of the wing's airfoil section(s). So here assume we have the following information available, obtained from Section 5.2 or elsewhere.

Section lift effectiveness, $c_{l_\alpha}(y)$
Section zero-lift angle of attack, $\alpha_0(y)$
Location of section aerodynamic center along the local chord, $x_{\text{ac}}(y)$
Section pitching moment (coefficient) about its aerodynamic center, $c_{m_{\text{ac}}}(y)$

Note that for each of the above parameters, we assume we have the span-wise (y) distribution of that parameter, since in general the airfoil section may vary with wing span.

5.3.1 Wing Lift

Wing lift-curve slope C_{L_α} in subsonic flow may be determined from modified lifting-line theory (Ref. 2), and an expression yielding good results for the *angle-of-attack lift effectiveness* over a wide range of aspect ratios is

$$\boxed{C_{L_\alpha} = \frac{2\pi A}{2 + \sqrt{\dfrac{A^2\beta^2}{k^2}\left(1 + \dfrac{\tan^2\Lambda_{c/2}}{\beta^2}\right) + 4}} \text{ per radian}} \tag{5.18}$$

where

$\beta = \sqrt{1 - M_\infty^2}$ is the Prandtl–Glauert subsonic compressibility factor

$k = \dfrac{c_{l_\alpha}}{2\pi}$ is the ratio of section lift effectiveness to the theoretical value of 2π

From the above expression we note that a wing's lift effectiveness generally decreases with wing sweep and increases with aspect ratio, especially at lower aspect ratio.

EXAMPLE 5.3

Subsonic 3-D Wing Lift

Consider a straight, unswept wing with aspect ratio of 4. Each wing cross section, constant with span, consists of an NACA 2412 airfoil shape. Ignoring compressibility effects (i.e., $M_\infty = 0$), determine the lift effectiveness of the 3-D wing. Also determine the wing lift coefficient at an angle of attack of 5 degrees if there is no span-wise wing twist.

■ Solution

Using Equation (5.18), we note that for $M_\infty = 0$, $\beta = 1$, and if the wing is not swept, $\Lambda_{c/2} = 0$ and $\tan\Lambda_{c/2} = 0$. For the NACA 2412 airfoil, the section lift-curve slope

$c_{l_\alpha} = 0.105$ per degree, and the zero-lift angle of attack $\alpha_0 = -2$ degrees. Consequently, $k = 0.105 \times (180/\pi)/2\pi = 0.9575$. From Equation (5.18) we find that the wing's lift effectiveness is

$$C_{L_\alpha} = \frac{2\pi(4)}{2 + \sqrt{\left(\frac{4 \times 1}{.9575}\right)^2 (1) + 4}} = \frac{2\pi(4)}{2 + 4.63} = 3.79 \text{ per rad}$$

Note that due primarily to the low wing aspect ratio this value is considerably less than the theoretical value for the 2-D section of 2π.

With no wing span-wise twist, all wing sections are at an angle of attack of 5 degrees. Therefore, the 3-D wing lift (coefficient) is

$$C_L\,|_{\alpha = 5\text{ deg}} = C_{L_\alpha}(\alpha - \alpha_0) = 3.79(5 + 2)\left(\frac{\pi}{180}\right) = 0.463$$

For supersonic flow ($M_\infty > 1.4$), supersonic thin-airfoil theory may be used to determine lift effectiveness. Theoretical results for straight, tapered, swept wings with sufficiently sharp leading edges are presented graphically in Figures 5.20*a–f*. This set of figures shows the wing *normal force coefficient* as a function of the sweep of the leading edge Λ_{LE}, taper ratio λ, aspect ratio A, and the Prandtl–Glauert supersonic compressibility factor, β, given by

$$\beta = \sqrt{M_\infty^2 - 1}$$

Normal force is similar to lift, but is the component of the total force normal to the wing chord. Recall that lift is normal to the free-stream flow $\mathbf{V}_\infty$. Referring to Figure 5.3, one can see that $C_L = C_N \cos\alpha$.

As an example of the use of Figures 5.20*a–f*, consider a straight swept wing with taper ratio $\lambda = 0.2$, aspect ratio $A = 3.5$, and a leading-edge sweep $\Lambda_{LE} =$ 52 deg. Let the Mach number $M_\infty = 2.0$. Therefore, $\beta = 1.73$, $\tan\Lambda_{LE}/\beta = 0.73$, and $A \tan \Lambda_{LE} = 4.42$. From Figure 5.20*b*, we find that $\beta C_{N_\alpha} = 4.24$ /rad, and so $C_{N_\alpha} = 2.45$ per radian.

5.3.2 Wing Zero-Lift Angle of Attack

The wing zero-lift angle of attack α_0 is primarily a function of the airfoil sections of the wing and the wing's span-wise angle of twist $\varepsilon(y)$. Twist is used, for example, to improve the stall characteristics of a wing. In that case, the twist angle used typically becomes more leading-edge down from wing root to tip, thus making the local angle of attack of the sections nearer the tip smaller than the local angle of attack near the root. As a result, the inboard sections will stall before the outboard sections, thus preserving smooth attached flow over any control surfaces located near the wing tip.

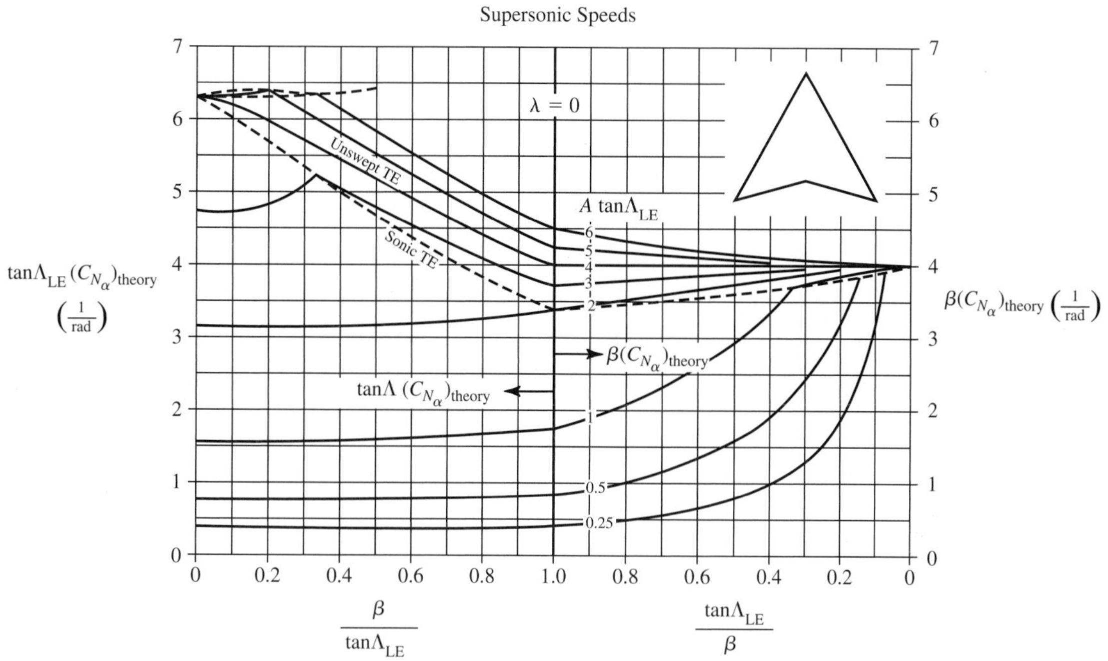

(*a*)

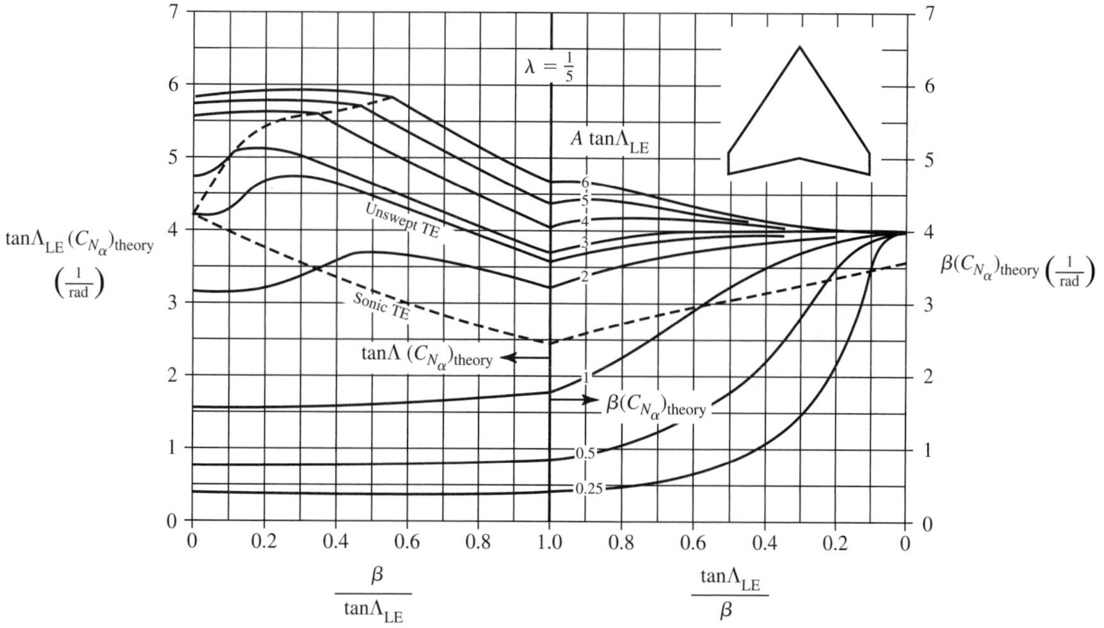

(*b*)

Figure 5.20*a–b* Normal-force coefficients for supersonic wings.

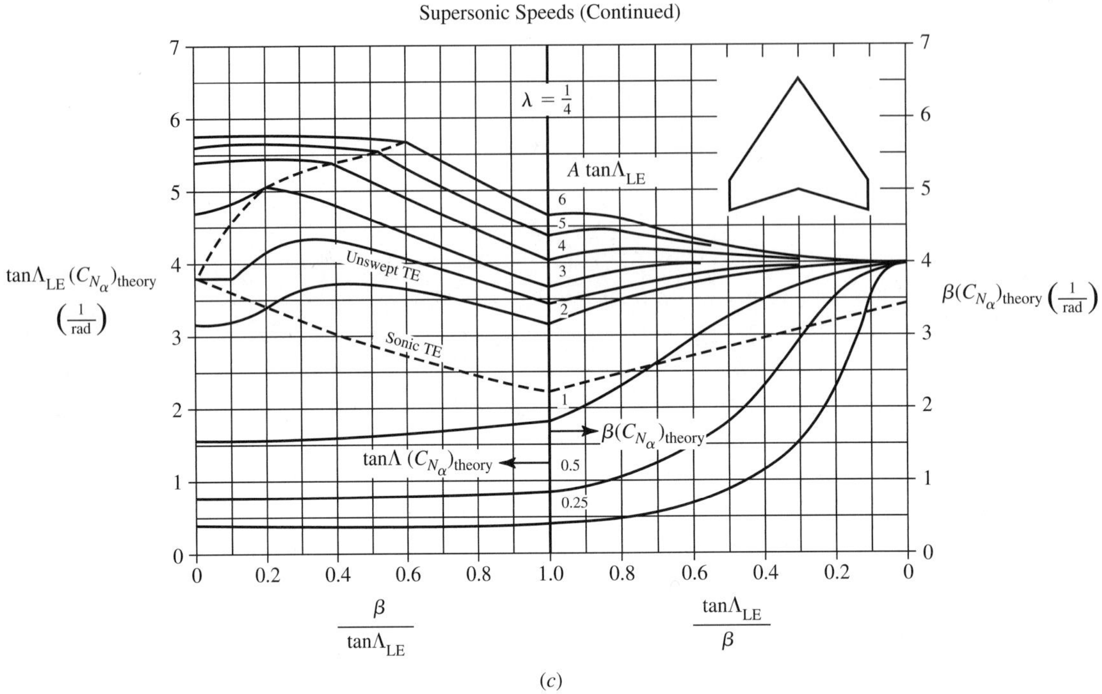

(*c*)

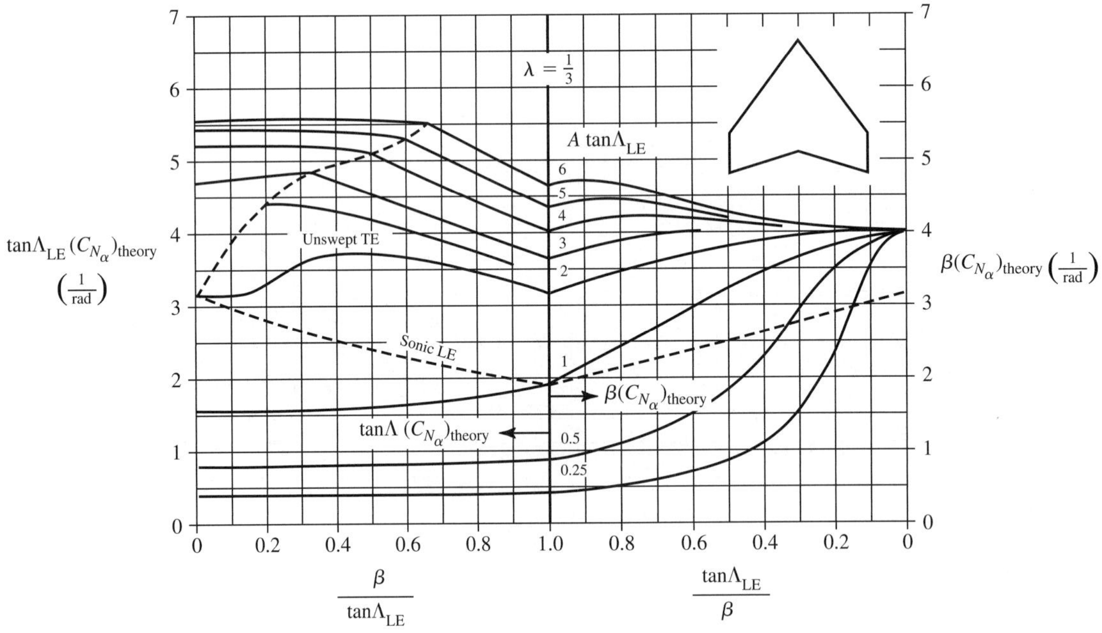

(*d*)

Figure 5.20*c–d* Normal-force coefficients for supersonic wings.

Supersonic Speeds (Continued)

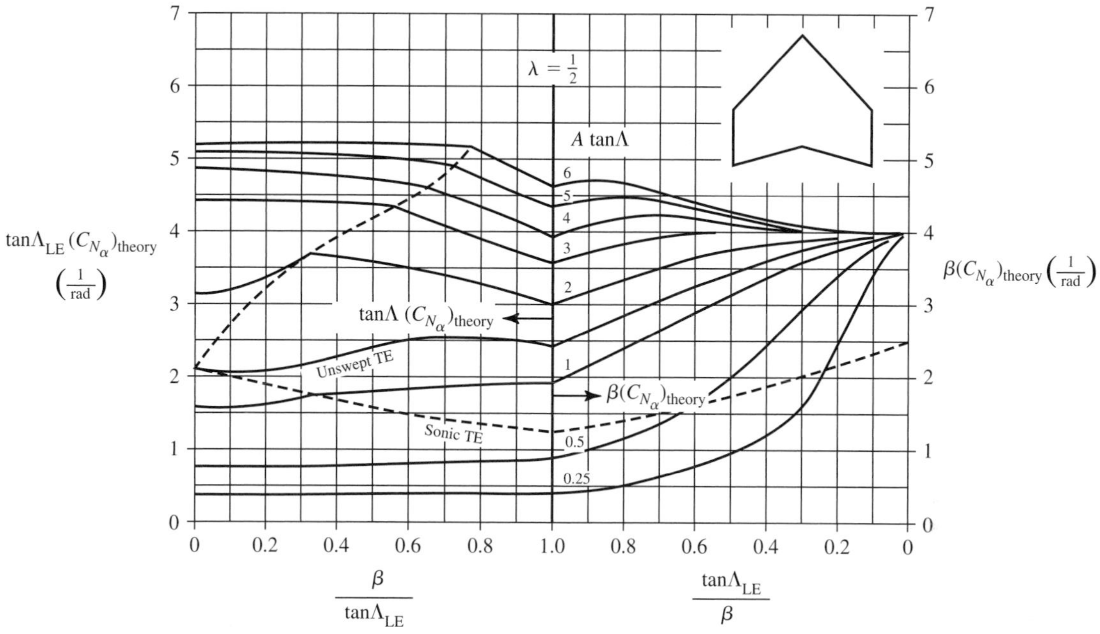

(*e*)

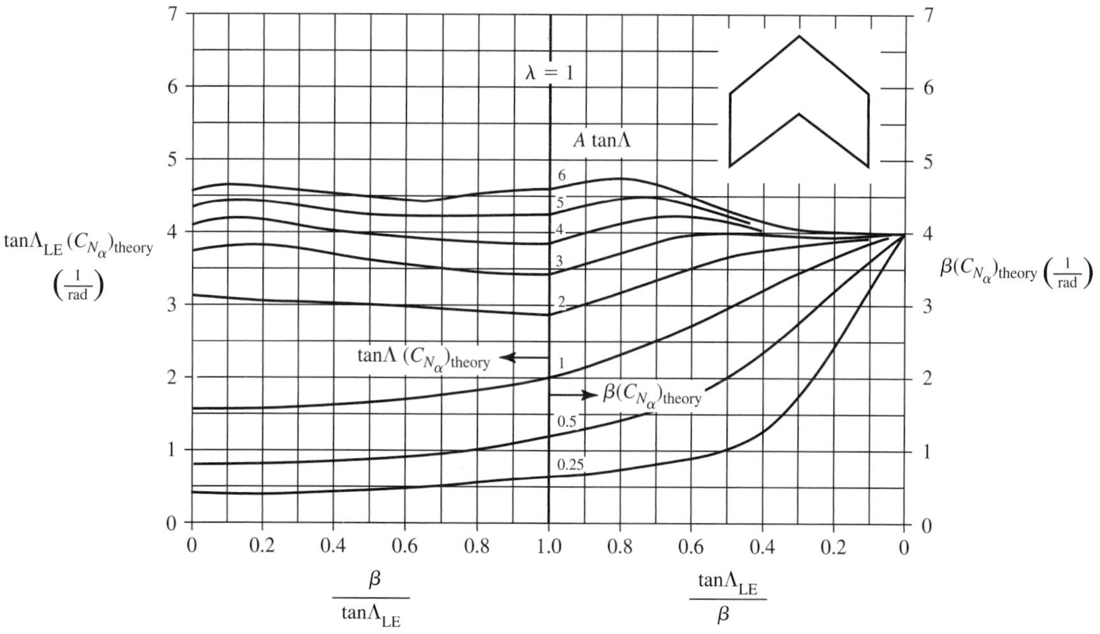

(*f*)

Figure 5.20*e–f* Normal-force coefficients for supersonic wings.

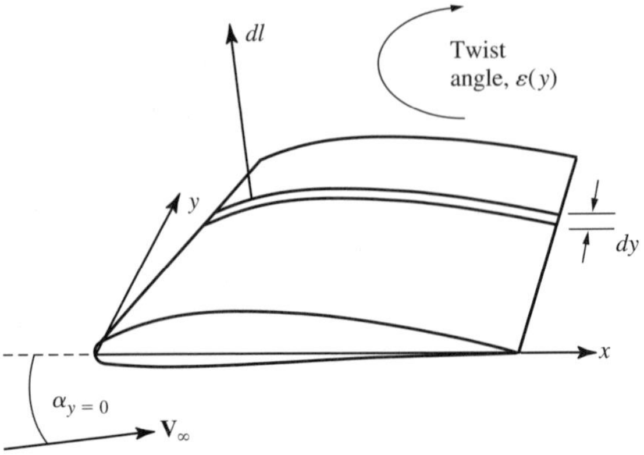

Figure 5.21 Right half of straight, swept wing with strip shown.

We will use strip theory to determine the zero-lift angle of attack. Figure 5.21 shows a sketch of the right half of a swept wing, looking along the span from root to tip. The usual wing coordinate system is used, with the y coordinate along the span and normal to the wing's plane of symmetry, and x positive aft from the leading edge of the wing root chord. Also shown is an infinitesimal (2-D) strip of the wing with thickness dy. The plane of this strip's airfoil cross section is parallel to the wing plane of symmetry, and this section's chord line may be twisted relative to the root chord an angle $\varepsilon(y)$, positive leading edge up. As a result of any such twist, the local angle of attack of the strip's airfoil section is $\alpha(y)$, where

$$\alpha(y) = \alpha_{y=0} + \varepsilon(y) \tag{5.19}$$

The wing strip generates an infinitesimal 2-D lift dl, where

$$\begin{aligned} dl &= c_l(y)q(y)dS = c_{l_\alpha}(y)\big(\alpha(y) - \alpha_0(y)\big)q(y)dS \\ &= c_{l_\alpha}(y)\big(\alpha_{y=0} + \varepsilon(y) - \alpha_0(y)\big)q(y)c(y)dy \end{aligned} \tag{5.20}$$

Note that $c_{l_\alpha}(y)$ is the lift-curve slope of the strip's airfoil section at span y, and $\alpha_0(y)$ is the zero-lift angle of attack of that same airfoil section. Therefore, integrating the above over the wing span yields the total lift of the wing, or

$$L = 2\int_0^{b/2} c_{l_\alpha}(y)\big(\alpha_{y=0} + \varepsilon(y) - \alpha_0(y)\big)q(y)c(y)dy \tag{5.21}$$

NOTE TO STUDENT

In Equation (5.21) we have integrated over the right wing ($y > 0$) and doubled the result to obtain the value for the entire wing. This is usually the easiest way to evaluate the integrals that arise in strip theory, especially due to the fact that for symmetric wings the functions in the integrand are frequently discontinuous at $y = 0$. For a vertical tail, which basically looks like a half wing, integrals similar to that in Equation (5.21) are obtained, but the factor of two is not required.

Now we will define a wing's angle of attack to be the angle of attack of its root chord. Or

$$\alpha_{\text{Wing}} \triangleq \alpha_W \triangleq \alpha_{y=0} \tag{5.22}$$

Furthermore, the angle of attack at zero lift α_0 is defined as that wing angle of attack that results in zero lift. Therefore, at $L = 0$ we have from Equation (5.21)

$$\int_0^{b/2} c_{l_\alpha}(y)\alpha_{0_{\text{wing}}} q(y)c(y)dy = \int_0^{b/2} c_{l_\alpha}(y)\big(\alpha_0(y) - \varepsilon(y)\big)q(y)c(y)dy$$

where $\alpha_{0_{\text{wing}}}$ is the wing's angle of attack for zero lift. Noting that $\alpha_{0_{\text{wing}}}$ is not a function of span y, and assuming the dynamic pressure equals the free-stream ($q(y) = q_\infty$) independent of span, we have the wing's zero-lift angle of attack given by

$$\alpha_{0_{\text{wing}}} = \frac{\displaystyle\int_0^{b/2} c_{l_\alpha}(y)\big(\alpha_0(y) - \varepsilon(y)\big)c(y)dy}{\displaystyle\int_0^{b/2} c_{l_\alpha}(y)c(y)dy} \approx \frac{2}{S}\int_0^{b/2} \big(\alpha_0(y) - \varepsilon(y)\big)c(y)dy \tag{5.23}$$

The approximate result on the far right in Equation (5.23) is valid when section lift effectiveness $c_{l_\alpha}(y)$ is approximately constant with span, which is frequently a good approximation. Note that in obtaining Equation (5.23) we have assumed the wing is symmetric about its root-chord plane, and we have defined the reference area S used with all wing aerodynamic coefficients to be

$$S = \int_{-b/2}^{b/2} c(y)dy = 2\int_0^{b/2} c(y)dy \tag{5.24}$$

which is the planform area of the wing. Finally, note that with no wing twist and with constant section $\alpha_0(y) = \alpha_0$, the wing zero-lift angle of attack is simply α_0 of the section.

5.3.3 Wing Pitching Moment and Aerodynamic Center

We will also use strip theory to express the pitching moment (coefficient) of the 3-D wing about its aerodynamic center. Figure 5.22, similar to Figure 5.21, again shows the sketch of the right half of a swept wing looking along the span. As in the discussion of wing zero-lift angle of attack, consider a wing strip of width dy, located at span location y, as shown. This strip generates an infinitesimal lift dl, and an infinitesimal pitching moment dm about the x_{ac} location of its

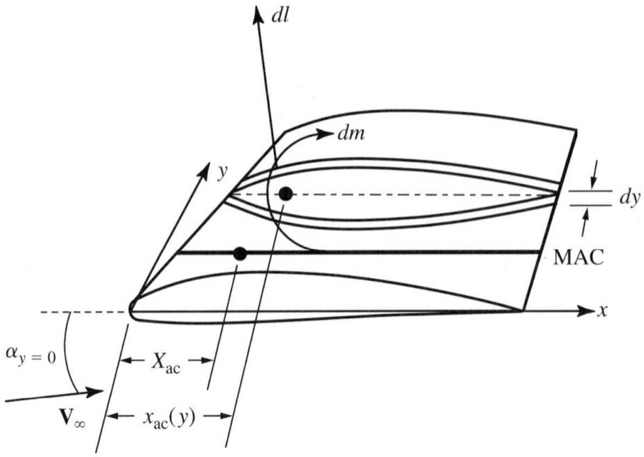

Figure 5.22 Right half of straight wing with aerodynamic center(s).

aerodynamic center on its chord. The location of the *wing aerodynamic center* (*ac*) on the wing's MAC is denoted as X_{ac}, which must also be determined.

We may now develop the expression for the wing's pitching moment taken about the wing's aerodynamic center (or about the line through the wing's *ac* and parallel to the *y* axis). We will use this result to both locate the position of the wing aerodynamic center as well as to determine the wing's moment (coefficient) about its *ac*.

The moment about a line through the wing aerodynamic center generated by the infinitesimal strip at span location *y* is

$$dm_{ac} = dm - dl\left(x_{ac}(y) - X_{ac}\right) \tag{5.25}$$

where *dm* is the local section's pitching moment about its *ac*, or

$$dm = c_{m_{ac}}(y)q(y)dSc(y) = c_{m_{ac}}(y)q(y)c^2(y)dy$$

and the local section lift *dl* is given by Equation (5.20). Integrating Equation (5.25) over the span of the wing yields the expression for the wing pitching moment about its *ac*, or

$$M_{ac} = 2\left(\int_0^{b/2} c_{m_{ac}} q(y)c^2(y)dy - \int_0^{b/2} c_{l_\alpha}(y)\left(\alpha_{\text{wing}} + \varepsilon(y) - \alpha_0(y)\right)\left(x_{ac}(y) - X_{ac}\right)q(y)c(y)dy\right)$$
$$= C_{M_{ac}} q_\infty S\bar{c} \tag{5.26}$$

The wing's moment coefficient $C_{M_{ac}}$ is defined by the above expression, and the characteristic length used is the wing's MAC.

Dividing Equation (5.26) by $q_\infty S\bar{c}$ yields the following expression for the wing's moment coefficient, which assumes the dynamic pressure equals the freestream ($q(y) = q_\infty$) independent of wing span.

$$C_{M_{ac}} = \frac{2}{S\bar{c}}\left(\int_0^{b/2} c_{m_{ac}} c^2(y)dy - \int_0^{b/2} c_{l_\alpha}(y)\left(\alpha_{\text{wing}} + \varepsilon(y) - \alpha_0(y)\right)\left(x_{ac}(y) - X_{ac}\right)c(y)dy\right) \quad (5.27)$$

However, since by definition the wing's moment about its aerodynamic center is invariant with (wing) angle of attack, we may choose the wing's angle of attack in the above expression to be its zero-lift angle of attack $\alpha_{0_{\text{wing}}}$. After making this substitution the wing's pitching-moment coefficient becomes

$$C_{M_{ac}} = \frac{2}{S\bar{c}}\left(\int_0^{b/2} c_{m_{ac}}(y)c^2(y)dy - \int_0^{b/2} c_{l_\alpha}(y)\left(\alpha_{0_{\text{wing}}} + \varepsilon(y) - \alpha_0(y)\right)\left(x_{ac}(y) - X_{ac}\right)c(y)dy\right) \quad (5.28)$$

Clearly, to determine the moment coefficient from Equation (5.28), one must know the wing's zero-lift angle of attack, given by Equation (5.23), and the location of the wing's *ac*, X_{ac}. For swept wings with straight leading and trailing edges, Figures 5.23*a*–*f* may be used to locate the wing's *ac*. These figures show the location of the *ac* for both subsonic and supersonic flows, as a function of aspect ratio A and Prandtl–Glauert compressibility factor β.

But it is instructive to also use Equation (5.27) to locate the wing's *ac*. Recall that from the definition of the aerodynamic center we have for the wing

$$C_{M_\alpha}\big|_{ac} = \frac{\partial C_{M_{ac}}}{\partial \alpha_{\text{wing}}} = 0 \quad (5.29)$$

Differentiating Equation (5.27) with respect to wing angle of attack, and setting the result to zero yields

$$\frac{\partial C_{M_{ac}}}{\partial \alpha_{\text{wing}}} = \frac{2}{S\bar{c}}\int_0^{b/2} c_{l_\alpha}(y)\left(x_{ac}(y) - X_{ac}\right)c(y)dy = 0$$

Since X_{ac} is invariant with respect to the span-wise integration, we may bring it outside the integral and solve for it yielding

$$X_{ac} = \frac{\int_0^{b/2} c_{l_\alpha}(y)x_{ac}(y)c(y)dy}{\int_0^{b/2} c_{l_\alpha}(y)c(y)dy} \quad (5.30)$$

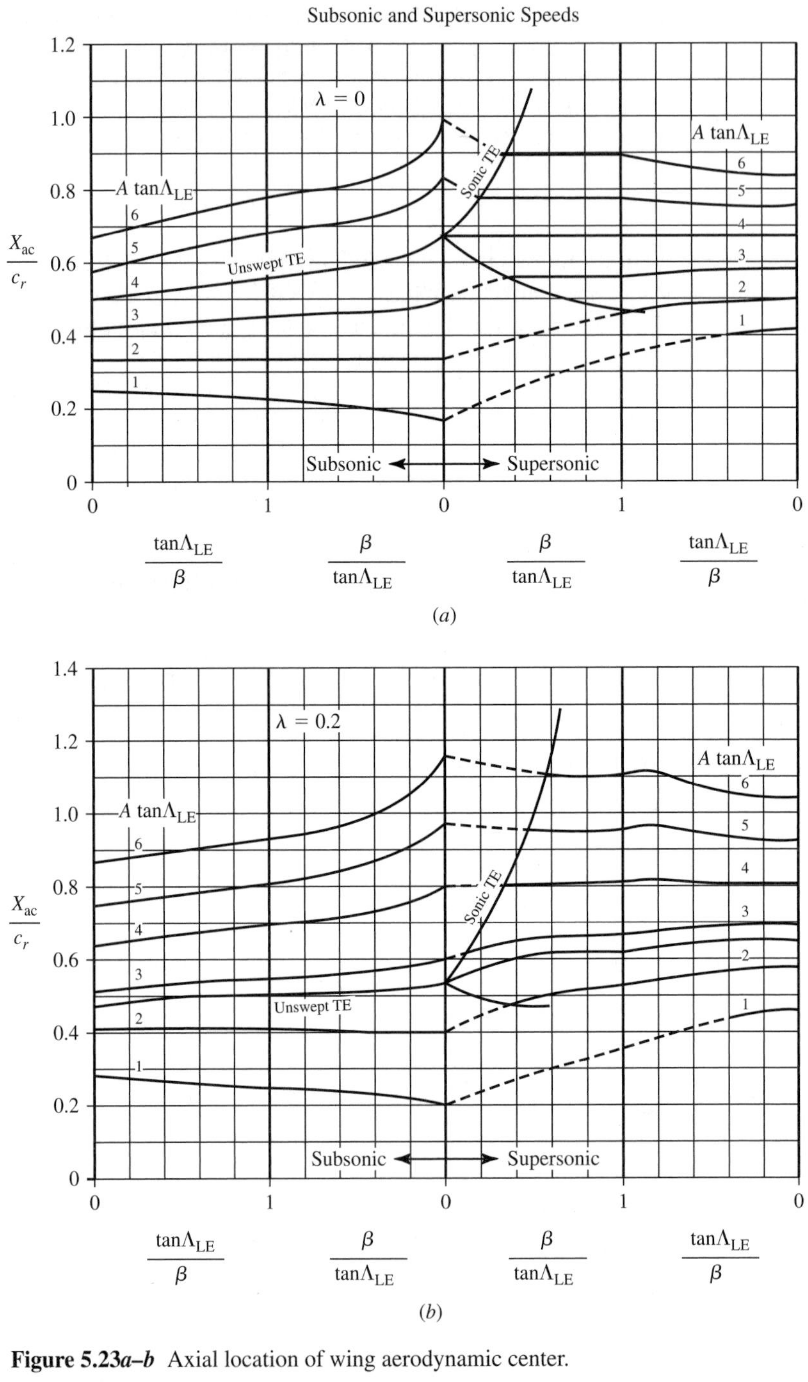

Figure 5.23*a–b* Axial location of wing aerodynamic center.

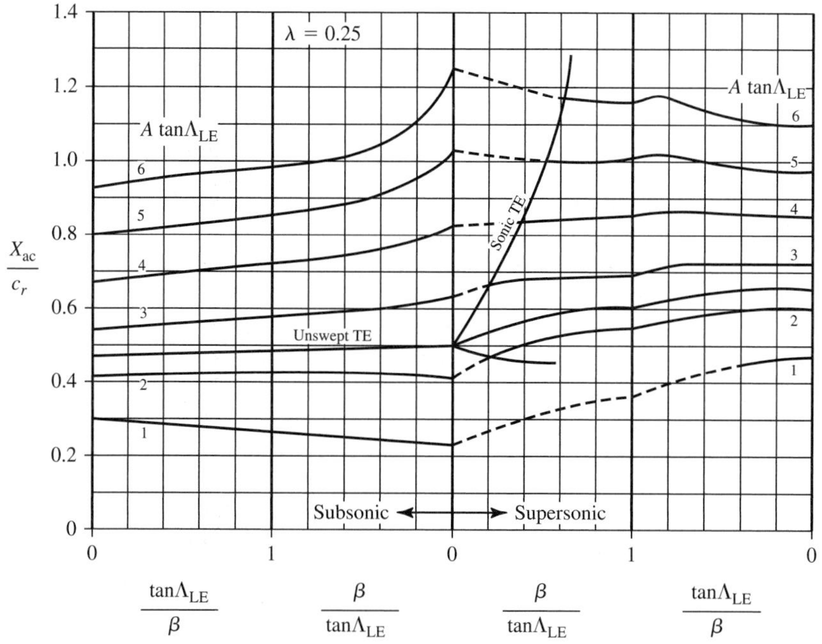

(c)

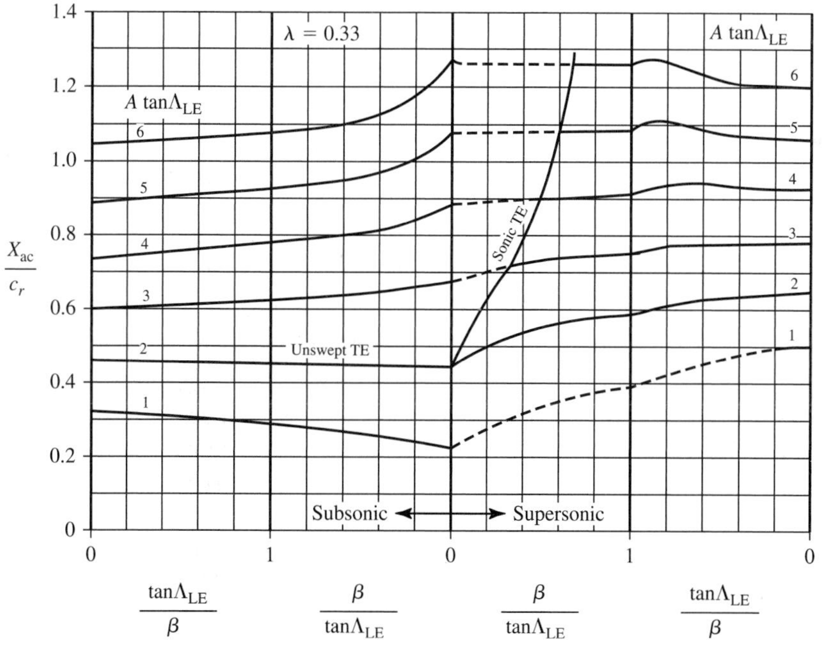

(d)

Figure 5.23c–d Axial location of wing aerodynamic center.

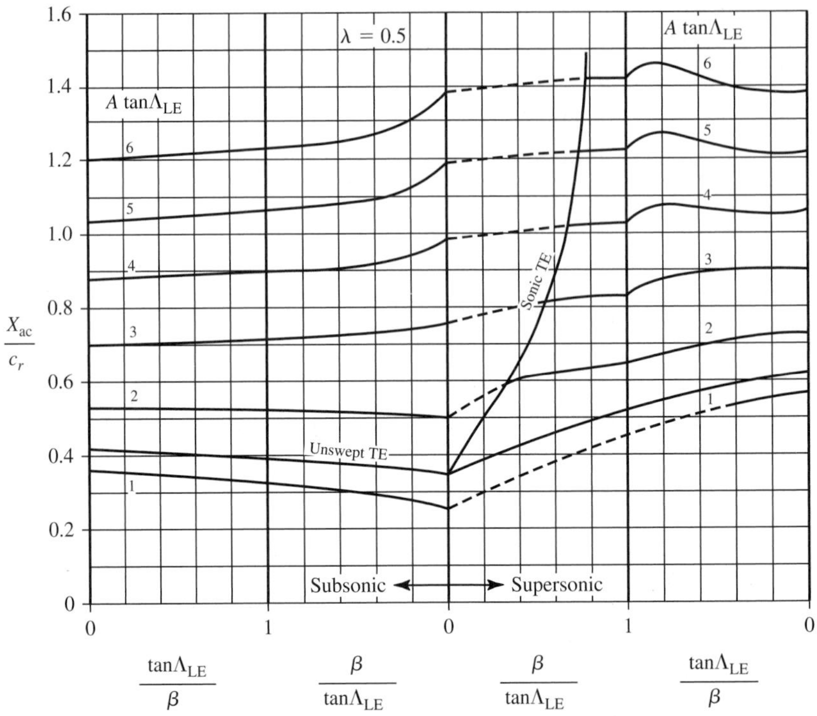

(e)

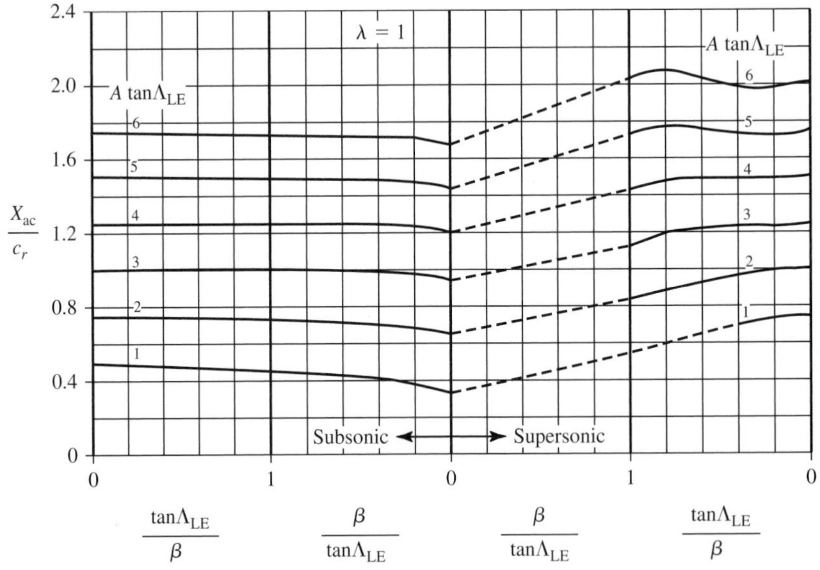

(f)

Figure 5.23e–f Axial location of wing aerodynamic center.

If the section lift-curve slope is approximately constant with span (a good approximation), Equation (5.30) may be simplified to yield

$$X_{ac} = \frac{2}{S}\int_0^{b/2} x_{ac}(y)c(y)dy \qquad (5.31)$$

But Equation (5.31) is simply the location of the aerodynamic center of the wing's MAC. Hence, knowing the x location and length of the wing's MAC locates the wing's *ac*. Finally, note from Equation (5.28) that for no wing twist and constant section characteristics with span, $C_{M_{ac}} = c_{m_{ac}}$, or the wing moment coefficient is the same as the section moment coefficient taken about their respective aerodynamic centers.

5.3.4 Wing Rolling Moment

A wing may also generate a rolling moment about its root (or the wing's x axis), even though the wing is symmetric. A rolling moment typically occurs when the wing is experiencing a sideslip—that is, the airflow over the wing has a spanwise component as shown in Figure 5.24. Defined in this figure is the sideslip angle β, which is analogous to the angle of attack α. A positive sideslip angle is as shown. (The student should not confuse sideslip angle with the Prandtl–Glauert compressibility correction factor, also denoted as β. The usage should be clear from the context.)

The rolling moment on a wing arises primarily due to its sweep and dihedral angles. The effect of the sweep angle is depicted schematically in Figure 5.25. (The dihedral angle is discussed below.) Shown in this figure are two sets of velocity vectors. The set on the left depicts the case with no sideslip angle, while the set on the right includes sideslip. In each case, the free-stream velocity is decomposed into a component normal to the wing leading edge and a component parallel to the leading edge.

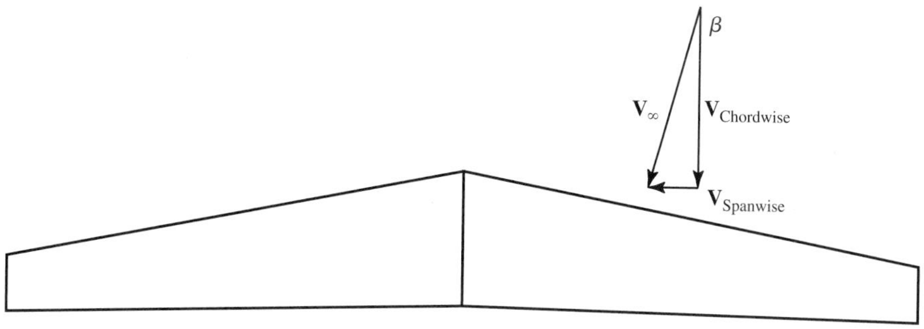

Figure 5.24 Swept wing in sideslip.

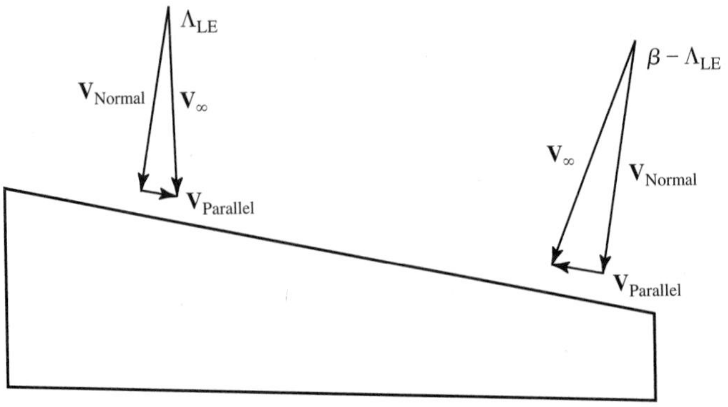

Figure 5.25 Normal velocity component with wing sweep.

Note that for the right wing shown in the figure, the normal velocity component is

$$V_{\text{Normal}} = V_\infty \cos(\beta - \Lambda_{\text{LE}}) = V_\infty \cos(\Lambda_{\text{LE}} - \beta) \tag{5.32}$$

And the parallel velocity component is

$$V_{\text{Parallel}} = -V_\infty \sin(\beta - \Lambda_{\text{LE}}) = V_\infty \sin(\Lambda_{\text{LE}} - \beta) \tag{5.33}$$

(A negative parallel velocity is directed in the negative y direction.) Conversely, for the left wing, one can show that the normal velocity component is

$$V_{\text{Normal}} = V_\infty \cos(\Lambda_{\text{LE}} + \beta) \tag{5.34}$$

And the parallel velocity component is

$$V_{\text{Parallel}} = -V_\infty \sin(\Lambda_{\text{LE}} + \beta) \tag{5.35}$$

Assume that the 2-D lift generated by a strip at span location y on the wing is proportional to V^2_{Normal} (from dynamic pressure), and that the flow component parallel to the wing leading edge does not affect the lift on that strip. In other words, let the 2-D lift per unit span be expressed as

$$l = c_l q_{\text{Normal}} c \tag{5.36}$$

where

$$q_{\text{Normal}} = \frac{1}{2}\rho_\infty V^2_{\text{Normal}}$$

Therefore, for a strip on the right wing, the 2-D lift is

$$l_{\text{Right}} = c_l\left(\frac{1}{2}\rho_\infty V_\infty^2 \cos^2(\Lambda_{\text{LE}} - \beta)\right)c = c_l q_\infty c \cos^2(\Lambda_{\text{LE}} - \beta) \tag{5.37}$$

That is, the lift on the right wing is proportional to $\cos^2(\Lambda_{\text{LE}} - \beta)$. For a strip on the left wing the 2-D lift is

$$l_{\text{Left}} = c_l\left(\frac{1}{2}\rho_\infty V_\infty^2 \cos^2\left(\Lambda_{\text{LE}} + \beta\right)\right)c = c_l q_\infty c \cos^2\left(\Lambda_{\text{LE}} + \beta\right) \tag{5.38}$$

So the lift on the left wing is proportional to $\cos^2(\Lambda_{\text{LE}} + \beta)$. Clearly, with no sideslip ($\beta = 0$), the lift on the right wing is the same as the lift on the left wing (assuming symmetric span-wise locations), and no rolling moment results. But a rolling moment is generated when sideslip β is not zero. For the wing, the 2-D rolling moment due to a strip at span-wise location y is

$$l_{\text{Roll}}(y) = -l(y)y \tag{5.39}$$

NOTE TO STUDENT

By convention, rolling moment is denoted as l for the 2-D section, and L for the 3-D wing. This can cause confusion since lift is also denoted as l and L. Although the usage should be clear from the context, we will include the subscript "Roll" to avoid confusion. Also by convention, roll is positive right-wing down, and recall that the origin of the wing coordinate system is at the wing apex. Hence, y is negative along the left wing and lift on that wing produces a positive rolling moment.

We may now integrate the 2-D rolling moment over the entire wingspan to obtain an expression for the 3-D wing rolling moment due to wing sweep. Or

$$\begin{aligned}
L_{\text{Roll}} &= \int_{-b/2}^{b/2} -l(y)y\,dy = \int_{-b/2}^{0} -l_{\text{Left}}(y)y\,dy + \int_{0}^{b/2} -l_{\text{Right}}(y)y\,dy \\
&= \int_{0}^{b/2} l_{\text{Left}}(y)y\,dy - \int_{0}^{b/2} l_{\text{Right}}(y)y\,dy \\
&= \int_{0}^{b/2} \left(c_l(y)q(y)c(y)\right)y\,dy \cos^2\left(\Lambda_{\text{LE}} + \beta\right) \\
&\quad - \int_{0}^{b/2} \left(c_l(y)q(y)c(y)\right)y\,dy \cos^2\left(\Lambda_{\text{LE}} - \beta\right) \\
&= \left(\int_{0}^{b/2} \left(c_l(y)q(y)c(y)\right)y\,dy\right)\left(\cos^2\left(\Lambda_{\text{LE}} + \beta\right) - \cos^2\left(\Lambda_{\text{LE}} - \beta\right)\right)
\end{aligned} \tag{5.40}$$

Noting the trigonometric identities

$$\cos^2 a = \frac{1}{2}\left(\cos 2a + 1\right)$$

and

$$\cos(a + b) = \cos a \cos b - \sin a \sin b \tag{5.41}$$

Equation (5.40) may be written as

$$L_{\text{Roll}} = \left(\int_0^{b/2} \big(c_l(y) q(y) c(y)\big) y\,dy \right) \left(-\sin 2\Lambda_{\text{LE}} \sin 2\beta \right) = C_{L_{\text{Roll}}} q_\infty S b \tag{5.42}$$

assuming the dynamic pressure $q(y) = q_\infty$ is constant with span. The wing's rolling-moment coefficient $C_{L_{\text{Roll}}}$ is defined here, and the characteristic length is the wing span b, by convention. Note that this rolling moment is proportional to both the sine of twice the leading-edge sweep angle and the sine of twice the sideslip angle.

The integral in Equation (5.42) should be recognized as just the rolling moment due to the lift distribution on the right wing. One can show that for constant section characteristics and dynamic pressure with span, a reasonable approximation for this integral is

$$\int_0^{b/2} \big(c_l(y) q(y) c(y)\big) y\,dy \approx C_{L_\alpha} \left(\alpha_{\text{Wing}} - \alpha_{0_{\text{Wing}}} \right) q_\infty Y_{\text{MAC}} \frac{S}{2} \tag{5.43}$$

where C_{L_α} is the 3-D wing lift effectiveness and Y_{MAC} is the span location of the wing's MAC. This yields a rolling-moment coefficient equal to

$$\begin{aligned} C_{L_{\text{Roll}}} &= -C_{L_\alpha} \left(\alpha_{\text{Wing}} - \alpha_{0_{\text{Wing}}} \right) \frac{Y_{\text{MAC}}}{2b} \left(\sin 2\Lambda_{\text{LE}} \sin 2\beta \right) \\ &\approx -C_{L_{\text{Wing}}} \overline{Y}_{\text{MAC}} \left(\sin 2\Lambda_{\text{LE}} \right) \beta \end{aligned} \tag{5.44}$$

which is the coefficient of rolling moment associated with wing sweep.

The wing *dihedral angle* Γ is defined in Figure 5.26. A positive dihedral angle results in the left and right wing each being canted up at the angle Γ about the wing root. At a typical span-wise location y, and without dihedral, the local

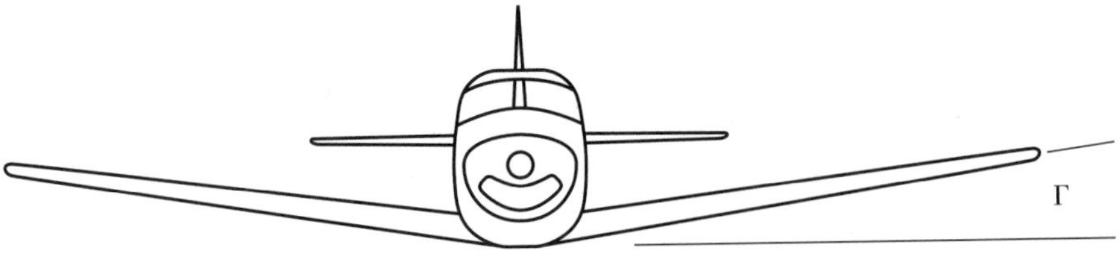

Figure 5.26 Wing dihedral angle Γ.

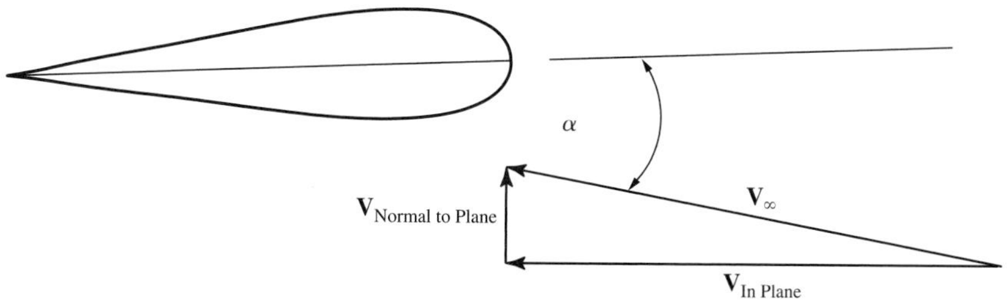

Figure 5.27 In-plane and vertical components of free-stream velocity vector -no dihedral.

angle of attack α is as shown in Figure 5.27, in which the velocity component normal to the plane of the wing, $V_{\text{Normal to Plane}}$ is also indicated. This view is along the span looking from the right wing tip. Clearly, the velocity component normal to the plane is

$$V_{\text{Normal to Plane}} = V_\infty \sin\alpha \tag{5.45}$$

Now with dihedral, $V_{\text{Normal to Plane}}$ can be further decomposed into a component normal to the wing V_{Normal}, and a component along the wing span V_{Span}, as shown in Figure 5.28. This view is a close up of a portion of the right wing, looking forward from behind the wing. Clearly, from this figure the local-flow-velocity component normal to the right wing is

$$V_{\text{Normal}} = V_{\text{Normal to Plane}} \cos\Gamma = V_\infty \sin\alpha \cos\Gamma. \tag{5.46}$$

A similar analysis would reveal that the flow component normal to the left wing is the same as that given above.

Defining the sine of the effective local angle of attack at a given section of the wing to be

$$\sin\alpha_{\text{eff}} \stackrel{\Delta}{=} \frac{V_{\text{Normal}}}{V_\infty} = \sin\alpha \cos\Gamma \tag{5.47}$$

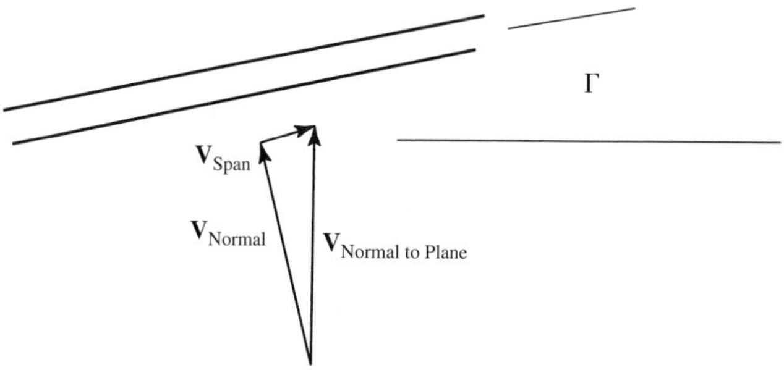

Figure 5.28 Normal velocity component with dihedral—right wing.

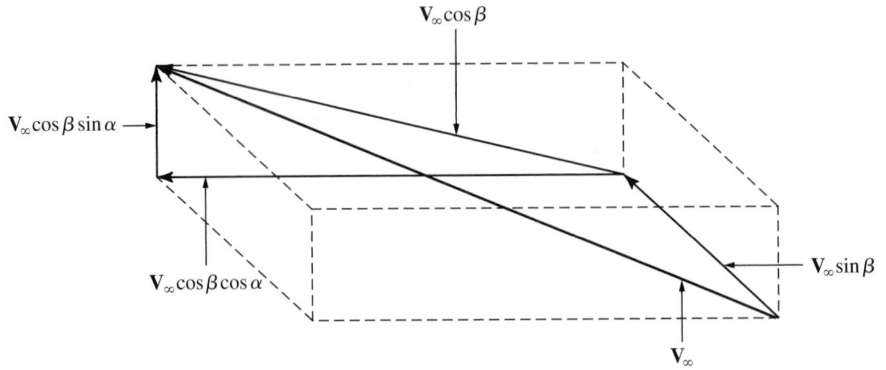

Figure 5.29 Components of free-stream velocity vector.

we see that with no sideslip the effect of a small dihedral angle on the local section angle of attack is negligible (since the cosine of a small angle is approximately unity.) In addition, since the effective angle of attack on each side of the wing is identical, the lift is symmetrical, and no rolling moment is generated.

But with sideslip the situation changes. Shown in Figure 5.29 is the flow velocity vector with an angle of attack α and sideslip angle β. Now the free-stream velocity has two components in the horizontal plane, namely $V_\infty \cos\beta\cos\alpha$ and $V_\infty \sin\beta$. The first of these two velocity components may be further decomposed, as discussed previously, to find its component normal to the surface of the wing. Plus we need to account for the second component, $V_\infty \sin\beta$.

So with sideslip, Figure 5.28 now becomes as shown in Figure 5.30. The total component of velocity normal to the surface of the right wing is

$$V_{\text{Normal Right}} = V_{\text{Normal 1}} + V_{\text{Normal 2}} = V_\infty \sin\alpha\cos\beta\cos\Gamma + V_\infty \sin\beta\sin\Gamma \quad (5.48)$$

But the component normal to the surface of the left wing will be

$$V_{\text{Normal Left}} = V_\infty \sin\alpha\cos\beta\cos\Gamma - V_\infty \sin\beta\sin\Gamma \quad (5.49)$$

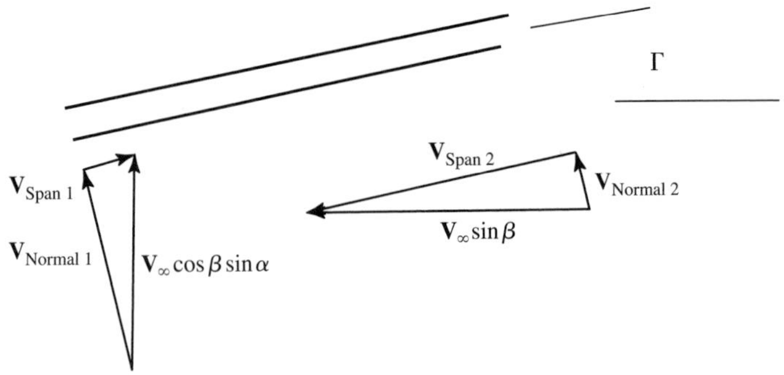

Figure 5.30 Normal velocity components with dihedral and sideslip—right wing.

Now, similar to the case without sideslip, define the sine of the effective local angle of attack to be

$$\sin\alpha_{\text{eff}} \triangleq \frac{V_{\text{Normal}}}{V_\infty \cos\beta} \tag{5.50}$$

For the right wing this becomes

$$\begin{aligned}\sin\alpha_{\text{eff Right}} \triangleq \frac{V_{\text{Normal}}}{V_\infty \cos\beta} &= \frac{V_\infty \sin\alpha\cos\beta\cos\Gamma + V_\infty \sin\beta\sin\Gamma}{V_\infty \cos\beta} \\ &= \sin\alpha\cos\Gamma + \tan\beta\sin\Gamma\end{aligned} \tag{5.51}$$

Or, for small angles,

$$\alpha_{\text{eff Right}} \approx \alpha + \beta\Gamma \tag{5.52}$$

And for the left wing we have

$$\begin{aligned}\sin\alpha_{\text{eff Left}} \triangleq \frac{V_{\text{Normal}}}{V_\infty \cos\beta} &= \frac{V_\infty \sin\alpha\cos\beta\cos\Gamma - V_\infty \sin\beta\sin\Gamma}{V_\infty \cos\beta} \\ &= \sin\alpha\cos\Gamma - \tan\beta\sin\Gamma\end{aligned} \tag{5.53}$$

Or, for small angles,

$$\alpha_{\text{eff Left}} = \alpha - \beta\Gamma \tag{5.54}$$

Therefore with sideslip, the (sine of the) effective angle of attack on the right wing is <u>increased</u> by the amount $\tan\beta\sin\Gamma$, and on the left wing it is <u>decreased</u> by the same amount. This creates an antisymmetric lift distribution, and therefore a rolling moment.

As in the analysis of wing sweep, the rolling moment due to a 2-D strip of the wing at span-wise location y may be written

$$l_{\text{Roll}} = -c_l q(y) y c(y) = -c_{l_\alpha}(\alpha - \alpha_0) q(y) y c(y) \tag{5.55}$$

Using Equation (5.40), and substituting the appropriate effective angles of attack for the left and right wing, the total rolling moment associated with dihedral may be expressed as

$$\begin{aligned}L_{\text{Roll}} &= \int_0^{b/2} l_{\text{Left}}(y) y\,dy - \int_0^{b/2} l_{\text{Right}}(y) y\,dy \\ &= \int_0^{b/2} c_{l_\alpha}(\alpha_{\text{Left}} - \alpha_0) q(y) c(y) y\,dy - \int_0^{b/2} c_{l_\alpha}(\alpha_{\text{Right}} - \alpha_0) q(y) c(y) y\,dy \\ &= \int_0^{b/2} c_{l_\alpha}(\alpha - \beta\Gamma - \alpha_0) q(y) c(y) y\,dy - \int_0^{b/2} c_{l_\alpha}(\alpha + \beta\Gamma - \alpha_0) q(y) c(y) y\,dy\end{aligned}$$

$$= \int_0^{b/2} c_{l_\alpha}(-\beta\Gamma)q(y)c(y)y\,dy - \int_0^{b/2} c_{l_\alpha}(\beta\Gamma)q(y)c(y)y\,dy$$

$$= -2\Gamma \int_0^{b/2} c_{l_\alpha}(\beta)q(y)c(y)y\,dy \tag{5.56}$$

Now assuming c_{l_α}, $q(y)$ $(= q_\infty)$, and β are constant with span, the above rolling moment due to dihedral becomes

$$\begin{aligned} L_{\text{Roll}} &= -2\Gamma c_{l_\alpha}\beta q_\infty \int_0^{b/2} c(y)y\,dy \approx -2\Gamma C_{L_\alpha}\beta q_\infty\left(Y_{\text{MAC}}\frac{S}{2}\right) \\ &= -\Gamma C_{L_\alpha}\beta q_\infty Y_{\text{MAC}} S = C_{L_{\text{Roll}}} q_\infty S b \end{aligned} \tag{5.57}$$

Here we have again note that the integral equals the term in parentheses, and the wing 3-D lift effectiveness has been used instead of section lift effectiveness. Consequently, the rolling moment coefficient associated with dihedral becomes

$$C_{L_{\text{Roll}}} = -\Gamma C_{L_\alpha}\beta \frac{Y_{\text{MAC}}}{b} = -\Gamma C_{L_\alpha}\beta \overline{Y}_{\text{MAC}} \tag{5.58}$$

Finally, if both wing sweep and dihedral are present, the total rolling moment coefficient is the sum of Equation (5.44) plus Equation (5.58), or

$$\begin{aligned} C_{L_{\text{Roll}}} &= -C_{L_{\text{Wing}}}\overline{Y}_{\text{MAC}}(\sin 2\Lambda_{\text{LE}})\beta - \Gamma C_{L_\alpha}\beta\overline{Y}_{\text{MAC}} \\ &= -(C_{L_{\text{Wing}}}\sin 2\Lambda_{\text{LE}} + \Gamma C_{L_\alpha})\overline{Y}_{\text{MAC}}\beta \end{aligned} \tag{5.59}$$

So a wing's *dihedral effect,* or its *sideslip rolling-moment effectiveness,* is

$$C_{L_\beta} \triangleq \frac{\partial C_{L_{\text{Roll}}}}{\partial \beta} = -(C_{L_{\text{Wing}}}\sin 2\Lambda_{\text{LE}} + \Gamma C_{L_\alpha})\overline{Y}_{\text{MAC}} \tag{5.60}$$

NOTE TO STUDENT

Although we have added the subscript "Roll" to the rolling-moment coefficient $C_{L_{\text{Roll}}}$ to avoid confusion with the lift coefficient, we will not add the subscript to the rolling-moment effectiveness such as C_{L_β}. There is little chance this would be confused with sideslip lift-effectiveness, since such an effectiveness is much less common.

A final comment on rolling moment due to dihedral is warranted. The analysis above focused on the components of velocity normal to the surface of the wing. But as shown in Figure 5.30, two span-wise components of velocity are also present. It can be shown that the first such component, $V_{\text{Span 1}}$ is symmetric about the wing centerline, and hence will produce no rolling moment. But the second component $V_{\text{Span }2}$ associated with the sideslip is not symmetric. This second span-wise component can create an additional rolling moment if something blocks this span-wise flow, such as the fuselage of the vehicle protruding below the wing. Hence, such a blockage will increase the wing's dihedral effect.

5.3.5 Wing Drag

The aerodynamic drag on a 3-D wing is composed of two components—*parasite* (or zero-lift) drag and *induced* drag (or drag due to lift). That is,

$$C_{D_{\text{wing}}} = C_{D_P} + C_{D_I} \tag{5.61}$$

Parasite drag includes friction drag and pressure drag due to flow separation. We will usually assume that no separation occurs on a smooth wing (i.e., no protuberances or openings). Friction drag arises from the friction associated with the flow passing over the wing surface, and is proportional to the wetted area (i.e, the entire surface area) of the wing S_{wet}. The friction drag coefficient of a wing may be estimated from the following expression.

$$\boxed{C_{D_f} = C_f\left(1 + 2\left(\frac{t}{c}\right) + 100\left(\frac{t}{c}\right)^4\right)\frac{S_{\text{wet}}}{S}} \tag{5.62}$$

where

C_f is the skin-friction coefficient, which depends on the flow Mach number and Reynolds number R_l. ($R_l = \dfrac{\rho_\infty V_\infty l}{\mu}$, where ρ_∞ = air density, μ = coefficient of viscosity of air, l = characteristic length)

t/c is the thickness ratio of the wing's airfoil section

The skin-friction coefficient C_f for fully turbulent flow may be determined from Figure 5.31, which gives friction coefficient as a function of Reynolds number (based on wing chord length) and Mach number for a smooth surface (low surface roughness).

The induced drag of a wing, C_{D_I}, is theoretically proportional to the square of the wing's lift coefficient C_L, for both subsonic and supersonic flow. The subsonic induced drag of a wing is given by

$$\boxed{C_{D_I} = \frac{C_L^2}{\pi A e}} \tag{5.63}$$

where e is Oswald's span-efficient factor ($0 < e < 1$).

The efficiency factor e is a function of wing geometry and wing loading, with typical values of about 0.8–0.9 for high-aspect ratio, unswept wings. For highly swept wings, or wings with low aspect ratio, the value of e can decrease to as low as around 0.4. For supersonic drag estimation, the reader is referred to Ref. 1 at the end of this chapter.

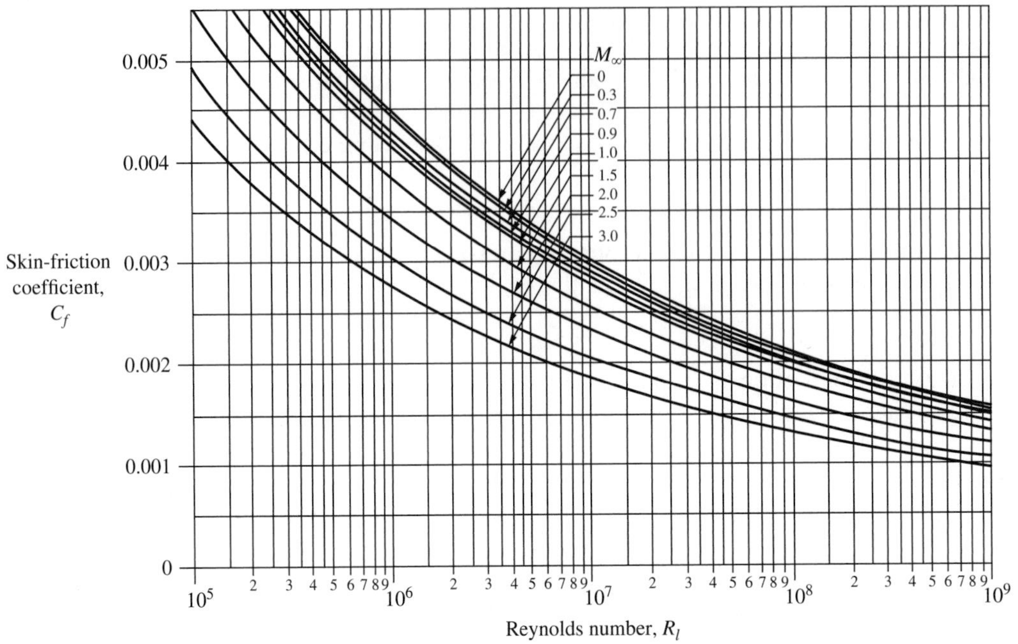

Figure 5.31 Skin friction coefficient as a function of Reynolds and Mach numbers.

EXAMPLE 5.4

Wing 3-D Aerodynamic Analysis

Consider a swept wing with the following characteristics:

1. Straight leading and trailing edges.
2. Leading-edge sweep angle $\Lambda_{\mathrm{LE}} = 26.6$ deg.
3. Taper ratio $\lambda = 0.5$.
4. Root chord length $c_r = 7.5$ ft.
5. Span $b = 30$ ft.
6. Linear twist distribution with span, and $\varepsilon_{\mathrm{Tip}} = -3$ deg.
7. Dihedral $\Gamma = 5$ deg.
8. Constant airfoil section with span—NACA 0009.

Sketch the wing planform and determine the following:

1. Wing planform area S and aspect ratio A.
2. Sweep angle of the mid-chord line, $\Lambda_{c/2}$.
3. Wing lift effectiveness C_{L_α} at Mach number $M_\infty = 0.2$.
4. Wing zero-lift angle of attack, $\alpha_{0_{\mathrm{Wing}}}$.
5. Length and position of the mean aerodynamic (geometric) chord, $\bar{c}$.

6. Axial location of the wing aerodynamic center, $X_{AC_{Wing}}$.
7. Wing pitching-moment coefficient about its aerodynamic center, $C_{m_{AC}}$.
8. Wing dihedral effect and rolling-moment coefficient at 2 deg angle of attack and 2 deg angle of sideslip.
9. Wing drag coefficient at 2 deg angle of attack, C_D.

■ Solution

Knowing that the wing taper ratio is 0.5, the length of the root chord is 7.5 ft, the span is 30 ft, and the leading-edge sweep is 26.6 deg, we can sketch the wing planform as shown in Figure 5.32 (not to scale). Note that from the geometry given, the location of the leading edge of the wing tip $X_{LE_{Tip}}$ is 7.5 ft aft of the leading edge of the root chord (or wing apex), which is taken as the origin of the wing x,y coordinate frame. The sweep of the mid-chord line, again from the geometry, is found from

$$\tan\Lambda_{c/2} = \frac{7.5 + \left(\dfrac{3.75}{2}\right) - \left(\dfrac{7.5}{2}\right)}{15} = \frac{5.625}{15}$$

So $\Lambda_{c/2} \approx 20.5$ deg.

With straight leading and trailing edges, the chord length is a linear function of span y, so for the right wing the expression for the chord length is given by

$$c(y) = 7.5 + \frac{(3.75 - 7.5)}{15} y = \left(7.5 - \frac{y}{4}\right) \text{ ft}$$

Using the expressions from Table 5.3 we find that the planform area is

$$S = 2\int_0^{15} c(y)dy = 2\int_0^{15}\left(7.5 - \frac{y}{4}\right)dy = 2\left(7.5y - \frac{y^2}{8}\right)\Bigg|_0^{15} \approx 169 \text{ ft}^2$$

the aspect ratio is

$$A = \frac{b^2}{S} = \frac{(30)^2}{169} = 5.33$$

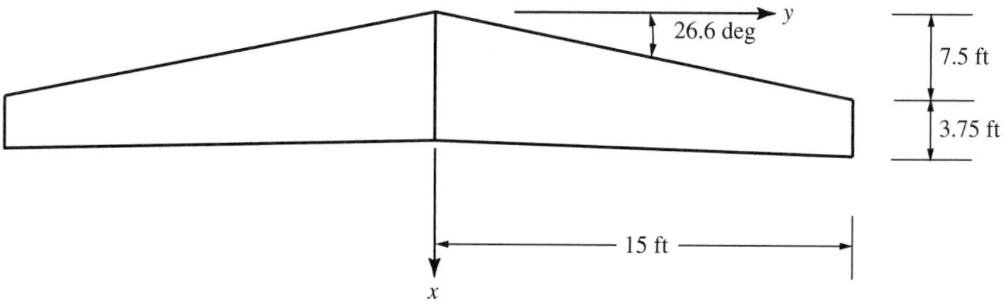

Figure 5.32 Wing planform.

and the length of the mean aerodynamic (geometric) chord is

$$\bar{c} = \frac{2}{S}\int_0^{15} c^2 dy = \frac{2}{169}\int_0^{15}\left(7.5 - \frac{y}{4}\right)^2 dy = \frac{2}{169}\left((7.5)^2 y - 7.5\frac{y^2}{(2)(2)} + \frac{y^3}{(16)(3)}\right)\Bigg|_0^{15} = 5.825 \text{ ft}$$

Using expressions from Tables 5.3 and 5.4 we find the x,y location of the mean aerodynamic chord to be

$$X_{\text{LE}_{\text{MAC}}} = \frac{2}{S}\int_0^{15} x_{\text{LE}}(y)c(y)dy = \frac{2}{169}\int_0^{15}\left(\frac{y}{2}\right)\left(7.5 - \frac{y}{4}\right)dy = \frac{2}{169}\left(7.5\frac{y^2}{4} - \frac{y^3}{24}\right)\Bigg|_0^{15} = 3.33 \text{ ft}$$

and

$$Y_{\text{MAC}} = \frac{2}{S}\int_0^{15} yc(y)dy = \frac{2}{169}\int_0^{15} y\left(7.5 - \frac{y}{4}\right)dy = \frac{2}{169}\left(7.5\frac{y^2}{2} - \frac{y^3}{12}\right)\Bigg|_0^{15} = 6.66 \text{ ft}$$

The wing's aerodynamic center (*ac*) is located at the aerodynamic center of its mean aerodynamic chord (MAC). And from the above we found that the leading edge of the MAC is 3.33 ft aft of the wing apex. From Table 5.2 we note that the aerodynamic center of the NACA 0009 airfoil is located at 25 percent of the chord aft of its leading edge. Consequently the axial location of the wing's *ac* is

$$X_{\text{AC}_{\text{Wing}}} = 3.33 + 0.25\bar{c} = 4.79 \text{ ft}$$

aft of the leading edge of the root chord, or wing apex.

As in Example 5.3, the wing lift effectiveness may be determined by using Equation (5.18), which is

$$C_{L_\alpha} = \frac{2\pi A}{2 + \sqrt{\frac{A^2\beta^2}{k^2}\left(1 + \frac{\tan^2\Lambda_{c/2}}{\beta^2}\right) + 4}}$$

At $M_\infty = 0.2$, $\beta = \sqrt{0.96}$, and using the section characteristics in Table 5.2 we see that $k = (0.109)(180/\pi)/2\pi = 0.99$. Consequently, the wing lift effectiveness is

$$C_{L_\alpha} = \frac{2\pi(5.33)}{2 + \sqrt{\frac{(5.33)^2(0.96)}{(0.99)^2}\left(1 + \frac{(5.625/15)^2}{0.96}\right) + 4}} = 2\pi\frac{5.33}{7.99} = 4.19 \text{ /rad}$$

The wing zero-lift angle of attack may be determined from Equation (5.23), or

$$\alpha_{0_{\text{Wing}}} = \frac{2}{S}\int_0^{b/2}(\alpha_0(y) - \varepsilon(y))c(y)dy = \frac{2}{169}\int_0^{15}\left(\frac{y}{5}\right)\left(7.5 - \frac{y}{4}\right)dy$$

$$= \frac{2}{169}\left(7.5\frac{y^2}{10} - \frac{y^3}{60}\right)\Bigg|_0^{15} = 1.33 \text{ deg}$$

So even though the NACA 0009 airfoil cross section has an $\alpha_0 = 0$, the wing zero-lift angle of attack is not zero, due to wing twist. And note that the wing-twist distribution with span for the right wing is

$$\varepsilon(y) = -\frac{3}{15}y \text{ deg}, \; y > 0$$

The sign is reversed for the left wing, since y would be negative.

We may now determine the wing's pitching-moment coefficient taken about its aerodynamic center. From Equation (5.28) we have

$$C_{M_{ac}} = \frac{2}{S\bar{c}}\left(\int_0^{b/2} c_{m_{ac}}(y)c^2(y)dy - \int_0^{b/2} c_{l_\alpha}(y)\left(\alpha_{0_{\text{Wing}}} + \varepsilon(y) - \alpha_0(y)\right)\left(x_{ac}(y) - X_{ac}\right)c(y)dy\right)$$

which here simplifies considerably since for the symmetric NACA 0009 airfoil $\alpha_0 = c_{m_{ac}} = 0$. In this case then

$$C_{M_{ac}} = \frac{2}{S\bar{c}}\int_0^{b/2} c_{l_\alpha}(y)\left(\alpha_{0_{\text{Wing}}} + \varepsilon(y)\right)\left(X_{ac} - x_{ac}(y)\right)c(y)dy$$

All the quantities in the above expression are known, except for the span-wise function locating the aerodynamic centers of the airfoil sections, or $x_{ac}(y)$. Here this function takes the form

$$x_{ac}(y) = x_{LE}(y) + 0.25c(y) = \frac{y}{2} + \frac{1}{4}\left(7.5 - \frac{y}{4}\right) = \frac{7.5}{4} + \frac{7}{16}y$$

Consequently, the wing pitching-moment coefficient is

$$C_{M_{ac}} = \frac{2(0.109)}{(169)(5.825)}\int_0^{15}\left(1.33 - \frac{y}{5}\right)\left(4.79 - \left(\frac{7.5}{4} + \frac{7}{16}y\right)\right)\left(7.5 - \frac{y}{4}\right)dy$$

$$= \frac{2(0.109)}{(984.425)}\left(29.077y - 9.741\frac{y^2}{2} + 0.954\frac{y^3}{3} - 0.022\frac{y^4}{4}\right)\Bigg|_0^{15} = 0.03$$

The wing's rolling-moment coefficient may be determined from Equations (5.44) and (5.58), or

$$C_{L_{\text{Roll}}} = -0.5C_{L_\alpha}\left(\alpha_{\text{Wing}} - \alpha_{0_{\text{Wing}}}\right)\bar{Y}_{\text{MAC}}\left(\sin 2\Lambda_{\text{LE}}\sin 2\beta\right) - C_{L_\alpha}\Gamma\beta\bar{Y}_{\text{MAC}}$$

and from Equation (5.60), the wing's dihedral effect is given by

$$C_{L_\beta} \triangleq \frac{\partial C_{L_{\text{Roll}}}}{\partial\beta} = -\left(C_{L_{\text{Wing}}}\sin 2\Lambda_{\text{LE}} + \Gamma C_{L_\alpha}\right)\bar{Y}_{\text{MAC}}$$

Since everything in the above two expressions is known, we have for the rolling-moment coefficient

$$C_{L_{\text{Roll}}} = -\frac{4.19}{2}\left(\frac{6.66}{30}\right)\left((2 - 1.33)\left(\frac{\pi}{180}\right)(\sin 53.2° \times \sin 4°) + 5\left(\frac{\pi}{180}\right)2\left(\frac{\pi}{180}\right)\right)$$

$$= -\frac{4.19}{2}\left(\frac{6.66}{30}\right)(0.0007 + 0.0030) = -0.0017$$

And the dihedral effect is

$$C_{L_\beta} = -(4.19)((2 - 1.33)\sin 53.2° + 5)\left(\frac{\pi}{180}\right)\left(\frac{6.66}{30}\right) = -0.09\,/\text{rad}$$

Finally, we will determine the wing's drag coefficient at $M_\infty = 0.2$. We will assume that the wing is sufficiently smooth to preclude pressure drag due to separation, so the parasite drag is only due to friction. Since the altitude is not specified we will assume sea-level conditions. Using a standard atmospheric model (e.g., Appendix A), we note the following:

Air density (sea level) $\rho_\infty = 0.002376$ sl/ft^3

Sonic velocity (sea level) $a = 1{,}116$ fps

Coefficient of viscosity (sea level) $\mu = 3.74 \times 10^{-7}$ lb-s/ft^2

With a wing MAC of 5.825 ft, we have a Reynolds number of

$$R_l = \frac{\rho V l}{\mu} = \frac{(0.002376)\,(0.2 \times 1{,}116)\,(5.825)}{3.74 \times 10^{-7}} = 8.4 \times 10^6$$

From Figure 5.31 we see that the skin friction coefficient is $C_f = 0.0031$. This yields a parasite (friction only) drag coefficient of

$$C_{D_p} = C_f\left(1 + 2\left(\frac{t}{c}\right) + 100\left(\frac{t}{c}\right)^4\right)\frac{S_{\text{wet}}}{S} = 0.0031(1 + 2(.09) + 100(.09)^4)(2) = 0.0074$$

The induced drag is simply

$$C_{D_I} = \frac{C_L^2}{\pi A e}$$

At an angle of attack of 2 deg, the wing's lift coefficient is

$$C_{L_{\alpha=2}} = C_{L_\alpha}(\alpha - \alpha_0) = 4.19(2 - 1.33)\left(\frac{\pi}{180}\right) = 0.049$$

Therefore the wing's induced drag (assuming an Oswald's efficiency factor $e = 0.85$) is

$$C_{D_I} = \frac{C_L^2}{\pi A e} = \frac{(0.049)^2}{\pi(5.33)(0.85)} = 1.7 \times 10^{-4}$$

So the total drag coefficient of the wing at 2 deg angle of attack is

$$C_D = C_{D_p} + C_{D_I} = 0.0074 + 0.0002 = 0.0076$$

5.4 EFFECTS OF FLAPS ON WING AERODYNAMIC CHARACTERISTICS

High-lift devices, such as trailing-edge flaps or control surfaces, affect the following characteristics of the 3-D wing.

α_δ—flap lift effectiveness for the wing
α_0—zero-lift wing angle of attack
$C_{L_{\max}}$—maximum wing lift coefficient
$C_{M_{ac}}$—moment (coefficient) about the wing's *ac*
C_D—wing drag

To estimate these parameters, we will make maximum use of the characteristics of the 2-D airfoil sections making up the wing. These section characteristics may be obtained from Sections 5.2 and 5.3, or elsewhere. It is also important to recall, as discussed when dealing with flapped 2-D sections, that under linear wing or lifting-line theory, trailing-edge high-lift devices have little appreciable effect on the wing's lift-curve slope C_{L_α} or location of its aerodynamic center, X_{ac}. Both of these results assume that flap deflection does not lead to flow separation on the upper wing surface; that is, flap deflections are not too large.

5.4.1 Flaps and Control Surfaces

As we discussed when dealing with section characteristics, a control surface is a trailing-edge flap, so except for ailerons (treated in Section 5.4.2) no distinction will be made here between the effects of flaps or control surfaces. We will simply refer to "flaps" here when we could just as well be dealing with a control surface such as an elevator. Finally, strip theory, or area-weighted averaging, will be used frequently to determine wing characteristics from section characteristics.

To first address the flap lift effectiveness C_{L_δ} we refer back to Figure 5.21, showing the 2-D lift of an infinitesimal wing section *dl*. If that section also included a trailing-edge flap, Equation (5.20) may be modified to account for the additional section lift from the flap, yielding

$$dl = \left(c_{l_\alpha}(y)\left(\alpha_{y=0} + \varepsilon(y) - \alpha_0(y)\right) + c_{l_\delta}(y)\delta_{\text{flap}}\right)q(y)c(y)dy \qquad (5.64)$$

Equation (5.64) may now be integrated over the wing span to yield an expression for the 3-D wing lift. However, not all airfoil sections along the wing's span include the trailing-edge flap, so care must be taken in performing this integration. Referring to Figure 5.19, let us assume that a single flap is present on each side of the wing. We will here address this single flap, and the same approach may be used to develop similar expressions for each trailing-edge device present.

Note that the wingspan location of the inboard end of the flap is denoted in the figure as η_i, and the span location of the outboard end is denoted η_o. As a result, the additional lift of the wing due to flap deflection may be expressed as

$$\Delta L_{\text{Flap}} = 2\int_{\eta_i}^{\eta_o} c_{l_\delta}(y)\delta_{\text{flap}}q(y)c(y)dy = C_{L_{\delta_F}}\delta_{\text{flap}}q_\infty S \qquad (5.65)$$

Without any flap deflection, the total wing lift was given by Equation (5.21). Note that in Equation (5.65) we have also expressed the additional lift in terms of the total planform area of the wing S, and a wing flap-effectiveness $C_{L_{\delta_F}}$, thus defining this parameter. Assuming the flap deflection δ_{flap} and free-stream dynamic pressure are invariant with respect to the span-wise integration, the flap lift effectiveness becomes

$$\boxed{C_{L_{\delta_F}} = \frac{2}{S}\int_{\eta_i}^{\eta_o} c_{l_\delta}(y)c(y)dy} \tag{5.66}$$

Or, if the section flap effectiveness c_{l_δ} is constant with span, Equation (5.66) reduces to

$$C_{L_{\delta_F}} = c_{l_\delta}\left(\frac{2S_{W_f}}{S}\right) \tag{5.67}$$

where S_{W_f}, shown in Figure 5.19, is the planform area of one side of the wing with the flap on its trailing edge.

The alternate flap-effectiveness parameter for the wing $\alpha_{\delta_{\text{wing}}}$ is defined as

$$\boxed{\alpha_{\delta_{\text{wing}}} \triangleq \frac{C_{L_\delta}}{C_{L_\alpha}} = \frac{\partial \alpha_{\text{wing}}}{\partial \delta_{\text{flap}}}\bigg|_{C_L=\text{constant}}} \tag{5.68}$$

or the ratio of $C_{L_{\delta_F}}$, found from Equation (5.67), and the wing lift-curve slope C_{L_α}, either found from Equation (5.18) or from Figures 5.20*a–f*. Equation (5.68) is consistent with the similar definition used for 2-D airfoil sections in Section 5.2.

Knowing this wing flap effectiveness $\alpha_{\delta_{\text{wing}}}$, one may easily obtain the change in the wing's zero-lift angle of attack due to flap deflection. From Equation (5.68), and referring to Figure 5.12 for reference, we can see that for the wing with flaps

$$\Delta\alpha_0 = \alpha_{\delta_{\text{wing}}}\delta_{\text{flap}} \tag{5.69}$$

The increment in maximum wing lift due to flap deflection may be estimated from the following expression (Ref. 1).

$$\boxed{\Delta C_{L_{\max}} = \Delta c_{l_{\max}} \frac{2S_{W_f}}{S} K_{\text{sweep}}} \tag{5.70}$$

where

$\Delta c_{l_{\max}}$ is the increment in 2-D airfoil section maximum lift due to flap deflection

S_{W_f} is the planform area of portion of one side of wing with trailing edge flap

$K_{\text{sweep}} = \left(1 - 0.08\cos^2\Lambda_{c/4}\right)\cos^{3/4}\Lambda_{c/4}$ is the sweep correction factor

The change in wing pitching moment due to flap deflection $\Delta C_{M_{\text{ac}}}$ may also be determined from strip theory. Referring to Figure 5.22, assume that the infinitesimal 2-D section shown now includes a trailing edge flap. Consistent with Equation (5.25), the pitching moment about the wing's aerodynamic center due to this 2-D section may be written as

$$dm = \left(c_{m_{\text{ac}}}(y) + c_{m_\delta}(y)\delta_{\text{flap}}\right)q(y)c^2(y)dy$$
$$- \left(c_{l_\alpha}(y)\left(\alpha_{\text{wing}} + \varepsilon(y) - \alpha_0(y)\right) + c_{l_\delta}(y)\delta_{\text{flap}}\right)\left(x_{\text{ac}}(y) - X_{\text{ac}}\right)q(y)c(y)dy \quad (5.71)$$

where the above expression now includes the effect of flap deflection δ_{flap} on both the section 2-D lift and pitching moment.

Integrating the above expression over the wing span would yield the total wing pitching moment about its *ac*. Instead we will determine the incremental moment due only to flap deflection ΔM, which is simply

$$\Delta M_{\text{Flap}} = 2\left(\int_{\eta_i}^{\eta_o} c_{m_\delta}(y)\delta_{\text{flap}}q(y)c^2(y)dy - \int_{\eta_i}^{\eta_o} c_{l_\delta}(y)\delta_{\text{flap}}\left(x_{\text{ac}}(y) - X_{\text{ac}}\right)q(y)c(y)dy\right)$$
$$= C_{M_{\delta_F}}\delta_{\text{flap}}q_\infty S\bar{c} \quad (5.72)$$

Here the limits on the integrals are again defined by the span locations of the inboard and outboard ends of the flaps, respectively. Without flap deflection, the wing's pitching moment was given by Equation (5.26).

In Equation (5.72), the incremental pitching moment ΔM is also expressed in terms of a wing moment coefficient due to flaps, or $C_{M_{\delta_F}}$ (thus defining this coefficient). Again assuming the dynamic pressure equals free-stream $q = q_\infty$ and the flap deflection δ_{flap} is invariant with respect to the span-wise integration, we find that the wing's flap pitching-moment effectiveness is given by

$$C_{M_{\delta_F}} = \frac{2}{S\bar{c}}\left(\int_{\eta_i}^{\eta_o} c_{m_\delta}(y)c^2(y)dy - \int_{\eta_i}^{\eta_o} c_{l_\delta}(y)\left(x_{\text{ac}}(y) - X_{\text{ac}}\right)c(y)dy\right) \quad (5.73)$$

The incremental section drag coefficient due to flap deflection may also be estimated from strip theory. Following the approach just presented, let us first refer to Figure 5.21. Let the infinitesimal drag force (not shown in the figure) of the 2-D flapped section be denoted *dd,* and recall that drag by definition acts parallel to the free-stream velocity vector. The component of this section drag due only to flap deflection is

$$dd = c_{d_\delta}(y)\delta_{\text{flap}}q(y)c(y)dy = \frac{\Delta c_{d_{\text{flap}}}}{\Delta\delta_{\text{flap}}}(y)\delta_{\text{flap}}q(y)c(y)dy \quad (5.74)$$

where $\Delta c_{d_{\text{flap}}}$ is the incremental section drag coefficient due to a given flap deflection $\Delta\delta_{\text{flap}}$ obtained from Figure 5.17. Integrating the above expression over the flap span for the right wing and doubling the results to account for the left wing yields the incremental wing drag due to flap deflection, or

$$\Delta D_{\text{Flap}} = 2\int_{\eta_i}^{\eta_o} \frac{\Delta c_{d_{\text{flap}}}}{\Delta\delta_{\text{flap}}}(y)\delta_{\text{flap}} q(y)c(y)dy = C_{D_{\delta_F}}\delta_{\text{flap}} q_\infty S \tag{5.75}$$

So the wing's flap drag effectiveness is given by

$$C_{D_{\delta_F}} = \frac{2}{S}\int_{\eta_i}^{\eta_o} \frac{\Delta c_{d_{\text{flap}}}}{\Delta\delta_{\text{flap}}}(y)c(y)dy = \frac{\Delta c_{d_{\text{flap}}}}{\Delta\delta_{\text{flap}}}\frac{2S_{W_f}}{S} \tag{5.76}$$

The last expression on the right assumes that the section drag due to flap deflection and local dynamic pressure are constant with span.

5.4.2 Ailerons

Typically trailing-edge control surfaces are installed near the tips of the wings to induce a rolling moment on the vehicle. This moment is generated as a result of antisymmetric deflections of these control surfaces called *ailerons*. Ailerons are treated separately here because they deflect antisymmetrically. We will now estimate the rolling and yawing effectiveness of ailerons using a technique similar to that used to estimate the lift effectiveness of flaps (Equations (5.66) and (5.67)).

Again referring to Figure 5.21, note that the lift generated by the infinitesimal 2-D section dl generates a rolling moment about a line at the wing root ($y = 0$) that is parallel to the free-stream velocity vector. (Recall that lift dl is orthogonal to the free-stream velocity vector.) We will denote this moment dl_{roll}, and it may be expressed as

$$dl_{\text{roll}} = -y dl \tag{5.77}$$

Here the negative sign arises since we define a positive rolling moment to be right wing down.

Referring to Equation (5.64), we may write the section lift due only to a "flap" deflection as

$$dl_{\text{flap}} = c_{l_\delta}(y)\delta_{\text{flap}} q(y)c(y)dy \tag{5.78}$$

So the rolling moment associated with this component of lift is

$$dl_{\text{roll}} = -c_{l_\delta}(y)\delta_{\text{flap}} q(y)c(y)y dy \tag{5.79}$$

(To attempt to clarify the notation here we have again included a subscript, and used l_{roll} for section rolling moment.) But the above is the infinitesimal rolling moment due to deflecting the "flap" trailing-edge down on the right wing. If the left flap is deflected trailing-edge up, opposite to the right flap deflection, the rolling moment associated with a 2-D section on the left wing would be

$$dl_{\text{roll}} = -c_{l_\delta}(y)(-\delta_{\text{flap}})q(y)c(y)y\,dy \tag{5.80}$$

Along the left wing, span location y is negative, so this rolling moment will be negative. But that is due to the fact that the left flap was deflected in a negative direction (trailing edge up).

Assuming the right and left wing sections are equidistant from the wing root, defining positive aileron deflection as trailing-edge up on the right wing, and adding the two moments above together yields the infinitesimal 2-D rolling moment due to the antisymmetric aileron deflection, or

$$dl_{\text{roll}} = 2c_{l_\delta}(y)\delta_{\text{aileron}}q(y)c(y)y\,dy \tag{5.81}$$

We have denoted the deflection as "aileron" here to emphasize that the two trailing-edge surfaces are deflected antisymmetrically.

The above expression may now be integrated over the span locations of the inboard and outboard ends of the ailerons, η_i and η_o, respectively, to obtain the change in wing rolling moment due to aileron deflection. The result is

$$\boxed{\Delta L_{\text{Roll Aileron}} = 2\int_{\eta_i}^{\eta_o} c_{l_\delta}(y)\delta_{\text{aileron}}q(y)c(y)y\,dy = C_{L_{\delta_A}}\delta_{\text{aileron}}q_\infty Sb} \tag{5.82}$$

where a wing aileron rolling effectiveness $C_{L_{\delta_A}}$ has been introduced. Note also that the characteristic length used to nondimensionalize the coefficient is the wing span b. If the strip dynamic pressure $q(y)$ equals the free-stream q_∞, and the aileron deflection is invariant with respect to the span-wise integration we have the aileron rolling-moment effectiveness given by

$$\boxed{C_{L_{\delta_A}} = \frac{2}{Sb}\int_{\eta_i}^{\eta_o} c_{l_\delta}(y)c(y)y\,dy} \tag{5.83}$$

Just as aileron deflection generates a rolling moment due to the antisymmetric change in lift, it also generates a yawing moment due to the antisymmetric change in drag. Following the approach presented above, let us refer again to Figure 5.21. Let the infinitesimal drag force (not shown in the figure) of the 2-D flapped section be denoted dd, and recall that drag by definition acts parallel to the free-stream velocity vector. The increase in this section drag due only to a (positive trailing-edge down) flap deflection may be expressed as

$$dd = c_{d_\delta}(y)\delta_{\text{flap}}q(y)c(y)dy = \frac{\Delta c_{d_{\text{flap}}}}{\Delta\delta_{\text{flap}}}(y)\delta_{\text{flap}}q(y)c(y)dy \tag{5.84}$$

where $\Delta c_{d_{\text{flap}}}$ is the incremental section drag coefficient due to a given flap deflection $\Delta\delta_{\text{flap}}$, obtained from Figure 5.17.

This incremental section drag force dd on the right wing generates a positive (nose-right) yawing moment n about a line at the wing root ($y = 0$) that is mutually orthogonal to the wing's y axis and the free-steam velocity vector. This infinitesimal yawing moment (positive nose right) may be expressed as

$$dn = \frac{\Delta c_{d_{\text{flap}}}}{\Delta\delta_{\text{flap}}}(y)\delta_{\text{flap}}q(y)c(y)y\,dy \tag{5.85}$$

(By convention, section yawing moment is denoted by n, and 3-D wing yawing moment is denoted by N.)

Likewise, an antisymmetric (negative trailing-edge up) "flap" deflection on the left wing would also generate a change in section drag dd, but this change in drag would generate the following change in yawing moment

$$dn = \frac{\Delta c_{d_{\text{flap}}}}{\Delta\delta_{\text{flap}}}(y)(-\delta_{\text{flap}})q(y)c(y)y\,dy \tag{5.86}$$

Note here that for the left wing a negative (trailing-edge up) "flap" deflection is indicated. Plus, along the left wing the span location y is negative. Therefore, a reduction in flap deflection near the left wing tip induces a reduction in section drag, which in turn also induces a positive (nose-right) change in yawing moment.

Assuming the two sections above are equidistant from the wing root, defining positive aileron deflection as trailing-edge up on the right wing, and adding together the infinitesimal yawing moments from the right and left wing sections, yields the total 2-D section yawing moment associated with aileron deflection, or

$$dn = -2\frac{\Delta c_{d_{\text{flap}}}}{\Delta\delta_{\text{flap}}}(y)\delta_{\text{aileron}}q(y)c(y)y\,dy \tag{5.87}$$

We have denoted the deflection as "aileron" here to emphasize that the two trailing-edge surfaces are deflected antisymmetrically.

Integrating the above change in section yawing moment over the span locations of the inboard and outboard ends of the ailerons yields the total change in yawing moment due to aileron deflection (called *adverse yaw*):

$$\boxed{\Delta N_{\text{Aileron}} = -2\int_{\eta_i}^{\eta_o}\frac{\Delta c_{d_{\text{flap}}}}{\Delta\delta_{\text{flap}}}(y)\delta_{\text{aileron}}q(y)c(y)y\,dy = C_{N_{\delta_A}}\delta_{\text{aileron}}q_\infty Sb} \tag{5.88}$$

An aileron yawing-moment effectiveness $C_{N_{\delta_A}}$ has here been defined. Again assuming that aileron deflection and dynamic pressure are invariant with respect to the span-wise integration, the aileron yawing-moment effectiveness is given by

$$C_{N_{\delta_A}} = \frac{-2}{Sb}\int_{\eta_i}^{\eta_o}\frac{\Delta c_{d_{\text{flap}}}}{\Delta\delta_{\text{flap}}}(y)c(y)y\,dy \tag{5.89}$$

EXAMPLE 5.5

Effects of Flaps and Ailerons on Wing Aerodynamic Characteristics

Consider the wing analyzed in Example 5.4, and assume that plain trailing-edge flaps are added to that wing, one on each side. The size of the flaps is given by:

Inboard end at span location $y = 5$ ft; outboard end at span location $y = 10$ ft

Ratio of flap chord to wing chord c_f/c is 25 percent, constant with span

Find the flap lift effectiveness $C_{L_{\delta_F}}$ and α_δ, pitching-moment effectiveness $C_{m_{\delta_F}}$, and drag effectiveness $C_{D_{\delta_F}}$. Also assume that the wing is equipped with three-foot-span ailerons, with the outboard section at the wing tips. If these ailerons also have a 25 percent chord ratio, find the aileron rolling-moment and yawing-moment effectiveness, $C_{L_{\delta_A}}$ and $C_{N_{\delta_A}}$, for the wing.

■ Solution

For the NACA 0009 airfoil and the flap geometry given, Equation (5.14) and Figures 5.14 and 5.15 may be used to determine the section flap lift effectiveness c_{l_δ}. For this airfoil section

$$\frac{c_{l_\alpha}}{c_{l_\alpha\,\text{Theory}}} = \frac{0.109(180/\pi)}{2\pi} = 0.994$$

So, for a 25 percent flap-to-wing chord ratio, Figures 5.14 and 5.15 indicate that $c_{l_\delta\,\text{Theory}} = 4\,/\text{rad}$ and that

$$\frac{c_{l_\delta}}{c_{l_\delta\,\text{Theory}}} = 0.99$$

So with the Prandtl–Glauert factor $\beta = \sqrt{0.96}$,

$$c_{l_\delta} = \frac{1}{0.98}(4 \times 0.99) = 4.04\,/\text{rad} = 0.071\,/\text{deg}$$

The section flap pitching-moment effectiveness may be estimated using Equation (5.15), or

$$c_{m_\delta} = -2\sqrt{\frac{c_f}{c}\left(1-\frac{c_f}{c}\right)^3} = -2\sqrt{0.25(1-0.25)^3} = -0.65\,/\text{rad} = -0.0113\,/\text{deg}$$

Figure 5.17 may be used to estimate the section flap drag effectiveness. From the figure we find that for $\delta_{\text{flap}} = 10$ deg, for example, $\Delta c_d = 0.007$. So the section flap drag effectiveness is

$$c_{d_\delta} \approx \frac{\Delta c_d}{\Delta \delta_{\text{flap}}} = \frac{.007}{10} = 0.0007\,/\text{deg} = 0.040\,/\text{rad}$$

Now the flap lift effectiveness for the wing may be estimated from Equation (5.66), or

$$C_{L_{\delta_F}} = \frac{2}{S}\int_{\eta_i}^{\eta_o} c_{l_\delta}(y)c(y)dy = c_{l_\delta}\frac{2S_{W_F}}{S}$$

since c_{l_δ} is constant with span here, and S_{W_F} is as shown in Figure 5.19. From the geometry of the wing, we can write

$$S_{W_F} = \left(5c(\eta_i) - 0.5(5)(2.5) + 0.5(5)(1.25)\right)\ \text{ft}^2$$

where $c(\eta_i)$ is the wing chord length at the inboard end of the flap. Since for this wing

$$c(y) = 7.5 - \frac{y}{4}\ \text{ft}$$

we have $c(\eta_i) = c(5) = 6.25$ ft. Therefore,

$$S_{W_F} = (31.25 - 6.25 + 3.125) = 28.125\ \text{ft}^2$$

So the flap lift effectiveness for the wing is

$$C_{L_{\delta_F}} = c_{l_\delta}\frac{2S_{W_F}}{S} = 0.071\frac{56.25}{169} = 0.024\,/\text{deg} = 1.35\,/\text{rad}$$

And the alternate form of flap lift effectiveness is

$$\alpha_\delta \triangleq \frac{C_{L_\delta}}{C_{L_\alpha}} = \frac{.024(180/\pi)}{4.19} = 0.328$$

The flap moment effectiveness may be estimated using Equation (5.73), or

$$C_{M_{\delta_F}} = \frac{2}{S\bar{c}}\left(\int_{\eta_i}^{\eta_o} c_{m_\delta}(y)c^2(y)dy - \int_{\eta_i}^{\eta_o} c_{l_\delta}(y)(x_{\text{ac}}(y) - X_{\text{ac}})c(y)dy\right)$$

We have the chord given as

$$c(y) = 7.5 - \frac{y}{4}\ \text{ft}$$

And in Example 5.4 we had

$$x_{\text{ac}}(y) = \frac{7.5}{4} + \frac{7}{16}y\ \text{ft}$$

and

$$X_{\text{AC}_{\text{wing}}} = 4.79\ \text{ft aft of wing apex}$$

Consequently, we have the flap pitching-moment effectiveness given by

$$C_{M_{\delta_F}} = \frac{2}{(169)(5.825)}\left(\int_5^{10} -0.65\left(7.5-\frac{y}{4}\right)^2 dy - \int_5^{10} 4.04\left(\frac{7.5}{4}+\frac{7}{16}y-4.79\right)\left(7.5-\frac{y}{4}\right)dy\right)$$

$$= \frac{2}{(169)(5.825)}\left(-0.65\left(7.5^2 y - 7.5\frac{y^2}{4}+\frac{y^3}{48}\right)\Bigg|_5^{10}\right.$$

$$\left. + 4.04\left(21.8625y - 4.01\frac{y^2}{2}+0.109\frac{y^3}{3}\right)\Bigg|_5^{10}\right)$$

$$= 0.00203(-103.26-37.46) = -0.286\ /\text{rad}$$

Finally, the flap drag effectiveness $C_{D_{\delta_F}}$ may be estimated from Equation (5.76) and Figure 5.17, or

$$C_{D_{\delta_F}} = \frac{2}{S}\int_{\eta_i}^{\eta_o}\frac{\Delta c_{d_{\text{flap}}}}{\Delta\delta_{\text{flap}}}(y)c(y)dy = 2\frac{\Delta c_{d_{\text{flap}}}}{\Delta\delta_{\text{flap}}}\frac{S_{W_F}}{S} = 2(0.04)\frac{56.25}{169} = 0.0266\ /\text{rad}$$

To find the aileron rolling-moment effectiveness we may refer to Equation (5.83), which in this case is

$$C_{L_{\delta_A}} = \frac{2c_{l_\delta}}{Sb}\int_{\eta_i}^{\eta_o} c(y)y dy = \frac{2(4.04)}{(169)(30)}\int_{12}^{15} c(y)y dy$$

since the section lift effectiveness is constant due to the constant chord ratio. For this wing we have the section chord lengths given by

$$c(y) = 7.5 - \frac{y}{4}\ \text{ft}$$

so

$$\int_{12}^{15} c(y)y dy = \int_{12}^{15}\left(7.5y - \frac{y^2}{4}\right)dy = \left(3.75y^2 - \frac{y^3}{12}\right)\Bigg|_{12}^{15} = (303.75-137.25) = 166.5\ \text{ft}^3$$

Therefore, we have

$$C_{L_{\delta_A}} = \frac{2(4.04)}{(169)(30)}(166.5) = 0.265\ /\text{rad}$$

To address the aileron yawing-moment effectiveness we refer to Equation (5.89), which becomes

$$C_{N_{\delta_A}} = \frac{-2}{Sb}\int_{\eta_i}^{\eta_o}\frac{\Delta c_{d_{\text{flap}}}}{\Delta\delta_{\text{flap}}}(y)c(y)y dy = \frac{-2}{(169)(30)}\int_{12}^{15}(0.04)c(y)y dy = \frac{-2(0.04)(166.5)}{(169)(30)}$$

$$= -0.0026\ /\text{rad}$$

EXAMPLE 5.6

Analysis of a Horizontal Tail

Consider a horizontal tail with the same airfoil section, same aspect and taper ratios, and same leading-edge sweep angle as the wing discussed in Examples 5.4 and 5.5, but scaled such that the tail span is 15 ft. The tail has no span-wise twist. If this tail surface is equipped with a full span trailing-edge elevator with an elevator chord ratio $c_\delta / c_H = 0.25$ constant with span, estimate the axial location of the tail's aerodynamic center X_{AC_H} from its apex, as well as the elevator effectiveness parameters $C_{L_{\delta_E}}$, α_{δ_E}, and $C_{M_{\delta_E}}$ for this tail surface.

■ **Solution**

Though scaled, the tail has the same planform geometry and airfoil section as the wing analyzed in Example 5.5. Therefore, we again have for this airfoil section

$$c_{l_\delta} = \frac{1}{0.98}(4 \times 0.99) = \frac{1}{0.98}3.96 = 4.04\ /\text{rad} = 0.071\ /\text{deg}$$

This yields an elevator lift effectiveness α_δ for the section of

$$\alpha_\delta\big|_{\text{section}} \triangleq \frac{c_{l_{\delta_E}}}{c_{l_\alpha}} = \frac{4.04}{0.109(180/\pi)} = 0.647$$

Plus, the section flap pitching-moment effectiveness is again

$$c_{m_\delta} = -0.65\ /\text{rad} = -0.0113\ /\text{deg}$$

For the 3-D horizontal tail, the angle-of-attack lift effectiveness is the same as the wing, or $C_{L_{\alpha_H}} = 4.19\ /\text{rad}$. Also, Equation (5.66) reveals that for full-span elevators, the 3-D elevator lift effectiveness equals the 2-D effectiveness. Therefore we will take the 3-D α_{δ_H} to be the same as the section α_δ, or

$$\alpha_{\delta_H} = 0.647$$

But since

$$\alpha_{\delta_H} \triangleq \frac{C_{L_{\delta_E}}}{C_{L_{\alpha_H}}}$$

we have

$$C_{L_{\delta_E}} = 0.647 \times 4.19 = 2.71\ /\text{rad} = 0.047\ /\text{deg}$$

(Note that this approach tends to lead to better estimates for full-span elevators than just setting the 3-D elevator lift effectiveness $C_{L_{\delta_E}}$ equal to the 2-D section elevator lift effectiveness $c_{l_{\delta_E}}$.)

Now the area may be found from the span and aspect ratio, or

$$S_H = \frac{b_H^2}{A} = \frac{15^2}{5.33} = 42.2\ \text{ft}^2$$

Plus, the chord as a function of span is

$$c(y) = \frac{7.5}{2} - \frac{y}{4}\ \text{ft}$$

so the mean aerodynamic (geometric) chord for the horizontal tail becomes

$$\bar{c}_H = \frac{2}{S_H}\int_0^{b_H/2} c^2 dy = \frac{2}{42.2}\int_0^{7.5}\left(\frac{7.5}{2}-\frac{y}{4}\right)^2 dy = \frac{2}{42.2}\int_0^{7.5}\left(\frac{7.5^2}{4}-\frac{7.5}{4}y+\frac{y^2}{16}\right)dy$$

$$= \frac{2}{42.2}\left(\frac{7.5^2}{4}y-\frac{7.5}{8}y^2+\frac{y^3}{48}\right)\Bigg|_0^{7.5} = \frac{2}{42.2}(105.47-52.73+8.79) = 2.92 \text{ ft}$$

Furthermore, the section aerodynamic centers for the tail are located at $0.25c(y) + x_{LE}(y)$, and so

$$x_{ac}(y) = \frac{7.5}{8} - \frac{y}{16} + \frac{1}{2}y = \frac{7.5}{8} + \frac{7}{16}y \text{ ft}$$

The position of the tail's aerodynamic center may be found using Equation (5.17), which states that

$$X_{LE_{MAC}} = \frac{2}{S_H}\int_0^{b_H/2} x_{LE}(y)c(y)dy = \frac{2}{(42.2)}\int_0^{15/2}(0.5y)\left(\frac{7.5}{2}-\frac{y}{4}\right)dy$$

$$= \frac{0.5}{(42.2)}\int_0^{15/2}\left(7.5y-\frac{y^2}{2}\right)dy = \frac{0.5}{(42.2)}\left(7.5\frac{y^2}{2}-\frac{y^3}{6}\right)\Bigg|_0^{7.5}$$

$$= 0.0118(210.94-70.31) = 1.659 \text{ ft}$$

Consequently, the position of the aerodynamic center aft of the apex of the tail is

$$X_{AC_H} = X_{LE_{MAC_H}} + 0.25\bar{c}_H = 1.659 + 0.25(2.92) = 2.39 \text{ ft}$$

The elevator moment effectiveness may again be estimated using Equation (5.73), or

$$C_{M_{\delta_E}} = \frac{2}{S\bar{c}}\left(\int_{\eta_i}^{\eta_o} c_{m_\delta}(y)c^2(y)dy - \int_{\eta_i}^{\eta_o} c_{l_\delta}(y)(x_{ac}(y)-X_{ac})c(y)dy\right)$$

Consequently, the tail's pitching-moment effectiveness about its aerodynamic center is given by

$$C_{M_{\delta_E}} = \frac{2}{(42.2)(2.92)}\left(\int_0^{15/2} -0.65\left(\frac{7.5}{2}-\frac{y}{4}\right)^2 dy - \int_0^{15/2} 4.04\left(\frac{7.5}{8}+\frac{7}{16}y-2.39\right)\left(\frac{7.5}{2}-\frac{y}{4}\right)dy\right)$$

$$= \frac{2}{(42.2)(2.92)}\left(-0.65\left(\frac{7.5^2}{4}y-\frac{7.5}{4}\frac{y^2}{2}+\frac{y^3}{48}\right)\Bigg|_0^{7.5}\right.$$

$$\left.+ 4.04\left(5.447y-2.004\frac{y^2}{2}+0.109\frac{y^3}{3}\right)\Bigg|_0^{7.5}\right)$$

$$= 0.0162(-39.99-0.74) = -0.660 \text{ /rad}$$

5.5 DOWNWASH

All lifting surfaces affect the flow aft of those surfaces. In subsonic flow, a lifting surface also affects the flow forward of that surface. This effect is referred to as *downwash.* The downwash behind a wing in subsonic flow is a consequence of the wing's trailing-vortex system, shown in Figure 5.33. A *vortex sheet* is shed from a lifting wing, and the sides of the sheet roll up into *tip vortices.* For high-aspect-ratio wings the sheet is relatively flat, but for low-aspect-ratio and/or highly swept wings the sheet may be bowed up from wing root to tip.

The main effect of this trailing-vortex system is to deflect the airflow (behind the wing) downward relative to the free-stream flow $\mathbf{V}_\infty$. This flow deflection reduces the local angle of attack on any lifting surfaces located behind the wing, such as a horizontal stabilizing surface as shown in Figures 5.33. As shown in Figure 5.34, the local angle of attack on the aft lifting surface is reduced by the *downwash angle* ε, which in turn depends on the location of the aft surface

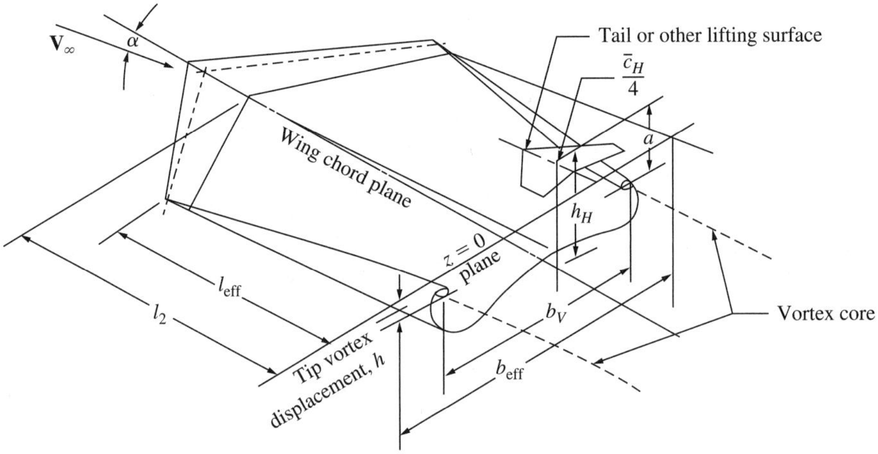

Figure 5.33 Wing trailing-vortex system and aft lifting surface.

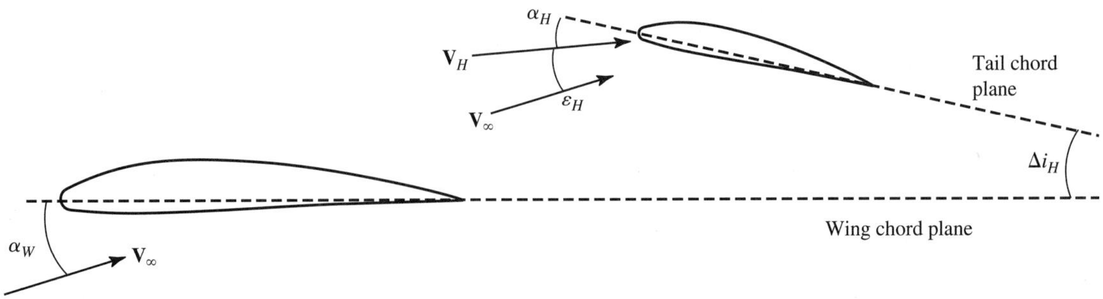

Figure 5.34 Influence of wing downwash on tail angle of attack.

relative to the wing, namely h_H and l_2 shown in Figure 5.33. In terms of the downwash angle, the angle of attack on the aft lifting surface α_H may be written as

$$\alpha_H = \alpha_W + \Delta i_H - \varepsilon_H \tag{5.90}$$

where

α_W is the wing angle of attack (= angle of attack of wing root chord)
Δi_H is the tail incidence angle (= angle between the wing and tail root chords)
ε_H is the downwash angle at the aft lifting surface

Equal in importance to the downwash angle is the change in downwash angle with wing angle of attack, or $d\varepsilon/d\alpha_W$. This parameter, known as the *downwash gradient,* is also a function of location aft of the wing. Theoretically, the downwash gradient is

$$\left.\frac{d\varepsilon}{d\alpha_W}\right|_{\text{theory}} = \begin{cases} 1 & \text{At the wing's trailing edge} \\ \dfrac{2C_{L_{\alpha_W}}}{\pi A} & \text{At infinity behind the wing} \end{cases} \tag{5.91}$$

Note that the angle of attack of the aft lifting surface may now be written in terms of the following linear expression

$$\alpha_H = \left(1 - \frac{d\varepsilon_H}{d\alpha_W}\right)\alpha_W + \Delta i_H \tag{5.92}$$

In addition, when the wing angle of attack is at α_0, the wing lift is zero by definition. Therefore, assuming the downwash angle varies linearly with wing angle of attack, the downwash angle aft of the wing may be written as

$$\boxed{\varepsilon = \frac{d\varepsilon}{d\alpha_W}\left(\alpha_W - \alpha_{0_W}\right)} \tag{5.93}$$

At intermediate distances behind the wing, $d\varepsilon/d\alpha_W$ may be estimated from Figure 5.35. This figure shows the downwash gradient $d\varepsilon/d\alpha_W$ in the wing's plane of symmetry and at the height of the vortex core. The gradient is given as a function of position of the aft lifting surface behind the wing in wing semispans $l_2/(b/2)$, effective wing aspect ratio A_{eff}, (which equals Ae), sweep of the wing's quarter chord line $\Lambda_{c/4}$, and the downwash gradient at infinity, found from Equation (5.91). The dashed lines show an example for $\Lambda_{c/4} = 45°$, $A_{\text{eff}} = 3$, $l_2/(b/2) = 0.8$, and a downwash gradient at infinity of 0.5. If the aft surface is located well above or below the height of the vortex core, as shown in Figure 5.33, the downwash gradient will be somewhat reduced.

As noted previously, ahead of a wing in subsonic flow, the flow is deflected upward relative to the free-stream flow, and this upward angle of deflection is referred to as the *upwash angle* ε_u. Figure 5.36 may be used to estimate the

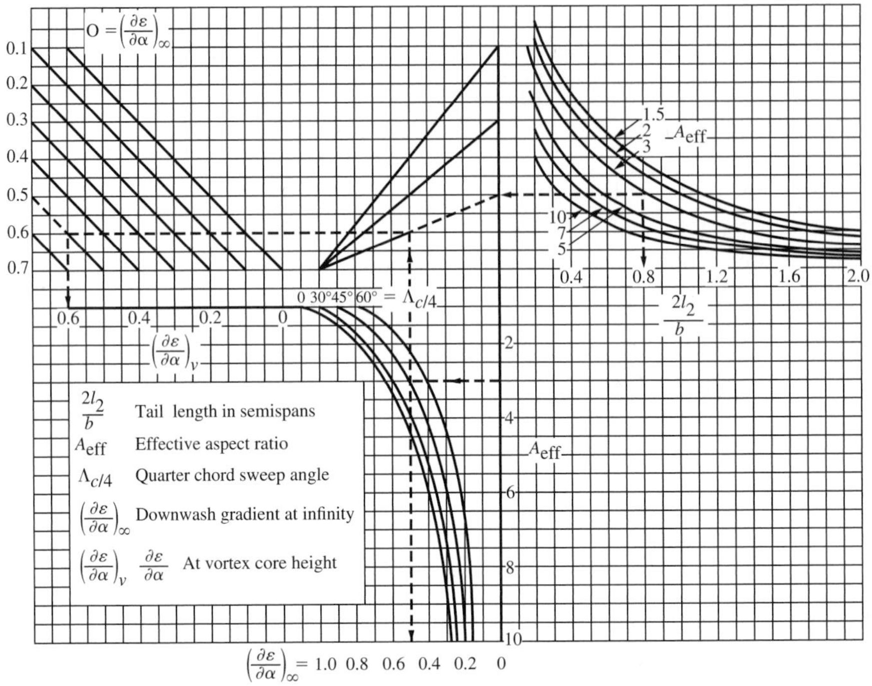

Figure 5.35 Downwash gradient aft of a wing.

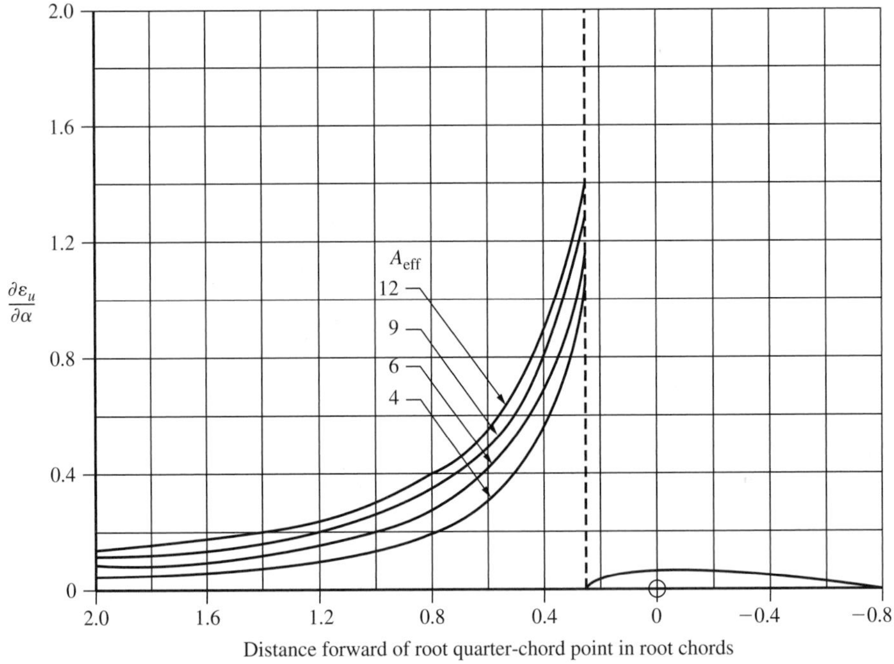

Figure 5.36 Upwash gradient forward of a wing.

upwash gradient $d\varepsilon_u/d\alpha_W$ in the wing's plane of symmetry, at various distances ahead of a wing with effective aspect ratio Ae. Analogous to the case with downwash, the local angle of attack forward of a wing may be expressed as

$$\alpha_{\text{forward}} = \left(1 + \frac{d\varepsilon_u}{d\alpha_W}\right)\alpha_W + \Delta i_{\text{forward}} \tag{5.94}$$

where $\Delta i_{\text{forward}}$ is the incidence angle between the root chord of the wing and the root chord of any lifting surface located forward of the wing (such as a canard).

EXAMPLE 5.7

Downwash Aft of Wing

Again consider the wing analyzed in Example 5.4. If the wing's angle of attack is 2 deg, determine the downwash angle at a point located at the height of the vortex core, and 2.5 root-chord lengths aft of the wing apex. Also, if the aerodynamic center of a horizontal tail is located at the above point, and the tail's geometric incidence angle $\Delta i_H = 2$ deg, find the local angle of attack of the horizontal tail (see Figure 5.34).

■ Solution

For the given wing, the aspect ratio is $A = 5.33$, the sweep of the quarter-chord line is $\Lambda_{c/4} \approx 25$ deg, and the lift effectiveness is $C_{L_\alpha} = 4.19$ /rad. Consequently, from Equation (5.91), the downwash gradient an infinite distance behind the wing is

$$\frac{d\varepsilon}{d\alpha_W}\Big|_\infty = \frac{2C_{L_\alpha}}{\pi A} = \frac{2(4.19)}{\pi(5.33)} = 0.5$$

The point of interest is located 2.5 wing-root-chord lengths behind the wing apex. Hence the distance l_2 in Figure 5.33 is 1.5 root-chord lengths, and the nondimensionalized distance is

$$\frac{2l_2}{b} = \frac{(7.5 \times 1.5)}{(15)} = 0.75$$

From Figure 5.35 (starting from the bottom axes in the upper-right and lower-left quadrants), we find that the downwash gradient at the location of the tail is

$$\frac{d\varepsilon}{d\alpha_W} = 0.57$$

The downwash angle aft of the wing may be estimated from Equation (5.93), or

$$\varepsilon = \frac{d\varepsilon}{d\alpha_W}(\alpha_W - \alpha_{0_W})$$

So for the wing in question, the downwash angle at the tail when the wing angle of attack is 2 deg is

$$\varepsilon_H = 0.57(2 - 1.33) = 0.38 \text{ deg}$$

For the tail incidence angle given, using Equation (5.90) we find that the local angle of attack at the horizontal tail is then

$$\alpha_H = \alpha_W + \Delta i_H - \varepsilon_H = 2 + 2 - 0.38 = 3.62 \text{ deg}$$

5.6 Summary

As noted at the beginning of this chapter, simple, empirically based methods for estimating the aerodynamic characteristics of flight vehicles can yield quite accurate results, and are quick and easy to use. In this chapter we presented several such empirical or semi-empirical methods. The overall approach used for estimating these aerodynamic characteristics is referred to as the *component build-up method.* That is, key characteristics of components of a wing, for example, are first estimated, and then these results are aggregated to estimate the characteristics of the complete wing.

Fundamental to the approach is the use of strip theory, along with the 2-D section lift, moment, and drag characteristics of airfoils. We considered the effects of flaps, along with the effect of downwash ahead and aft of a lifting surface. We defined the aerodynamic center of a 2-D section and of a 3-D wing, and presented methods for determining the location of the aerodynamic center of a lifting surface.

5.7 Problems

5.1 Starting with Equations (5.23) and (5.28), show that under strip theory for a wing with constant airfoil section with span and with no span-wise twist, the wing's zero-lift angle of attack and the wing's pitching-moment coefficient (about its *ac*) equals the wing-section zero-lift angle of attack and the wing-section pitching-moment coefficient, respectively.

5.2 Using MATLAB, calculate and plot 3-D wing lift effectiveness C_{L_α} versus aspect ratio A for wing mid-chord sweep angles of 10 and 20 degrees, and at Mach numbers of 0.25 and 0.5. Assume $k = 0.95$ in all cases.

5.3 Calculate the dihedral effect, or C_{L_β}, for the wing analyzed in Examples 5.4 and 5.5, and compare it to the aileron's rolling-moment effectiveness $C_{L_{\delta_A}}$. In particular, determine the approximate aileron deflection required to overcome the dihedral effect at a given sideslip angle.

5.4 Extra-credit project: Using the MATAERO wing-analysis code available on the book's website at www.mhhe.com/schmidt, determine the aerodynamic characteristics of the wing and horizontal tail considered in Examples 5.4 and 5.6, and compare the numerical results with those obtained in the examples.

References

1. "USAF Stability and Control DATCOM," prepared by the Douglas Aircraft Div., McDonnell Douglas Corp., for the USAF Flight Dynamics Laboratory, Wright Patterson AFB, Ohio, October 1960 (revised April 1978).
2. Anderson, John D., Jr.: *Fundamentals of Aerodynamics,* 5th ed., McGraw-Hill, New York, 2011.
3. Bertin, John J. and Michael L. Smith: *Aerodynamics for Engineers,* 2nd ed., Prentice-Hall, Upper Saddle River, NJ, 1989.

CHAPTER 6

Modeling the Forces and Moments on the Vehicle

Chapter Roadmap: *The material in this chapter would typically be covered in a first course in flight dynamics (especially the material associated with the longitudinal forces and moments). The modeling framework and all the aerodynamic and propulsive effectiveness coefficients, or stability derivatives, are defined here. Plus techniques are presented to estimate these important parameters, and the models for the forces and moments are integrated into the equations of motion.*

In this chapter we will address two major topics. The first topic involves the development of a framework for modeling the forces and moments acting on the flight vehicle. This framework will be based on the concept of expanding the forces and moments in Taylor series. The second topic involves the development of techniques for estimating the parameters in these Taylor-series expansions, given the geometry of the vehicle. These techniques are especially useful in preliminary design, and for gaining insight into a vehicle's aerodynamic characteristics and ultimately its dynamic characteristics. In fact, the concepts presented in this chapter provide the means by which the student may ultimately relate the vehicle's geometry to its dynamic characteristics.

Several techniques exist for estimating the forces and moments. As noted in Chapter 5, we may use analytical, computation, and experiment techniques to characterize the aerodynamic properties of airfoils. We also use these three methods to characterize the forces and moments on complete vehicles. But a vehicle's aerodynamic properties determined by any of these methods can be incorporated into the modeling framework developed here. For instance, as additional estimates for the aerodynamic characteristics become available, parameters in the model may be updated, but the modeling framework remains the same.

As an example, consider the data plotted in Figure 6.1, which is the drag polar (drag vs. lift coefficients) for a modern fighter aircraft. The coefficient data may be initially estimated using computational techniques, and then compared to

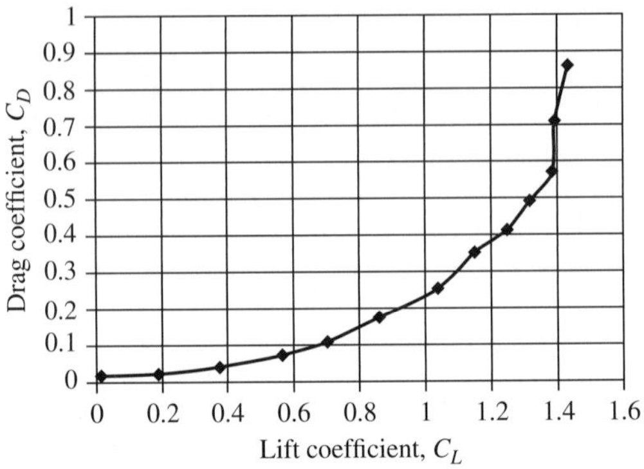

Figure 6.1 Drag polar for a modern fighter aircraft.

wind tunnel and flight test results at a later time. This data may be manipulated graphically and used in the modeling framework presented.

As discussed in Chapter 2, a resultant force vector **F** and a moment vector **M** are acting on the flight vehicle, as shown in Figure 6.2. These force and moment vectors arise from aerodynamic and propulsive effects. For now, we will consider this vehicle to be rigid. (The effects of flexibility will be considered in Chapter 7.) Also shown in Figure 6.2 is a vehicle-fixed reference frame to be discussed in more detail in Section 6.4. Consistent with Chapter 2, we have the aerodynamic and propulsive contributions to these forces and moments expressed as vector sums, or

$$\mathbf{F} = \mathbf{F}_A + \mathbf{F}_P \quad \text{and} \quad \mathbf{M} = \mathbf{M}_A + \mathbf{M}_P \tag{6.1}$$

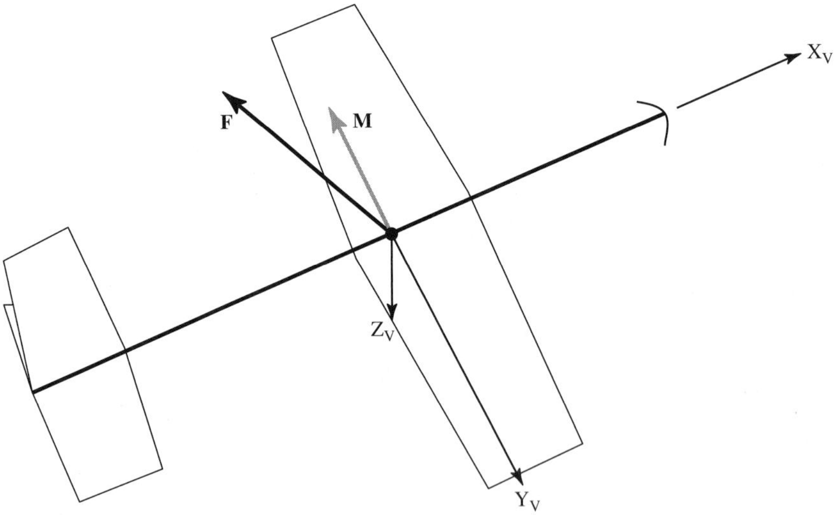

Figure 6.2 Resultant force and moment vectors acting on vehicle.

These force and moment vectors have components in the vehicle-fixed frame given by

$$\begin{aligned} \mathbf{F} &= \left(F_{A_X} + F_{P_X}\right)\mathbf{i}_V + \left(F_{A_Y} + F_{P_Y}\right)\mathbf{j}_V + \left(F_{A_Z} + F_{P_Z}\right)\mathbf{k}_V \\ \mathbf{M} &= \left(L_A + L_P\right)\mathbf{i}_V + \left(M_A + M_P\right)\mathbf{j}_V + \left(N_A + N_P\right)\mathbf{k}_V \end{aligned} \tag{6.2}$$

where the definitions are obvious for the components of $\mathbf{F}_A$, $\mathbf{F}_P$, $\mathbf{M}_A$, and $\mathbf{M}_P$.

Now, as also discussed in Chapters 1 and 2, we will eventually perform a perturbation analysis on the model of the vehicle's dynamics (or the dynamic system). In perturbation analysis the forces and moments must be expressed in terms of those associated with the reference condition, plus those perturbation forces and moments relative to the reference condition. That is, consistent with Equations (2.41), the force and moment vectors will be expressed in terms of reference values plus perturbations, or

$$\begin{aligned} \mathbf{F}_A &= \mathbf{F}_{A_0} + \boldsymbol{f}_A \qquad \mathbf{M}_A = \mathbf{M}_{A_0} + \boldsymbol{m}_A \\ \mathbf{F}_P &= \mathbf{F}_{P_0} + \boldsymbol{f}_P \qquad \mathbf{M}_P = \mathbf{M}_{P_0} + \boldsymbol{m}_P \end{aligned} \tag{6.3}$$

Here, the quantities with the zero subscript are associated with a particular <u>reference</u> flight condition, and the quantities denoted in lower-case italics are the <u>perturbation</u> quantities.

6.1 TAYLOR-SERIES EXPANSION OF AERODYNAMIC FORCES AND MOMENTS

We know that the aerodynamic forces and moments acting on a vehicle are functions of many variables. Among these are:

- Vehicle velocity relative to the air mass—$\mathbf{V}_\infty$ (or U, V, and W).
- Atmospheric density ρ_∞ (or altitude h).
- Vehicle geometry—for example, tail incidence angle i_H, flap and control surface deflections δ.
- Vehicle orientation relative to the free-stream velocity—angles of attack α and sideslip β.
- Rates of change of vehicle orientation—for example, rate of change of angle of attack $\dot{\alpha}$ or pitch rate Q.

So define a parameter vector $\mathbf{p}$ that includes all the above variables. That is, let

$$\mathbf{p} = \begin{bmatrix} U & V & W & \mathrm{h} & i_H & \delta_\mathrm{E} & \cdots & \alpha & \beta & \dot{\alpha} & Q & \cdots \end{bmatrix}^T \tag{6.4}$$

Also note that perturbations in these parameters are then used to define the perturbation parameter vector $\delta\mathbf{p}$.

NOTE TO STUDENT

In some texts and reports, the parameters in $\delta\mathbf{p}$ are nondimensionalized, such as u/U_0. We will not do that in this book because it adds notational complexity, and in the experience of the author is frequently a source of confusion. Plus, the final results for the perturbation forces and moments will be identical, regardless of whether or not the parameters were nondimensionalized. However, remember that the elements of $\mathbf{p}$ and $\delta\mathbf{p}$ used here are not nondimensionalized. This is especially important when using data provided from sources other than this book. Further discussion of this issue is presented in Section 6.9.

Let us now consider the aerodynamic force and moment vectors given in Equations (6.1), and let each component be expanded in a Taylor series about a reference condition (as yet undefined). That is, let us write

$$\begin{aligned}\{\mathbf{F}_A\} &= \{\mathbf{F}_{A_0}\} + \left[\frac{\partial \mathbf{F}_A}{\partial \mathbf{p}}\Big|_0\right]\delta\mathbf{p} + \frac{1}{2}\delta\mathbf{p}^T\left[\frac{\partial^2 \mathbf{F}_A}{\partial \mathbf{p}^2}\Big|_0\right]\delta\mathbf{p} + \cdots \\ \{\mathbf{M}_A\} &= \{\mathbf{M}_{A_0}\} + \left[\frac{\partial \mathbf{M}_A}{\partial \mathbf{p}}\Big|_0\right]\delta\mathbf{p} + \frac{1}{2}\delta\mathbf{p}^T\left[\frac{\partial^2 \mathbf{M}_A}{\partial \mathbf{p}^2}\Big|_0\right]\delta\mathbf{p} + \cdots\end{aligned} \tag{6.5}$$

where, for example, consistent with Equation (6.2)

$$\{\mathbf{F}_A\} \triangleq \begin{Bmatrix} F_{A_X} \\ F_{A_Y} \\ F_{A_Z} \end{Bmatrix} \quad \text{and} \quad \{\mathbf{M}_A\} \triangleq \begin{Bmatrix} L_A \\ M_A \\ N_A \end{Bmatrix}$$

and the terms in square brackets are matrices with elements consisting of partial derivatives. Note that all these partial derivatives are evaluated at the reference condition, as indicated in the above expressions. Finally, two similar Taylor-series expansions may be used to model the propulsive force and moment vectors $\mathbf{F}_P$ and $\mathbf{M}_P$.

In Equations (6.5) above, the first terms on the right sides represent the components of the aerodynamic force and moment vectors, respectively, acting on the vehicle in the chosen reference flight condition. Also in Equations (6.5), the vector $\delta\mathbf{p}$ includes perturbations in the variables in the parameter vector $\mathbf{p}$. For example, if surge velocity is expressed as the surge velocity at the chosen reference condition plus a perturbation in surge velocity, or

$$U = \mathrm{U}_0 + u$$

then u is an element of $\delta\mathbf{p}$.

The concept of using the Taylor-series expansion is the key to the modeling framework, and it is important to note that this concept is independent of the method used to estimate the force $\mathbf{F}_A$ or moment $\mathbf{M}_A$. This force and moment could be estimated analytically, computationally, or experimentally. For example, imagine if the forces and moments were measured in wind tunnel tests, and

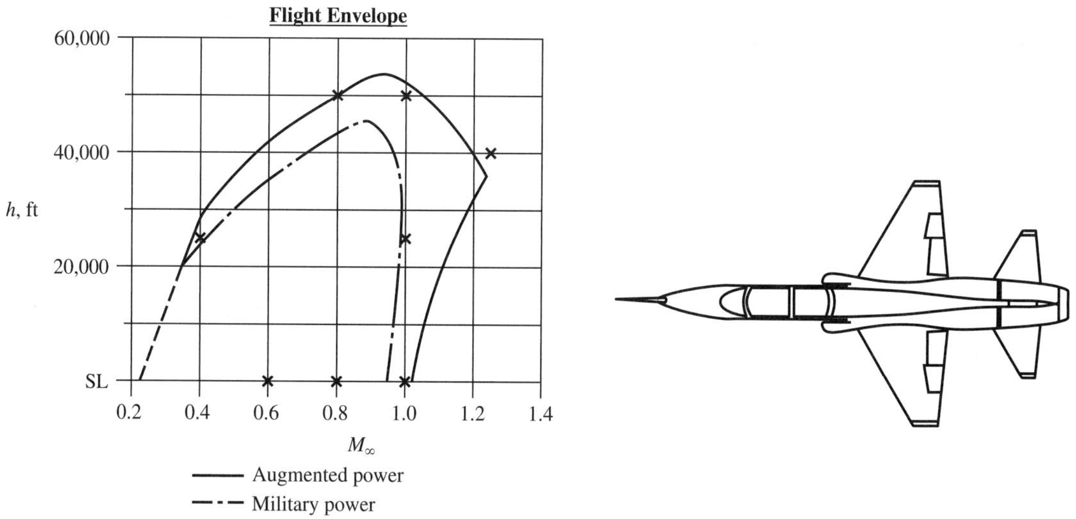

	Reference Flight Condition (all level flight)							
Parameter	**1**	**2**	**3**	**4**	**5**	**6**	**7**	**8**
Altitude, h ft	0	0	0	25,000	25,000	50,000	50,000	40,000
Mach number, M_∞	0.6	0.8	1.0	0.4	1.0	0.8	1.0	1.25
Velocity, V_∞ fps	670	893	1,117	406	1,016	774	969	1,210
Dynamic pressure, q_∞ psf	535	950	1,482	88	550	109	170	424

Figure 6.3 Some reference flight conditions for the T-38 (Ref. 1).

consider all this data plotted against each of the variables in **p**. Then, for example, the first-partial derivatives in Equations (6.5) could be extracted graphically (i.e., slopes) from this data.

The reference conditions correspond to specific flight conditions (i.e., altitude, velocity, orientation) for the vehicle. For example, the X's in the flight envelope in Figure 6.3 indicate eight reference flight conditions for a T-38 turbojet trainer. Each point in the flight envelope indicates an altitude-velocity (Mach number) pair at which the vehicle can sustain level flight. Also included in the tabulated data in the figure are details defining these reference flight conditions. For example, flight tests conducted at these flight conditions could be used to check and improve the accuracy of previously obtained wind tunnel data.

Now consider Equations (6.3) and (6.5), and recall that the first terms in the two Taylor-series expansions in Equations (6.5) correspond to the force and moment vectors, respectively, acting on the vehicle in the chosen reference condition. Therefore, the perturbation quantities $\boldsymbol{f}_A$ and $\boldsymbol{m}_A$ in Equations (6.3) correspond to the terms in the expansions in Equations (6.5) not including the first terms. That is, the perturbation aerodynamic force and moment vectors may be expressed as

$$
\begin{aligned}
f_A &= \left[\frac{\partial \mathbf{F}_A}{\partial \mathbf{p}} |_0 \right] \delta\mathbf{p} + \frac{1}{2}\delta\mathbf{p}^T \left[\frac{\partial^2 \mathbf{F}_A}{\partial \mathbf{p}^2} |_0 \right] \delta\mathbf{p} + \cdots \\
m_A &= \left[\frac{\partial \mathbf{M}_A}{\partial \mathbf{p}} |_0 \right] \delta\mathbf{p} + \frac{1}{2}\delta\mathbf{p}^T \left[\frac{\partial^2 \mathbf{M}_A}{\partial \mathbf{p}^2} |_0 \right] \delta\mathbf{p} + \cdots
\end{aligned}
\tag{6.6}
$$

Two similar expressions may be used to model the perturbation force and moment vectors arising from propulsive effects, or $\boldsymbol{f}_P$ and $\boldsymbol{m}_P$.

The components of the force and moment vectors in the vehicle-fixed frame introduced in Equations (6.2) may now be addressed. Consistent with Equations (6.3) we have the components of the aerodynamic force vector expressed as

$$
\begin{aligned}
F_{A_X} &= \mathrm{F}_{A_{X_0}} + \left[\frac{\partial F_{A_X}}{\partial \mathbf{p}} |_0 \right] \delta\mathbf{p} + \frac{1}{2}\delta\mathbf{p}^T \left[\frac{\partial^2 F_{A_X}}{\partial \mathbf{p}^2} |_0 \right] \delta\mathbf{p} + \cdots \\
f_{A_X} &= \left[\frac{\partial F_{A_X}}{\partial \mathbf{p}} |_0 \right] \delta\mathbf{p} + \frac{1}{2}\delta\mathbf{p}^T \left[\frac{\partial^2 F_{A_X}}{\partial \mathbf{p}^2} |_0 \right] \delta\mathbf{p} + \cdots \\
F_{A_Y} &= \mathrm{F}_{A_{Y_0}} + \left[\frac{\partial F_{A_Y}}{\partial \mathbf{p}} |_0 \right] \delta\mathbf{p} + \frac{1}{2}\delta\mathbf{p}^T \left[\frac{\partial^2 F_{A_Y}}{\partial \mathbf{p}^2} |_0 \right] \delta\mathbf{p} + \cdots \\
f_{A_Y} &= \left[\frac{\partial F_{A_Y}}{\partial \mathbf{p}} |_0 \right] \delta\mathbf{p} + \frac{1}{2}\delta\mathbf{p}^T \left[\frac{\partial^2 F_{A_Y}}{\partial \mathbf{p}^2} |_0 \right] \delta\mathbf{p} + \cdots \\
F_{A_Z} &= \mathrm{F}_{A_{Z_0}} + \left[\frac{\partial F_{A_Z}}{\partial \mathbf{p}} |_0 \right] \delta\mathbf{p} + \frac{1}{2}\delta\mathbf{p}^T \left[\frac{\partial^2 F_{A_Z}}{\partial \mathbf{p}^2} |_0 \right] \delta\mathbf{p} + \cdots \\
f_{A_Z} &= \left[\frac{\partial F_{A_Z}}{\partial \mathbf{p}} |_0 \right] \delta\mathbf{p} + \frac{1}{2}\delta\mathbf{p}^T \left[\frac{\partial^2 F_{A_Z}}{\partial \mathbf{p}^2} |_0 \right] \delta\mathbf{p} + \cdots
\end{aligned}
\tag{6.7}
$$

Likewise, for the components of the aerodynamic moment vector we have

$$
\begin{aligned}
L_A &= \mathrm{L}_{A_0} + \left[\frac{\partial L_A}{\partial \mathbf{p}} |_0 \right] \delta\mathbf{p} + \frac{1}{2}\delta\mathbf{p}^T \left[\frac{\partial^2 L_A}{\partial \mathbf{p}^2} |_0 \right] \delta\mathbf{p} + \cdots \\
l_A &= \left[\frac{\partial L_A}{\partial \mathbf{p}} |_0 \right] \delta\mathbf{p} + \frac{1}{2}\delta\mathbf{p}^T \left[\frac{\partial^2 L_A}{\partial \mathbf{p}^2} |_0 \right] \delta\mathbf{p} + \cdots \\
M_A &= \mathrm{M}_{A_0} + \left[\frac{\partial M_A}{\partial \mathbf{p}} |_0 \right] \delta\mathbf{p} + \frac{1}{2}\delta\mathbf{p}^T \left[\frac{\partial^2 M_A}{\partial \mathbf{p}^2} |_0 \right] \delta\mathbf{p} + \cdots \\
m_A &= \left[\frac{\partial M_A}{\partial \mathbf{p}} |_0 \right] \delta\mathbf{p} + \frac{1}{2}\delta\mathbf{p}^T \left[\frac{\partial^2 M_A}{\partial \mathbf{p}^2} |_0 \right] \delta\mathbf{p} + \cdots
\end{aligned}
\tag{6.8}
$$

$$N_A = N_{A_0} + \left[\frac{\partial N_A}{\partial \mathbf{p}} \Big|_0 \right] \delta\mathbf{p} + \frac{1}{2} \delta\mathbf{p}^T \left[\frac{\partial^2 N_A}{\partial \mathbf{p}^2} \Big|_0 \right] \delta\mathbf{p} + \cdots$$

$$n_A = \left[\frac{\partial N_A}{\partial \mathbf{p}} \Big|_0 \right] \delta\mathbf{p} + \frac{1}{2} \delta\mathbf{p}^T \left[\frac{\partial^2 N_A}{\partial \mathbf{p}^2} \Big|_0 \right] \delta\mathbf{p} + \cdots$$

Again, similar expressions may be developed for the components of the propulsive force and moment vectors $\mathbf{F}_P$ and $\mathbf{M}_P$.

Now recall that in perturbation analysis, as discussed in Chapter 1, a small-perturbation assumption is made which yields <u>linear</u> equations governing the perturbation quantities. Under such a small-perturbation assumption the expressions for the perturbation forces and moments in Equations (6.7) and (6.8) are simplified since higher-order terms in $\delta\mathbf{p}$ are neglected, leaving only terms that are linear in the parameter-vector $\delta\mathbf{p}$. Or, <u>under the small-perturbation assumption</u>, we have the perturbations in the aerodynamic forces and moments given by

$$\begin{aligned} f_{A_X} &= \frac{\partial F_{A_X}}{\partial \mathbf{p}} \Big|_0 \delta\mathbf{p} & l_A &= \frac{\partial L_A}{\partial \mathbf{p}} \Big|_0 \delta\mathbf{p} \\ f_{A_Y} &= \frac{\partial F_{A_Y}}{\partial \mathbf{p}} \Big|_0 \delta\mathbf{p} & m_A &= \frac{\partial M_A}{\partial \mathbf{p}} \Big|_0 \delta\mathbf{p} \\ f_{A_Z} &= \frac{\partial F_{A_Z}}{\partial \mathbf{p}} \Big|_0 \delta\mathbf{p} & n_A &= \frac{\partial N_A}{\partial \mathbf{p}} \Big|_0 \delta\mathbf{p} \end{aligned} \tag{6.9}$$

with similar expressions developed for the propulsive forces and moments. Note that the above expressions depend explicitly on the partial derivatives of the aerodynamic forces and moments with respect to the parameters in $\mathbf{p}$ (Equation (6.4)). Consequently, we would expect the vehicle's force- and moment-effectiveness coefficients, such as angle-of-attack lift and pitching-moment effectiveness, C_{L_α} and C_{M_α}, respectively, to play key roles in the aerodynamic modeling. We shall see that this is the case.

NOTE TO STUDENT

Historically, the partial derivatives in Equation (6.9), and those that appear in similar expressions for the propulsive forces and moments, were referred to as *stability derivatives.* The phrase was also used to describe the effectiveness coefficients such as C_{L_α}, C_{M_α}, and so on, since they are also partial derivatives. We prefer not use that terminology in this book.

In this section, we have developed a framework for modeling the aerodynamic and propulsive forces and moments acting on the vehicle. This framework, based on Taylor-series expansions, is particularly compatible with small-perturbation analysis of the vehicle's dynamics, but will be useful in other analyses as well.

6.2 AERODYNAMIC FORCES AND MOMENTS ACTING ON THE VEHICLE

We now seek to develop analytical expressions for the aerodynamic forces and moments acting on the vehicle. Then, using these analytical expressions, we will derive expressions for the various force and moment effectivenesses for the vehicle, such as angle-of-attack lift effectiveness C_{L_α}.

As noted in Chapter 5, various methods may be used to estimate these aerodynamic forces and moments. But in this chapter we will use the *component build-up method,* aggregating the forces or moments acting on the various components of the vehicle, and drawing on the results from Chapter 5. We will assume conventional aircraft geometry, such as the T-38 aircraft shown in Figure 6.3, consisting of a fuselage, wing, and horizontal and vertical tail surfaces, with elevator, rudder, and ailerons as control devices. We will also assume that the aerodynamic lift and drag characteristics for all the vehicle's lifting surfaces are known (e.g., from techniques presented in the previous chapter or from experimental data). For vehicles with other geometries, the methods presented in this chapter may be applied to develop expressions similar to those developed here. Other vehicle geometries are considered in the problems at the end of this chapter and in Appendix B.

Three familiar components of the total aerodynamic force vector $\mathbf{F}_A$ acting on the vehicle are depicted in Figure 6.4. These force components are lift, L, drag, D, and side force S (not to be confused with S, the reference area used to nondimensionalize the aerodynamic coefficients). Fuselage-referenced axes, aligned with and fixed to the vehicle, are also shown, as are the definitions of the vehicle angle of attack α and sideslip angle β. The forces as shown are acting at the aerodynamic center of the vehicle, to be discussed in Section 6.2.5.

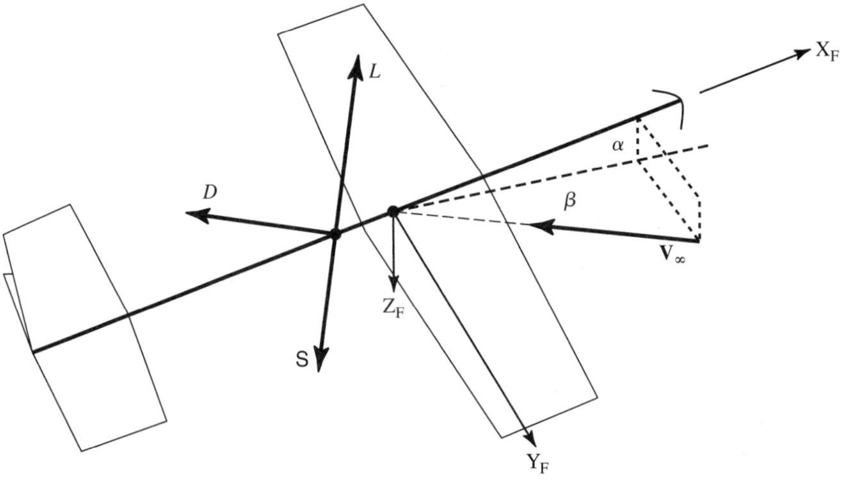

Figure 6.4 Aerodynamic forces on the vehicle.

By the convention used in aerodynamics, these force components are tied to the free-stream velocity vector. Drag D is defined as acting parallel to the free-stream velocity vector. Lift L acts normal to the free-stream velocity vector, and lies in the XZ plane of the vehicle. Side-force S acts orthogonally to both the lift and drag force components. Hence drag always acts at an angle β from the vehicle-fixed XZ plane, and side force always acts at an angle β from the vehicle's YZ plane.

NOTE TO STUDENT

Unless otherwise noted, the air mass through which the vehicle is flying is assumed fixed with respect to the earth's surface. Therefore, the velocity of the vehicle relative to a point on the earth's surface will be the same as the velocity of the vehicle relative to the air mass. Plus, the free-stream velocity of the air mass relative to the vehicle, $\mathbf{V}_\infty$, is equal and opposite in direction to the velocity of the vehicle relative to this air mass, or $\mathbf{V}_V$. The effects of wind (i.e., the air mass moving with respect to the earth's surface) will be discussed in Section 6.8.

6.2.1 Vehicle Lift

We will assume here that the vehicle lift is basically generated by a wing and horizontal tail. That is, we assume that the lift generated by the fuselage is negligible compared to that generated by the wing and tail surfaces. This assumption is valid for most conventional aircraft, but would not be valid for a wingless vehicle, for example. (See Problems 6.2 and 6.3 at the end of this chapter.) The lift (and drag) acting on the wing and tail are depicted in Figure 6.5. Note that the vehicle angle of attack α is also shown, defined in terms of a fuselage-referenced axis aligned with the fuselage, and that the lift and drag act at the aerodynamic center of the respective surface.

Under the above assumptions, the lift of the vehicle may be written as

$$\begin{aligned} L &= L_W + L_H \\ &= C_{L_W} q_\infty S_W + C_{L_H} q_H S_H = C_L q_\infty S \end{aligned} \tag{6.10}$$

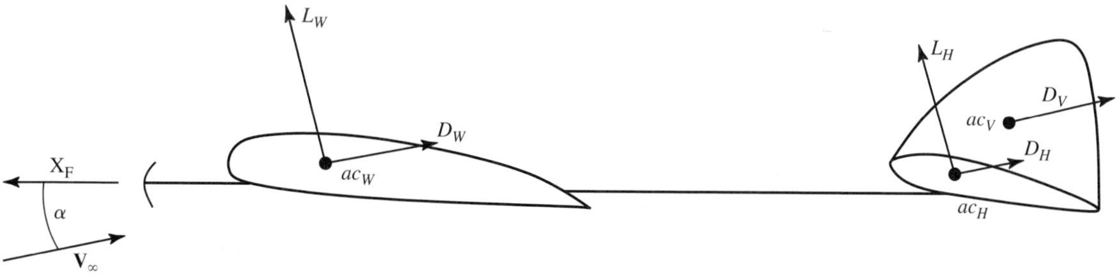

Figure 6.5 Wing and tail lift and drag.

Note that the lift coefficient for the vehicle C_L has been defined by Equation (6.10), and that the dynamic pressure at the tail q_H usually differs from the free-stream value q_∞ due to viscous effects in the boundary layer on the fuselage and wing ahead of the tail. From Equation (6.10), and defining the *vehicle reference area S* to be the wing planform area, we can write the vehicle lift coefficient C_L as

$$\boxed{C_L = C_{L_W} + C_{L_H}\frac{q_H}{q_\infty}\frac{S_H}{S_W}} \tag{6.11}$$

NOTE TO STUDENT

Again, regarding notation, we will typically not use any additional subscripts to indicate that a coefficient corresponds to the vehicle, such as C_L above. This is in contrast to the subscripts W and H used in the above equation to indicate coefficients corresponding to the wing or horizontal tail. The student should take care in this regard, but the notation should be clear from the context.

Now assume the wing is attached to the fuselage such that the wing root chord makes an angle i_W (positive leading-edge up) from the fuselage-referenced X axis. Therefore, the angle of attack of the wing root chord, defined as the wing angle of attack, is

$$\alpha_W = \alpha + i_W \tag{6.12}$$

Likewise, let the root chord of the horizontal tail make an angle i_H (positive leading-edge up) from the fuselage-referenced X axis. Therefore, the local angle of attack at the horizontal tail is

$$\alpha_H = \alpha + i_H - \varepsilon \tag{6.13}$$

where ε is the downwash angle at the tail. In terms of the downwash gradient, introduced in Chapter 5, the downwash angle may be expressed as

$$\varepsilon = \frac{d\varepsilon}{d\alpha}\left(\alpha_W - \alpha_{0_W}\right) = \frac{d\varepsilon}{d\alpha}\left(\alpha + i_W - \alpha_{0_W}\right) \tag{6.14}$$

where α_{0_W} is the wing's zero-lift angle of attack. Therefore, the local angle of attack at the tail may be written as

$$\alpha_H = \alpha + i_H - \frac{d\varepsilon}{d\alpha}\left(\alpha + i_W - \alpha_{0_W}\right) = \left(1 - \frac{d\varepsilon}{d\alpha}\right)\alpha - \frac{d\varepsilon}{d\alpha}\left(i_W - \alpha_{0_W}\right) + i_H \tag{6.15}$$

With the geometry defined as above, and assuming the lift coefficients lie in their linear ranges, the wing and tail lift coefficients may be written as

$$\boxed{\begin{aligned} C_{L_W} &= C_{L_{\alpha_W}}\left(\alpha_W - \alpha_{0_W}\right) = C_{L_{\alpha_W}}\left(\alpha + i_W - \alpha_{0_W}\right) \\ C_{L_H} &= C_{L_{\alpha_H}}\left(\alpha_H - \alpha_{0_H} + \alpha_\delta \delta_E\right) \\ &= C_{L_{\alpha_H}}\left(\left(1 - \frac{d\varepsilon}{d\alpha}\right)\alpha - \frac{d\varepsilon}{d\alpha}\left(i_W - \alpha_{0_W}\right) + i_H - \alpha_{0_H} + \alpha_\delta \delta_E\right) \end{aligned}} \tag{6.16}$$

In these expressions, wing and tail zero-lift angles of attack α_0 and elevator lift effectiveness α_δ are included. These parameters, as well as the downwash gradient $d\varepsilon/d\alpha$, were all discussed in Chapter 5. (Note that typically the zero-lift angle of attack for the tail α_{0_H} is zero since a symmetric airfoil cross section is usually used for that surface.)

NOTE TO STUDENT

In this section, frequent mention is made of aerodynamic coefficients lying in their linear ranges. Note, however, that the all-important force- and moment-effectiveness coefficients found in this chapter correspond to partial derivatives (i.e., local slopes). Hence, the expressions for the effectiveness coefficients are valid regardless of whether the aerodynamics are linear or not, as long as local slopes are used.

For example, consider the data presented in Figure 6.6, which shows the lift coefficient for a modern fighter aircraft plotted versus vehicle angle of attack. Also shown is the angle-of-attack lift effectiveness C_{L_α}, represented by the local slope evaluated at some reference angle of attack α_0. Note that this lift effectiveness is therefore a function of the reference angle of attack.

Inserting Equations (6.16) into Equations (6.10), and letting the vehicle reference area S equal the wing planform area S_W, we have the vehicle lift coefficient given by

$$C_L = C_{L_{\alpha_W}}\left(\alpha + i_W - \alpha_{0_W}\right) + C_{L_{\alpha_H}}\frac{q_H}{q_\infty}\frac{S_H}{S_W}\left(\left(1 - \frac{d\varepsilon}{d\alpha}\right)\alpha - \frac{d\varepsilon}{d\alpha}\left(i_W - \alpha_{0_W}\right) + i_H - \alpha_{0_H} + \alpha_\delta \delta_E\right) \tag{6.17}$$

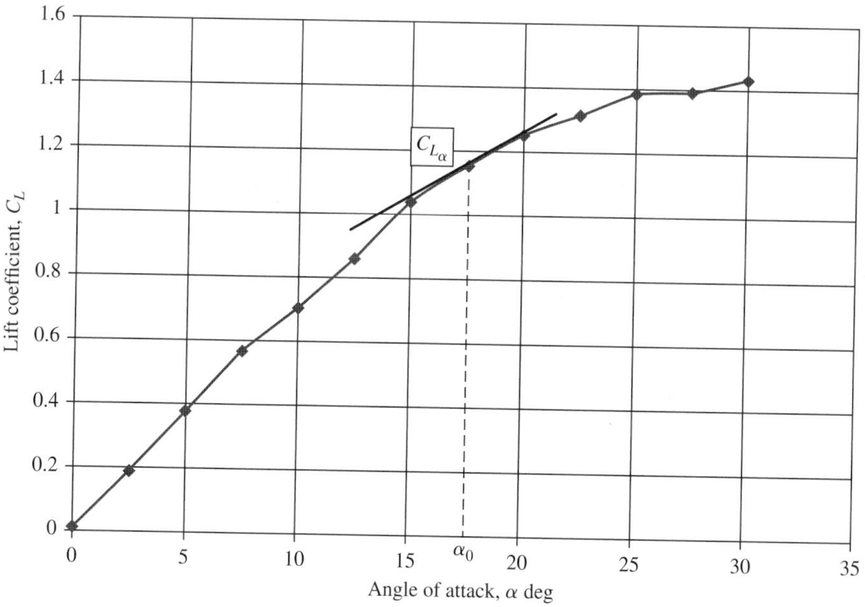

Figure 6.6 Lift coefficient for a modern fighter aircraft.

Note that the lift-effectiveness coefficients may be obtained by differentiating this expression with respect to the parameter in question. Specifically, differentiating this expression with respect to vehicle angle of attack α, we can deduce that the vehicle angle-of-attack lift effectiveness is

$$C_{L_\alpha} = C_{L_{\alpha_W}} + C_{L_{\alpha_H}} \frac{q_H}{q_\infty} \frac{S_H}{S_W} \left(1 - \frac{d\varepsilon}{d\alpha}\right) \tag{6.18}$$

Furthermore, differentiating Equation (6.17) with respect to elevator deflection δ_E yields the following elevator lift effectiveness for the vehicle.

$$C_{L_{\delta_E}} = C_{L_{\alpha_H}} \alpha_\delta \frac{q_H}{q_\infty} \frac{S_H}{S_W} = C_{L_{\delta_H}} \frac{q_H}{q_\infty} \frac{S_H}{S_W} \tag{6.19}$$

And finally, differentiating Equation (6.17) with respect to tail incidence angle i_H yields the following tail-incidence lift effectiveness for the vehicle.

$$C_{L_{i_H}} = C_{L_{\alpha_H}} \frac{q_H}{q_\infty} \frac{S_H}{S_W} \tag{6.20}$$

This tail-incidence effectiveness is especially important when tail incidence may be varied, to trim the pilot's stick forces, for example. Wing incidence is usually not variable.

Lastly, when $\alpha = \delta_E = i_H = 0$, the vehicle's lift coefficient is typically not zero. In fact, from Equation (6.17) we find that

$$C_L\big|_{\alpha=\delta_E=i_H=0} = C_{L_{\alpha_W}}\left(i_W - \alpha_{0_W}\right) + C_{L_{\alpha_H}} \frac{q_H}{q_\infty} \frac{S_H}{S_W} \left(\frac{d\varepsilon}{d\alpha}\left(\alpha_{0_W} - i_W\right) - \alpha_{0_H}\right) \tag{6.21}$$

EXAMPLE 6.1

Vehicle Lift Analysis

Consider an aircraft with a wing as described in Example 5.4 and a horizontal tail as described in Examples 5.6 and 5.7. The wing is joined to the fuselage such that the wing's root chord makes an angle of 2 deg (leading-edge up) relative to the fuselage reference axis. The horizontal tail, furthermore, is located such that its aerodynamic center is 2.5 wing root-chord lengths aft of the wing's apex. Find the vehicle's angle-of-attack, elevator, and tail-incidence lift effectiveness C_{L_α}, $C_{L_{\delta_E}}$, and $C_{L_{i_H}}$. Also find the vehicle's lift coefficient when $\alpha = \delta_E = i_H = 0$.

■ Solution

From Example 5.4 we found that for this wing, $S_W = 169$ ft^2, $\bar{c}_W = 5.825$ ft, $c_r = 7.5$ ft, $C_{L_{\alpha_W}} = 4.19$ /rad, $\alpha_{0_W} = 1.33$ deg, $X_{AC_W} = 4.79$ ft aft of the wing apex. And from Examples 5.6 and 5.7 we have for the horizontal tail, $S_H = 42$ ft^2, $\bar{c}_H = 2.92$ ft, $C_{L_{\alpha_H}} = C_{L_{\alpha_W}} = 4.19$ /rad, $X_{AC_H} = 2.39$ ft aft of the tail apex, $\alpha_\delta = 0.65$, and $\dfrac{d\varepsilon_H}{d\alpha_W} = 0.57$.

Now from Equation (6.20) we have the vehicle's tail-incidence lift effectiveness given by

$$C_{L_{i_H}} = C_{L_{\alpha_H}} \frac{q_H}{q_\infty} \frac{S_H}{S_W} = 4.19(0.9)\frac{42}{169} = 0.94 \text{ /rad}$$

Note that we have assumed that the dynamic pressure ratio $q_H/q_\infty \approx 0.9$, which is typically a reasonable first approximation. Then, from Equation (6.19), the vehicle's elevator lift effectiveness is

$$C_{L_{\delta_E}} = C_{L_{\alpha_H}} \alpha_\delta \frac{q_H}{q_\infty} \frac{S_H}{S_W} = \alpha_\delta C_{L_{i_H}} = 0.65(0.94) = 0.61 \text{ /rad}$$

And from Equation (6.18) we have the vehicle's angle-of-attack lift effectiveness given by

$$C_{L_\alpha} = C_{L_{\alpha_W}} + C_{L_{\alpha_H}} \frac{q_H}{q_\infty} \frac{S_H}{S_W}\left(1 - \frac{d\varepsilon}{d\alpha}\right) = C_{L_{\alpha_W}} + C_{L_{i_H}}\left(1 - \frac{d\varepsilon}{d\alpha}\right)$$

$$= 4.19 + 0.94(0.43) = 4.59 \text{ /rad}$$

Finally, from Equation (6.21), we have

$$C_L|_{\alpha=\delta_E=i_H=0} = C_{L_{\alpha_W}}(i_W - \alpha_{0_W}) + C_{L_{\alpha_H}} \frac{q_H}{q_\infty} \frac{S_H}{S_W}\left(\frac{d\varepsilon}{d\alpha}(\alpha_{0_W} - i_W) - \alpha_{0_H}\right)$$

$$= C_{L_{\alpha_W}}(i_W - \alpha_{0_W}) + C_{L_{i_H}}\left(\frac{d\varepsilon}{d\alpha}(\alpha_{0_W} - i_W) - \alpha_{0_H}\right)$$

Since the horizontal tail's airfoil section is symmetric, and the tail has no span-wise twist, its zero-lift angle of attack α_{0_H} will be zero. Therefore,

$$C_L|_{\alpha=\delta_E=i_H=0} = 4.19(2 - 1.33)(\pi/180) + 0.94(0.57(1.33 - 2))(\pi/180) = 0.072$$

Note in this example that the largest contributor to the vehicle's aerodynamic lift, or lift effectiveness, is its angle of attack, compared to elevator deflection and tail incidence. Also note the large magnitude of wing lift compared to that for the horizontal tail, when both are referenced to the same reference area. These results are typical since the tail surface is much smaller than the wing.

6.2.2 Vehicle Side Force

If a vehicle is experiencing a sideslip angle β in addition to an angle of attack α, the vehicle will generate a side force S. Also, a side force will be generated if the rudder, the control surface on the vertical tail, is deflected, or $\delta_R \neq 0$. This side force is primarily due to the force on the vertical tail S_V. This force S_V is analogous to the lift generated by a horizontal tail at an angle of attack α and/or with elevator deflection δ_E, and may be written in terms of the sideslip side-force effectiveness of the vertical tail $C_{S_{\beta_V}}$. Specifically, we may write

$$\mathsf{S}_V = C_{S_V} q_H S_V = C_{S_{\beta_V}}(\beta + \beta_\delta \delta_R) q_H S_V = C_S q_\infty S_W \tag{6.22}$$

which is valid as long as the side-force effectiveness $C_{S_{\beta_V}}$ lies in a linear range. Here the <u>vehicle</u> side-force coefficient C_S and a rudder sideslip effectiveness β_δ

have been introduced, and the dynamic pressure at the vertical tail is taken to be the same as the dynamic pressure at the horizontal tail q_H. The rudder sideslip effectiveness is defined, analogous to the elevator angle-of-attack effectiveness α_δ, as

$$\beta_\delta \triangleq \frac{C_{S_{\delta_V}}}{C_{S_{\beta_V}}} \tag{6.23}$$

and will be negative due to the sign convention used for sideslip angle and rudder deflection. (A positive rudder deflection produces a side force in the positive Y direction.) Finally, a symmetric airfoil with no span-wise twist has been also assumed for the vertical tail, and hence the zero-side-force sideslip angle β_0 will be zero. Note finally that the vertical tail's sideslip side-force effectiveness $C_{S_{\beta_V}}$ will be negative, given the fact that a positive sideslip β generates a negative side force S, in light of the definitions of positive sideslip angle and side force indicated in Figure 6.4.

Consequently, the vehicle's side-force coefficient may be written as

$$\boxed{C_S = C_{S_V}\frac{q_H}{q_\infty}\frac{S_V}{S_W} = C_{S_{\beta_V}}(\beta + \beta_\delta \delta_R)\frac{q_H}{q_\infty}\frac{S_V}{S_W}} \tag{6.24}$$

Differentiating the above with respect to β yields the vehicle sideslip side-force effectiveness, or

$$\boxed{C_{S_\beta} = C_{S_{\beta_V}}\frac{q_H}{q_\infty}\frac{S_V}{S_W}} \tag{6.25}$$

Likewise, differentiating Equation (6.24) with respect to rudder deflection δ_R yields the vehicle rudder side-force effectiveness, or

$$\boxed{C_{S_{\delta_R}} = C_{S_{\beta_V}}\beta_\delta\frac{q_H}{q_\infty}\frac{S_V}{S_W} = C_{S_{\delta_V}}\frac{q_H}{q_\infty}\frac{S_V}{S_W}} \tag{6.26}$$

6.2.3 Vehicle Drag

Again with reference to Figure 6.5, the total drag on the vehicle, including the fuselage and a vertical tail, may be expressed as

$$\begin{aligned} D &= D_W + D_H + D_V + D_F \\ &= C_{D_W}q_\infty S_W + C_{D_H}q_H S_H + C_{D_V}q_H S_V + C_{D_F}q_\infty S_F = C_D q_\infty S \end{aligned} \tag{6.27}$$

Here the drag coefficient for the vehicle C_D has been introduced, as well as the drag coefficient for the fuselage C_{D_F}.

We will assume that the drag polars, discussed in Chapter 5, are available for all three lifting surfaces, wing and horizontal and vertical tails. Or we have

$$
\begin{aligned}
C_{D_W} &= C_{D_{0_W}} + \frac{C_{L_W}^2}{\pi A_W e_W} \\
C_{D_H} &= C_{D_{0_H}} + \frac{C_{L_H}^2}{\pi A_H e_H} \\
C_{D_V} &= C_{D_{0_V}} + \frac{C_{S_V}^2}{\pi A_V e_V}
\end{aligned} \tag{6.28}
$$

In these expressions, the parasite drag on all three surfaces has been included, as well as the induced drag. Note that the induced drag on the vertical tail is a function of the side-force coefficient C_{S_V} for that tail surface. This coefficient was introduced in Equation (6.22), while the wing and horizontal-tail lift coefficients in Equations (6.28) are as given in Equations (6.16).

Collecting all terms, and letting the vehicle reference area S equal the wing planform area S_W, we may now write the vehicle's drag coefficient as

$$
\begin{aligned}
C_D &= \left(C_{D_{0_W}} + C_{D_F}\frac{S_F}{S} + C_{D_{0_H}}\frac{q_H}{q_\infty}\frac{S_H}{S_W} + C_{D_{0_V}}\frac{q_H}{q_\infty}\frac{S_V}{S_W} \right) \\
&\quad + \left(\frac{C_{L_W}^2}{\pi A_W e_W} + \frac{C_{L_H}^2}{\pi A_H e_H}\frac{q_H}{q_\infty}\frac{S_H}{S_W} + \frac{C_{S_V}^2}{\pi A_V e_V}\frac{q_H}{q_\infty}\frac{S_V}{S_W} \right) \\
&\triangleq C_{D_0} + \left(\frac{C_{L_W}^2}{\pi A_W e_W} + \frac{C_{L_H}^2}{\pi A_H e_H}\frac{q_H}{q_\infty}\frac{S_H}{S_W} + \frac{C_{S_V}^2}{\pi A_V e_V}\frac{q_H}{q_\infty}\frac{S_V}{S_W} \right)
\end{aligned} \tag{6.29}
$$

Here we have introduced the *parasite drag coefficient of the vehicle* C_{D_0}, with obvious definition. If there is no side force on the vehicle (or vertical tail), the vehicle's drag coefficient above is frequently expressed as

$$
C_D = C_{D_0} + \frac{1}{\pi A_W e_W}\left(C_{L_W}^2 + C_{L_H}^2 \frac{A_W e_W}{A_H e_H}\frac{q_H}{q_\infty}\frac{S_H}{S_W} \right) = C_{D_0} + \frac{C_L^2}{\pi A_W e_{\mathrm{Eff}}} \tag{6.30}
$$

Here we have expressed the induced drag of the vehicle in terms of the vehicle's lift coefficient C_L and an effective Oswald's efficiency factor e_{Eff} for the vehicle. This effective factor will typically be about the same as that for the wing alone.

Differentiating Equation (6.29) or (6.30) with respect to vehicle angle of attack yields the vehicle's angle-of-attack drag effectiveness, or

$$
C_{D_\alpha} = \frac{2}{\pi}\left(\frac{C_{L_W}}{A_W e_W} C_{L_{\alpha_W}} + \frac{C_{L_H}}{A_H e_H} C_{L_{\alpha_H}} \frac{q_H}{q_\infty}\frac{S_H}{S_W}\left(1 - \frac{d\varepsilon}{d\alpha}\right)\right) \tag{6.31}
$$

Likewise, differentiating Equation (6.29) or (6.30) with respect to elevator deflection δ_E yields the vehicle elevator drag effectiveness, or

$$C_{D_{\delta_E}} = \frac{2C_{L_H}}{\pi A_H e_H} C_{L_{\alpha_H}} \alpha_\delta \frac{q_H}{q_\infty} \frac{S_H}{S_W} = \frac{2C_{L_H}}{\pi A_H e_H} C_{L_{\delta_H}} \frac{q_H}{q_\infty} \frac{S_H}{S_W} \tag{6.32}$$

Furthermore, differentiating Equation (6.29) or (6.30) with respect to horizontal tail incidence angle i_H yields the vehicle tail-incidence-angle drag effectiveness, or

$$C_{D_{i_H}} = \frac{2C_{L_H}}{\pi A_H e_H} C_{L_{\alpha_H}} \frac{q_H}{q_\infty} \frac{S_H}{S_W} \tag{6.33}$$

Differentiating Equation (6.29) with respect to sideslip angle β yields the vehicle sideslip drag effectiveness, or

$$C_{D_\beta} = \frac{2C_{S_V}}{\pi A_V e_V} C_{S_{\beta_V}} \frac{q_H}{q_\infty} \frac{S_V}{S_W} \tag{6.34}$$

And finally, differentiating Equation (6.29) with respect to rudder deflection δ_R yields the vehicle rudder drag effectiveness, or

$$C_{D_{\delta_R}} = \frac{2C_{S_V}}{\pi A_V e_V} C_{S_{\beta_V}} \beta_\delta \frac{q_H}{q_\infty} \frac{S_V}{S_W} = \frac{2C_{S_V}}{\pi A_V e_V} C_{S_{\delta_V}} \frac{q_H}{q_\infty} \frac{S_V}{S_W} \tag{6.35}$$

Note that when $\alpha = \beta = \delta_E = i_H = \delta_R = 0$, the vehicle's drag coefficient is

$$C_D \big|_{\alpha=\beta=\delta_E=i_H=\delta_R=0} = C_{D_0} + \frac{1}{\pi A_W e_W} \left(C_{L_W}^2 + C_{L_H}^2 \frac{A_W e_W}{A_H e_H} \frac{q_H}{q_\infty} \frac{S_H}{S_W} \right) \tag{6.36}$$

where C_{L_W} and C_{L_H} are given by Equations (6.16) with $\alpha = i_H = 0$, and C_{D_0} is defined in Equation (6.29). Also note that the drag coefficient with $\alpha = \beta = \delta_E = i_H = \delta_R = 0$ is not the same as C_{D_0} given in Equations (6.29) and (6.30). The difference should be obvious from Equation (6.36).

6.2.4 Vehicle Rolling Moment

The three components of the aerodynamic moment vector $\mathbf{M}_A$ acting on the vehicle are shown in Figure 6.7. These three components act along the three axes of the vehicle-fixed reference frame. By convention, the components of aerodynamic moment are given as

$$\mathbf{M}_A = L_A \mathbf{i}_V + M_A \mathbf{j}_V + N_A \mathbf{k}_V \tag{6.37}$$

A rolling moment L_A may be generated if the vehicle is experiencing a sideslip angle β, or if the ailerons and/or rudder are deflected. If none of these conditions

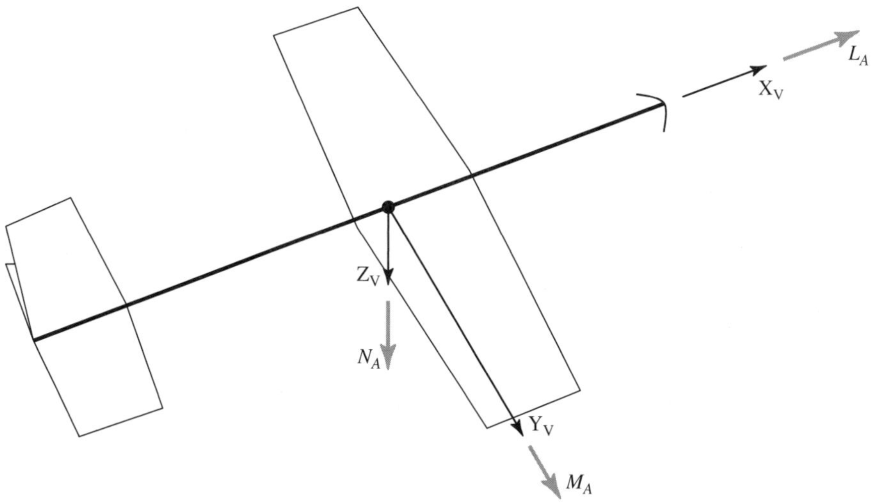

Figure 6.7 Components of aerodynamic moment on the vehicle.

are present, the symmetry of the vehicle is typically such that no rolling moment is generated.

All the lifting surfaces (e.g., wing and horizontal and vertical tails) may contribute to the rolling moment. Therefore, we write the vehicle's aerodynamic rolling moment as

$$L_A = L_{A_W} + L_{A_{H\,\&\,V}} = C_{L_{\text{Roll}}} q_\infty S_W b_W \tag{6.38}$$

Here we have introduced the vehicle's rolling-moment coefficient $C_{L_{\text{Roll}}}$, and we include the subscript Roll to distinguish this coefficient from the lift coefficient C_L. The wing span b_W is defined as the characteristic length associated with this coefficient.

In Chapter 5 we found that the rolling moment generated by a wing may be expressed in terms of the wing's dihedral effect $C_{L_{\beta_W}}$, plus the aileron rolling-moment effectiveness $C_{L_{\delta_A}}$. Or

$$L_{A_W} = \left(C_{L_{\beta_W}} \beta + C_{L_{\delta_A}} \delta_A\right) q_\infty S_W b_W \tag{6.39}$$

Similarly, assuming the elevators on the horizontal tail can only be deflected symmetrically, the rolling moment generated by the horizontal and vertical tail may be expressed as

$$L_{A_{H\,\&\,V}} = C_{L_{\beta_H}} \beta q_H S_H b_H + \left(C_{L_{\beta_V}} \beta + C_{L_{\delta_V}} \delta_R\right) q_H S_V b_V \tag{6.40}$$

Here we have included the dihedral effect C_{L_β} of both the horizontal and vertical tails, along with the rudder rolling-moment effectiveness $C_{L_{\delta_V}}$ of the vertical tail.

Both the dihedral effect and the rudder rolling-moment effectiveness of the vertical tail arise due to the fact that the side force generated by the vertical tail (discussed earlier in this chapter) is taken to act at its aerodynamic center. And since the span-wise location of this aerodynamic center lies off the vehicle-fixed

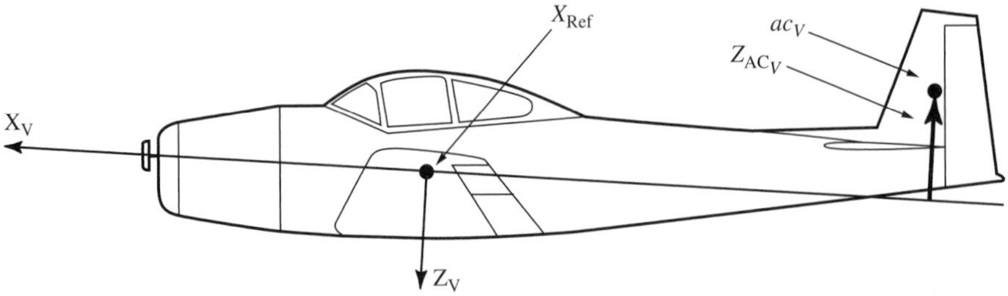

Figure 6.8 Location of the aerodynamic center of the vertical tail.

X axis of the vehicle (see Figure 6.8), a moment arm exists for the side force to generate a rolling moment. The location of this aerodynamic center is typically above the vehicle's X axis. Therefore, if the aerodynamic center of the vertical tail is located a distance Z_{AC_V} above the vehicle-fixed X axis, the dihedral effect of the vertical tail may be written in terms of the sideslip side-force effectiveness of the vertical tail, or

$$\boxed{C_{L_{\beta_V}} = C_{S_{\beta_V}} \frac{Z_{AC_V}}{b_V}} \tag{6.41}$$

So the side force from the vertical tail S_V generates the rolling moment L_{A_V}. Note, finally, that the sideslip rolling-moment effectiveness $C_{L_{\beta_V}}$ will typically be negative, due to the sign convention adopted for side force and sideslip (see Figure 6.4), except when Z_{AC_V} is negative.

As for the rudder rolling-moment effectiveness, we may also express it in terms of the rudder side force acting over the same moment arm discussed above. Or, if the side force coefficient is expressed in terms of the rudder sideslip effectiveness β_δ, we may write

$$\boxed{C_{L_{\delta_V}} = C_{S_{\beta_V}} \beta_\delta \frac{Z_{AC_V}}{b_V}} \tag{6.42}$$

Note that this coefficient is positive if Z_{AC_V} (shown in Figure 6.8) is positive, since both $C_{S_{\beta_V}}$ and β_δ will be negative due to the sign conventions noted above.

From Equations (6.38–6.40), and again defining the wing-planform area as the vehicle reference area, we find that the vehicle's aerodynamic rolling-moment coefficient may be expressed as

$$\boxed{\begin{aligned} C_{L_{\text{Roll}}} = &\left(C_{L_{\beta_W}} + C_{L_{\beta_H}} \frac{q_H}{q_\infty} \frac{S_H}{S_W} \frac{b_H}{b_W} + C_{L_{\beta_V}} \frac{q_H}{q_\infty} \frac{S_V}{S_W} \frac{b_V}{b_W}\right)\beta \\ &+ C_{L_{\delta_A}} \delta_A + C_{L_{\delta_V}} \frac{q_H}{q_\infty} \frac{S_V}{S_W} \frac{b_V}{b_W} \delta_R \end{aligned}} \tag{6.43}$$

This expression is valid as long as the rolling-moment coefficients lie in their locally linear ranges. Differentiating Equation (6.43) with respect to vehicle sideslip angle yields the vehicle's sideslip rolling-moment effectiveness, or

$$C_{L_\beta} = C_{L_{\beta_W}} + C_{L_{\beta_H}} \frac{q_H}{q_\infty} \frac{S_H}{S_W} \frac{b_H}{b_W} + C_{L_{\beta_V}} \frac{q_H}{q_\infty} \frac{S_V}{S_W} \frac{b_V}{b_W} \tag{6.44}$$

Differentiating Equation (6.43) with respect to rudder deflection δ_R yields the vehicle's rudder rolling-moment effectiveness, or

$$C_{L_{\delta_R}} = C_{L_{\delta_V}} \frac{q_H}{q_\infty} \frac{S_V}{S_W} \frac{b_V}{b_W} \tag{6.45}$$

Finally, differentiating Equation (6.43) with respect to aileron deflection δ_A reveals that the vehicle's aileron rolling-moment effectiveness is identical to the aileron rolling-moment effectiveness of the wing $C_{L_{\delta_A}}$, which appears in Equation (6.39).

6.2.5 Vehicle Pitching Moment

The lift-generating surfaces (e.g., the wing and the horizontal tail) plus the fuselage all contribute to the pitching moment acting on the vehicle. The effect of the wing and tail will be discussed first. Figure 6.9, similar to Figure 6.5, shows the forces and pitching moments acting on the wing and tail. Also shown are the locations, along the vehicle's X axis, of the wing and tail aerodynamic centers, plus a reference point located at X_{Ref}.

Summing pitching moments about this reference point (taking X locations as positive forward) yields the following expression for the pitching moment generated by the wing and horizontal tail.

$$\begin{aligned} M_{A_{W\&H}} = M_{\text{AC}_W} &+ (L_W \cos\alpha + D_W \sin\alpha)(X_{\text{AC}_W} - X_{\text{Ref}}) \\ &+ (-L_W \sin\alpha + D_W \cos\alpha) Z_{\text{AC}_W} + M_{\text{AC}_H} \\ &- (L_H \cos\alpha + D_H \sin\alpha)(X_{\text{Ref}} - X_{\text{AC}_H}) + (-L_H \sin\alpha + D_H \cos\alpha) Z_{\text{AC}_H} \end{aligned}$$

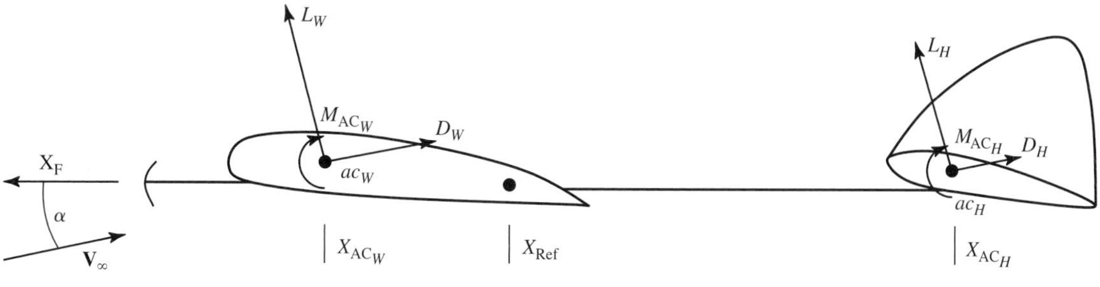

Figure 6.9 Wing and tail aerodynamic forces and pitching moments.

$$= (M_{AC_W} + M_{AC_H}) + (L_W \cos\alpha + D_W \sin\alpha)\Delta X_W + (-L_W \sin\alpha + D_W \cos\alpha) Z_{AC_W}$$
$$- (L_H \cos\alpha + D_H \sin\alpha)\Delta X_H + (-L_H \sin\alpha + D_H \cos\alpha) Z_{AC_H}$$

Here $Z_{AC\bullet}$ (not shown in Figure 6.9 to avoid clutter) is the vertical distance of the wing or tail aerodynamic center above the X axis of the vehicle. However, Z_{AC_V} was defined in Figure 6.8.

For conventional-geometry vehicles, typically $D_\bullet \ll L_\bullet$, and $|Z_{AC\bullet}| \ll |\Delta X_\bullet|$. In addition, $\cos\alpha \gg \sin\alpha$, and for a symmetric tail airfoil section $M_{AC_H} = \alpha_{0_H} = 0$. So the above equation for the pitching moment can usually be approximated by the following expression;

$$M_{A_{W\,\&\,H}} = M_{AC_W} + L_W(X_{AC_W} - X_{Ref}) - L_H(X_{Ref} - X_{AC_H}) \tag{6.46}$$

Now using Equations (6.16), the lift generated by the wing and tail, respectively, may be written as

$$L_W = C_{L_W} q_\infty S_W = C_{L_{\alpha_W}}(\alpha + i_W - \alpha_{0_W}) q_\infty S_W \tag{6.47}$$

and

$$L_H = C_{L_H} q_H S_H = C_{L_{\alpha_H}}\left(\left(1 - \frac{d\varepsilon}{d\alpha}\right)\alpha + \frac{d\varepsilon}{d\alpha}(\alpha_{0_W} - i_W) + (i_H - \alpha_{0_H}) + \alpha_\delta \delta_E\right) q_H S_H \tag{6.48}$$

Again, the above expressions are valid as long as the effectiveness coefficients lie in their locally linear ranges. Consequently, the pitching moment due to the wing and horizontal tail, under the local-linearity assumption, may be expressed as

$$\begin{aligned} M_{A_{W\,\&\,H}} &= C_{M_{AC_W}} q_\infty S_W \bar{c}_W + C_{L_{\alpha_W}}(\alpha + i_W - \alpha_{0_W})(X_{AC_W} - X_{Ref}) q_\infty S_W \\ &\quad - C_{L_{\alpha_H}}\left(\left(1 - \frac{d\varepsilon}{d\alpha}\right)\alpha + \frac{d\varepsilon}{d\alpha}(\alpha_{0_W} - i_W) + (i_H - \alpha_{0_H}) + \alpha_\delta \delta_E\right)(X_{Ref} - X_{AC_H}) q_H S_H \end{aligned} \tag{6.49}$$

Though the effect of the fuselage on lift is typically negligible compared to that of the wing for conventional vehicle configurations, the fuselage can appreciably affect the vehicle pitching moment due to wing upwash and downwash. Specifically, the fuselage both influences the pitching-moment of the wing-fuselage combination, as well as shifts the location of the aerodynamic center of that combination.

The schematic in Figure 6.10, similar to that in Figure 6.9, shows the upwash ahead of the wing and the downwash aft. In the situation shown, the fuselage section forward of the wing experiences upwash, and the fuselage section aft of the wing experiences downwash. The net effect is an increase (nose-up) in the pitching moment acting on the vehicle.

Based on momentum considerations, and for subsonic flow over a straight fuselage in the presence of a wing, the fuselage-generated zero-lift pitching moment M_{0_F} may be quantified in terms of Equation (6.50), which involves breaking the

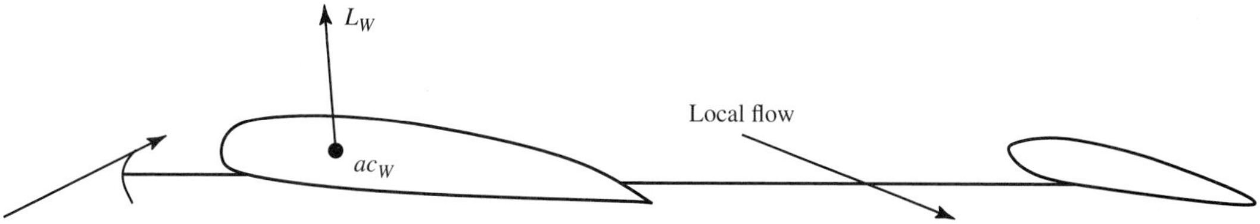

Figure 6.10 Upwash and downwash acting on the fuselage.

fuselage up into sections along its length (Refs. 2, 3). This procedure will be demonstrated in Example 6.2.

$$M_{0_F} = C_{M_{0_F}} q_\infty S_W \bar{c}_W = \frac{k q_\infty \alpha_{0_W}}{36.5} \sum_{x_i=0}^{x_i=l_F} w_{F_i}^2 \Delta x_i \tag{6.50}$$

Here,

k = Munk apparent-mass term (from Figure 6.11)
x_i = distance of i'th fuselage section aft of nose (fuselage broken into multiple sections for summation)
Δx_i = length of i'th fuselage section
l_F = total fuselage length
w_{F_i} = average width of i'th fuselage section
$\alpha_{0_W} = \alpha_0 + i_w$ = wing zero-lift angle of attack (deg) measured from the fuselage-referenced X axis (see Equation (6.12))

Note that this moment may be positive or negative, depending on the sign of α_{0_W}.

Furthermore, the fuselage changes the angle-of-attack pitching-moment effectiveness of the wing-fuselage combination, or $C_{M_{\alpha_{W\&F}}}$. Similar to Equation (6.50), the expression for the angle-of-attack pitching-moment effectiveness due to the fuselage is (Ref. 3)

$$M_{\alpha_F} = C_{M_{\alpha_F}} q_\infty S_W \bar{c}_W \approx \frac{q_\infty}{36.5} \sum_{x_i=0}^{x_i=l_F} w_{F_i}^2 \frac{d\alpha_{\text{local}}}{d\alpha_W}(x_i) \Delta x_i \text{ /deg} \tag{6.51}$$

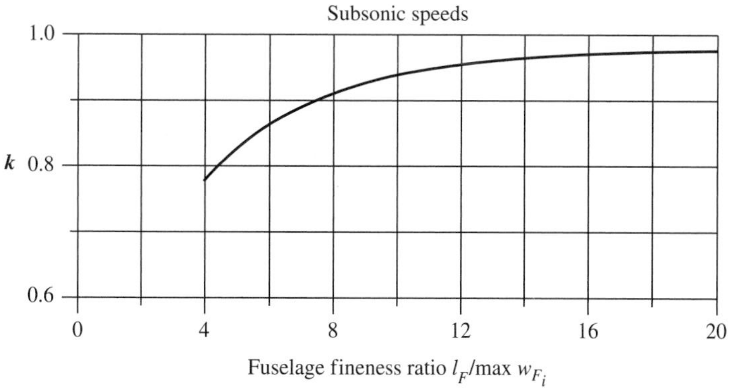

Figure 6.11 Munk apparent-mass term.

where x_i, Δx_i, and w_{F_i} are as defined with Equation (6.50), and

$$\frac{d\alpha_{\text{local}}}{d\alpha_W}(x_i) = \text{local flow-angle gradient at fuselage location } x_i \text{, with}$$

$$\alpha_{\text{local}} = \alpha_W \pm \varepsilon = \text{local flow-incidence angle at the section}$$

Note then that

$$\frac{d\alpha_{\text{local}}}{d\alpha_W} = 1 \pm \frac{d\varepsilon}{d\alpha_W}$$

where ε is the upwash (downwash) angle ahead of (behind) the wing. The upwash and downwash gradients may be determined from Figures 5.36 and 5.35, respectively. This procedure will also be demonstrated in Example 6.2.

The fuselage pitching-moment effectiveness in Equation (6.51) induces a shift in the location of the aerodynamic center of the wing-fuselage combination relative to that of the wing alone. Specifically, this shift may be expressed as

$$\Delta \bar{X}_{\text{AC}_F} = \frac{C_{M_{\alpha_F}}}{C_{L_{\alpha_W}}} \tag{6.52}$$

and so the location of the aerodynamic center of the wing-fuselage combination becomes

$$X_{\text{AC}_{W\,\&\,F}} = X_{\text{AC}_W} + \Delta X_{\text{AC}_F} \tag{6.53}$$

Note that a positive $C_{M_{\alpha_F}}$ leads to a <u>forward</u> shift in the location of the aerodynamic center.

Consequently, using Equations (6.49–6.51), the total aerodynamic pitching moment on the vehicle, generated by the wing, fuselage, and horizontal tail is now

$$M_A = M_{0_F} + M_{A_{W\,\&\,H}} = C_M q_\infty S_W \bar{c}_W \tag{6.54}$$

Here we have introduced the vehicle's pitching-moment coefficient C_M, and defined the length of the wing's mean aerodynamic chord (MAC) $\bar{c}_W$ as the vehicle reference length associated with this coefficient.

From Equations (6.49–6.54) we can now solve for the <u>vehicle's pitching-moment coefficient</u>. Or, under the local-linearity assumption, we have

$$\begin{aligned} C_M = {} & \left(C_{M_{\text{AC}_W}} + C_{M_{0_F}}\right) + C_{L_{\alpha_W}}\left(\alpha + i_W - \alpha_{0_W}\right)\left(\frac{X_{\text{AC}_{W\,\&\,F}} - X_{\text{Ref}}}{\bar{c}_W}\right) \\ & - C_{L_{\alpha_H}}\left(\left(1 - \frac{d\varepsilon}{d\alpha}\right)\alpha + \frac{d\varepsilon}{d\alpha}\left(\alpha_{0_W} - i_W\right) + \left(i_H - \alpha_{0_H}\right) + \alpha_\delta \delta_E\right)\left(\frac{X_{\text{Ref}} - X_{\text{AC}_H}}{\bar{c}_W}\right)\frac{q_H}{q_\infty}\frac{S_H}{S_W} \end{aligned} \tag{6.55}$$

Differentiating Equation (6.55) with respect to vehicle angle of attack α yields the vehicle's angle-of-attack pitching-moment effectiveness, or

$$\begin{aligned} C_{M_\alpha} &= C_{L_{\alpha_W}}\left(\frac{X_{\mathrm{AC}_{W\,\&\,F}} - X_{\mathrm{Ref}}}{\bar{c}_W}\right) - C_{L_{\alpha_H}}\left(1 - \frac{d\varepsilon}{d\alpha}\right)\left(\frac{X_{\mathrm{Ref}} - X_{\mathrm{AC}_H}}{\bar{c}_W}\right)\frac{q_H}{q_\infty}\frac{S_H}{S_W} \\ &= C_{L_{\alpha_W}}\Delta\bar{X}_{\mathrm{AC}_{W\,\&\,F}} - C_{L_{\alpha_H}}\left(1 - \frac{d\varepsilon}{d\alpha}\right)\Delta\bar{X}_{\mathrm{AC}_H}\frac{q_H}{q_\infty}\frac{S_H}{S_W} \end{aligned} \tag{6.56}$$

(Note that some authors define a term called the *tail volume coefficient* V_H, where

$$V_H = \Delta\bar{X}_{\mathrm{AC}_H}\frac{S_H}{S_W} \tag{6.57}$$

but we choose not to use this term.) As we shall see in Chapter 9, the pitching-moment effectiveness given in Equation (6.56) is a very important design parameter for the vehicle. Note that in the above two expressions we have introduced the notation $\Delta\bar{X}_{\cdot}$ for the nondimensionalized distances between the aerodynamic centers of the fuselage-wing or of the tail and the reference point about which we defined the pitching moment.

NOTE TO STUDENT

In the derivation of Equation (6.55) and in all other expressions that follow from that equation, such as (6.56), we have taken X locations on the vehicle to be defined as positive forward. However, note that many times in practice, X locations are provided measured aft from the wing apex, or aft in terms of percent mean aerodynamic chord (MAC) of the wing. So care must be exercised in terms of the signs of these X locations when using Equation (6.55) and related expressions. When we use these equations in this book, we will try to remind the reader whether X is measured positive forward or aft. But a quick check of each equation as to whether positive forward or aft makes sense is always a good practice.

Furthermore, differentiating Equation (6.55) with respect to elevator deflection δ_E yields the elevator pitching-moment effectiveness for the vehicle, or

$$C_{M_{\delta_E}} = -C_{L_{\alpha_H}}\alpha_\delta\Delta\bar{X}_{\mathrm{AC}_H}\frac{q_H}{q_\infty}\frac{S_H}{S_W} = -C_{L_{\delta_H}}\Delta\bar{X}_{\mathrm{AC}_H}\frac{q_H}{q_\infty}\frac{S_H}{S_W} \tag{6.58}$$

And differentiating Equation (6.55) with respect to tail incidence angle i_H yields the vehicle's tail-incidence pitching-moment effectiveness, or

$$C_{M_{i_H}} = -C_{L_{\alpha_H}}\Delta\bar{X}_{\mathrm{AC}_H}\frac{q_H}{q_\infty}\frac{S_H}{S_W} \tag{6.59}$$

Finally, note that when $\alpha = \delta_E = i_H = 0$, the pitching-moment coefficient is

$$C_M\big|_{\alpha=\delta_E=i_H=0} = \left(C_{M_{\mathrm{AC}_W}} + C_{M_{0_F}}\right) + C_{L_{\alpha_W}}\left(i_W - \alpha_{0_W}\right)\Delta\overline{X}_{\mathrm{AC}_{W\&F}} - C_{L_{\alpha_H}}\left(\frac{d\varepsilon}{d\alpha}\left(\alpha_{0_W} - i_W\right) - \alpha_{0_H}\right)\Delta\overline{X}_{\mathrm{AC}_H}\frac{q_H}{q_\infty}\frac{S_H}{S_W} \tag{6.60}$$

In Equations (6.59) and (6.60), X locations are again defined to be positive forward on the vehicle.

As an important application of Equation (6.56), we may now solve for the axial location of the *vehicle's aerodynamic center*, $X_{\mathrm{AC}_{\mathrm{Veh}}}$. Until this point the location of X_{Ref} has been arbitrary. By definition the aerodynamic center is that point about which the pitching moment is invariant with angle of attack. Therefore, we may solve for the location of the aerodynamic center by finding that X_{Ref} for which $C_{M_\alpha} = 0$. Setting Equation (6.56) equal to zero, and solving for the X_{Ref} in that case yields the following location for the vehicle's aerodynamic center.

$$\overline{X}_{\mathrm{AC}_{\mathrm{Veh}}} \triangleq \frac{X_{\mathrm{AC}_{\mathrm{Veh}}}}{\bar{c}_W} = \frac{C_{L_{\alpha_W}}\overline{X}_{\mathrm{AC}_{W\&F}} + C_{L_{\alpha_H}}\left(1 - \frac{d\varepsilon}{d\alpha}\right)\overline{X}_{\mathrm{AC}_H}\frac{q_H}{q_\infty}\frac{S_H}{S_W}}{C_{L_{\alpha_W}} + C_{L_{\alpha_H}}\left(1 - \frac{d\varepsilon}{d\alpha}\right)\frac{q_H}{q_\infty}\frac{S_H}{S_W}} \tag{6.61}$$

EXAMPLE 6.2

Vehicle Pitching-Moment Analysis

Again consider the vehicle described in Example 6.1 and the tail analyzed in Example 5.7. Let the fuselage, shown in Figure 6.12, be described in terms of sections along its length, with dimensions given in Table 6.1. If the axial location of the reference point X_{Ref} is at $0.5\bar{c}_W$ aft of the wing apex, and the wing's aerodynamic center is located at $X_{\mathrm{AC}_W} = x_9$ in Table 6.1, determine the pitching-moment effectivenesses C_{M_α}, $C_{M_{\delta_E}}$, and $C_{M_{i_H}}$ for the vehicle. Also, find the location of the vehicle's aerodynamic center and determine the pitching-moment coefficient when $\alpha = \delta_E = i_H = 0$.

■ Solution

The relevant data for the wing and tail are as follows:

Wing: $C_{M_{\mathrm{AC}_W}} = 0.03$, $\alpha_{0_W} = 1.33$ deg, $i_W = 2$ deg, $X_{\mathrm{AC}_W} = 4.79$ ft aft of wing apex,
$A_W = 5.33$, $S_W = 169$ ft^2, $\bar{c}_W = 5.825$ ft, $c_{r_W} = 7.5$ ft, $b_W = 30$ ft,
$C_{L_{\alpha_W}} = C_{L_{\alpha_H}} = 4.19$ /rad, $\dfrac{d\varepsilon}{d\alpha_W}\Big|_\infty = 0.5$

Tail: $\alpha_\delta = 0.65$, $S_H = 42.2$ ft^2, $\bar{c}_H = 2.92$ ft, $X_{\mathrm{AC}_H} = 2.5c_{r_W}$ aft of wing apex,
$\dfrac{d\varepsilon_H}{d\alpha_W} = 0.57$

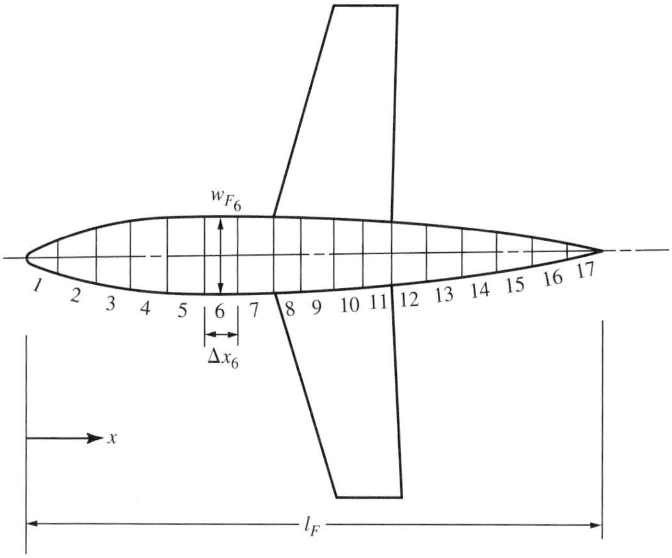

Figure 6.12 Fuselage broken into sections (parameters shown for Section 6 are typical).

Table 6.1 Data for Fuselage Sections

Section, i	Average Width, w_{F_i} ft	Section Length, Δx_i ft	Axial Location of Centroid of Section, x_i ft	Centroid Distance from X_{AC_W}, ft	Local Flow-Angle Gradient, $\frac{d\alpha_{local}}{d\alpha_W}(x_i)$
1	1.34	2.49	1.25	19.31	1.09
2	3.12	2.49	3.74	16.82	1.09
3	4.33	2.49	6.23	14.33	1.09
4	4.92	2.49	8.72	11.74	1.10
5	5.18	2.49	11.21	9.35	1.13
6	5.28	2.49	13.7	6.86	1.22
7	5.22	2.49	16.19	4.37	1.42
8	5.03	2.08	18.47	2.09	2.2
9	4.79	2.08	20.56	0	–
10	4.40	2.08	22.64	2.08	–
11	4.12	2.08	24.72	4.16	–
12	3.90	2.49	27	6.44	0
13	3.61	2.49	29.49	8.93	0.38
14	3.12	2.49	31.98	11.42	0.42
15	2.41	2.49	34.47	13.91	0.43
16	1.59	2.49	36.96	16.40	0.45
17	0.57	2.49	39.45	18.89	0.47

(Note that Sections 9–11 are between the wings, and assumed unaffected by wing downwash.)

To address the effects of the fuselage, the data needed to apply Equations (6.50) and (6.51) are summarized in Table 6.1. The last column of the table lists the quantity $1 \pm \frac{d\varepsilon}{d\alpha_W}$, in terms of the downwash or upwash gradient at the fuselage section obtained from Figures 5.35 and 5.36, respectively. Using the given data in Equation (6.50), and the fact that the fuselage fineness ratio is 7.6, we find that the fuselage's contribution to the zero-lift pitching moment coefficient is

$$C_{M_{0_F}} = \frac{k\alpha_{0_W}}{36.5 S_W \bar{c}_W} \sum_{x_i=0}^{x_i=l_F} w_{F_i}^2 \Delta x_i = \frac{0.9(1.33+2)}{36.5(169)(5.825)}(631.8) = 0.053$$

Note that this coefficient is positive, corresponding to a small nose-up moment. This result is typical.

Using the given data in Equation (6.51) yields the following fuselage contribution to the angle-of-attack pitching-moment effectiveness

$$C_{M_{\alpha_F}} = \frac{1}{36.5 S_W \bar{c}_W} \sum_{x_i=0}^{x_i=l_F} w_{F_i}^2 \frac{d\varepsilon}{d\alpha_W}(x_i)\Delta x_i = \frac{552.7}{36.5(169)(5.825)} = 0.015 \text{ /deg}$$

Consequently, the concomitant change in the nondimensional location of the aerodynamic center of the wing and fuselage in combination is

$$\Delta\bar{X}_{\text{AC}_F} = \frac{C_{M_{\alpha_F}}}{C_{L_{\alpha_W}}} = \frac{0.015(57.3)}{4.19} = 0.205 \text{ forward}$$

corresponding to a change in the absolute location of the aerodynamic center of wing and fuselage of

$$\Delta X_{\text{AC}_F} = \Delta\bar{X}_{\text{AC}_F}\bar{c}_W = 0.205(5.825) = 1.19 \text{ ft forward}$$

So the effect of the fuselage here is to shift the location of the aerodyamic center of the wing-fuselage combination forward of the location of the aerodynamic center of the wing alone. But it is possible, depending on wing and/or fuselage geometry, to have an aft shift as well.

We may now determine the pitching-moment effectivenesses for the vehicle. From Equation (6.59), the tail-incidence moment effectiveness is given by

$$C_{M_{i_H}} = -C_{L_{\alpha_H}}\Delta\bar{X}_{\text{AC}_H}\frac{q_H}{q_\infty}\frac{S_H}{S_W} = -4.19\left(\frac{2.5(7.5)-0.5(5.825)}{5.825}\right)(0.9)\left(\frac{42.2}{169}\right)$$
$$= -2.54 \text{ /rad}$$

(Note that Equation (6.59) was derived assuming X locations were measured positive forward on the vehicle. And since X_{Ref} and X_{AC_H} were given in this problem as distances aft of the wing apex, their signs were reversed from those in the given equation.)

Then from Equation (6.58) we have the elevator moment effectiveness given by

$$C_{M_{\delta_E}} = -C_{L_{\alpha_H}}\alpha_\delta\Delta\bar{X}_{\text{AC}_H}\frac{q_H}{q_\infty}\frac{S_H}{S_W} = \alpha_\delta C_{M_{i_H}} = 0.65(-2.54) = -1.65 \text{ /rad}$$

and from Equation (6.56) we have the vehicle's angle-of-attack moment effectiveness given by

$$C_{M_\alpha} = C_{L_{\alpha_W}}\left(\frac{X_{AC_{W\&F}} - X_{Ref}}{\bar{c}_W}\right) + C_{M_{i_H}}\left(1 - \frac{d\varepsilon}{d\alpha}\right)$$

$$= 4.19\left(\frac{-(4.79 - 1.19) + 0.5(5.825)}{5.825}\right) - 2.54(1 - 0.57) = -1.59\ /\text{rad}$$

(Again note the reversal of signs on X_{Ref} and $X_{AC_{W\&F}}$.) Note that this coefficient is negative, which, as we will see in Chapters 9 and 10, has important implications with regards to the stability of the vehicle. Again note the relative contributions of the wing and tail to this coefficient.

Applying Equation (6.61) we find that the nondimensionalized location of the vehicle's aerodynamic center aft of the wing apex is

$$\bar{X}_{AC_{Veh}} \triangleq \frac{X_{AC_{Veh}}}{\bar{c}_W} = \frac{C_{L_{\alpha_W}}\bar{X}_{AC_{W\&F}} + C_{L_{\alpha_H}}\left(1 - \frac{d\varepsilon}{d\alpha}\right)\bar{X}_{AC_H}\frac{q_H}{q_\infty}\frac{S_H}{S_W}}{C_{L_{\alpha_W}} + C_{L_{\alpha_H}}\left(1 - \frac{d\varepsilon}{d\alpha}\right)\frac{q_H}{q_\infty}\frac{S_H}{S_W}}$$

$$= \frac{4.19\left(\frac{4.79 - 1.19}{5.825}\right) + 4.19(1 - 0.57)\left(\frac{2.5(7.5)}{5.825}\right)(0.9)\left(\frac{42}{169}\right)}{4.19 + 4.19(1 - 0.57)(0.9)\left(\frac{42}{169}\right)} = 0.846$$

And so the aerodynamic center is located 0.846(5.825) = 4.928 ft aft of the wing apex. Note that, as expected, the vehicle's aerodynamic center is located aft of the aerodynamic center of the wing and fuselage alone, which is located at 4.79 − 1.19 = 3.6 ft aft of the wing apex.

Finally, from Equation (6.60) we have the vehicle's pitching-moment coefficient with $\alpha = \delta = i_H = 0$ given by

$$C_M\big|_{\alpha=\delta_E=i_H=0} = \left(C_{M_{AC_W}} + C_{M_{0_F}}\right) + C_{L_{\alpha_W}}\left(i_W - \alpha_{0_W}\right)\Delta\bar{X}_{AC_{W\&F}} + C_{M_{i_H}}\left(\frac{d\varepsilon}{d\alpha}\left(\alpha_{0_W} - i_W\right) - \alpha_{0_H}\right)$$

$$= (0.03 + 0.053) + 4.19\left(\frac{2 - 1.33}{57.3}\right)(-0.118) - 2.54\left(0.57\left(\frac{1.33 - 2}{57.3}\right) - 0\right)$$

$$= 0.083 - 0.006 - 2.54(-0.007) = 0.095$$

Again, this moment is small and positive, another typical result.

6.2.6 Vehicle Yawing Moment

A yawing moment on the vehicle may be generated by any surface generating a side force, such as the vertical tail in sideslip or with its rudder deflected. Plus, as discussed in Chapter 5, since aileron deflection generates antisymmetric drag a yawing moment would be generated as well. Consequently, let us write the vehicle's aerodynamic yawing moment N_A as

$$N_A = N_V + N_{\delta_A} \tag{6.62}$$

where the yawing moment due to the vertical tail is

$$N_V = N_{\beta_V} + N_{\delta_V} \tag{6.63}$$

Letting the X location of the rudder's aerodynamic center be denoted as X_{AC_V}, writing side forces in terms of (locally linear) sideslip and rudder effectiveness, and, with reference to Figure 6.9, summing yawing moments about X_{Ref} yields

$$\begin{aligned} N_{\beta_V} &= -C_{S_{\beta_V}} \beta (X_{\mathrm{Ref}} - X_{\mathrm{AC}_V}) q_H S_V \\ N_{\delta_V} &= -C_{S_{\delta_V}} \delta_R (X_{\mathrm{Ref}} - X_{\mathrm{AC}_V}) q_H S_V \end{aligned} \tag{6.64}$$

The negative signs arise from the fact that a positive side force (positive Y direction) on the aft vertical tail generates a negative yawing moment. Finally, consistent with Chapter 5, we have the yawing moment due to aileron deflection given by

$$N_{\delta_A} = C_{N_{\delta_{A_W}}} \delta_A q_\infty S_W b_W \tag{6.65}$$

where $C_{N_{\delta_{A_W}}}$ is the wing's aileron yawing-moment effectiveness as found in Chapter 5.

Collecting terms yields the expression for the vehicle's aerodynamic yawing moment, or

$$N_A = -(C_{S_{\beta_V}} \beta + C_{S_{\delta_V}} \delta_R)(X_{\mathrm{Ref}} - X_{\mathrm{AC}_V}) q_H S_V + C_{N_{\delta_{A_W}}} \delta_A q_\infty S_W b_W = C_N q_\infty S_W b_W \tag{6.66}$$

Here we have introduced the vehicle's yawing-moment coefficient C_N, with the wingspan b_W defined as the reference length and the wing-planform area defined as the vehicle reference area.

Solving for the <u>vehicle yawing-moment coefficient</u> we have

$$\boxed{\begin{aligned} C_N &= C_{N_{\delta_{A_W}}} \delta_A - (C_{S_{\beta_V}} \beta + C_{S_{\delta_V}} \delta_R)\left(\frac{X_{\mathrm{Ref}} - X_{\mathrm{AC}_V}}{b_W}\right) \frac{q_H}{q_\infty} \frac{S_V}{S_W} \\ &= C_{N_{\delta_{A_W}}} \delta_A - (C_{S_{\beta_V}} \beta + C_{S_{\delta_V}} \delta_R) \Delta \overline{X}_{\mathrm{AC}_V} \frac{q_H}{q_\infty} \frac{S_V}{S_W} \end{aligned}} \tag{6.67}$$

Differentiating the above with respect to sideslip angle β yields the <u>vehicle's sideslip yawing-moment effectiveness</u>, or

$$\boxed{C_{N_\beta} = -C_{S_{\beta_V}} \Delta \overline{X}_{\mathrm{AC}_V} \frac{q_H}{q_\infty} \frac{S_V}{S_W}} \tag{6.68}$$

Differentiating Equation (6.67) with respect to rudder deflection δ_R yields the vehicle's rudder yawing-moment effectiveness, or

$$C_{N_{\delta_R}} = -C_{S_{\delta_V}} \Delta \overline{X}_{AC_V} \frac{q_H}{q_\infty} \frac{S_V}{S_W} \tag{6.69}$$

Differentiating Equation (6.67) with respect to aileron deflection δ_A reveals that the vehicle's aileron yawing-moment effectiveness is identical to that of the wing, or

$$C_{N_{\delta_A}} = C_{N_{\delta_{A_W}}} \tag{6.70}$$

EXAMPLE 6.3

Vehicle Side Force, Rolling and Yawing Moment Analysis

Again consider the vehicle addressed in Examples 6.1 and 6.2, and assume the vehicle's vertical tail has the same geometry (sweep, span, taper, etc.) as one half of the horizontal tail. In addition, assume the aerodynamic center of the vertical tail is also located 2.5 wing-root-chord lengths aft of the wing apex, and let the vertical tail have a full-span rudder with a rudder-to-tail chord ratio c_R/c_V of 0.25, constant with span. Also assume the horizontal tail has no dihedral. For this vehicle, at an angle of attack of zero, find the following:

1. Dihedral effect, C_{L_β}.
2. Aileron and rudder rolling-moment effectiveness, $C_{L_{\delta_A}}$ and $C_{L_{\delta_R}}$.
3. Sideslip side-force and yawing-moment effectiveness, C_{S_β} and C_{N_β}.
4. Aileron and rudder side-force and yawing-moment effectiveness, $C_{S_{\delta_R}}$, $C_{N_{\delta_A}}$, and $C_{N_{\delta_R}}$.

■ Solution

From the given information, and due to the similarity between the geometries of the horizontal and vertical tails, we find that for the vertical tail, $S_V = 21.1$ ft^2, $\overline{c}_V = 2.92$ ft, $C_{L_{\alpha_V}} = 4.19$ /rad, $\beta_\delta = -0.65$, and $X_{AC_V} = 2.5c_{r_W}$ aft of wing apex. Also, from Example 5.6 we have the chord of the vertical tail given by

$$c(z) = \frac{7.5}{2} - \frac{z}{4} \text{ ft, } z \text{ positive along vertical tail}$$

So the span-wise location of the MAC of the vertical tail is given by

$$Z_{MAC_V} = \frac{1}{S_V}\int_0^{b_V} zc(z)dz = \frac{1}{21.1}\int_0^{7.5}\left(\frac{7.5z}{2} - \frac{z^2}{4}\right)dz = \frac{1}{21.1}\left(\frac{7.5z^2}{4} - \frac{z^3}{12}\right)\Bigg|_0^{7.5}$$

$$= \frac{1}{21.1}(105.5 - 35.16) = 3.33 \text{ ft}$$

From Example 5.5 we found that the aileron rolling-moment and yawing-moment effectiveness for the wing alone were $C_{L_{\delta_A}} = 0.265$ /rad and $C_{N_{\delta_A}} = -0.0026$ /rad, respectively. And from Example 5.4 we found that the wing's dihedral effect was $C_{L_{\beta_W}} = -0.09$ /rad.

Now from Equation (6.44) we have the dihedral effect for the vehicle given by

$$C_{L_\beta} = C_{L_{\beta_W}} + C_{L_{\beta_H}} \frac{q_H}{q_\infty} \frac{S_H}{S_W} \frac{b_H}{b_W} + C_{L_{\beta_V}} \frac{q_H}{q_\infty} \frac{S_V}{S_W} \frac{b_V}{b_W}$$

Since the dihedral angle is zero and the leading-edge sweep is small for the horizontal tail, let $C_{L_{\beta_H}} \approx 0$. And from Equation (6.41) we have

$$C_{L_{\beta_V}} = C_{S_{\beta_V}} \frac{Z_{AC_V}}{b_V} = -C_{L_{\alpha_V}} \frac{Z_{AC_V}}{b_V} = -4.19 \frac{3.33}{7.5} = -1.86 \text{ /rad}$$

where Z_{AC_V} is the vertical location of the tail's aerodynamic center above the vehicle's X axis. Since the vehicle's angle of attack is given as zero, the free-stream velocity is parallel to the fuselage-referenced X axis, so $Z_{AC_V} = Z_{MAC_V} = 3.33$ ft. Consequently, the vehicle's dihedral effect is

$$C_{L_\beta} = -0.09 + 0 - 1.86(0.9)\left(\frac{21.1}{169}\right)\left(\frac{7.5}{30}\right) = -0.142 \text{ /rad}$$

Note that approximately one third of this dihedral effect is due to the vertical tail, and that this parameter is typically negative.

In prior discussions of rolling moment it was noted that the aileron rolling-moment effectiveness for the vehicle is the same as that for the wing. And, as given above, this wing's aileron rolling-moment effectiveness is $C_{L_{\delta_{A_W}}} = 0.265$ /rad. Hence for the vehicle we have

$$C_{L_{\delta_A}} = 0.265 \text{ /rad}$$

This parameter is typically positive, due to the convention for positive aileron deflection and positive roll.

From Equation (6.45) we have the vehicle's rudder rolling-moment effectiveness given by

$$C_{L_{\delta_R}} = C_{L_{\delta_V}} \frac{q_H}{q_\infty} \frac{S_V}{S_W} \frac{b_V}{b_W}$$

where, from Equation (6.42), the rudder rolling-moment effectiveness of the vertical tail is given by

$$C_{L_{\delta_V}} = C_{S_{\beta_V}} \beta_\delta \frac{Z_{AC_V}}{b_V}$$

Therefore, the vehicle's rudder rolling-moment effectiveness is

$$C_{L_{\delta_R}} = \left(C_{S_{\beta_V}} \beta_\delta \frac{Z_{AC_V}}{b_V}\right) \frac{q_H}{q_\infty} \frac{S_V}{S_W} \frac{b_V}{b_W} = \left(-4.19(-0.65)\left(\frac{3.33}{7.5}\right)\right)(0.9)\left(\frac{21.1}{169}\right)\left(\frac{7.5}{30}\right) = 0.034 \text{ /rad}$$

Note that this parameter is typically positive, again due to the convention for positive roll and positive rudder deflection.

From Equation (6.25) we have the vehicle sideslip side-force effectiveness given by

$$C_{S_\beta} = C_{S_{\beta_V}} \frac{q_H}{q_\infty} \frac{S_V}{S_W} = -C_{L_{\alpha_V}} \frac{q_H}{q_\infty} \frac{S_V}{S_W} = -4.19(0.9)\left(\frac{21.1}{169}\right) = -0.471 \text{ /rad}$$

This parameter is also typically negative, again due to the sign conventions on side force and sideslip. From Equation (6.68) we have the vehicle's sideslip yawing-moment effectiveness given by

$$C_{N_\beta} = -C_{S_{\beta_V}} \Delta \overline{X}_{AC_V} \frac{q_H}{q_\infty} \frac{S_V}{S_W} = 4.19\left(\frac{2.5(7.5) - 0.5(5.825)}{5.825}\right)(0.9)\left(\frac{21.1}{169}\right) = 1.28 \text{ /rad}$$

This parameter is typically positive (why?). From Equation (6.26) we have the vehicle's rudder side-force effectiveness given by

$$C_{S_{\delta_R}} = C_{S_{\beta_V}} \beta_\delta \frac{q_H}{q_\infty} \frac{S_V}{S_W} = (-4.19)(-0.65)(0.9)\left(\frac{21.1}{169}\right) = 0.306 \text{ /rad}$$

This parameter is also typically positive, again due to the sign conventions. From Equation (6.69) we have the vehicle's rudder yawing-moment effectiveness given by

$$C_{N_{\delta_R}} = -C_{S_{\delta_V}} \Delta \overline{X}_{AC_V} \frac{q_H}{q_\infty} \frac{S_V}{S_W} = -C_{S_{\beta_V}} \beta_\delta \Delta \overline{X}_{AC_V} \frac{q_H}{q_\infty} \frac{S_V}{S_W}$$

$$= -(-4.19)(-0.65)\left(\frac{2.5(7.5) - 0.5(5.825)}{5.825}\right)(0.9)\left(\frac{21.1}{169}\right) = -0.832 \text{ /rad}$$

This parameter will always be negative for a rudder on an aft vertical tail, given the sign convention on rudder deflection. Finally, from Equation (6.70) we have the vehicle's aileron yawing-moment effectiveness given by

$$C_{N_{\delta_A}} = C_{N_{\delta_{A_W}}} = -0.0026 \text{ /rad}$$

Since this parameter is negative, the ailerons produce adverse yaw. If the ailerons are deflected positively to bank the vehicle to the right (positive roll) to initiate a right turn, a <u>negative</u> yawing moment is produced, which is in the direction opposite that of the turn.

6.3 PROPULSIVE FORCES AND MOMENTS ACTING ON THE VEHICLE

All aircraft propulsion systems generate thrust by accelerating a column (mass) of air. Propeller-driven systems, including rotorcraft, accelerate the air mechanically, while turbojet engines do so thermodynamically. But regardless of how it is accomplished, the propulsion system generates a thrust force acting on the vehicle.

To address the propulsive forces and moments, consider the schematic depicted in Figure 6.13. (If there are multiple propulsive devices, the analysis presented here may be applied to each device.) We will assume that the propulsive

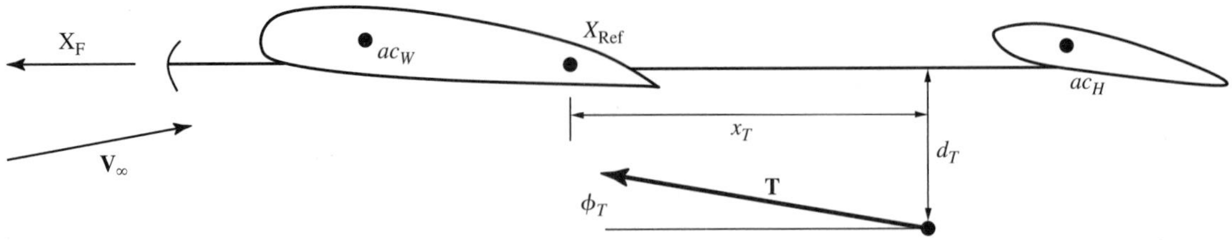

Figure 6.13 Position and line of action of propulsive thrust.

device(s) and the total thrust T are symmetric with respect to the XZ plane of the vehicle. Due to the manner by which the propulsive device is mounted on the vehicle, this force may act at an angle ϕ_T (positive up) relative to the vehicle's (fuselage-referenced) X axis. Thus, there may be both an axial (X) and a vertical (Z) component of this thrust.

If the thrust is controlled, such as with a throttle setting π, then the expression for the magnitude of the thrust may take on the following form.

$$T = T\left(V_\infty, \text{h}, \pi\right) \tag{6.71}$$

The thrust at a particular flight velocity and altitude is typically available from an engine model or tabulated data.

The point of action of $\mathbf{T}$ is assumed to be located a distance d_T below the (fuselage-referenced) X axis, and a distance x_T aft of the selected reference location X_{Ref}. Thus, this thrust T also generates a propulsive pitching moment M_P about X_{Ref} equal to

$$M_P = T\left(V_\infty, \text{h}, \pi\right)\left(d_T \cos\phi_T - x_T \sin\phi_T\right) \tag{6.72}$$

If the vehicle has multiple propulsive devices and the thrust forces are not symmetric with respect to the XZ plane of the vehicle, then those thrust forces would also generate a yawing moment that could be expressed in a fashion analogous to Equation (6.72).

NOTE TO STUDENT

If the line of action of $\mathbf{T}$ passes close to X_{Ref} then

$$\left(d_T \cos\phi_T - x_T \sin\phi_T\right) \approx 0 \tag{6.73}$$

in which case $M_P \approx 0$. This fact must be kept in mind in all later discussions of propulsive pitching moment.

The propulsive forces and moments acting on the vehicle are influenced by changes in the axial and transverse inlet-flow velocities of the working fluid, although turbojet inlets are frequently designed to minimize these effects. The sketches in Figure 6.14 represent a turbojet and a propeller propulsive device, including the column of working fluid being accelerated by the device. The thrust

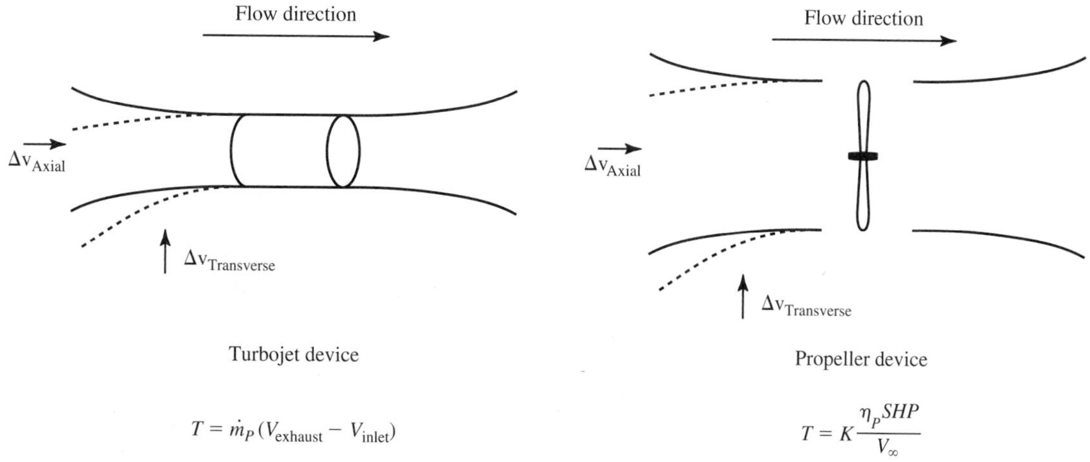

Figure 6.14 Changes in incoming-flow velocity.

axes of the devices are nominally aligned with the centerline of the column of working fluid. Also shown are changes in axial and transverse flow velocities superimposed on the oncoming velocity of the working fluid. A change in axial-flow velocity affects the thrust, while transverse flow induces a force normal to the centerline of the column of working fluid. Finally, the equations governing the thrust of the two types of devices are also listed, where $\dot{m}_P$ is the mass-flow rate through the turbojet, *SHP* is the power delivered to the propeller shaft, η_P is the propeller efficiency, and K is the appropriate unit-conversion factor.

Consider first the effect of a change in axial flow velocity, Δv_{Axial}. For a turbojet, this effect may be quantified in terms of the change in thrust due to a change in free-stream axial Mach number, or $\partial T/\partial M_\infty$, which must be obtained from engine test data. So for the turbojet, the change in thrust due to a change in axial flow velocity is

$$\frac{\partial T}{\partial v_{\text{Axial}}} = \frac{1}{a}\frac{\partial T}{\partial M_\infty} \tag{6.74}$$

where a is the sonic velocity in the oncoming flow.

For a fixed-pitch propeller, the change in axial velocity typically affects the propeller efficiency η_P, and hence the thrust. This change in efficiency $\Delta\eta_P$, due to a small change in axial velocity Δv_{Axial}, must be determined from propeller data (i.e., propeller map), and would depend on the particular flight condition being investigated. Also, the thrust equation for the propeller-driven propulsive device is an explicit function of nominal axial velocity V_∞. Therefore, the change in thrust due to a change in axial flow velocity for a fixed-pitch propeller device may be expressed as

$$\frac{\partial T}{\partial v_{\text{Axial}}} = -K\frac{SHP\eta_P}{V_\infty^2} + K\frac{SHP}{V_\infty}\frac{\Delta\eta_P}{\Delta v_{\text{Axial}}} = -\frac{T}{V_\infty} + \frac{T}{\eta_P}\frac{\partial\eta_P}{\partial v_{\text{Axial}}} \tag{6.75}$$

Here it is assumed that the throttle setting is fixed, and hence the shaft power is constant.

However, for a propulsive device with a variable-pitch propeller, the propeller efficiency is kept approximately constant, and thus the thrust power is approximately constant. In such a case,

$$T\left(V_\infty + \Delta v_{\text{Axial}}\right) = \text{constant} \tag{6.76}$$

Differentiating this with respect to Δv_{Axial} yields the change in thrust due to a change in axial velocity for a variable-pitch propeller device, or

$$\frac{\partial T}{\partial v_{\text{Axial}}} = -T \tag{6.77}$$

Next we will consider the effects of a change in transverse velocity $\Delta v_{\text{Transverse}}$ on the normal forces generated by the propulsive device. As indicated in Figure 6.14, any transverse velocity present in the incoming flow will involve a turning of the flow (or the flow momentum) as it passes through the engine or propeller disk. This change in momentum, of course, creates a transverse or normal force F_N acting on the engine or propeller shaft in the direction of the transverse velocity $\Delta v_{\text{Transverse}}$.

Frequently the change in this normal force is expressed in terms of the turning angle of the flow α_P, or

$$\frac{\partial F_N}{\partial \alpha_P}$$

where

$$\tan \alpha_P \approx \alpha_P = \frac{\Delta v_{\text{Transverse}}}{V_\infty} \tag{6.78}$$

Note then that

$$\frac{\partial F_N}{\partial v_{\text{Transverse}}} = \frac{1}{v_{\text{Axial}}} \frac{\partial F_N}{\partial \alpha_P} \tag{6.79}$$

Methods for estimating this effect for propeller-driven devices are presented in Ref. 4. For a turbojet engine, the normal force F_N may be written in terms of the engine mass-flow rate. That is,

$$F_N = \dot{m}_P v_{\text{Axial}} \sin \alpha_P \approx \dot{m}_P v_{\text{Axial}} \alpha_P \tag{6.80}$$

where v_{Axial} is the axial flow velocity at the engine inlet. This inlet velocity, in turn, my be expressed as

$$v_{\text{Axial}} = \frac{\dot{m}_P}{\rho_\infty A_{\text{in}}} \tag{6.81}$$

where

A_{in} = cross-sectional area of engine inlet

Consequently, the normal force on the engine is given by

$$F_N = \frac{\dot{m}_P^2}{\rho_\infty A_{\text{in}}} \alpha_P \tag{6.82}$$

and the change in normal force due to a change in transverse velocity for a turbojet engine may be expressed as

$$\boxed{\frac{\partial F_N}{\partial v_{\text{Transverse}}} = \frac{1}{V_\infty} \frac{\partial F_N}{\partial \alpha_P} = \frac{1}{V_\infty} \frac{\dot{m}_P^2}{\rho_\infty A_{\text{in}}}} \tag{6.83}$$

Again, the mass-flow rate must be obtained from the engine data.

Based on the above discussion, we may now express the propulsive forces and moments on the vehicle in terms of vehicle angles of attack and sideslip. Note that in terms of these angles

$$\begin{aligned} \Delta v_{\text{Axial}} &= V_\infty(\cos\alpha - 1) + V_\infty(\cos\beta - 1) \approx 0 \\ \Delta v_{Z_{\text{Transverse}}} &= -V_\infty \sin\alpha \\ \Delta v_{Y_{\text{Transverse}}} &= -V_\infty \sin\beta \end{aligned} \tag{6.84}$$

Assuming the thrust T and propulsive normal force F_N are symmetric with respect to the XZ plane of the vehicle, the components of the propulsive force may now be written in the fuselage-referenced axes shown in Figure 6.13, for example, as

$$\boxed{\begin{aligned} F_{P_X} &= T\cos\phi_T - \frac{\partial F_N}{\partial v_{\text{Transverse}}} V_\infty \sin\alpha \sin\phi_T \\ F_{P_Y} &= -\frac{\partial F_N}{\partial v_{\text{Transverse}}} V_\infty \sin\beta \\ F_{P_Z} &= -T\sin\phi_T - \frac{\partial F_N}{\partial v_{\text{Transverse}}} V_\infty \sin\alpha \cos\phi_T \end{aligned}} \tag{6.85}$$

And the components of the propulsive moment about X_{Ref}, in the same axes, are

$$\boxed{\begin{aligned} L_P &= \frac{\partial F_N}{\partial v_{\text{Transverse}}} V_\infty d_T \sin\beta \\ M_P &= T(d_T\cos\phi_T - x_T\sin\phi_T) - \frac{\partial F_N}{\partial v_{\text{Transverse}}} V_\infty \sin\alpha(x_T\cos\phi_T - d_T\sin\phi_T) \\ N_P &= \frac{\partial F_N}{\partial v_{\text{Transverse}}} V_\infty x_T \sin\beta \end{aligned}} \tag{6.86}$$

6.4 FUSELAGE-REFERENCE AND STABILITY AXES

Before proceeding further with modeling the forces and moments, we must more precisely define two vehicle-fixed coordinate systems that are frequently used. In truth, any vehicle-fixed frame could be used for a rigid vehicle. For a flexible vehicle, any vehicle-fixed frame that satisfies the mean-axis constraints (Equations (4.13) or (4.14)) would be valid. So, practically speaking, the choice of the vehicle-fixed frame is made for convenience—to simplify the analysis as much as possible.

Sometimes a frame referenced to the vehicle's structure is used, with the X axis of the frame aligned with a fuselage reference line such as its centerline. We refer to such a frame as *fuselage-reference axes;* we have used this frame in previous sections without defining it precisely. Fuselage-reference axes are frequently used when analyzing flexible vehicles, and are also commonly used in nonlinear simulations.

But frequently a frame known as the *stability axes* is selected for perturbation analysis, since it simplifies the expressions for the aerodynamic forces and moments. In the rest of this chapter we will also use the stability axes, and components of the forces and moments will frequently be expressed in terms of these axes. However, if for some reason the analyst chooses to use other axes, the stability-axes components of the forces and moments developed in this chapter may be transformed into components expressed in the new coordinate frame via coordinate transformation and direction cosine matrices. (See Chapter 1.)

The stability axes are special vehicle-fixed axes that are associated with a particular reference flight condition, such as steady level flight, and are tied to the reference free-stream velocity vector $\mathbf{V}_{\infty_0}$ associated with that flight condition. The axes are vehicle-fixed in the sense that once the stability axes are defined for the chosen reference flight condition, the axes are then fixed to the vehicle and translate and rotate with the vehicle as the vehicle is perturbed from the reference condition.[1]

Consider the vehicle depicted in Figure 6.15. Shown in this figure are the reference free-stream velocity vector $\mathbf{V}_{\infty_0}$, the fuselage-reference axes identified as X_F, Y_F, and Z_F, and the stability axes denoted as X_S, Y_S, Z_S. Of course, both sets of axes have their origins at the vehicle center of mass (gravity). Also shown are the vehicle's aerodynamic angle of attack α_0 and angle of sideslip β_0 at the chosen reference condition.

These aerodynamic angles are defined in terms of the fuselage-reference axes and the reference free-stream velocity vector as shown. The sideslip angle β_0 lies in the X_SY_S plane, and is the angle between $\mathbf{V}_{\infty_0}$ and the X_S axis. The angle of attack α_0 lies in the X_SZ_S plane as well as the X_FZ_F plane, and is the angle between the X_F and X_S axes (as well as the angle between the Z_F and Z_S axes).

[1] This is in contrast to the wind axes introduced in Chapter 2, with the X_W axis fixed with respect to the instantaneous free-stream velocity vector $\mathbf{V}_{\infty}$.

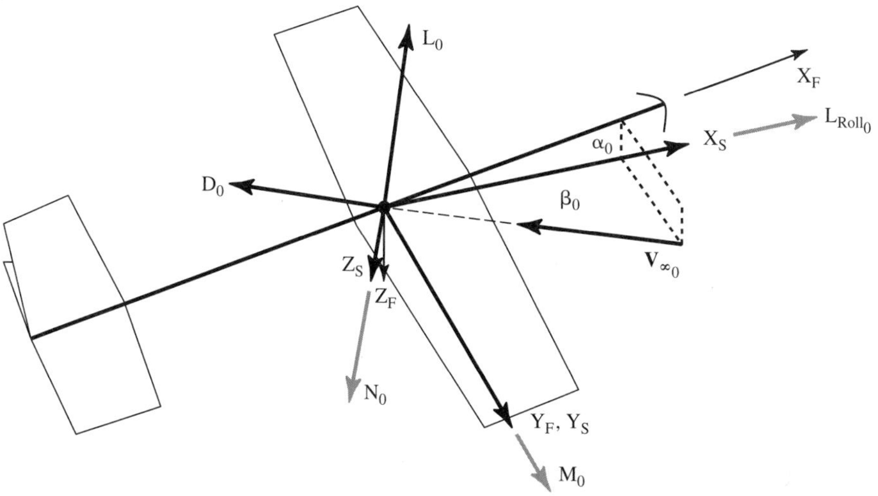

Figure 6.15 Stability axes.

Now recall that once the stability axis is defined in terms of the reference flight condition, it is then fixed to the vehicle and allowed to rotate with the vehicle as the vehicle is perturbed from this reference condition.

6.5 AERODYNAMIC AND PROPULSIVE FORCES AND MOMENTS AT THE REFERENCE CONDITION

Shown in Figure 6.15 were two of the three aerodynamic forces, at their reference values, acting on the vehicle—the lift L_0 and drag D_0. (Not shown for clarity is the reference aerodynamic side force S_0 and the reference propulsive thrust force T_0.) Using Figure 6.15, recalling the definitions of vehicle lift, drag, and side force, and consistent with Equations (6.1) and (6.2), we can deduce that at the chosen reference condition the components of the aerodynamic force in the stability axes are

$$\boxed{\begin{aligned} F_{A_{X_0}} &= -D_0 \cos\beta_0 - S_0 \sin\beta_0 \\ F_{A_{Y_0}} &= -D_0 \sin\beta_0 + S_0 \cos\beta_0 \\ F_{A_{Z_0}} &= -L_0 \end{aligned}} \tag{6.87}$$

The lift, drag, and side forces in these equations may be determined from the expressions developed in Section 6.2.

Also shown in Figure 6.15 were the stability-axis components of the total (aerodynamic plus propulsive) moment acting on the vehicle at the reference condition. The aerodynamic moment vector is

$$\mathbf{M}_A = L_A \mathbf{i}_S + M_A \mathbf{j}_S + N_A \mathbf{k}_S \tag{6.88}$$

consistent with Equations (6.1) and (6.2). At the chosen reference condition the components of the reference value of the aerodynamic moment in the stability axes are simply L_{A_0}, M_{A_0} and N_{A_0}, where these rolling, pitching, and yawing moments may also be determined from the expressions developed in Section 6.2. In developing the expressions in Section 6.2, note that we simply specified the components were in some vehicle-fixed axes. Here, we are specifying the stability axes as the vehicle-fixed axis being used.

Based on the geometry defined in Figure 6.13, in which the fuselage-reference axes were used, and the definition of the stability axes, the reference values of the components of the propulsive force in the stability axes are

$$\boxed{\begin{aligned} F_{P_{X_0}} &= \mathrm{T}_0 \cos(\phi_T + \alpha_0) \\ F_{P_{Y_0}} &= 0 \\ F_{P_{Z_0}} &= -\mathrm{T}_0 \sin(\phi_T + \alpha_0) \end{aligned}} \tag{6.89}$$

Finally, let the components of the reference value of the propulsive moment $\mathbf{M}_{P_0}$ in the stability axes be defined by

$$\mathbf{M}_{P_0} = \mathrm{L}_{P_0}\mathbf{i}_S + \mathrm{M}_{P_0}\mathbf{j}_S + \mathrm{N}_{P_0}\mathbf{k}_S \tag{6.90}$$

So for the notional vehicle configuration depicted in Figure 6.15, and from the definition of the stability axes, the components of the reference value of the propulsive moment in the stability axis are

$$\boxed{\begin{aligned} \mathrm{L}_{P_0} &= 0 \\ \mathrm{M}_{P_0} &= \mathrm{T}_0(d_T \cos\phi_T - x_T \sin\phi_T) \\ \mathrm{N}_{P_0} &= 0 \end{aligned}} \tag{6.91}$$

EXAMPLE 6.4

Reference-Values of Forces and Moments

Again consider the vehicle described in Examples 6.1–6.3. Assume the vehicle is equipped with two turbojet engines, mounted on either side of the fuselage just ahead of the horizontal tail. (Here the horizontal tail may be assumed to be a "T-tail," attached to the top of the vertical tail.) Let the engines produce equal thrust $T/2$, with the thrust axis aligned with the fuselage-referenced X axis, and with the engines mounted along the (fuselage) X axis a distance $2\bar{c}_W$ aft of the reference location X_{Ref}. If the vehicle is in steady, unaccelerating, level flight at an angle of attack $\alpha_0 = 2$ deg with no sideslip, find the reference values of the aerodynamic and propulsive forces and moments acting on the vehicle.

■ Solution

Note here that the reference flight condition is steady, level flight with no sideslip angle. So from Equations (6.87), we have the reference values of the aerodynamic forces in the stability axes given by

$$\mathrm{F}_{A_{X_0}} = -\mathrm{D}_0\cos\beta_0 - \mathrm{S}_0\sin\beta_0 = -\mathrm{D}_0 = -C_{D_0}q_\infty S$$

$$\mathrm{F}_{A_{Y_0}} = -\mathrm{D}_0\sin\beta_0 + \mathrm{S}_0\cos\beta_0 = \mathrm{S}_0 = C_{S_0}q_\infty S$$

$$\mathrm{F}_{A_{Z_0}} = -\mathrm{L}_0 = -C_{L_0}q_\infty S$$

(Note that C_{D_0} in the first of the above expressions refers to the reference value of drag coefficient, and not the parasite drag coefficient.)

For the data given, and with reference to Figure 6.13, we have $d_T = \phi_T = 0$. So from Equations (6.89), with $\alpha_0 = 2$ deg, we find that the reference-values of the propulsive forces in the stability axes are

$$F_{P_{X_0}} = \mathrm{T}_0\cos(\phi_T + \alpha_0) \approx \mathrm{T}_0$$

$$F_{P_{Y_0}} = 0$$

$$F_{P_{Z_0}} = -\mathrm{T}_0\sin(\phi_T + \alpha_0) = -0.03\mathrm{T}_0$$

Then from Equations (6.91) we have for the reference values of the propulsive moments

$$\mathrm{L}_{P_0} = 0$$

$$\mathrm{M}_{P_0} = \mathrm{T}_0(d_T\cos\phi_T - x_T\sin\phi_T) = 0$$

$$\mathrm{N}_{P_0} = 0$$

Note that the line of action of **T** passes through the X_{Ref} location and so

$$(d_T\cos\phi_T - x_T\sin\phi_T) = 0$$

Consistent with Equations (6.2), the components of the aerodynamic moment in the stability axes at the reference condition are

$$\mathrm{L}_{A_0} = C_{L\mathrm{roll}_0}q_\infty S_W b_W$$

$$\mathrm{M}_{A_0} = C_{M_0}q_\infty S_W \bar{c}_W$$

$$\mathrm{N}_{A_0} = C_{N_0}q_\infty S_W b_W$$

Therefore, for the reference flight condition to be in steady unaccelerated flight, the following must hold.

$$F_{A_{X_0}} + F_{P_{X_0}} = 0 \Rightarrow C_{D_0} = \frac{\mathrm{T}_0}{q_\infty S_W}$$

$$F_{A_{Y_0}} + F_{P_{Y_0}} = 0 \Rightarrow C_{S_0} = 0$$

$$F_{A_{Z_0}} + F_{P_{Z_0}} + mg = 0 \Rightarrow C_{L_0} = \frac{mg - 0.03\mathrm{T}_0}{q_\infty S_W}$$

$$\mathrm{L}_{A_0} + \mathrm{L}_{P_0} = 0 \Rightarrow C_{L\mathrm{roll}_0} = 0$$

$$\mathrm{M}_{A_0} + \mathrm{M}_{P_0} = 0 \Rightarrow C_{M_0} = 0$$

$$\mathrm{N}_{A_0} + \mathrm{N}_{P_0} = 0 \Rightarrow C_{N_0} = 0$$

Here mg is the vehicle weight, and $S_W = 169\ \text{ft}^2$. Note that since the reference value of the propulsive moment is zero, the reference value of the aerodynamic moment must be zero as well. So the reference values of all aerodynamic moments (and side force) are zero. This is a normal result for straight and level flight if the propulsive moment is zero.

6.6 FORCES AND MOMENTS DUE TO TRANSLATIONAL VELOCITY PERTURBATIONS

The forces and moments acting on the vehicle are affected when the vehicle undergoes motion. That is, changes in the vehicle's translational and rotational velocities relative to the air mass affect both the aerodynamic and propulsive forces and moments. In Sections 6.6 and 6.7, we will address the effects of three translations and three rotations, plus the rate of change of angle of attack. In each case, the analysis procedure will involve:

- Describing the mechanisms by which the motion affects the forces and moments.
- Expressing the forces and moments in appropriate terms, including their nondimensional coefficients.
- Deriving expressions that capture the effects of the particular motion on these forces and moments.
- Determining the partial derivatives of the forces and moments with respect to the motion variables.

Let's first address, in general, the effects of perturbations in translational velocity. Figure 6.16, which is critical to understanding the effects of these motion perturbations, shows the vehicle's stability and fuselage-reference axes,

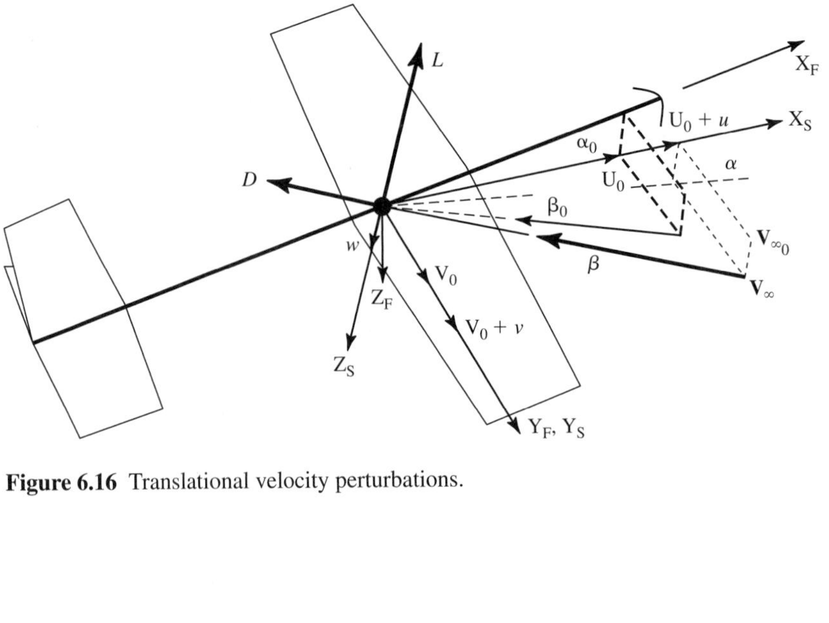

Figure 6.16 Translational velocity perturbations.

along with the reference and the instantaneous free-stream velocity vectors, $\mathbf{V}_{\infty_0}$ and $\mathbf{V}_{\infty}$, respectively. Also shown are the lift L and drag D at the perturbed flight condition. The side-force S is not shown for clarity.

Recall that the components of $\mathbf{V}_{\infty_0}$ in the stability axes are

$$\mathbf{V}_{\infty_0} = \mathrm{U}_0\mathbf{i}_S + \mathrm{V}_0\mathbf{j}_S + \mathrm{W}_0\mathbf{k}_S = \mathrm{U}_0\mathbf{i}_S + \mathrm{V}_0\mathbf{j}_S \tag{6.92}$$

since due to the definition of the stability axes W_0 will always be zero. With no sideslip V_0 will also be zero, so then

$$V_{\infty_0} = \mathrm{U}_0. \tag{6.93}$$

Also, the components of the perturbed velocity vector $\mathbf{V}_{\infty}$ in the same axis system are defined by

$$\boxed{\mathbf{V}_{\infty} = (\mathrm{U}_0 + u)\mathbf{i}_S + (\mathrm{V}_0 + v)\mathbf{j}_S + (\mathrm{W}_0 + w)\mathbf{k}_S} \tag{6.94}$$

In the perturbed flight condition, the vehicle's angle of attack becomes $\alpha_0 + \alpha$, while the sideslip angle becomes $\beta_0 + \beta$.

From Figure 6.16 one can see that the perturbation angle of attack α can be expressed in terms of the perturbation plunge velocity w, or

$$\tan\alpha = \frac{w}{\mathrm{U}_0 + u} \tag{6.95}$$

Assuming α is sufficiently small, and u is small compared to U_0, the above may be approximated as

$$\boxed{\alpha \approx \frac{w}{\mathrm{U}_0}} \tag{6.96}$$

Likewise, the sideslip angle may be expressed in terms of the lateral component of velocity, or

$$\tan(\beta_0 + \beta) = \frac{\mathrm{V}_0 + v}{\mathrm{U}_0 + u} \tag{6.97}$$

Again assuming $\beta_0 + \beta$ is sufficiently small, and u is small compared to U_0, we can write

$$\beta_0 + \beta \approx \frac{\mathrm{V}_0 + v}{\mathrm{U}_0} \tag{6.98}$$

and thus the reference and perturbation sideslip angles may be expressed in terms of the lateral velocities as

$$\boxed{\beta_0 \approx \frac{\mathrm{V}_0}{\mathrm{U}_0} \quad \beta \approx \frac{v}{\mathrm{U}_0}} \tag{6.99}$$

Perturbations in the magnitude of free-stream velocity affect the dynamic pressure q_∞, and thus the magnitudes of the aerodynamic forces and moments as well. The dynamic pressure, written in terms of the reference and perturbation velocities, is given by

$$q_\infty = \frac{1}{2}\rho_\infty V_\infty^2 = \frac{1}{2}\rho_\infty\left((\mathrm{U}_0 + u)^2 + (\mathrm{V}_0 + v)^2 + (\mathrm{W}_0 + w)^2\right) \quad (6.100)$$

Perturbations in the direction of the free-stream velocity vector, or perturbations α and β, affect both the magnitudes as well as the orientation of the lift, drag, and side force with respect to the stability axis. Recall that lift is always defined as acting normal to the free-stream velocity vector and lying in the XZ plane of the vehicle (and thus the $X_S Z_S$ plane). Likewise, drag always acts along the free-stream velocity vector, and side force is mutually orthogonal to both lift and drag. Although the lift and drag shown in Figure 6.16 may appear to be acting in the same directions as L_0 and D_0 in Figure 6.15, that will not be the case in general due to the velocity perturbations.

To aid in the following analysis, let us recall the wind axes introduced in Chapter 2, with unit vectors $\boldsymbol{i}_W, \boldsymbol{j}_W$, and $\boldsymbol{k}_W$. And let $\boldsymbol{i}_W$ act in the direction opposite to $\mathbf{V}_\infty$, $\boldsymbol{j}_W$ in the same direction as the side force S, and $\boldsymbol{k}_W$ in the direction opposite to lift L. From Figure 6.16 we can see that the orientation of this velocity-fixed frame relative to the stability axis will be defined by the perturbation angle of attack α and the total sideslip angle $(\beta_0 + \beta)$. Note that α is a rotation about the $\boldsymbol{j}_S$ axis, while $(\beta_0 + \beta)$ is a rotation about the $\boldsymbol{k}_W$ axis. Therefore, following the techniques presented in Chapter 1, the unit vectors of these two frames may be related through a direction-cosine matrix $\mathbf{T}_{W,S}$. Or

$$\begin{Bmatrix} \mathbf{i}_S \\ \mathbf{j}_S \\ \mathbf{k}_S \end{Bmatrix} = \mathbf{T}_{W,S} \begin{Bmatrix} \mathbf{i}_W \\ \mathbf{j}_W \\ \mathbf{k}_W \end{Bmatrix} \quad (6.101)$$

where

$$\begin{aligned} \mathbf{T}_{W,S} &= \begin{bmatrix} \cos(\beta_0+\beta) & \sin(\beta_0+\beta) & 0 \\ -\sin(\beta_0+\beta) & \cos(\beta_0+\beta) & 0 \\ 0 & 0 & 1 \end{bmatrix} \begin{bmatrix} \cos\alpha & 0 & \sin\alpha \\ 0 & 1 & 0 \\ -\sin\alpha & 0 & \cos\alpha \end{bmatrix} \\ &= \begin{bmatrix} \cos\alpha\cos(\beta_0+\beta) & \sin(\beta_0+\beta) & \sin\alpha\cos(\beta_0+\beta) \\ -\cos\alpha\sin(\beta_0+\beta) & \cos(\beta_0+\beta) & -\sin\alpha\sin(\beta_0+\beta) \\ -\sin\alpha & 0 & \cos\alpha \end{bmatrix} \end{aligned} \quad (6.102)$$

Now the total aerodynamic force vector $\mathbf{F_A}$ acting on the vehicle may be expressed as

$$\mathbf{F_A} = F_{A_X}\mathbf{i}_S + F_{A_Y}\mathbf{j}_S + F_{A_Z}\mathbf{k}_S = -D\mathbf{i}_W + \mathsf{S}\mathbf{j}_W - L\mathbf{k}_W \quad (6.103)$$

Or

$$\begin{bmatrix} F_{A_X} & F_{A_Y} & F_{A_Z} \end{bmatrix} \begin{Bmatrix} \mathbf{i}_S \\ \mathbf{j}_S \\ \mathbf{k}_S \end{Bmatrix} = \begin{bmatrix} -D & \mathsf{S} & -L \end{bmatrix} \begin{Bmatrix} \mathbf{i}_W \\ \mathbf{j}_W \\ \mathbf{k}_W \end{Bmatrix} \tag{6.104}$$

So, using Equation 6.101, we have

$$\begin{bmatrix} F_{A_X} & F_{A_Y} & F_{A_Z} \end{bmatrix} \begin{Bmatrix} \mathbf{i}_S \\ \mathbf{j}_S \\ \mathbf{k}_S \end{Bmatrix} = \begin{bmatrix} -D & \mathsf{S} & -L \end{bmatrix} \mathbf{T}_{W,S}^T \begin{Bmatrix} \mathbf{i}_S \\ \mathbf{j}_S \\ \mathbf{k}_S \end{Bmatrix} \tag{6.105}$$

Therefore,

$$\begin{Bmatrix} F_{A_X} \\ F_{A_Y} \\ F_{A_Z} \end{Bmatrix} = \mathbf{T}_{W,S} \begin{Bmatrix} -D \\ \mathsf{S} \\ -L \end{Bmatrix} \tag{6.106}$$

and the components of the total aerodynamic force in the stability axes are

$$\boxed{\begin{aligned} F_{A_X} &= C_X q_\infty S = -D\cos\alpha\cos(\beta_0+\beta) - \mathsf{S}\cos\alpha\sin(\beta_0+\beta) + L\sin\alpha \\ F_{A_Y} &= C_Y q_\infty S = -D\sin(\beta_0+\beta) + \mathsf{S}\cos(\beta_0+\beta) \\ F_{A_Z} &= C_Z q_\infty S = -D\sin\alpha\cos(\beta_0+\beta) - \mathsf{S}\sin\alpha\sin(\beta_0+\beta) - L\cos\alpha \end{aligned}} \tag{6.107}$$

By total forces we mean reference plus perturbation. For example, for the total drag we have $D = D_0 + \delta D$. Also note that by using the appropriate direction-cosine matrix, the aerodynamic-force components in any coordinate frame may be determined by performing an analysis similar to that above.

In Equations (6.107) we have also expressed these force components in coefficient form, introducing the coefficients C_X, C_Y, and C_Z. We will use Equations (6.107) frequently in the following sections. Note also that setting both perturbation angles α and β to zero in these equations leads to the results given in Equations (6.87) for the force components in the reference condition.

Perturbations in translational velocity may also affect the propulsive forces and moments acting on the vehicle. Referring back to Section 6.3 and the sketches in Figure 6.14, we found that for the turbojet, the change in thrust due to a perturbation in axial velocity is

$$\frac{\partial T}{\partial v_{\text{Axial}}} = \frac{1}{a}\frac{\partial T}{\partial M_\infty} \tag{6.108}$$

where a is the sonic velocity in the oncoming flow. We also found that the change in thrust due to a perturbation in axial velocity for a fixed-pitch propeller device may be expressed as

$$\frac{\partial T}{\partial v_{\text{Axial}}} = -K\frac{SHP\eta_P}{V_\infty^2} + K\frac{SHP}{V_\infty}\frac{\Delta\eta_P}{\Delta v_{\text{Axial}}} = -\frac{T}{V_\infty} + \frac{T}{\eta_P}\frac{\partial\eta_P}{\partial v_{\text{Axial}}} \tag{6.109}$$

where it is assumed that the throttle setting is fixed, and hence the shaft power is constant. Finally, we found that the change in thrust due to a perturbation in axial velocity for a variable-pitch propeller device was

$$\frac{\partial T}{\partial v_{\text{Axial}}} = -T \tag{6.110}$$

Considering the effects of a perturbation in transverse velocity $\Delta v_{\text{Transverse}}$ we noted that in terms of the flow turning angle α_P

$$\frac{\partial F_N}{\partial v_{\text{Transverse}}} = \frac{1}{v_{\text{Axial}}} \frac{\partial F_N}{\partial \alpha_P} \tag{6.111}$$

where methods for estimating this effect for propeller-driven devices are presented in Ref. 4. And we had the change in normal force due to a change in transverse velocity for a turbojet engine given by

$$\frac{\partial F_N}{\partial v_{\text{Transverse}}} = \frac{1}{V_\infty} \frac{\partial F_N}{\partial \alpha_P} = \frac{1}{V_\infty} \frac{\dot{m}_P^2}{\rho_\infty A_{\text{in}}} \tag{6.112}$$

where the mass-flow rate must be obtained from the engine data. We will use Equations (6.108–6.112) in the sections to follow.

6.6.1 Surge-Velocity Perturbation *u*

Perturbations in the surge velocity u affect the dynamic pressure q_∞ and the flight Mach number M_∞, both of which affect the aerodynamic forces and moments in general.

From Equations (6.107) we have

$$\begin{aligned} F_{A_X} &= C_X q_\infty S \\ F_{A_Y} &= C_Y q_\infty S \\ F_{A_Z} &= C_Z q_\infty S \end{aligned} \tag{6.113}$$

where

$$\begin{aligned} C_X &= -C_D \cos\alpha \cos(\beta_0 + \beta) - C_S \cos\alpha \sin(\beta_0 + \beta) + C_L \sin\alpha \\ C_Y &= -C_D \sin(\beta_0 + \beta) + C_S \cos(\beta_0 + \beta) \\ C_Z &= -C_D \sin\alpha \cos(\beta_0 + \beta) - C_S \sin\alpha \sin(\beta_0 + \beta) - C_L \cos\alpha \end{aligned} \tag{6.114}$$

Now let the effect of surge perturbations on these three force components be quantified in terms of the partial derivative $\partial F_{A_\bullet}/\partial u$, evaluated at the reference condition. So then we have

$$\frac{\partial F_{A_\bullet}}{\partial u}\Big|_0 = \left(\frac{\partial C_\bullet}{\partial u} q_\infty S_W + C_\bullet S_W \frac{\partial q_\infty}{\partial u} \right)\Big|_0, \; \bullet = X, Y, \text{or } Z \tag{6.115}$$

And from Equation (6.100) we have

$$\frac{\partial q_\infty}{\partial u}\Big|_0 = \rho_\infty(U_0 + u)\big|_0 = \rho_\infty U_0 \tag{6.116}$$

Note that we can also express $\partial C_\bullet/\partial u$, the effect of surge velocity on the three force coefficients, in terms of the effect of Mach number M_∞ on these coefficients. That is, we may write

$$\frac{\partial C_\bullet}{\partial u} = \frac{1}{a}\frac{\partial C_\bullet}{\partial\left(\frac{u}{a}\right)} = \frac{1}{a}\frac{\partial C_\bullet}{\partial m_\infty} = \frac{1}{a}\frac{\partial C_\bullet}{\partial M_\infty}, \ \bullet = X,\ Y, \text{ or } Z \tag{6.117}$$

where a is the sonic velocity in air, and m_∞ is the perturbation in Mach number due to the perturbation in surge velocity. The last equality in the above equation reflects the fact that a derivative is defined in terms of infinitesimal variations in the variables.

Interpreting u/a as the perturbation Mach number m_∞ in Equation (6.117) is justified by the following analysis. First, we write the Mach number in terms of its reference value plus the perturbation, or

$$M_\infty = M_{\infty_0} + m_\infty$$

Then using a Taylor-series expansion, we can write the perturbation Mach number as

$$m_\infty = \frac{\partial M_\infty}{\partial u}\Big|_0 u + \frac{\partial M_\infty}{\partial v}\Big|_0 v + \frac{\partial M_\infty}{\partial w}\Big|_0 w$$

where Mach number is defined as

$$M_\infty \triangleq \frac{V_\infty}{a} = \frac{1}{a}\sqrt{(U_0 + u)^2 + (V_0 + v)^2 + (W_0 + w)^2} \tag{6.118}$$

Therefore, we have the perturbation in Mach number due to the perturbation in surge velocity given by

$$\frac{\partial M_\infty}{\partial u}\Big|_0 u = \left(\frac{1}{a}\frac{U_0 + u}{V_\infty}\right)\Big|_0 u = \frac{1}{a}\frac{U_0}{V_\infty}u \approx \frac{u}{a} \tag{6.119}$$

thus proving the original assertion. It may also be noted that by following a similar analysis one finds that

$$\frac{\partial M_\infty}{\partial v}\Big|_0 v \approx \beta_0\frac{v}{a} \tag{6.120}$$

and

$$\frac{\partial M_\infty}{\partial w}\Big|_0 w = 0$$

since W_0 is always zero in the stability axes.

Therefore we have the effect of surge-velocity perturbation on the aerodynamic force components given as

$$\begin{aligned}\frac{\partial F_{A\bullet}}{\partial u}\Big|_0 &= \left(\frac{1}{a}\frac{\partial C_\bullet}{\partial M_\infty}q_\infty S_W + C_\bullet S_W \rho_\infty (U_0 + u)\right)\Big|_0 \\ &= C_{\bullet_u} q_\infty S_W + C_{\bullet_0} S_W \rho_\infty U_0,\ \bullet = X, Y, \text{or } Z\end{aligned} \tag{6.121}$$

If β_0, and therefore V_0, equals zero, this expression reduces to

$$\frac{\partial F_{A\bullet}}{\partial u}\Big|_0 = \left(\frac{1}{a}\frac{\partial C_\bullet}{\partial M_\infty}\Big|_0 + \frac{2C_{\bullet_0}}{U_0}\right)q_\infty S_W = \left(C_{\bullet_u} + \frac{2C_{\bullet_0}}{U_0}\right)q_\infty S_W,\ \bullet = X, Y, \text{or } Z \tag{6.122}$$

To evaluate Equation (6.121), we can again refer back to Equations (6.114), from which we find that

$$\begin{aligned}C_{X_0} &= -C_{D_0}\cos\beta_0 - C_{S_0}\sin\beta_0 \\ C_{Y_0} &= -C_{D_0}\sin\beta_0 + C_{S_0}\cos\beta_0 \\ C_{Z_0} &= -C_{L_0}\end{aligned} \tag{6.123}$$

and that

$$\begin{aligned}\frac{\partial C_X}{\partial M_\infty}\Big|_0 &= -C_{D_{M_\infty}}\cos\beta_0 - C_{S_{M_\infty}}\sin\beta_0 \\ \frac{\partial C_Y}{\partial M_\infty}\Big|_0 &= -C_{D_{M_\infty}}\sin\beta_0 + C_{S_{M_\infty}}\cos\beta_0 \\ \frac{\partial C_Z}{\partial M_\infty}\Big|_0 &= -C_{L_{M_\infty}}\end{aligned} \tag{6.124}$$

where

$$C_{\bullet_{M_\infty}} \triangleq \frac{\partial C_\bullet}{\partial M_\infty}\Big|_0 = a\frac{\partial C_\bullet}{\partial U}\Big|_0 \triangleq aC_{\bullet_u},\ C_\bullet = C_L, C_D, C_S, C_X, C_Y, \text{or } C_Z \tag{6.125}$$

Note that all these partial derivatives are evaluated at the reference condition, at which perturbation α and β are zero. Therefore, Equations (6.121–6.125) quantify the effects of surge-velocity (u) perturbation on the aerodynamic forces.

It is worth noting that since the effect of surge velocity on the force coefficients is inversely proportional to sonic velocity a, this effect will be small unless the effect of Mach number on the lift, drag, or side force coefficient is large. But the effect of Mach number may, in fact, be quite large in the transonic speed range, especially in the case of drag. Shown in the sketch in Figure 6.17 is

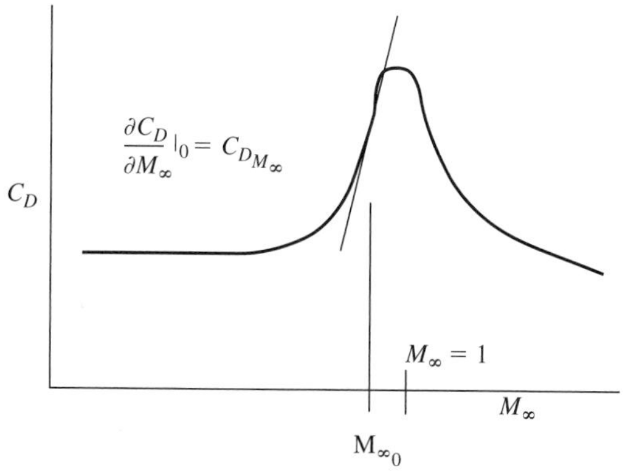

Figure 6.17 Effect of Mach number on drag coefficient.

the qualitative effect of Mach number on vehicle drag coefficient. It is clear that when the reference Mach number is near $M_{\infty_0} = 1$, not only may the magnitude of $C_{D_{M_\infty}}$ be large, but the sign of $C_{D_{M_\infty}}$ may also quickly change if the reference Mach number changes only slightly.

Ideally, wind-tunnel data would be available to evaluate the effects of Mach number on the lift, drag, and side-force coefficients. But if such data is not available, one may develop simple estimates of the Mach effect by recalling the Prandtl–Glauert compressibility factor from Chapter 5. This factor reflects the effect of compressibility on the pressure distribution over a 2-D airfoil. Applying this factor to the vehicle lift coefficient, we may write

$$C_L = \frac{C_L|_{M_\infty=0}}{\sqrt{1-M_\infty^2}}, \quad M_\infty < 1 \tag{6.126}$$

Therefore, for subsonic free-stream velocities we have

$$\boxed{\frac{\partial C_L}{\partial M_\infty}\bigg|_0 = \frac{M_{\infty_0}}{1-M_{\infty_0}^2} C_{L_0}} \tag{6.127}$$

where C_{L_0} is the reference value of the vehicle's lift coefficient. For vehicle drag and side-force coefficients we may likewise use (for subsonic velocities)

$$\boxed{\frac{\partial C_S}{\partial M_\infty}\bigg|_0 = \frac{M_{\infty_0}}{1-M_{\infty_0}^2} C_{S_0} \qquad \frac{\partial C_D}{\partial M_\infty}\bigg|_0 = \frac{M_{\infty_0}}{1-M_{\infty_0}^2} C_{D_0}} \tag{6.128}$$

Attention now turns to the effect of perturbation surge velocity on the aerodynamic moments acting on the vehicle. The components of the total aerodynamic moment $\mathbf{M}_A$ in the stability axes are defined by Equation (6.88), or

$$\mathbf{M}_A = L_A \mathbf{i}_S + M_A \mathbf{j}_S + N_A \mathbf{k}_S$$

And in terms of nondimensional coefficients we have

$$
\begin{aligned}
L_A &= C_{L_{\text{Roll}}} q_\infty S_W b_W \\
M_A &= C_M q_\infty S_W \bar{c}_W \\
N_A &= C_N q_\infty S_W b_W
\end{aligned}
\tag{6.129}
$$

The effects of perturbation in surge velocity u on these moments are quantified by the partial derivatives

$$
\frac{\partial L_A}{\partial u}\Big|_0, \quad \frac{\partial M_A}{\partial u}\Big|_0, \text{ and } \frac{\partial N_A}{\partial u}\Big|_0
$$

all evaluated at the reference flight condition. And just as in the case with the aerodynamic forces, the effects of perturbations in surge velocity u on the moments arise due to the effect of u on dynamic pressure and the effect of Mach number on the nondimensional moment coefficients. Therefore, following the approach taken previously when dealing with the aerodynamic forces, we find that

$$
\boxed{
\begin{aligned}
\frac{\partial L_A}{\partial u}\Big|_0 &= \frac{1}{a}\frac{\partial C_{L_{\text{Roll}}}}{\partial M_\infty}\Big|_0 q_\infty S_W b_W + C_{L\text{roll}_0} S_W b_W \rho_\infty \mathrm{U}_0 \\
&\triangleq \left(C_{L\text{roll}_u} + \frac{2C_{L\text{roll}_0}}{\mathrm{U}_0}\right) q_\infty S_W b_W \approx 0 \\
\frac{\partial M_A}{\partial u}\Big|_0 &= \frac{1}{a}\frac{\partial C_M}{\partial M_\infty}\Big|_0 q_\infty S_W \bar{c}_W + C_{M_0} S_W \bar{c}_W \rho_\infty \mathrm{U}_0 \\
&\triangleq \left(C_{M_u} + \frac{2C_{M_0}}{\mathrm{U}_0}\right) q_\infty S_W \bar{c}_W \\
\frac{\partial N_A}{\partial u}\Big|_0 &= \frac{1}{a}\frac{\partial C_N}{\partial M_\infty}\Big|_0 q_\infty S_W b_W + C_{N_0} S_W b_W \rho_\infty \mathrm{U}_0 \\
&\triangleq \left(C_{N_u} + \frac{2C_{N_0}}{\mathrm{U}_0}\right) q_\infty S_W b_W \approx 0
\end{aligned}
}
\tag{6.130}
$$

The above expressions, therefore, quantify the effect of perturbation in surge velocity u on the aerodynamic moments acting on the vehicle.

Note that frequently reference conditions will result in the reference values of rolling-, pitching-, and yawing-moment coefficients being equal to zero. In such cases the second terms in parentheses on the right-hand sides of Equations (6.130) will vanish. Also, since the effects of surge velocity u on the moment coefficients (the first terms in parentheses on the right-hand sides of Equations (6.130)) are inversely proportional to sonic velocity a, these first terms will be small unless the moment coefficients are very sensitive to Mach number. Such may be the case with pitching-moment coefficient when reference flight velocities are in the transonic speed range. The reason for this is the shift in location of the aerodynamic center of the lifting surfaces, and hence the shift in location of the vehicle's aerodynamic center. As discussed in Chapter 5, as the free-stream velocity increases from subsonic to supersonic, the aerodynamic center of an airfoil shifts from around the quarter-chord to near the mid-chord. As a result, though difficult to estimate, in the transonic speed range the change in pitching moment with Mach number, that is, $C_{M_{M_\infty}}$, may become quite large in magnitude and either positive or negative. When a large negative value occurs, it is referred to as a *Mach tuck* condition since the vehicle tends to pitch down.

Attention now turns to the effects of perturbations in surge velocity on the propulsive forces and moments. Consistent with Equations (6.2), the components of the propulsive force vector in the stability axes are defined by

$$\mathbf{F}_P = F_{P_X}\mathbf{i}_S + F_{P_Y}\mathbf{j}_S + F_{P_Z}\mathbf{k}_S \tag{6.131}$$

And, as with the aerodynamic forces, we may express these forces in terms of nondimensional force coefficients defined here.

$$\begin{aligned} F_{P_X} &= C_{P_X} q_\infty S_W \\ F_{P_Y} &= C_{P_Y} q_\infty S_W \\ F_{P_Z} &= C_{P_Z} q_\infty S_W \end{aligned} \tag{6.132}$$

Consistent with Equations (6.115) and (6.116), the effects of surge velocity on these propulsive forces are then expressed as

$$\boxed{\begin{aligned} \frac{\partial F_{P_\bullet}}{\partial u}\Big|_0 &= \left(\frac{\partial C_{P_\bullet}}{\partial u} q_\infty S_W + C_{P_\bullet} S_W \rho_\infty \mathrm{U}\right)\Big|_0 \\ &\triangleq \left(C_{P_{\bullet_u}} + \frac{2C_{P_{\bullet_0}}}{\mathrm{U}_0}\right) q_\infty S_W, \bullet = X, Y, \text{ or } Z \end{aligned}} \tag{6.133}$$

Now by referring to Figure 6.13, and noting that it has been assumed that the thrust force(s) are symmetric with respect to the XZ plane of the vehicle, we will write the components of the total propulsive force (Equation (6.131)) in terms of their reference values plus perturbations in surge velocity u, thrust δT, and normal force on the engine δF_N. Or

$$\begin{aligned} F_{P_X} &= F_{P_{X_0}} + C_{P_{X_0}} S_W \frac{\partial q_\infty}{\partial u}\Big|_0 u + \delta T\cos(\phi_T + \alpha_0) - \delta F_N \sin(\phi_T + \alpha_0) \\ F_{P_Y} &= 0 \\ F_{P_Z} &= F_{P_{Z_0}} + C_{P_{Z_0}} S_W \frac{\partial q_\infty}{\partial u}\Big|_0 u - \delta T\sin(\phi_T + \alpha_0) - \delta F_N \cos(\phi_T + \alpha_0) \end{aligned} \tag{6.134}$$

Note that since we are considering surge-velocity perturbations, any propulsive normal force F_N will lie in the vehicle's XZ plane; hence the side force F_{P_Y} will remain zero.

The force perturbations δT and δF_N in Equations (6.134) may now be expressed as

$$\delta T = \frac{\partial T}{\partial u} u = \frac{\partial T}{\partial v_{\text{Axial}}} \frac{\partial v_{\text{Axial}}}{\partial u} u$$
$$\delta F_N = \frac{\partial F_N}{\partial u} u = \frac{\partial F_N}{\partial v_{\text{Transverse}}} \frac{\partial v_{\text{Transverse}}}{\partial u} u \tag{6.135}$$

To address the partial derivatives of thrust and normal force in the above equations, we may refer to the earlier discussion of the effects of changes in axial and transverse flow velocities on the thrust and normal force associated with the propulsive devices. Then we must deal with the geometry of the flow—the orientation of the axial and transverse flow perturbations arising due to the surge perturbation.

From Figures 6.14 and 6.16 we can write

$$\Delta v_{\text{Axial}} = u\cos(\phi_T + \alpha_0)$$
$$\Delta v_{\text{Transverse}} = u\sin(\phi_T + \alpha_0) \tag{6.136}$$

Therefore,

$$\delta T = \frac{\partial T}{\partial v_{\text{Axial}}} \cos(\phi_T + \alpha_0) u$$
$$\delta F_N = \frac{\partial F_N}{\partial v_{\text{Transverse}}} \sin(\phi_T + \alpha_0) u \tag{6.137}$$

Consequently, the partial derivatives of the propulsive force components with respect to surge velocity are given by

$$\begin{aligned}
\frac{\partial F_{P_X}}{\partial u}\Big|_0 &= C_{P_{X_0}} \rho_\infty U_0 S_W + \frac{\partial T}{\partial v_{\text{Axial}}}\Big|_0 \cos^2(\phi_T + \alpha_0) \\
&\quad - \frac{\partial F_N}{\partial v_{\text{Transverse}}}\Big|_0 \sin^2(\phi_T + \alpha_0) \\
\frac{\partial F_{P_Y}}{\partial u}\Big|_0 &= 0 \\
\frac{\partial F_{P_Z}}{\partial u}\Big|_0 &= C_{P_{Z_0}} \rho_\infty U_0 S_W - \frac{\partial T}{\partial v_{\text{Axial}}}\Big|_0 \cos(\phi_T + \alpha_0)\sin(\phi_T + \alpha_0) \\
&\quad - \frac{\partial F_N}{\partial v_{\text{Transverse}}}\Big|_0 \sin(\phi_T + \alpha_0)\cos(\phi_T + \alpha_0)
\end{aligned} \tag{6.138}$$

Typically the angle $(\phi_T + \alpha_0)$ is small, and the above equations may be simplified accordingly.

The partial derivatives of thrust and normal force in the above expressions may be evaluated using Equations (6.108–6.112), depending on the type of propulsion system(s) installed on the vehicle. Also, by comparing Equations (6.133) with Equations (6.138), and assuming the angle $(\phi_T + \alpha_0)$ is sufficiently small, we find that in terms of coefficients we have

$$\boxed{\begin{aligned} C_{P_{X_u}} &= \frac{1}{q_\infty S_W} \frac{\partial T}{\partial v_{\text{Axial}}}\Big|_0 \\ C_{P_{Y_u}} &= 0 \\ C_{P_{Z_u}} &= -\frac{1}{q_\infty S_W}\left(\frac{\partial T}{\partial v_{\text{Axial}}}\Big|_0 + \frac{\partial F_N}{\partial v_{\text{Transverse}}}\Big|_0\right)(\phi_T + \alpha_0) \end{aligned}} \tag{6.139}$$

The changes in propulsive pitching moment due to velocity perturbations can be important if the engines are mounted far forward or far aft on the vehicle. The changes in propulsive rolling and yawing moments will be negligible if the thrust and propulsive normal force is symmetric with respect to the vehicle's XZ plane and with the engine-installation geometry assumed here.

To address the pitching moment, we draw from Equations (6.134) and Figure 6.13 to write the propulsive moment M_P as

$$\begin{aligned} M_P = \mathrm{M}_{P_0} &+ C_{P_{M_0}} S_W \bar{c}_W \frac{\partial q_\infty}{\partial u}\Big|_0 u + \delta T(d_T \cos\phi_T - x_T \sin\phi_T) \\ &- \delta F_N(x_T \cos\phi_T + d_T \sin\phi_T) \\ &= C_{P_M} q_\infty S_W \bar{c}_W \end{aligned} \tag{6.140}$$

So, consistent with Equations (6.130), we have the <u>effect of surge-velocity perturbations on the propulsive pitching moment</u> given by

$$\boxed{\frac{\partial M_P}{\partial u}\Big|_0 = \frac{1}{a}\frac{\partial C_{P_M}}{\partial M_\infty}\Big|_0 q_\infty S_W \bar{c}_W + C_{P_{M_0}} S_W \bar{c}_W \rho_\infty \mathrm{U}_0 \triangleq \left(C_{P_{M_u}} + \frac{2C_{P_{M_0}}}{\mathrm{U}_0}\right) q_\infty S_W \bar{c}_W} \tag{6.141}$$

Again, all the zero subscripts indicate the reference values of the corresponding parameters.

Following the approach taken above in analyzing the perturbation propulsive forces, we find that the <u>effect of surge-velocity perturbation on the propulsive pitching moment</u> may also be expressed as

$$\boxed{\begin{aligned} \frac{\partial M_P}{\partial u}\Big|_0 &= C_{P_{M_0}} S_W \bar{c}_W \frac{\partial q_\infty}{\partial u}\Big|_0 + (d_T \cos\phi_T - x_T \sin\phi_T)\frac{\partial T}{\partial u}\Big|_0 \\ &\quad - (x_T \cos\phi_T + d_T \sin\phi_T)\frac{\partial F_N}{\partial u}\Big|_0 \\ &= C_{P_{M_0}} S_W \bar{c}_W \rho_\infty \mathrm{U}_0 + (d_T \cos\phi_T - x_T \sin\phi_T)\frac{\partial T}{\partial v_{\text{Axial}}}\Big|_0 \cos(\phi_T + \alpha_0) \\ &\quad - (x_T \cos\phi_T + d_T \sin\phi_T)\frac{\partial F_N}{\partial v_{\text{Transverse}}}\Big|_0 \sin(\phi_T + \alpha_0) \end{aligned}} \tag{6.142}$$

Typically the angle $(\phi_T + \alpha_0)$ is small, and the reference value of propulsive pitching-moment coefficient is also frequently small or zero. Consequently, Equation (6.142) can often be significantly simplified. Again, the partial derivatives of thrust and normal force in Equation (6.142) may be evaluated using Equations (6.108–6.112), depending on the type of propulsion system(s) installed on the vehicle. By comparing Equations (6.141) and (6.142), and assuming angles ϕ_T and $(\phi_T + \alpha_0)$ are sufficiently small, we find that in terms of the propulsive-moment coefficient we have

$$C_{P_{M_u}} = \frac{1}{q_\infty S_W \bar{c}_W}\left((d_T - x_T\phi_T)\frac{\partial T}{\partial v_{\text{Axial}}}\Big|_0 - (x_T + d_T\phi_T)\frac{\partial F_N}{\partial v_{\text{Transverse}}}\Big|_0 (\phi_T + \alpha_0)\right) \tag{6.143}$$

EXAMPLE 6.5

Effects of Surge-Velocity Perturbations on the Forces and Moments

Again consider the vehicle described in Examples 6.1–6.4. Determine the effects of surge-velocity perturbations on the aerodynamic and propulsive forces and moments acting on the vehicle. Assume the flight Mach number $M_\infty = 0.2$, as in the previous examples.

■ **Solution**

From Equations (6.121), with no sideslip we have

$$\frac{\partial F_{A_\bullet}}{\partial u}\Big|_0 = \left(\frac{1}{a}\frac{\partial C_\bullet}{\partial M_\infty}\Big|_0 + \frac{2C_{\bullet_0}}{U_0}\right)q_\infty S_W = \left(C_{\bullet_u} + \frac{2C_{\bullet_0}}{U_0}\right)q_\infty S_W, \quad \bullet = X, Y, \text{ or } Z$$

From Example 6.4 and Equations (6.123) we find that the reference values of the force coefficients are

$$C_{X_0} = -C_{D_0} = -\frac{T_0}{q_\infty S_W}, \quad C_{Y_0} = C_{S_0} = 0, \text{ and } C_{Z_0} = -C_{L_0} = \frac{-mg + T_0 \sin 2^\circ}{q_\infty S_W}$$

And from Equations (6.124) with no sideslip angle we have

$$\frac{\partial C_X}{\partial M_\infty}\Big|_0 = -C_{D_{M_\infty}}\cos\beta_0 - C_{S_{M_\infty}}\sin\beta_0 = -C_{D_{M_\infty}}$$

$$\frac{\partial C_Y}{\partial M_\infty}\Big|_0 = -C_{D_{M_\infty}}\sin\beta_0 + C_{S_{M_\infty}}\cos\beta_0 = C_{S_{M_\infty}} \approx 0$$

$$\frac{\partial C_Z}{\partial M_\infty}\Big|_0 = -C_{L_{M_\infty}}$$

But from Equations (6.127) and (6.128) we have

$$\frac{\partial C_L}{\partial M_\infty}\Big|_0 \triangleq C_{L_{M_\infty}} = \frac{M_{\infty_0}}{1 - M_{\infty_0}^2}C_{L_0} = \frac{0.2}{1 - 0.2^2}C_{L_0} = 0.21C_{L_0}$$

$$\frac{\partial C_D}{\partial M_\infty}\Big|_0 \triangleq C_{D_{M_\infty}} = \frac{M_{\infty_0}}{1 - M_{\infty_0}^2}C_{D_0} = \frac{0.2}{1 - 0.2^2}C_{D_0} = 0.21C_{D_0}$$

Therefore, with $U_0 = 0.2a$ the effects of surge-velocity perturbations on the aerodynamic forces are given by

$$\frac{\partial F_{A_X}}{\partial u}\Big|_0 = \left(\frac{1}{a}\frac{\partial C_X}{\partial M_\infty}\Big|_0 + \frac{2C_{X_0}}{U_0}\right)q_\infty S_W = -\frac{1}{a}(0.21 + 10)C_{D_0}q_\infty S_W = -\frac{10.21}{a}C_{D_0}q_\infty S_W$$

$$\frac{\partial F_{A_Y}}{\partial u}\Big|_0 = \left(\frac{1}{a}\frac{\partial C_Y}{\partial M_\infty}\Big|_0 + \frac{2C_{Y_0}}{U_0}\right)q_\infty S_W = 0$$

$$\frac{\partial F_{A_Z}}{\partial u}\Big|_0 = \left(\frac{1}{a}\frac{\partial C_Z}{\partial M_\infty}\Big|_0 + \frac{2C_{Z_0}}{U_0}\right)q_\infty S_W = -\frac{10.21}{a}C_{L_0}q_\infty S_W$$

where C_{L_0} and C_{D_0} were determined as a function of thrust and weight in Example 6.4.

With regard to the effects on the aerodynamic moments, Equations (6.130) state that

$$\frac{\partial M_A}{\partial u}\Big|_0 = \frac{1}{a}\frac{\partial C_M}{\partial M_\infty}\Big|_0 q_\infty S_W\bar{c}_W + C_{M_0}S_W\bar{c}_W\rho_\infty U_0 = \left(C_{M_u} + \frac{2C_{M_0}}{U_0}\right)q_\infty S_W\bar{c}_W$$

and that

$$\frac{\partial L_A}{\partial u}\Big|_0 \approx 0 \text{ and } \frac{\partial N_A}{\partial u}\Big|_0 \approx 0$$

Furthermore, since the flight Mach number of 0.2 is likely well below the critical Mach number for the lifting surfaces, little or no shift in the aerodynamic center is expected. Consequently,

$$C_{M_{M_\infty}} \approx 0$$

Therefore we have the effect of surge-velocity perturbations on the aerodynamic pitching moment given by

$$\frac{\partial M_A}{\partial u}\Big|_0 = \frac{10}{a}C_{M_0}q_\infty S_W\bar{c}_W = 0$$

where C_{M_0} was found to be zero in Example 6.4.

The effects of surge-velocity perturbations on the propulsive forces are given by Equations (6.138), or, including the more significant terms, we have

$$\frac{\partial F_{P_X}}{\partial u}\Big|_0 = \frac{\partial T}{\partial v_{\text{Axial}}}\Big|_0$$

$$\frac{\partial F_{P_Y}}{\partial u}\Big|_0 = 0 \quad \text{and} \quad \frac{\partial F_{P_Z}}{\partial u}\Big|_0 \approx 0$$

But from Equation (6.108) we have

$$\frac{\partial T}{\partial v_{\text{Axial}}}\Big|_0 = \frac{1}{a}\frac{\partial T}{\partial M_\infty}\Big|_0$$

where the change in engine thrust with respect to flight Mach number would be evaluated from the engine test data.

Finally, the effect of surge-velocity perturbations on the propulsive pitching moment is given by Equation (6.141), which is

$$\frac{\partial M_P}{\partial u}\Big|_0 = \left(C_{P_{M_u}} + \frac{2C_{P_{M_0}}}{U_0}\right)q_\infty S_W\bar{c}_W$$

while the effects on propulsive rolling and yawing moments are negligible. And from Equation (6.143) we have

$$C_{P_{M_u}} = \frac{1}{q_\infty S_W \bar{c}_W}\left((d_T - x_T\phi_T)\frac{\partial T}{\partial v_{\text{Axial}}}\Big|_0 - (x_T + d_T\phi_T)\frac{\partial F_N}{\partial v_{\text{Transverse}}}\Big|_0(\phi_T + \alpha_0)\right)$$

which with $\phi_T = d_T = 0$, $\alpha_0 = 2$ deg, and $x_T = 2\bar{c}_W$ reduces to

$$C_{P_{M_u}} = \frac{-2}{q_\infty S_W}\left(\frac{\partial F_N}{\partial v_{\text{Transverse}}}\Big|_0 (2/57.3)\right)$$

Since in Example 6.4 we found that $C_{P_{M_0}} = 0$, the effect of surge velocity on the propulsive pitching moment is given by

$$\frac{\partial M_P}{\partial u}\Big|_0 = -2\bar{c}\frac{\partial F_N}{\partial v_{\text{Transverse}}}\Big|_0 (2/57.3)$$

6.6.2 Plunge-Velocity Perturbation *w*

The effect of plunge-velocity perturbation w on the aerodynamic and propulsive forces and moments arises due to the concomitant change in vehicle perturbation angle of attack α (see Equation (6.96)).

As in the previous analysis dealing with surge velocity u, we will quantify the effects of plunge-velocity perturbations on the aerodynamic forces in terms of the partial derivatives $\partial F_{A_\bullet}/\partial w$, evaluated at the reference flight condition. From Equation (6.96) we found that the perturbation angle of attack may be expressed as

$$\alpha \approx \frac{w}{U_0}$$

And from Equations (6.113) we have the components of the aerodynamic force expressed as

$$F_{A_X} = C_X q_\infty S$$
$$F_{A_Y} = C_Y q_\infty S$$
$$F_{A_Z} = C_Z q_\infty S$$

where, from Equations (6.114), we have the force coefficients in the stability axes given by

$$C_X = -C_D \cos\alpha \cos(\beta_0 + \beta) - C_S \cos\alpha \sin(\beta_0 + \beta) + C_L \sin\alpha$$
$$C_Y = -C_D \sin(\beta_0 + \beta) + C_S \cos(\beta_0 + \beta)$$
$$C_Z = -C_D \sin\alpha \cos(\beta_0 + \beta) - C_S \sin\alpha \sin(\beta_0 + \beta) - C_L \cos\alpha$$

Therefore, for each of the three components in Equations (6.113), we have the effect of plunge-velocity perturbations given by

$$\frac{\partial F_{A\bullet}}{\partial w}|_0 = \frac{\partial C_\bullet}{\partial w}|_0\, q_\infty S_W + C_{\bullet_0} S_W \frac{\partial q_\infty}{\partial w}|_0, \bullet = X, Y, \text{ or } Z \tag{6.144}$$

From Equation (6.100), we find that

$$\frac{\partial q}{\partial w}|_0 = \rho_\infty (W_0 + w)|_0 = 0 \tag{6.145}$$

This result follows from the facts that in the stability axis $W_0 = 0$, plus all perturbations vanish at the reference condition. Therefore, Equation (6.144) reduces to

$$\boxed{\frac{\partial F_{A\bullet}}{\partial w}|_0 = \frac{\partial C_\bullet}{\partial w}|_0\, q_\infty S_W = C_{\bullet_w} q_\infty S_W = \frac{1}{U_0} C_{\bullet_\alpha} q_\infty S_W, \bullet = X, Y, \text{ or } Z} \tag{6.146}$$

where, from Equations (6.114), we find that

$$\boxed{\begin{aligned} \frac{\partial C_X}{\partial \alpha}|_0 &= -\frac{\partial C_D}{\partial \alpha}|_0 \cos\beta_0 + C_L|_0 = -C_{D_\alpha}\cos\beta_0 + C_{L_0} \\ \frac{\partial C_Y}{\partial \alpha}|_0 &= -\frac{\partial C_D}{\partial \alpha}|_0 \sin\beta_0 = -C_{D_\alpha}\sin\beta_0 \\ \frac{\partial C_Z}{\partial \alpha}|_0 &= -\frac{\partial C_L}{\partial \alpha}|_0 - C_D|_0 \cos\beta_0 = -C_{L_\alpha} - C_{D_0}\cos\beta_0 \end{aligned}} \tag{6.147}$$

Note here that we have assumed that $\partial C_S/\partial\alpha = 0$. So Equations (6.146–6.147) quantify the effects of plunge-velocity perturbations on the aerodynamic force components in the stability axes. And Equations (6.18) and (6.31) may be used to evaluate C_{L_α} and C_{D_α}, respectively.

In a similar fashion we will quantify the effects of plunge-velocity perturbation on the aerodynamic moments acting on the vehicle. The components of the aerodynamic moment $\mathbf{M}_A$ are again defined by Equation (6.88), or

$$\mathbf{M}_A = L_A \mathbf{i}_S + M_A \mathbf{j}_S + N_A \mathbf{k}_S$$

And in terms of nondimensional coefficients, we have from Equations (6.129)

$$L_A = C_{L_{\text{Roll}}} q_\infty S_W b_W$$

$$M_A = C_M q_\infty S_W \bar{c}_W$$

$$N_A = C_N q_\infty S_W b_W$$

The effects of perturbation in plunge velocity w on the aerodynamic moments are quantified by the partial derivatives

$$\frac{\partial L_A}{\partial w}|_0, \frac{\partial M_A}{\partial w}|_0, \text{ and } \frac{\partial N_A}{\partial w}|_0$$

all evaluated at the reference flight condition. And just as in the case with the aerodynamic forces, the effects of perturbations in plunge velocity w on the moments arise due to the change in angle of attack α.

Typically, due to the symmetry of the vehicle about its XZ plane, a change in vehicle angle of attack has little effect on its rolling or yawing moment. Therefore, we will focus attention on evaluating the effect of angle of attack on pitching moment. In terms of the pitching moment coefficient, we may write the effect of plunge-velocity perturbation on pitching moment as

$$\boxed{\frac{\partial M_A}{\partial w}|_0 = \frac{1}{U_0}\frac{\partial M_A}{\partial \alpha}|_0 = \frac{1}{U_0} C_{M_\alpha} q_\infty S_W \bar{c}_W} \tag{6.148}$$

where Equation (6.56) may be used to evaluate the vehicle's angle-of-attack pitching-moment effectiveness C_{M_α}.

Attention now turns to the effects of perturbations in plunge velocity on the propulsive forces and moments. Recall that the total propulsive force expressed in terms of its components in the stability axes was given in Equation (6.131). Referring to Figure 6.13, and noting that it has been assumed that the thrust force(s) are symmetric with respect to the XZ plane of the vehicle, we will write the components of the total propulsive force in terms of their reference values, plus perturbations in the thrust δT and the normal force on the engine δF_N. Or, consistent with Equation (6.134), we now have

$$\begin{aligned} F_{P_X} &= F_{P_{X_0}} + \delta T \cos(\phi_T + \alpha_0) - \delta F_N \sin(\phi_T + \alpha_0) \\ F_{P_Y} &= 0 \\ F_{P_Z} &= F_{P_{X_0}} - \delta T \sin(\phi_T + \alpha_0) - \delta F_N \cos(\phi_T + \alpha_0) \end{aligned} \tag{6.149}$$

Note that since we are only considering plunge-velocity perturbations, any propulsive normal force F_N will also lie in the XZ plane; hence the side force F_{P_Y} will remain zero. Consistent with Equation (6.146), the effects of plunge velocity on these force components are expressed as

$$\boxed{\frac{\partial F_{P_\bullet}}{\partial w}|_0 = \frac{\partial C_{P_\bullet}}{\partial w}|_0 q_\infty S_W = C_{P_{\bullet_w}} q_\infty S_W = \frac{1}{U_0} C_{P_{\bullet_\alpha}} q_\infty S_W, \bullet = X, Y, \text{or } Z} \tag{6.150}$$

The force perturbations δT and δF_N in Equations (6.149) may now be expressed as

$$\begin{aligned} \delta T &= \frac{\partial T}{\partial w} w = \frac{\partial T}{\partial v_{\text{Axial}}} \frac{\partial v_{\text{Axial}}}{\partial w} w \\ \delta F_N &= \frac{\partial F_N}{\partial w} w = \frac{\partial F_N}{\partial v_{\text{Transverse}}} \frac{\partial v_{\text{Transverse}}}{\partial w} w \end{aligned} \tag{6.151}$$

To address the partial derivatives of thrust and normal force in Equations (6.151), we may refer to Equations (6.108–6.112). Here, then we simply must deal with the geometry of the flow—that is, the orientation of the axial and transverse flows due to the plunge velocity perturbation.

From Figures 6.14 and 6.16, and including the effect of wing upwash or downwash at the engine inlet or propeller shaft on the local plunge velocity, we can write

$$\begin{aligned}\Delta v_{\text{Axial}} &= -w_{\text{inlet}} \sin(\phi_T + \alpha_0) = -w\left(1 \pm \frac{d\varepsilon_{\text{inlet}}}{d\alpha_W}\right)\sin(\phi_T + \alpha_0)\\ \Delta v_{\text{Transverse}} &= w_{\text{inlet}} \cos(\phi_T + \alpha_0) = w\left(1 \pm \frac{d\varepsilon_{\text{inlet}}}{d\alpha_W}\right)\cos(\phi_T + \alpha_0)\end{aligned} \tag{6.152}$$

In Equations (6.152), the positive signs on the downwash gradient are used if the propulsive device is located ahead of the wing, and the negative signs if located aft of the wing. Therefore, we may write the perturbation forces as

$$\begin{aligned}\delta T &= -\frac{\partial T}{\partial v_{\text{Axial}}}\sin(\phi_T + \alpha_0)\left(1 \pm \frac{d\varepsilon_{\text{inlet}}}{d\alpha}\right)w\\ \delta F_N &= \frac{\partial F_N}{\partial v_{\text{Transverse}}}\cos(\phi_T + \alpha_0)\left(1 \pm \frac{d\varepsilon_{\text{inlet}}}{d\alpha}\right)w\end{aligned} \tag{6.153}$$

As a result, we have the <u>effects of plunge velocity on the propulsive force components</u> given in terms of the following partial derivatives, evaluated at the reference fight condition.

$$\begin{aligned}\frac{\partial F_{P_X}}{\partial w}\Big|_0 &= -\left(\frac{\partial T}{\partial v_{\text{Axial}}}\Big|_0 + \frac{\partial F_N}{\partial v_{\text{Transverse}}}\Big|_0\right)\cos(\phi_T + \alpha_0)\sin(\phi_T + \alpha_0)\left(1 \pm \frac{d\varepsilon_{\text{inlet}}}{d\alpha}\right)\\ \frac{\partial F_{P_Y}}{\partial w}\Big|_0 &= 0\\ \frac{\partial F_{P_Z}}{\partial w}\Big|_0 &= \left(\frac{\partial T}{\partial v_{\text{Axial}}}\Big|_0 \sin^2(\phi_T + \alpha_0) - \frac{\partial F_N}{\partial v_{\text{Transverse}}}\Big|_0 \cos^2(\phi_T + \alpha_0)\right)\left(1 \pm \frac{d\varepsilon_{\text{inlet}}}{d\alpha}\right)\end{aligned} \tag{6.154}$$

Note that typically the angle $(\phi_T + \alpha_0)$ is small, and the above expressions may be simplified accordingly. Finally, by comparing Equations (6.150) and (6.154), and assuming the angle $(\phi_T + \alpha_0)$ is sufficiently small, we find that

$$\begin{aligned}C_{P_{X_w}} &= -\frac{1}{q_\infty S_W}\left(\frac{\partial T}{\partial v_{\text{Axial}}}\Big|_0 + \frac{\partial F_N}{\partial v_{\text{Transverse}}}\Big|_0\right)\left(1 \pm \frac{d\varepsilon_{\text{inlet}}}{d\alpha}\right)(\phi_T + \alpha_0)\\ C_{P_{Y_w}} &= 0\\ C_{P_{Z_w}} &\approx -\frac{1}{q_\infty S_W}\frac{\partial F_N}{\partial v_{\text{Transverse}}}\Big|_0\left(1 \pm \frac{d\varepsilon_{\text{inlet}}}{d\alpha}\right)\end{aligned} \tag{6.155}$$

Typically, the effects of plunge-velocity perturbations on the propulsive forces are not large. However, again, if the propulsion devices are located on the vehicle such that they have a sufficiently large moment arm relative to the moment-reference point, the changes in pitching moment due to these forces can be significant.

Under the assumed installation geometry of the propulsive devices, and since here we are dealing with only plunge velocity perturbation, the perturbation forces δT and δF_N both lie in the XZ plane of the vehicle. Hence they will generate no yawing or rolling moments. To address the pitching moment, we again draw from Figure 6.13 and note that the equation for the propulsive pitching moment, consistent with Equation (6.140), is

$$\begin{aligned} M_P &= \mathrm{M}_{P_0} + \delta T\big(d_T \cos\phi_T - x_T \sin\phi_T\big) - \delta F_N\big(x_T \cos\phi_T + d_T \sin\phi_T\big) \\ &= C_{P_M} q_\infty S_W \bar{c}_W \end{aligned} \tag{6.156}$$

Therefore, consistent with Equation (6.148), we find that the effect of perturbations in plunge velocity on the propulsive pitching moment is

$$\boxed{\frac{\partial M_P}{\partial w}\Big|_0 = \frac{1}{\mathrm{U}_0} C_{P_{M_\alpha}} q_\infty S_W \bar{c}_W} \tag{6.157}$$

Furthermore, following the approach taken above in analyzing the propulsive forces, we find that the effect of plunge-velocity perturbations on the propulsive pitching moment may also be expressed as

$$\boxed{\begin{aligned} \frac{\partial M_P}{\partial w}\Big|_0 &= \left(\big(d_T \cos\phi_T - x_T \sin\phi_T\big)\frac{\partial T}{\partial w}\Big|_0 - \big(x_T \cos\phi_T + d_T \sin\phi_T\big)\frac{\partial F_N}{\partial w}\Big|_0\right) \\ &= -\left(\begin{aligned} &\big(d_T \cos\phi_T - x_T \sin\phi_T\big)\frac{\partial T}{\partial v_{\text{Axial}}}\Big|_0 \sin(\phi_T + \alpha_0) \\ &+\big(x_T \cos\phi_T + d_T \sin\phi_T\big)\frac{\partial F_N}{\partial v_{\text{Transverse}}}\Big|_0 \cos(\phi_T + \alpha_0) \end{aligned}\right)\left(1 \pm \frac{d\varepsilon_{\text{inlet}}}{d\alpha}\right) \end{aligned}} \tag{6.158}$$

Again, the partial derivatives of thrust and normal force in Equation (6.158) may be evaluated using Equations (6.108–6.112), depending on the type of propulsion system(s) installed on the vehicle. Also, recall that if the line of action of the thrust passes close to X_{Ref}, then

$$\big(d_T \cos\phi_T - x_T \sin\phi_T\big) \approx 0$$

Finally, by comparing Equations (6.157) and (6.158), and assuming that ϕ_T and α_0 are sufficiently small, we find that

$$\boxed{\begin{aligned} C_{P_{M_w}} = -\frac{1}{q_{\infty_0} S_W \bar{c}_W}\Bigg(&\big(d_T - x_T\phi_T\big)\frac{\partial T}{\partial v_{\text{Axial}}}\Big|_0(\phi_T + \alpha_0) \\ &+ \big(x_T + d_T\phi_T\big)\frac{\partial F_N}{\partial v_{\text{Transverse}}}\Big|_0\Bigg)\left(1 \pm \frac{d\varepsilon_{\text{inlet}}}{d\alpha}\right) \end{aligned}} \tag{6.159}$$

EXAMPLE 6.6

Effects of Plunge-Velocity Perturbations on the Forces and Moments

Again consider the vehicle described in Examples 6.1–6.5. Determine the effects of plunge-velocity perturbations on the aerodynamic and propulsive forces and moments acting on the vehicle.

■ Solution

From Equations (6.146) we have the effects on the aerodynamic forces given by

$$\frac{\partial F_{A\bullet}}{\partial w}\Big|_0 = \frac{\partial C_\bullet}{\partial w}\Big|_0 q_\infty S_W = C_{\bullet_w} q_\infty S_W = \frac{1}{U_0} C_{\bullet_\alpha} q_\infty S_W,\ \bullet = X,\ Y, \text{ or } Z$$

where from Equations (6.147) with no sideslip angle

$$\frac{\partial C_X}{\partial \alpha}\Big|_0 = -\frac{\partial C_D}{\partial \alpha}\Big|_0 \cos\beta_0 + C_L|_0 = -C_{D_\alpha} + C_{L0}$$

$$\frac{\partial C_Y}{\partial \alpha}\Big|_0 = -\frac{\partial C_D}{\partial \alpha}\Big|_0 \sin\beta_0 = 0$$

$$\frac{\partial C_Z}{\partial \alpha}\Big|_0 = -\frac{\partial C_L}{\partial \alpha}\Big|_0 - C_D|_0 \cos\beta_0 = -C_{L_\alpha} - C_{D_0}$$

In Example 6.4 we determined the vehicle's reference lift and drag coefficients C_{L_0} and C_{D_0} in terms of the reference propulsive thrust and vehicle weight, while the vehicle's angle-of-attack lift effectiveness was determined in Example 6.1 to be $C_{L_\alpha} = 4.59$ /rad. From Equation (6.31), the vehicle's angle-of-attack drag effectiveness is given by

$$C_{D_\alpha} = \frac{2}{\pi}\left(\frac{C_{L_{W_0}}}{A_W e_W} C_{L_{\alpha_W}} + \frac{C_{L_{H_0}}}{A_H e_H} C_{L_{\alpha_H}} \frac{q_H}{q_\infty}\frac{S_H}{S_W}\left(1 - \frac{d\varepsilon}{d\alpha}\right)\right)$$

$$= \frac{2(4.19)}{\pi(5.33)(0.85)}\left(C_{L_{W_0}} + C_{L_{H_0}}(0.9)\frac{42}{169}(0.43)\right) \text{ per radian}$$

Now for the wing and horizontal tail we may write

$$C_{L_W} = C_{L_{\alpha_W}}\left(\alpha_W - \alpha_{0_W}\right)$$

$$C_{L_H} = C_{L_{\alpha_H}}\left(\left(1 - \frac{d\varepsilon}{d\alpha}\right)\alpha - \frac{d\varepsilon}{d\alpha}\left(i_W - \alpha_{0_W}\right) + i_H - \alpha_{0_H}\right)$$

So for a reference angle of attack of 2 deg, a wing $i_W = 2$ deg, and a wing zero-lift angle of attack $\alpha_0 = 1.33$ deg, we have

$$C_{L_{W_0}} = 4.19\left(\frac{2 + 2 - 1.33}{57.3}\right) = 0.195$$

And with a tail zero-lift angle of attack $\alpha_0 = 0$ we have

$$C_{L_H} = \frac{4.19}{57.3}\left((1 - 0.57)2 - 0.57(2 - 1.33) + 0\right) = 0.035$$

(Note that we have assumed that tail incidence $i_H = 0$ here.)

Consequently, $C_{D_\alpha} = \dfrac{2(4.19)}{\pi(5.33)(0.85)}\left(0.195 + 0.035(0.9)\left(\dfrac{42}{169}\right)(0.43)\right) = 0.117$ /rad

Therefore, from Equations (6.146), the effects of plunge-velocity perturbations on the aerodynamic forces are

$$\frac{\partial F_{A_X}}{\partial w}\Big|_0 = \frac{1}{U_0} C_{X_\alpha} q_\infty S_W = -\frac{5}{a}\left(C_{D_\alpha} - C_{L_0}\right) q_\infty S_W = -\frac{5}{a}\left(0.117 - \frac{mg - 0.03T_0}{q_\infty S_W}\right) q_\infty S_W$$

$$\frac{\partial F_{A_Y}}{\partial w}\Big|_0 = \frac{1}{U_0} C_{Y_\alpha} q_\infty S_W = 0$$

$$\frac{\partial F_{A_Z}}{\partial w}\Big|_0 = \frac{1}{U_0} C_{Z_\alpha} q_\infty S_W = -\frac{1}{U_0}\left(C_{L_\alpha} + C_{D_0}\right) q_\infty S_W = -\frac{5}{a}\left(4.59 + \frac{T_0}{q_\infty S_W}\right) q_\infty S_W$$

Regarding the effect of plunge-velocity perturbations on the aerodynamic pitching moment, Equation (6.148) indicates that

$$\frac{\partial M_A}{\partial w}\Big|_0 = \frac{1}{U_0}\frac{\partial M_A}{\partial \alpha}\Big|_0 = \frac{1}{U_0} C_{M_\alpha} q_\infty S_W \bar{c}_W$$

while the effects on rolling and yawing moments are negligible. From Example 6.2 we found that the vehicle's angle-of-attack pitching-moment effectiveness was $C_{M_\alpha} = -1.59$ /rad. Therefore, the effect of plunge velocity on the aerodynamic pitching moment is

$$\frac{\partial M_A}{\partial w}\Big|_0 = \frac{1}{U_0} C_{M_\alpha} q_\infty S_W \bar{c}_W = -\frac{5}{a}(1.59) q_\infty S_W \bar{c}_W$$

The effects of plunge-velocity perturbations on the propulsive forces are indicted by Equations (6.154), which for this vehicle become

$$\frac{\partial F_{P_X}}{\partial w}\Big|_0 \approx 0, \quad \frac{\partial F_{P_Y}}{\partial w}\Big|_0 = 0$$

$$\frac{\partial F_{P_Z}}{\partial w}\Big|_0 = -\frac{\partial F_N}{\partial v_{\text{Transverse}}}\Big|_0 \left(1 - \frac{d\varepsilon_{\text{inlet}}}{d\alpha}\right)$$

If the engines are mounted $2\bar{c}_W$ aft of X_{Ref}, and $X_{\text{Ref}} = 0.5\bar{c}_W$ aft of the wing apex, then the engine inlets are approximately $2.5\bar{c}_W - c_r$ aft of the wing's trailing edge. Figure 5.35 shows the downwash gradient at a normalized distance $l/(b/2)$ aft of the wing's trailing edge. And here

$$\frac{l}{b/2} = \frac{2.5\bar{c}_W - c_r}{15} = \frac{2.5(5.825) - 7.5}{15} = 0.47$$

From Figure 5.35 we find that

$$\frac{d\varepsilon_{\text{inlet}}}{d\alpha} \approx 0.62$$

And from Equation (6.112) we have

$$\frac{\partial F_N}{\partial v_{\text{Transverse}}} = \frac{1}{U_0}\frac{\dot{m}_P^2}{\rho_\infty A_{\text{in}}}$$

where the inlet area and mass-flow rate must be obtained from engine data. Therefore, the effects of plunge-velocity perturbations on propulsive forces are

$$\frac{\partial F_{P_X}}{\partial w}|_0 \approx 0, \quad \frac{\partial F_{P_Y}}{\partial w}|_0 = 0$$

$$\frac{\partial F_{P_Z}}{\partial w}|_0 = -\frac{5}{a}\frac{\dot{m}_P^2}{\rho_\infty A_{\text{in}}}(1-0.62) = -\frac{1.9}{a}\frac{\dot{m}_P^2}{\rho_\infty A_{\text{in}}}$$

The effect of plunge-velocity perturbation on the propulsive pitching moment is indicated by Equation (6.158), while the effects on propulsive rolling and yawing moments are negligible. Here, Equation (6.158) becomes

$$\frac{\partial M_P}{\partial w}|_0 = -x_T\frac{\partial F_N}{\partial v_{\text{Transverse}}}|_0\left(1-\frac{d\varepsilon_{\text{inlet}}}{d\alpha}\right) = -2\bar{c}_W\left(\frac{1.9}{a}\frac{\dot{m}_P^2}{\rho_\infty A_{\text{in}}}\right)$$

6.6.3 Sideslip-Velocity Perturbation *v*

The effects of sideslip-velocity perturbation v on the aerodynamic and propulsive forces and moments arise due to the concomitant change in vehicle perturbation sideslip angle β.

Similar to before, we will quantify the effect of sideslip-velocity perturbation on the aerodynamic forces in terms of the partial derivatives $\partial F_{A_\bullet}/\partial v$, evaluated at the reference flight condition. From Equation (6.99) we have

$$\beta \approx \frac{v}{U_0}$$

And from Equations (6.113) we have the components of aerodynamic force in the stability axes expressed as

$$F_{A_X} = C_X q_\infty S$$
$$F_{A_Y} = C_Y q_\infty S$$
$$F_{A_Z} = C_Z q_\infty S$$

where, from Equations (6.114), we have the aerodynamic force coefficients given by

$$C_X = -C_D\cos\alpha\cos(\beta_0+\beta) - C_S\cos\alpha\sin(\beta_0+\beta) + C_L\sin\alpha$$
$$C_Y = -C_D\sin(\beta_0+\beta) + C_S\cos(\beta_0+\beta)$$
$$C_Z = -C_D\sin\alpha\cos(\beta_0+\beta) - C_S\sin\alpha\sin(\beta_0+\beta) - C_L\cos\alpha$$

Therefore, for each of the three force components in Equations (6.113), we have the effect of lateral-velocity perturbations given by

$$\boxed{\begin{aligned}\frac{\partial F_{A_\bullet}}{\partial v}|_0 &= \frac{\partial C_\bullet}{\partial v}|_0\, q_\infty S_W + C_{\bullet_0} S_W \frac{\partial q_\infty}{\partial v}|_0 \\ &= \frac{1}{U_0} C_{\bullet_\beta} q_\infty S_W + C_{\bullet_0} S_W \frac{\partial q_\infty}{\partial v}|_0, \bullet = X, Y, \text{ or } Z\end{aligned}} \qquad (6.160)$$

where, from Equation (6.100), we have

$$\boxed{\frac{\partial q_\infty}{\partial v}\Big|_0 = \rho_\infty(V_0 + v)\big|_0 = \rho_\infty V_0 = \rho_\infty U_0 \beta_0} \tag{6.161}$$

Plus, from Equations (6.114), we have sideslip effectivenesses given by

$$\boxed{\begin{aligned} C_{X_\beta} &\triangleq \frac{\partial C_X}{\partial \beta}\Big|_0 = -\frac{\partial C_S}{\partial \beta}\Big|_0 \sin\beta_0 = -C_{S_\beta}\sin\beta_0 \\ C_{Y_\beta} &\triangleq \frac{\partial C_Y}{\partial \beta}\Big|_0 = C_{S_\beta}\cos\beta_0 \\ C_{Z_\beta} &\triangleq \frac{\partial C_Z}{\partial \beta}\Big|_0 = 0 \end{aligned}} \tag{6.162}$$

In obtaining the above three expressions we have assumed that perturbation sideslip angle has no effect on lift or drag coefficient. Therefore Equations (6.160–6.162) quantify the effects of sideslip-velocity perturbations on the aerodynamic force components in the stability axes. Finally, Equation (6.25) may be used to evaluate C_{S_β}, while Equations (6.123) may be used to evaluate C_{X_0}, C_{Y_0}, and C_{Z_0}.

Similarly, we can quantify the effects of sideslip-velocity perturbation on the aerodynamic moments acting on the vehicle. The components of the total aerodynamic moment $\mathbf{M}_A$ are defined in Equation (6.88). And in terms of nondimensional coefficients, from Equations (6.129), these components are

$$\begin{aligned} L_A &= C_{L_{\text{Roll}}} q_\infty S_W b_W \\ M_A &= C_M q_\infty S_W \bar{c}_W \\ N_A &= C_N q_\infty S_W b_W \end{aligned}$$

The effects of perturbations in sideslip velocity v on these moments are quantified by the partial derivatives

$$\frac{\partial L_A}{\partial v}\Big|_0, \quad \frac{\partial M_A}{\partial v}\Big|_0, \text{ and } \frac{\partial N_A}{\partial v}\Big|_0$$

all evaluated at the reference condition. And just as in the case with the aerodynamic forces, the effects of perturbations in sideslip velocity v on these moments arise due to the change in sideslip angle β.

Typically, a change in a vehicle's sideslip angle has little effect on its pitching moment. Therefore, we will focus attention on evaluating the effects of sideslip angle on the rolling and yawing moments. In terms of the moment coefficients, we may write

$$\frac{\partial L_A}{\partial v}\bigg|_0 = \frac{\partial C_{L_{\text{Roll}}}}{\partial v}\bigg|_0 q_\infty S_W b_W + C_{L\text{roll}_0} S_W b_W \frac{\partial q_\infty}{\partial v}\bigg|_0 = \frac{1}{U_0} C_{L_\beta} q_\infty S_W b_W + C_{L\text{roll}_0} S_W b_W \frac{\partial q_\infty}{\partial v}\bigg|_0$$

and (6.163)

$$\frac{\partial N_A}{\partial v}\bigg|_0 = \frac{\partial C_N}{\partial v}\bigg|_0 q_\infty S_W b_W + C_{N_0} S_W b_W \frac{\partial q_\infty}{\partial v}\bigg|_0 = \frac{1}{U_0} C_{N_\beta} q_\infty S_W b_W + C_{N_0} S_W b_W \frac{\partial q_\infty}{\partial v}\bigg|_0$$

We had previously determined in Equation (6.161) that

$$\frac{\partial q_\infty}{\partial v}\bigg|_0 = \rho_\infty(V_0 + v)\big|_0 = \rho_\infty V_0 = \rho_\infty U_0 \beta_0$$

So then we have

$$\boxed{\begin{aligned} \frac{\partial L_A}{\partial v}\bigg|_0 &= \frac{1}{U_0}\left(C_{L_\beta} + 2C_{L\text{roll}_0}\beta_0\right) q_\infty S_W b_W \\ \text{and} \qquad & \\ \frac{\partial N_A}{\partial v}\bigg|_0 &= \frac{1}{U_0}\left(C_{N_\beta} + 2C_{N_0}\beta_0\right) q_\infty S_W b_W \end{aligned}} \qquad (6.164)$$

Furthermore, Equations (6.44) and (6.68) may be used to evaluate C_{L_β} and C_{N_β}, respectively.

So Equations (6.163) and (6.164) quantify the effects of sideslip-velocity perturbation v on the aerodynamic rolling and yawing moment, while the effect of this velocity perturbation on pitching moment is usually negligible. Note that frequently the second terms on the right sides of Equations (6.164) are small compared to the other terms, since β_0 will be small and/or the reference values of rolling and yawing moment coefficients will be zero.

Attention now turns to the effects of perturbations in lateral velocity on the propulsive forces and moments. Recall that the total propulsive force expressed in terms of its components in the stability axes was given in Equation (6.131). Referring to Figure 6.14, and noting that it has been assumed that the thrust force(s) are symmetric with respect to the XZ plane of the vehicle, we will write the components of the total propulsive force in terms of their reference values, plus perturbations in the normal force on the engine δF_N. Since we're dealing with only lateral-velocity perturbations here, instead of the results given in Equation (6.134) we now have

$$\begin{aligned} F_{P_X} &= F_{P_{X_0}} + C_{P_{X_0}} S_W \frac{\partial q_\infty}{\partial v}\bigg|_0 v \\ F_{P_Y} &= C_{P_{Y_0}} S_W \frac{\partial q_\infty}{\partial v}\bigg|_0 v - \delta F_N \\ F_{P_Z} &= F_{P_{Z_0}} + C_{P_{Z_0}} S_W \frac{\partial q_\infty}{\partial v}\bigg|_0 v \end{aligned} \qquad (6.165)$$

where the normal-force perturbation may be expressed as

$$\delta F_N = \frac{\partial F_N}{\partial v} v = \frac{\partial F_N}{\partial v_{\text{Transverse}}} \frac{\partial v_{\text{Transverse}}}{\partial v} v \tag{6.166}$$

Note that this normal force F_N is now acting laterally on the vehicle, rather than vertically.

To address the partial derivative of propulsive normal force in Equation (6.166), we may refer to Equations (6.111) or (6.112), depending on the type of propulsion system. Plus, from Figures 6.14 and 6.16 we can see that here

$$\Delta v_{\text{Transeverse}} \approx v \tag{6.167}$$

where the effect of side wash (analogous to downwash) has been ignored. Therefore, we have

$$\delta F_N = \frac{\partial F_N}{\partial v_{\text{Transverse}}} v \tag{6.168}$$

Consistent with Equations (6.160), the effects of lateral-velocity perturbation on the propulsive forces may be expressed as

$$\boxed{\begin{aligned} \frac{\partial F_{P_X}}{\partial v}\Big|_0 &= C_{P_{X_0}} S_W \frac{\partial q_\infty}{\partial v}\Big|_0 = C_{P_{X_0}} \rho_\infty S_W \mathrm{U}_0 \beta_0 \\ \frac{\partial F_{P_Y}}{\partial v}\Big|_0 &= \frac{\partial C_{P_Y}}{\partial v}\Big|_0 q_\infty S_W + C_{P_{Y_0}} S_W \frac{\partial q_\infty}{\partial v}\Big|_0 = \frac{1}{\mathrm{U}_0}(C_{P_{Y_\beta}} + 2C_{P_{Y_0}}\beta_0) q_\infty S_W \\ \frac{\partial F_{P_Z}}{\partial v}\Big|_0 &= C_{P_{Z_0}} S_W \frac{\partial q_\infty}{\partial v}\Big|_0 = C_{P_{Z_0}} \rho_\infty S_W \mathrm{U}_0 \beta_0 \end{aligned}} \tag{6.169}$$

Finally, using Equations (6.165), (6.168),and (6.169), we find that

$$\boxed{C_{P_{Y_\beta}} = -\frac{\mathrm{U}_0}{q_\infty S_W} \frac{\partial F_N}{\partial v_{\text{Transverse}}}\Big|_0} \tag{6.170}$$

Typically, the effects of sideslip-velocity perturbations on the propulsive forces are not large. However, if the propulsion devices are installed such that they have a large moment arm relative to the moment reference point, the changes in propulsive rolling or yawing moments due to this force can again be significant. Under the assumed installation geometry of the propulsive devices, and since here we are dealing with only lateral velocity perturbation, the perturbation force δF_N acts along the negative Y axis of the vehicle for positive sideslip. Hence it generates no pitching moment.

To address the rolling and yawing propulsive moments, we draw from Equations (6.165) and Figure 6.13 to write the equations for the perturbations in propulsive rolling and yawing moments about the X_S and Z_S stability axes, respectively. These are

$$\delta L_P = C_{P_{L\text{roll}_0}} S_W \bar{c}_W \frac{\partial q_\infty}{\partial v}\Big|_0 v + (d_T \cos\alpha_0 + x_T \sin\alpha_0)\delta F_N$$
$$\delta N_P = C_{P_{N_0}} S_W \bar{c}_W \frac{\partial q_\infty}{\partial v}\Big|_0 v + (x_T \cos\alpha_0 - d_T \sin\alpha_0)\delta F_N \tag{6.171}$$

In terms of nondimensional coefficients, these moments may be expressed as

$$L_P = C_{P_{L\text{Roll}}} q_\infty S_W b_W$$
$$N_P = C_{P_N} q_\infty S_W b_W \tag{6.172}$$

Consistent with Equations (6.164), we may therefore express the effects of perturbations in lateral velocity on the propulsive moments as

$$\boxed{\begin{aligned} \frac{\partial L_P}{\partial v}\Big|_0 &= \frac{1}{U_0}(C_{P_{L_\beta}} + 2C_{P_{L\text{roll}_0}}\beta_0) q_\infty S_W b_W \\ \frac{\partial N_P}{\partial v}\Big|_0 &= \frac{1}{U_0}(C_{P_{N_\beta}} + 2C_{P_{N_0}}\beta_0) q_\infty S_W b_W \end{aligned}} \tag{6.173}$$

Following the approach taken above in analyzing the propulsive forces, we find that the effects of lateral-velocity perturbation on the propulsive moments may also be expressed as

$$\boxed{\begin{aligned} \frac{\partial L_P}{\partial v} &= C_{P_{L\text{roll}_0}} S_W b_W \frac{\partial q_\infty}{\partial v}\Big|_0 + (d_T \cos\alpha_0 + x_T \sin\alpha_0)\frac{\partial F_N}{\partial v} \\ &= C_{P_{L\text{roll}_0}} \rho_\infty U_0 \beta_0 S_W b_W + (d_T \cos\alpha_0 + x_T \sin\alpha_0)\frac{\partial F_N}{\partial v_{\text{Transverse}}} \\ \frac{\partial N_P}{\partial v} &= C_{P_{N_0}} S_W b_W \frac{\partial q_\infty}{\partial v}\Big|_0 + (x_T \cos\alpha_0 - d_T \sin\alpha_0)\frac{\partial F_N}{\partial v} \\ &= C_{P_{N_0}} \rho_\infty U_0 \beta_0 S_W b_W + (x_T \cos\alpha_0 - d_T \sin\alpha_0)\frac{\partial F_N}{\partial v_{\text{Transverse}}} \end{aligned}} \tag{6.174}$$

Note that if the reference sideslip angle β_0 is small, the reference propulsion rolling and yawing moments are typically both zero, in which case the first terms on the right in Equations (6.174) are both zero. The partial derivative of normal force in Equations (6.174) may be evaluated using Equation (6.111) or (6.112). Finally, by comparing Equations (6.173) and (6.174), and assuming α_0 is sufficiently small, we find that sideslip effectiveness coefficients are

$$\boxed{\begin{aligned} C_{P_{L_\beta}} &= \frac{U_0}{q_\infty S_W b_W}(d_T + x_T \alpha_0)\frac{\partial F_N}{\partial v_{\text{Transverse}}} \\ C_{P_{N_\beta}} &= \frac{U_0}{q_\infty S_W b_W}(x_T - d_T \alpha_0)\frac{\partial F_N}{\partial v_{\text{Transverse}}} \end{aligned}} \tag{6.175}$$

6.7 FORCES AND MOMENTS DUE TO ANGULAR-VELOCITY PERTURBATIONS

As noted at the beginning of Section 6.6, changes in the vehicle's translational and rotational velocities relative to the air mass affect both the aerodynamic and propulsive forces and moments. We now will address the effects of perturbations in the vehicle's angular velocity on these forces and moments.

Angular velocities influence the aerodynamic forces and moments due to the fact that the vehicle's rotations affect the local angles of attack and sideslip of the various lifting surfaces. Figure 6.18 reveals how a perturbation in pitch rate q, for example, induces a perturbation in the local angle of attack of the horizontal tail, or $\Delta\alpha_H$. Clearly, if the X locations along the vehicle are taken to be positive forward, this change in local angle of attack is

$$\Delta\alpha_H \approx \frac{w_H}{U_0} = \frac{q}{U_0}(X_{\text{Ref}} - X_{\text{AC}_H}) \qquad (6.176)$$

Note that a positive pitch rate induces a downward velocity of an aft tail surface w_H, which in turn induces a positive change in the local angle of attack of that surface.

Similarly, as shown in Figure 6.19, a perturbation in the vehicle's yaw rate r (rotation about the Z_S axis) will induce a perturbation in the local angle of attack of an aft vertical tail by the amount $\Delta\alpha_V$, where

$$\Delta\alpha_V \approx \frac{r}{U_0}(X_{\text{Ref}} - X_{\text{AC}_V}) \qquad (6.177)$$

Recall that we are dealing with the yaw rate about the vehicle's stability axis while the X locations are taken along the fuselage-reference X axis. (See Figure 6.13.) Note that a positive $\Delta\alpha_V$ corresponds to an increase in the aerodynamic "lift" generated by the vertical tail, which acts in the positive Y_S direction.

Additionally, a yaw-rate perturbation induces a perturbation in the local surge velocity $\Delta u(y)$ at a given span location on both the wing and horizontal tail. This change in local velocity may be written as

$$\Delta u(y) = -ry \qquad (6.178)$$

or for a positive yaw rate this velocity change is negative along the right span, and positive along the left.

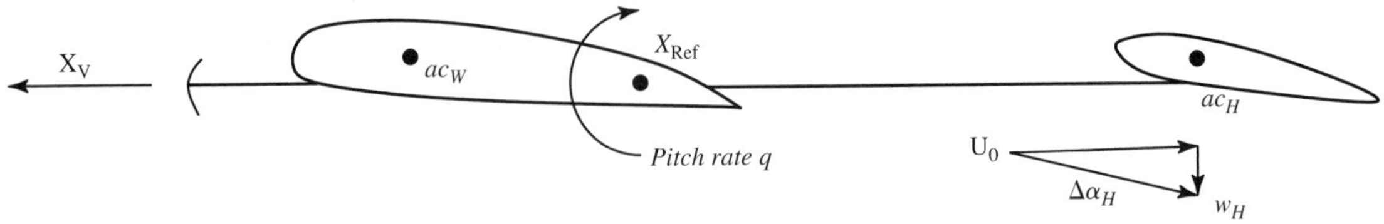

Figure 6.18 Induced angle of attack due to pitch rate.

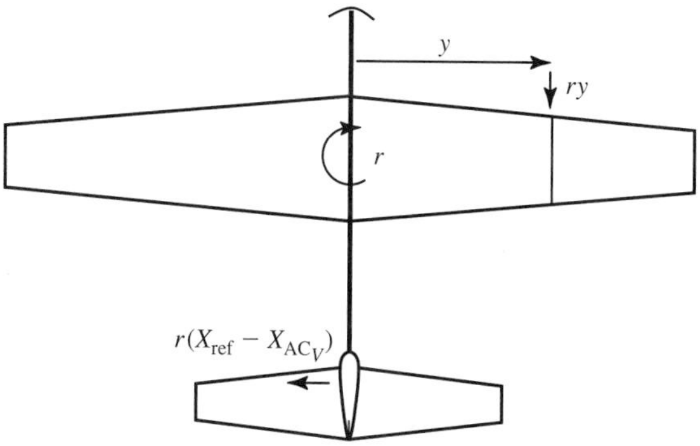

Figure 6.19 Induced local velocities due to yaw rate.

Finally, a perturbation in the vehicle's roll rate p about the X_S axis will induce a perturbation in the local angle of attack $\Delta\alpha(y)$ of the airfoil section at a span location y of a wing or horizontal tail surface equal to

$$\Delta\alpha_{W\text{ or }H}(y) = p\frac{y}{U_0} \tag{6.179}$$

Note that span y is defined to be positive along the right wing or horizontal tail. Therefore along the left wing or left horizontal tail, where y is negative, a positive roll rate induces a reduction in local angle of attack.

Perturbation roll rate also affects the local angle of attack of the airfoil sections of the vertical tail. Since we are addressing roll rate about the stability axis, the key dimension is the vertical distance between the X_S axis and the aerodynamic center of a particular airfoil section on the vertical tail. (Again, see Figure 6.8.) Denoting this distance as z_V, the change in local angle of attack due to roll rate is

$$\Delta\alpha_V(z) = -p\frac{z_V}{U_0} \tag{6.180}$$

6.7.1 Pitch-Rate Perturbation q

As noted above, a perturbation in pitch rate will induce a change in the local angle of attack of any lifting surface that is not located at the point of rotation, or X_{Ref}. This typically would correspond to a horizontal tail and/or a canard (forward surface). A pitch rate usually has little effect on the wing's angle of attack because the wing is typically located close to X_{Ref}. We will assume here that the vehicle has a horizontal tail and no canard.

Again, we have the components of the aerodynamic force expressed in terms of nondimensional coefficients, given in Equations (6.113), repeated here.

$$F_{A_X} = C_X q_\infty S$$

$$F_{A_Y} = C_Y q_\infty S$$

$$F_{A_Z} = C_Z q_\infty S$$

where from Equations (6.114) we have

$$C_X = -C_D \cos\alpha \cos(\beta_0 + \beta) - C_S \cos\alpha \sin(\beta_0 + \beta) + C_L \sin\alpha$$
$$C_Y = -C_D \sin(\beta_0 + \beta) + C_S \cos(\beta_0 + \beta)$$
$$C_Z = -C_D \sin\alpha \cos(\beta_0 + \beta) - C_S \sin\alpha \sin(\beta_0 + \beta) - C_L \cos\alpha$$

As done previously, we will quantify the effects of pitch-rate perturbation q on these forces in terms of partial derivatives, which are

$$\boxed{\frac{\partial F_{A\bullet}}{\partial q}\Big|_0 = \frac{\partial C_\bullet}{\partial q}\Big|_0 q_\infty S_W \triangleq C_{\bullet_q} q_\infty S_W, \quad C_\bullet = C_X, C_Y, \text{and } C_Z} \tag{6.181}$$

Applying Equations (6.114), and noting that side force will not be significantly influenced by pitch rate, we have

$$\boxed{\begin{aligned}
\frac{\partial C_X}{\partial q}\Big|_0 &= -\frac{\partial C_D}{\partial q}\Big|_0 \cos\beta_0 \triangleq -C_{D_q} \cos\beta_0 \\
\frac{\partial C_Y}{\partial q}\Big|_0 &= -\frac{\partial C_D}{\partial q}\Big|_0 \sin\beta_0 \triangleq -C_{D_q} \sin\beta_0 \\
\frac{\partial C_Z}{\partial q}\Big|_0 &= -\frac{\partial C_L}{\partial q}\Big|_0 \triangleq -C_{L_q}
\end{aligned}} \tag{6.182}$$

With reference to Figure 6.18, the change in lift and drag acting on the horizontal tail due to the change in local angle of attack $\Delta\alpha_H$ may be written as

$$\begin{aligned}
\Delta L_H &= \Delta C_{L_H} q_H S_H = C_{L_{\alpha_H}} \Delta\alpha_H q_H S_H \\
\Delta D_H &= \Delta C_{D_H} q_H S_H = \left(\frac{\partial C_{D_H}}{\partial C_{L_H}} \Delta C_{L_H}\right) q_H S_H = \left(\frac{2C_{L_H}}{\pi A_H e_H}\right) \Delta C_{L_H} q_H S_H
\end{aligned} \tag{6.183}$$

Assuming the change in lift and drag on the tail constitutes the only change in lift and drag for the vehicle, we have

$$\begin{aligned}
\Delta L &= \Delta C_L q_\infty S_W = \Delta L_H \\
\Delta D &= \Delta C_D q_\infty S_W = \Delta D_H
\end{aligned} \tag{6.184}$$

Therefore, from Equations (6.176), (6.183), and (6.184), we find that

$$\boxed{\begin{aligned}
C_{L_q} &= C_{L_{\alpha_H}} \frac{q_H}{q_\infty} \frac{S_H}{S_W} \frac{\partial \alpha_H}{\partial q} = C_{L_{\alpha_H}} \frac{q_H}{q_\infty} \frac{S_H}{S_W} \frac{(X_{\text{Ref}} - X_{\text{AC}_H})}{U_0} \\
C_{D_q} &= \left(\frac{2C_{L_{H_0}}}{\pi A_H e_H}\right) C_{L_{\alpha_H}} \frac{q_H}{q_\infty} \frac{S_H}{S_W} \frac{\partial \alpha_H}{\partial q} = \left(\frac{2C_{L_{H_0}}}{\pi A_H e_H}\right) C_{L_q}
\end{aligned}} \tag{6.185}$$

Substituting Equations (6.185) into Equations (6.182), and applying Equations (6.181), yields the effects of pitch-rate perturbation on the aerodynamic forces. Finally, it should be noted that typically the pitch-rate drag effectiveness is small compared to the lift effectiveness, and may be neglected.

With regard to the effects of pitch-rate perturbations on the aerodynamic moments, the effects of perturbation pitch rate on the rolling and yawing moments are typically negligible. The effect of a perturbation in pitch rate on the pitching moment is quantified in terms of the partial derivative

$$\frac{\partial M_A}{\partial q}\Big|_0 = \frac{\partial C_M}{\partial q}\Big|_0 q_\infty S_W \bar{c}_W \triangleq C_{M_q} q_\infty S_W \bar{c}_W \tag{6.186}$$

With reference to Equations (6.183) and Figure 6.18, the change in vehicle aerodynamic pitching moment M_A due to a perturbation in local angle of attack on the horizontal tail $\Delta\alpha_H$ may be written as

$$\Delta M_A = -\left(\Delta L_H \cos\alpha_0 + \Delta D_H \sin\alpha_0\right)\left(X_{\text{Ref}} - X_{\text{AC}_H}\right) = \Delta C_M q_\infty S_W \bar{c}_W \tag{6.187}$$

Note that we have also expressed this change in pitching moment in terms of a change in vehicle pitching-moment coefficients ΔC_M. Therefore, using Equations (6.176), (6.183), and (6.187), we obtain

$$\begin{aligned} C_{M_q} &= -\left(\cos\alpha_0 + \left(\frac{2C_{L_{H_0}}}{\pi A_H e_H}\right)\sin\alpha_0\right) C_{L_{\alpha_H}} \frac{\left(X_{\text{Ref}} - X_{\text{AC}_H}\right)^2}{U_0 \bar{c}_W} \frac{q_H}{q_\infty} \frac{S_H}{S_W} \\ &\approx -C_{L_{\alpha_H}} \frac{\left(X_{\text{Ref}} - X_{\text{AC}_H}\right)^2}{U_0 \bar{c}_W} \frac{q_H}{q_\infty} \frac{S_H}{S_W} \end{aligned} \tag{6.188}$$

Equations (6.186) and (6.188) then quantify the effect of a perturbation in pitch rate on the vehicle's aerodynamic pitching moment. Note that C_{M_q} is always negative and is called the *pitch-damping coefficient.* We shall see in Chapter 10 that this is an important parameter with regard to the vehicle's dynamics.

A perturbation in pitch rate may also affect the propulsive forces and moments, since pitch rate affects the local flow angle aft or forward of X_{Ref}. A modest change in local angle of attack has little effect on the magnitude of propulsive thrust, but can change the normal force F_N generated on a turbojet engine or propeller hub. This change in normal force can be appreciable if the propulsion devices are located far forward or aft on the vehicle. The presentation here will closely follow that dealing with the effect of plunging velocity w on the propulsive forces and moments.

To address the propulsive forces, from Equations (6.149) we have

$$\begin{aligned} F_{P_X} &= F_{P_{X_0}} + \delta T \cos(\phi_T + \alpha_0) - \delta F_N \sin(\phi_T + \alpha_0) = C_{P_X} q_\infty S_W \\ F_{P_Y} &= 0 = C_{P_Y} q_\infty S_W \\ F_{P_Z} &= F_{P_{X_0}} - \delta T \sin(\phi_T + \alpha_0) - \delta F_N \cos(\phi_T + \alpha_0) = C_{P_Z} q_\infty S_W \end{aligned}$$

Note that since we are now just considering the change in local angle of attack associated with pitch-rate perturbations, any propulsive normal force F_N will lie in the XZ plane. Hence the side force F_{P_Y} remains zero.

Consistent with Equation (6.181), the effects of pitch-rate perturbation of the propulsive forces are expressed in terms of partial derivatives, or

$$\frac{\partial F_{P_\bullet}}{\partial q}\Big|_0 = \frac{\partial C_{P_\bullet}}{\partial q}\Big|_0 q_\infty S_W \triangleq C_{P_{\bullet_q}} q_\infty S_W, \ \bullet = X,\ Y,\ \text{or } Z \tag{6.189}$$

Also, the propulsive force perturbations δT and δF_N in Equations (6.149) may now be expressed in terms of pitch-rate perturbation, or

$$\begin{aligned}\delta T &= \frac{\partial T}{\partial q} q = \frac{\partial T}{\partial v_{\text{Axial}}} \frac{\partial v_{\text{Axial}}}{\partial w_{\text{inlet}}} \frac{\partial w_{\text{inlet}}}{\partial q} q \\ \delta F_N &= \frac{\partial F_N}{\partial q} q = \frac{\partial F_N}{\partial v_{\text{Transverse}}} \frac{\partial v_{\text{Transverse}}}{\partial w_{\text{inlet}}} \frac{\partial w_{\text{inlet}}}{\partial q} q\end{aligned} \tag{6.190}$$

With reference to Figures 6.14, 6.16, and 6.18, we can write

$$\begin{aligned}\Delta v_{\text{Axial}} &= -w_{\text{inlet}} \sin(\phi_T + \alpha_0) = -x_T q \sin(\phi_T + \alpha_0) \\ \Delta v_{\text{Transverse}} &= w_{\text{inlet}} \cos(\phi_T + \alpha_0) = x_T q \cos(\phi_T + \alpha_0)\end{aligned} \tag{6.191}$$

Consequently, the effects of perturbation in pitch rate on the propulsive force components may be expressed in terms of the following partial derivatives, all evaluated at the reference flight condition.

$$\begin{aligned}\frac{\partial F_{P_X}}{\partial q}\Big|_0 &= -\left(\frac{\partial T}{\partial v_{\text{Axial}}}\Big|_0 + \frac{\partial F_N}{\partial v_{\text{Transverse}}}\Big|_0\right) x_T \cos(\phi_T + \alpha_0) \sin(\phi_T + \alpha_0) \\ \frac{\partial F_{P_Y}}{\partial q}\Big|_0 &= 0 \\ \frac{\partial F_{P_Z}}{\partial q}\Big|_0 &= x_T\left(\frac{\partial T}{\partial v_{\text{Axial}}}\Big|_0 \sin^2(\phi_T + \alpha_0) - \frac{\partial F_N}{\partial v_{\text{Transverse}}}\Big|_0 \cos^2(\phi_T + \alpha_0)\right)\end{aligned} \tag{6.192}$$

Typically the angle $(\phi_T + \alpha_0)$ is small, and the above expressions may be appropriately simplified. The partial derivatives of thrust and normal force in Equations (6.192) may be evaluated using Equations (6.108–6.112), depending on the type of propulsion system(s) on the vehicle. Finally, by comparing Equations (6.189) and (6.192), and assuming $(\phi_T + \alpha_0)$ is sufficiently small, we find that

$$
\begin{aligned}
C_{P_{X_q}} &\approx -\frac{x_T}{q_\infty S_W}\left(\frac{\partial T}{\partial v_{\text{Axial}}}\Big|_0 + \frac{\partial F_N}{\partial v_{\text{Transverse}}}\Big|_0\right)(\phi_T + \alpha_0) \\
C_{P_{Z_q}} &\approx -\frac{x_T}{q_\infty S_W}\frac{\partial F_N}{\partial v_{\text{Transverse}}}\Big|_0
\end{aligned}
\qquad (6.193)
$$

Normally the effect of pitch-rate perturbations on the above propulsive forces is not large. However, if the propulsive devices are installed such that they have a significant moment arm with respect to the moment reference point, the change in pitching moment due to these forces can again be significant. Under the geometry of the propulsive devices we have assumed in this chapter, and since here we are dealing with only local plunge velocity associated with a pitch-rate perturbation, the perturbation forces δT and δF_N both lie in the XZ plane of the vehicle. Hence they would generate no yawing or rolling moments.

To address the effect on pitching moment, we again draw from Equations (6.149) and Figures 6.13 and 6.16 to note that the equation for the propulsive pitching moment is given in Equation (6.156), repeated here.

$$
M_P = \mathrm{M}_{P_0} + \delta T(d_T\cos\phi_T - x_T\sin\phi_T) - \delta F_N(x_T\cos\phi_T + d_T\sin\phi_T) = C_{P_M} q_\infty S_W \bar{c}_W
$$

Note we have again used the nondimensional propulsive pitching-moment coefficient C_{P_M}.

Consistent with Equation (6.186), the <u>effect of pitch-rate perturbations on the propulsive pitching moment</u> may be expressed as

$$
\frac{\partial M_P}{\partial q}\Big|_0 = \frac{\partial C_{P_M}}{\partial q}\Big|_0 q_\infty S_W \bar{c}_W \triangleq C_{P_{M_q}} q_\infty S_W \bar{c}_W \qquad (6.194)
$$

Following the approach taken in analyzing the perturbation propulsive forces, we find that the <u>effect of pitch-rate perturbation on the propulsive pitching moment</u> may be expressed as

$$
\begin{aligned}
\frac{\partial M_P}{\partial q}\Big|_0 &= \left((d_T\cos\phi_T - x_T\sin\phi_T)\frac{\partial T}{\partial q}\Big|_0 - (x_T\cos\phi_T + d_T\sin\phi_T)\frac{\partial F_N}{\partial q}\Big|_0\right) \\
&= -x_T\begin{pmatrix} (d_T\cos\phi_T - x_T\sin\phi_T)\dfrac{\partial T}{\partial v_{\text{Axial}}}\Big|_0 \sin(\phi_T + \alpha_0) \\ +(x_T\cos\phi_T + d_T\sin\phi_T)\dfrac{\partial F_N}{\partial v_{\text{Transverse}}}\Big|_0 \cos(\phi_T + \alpha_0) \end{pmatrix}
\end{aligned}
\qquad (6.195)
$$

Typically the thrust angle ϕ_T and the reference angle of attack α_0 are small, and so simplifications of the above expressions are usually possible. Again, the partial derivatives of thrust and normal force in Equation (6.195) may be evaluated using Equations (6.108–6.112), depending on the type of propulsion system(s) installed on the vehicle. Finally, by comparing Equations (6.194) and (6.195), and assuming the angles ϕ_T and α_0 are sufficiently small, we find that

$$C_{P_{M_q}} \approx -\frac{x_T}{q_\infty S_W \bar{c}_W}\left(\frac{\partial T}{\partial v_{\text{Axial}}}\bigg|_0 d_T(\phi_T + \alpha_0) + \frac{\partial F_N}{\partial v_{\text{Transverse}}}\bigg|_0 (x_T + d_T\phi_T)\right) \quad (6.196)$$

EXAMPLE 6.7

Effects of Pitch-Rate Perturbations on the Vehicle Forces and Moments

Again consider the vehicle analyzed in Examples 6.1–6.6. Determine the effects of pitch-rate perturbations on the aerodynamic and propulsive forces and moments acting on the vehicle.

■ Solution

From Equations (6.181) we have the effect of pitch-rate perturbations on the aerodynamic forces given by

$$\frac{\partial F_{A\bullet}}{\partial q}\bigg|_0 = \frac{\partial C_\bullet}{\partial q}\bigg|_0 q_\infty S_W \triangleq C_{\bullet_q} q_\infty S_W, \; \bullet = X, Y, \text{or } Z$$

And from Equations (6.182) we have, since $\beta_0 = 0$,

$$\frac{\partial C_X}{\partial q}\bigg|_0 = -\frac{\partial C_D}{\partial q}\bigg|_0 \cos\beta_0 = -C_{D_q}$$

$$\frac{\partial C_Y}{\partial q}\bigg|_0 = -\frac{\partial C_D}{\partial q}\bigg|_0 \sin\beta_0 = 0$$

$$\frac{\partial C_Z}{\partial q}\bigg|_0 = -\frac{\partial C_L}{\partial q}\bigg|_0 = -C_{L_q}$$

Now from Equations (6.185) we have

$$C_{L_q} = C_{L_{\alpha_H}} \frac{q_H}{q_\infty}\frac{S_H}{S_W}\frac{(X_{\text{Ref}} - X_{\text{AC}_H})}{U_0} = 4.19(0.9)\left(\frac{42}{169}\right)\frac{(2.5(7.5) - 0.5(5.825))}{0.2a} = \frac{74.2}{a}$$

$$C_{D_q} = \left(\frac{2C_{L_{H_0}}}{\pi A_H e_H}\right) C_{L_q} \approx 0$$

(Again note that in the above expression, X locations are defined as positive forward, and since X_{Ref} and X_{AC_H} were given in the previous examples as distances aft of the wing apex, their signs were reversed.) Therefore, the effects of pitch-rate perturbations on the aerodynamic forces are given by

$$\frac{\partial F_{A_X}}{\partial q}\bigg|_0 = C_{X_q} q_\infty S_W \approx 0, \quad \frac{\partial F_{A_Y}}{\partial q}\bigg|_0 = C_{Y_q} q_\infty S_W = 0$$

$$\frac{\partial F_{A_Z}}{\partial q}\bigg|_0 = C_{Z_q} q_\infty S_W = -C_{L_q} q_\infty S_W = -\frac{74.2}{a} q_\infty S_W$$

Concerning the effects of pitch-rate perturbations on the aerodynamic pitching moment, Equation (6.186) states that

$$\frac{\partial M_A}{\partial q}\Big|_0 = \frac{\partial C_M}{\partial q}\Big|_0 q_\infty S_W \bar{c}_W \triangleq C_{M_q} q_\infty S_W \bar{c}_W$$

while the effects on rolling and yawing moments are negligible. But from Equation (6.188) we have

$$C_{M_q} = -C_{L_{\alpha_H}} \frac{\left(X_{\text{Ref}} - X_{\text{AC}_H}\right)^2}{U_0 \bar{c}_W} \frac{q_H}{q_\infty} \frac{S_H}{S_W}$$

$$\approx -(4.19)\frac{\left(2.5(7.5) - 0.5(5.825)\right)^2}{0.2a(5.825)}(0.9)\left(\frac{42}{169}\right)$$

$$= -(1.02)(4.19)\frac{250.8}{1.17a}(0.9)\left(\frac{42}{169}\right) = -\frac{204.9}{a}$$

(Again note the sign reversals on X_{Ref} and X_{AC_H}.) Therefore, the effect of pitch-rate perturbations on the aerodynamic pitching moment is

$$\frac{\partial M_A}{\partial q}\Big|_0 = C_{M_q} q_\infty S_W \bar{c}_W = -\frac{204.9}{a} q_\infty S_W \bar{c}_W$$

Note that the effect of q perturbation on pitching moment, or pitch damping, is significantly larger than the effect on the vertical force.

Turning now to the effects of pitch-rate perturbations on the propulsive forces, from Equations (6.192) we have (with $\phi_T = d_T = 0$ and $\alpha_0 = 2$ deg)

$$\frac{\partial F_{P_X}}{\partial q}\Big|_0 \approx 0, \qquad \frac{\partial F_{P_Y}}{\partial q}\Big|_0 = 0$$

$$\frac{\partial F_{P_Z}}{\partial q}\Big|_0 \approx -\frac{\partial F_N}{\partial v_{\text{Transverse}}}\Big|_0 x_T$$

where only the more significant terms have been included. Here, x_T is $2\bar{c}_W$, and the propulsive normal-force gradient with respect to transverse flow velocity must be determined from the engine test data.

Finally, with respect to the propulsive moments, the effect of pitch-rate perturbations on pitching moment is given by Equation (6.195), or

$$\frac{\partial M_P}{\partial q}\Big|_0 \approx -\frac{\partial F_N}{\partial v_{\text{Transverse}}}\Big|_0 x_T^2 = x_T \frac{\partial F_{P_Z}}{\partial q}\Big|_0 = 2\bar{c}_W \frac{\partial F_{P_Z}}{\partial q}\Big|_0$$

So the effect of pitch-rate perturbations on propulsive pitching moment is also significantly larger (by approximately an order of magnitude) than the effect on the propulsive force. The effects of pitch-rate perturbations on rolling and yawing moments are negligible, given vehicle symmetry about its XZ plane.

6.7.2 Roll-Rate Perturbation *p*

As noted in Section 6.7, a perturbation in the vehicle's roll rate induces a change in the local angle of attack of a 2-D airfoil section at a given span location on all lifting surfaces. This change in local angle of attack arises due to the change in local velocity depicted schematically in Figure 6.20. Here, the wing or horizontal tail is viewed from behind, and a positive roll rate is shown. This induced local velocity leads to an increase in local angle of attack along the right span as expressed in Equation (6.179), repeated here.

$$\Delta\alpha(y) = p\frac{y}{U_0}$$

Along the left span, local angle of attack would be decreased. For the vertical tail, the change in local angle of attack of a section is given by Equation (6.180), also repeated here.

$$\Delta\alpha(z) = -p\frac{z_V}{U_0}$$

The change in local angle of attack given in Equation (6.179) or (6.180) leads to a change in 2-D section lift at span location y for wing or horizontal tail, or z_V for the vertical tail. Expressed in terms of the local angle-of-attack 2-D lift effectiveness c_{l_α}, this change in lift is

$$\Delta c_l(\bullet) = c_{l_\alpha}(\bullet)\Delta\alpha(\bullet)q(\bullet)c(\bullet),\ \bullet = y \text{ or } z_V \tag{6.197}$$

For the wing and horizontal and vertical tails, respectively, let us write this change in 2-D section lift as

$$\begin{aligned}
\Delta c_{l_W}(y) &= c_{l_{\alpha_W}}(y)\Delta\alpha(y)q_W c_W(y) = \frac{py}{U_0}c_{l_{\alpha_W}}(y)q_\infty c_W(y)\\
\Delta c_{l_H}(y) &= c_{l_{\alpha_H}}(y)\Delta\alpha(y)q_H c_H(y) = \frac{py}{U_0}c_{l_{\alpha_H}}(y)q_H c_H(y)\\
\Delta c_{l_V}(z_V) &= c_{l_{\alpha_V}}(z_V)\Delta\alpha(z_V)q_H c_V(z_V) = -\frac{pz_V}{U_0}c_{l_{\alpha_V}}(z_V)q_H c_V(z_V)
\end{aligned} \tag{6.198}$$

where we have used the free-stream dynamic pressure for the change in wing section lift. In the third equation above, we have taken "lift" to be the same as side force.

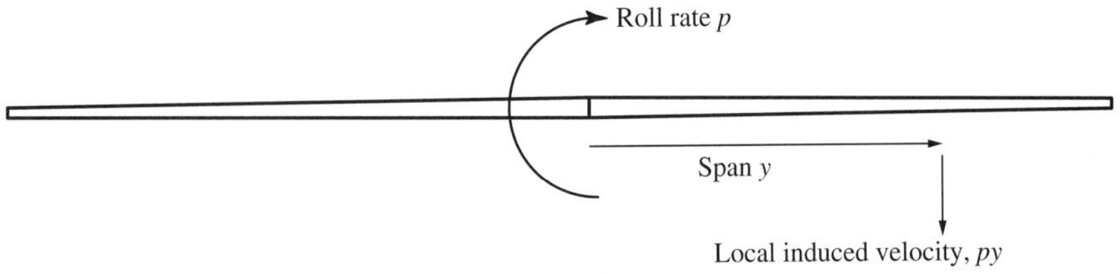

Figure 6.20 Induced local velocity along wing due to roll rate.

Due to vehicle symmetry about its XZ plane, the net change in lift for the entire wing or horizontal tail will be effectively zero. However, the change in lift on the wing and both tails will generate a change in aerodynamic rolling moment, and the vertical tail will also generate a side force. These force and moments may be estimated by appropriately integrating the force or moment due to the changes in lift given in Equations (6.198) over the wing and tail spans.

Regarding the side force generated by the vertical tail, we may write this side force as

$$\begin{aligned}\Delta S_V &= \int_{z_{V_{\text{Root}}}}^{z_{V_b}} c_{l_{\alpha_V}}(z_V)\Delta\alpha(z_V)q_H c_V(z_V)dz_V \\ &= -\frac{q_H p}{U_0}\int_{z_{V_{\text{Root}}}}^{z_{V_b}} c_{l_{\alpha_V}}(z_V)z_V c_V(z_V)dz_V = \Delta C_S q_H S_V\end{aligned} \tag{6.199}$$

But we will express the last integral term as

$$\int_{z_{V_{\text{Root}}}}^{z_{V_b}} c_{l_{\alpha_V}}(z_V)z_V c_V(z_V)dz_V \approx C_{L_{\alpha_V}}\int_{z_{V_{\text{Root}}}}^{z_{V_b}} z_V c_V(z_V)dz_V = C_{L_{\alpha_V}} Z_{S_{\text{MAC}}} S_V \tag{6.200}$$

where

$$Z_{S_{\text{MAC}}} \triangleq \frac{1}{S_V}\int_{z_{V_{\text{Root}}}}^{z_{V_b}} z_V c_V(z_V)dz_V$$

is the vertical distance (in the negative Z_S direction) from the X_S (stability) axis to the span-wise location of the MAC of the vertical tail. Consequently, using Equations (6.113), (6.114), (6.199), and (6.200), and assuming small reference sideslip β_0, we find that the effects of roll-rate perturbation on the aerodynamic forces are given as

$$\begin{aligned}\frac{\partial F_{A_X}}{\partial p}\Big|_0 &= \frac{\partial C_X}{\partial p}\Big|_0 q_\infty S_W = -\frac{\partial C_S}{\partial p}\Big|_0 \sin\beta_0 q_\infty S_W \triangleq -C_{S_p}\sin\beta_0 q_\infty S_W \approx 0 \\ \frac{\partial F_{A_Y}}{\partial p}\Big|_0 &= \frac{\partial C_Y}{\partial p}\Big|_0 q_\infty S_W = \frac{\partial C_S}{\partial p}\Big|_0 \cos\beta_0 q_\infty S_W \triangleq C_{S_p}\cos\beta_0 q_\infty S_W \\ &\approx -\left(C_{L_{\alpha_V}}\frac{Z_{S_{\text{MAC}}}}{U_0}\frac{q_H}{q_\infty}\frac{S_V}{S_W}\right)q_\infty S_W\end{aligned} \tag{6.201}$$

with obvious definition for C_{S_p}.

Concerning the effect of roll rate on the vehicle's aerodynamic rolling and yawing moments, let the changes in moments due to roll rate generated by the three lifting surfaces be denoted as ΔL_{A_W}, ΔL_{A_H}, ΔL_{A_V}, and ΔN_{A_V}. Using Equations (6.198) and integrating the moments generated by the section lift over the spans of each lifting surface we have

$$
\begin{aligned}
\Delta L_{A_W} &= -\int_{-b_W/2}^{b_W/2} c_{l_{\alpha_W}}(y)\Delta\alpha_W(y)q_\infty c_W(y)y\,dy = -\frac{q_\infty p}{U_0}\int_{-b_W/2}^{b_W/2} c_{l_{\alpha_W}}(y)y^2 c_W(y)\,dy \\
\Delta L_{A_H} &= -\int_{-b_H/2}^{b_H/2} c_{l_{\alpha_H}}(y)\Delta\alpha_H(y)q_H c_H(y)y\,dy = -\frac{q_H p}{U_0}\int_{-b_H/2}^{b_H/2} c_{l_{\alpha_H}}(y)y^2 c_H(y)\,dy \\
\Delta L_{A_V} &= \int_{z_{V_{\text{Root}}}}^{z_{V_b}} c_{l_{\alpha_V}}(z_V)\Delta\alpha_V(z_V)q_H c_V(z_V)z_V\,dz_V = -\frac{q_H p}{U_0}\int_{z_{V_{\text{Root}}}}^{z_{V_b}} c_{l_{\alpha_V}}(z_V)z_V^2 c_V(z_V)\,dz_V \\
\Delta N_{A_V} &= -\int_{z_{V_{\text{Root}}}}^{z_{b_V}} c_{l_{\alpha_V}}(z_V)\Delta\alpha_V(z_V)q_H c_V(z_V)\left(X_{\text{Ref}} - x_{\text{AC}_V}(z_V)\right)dz_V \\
&\approx \frac{q_H p}{U_0}\left(X_{\text{Ref}} - X_{\text{AC}_V}\right)\int_{z_{V_{\text{Root}}}}^{z_{b_V}} c_{l_{\alpha_V}}(z_V)z_V c_V(z_V)\,dz_V
\end{aligned}
\tag{6.202}
$$

So the total change in rolling and yawing moments for the vehicle are

$$
\begin{aligned}
\Delta L_A &= \Delta L_{A_W} + \Delta L_{A_H} + \Delta L_{A_V} = \Delta C_{L_{\text{Roll}}} q_\infty S_W b_W \\
\Delta N_A &= \Delta N_{A_V} = \Delta C_N q_\infty S_W b_W
\end{aligned}
\tag{6.203}
$$

where we have introduced the changes in rolling- and yawing-moment coefficients $\Delta C_{L_{\text{Roll}}}$ and ΔC_N.

Note that for the wing and horizontal tail surfaces, the integral terms in Equation (6.202) may be expressed as

$$
\int_{-b_\bullet/2}^{b_\bullet/2} c_{l_{\alpha_\bullet}}(y)y^2 c_\bullet(y)\,dy \approx C_{L_{\alpha_\bullet}}\int_{-b_\bullet/2}^{b_\bullet/2} y^2 c_\bullet(y)\,dy = C_{L_{\alpha_\bullet}}\widetilde{Y}_\bullet S_\bullet, \quad \bullet = W \text{ or } H \tag{6.204}
$$

with

$$
\widetilde{Y}_\bullet \triangleq \frac{1}{S_\bullet}\int_{-b_\bullet/2}^{b_\bullet/2} y^2 c_\bullet(y)\,dy = \frac{2}{S_\bullet}\int_0^{b_\bullet/2} y^2 c_\bullet(y)\,dy
$$

Additionally, $C_{L_{\alpha_\bullet}}$ is the 3-D angle-of-attack lift effectiveness and $c_\bullet(y)$ is the span-wise chord distribution along the wing or tail. Likewise, for the vertical tail we can write

$$\int_{z_{V_{\mathrm{Root}}}}^{z_{V_b}} c_{l_{\alpha_V}}(z_V) z_V^2 c_V(z_V) dz_V \approx C_{L_{\alpha_V}} \int_{z_{V_{\mathrm{Root}}}}^{z_{V_b}} z_V^2 c_V(z_V) dz_V = C_{L_{\alpha_V}} \widetilde{Z}_S S_V \tag{6.205}$$

with

$$\widetilde{Z}_S \triangleq \frac{1}{S_V} \int_{z_{V_{\mathrm{Root}}}}^{z_{V_b}} z_V^2 c_V(z_V) dz_V$$

Here note that the term $\widetilde{Z}_S$ is a function of the height of the vertical-tail sections above the X_S axis. For sections nearer the tail's root chord this height can become negative when the vehicle reference flight condition involves a large reference angle of attack α_0.

Using these facts, and by substituting Equations (6.200) and (6.202) into Equations (6.203) and solving for the changes in the coefficients due to roll rate, we find that

$$\begin{aligned}
\Delta C_{L_{\mathrm{Roll}}} &= -\left(C_{L_{\alpha_W}} \widetilde{Y}_{\mathrm{MAC}_W} + C_{L_{\alpha_H}} \widetilde{Y}_{\mathrm{MAC}_H} \frac{q_H}{q_\infty} \frac{S_H}{S_W} + C_{L_{\alpha_V}} \widetilde{Z}_{S_{\mathrm{MAC}}} \frac{q_H}{q_\infty} \frac{S_V}{S_W} \right) \frac{p}{\mathrm{U}_0 b_W} \\
\Delta C_N &= \frac{p}{\mathrm{U}_0} \frac{(X_{\mathrm{Ref}} - X_{\mathrm{AC}_V})}{b_W} C_{L_{\alpha_V}} Z_{S_{\mathrm{MAC}}} \frac{q_H}{q_\infty} \frac{S_V}{S_W}
\end{aligned} \tag{6.206}$$

So the effects of roll-rate perturbation on aerodynamic rolling and yawing moments may be quantified by the following partial derivatives

$$\begin{aligned}
\frac{\partial L_A}{\partial p}\Big|_0 &= \frac{\partial C_{L_{\mathrm{Roll}}}}{\partial p}\Big|_0 q_\infty S_W b_W \triangleq C_{L_p} q_\infty S_W b_W \\
&= \left(-\frac{1}{\mathrm{U}_0 b_W} \left(C_{L_{\alpha_W}} \widetilde{Y}_{\mathrm{MAC}_W} + C_{L_{\alpha_H}} \widetilde{Y}_{\mathrm{MAC}_H} \frac{q_H}{q_\infty} \frac{S_H}{S_W} \right.\right. \\
&\qquad \left.\left. + C_{L_{\alpha_V}} \widetilde{Z}_{S_{\mathrm{MAC}}} \frac{q_H}{q_\infty} \frac{S_V}{S_W} \right) \right) q_\infty S_W b_W \\
\frac{\partial N_A}{\partial p}\Big|_0 &= \frac{\partial C_N}{\partial p}\Big|_0 q_\infty S_W b_W \triangleq C_{N_p} q_\infty S_W b_W \\
&= \left(\frac{(X_{\mathrm{Ref}} - X_{\mathrm{AC}_V})}{\mathrm{U}_0 b_W} C_{L_{\alpha_V}} Z_{S_{\mathrm{MAC}}} \frac{q_H}{q_\infty} \frac{S_V}{S_W} \right) q_\infty S_W b_W
\end{aligned} \tag{6.207}$$

with obvious definitions for C_{L_p} and C_{N_p}. Similar to the case with pitching moment, C_{L_p} is called the *roll-damping coefficient.* We will see in Chapter 10 that it is also an important parameter with regard to the vehicle dynamics.

The roll-rate perturbations may also affect the rolling moment generated by the propulsive devices if these devices are not located near the XZ plane of the vehicle. The propulsive forces and other moments are not appreciably affected, due to the assumed symmetry of the installed propulsive devices with respect to the XZ plane. Here we will only address the case in which the vehicle has engines located at span-wise locations y off the XY plane of the vehicle, and we will only address the effect of roll-rate perturbations on propulsive rolling moment.

We again draw from Equations (6.149) and Figures 6.13, 6.14, and 6.15 to find that the propulsive rolling moment may be expressed as

$$L_P = -\sum_{i=1}^{n_P} y_i\left(F_{N_i}\cos(\phi_{T_i} + \alpha_0) + T_i\sin(\phi_{T_i} + \alpha_0)\right) = C_{P_{L_{\text{Roll}}}} q_\infty S_W b_W \quad (6.208)$$

where

n_P = number of installed propulsive devices
y_i = distance of the i'th propulsive device from the XZ plane (positive along right wing)

Consistent with Equation (6.207), we may express the effect of roll-rate perturbations on propulsive rolling moment as

$$\boxed{\frac{\partial L_P}{\partial p}\Big|_0 = \frac{\partial C_{P_{L_{\text{Roll}}}}}{\partial p}\Big|_0 q_\infty S_W b_W \triangleq C_{P_{L_p}} q_\infty S_W b_W} \quad (6.209)$$

Now from Equation (6.179) the change in local angle of attack due to roll rate was given as

$$\Delta\alpha(y) = p\frac{y}{\mathrm{U}_0}$$

And this change in local angle of attack gives rise to the following changes in the propulsion-inlet flow velocity and direction.

$$\begin{aligned} \Delta v_{\text{Axial}}(y) &= -\Delta\alpha(y)\mathrm{U}_0\sin(\phi_T + \alpha_0) = -py\sin(\phi_T + \alpha_0) \\ \Delta v_{\text{Transverse}}(y) &= \Delta\alpha(y)\mathrm{U}_0\cos(\phi_T + \alpha_0) = py\cos(\phi_T + \alpha_0) \end{aligned} \quad (6.210)$$

So from Equation (6.208) we can see that the change in propulsive rolling moment due to roll rate may be written as

$$\frac{\partial L_P}{\partial p} = -\sum_{i=1}^{n_P} y_i\left(\frac{\partial F_{N_i}}{\partial p}\cos(\phi_{T_i} + \alpha_0) + \frac{\partial T_i}{\partial p}\sin(\phi_{T_i} + \alpha_0)\right) \quad (6.211)$$

But from Equation (6.210) we can write

$$\begin{aligned} \frac{\partial F_{N_i}}{\partial p} &= \frac{\partial F_{N_i}}{\partial v_{\text{Transverse}}}\frac{\partial v_{\text{Transverse}}}{\partial p} = \frac{\partial F_{N_i}}{\partial v_{\text{Transverse}}} y_i\cos(\phi_{T_i} + \alpha_0) \\ \frac{\partial T_i}{\partial p} &= \frac{\partial T_i}{\partial v_{\text{Axial}}}\frac{\partial v_{\text{Axial}}}{\partial p} = -\frac{\partial T_i}{\partial v_{\text{Axial}}} y_i\sin(\phi_{T_i} + \alpha_0) \end{aligned} \quad (6.212)$$

Therefore, we may express the effect of roll rate on the propulsive rolling moment as

$$\begin{aligned}\frac{\partial L_P}{\partial p} &= -\sum_{i=1}^{n_P} y_i^2\left(\frac{\partial F_{N_i}}{\partial v_{\text{Transverse}}}\cos^2(\phi_{T_i}+\alpha_0) - \frac{\partial T_i}{\partial v_{\text{Axial}}}\sin^2(\phi_{T_i}+\alpha_0)\right)\\ &\approx -\sum_{i=1}^{n_P} y_i^2\frac{\partial F_{N_i}}{\partial v_{\text{Transverse}}}\end{aligned} \tag{6.213}$$

Again, the partial derivative of normal force in the above equation may be evaluated using Equations (6.108–6.112), depending on the type of propulsion system(s) installed on the vehicle. Finally, by comparing Equations (6.209) and (6.213), and assuming the angle $(\phi_T + \alpha_0)$ is sufficiently small, we find that

$$C_{P_{L_p}} \approx -\frac{1}{q_\infty S_W b_W}\sum_{i=1}^{n_P}\frac{\partial F_{N_i}}{\partial v_{\text{Transverse}}} y_i^2 \tag{6.214}$$

6.7.3 Yaw-Rate Perturbation *r*

As noted in Section 6.7, a perturbation in the vehicle's yaw rate induces a change in the local angle of attack at the vertical tail, as well as a change in the local dynamic pressure on the wing and horizontal tail. The change in local angle of attack on the vertical tail arises due to the change in local velocity depicted schematically in Figure 6.19, and this change in angle of attack, given by Equation (6.177) repeated here, is

$$\Delta\alpha_V \approx \frac{r}{U_0}(X_{\text{Ref}} - X_{AC_V})$$

In addition, a yaw rate induces a change in the local surge velocity at the 2-D airfoil section of the wing or horizontal tail located at span y. For a positive yaw rate this change in velocity is negative along the right span, and positive along the left span, and is given in Equation (6.178) repeated here.

$$\Delta u_{W \text{ or } H}(y) = -ry$$

This change in velocity, of course, leads to a change in the local dynamic pressure, which may be written as

$$\Delta q(y) = \frac{\partial q_\infty}{\partial u}\Big|_0 \Delta u(y) = -\rho_\infty U_0 r y \tag{6.215}$$

Following an approach similar to that above when dealing with roll rate, the change in side force on the vertical tail due to yaw rate may be expressed as

$$\Delta S_V = C_{L_{\alpha_V}}\Delta\alpha_V q_H S_V = C_{L_{\alpha_V}}\frac{r}{U_0}(X_{\text{Ref}} - X_{AC_V})q_H S_V \tag{6.216}$$

Therefore, we can expressed the effects of yaw rate on the aerodynamic forces for small β_0 as

$$\frac{\partial F_{A_X}}{\partial r}\Big|_0 = \frac{\partial C_X}{\partial r}\Big|_0 q_\infty S_W = -\frac{\partial C_S}{\partial r}\Big|_0 \sin\beta_0 q_\infty S_W \triangleq -C_{S_r}\sin\beta_0 q_\infty S_W \approx 0$$
$$\frac{\partial F_{A_Y}}{\partial r}\Big|_0 = \frac{\partial C_Y}{\partial r}\Big|_0 q_\infty S_W = C_{S_r}\cos\beta_0 q_\infty S_W \approx \left(C_{L_{\alpha_V}}\frac{(X_{\text{Ref}} - X_{\text{AC}_V})}{U_0}\frac{q_H}{q_\infty}\frac{S_V}{S_W}\right) q_\infty S_W \tag{6.217}$$

with obvious definition for C_{S_r}.

The change in side force generated by the vertical tail, discussed above, also generates a change in yawing moment. This change in yawing moment may be expressed as

$$\Delta N_V = -\Delta S_V(X_{\text{Ref}} - X_{\text{AC}_V}) = -C_{L_{\alpha_V}}\frac{r}{U_0}(X_{\text{Ref}} - X_{\text{AC}_V})^2 q_H S_V \tag{6.218}$$

Consequently, we may write the effect of yaw rate on vehicle aerodynamic yawing moment as

$$\frac{\partial N_A}{\partial r}\Big|_0 = \frac{\partial C_N}{\partial r}\Big|_0 q_\infty S_W b_W \triangleq C_{N_r} q_\infty S_W b_W$$
$$= \left(-C_{L_{\alpha_V}}\frac{(X_{\text{Ref}} - X_{\text{AC}_V})^2}{U_0 b_W}\frac{q_H}{q_\infty}\frac{S_V}{S_W}\right) q_\infty S_W b_W \tag{6.219}$$

with obvious definition for C_{N_r}. This coefficient is called the *yaw-damping coefficient,* and like the pitch-damping and roll-damping coefficients it is an important parameter with regard to the vehicle's dynamics.

In addition, the antisymmetric changes in local dynamic pressure along the spans of the wing and horizontal tail due to yaw-rate perturbations induce a change in the local lift and drag at each 2-D section. The antisymmetric changes in lift generate a vehicle rolling moment, while the changes in drag generate a yawing moment. But we will assume here that this yawing moment is small compared to that generated by the vertical tail, as expressed in Equation (6.218), and it will be neglected.

Let us write the change in rolling moment in terms of the local 2-D section lift, and integrate over the spans of the wing and tail. That is, for the wing or horizontal tail we may write

$$
\begin{aligned}
\Delta L_{A\bullet} &= -\int_{-b_\bullet/2}^{b_\bullet/2} c_l(y)\Delta q(y) c_\bullet(y) y dy \\
&= \rho_\infty U_0 r \int_{-b_\bullet/2}^{b_\bullet/2} c_{l_\alpha}(y)\big(\alpha_\bullet + \varepsilon_{T_\bullet}(y) - \alpha_{0_\bullet}(y)\big)c_\bullet(y) y^2 dy \\
&\approx \rho_\infty U_0 r C_{L_{\alpha_\bullet}}\left(\alpha_\bullet \int_{-b_\bullet/2}^{b_\bullet/2} c_\bullet(y) y^2 dy + \int_{-b_\bullet/2}^{b_\bullet/2} \big(\varepsilon_{T_\bullet}(y) - \alpha_{0_\bullet}(y)\big)c_\bullet(y) y^2 dy\right) \\
&= \rho_\infty U_0 r C_{L_{\alpha_\bullet}}(\alpha_\bullet \widetilde{Y}_\bullet + \widetilde{E}_\bullet) S_\bullet = \frac{\rho_\infty U_0}{q_\infty} r C_{L_{\alpha_\bullet}} \frac{(\alpha_\bullet \widetilde{Y}_\bullet + \widetilde{E}_\bullet)}{b_\bullet} q_\infty S_\bullet b_\bullet, \bullet = W \text{ or } H
\end{aligned}
\tag{6.220}
$$

where

$$
\widetilde{E}_\bullet \triangleq \frac{1}{S_\bullet} \int_{-b_\bullet/2}^{b_\bullet/2} \big(\varepsilon_{T_\bullet}(y) - \alpha_{0_\bullet}(y)\big)c_\bullet(y) y^2 dy, \bullet = W \text{ or } H
$$

$\widetilde{Y}_\bullet$ as defined in Equation (6.204)

Now let the change in vehicle aerodynamic rolling moment be written as the following sum

$$
\Delta L_A = \Delta L_{A_W} + \Delta L_{A_H} = \Delta C_{L_{\text{Roll}}} q_\infty S_W b_W \tag{6.221}
$$

and assuming the horizontal tail has no span-wise twist and that $\alpha_{0_H}(y) = 0$, we have the effect of yaw-rate perturbation on vehicle aerodynamic rolling moment given by

$$
\begin{aligned}
\frac{\partial L_A}{\partial r}\Big|_0 &= \frac{\partial C_{L_{\text{Roll}}}}{\partial r}\Big|_0 q_\infty S_W b_W \triangleq C_{L_r} q_\infty S_W b_W \\
&= \frac{\rho_\infty U_0}{q_\infty}\left(C_{L_{\alpha_W}} \frac{(\alpha_0 + i_W)\widetilde{Y}_W + \widetilde{E}_W}{b_W} + C_{L_{\alpha_H}} \frac{\left(\alpha_0\left(1 - \frac{d\varepsilon}{d\alpha}\right) + i_H - \frac{d\varepsilon}{d\alpha}(i_W - \alpha_{0_W})\right)\widetilde{Y}_H}{b_W} \frac{S_H}{S_W}\right) q_\infty S_W b_W
\end{aligned}
\tag{6.222}
$$

with obvious definition for C_{L_r}. Here α_0 is the vehicle angle of attack at the reference condition.

Attention now turns to the effects of yaw-rate perturbations on the propulsive forces and moments. Recall that the total propulsive force expressed in terms of its components in the stability axes was given in Equation (6.131). Referring to Figure 6.13, note that it has been assumed that the propulsive devices are symmetric with respect to the XZ plane of the vehicle. Following the development in Section 6.6.3 dealing with lateral-velocity perturbations v, we can write the components of the total propulsive force in terms of their reference values plus perturbations in the thrust δT and normal force on the engine δF_N.

If the engines happen to be located far from the XZ plane of the vehicle, along the wing, for example, a yaw rate could also affect the axial-flow velocity into the engine and therefore the thrust. But since this effect would be symmetric with respect to the vehicle XZ plane, the net effect on the propulsive forces acting on the vehicle will be negligible.

Since we're dealing with only yaw rate, we have from Equations (6.165),

$$F_{P_X} = F_{P_{X_0}}$$

$$F_{P_Y} = -\delta F_N = C_{P_Y} q_\infty S_W$$

$$F_{P_Z} = F_{P_{X_0}}$$

So the effect of yaw-rate perturbation on the lateral propulsive force is expressed as

$$\boxed{\frac{\partial F_{P_Y}}{\partial r}\Big|_0 = \frac{\partial C_{P_Y}}{\partial r}\Big|_0 q_\infty S_W \triangleq C_{P_{Y_r}} q_\infty S_W} \tag{6.223}$$

The propulsive force perturbations may now be expressed in terms of yaw rate, or

$$\begin{aligned}\delta F_N &= \frac{\partial F_N}{\partial r} r = \frac{\partial F_N}{\partial v_{\text{Transverse}}} \frac{\partial v_{\text{Transverse}}}{\partial r} r \\ \delta T &= \frac{\partial T}{\partial r} r = \frac{\partial T}{\partial v_{\text{Axial}}} \frac{\partial v_{\text{Axial}}}{\partial r} r\end{aligned} \tag{6.224}$$

The perturbation thrust will be used later when addressing propulsive moments. Now from Figures 6.14 and 6.19 we can see that

$$\Delta v_{\text{Transverse}} = r x_T \cos\alpha_0$$

Therefore, we have

$$\frac{\partial v_{\text{Transverse}}}{\partial r} = x_T \cos\alpha_0 \tag{6.225}$$

Also, for an engine located a span-wise distance y_T from the XZ plane of the vehicle we have

$$\Delta v_{\text{Axial}} = -r y_T$$

and so

$$\frac{\partial v_{\text{Axial}}}{\partial r} = -y_T \tag{6.226}$$

Consequently, the effect of yaw rate on the lateral propulsive force is given as

$$\boxed{\frac{\partial F_{P_Y}}{\partial r}\Big|_0 = \frac{\partial F_N}{\partial v_{\text{Transverse}}}\Big|_0 x_T \cos\alpha_0} \tag{6.227}$$

The partial derivative of the normal force above may be evaluated using Equations (6.111) or (6.112), depending on the type of propulsion system(s) on the vehicle. Finally, by comparing Equations (6.223) and (6.227), and assuming α_0 is sufficiently small, we find that

$$\boxed{C_{P_{Y_r}} = \frac{x_T}{q_\infty S_W}\frac{\partial F_N}{\partial v_{\text{Transverse}}}\Big|_0} \tag{6.228}$$

Typically, the effect of yaw-rate perturbations on the lateral propulsive force is not large. However, the changes in moments due to this force can again be significant if the moment arm for the propulsive devices is large, or if the engines are located far off the XZ plane of the vehicle. Under the assumed geometry of the propulsive devices, a positive yaw-rate perturbation induces a perturbation normal force δF_N along the positive Y_S axis of the vehicle. Plus, the perturbation thrust δT acts asymmetrically with respect to the XZ plane of the vehicle. Hence δT can generate a yawing moment but little or no pitching moment.

To address these propulsive rolling and yawing moments, we again draw from Equations (6.149) and Figure 6.13 to write the equations for the propulsive rolling and yawing moments about the X_S and Z_S stability axes, respectively. Or, similar to that given in Equations (6.171), we have

$$\begin{aligned} L_P &= -(d_T\cos\alpha_0 + x_T\sin\alpha_0)F_N = C_{P_{L_{\text{Roll}}}} q_\infty S_W b_W \\ N_P &= -(x_T\cos\alpha_0 - d_T\sin\alpha_0)F_N - \sum_{i=1}^{n_P} y_{T_i}\delta T_i = C_{P_N} q_\infty S_W b_W \end{aligned} \tag{6.229}$$

where

i is the i'th propulsive device mounted off the XZ plane of the vehicle
n_P is the number of propulsive devices on the vehicle mounted off the XZ plane of the vehicle
y_{T_i} is the span-wise location of i'th propulsive device

Consistent with Equations (6.219) and (6.222), we may now express the effects of yaw-rate perturbations on the propulsive rolling and yawing moments as

$$\boxed{\begin{aligned} \frac{\partial L_P}{\partial r}\Big|_0 &= \frac{\partial C_{P_{L_{\text{Roll}}}}}{\partial r}\Big|_0 q_\infty S_W b_W \triangleq C_{P_{L_r}} q_\infty S_W b_W \\ \frac{\partial N_P}{\partial r}\Big|_0 &= \frac{\partial C_{P_N}}{\partial r}\Big|_0 q_\infty S_W b_W \triangleq C_{P_{N_r}} q_\infty S_W b_W \end{aligned}} \tag{6.230}$$

Following the approach taken previously in analyzing the propulsive forces, we find that the effects of yaw-rate perturbation on the propulsive moments may be expressed as

$$\begin{aligned}
\frac{\partial L_P}{\partial r}\Big|_0 &= -(d_T\cos\alpha_0 + x_T\sin\alpha_0)\frac{\partial F_N}{\partial r}\Big|_0 \\
&= -(d_T\cos\alpha_0 + x_T\sin\alpha_0)\frac{\partial F_N}{\partial v_{\text{Transverse}}}x_T\cos\alpha_0 \\
\frac{\partial N_P}{\partial r}\Big|_0 &= -(x_T\cos\alpha_0 - d_T\sin\alpha_0)\frac{\partial F_N}{\partial r}\Big|_0 - \sum_{i=1}^{n_P} y_{T_i}\frac{\partial T_i}{\partial r} \\
&= -(x_T\cos\alpha_0 - d_T\sin\alpha_0)\frac{\partial F_N}{\partial v_{\text{Transverse}}}x_T\cos\alpha_0 + \sum_{i=1}^{n_P} y_{T_i}^2\frac{\partial T_i}{\partial v_{\text{Axial}}}
\end{aligned} \tag{6.231}$$

Typically the reference angle of attack α_0 is small, and the expressions in Equations (6.231) may be simplified accordingly. And if the engines are mounted close to the vehicle's XY plane, the summation in the second of the above equations is negligible. The partial derivatives of thrust and normal force in Equations (6.231) may be evaluated using Equations (6.108–6.112), depending on the type of propulsion system(s) installed on the vehicle. Finally, by comparing Equations (6.230) and (6.231), and assuming α_0 is sufficiently small, we find that the yaw-rate propulsive-moment effectivenesses are

$$\begin{aligned}
C_{P_{L_r}} &= -\frac{x_T(d_T + x_T\alpha_0)}{q_\infty S_W b_W}\frac{\partial F_N}{\partial v_{\text{Transverse}}} \\
C_{P_{N_r}} &= -\frac{1}{q_\infty S_W b_W}\left(x_T(x_T - d_T\alpha_0)\frac{\partial F_N}{\partial v_{\text{Transverse}}} + \sum_{i=1}^{n_P} y_{T_i}^2\frac{\partial T_i}{\partial v_{\text{Axial}}}\right)
\end{aligned} \tag{6.232}$$

6.7.4 Perturbation in Rate of Change of Angle of Attack $\dot{\alpha}$

We will consider next the effects of the rate of change of free-stream flow direction, which may expressed in terms of the rates of change of angle of attack and sideslip angle, or $\dot{\alpha}$ and $\dot{\beta}$, respectively. These effects are usually related to, but fundamentally different from, the effects of pitch rate and yaw rates. A change in flow direction may arise even though the vehicle is not initially undergoing an angular rate. For example, imagine the effects of a sudden vertical or lateral wind gust on a vehicle in straight and level flight. Such gusts would change the local flow direction over the vehicle's lifting surfaces.

If the rate of change of angle of attack is slow, compared to the rate at which the pressure distribution over the lifting surfaces stabilizes, the primary effect of a change in flow direction is due to a lag of downwash ε. (Since side wash has been ignored in this chapter, due to its typically small effect, here we will only address downwash.) A lag in downwash occurs when the angle of attack of a forward lifting surface (e.g., wing) changes. This lift change leads to a change in downwash at the trailing edge of the wing. But the effect of this change in downwash on a trailing lifting surface (e.g., horizontal tail) will be encountered some time Δt later, when the flow from the wing reaches the tail surface.

The time lag Δt between a change in wing angle of attack and the corresponding change in downwash encountered at the horizontal tail may be expressed in terms of the distance between the wing and tail and the vehicle's velocity relative to the air mass. Or, assuming X locations are positive forward on the vehicle, we have

$$\Delta t \approx \frac{\left(X_{\mathrm{AC}_W} - X_{\mathrm{AC}_H}\right)}{\mathrm{U}_0} \tag{6.233}$$

As a result of this change in downwash at the tail, let us write the angle of attack at the tail at time $(t + \Delta t)$ as

$$\begin{aligned} \alpha_H(t+\Delta t) &= \alpha(t+\Delta t) + i_H - \varepsilon_H(t) \\ &= \alpha(t+\Delta t) + i_H - \Big(\varepsilon_H(t+\Delta t) - \big(\varepsilon_H(t+\Delta t) - \varepsilon_H(t)\big)\Big) \\ &= \alpha(t+\Delta t) + i_H - \big(\varepsilon_H(t+\Delta t) - \Delta\varepsilon_H\big) \end{aligned} \tag{6.234}$$

or

$$\alpha_H = \alpha + i_H - \varepsilon_H + \Delta\varepsilon_H$$

where $\varepsilon_H(t)$ is the downwash at the tail prior to experiencing the effect of the wing's change in angle of attack. But the change in downwash $\Delta\varepsilon_H$ in Equation (6.234) may be expressed as

$$\Delta\varepsilon_H \approx \frac{d\varepsilon_H}{d\alpha_W}\Delta\alpha_W = \frac{d\varepsilon_H}{d\alpha_W}\dot{\alpha}\Delta t = \frac{d\varepsilon_H}{d\alpha_W}\dot{\alpha}\frac{\left(X_{\mathrm{AC}_W} - X_{\mathrm{AC}_H}\right)}{\mathrm{U}_0} \tag{6.235}$$

So the change in the angle of attack at the tail $\Delta\alpha_H$ due to a lag of downwash is given by Equation (6.235). Further, note from Equation (6.234) that a positive $\Delta\varepsilon_H$ corresponds to an increase in the angle of attack of the horizontal tail at $(t + \Delta t)$.

The change in the horizontal-tail lift corresponding to this change in downwash is therefore

$$\begin{aligned} \Delta L_H &= \Delta C_{L_H} q_H S_H = C_{L_{\alpha_H}} \Delta\alpha_H q_H S_H \\ &= C_{L_{\alpha_H}} \Delta\varepsilon_H q_H S_H = C_{L_{\alpha_H}} \frac{d\varepsilon_H}{d\alpha_W}\dot{\alpha}\frac{\left(X_{\mathrm{AC}_W} - X_{\mathrm{AC}_H}\right)}{\mathrm{U}_0} q_H S_H \end{aligned} \tag{6.236}$$

So the change in vehicle lift, assuming $\Delta L = \Delta L_H$, is given by

$$\Delta L = \frac{\partial C_L}{\partial\dot{\alpha}}\dot{\alpha}q_\infty S_W \triangleq C_{L_{\dot{\alpha}}}\dot{\alpha}q_\infty S_W = \left(C_{L_{\alpha_H}}\frac{d\varepsilon_H}{d\alpha_W}\frac{\left(X_{\mathrm{AC}_W} - X_{\mathrm{AC}_H}\right)}{\mathrm{U}_0}\frac{q_H}{q_\infty}\frac{S_H}{S_W}\right)\dot{\alpha}q_\infty S_W \tag{6.237}$$

with obvious definition for $C_{L_{\dot{\alpha}}}$.

Using Equations (6.113) and (6.114), and assuming the effects of $\dot{\alpha}$ on drag and side force are negligible, we may then write the effects of the rate of change of angle of attack on the aerodynamic forces as

$$\boxed{\begin{aligned} &\frac{\partial F_{A_X}}{\partial \dot{\alpha}}\bigg|_0 = 0, \qquad \frac{\partial F_{A_Y}}{\partial \dot{\alpha}}\bigg|_0 = 0 \\ &\frac{\partial F_{A_Z}}{\partial \dot{\alpha}}\bigg|_0 = -\frac{\partial L}{\partial \dot{\alpha}}\bigg|_0 = -C_{L_{\dot{\alpha}}} q_\infty S_W \\ &\qquad = -\left(C_{L_{\alpha_H}} \frac{d\varepsilon_H}{d\alpha_W} \frac{(X_{AC_W} - X_{AC_H})}{U_0} \frac{q_H}{q_\infty} \frac{S_H}{S_W}\right) q_\infty S_W \end{aligned}} \tag{6.238}$$

More significantly, the lag of downwash also affects the vehicle's aerodynamic pitching moment. Using Equation (6.236) for the change in the lift generated by the horizontal tail, and again assuming that X locations are measured positive forward, we may write the change in vehicle pitching moment due to the lag of downwash as

$$\begin{aligned} \Delta M_A &= \Delta C_m q_\infty S_W \bar{c}_W = -\Delta L_H (X_{Ref} - X_{AC_H}) \\ &= -\left(C_{L_{\alpha_H}} \frac{d\varepsilon_H}{d\alpha_W} \dot{\alpha} \frac{(X_{AC_W} - X_{AC_H})(X_{Ref} - X_{AC_H})}{U_0 \bar{c}_W} \frac{q_H}{q_\infty} \frac{S_H}{S_W}\right) q_\infty S_W \bar{c}_W \end{aligned} \tag{6.239}$$

Consequently, the effect of rate of change of angle on the vehicle's aerodynamic pitching moment is

$$\boxed{\begin{aligned} &\frac{\partial M_A}{\partial \dot{\alpha}}\bigg|_0 = \frac{\partial C_M}{\partial \dot{\alpha}}\bigg|_0 q_\infty S_W \bar{c}_W \triangleq C_{M_{\dot{\alpha}}} q_\infty S_W \bar{c}_W \\ &\quad = -\left(C_{L_{\alpha_H}} \frac{d\varepsilon_H}{d\alpha_W} \frac{(X_{AC_W} - X_{AC_H})(X_{Ref} - X_{AC_H})}{U_0 \bar{c}_W} \frac{q_H}{q_\infty} \frac{S_H}{S_W}\right) q_\infty S_W \bar{c}_W \end{aligned}} \tag{6.240}$$

with obvious definition for $C_{M_{\dot{\alpha}}}$. This effectiveness is always negative for vehicles with an aft tail. And, as noted previously, the effect of rate of change of angle of attack on the rolling and yawing moments are taken to be negligible.

The lag of downwash can also affect the propulsive forces and moments, since the magnitude and direction of inlet flow is being affected. Previously we noted that the change in local angle of attack due to the lag in downwash was given by the change in downwash $\Delta\varepsilon(x)$, from Equation (6.235), where in this equation, $x = X_{AC_H}$. Consequently, we can write the local change in angle of attack at the inlet of a propulsive device as

$$\Delta\alpha_{\text{inlet}} \approx \frac{d\varepsilon_{\text{inlet}}}{d\alpha_W}\dot{\alpha}\frac{\left(X_{\text{AC}_W} - X_{\text{Ref}} + x_T\right)}{\text{U}_0} \tag{6.241}$$

From Equations (6.149) we have the propulsive forces given by

$$F_{P_X} = F_{P_{X_0}} + \delta T\cos(\phi_T + \alpha_0) - \delta F_N \sin(\phi_T + \alpha_0) = C_{P_X} q_\infty S_W$$

$$F_{P_Y} = 0$$

$$F_{P_Z} = F_{P_{X_0}} - \delta T\sin(\phi_T + \alpha_0) - \delta F_N \cos(\phi_T + \alpha_0) = C_{P_Z} q_\infty S_W$$

where the perturbation forces δT and δF_N are symmetric with respect to the XZ plane of the vehicle.

Consistent with Equation (6.238), we may write the effects of perturbations in angle-of-attack rate on the propulsive forces in terms of coefficients as

$$\boxed{\begin{aligned} \frac{\partial F_{P_X}}{\partial \dot{\alpha}}\Big|_0 &= C_{P_{X_{\dot{\alpha}}}} q_\infty S_W \\ \frac{\partial F_{P_Z}}{\partial \dot{\alpha}}\Big|_0 &= C_{P_{Z_{\dot{\alpha}}}} q_\infty S_W \end{aligned}} \tag{6.242}$$

And we may now write the changes in thrust and normal force in terms of angle-of-attack rate, or

$$\begin{aligned} \delta T &= \frac{\partial T}{\partial \dot{\alpha}}\dot{\alpha} = \frac{\partial T}{\partial v_{\text{Axial}}}\frac{\partial v_{\text{Axial}}}{\partial \dot{\alpha}}\dot{\alpha} \\ \delta F_N &= \frac{\partial F_N}{\partial \dot{\alpha}}\dot{\alpha} = \frac{\partial F_N}{\partial v_{\text{Transverse}}}\frac{\partial v_{\text{Transverse}}}{\partial \dot{\alpha}}\dot{\alpha} \end{aligned} \tag{6.243}$$

From Equation (6.241) and Figures 6.13, 6.14, 6.16, and 6.18, we have the changes in inlet flow velocity given by

$$\begin{aligned} \Delta v_{\text{Axial}} &= -\Delta\alpha_{\text{inlet}}\text{U}_0\sin(\phi_T + \alpha_0) \\ &= -\frac{d\varepsilon_{\text{inlet}}}{d\alpha_W}\dot{\alpha}\left(X_{\text{AC}_W} - X_{\text{Ref}} + x_T\right)\sin(\phi_T + \alpha_0) \\ \Delta v_{\text{Transverse}} &= \Delta\alpha_{\text{inlet}}\text{U}_0\cos(\phi_T + \alpha_0) \\ &= \frac{d\varepsilon_{\text{inlet}}}{d\alpha_W}\dot{\alpha}\left(X_{\text{AC}_W} - X_{\text{Ref}} + x_T\right)\cos(\phi_T + \alpha_0) \end{aligned} \tag{6.244}$$

Consequently, the <u>effects of perturbations in the rate of change of angle of attack on the propulsive forces</u> are given by

$$
\begin{aligned}
\frac{\partial F_{P_X}}{\partial \dot{\alpha}} &= \frac{\partial T}{\partial \dot{\alpha}}\cos(\phi_T+\alpha_0) - \frac{\partial F_N}{\partial \dot{\alpha}}\sin(\phi_T+\alpha_0) \\
&= -\left(\frac{\partial T}{\partial v_{\text{Axial}}} + \frac{\partial F_N}{\partial v_{\text{Transverse}}}\right)\frac{d\varepsilon_{\text{inlet}}}{d\alpha_W}\left(X_{AC_W} - X_{\text{Ref}} + x_T\right)\cos(\phi_T+\alpha_0)\sin(\phi_T+\alpha_0) \\
\frac{\partial F_{P_Y}}{\partial \dot{\alpha}} &= 0 \qquad (6.245) \\
\frac{\partial F_{P_Z}}{\partial \dot{\alpha}} &= -\frac{\partial T}{\partial \dot{\alpha}}\sin(\phi_T+\alpha_0) - \frac{\partial F_N}{\partial \dot{\alpha}}\cos(\phi_T+\alpha_0) \\
&= \left(\frac{\partial T}{\partial v_{\text{Axial}}}\sin^2(\phi_T+\alpha_0) - \frac{\partial F_N}{\partial v_{\text{Transverse}}}\cos^2(\phi_T+\alpha_0)\right)\frac{d\varepsilon_{\text{inlet}}}{d\alpha_W}\left(X_{AC_W} - X_{\text{Ref}} + x_T\right)
\end{aligned}
$$

Typically, the angle $(\phi_T + \alpha_0)$ is small, and the above expressions may be simplified accordingly. The partial derivatives of thrust and normal force in Equations (6.245) may be evaluated using Equations (6.108–6.112), depending on the type of propulsion system(s) on the vehicle. Finally, by comparing Equations (6.242) with Equations (6.245), we find that

$$
\begin{aligned}
C_{P_{X_{\dot{\alpha}}}} &\approx -\frac{1}{q_\infty S_W}\left(\frac{\partial T}{\partial v_{\text{Axial}}} + \frac{\partial F_N}{\partial v_{\text{Transverse}}}\right)\frac{d\varepsilon_{\text{inlet}}}{d\alpha_W}\left(X_{AC_W} - X_{\text{Ref}} + x_T\right)(\phi_T+\alpha_0) \approx 0 \\
C_{P_{Z_{\dot{\alpha}}}} &\approx -\frac{1}{q_\infty S_W}\frac{\partial F_N}{\partial v_{\text{Transverse}}}\frac{d\varepsilon_{\text{inlet}}}{d\alpha_W}\left(X_{AC_W} - X_{\text{Ref}} + x_T\right) \qquad (6.246)
\end{aligned}
$$

For typical vehicle geometries, the effect of angle-or-attack rate perturbations on the above propulsive forces is not large. However, if the propulsive devices are installed such that they have a significant moment arm relative to the moment reference point, the change in pitching moment due to these forces can again be significant. Under the assumed geometry of the propulsive devices, and since here we are dealing with only a change in local angle of attack, the perturbation forces δT and δF_N are symmetric with respect to the XZ plane of the vehicle. Hence, they generate no yawing or rolling moments.

To address the change in propulsive pitching moment, we again draw from Equations (6.149) and Figures 6.13, 6.14, 6.16, and 6.18 to note that the equation for the pitching moment is given as in Equation (6.156), repeated here.

$$
M_P = C_{P_M} q_\infty S_W \bar{c}_W = \mathrm{M}_{P_0} + \delta T\left(d_T\cos\phi_T - x_T\sin\phi_T\right) - \delta F_N\left(x_T\cos\phi_T + d_T\sin\phi_T\right)
$$

Consistent with Equation (6.240), the effect of rate of change of angle of attack on the propulsive pitching moment may be expressed as

$$\boxed{\frac{\partial M_P}{\partial \dot{\alpha}}\Big|_0 = C_{P_{M_{\dot{\alpha}}}} q_\infty S_W \bar{c}_W} \tag{6.247}$$

And by following the approach taken previously in analyzing the propulsive forces, we find that the effect of perturbations in the rate of change of angle of attack on the propulsive pitching moment may be expressed as

$$\begin{aligned}\frac{\partial M_P}{\partial \dot{\alpha}}\Big|_0 &= \frac{\partial T}{\partial \dot{\alpha}}\Big|_0 \left(d_T \cos\phi_T - x_T \sin\phi_T\right) - \frac{\partial F_N}{\partial \dot{\alpha}}\Big|_0 \left(x_T \cos\phi_T + d_T \sin\phi_T\right) \\ &= -\begin{pmatrix} \left(d_T \cos\phi_T - x_T \sin\phi_T\right) \dfrac{\partial T}{\partial v_{\text{Axial}}}\Big|_0 \sin\left(\phi_T + \alpha_0\right) \\ +\left(x_T \cos\phi_T + d_T \sin\phi_T\right) \dfrac{\partial F_N}{\partial v_{\text{Transverse}}}\Big|_0 \cos\left(\phi_T + \alpha_0\right) \end{pmatrix} \frac{d\varepsilon_{\text{inlet}}}{d\alpha_W}\left(X_{\text{AC}_W} - X_{\text{Ref}} + x_T\right)\end{aligned} \tag{6.248}$$

Typically the thrust angle ϕ_T and the reference angle of attack α_0 are small, and the above expression may be simplified accordingly. Again, the partial derivatives of thrust and normal force in Equation (6.248) may be evaluated using Equations (6.108–6.112), depending on the type of propulsion system(s) installed on the vehicle. Finally, by comparing Equation (6.247) with (6.248), and assuming ϕ_T and α_0 are sufficiently small, we find that

$$\begin{aligned} C_{P_{M_{\dot{\alpha}}}} \approx -\frac{1}{q_\infty S_W \bar{c}_W}\Bigg(&\frac{\partial T}{\partial v_{\text{Axial}}}\Big|_0 d_T\left(\phi_T + \alpha_0\right) \\ &+ \frac{\partial F_N}{\partial v_{\text{Transverse}}}\Big|_0 \left(x_T + d_T\phi_T\right)\Bigg) \frac{d\varepsilon_{\text{inlet}}}{d\alpha_W}\left(X_{\text{AC}_W} - X_{\text{Ref}} + x_T\right)\end{aligned} \tag{6.249}$$

6.8 EFFECTS OF ATMOSPHERIC TURBULENCE ON THE FORCES AND MOMENTS

Although models for atmospheric turbulence are presented in Appendix C, here we will consider techniques for modeling the effects of gusts on the aerodynamic forces and moments acting on the vehicle. The schematics in Figures 6.4 and 6.5, showing the aerodynamic forces and the angles of attack and sideslip, are still appropriate. But wind gusts affect the total angle of attack of the vehicle's lifting surfaces, as well as the free-stream axial velocity.

To illustrate the concepts, consider a vehicle flying at a velocity U_0 relative to the surface of the earth, and assume the air mass through which the vehicle is

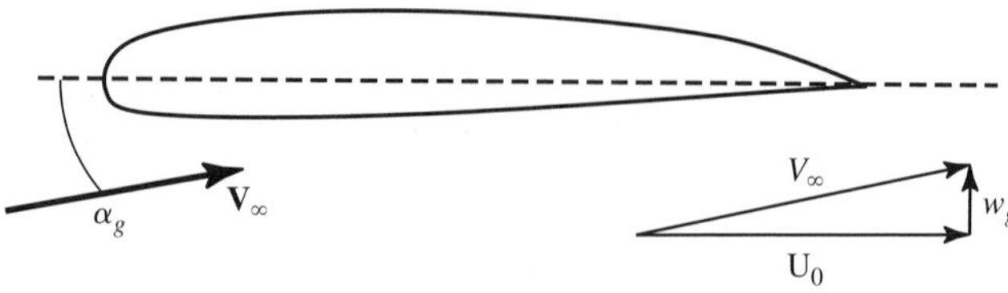

Figure 6.21 Effect of vertical gust on total angle of attack.

flying is fixed to the surface of the earth. Consequently, the free-stream velocity of the air V_∞, shown in Figures 6.4 and 6.5, is simply U_0.

Next, define wind gusts to be localized movement of the air relative to a point on the earth's surface. With these concepts in mind, consider at first only a vertical gust with earth-relative velocity w_g (positive up). This gust produces a perturbation in the free-stream velocity experienced by the lifting surfaces on the vehicle. For example, consider a horizontal lifting surface as shown in Figure 6.21. With a vertical wind gust present, the angle of attack of the surface is perturbed by an amount α_g:

$$\alpha_g = \frac{w_g}{U_0} \tag{6.250}$$

If the vehicle is also moving with perturbation plunge velocity w (positive down) relative to the earth's surface, the total perturbation angle of attack consists of the sum of two angles, or

$$\alpha_{\text{total}} = \alpha + \alpha_g \approx \frac{w}{U_0} + \frac{w_g}{U_0} \tag{6.251}$$

where α is the perturbation angle of attack in still air (no gust) as given in Equation (6.96).

In general, wind gusts not only have a vertical, or plunge component w_g, but also an axial component u_g and a lateral component v_g. Or the components of the gust velocity in the vehicle's stability axes are given as

$$\mathbf{v}_g = -\begin{bmatrix} u_g & v_g & w_g \end{bmatrix} \begin{Bmatrix} \mathbf{i}_S \\ \mathbf{j}_S \\ \mathbf{k}_S \end{Bmatrix} \tag{6.252}$$

In a fashion similar to the effect of vertical gusts, lateral gusts induce a perturbation in the total angle of attack on vertical lifting surfaces, or a perturbation in the angle of sideslip β. Finally, axial gusts simply produce a perturbation in the surge component of free-stream velocity, or u. In other words, if the vehicle is experiencing a lateral gust we may write (consistent with Equation (6.99))

$$\beta_{\text{total}} = \beta + \beta_g \approx \frac{v}{U_0} + \frac{v_g}{U_0} \tag{6.253}$$

while

$$u_{\text{total}} = u + u_g \tag{6.254}$$

In addition to the effects of gusts on surge velocity and angles of attack and sideslip, they may also influence the rate of change of angle of attack, or $\dot{\alpha}_{\text{total}}$. (This effect is only defined when the gust velocity has a finite first derivative with respect to time.) In such a case we may also write

$$\dot{\alpha}_{\text{total}} = \dot{\alpha} + \dot{\alpha}_g \approx \dot{\alpha} + \frac{\dot{w}_g}{U_0} \tag{6.255}$$

With a simplifying assumption, it is now quite straightforward to model the effects of these gust components. The assumption is that although the axial and lateral gust velocities may vary temporally with respect to a point on the earth (i.e., the wind may not be uniform in these directions), they do not vary spatially over the dimensions of the vehicle. That is, the spatial variations in axial and lateral wind gusts over the vehicle are small, compared to the global variations in these gusts. This assumption states nothing about the vertical gusts.

Recall that in Section 6.6 we modeled the effects of perturbations in the translational velocity of the vehicle (relative to a still air mass) on the aerodynamic forces and moments acting on the vehicle. And, as noted above, wind gusts are modeled as perturbations in the velocity of the vehicle relative to the air. In addition, in Section 6.7.4 we modeled the effects of rate of change of angle of attack on those same forces and moments.

For example, consistent with Equations (6.9), let us write the perturbations in the aerodynamic forces associated only with <u>perturbations in translational velocity and rate of change of angle of attack</u> (with no gusts present) as

$$f_A = \begin{bmatrix} f_{A_x} & f_{A_Y} & f_{A_Z} \end{bmatrix} \begin{Bmatrix} \mathbf{i}_S \\ \mathbf{j}_S \\ \mathbf{k}_S \end{Bmatrix} = \begin{bmatrix} \dfrac{\partial \mathbf{F}_A}{\partial u} & \dfrac{\partial \mathbf{F}_A}{\partial v} & \dfrac{\partial \mathbf{F}_A}{\partial w} & \dfrac{\partial \mathbf{F}_A}{\partial \dot{\alpha}} \end{bmatrix} \begin{Bmatrix} u \\ v \\ w \\ \dot{\alpha}\left(= \dfrac{\dot{w}}{U_0}\right) \end{Bmatrix} \tag{6.256}$$

and recall that

$$\mathbf{F}_A = \begin{bmatrix} F_{A_X} & F_{A_Y} & F_{A_Z} \end{bmatrix} \begin{Bmatrix} \mathbf{i}_S \\ \mathbf{j}_S \\ \mathbf{k}_S \end{Bmatrix}$$

But as in Section 6.6, we find from Equations (6.121), (6.160), and (6.146) that

$$\begin{aligned} \frac{\partial F_{A\bullet}}{\partial u}\Big|_0 &= \left(\frac{1}{a}\frac{\partial C_\bullet}{\partial M_\infty} q_\infty S_W + C_{\bullet_0} S_W \rho_\infty (U_0 + u)\right)\Big|_0 \\ &= \frac{1}{a}\frac{\partial C_\bullet}{\partial M_\infty}\Big|_0 q_\infty S_W + C_{\bullet_0} S_W \rho_\infty U_0, \ \bullet = X, Y, \text{ or } Z \end{aligned}$$

$$\frac{\partial F_{A\bullet}}{\partial v}\Big|_0 = \frac{\partial C_\bullet}{\partial v}\Big|_0 q_\infty S_W + C_{\bullet_0} S_W \frac{\partial q_\infty}{\partial v}\Big|_0$$

$$= \frac{1}{U_0} C_{\bullet_\beta} q_\infty S_W + C_{\bullet_0} S_W \frac{\partial q_\infty}{\partial v}\Big|_0, \bullet = X, Y, \text{ or } Z$$

$$\frac{\partial F_{A\bullet}}{\partial w}\Big|_0 = \frac{\partial C_\bullet}{\partial w}\Big|_0 q_\infty S_W = \frac{1}{U_0}\frac{\partial C_\bullet}{\partial \alpha}\Big|_0 q_\infty S_W, \bullet = X, Y, \text{ or } Z$$

Additionally, as in Section 6.7 we find from Equations (6.238) that

$$\frac{\partial F_{A_X}}{\partial \dot{\alpha}}\Big|_0 = 0, \quad \frac{\partial F_{A_Y}}{\partial \dot{\alpha}}\Big|_0 = 0$$

$$\frac{\partial F_{A_Z}}{\partial \dot{\alpha}}\Big|_0 = -\frac{\partial L}{\partial \dot{\alpha}}\Big|_0 = -C_{L_{\dot{\alpha}}} q_\infty S_W = -\left(C_{L_{\alpha_H}} \frac{d\varepsilon_H}{d\alpha_W} \frac{\left(X_{AC_W} - X_{AC_H}\right)}{U_0} \frac{q_H}{q_\infty} \frac{S_H}{S_W}\right) q_\infty S_W$$

Therefore, with gusts present we may now write Equations (6.256), the perturbations in aerodynamic forces, as

$$f_A = \begin{bmatrix} f_{A_X} & f_{A_Y} & f_{A_Z} \end{bmatrix} \begin{Bmatrix} \mathbf{i}_S \\ \mathbf{j}_S \\ \mathbf{k}_S \end{Bmatrix} = \begin{bmatrix} \dfrac{\partial \mathbf{F}_A}{\partial u} & \dfrac{\partial \mathbf{F}_A}{\partial v} & \dfrac{\partial \mathbf{F}_A}{\partial w} & \dfrac{\partial \mathbf{F}_A}{\partial \dot{\alpha}} \end{bmatrix} \begin{Bmatrix} u_{\text{total}} \\ v_{\text{total}} \\ w_{\text{total}} \\ \dot{\alpha}_{\text{total}} \end{Bmatrix} \tag{6.257}$$

where, from Equations (6.251) and (6.253–6.255), we have

$$u_{\text{total}} = u + u_g$$
$$v_{\text{total}} = v + v_g$$
$$w_{\text{total}} = w + w_g$$
$$\dot{\alpha}_{\text{total}} = \dot{\alpha} + \dot{\alpha}_g$$

And we may use Equations (6.121), (6.146), (6.160), and (6.238) to model the effects of the perturbations in translational velocities and rate of change of angle of attack on these forces. So one can see that the mechanism generating the perturbation forces is the same, except the total perturbations (including the gust terms) in translational velocities and angle-of-attack rate are substituted for their counterparts without gusts.

The techniques presented above may also be used to model the effects of gusts on the aerodynamic moments (e.g., Problem 6.9 at the end of this chapter), as well as the propulsive forces and moments, all of which were discussed in the absence of gusts in Sections 6.6 and 6.7.

6.9 DIMENSIONAL VERSUS NONDIMENSIONAL DERIVATIVES

It is now useful to compare the results developed in this chapter to those based on using nondimensional parameters, as discussed in the *Note to Student* below Equation (6.4). It was previously shown that the perturbation forces and moments acting on the vehicle come from Taylor-series expansions, in which the first-order terms are

$$f_{\bullet_{A \text{ or } P}} = \left[\frac{\partial F_{\bullet_{A \text{ or } P}}}{\partial \mathbf{p}}\right]\delta\mathbf{p}, \quad \bullet = X, Y, \text{ or } Z$$
$$m_{\bullet_{A \text{ or } P}} = \left[\frac{\partial M_{\bullet_{A \text{ or } P}}}{\partial \mathbf{p}}\right]\delta\mathbf{p}, \quad \bullet = X, Y, \text{ or } Z \tag{6.258}$$

where the parameter-vector $\mathbf{p}$ includes the components of translational and rotational velocities, angle-of-attack rate, and control-surface deflections. So the key aspects of this model consist of the partial derivatives in these expansions.

The aerodynamic lift, drag, and side force and pitching, rolling, and yawing moments were also expressed in terms of similar partial derivatives. Consequently, results from this chapter may be succinctly summarized in tables containing all the partial derivatives. In the tables discussed below, please note that the following notation has been adopted for conciseness.

$$\overline{qS_H} \triangleq \frac{q_H}{q_\infty}\frac{S_H}{S_W}, \quad \overline{qS_V} \triangleq \frac{q_H}{q_\infty}\frac{S_V}{S_W}, \quad \bar{\varepsilon} \triangleq \left(1 - \frac{d\varepsilon}{d\alpha}\right) \tag{6.259}$$

Finally, note that a conventional vehicle geometry has been assumed throughout this chapter, unless otherwise noted. Such a vehicle consists of a wing and horizontal and vertical tail surfaces, along with elevator, aileron, and rudder control surfaces. However, using the techniques presented in this chapter, similar expressions for the partial derivatives for nonconventional vehicles may also be derived. (See, for example, Problems 6.2 and 6.3 at the end of this chapter, and especially Reference 5, in which a hypersonic vehicle is modeled and analyzed.)

Table 6.2 summarizes the partial derivatives associated with lift, drag, and pitching-moment coefficients. Table 6.3 summarizes the partial derivatives associated with side-force and rolling- and yawing-moment coefficients, and the remaining derivatives associated with drag coefficient. To help understand the contents of these two tables, let's consider the lift coefficient in Table 6.2. Expressed in terms of the Taylor series to first order, and in the absence of any perturbations in translational or rotational velocity, the lift coefficient is given as

$$C_L = C_{L_{\alpha=\delta_e=i_H=0}} + C_{L_\alpha}\alpha + C_{L_{\delta_E}}\delta_E + C_{L_{i_H}}i_H \tag{6.260}$$

and expressions for the four coefficients on the right side of this expression are listed in the first row in the table below the headings. If angle of attack, elevator deflection, and tail incidence angle are all assumed to be in radians, all four of these coefficients are per radian, or dimensionless. Finally, the defining equation number is listed below each entry in all the tables.

Table 6.2 Summary for Vehicle Lift, Drag, and Pitching Moment

	$\alpha = \delta_E = i_H = 0$ (plus $\beta = \delta_R = 0$ for drag)	Angle of Attack, α	Elevator Deflection, δ_E	Tail Incidence, i_H
Lift, C_L	$C_{L_{\alpha_W}}(i_W - \alpha_{0_W}) +$ $C_{L_{\alpha_H}}\overline{q}S_H\left(\frac{d\varepsilon}{d\alpha}(\alpha_{0_W} - i_W) - \alpha_{0_H}\right)$ Equation (6.21)	$C_{L_{\alpha_W}} + C_{L_{\alpha_H}}\overline{q}S_H\overline{\varepsilon}$ Equation (6.18)	$C_{L_{\alpha_H}}\alpha_\delta\overline{q}S_H$ $= C_{L_{\delta_H}}\overline{q}S_H$ Equation (6.19)	$C_{L_{\alpha_H}}\overline{q}S_H$ Equation (6.20)
Drag, C_D	$C_{D_0} + \frac{1}{\pi A_W e_W}\left(C_{L_W}^2 + C_{L_H}^2\frac{A_W e_W}{A_H e_H}\overline{q}S_H\right)$ Equation (6.36)	$\frac{2C_{L_W}C_{L_{\alpha_W}}}{\pi A_W e_W}\left(1 + \frac{C_{L_H}C_{L_{\alpha_H}}}{C_{L_W}C_{L_{\alpha_W}}}\frac{A_W e_W}{A_H e_H}\overline{q}S_H\overline{\varepsilon}\right)$ Equation (6.31)	$C_{L_{\alpha_H}}\alpha_\delta\frac{2C_{L_H}}{\pi A_H e_H}\overline{q}S_H$ $= C_{L_{\delta_H}}\frac{2C_{L_H}}{\pi A_H e_H}\overline{q}S_H$ Equation (6.32)	$C_{L_{\alpha_H}}\frac{2C_{L_H}}{\pi A_H e_H}\overline{q}S_H$ Equation (6.33)
Pitching moment, C_m	$C_{M_{AC_W}} + C_{M_{0_F}} + C_{L_{\alpha_W}}(i_W - \alpha_{0_W})\Delta\overline{X}_{AC_{W\&F}}$ $- C_{L_{\alpha_H}}\left(\frac{d\varepsilon}{d\alpha}(\alpha_{0_W} - i_W) - \alpha_{0_H}\right)\Delta\overline{X}_{AC_H}\overline{q}S_H$ Equation (6.60)	$C_{L_{\alpha_W}}\Delta\overline{X}_{AC_{W\&F}}$ $- C_{L_{\alpha_H}}\overline{\varepsilon}\Delta\overline{X}_{AC_H}\overline{q}S_H$ Equation (6.56)	$-C_{L_{\alpha_H}}\alpha_\delta\Delta\overline{X}_{AC_H}\overline{q}S_H$ $= -C_{L_{\delta_H}}\Delta\overline{X}_{AC_H}\overline{q}S_H$ Equation (6.58)	$-C_{L_{\alpha_H}}\Delta\overline{X}_{AC_H}\overline{q}S_H$ Equation (6.59)

Table 6.3 Summary for Vehicle Side Force, Rolling and Yawing Moments, and Drag

Coefficient	Sideslip Angle, β	Rudder Deflection, δ_R	Aileron Deflection, δ_A
Side force, C_S	$C_{S_{\beta_V}}\overline{q}S_V$ Equation (6.25)	$C_{S_{\beta_V}}\beta_\delta\overline{q}S_V = C_{S_{\delta_V}}\overline{q}S_V$ Equation (6.26)	≈ 0
Rolling moment, $C_{L_{Roll}}$	$C_{L_{\beta_W}} + C_{L_{\beta_H}}\overline{q}S_H\frac{b_H}{b_W} + C_{L_{\beta_V}}\overline{q}S_V\frac{b_V}{b_W}$ Equation (6.44)	$C_{L_{\delta_V}}\overline{q}S_V\frac{b_V}{b_W}$ Equation (6.45)	$C_{L_{\delta_{A_{Wing}}}}$ Equation (6.39)
Yawing moment, C_N	$-C_{S_{\beta_V}}\Delta\overline{X}_{AC_V}\overline{q}S_V$ Equation (6.68)	$-C_{S_{\delta_V}}\Delta\overline{X}_{AC_V}\overline{q}S_V$ Equation (6.69)	$C_{N_{\delta_{A_{Wing}}}}$ Equation (6.65)
Drag, C_D	$\frac{2C_{S_V}}{\pi A_V e_V}C_{S_{\beta_V}}\overline{q}S_V$ Equation (6.34)	$C_{S_{\beta_V}}\beta_\delta\frac{2C_{S_V}}{\pi A_V e_V}\overline{q}S_V = C_{S_{\delta_V}}\frac{2C_{S_V}}{\pi A_V e_V}\overline{q}S_V$ Equation (6.35)	≈ 0

Table 6.4 Dimensional Versus Nondimensional Aerodynamic and Propulsive Force Derivatives

Partial Derivative (• = X, Y, or Z)	Dimensional Derivative	Nondimensional Derivative	Nondimensional Parameter
$\left.\frac{\partial F_{A\text{ or }P\bullet}}{\partial u}\right\|_0$	$\left(C_{\bullet_u} + \frac{2C_{\bullet_0}}{U_0}\right) q_\infty S_W$ Equations (6.122) and (6.133)	$(C_{\bullet_{u'}} + 2C_{\bullet_0}) q_\infty S_W$ $C_{\bullet_{u'}} = U_0 C_{\bullet_u}$	$\left(\frac{u}{U_0} = u'\right)$
$\left.\frac{\partial F_{A\text{ or }P\bullet}}{\partial w}\right\|_0$	$\frac{1}{U_0} C_{\bullet_\alpha} q_\infty S_W$ Equations (6.146) and (6.150)	$C_{\bullet_\alpha} q_\infty S_W$ $C_{\bullet_\alpha} = U_0 C_{\bullet_w}$	$\left(\frac{w}{U_0} = \alpha\right)$
$\left.\frac{\partial F_{A\text{ or }P\bullet}}{\partial v}\right\|_0$	$\frac{1}{U_0}(C_{\bullet_\beta} + 2C_{\bullet_0}\beta_0) q_\infty S_W$ Equations (6.160) and (6.169)	$(C_{\bullet_\beta} + 2C_{\bullet_0}\beta_0) q_\infty S_W$ $C_{\bullet_\beta} = U_0 C_{\bullet_v}$	$\left(\frac{v}{U_0} = \beta\right)$
$\left.\frac{\partial F_{A\text{ or }P\bullet}}{\partial q}\right\|_0$	$C_{\bullet_q} q_\infty S_W$ Equation (6.181) or (6.189)	$C_{\bullet_{q'}} q_\infty S_W$ $C_{\bullet_{q'}} = \frac{2U_0}{\bar{c}_W} C_{\bullet_q}$	$\left(\frac{q\bar{c}_W}{2U_0} = q'\right)$
$\left.\frac{\partial F_{A_Y}}{\partial p}\right\|_0 \left(\left.\frac{\partial F_{P_Y}}{\partial p}\right\|_0 \approx 0\right)$	$C_{Y_p} q_\infty S_W$ Equation (6.201)	$C_{Y_{p'}} q_\infty S_W$ $C_{Y_{p'}} = \frac{2U_0}{b_W} C_{Y_p}$	$\left(\frac{pb_W}{2U_0} = p'\right)$
$\left.\frac{\partial F_{A\text{ or }P\bullet}}{\partial r}\right\|_0$	$C_{\bullet_r} q_\infty S_W$ Equation (6.217) or (6.223)	$C_{\bullet_{r'}} q_\infty S_W$ $C_{\bullet_{r'}} = \frac{2U_0}{b_W} C_{\bullet_r}$	$\left(\frac{rb_W}{2U_0} = r'\right)$
$\left.\frac{\partial F_{A\text{ or }P\bullet}}{\partial \dot{\alpha}}\right\|_0$	$C_{\bullet_{\dot{\alpha}}} q_\infty S_W$ Equation (6.238) or (6.245)	$C_{\bullet_{\dot{\alpha}'}} q_\infty S_W$ $C_{\bullet_{\dot{\alpha}'}} = \frac{2U_0}{\bar{c}_W} C_{\bullet_{\dot{\alpha}}}$	$\left(\frac{\dot{\alpha}\bar{c}_W}{2U_0} = \dot{\alpha}'\right)$

Table 6.4 summarizes additional partial derivatives associated with the aerodynamic and propulsive forces (the form of the derivatives are the same, whether aerodynamic or propulsive). These derivatives are associated with velocity perturbations. The second column contains the derivatives derived in this chapter. The third and fourth columns, respectively, lists the companion nondimensional derivatives and defines the nondimensionalized parameters in the parameter vector $\delta\mathbf{p}$ in Equations (6.258). Also listed in the third column is the relationship between the dimensional and the nondimensional derivatives.

As a key example, the perturbation force f_{A_X} associated with velocity perturbations, expressed in terms of dimensional derivatives defined in this chapter, is

$$f_{A_X} = \frac{\partial F_{A_X}}{\partial u} u + \frac{\partial F_{A_X}}{\partial v} v + \frac{\partial F_{A_X}}{\partial w} w + \frac{\partial F_{A_X}}{\partial q} q \qquad (6.261)$$

But in terms of the nondimensional derivatives, this same perturbation force is

$$f_{A_X} = \frac{\partial F_{A_X}}{\partial u'} u' + \frac{\partial F_{A_X}}{\partial \beta} \beta + \frac{\partial F_{A_X}}{\partial \alpha} \alpha + \frac{\partial F_{A_X}}{\partial q'} q' \tag{6.262}$$

A little algebra reveals that the final results (i.e., f_{A_X}) from Equations (6.261) and (6.262) are the same, but the partial derivatives themselves are not! Hence, when using numerical results from other sources, it is important to clarify the definition of the derivatives being used.

Finally, Table 6.5 summarizes the partial derivatives associated with the aerodynamic and propulsive moments. As with Table 6.4, the second column of Table 6.5 lists the dimensional derivatives as derived in this chapter. The third and fourth columns, respectively, lists the companion nondimensional derivatives and defines the nondimensionalized parameters in $\delta\mathbf{p}$.

Table 6.5 Dimensional Versus Nondimensional Aerodynamic and Propulsive Moment Derivatives

Partial Derivative (• = $L, M,$ or N)	Dimensional Derivative	Nondimensional Derivative	Nondimensional Parameter
$\left.\frac{\partial \bullet_{A \text{ or } P}}{\partial u}\right\vert_0$	$\left(C_{\bullet_u} + \frac{2C_{\bullet_0}}{U_0}\right) q_\infty S_W (\bar{c}_W \text{ or } b_W)$ Equation (6.130) or (6.141)	$(C_{\bullet_{u'}} + 2C_{\bullet_0}) q_\infty S_W (\bar{c}_W \text{ or } b_W)$ $C_{\bullet_{u'}} = U_0 C_{\bullet_u}$	$\left(\frac{u}{U_0} = u'\right)$
$\left.\frac{\partial M_{A \text{ or } P}}{\partial w}\right\vert_0$	$\frac{1}{U_0} C_{M_\alpha} q_\infty S_W \bar{c}_W$ Equation (6.148) or (6.157)	$C_{M_\alpha} q_\infty S_W \bar{c}_W$ $C_{M_\alpha} = U_0 C_{M_w}$	$\left(\frac{w}{U_0} = \alpha\right)$
$\left.\frac{\partial \bullet_{A \text{ or } P}}{\partial v}\right\vert_0$	$\frac{1}{U_0} (C_{\bullet_\beta} + 2C_{\bullet_0} \beta_0) q_\infty S_W b_W$ Equation (6.164) or (6.173)	$(C_{\bullet_\beta} + 2C_{\bullet_0} \beta_0) q_\infty S_W b_W$ $C_{\bullet_\beta} = U_0 C_{\bullet_v}$	$\left(\frac{v}{U_0} = \beta\right)$
$\left.\frac{\partial M_{A \text{ or } P}}{\partial q}\right\vert_0$	$C_{M_q} q_\infty S_W \bar{c}_W$ Equation (6.186) or (6.194)	$C_{M_{q'}} q_\infty S_W \bar{c}_W$ $C_{M_{q'}} = \frac{2U_0}{\bar{c}_W} C_{M_q}$	$\left(\frac{q\bar{c}_W}{2U_0} = q'\right)$
$\left.\frac{\partial L_{A \text{ or } P}}{\partial p}\right\vert_0$	$C_{L_p} q_\infty S_W b_W$ Equation (6.207) or (6.209)	$C_{L_{p'}} q_\infty S_W b_W$ $C_{L_{p'}} = \frac{2U_0}{b_W} C_{L_p}$	$\left(\frac{pb_W}{2U_0} = p'\right)$
$\left.\frac{\partial N_{A \text{ or } P}}{\partial r}\right\vert_0$	$C_{N_r} q_\infty S_W b_W$ Equation (6.219) or (6.230)	$C_{N_{r'}} q_\infty S_W b_W$ $C_{N_{r'}} = \frac{2U_0}{b_W} C_{N_r}$	$\left(\frac{rb_W}{2U_0} = r'\right)$
$\left.\frac{\partial L_{A \text{ or } P}}{\partial r}\right\vert_0$	$C_{L_r} q_\infty S_W b_W$ Equation (6.222) or (6.230)	$C_{L_{r'}} q_\infty S_W b_W$ $C_{L_{r'}} = \frac{2U_0}{b_W} C_{L_r}$	$\left(\frac{rb_W}{2U_0} = r'\right)$
$\left.\frac{\partial M_{A \text{ or } P}}{\partial \dot{\alpha}}\right\vert_0$	$C_{M_{\dot{\alpha}}} q_\infty S_W \bar{c}_W$ Equation (6.240) or (6.247)	$C_{M_{\dot{\alpha}'}} q_\infty S_W \bar{c}_W$ $C_{M_{\dot{\alpha}'}} = \frac{2U_0}{\bar{c}_W} C_{M_{\dot{\alpha}}}$	$\left(\frac{\dot{\alpha}\bar{c}_W}{2U_0} = \dot{\alpha}'\right)$

6.10 INTEGRATION OF FORCES AND MOMENTS INTO THE EQUATIONS OF MOTION

The integration of the forces and moments into the nonlinear equations of motion (Equations (2.22) and (2.27), in the case of a flat, nonrotating earth) will be discussed in more detail in Chapter 8, which addresses simulation. So here we will only highlight key points. However, we will address in more detail the integration of the forces and moments into the linearized equations of motion (Equations (2.45) and (2.49)).

6.10.1 Integration into the Nonlinear Equations of Motion

The forces appearing in the nonlinear Equations (2.22) consist of the components of the resultant aerodynamic and propulsive forces, resolved in a vehicle-fixed frame. That is,

$$\mathbf{F}_A + \mathbf{F}_P = \left(F_{A_X} + F_{P_X}\right)\mathbf{i}_V + \left(F_{A_Y} + F_{P_Y}\right)\mathbf{j}_V + \left(F_{A_Z} + F_{P_Z}\right)\mathbf{k}_V \tag{6.263}$$

When using the nonlinear equations of motion, it is usually appropriate to use fuselage-referenced axes as that vehicle-fixed frame. Now, consistent with Equations (6.107), the three aerodynamic force components in the fuselage-referenced axes, expressed in terms of lift, drag, and side force are

$$\boxed{\begin{aligned} F_{A_X} &= C_X q_\infty S_W = -D\cos\alpha\cos\beta - \mathsf{S}\cos\alpha\sin\beta + L\sin\alpha \\ F_{A_Y} &= C_Y q_\infty S_W = -D\sin\beta + \mathsf{S}\cos\beta \\ F_{A_Z} &= C_Z q_\infty S_W = -D\sin\alpha\cos\beta - \mathsf{S}\sin\alpha\sin\beta - L\cos\alpha \end{aligned}} \tag{6.264}$$

where now α and β are the <u>total</u> (i.e., reference plus perturbation) angles of attack and sideslip of the fuselage-referenced frame and D, L, and S are the total drag, lift, and side force, respectively. The lift, drag, and side force, and their nondimensional coefficients, were discussed in Sections 6.2.1–6.2.3, in which expressions were developed for these forces and coefficients.

The propulsive forces were discussed in Section 6.3. Under the assumptions cited in that section, and from Figure 6.13, we may write the propulsive-force components listed in Equation (6.263) as

$$\boxed{\begin{aligned} F_{P_X} &= C_{P_X} q_\infty S_W = T\left(V_{\text{inlet}}, \text{h}, \pi\right)\cos\phi_T \\ F_{P_Y} &= C_{P_Y} q_\infty S_W = F_{N_Y} \\ F_{P_Z} &= C_{P_Z} q_\infty S_W = -T\left(V_{\text{inlet}}, \text{h}, \pi\right)\sin\phi_T + F_{N_Z} \end{aligned}} \tag{6.265}$$

where fuselage-referenced axes are again assumed. So if the thrust and normal forces are known from an engine model as a function of inlet velocity V_{inlet}, altitude h, and power setting π, for example, the above three force components, or their associated force coefficients, are also known.

The moments appearing in nonlinear Equations (2.27) consist of the components of the resultant aerodynamic and propulsive moments again resolved in a vehicle-fixed frame, here assumed to be fuselage-referenced. That is,

$$\mathbf{M}_A + \mathbf{M}_P = \left(L_A + L_P\right)\mathbf{i}_V + \left(M_A + M_P\right)\mathbf{j}_V + \left(N_A + N_P\right)\mathbf{k}_V \tag{6.266}$$

And consistent with Equation (6.37) we have

$$\boxed{\begin{aligned} L_A &= C_{L_{\text{Roll}}} q_\infty S_W b_W && \text{aerodynamic rolling moment} \\ M_A &= C_M q_\infty S_W \bar{c}_W && \text{aerodynamic pitching moment} \\ N_A &= C_N q_\infty S_W b_W && \text{aerodynamic yawing moment} \end{aligned}} \tag{6.267}$$

where these three aerodynamic moments and their coefficients were discussed in Sections 6.2.4–6.2.6. Expressions were also developed in these sections that may be used to estimate the three moments in Equation (6.267).

For the propulsive moments we have

$$\boxed{\begin{aligned} L_P &= C_{P_{L_{\text{Roll}}}} q_\infty S_W b_W && \text{propulsive rolling moment} \\ M_P &= C_{P_M} q_\infty S_W \bar{c}_W && \text{propulsive pitching moment} \\ N_P &= C_{P_N} q_\infty S_W b_W && \text{propulsive yawing moment} \end{aligned}} \tag{6.268}$$

where consistent with Equations (6.156), (6.208), and (6.229),

$$L_P = \sum_{i=1}^{n_P} \Big(y_{T_i} \big(F_{N_{Z_i}} \cos\phi_{T_i} - T_i(V_{\text{inlet}}, \text{h}, \pi) \sin\phi_{T_i} \big) - d_{T_i} F_{N_{Y_i}} \Big)$$

$$M_P = \sum_{i=1}^{n_P} \Big(T_i(V_{\text{inlet}}, \text{h}, \pi) \big(d_{T_i} \cos\phi_{T_i} - x_{T_i} \sin\phi_{T_i} \big) - F_{N_{Z_i}} \big(x_{T_i} \cos\phi_{T_i} + d_{T_i} \sin\phi_{T_i} \big) \Big)$$

$$N_P = -\sum_{i=1}^{n_P} \Big(y_{T_i} T_i(V_{\text{inlet}}, \text{h}, \pi) \cos\phi_{T_i} + x_{T_i} F_{N_{Y_i}} \Big)$$

Again, from an engine model and the vehicle geometry, these moments are known.

6.10.2 Integration into the Linearized Equations of Motion

The linearized equations of motion (flat earth) were given in Equations (2.45) and (2.49). The perturbation forces appearing in Equations (2.45) were the three components of the perturbation aerodynamic plus propulsive forces, expressed now in the vehicle-fixed stability axes. Or

$$\boldsymbol{f}_A + \boldsymbol{f}_P = \big(f_{A_X} + f_{P_X}\big)\mathbf{i}_S + \big(f_{A_Y} + f_{P_Y}\big)\mathbf{j}_S + \big(f_{A_Z} + f_{P_Z}\big)\mathbf{k}_S \tag{6.269}$$

But the perturbation aerodynamic forces arise from the linear terms in the Taylor-series expansion. That is,

$$f_{A\bullet} = \frac{\partial F_{A\bullet}}{\partial \delta \mathbf{p}}\Big|_0 \delta\mathbf{p}, \bullet = X, Y, \text{ or } Z$$

Based on the analysis in this chapter, and assuming conventional vehicle geometry, we determined that the appropriate perturbation parameter vector upon which the perturbation aerodynamic forces depend is

$$\delta\mathbf{p}^T = \begin{bmatrix} u & v & w & p & q & r & \dot{\alpha} & \delta_E & \delta_A & \delta_R \end{bmatrix}$$

or, equivalently,

$$\delta \mathbf{p}^T = \begin{bmatrix} u & \beta & \alpha & p & q & r & \dot{\alpha} & \delta_E & \delta_A & \delta_R \end{bmatrix}$$

since

$$\alpha \approx \frac{w}{\mathrm{U}_0}$$

and

$$\beta \approx \frac{v}{\mathrm{U}_0}$$

And so, consistent with the development in Sections 6.6–6.8, we may write

$$\boxed{\begin{aligned} f_{A_Z} &= \frac{\partial F_{A_Z}}{\partial u}\Big|_0 u + \frac{\partial F_{A_Z}}{\partial \alpha}\Big|_0 \alpha + \frac{\partial F_{A_Z}}{\partial \dot{\alpha}}\Big|_0 \dot{\alpha} + \frac{\partial F_{A_Z}}{\partial q}\Big|_0 q + \frac{\partial F_{A_Z}}{\partial \delta_E}\Big|_0 \delta_E \\ f_{A_X} &= \frac{\partial F_{A_X}}{\partial u}\Big|_0 u + \frac{\partial F_{A_X}}{\partial \alpha}\Big|_0 \alpha + \frac{\partial F_{A_X}}{\partial \dot{\alpha}}\Big|_0 \dot{\alpha} + \frac{\partial F_{A_X}}{\partial q}\Big|_0 q + \frac{\partial F_{A_X}}{\partial \delta_E}\Big|_0 \delta_E \\ f_{A_Y} &= \frac{\partial F_{A_Y}}{\partial \beta}\Big|_0 \beta + \frac{\partial F_{A_Y}}{\partial p}\Big|_0 p + \frac{\partial F_{A_Y}}{\partial r}\Big|_0 r + \frac{\partial F_{A_Y}}{\partial \delta_A}\Big|_0 \delta_A + \frac{\partial F_{A_Y}}{\partial \delta_R}\Big|_0 \delta_R \end{aligned}} \tag{6.270}$$

where the remaining terms not appearing in the above expressions typically have a negligible effect. Now Equations (6.107) give the aerodynamic force components in terms of lift, drag, and side force, and so the partial derivatives in Equations (6.270) may also be expressed in terms of partial derivatives of lift, drag, and side force. These derivatives were all discussed in this chapter.

Applying Equations (6.107) and working through the details, the partial derivatives of the aerodynamic forces and moments listed in Table 6.6 may be derived, and these results are consistent with those presented in Sections 6.6–6.8. As an example of such a derivation, consider the derivative $\frac{\partial F_{A_Z}}{\partial \alpha}\Big|_0$.

$$\begin{aligned} \frac{\partial F_{A_Z}}{\partial \alpha}\Big|_0 &= \frac{\partial}{\partial \alpha}\Big|_0 \left(-D\sin\alpha\cos(\beta_0+\beta) - \mathrm{S}\sin\alpha\sin(\beta_0+\beta) - L\cos\alpha\right) \\ &= -\left(\left(\frac{\partial D}{\partial \alpha}\sin\alpha + D\cos\alpha\right)\cos(\beta_0+\beta) - \mathrm{S}\cos\alpha\sin(\beta_0+\beta)\right. \\ &\quad \left. + \frac{\partial L}{\partial \alpha}\cos\alpha - L\sin\alpha\right)\Big|_0 \\ &= -\mathrm{D}_0\cos\beta_0 - \mathrm{S}_0\sin\beta_0 - \frac{\partial L}{\partial \alpha} \approx -\mathrm{D}_0 - \frac{\partial L}{\partial \alpha} = -(C_{D_0} + C_{L_\alpha})q_\infty S_W \end{aligned} \tag{6.271}$$

In obtaining the above result, first note that α and β in Equations (6.107) are the perturbation angles of attack and sideslip, respectively, which are both zero at the reference condition. And, secondly, note that the reference value of the sideslip angle β_0 is assumed small or zero.

Table 6.6 Partial Derivatives of Aerodynamic Forces and Moments

	u	α	q	$\dot{\alpha}$	δ_E
$\frac{1}{q_\infty S_W}\frac{\partial F_{A_X}}{\partial \bullet}$	$-\left(C_{D_u}+\frac{2}{U_0}C_{D_0}\right)$	$(-C_{D_\alpha}+C_{L_0})$	$-C_{D_q}\approx 0$	$-C_{D_{\dot{\alpha}}}\approx 0$	$-C_{D_{\delta_E}}$
$\frac{1}{q_\infty S_W}\frac{\partial F_{A_Z}}{\partial \bullet}$	$-\left(C_{L_u}+\frac{2}{U_0}C_{L_0}\right)$	$(-C_{L_\alpha}-C_{D_0})$	$-C_{L_q}$	$-C_{L_{\dot{\alpha}}}$	$-C_{L_{\delta_E}}$
$\frac{1}{q_\infty S_W \bar{c}_W}\frac{\partial M_A}{\partial \bullet}$	$\left(C_{M_u}+\frac{2}{U_0}C_{M_0}\right)$	C_{M_α}	C_{M_q}	$C_{M_{\dot{\alpha}}}$	$C_{M_{\delta_E}}$

	β	p	r	δ_A	δ_R
$\frac{1}{q_\infty S_W}\frac{\partial F_{A_Y}}{\partial \bullet}$	C_{S_β}	C_{S_p}	C_{S_r}	$C_{S_{\delta_A}}$	$C_{S_{\delta_R}}$
$\frac{1}{q_\infty S_W b_W}\frac{\partial L_A}{\partial \bullet}$	C_{L_β}	C_{L_p}	C_{L_r}	$C_{L_{\delta_A}}$	$C_{L_{\delta_R}}$
$\frac{1}{q_\infty S_W b_W}\frac{\partial N_A}{\partial \bullet}$	C_{N_β}	C_{N_p}	C_{N_r}	$C_{N_{\delta_A}}$	C_{N_r}

Based on the above analysis, the perturbation aerodynamic forces may now be expressed as follows:

$$\begin{aligned}
f_{A_X} &= q_\infty S_W\left(-\left(C_{D_u}+\frac{2}{U_0}C_{D_0}\right)u+(-C_{D_\alpha}+C_{L_0})\alpha - C_{D_{\dot{\alpha}}}\dot{\alpha} - C_{D_q}q - C_{D_{\delta_E}}\delta_E\right)\\
f_{A_Y} &= q_\infty S_W(C_{S_\beta}\beta + C_{S_p}p + C_{S_r}r + C_{S_{\delta_A}}\delta_A + C_{S_{\delta_R}}\delta_R)\\
f_{A_Z} &= q_\infty S_W\left(-\left(C_{L_u}+\frac{2}{U_0}C_{L_0}\right)u-(C_{L_\alpha}+C_{D_0})\alpha - C_{L_{\dot{\alpha}}}\dot{\alpha} - C_{L_q}q - C_{L_{\delta_E}}\delta_E\right)
\end{aligned} \quad (6.272)$$

All the coefficients in these expressions have been discussed in this chapter, except $C_{D_{\dot{\alpha}}}$ and C_{D_q}, which are typically negligible.

The aerodynamic and propulsive moments are to be integrated into Equations (2.49). The expressions for the perturbation aerodynamic moments are a little more straightforward, and are

$$\begin{aligned}
l_A &= q_\infty S_W b_W(C_{L_\beta}\beta + C_{L_p}p + C_{L_r}r + C_{L_{\delta_A}}\delta_A + C_{L_{\delta_R}}\delta_R)\\
m_A &= q_\infty S_W \bar{c}_W\left(\left(C_{M_u}+\frac{2}{U_0}C_{M_0}\right)u + C_{M_\alpha}\alpha + C_{M_{\dot{\alpha}}}\dot{\alpha} + C_{M_q}q + C_{M_{\delta_E}}\delta_E\right)\\
n_A &= q_\infty S_W b_W(C_{N_\beta}\beta + C_{N_p}p + C_{N_r}r + C_{N_{\delta_A}}\delta_A + C_{N_{\delta_R}}\delta_R)
\end{aligned} \quad (6.273)$$

In a similar fashion, the expressions for the perturbation propulsive forces and moments are found to be

$$
\begin{aligned}
f_{P_X} &= q_\infty S_W\left(C_{P_{X_u}} + \frac{2}{U_0}C_{P_{X_0}}\right)u + \delta T\cos(\phi_T + \alpha_0)\\
f_{P_Z} &= q_\infty S_W\left(C_{P_{Z_u}} + \frac{2}{U_0}C_{P_{Z_0}}\right)u - \delta T\sin(\phi_T + \alpha_0) \qquad (6.274)\\
m_P &= q_\infty S_W\bar{c}_W\left(\left(C_{P_{M_u}} + \frac{2}{U_0}C_{P_{M_0}}\right)u + C_{P_{M_\alpha}}\alpha\right) + \delta T\left(d_T\cos\phi_T - x_T\sin\phi_T\right)
\end{aligned}
$$

Here, δT is the perturbation in thrust due to a perturbation in throttle setting π. Or

$$
\delta T = \frac{\partial T}{\partial \pi}\Big|_0\,\delta\pi
$$

All the coefficients in Equations (6.273) and (6.274) have also been discussed in this chapter.

Finally, the development above assumed a constant atmospheric density, which is typical for linear models of conventional flight vehicles. If, however, the atmospheric density is not assumed constant with altitude, additional altitude-dependent perturbation forces must be added to Equations (6.272). Density variations can be important for hypersonic vehicles, for example, which due to their high velocity can traverse large altitude differences even when their angles of attack and flight-path are small.

The additional aerodynamic forces due to density variations may be expressed as

$$
\begin{aligned}
f_{A_{X_h}} &\triangleq \frac{\partial F_{A_X}}{\partial h}\Big|_0 h = C_{X_0}\frac{q_\infty S_W}{\rho_\infty}\frac{\partial \rho_\infty}{\partial h}\Big|_0 = -\left(C_{D_0}\cos\beta_0 + C_{S_0}\sin\beta_0\right)\frac{q_\infty S_W}{\rho_\infty}\frac{\partial \rho_\infty}{\partial h}\Big|_0\\
f_{A_{Y_h}} &\triangleq \frac{\partial F_{A_Y}}{\partial h}\Big|_0 h = C_{Y_0}\frac{q_\infty S_W}{\rho_\infty}\frac{\partial \rho_\infty}{\partial h}\Big|_0 = \left(C_{S_0}\cos\beta_0 - C_{D_0}\sin\beta_0\right)\frac{q_\infty S_W}{\rho_\infty}\frac{\partial \rho_\infty}{\partial h}\Big|_0\\
f_{A_{Z_h}} &\triangleq \frac{\partial F_{A_Z}}{\partial h}\Big|_0 h = C_{Z_0}\frac{q_\infty S_W}{\rho_\infty}\frac{\partial \rho_\infty}{\partial h}\Big|_0 = -C_{L_0}\frac{q_\infty S_W}{\rho_\infty}\frac{\partial \rho_\infty}{\partial h}\Big|_0 \qquad (6.275)
\end{aligned}
$$

where the density-altitude gradient $\partial\rho_\infty/\partial h$ is obtained from an atmospheric model such as that given in Appendix A. Note that these perturbation forces are functions of the force coefficients C_{X_0}, C_{Y_0}, and C_{Z_0}, which are the force coefficients evaluated at the reference flight condition. Since the aerodynamic moment coefficients evaluated at this condition are often either small or zero, there are frequently no perturbation moments associated with density variations.

6.11 Summary

A great deal of material has been introduced in this chapter, including some initially confusing notation perhaps. But as mentioned, in Chapters 9 and 10 we will find how the effectiveness coefficients, or stability derivatives, defined in this chapter influence the dynamics of the vehicle, including its stability. Thus, being able to estimate these coefficients for a given vehicle is crucial.

We took the basic aerodynamic concepts introduced in Chapter 5 and developed models for the aerodynamic and propulsive forces acting on the flight vehicle. We also noted that these models could be updated as additional aerodynamic data became available, without changing the overall structure of the models. Then we demonstrated how to incorporate these forces and moments into the equations of motion.

We introduced several important concepts, including the definition and use of the stability and fuselage-reference axes, the determination of the vehicle's aerodynamic center, and the modeling of the effects of atmospheric turbulence on the forces and moments. These concepts will be applied further in Chapters 8–10.

6.12 Problems

6.1 Consider a vehicle with geometry similar to that shown in Figure 6.9, but with forward pitch-control surfaces, or canards, added to the vehicle. Canards are horizontal lifting surfaces mounted forward of the wing, and their local angle of attack is sometimes controlled by deflecting the entire surface in pitch relative to the fuselage-referenced axes. Let this canard pitch deflection be denoted δ_c. Sketch the vehicle geometry similar to that shown in Figure 6.9, showing all relevant aerodynamic forces and moments. Using an approach similar to that used in Section 6.2.5, derive an expression for the vehicle's pitching-moment coefficient that includes the effects of the canards. Finally, find the canard-deflection pitching-moment effectiveness for this vehicle, or $C_{M_{\delta_c}}$, and compare this result to the elevator-deflection pitching-moment effectiveness. (Ignore the effects of downwash from the canard.)

6.2 Consider a wingless missile as depicted in Figure 6.22, for which the nose section generates a considerable portion of the vehicle's aerodynamic lift. Assume this force acts at the axial location of the center of pressure of the nose section, X_{AC_N}, which is also the location of its aerodynamic center. Sketch the vehicle geometry similar to Figure 6.9, showing all relevant component aerodynamic forces and moments. Then, using an approach similar to that used in Section 6.2.5, derive an expression for the vehicle's pitching-moment coefficient similar to Equation (6.55). Assume the tail fins have a symmetric airfoil section and can be deflected relative to the body-referenced X axis, denoted as δ_F. By convention, use the body cross-sectional area A as the reference area and body diameter d as the reference length in nondimensionalizing aerodynamic forces and moments. Ignore body lift other than that of the nose, as well as any downwash effects.

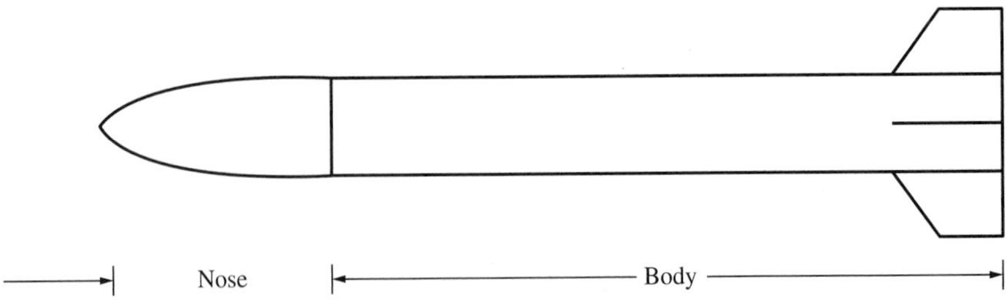

Figure 6.22 Wingless missile.

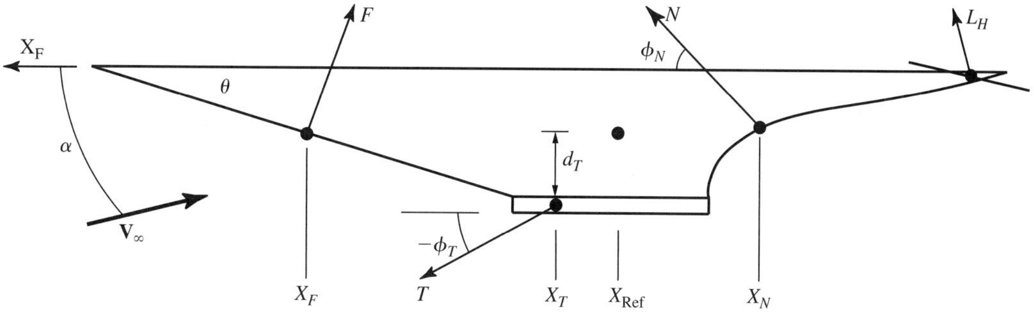

Figure 6.23 Hypersonic vehicle.

6.3 Consider a hypersonic vehicle, similar to NASA's Hyper-X or X-43, as shown schematically in Figure 6.23. The forebody generates an aerodynamic normal force F, plus compresses the oncoming air prior to ingestion into the under slung SCRAMjet propulsion system. This engine produces a thrust T, but due to the turning of the ingested air this resultant force T acts at an $-\phi_T$ relative to the vehicle's X axis. The engine exhaust plume creates a pressure distribution on the aft external nozzle generating a resultant force N as shown. Assume the points of action of forces F and N lie on a horizontal line that passes through the reference point shown, but T acts at a point d_T below the reference point (consistent with Figure 6.13). Also assume the all-moving horizontal tail has a symmetric airfoil section and can be deflected relative to the vehicle, with deflection angle denoted as δ_H.

Using an approach similar to that used in Section 6.2.5, derive an expression for the vehicle's pitching-moment about the reference location and then develop an expression for the vehicle's pitching-moment coefficient similar to Equation (6.55). Use S_{Ref} as the reference area and l_{Ref} as the reference length in nondimensionalizing aerodynamic forces and moments. Assume that aerodynamic lift is generated only by the forebody and horizontal tail, and ignore any downwash effects.

6.4 Consideration is being given to stretching a commercial transport aircraft by extending the length of its fuselage. This would result in the horizontal tail's aerodynamic center being $2\bar{c}_W$ further aft, and the location of X_{Ref} (the reference location about which pitching moment is defined) being $0.25\bar{c}_W$ further aft of the wing's aerodynamic center. The wing and tail geometry would remain the same. If wind-tunnel and flight tests of the original design indicate that the original vehicle's angle-of-attack lift and pitching-moment effectivenesses were C'_{L_α} and C'_{M_α}, respectively, derive expressions in terms of C'_{L_α} and C'_{M_α} that predict the change in the values in these coefficients, or ΔC_{L_α} and ΔC_{M_α}, due to the stretched fuselage.

6.5 Consider the vehicle addressed in Examples 6.1–6.6. Determine the effects of lateral-velocity perturbations on the aerodynamic and propulsive forces and moments on this vehicle.

6.6 Consider the vehicle addressed in Examples 6.1–6.7. Determine the effects of roll-rate perturbations on the aerodynamic and propulsive forces and moments on this vehicle.

6.7 Consider the vehicle addressed in Examples 6.1–6.7. Determine the effects of yaw-rate perturbations on the aerodynamic and propulsive forces and moments on this vehicle.

6.8 Consider the vehicle addressed in Examples 6.1–6.7. Determine the effects of perturbations in rate of change of angle of attack on the aerodynamic and propulsive forces and moments on this vehicle. Compare your results to the effects of pitch-rate perturbations on these forces and moments.

6.9 Following the approach presented in Section 6.8 leading to Equation (6.257), develop similar expressions that capture the effects of gusts on the aerodynamic moments acting on a vehicle.

6.10 Starting with Equations (6.107), verify the entries for $\frac{\partial F_{A_X}}{\partial u}$ and $\frac{\partial F_{A_X}}{\partial \alpha}$ in Table 6.6.

References

1. Teper, Gary L.: "Aircraft Stability and Control Data," STI TR-176-1, Systems Technology, Inc., Hawthorne, CA, April 1969.
2. "USAF Stability and Control DATCOM," prepared by the Douglas Aircraft Div., McDonnell Douglas Corp. for the USAF Flight Dynamics Laboratory, Wright Patterson AFB, Ohio, October 1960 (revised April 1978).
3. Multhopp, H.: "Aerodynamics of the Fuselage," NACA TM-1036, 1942.
4. Perkins, C. D. and R. E. Hage: *Airplane Performance Stability and Control,* Wiley, New York, 1949.
5. Chavez, F. R. and D. K. Schmidt: "An Analytical Model and Dynamic Analysis of an Aeropropulsive/Aeroelastic Hypersonic Vehicle," *Journal of Guidance, Control, and Dynamics,* vol. 17, no. 6, 1994.

CHAPTER 7

Effects of Elastic Deformation on the Forces and Moments

Chapter Roadmap: *The material in this chapter would not normally be included in a first course in flight dynamics, but in later courses. Before beginning the study of the chapter, the student should be very familiar with the material in Chapters 3–6.*

In this chapter we will continue the discussion we began in Chapter 6 on modeling the aerodynamic forces and moments acting on the vehicle, but we will now focus on the effects of elastic deformation of the vehicle's structure on these forces and moments. We will continue to draw from techniques presented in Chapters 5 and 6, including the component build-up method and strip theory, which has been successfully applied to aeroelastic modeling for quite a while (Ref. 1). Conventional vehicle geometry will be assumed—the vehicle consists of a fuselage, wing, and aft horizontal and vertical tails—although other geometries may be treated using the approaches presented here.

Given that our overall topic is atmospheric flight dynamics, rather than purely structural dynamics, we will be especially interested in possible interactions between the flexible structure and the "rigid-body" motion (i.e., rigid-body degrees of freedom). That is, we will be interested in determining whether the dynamics of the rigid-body degrees of freedom are significantly affected by the elastic motion or by the static-elastic deflections of the structure. We are not, for example, focused on flutter, which involves the interactions of two or more elastic degrees of freedom. As a result, the lower-frequency elastic modes of the structure are of particular interest. And, therefore, unsteady aerodynamic effects will not be considered, which is justified whenever the elastic modal periods are long compared to the rate of change of the pressure distributions over the lifting surfaces.

7.1 A MOTIVATIONAL AEROELASTIC EXAMPLE

To motivate the study of elastic effects on the aerodynamic forces and moments, consider an idealized 2-D airfoil as shown in Figure 7.1. The airfoil shape and control surface are those considered in Example 5.2. The airfoil here is suspended by two springs, one translational and one rotational, as in a wind tunnel, for example. The center of mass (gravity), the aerodynamic center (ac) of the section, and a trailing-edge control surface are also shown, along with the 2-D lift l and moment about the aerodynamic center m_{ac}. The section can both pitch and plunge with displacements given by θ and z_{cm}, respectively. Using the airfoil data from Example 5.2, along with the additional data below, we will compare the lift and moment acting on the elastic airfoil to that of the airfoil as if it were rigid.

$$V_\infty = 75 \text{ fps}, c\left(\text{chord}\right) = 2 \text{ ft}, \Delta x = 0.2 \text{ ft}$$

$$m\left(\text{mass}\right) = 1 \text{ sl/ft}, I_{cm}\left(\text{inertia}\right) = 1 \text{ sl-ft}^2\text{/ft}$$

$$k_p\left(\text{spring constant}\right) = 35 \text{ lb/ft}, k_t\left(\text{spring constant}\right) = 144 \text{ ft-lb/ft}$$

First, note that in the absence of any aerodynamic forces and moments, the equations governing the free-vibration of the suspended airfoil section are

$$m\ddot{z}_{cm} + k_p z_{cm} = 0$$

$$I_{cm}\ddot{\theta} + k_t\theta = 0$$

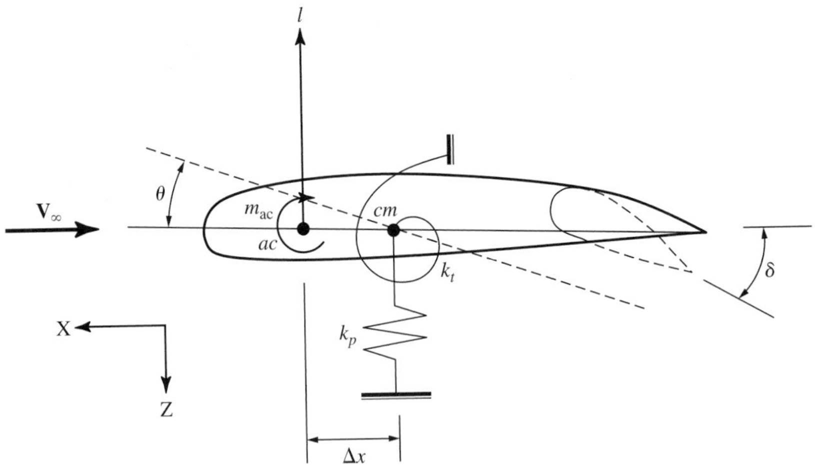

Figure 7.1 Aeroelastic 2-D airfoil schematic.

Or in matrix notation, with the mass and stiffness matrices **M** and **K** introduced in Chapter 3, we have

$$\begin{bmatrix} m & 0 \\ 0 & I_{\text{cm}} \end{bmatrix} \begin{Bmatrix} \ddot{z}_{\text{cm}} \\ \ddot{\theta} \end{Bmatrix} + \begin{bmatrix} k_p & 0 \\ 0 & k_t \end{bmatrix} \begin{Bmatrix} z_{\text{cm}} \\ \theta \end{Bmatrix} = \begin{Bmatrix} 0 \\ 0 \end{Bmatrix}$$

$$[\mathbf{M}]\{\ddot{\mathbf{q}}\} + [\mathbf{K}]\{\mathbf{q}\} = \{\mathbf{0}\}$$

The equations are decoupled, and the two vibration frequencies are simply

$$\omega_1 = \sqrt{\frac{k_p}{m}} = \sqrt{35} \text{ rad/sec} \qquad \omega_2 = \sqrt{\frac{k_t}{I_{\text{cm}}}} = \sqrt{144} \text{ rad/sec}$$

Now with the aerodynamic lift and moment (taken about the *cm*), and adding structural viscous damping ζ, the equations of motion become

$$\ddot{z}_{\text{cm}} + \left(2\zeta\sqrt{\frac{k_p}{m}}\right)\dot{z}_{\text{cm}} + (k_p/m)z_{\text{cm}} + l/m = 0$$

$$\ddot{\theta} + \left(2\zeta\sqrt{\frac{k_t}{I_{\text{cm}}}}\right)\dot{\theta} + (k_t/I_{\text{cm}})\theta - m_{\text{cm}}/I_{\text{cm}} = 0$$

where, in terms of the 2-D aerodynamic coefficients for the airfoil, the lift and moment (per unit span) may be expressed as

$$l = \left(c_{l_\alpha}(\alpha - \alpha_0) + c_{l_\delta}\delta\right)q_\infty c$$

$$m_{\text{cm}} = m_{\text{ac}} + l\Delta x = \left(\left(c_{m_{\text{ac}}} + c_{m_\delta}\delta\right) + c_{l_\alpha}(\alpha - \alpha_0)\frac{\Delta x}{c}\right)q_\infty c^2$$

And from Example 5.2 we know that

$$\alpha_0 = -2 \text{ deg } (-0.035 \text{ rad}),\ c_{l_\alpha} = 0.108 \text{ /deg } (6.19 \text{ /rad}),\ c_{l_\delta} = 0.092 \text{ /deg } (5.27 \text{ /rad}),$$

$$c_{m_{\text{ac}}} = -0.047,\ c_{m_\delta} = -0.012 \text{ /deg } (-0.688 \text{ /rad})$$

But note that the angle of attack of the airfoil section may be expressed in terms of the motion variables, or

$$\boxed{\alpha = \theta + \frac{\dot{z}_{\text{cm}}}{V_\infty}}$$

Consequently, the lift and moment may also be expressed in terms of these motion variables, or

$$l = \left(c_{l_\alpha}\left(\theta + \frac{\dot{z}_{cm}}{V_\infty} - \alpha_0\right) + c_{l_\delta}\delta\right) q_\infty c \triangleq \left(c_{l_0} + c_{l_\theta}\theta + c_{l_{\dot{z}_{cm}}}\dot{z}_{cm} + c_{l_\delta}\delta\right) q_\infty c$$

$$m_{cm} = \left(\left(c_{m_{ac}} + c_{m_\delta}\delta\right) + c_{l_\alpha}\left(\theta + \frac{\dot{z}_{cm}}{V_\infty} - \alpha_0\right)\frac{\Delta x}{c}\right) q_\infty c^2 \triangleq \left(c_{m_0} + c_{m_\theta}\theta + c_{m_{\dot{z}_{cm}}}\dot{z}_{cm} + c_{m_\delta}\delta\right) q_\infty c^2$$

where

$$c_{l_0} = -c_{l_\alpha}\alpha_0, \quad c_{l_\theta} = c_{l_\alpha}, \quad c_{l_\delta} = c_{l_\delta}, \quad c_{l_{\dot{z}_{cm}}} = \frac{c_{l_\alpha}}{V_\infty}$$

$$c_{m_0} = \left(c_{m_{ac}} - c_{l_\alpha}\alpha_0\frac{\Delta x}{c}\right), \quad c_{m_\delta} = c_{m_\delta}, \quad c_{m_\theta} = c_{l_\alpha}\frac{\Delta x}{c}, \quad c_{m_{\dot{z}_{cm}}} = \frac{c_{l_\alpha}}{V_\infty}\frac{\Delta x}{c}$$

The above eight coefficients are the 2-D aeroelastic effectiveness coefficients describing how the motion variables (pitch and plunge) influence the lift and moment on the section. The development of similar coefficients, but for a vehicle, will be a main focus of this chapter.

The equations of motion for the airfoil section now become

$$\ddot{z}_{cm} + \left(2\zeta\sqrt{\frac{k_p}{m}} + \frac{1}{m}c_{l_{\dot{z}_{cm}}}q_\infty c\right)\dot{z}_{cm} + \frac{k_p}{m}z_{cm} + \frac{1}{m}\left(c_{l_\theta}q_\infty c\right)\theta = -\frac{1}{m}\left(c_{l_0}q_\infty c\right) - \frac{1}{m}\left(c_{l_\delta}q_\infty c\right)\delta$$

$$\ddot{\theta} + \left(2\zeta\sqrt{\frac{k_t}{I_{cm}}}\right)\dot{\theta} + \frac{1}{I_{cm}}\left(k_t - c_{m_\theta}q_\infty c^2\right)\theta - \frac{1}{I_{cm}}\left(c_{m_{\dot{z}_{cm}}}q_\infty c^2\right)\dot{z}_{cm} = \frac{1}{I_{cm}}\left(c_{m_0}q_\infty c^2\right) + \frac{1}{I_{cm}}\left(c_{m_\delta}q_\infty c^2\right)\delta$$

which, for a given dynamic pressure, may be solved via simulation techniques discussed in Chapter 8. Taking the atmospheric density ρ_∞ to be that for sea level, or 0.002377 sl/ft^3 in the calculation of dynamic pressure q_∞, and letting the structural viscous damping be a nominal $\zeta = 0.02$, the time histories of the section pitch and plunge displacements for a step input of the control surface of 1 deg are shown in Figure 7.2.

To compare the lift and moment between the rigid and elastic section, we have the lift and moment for the rigid section given by

$$l = \left(c_{l_\alpha}(-\alpha_0) + c_{l_\delta}\delta\right) q_\infty c = 4.12 \text{ lb/ft}$$

$$m_{cm} = \left(\left(c_{m_{ac}} + c_{m_\delta}\delta\right) + c_{l_\alpha}(-\alpha_0)\frac{\Delta x}{c}\right) q_\infty c^2 = -1.00 \text{ ft-lb/ft}$$

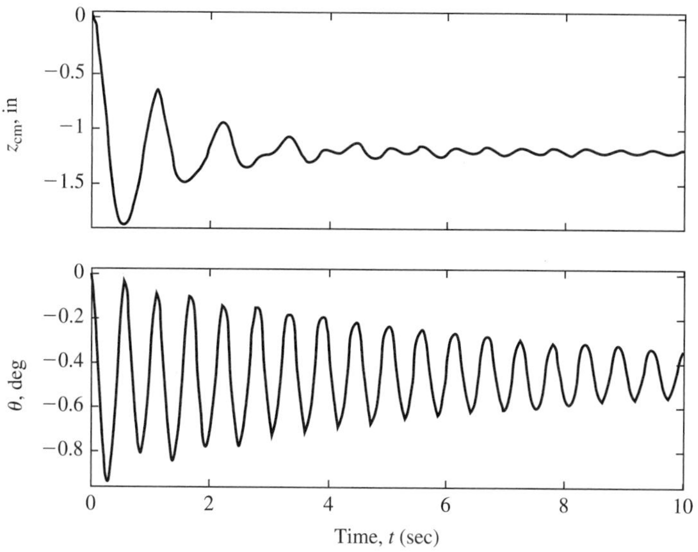

Figure 7.2 Time histories of aeroelastic wing section.

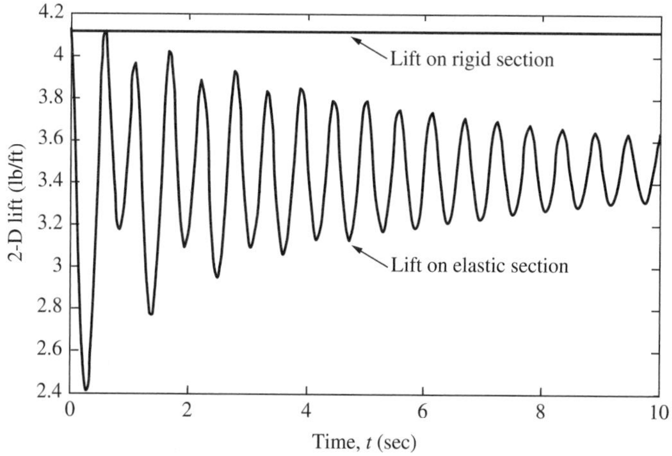

Figure 7.3 Lift on rigid and elastic airfoil section.

where here the angle of attack remains zero. For the elastic section, with the displacements given as shown in Figures 7.2, the lift and moment are as shown in Figures 7.3 and 7.4. Clearly, the results for the rigid and elastic section are quite different, thus demonstrating that the elastic effects on the aerodynamic forces and moments can be significant. The steady-state value of lift on the elastic section is approximately 3.5 lb/ft, which corresponds to a 15 percent reduction

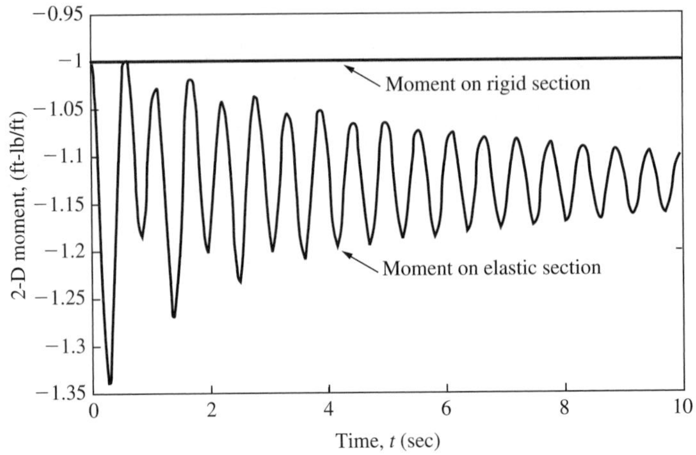

Figure 7.4 Pitching moment on rigid and elastic airfoil section.

in lift. Also, the steady-state value of pitching moment for the elastic section is about 1.13 ft-lb/ft, which corresponds to a 13 percent increase in moment. The peak differences in lift and moment are much larger. Such differences in forces and moments can, in turn, significantly affect the vehicle's dynamics.

7.2 ELASTIC DEFORMATION REVISITED

In Chapter 6 we noted that the total aerodynamic force and moment vectors acting on the vehicle, with components given in vehicle-fixed axes, were

$$\begin{aligned} \mathbf{F}_A &= F_{A_X}\mathbf{i}_V + F_{A_Y}\mathbf{j}_V + F_{A_Z}\mathbf{k}_V \\ \mathbf{M}_A &= L_A\mathbf{i}_V + M_A\mathbf{j}_V + N_A\mathbf{k}_V \end{aligned} \tag{7.1}$$

In this chapter we seek to determine how elastic deformation affects the components of these two vectors. Consistent with the approach in Chapter 6, we will develop Taylor-series expansions of these forces and moments in terms of a parameter vector **p**. Or

$$\begin{aligned} \mathbf{F}_A &= \mathbf{F}_A(\mathbf{p}) \\ \mathbf{M}_A &= \mathbf{M}_A(\mathbf{p}) \end{aligned}$$

In Chapter 6, **p** included only the variables associated with rigid-body motion (e.g., U, α, β, P, Q, R, δ_E, i_H, δ_R, δ_A). Here we will expand **p** to include the variables associated with elastic deformation. Specifically, we will include the effects of the modal coordinates η_i. Note that consistent with our notational convention, the total modal-coordinate displacement may also be expressed in terms of a reference value plus a perturbation from that reference, or

$$\eta_i(t) = H_{i_0}(t) + \eta_i(t) \tag{7.2}$$

Also in Chapter 6 we showed that the force components in vehicle-fixed axes could be written as

$$\begin{aligned} F_{A_X} &= C_X q_\infty S_W = -D\cos\alpha\cos\beta - \mathsf{S}\cos\alpha\sin\beta + L\sin\alpha \\ F_{A_Y} &= C_Y q_\infty S_W = -D\sin\beta + \mathsf{S}\cos\beta \\ F_{A_Z} &= C_Z q_\infty S_W = -D\sin\alpha\cos\beta - \mathsf{S}\sin\alpha\sin\beta - L\cos\alpha \end{aligned} \tag{7.3}$$

where L is the total vehicle lift, D the total drag, S the total side force, and α and β define the direction of the free-stream velocity vector $\mathbf{V}_\infty$ relative to the chosen vehicle-fixed frame. We will determine here the effects of elastic deformation on these force components by focusing on lift and side force. The elastic effects on drag will be assumed negligible, which is a good approximation when the elastic deformations are small compared to the dimensions of the overall vehicle.

Regarding the components of the moments appearing in Equations (7.1), L_A is the aerodynamic rolling moment, M_A the aerodynamic pitching moment, and N_A the aerodynamic yawing moment acting on the vehicle. We will also find the effects of elastic deformation on these moments.

Finally, in Chapter 4, Equation (4.84), we determined that the equations of motion governing the elastic degrees of freedom are of the form

$$\ddot{\eta}_i + \omega_i^2 \eta_i = \frac{Q_i}{\mathcal{M}_i}, \; i = 1 \ldots n \tag{7.4}$$

where Q_i is the generalized force and $\mathcal{M}_i$ the generalized mass, each associated with the i'th free-vibration mode. Here we will also derive expressions for estimating these generalized forces.

Regarding the elastic motion itself, we had shown in Chapter 3, and applied in Chapter 4, the fact that the displacement $\mathbf{d}_E(x,y,z)$ of any point on a structure due to elastic deformation could be represented as an infinite sum of modal displacements, each written in terms of the mode shapes $\boldsymbol{\nu}_i(x,y,z)$ and modal coordinates $\eta_i(t)$ of the vibration modes of the structure. That is, at any time t the elastic displacement (vector) $\mathbf{d}_E$ of any point (x,y,z) on the undeformed structure may be written as

$$\mathbf{d}_E(x,y,z,t) = \sum_{i=1}^{\infty} \boldsymbol{\nu}_i(x,y,z)\eta_i(t)_i \tag{7.5}$$

Also, recall from Equation (4.15) that $\boldsymbol{\nu}_i(x,y,z)$ is the (vector) value of the mode shape associated with mode i, evaluated at the location (x,y,z) on the structure, and $\eta_i(t)$ is the value of the i'th modal coordinate at time t. Consistent with Chapter 4, it is assumed throughout this chapter that the mode shapes are available from a previous free-vibration analysis of the structure (e.g., finite-element analysis).

The mode shape $\boldsymbol{\nu}_i$ includes displacements in each of the X, Y, and Z directions of the chosen vehicle-fixed reference frame, That is,

$$\boldsymbol{\nu}_i = \begin{bmatrix} \nu_{X_i} & \nu_{Y_i} & \nu_{Z_i} \end{bmatrix} \begin{Bmatrix} \mathbf{i}_V \\ \mathbf{j}_V \\ \mathbf{k}_V \end{Bmatrix} \tag{7.6}$$

where $\mathbf{i}_V$, $\mathbf{j}_V$, and $\mathbf{k}_V$ are unit vectors in that chosen frame (e.g., fuselage-referenced axes). For vehicles that are symmetric with respect to their XZ plane, the elastic modes may be grouped into symmetric and antisymmetric modes. As the name implies, the symmetric mode shapes will be symmetric with respect to the plane of symmetry, and these modes will contain symmetric wing bending about the X_V axis as well as fuselage bending about the Y_V axis, for example. The antisymmetric modes have mode shapes that are asymmetric with respect to the plane of symmetry, and will contain asymmetric wing bending and fuselage bending about the Z_V axis, for example. We will make note of this modal grouping throughout this chapter.

In Sections 7.3–7.8 that follow, the same basic approach will be repeatedly taken in the modeling of elastic effects on the forces and moment. This approach will consist of four steps:

1. Characterize the mechanisms by which elastic deformation influences the local 2-D airfoil section lift and moment for all relevant lifting surfaces.
2. Express the elastic deformation in terms of the mode shapes and modal coordinates of the vibration modes of the structure.
3. Integrate the resulting expressions over the spans of the relevant lifting surfaces.
4. Extract closed-form expressions for the partial derivatives of the forces and moments with respect to the modal coordinates of the vibration modes.

7.3 ELASTIC EFFECTS ON LIFT

In Section 6.2.1, assuming the vehicle lift is essentially generated by the lift of a wing and horizontal tail, we wrote the vehicle lift as

$$L = L_W + L_H$$

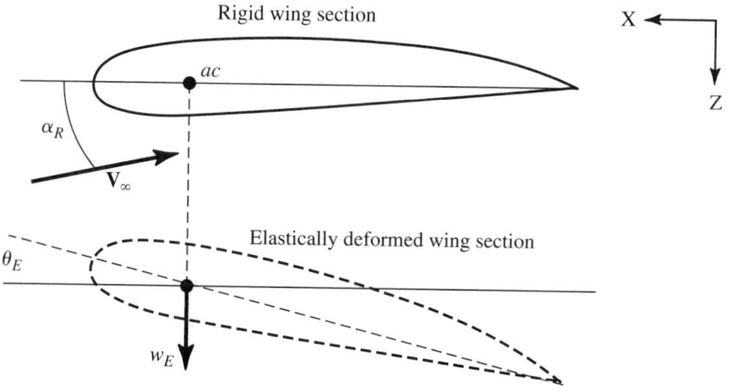

Figure 7.5 Rigid and deformed wing section.

And we proceeded to write the wing and tail lift in terms of the local angles of attack of the wing and tail sections.

But elastic deformation affects those local angles of attack, and hence the lift. To model this effect, consider a wing or tail 2-D section as depicted in Figure 7.5. Let the elastic motion of the section, relative to its undeformed position, include local plunge velocity of the aerodynamic center $w_E(y)$ and local rotational $\theta_E(y)$ of the chord, where y is the span location of the section. (Note that in Chapter 5 we denoted the span-wise twist angle of a rigid wing as $\varepsilon(y)$. The elastic twist here is similar geometrically, but we have denoted it as $\theta_E(y)$). Therefore, writing the angle of attack due to plunge velocity as $w_E(y)/V_\infty$, where V_∞ is the free-stream velocity, we may write the local angle of attack of a wing section as

$$\alpha_W(y) = \alpha_{R_W}(y) + \alpha_{E_W}(y) = \alpha_{R_W}(y) + \left(\theta_E(y) + \frac{w_E(y)}{V_\infty}\right) \tag{7.7}$$

where α_{R_W} is the angle of attack of the undeformed or rigid wing and α_{E_W} is the contribution to the section's angle of attack due to the elastic deformation. Note that the rigid angle of attack α_{R_W} will include the effects of vehicle rigid-body plunge velocity W, rigid-body roll rate P, rigid-wing twist $\varepsilon(y)$, and so on.

For a tail section we have the local angle of attack given by

$$\alpha_H(y) = \alpha_{R_H}(y) + \alpha_{E_H}(y) = \alpha_{R_H}(y) + \left(\theta_{E_H}(y) + \frac{w_{E_H}(y)}{V_\infty} - \varepsilon_{E_H}(y)\right) \tag{7.8}$$

Here we have included an additional downwash angle ε_{E_H} at the tail due to elastic deformation of the wing. Note that α_{R_H} will include the downwash at the tail associated with the angle of attack α_{R_W} of the rigid wing.

Ignoring any change in section camber due to elastic deformation, and assuming locally linear aerodynamics, the 2-D section wing- or tail-section lift may be written as

$$l(y) = c_{l_\alpha}(y)\big(\alpha(y) - \alpha_0(y)\big)q(y)c(y) \tag{7.9}$$

where the angle of attack α is given by either Equation (7.7) or (7.8). Integrating this section lift over the wing span b_W and tail span b_H, respectively, and assuming the wing and tail are symmetric with respect to their root-chord planes, yields the following expressions for the lift on the wing and tail.

$$\begin{aligned} L_W &= 2\int_0^{b_W/2} c_{l_{\alpha_W}}(y)\big(\alpha_W(y) - \alpha_{0_W}(y)\big)q_W(y)c_W(y)\,dy \\ L_H &= 2\int_0^{b_H/2} c_{l_{\alpha_H}}(y)\big(\alpha_H(y) - \alpha_{0_H}(y)\big)q_H(y)c_H(y)\,dy \end{aligned} \tag{7.10}$$

But by substituting Equations (7.7) and (7.8) into the above expressions, we may extract only that part of the wing and tail lift associated with elastic deformation. So the change in wing lift <u>due to elastic effects</u> is simply

$$\boxed{\begin{aligned} L_{E_W} &= 2\int_0^{b_W/2} c_{l_{\alpha_W}}(y)\left(\theta_E(y) + \frac{w_E(y)}{V_\infty}\right)q_W(y)c_W(y)\,dy \\ &\approx 2q_\infty\int_0^{b_W/2} c_{l_{\alpha_W}}(y)\left(\theta_E(y) + \frac{w_E(y)}{V_\infty}\right)c_W(y)\,dy \end{aligned}} \tag{7.11}$$

where in obtaining the result we have assumed that local dynamic pressure $q(y)$ is constant with span and equal to the free-stream dynamic pressure. In making this assumption, we are in effect ignoring any surge (X) elastic displacement of the wing. This is justified due to the fact that wing sections are normally stiffer in surge bending (bending about the Z axis) than in plunge bending (bending about the X axis).

Similar to the wing, the change in lift <u>due to elastic effects</u> on the horizontal tail may be written as

$$L_{E_H} = 2\int_0^{b_H/2} c_{l_{\alpha_H}}(y)\left(\theta_{E_H}(y) + \frac{w_{E_H}(y)}{V_\infty} - \varepsilon_{E_H}(y)\right)q_H(y)c_H(y)dy$$
$$\approx 2q_H\int_0^{b_H/2} c_{l_{\alpha_H}}(y)\left(\theta_{E_H}(y) + \frac{w_{E_H}(y)}{V_\infty} - \varepsilon_{E_H}(y)\right)c_H(y)dy \tag{7.12}$$

where the additional downwash at the tail due to elastic effects ε_{E_H} may be found as follows. Since

$$\varepsilon_H \overset{\Delta}{=} \varepsilon_{R_H} + \varepsilon_{E_H} = \frac{d\varepsilon_H}{d\alpha_W}(\alpha_W - \alpha_{0_W})$$
$$= \frac{d\varepsilon_H}{d\alpha_W}(\alpha_{R_W} + \alpha_{E_W} - \alpha_{0_W}) = \frac{d\varepsilon_H}{d\alpha_W}((\alpha_{R_W} - \alpha_{0_W}) + \alpha_{E_W}) \tag{7.13}$$

we have

$$\varepsilon_{E_H} = \frac{d\varepsilon_H}{d\alpha_W}\alpha_{E_W} = \frac{d\varepsilon_H}{d\alpha_W}\left(\theta_{E_W}(y) + \frac{w_{E_W}(y)}{V_\infty}\right) \tag{7.14}$$

And therefore we have the change in tail lift due to elastic effects given by

$$L_{E_H} \approx 2q_H\int_0^{b_H/2} c_{l_{\alpha_H}}(y)\left(\left(\theta_{E_H}(y) + \frac{w_{E_H}(y)}{V_\infty}\right) - \frac{d\varepsilon_H}{d\alpha_W}\left(\theta_{E_W}(y) + \frac{w_{E_W}(y)}{V_\infty}\right)\right)c_H(y)dy \tag{7.15}$$

Note here that due to the downwash, the elastic contribution to tail lift is a function of the wing's elastic deformation, and that the lag of downwash discussed in Section 6.7.4 has been ignored.

7.3.1 Effects of Modal Displacement

With the wing and tail lift associated with elastic effects given by Equations (7.11) and (7.15), respectively, attention now turns to finding the modal contribution to the lift. Recall from Equation (7.5) that elastic displacement $\mathbf{d}_E$ of a point on the undeformed structure (x,y,z) may be expressed in terms of the mode shapes and modal coordinates, or

$$\mathbf{d}_E(x,y,z,t) = \sum_{i=1}^{\infty} \boldsymbol{\nu}_i(x,y,z)\eta_i(t)$$

Therefore, one should be able to express the relevant elastic deformations in terms of the modal parameters. Also, due to the symmetry of the vehicle, only the symmetric vibration modes will contribute significantly to the vehicle lift, so only these modes need be considered here.

We may now write the elastic wing twist $\theta_E(\mathrm{y})$ in terms of the local slopes of the wing's vibration mode shapes. Consider the elastic deformation of the 2-D section shown in Figure 7.6, similar to Figure 7.5. Recall that we are ignoring any deformation of the airfoil section such as a change in camber. Note that the plunge (Z) displacement of two points A and B on the airfoil section, with locations (x_A,y,z_A) and (x_B,y,z_B), may be written as

$$\begin{aligned} z_A(y,t) &= \sum_{i=1}^{\infty} \nu_{Z_i}(x_A,y,z_A)\eta_i(t) \\ z_B(y,t) &= \sum_{i=1}^{\infty} \nu_{Z_i}(x_B,y,z_B)\eta_i(t) \end{aligned} \tag{7.16}$$

where $\nu_{Z_i}(x,y,z)$ is the Z (plunge) component of the i'th mode shape evaluated at the location (x,y,z) on the (undeformed) structure.

Now note that the elastic wing twist of the section $\theta_E(y)$ may be written in terms of the plunge displacement of these two points A and B. Or at time t we have

$$\theta_E(y,t) \approx \tan\theta_E(y,t) = \frac{1}{x_A - x_B}\sum_{i=1}^{\infty}\left(\nu_{Z_i}(x_B,y,z_B) - \nu_{Z_i}(x_A,y,z_A)\right)\eta_i(t) \tag{7.17}$$

But also note that

$$\lim_{(x_A - x_B)\to 0}\left(\frac{\nu_{Z_i}(x_B,y,z_B) - \nu_{Z_i}(x_A,y,z_A)}{x_A - x_B}\right) \triangleq \nu'_{Z_i}(y) \tag{7.18}$$

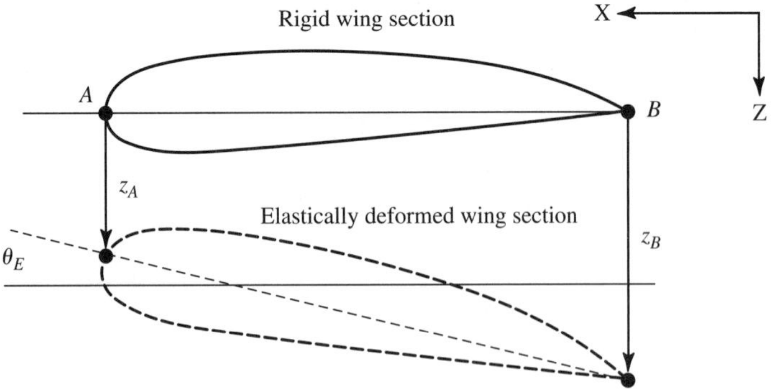

Figure 7.6 Elastic wing twist in terms of plunge displacements.

is just the slope (leading-edge up) of the i'th mode shape evaluated at span location y, as shown in Figure 7.7. Consequently, we may write the elastic wing or tail twist angle at span y as

$$\boxed{\theta_E(y,t) = \sum_{i=1}^{\infty} \nu'_{Z_i}(y)\eta_i(t)} \tag{7.19}$$

where $\nu'_{Z_i}(y)$ is the slope of the i'th mode shape evaluated at span y on the wing or tail.

With the elastic twist (or torsion) angle expressed in terms of modal parameters, we may now determine the contribution of modal displacements to vehicle lift. Or, consistent with Chapter 6, we seek to determine the partial derivative of the lift with respect to the modal coordinate $\dfrac{\partial L}{\partial \eta_i}$. Substituting Equation (7.19) into Equations (7.11) and (7.15), we may write the changes in wing and tail lift, respectively, due to elastic deformation as

$$\begin{aligned}
L_{E_W} &= 2q_\infty \int_0^{b_W/2} c_{l_{\alpha_W}}(y)\left(\sum_{i=1}^{\infty} \nu'_{Z_{i_W}}(y)\eta_i(t) + \frac{w_E(y)}{V_\infty}\right)c_W(y)\,dy \\
&= 2q_\infty\left(\sum_{i=1}^{\infty}\left(\underline{\int_0^{b_W/2} c_{l_{\alpha_W}}(y)\nu'_{Z_{i_W}}(y)c_W(y)\,dy}\right)\eta_i(t)\right. \\
&\quad \left. + \frac{1}{V_\infty}\int_0^{b_W/2} c_{l_{\alpha_W}}(y)w_E(y)c_W(y)\,dy\right) \\
L_{E_H} &= 2q_H \int_0^{b_H/2} c_{l_{\alpha_H}}(y)\left(\left(\sum_{i=1}^{\infty} \nu'_{Z_{i_H}}(y)\eta_i(t) + \frac{w_{E_H}(y)}{V_\infty}\right)\right. \\
&\quad \left. - \frac{d\varepsilon_H}{d\alpha_W}\left(\sum_{i=1}^{\infty} \nu'_{Z_{i_W}}(y)\eta_i(t) + \frac{w_{E_W}(y)}{V_\infty}\right)\right)c_H(y)\,dy \\
&= 2q_H\left(\sum_{i=1}^{\infty}\left(\underline{\int_0^{b_H/2} c_{l_{\alpha_H}}(y)\left(\nu'_{Z_{i_H}}(y) - \frac{d\varepsilon_H}{d\alpha_W}\nu'_{Z_{i_W}}(y)\right)c_H(y)\,dy}\right)\eta_i(t)\right. \\
&\quad \left. + \frac{1}{V_\infty}\int_0^{b_H/2} c_{l_{\alpha_H}}(y)\left(w_{E_H}(y) - \frac{d\varepsilon_H}{d\alpha_W}w_{E_W}(y)\right)c_H(y)\,dy\right)
\end{aligned} \tag{7.20}$$

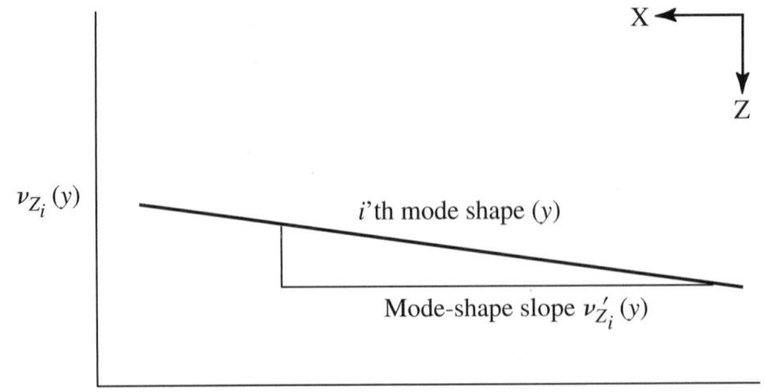

Figure 7.7 Mode-shape slope at span y.

In obtaining the two expressions in Equations (7.20) for the wing and tail, we have replaced the integrals over the summations with summations of integrals. Also note that the underlined terms in the summations are not explicit functions of time, since the time dependence of the elastic displacement is contained only in the modal coordinates $\eta_i(t)$.

So ignoring any fuselage lift, Equations (7.20) explicitly include the effects of the modal displacements η_i on vehicle lift. Now we wish to find the partial derivative of vehicle lift with respect to the modal displacement. Combining the lift on the wing and tail, differentiation of Equations (7.20) yields

$$\begin{aligned}\frac{\partial L}{\partial \eta_i} &\triangleq C_{L_{\eta_i}} q_\infty S_W = \frac{\partial L_{E_W}}{\partial \eta_i} + \frac{\partial L_{E_H}}{\partial \eta_i} \\ &= 2q_\infty \int_0^{b_W/2} c_{l_{\alpha_W}}(y)\nu'_{Z_{i_W}}(y) c_W(y)\,dy \\ &+ 2q_H \int_0^{b_H/2} c_{l_{\alpha_H}}(y)\left(\nu'_{Z_{i_H}}(y) - \frac{d\varepsilon_H}{d\alpha_W}\nu'_{Z_{i_W}}(y)\right) c_H(y)\,dy\end{aligned} \tag{7.21}$$

Here we have also introduced the modal-displacement lift effectiveness $C_{L_{\eta_i}}$, which from the above expression is found to be

$$\begin{aligned}C_{L_{\eta_i}} = \frac{2}{S_W}\Bigg(&\int_0^{b_W/2} c_{l_{\alpha_W}}(y)\nu'_{Z_{i_W}}(y) c_W(y)\,dy \\ &+ \frac{q_H}{q_\infty}\int_0^{b_H/2} c_{l_{\alpha_H}}(y)\left(\nu'_{Z_{i_H}}(y) - \frac{d\varepsilon_H}{d\alpha_W}\nu'_{Z_{i_W}}(y)\right) c_H(y)\,dy\Bigg)\end{aligned} \tag{7.22}$$

Note that this coefficient is a function of wing and tail geometry, section lift-curve slopes, and slopes of the vibration mode shapes. The integral terms are usually evaluated numerically from tabulated mode-shape data, and only symmetric modes need be considered. This effectiveness is analogous to the vehicle's angle-of-attack lift effectiveness C_{L_α}, for example.

7.3.2 Effects of Modal Velocity

We will show in this section how the vehicle lift is also a function of modal velocity, or $\dot{\eta}_i(t)$, and we will determine the partial derivative of lift with respect to this velocity, or $\dfrac{\partial L}{\partial \dot{\eta}_i}$.

From Equations (7.11) and (7.15) we can see that the wing and tail lift depend on the local elastic plunge velocities $w_{E_W}(y)$ and $w_{E_H}(y)$. But these velocities are expressible in terms of modal parameters. In fact, from Figures 7.5 and 7.6 and Equations (7.16) we can write for the local plunge velocity for either a wing or tail section

$$w_E(y,t) = \dot{z}_{\mathrm{ac}}(y,t) = \sum_{i=1}^{\infty} \nu_{Z_i}(x_{\mathrm{ac}},y,z_{\mathrm{ac}})\dot{\eta}_i\,(t) \triangleq \sum_{i=1}^{\infty} \nu_{Z_i}(y)\dot{\eta}_i(t) \tag{7.23}$$

And, therefore, we can write Equations (7.11) and (7.15) for the wing and tail, respectively, as

$$\begin{aligned}
L_{E_W} &= 2q_\infty \int_0^{b_W/2} c_{l_{\alpha_W}}(y)\left(\theta_E(y) + \frac{1}{V_\infty}\sum_{i=1}^{\infty} \nu_{Z_{i_W}}(y)\dot{\eta}_i(t)\right)c_W(y)\,dy \\
&= 2q_\infty\left(\int_0^{b_W/2} c_{l_{\alpha_W}}(y)\theta_E(y)c_W(y)\,dy + \frac{1}{V_\infty}\sum_{i=1}^{\infty}\left(\underline{\int_0^{b_W/2} c_{l_{\alpha_W}}(y)\nu_{Z_{i_W}}(y)c_W(y)\,dy}\right)\dot{\eta}_i(t)\right)
\end{aligned} \tag{7.24}$$

$$\begin{aligned}
L_{E_H} &= 2q_H \int_0^{b_H/2} c_{l_{\alpha_H}}(y)\left(\begin{aligned}&\left(\theta_{E_H}(y) + \frac{1}{V_\infty}\sum_{i=1}^{\infty}\nu_{Z_{i_H}}(y)\dot{\eta}_i(t)\right)\\ &-\frac{d\varepsilon_H}{d\alpha_W}\left(\theta_{E_W}(y) + \frac{1}{V_\infty}\sum_{i=1}^{\infty}\nu_{Z_{i_W}}(y)\dot{\eta}_i(t)\right)\end{aligned}\right)c_H(y)\,dy \\
&= 2q_H\left(\int_0^{b_H/2} c_{l_{\alpha_H}}(y)\left(\theta_{E_H}(y) - \frac{d\varepsilon_H}{d\alpha_W}\theta_{E_W}(y)\right)c_H(y)\,dy\right. \\
&\quad\left.+\frac{1}{V_\infty}\sum_{i=1}^{\infty}\left(\underline{\int_0^{b_H/2} c_{l_{\alpha_H}}\left(\nu_{Z_{i_H}}(y) - \frac{d\varepsilon_H}{d\alpha_W}\nu_{Z_{i_W}}(y)\right)c_H(y)\,dy}\right)\dot{\eta}_i(t)\right)
\end{aligned}$$

In Equations (7.24), we have again replaced the integrals over summations with summations of integrals. Note that the underlined terms are not explicit functions of time. These two equations then capture the effects of modal velocity on vehicle lift, ignoring any fuselage lift.

Differentiating Equations (7.24) with respect to modal velocity and combining the effect of the wing and tail yields the partial derivative of vehicle lift with respect to modal velocities. Or

$$\frac{\partial L}{\partial \dot{\eta}_i} \triangleq C_{L_{\dot{\eta}_i}} q_\infty S_W = \frac{\partial L_{E_W}}{\partial \dot{\eta}_i} + \frac{\partial L_{E_H}}{\partial \dot{\eta}_i}$$
$$= \frac{2q_\infty}{V_\infty}\left(\int_0^{b_W/2} c_{l_{\alpha_W}}(y)\nu_{Z_{i_W}}(y)c_W(y)\,dy + \frac{q_H}{q_\infty}\int_0^{b_H/2} c_{l_{\alpha_H}}\left(\nu_{Z_{i_H}}(y) - \frac{d\varepsilon_H}{d\alpha_W}\nu_{Z_{i_W}}(y)\right)c_H(y)\,dy\right) \tag{7.25}$$

From this expression we find that the modal-velocity lift effectiveness $C_{L_{\dot{\eta}}}$ for the i'th mode is

$$C_{L_{\dot{\eta}_i}} = \frac{2}{V_\infty S_W}\left(\int_0^{b_W/2} c_{l_{\alpha_W}}(y)\nu_{Z_{i_W}}(y)c_W(y)\,dy + \frac{q_H}{q_\infty}\int_0^{b_H/2} c_{l_{\alpha_H}}\left(\nu_{Z_{i_H}}(y) - \frac{d\varepsilon_H}{d\alpha_W}\nu_{Z_{i_W}}(y)\right)c_H(y)\,dy\right) \tag{7.26}$$

Again note that only symmetric modes need be considered here. Except for the mode-shape data used in the two integrals above and in Equation (7.22), the two equations are very similar.

EXAMPLE 7.1

Elastic Effects on Vehicle Lift

Consider a vehicle with the wing and horizontal tail as analyzed in Example 6.1. An analysis of the vehicle's structural vibration yielded the wing and horizontal tail (symmetric) bending and torsion (mode slope) mode shapes as shown in Figures 7.8 and 7.9. Evaluate the partial derivatives of the vehicle lift with respect to the first mode's displacement and velocity. That is, evaluate $\frac{\partial L}{\partial \eta_1}$ and $\frac{\partial L}{\partial \dot{\eta}_1}$, and the associated effectiveness coefficients.

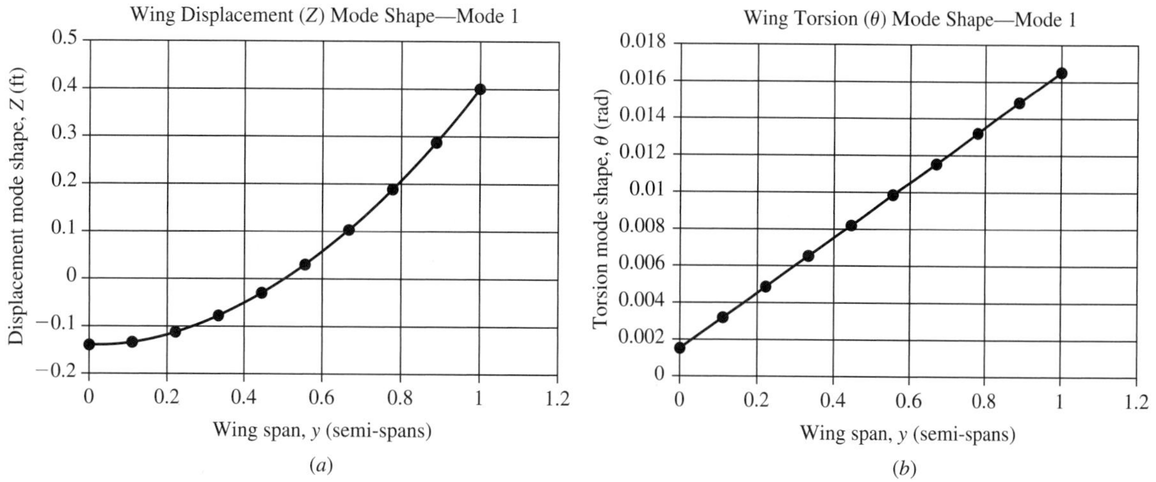

Figure 7.8 Wing mode shapes (along quarter chord).

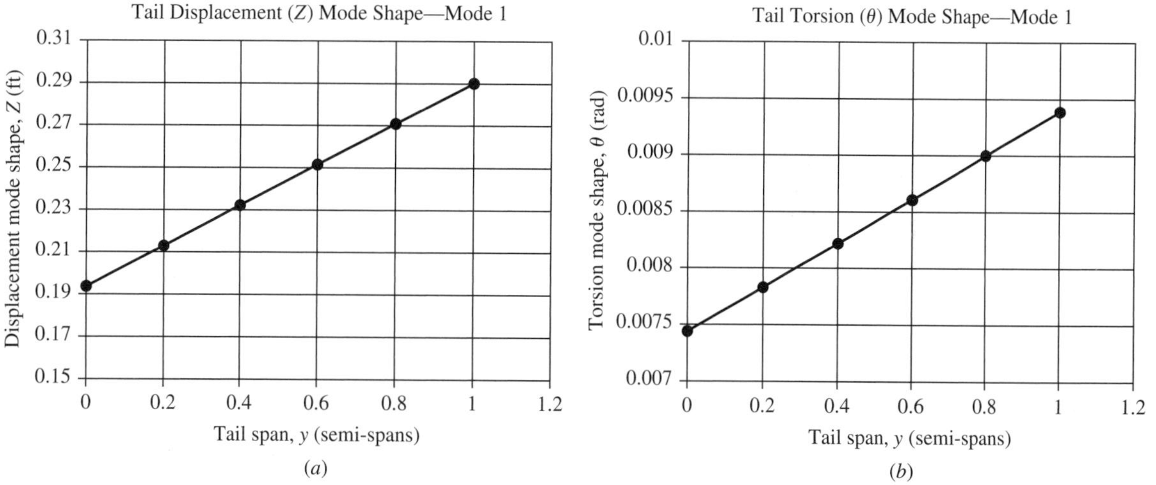

Figure 7.9 Horizontal tail mode shapes (along quarter chord).

Solution

The relevant data for the wing and tail surfaces are as follows:

Wing: $S_W = 169\ \text{ft}^2$, $\bar{c}_W = 5.825$, $b_W = 30$ ft, $c_{l_\alpha} = 0.107$ /deg, $c(y) = \left(7.5 - \dfrac{y}{4}\right)$ ft

Tail: $S_H = 42\ \text{ft}^2$, $\bar{c}_H = 2.92$ ft, $b_H = 15$ ft, $c_{l_{\alpha_H}} = 0.107$ /deg, $c_H(y) = \dfrac{7.5}{2} - \dfrac{y}{4}$ ft,

$$\frac{d\varepsilon_H}{d\alpha_W} = 0.57$$

Fitting polynomials to the mode-shape data allows for analytical evaluation of integral terms, although they may also be evaluated numerically. The fitted polynomials for the wing mode shapes, with y given in wing semi-spans (i.e., $y/(b/2)$) are

$$\begin{aligned} \nu_{Z_{1_W}}(y) &= \left(0.5265\left(\frac{y}{15}\right)^2 + 0.0135\left(\frac{y}{15}\right) - 0.14\right)\text{ ft} \\ \nu'_{Z_{1_W}}(y) &= \left(0.015\left(\frac{y}{15}\right) + 0.00156\right) \end{aligned} \tag{7.27}$$

For the tail mode shapes, the polynomials are

$$\begin{aligned} \nu_{Z_{1_H}}(y) &= \left(0.09625\left(\frac{y}{7.5}\right) + 0.194\right)\text{ ft} \\ \nu'_{Z_{1_H}}(y) &= \left(0.00195\left(\frac{y}{7.5}\right) + 0.00744\right) \end{aligned} \tag{7.28}$$

From Equation (7.21) we have the partial derivative of lift with respect to modal displacement given as

$$\begin{aligned} \frac{\partial L}{\partial \eta_1} = C_{L_{\eta_1}} q_\infty S_W &= \frac{\partial L_W}{\partial \eta_1} + \frac{\partial L_H}{\partial \eta_1} \\ &= 2q_\infty \int_0^{b_W/2} c_{l_{\alpha_W}}(y)\nu'_{Z_{1_W}}(y)c_W(y)dy \\ &\quad + 2q_H \int_0^{b_H/2} c_{l_{\alpha_H}}(y)\left(\nu'_{Z_{1_H}}(y) - \frac{d\varepsilon_H}{d\alpha_W}\nu'_{Z_{1_W}}(y)\right)c_H(y)dy \end{aligned}$$

The first integral in the above expression becomes

$$\begin{aligned} \int_0^{b_W/2} c_{l_{\alpha_W}}(y)\nu'_{Z_{1_W}}(y)c_W(y)dy &= \int_0^{15} (0.107 \times 57.3)(0.015(y/15) + 0.00156)\left(7.5 - \frac{y}{4}\right)dy \\ &= (0.107 \times 57.3)(0.001)\int_0^{15} (y + 1.56)\left(7.5 - \frac{y}{4}\right)dy = 4.256\ \text{ft}^2 \end{aligned}$$

While the second integral is

$$\int_0^{b_H/2} c_{l_{\alpha_H}}(y)\left(\nu'_{Z_{1_H}}(y) - \frac{d\varepsilon_H}{d\alpha_W}\nu'_{Z_{1_W}}(y)\right)c_H(y)\,dy$$

$$= \int_0^{7.5} (0.107\times 57.3)\Big(\big(0.00195(y/7.5) + 0.00744\big)$$

$$- 0.57\big(0.015(y/15) + 0.00156\big)\Big)\left(\frac{7.5}{2} - \frac{y}{4}\right)dy$$

$$= (0.107\times 57.3)(0.001)\int_0^{7.5}(-0.31y + 6.55)\left(\frac{7.5}{2} - \frac{y}{4}\right)dy = 0.713\ \text{ft}^2$$

Therefore, the partial derivative of lift with respect to modal displacement is

$$\frac{\partial L}{\partial \eta_1} = 2q_\infty\left(4.256 + \frac{q_H}{q_\infty}0.713\right) = 9.8q_\infty\ \text{lb}$$

assuming $\frac{q_H}{q_\infty} = 0.9$. And the modal-displacement lift effectiveness for the first mode becomes

$$C_{L_{\eta_1}} = \frac{9.8}{S_W} = 0.058$$

Note here that the modal coordinate η_1 is dimensionless, as is the above effectiveness coefficient. This coefficient is also positive, which indicates that a positive modal displacement $\eta_1(t)$ leads to an increase in lift.

From Equation (7.25), the partial derivative of lift with respect to modal velocity is

$$\frac{\partial L}{\partial \dot\eta_1} = C_{L_{\dot\eta_1}}q_\infty S_W = \frac{\partial L_{E_W}}{\partial \dot\eta_1} + \frac{\partial L_{E_H}}{\partial \dot\eta_1}$$

$$= \frac{2q_\infty}{V_\infty}\left(\int_0^{b_W/2} c_{l_{\alpha_W}}(y)\nu_{Z_{1_W}}(y)c_W(y)\,dy\right.$$

$$\left. + \frac{q_H}{q_\infty}\int_0^{b_H/2} c_{l_{\alpha_H}}\left(\nu_{Z_{1_H}}(y) - \frac{d\varepsilon_H}{d\alpha_W}\nu_{Z_{1_W}}(y)\right)c_H(y)\,dy\right)$$

The first integral above is now

$$\int_0^{b_W/2} c_{l_{\alpha_W}}(y)\nu_{Z_{1_W}}(y)c_W(y)\,dy = \int_0^{15}(0.107\times 57.3)\left(0.5265\left(\frac{y}{15}\right)^2 + 0.0135\left(\frac{y}{15}\right) - 0.14\right)\left(7.5 - \frac{y}{4}\right)dy$$

$$= (0.107\times 57.3)(0.00234)\int_0^{15}(y^2 + 0.385y - 59.83)\left(7.5 - \frac{y}{4}\right)dy = 6.34\ \text{ft}^3$$

and the second integral is

$$\int_0^{b_H/2} c_{l_{\alpha_H}}\left(\nu_{Z_{1_H}}(y) - \frac{d\varepsilon_H}{d\alpha_W}\nu_{Z_{1_W}}(y)\right)c_H(y)dy$$

$$= \int_0^{7.5}(0.107\times 57.3)\left(\left(0.09625\left(\frac{y}{7.5}\right)+0.194\right)\right.$$

$$\left.-0.57\left(0.5265\left(\frac{y}{15}\right)^2+0.0135\left(\frac{y}{15}\right)-0.14\right)\right)\left(\frac{7.5}{2}-\frac{y}{4}\right)dy$$

$$=(0.107\times 57.3)(0.00133)\int_0^{7.5}(-y^2+9.235y+205.9)\left(\frac{7.5}{2}-\frac{y}{4}\right)dy = 38.0\ \text{ft}^3$$

Consequently, the sought partial derivative is

$$\frac{\partial L}{\partial \dot{\eta}_1} = \frac{2q_\infty}{V_\infty}\left(6.34+\frac{q_H}{q_\infty}38.0\right) = 81.1\frac{q_\infty}{V_\infty}\ \text{lb-sec}$$

assuming $\frac{q_H}{q_\infty} = 0.9$, and the modal-velocity lift effectiveness for the first mode is

$$C_{L_{\dot{\eta}_1}} = \frac{81.1}{V_\infty S_W} = \frac{0.480}{V_\infty}\ \text{sec}$$

Note here that this effectiveness coefficient is not dimensionless. This coefficient is also positive, which indicates that a positive modal rate $\dot{\eta}_1(t)$ leads to an increase in lift.

NOTE TO STUDENT

Care must be taken when comparing effectiveness coefficients among the various vibration modes. The coefficient magnitudes are implicitly a function of the generalized masses of the vibration modes, which in turn are a function of the technique used to normalize the vibration mode shapes. Some mode shapes are normalized to unity generalized mass, others are normalized to some maximum mode-shape displacement, and still others are normalized to unity norm of the mode shape. While this may seem somewhat obscure at this point, just be careful when comparing modal effectiveness coefficients associated with different modes.

7.4 ELASTIC EFFECTS ON SIDE FORCE

Now that we have presented the methodology in detail in addressing the lift, we may move through the remaining analysis more quickly. We want to assess the effects of elastic deformation on vehicle side force. We will assume that the side force is generated only by the vehicle's vertical tail, and only antisymmetric vibration modes need be considered.

For an airfoil section on the vertical tail at span location z, and ignoring any side wash effects, the local angle of attack may be written in terms of its rigid and elastic components as

$$\alpha_V(z) = \alpha_{R_V}(z) + \alpha_{E_V}(z) = \alpha_{R_V}(z) + \left(\theta_{E_V}(z) - \frac{v_E(z)}{V_\infty}\right) \tag{7.29}$$

Here v_E is the elastic lateral velocity of the section's aerodynamic center, and θ_{E_V} is the elastic twist angle of the section. These two quantities are analogous to w_E and θ_E for a wing section as discussed in Section 7.3. The rigid angle of attack α_{R_V} is the angle of attack for a rigid vehicle, and angles of attack and elastic twist are defined here as positive when they generate an aerodynamic force in the positive Y direction (towards the right wing tip). This sign convention leads to the negative sign in Equation (7.29).

The 2-D side force generated by the tail section may now be written as

$$\mathsf{s}(z) = c_{l_{\alpha_V}}\left(\alpha_V - \alpha_{0_V}\right)q_H(z)c_V(z) \tag{7.30}$$

Integrating this expression over the tail span yields the side force generated by the vertical tail, or

$$\mathsf{S}_V = \int_0^{b_V} c_{l_{\alpha_V}}(z)\left(\alpha_V(z) - \alpha_{0_V}(z)\right)q_H(z)c_V(z)\,dz \tag{7.31}$$

Substituting Equation (7.29) into the above expression, we may extract only that portion of side force associated with elastic deformation. So the change in side force generated by the vertical tail (which we have assumed to be the only side force on the vehicle) <u>due to elastic effects</u> is given by

$$\boxed{\mathsf{S}_{E_V} = \int_0^{b_V} c_{l_{\alpha_V}}(z)\left(\theta_{E_V}(z) - \frac{v_E(z)}{V_\infty}\right)q_H(z)c_V(z)\,dz} \tag{7.32}$$

7.4.1 Effects of Modal Displacement

The elastic deformations in Equation (7.32) may now be written in terms of the modal parameters. Specifically, as with the wing, the tail-section elastic twist angle $\theta_{E_V}(y)$ may be expressed in terms of slopes of the vibration mode shapes at the tail. If the lateral component of the i'th antisymmetric mode shape at a location (x,y,z) on the tail is $\nu_{Y_i}(x,y,z)$, then the elastic twist angle is

$$\theta_{E_V}(z,t) \approx \tan\theta_{E_V}(z,t) = \frac{1}{x_A - x_B}\sum_{i=1}^{\infty}\left(\nu_{Y_i}(x_A,y_A,z) - \nu_{Y_i}(x_B,y_B,z)\right)\eta_i(t) \tag{7.33}$$

where, similar to Figure 7.6, two points A and B on the tail section (e.g., leading and trailing edges) have been considered. But each of the coefficients of $\eta_i(t)$ in the summation in Equation (7.33) is just an approximation for the local slope of the mode shape $\nu'_{Y_i}(z)$, where, similar to Figure 7.7,

$$\nu'_{Y_i}(z) \triangleq \lim_{(x_A - x_B)\to 0} \frac{\nu_{Y_i}(x_A, y_A, z) - \nu_{Y_i}(x_B, y_B, z)}{x_A - x_B} \tag{7.34}$$

Consequently, the elastic tail-twist angle may be written as

$$\theta_{E_V}(z,t) \approx \tan\theta_{E_V}(z,t) = \sum_{i=1}^{\infty} \nu'_{Y_i}(z)\eta_i(t) \tag{7.35}$$

Substituting Equation (7.35) into Equation (7.32), we have the change in the side force on the vertical tail due to elastic deformation given by

$$\begin{aligned} \mathsf{S}_{E_V} &= \int_0^{b_V} c_{l_{\alpha_V}}(z)\left(\sum_{i=1}^{\infty} \nu'_{Y_i}(z)\eta_i(t) - \frac{\nu_E(z)}{V_\infty}\right) q_H(z) c_V(z)\,dz \\ &\approx q_H\left(\sum_{i=1}^{\infty}\left(\underline{\int_0^{b_V} c_{l_{\alpha_V}}(z)\nu'_{Y_i}(z) c_V(z)\,dz}\right)\eta_i(t) - \int_0^{b_V} c_{l_{\alpha_V}}(z)\frac{\nu_E(z)}{V_\infty} c_V(z)\,dz\right) \end{aligned} \tag{7.36}$$

Here we have again assumed that the dynamic pressure is constant with respect to span, and we have replaced the integral over a summation with a summation of integrals.

We can now proceed in a straightforward manner to determine the effect of modal displacement $\eta_i(t)$ on side force, or $\dfrac{\partial \mathsf{S}}{\partial \eta_i}$. Still assuming that vehicle side force is generated only by the vertical tail, and taking the partial derivative of Equation (7.36), we find that the <u>effect of modal displacement on vehicle side force</u> is given by

$$\boxed{\frac{\partial \mathsf{S}}{\partial \eta_i} \triangleq C_{S_{\eta_i}} q_\infty S_W = \frac{\partial \mathsf{S}_{E_V}}{\partial \eta_i} = q_H \int_0^{b_V} c_{l_{\alpha_V}}(z)\nu'_{Y_i}(z) c_V(z)\,dz} \tag{7.37}$$

The modal-displacement side-force effectiveness $C_{S_{\eta_i}}$ has been introduced here, and from the above equation we find that it may be expressed as

$$\boxed{C_{S_{\eta_i}} = \frac{1}{S_W}\frac{q_H}{q_\infty}\int_0^{b_V} c_{l_{\alpha_V}}(z)\nu'_{Y_i}(z) c_V(z)\,dz} \tag{7.38}$$

As with the results for lift, this coefficient is a function of vertical-tail planform geometry, aerodynamic characteristics, and vibration mode shape for the i'th vibration mode. Again note that only antisymmetric modes need be considered here.

7.4.2 Effects of Modal Velocity

The vehicle's side force due to elastic deformation is also a function of the modal velocities, or $\dot{\eta}_i(t)$. Here we seek to determine the partial derivative $\dfrac{\partial \mathsf{S}}{\partial \dot{\eta}_i}$. As in Section 7.4.1, we assume that the side force is essentially generated by the vertical tail, and that the tail's side force due only to elastic deformation was given by Equation (7.32).

But as with the elastic plunge velocity addressed in Section 7.3.2, the lateral velocity associated with the elastic deformation of the aerodynamic center of a tail section at span z, or $v_E(z)$, is expressible in terms of the modal parameters. That is, we may write this velocity in terms of the mode shapes and rates of change of the modal coordinates. Or

$$v_E(z,t) = \sum_{i=1}^{\infty} \nu_{Y_i}(x_{\mathrm{AC}}, y_{\mathrm{AC}}, z)\dot{\eta}_i(t) \triangleq \sum_{i=1}^{\infty} \nu_{Y_i}(z)\dot{\eta}_i(t) \tag{7.39}$$

Hence, Equation (7.32) may now be written as

$$\begin{aligned}
\mathsf{S}_{E_V} &= \int_0^{b_V} c_{l_{\alpha_V}}(z)\left(\theta_{E_V}(z) - \frac{1}{V_\infty}\sum_{i=1}^{\infty}\nu_{Y_i}(z)\dot{\eta}_i(t)\right)q_H(z)c_V(z)\,dz \\
&= \int_0^{b_V} c_{l_{\alpha_V}}(z)\theta_{E_V}(z)q_H(z)c_V(z)\,dz \\
&\quad - \frac{1}{V_\infty}\sum_{i=1}^{\infty}\left(\underline{\int_0^{b_V} c_{l_{\alpha_V}}(z)\nu_{Y_i}(z)q_H(z)c_V(z)\,dz}\right)\dot{\eta}_i(t)
\end{aligned} \tag{7.40}$$

Again note that the underlined term is not an explicit function of time, and we have replaced an integral over a summation with a summation of integrals.

Differentiating Equation (7.40) with respect to modal velocity $\dot{\eta}_i$, and recalling that we have assumed that the vehicle side force is only generated by the vertical tail, we have the <u>effect of modal rate on vehicle side force</u> given by

$$\boxed{\frac{\partial \mathsf{S}}{\partial \dot{\eta}_i} \triangleq C_{S_{\dot{\eta}_i}} q_\infty S_W = \frac{\partial \mathsf{S}_{E_V}}{\partial \dot{\eta}_i} \approx -\frac{q_H}{V_\infty}\int_0^{b_V} c_{l_{\alpha_V}}(z)\nu_{Y_i}(z)c_V(z)\,dz} \tag{7.41}$$

We have assumed that the dynamic pressure at the tail, q_H, is constant with respect to the span-wise integration. The modal-rate side-force effectiveness $C_{S_{\dot{\eta}_i}}$ has also been introduced, and we can see from Equation (7.41) that

$$C_{S_{\dot{\eta}_i}} = -\frac{1}{V_\infty S_W}\frac{q_H}{q_\infty}\int_0^{b_V} c_{l_{\alpha_V}}(z)\nu_{Y_i}(z)c_V(z)\,dz \tag{7.42}$$

Note the similarity between this result and that given by Equation (7.38). The main differences are the specific mode-shape data appearing in the integrals and the negative sign in the above expression. Finally, only antisymmetric modes need be considered here.

7.5 ELASTIC EFFECTS ON PITCHING MOMENT

We will now turn to the elastic effects on the moments acting on the vehicle. In this section we will focus on the pitching moment, which is assumed to be generated by the lift on the wing and horizontal tail. The direct effects of fuselage bending on this moment may usually be neglected, except, of course, for the displacements of the roots of the wing and tail.

Consider the geometry depicted in Figure 7.10, in which a wing semi-span is shown. From the figure, and assuming the X dimension is measured positive forward, we can see that the pitching moment due to a 2-D lift force $l(y)$ acting at the aerodynamic center of the wing section at span y may be written as

$$m_W(y) = l(y)\cos\alpha\left(x_{AC}(y) - X_{Ref}\right) \approx l(y)\left(x_{AC}(y) - X_{Ref}\right) \tag{7.43}$$

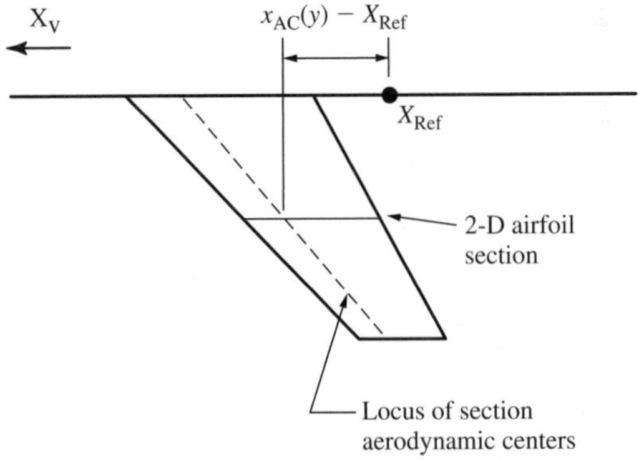

Figure 7.10 Wing geometry and section moment arm.

where the section lift $l(y)$, given by Equation (7.9), is repeated here for convenience.

$$l(y) = c_{l_{\alpha_W}}(y)\big(\alpha(y) - \alpha_{0_W}(y)\big)q(y)c_W(y)$$

Also recall that the angle of attack of the sections may be written in terms of rigid and elastic components, as in Equation (7.7) for the wing and Equation (7.8) for the horizontal tail. Therefore, ignoring the effects of section drag and lag of downwash, we can write the pitching moment from the wing and tail sections, respectively, due only to the elastic effects as

$$m_{E_W}(y) = c_{l_{\alpha_W}}(y)\left(\theta_{E_W}(y) + \frac{w_{E_W}(y)}{V_\infty}\right)\left(x_{AC_W}(y) - X_{Ref}\right)q_W(y)c_W(y)$$

$$\begin{aligned} m_{E_H}(y) = -c_{l_{\alpha_H}}(y)&\left(\left(\theta_{E_H}(y) + \frac{w_{E_H}(y)}{V_\infty}\right)\right. \\ &\left.- \frac{d\varepsilon_H}{d\alpha_W}\left(\theta_{E_W}(y) + \frac{w_{E_W}(y)}{V_\infty}\right)\right)\left(X_{Ref} - x_{AC_H}(y)\right)q_H(y)c_H(y) \end{aligned} \tag{7.44}$$

Integrating the above two expressions over the wing and tail spans, respectively, yields the change in pitching moment on the vehicle due to wing and tail elastic effects. Or, assuming the vehicle is symmetric with respect to its XZ plane, we have

$$\begin{aligned} M_{E_W} &= 2\int_0^{b_W/2} c_{l_{\alpha_W}}(y)\left(\theta_{E_W}(y) + \frac{w_{E_W}(y)}{V_\infty}\right)\left(x_{AC_W}(y) - X_{Ref}\right)q_W(y)c_W(y)\,dy \\ M_{E_H} &= -2\int_0^{b_H/2} c_{l_{\alpha_H}}(y)\left(\left(\theta_{E_H}(y) + \frac{w_{E_H}(y)}{V_\infty}\right)\right. \\ &\qquad\left.- \frac{d\varepsilon_H}{d\alpha_W}\left(\theta_{E_W}(y) + \frac{w_{E_W}(y)}{V_\infty}\right)\right)\left(X_{Ref} - x_{AC_H}(y)\right)q_H(y)c_H(y)\,dy \end{aligned} \tag{7.45}$$

Finally, the change in pitching moment on the vehicle due to elastic effects is then the sum of the above two expressions.

7.5.1 Effects of Modal Displacement

In Section 7.3.1 we expressed the elastic deformation of the wing and tail in terms of the modal parameters, that is, the mode shapes and modal coordinates. In particular, we showed in Equation (7.19) that the elastic twist of the wing and tail could be written as

$$\theta_E(y,t) \approx \tan\theta_E(y,t) = \sum_{i=1}^{\infty} \nu'_{Z_i}(y)\eta_i(t)$$

where $\nu'_{Z_i}(y)$ is the slope of the plunge component of mode shape evaluated at span location (y) on the wing or tail, and $\eta_i(t)$ is the i'th modal coordinate.

As a result, we may express Equations (7.45), the changes in moments due to elastic effects, as

$$M_{E_W} = 2\int_0^{b_W/2} c_{l_{\alpha_W}}(y)\left(\sum_{i=1}^{\infty}\nu'_{Z_{i_W}}(y)\eta_i(t) + \frac{w_{E_W}(y)}{V_\infty}\right)(x_{AC_W}(y) - X_{Ref})q_W(y)c_W(y)\,dy$$

$$\approx 2q_\infty\left(\begin{array}{l}\sum_{i=1}^{\infty}\left(\underline{\int_0^{b_W/2} c_{l_{\alpha_W}}(y)\,\nu'_{Z_{i_W}}(y)(x_{AC_W}(y) - X_{Ref})c_W(y)\,dy}\right)\eta_i(t) \\ + \underline{\int_0^{b_W/2} c_{l_{\alpha_W}}(y)\left(\frac{w_{E_W}(y)}{V_\infty}\right)(x_{AC_W}(y) - X_{Ref})c_W(y)\,dy}\end{array}\right) \tag{7.46}$$

$$M_{E_H} = -2\int_0^{b_H/2} c_{l_{\alpha_H}}(y)\left(\left(\sum_{i=1}^{\infty}\nu'_{Z_{i_H}}(y)\eta_i(t) + \frac{w_{E_H}(y)}{V_\infty}\right) - \frac{d\varepsilon_H}{d\alpha_W}\left(\sum_{i=1}^{\infty}\nu'_{Z_{i_W}}(y)\eta_i(t) + \frac{w_{E_W}(y)}{V_\infty}\right)\right)(X_{Ref} - x_{AC_H}(y))q_H(y)c_H(y)\,dy$$

$$\approx -2q_H\left(\begin{array}{l}\sum_{i=1}^{\infty}\left(\underline{\int_0^{b_H/2} c_{l_{\alpha_H}}(y)\left(\nu'_{Z_{i_H}}(y) - \frac{d\varepsilon_H}{d\alpha_W}\nu'_{Z_{i_W}}(y)\right)(X_{Ref} - x_{AC_H}(y))c_H(y)\,dy}\right)\eta_i(t) \\ + \underline{\frac{1}{V_\infty}\int_0^{b_H/2} c_{l_{\alpha_H}}(y)\left(w_{E_H}(y) - \frac{d\varepsilon_H}{d\alpha_W}w_{E_W}(y)\right)(X_{Ref} - x_{AC_H}(y))c_H(y)\,dy}\end{array}\right)$$

In obtaining the two expressions in Equations (7.46), note again that we have replaced the integral over a summation with a summation of integrals. Also note that the underlined terms are not explicit functions of time, since the time dependence of modal displacement is contained only in the modal coordinates $\eta_i(t)$.

Summing the two expressions, and defining the result as the change in vehicle pitching moment due to the elastic effects, we can determine the effects of modal displacement. By adding the above wing and tail moments, and differentiating them with respect to modal displacement $\eta_i(t)$, we have the effects of modal displacement on the vehicle pitching moment given by

$$\frac{\partial M}{\partial \eta_i} \triangleq C_{M_{\eta_i}} q_\infty S_W \bar{c}_W = \frac{\partial M_{E_W}}{\partial \eta_i} + \frac{\partial M_{E_H}}{\partial \eta_i}$$

$$= 2\left(q_\infty \int_0^{b_W/2} c_{l_{\alpha_W}}(y) \nu'_{Z_{i_W}}(y)\left(x_{\mathrm{AC}_W}(y) - X_{\mathrm{Ref}}\right) c_W(y)\, dy \right.$$

$$\left. - q_H \int_0^{b_H/2} c_{l_{\alpha_H}}(y) \left(\nu'_{Z_{i_H}}(y) - \frac{d\varepsilon_H}{d\alpha_W} \nu'_{Z_{i_W}}(y) \right)\left(X_{\mathrm{Ref}} - x_{\mathrm{AC}_H}(y)\right) c_H(y)\, dy \right) \tag{7.47}$$

Note the similarity between the integral terms in Equation (7.47) and those in Equation (7.21). The above integrals simply contain the section moment arms.

In addition, the modal-displacement pitching-moment effectiveness $C_{M_{\eta_i}}$ has been introduced. From the above expression, we can see that this effectiveness is given by

$$C_{M_{\eta_i}} = \frac{2}{S_W \bar{c}_W} \left(\int_0^{b_W/2} c_{l_{\alpha_W}}(y) \nu'_{Z_{i_W}}(y)\left(x_{\mathrm{AC}_W}(y) - X_{\mathrm{Ref}}\right) c_W(y)\, dy \right.$$

$$\left. - \frac{q_H}{q_\infty} \int_0^{b_H/2} c_{l_{\alpha_H}}(y) \left(\nu'_{Z_{i_H}}(y) - \frac{d\varepsilon_H}{d\alpha_W} \nu'_{Z_{i_W}}(y) \right)\left(X_{\mathrm{Ref}} - x_{\mathrm{AC}_H}(y)\right) c_H(y)\, dy \right) \tag{7.48}$$

This effectiveness is seen to be a function of wing, tail and vehicle geometry, section lift-curve slopes, and slopes of the mode shapes. The integral terms are usually evaluated numerically from tabulated mode-shape data, and only symmetric modes need be considered here.

7.5.2 Effects of Modal Velocity

Similar to the discussion in Section 7.3.2 involving the elastic effects on vehicle lift, the change in pitching moments due to elastic effects are also a function of the rate of change of the modal coordinates, or modal velocities.

From Equation (7.23) we have the plunge velocity of the 2-D wing or tail section given in terms of modal parameters. Or

$$w_E(y,t) = \dot{z}_{\mathrm{AC}}(y,t) = \sum_{i=1}^{\infty} \nu_{Z_i}(x_{\mathrm{AC}}, y, z_{\mathrm{AC}}) \dot{\eta}_i(t) \triangleq \sum_{i=1}^{\infty} \nu_{Z_i}(y) \dot{\eta}_i(t)$$

Therefore, inserting the above into Equations (7.45), we have the changes in wing and tail pitching moments due to elastic effects given by

$$M_{E_W} = 2\int_0^{b_W/2} c_{l_{\alpha_W}}(y)\left(\theta_{E_W}(y) + \frac{1}{V_\infty}\sum_{i=1}^{\infty}\nu_{Z_{i_W}}(y)\dot{\eta}_i(t)\right)\left(x_{\mathrm{AC}_W}(y) - X_{\mathrm{Ref}}\right)q_W(y)c_W(y)\,dy$$

$$\approx 2q_\infty\left(\frac{1}{V_\infty}\sum_{i=1}^{\infty}\left(\int_0^{b_W/2} c_{l_{\alpha_W}}(y)\nu_{Z_{i_W}}(y)\left(x_{\mathrm{AC}_W}(y) - X_{\mathrm{Ref}}\right)c_W(y)\,dy\right)\dot{\eta}_i(t) + \int_0^{b_W/2} c_{l_{\alpha_W}}(y)\theta_{E_W}(y)\left(x_{\mathrm{AC}_W}(y) - X_{\mathrm{Ref}}\right)c_W(y)\,dy\right)$$

$$M_{E_H} = -2\int_0^{b_H/2} c_{l_{\alpha_H}}(y)\left(\left(\theta_{E_H}(y) + \frac{1}{V_\infty}\sum_{i=1}^{\infty}\nu_{Z_{i_H}}(y)\dot{\eta}_i(t)\right) - \frac{d\varepsilon_H}{d\alpha_W}\left(\theta_{E_W}(y) + \frac{1}{V_\infty}\sum_{i=1}^{\infty}\nu_{Z_{i_W}}(y)\dot{\eta}_i(t)\right)\right)$$

$$\left(X_{\mathrm{Ref}} - x_{\mathrm{AC}_H}(y)\right)q_H(y)c_H(y)\,dy$$

$$\approx -2q_H\left(\frac{1}{V_\infty}\sum_{i=1}^{\infty}\left(\int_0^{b_H/2} c_{l_{\alpha_H}}(y)\left(\nu_{Z_{i_H}}(y) - \frac{d\varepsilon_H}{d\alpha_W}\nu_{Z_{i_W}}(y)\right)\left(X_{\mathrm{Ref}} - x_{\mathrm{AC}_H}(y)\right)c_H(y)\,dy\right)\dot{\eta}_i(t) + \int_0^{b_H/2} c_{l_{\alpha_H}}(y)\left(\theta_{E_H}(y) - \frac{d\varepsilon_H}{d\alpha_W}\theta_{E_W}(y)\right)\left(X_{\mathrm{Ref}} - x_{\mathrm{AC}_H}(y)\right)c_H(y)\,dy\right)$$

(7.49)

So by summing the two expressions in Equations (7.49) and taking the partial derivatives with respect to modal velocity $\dot{\eta}_i(t)$, we find that the effect of the i'th modal velocity on the vehicle pitching moment is given by

$$\frac{\partial M}{\partial \dot{\eta}_i} \triangleq C_{M_{\dot{\eta}_i}} q_\infty S_W \bar{c}_W = \frac{\partial M_{E_W}}{\partial \dot{\eta}_i} + \frac{\partial M_{E_H}}{\partial \dot{\eta}_i}$$

$$= \frac{2q_\infty}{V_\infty}\left(\int_0^{b_W/2} c_{l_{\alpha_W}}(y)\nu_{Z_{i_W}}(y)\left(x_{\mathrm{AC}_W}(y) - X_{\mathrm{Ref}}\right)c_W(y)\,dy\right.$$

$$\left. - \frac{q_H}{q_\infty}\int_0^{b_H/2} c_{l_{\alpha_H}}(y)\left(\nu_{Z_{i_H}}(y) - \frac{d\varepsilon_H}{d\alpha_W}\nu_{Z_{i_W}}(y)\right)\left(X_{\mathrm{Ref}} - x_{\mathrm{AC}_H}(y)\right)c_H(y)\,dy\right) \quad (7.50)$$

We have introduced here the modal-velocity pitching-moment effectiveness $C_{M_{\dot{\eta}_i}}$, which from Equation (7.50) may be found to be

$$C_{M_{\dot{\eta}_i}} = \frac{2}{V_\infty S_W \bar{c}_W}\left(\int_0^{b_W/2} c_{l_{\alpha_W}}(y) \nu_{Z_{i_W}}(y)\big(x_{\mathrm{AC}_W}(y) - X_{\mathrm{Ref}}\big) c_W(y)\,dy \right.$$
$$\left. - \frac{q_H}{q_\infty}\int_0^{b_H/2} c_{l_{\alpha_H}}(y)\left(\nu_{Z_{i_H}}(y) - \frac{d\varepsilon_H}{d\alpha_W}\nu_{Z_{i_W}}(y)\right)\big(X_{\mathrm{Ref}} - x_{\mathrm{AC}_H}(y)\big)c_H(y)\,dy\right) \tag{7.51}$$

Again note the similarity between the above integrals and those in the lift-effectiveness given in Equation (7.26). The only difference is the appearance of the 2-D section moment arm in the integrals above. Finally, only symmetric modes need be considered in the evaluation of this effectiveness.

7.6 ELASTIC EFFECTS ON ROLLING MOMENT

Any side force on the vertical tail or any antisymmetric lift on the wing or horizontal tail induces a rolling moment on the vehicle. We will first address elastic effects on the wing and horizontal tail, and then on the vertical tail.

The 2-D section lift generated by an airfoil section on the wing or horizontal tail was given in Equation (7.9) as

$$l(y) = c_{l_\alpha}(y)\big(\alpha(y) - \alpha_0(y)\big)q(y)c(y)$$

where $\alpha(y)$ is the local angle of attack of the section. For the wing and tail, respectively, these angles of attack are given in Equations (7.7) and (7.8), or

$$\alpha_W(y) = \alpha_{R_W}(y) + \alpha_{E_W}(y) = \alpha_{R_W}(y) + \left(\theta_E(y) + \frac{w_E(y)}{V_\infty}\right)$$

$$\alpha_H(y) = \alpha_{R_H}(y) + \alpha_{E_H}(y) = \alpha_{R_H}(y) + \left(\theta_{E_H}(y) + \frac{w_{E_H}(y)}{V_\infty} - \varepsilon_{E_H}(y)\right)$$

Note again that these angles of attack are expressed in terms of rigid and elastic contributions. Hence, the contribution of the elastic deformation to the section lift can be isolated. In addition, the elastic contribution to the downwash angle at the tail $\varepsilon_{E_H}(y)$ was given by Equation (7.14).

By referring to Figure 7.11, we can see that the rolling moment generated by the 2-D section lift $l(y)$ may be written as

$$l_{\mathrm{Roll}}(y) = -y\,l(y) \tag{7.52}$$

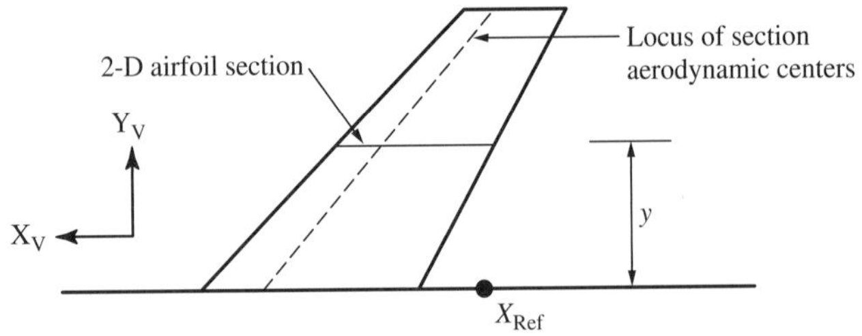

Figure 7.11 Wing geometry and rolling-moment arm.

where y is positive along the right wing or tail, and negative along the left. Consequently, by integrating Equation (7.52) over the spans, we have the change in rolling moment due to elastic deformation of the wing and tail, respectively, given by

$$L_{\text{Roll}_{E_W}} \approx -q_\infty \int_{-b_W/2}^{b_W/2} c_{l_{\alpha_W}}(y)\left(\theta_{E_W}(y)+\frac{w_{E_W}(y)}{V_\infty}\right)c_W(y)\,y\,dy \tag{7.53}$$

$$L_{\text{Roll}_{E_H}} \approx -q_H \int_{-b_H/2}^{b_H/2} c_{l_{\alpha_H}}(y)\left(\left(\theta_{E_H}(y)+\frac{w_{E_H}(y)}{V_\infty}\right)-\frac{d\varepsilon_H}{d\alpha_W}\left(\theta_{E_W}(y)+\frac{w_{E_W}(y)}{V_\infty}\right)\right)c_H(y)\,y\,dy$$

Note here that these integrals are over the entire spans. If the elastic deformation is symmetric with respect to the vehicle's XZ plane, the integrals will equal zero. So only asymmetric elastic deformations need be considered.

Now to address the rolling moment generated by the elastic deformation of the vertical tail, recall that the 2-D aerodynamic side force generated by a section at span location z was given by Equation (7.30). Or

$$\mathsf{s}(z) = c_{l_{\alpha_V}}(\alpha_V - \alpha_{0_V})q_H(z)c_V(z)$$

And the local angle of attack of the section, as given in Equation (7.29), is

$$\alpha_V(z) = \alpha_{R_V}(z) + \alpha_{E_V}(z) = \alpha_{R_V}(z) + \left(\theta_{E_V}(z) - \frac{v_E(z)}{V_\infty}\right)$$

where the angle of attack again consists of both rigid and elastic components. As with the other lifting surfaces, the elastic component of angle of attack includes the effect of both elastic twist and lateral displacement, and the change in side force associated only with these elastic effects $\mathsf{s}_E(z)$ may be identified.

By referring to Figure 7.12, we can see that the rolling moment generated by the 2-D section side force due only to elastic deformation may be written as

$$l_{\text{Roll}_{E_V}}(z) = \mathsf{s}_E(z)z_{AC}(z) \tag{7.54}$$

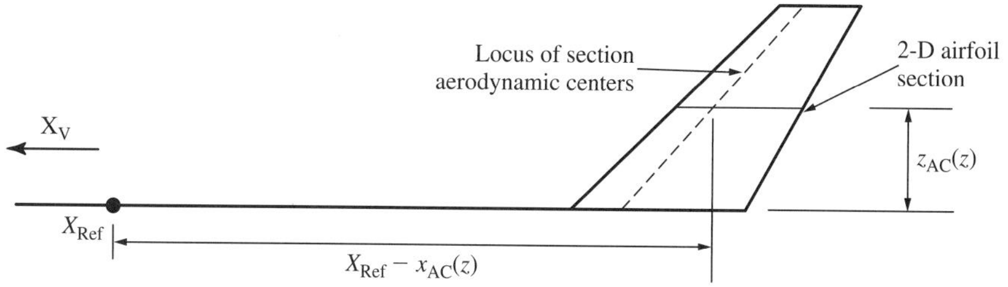

Figure 7.12 Vertical tail geometry and rolling-moment arm.

Here, $z_{AC}(z)$ is the height of the section aerodynamic center above the chosen vehicle-fixed X_V axis. For example, if fuselage-referenced axes are chosen, $z_{AC}(z) = z + z_{Root}$, but if stability axes are chosen

$$z_{AC}(z) = \left(z + z_{Root}\right)\cos\alpha_0 - \left(X_{Ref} - x_{AC}(z)\right)\sin\alpha_0 \tag{7.55}$$

where z is the span location of the airfoil section measured from the tail root plane. Finally, by integrating Equation (7.54) over the span of the vertical tail, we obtain the following expression for the change in tail's rolling moment due to elastic effects.

$$L_{E_V} = \int_0^{b_V} c_{l_{\alpha_V}}\left(\theta_{E_V}(z) - \frac{v_E(z)}{V_\infty}\right) z_{AC}(z)\, q_H(z)\, c_V(z)\, dz \tag{7.56}$$

Thus, the total change in rolling moment on the vehicle due to elastic deformation is then the sum of the wing and horizontal- and vertical-tail contributions, or the sum of Equations (7.53) plus (7.56).

7.6.1 Effects of Modal Displacement

As with the other aeroelastic effects on forces and moments, the change in rolling moment due to elastic deformation may of course be expressed in terms of the modal parameters. Specifically, the elastic twist θ_E for each of the lifting surfaces may be written in terms of the modal displacement $\eta_i(t)$ as in Equation (7.19) for the wing and horizontal tail, and in Equation (7.35) for the vertical tail, or

$$\theta_{E_{W \text{ or } H}}(y,t) = \sum_{i=1}^{\infty} \nu'_{Z_{i_{W \text{ or } H}}}(y)\eta_i(t)$$

$$\theta_{E_V}(z,t) = \sum_{i=1}^{\infty} \nu'_{Y_{i_V}}(z)\eta_i(t)$$

As discussed previously, $\nu'_{Z_i}(y)$ is the local slope of the Z-displacement mode shape evaluated at span location y on either the wing or horizontal tail, and $\nu'_{Y_{i_V}}(z)$

is the local slope of the Y-displacement mode shape evaluated at span location z on the vertical tail.

Therefore, the change in rolling moments due to the elastic deformation of the wing and tails, respectively, from Equations (7.53) and (7.56), may now be written as

$$L_{\text{Roll}_{E_W}} \approx -q_\infty \int_{-b_W/2}^{b_W/2} c_{l_{\alpha_W}}(y)\left(\sum_{i=1}^{\infty} \nu'_{Z_{i_W}}(y)\eta_i(t) + \frac{w_{E_W}(y)}{V_\infty}\right)c_W(y)\,y\,dy$$

$$L_{\text{Roll}_{E_H}} \approx -q_H \int_{-b_H/2}^{b_H/2} c_{l_{\alpha_H}}(y)\left(\left(\sum_{i=1}^{\infty} \nu'_{Z_{i_H}}(y)\eta_i(t) + \frac{w_{E_H}(y)}{V_\infty}\right) - \frac{d\varepsilon_H}{d\alpha_W}\left(\sum_{i=1}^{\infty} \nu'_{Z_{i_W}}(y)\eta_i(t) + \frac{w_{E_W}(y)}{V_\infty}\right)\right)c_H(y)\,y\,dy \tag{7.57}$$

$$L_{E_V} = q_H \int_0^{b_V} c_{l_{\alpha_V}}\left(\sum_{i=1}^{\infty} \nu'_{Y_{i_V}}(z)\eta_i(t) - \frac{v_{E_V}(z)}{V_\infty}\right)z_{\text{AC}}(z)c_V(z)\,dz$$

Rewriting the integrals over the summations as summations of integrals and rearranging, we have the change in rolling moments given by

$$L_{\text{Roll}_{E_W}} \approx -q_\infty\left(\sum_{i=1}^{\infty}\left(\underline{\int_{-b_W/2}^{b_W/2} c_{l_{\alpha_W}}(y)\nu'_{Z_{i_W}}(y)c_W(y)\,y\,dy}\right)\eta_i(t) + \int_{-b_W/2}^{b_W/2} c_{l_\alpha}(y)\left(\frac{w_{E_W}(y)}{V_\infty}\right)c_W(y)\,y\,dy\right)$$

$$L_{\text{Roll}_{E_H}} \approx -q_H\left(\sum_{i=1}^{\infty}\left(\underline{\int_{-b_H/2}^{b_H/2} c_{l_{\alpha_H}}(y)\left(\nu'_{Z_{i_H}}(y) - \frac{d\varepsilon_H}{d\alpha_W}\nu'_{Z_{i_W}}(y)\right)c_H(y)\,y\,dy}\right)\eta_i(t) + \int_{-b_H/2}^{b_H/2} c_{l_{\alpha_H}}(y)\left(\frac{w_{E_H}(y)}{V_\infty} - \frac{d\varepsilon_H}{d\alpha_W}\frac{w_{E_W}(y)}{V_\infty}\right)c_H(y)\,y\,dy\right) \tag{7.58}$$

$$L_{E_V} = q_H\left(\sum_{i=1}^{\infty}\left(\underline{\int_0^{b_V} c_{l_{\alpha_V}}(z)\nu'_{Y_{i_V}}(z)z_{\text{AC}}(z)c_V(z)\,dz}\right)\eta_i(t) - \int_0^{b_V} c_{l_{\alpha_V}}(z)\left(\frac{v_E(z)}{V_\infty}\right)z_{\text{AC}}(z)c_V(z)\,dz\right)$$

By differentiating the three expressions in Equations (7.58) with respect to the modal coordinate $\eta_i(t)$, we find that the effects of modal displacement on the rolling moments may be expressed as

$$\begin{aligned}
\frac{\partial L_{\text{Roll}_{E_W}}}{\partial \eta_i} &= -q_\infty \int_{-b_W/2}^{b_W/2} c_{l_{\alpha_W}}(y) \nu'_{Z_{i_W}}(y) c_W(y) y dy \\
\frac{\partial L_{\text{Roll}_{E_H}}}{\partial \eta_i} &= -q_H \int_{-b_H/2}^{b_H/2} c_{l_{\alpha_H}}(y) \left(\nu'_{Z_{i_H}}(y) - \frac{d\varepsilon_H}{d\alpha_W} \nu'_{Z_{i_W}}(y) \right) c_H(y) y dy \\
\frac{\partial L_{E_V}}{\partial \eta_i} &= q_H \int_0^{b_V} c_{l_{\alpha_V}}(z) \nu'_{Y_{i_V}}(z) z_{\text{AC}}(z) c_V(z) dz
\end{aligned} \qquad (7.59)$$

And so the effect of modal displacement on the vehicle's rolling moment is the sum of the wing and tails' contributions given by Equations (7.59), or

$$\frac{\partial L_{\text{Roll}}}{\partial \eta_i} \triangleq C_{L\text{roll}_{\eta_i}} q_\infty S_W b_W = \frac{\partial L_{\text{Roll}_{E_W}}}{\partial \eta_i} + \frac{\partial L_{\text{Roll}_{E_H}}}{\partial \eta_i} + \frac{\partial L_{E_V}}{\partial \eta_i} \qquad (7.60)$$

Here, the vehicle's modal-displacement rolling-moment effectiveness $C_{L\text{roll}_{\eta_i}}$ has been introduced, and we can see that this effectiveness is

$$\begin{aligned}
C_{L\text{roll}_{\eta_i}} = -\frac{1}{S_W b_W} \Bigg(& \int_{-b_W/2}^{b_W/2} c_{l_{\alpha_W}}(y) \nu'_{Z_{i_W}}(y) c_W(y) y dy + \frac{q_H}{q_\infty} \int_{-b_H/2}^{b_H/2} c_{l_{\alpha_H}}(y) \left(\nu'_{Z_{i_H}}(y) - \frac{d\varepsilon_H}{d\alpha_W} \nu'_{Z_{i_W}}(y) \right) c_H(y) y dy \\
& - \frac{q_H}{q_\infty} \int_0^{b_V} c_{l_{\alpha_V}}(z) \nu'_{Y_{i_V}}(z) z_{\text{AC}}(z) c_V(z) dz \Bigg)
\end{aligned} \qquad (7.61)$$

Due to the symmetry of the vehicle about its XZ plane, this effectiveness coefficient will be zero for the symmetric modes, so only antisymmetric modes need be considered here.[1]

7.6.2 Effects of Modal Velocity

The elastic rolling moment is also a function of the rate of change of the modal coordinate $\dot{\eta}_i$, or the modal velocity. From Equation (7.23), we have the (elastic) plunge velocity of the aerodynamic center of a 2-D section on the wing or tail given by

$$w_E(y,t) = \sum_{i=1}^{\infty} \nu_{Z_i}(x_{\text{AC}}, y, z_{\text{AC}}) \dot{\eta}_i(t) \triangleq \sum_{i=1}^{\infty} \nu_{Z_i}(y) \dot{\eta}_i(t)$$

---

[1] In the evaluation of the above integrals for the wing and horizontal tail, consider only integrating from the root to the tip and doubling the result, using only antisymmetric modes.

And from Equation (7.39), the (elastic) lateral velocity of the aerodynamic center of a 2-D section on the vertical tail is

$$v_E(z,t) = \sum_{i=1}^{\infty} \nu_{Y_i}(x_{AC}, y_{AC}, z)\dot{\eta}_i(t) \triangleq \sum_{i=1}^{\infty} \nu_{Y_i}(z)\dot{\eta}_i(t)$$

Consequently, we may write Equations (7.53), the change in rolling moments due to elastic effects for the wing and horizontal tail, respectively, as

$$L_{\text{Roll}_{E_W}} \approx -q_\infty \int_{-b_W/2}^{b_W/2} c_{l_{\alpha_W}}(y)\left(\theta_{E_W}(y) + \frac{1}{V_\infty}\sum_{i=1}^{\infty} \nu_{Z_{i_W}}(y)\dot{\eta}_i(t)\right)c_W(y)\,y\,dy$$

$$L_{\text{Roll}_{E_H}} \approx -q_H \int_{-b_H/2}^{b_H/2} c_{l_{\alpha_H}}(y)\left(\begin{matrix}\left(\theta_{E_H}(y) + \frac{1}{V_\infty}\sum_{i=1}^{\infty} \nu_{Z_{i_H}}(y)\dot{\eta}_i(t)\right)\\ -\frac{d\varepsilon_H}{d\alpha_W}\left(\theta_{E_W}(y) + \frac{1}{V_\infty}\sum_{i=1}^{\infty} \nu_{Z_{i_W}}(y)\dot{\eta}_i(t)\right)\end{matrix}\right)c_H(y)\,y\,dy \tag{7.62}$$

and Equation (7.56), the change in rolling moment from the vertical tail, as

$$L_{E_V} = q_H \int_0^{b_V} c_{l_{\alpha_V}}\left(\theta_{E_V}(z) - \frac{1}{V_\infty}\sum_{i=1}^{\infty} \nu_{Y_{i_V}}(z)\dot{\eta}_i(t)\right)z_{AC}(z)c_V(z)\,dz \tag{7.63}$$

In the usual fashion, we may rewrite the above three expressions as

$$L_{\text{Roll}_{E_W}} \approx -q_\infty\left(\frac{1}{V_\infty}\sum_{i=1}^{\infty}\left(\underline{\int_{-b_W/2}^{b_W/2} c_{l_{\alpha_W}}(y)\nu_{Z_{i_W}}(y)c_W(y)\,y\,dy}\right)\dot{\eta}_i(t) + \int_{-b_W/2}^{b_W/2} c_{l_{\alpha_W}}(y)\theta_E(y)c_W(y)\,y\,dy\right)$$

$$L_{\text{Roll}_{E_H}} \approx -q_H\left(\begin{matrix}\frac{1}{V_\infty}\sum_{i=1}^{\infty}\left(\underline{\int_{-b_H/2}^{b_H/2} c_{l_{\alpha_H}}(y)\left(\nu_{Z_{i_H}}(y) - \frac{d\varepsilon_H}{d\alpha_W}\nu_{Z_{i_W}}(y)\right)c_H(y)\,y\,dy}\right)\dot{\eta}_i(t)\\ +\int_{-b_H/2}^{b_H/2} c_{l_{\alpha_H}}(y)\left(\theta_{E_H}(y) - \frac{d\varepsilon_H}{d\alpha_W}\theta_{E_W}(y)\right)c_H(y)\,y\,dy\end{matrix}\right) \tag{7.64}$$

$$L_{E_V} = -q_H\left(\begin{matrix}\frac{1}{V_\infty}\sum_{i=1}^{\infty}\left(\underline{\int_0^{b_V} c_{l_{\alpha_V}}(z)\nu_{Y_{i_V}}(z)z_{AC}(z)c_V(z)\,dz}\right)\dot{\eta}_i(t)\\ -\int_0^{b_V} c_{l_{\alpha_V}}(z)\theta_{E_V}(z)z_{AC}(z)c_V(z)\,dz\end{matrix}\right)$$

where the underlined terms are not explicit functions of time. Differentiating Equations (7.64) with respect to modal velocity yields the effects of modal velocity on the rolling moments from the wing and tails. Or

$$
\begin{aligned}
\frac{\partial L_{\text{Roll}_{E_W}}}{\partial \dot{\eta}_i} &= -\frac{q_\infty}{V_\infty} \int_{-b_W/2}^{b_W/2} c_{l_{\alpha_W}}(y)\nu_{Z_{i_W}}(y)c_W(y)y\,dy \\
\frac{\partial L_{\text{Roll}_{E_H}}}{\partial \dot{\eta}_i} &= -\frac{q_H}{V_\infty} \int_{-b_H/2}^{b_H/2} c_{l_{\alpha_H}}(y)\left(\nu_{Z_{i_H}}(y) - \frac{d\varepsilon_H}{d\alpha_W}\nu_{Z_{i_W}}(y)\right)c_H(y)y\,dy \\
\frac{\partial L_{E_V}}{\partial \dot{\eta}_i} &= -\frac{q_H}{V_\infty} \int_{0}^{b_V} c_{l_{\alpha_V}}(z)\nu_{Y_{i_V}}(z)z_{AC}(z)c_V(z)\,dz
\end{aligned}
\tag{7.65}
$$

And so the effect of modal velocity on the vehicle's rolling moment, written in terms of the moments from the wing and tails, is then

$$
\frac{\partial L_{\text{Roll}}}{\partial \dot{\eta}_i} \triangleq C_{L\text{roll}_{\dot{\eta}_i}} q_\infty S_W b_W = \frac{\partial L_{\text{Roll}_{E_W}}}{\partial \dot{\eta}_i} + \frac{\partial L_{\text{Roll}_{E_H}}}{\partial \dot{\eta}_i} + \frac{\partial L_{E_V}}{\partial \dot{\eta}_i}
\tag{7.66}
$$

Finally, the modal-rate rolling-moment effectiveness for the vehicle $C_{L\text{roll}_{\dot{\eta}_i}}$ has been introduced in the above expression, and this effectiveness may be expressed as

$$
\begin{aligned}
C_{L\text{roll}_{\dot{\eta}_i}} = -\frac{1}{V_\infty S_W b_W}\Bigg(&\int_{-b_W/2}^{b_W/2} c_{l_{\alpha_W}}(y)\nu_{Z_{i_W}}(y)c_W(y)y\,dy \\
&+ \frac{q_H}{q_\infty} \int_{-b_H/2}^{b_H/2} c_{l_{\alpha_H}}(y)\left(\nu_{Z_{i_H}}(y) - \frac{d\varepsilon_H}{d\alpha_W}\nu_{Z_{i_W}}(y)\right)c_H(y)y\,dy \\
&+ \frac{q_H}{q_\infty} \int_{0}^{b_V} c_{l_{\alpha_V}}(z)\nu_{Y_{i_V}}(z)z_{AC}(z)c_V(z)\,dz\Bigg)
\end{aligned}
\tag{7.67}
$$

Due to the symmetry of the vehicle about its XZ plane, all the effectiveness coefficients in Equation (7.67) will be zero for the symmetric modes, so only antisymmetric modes need be considered here.[2]

---

[2] In the evaluation of the above integrals for the wing and horizontal tail, consider only integrating from the root to the tip and doubling the result, using only antisymmetric modes.

7.7 ELASTIC EFFECTS ON YAWING MOMENT

In Section 7.4, we discussed the elastic effects on the vehicle side force, considering only the side force on the vertical tail. This side force may also generate a yawing moment on the vehicle, and this moment will be considered here.

From Equation (7.30) we had the side force generated by the 2-D aerodynamic section on the vertical tail at span z given by

$$\mathsf{s}(z) = c_{l_{\alpha_V}}\left(\alpha_V(z) - \alpha_{0_V}\right)q_H(z)c_V(z)$$

And the local angle of attack of the section was given in Equation (7.29). Or

$$\alpha_V(z) = \alpha_{R_V}(z) + \alpha_{E_V}(z) = \alpha_{R_V}(z) + \left(\theta_{E_V}(z) - \frac{v_E(z)}{V_\infty}\right)$$

Note again that this angle of attack is expressed in terms of two components—rigid and elastic. Hence, the side force $\mathsf{s}_E(z)$ due only to elastic deformation may be identified.

From Figure 7.13, and assuming that the X direction is taken to be positive forward, we can see that the yawing moment associated with the above 2-D elastic side force may be written (essentially in either stability or fuselage-reference axes) as

$$n_E(z) = -\mathsf{s}_E(z)\left(X_{\text{Ref}} - x_{\text{AC}_V}(z)\right)\cos\alpha_0 \approx -\mathsf{s}_E(z)\left(X_{\text{Ref}} - x_{\text{AC}_V}(z)\right) \quad (7.68)$$

The negative sign reflects the fact that a positive Y side force at the tail induces a negative (nose left) yawing moment. By integrating Equation (7.68) over the span of the vertical tail, we have the change in yawing moment on the vehicle due to elastic deformation given by

$$\boxed{N_E = N_{E_V} = -\int_0^{b_V} c_{l_{\alpha_V}}\left(\theta_{E_V}(z) - \frac{v_E(z)}{V_\infty}\right)\left(X_{\text{Ref}} - x_{\text{AC}_V}(z)\right)q_H(z)c_V(z)\,dz} \quad (7.69)$$

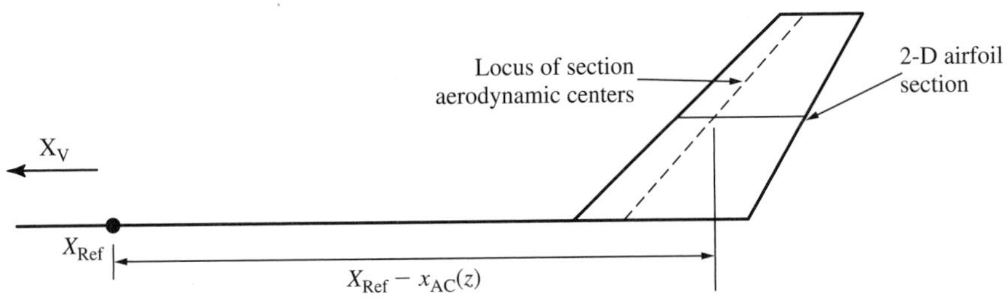

Figure 7.13 Vertical tail geometry and section yawing-moment arm.

7.7.1 Effects of Modal Displacement

The elastic yawing moment given in Equation (7.69) may be expressed in terms of the modal parameters, or the mode shapes and modal coordinates. Specifically, from Equation (7.35), the elastic tail twist $\theta_{E_V}(z)$ may be expressed as

$$\theta_{E_V}(z,t) \approx \tan\theta_{E_V}(z,t) = \sum_{i=1}^{\infty} \nu'_{Y_{i_V}}(z)\eta_i(t)$$

where $\nu'_{Y_{i_V}}(z)$ is the local slope of the Y-displacement component of the i'th mode shape, evaluated at span location z of the vertical tail. Substituting the above expression into Equation (7.69) yields the following result for the change in the vehicle's yawing moment.

$$\begin{aligned} N_E &= -\int_0^{b_V} c_{l_{\alpha_V}}\left(\sum_{i=1}^{\infty} \nu'_{Y_{i_V}}(z)\eta_i(t) - \frac{v_E(z)}{V_\infty}\right)(X_{\text{Ref}} - x_{\text{AC}_V}(z))q_H(z)c_V(z)dz \\ &= -\sum_{i=1}^{\infty}\left(\underline{\int_0^{b_V} c_{l_{\alpha_V}}\nu'_{Y_{i_V}}(z)(X_{\text{Ref}} - x_{\text{AC}_V}(z))q_H(z)c_V(z)dz}\right)\eta_i(t) \\ &\quad + \int_0^{b_V} c_{l_{\alpha_V}}\left(\frac{v_E(z)}{V_\infty}\right)(X_{\text{Ref}} - x_{\text{AC}_V}(z))q_H(z)c_V(z)dz \end{aligned} \tag{7.70}$$

Again note that the underlined term is not a function of time.

Taking the partial derivative of the above expression with respect to the modal coordinate $\eta_i(z)$ yields the effect of modal displacement on the vehicle's yawing moment. Or

$$\boxed{\frac{\partial N}{\partial \eta_i} \triangleq C_{N_{\eta_i}} q_\infty S_W b_W = \frac{\partial N_E}{\partial \eta_i} = -q_H \int_0^{b_V} c_{l_{\alpha_V}} \nu'_{Y_{i_V}}(z)(X_{\text{Ref}} - x_{\text{AC}_V}(z))c_V(z)dz} \tag{7.71}$$

The modal-displacement yawing-moment effectiveness $C_{N_{\eta_i}}$ has been introduced in Equation (7.71) and we can see that this effectiveness is given by

$$\boxed{C_{N_{\eta_i}} = -\frac{1}{S_W b_W}\frac{q_H}{q_\infty}\int_0^{b_V} c_{l_{\alpha_V}} \nu'_{Y_{i_V}}(z)(X_{\text{Ref}} - x_{\text{AC}_V}(z))c_V(z)dz} \tag{7.72}$$

Note the similarity between this result and that for the side-force effectiveness given in Equation (7.38). Only antisymmetric modes need be considered in evaluating the above effectiveness.

7.7.2 Effects of Modal Velocity

The elastic effects on the yawing moment on the vehicle, given in Equation (7.69), may also be expressed in terms of the modal velocities, or the rates of change of the modal coordinates $\dot{\eta}_i(t)$. Specifically, from Equation (7.39) we have the lateral velocity of the tail section due to elastic deformation given by

$$v_E(z,t) = \sum_{i=1}^{\infty} \nu_{Y_{i_V}}(x_{AC}, y_{AC}, z)\dot{\eta}_i(t) \triangleq \sum_{i=1}^{\infty} \nu_{Y_i}(z)\dot{\eta}_i(t)$$

Consequently, the change in yawing moment due to the elastic deformation may be written as

$$\begin{aligned} N_E &= -\int_0^{b_V} c_{l_{\alpha_V}}\left(\theta_{E_V}(z) - \frac{1}{V_\infty}\sum_{i=1}^{\infty} \nu_{Y_{i_V}}(z)\dot{\eta}_i(t)\right)\left(X_{\text{Ref}} - x_{AC_V}(z)\right)q_H(z)c_V(z)dz \\ &= \frac{1}{V_\infty}\sum_{i=1}^{\infty}\left(\underline{\int_0^{b_V} c_{l_{\alpha_V}}\nu_{Y_{i_V}}(z)\left(X_{\text{Ref}} - x_{AC_V}(z)\right)q_H(z)c_V(z)dz}\right)\dot{\eta}_i(t) \\ &\quad - \int_0^{b_V} c_{l_{\alpha_V}}\theta_{E_V}(z)\left(X_{\text{Ref}} - x_{AC_V}(z)\right)q_H(z)c_V(z)dz \end{aligned} \tag{7.73}$$

where the underlined term is not a function of time.

By differentiating the above expression with respect to the modal velocities $\dot{\eta}_i$, we find that the effect of the modal velocities on the vehicle yawing moment may be written as

$$\frac{\partial N}{\partial \dot{\eta}_i} \triangleq C_{N_{\dot{\eta}_i}} q_\infty S_W b_W \approx \frac{\partial N_E}{\partial \dot{\eta}_i} = \frac{1}{V_\infty}\int_0^{b_V} c_{l_{\alpha_V}}\nu_{Y_{i_V}}(z)\left(X_{\text{Ref}} - x_{AC_V}(z)\right)q_H(z)c_V(z)dz \tag{7.74}$$

The modal-velocity yawing-moment effectiveness $C_{N_{\dot{\eta}_i}}$ has been introduced in the above equation, and we see that this effectiveness can be expressed as

$$C_{N_{\dot{\eta}_i}} = \frac{1}{V_\infty S_W b_W}\frac{q_H}{q_\infty}\int_0^{b_V} c_{l_{\alpha_V}}\nu_{Y_{i_V}}(z)\left(X_{\text{Ref}} - x_{AC_V}(z)\right)c_V(z)dz \tag{7.75}$$

The similarity between the above result and that for the side-force effectiveness given by Equation (7.42) is noted. Only antisymmetric modes need be considered when evaluating the above effectiveness.

7.8 GENERALIZED FORCES ACTING ON THE ELASTIC DEGREES OF FREEDOM

In this section we will address the generalized forces Q_i that act as forcing functions on the elastic degrees of freedom. We noted at the beginning of this chapter that the forced response of the elastic degrees of freedom is governed by Equations (7.4), or

$$\ddot{\eta}_i + \omega_i^2 \eta_i = \frac{Q_i}{\mathcal{M}_i},\ i = 1 \dots n$$

Here, Q_i is the generalized force acting on the i'th elastic degree of freedom, and $\mathcal{M}_i$ is the i'th generalized mass. Like the mode shapes $\boldsymbol{\nu}_i$ and vibration frequencies ω_i, the generalized masses are assumed available from a free-vibration analysis of the structure.

From Chapter 4 we know that the generalized forces may be expressed in terms of virtual work δW. Or

$$Q_i = \frac{\partial(\delta W)}{\partial \eta_i} \tag{7.76}$$

Also from Chapter 4 we had expressed the virtual work associated with virtual elastic deformation in terms of a surface integral over the pressure distribution acting on the structure. Or

$$\delta W_E = \int_S \mathbf{P}(x,y,z) \cdot \sum_{i=1}^{\infty} \boldsymbol{\nu}_i(x,y,z) \delta\eta_i dS \tag{7.77}$$

But this integral can now be written more conveniently in terms of the 2-D lift and pitching moment of the airfoil sections of the lifting surfaces.

Consider the 2-D forces and moments shown in Figure 7.14, and note that the virtual work per unit span due to virtual elastic displacement, δw_E, may be

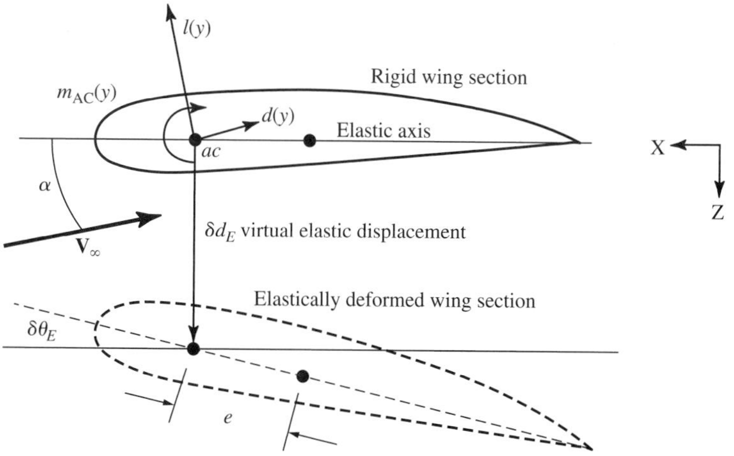

Figure 7.14 Forces, moments, and displacements of the 2-D airfoil section.

expressed in terms of virtual translational displacements and forces (per unit span), and rotational displacement and moments (per unit span).

That is, the virtual work done due to the virtual elastic deformation of an airfoil section of the wing or horizontal tail may be written as

$$\begin{aligned}\delta w_{E_{W\text{ or }H}}(y) &= -\big(l(y)\cos\alpha + d(y)\sin\alpha\big)\sum_{i=1}^{\infty}\nu_{Z_i}(y)\delta\eta_i \\ &+\big(l(y)\sin\alpha - d(y)\cos\alpha\big)\sum_{i=1}^{\infty}\nu_{X_i}(y)\delta\eta_i \\ &+\Big(m_{\text{AC}}(y) + e(y)\big(l(y)\cos\alpha + d(y)\sin\alpha\big)\Big)\sum_{i=1}^{\infty}\nu'_{Z_i}(y)\delta\eta_i\end{aligned} \tag{7.78}$$

Here $e(y)$ is the distance between the section aerodynamic center and the wing's *elastic axis.* The elastic axis is the locus of points along the span at which transverse (i.e., bending) loads applied to the wing will produce no elastic twist (Ref. 3).

Likewise with regard to a vertical tail 2-D section, we may write its virtual work as

$$\begin{aligned}\delta w_{E_V}(z) &= \big(\mathsf{s}(z)\cos\beta - d(z)\sin\beta\big)\sum_{i=1}^{\infty}\nu_{Y_{i_V}}(z)\delta\eta_i \\ &-\big(\mathsf{s}(z)\sin\beta + d(z)\cos\beta\big)\sum_{i=1}^{\infty}\nu_{X_{i_V}}(z)\delta\eta_i \\ &+\Big(m_{\text{AC}}(z) + e(z)\big(\mathsf{s}(z)\cos\beta - d(z)\sin\beta\big)\Big)\sum_{i=1}^{\infty}\nu'_{Y_{i_V}}(z)\delta\eta_i\end{aligned} \tag{7.79}$$

Note that in Equations (7.78) and (7.79), ν_{X_i}, ν_{Y_i} and ν_{Z_i} are the components of the i'th displacement mode shape evaluated at the aerodynamic center of the respective 2-D section, and $\nu'_{Z_i \text{ or } Y_i}$ is the slope of the Z- or Y-displacement mode shape evaluated at span location y or z of the section aerodynamic centers of the wing or tails, respectively.

Typically, Equations (7.78) and (7.79) can be considerably simplified under some reasonable assumptions. These assumptions include:

- α and β are small.
- Drag $d(y) \ll l(y)$ (lift) on the wing and horizontal tail, and $d(z) \ll \mathsf{s}(z)$ (side force) on the vertical tail.
- Wing and tail "surge stiffness" is much greater than plunge or lateral stiffness. Therefore, for the wing and horizontal tail $\nu_{Z_i}(y) \gg \nu_{X_i}(y)$, and for the vertical tail $\nu_{Y_i}(z) \gg \nu_{X_i}(z)$ for all i, y, and z.
- The tail sections consist of symmetric airfoils, and therefore $\alpha_{0_H}(y) = \alpha_{0_V}(z) = 0$, and $m_{\text{AC}_H}(y) = m_{\text{AC}_V}(z) = 0$.

Under these assumptions, Equations (7.78) and (7.79) become simply

$$\delta w_{E_W}(y) \approx -l_W(y) \sum_{i=1}^{\infty} \nu_{Z_{i_W}}(y)\,\delta\eta_i + \big(m_{AC_W}(y) + e_W(y)\,l_W(y)\big) \sum_{i=1}^{\infty} \nu'_{Z_{i_W}}(y)\,\delta\eta_i$$

$$\delta w_{E_H}(y) \approx -l_H(y)\left(\sum_{i=1}^{\infty} \big(\nu_{Z_{i_H}}(y) + e_H(y)\nu'_{Z_{i_H}}(y)\big)\delta\eta_i\right) \tag{7.80}$$

$$\delta w_{E_V}(z) \approx \mathsf{s}(z)\left(\sum_{i=1}^{\infty} \big(\nu_{Y_{i_V}}(z) + e_V(z)\nu'_{Y_{i_V}}(z)\big)\delta\eta_i\right)$$

We may now integrate Equations (7.80) over the spans of the wing and tails to obtain the total virtual work done due to virtual elastic deformation. Or

$$\delta W_E \approx \int_{-b_W/2}^{b_W/2} \delta w_{E_W}(y)\,dy + \int_{-b_H/2}^{b_H/2} \delta w_{E_H}(y)\,dy + \int_{0}^{b_V} \delta w_{E_V}(z)\,dz \tag{7.81}$$

By substituting Equations (7.80) into Equation (7.81), and differentiating the result with respect to virtual modal displacement $\delta\eta_i$, we have the generalized forces given by

$$\boxed{\begin{aligned} Q_i = &\int_{-b_W/2}^{b_W/2} \Big(-l_W(y)\nu_{Z_{i_W}}(y) + \big(m_{AC_W}(y) + e_W(y)\,l_W(y)\big)\nu'_{Z_{i_W}}(y)\Big)dy \\ &- \int_{-b_H/2}^{b_H/2} l_H(y)\big(\nu_{Z_{i_H}}(y) + e_H(y)\nu'_{Z_{i_H}}(y)\big)dy \\ &+ \int_{0}^{b_V} \mathsf{s}(z)\big(\nu_{Y_{i_V}}(z) + e_V(z)\nu'_{Y_{i_V}}(z)\big)dz \end{aligned}} \tag{7.82}$$

Note the limits on the integrals. If antisymmetric as well as symmetric modes are being considered, integration over the full spans may be required.

We must now stress that the section lift and side force in Equation (7.82) are functions of the total motion of the airfoil section, plus any control surface deflection. For example, using a familiar expression developed in Chapter 5, the section lift for the wing or horizontal tail may be written as

$$l(y) = c_{l_\alpha}(y)\big(\alpha(y) - \alpha_0(y) + \alpha_\delta(y)\delta(y)\big)q(y)c(y) \tag{7.83}$$

where δ corresponds to a control-surface or flap deflection and α_0 is the zero-lift angle of attack. Furthermore, the local section angle of attack $\alpha(y)$ in Equation (7.83) is a function of both the rigid-body and elastic motion of the section. Or, assuming the X direction is positive forward, the local angle of attack of a wing section is

$$\alpha_W(y) = \alpha_{R_W}(y) + \alpha_{E_W}(y) \tag{7.84}$$
$$= \frac{1}{V_\infty}\left(W + Py - Q\left(x_{\mathrm{AC}_W}(y) - X_{\mathrm{Ref}}\right)\right) + \left(i_W + \varepsilon_{\mathrm{Twist}}(y)\right) + \left(\theta_{E_W}(y) + \frac{w_{E_W}(y)}{V_\infty}\right)$$

where as defined in Chapter 2, P and Q are the rigid-body roll and pitch rate, respectively, and W is the rigid-body plunge velocity. The elastic motion consists of twist $\theta_E(y)$ and plunge $w_E(y)$, as discussed in Section 7.3.

For the local angle of attack of a horizontal tail section we have

$$\alpha_H(y) = \frac{1}{V_\infty}\left(W + Py + Q\left(X_{\mathrm{Ref}} - x_{\mathrm{AC}_H}(y)\right)\right) + \left(i_H + \varepsilon_{\mathrm{Twist}}(y) - \varepsilon_H(y)\right)$$
$$+ \left(\theta_{E_H}(y) + \frac{w_{E_H}(y)}{V_\infty}\right) \tag{7.85}$$

Here, ε_H is the local downwash angle, which may be expressed as

$$\varepsilon_H = \frac{d\varepsilon_H}{d\alpha_W}\left(\alpha_W - \alpha_{0_W}\right) \tag{7.86}$$

Hence, the angles of attack of the wing and tail are coupled.

And finally for the vertical tail we may write the local angle of attack as

$$\alpha_V(z) = \frac{-1}{V_\infty}\left(V + Pz_{\mathrm{AC}}(z) - R\left(X_{\mathrm{Ref}} - x_{\mathrm{AC}_V}(z)\right)\right) + \left(\theta_{E_V}(z) - \frac{v_{E_V}(z)}{V_\infty}\right) \tag{7.87}$$

As discussed in Section 7.6, z_{AC} is the height of the section aerodynamic center above the X_V axis of the chosen vehicle-fixed reference frame. Consequently, we can see that the generalized forces in Equation (7.82) are functions of virtually all the motion variables and control deflections.

Let us consider the generalized force (Equation (7.82)) now expanded in a Taylor series.[3] Or let

$$Q_i = Q_{i_{\mathbf{p}=\mathbf{0}}} + \frac{\partial Q_i}{\partial \mathbf{p}}\mathbf{p} \tag{7.88}$$

where the parameter vector $\mathbf{p}$ here includes the modal displacements and velocities. That is,

$$\mathbf{p}^T = \begin{bmatrix} U & V & W & P & Q & R & \eta_i, i = 1\cdots n & \dot{\eta}_i, i = 1\cdots n & i_H & \delta_E & \delta_A & \delta_R \end{bmatrix} \tag{7.89}$$

[3] Note that this Taylor series is expanded about the condition $\mathbf{p} = \mathbf{0}$, rather than a flight reference condition.

In addition, let us define the i'th *generalized force coefficient* C_{Q_i} from the following expression

$$Q_i \triangleq C_{Q_i} q_\infty S_W \bar{c}_W \tag{7.90}$$

And consistent with the above treatment of the generalized force, let us expand the generalized-force coefficient in a Taylor series. Or

$$C_{Q_i} = C_{Q_{i_{\mathbf{p}=\mathbf{0}}}} + \frac{\partial C_{Q_i}}{\partial \mathbf{p}} \mathbf{p} \tag{7.91}$$

where, for example,

$$C_{Q_{i_{\mathbf{p}=\mathbf{0}}}} = \frac{1}{q_\infty S_W \bar{c}_W} Q_{i_0} \tag{7.92}$$

and for the k'th parameter in $\mathbf{p}$

$$C_{Q_{i_{p_k}}} \triangleq \frac{\partial C_{Q_i}}{\partial p_k} = \frac{1}{q_\infty S_W \bar{c}_W} Q_{i_{p_k}} \tag{7.93}$$

We may now develop the expressions for the partial derivatives in Equation (7.88), and for the generalized-force coefficients. Consider a vehicle with both horizontal and vertical tails, and let us assume that the airfoils for the tails are symmetric and that the distance between the aerodynamic centers and the elastic axis for all wing and tail sections, or e, is negligible. Then for the partial derivatives with respect to the rigid-body degrees of freedom and the control surfaces, and for the corresponding generalized-force coefficients, one can show (see Problem 7.2) that

$$\begin{aligned}
Q_{i_{\mathbf{p}=\mathbf{0}}} &= C_{Q_{i_0}} q_\infty S_W \bar{c}_W = q_\infty \int_{-b_W/2}^{b_W/2} \begin{pmatrix} -c_{l_{\alpha_W}}(y)\big(i_W + \varepsilon_{\text{Twist}_W}(y) - a_{0_W}(y)\big)\nu_{Z_{i_W}}(y) \\ +c_{m_{\text{AC}_W}}(y) c_W(y) \nu'_{Z_{i_W}}(y) \end{pmatrix} c_W(y)\,dy \\
&\quad + q_H \int_{-b_H/2}^{b_H/2} c_{l_{\alpha_H}}(y) \frac{d\varepsilon_H}{d\alpha_W}\big(i_W + \varepsilon_{\text{Twist}_W}(y) - \alpha_{0_W}(y)\big)\nu_{Z_{i_H}}(y) c_H(y)\,dy \\
Q_{i_u} &= C_{Q_{i_u}} q_\infty S_W \bar{c}_W = \frac{2q_\infty}{V_\infty}\frac{\partial Q_i}{\partial q_\infty} = \frac{2q_\infty}{V_\infty} C_{Q_i} S_W \bar{c}_W \\
Q_{i_\alpha} &= V_\infty Q_{i_w} = C_{Q_{i_\alpha}} q_\infty S_W \bar{c}_W = -q_\infty \int_{-b_W/2}^{b_W/2} c_{l_{\alpha_W}}(y)\nu_{Z_{i_W}}(y) c_W(y)\,dy \\
&\quad - q_H \int_{-b_H/2}^{b_H/2} c_{l_{\alpha_H}}(y)\left(1 - \frac{d\varepsilon_H}{d\alpha_W}\right)\nu_{Z_{i_H}}(y) c_H(y)\,dy \\
Q_{i_\beta} &= V_\infty Q_{i_v} = C_{Q_{i_\beta}} q_\infty S_W \bar{c}_W = -q_H \int_0^{b_V} c_{l_{\alpha_V}}(z)\nu_{Y_{i_V}}(z) c_V(z)\,dz
\end{aligned} \tag{7.94}$$

$$Q_{i_p} = C_{Q_{i_p}} q_\infty S_W \bar{c}_W = -\frac{q_\infty}{V_\infty} \int_{-b_W/2}^{b_W/2} c_{l_{\alpha_W}}(y) \nu_{Z_{i_W}}(y) c_W(y) y dy$$
$$- \frac{q_H}{V_\infty} \int_{-b_H/2}^{b_H/2} c_{l_{\alpha_H}}(y) \left(1 - \frac{d\varepsilon_H}{d\alpha_W}\right) \nu_{Z_{i_H}}(y) c_H(y) y dy$$
$$- \frac{q_H}{V_\infty} \int_0^{b_V} c_{l_{\alpha_V}}(z) z_{AC}(z) \nu_{Y_{i_V}}(z) c_V(z) dz$$

$$Q_{i_q} = C_{Q_{i_q}} q_\infty S_W \bar{c}_W = \frac{q_\infty}{V_\infty} \int_{-b_W/2}^{b_W/2} c_{l_{\alpha_W}}(y) \left(x_{AC_W}(y) - X_{\text{Ref}}\right) \nu_{Z_{i_W}}(y) c_W(y) dy$$
$$+ \frac{q_H}{V_\infty} \int_{-b_H/2}^{b_H/2} c_{l_{\alpha_H}}(y) \left(\left(X_{\text{Ref}} - x_{AC_H}(y)\right) - \frac{d\varepsilon_H}{d\alpha_W}\left(x_{AC_W}(y) - X_{\text{Ref}}\right)\right) \nu_{Z_{i_H}}(y) c_H(y) dy$$

(7.94) cont'd

$$Q_{i_r} = C_{Q_{i_r}} q_\infty S_W \bar{c}_W = \frac{q_H}{V_\infty} \int_0^{b_V} c_{l_{\alpha_V}}(z) \left(X_{\text{Ref}} - x_{AC_V}(z)\right) \nu_{Y_{i_V}}(z) c_V(z) dz$$

$$Q_{i_{i_H}} = C_{Q_{i_{i_H}}} q_\infty S_W \bar{c}_W = -q_H \int_{-b_H/2}^{b_H/2} c_{l_{\alpha_H}}(y) \nu_{Z_{i_H}}(y) c_H(y) dy$$

$$Q_{i_{\delta_E}} = C_{Q_{i_{\delta_E}}} q_\infty S_W \bar{c}_W = -q_H \int_{-b_H/2}^{b_H/2} c_{l_{\alpha_H}}(y) \alpha_{\delta_E}(y) \nu_{Z_{i_H}}(y) c_H(y) dy$$

$$Q_{i_{\delta_R}} = C_{Q_{i_{\delta_R}}} q_\infty S_W \bar{c}_W = q_H \int_0^{b_V} c_{l_{\alpha_V}}(z) \alpha_{\delta_R}(z) \nu_{Y_{i_V}}(z) c_V(z) dz$$

$$Q_{i_{\delta_A}} = C_{Q_{i_{\delta_A}}} q_\infty S_W \bar{c}_W = q_\infty \left(\int_{-b_o/2}^{-b_i/2} c_{l_{\alpha_W}}(y) \alpha_{\delta_A}(y) \nu_{Zi_W}(y) c_W(y) dy - \int_{b_i/2}^{b_o/2} c_{l_{\alpha_W}}(y) \alpha_{\delta_A}(y) \nu_{Z_{i_W}}(y) c_W(y) dy \right)$$

Note that in the last of Equations (7.94) involving the aileron, or $Q_{i_{\delta_A}}$, the two integrals correspond to the left and right wing, respectively, and the limits on the integrals, b_i and b_o correspond to the inboard and outboard spans of the ailerons, respectively.

EXAMPLE 7.2

Calculating Generalized-Force Coefficients

Again consider the vehicle analyzed in Example 7.1. For this vehicle, find the generalized-force coefficients $C_{Q_{1_\alpha}}$, $C_{Q_{1_q}}$, and $C_{Q_{1_{\delta_E}}}$.

■ Solution

The data for the wing and horizontal tail was given in Example 7.1, including the plunge and twist mode shapes for the first symmetric vibration mode. The polynomial fits for the plunge-displacement mode shapes were

$$\nu_{Z_{1_W}}(y) = 0.5265\left(\frac{y}{15}\right)^2 + 0.0135\left(\frac{y}{15}\right) - 0.14$$

$$\nu_{Z_{1_H}}(y) = 0.09625\left(\frac{y}{7.5}\right) + 0.194$$

In addition, in Example 6.2 it was noted that the location of the tail's aerodynamic center was $X_{AC_H} = 2.5c_{r_W}$ aft of the wing apex, and that the reference point used to define the pitching moment X_{Ref} was located at 0.5 MAC aft of the apex as well. For this wing, $S_W = 169\ \text{ft}^2$ and MAC $= \bar{c}_W = 5.825$ ft. Finally, from Example 5.6 we had a constant section elevator effectiveness $\alpha_{\delta_E}(y) = 0.65$ for the tail.

Now from Equations (7.94) we have

$$C_{Q_{i_\alpha}} q_\infty S_W \bar{c}_W = -q_\infty \int_{-b_W/2}^{b_W/2} c_{l_{\alpha_W}}(y)\nu_{Z_{i_W}}(y)c_W(y)\,dy - q_H \int_{-b_H/2}^{b_H/2} c_{l_{\alpha_H}}(y)\left(1 - \frac{d\varepsilon_H}{d\alpha_W}\right)\nu_{Z_{i_H}}(y)c_H(y)\,dy$$

$$C_{Q_{i_q}} q_\infty S_W \bar{c}_W = \frac{q_\infty}{V_\infty} \int_{-b_W/2}^{b_W/2} c_{l_{\alpha_W}}(y)\left(x_{AC_W}(y) - X_{\text{Ref}}\right)\nu_{Z_{i_W}}(y)c_W(y)\,dy + \frac{q_H}{V_\infty} \int_{-b_H/2}^{b_H/2} c_{l_{\alpha_H}}(y)\left(\left(X_{\text{Ref}} - x_{AC_H}(y)\right) - \frac{d\varepsilon_H}{d\alpha_W}\left(x_{AC_W}(y) - X_{\text{Ref}}\right)\right)\nu_{Z_{i_H}}(y)c_H(y)\,dy$$

$$C_{Q_{i_{\delta_E}}} q_\infty S_W \bar{c}_W = -q_H \int_{-b_H/2}^{b_H/2} c_{l_{\alpha_H}}(y)\alpha_{\delta_E}(y)\nu_{Z_{i_H}}(y)c_H(y)\,dy$$

So the solution depends on the five integral expressions in these three equations.

To evaluate these integrals, first recall that in Example 7.1 we had found that

$$\int_0^{b_W/2} c_{l_{\alpha_W}}(y)\nu_{Z_{1_W}}(y)c_W(y)dy = 6.34 \text{ ft}^3$$

and since the wing and this mode shape are symmetric we have the first integral given by

$$\int_{-b_W/2}^{b_W/2} c_{l_{\alpha_W}}(y)\nu_{Z_{1_W}}(y)c_W(y)dy = 2(6.34) = 12.68 \text{ ft}^3$$

The second integral, again since the horizontal tail and the mode shape are symmetric, is

$$\int_{-b_H/2}^{b_H/2} c_{l_{\alpha_H}}(y)\left(1 - \frac{d\varepsilon_H}{d\alpha_W}\right)\nu_{Z_{i_H}}(y)c_H(y)dy$$

$$= 2\int_0^{7.5} (0.107 \times 57.3)(1 - 0.57)\left(0.09625\left(\frac{y}{7.5}\right) + 0.194\right)\left(\frac{7.5}{2} - \frac{y}{4}\right)dy$$

$$= 2(0.107 \times 57.3)(0.43)(0.0128)\int_0^{7.5} (y + 15.16)\left(\frac{7.5}{2} - \frac{y}{4}\right)dy = 26.4 \text{ ft}^3$$

The fifth integral, similar to the second, is

$$\int_{-b_H/2}^{b_H/2} c_{l_{\alpha_H}}(y)\alpha_{\delta_E}(y)\nu_{Z_{1_H}}(y)c_H(y)dy$$

$$= 2\int_0^{7.5} (0.107 \times 57.3)(0.65)\left(0.09625\left(\frac{y}{7.5}\right) + 0.194\right)\left(\frac{7.5}{2} - \frac{y}{4}\right)dy = 39.91 \text{ ft}^3$$

To evaluate the third integral, we note that measured <u>aft</u> from the wing apex the section aerodynamic centers are located at

$$x_{AC_W}(y) = x_{LE_W}(y) + 0.25c_W(y) = y\tan\Lambda_{LE} + 0.25c_W(y)$$

So then if X locations are measured positive forward, and with a leading-edge sweep angle of 26.6 deg,

$$x_{AC_W}(y) - X_{Ref} = 0.5\bar{c}_W - \left(y\tan 26.6° + 0.25\left(-\frac{y}{4} + 7.5\right)\right) = -0.438y + 1.0375 \text{ ft}$$

and the third integral becomes

$$\int_{-b_W/2}^{b_W/2} c_{l_{\alpha_W}}(y)\left(x_{\mathrm{AC}_W}(y)-X_{\mathrm{Ref}}\right)\nu_{Z_{1_W}}(y)c_W(y)dy$$

$$=2\int_0^{15}(0.107\times 57.3)(-0.438y+1.0375)$$

$$\left(0.5265\left(\frac{y}{15}\right)^2+0.0135\left(\frac{y}{15}\right)-0.14\right)\left(-\frac{y}{4}+7.5\right)dy=-282.75\ \mathrm{ft}^4$$

Finally, to evaluate the fourth integral, note from Examples 5.6 and 5.7 that measured aft of the wing apex, the location of the tail apex is

$$x_{\mathrm{Apex}_H}=2.5c_{r_W}-2.39=16.36\ \mathrm{ft}$$

And measured aft of the tail apex with a leading-edge sweep angle of 26.6 deg, the tail aerodynamic centers are located at

$$x_{\mathrm{AC}_H}(y)=x_{\mathrm{LE}_H}(y)+0.25c_H(y)=y\tan 26.6°+0.25\left(-\frac{y}{4}+\frac{7.5}{2}\right)$$

So then if X locations are measured positive forward,

$$X_{\mathrm{Ref}}-x_{\mathrm{AC}_H}(y)=16.36+y\tan 26.6°+0.25\left(-\frac{y}{4}+\frac{7.5}{2}\right)-0.5\bar{c}_W$$

$$=0.438y+14.385\ \mathrm{ft}$$

and the fourth integral becomes

$$\int_{-b_H/2}^{b_H/2} c_{l_{\alpha_H}}(y)\left(\left(X_{\mathrm{Ref}}-x_{\mathrm{AC}_H}(y)\right)-\frac{d\varepsilon_H}{d\alpha_W}\left(x_{\mathrm{AC}_W}(y)-X_{\mathrm{Ref}}\right)\right)\nu_{Z_{1_H}}(y)c_H(y)dy$$

$$=2\int_0^{7.5}(0.107\times 57.3)\left((0.438y+14.385)-0.57(-0.438y+1.0375)\right)$$

$$\left(0.09625\left(\frac{y}{7.5}\right)+0.194\right)\left(\frac{7.5}{2}-\frac{y}{4}\right)dy$$

$$=2(0.107\times 57.3)(0.0128)\int_0^{7.5}(0.688y+13.79)(y+15.16)\left(\frac{7.5}{2}-\frac{y}{4}\right)dy$$

$$=972.68\ \mathrm{ft}^4$$

Collecting terms, and assuming $q_H/q_\infty = 0.9$, we have the desired coefficients given by

$$C_{Q_{i_\alpha}} = \frac{1}{S_W \bar{c}_W}\left(-12.68 - \frac{q_H}{q_\infty}(26.4)\right) = \frac{-1}{(169)(5.825)}\left(12.68 + 0.9(26.4)\right) = -0.037$$

$$C_{Q_{i_q}} = \frac{1}{S_W \bar{c}_W V_\infty}\left(-282.75 + \frac{q_H}{q_\infty}(972.68)\right)$$
$$= \frac{1}{(169)(5.825)V_\infty}\left(-282.75 + 0.9(972.68)\right) = \frac{0.602}{V_\infty}\ \text{sec}$$

$$C_{Q_{i_{\delta_E}}} = -\frac{1}{S_W \bar{c}_W}\frac{q_H}{q_\infty}(39.91) = -\frac{0.9}{(169)(5.825)}(39.91) = 0.036$$

Note that not all these coefficients are dimensionless.

The procedure for determining the partial derivatives of the generalized forces with respect to the elastic degrees of freedom, or the modal displacement and rate, is the same as that presented in Section 7.3–7.7, but the algebra is a little more involved. First recall that the airfoil-section elastic twist angle θ_E may be expressed in terms of the slopes of the mode shapes. Or, for the wing and horizontal tail we wrote this elastic twist angle as in Equation (7.19), or

$$\theta_{E_{W\text{ or }H}}(y,t) = \sum_{i=1}^{\infty} \nu'_{Z_{i_{W\text{ or }H}}}(y)\eta_i(t)$$

and for the vertical tail we have, as in Equation (7.35),

$$\theta_{E_V}(z,t) = \sum_{i=1}^{\infty} \nu'_{Y_{i_V}}(z)\eta_i(t)$$

In addition, the plunge and lateral velocities of the aerodynamic centers of the wing and tail airfoil sections may be expressed in terms of the modal velocities. Or, for the wing or horizontal-tail sections the plunge velocities, as in Equations (7.23), are

$$w_{E_{W\text{ or }H}}(y,t) = \sum_{i=1}^{\infty} \nu_{Z_{i_{W\text{ or }H}}}(y)\dot{\eta}_i(t)$$

and for the vertical-tail sections the lateral velocities, as in Equations (7.39), are

$$v_{E_V}(z,t) = \sum_{i=1}^{\infty} \nu_{Y_{i_V}}(z)\dot{\eta}_i(t)$$

Finally, the above four equations may be substituted into the equations for the section angles of attack, or Equations (7.84–7.87).

After making the above substitutions, we find that the partial derivatives of the generalized forces with respect to modal displacement and velocity, and their associated generalized-force coefficients, are given by

$$Q_{i_{\eta_j}} = C_{Q_{i_{\eta_j}}} q_\infty S_W \bar{c}_W = -q_\infty \int_{-b_W/2}^{b_W/2} c_{l_{\alpha_W}}(y) \nu'_{Z_{j_W}}(y) \nu_{Z_{i_W}}(y) c_W(y)\,dy$$
$$-q_H \int_{-b_H/2}^{b_H/2} c_{l_{\alpha_H}}(y) c_H(y) \left(\nu'_{Z_{j_H}}(y) - \frac{d\varepsilon_H}{d\alpha_W} \nu'_{Z_{j_W}}(y) \right) \nu_{Z_{i_H}}(y)\,dy$$
$$+q_H \int_0^{b_V} c_{l_{\alpha_V}}(z) \nu'_{Y_{j_V}}(z) \nu_{Y_{i_V}}(z) c_V(z)\,dz$$

$$Q_{i_{\dot\eta_j}} = C_{Q_{i_{\dot\eta_j}}} q_\infty S_W \bar{c}_W = -\frac{q_\infty}{V_\infty} \int_{-b_W/2}^{b_W/2} c_{l_{\alpha_W}}(y) \nu_{Z_{j_W}}(y) \nu_{Z_{i_W}}(y) c_W(y)\,dy$$
$$-\frac{q_H}{V_\infty} \int_{-b_H/2}^{b_H/2} c_{l_{\alpha_H}}(y) \left(\nu_{Z_{j_H}}(y) - \frac{d\varepsilon_H}{d\alpha_W} \nu_{Z_{j_W}}(y) \right) \nu_{Z_{i_H}}(y) c_H(y)\,dy$$
$$-\frac{q_H}{V_\infty} \int_0^{b_V} c_{l_{\alpha_V}}(z) \nu_{Y_{j_V}}(z) \nu_{Y_{i_V}}(z) c_V(z)\,dz \tag{7.95}$$

Note here that two sets of mode shapes are involved in the above two expressions, the *i*'th corresponding to the *i*'th mode and generalized force, and the *j*'th corresponding to the modal coordinate with respect to which the partial derivative was taken.

7.9 ELASTIC EFFECTS ON THE FORCES AND MOMENTS FOR A LARGE HIGH-SPEED AIRCRAFT—A CASE STUDY

We will now consider a large high-speed aircraft, with geometry as shown in Figure 7.15 (Ref.2). The vehicle is representative of a large military vehicle or perhaps a supersonic transport. (Note the small control vanes near the cockpit added to control the vibrations experienced by the pilot.) The focus of the modeling is the elastic effects on the vehicle's flight dynamics, and a model of the rigid-vehicle dynamics is assumed available. Such a situation may arise, for example,

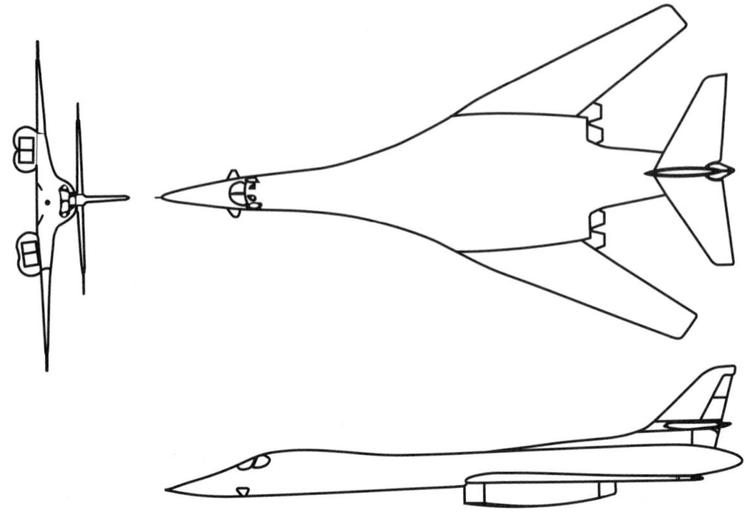

Figure 7.15 Large, high-speed aircraft.

Table 7.1 Data for Study Vehicle

Wing geometry	$S_W = 1{,}950\ \text{ft}^2$	**Inertias**	$I_{xx} = 9.5 \times 10^5\ \text{sl-ft}^2$
	$\bar{c}_W = 15.3\ \text{ft}$		$I_{yy} = 6.4 \times 10^6\ \text{sl-ft}^2$
	$b_W = 70\ \text{ft}$		$I_{zz} = 7.1 \times 10^6\ \text{sl-ft}^2$
	$\Lambda_{LE} = 65\ \text{deg}$		$I_{xz} = -52{,}700\ \text{sl-ft}^2$
Weight	$W = 288{,}000\ \text{lb}$	**Vehicle length**	143 ft
Modal generalized masses	$\mathcal{M}_1 = 184\ \text{sl-ft}^2$	**Modal frequencies**	$\omega_1 = 12.6\ \text{rad/sec}$
	$\mathcal{M}_2 = 9{,}587\ \text{sl-ft}^2$		$\omega_2 = 14.1\ \text{rad/sec}$
	$\mathcal{M}_3 = 1{,}334\ \text{sl-ft}^2$		$\omega_3 = 21.2\ \text{rad/sec}$
	$\mathcal{M}_4 = 436{,}000\ \text{sl-ft}^2$		$\omega_4 = 22.1\ \text{rad/sec}$

when a simulation model of the rigid vehicle has been developed, and the model is to be modified to capture the elastic effects. Such was the case in the investigation reported in Ref. 4.

Key geometric data and mass properties of the vehicle are given in Table 7.1. In addition, a vibration analysis of the structure has been performed, and the mode shapes for the first four symmetric vibration modes of the structure are sketched in Figure 7.16. Detailed tabulated data for the mode shapes are also assumed available (Ref. 5), so the sketches in Figure 7.16 are included just to help visualize the modes. For example, Mode 1 might be called the *first fuselage bending mode*, while Mode 2 might be called the *first wing bending mode*.

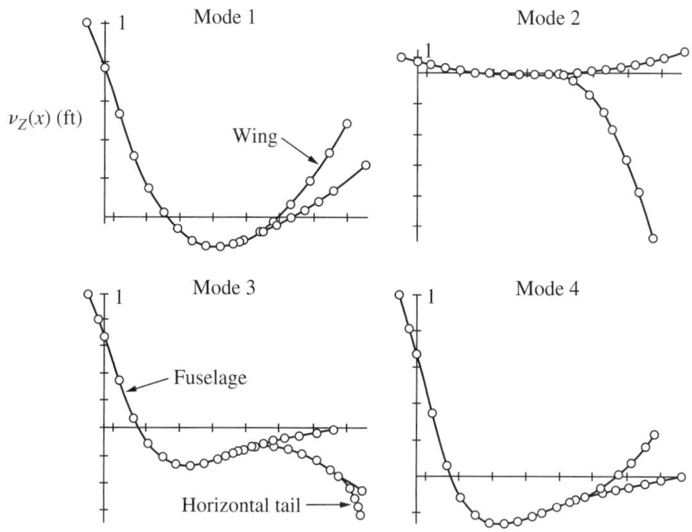

Figure 7.16 Sketches of the vibration mode shapes.

All four mode shapes have been normalized such that the magnitude of the mode shape is one foot at the nose of the aircraft. For this normalization, the modal generalized masses are tabulated in Table 7.1. Note that the magnitudes of these masses differ widely due to the normalization.

The natural frequencies for the four vibration modes are also listed in Table 7.1. Although we do not use these frequencies in developing the aerodynamic model, they are useful in helping select the number of modes to include in the model. For example, if we are interested in the motion of the vehicle due to pilot inputs, we do not need to include modes with modal frequencies much above 20 rad/sec, as the pilot would not be able to excite the higher frequency modes.

In this study we will model the elastic effects on the aerodynamic force F_{A_Z} and the pitching moment M_A, as well as develop the model for the generalized forces associated with the four modal coordinates η_{1-4}. Similar to the examples in this chapter, and noting that the wing root chord is on the centerline of the vehicle, each coefficient was determined by graphically evaluating the necessary integrals appearing in the equation corresponding to that coefficient. The results are listed in Tables 7.2 and 7.3. (All angles are in radians.) Also listed in the table are the equations used. For example, using Equation (7.22) we find that $C_{L_{\eta_1}} = 0.029$, $C_{L_{\eta_2}} = -0.306$, $C_{L_{\eta_3}} = -0.015$, and $C_{L_{\eta_4}} = 0.014$. In obtaining these results, we have assumed $c_{l_\alpha}(y) = 5.73$ /rad and $m_{AC}(y) = 0$ for both the wing and horizontal tail, and that $\dfrac{d\varepsilon_H}{d\alpha_W} = 0.5$.

Table 7.2 Elastic Coefficients for Lift and Pitching Moment

	Coefficient	Equation	Mode 1	Mode 2	Mode 3	Mode 4
Lift	$C_{L_{\eta_i}}$	7.22	0.029	−0.306	−0.015	0.014
	$C_{L_{\dot{\eta}_i}}$ (sec)	7.26	$0.658/V_\infty$	$-7.896/V_\infty$	$-0.461/V_\infty$	$0.132/V_\infty$
Pitching moment	$C_{M_{\eta_i}}$	7.48	−0.032	−0.025	0.041	−0.018
	$C_{M_{\dot{\eta}_i}}$ (sec)	7.51	$-1.184/V_\infty$	$9.409/V_\infty$	$1.316/V_\infty$	$-0.395/V_\infty$

Table 7.3 Coefficients for the Modal Generalized Forces

Coefficient	Equation	Mode 1	Mode 2	Mode 3	Mode 4
$C_{Q_{i_0}}$	7.94	0	0	0	0
$C_{Q_{i_\alpha}}$	same	-1.49×10^{-2}	2.58×10^{-2}	1.49×10^{-2}	3.35×10^{-5}
$C_{Q_{i_q}}$ (sec)	same	$-0.726/V_\infty$	$0.089/V_\infty$	$0.304/V_\infty$	~0
$C_{Q_{i_{\delta_E}}}$	same	-1.28×10^{-2}	-6.42×10^{-2}	2.56×10^{-2}	1.50×10^{-4}
$C_{Q_{i_{\eta_1}}}$	7.95	5.85×10^{-5}	4.21×10^{-3}	2.91×10^{-4}	2.21×10^{-5}
$C_{Q_{i_{\eta_2}}}$	same	-9.0×10^{-5}	-9.22×10^{-2}	1.44×10^{-3}	-1.32×10^{-4}
$C_{Q_{i_{\eta_3}}}$	same	3.55×10^{-4}	1.97×10^{-3}	-3.46×10^{-4}	9.68×10^{-6}
$C_{Q_{i_{\eta_4}}}$	same	1.20×10^{-4}	3.37×10^{-3}	1.44×10^{-4}	1.77×10^{-3}
$C_{Q_{i_{\dot{\eta}_1}}}$ (sec)	same	$-0.0032/V_\infty$	$0.0665/V_\infty$	$-0.0048/V_\infty$	$-0.0004/V_\infty$
$C_{Q_{i_{\dot{\eta}_2}}}$ (sec)	same	$-0.0015/V_\infty$	$-2.277/V_\infty$	$0.1494/V_\infty$	$0.0031/V_\infty$
$C_{Q_{i_{\dot{\eta}_3}}}$ (sec)	same	$0.0050/V_\infty$	$0.0320/V_\infty$	$-0.0001/V_\infty$	$-0.0004/V_\infty$
$C_{Q_{i_{\dot{\eta}_4}}}$ (sec)	same	$-0.0011/V_\infty$	$0.0317/V_\infty$	$-0.0100/V_\infty$	$0.6112/V_\infty$

Now with regard to the surge-velocity modal effectiveness $C_{Q_{i_u}}$, Equations (7.94) indicate that

$$C_{Q_{i_u}} = \frac{2}{V_\infty} C_{Q_i} = \frac{2}{V_\infty}\left(C_{Q_{i_0}} + \frac{\partial C_{Q_i}}{\partial \mathbf{p}}\mathbf{p}\right)$$

where C_{Q_i} is the total generalized force coefficient. The partial derivatives with respect to the various parameters in $\mathbf{p}$ (= α, Q, δ_E, etc.) in the above expression correspond to the effectiveness coefficients given in Equations (7.94) and (7.95). However, the resulting value for $C_{Q_{i_u}}$ will be quite small, and therefore has been ignored.

We may now use the results in Tables 7.1–7.3 to complete the model. Specifically, let us assemble a model for the elastic effects on the aerodynamic force F_{A_Z}, or $F_{A_{Z_E}}$, the pitching moment M_E, and the four generalized forces Q_i, i = 1-4. Assuming α_0 is small, we have the following model for the changes in the aerodynamic force and pitching moment due to elastic effects.

$$
\begin{aligned}
F_{A_{Z_E}} &= q_\infty S_W\Bigg(-0.029\eta_1 + 0.306\eta_2 + 0.015\eta_3 - 0.014\eta_4 \\
&\quad - \frac{0.658}{V_\infty}\dot{\eta}_1 + \frac{7.896}{V_\infty}\dot{\eta}_2 + \frac{0.461}{V_\infty}\dot{\eta}_3 - \frac{0.132}{V_\infty}\dot{\eta}_4\Bigg) \\
M_{A_E} &= q_\infty S_W \bar{c}_W\Bigg(-0.032\eta_1 - 0.025\eta_2 + 0.041\eta_3 - 0.018\eta_4 \\
&\quad - \frac{1.184}{V_\infty}\dot{\eta}_1 + \frac{9.409}{V_\infty}\dot{\eta}_2 + \frac{1.316}{V_\infty}\dot{\eta}_3 - \frac{0.395}{V_\infty}\dot{\eta}_4\Bigg)
\end{aligned}
\tag{7.96}
$$

Plus, we have the model for the four generalized forces Q_i, $i = 1$-4 given by

$$
\begin{aligned}
Q_1 &= q_\infty S_W \bar{c}_W\left(\begin{aligned}&-0.0149\alpha - \frac{0.726}{V_\infty}Q - 0.0128\delta_E + 5.85\times10^{-5}\eta_1 - 9.0\times10^{-5}\eta_2 + 3.55\times10^{-4}\eta_3 \\ &+1.2\times10^{-4}\eta_4 - \frac{0.0032}{V_\infty}\dot{\eta}_1 - \frac{0.0015}{V_\infty}\dot{\eta}_2 + \frac{0.0050}{V_\infty}\dot{\eta}_3 - \frac{0.0011}{V_\infty}\dot{\eta}_4\end{aligned}\right) \\
Q_2 &= q_\infty S_W \bar{c}_W\left(\begin{aligned}&0.0258\alpha + \frac{0.089}{V_\infty}Q - 0.0642\delta_E + 4.21\times10^{-3}\eta_1 - 9.22\times10^{-2}\eta_2 + 1.97\times10^{-3}\eta_3 \\ &+ 3.37\times10^{-3}\eta_4 + \frac{0.0665}{V_\infty}\dot{\eta}_1 - \frac{2.277}{V_\infty}\dot{\eta}_2 + \frac{0.0320}{V_\infty}\dot{\eta}_3 + \frac{0.0317}{V_\infty}\dot{\eta}_4\end{aligned}\right) \\
Q_3 &= q_\infty S_W \bar{c}_W\left(\begin{aligned}&0.0149\alpha + \frac{0.304}{V_\infty}Q + 0.0256\delta_E + 2.91\times10^{-4}\eta_1 + 1.44\times10^{-3}\eta_2 - 3.46\times10^{-4}\eta_3 \\ &+1.44\times10^{-4}\eta_4 - \frac{0.0048}{V_\infty}\dot{\eta}_1 + \frac{0.1494}{V_\infty}\dot{\eta}_2 - \frac{0.0001}{V_\infty}\dot{\eta}_3 - \frac{0.0100}{V_\infty}\dot{\eta}_4\end{aligned}\right) \\
Q_4 &= q_\infty S_W \bar{c}_W\left(\begin{aligned}&3.35\times10^{-5}\alpha + 0.0Q + 1.5\times10^{-4}\delta_E + 2.21\times10^{-5}\eta_1 - 1.32\times10^{-4}\eta_2 \\ &+ 9.68\times10^{-6}\eta_3 + 1.77\times10^{-3}\eta_4 - \frac{0.0004}{V_\infty}\dot{\eta}_1 + \frac{0.0031}{V_\infty}\dot{\eta}_2 - \frac{0.0004}{V_\infty}\dot{\eta}_3 + \frac{0.6112}{V_\infty}\dot{\eta}_4\end{aligned}\right)
\end{aligned}
\tag{7.97}
$$

Note that the above generalized forces are functions of both rigid-body plus elastic modal coordinates.

7.10 INTEGRATING ELASTIC EFFECTS INTO THE EQUATIONS OF MOTION

Having determined the effects of modal displacement and velocity on the vehicle's lift, side force, moments, and modal generalized forces, we now discuss how to integrate these elastic effects into the equations of motion. The complete nonlinear equations of motion will be discussed in more detail in Chapter 8, when discussing simulation. So here we will just cover the highlights. We will, however, more completely address the integration of elastic effects into the linearized equations of motion.

7.10.1 Integrating into the Nonlinear Equations

The nonlinear equations of motion for an elastic vehicle, under the assumption of a flat, nonrotating earth, were given in Equations (4.65) (same as Equations (2.22)), (4.82) (same as Equations (2.27)), and (4.84). Consistent with Equations (4.65), the equations governing the rigid-body translational degrees of freedom are

$$\begin{aligned} m(\dot{U} - VR + WQ) &= -mg\sin\theta + (F_{A_{X_R}} + F_{A_{X_E}}) + F_{P_X} \\ m(\dot{V} + UR - WP) &= mg\cos\theta\sin\phi + (F_{A_{Y_R}} + F_{A_{Y_E}}) + F_{P_Y} \\ m(\dot{W} - UQ + VP) &= mg\cos\theta\cos\phi + (F_{A_{Z_R}} + F_{A_{Z_E}}) + F_{P_Z} \end{aligned} \tag{7.98}$$

The aerodynamic forces acting on the vehicle appear here, and these components have now been written in terms of the rigid-body plus elastic contributions.

From Equations (7.3) we have the aerodynamic force components in a vehicle-fixed frame given by

$$\begin{aligned} F_{A_X} &= C_X q_\infty S_W = -D\cos\alpha\cos\beta - \mathsf{S}\cos\alpha\sin\beta + L\sin\alpha \\ F_{A_Y} &= C_Y q_\infty S_W = -D\sin\beta + \mathsf{S}\cos\beta \\ F_{A_Z} &= C_Z q_\infty S_W = -D\sin\alpha\cos\beta - \mathsf{S}\sin\alpha\sin\beta - L\cos\alpha \end{aligned}$$

(As noted in Section 6.10.1, fuselage-referenced axes are typically used here.) Now in Equations (7.11) and (7.15) we have the expressions for the change in lift L on the vehicle due to the elastic effects. Likewise, Equation (7.32) is the expression for the elastic effect on the vehicle's side force S. These changes in lift and side force must be added to the forces as modeled in Chapter 6. Consequently, using these results for the changes in lift and side force, along with Equations (7.3), one may clearly identify the contributions to F_{A_X}, F_{A_Y}, and F_{A_Z} from elastic effects.

Furthermore, if we now write these elastic contributions to these forces as

$$\boxed{\begin{aligned} F_{A_{X_E}} &= \sum_{i=1}^{n}\left(\frac{\partial F_{A_X}}{\partial \eta_i}\eta_i + \frac{\partial F_{A_X}}{\partial \dot{\eta}_i}\dot{\eta}_i\right) \\ F_{A_{Y_E}} &= \sum_{i=1}^{n}\left(\frac{\partial F_{A_Y}}{\partial \eta_i}\eta_i + \frac{\partial F_{A_Y}}{\partial \dot{\eta}_i}\dot{\eta}_i\right) \\ F_{A_{Z_E}} &= \sum_{i=1}^{n}\left(\frac{\partial F_{A_Z}}{\partial \eta_i}\eta_i + \frac{\partial F_{A_Z}}{\partial \dot{\eta}_i}\dot{\eta}_i\right) \end{aligned}} \tag{7.99}$$

we may then use Equations (7.21) and (7.37) to find the partial derivatives with respect to modal displacement, and Equations (7.25) and (7.41) to find the partial derivatives with respect to modal velocities. Assuming the forces in Section 6.10.1 were expressed in a fuselage-referenced axes, the three expressions given in Equations (7.99) may then simply be added to the forces described in Section 6.10.1, and all will appear on the right-hand sides of the three equations that govern rigid-body translations (Equations (7.98)).

Regarding the moments on the vehicle, consistent with Equations (4.82) we have

$$
\begin{aligned}
&I_{xx}\dot{P} - (I_{yy} - I_{zz})QR - I_{xy}(\dot{Q} - PR) \\
&\qquad - I_{yz}(Q^2 - R^2) - I_{xz}(\dot{R} + PQ) = (L_{A_R} + L_{A_E}) + L_P \\
&I_{yy}\dot{Q} + (I_{xx} - I_{zz})PR - I_{xy}(\dot{P} + QR) \\
&\qquad - I_{yz}(\dot{R} - PQ) + I_{xz}(P^2 - R^2) = (M_{A_R} + M_{A_E}) + M_P \\
&I_{zz}\dot{R} + (I_{yy} - I_{xx})PQ + I_{xy}(Q^2 - P^2) \\
&\qquad - I_{yz}(\dot{Q} + PR) - I_{xz}(\dot{P} - QR) = (N_{A_R} + N_{A_E}) + N_P
\end{aligned}
\tag{7.100}
$$

in which the aerodynamic rolling L_A, pitching M_A, and yawing N_A moments are now expressed in terms of the rigid-body plus the elastic contributions.

Equations (7.45), (7.53), (7.56), and (7.69) were derived to determine the elastic effects on aerodynamic pitching moment M_A, rolling moment L_A, and yawing moment N_A, respectively. Then Equations (7.47), (7.60), and (7.71) were developed to express the effects due to modal displacement, and Equations (7.50), (7.66), and (7.74) to express the effects due to modal velocities on these same moments. These latter six equations may then be used to obtain the partial derivatives appearing below in the expressions for the contributions to the moments from elastic effects.

$$
\boxed{
\begin{aligned}
L_{A_E} &= \sum_{i=1}^{n}\left(\frac{\partial L_A}{\partial \eta_i}\eta_i + \frac{\partial L_A}{\partial \dot{\eta}_i}\dot{\eta}_i\right) \\
M_{A_E} &= \sum_{i=1}^{n}\left(\frac{\partial M_A}{\partial \eta_i}\eta_i + \frac{\partial M_A}{\partial \dot{\eta}_i}\dot{\eta}_i\right) \\
N_{A_E} &= \sum_{i=1}^{n}\left(\frac{\partial N_A}{\partial \eta_i}\eta_i + \frac{\partial N_A}{\partial \dot{\eta}_i}\dot{\eta}_i\right)
\end{aligned}
}
\tag{7.101}
$$

Again assuming the moments developed in Section 6.10.1 were expressed in fuselage-referenced axes, the three expressions above for the elastic effects on the moments may then simply be added to the rigid-body contributions to these moments discussed in Section 6.10.1, and all will appear on the right-hand sides of the three equations of motion governing the rigid-body rotations (Equations (7.100)).

To complete the model of the elastic effects, Equations (4.84) govern the generalized coordinates associated with the elastic degrees of freedom. Rewriting these equations we have

$$\ddot{\eta}_i + \omega_i^2 \eta_i = \frac{1}{\mathcal{M}_i}(Q_{i_R} + Q_{i_E}), \quad i = 1 \cdots n \tag{7.102}$$

where the generalized forces Q_i have now been written in terms of the rigid-body plus elastic contributions. These generalized forces acting on the elastic degrees of freedom themselves were also developed in this chapter. The model for these generalized forces consists of Equations (7.88), (7.94), and (7.95). But consistent with Equations (7.102), let us here write these forces in the following form.

$$Q_i = \left(Q_{i_{\mathbf{p}=0}} + \frac{\partial Q_i}{\partial \mathbf{p}_{\text{Rigid}}} \mathbf{p}_{\text{Rigid}} \right) + \sum_{j=1}^{n} \left(\frac{\partial Q_i}{\partial \eta_j} \eta_j + \frac{\partial Q_i}{\partial \dot{\eta}_j} \dot{\eta}_j \right) \tag{7.103}$$

where

$$\mathbf{p}_{\text{Rigid}}^T = \left[U \quad \beta\left(\approx \frac{V}{U}\right) \quad \alpha\left(\approx \frac{W}{U}\right) \quad P \quad Q \quad R \quad i_H \quad \delta_E \quad \delta_A \quad \delta_R \right]$$

and n is the number of elastic modes to be included in the model. Then Equations (7.94) correspond to the partial derivatives in the first parentheses on the right of Equation (7.103), and Equations (7.95) correspond to the partial derivatives in the second parentheses. And the n generalized forces Q_i given by Equations (7.103) would appear in the right-hand sides of the corresponding n equations of motion governing the elastic degrees of freedom (Equations (7.102)).

7.10.2 Integrating into the Linearized Equations

We now need to address incorporating the changes in the perturbation forces and moments into the linear small-perturbation equations of motion (Equations (2.45) and (2.49)). When dealing with a rigid vehicle, as in Section 6.10.2, it was convenient to use the stability axes as the selected vehicle-fixed axes. We may also use the stability axes when dealing with a flexible vehicle, assuming these axes satisfy the mean-axis constraints. Typically, however, one selects fuselage-referenced axes. In both cases care must be taken to develop the forces and moments, and to use force and moment components consistent with the vehicle-fixed axes selected. Here we will use fuselage-referenced axes, since these axes would typically be used in the development of the effectiveness coefficients in this chapter.

With regard to the forces, from Equations (7.3) we first note that the three components of the total aerodynamic force acting on the vehicle are given by

$$F_{A_X} = C_X q_\infty S_W = -D\cos\alpha\cos\beta - S\cos\alpha\sin\beta + L\sin\alpha$$
$$F_{A_Y} = C_Y q_\infty S_W = -D\sin\beta + S\cos\beta$$
$$F_{A_Z} = C_Z q_\infty S_W = -D\sin\alpha\cos\beta - S\sin\alpha\sin\beta - L\cos\alpha$$

where α and β are the total vehicle angles of attack and sideslip (reference plus perturbation). But consistent with Equations (2.45), the equations governing perturbations in the rigid-body translational degrees of freedom may be written as

$$\begin{aligned} m\left(\dot{u} + (Q_0 w + W_0 q) - (V_0 r + R_0 v)\right) &= -mg\cos\Theta_0\theta + \left(f_{A_{X_R}} + f_{A_{X_E}}\right) + f_{P_X} \\ m\left(\dot{v} + (R_0 u + U_0 r) - (P_0 w + W_0 p)\right) &= mg\left(\cos\Theta_0\cos\Phi_0\phi - \sin\Theta_0\sin\Phi_0\theta\right) \\ &\quad + \left(f_{A_{Y_R}} + f_{A_{Y_E}}\right) + f_{P_Y} \\ m\left(\dot{w} + (P_0 v + V_0 p) - (Q_0 u + U_0 q)\right) &= -mg\left(\cos\Theta_0\sin\Phi_0\phi + \sin\Theta_0\cos\Phi_0\theta\right) \\ &\quad + \left(f_{A_{Z_R}} + f_{A_{Z_E}}\right) + f_{P_Z} \end{aligned} \tag{7.104}$$

where the perturbation aerodynamic-force components have been expressed in terms of rigid-body plus elastic contributions. Let us write the perturbation force component in the Z direction, for example, as

$$f_{A_Z} \triangleq \frac{\partial F_{A_Z}}{\partial \delta\mathbf{p}}\Big|_0 \delta\mathbf{p} = f_{A_{Z_R}} + f_{A_{Z_E}} = \left(\frac{\partial F_{A_Z}}{\partial \delta\mathbf{p}_{\text{Rigid}}}\Big|_0 \delta\mathbf{p}_{\text{Rigid}}\right) + \sum_{i=1}^{n}\left(\frac{\partial F_{A_Z}}{\partial \eta_i}\Big|_0 \eta_i + \frac{\partial F_{A_Z}}{\partial \dot{\eta}_i}\Big|_0 \dot{\eta}_i\right) \tag{7.105}$$

where, consistent with Equation (6.9), the perturbation parameter vector for this force is

$$\delta\mathbf{p}_{\text{Rigid}}^T = \begin{bmatrix} u & \alpha & \dot{\alpha} & q & \delta_E \end{bmatrix} \tag{7.106}$$

Note that Equation (7.105), the expression for the perturbation force, includes terms associated with the rigid-body degrees of freedom plus those associated with the elastic degrees of freedom.

Then, continuing the example, writing Equation (7.105) in terms of effectiveness coefficients, we have

$$f_{A_Z} = q_\infty S_W\left(\left(C_{Z_u} u + C_{Z_\alpha}\alpha + C_{Z_{\dot{\alpha}}}\dot{\alpha} + C_{Z_q} q + C_{Z_{\delta_E}}\delta_E\right) + \left(\sum_{i=1}^{n}\left(C_{Z_{\eta_i}}\eta_i + C_{Z_{\dot{\eta}_i}}\dot{\eta}_i\right)\right)\right) \tag{7.107}$$

So the effect of elastic deformation on this force is seen to be

$$f_{A_{Z_E}} = q_\infty S_W \sum_{i=1}^{n}\left(C_{Z_{\eta_i}}\eta_i + C_{Z_{\dot{\eta}_i}}\dot{\eta}_i\right) \tag{7.108}$$

where the coefficients in this expression may be found using Equations (7.3) above, or

$$C_{Z_\bullet} = \left(-C_{D_\bullet}\sin\alpha_0\cos\beta_0 - C_{S_\bullet}\sin\alpha_0\sin\beta_0 - C_{L_\bullet}\cos\alpha_0\right),\ \bullet = \eta_i \text{ or } \dot{\eta}_i \tag{7.109}$$

Consequently, consistent with Equations (7.105) and (7.108), the elastic effects on the three perturbation-force components may be written as

$$\boxed{\begin{aligned} f_{A_{X_E}} &= q_\infty S_W \sum_{i=1}^{n} \left(C_{X_{\eta_i}} \eta_i + C_{X_{\dot{\eta}_i}} \dot{\eta}_i\right) \\ f_{A_{Y_E}} &= q_\infty S_W \sum_{i=1}^{n} \left(C_{Y_{\eta_i}} \eta_i + C_{Y_{\dot{\eta}_i}} \dot{\eta}_i\right) \\ f_{A_{Z_E}} &= q_\infty S_W \sum_{i=1}^{n} \left(C_{Z_{\eta_i}} \eta_i + C_{Z_{\dot{\eta}_i}} \dot{\eta}_i\right) \end{aligned}} \tag{7.110}$$

where the six coefficients in the above expressions are given in Table 7.4. In obtaining the results in Table 7.4, we have ignored any elastic effects on the vehicle drag, and the four coefficients appearing as entries in the table are the same as those discussed in Sections 7.3 and 7.4, with one exception. That exception involves the substitution of the velocity at the reference condition V_{∞_0} for V_∞ in the modal-rate effectiveness coefficients.

One might think that the three expressions in Equations (7.110) may simply be added to the forces (discussed in Section 6.10.2) appearing on the right-hand sides of the equations governing rigid-body translations (Equations (7.104)). This would be true if the coefficients in Section 6.10.2 were developed in the same fuselage-referenced axes being used here. But in Section 6.10.2 stability axes were being used. So adjustments must be made to the expressions developed in Section 6.10.2 to account for the different vehicle-fixed axes. Let us take a moment here to discuss the necessary adjustments.

The three unit vectors defining stability and fuselage-referenced axes are related through a direction-cosine matrix, or

$$\begin{Bmatrix} \mathbf{i}_S \\ \mathbf{j}_S \\ \mathbf{k}_S \end{Bmatrix} = \begin{bmatrix} \cos\alpha_0 & 0 & \sin\alpha_0 \\ 0 & 1 & 0 \\ -\sin\alpha_0 & 0 & \cos\alpha_0 \end{bmatrix} \begin{Bmatrix} \mathbf{i}_F \\ \mathbf{j}_F \\ \mathbf{k}_F \end{Bmatrix} \tag{7.111}$$

Therefore, the components of the aerodynamic plus propulsive force vectors in the two axes are related by

$$\begin{aligned} \mathbf{F}_A + \mathbf{F}_P &= \begin{bmatrix} F_{X_S} & F_{Y_S} & F_{Z_S} \end{bmatrix} \begin{Bmatrix} \mathbf{i}_S \\ \mathbf{j}_S \\ \mathbf{k}_S \end{Bmatrix} = \begin{bmatrix} F_{X_S} & F_{Y_S} & F_{Z_S} \end{bmatrix} \begin{bmatrix} \cos\alpha_0 & 0 & \sin\alpha_0 \\ 0 & 1 & 0 \\ -\sin\alpha_0 & 0 & \cos\alpha_0 \end{bmatrix} \begin{Bmatrix} \mathbf{i}_F \\ \mathbf{j}_F \\ \mathbf{k}_F \end{Bmatrix} \\ &= \begin{bmatrix} \left(F_{X_S}\cos\alpha_0 - F_{Z_S}\sin\alpha_0\right) & F_{Y_S} & \left(F_{X_S}\sin\alpha_0 + F_{Z_S}\cos\alpha_0\right) \end{bmatrix} \begin{Bmatrix} \mathbf{i}_F \\ \mathbf{j}_F \\ \mathbf{k}_F \end{Bmatrix} \end{aligned} \tag{7.112}$$

Table 7.4 Force Coefficients

	$C_{X.}$	$C_{Y.}$	$C_{Z.}$
η_i	$-C_{S_{\eta_i}}\cos\alpha_0\sin\beta_0 + C_{L_{\eta_i}}\sin\alpha_0$	$C_{S_{\eta_i}}\cos\beta_0$	$-(C_{S_{\eta_i}}\sin\alpha_0\sin\beta_0 + C_{L_{\eta_i}}\cos\alpha_0)$
$\dot{\eta}_i$	$-C_{S_{\dot{\eta}_i}}\cos\alpha_0\sin\beta_0 + C_{L_{\dot{\eta}_i}}\sin\alpha_0$	$C_{S_{\dot{\eta}_i}}\cos\beta_0$	$-(C_{S_{\dot{\eta}_i}}\sin\alpha_0\sin\beta_0 + C_{L_{\dot{\eta}_i}}\cos\alpha_0)$

And, for example, since the components of the aerodynamic force vector in the stability axes were

$$
\begin{aligned}
F_{A_{X_S}} &= -D\cos\beta_0 - \mathsf{S}\sin\beta_0 \\
F_{A_{Y_S}} &= -D\sin\beta_0 + \mathsf{S}\cos\beta_0 \\
F_{A_{Z_S}} &= -L
\end{aligned}
$$

then the components of that vector in the fuselage-referenced axes are

$$
\begin{aligned}
F_{A_{X_F}} &= -D\cos\alpha_0\cos\beta_0 - \mathsf{S}\cos\alpha_0\sin\beta_0 + L\sin\alpha_0 \\
F_{A_{Y_F}} &= -D\sin\beta_0 + \mathsf{S}\cos\beta_0 \\
F_{A_{Z_F}} &= -D\sin\alpha_0\cos\beta_0 - \mathsf{S}\sin\alpha_0\sin\beta_0 - L\cos\alpha_0
\end{aligned}
\tag{7.113}
$$

Clearly, if α_0 is small, there is little difference between the magnitudes of the force components in the stability axes and those of the force components in the fuselage-referenced axes. But they are not identical.

Therefore, under the assumption that perturbation forces expressed in stability axes were used to develop the rigid-body model for these forces (e.g., $f_{A_{Z_R}}$ in Equation (7.105)), the components of the rigid-body model in the fuselage-referenced axes become

$$
\boxed{
\begin{aligned}
\left(f_{A_{X_{R_F}}} + f_{P_{X_{R_F}}}\right) &= \left(f_{A_{X_R}} + f_{P_{X_R}}\right)\cos\alpha_0 - \left(f_{A_{Z_R}} + f_{P_{Z_R}}\right)\sin\alpha_0 \\
\left(f_{A_{Y_{R_F}}} + f_{P_{Y_{R_F}}}\right) &= \left(f_{A_{Y_R}} + f_{P_{Y_R}}\right) \\
\left(f_{A_{Z_{R_F}}} + f_{P_{Z_{R_F}}}\right) &= \left(f_{A_{X_R}} + f_{P_{X_R}}\right)\sin\alpha_0 + \left(f_{A_{Z_R}} + f_{P_{Z_R}}\right)\cos\alpha_0
\end{aligned}
}
\tag{7.114}
$$

So the three components of the perturbation force vector appearing on the right-hand sides of Equations (7.104) that govern rigid-body translation would be the three expressions in Equations (7.114) plus the three expressions in Equations (7.110).

The perturbations in the rigid-body rotational degrees of freedom are governed by Equations (2.49), which, assuming the vehicle is symmetric with respect to it's XZ plane, may now be written as

$$I_{xx}\dot{p} - I_{xz}\left(\dot{r} + (\mathrm{Q}_0 p + \mathrm{P}_0 q)\right) + \left(I_{zz} - I_{yy}\right)\left(\mathrm{R}_0 q + \mathrm{Q}_0 r\right) = \left(l_{A_R} + l_{A_E}\right) + l_P$$
$$I_{yy}\dot{q} + \left(I_{xx} - I_{zz}\right)\left(\mathrm{R}_0 p + \mathrm{P}_0 r\right) + 2I_{xz}\left(\mathrm{P}_0 p - \mathrm{R}_0 r\right) = \left(m_{A_R} + m_{A_E}\right) + m_p \tag{7.115}$$
$$I_{zz}\dot{r} - I_{xz}\left(\dot{p} - (\mathrm{R}_0 q + \mathrm{Q}_0 r)\right) + \left(I_{yy} - I_{xx}\right)\left(\mathrm{Q}_0 p + \mathrm{P}_0 q\right) = \left(n_{A_R} + n_{A_E}\right) + n_P$$

Again the components of the aerodynamic moment, or l_A (rolling moment), m_A (pitching moment), and n_A (yawing moment), have been expressed in terms of rigid-body plus elastic contributions.

Analogous to Equation (7.105), which expressed a perturbation force in terms of the rigid-body and elastic contributions, let us write the perturbation aerodynamic pitching moment, for example, in the following form.

$$m_A \triangleq \frac{\partial M_A}{\partial \delta \mathbf{p}}\Big|_0 \delta \mathbf{p} = m_{A_R} + m_{A_E}$$
$$= \left(\frac{\partial M_A}{\partial \delta \mathbf{p}_{\mathrm{Rigid}}}\Big|_0 \delta \mathbf{p}_{\mathrm{Rigid}}\right) + \sum_{i=1}^{n}\left(\frac{\partial M_A}{\partial \eta_i}\Big|_0 \eta_i + \frac{\partial M_A}{\partial \dot{\eta}_i}\Big|_0 \dot{\eta}_i\right) \tag{7.116}$$

We again see that the perturbation moments will consist of rigid-body terms (here, m_{A_R}) and elastic terms (here, m_{A_E}). And so the elastic contributions to the three perturbation moments may be written as

$$l_{A_E} = \sum_{i=1}^{n}\left(\frac{\partial L_A}{\partial \eta_i}\Big|_0 \eta_i + \frac{\partial L_A}{\partial \dot{\eta}_i}\Big|_0 \dot{\eta}_i\right) = q_\infty S_W b_W \sum_{i=1}^{n}\left(C_{L\mathrm{roll}_{\eta_i}}\eta_i + C_{L\mathrm{roll}_{\dot{\eta}_i}}\dot{\eta}_i\right)$$
$$m_{A_E} = \sum_{i=1}^{n}\left(\frac{\partial M_A}{\partial \eta_i}\Big|_0 \eta_i + \frac{\partial M_A}{\partial \dot{\eta}_i}\Big|_0 \dot{\eta}_i\right) = q_\infty S_W \bar{c}_W \sum_{i=1}^{n}\left(C_{M_{\eta_i}}\eta_i + C_{M_{\dot{\eta}_i}}\dot{\eta}_i\right) \tag{7.117}$$
$$n_{A_E} = \sum_{i=1}^{n}\left(\frac{\partial N_A}{\partial \eta_i}\Big|_0 \eta_i + \frac{\partial N_A}{\partial \dot{\eta}_i}\Big|_0 \dot{\eta}_i\right) = q_\infty S_W b_W \sum_{i=1}^{n}\left(C_{N_{\eta_i}}\eta_i + C_{N_{\dot{\eta}_i}}\dot{\eta}_i\right)$$

All six aeroelastic coefficients appearing in the above expressions are the same as those discussed in Sections 7.5, 7.6, and 7.7, again with one exception. In the equations for the modal-rate coefficients, the reference velocity V_{∞_0} must be substituted for V_∞.

Since the perturbation moments for the rigid-body degrees of freedom discussed in Section 6.10.2, or l_{A_R}, m_{A_R}, and n_{A_R}, were developed in stability axes, then as with the perturbation forces, these rigid-body contributions must be converted to moments corresponding to the fuselage-referenced axes. And so the rigid-body contributions to the three moments in the fuselage-referenced axes become

$$\boxed{\begin{aligned} l_{A_{R_F}} &= l_{A_R}\cos\alpha_0 - n_{A_R}\sin\alpha_0 \\ m_{A_{R_F}} &= m_{A_R} \\ n_{A_{R_F}} &= l_{A_R}\sin\alpha_0 + n_{A_R}\cos\alpha_0 \end{aligned}} \tag{7.118}$$

Now the right-hand sides of Equations (7.115), which govern the rigid-body rotational degrees of freedom, will consist of the three expressions in Equations (7.118), plus the terms on the right-hand sides of Equations (7.117).

To complete the incorporation of the forces and moments into the linearized equations of motion, we must now consider the perturbation generalized forces acting on the perturbation modal coordinates η_i. Equations (4.84) gave the equations of motion governing the (total) modal coordinates η_i, or

$$\ddot{\eta}_i + \omega_i^2\eta_i = \frac{Q_i}{\mathcal{M}_i}$$

Since this equation is linear, the equations of motion governing the <u>perturbation</u> modal coordinates are simply

$$\ddot{\eta}_i + \omega_i^2\eta_i = \frac{1}{\mathcal{M}_i}(q_{i_R} + q_{i_E}) \tag{7.119}$$

where η_i is the perturbation modal coordinate and the perturbation generalized force q_i has been expressed in terms of the rigid-body plus elastic contributions.

Again consider Equation (7.103), repeated here, which is the expression for the total generalized force Q_i, or

$$Q_i = \left(Q_{i_{\mathbf{p}=0}} + \frac{\partial Q_i}{\partial \mathbf{p}_{\text{Rigid}}}\mathbf{p}_{\text{Rigid}}\right) + \sum_{i=1}^{n}\left(\frac{\partial Q_i}{\partial \eta_j}\eta_j + \frac{\partial Q_i}{\partial \dot{\eta}_j}\dot{\eta}_j\right)$$

with

$$\mathbf{p}_{\text{Rigid}}^T = \left[U \quad \beta\left(\approx\frac{V}{U}\right) \quad \alpha\left(\approx\frac{W}{U}\right) \quad P \quad Q \quad R \quad i_H \quad \delta_E \quad \delta_A \quad \delta_R\right]$$

Since the above expression gives the <u>total</u> generalized force, the <u>perturbation</u> generalized force must then be

$$\boxed{q_i \triangleq \frac{\partial Q_i}{\partial \delta\mathbf{p}}\Big|_0 \delta\mathbf{p} = q_{i_R} + q_{i_E}} \tag{7.120}$$

with

$$\delta\mathbf{p}^T = \left[\delta\mathbf{p}_{\text{Rigid}}^T \quad \delta\boldsymbol{\eta}^T \quad \delta\dot{\boldsymbol{\eta}}^T\right]$$

and, consistent with the development in Section 6.10.2,

$$\delta \mathbf{p}_{\text{Rigid}}^T = \left[u \;\; \beta\left(\approx \frac{v}{U_0}\right) \;\; \alpha\left(\approx \frac{w}{U_0}\right) \;\; \dot{\alpha} \;\; p \;\; q \;\; r \;\; \delta_E \;\; \delta_A \;\; \delta_R \right] \tag{7.121}$$

Therefore, the rigid-body contributions to the perturbation generalized force are given by

$$\boxed{q_{i_R} = \frac{\partial Q_i}{\partial \delta \mathbf{p}_{\text{Rigid}}}\Big|_0 \delta \mathbf{p}_{\text{Rigid}}} \tag{7.122}$$

and the elastic contributions are given by

$$\boxed{q_{i_E} = q_\infty S_W \bar{c}_W \sum_{j=1}^{n} \left(C_{Q_{i_{\eta_j}}} \eta_j + C_{Q_{i_{\dot{\eta}_j}}} \dot{\eta}_j \right)} \tag{7.123}$$

Note that the generalized-force coefficients that will appear when Equation (7.122) is expanded will be the same as those given in Equations (7.94), while, with one exception, the generalized-force coefficients in Equations (7.123) are the same as those given in Equations (7.95). The exception involves substituting V_{∞_0} for V_∞. Finally, the perturbation generalized forces on the right-hand sides of Equations (7.119) will consist of the expressions given by Equations (7.122) plus those given by Equations (7.123).

7.11 STATIC-ELASTIC EFFECTS ON A VEHICLE'S AERODYNAMICS

There are two classes of mathematical models of the dynamics of elastic vehicles—dynamic and static. In Section 7.10, we discussed the assembly of *dynamic-elastic models,* in which the dynamics of the elastic degrees of freedom, as well as the dynamics of the rigid-body degrees of freedom, are all included in the model. The second class of model, or *static-elastic models,* includes only the dynamics of the rigid-body degrees of freedom, but with the effects of static-elastic deformation on the aerodynamic forces and moments.

These static-elastic effects may be accounted for in wind-tunnel tests, for example, using a wind-tunnel model of the vehicle based on its in-flight shape, as opposed to its manufactured shape, or "jig shape." Of course, this requires prediction of the in-flight shape including the elastic deformations, which is one key aspect of the material presented in this section. The second aspect is the actual prediction of the static-elastic effects on the aerodynamics of the rigid body.

7.11.1 Static-Elastic Deformations

To explore the static-elastic deformations of a vehicle's structure, recall that in this chapter we previously developed expressions capturing the effects of elastic deformation on the forces and moments acting on the rigid-body degrees of freedom. We expressed these forces and moments in terms of the rigid-body plus elastic contributions. That is, using Equations (7.99) and (7.101), we have

$$
\begin{aligned}
F_{A_X} &= F_{A_{X_R}} + F_{A_{X_E}} = F_{A_{X_R}} + \sum_{i=1}^{n}\left(\frac{\partial F_{A_X}}{\partial \eta_i}\eta_i + \frac{\partial F_{A_X}}{\partial \dot{\eta}_i}\dot{\eta}_i\right)\\
F_{A_Y} &= F_{A_{Y_R}} + F_{A_{Y_E}} = F_{A_{Y_R}} + \sum_{i=1}^{n}\left(\frac{\partial F_{A_Y}}{\partial \eta_i}\eta_i + \frac{\partial F_{A_Y}}{\partial \dot{\eta}_i}\dot{\eta}_i\right)\\
F_{A_Z} &= F_{A_{Z_R}} + F_{A_{Z_E}} = F_{A_{Z_R}} + \sum_{i=1}^{n}\left(\frac{\partial F_{A_Z}}{\partial \eta_i}\eta_i + \frac{\partial F_{A_Z}}{\partial \dot{\eta}_i}\dot{\eta}_i\right)\\
L_A &= L_{A_R} + L_{A_E} = L_{A_R} + \sum_{i=1}^{n}\left(\frac{\partial L_A}{\partial \eta_i}\eta_i + \frac{\partial L_A}{\partial \dot{\eta}_i}\dot{\eta}_i\right)\\
M_A &= M_{A_R} + M_{A_E} = M_{A_R} + \sum_{i=1}^{n}\left(\frac{\partial M_A}{\partial \eta_i}\eta_i + \frac{\partial M_A}{\partial \dot{\eta}_i}\dot{\eta}_i\right)\\
N_A &= N_{A_R} + N_{A_E} = N_{A_R} + \sum_{i=1}^{n}\left(\frac{\partial N_A}{\partial \eta_i}\eta_i + \frac{\partial N_A}{\partial \dot{\eta}_i}\dot{\eta}_i\right)
\end{aligned}
\tag{7.124}
$$

And, of course, the rigid-body contributions to these forces and moments are all functions of the rigid-body motion and control-surface deflections, such as U, β, α, P, Q, R, $\dot{\alpha}$, i_H, δ_E, δ_A, and δ_R, as developed in Chapter 6.

Similarly, we have also developed expressions for the generalized forces acting on the elastic degrees of freedom. These generalized forces are also functions of both the rigid-body motion, control deflections, and the modal coordinates associated with the elastic degrees of freedom included in the model η_i, $i = 1 \ldots$ n. That is, from Equations (7.103) these generalized forces were expressed as

$$
Q_i = \left(Q_{i_{\mathbf{p}=\mathbf{0}}} + \frac{\partial Q_i}{\partial \mathbf{p}_{\text{Rigid}}}\mathbf{p}_{\text{Rigid}}\right) + \sum_{j=1}^{n}\left(\frac{\partial Q_i}{\partial \eta_j}\eta_j + \frac{\partial Q_i}{\partial \dot{\eta}_j}\dot{\eta}_j\right) \tag{7.125}
$$

where the vector of rigid-body variables $\mathbf{p}_{\text{Rigid}}$ includes U, β, α, P, Q, R, $\dot{\alpha}$, i_H, δ_E, δ_A, and δ_R.

For the purposes of illustration, let's assume that all the forces and moments appearing in Equations (7.124) and (7.125) are linear in the rigid-body variables, elastic degrees of freedom, and control deflections. Hence the nonlinear equations of motion governing these rigid-body and elastic degrees of freedom, or Equations (7.98), (7.100), and (7.102), may be written in the following general state-variable format. (This will be demonstrated in the discussion to follow.)

$$
\begin{aligned}
\mathbf{M}\dot{\mathbf{x}}_R &= \mathbf{f}_R(\mathbf{x}_R, T) + \mathbf{A}_R\mathbf{x}_R + \begin{bmatrix}\mathbf{A}_{R\eta} & \mathbf{A}_{R\dot{\eta}}\end{bmatrix}\mathbf{x}_E + \mathbf{B}_R\mathbf{u}\\
\dot{\mathbf{x}}_E &= \begin{bmatrix}\mathbf{0}\\ \mathbf{A}_{ER}\end{bmatrix}\mathbf{x}_R + \begin{bmatrix}\mathbf{0} & \mathbf{I}\\ \mathbf{A}_\eta & \mathbf{A}_{\dot{\eta}}\end{bmatrix}\mathbf{x}_E + \begin{bmatrix}\mathbf{0}\\ \mathbf{B}_E\end{bmatrix}\mathbf{u}
\end{aligned}
\tag{7.126}
$$

where

$$\mathbf{x}_R^T = \begin{bmatrix} U & \alpha & Q & \beta & P & R \end{bmatrix}$$

$$\mathbf{x}_E^T = \begin{bmatrix} \eta_1 & \cdots & \eta_n & \dot{\eta}_1 & \cdots & \dot{\eta}_n \end{bmatrix}$$

$$\mathbf{u}^T = \begin{bmatrix} i_H & \delta_E & \delta_A & \delta_R \end{bmatrix}$$

and **M** is the matrix containing the mass and inertias, plus any terms arising from the $\dot{\alpha}$ terms in the aerodynamic forces and moments that may be moved to the left-hand side of the matrix equation. The matrices **A.** and **B.** are seen to contain dimensional stability derivatives such as M_α and Q_{i_α} and so forth.

The functions $\mathbf{f}_R(\mathbf{x}_R,T)$ in Equations (7.126) contain all the nonlinear terms in the equations of motion, but do not include the (linear) models for the aerodynamic forces and moments. For example, if we consider just the first of Equations (7.98) we have

$$f_R(\mathbf{x}_R,T) = m(VR - WQ) - mg\sin\theta + T\cos\phi_T$$

Since we are interested in static-elastic effects, we set all $\dot{\eta}_i$ and $\ddot{\eta}_i = 0$ in the second of Equations (7.126), and solve for the vector containing the static modal displacements, or $\boldsymbol{\eta}_0$, in terms of the rigid-body variables $\mathbf{x}_R$ and the control-surface deflections **u** yielding

$$\boxed{\boldsymbol{\eta}_0 = -\mathbf{A}_\eta^{-1}(\mathbf{A}_{ER}\mathbf{x}_R + \mathbf{B}_E\mathbf{u})} \tag{7.127}$$

This result is the *static-elastic constraint.* Substituting this result back into the first of the equations in Equations (7.126), we find that the equations governing only the rigid-body degrees of freedom of the elastic vehicle now become

$$\boxed{\begin{aligned}\mathbf{M}\dot{\mathbf{x}}_R &= \mathbf{f}_R(\mathbf{x}_R,T) + \mathbf{A}_R\mathbf{x}_R - \mathbf{A}_{R\eta}\mathbf{A}_\eta^{-1}(\mathbf{A}_{ER}\mathbf{x}_R + \mathbf{B}_E\mathbf{u}) + \mathbf{B}_R\mathbf{u}\\ &= \mathbf{f}_R(\mathbf{x}_R,T) + (\mathbf{A}_R - \mathbf{A}_{R\eta}\mathbf{A}_\eta^{-1}\mathbf{A}_{ER})\mathbf{x}_R + (\mathbf{B}_R - \mathbf{A}_{R\eta}\mathbf{A}_\eta^{-1}\mathbf{B}_E)\mathbf{u}\end{aligned}} \tag{7.128}$$

The above process is known as *residualization* of the elastic degrees of freedom, and the model for the aerodynamic forces and moments including static-elastic effects is

$$\boxed{(\mathbf{A}_R - \mathbf{A}_{R\eta}\mathbf{A}_\eta^{-1}\mathbf{A}_{ER})\mathbf{x}_R + (\mathbf{B}_R - \mathbf{A}_{R\eta}\mathbf{A}_\eta^{-1}\mathbf{B}_E)\mathbf{u}} \tag{7.129}$$

One can see that both matrices of coefficients $\mathbf{A}_R$ and $\mathbf{B}_R$ will be modified as a result of the static-elastic deformations.

Equations (7.128) then constitute the static-elastic model for the vehicle's rigid-body dynamics, and the solution, or the rigid-body state history $\mathbf{x}_R(t)$, given the thrust and control-surface input $\mathbf{u}(t)$, may be substituted into Equation (7.127) to determine the static-elastic deformations of the structure. Note that in general these static-elastic deformations are not constant, but vary as the flight condition affects the load on the structure. Finally, if the aerodynamic forces and moments are not linear in the rigid-body and elastic variables and control deflections, as originally assumed here, then numerical techniques must be employed to find the static-elastic deflections.

7.11.2 Effects on the Aerodynamics

We will continue to assume that the aerodynamic forces and moments are linear in the rigid-body and elastic degrees of freedom and control-surface deflections, such that the nonlinear equations of motion may be expressed in the forms given in Equations (7.126). And recall that the nonlinear functions $\mathbf{f}_R(\mathbf{x}_R,T)$ contain the right-hand sides of the equations of motions excluding the linear models for the aerodynamic forces and moments.

For simplicity of presentation we will continue this discussion using only the longitudinal equations of motion and ignore $\dot{\alpha}$ effects. The approach taken here may be applied directly to the lateral-directional equations with little modification. Consistent with Equations (7.124), let us express the aerodynamic axial and vertical forces and pitching moment as

$$\begin{aligned}
F_{A_X} &= \Big(C_{X_{\alpha=\delta=i_H=\eta=0}} + C_{X_u}U + C_{X_\alpha}\alpha + C_{X_q}Q + C_{X_{\delta_E}}\delta_E + C_{X_{i_H}}i_H \\
&\quad + \sum_i \left(C_{X_{\eta_i}}\eta_i + C_{X_{\dot{\eta}_i}}\dot{\eta}_i\right)\Big)q_\infty S_W \\
F_{A_Z} &= \Big(C_{Z_{\alpha=\delta=i_H=\eta=0}} + C_{Z_u}U + C_{Z_\alpha}\alpha + C_{Z_q}Q + C_{Z_{\delta_E}}\delta_E + C_{Z_{i_H}}i_H \\
&\quad + \sum_i \left(C_{Z_{\eta_i}}\eta_i + C_{Z_{\dot{\eta}_i}}\dot{\eta}_i\right)\Big)q_\infty S_W \\
M_A &= \Big(C_{M_0} + C_{M_u}U + C_{M_\alpha}\alpha + C_{M_q}Q + C_{M_{\delta_E}}\delta_E + C_{M_{i_H}}i_H \\
&\quad + \sum_i \left(C_{M_{\eta_i}}\eta_i + C_{M_{\dot{\eta}_i}}\dot{\eta}_i\right)\Big)q_\infty S_W \bar{c}_W
\end{aligned} \tag{7.130}$$

With $\mathbf{x}_R = \left[U\ \alpha\ Q\right]^T$ and $\mathbf{u} = \left[i_H\ \delta_E\right]^T$, we may also define the following vectors (arrays) to be

$$\begin{aligned}
\mathbf{C}_{X_R} &= \left[C_{X_u}\quad C_{X_\alpha}\quad C_{X_q}\right] & \mathbf{C}_{X_{\boldsymbol{\eta}}} &= \left[C_{X_{\eta_1}}\quad \cdots \quad C_{X_{\eta_n}}\right] \\
\mathbf{C}_{Z_R} &= \left[C_{Z_u}\quad C_{Z_\alpha}\quad C_{Z_q}\right] & \mathbf{C}_{Z_{\boldsymbol{\eta}}} &= \left[C_{Z_{\eta_1}}\quad \cdots \quad C_{Z_{\eta_n}}\right] \\
\mathbf{C}_{M_R} &= \left[C_{M_u}\quad C_{M_\alpha}\quad C_{M_q}\right] & \mathbf{C}_{M_{\boldsymbol{\eta}}} &= \left[C_{M_{\eta_1}}\quad \cdots \quad C_{M_{\eta_n}}\right]
\end{aligned} \tag{7.131}$$

and

$$
\begin{aligned}
&\mathbf{C}_{X_{\dot{\boldsymbol{\eta}}}} = \begin{bmatrix} C_{X_{\dot{\eta}_1}} & \cdots & C_{X_{\dot{\eta}_n}} \end{bmatrix} \qquad && \mathbf{C}_{X_{\mathbf{u}}} = \begin{bmatrix} C_{X_{i_H}} & C_{X_{\delta_E}} \end{bmatrix} \\
&\mathbf{C}_{Z_{\dot{\boldsymbol{\eta}}}} = \begin{bmatrix} C_{Z_{\dot{\eta}_1}} & \cdots & C_{Z_{\dot{\eta}_n}} \end{bmatrix} && \mathbf{C}_{Z_{\mathbf{u}}} = \begin{bmatrix} C_{Z_{i_H}} & C_{Z_{\delta_E}} \end{bmatrix} \\
&\mathbf{C}_{M_{\dot{\boldsymbol{\eta}}}} = \begin{bmatrix} C_{M_{\dot{\eta}_1}} & \cdots & C_{M_{\dot{\eta}_n}} \end{bmatrix} && \mathbf{C}_{M_{\mathbf{u}}} = \begin{bmatrix} C_{M_{i_H}} & C_{M_{\delta_E}} \end{bmatrix}
\end{aligned}
\tag{7.132}
$$

These are all matrices with elements consisting of aerodynamic effectiveness coefficients, and it is these coefficients for which we seek the static-elastic corrections.

Equations (7.130) may now be written as

$$
\begin{aligned}
F_{A_X} &= F_{A_{X_0}} + \left(\mathbf{C}_{X_R}\mathbf{x}_R + \begin{bmatrix}\mathbf{C}_{X_{\boldsymbol{\eta}}} & \mathbf{C}_{X_{\dot{\boldsymbol{\eta}}}}\end{bmatrix}\mathbf{x}_E + \mathbf{C}_{X_{\mathbf{u}}}\mathbf{u}\right)q_\infty S_W \\
F_{A_Z} &= F_{A_{Z_0}} + \left(\mathbf{C}_{Z_R}\mathbf{x}_R + \begin{bmatrix}\mathbf{C}_{Z_{\boldsymbol{\eta}}} & \mathbf{C}_{Z_{\dot{\boldsymbol{\eta}}}}\end{bmatrix}\mathbf{x}_E + \mathbf{C}_{Z_{\mathbf{u}}}\mathbf{u}\right)q_\infty S_W \\
M_A &= M_{A_0} + \left(\mathbf{C}_{M_R}\mathbf{x}_R + \begin{bmatrix}\mathbf{C}_{M_{\boldsymbol{\eta}}} & \mathbf{C}_{M_{\dot{\boldsymbol{\eta}}}}\end{bmatrix}\mathbf{x}_E + \mathbf{C}_{M_{\mathbf{u}}}\mathbf{u}\right)q_\infty S_W \bar{c}_W
\end{aligned}
\tag{7.133}
$$

with obvious definitions for $F_{A_{X_0}}$, $F_{A_{Z_0}}$ and M_{A_0}. And so, from Equations (7.98) and (7.100), the longitudinal equations governing the rigid-body degrees of freedom may be written as

$$
\begin{aligned}
\begin{bmatrix} m & 0 & 0 \\ 0 & mU & 0 \\ 0 & 0 & I_{yy} \end{bmatrix}\dot{\mathbf{x}}_R &= \begin{Bmatrix} -mWQ - mg\sin\theta + F_{A_{X_0}} + T\cos\phi_T \\ mUQ + mg\cos\theta + F_{A_{Z_0}} - T\sin\phi_T \\ M_{A_0} + T(d_T\cos\phi_T - x_T\sin\phi_T) \end{Bmatrix} + \begin{bmatrix} q_\infty S_W \mathbf{C}_{X_R} \\ q_\infty S_W \mathbf{C}_{Z_R} \\ q_\infty S_W \bar{c}_W \mathbf{C}_{M_R} \end{bmatrix}\mathbf{x}_R \\
&+ \begin{bmatrix} q_\infty S_W \mathbf{C}_{X_{\boldsymbol{\eta}}} & q_\infty S_W \mathbf{C}_{X_{\dot{\boldsymbol{\eta}}}} \\ q_\infty S_W \mathbf{C}_{Z_{\boldsymbol{\eta}}} & q_\infty S_W \mathbf{C}_{Z_{\dot{\boldsymbol{\eta}}}} \\ q_\infty S_W \bar{c}_W C_{M_{\boldsymbol{\eta}}} & q_\infty S_W \bar{c}_W \mathbf{C}_{M\dot{\boldsymbol{\eta}}} \end{bmatrix}\mathbf{x}_E + \begin{bmatrix} q_\infty S_W \mathbf{C}_{X_{\mathbf{u}}} \\ q_\infty S_W \mathbf{C}_{Z_{\mathbf{u}}} \\ q_\infty S_W \bar{c}_W C_{M_{\mathbf{u}}} \end{bmatrix}\mathbf{u}
\end{aligned}
\tag{7.134}
$$

Comparing the above expression with Equations (7.126) we can see that

$$
\begin{aligned}
&\mathbf{f}_R(\mathbf{x}_R, T) = \begin{Bmatrix} -mWQ - mg\sin\theta + F_{A_{X_0}} + T\cos\phi_T \\ mUQ + mg\cos\theta + F_{A_{Z_0}} - T\sin\phi_T \\ M_{A_0} + T(d_T\cos\phi_T - x_T\sin\phi_T) \end{Bmatrix}, \quad \mathbf{A}_R = \begin{bmatrix} q_\infty S_W \mathbf{C}_{X_R} \\ q_\infty S_W \mathbf{C}_{Z_R} \\ q_\infty S_W \bar{c}_W \mathbf{C}_{M_R} \end{bmatrix} \\
&\mathbf{A}_{R_\eta} = \begin{bmatrix} q_\infty S_W \mathbf{C}_{X_{\boldsymbol{\eta}}} \\ q_\infty S_W \mathbf{C}_{Z_{\boldsymbol{\eta}}} \\ q_\infty S_W \bar{c}_W \mathbf{C}_{M_{\boldsymbol{\eta}}} \end{bmatrix}, \quad \mathbf{A}_{R_{\dot{\eta}}} = \begin{bmatrix} q_\infty S_W \mathbf{C}_{X_{\dot{\boldsymbol{\eta}}}} \\ q_\infty S_W \mathbf{C}_{Z_{\dot{\boldsymbol{\eta}}}} \\ q_\infty S_W \bar{c}_W \mathbf{C}_{M_{\dot{\boldsymbol{\eta}}}} \end{bmatrix}, \quad \mathbf{B}_R = \begin{bmatrix} q_\infty S_W \mathbf{C}_{X_{\mathbf{u}}} \\ q_\infty S_W \mathbf{C}_{Z_{\mathbf{u}}} \\ q_\infty S_W \bar{c}_W \mathbf{C}_{M_{\mathbf{u}}} \end{bmatrix}
\end{aligned}
\tag{7.135}
$$

Similar expressions may be derived for the three lateral-directional equations of motion.

Likewise, using Equations (7.103), (7.94), and (7.95), we may write the generalized forces acting on the elastic degrees of freedom as

$$Q_i = \Big(C_{Q_{i_0}} + C_{Q_{i_u}} U + C_{Q_{i_\alpha}} \alpha + C_{Q_{i_q}} Q + C_{Q_{i_{i_H}}} i_H + C_{Q_{i_{\delta_E}}} \delta_E + \sum_j \big(C_{Q_{i_{\eta_j}}} \eta_j + C_{Q_{i_{\dot{\eta}_j}}} \dot{\eta}_j\big)\Big) q_\infty S_W \bar{c}_W \tag{7.136}$$

So by defining the aeroelastic-coefficient matrices as

$$\begin{aligned} \mathbf{C}_{Q_R} &= \begin{bmatrix} C_{Q_{i_u}} & C_{Q_{i_\alpha}} & C_{Q_{i_q}} \end{bmatrix} \quad \mathbf{C}_{Q_{\dot{\boldsymbol{\eta}}}} = \begin{bmatrix} C_{Q_{i_{\dot{\eta}_j}}} \end{bmatrix},\ j = 1, \ldots, n \\ \mathbf{C}_{Q_{\boldsymbol{\eta}}} &= \begin{bmatrix} C_{Q_{i_{\eta_j}}} \end{bmatrix},\ j = 1, \ldots, n \quad \mathbf{C}_{Q_{\mathbf{u}}} = \begin{bmatrix} C_{Q_{i_{u_j}}} \end{bmatrix},\ j = 1, \ldots, m \end{aligned} \tag{7.137}$$

along with

$$\boldsymbol{\Omega} = \mathrm{diag}\big[\omega_i^2\big], \quad [\mathcal{M}] = diag[\mathcal{M}_i] \tag{7.138}$$

we can see that the equations of motion governing the elastic degrees of freedom may now be written as

$$\dot{\mathbf{x}}_E = \begin{bmatrix} \mathbf{0}_n \\ q_\infty S_W \bar{c}_W [\mathcal{M}]^{-1} \mathbf{C}_{Q_R} \end{bmatrix} \mathbf{x}_R + \begin{bmatrix} \mathbf{0}_n & \mathbf{I}_n \\ \big(q_\infty S_W \bar{c}_W [\mathcal{M}]^{-1} \mathbf{C}_{Q_{\boldsymbol{\eta}}} - \boldsymbol{\Omega}\big) & q_\infty S_W \bar{c}_W [\mathcal{M}]^{-1} \mathbf{C}_{Q_{\dot{\boldsymbol{\eta}}}} \end{bmatrix} \mathbf{x}_E + \begin{bmatrix} \mathbf{0}_n \\ q_\infty S_W \bar{c}_W [\mathcal{M}]^{-1} \mathbf{C}_{Q_{\mathbf{u}}} \end{bmatrix} \mathbf{u} \tag{7.139}$$

Here, $\mathbf{0}_n$ and $\mathbf{I}_n$ are n-by-n null and identity matrices, respectively.

Comparing this equation to the second of Equations (7.126) we find that

$$\begin{aligned} \mathbf{A}_{ER} &= q_\infty S_W \bar{c}_W [\mathcal{M}]^{-1} \mathbf{C}_{Q_R} & \mathbf{A}_{\dot{\eta}} &= q_\infty S_W \bar{c}_W [\mathcal{M}]^{-1} \mathbf{C}_{Q_{\dot{\boldsymbol{\eta}}}} \\ \mathbf{A}_\eta &= \big(q_\infty S_W \bar{c}_W [\mathcal{M}]^{-1} \mathbf{C}_{Q_{\boldsymbol{\eta}}} - \boldsymbol{\Omega}\big) & \mathbf{B}_E &= q_\infty S_W \bar{c}_W [\mathcal{M}]^{-1} \mathbf{C}_{Q_{\mathbf{u}}} \end{aligned} \tag{7.140}$$

And so from Equation (7.127), the static-elastic constraint becomes

$$\boxed{\begin{aligned} \boldsymbol{\eta}_0 &= -\mathbf{A}_\eta^{-1}\big(\mathbf{A}_{ER}\mathbf{x}_R + \mathbf{B}_E \mathbf{u}\big) \\ &= -\big[q_\infty S_W \bar{c}_W [\mathcal{M}]^{-1} \mathbf{C}_{Q_{\boldsymbol{\eta}}} - \boldsymbol{\Omega}\big]^{-1}\Big(\big[q_\infty S_W \bar{c}_W [\mathcal{M}]^{-1} \mathbf{C}_{Q_R}\big]\mathbf{x}_R \\ &\qquad + \big[q_\infty S_W \bar{c}_W [\mathcal{M}]^{-1} \mathbf{C}_{Q_{\mathbf{u}}}\big]\mathbf{u}\Big) \end{aligned}} \tag{7.141}$$

Finally, in Equation (7.129) the model for the aerodynamic forces and moments in the rigid-body equations of motion were given as

$$\big(\mathbf{A}_R - \mathbf{A}_{R\eta}\mathbf{A}_\eta^{-1}\mathbf{A}_{ER}\big)\mathbf{x}_R + \big(\mathbf{B}_R - \mathbf{A}_{R\eta}\mathbf{A}_\eta^{-1}\mathbf{B}_E\big)\mathbf{u}$$

So the matrix of corrected aerodynamic coefficients corresponding to the rigid-body variables $\mathbf{x}_R$ is

$$\mathbf{A}_R - \mathbf{A}_{R\eta}\mathbf{A}_\eta^{-1}\mathbf{A}_{ER}$$

$$= \begin{bmatrix} q_\infty S_W \mathbf{C}_{X_R} \\ q_\infty S_W \mathbf{C}_{Z_R} \\ q_\infty S_W \bar{c}_W \mathbf{C}_{M_R} \end{bmatrix} - \begin{bmatrix} q_\infty S_W \mathbf{C}_{X_\eta} \\ q_\infty S_W \mathbf{C}_{Z_\eta} \\ q_\infty S_W \bar{c}_W \mathbf{C}_{M_\eta} \end{bmatrix} \left[q_\infty S_W \bar{c}_W [\mathcal{M}]^{-1} \mathbf{C}_{Q_\eta} - \boldsymbol{\Omega}\right]^{-1} \left[q_\infty S_W \bar{c}_W [\mathcal{M}]^{-1} \mathbf{C}_{Q_R}\right] \tag{7.142}$$

while the matrix of <u>corrected</u> aerodynamic coefficients corresponding to the control-surface deflections is

$$\mathbf{B}_R - \mathbf{A}_{R\eta}\mathbf{A}_\eta^{-1}\mathbf{B}_E$$

$$= \begin{bmatrix} q_\infty S_W \mathbf{C}_{X_\mathbf{u}} \\ q_\infty S_W \mathbf{C}_{Z_\mathbf{u}} \\ q_\infty S_W \bar{c}_W \mathbf{C}_{M_\mathbf{u}} \end{bmatrix} - \begin{bmatrix} q_\infty S_W \mathbf{C}_{X_\eta} \\ q_\infty S_W \mathbf{C}_{Z_\eta} \\ q_\infty S_W \bar{c}_W \mathbf{C}_{M_\eta} \end{bmatrix} \left[q_\infty S_W \bar{c}_W [\mathcal{M}]^{-1} \mathbf{C}_{Q_\eta} - \boldsymbol{\Omega}\right]^{-1} \left[q_\infty S_W \bar{c}_W [\mathcal{M}]^{-1} \mathbf{C}_{Q_\mathbf{u}}\right] \tag{7.143}$$

Finally, by considering Equations (7.142) we see that the matrix of corrected aerodynamic coefficients analogous to Equations (7.131), denoted here with primes, is

$$\begin{bmatrix} \mathbf{C}'_{X_R} \\ \mathbf{C}'_{Z_R} \\ \mathbf{C}'_{M_R} \end{bmatrix} = \begin{bmatrix} \mathbf{C}_{X_R} \\ \mathbf{C}_{Z_R} \\ \mathbf{C}_{M_R} \end{bmatrix} - \begin{bmatrix} \mathbf{C}_{X_\eta} \\ \mathbf{C}_{Z_\eta} \\ \mathbf{C}_{M_\eta} \end{bmatrix} \left[q_\infty S_W \bar{c}_W [\mathcal{M}]^{-1} \mathbf{C}_{Q_\eta} - \boldsymbol{\Omega}\right]^{-1} \left[q_\infty S_W \bar{c}_W [\mathcal{M}]^{-1} \mathbf{C}_{Q_R}\right]$$

[Corrected Coefficients] = [Original Coefficients]
+ [Static-Elastic Corrections] (7.144)

And by considering Equations (7.143), we see that the matrix of corrected aerodynamic coefficients analogous to Equations (7.132), also denoted with primes, is

$$\begin{bmatrix} \mathbf{C}'_{X_\mathbf{u}} \\ \mathbf{C}'_{Z_\mathbf{u}} \\ \mathbf{C}'_{M_\mathbf{u}} \end{bmatrix} = \begin{bmatrix} \mathbf{C}_{X_\mathbf{u}} \\ \mathbf{C}_{Z_\mathbf{u}} \\ \mathbf{C}_{M_\mathbf{u}} \end{bmatrix} - \begin{bmatrix} \mathbf{C}_{X_\eta} \\ \mathbf{C}_{Z_\eta} \\ \mathbf{C}_{M_\eta} \end{bmatrix} \left[q_\infty S_W \bar{c}_W [\mathcal{M}]^{-1} \mathbf{C}_{Q_\eta} - \boldsymbol{\Omega}\right]^{-1} \left[q_\infty S_W \bar{c}_W [\mathcal{M}]^{-1} \mathbf{C}_{Q_\mathbf{u}}\right]$$

[Corrected Coefficients] = [Original Coefficients]
+ [Static-Elastic Corrections] (7.145)

Note that these static-elastic corrections are all inversely proportional to the squares of the vibration frequencies (in $\boldsymbol{\Omega}$). And so if these frequencies are high, corresponding to a stiff structure, the corrections will be small, as expected. Also, note that these corrections are in general a function of flight dynamic pressure.

EXAMPLE 7.3

Static-Elastic Corrections for a Large High-Speed Aircraft

Determine the static-elastic corrections for the coefficients C_{M_u}, C_{M_α}, and C_{M_q} for the large, high-speed aircraft considered in the case study in Section 7.9. Assume the flight condition of interest is level flight at 5,000 ft altitude at a Mach number of 0.6. Determine these corrections using only one elastic mode and compare to the results using all four elastic modes.

■ Solution

From Equations (7.131) we see that C_{M_u}, C_{M_α}, and C_{M_q} are elements of the coefficient matrix $\mathbf{C}_{M_R}$, or

$$\mathbf{C}_{M_R} = \begin{bmatrix} C_{M_u} & C_{M_\alpha} & C_{M_q} \end{bmatrix}$$

And from Equation (7.144) we see that the static-elastic corrections to the elements of this coefficient matrix are given by

$$\Delta\mathbf{C}_{M_R} = -\mathbf{C}_{M_\eta}\left[q_\infty S_W \bar{c}_W [\mathcal{M}]^{-1}\mathbf{C}_{Q_\eta} - \mathbf{\Omega}\right]^{-1}\left[q_\infty S_W \bar{c}_W [\mathcal{M}]^{-1}\mathbf{C}_{Q_R}\right] \tag{7.146}$$

So we need to assemble the matrices in the above expression to compute the corrections.

From Table 7.2, the results from the case study, we have

$$\begin{aligned}\mathbf{C}_{M_\eta} &= \begin{bmatrix} C_{M_{\eta_1}} & \cdots & C_{M_{\eta_n}} \end{bmatrix} \\ &= \begin{bmatrix} -0.032 & -0.025 & 0.041 & -0.018 \end{bmatrix}\end{aligned}$$

Also from Table 7.3 (the case study) we have

$$\mathbf{C}_{Q_R} = \begin{bmatrix} C_{Q_{i_u}} & C_{Q_{i_\alpha}} & C_{Q_{i_q}} \end{bmatrix} = \begin{bmatrix} 0 & -0.0149 & -0.726/V_\infty \\ 0 & 0.0258 & 0.089/V_\infty \\ 0 & 0.0149 & 0.304/V_\infty \\ 0 & \sim 0 & \sim 0 \end{bmatrix}$$

$$\mathbf{C}_{Q_\eta} = \begin{bmatrix} C_{Q_{i_{\eta_j}}} \end{bmatrix} = \begin{bmatrix} 5.85\times10^{-5} & -9.00\times10^{-5} & 3.55\times10^{-4} & 1.20\times10^{-4} \\ 4.21\times10^{-3} & -9.22\times10^{-2} & 1.97\times10^{-3} & 3.37\times10^{-3} \\ 2.91\times10^{-4} & 1.44\times10^{-3} & -3.46\times10^{-4} & 1.44\times10^{-4} \\ 2.21\times10^{-5} & -1.32\times10^{-4} & 9.68\times10^{-6} & 1.77\times10^{-3} \end{bmatrix}$$

Furthermore,

$$\mathbf{\Omega} = diag\left[(12.6)^2 \quad (14.1)^2 \quad (21.2)^2 \quad (22.1)^2\right](\text{rad/sec})^2$$

and

$$[\mathcal{M}] = diag\left[184 \quad 9{,}587 \quad 1{,}334 \quad 436{,}000\right] \text{ sl-ft}^2$$

Finally, we have

$$q_\infty = \frac{1}{2}\rho_\infty V_\infty^2 = \frac{1}{2}(0.002048)(0.6 \times 1097)^2 = 444 \text{ psf}$$

$$S_W = 1{,}950 \text{ ft}^2$$

$$\bar{c}_W = 15.3 \text{ ft}$$

Using MATLAB, we find from Equation (7.146) that the static-elastic corrections using all four modes are

$$\Delta \mathbf{C}_{M_R} = \begin{bmatrix} 0 & 0.2325 & 0.0169 \end{bmatrix}$$

Or

$$\Delta C_{M_u} = 0$$

$$\Delta C_{M_\alpha} = 0.2325 \text{ /rad}$$

$$\Delta C_{M_q} = 0.0169 \text{ sec}$$

Repeating the process using only the first elastic mode ($i = 1$) we find that the static-elastic corrections are

$$\Delta C_{M_u} = 0$$

$$\Delta C_{M_\alpha} = 0.2221 \text{ /rad}$$

$$\Delta C_{M_q} = 0.0164 \text{ sec}$$

So the first elastic mode is the major contributor to the static-elastic corrections. This is typical, since the corrections are inversely proportional to the squares of the vibration frequencies.

The above static-elastic corrections for C_{M_α} and C_{M_q} are significant when compared to the typical values for these coefficients—especially for C_{M_α}. For example, for this vehicle C_{M_q} will be on the order of –0.4 sec and C_{M_α} will be on the order of –1.5 /rad. But more important than the magnitudes of the corrections are the signs. Both of the corrections are positive, while the coefficients themselves are both negative. So both coefficients will be less negative after the corrections are applied. As we shall see in Chapters 9 and 10, negative values for C_{M_α} and C_{M_q} are required for stability of the vehicle. So the effect of static-elastic deformations in this case is to make the vehicle less stable.

7.12 Summary

In this chapter we continued to address the modeling of the aerodynamic forces and moments acting on the vehicle, but we focused on the changes in those forces and moments arising due to elastic deformation of the vehicle's structure. Strip theory and the component build-up methods were the basic tools, and all the analysis in this chapter assumed a conventional vehicle geometry. But the techniques we presented may obviously be applied to other vehicle geometries.

We derived closed-form expressions for the effects of elastic deformation on all three components of the aerodynamic force acting on the vehicle, as well as the three components of the aerodynamic moment. And we derived similar expressions for the generalized forces acting on the elastic degrees of freedom. We ignored unsteady aerodynamic effects, due to the assumption that only the lower-frequency vibration modes would be of interest here.

We defined generalized-force coefficients, or modal-displacement and modal-velocity effectiveness coefficients, and derived expressions for all these coefficients. Then we expressed the effects of elastic deformation on the forces and moments in terms of these coefficients. We presented a case study involving a large high-speed aircraft to solidify the understanding of the concepts presented.

Finally, we addressed the effects of static-elastic deflections on the vehicle's aerodynamic forces and moments. We derived a static-elastic constraint that yields the static deflections of the elastic modal coordinates as a function of the rigid-body coordinates and control-surface deflections. Then we obtained the static-elastic corrections to the "rigid-body" aerodynamic coefficients. We demonstrated by example that such corrections may be significant.

7.13 Problems

7.1 Using a spreadsheet, rework Example 7.1 using a graphical technique (e.g., trapezoidal integration) to evaluate the integral terms. (You may use the polynomial fits of the mode shapes to generate the graph of the mode-shape "data.")

7.2 Verify Equations (7.94).

7.3 Assume an aircraft has a canard control surface in front. Develop an expression, given in terms of modal displacements η_i and modal velocities $\dot{\eta}_i$, for the change in the vehicle's pitching moment due to elastic deformation of the structure. That is, find an expression for M_E.

7.4 Consider the case study in Section 7.9. Assume only Mode 1 is to be included in the model for the flexible effects on F_{A_Z}, M_A, and the generalized forces Q_i. Assemble the model for these effects in this case.

References

1. Yates, E. C.: "Calculation of Flutter Characteristics for Finite-Span Swept or Unswept Wings at Subsonic and Supersonic Speeds by a Modified Strip Analysis," NACA RM L57L10, March 1958.
2. Waszak, M. R., and D. K. Schmidt: "Flight Dynamics of Aeroelastic Vehicles," *Journal of Aircraft,* Vol. 25, no. 6, June 1988.
3. Curtis, Howard D.: *Fundamentals of Aircraft Structural Analysis,* McGraw-Hill, New York, 1996.
4. Waszak, M. R. and D. K. Schmidt: "Analysis of Flexible Aircraft Longitudinal Dynamics and Handling Qualities," vols. I and II, NASA Contractor Report 177943, School of Aeronautics and Astronautics, Purdue University, West Lafayette, IN, June 1985.
5. Freeman, R. C. and T. I. Rozsa: "Basic Modal Data Package for -55B Mid-Penetration Weight 65 Degree Wing Sweep," North American Rockwell Corp., VDD-71-4, November 1971.
6. Bisplinghoff, R. L., H. Ashley, and H. Halfman: *Aeroelasticity,* Dover Science, Mineola, NY, 1996.

8 CHAPTER

Math Model Assembly and Flight Simulation

Chapter Roadmap: *The material presented in Sections 8.1.1–8.1.4 should be covered in a first undergraduate course in flight dynamics because here the linear models for the vehicle dynamics are assembled. The remainder of the chapter is frequently not included in a first course due to time limitations. Those wishing to include the most basic topics may skip Sections 8.1.5 and 8.2.4, dealing with flexible vehicles, Sections 8.1.7 and 8.2.6 dealing with atmospheric turbulence, and the Just-In-Time Tutorials in Sections 8.1.8 and 8.2.7.*

This chapter marks a shift in emphasis in the book—from that of building the mathematical model of the vehicle's dynamics to that of using the model for the analysis of those dynamics. Computer simulation is a powerful tool for studying the dynamics of flight vehicles. Such simulations are frequently referred to as *flight simulations,* since they simulate the motion, or flight, of the vehicle.

All flight simulations may be divided into two types, real time and nonreal time. Real-time simulations synchronize simulated and real time, using a clock in the simulation computer, such that inputs to the simulated vehicle (model) and its responses occur in real time. Such simulations are required if the simulation will interact with a physical component for testing, such as a flight-control computer or with a human pilot. Piloted real-time simulations range from simple laboratory simulations at a university, to complex systems with a realistic simulated visual scene as well as cockpit motion (Ref. 1). Nonreal-time simulations lack the synchronization of real and simulated time, and are used for engineering analysis.

Flight simulations may also be divided into two other types, nonlinear and linear, depending on whether the mathematical model of the dynamics in the simulation is linear or nonlinear. Usually, real-time simulations are nonlinear because of the applications of such simulations. But nonreal-time simulations may be either linear or nonlinear, again depending on the intended use of the simulation.

In this chapter we will discuss tools necessary to develop a flight simulation, and we will focus on nonreal-time simulations. We will emphasize the assembly of mathematical models for the vehicle's dynamics, numerical integration, and examples using the simulation tools in MATLAB. However, almost all the topics discussed in the chapter also apply to real-time simulations, and to other software simulation environments.

8.1 LINEAR MODEL ASSEMBLY AND SIMULATION

We will first address linear simulation, in part because it is the first simulation frequently developed in an engineering analysis. We will assemble linear models of the vehicle's dynamics, as well as discuss various linear-simulation techniques and tools.

8.1.1 Linear Equations of Motion

The major pieces of the mathematical model of the vehicle's dynamics have already been discussed in earlier chapters. We will now address assembling these pieces. The core of the models consists of the perturbation equations of motion developed in Chapter 2 for rigid vehicles, plus Chapter 4 for flexible vehicles.

We seek to express the linear models in state-variable format, the most general form for linear models. Let us first consider the equations developed in Chapter 2, and consider the flat-earth case specifically. We assume that the vehicle has constant mass and is symmetric with respect to its XZ plane, and we will ignore any effects of rotating machinery. (The procedures presented here may be applied directly to other cases by starting with the appropriate equations of motion in Chapter 2.)

From Equations (2.45), we have the three linear equations governing perturbations in rigid-body translation, or

$$
\begin{aligned}
m\big(\dot{u} + (\mathrm{Q}_0 w + \mathrm{W}_0 q) - (\mathrm{V}_0 r + \mathrm{R}_0 v)\big) &= -mg\cos\Theta_0\theta + f_{A_X} + f_{P_X}\\
m\big(\dot{v} + (\mathrm{R}_0 u + \mathrm{U}_0 r) - (\mathrm{P}_0 w + \mathrm{W}_0 p)\big) &= mg\big(\cos\Theta_0\cos\Phi_0\phi - \sin\Theta_0\sin\Phi_0\theta\big)\\
&\quad + f_{A_Y} + f_{P_Y}\\
m\big(\dot{w} + (\mathrm{P}_0 v + \mathrm{V}_0 p) - (\mathrm{Q}_0 u + \mathrm{U}_0 q)\big) &= -mg\big(\cos\Theta_0\sin\Phi_0\phi + \sin\Theta_0\cos\Phi_0\theta\big)\\
&\quad + f_{A_Z} + f_{P_Z}
\end{aligned}
$$

Dividing through by the vehicle mass m and rearranging we have

$$
\boxed{
\begin{aligned}
\dot{u} &= (\mathrm{V}_0 r + \mathrm{R}_0 v) - (\mathrm{Q}_0 w + \mathrm{W}_0 q) - g\cos\Theta_0\theta + (f_{A_X} + f_{P_X})/m\\
\dot{v} &= (\mathrm{P}_0 w + \mathrm{W}_0 p) - (\mathrm{R}_0 u + \mathrm{U}_0 r)\\
&\quad + g\big(\cos\Theta_0\cos\Phi_0\phi - \sin\Theta_0\sin\Phi_0\theta\big) + (f_{A_Y} + f_{P_Y})/m\\
\dot{w} &= (\mathrm{Q}_0 u + \mathrm{U}_0 q) - (\mathrm{P}_0 v + \mathrm{V}_0 p)\\
&\quad - g\big(\cos\Theta_0\sin\Phi_0\phi + \sin\Theta_0\cos\Phi_0\theta\big) + (f_{A_Z} + f_{P_Z})/m
\end{aligned}
}
\tag{8.1}
$$

Recall that u, v, and w are the three components of the vehicle's perturbation translational velocity, expressed in the selected vehicle-fixed axes. (variables with the zero subscript correspond to the reference flight condition.) Or

$$\delta \mathbf{V}_V = u\mathbf{i}_V + v\mathbf{j}_V + w\mathbf{k}_V = \begin{bmatrix} u & v & w \end{bmatrix} \begin{Bmatrix} \mathbf{i}_V \\ \mathbf{j}_V \\ \mathbf{k}_V \end{Bmatrix} \tag{8.2}$$

Then from Equations (2.49) we have the three linear equations governing perturbations in rigid-body rotation given by

$$\begin{aligned}
\dot{p} - \frac{I_{xz}}{I_{xx}}\dot{r} &= \frac{1}{I_{xx}}\left(I_{xz}(\mathrm{Q}_0 p + \mathrm{P}_0 q) + (I_{yy} - I_{zz})(\mathrm{R}_0 q + \mathrm{Q}_0 r) + (l_A + l_P)\right) \\
\dot{q} &= \frac{1}{I_{yy}}\left((I_{zz} - I_{xx})(\mathrm{R}_0 p + \mathrm{P}_0 r) + 2I_{xz}(\mathrm{R}_0 r - \mathrm{P}_0 p) + (m_A + m_p)\right) \\
\dot{r} - \frac{I_{xz}}{I_{zz}}\dot{p} &= \frac{1}{I_{zz}}\left(-I_{xz}(\mathrm{R}_0 q + \mathrm{Q}_0 r) + (I_{xx} - I_{yy})(\mathrm{Q}_0 p + \mathrm{P}_0 q) + (n_A + n_P)\right)
\end{aligned} \tag{8.3}$$

where we have rearranged the equations and divided each equation by the appropriate moment of inertia. Here p, q, and r are the three components of the vehicle's perturbation rotational velocity expressed in the same vehicle-fixed axes. Or

$$\delta \boldsymbol{\omega}_{V,I} = p\mathbf{i}_V + q\mathbf{j}_V + r\mathbf{k}_V = \begin{bmatrix} p & q & r \end{bmatrix} \begin{Bmatrix} \mathbf{i}_V \\ \mathbf{j}_V \\ \mathbf{k}_V \end{Bmatrix} \tag{8.4}$$

In addition, m is the vehicle mass and $I_{\bullet\bullet}$ are elements of the vehicle's inertia tensor, consisting of the moments and products of inertia. Also recall that in the development of Equations (2.49), the vehicle was assumed to be symmetric about its XZ plane, and therefore the products of inertia $I_{xy} = I_{yz} = 0$. If this is not the case, the derivation of Equations (2.49) should be revisited. Finally, the perturbation forces and moments appear on the right-hand sides of Equations (8.1) and (8.3).

Now note that the first and third equations in Equations (8.3) are coupled through the inertia matrix. Rewriting these two coupled equations in matrix form yields

$$\begin{bmatrix} 1 & -\dfrac{I_{xz}}{I_{xx}} \\ -\dfrac{I_{xz}}{I_{zz}} & 1 \end{bmatrix} \begin{Bmatrix} \dot{p} \\ \dot{r} \end{Bmatrix} = \begin{Bmatrix} \dfrac{1}{I_{xx}}\left(I_{xz}(\mathrm{Q}_0 p + \mathrm{P}_0 q) + (I_{yy} - I_{zz})(\mathrm{R}_0 q + \mathrm{Q}_0 r) + (l_A + l_P)\right) \\ \dfrac{1}{I_{zz}}\left(-I_{xz}(\mathrm{R}_0 q + \mathrm{Q}_0 r) + (I_{xx} - I_{yy})(\mathrm{Q}_0 p + \mathrm{P}_0 q) + (n_A + n_P)\right) \end{Bmatrix} \tag{8.5}$$

By inverting the leading coefficient matrix in Equation (8.5), and with the remaining equation from Equations (8.3), we have the decoupled equations governing rotation given by

$$\dot{q} = \frac{1}{I_{yy}}\Big((I_{zz} - I_{xx})(\mathrm{R}_0 p + \mathrm{P}_0 r) + 2I_{xz}(\mathrm{R}_0 r - \mathrm{P}_0 p) + (m_A + m_p)\Big)$$

$$\begin{Bmatrix} \dot{p} \\ \dot{r} \end{Bmatrix} = \frac{1}{1 - \left(\dfrac{I_{xz}^2}{I_{xx}I_{zz}}\right)} \begin{bmatrix} 1 & \dfrac{I_{xz}}{I_{xx}} \\ \dfrac{I_{xz}}{I_{zz}} & 1 \end{bmatrix} \begin{Bmatrix} \dfrac{1}{I_{xx}}\Big(I_{xz}(\mathrm{Q}_0 p + \mathrm{P}_0 q) + (I_{yy} - I_{zz})(\mathrm{R}_0 q + \mathrm{Q}_0 r) + (l_A + l_P)\Big) \\ \dfrac{1}{I_{zz}}\Big(-I_{xz}(\mathrm{R}_0 q + \mathrm{Q}_0 r) + (I_{xx} - I_{yy})(\mathrm{Q}_0 p + \mathrm{P}_0 q) + (n_A + n_P)\Big) \end{Bmatrix} \quad (8.6)$$

In addition to the above six equations governing translations and rotations, we also have three equations governing the kinematic relationships between the angular rates. Or, consistent with Equations (2.55), we have

$$\begin{aligned} \dot{\phi} &= p + \tan\Theta_0\big(\sin\Phi_0 q + \cos\Phi_0 r + (\mathrm{Q}_0\cos\Phi_0 - \mathrm{R}_0\sin\Phi_0)\phi\big) \\ &\quad + (\mathrm{Q}_0\sin\Phi_0 + \mathrm{R}_0\cos\Phi_0 + \dot{\Psi}_0\sin\Theta_0\tan\Theta_0)\theta \\ \dot{\theta} &= \cos\Phi_0 q - \sin\Phi_0 r - (\mathrm{Q}_0\sin\Phi_0 + \mathrm{R}_0\cos\Phi_0)\phi \\ \dot{\psi} &= \dot{\Psi}_0\tan\Theta_0\theta + \big(\sin\Phi_0 q + \cos\Phi_0 r - (\mathrm{R}_0\sin\Phi_0 - \mathrm{Q}_0\cos\Phi_0)\phi\big)/\cos\Theta_0 \end{aligned} \quad (8.7)$$

Here ψ, θ, and ϕ are perturbations in the 3-2-1 Euler angles defining the orientation of the vehicle-fixed frame with respect to the inertial frame. These three equations are greatly simplified when a particular reference condition is considered.

Finally, if we are going to track the vehicle's position in an earth-fixed reference frame, we have three more equations governing the kinematic relationships between the vehicle position and its velocity. Assuming the flat-earth case introduced in Chapter 2, Equations (2.40) give the nonlinear (not perturbation) kinematic relationships. Or

$$\begin{aligned} \dot{X}_E &= U\cos\theta\cos\psi + V(\sin\phi\sin\theta\cos\psi - \cos\phi\sin\psi) \\ &\quad + W(\cos\phi\sin\theta\cos\psi + \sin\phi\sin\psi) \\ \dot{Y}_E &= U\cos\theta\sin\psi + V(\sin\phi\sin\theta\sin\psi + \cos\phi\cos\psi) \\ &\quad + W(\cos\phi\sin\theta\sin\psi - \sin\phi\cos\psi) \\ \dot{\mathrm{h}} &= U\sin\theta - V\sin\phi\cos\theta - W\cos\phi\cos\theta \end{aligned}$$

In terms of the direction-cosine matrices, this expression may also be written as

$$\begin{Bmatrix} \dot{X}_E \\ \dot{Y}_E \\ -\dot{\mathrm{h}} \end{Bmatrix} = \begin{bmatrix} \cos\psi & -\sin\psi & 0 \\ \sin\psi & \cos\psi & 0 \\ 0 & 0 & 1 \end{bmatrix} \begin{bmatrix} \cos\theta & 0 & \sin\theta \\ 0 & 1 & 0 \\ -\sin\theta & 0 & \cos\theta \end{bmatrix} \begin{bmatrix} 1 & 0 & 0 \\ 0 & \cos\phi & -\sin\phi \\ 0 & \sin\phi & \cos\phi \end{bmatrix} \begin{Bmatrix} U \\ V \\ W \end{Bmatrix} \quad (8.8)$$

where X_E, and Y_E are the north and east coordinates of the vehicle, respectively, and h is its altitude. To find the linearized relationships, we must perform a perturbation analysis on Equations (8.8).

Making the necessary substitutions, using the trigonometric identities for the sum of two angles, and making the small-perturbation assumptions, we have the linear perturbation kinematic equations given by

$$
\begin{Bmatrix} \dot{x}_E \\ \dot{y}_E \\ -\dot{h} \end{Bmatrix} = \left(\begin{bmatrix} -S_{\Psi_0}\psi & -C_{\Psi_0}\psi & 0 \\ C_{\Psi_0}\psi & -S_{\Psi_0}\psi & 0 \\ 0 & 0 & 0 \end{bmatrix} \begin{bmatrix} C_{\Theta_0} & 0 & S_{\Theta_0} \\ 0 & 1 & 0 \\ -S_{\Theta_0} & 0 & C_{\Theta_0} \end{bmatrix} \begin{bmatrix} 1 & 0 & 0 \\ 0 & C_{\Phi_0} & -S_{\Phi_0} \\ 0 & S_{\Phi_0} & C_{\Phi_0} \end{bmatrix} \right.
$$
$$
+ \begin{bmatrix} C_{\Psi_0} & -S_{\Psi_0} & 0 \\ S_{\Psi_0} & C_{\Psi_0} & 0 \\ 0 & 0 & 1 \end{bmatrix} \begin{bmatrix} -S_{\Theta_0}\theta & 0 & C_{\Theta_0}\theta \\ 0 & 0 & 0 \\ -C_{\Theta_0}\theta & 0 & -S_{\Theta_0}\theta \end{bmatrix} \begin{bmatrix} 1 & 0 & 0 \\ 0 & C_{\Phi_0} & -S_{\Phi_0} \\ 0 & S_{\Phi_0} & C_{\Phi_0} \end{bmatrix}
$$
$$
\left. + \begin{bmatrix} C_{\Psi_0} & -S_{\Psi_0} & 0 \\ S_{\Psi_0} & C_{\Psi_0} & 0 \\ 0 & 0 & 1 \end{bmatrix} \begin{bmatrix} C_{\Theta_0} & 0 & S_{\Theta_0} \\ 0 & 1 & 0 \\ -S_{\Theta_0} & 0 & C_{\Theta_0} \end{bmatrix} \begin{bmatrix} 0 & 0 & 0 \\ 0 & -S_{\Phi_0}\phi & -C_{\Phi_0}\phi \\ 0 & C_{\Phi_0}\phi & -S_{\Phi_0}\phi \end{bmatrix} \right) \begin{Bmatrix} U_0 \\ V_0 \\ W_0 \end{Bmatrix}
$$
$$
+ \begin{bmatrix} C_{\Psi_0} & -S_{\Psi_0} & 0 \\ S_{\Psi_0} & C_{\Psi_0} & 0 \\ 0 & 0 & 1 \end{bmatrix} \begin{bmatrix} C_{\Theta_0} & 0 & S_{\Theta_0} \\ 0 & 1 & 0 \\ -S_{\Theta_0} & 0 & C_{\Theta_0} \end{bmatrix} \begin{bmatrix} 1 & 0 & 0 \\ 0 & C_{\Phi_0} & -S_{\Phi_0} \\ 0 & S_{\Phi_0} & C_{\Phi_0} \end{bmatrix} \begin{Bmatrix} u \\ v \\ w \end{Bmatrix} \tag{8.9}
$$

Note that in the above equation we have introduced the shorthand notation $S_\bullet = \sin(\bullet)$ and $C_\bullet = \cos(\bullet)$. Again, the above equations simplify greatly for a given reference condition.

NOTE TO STUDENT

The matrix perturbation equation above, as well as the nine previous equations of motion, are all linear in the perturbation variables, although they do not appear so at first glance. But looking more carefully, we see that all the coefficients of the perturbation variables involve the vehicle's mass and inertia, plus terms evaluated at the reference flight condition. Therefore, since we are using a vehicle-fixed coordinate system, and under a constant-mass assumption, these coefficients are simply constants depending only on the reference flight condition!

The 12 linear equations of motion we've just assembled make up the complete set. But in the vast majority of cases, fewer and much simpler equations will result. We will demonstrate this fact in the following example.

EXAMPLE 8.1

Linear Equations of Motion for Straight-and-Level Flight

Assume that the reference flight condition of interest is straight-and-level flight in an easterly direction, with zero bank and sideslip angles. Using stability axes corresponding to this flight condition, determine the linear equations of motion.

■ Solution

Since the reference flight condition involves zero bank and sideslip angles, then $\Phi_0 = \beta_0 = 0$. And if the flight condition also involves straight-and-level flight, then for stability axes $\Theta_0 = 0$. Recall also that for stability axes, $W_0 = 0$ and $\alpha_0 = 0$, so the X_S axis always lies in the same plane as the reference velocity vector $\mathbf{V}_{\infty_0}$. Plus if $\beta_0 = 0$ (and therefore $V_0 = 0$) as well, then the X_S axis is co-linear with the reference velocity vector, and $U_0 = |\mathbf{V}_{\infty_0}|$. Additionally, straight-and-level-flight implies that all angular rates are zero for the reference condition. Or $P_0 = Q_0 = R_0 = 0$. Finally, for an easterly heading, $\Psi_0 = 90$ deg.

Under these stated conditions, the equations governing translation and rotation (Equations (8.1) and (8.6)) become simply

$$\boxed{\begin{aligned} \dot{u} &= -g\theta + \left(f_{A_X} + f_{P_X}\right)/m \\ \dot{v} &= -\mathrm{U}_0 r + g\phi + \left(f_{A_Y} + f_{P_Y}\right)/m \\ \dot{w} &= \mathrm{U}_0 q + \left(f_{A_Z} + f_{P_Z}\right)/m \end{aligned}} \tag{8.10}$$

and

$$\boxed{\begin{aligned} &\dot{q} = \left(m_A + m_p\right)/I_{yy} \\ &\begin{Bmatrix} \dot{p} \\ \dot{r} \end{Bmatrix} = \frac{1}{I_{xx}I_{zz} - I_{xz}^2}\begin{bmatrix} I_{zz} & I_{xz} \\ I_{xz} & I_{xx} \end{bmatrix}\begin{Bmatrix} \left(l_A + l_P\right) \\ \left(n_A + n_P\right) \end{Bmatrix} \end{aligned}} \tag{8.11}$$

In addition, the kinematic equations relating angular rates, Equations (8.7), become simply

$$\boxed{\begin{aligned} \dot{\phi} &= p \\ \dot{\theta} &= q \\ \dot{\psi} &= r \end{aligned}} \tag{8.12}$$

(Note that students frequently think that the three equations above are always valid, but that is clearly not the case!) And finally, the three kinematic equations relating translational rates (Equations (8.9)) become simply

$$\boxed{\begin{aligned} \dot{x}_E &= -v - \mathrm{U}_0\psi \\ \dot{y}_E &= u \\ \dot{h} &= -w + \mathrm{U}_0\theta \end{aligned}} \tag{8.13}$$

The first two of these equations may look odd at first. But recall that x_E and y_E are defined as <u>north</u> and <u>east</u> inertial positions, respectively.

Table 8.1 Linear Models of Translational and Rotational Equations of Motion

Case	Translational Equations	Rotational Equations
Rotating machinery	Equations (2.45) (unchanged, same as Equations (8.1))	Equations (2.49) plus add'l underlined terms in Equations (2.75)
Variable mass	Equations (2.45) plus linearized results from underlined terms in Equations (2.89) (see Example 2.4)	Equations (2.49) plus linearized results from underlined terms in Equations (2.100) (see Example 2.4)
Rotating, spherical earth	Equations (2.45) plus linearized results from underlined terms in Equations (2.123)	Equations (2.49) *Note:* also need linearized results from Equations (2.114), (2.115), and (2.125) for new kinematic equations

Up to this point we have only considered the linear models corresponding to a flat-earth case, ignoring the effects of any rotating machinery on the vehicle. If, however, we need to consider rotating machinery, a rotating spherical earth, or the effects of variable mass, we must begin with the appropriate set of equations governing translation and rotation. In Table 8.1, these other three cases appear along with the equation numbers for the corresponding linearized equations of motion. After identifying the appropriate set of linearized equations, one may proceed in a fashion similar to that presented in the remainder of Section 8.1.

8.1.2 Linear Models of the Forces and Moments

Locally linear models for the aerodynamic and propulsive forces and moments were developed in Chapter 6 for the rigid vehicle. Then in Chapter 7 we developed models for the changes in the forces and moments due to elastic deformations. Now we will assemble the models for a rigid vehicle; the flexible vehicle will be addressed in Section 8.1.5.

From Equations (6.272), repeated here, we have the following locally linear model for the perturbation aerodynamic forces acting on the vehicle.

$$f_{A_X} = q_\infty S_W\left(-\left(C_{D_u} + \frac{2}{U_0}C_{D_0}\right)u + (-C_{D_\alpha} + C_{L_0})\alpha - C_{D_{\dot{\alpha}}}\dot{\alpha} - C_{D_q}q - C_{D_{\delta_E}}\delta_E\right)$$

$$f_{A_Y} = q_\infty S_W(C_{S_\beta}\beta + C_{S_p}p + C_{S_r}r + C_{S_{\delta_A}}\delta_A + C_{S_{\delta_R}}\delta_R)$$

$$f_{A_Z} = q_\infty S_W\left(-\left(C_{L_u} + \frac{2}{U_0}C_{L_0}\right)u - (C_{L_\alpha} + C_{D_0})\alpha - C_{L_{\dot{\alpha}}}\dot{\alpha} - C_{L_q}q - C_{L_{\delta_E}}\delta_E\right)$$

Also, from Equations (6.274), the perturbation propulsive forces are given as

$$f_{P_X} = q_\infty S_W\left(C_{P_{X_u}} + \frac{2}{U_0}C_{P_{X_0}}\right)u + \delta T\cos(\phi_T + \alpha_0)$$

$$f_{P_Y} = 0$$

$$f_{P_Z} = q_\infty S_W\left(C_{P_{Z_u}} + \frac{2}{U_0}C_{P_{Z_0}}\right)u - \delta T\sin(\phi_T + \alpha_0)$$

These two sets of perturbation forces when combined become

$$f_{A_X} + f_{P_X} = q_\infty S_W\begin{pmatrix}\left(-\left(C_{D_u} + \frac{2}{U_0}C_{D_0}\right) + \left(C_{P_{X_u}} + \frac{2}{U_0}C_{P_{X_0}}\right)\right)u \\ +(-C_{D_\alpha} + C_{L_0})\alpha - C_{D_{\dot\alpha}}\dot\alpha - C_{D_q}q - C_{D_{\delta_E}}\delta_E\end{pmatrix} + \delta T\cos(\phi_T + \alpha_0)$$

$$f_{A_Y} + f_{P_Y} = q_\infty S_W\left(C_{S_\beta}\beta + C_{S_p}p + C_{S_r} + C_{S_{\delta_A}}\delta_A + C_{S_{\delta_R}}\delta_R\right) \tag{8.14}$$

$$f_{A_Z} + f_{P_Z} = q_\infty S_W\begin{pmatrix}\left(-\left(C_{L_u} + \frac{2}{U_0}C_{L_0}\right) + \left(C_{P_{Z_u}} + \frac{2}{U_0}C_{P_{Z_0}}\right)\right)u \\ -(C_{L_\alpha} + C_{D_0})\alpha - C_{L_{\dot\alpha}}\dot\alpha - C_{L_q}q - C_{L_{\delta_E}}\delta_E\end{pmatrix} - \delta T\sin(\phi_T + \alpha_0)$$

Noting that these three sets of forces will each be divided by the vehicle mass m in the equations of motion. The forces per unit mass, or accelerations, may be written as

$$\frac{f_{A_X} + f_{P_X}}{m} = \frac{q_\infty S_W}{m}\begin{pmatrix}\left(-\left(C_{D_u} + \frac{2}{U_0}C_{D_0}\right) + \left(C_{P_{X_u}} + \frac{2}{U_0}C_{P_{X_0}}\right)\right)u \\ +(-C_{D_\alpha} + C_{L_0})\alpha - C_{D_{\dot\alpha}}\dot\alpha - C_{D_q}q - C_{D_{\delta_E}}\delta_E\end{pmatrix} + \frac{\delta T\cos(\phi_T + \alpha_0)}{m}$$

$$\frac{f_{A_Y} + f_{P_Y}}{m} = \frac{q_\infty S_W}{m}\left(C_{S_\beta}\beta + C_{S_p}p + C_{S_r}r + C_{S_{\delta_A}}\delta_A + C_{S_{\delta_R}}\delta_R\right) \tag{8.15}$$

$$\frac{f_{A_Z} + f_{P_Z}}{m} = \frac{q_\infty S_W}{m}\begin{pmatrix}\left(-\left(C_{L_u} + \frac{2}{U_0}C_{L_0}\right) + \left(C_{P_{Z_u}} + \frac{2}{U_0}C_{P_{Z_0}}\right)\right)u \\ -(C_{L_\alpha} + C_{D_0})\alpha - C_{L_{\dot\alpha}}\dot\alpha - C_{L_q}q - C_{L_{\delta_E}}\delta_E\end{pmatrix} - \frac{\delta T\sin(\phi_T + \alpha_0)}{m}$$

or, in terms of a standard shorthand notation,

$$\begin{aligned}(f_{A_X} + f_{P_X})/m &= X_u u + X_{P_u}u + X_\alpha\alpha + X_{\dot\alpha}\dot\alpha + X_q q + X_{\delta_E}\delta_E + X_T\delta T \\ (f_{A_Y} + f_{P_Y})/m &= Y_\beta\beta + Y_p p + Y_r r + Y_{\delta_A}\delta_A + Y_{\delta_R}\delta_R \\ (f_{A_Z} + f_{P_Z})/m &= Z_u u + Z_{P_u}u + Z_\alpha\alpha + Z_{\dot\alpha}\dot\alpha + Z_q q + Z_{\delta_E}\delta_E + Z_T\delta T\end{aligned} \tag{8.16}$$

The coefficients in Equations (8.16) are called *dimensional stability derivatives*, and their definitions are obvious from comparing Equations (8.15) and (8.16). Finally, the translational equations of motion are complete when Equations (8.16) are incorporated into Equations (8.1).

One note regarding the coefficients of u is in order here. Notice that X_u and X_{P_u} contain two similar terms involving the reference drag and propulsive thrust coefficients, or C_{D_0} and $C_{P_{X_0}}$. When a conventional vehicle is not axially accelerating, thrust equals drag, and these two terms cancel each other. Similarly, Z_u and Z_{P_u} contain two similar terms involving the reference lift and propulsive normal-force coefficients, or C_{L_0} and $C_{P_{Z_0}}$. If a conventional vehicle is in level flight, the sum of these two forces will equal the vehicle weight, or

$$mg = q_\infty S_W\left(-C_{L_0} + C_{P_{Z_0}}\right) \tag{8.17}$$

Also, the development above assumed a constant atmospheric density, which is typical for linear simulations of conventional flight vehicles. If, however, the atmospheric density is not assumed constant with altitude, additional altitude-dependent perturbation forces must be added to the above model. From Equations (6.275) these additional forces due to density variations may be expressed as

$$\boxed{\begin{aligned}
f_{A_{X_h}} &\triangleq \frac{\partial F_{A_X}}{\partial h}\Big|_0 h = C_{X_0}\frac{q_\infty S_W}{\rho_\infty}\frac{\partial \rho_\infty}{\partial h}\Big|_0 h \\
&= -\left(C_{D_0}\cos\beta_0 + C_{S_0}\sin\beta_0\right)\frac{q_\infty S_W}{\rho_\infty}\frac{\partial \rho_\infty}{\partial h}\Big|_0 h \\
f_{A_{Y_h}} &\triangleq \frac{\partial F_{A_Y}}{\partial h}\Big|_0 h = C_{Y_0}\frac{q_\infty S_W}{\rho_\infty}\frac{\partial \rho_\infty}{\partial h}\Big|_0 h \\
&= \left(C_{S_0}\cos\beta_0 - C_{D_0}\sin\beta_0\right)\frac{q_\infty S_W}{\rho_\infty}\frac{\partial \rho_\infty}{\partial h}\Big|_0 h \\
f_{A_{Z_h}} &\triangleq \frac{\partial F_{A_Z}}{\partial h}\Big|_0 h = C_{Z_0}\frac{q_\infty S_W}{\rho_\infty}\frac{\partial \rho_\infty}{\partial h}\Big|_0 h = -C_{L_0}\frac{q_\infty S_W}{\rho_\infty}\frac{\partial \rho_\infty}{\partial h}\Big|_0 h
\end{aligned}} \tag{8.18}$$

where the density-altitude gradient $\partial\rho_\infty/\partial h$ is obtained from an atmospheric model such as that given in Appendix A. The above forces would be added to Equations (8.15), giving rise to additional dimensional derivatives X_h, Y_h, and Z_h in Equations (8.16).

From Equations (6.273), the locally linear model for the perturbation aerodynamic moments is

$$\begin{aligned}
l_A &= q_\infty S_W b_W\left(C_{L_\beta}\beta + C_{L_p}p + C_{L_r}r + C_{L_{\delta_A}}\delta_A + C_{L_{\delta_R}}\delta_R\right) \\
m_A &= q_\infty S_W \bar{c}_W\left(\left(C_{M_u} + \frac{2}{U_0}C_{M_0}\right)u + C_{M_\alpha}\alpha + C_{M_{\dot\alpha}}\dot\alpha + C_{M_q}q + C_{M_{\delta_E}}\delta_E\right) \\
n_A &= q_\infty S_W b_W\left(C_{N_\beta}\beta + C_{N_p}p + C_{N_r}r + C_{N_{\delta_A}}\delta_A + C_{N_{\delta_R}}\delta_R\right)
\end{aligned}$$

And from Equations (6.274), the linear model for the perturbation propulsive pitching moment is

$$m_P = q_\infty S_W \bar{c}_W \left(\left(C_{P_{M_u}} + \frac{2}{U_0} C_{P_{M_0}} \right) u + C_{M_{P_\alpha}} \alpha \right) + \delta T \left(d_T \cos\phi_T - x_T \sin\phi_T \right)$$

Combining the above moments we have

$$\boxed{\begin{aligned}
l_A + l_P &= q_\infty S_W b_W \left(C_{L_\beta}\beta + C_{L_p} p + C_{L_r} r + C_{L_{\delta_A}} \delta_A + C_{L_{\delta_R}} \delta_R \right)\\
m_A + m_P &= q_\infty S_W \bar{c}_W \left(\begin{aligned} &\left(\left(C_{M_u} + \frac{2}{U_0} C_{M_0} \right) + \left(C_{P_{M_u}} + \frac{2}{U_0} C_{P_{M_0}} \right) \right) u \\ &+ \left(C_{M_\alpha} + C_{M_{P_\alpha}} \right)\alpha + C_{M_{\dot\alpha}}\dot\alpha + C_{M_q} q + C_{M_{\delta_E}} \delta_E \end{aligned} \right)\\
&\quad + \delta T \left(d_T \cos\phi_T - x_T \sin\phi_T \right)\\
n_A + n_P &= q_\infty S_W b_W \left(C_{N_\beta}\beta + C_{N_p} p + C_{N_r} r + C_{N_{\delta_A}} \delta_A + C_{N_{\delta_R}} \delta_R \right)
\end{aligned}} \tag{8.19}$$

If the above equation involving the rolling moments is divided by I_{xx}, the equation involving the pitching moments is divided by I_{yy}, and the equation involving the yawing moments is divided by I_{zz}, the three equations become

$$\boxed{\begin{aligned}
\frac{l_A + l_P}{I_{xx}} &= \frac{q_\infty S_W b_W}{I_{xx}} \left(C_{L_\beta}\beta + C_{L_p} p + C_{L_r} r + C_{L_{\delta_A}} \delta_A + C_{L_{\delta_R}} \delta_R \right)\\
\frac{m_A + m_P}{I_{yy}} &= \frac{q_\infty S_W \bar{c}_W}{I_{yy}} \left(\begin{aligned} &\left(\left(C_{M_u} + \frac{2}{U_0} C_{M_0} \right) + \left(C_{P_{M_u}} + \frac{2}{U_0} C_{P_{M_0}} \right) \right) u \\ &+ \left(C_{M_\alpha} + C_{M_{P_\alpha}} \right)\alpha + C_{M_{\dot\alpha}}\dot\alpha + C_{M_q} q + C_{M_{\delta_E}} \delta_E \end{aligned} \right)\\
&\quad + \delta T \frac{\left(d_T \cos\phi_T - x_T \sin\phi_T \right)}{I_{yy}}\\
\frac{n_A + n_P}{I_{zz}} &= \frac{q_\infty S_W b_W}{I_{zz}} \left(C_{N_\beta}\beta + C_{N_p} p + C_{N_r} r + C_{N_{\delta_A}} \delta_A + C_{N_{\delta_R}} \delta_R \right)
\end{aligned}} \tag{8.20}$$

These three equations may also be written in a simpler form by again introducing standard shorthand notation involving dimensional stability derivatives. So Equations (8.20), the model for the perturbation moments, become

$$\boxed{\begin{aligned}
&\left(l_A + l_P \right)/I_{xx} = L_\beta \beta + L_p p + L_r r + L_{\delta_A}\delta_A + L_{\delta_R}\delta_R\\
&\left(m_A + m_P \right)/I_{yy} = M_u u + M_{P_u} u + M_\alpha \alpha + M_{P_\alpha}\alpha + M_{\dot\alpha}\dot\alpha + M_q q + M_{\delta_E}\delta_E + M_T \delta T\\
&\left(n_A + n_P \right)/I_{zz} = N_\beta \beta + N_p p + N_r r + N_{\delta_A}\delta_A + N_{\delta_R}\delta_R
\end{aligned}} \tag{8.21}$$

The definitions for these dimensional derivatives are obvious by comparing Equations (8.20) and (8.21). Finally, Equations (8.21) may be inserted into Equations (8.3) to complete the model governing the perturbation angular rates.

However, as noted previously, the first and third of Equations (8.3) are coupled through the inertia matrix. So if instead of inserting Equations (8.21), we substituted Equations (8.19) into the decoupled equations, or Equations (8.6), the model for the perturbation rolling and yawing moments may now be expressed in terms of an alternate set of dimensional stability derivatives, or

$$\boxed{\begin{aligned} &(m_A + m_P)/I_{yy} = M_u u + M_{P_u} u + M_\alpha \alpha + M_{P_\alpha}\alpha \\ &\qquad + M_{\dot\alpha}\dot\alpha + M_q q + M_{\delta_E}\delta_E + M_T \delta T \\ &\left(\frac{1}{1 - I_{xz}^2/(I_{xx}I_{zz})}\right)\begin{bmatrix} 1 & I_{xz}/I_{xx} \\ I_{xz}/I_{zz} & 1 \end{bmatrix}\begin{Bmatrix} (l_A + l_P)/I_{xx} \\ (n_A + n_P)/I_{zz} \end{Bmatrix} \\ &\triangleq \begin{Bmatrix} L'_\beta \beta + L'_p p + L'_r r + L'_{\delta_A}\delta_A + L'_{\delta_R}\delta_R \\ N'_\beta \beta + N'_p p + N'_r r + N'_{\delta_A}\delta_A + N'_{\delta_R}\delta_R \end{Bmatrix} \end{aligned}} \tag{8.22}$$

Also, since

$$\begin{gathered} \begin{Bmatrix} L'_\beta \beta + L'_p p + L'_r r + L'_{\delta_A}\delta_A + L'_{\delta_R}\delta_R \\ N'_\beta \beta + N'_p p + N'_r r + N'_{\delta_A}\delta_A + N'_{\delta_R}\delta_R \end{Bmatrix} \\ = \left(\frac{1}{1 - I_{xz}^2/(I_{xx}I_{zz})}\right)\begin{bmatrix} 1 & I_{xz}/I_{xx} \\ I_{xz}/I_{zz} & 1 \end{bmatrix}\begin{Bmatrix} L_\beta \beta + L_p p + L_r r + L_{\delta_A}\delta_A + L_{\delta_R}\delta_R \\ N_\beta \beta + N_p p + N_r r + N_{\delta_A}\delta_A + N_{\delta_R}\delta_R \end{Bmatrix} \end{gathered} \tag{8.23}$$

almost by inspection we can see that

$$\boxed{\begin{aligned} L'_\beta &= (L_\beta + N_\beta I_{xz}/I_{xx})D \\ L'_p &= (L_p + N_p I_{xz}/I_{xx})D \\ L'_r &= (L_r + N_r I_{xz}/I_{xx})D \\ L'_{\delta_A} &= (L_{\delta_A} + N_{\delta_A} I_{xz}/I_{xx})D \\ L'_{\delta_R} &= (L_{\delta_R} + N_{\delta_R} I_{xz}/I_{xx})D \end{aligned}} \quad \text{and} \quad \boxed{\begin{aligned} N'_\beta &= (N_\beta + L_\beta I_{xz}/I_{zz})D \\ N'_p &= (N_p + L_p I_{xz}/I_{zz})D \\ N'_r &= (N_r + L_r I_{xz}/I_{zz})D \\ N'_{\delta_A} &= (N_{\delta_A} + L_{\delta_A} I_{xz}/I_{zz})D \\ N'_{\delta_R} &= (N_{\delta_R} + L_{\delta_R} I_{xz}/I_{zz})D \end{aligned}} \tag{8.24}$$

with

$$\boxed{D = \frac{1}{1 - I_{xz}^2/(I_{xx}I_{zz})}}$$

Now Equations (8.6) and (8.22) constitute the desired linear model governing perturbation angular rates.

NOTE TO STUDENT

If the numerical values for the moments and products of inertia in Equations (8.6) were developed in fuselage-referenced axes rather than stability axes, then strictly speaking they should be adjusted to the moments and products of inertia corresponding to the stability axes.

The final additions to the linear simulation model are the accelerations experienced at a particular location on the vehicle. Let this location be defined by the vector **p** relative to the vehicle's center of mass, with components in the stability axes given by (x,y,z). Then the acceleration at **p** is the inertial acceleration due to the rigid-body motion, which, consistent with the results from Example 1.2, is

$$\mathbf{a}_R(x,y,z,t) = \mathbf{a}_R(\mathbf{p},t) = \frac{d\mathbf{V}_V}{dt}\Big|_V + \left(\boldsymbol{\omega}_{V,I} \times \mathbf{V}_V\right) + \left(\boldsymbol{\omega}_{V,I} \times \left(\boldsymbol{\omega}_{V,I} \times \mathbf{p}\right)\right) + \left(\frac{d\boldsymbol{\omega}_{V,I}}{dt}\Big|_V \times \mathbf{p}\right) \quad (8.25)$$

where

$$\frac{d\mathbf{V}_V}{dt}\Big|_V = \dot{U}\mathbf{i}_V + \dot{V}\mathbf{j}_V + \dot{W}\mathbf{k}_V$$

and

$$\frac{d\boldsymbol{\omega}_{V,I}}{dt}\Big|_V = \dot{P}\mathbf{i}_V + \dot{Q}\mathbf{j}_V + \dot{R}\mathbf{k}_V$$

After expanding Equations (8.25), and using the small-perturbation assumption to linearize these results, the components of perturbation acceleration may be found. (See Section 8.2.3 and Problem 8.4.) Although in the general case the above expressions are rather complex, for a reference condition involving steady rectilinear flight with zero bank and sideslip angles, they simply become

$$\boxed{\begin{aligned} a_{X_R}(x,y,z) &= \dot{u} + \mathrm{W}_0 q + z\dot{q} - y\dot{r} \\ a_{Y_R}(x,y,z) &= \dot{v} + \mathrm{U}_0 r - \mathrm{W}_0 p + x\dot{r} - z\dot{p} \\ a_{Z_R}(x,y,z) &= \dot{w} - \mathrm{U}_0 q + y\dot{p} - x\dot{q} \end{aligned}} \quad (8.26)$$

These three components of acceleration at a particular location on the vehicle are additional responses frequently included in linear models for simulation.

8.1.3 Decoupling the Equations of Motion in Level Flight

Consider the case investigated in Example 8.1 involving a reference flight condition consisting of straight-and-level flight, with bank and sideslip angles equal to zero. Also assume that the atmospheric density is constant. We will now show that the equations of motion in this case decouple into two independent sets of equations.

We found in Example 8.1 (Equations (8.10)) that for a reference flight condition of steady, level flight the linear equations governing translation were

$$\dot{u} = -g\theta + \left(f_{A_X} + f_{P_X}\right)/m$$
$$\dot{v} = -\mathrm{U}_0 r + g\phi + \left(f_{A_Y} + f_{P_Y}\right)/m$$
$$\dot{w} = \mathrm{U}_0 q + \left(f_{A_Z} + f_{P_Z}\right)/m$$

Substituting Equations (8.16) into the above yields the translational equations of motion given by

$$\boxed{\begin{aligned}
\dot{u} &= -g\theta + X_u u + X_{P_u} u + X_\alpha \alpha + X_{\dot{\alpha}}\dot{\alpha} + X_q q + X_{\delta_E}\delta_E + X_T \delta T\\
\dot{v} &= g\phi + Y_\beta \beta + Y_p p + \left(Y_r - \mathrm{U}_0\right) r + Y_{\delta_A}\delta_A + Y_{\delta_R}\delta_R\\
\dot{w} &= Z_u u + Z_{P_u} u + Z_\alpha \alpha + Z_{\dot{\alpha}}\dot{\alpha} + \left(Z_q + \mathrm{U}_0\right) q + Z_{\delta_E}\delta_E + Z_T \delta T
\end{aligned}} \tag{8.27}$$

Also from Example 8.1 (Equations (8.11)), we found the three equations governing the rotational degrees of freedom are given by

$$\dot{q} = \left(m_A + m_p\right)/I_{yy}$$
$$\begin{Bmatrix}\dot{p}\\ \dot{r}\end{Bmatrix} = \frac{1}{I_{xx}I_{zz} - I_{xz}^2}\begin{bmatrix} I_{zz} & I_{xz}\\ I_{xz} & I_{xx}\end{bmatrix}\begin{Bmatrix}\left(l_A + l_P\right)\\ \left(n_A + n_P\right)\end{Bmatrix}$$

From Equations (8.22) and the above we obtain the following equations of motion governing the vehicle's rotation.

$$\boxed{\begin{aligned}
\dot{p} &= L'_\beta \beta + L'_p p + L'_r r + L'_{\delta_A}\delta_A + L'_{\delta_R}\delta_R\\
\dot{q} &= M_u u + M_{P_u} u + M_\alpha \alpha + M_{P_\alpha}\alpha + M_{\dot{\alpha}}\dot{\alpha} + M_q q + M_{\delta_E}\delta_E + M_T \delta T\\
\dot{r} &= N'_\beta \beta + N'_p p + N'_r r + N'_{\delta_A}\delta_A + N'_{\delta_R}\delta_R
\end{aligned}} \tag{8.28}$$

Observe that Equations (8.27), (8.28), and (8.12) may now be regrouped as follows:

$$\boxed{\begin{aligned}
\dot{u} &= -g\theta + X_u u + X_{P_u} u + X_\alpha \alpha + X_{\dot{\alpha}}\dot{\alpha} + X_q q + X_{\delta_E}\delta_E + X_T \delta T\\
\dot{w} &= Z_u u + Z_{P_u} u + Z_\alpha \alpha + Z_{\dot{\alpha}}\dot{\alpha} + \left(Z_q + \mathrm{U}_0\right) q + Z_{\delta_E}\delta_E + Z_T \delta T\\
\dot{q} &= M_u u + M_{P_u} u + M_\alpha \alpha + M_{P_\alpha}\alpha + M_{\dot{\alpha}}\dot{\alpha} + M_q q + M_{\delta_E}\delta_E + M_T \delta T\\
\dot{\theta} &= q
\end{aligned}} \tag{8.29}$$

and

$$\boxed{\begin{aligned}\dot{v} &= g\phi + Y_\beta\beta + Y_p p + (Y_r - U_0)r + Y_{\delta_A}\delta_A + Y_{\delta_R}\delta_R\\ \dot{p} &= L'_\beta\beta + L'_p p + L'_r r + L'_{\delta_A}\delta_A + L'_{\delta_R}\delta_R\\ \dot{r} &= N'_\beta\beta + N'_p p + N'_r r + N'_{\delta_A}\delta_A + N'_{\delta_R}\delta_R\\ \dot{\phi} &= p\end{aligned}} \tag{8.30}$$

Now noting that $\alpha = w/U_0$ and that $\beta = v/U_0$, observe that these two latter sets are decoupled. The first set, Equations (8.29), are referred to as the linear *longitudinal equations of motion,* while the second set, Equations (8.30), are referred to as the linear *lateral-directional equations of motion.* Note, however, that the decoupling is valid only for zero reference bank and sideslip angles!

But Equations (8.9) do not generally allow a grouping into the longitudinal and lateral-directional sets. The grouping may work nicely for certain reference flight conditions, but not at all for other flight conditions. This is usually not a problem because careful examination of Equations (8.29) and (8.30) reveals that none of the right-hand sides of the equations depend on the inertial positions x_E and y_E (or ψ), and so the equations governing these variables are frequently regarded separately. The equation for altitude rate $\dot{h}$, however, is frequently retained and grouped with the longitudinal set. (The equation for $\dot{\psi}$, if needed, is grouped with the lateral-directional set.)

8.1.4 Decoupled Models in State-Variable Format

Assuming the same reference flight condition considered in Example 8.1 (including constant atmospheric density), we will now assemble the models for the vehicle's longitudinal and lateral-directional dynamics in state-variable format. That is, we will write the models for the dynamics in the form

$$\dot{\mathbf{x}} = \mathbf{A}\mathbf{x} + \mathbf{B}\mathbf{u}$$

$$\mathbf{y} = \mathbf{C}\mathbf{x} + \mathbf{D}\mathbf{u}$$

where $\mathbf{x}$ is a selected state vector, $\mathbf{u}$ is the vector of control-input variables, and $\mathbf{y}$ is the vector of response variables, and we will identify the matrices $\mathbf{A}$, $\mathbf{B}$, $\mathbf{C}$, and $\mathbf{D}$.

First, for the longitudinal set, the first three equations of motion are as given in Equations (8.29), or

$$\begin{aligned}\dot{u} &= -g\theta + X_u u + X_{P_u} u + X_\alpha\alpha + X_{\dot{\alpha}}\dot{\alpha} + X_q q + X_{\delta_E}\delta_E + X_T\delta T\\ \dot{w} &= Z_u u + Z_{P_u} u + Z_\alpha\alpha + Z_{\dot{\alpha}}\dot{\alpha} + (Z_q + U_0)q + Z_{\delta_E}\delta_E + Z_T\delta T\\ \dot{q} &= M_u u + M_{P_u} u + M_\alpha\alpha + M_{P_\alpha}\alpha + M_{\dot{\alpha}}\dot{\alpha} + M_q q + M_{\delta_E}\delta_E + M_T\delta T\end{aligned}$$

To obtain the desired results we must first eliminate $\dot{\alpha}$ from the right-hand sides of Equations (8.29). Dividing the second of the equations by U_0 will lead to the equation governing the rate of change of angle of attack, or $\dot{\alpha}$. Or, since

$$\dot{\alpha} = \frac{\dot{w}}{U_0} = \frac{1}{U_0}\left(Z_u u + Z_{P_u} u + Z_\alpha \alpha + Z_{\dot{\alpha}}\dot{\alpha} + \left(Z_q + U_0\right)q + Z_{\delta_E}\delta_E + Z_T \delta T\right)$$

we have

$$\dot{\alpha} = \left(\frac{1}{U_0 - Z_{\dot{\alpha}}}\right)\left(Z_u u + Z_{P_u} u + Z_\alpha \alpha + \left(Z_q + U_0\right)q + Z_{\delta_E}\delta_E + Z_T \delta T\right) \quad (8.31)$$

Now substituting Equation (8.31) for $\dot{\alpha}$ into the first and third of Equations (8.29) yields

$$\begin{aligned}
\dot{u} &= \left(X_u + X_{P_u} + \frac{X_{\dot{\alpha}}\left(Z_u + Z_{P_u}\right)}{U_0 - Z_{\dot{\alpha}}}\right)u + \left(X_\alpha + \frac{X_{\dot{\alpha}} Z_\alpha}{U_0 - Z_{\dot{\alpha}}}\right)\alpha - g\theta \\
&\quad + \left(X_q + X_{\dot{\alpha}}\left(\frac{U_0 + Z_q}{U_0 - Z_{\dot{\alpha}}}\right)\right)q + \left(X_{\delta_E} + \frac{X_{\dot{\alpha}} Z_{\delta_E}}{U_0 - Z_{\dot{\alpha}}}\right)\delta_E + \left(X_T + \frac{X_{\dot{\alpha}} Z_T}{U_0 - Z_{\dot{\alpha}}}\right)\delta T \\
\dot{\alpha} &= \left(\frac{1}{U_0 - Z_{\dot{\alpha}}}\right)\left(Z_u u + Z_{P_u} u + Z_\alpha \alpha + \left(Z_q + U_0\right)q + Z_{\delta_E}\delta_E + Z_T \delta T\right) \\
\dot{q} &= \left(M_u + M_{P_u} + \frac{M_{\dot{\alpha}}\left(Z_u + Z_{P_u}\right)}{U_0 - Z_{\dot{\alpha}}}\right)u + \left(M_\alpha + M_{P_\alpha} + \frac{M_{\dot{\alpha}} Z_\alpha}{U_0 - Z_{\dot{\alpha}}}\right)\alpha \\
&\quad + \left(M_q + M_{\dot{\alpha}}\left(\frac{U_0 + Z_q}{U_0 - Z_{\dot{\alpha}}}\right)\right)q + \left(M_{\delta_E} + \frac{M_{\dot{\alpha}} Z_{\delta_E}}{U_0 - Z_{\dot{\alpha}}}\right)\delta_E + \left(M_T + \frac{M_{\dot{\alpha}} Z_T}{U_0 - Z_{\dot{\alpha}}}\right)\delta T
\end{aligned} \quad (8.32)$$

(Note that the above equations simplify significantly when $X_{\dot{\alpha}} = Z_{\dot{\alpha}} = 0$, which is frequently the case.) Also, from Equations (8.29) and (8.13) we have the two kinematic equations

$$\dot{\theta} = q$$

and

$$\dot{h} = -w + U_0\theta \; \left(\text{with } w = U_0\alpha\right). \quad (8.33)$$

So the longitudinal model consists of these five differential equations.

The five responses of interest corresponding to this model usually consist of perturbation surge velocity u, angle of attack α, pitch-attitude angle θ, pitch rate q, and altitude h. And the typical control-input vector includes perturbation elevator deflection δ_E and thrust δT. Therefore we take the response vector **y** and the control input vector **u** in the state-variable model to be

$$\mathbf{y} = \begin{Bmatrix} u \\ \alpha \\ \theta \\ q \\ h \end{Bmatrix} \text{ and } \mathbf{u} = \begin{Bmatrix} \delta_E \\ \delta T \end{Bmatrix} \tag{8.34}$$

Loosely speaking, the state variables are the variables governed by the equations of motion, but mathematically the selection of the state vector **x** is not unique in that several selections are possible. Under the assumption of constant atmospheric density, the equations do not depend on altitude, but if we want altitude to be one of the responses we have to include it in the state vector. So we may simply choose the state vector equal to the response vector, or

$$\mathbf{x} = \mathbf{y} = \begin{Bmatrix} u \\ \alpha \\ \theta \\ q \\ h \end{Bmatrix} \tag{8.35}$$

Under this selection of state, response, and input vectors, the four matrices in the state-variable model for the longitudinal dynamics become

$$\mathbf{A} = \begin{bmatrix} \left(X_u + X_{P_u} + \dfrac{X_{\dot\alpha}(Z_u + Z_{P_u})}{U_0 - Z_{\dot\alpha}}\right) & \left(X_\alpha + \dfrac{X_{\dot\alpha} Z_\alpha}{U_0 - Z_{\dot\alpha}}\right) & -g & \left(X_q + X_{\dot\alpha}\left(\dfrac{U_0 + Z_q}{U_0 - Z_{\dot\alpha}}\right)\right) & 0 \\ \left(\dfrac{Z_u + Z_{P_u}}{U_0 - Z_{\dot\alpha}}\right) & \left(\dfrac{Z_\alpha}{U_0 - Z_{\dot\alpha}}\right) & 0 & \left(\dfrac{U_0 + Z_q}{U_0 - Z_{\dot\alpha}}\right) & 0 \\ 0 & 0 & 0 & 1 & 0 \\ \left(M_u + M_{P_u} + \dfrac{M_{\dot\alpha}(Z_u + Z_{P_u})}{U_0 - Z_{\dot\alpha}}\right) & \left(M_\alpha + M_{P_\alpha} + \dfrac{M_{\dot\alpha} Z_\alpha}{U_0 - Z_{\dot\alpha}}\right) & 0 & \left(M_q + M_{\dot\alpha}\left(\dfrac{U_0 + Z_q}{U_0 - Z_{\dot\alpha}}\right)\right) & 0 \\ 0 & -U_0 & U_0 & 0 & 0 \end{bmatrix}$$

$$\mathbf{B} = \begin{bmatrix} \left(X_{\delta_E} + \dfrac{X_{\dot\alpha} Z_{\delta_E}}{U_0 - Z_{\dot\alpha}}\right) & \left(X_T + \dfrac{X_{\dot\alpha} Z_T}{U_0 - Z_{\dot\alpha}}\right) \\ \left(\dfrac{Z_{\delta_E}}{U_0 - Z_{\dot\alpha}}\right) & \left(\dfrac{Z_T}{U_0 - Z_{\dot\alpha}}\right) \\ 0 & 0 \\ \left(M_{\delta_E} + \dfrac{M_{\dot\alpha} Z_{\delta_E}}{U_0 - Z_{\dot\alpha}}\right) & \left(M_T + \dfrac{M_{\dot\alpha} Z_T}{U_0 - Z_{\dot\alpha}}\right) \\ 0 & 0 \end{bmatrix}, \mathbf{C} = \begin{bmatrix} 1 & 0 & 0 & 0 & 0 \\ 0 & 1 & 0 & 0 & 0 \\ 0 & 0 & 1 & 0 & 0 \\ 0 & 0 & 0 & 1 & 0 \\ 0 & 0 & 0 & 0 & 1 \end{bmatrix}, \mathbf{D} = \begin{bmatrix} 0 & 0 \\ 0 & 0 \\ 0 & 0 \\ 0 & 0 \\ 0 & 0 \end{bmatrix} \tag{8.36}$$

Note that the last column in the matrix **A** consists only of zeros. This corresponds to the fact that we assumed a constant atmospheric density, and hence none of the equations of motion depend on altitude. If we chose not to include altitude h in the response vector **y**, we could choose the state vector to be just the first four responses in Equation (8.35). In this case the **A** and **C** matrices could be reduced to 4×4 matrices, and the **B** and **D** matrices could be reduced to 4×2 matrices.

EXAMPLE 8.2

State-Variable Model of the Navion's Longitudinal Dynamics

Consider the Navion general-aviation aircraft shown in Figure 8.1. Using the data for this aircraft from Appendix B, assemble the state-variable model for the vehicle's perturbation longitudinal dynamics. Include altitude h in the vector of model responses. (Simulation of these dynamics will be addressed in additional examples that follow.)

■ Solution

Since altitude is to be included in the model's responses, the form of the state-variable model is that given in Equations (8.34–8.36) (with $w = U_0\alpha$). The data in Appendix B for the Navion is defined for vehicle-fixed stability axes corresponding to the following reference flight condition.

$$U_0 = 176 \text{ fps}, \Theta_0 = 0, h_0 = 0 \text{ ft}, \alpha_0 = 0.6 \text{ deg, and } \Phi_0 = 0 \text{ (implied)}$$

Also from Appendix B, the longitudinal dimensional derivatives for this flight condition are given in Table 8.2. The derivative X_T (= 1/m) was calculated based on the data provided. All angles are in radians, forces in pounds, and velocities in feet per second.

Now note from the table that $X_{\dot{\alpha}} = Z_{\dot{\alpha}} = 0$, so many of the elements of the matrices in Equations (8.36) may be considerably simplified. Inserting the data from the table into these equations yields the desired matrices

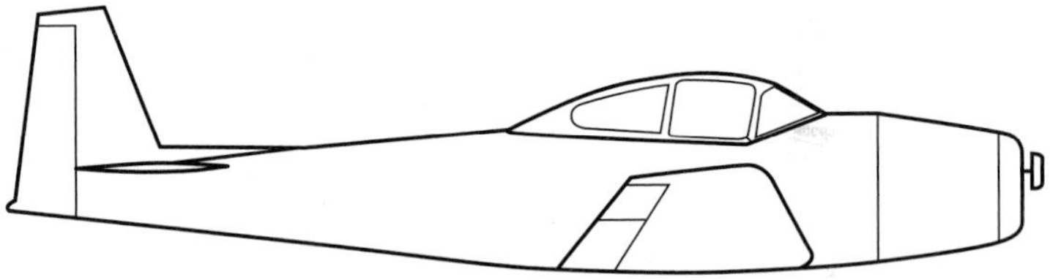

Figure 8.1 Sketch of Navion aircraft.

Table 8.2 Longitudinal Dimensional Derivatives for the Navion

Stability Derivative	u	α	$\dot{\alpha}$	q	δ_E	δT
$X.$	−0.0451	6.348	0	0	0	0.0117
$Z.$	−0.3697	−356.29	0	0	−28.17	0
$M.$	0	−8.80	−0.9090	−2.0767	−11.189	0

$$\mathbf{A} = \begin{bmatrix} -0.0451 & 6.348 & -32.2 & 0 & 0 \\ -0.0021 & -2.0244 & 0 & 1 & 0 \\ 0 & 0 & 0 & 1 & 0 \\ 0.0021 & -6.958 & 0 & -3.0757 & 0 \\ 0 & -176 & 176 & 0 & 0 \end{bmatrix}, \mathbf{B} = \begin{bmatrix} 0 & 0.0117 \\ -0.160 & 0 \\ 0 & 0 \\ -11.029 & 0 \\ 0 & 0 \end{bmatrix}$$

$$\mathbf{C} = \begin{bmatrix} 1 & 0 & 0 & 0 & 0 \\ 0 & 1 & 0 & 0 & 0 \\ 0 & 0 & 1 & 0 & 0 \\ 0 & 0 & 0 & 1 & 0 \\ 0 & 0 & 0 & 0 & 1 \end{bmatrix}, \mathbf{D} = \begin{bmatrix} 0 & 0 \\ 0 & 0 \\ 0 & 0 \\ 0 & 0 \\ 0 & 0 \end{bmatrix} \tag{8.37}$$

And the state, response, and control-input vectors for the model are as given in Equations (8.34) and (8.35).

Turning now to the lateral-directional set of equations, Equations (8.30) give the equations of motion when atmospheric density is assumed constant. The first three of these equations are

$$\begin{aligned} \dot{v} &= g\phi + Y_\beta \beta + Y_p p + (Y_r - \mathrm{U}_0) r + Y_{\delta_A}\delta_A + Y_{\delta_R}\delta_R \\ \dot{p} &= L'_\beta \beta + L'_p p + L'_r r + L'_{\delta_A}\delta_A + L'_{\delta_R}\delta_R \\ \dot{r} &= N'_\beta \beta + N'_p p + N'_r r + N'_{\delta_A}\delta_A + N'_{\delta_R}\delta_R \end{aligned}$$

Or, in terms of the sideslip angle, the first of the above equations may also be written as

$$\dot{\beta} = \frac{g}{\mathrm{U}_0}\phi + \frac{Y_\beta}{\mathrm{U}_0}\beta + \frac{Y_p}{\mathrm{U}_0}p + \left(\frac{Y_r}{\mathrm{U}_0} - 1\right) r + \frac{Y_{\delta_A}}{\mathrm{U}_0}\delta_A + \frac{Y_{\delta_R}}{\mathrm{U}_0}\delta_R \tag{8.38}$$

Adding the kinematic equations $\dot{\phi} = p$ and $\dot{\psi} = r$ yields the five equations making up the linear lateral-directional model.

The five responses for this model are perturbation sideslip angle β, bank angle ϕ, roll rate p, yaw rate r, and heading angle ψ, while the control inputs are perturbation aileron deflection δ_A and rudder deflection δ_R. Therefore we can take the response vector $\mathbf{y}$ and input vector $\mathbf{u}$ to be

$$\mathbf{y} = \begin{Bmatrix} \beta \\ \phi \\ p \\ r \\ \psi \end{Bmatrix} \text{ and } \mathbf{u} = \begin{Bmatrix} \delta_A \\ \delta_R \end{Bmatrix} \tag{8.39}$$

Again, the state vector is not unique, but we typically select this vector to be the same as the response vector, or

$$\mathbf{x} = \mathbf{y} = \begin{Bmatrix} \beta \\ \phi \\ p \\ r \\ \psi \end{Bmatrix} \tag{8.40}$$

With these selections for state, response, and input vectors, the four matrices completing the model for the linear lateral-directional dynamics are

$$\mathbf{A} = \begin{bmatrix} \dfrac{Y_\beta}{U_0} & \dfrac{g}{U_0} & \dfrac{Y_p}{U_0} & \left(\dfrac{Y_r}{U_0} - 1\right) & 0 \\ 0 & 0 & 1 & 0 & 0 \\ L'_\beta & 0 & L'_p & L'_r & 0 \\ N'_\beta & 0 & N'_p & N'_r & 0 \\ 0 & 0 & 0 & 1 & 0 \end{bmatrix}$$

$$\mathbf{B} = \begin{bmatrix} \dfrac{Y_{\delta_A}}{U_0} & \dfrac{Y_{\delta_R}}{U_0} \\ 0 & 0 \\ L'_{\delta_A} & L'_{\delta_R} \\ N'_{\delta_A} & N'_{\delta_R} \\ 0 & 0 \end{bmatrix}, \mathbf{C} = \begin{bmatrix} 1 & 0 & 0 & 0 & 0 \\ 0 & 1 & 0 & 0 & 0 \\ 0 & 0 & 1 & 0 & 0 \\ 0 & 0 & 0 & 1 & 0 \\ 0 & 0 & 0 & 0 & 1 \end{bmatrix}, \mathbf{D} = \begin{bmatrix} 0 & 0 \\ 0 & 0 \\ 0 & 0 \\ 0 & 0 \\ 0 & 0 \end{bmatrix} \tag{8.41}$$

Again note that the last column of the matrix **A** consists entirely of zeros. This corresponds to the fact that none of the equations of motion depend on heading angle ψ. Similar to the case for the longitudinal dynamics, if we chose to not include ψ in the model's response vector, then it need not be included in the state vector. In this case the **A** and **C** matrices could be reduced to 4×4 matrices, and the **B** and **D** matrices could be reduced to 4×2 matrices.

Finally, as noted previously, the equations decouple only for reference flight conditions involving zero reference bank and sideslip angles. Therefore if other flight conditions are being investigated, the complete coupled set of equations should be used. That coupled set may also be expressed in state-variable form, although it will obviously be of larger dimension.

8.1.5 Linear Models for Flexible Vehicles

When developing a simulation of a flexible vehicle, we must include the equations of motion governing the elastic degrees of freedom in the mathematical model of the vehicle dynamics. These equations were developed in Chapter 4.

The equations governing rigid-body translations and rotations remain unchanged except for necessary adjustments in the forces and moments acting on the vehicle. Models for these forces and moments were developed in Chapter 7.

NOTE TO STUDENT

Since we have assured in the derivations of the equations of motion that the selected vehicle-fixed axes satisfy the mean-axis constraints discussed in Chapters 3 and 4, the model includes a very nice feature: the equations governing the rigid-body degrees of freedom are identical to those used for a rigid vehicle. That is, for a flexible vehicle we also use Equations (8.1), (8.6), (8.7), and (8.9), all of which remain unchanged in form.

Consistent with Equations (4.84), the equations governing the perturbation dynamics of the elastic degrees of freedom are

$$\ddot{\eta}_i + \omega_i^2 \eta_i = \frac{q_i}{\mathcal{M}_i}, \; i = 1, \ldots, n \tag{8.42}$$

Recall here that η_i is the perturbation modal coordinate associated with the i'th vibration mode included in the model, q_i is the perturbation generalized force associated with this degree of freedom, ω_i is the in-vacuo vibration frequency of the vibration mode, and $\mathcal{M}_i$ is the modal generalized mass. The latter two quantities are assumed available from a prior free-vibration analysis of the vehicle's structure.

Note that frequently a small modal damping ζ_i of approximately 0.02 is added to each of Equations (8.42), to improve agreement between the analytical and experimental results from the vibration analysis. With damping included, Equations (8.42) become

$$\boxed{\ddot{\eta}_i + 2\zeta_i \omega_i \dot{\eta}_i + \omega_i^2 \eta_i = \frac{q_i}{\mathcal{M}_i}, \; i = 1, \ldots, n} \tag{8.43}$$

The forces and moments acting on a flexible vehicle were modeled in Chapter 7, and since vibration mode shapes are defined in terms of <u>fuselage-reference axes</u>, those axes are usually used as the vehicle-fixed axes in the development of the models for the forces and moments. But the components of the forces acting on the rigid vehicle, given in Equations (8.14), were developed assuming <u>stability axes</u>. So if we are to use fuselage-referenced axes, we should apply Equations (7.114) and obtain the forces and moments acting on the rigid-body degrees of freedom in these axes, or

$$\begin{aligned}
\left(f_{A_{X_{R_F}}} + f_{P_{X_{R_F}}}\right) &= \left(f_{A_{X_R}} + f_{P_{X_R}}\right)\cos\alpha_0 - \left(f_{A_{Z_R}} + f_{P_{Z_R}}\right)\sin\alpha_0 \\
\left(f_{A_{Y_{R_F}}} + f_{P_{Y_{R_F}}}\right) &= \left(f_{A_{Y_R}} + f_{P_{Y_R}}\right) \\
\left(f_{A_{Z_{R_F}}} + f_{P_{Z_{R_F}}}\right) &= \left(f_{A_{X_R}} + f_{P_{X_R}}\right)\sin\alpha_0 + \left(f_{A_{Z_R}} + f_{P_{Z_R}}\right)\cos\alpha_0
\end{aligned} \tag{8.44}$$

Finally, note that frequently the angle of attack at the reference flight condition α_0 is sufficiently small such that the difference between the force components in the stability and fuselage-referenced axes is small.

Regarding the elastic effects on the forces, Equations (7.110), or

$$f_{A_{X_E}} = q_\infty S_W \sum_{i=1}^{n} \left(C_{X\eta_i}\eta_i + C_{X\dot{\eta}_i}\dot{\eta}_i\right)$$

$$f_{A_{Y_E}} = q_\infty S_W \sum_{i=1}^{n} \left(C_{Y\eta_i}\eta_i + C_{Y\dot{\eta}_i}\dot{\eta}_i\right)$$

$$f_{A_{Z_E}} = q_\infty S_W \sum_{i=1}^{n} \left(C_{Z\eta_i}\eta_i + C_{Z\dot{\eta}_i}\dot{\eta}_i\right)$$

along with Table 7.4 model the elastic effects on the three aerodynamic force components in Equations (7.104). So the force components to be incorporated into the linear equations of motion governing rigid-body translations (Equations (8.1)) are the sums of the "rigid-body" force components plus the elastic contribution to these force components. Or, the terms to be incorporated into Equations (8.1) are

$$\begin{aligned}
\left(f_{A_X} + f_{P_X}\right) &= \left(f_{A_{X_{R_F}}} + f_{P_{X_{R_F}}}\right) + f_{A_{X_E}} \\
\left(f_{A_Y} + f_{P_Y}\right) &= \left(f_{A_{Y_{R_F}}} + f_{P_{Y_{R_F}}}\right) + f_{A_{Y_E}} \\
\left(f_{A_Z} + f_{P_Z}\right) &= \left(f_{A_{Z_{R_F}}} + f_{P_{Z_{R_F}}}\right) + f_{A_{Z_E}}
\end{aligned} \tag{8.45}$$

Therefore, in terms of dimensional stability derivatives, and assuming a constant atmospheric density, Equations (8.16) now become

$$\begin{aligned}
\left(f_{A_X} + f_{P_X}\right)/m &= X_{u_F}u + X_{P_{u_F}}u + X_{\alpha_F}\alpha + X_{\dot{\alpha}_F}\dot{\alpha} + X_{q_F}q + X_{\delta_{E_F}}\delta_E + X_{T_F}\delta T \\
&\quad + \sum_{i=1}^{n}\left(X_{\eta_i}\eta_i + X_{\dot{\eta}_i}\dot{\eta}_i\right) \\
\left(f_{A_Y} + f_{P_Y}\right)/m &= Y_{\beta_F}\beta + Y_{p_F}p + Y_{r_F}r + Y_{\delta_{A_F}}\delta_A + Y_{\delta_{R_F}}\delta_R + \sum_{i=1}^{n}\left(Y_{\eta_i}\eta_i + Y_{\dot{\eta}_i}\dot{\eta}_i\right) \\
\left(f_{A_Z} + f_{P_Z}\right)/m &= Z_{u_F}u + Z_{P_{u_F}}u + Z_{\alpha_F}\alpha + Z_{\dot{\alpha}_F}\dot{\alpha} + Z_{q_F}q + Z_{\delta_{E_F}}\delta_E + Z_{T_F}\delta T \\
&\quad + \sum_{i=1}^{n}\left(Z_{\eta_i}\eta_i + Z_{\dot{\eta}_i}\dot{\eta}_i\right)
\end{aligned} \tag{8.46}$$

where the subscript F has been added to remind us that the dimensional stability derivatives must be developed using fuselage-reference axes. (See Appendix B.) In addition, we have the dimensional stability derivatives associated with the vibration-modal coordinates given by

$$
\begin{aligned}
X_{\eta_i} &\triangleq \frac{q_\infty S_W}{m} C_{X_{\eta_i}} & X_{\dot{\eta}_i} &\triangleq \frac{q_\infty S_W}{m} C_{X_{\dot{\eta}_i}} \\
Y_{\eta_i} &\triangleq \frac{q_\infty S_W}{m} C_{Y_{\eta_i}} & Y_{\dot{\eta}_i} &\triangleq \frac{q_\infty S_W}{m} C_{Y_{\dot{\eta}_i}} \\
Z_{\eta_i} &\triangleq \frac{q_\infty S_W}{m} C_{Z_{\eta_i}} & Z_{\dot{\eta}_i} &\triangleq \frac{q_\infty S_W}{m} C_{Z_{\dot{\eta}_i}}
\end{aligned}
\tag{8.47}
$$

with the coefficients $C_{X\bullet}$, $C_{Y\bullet}$, and $C_{Z\bullet}$ all defined in Table 7.4. Equations (8.46) may now be incorporated into Equations (8.1) to complete the equations governing perturbations in rigid-body translations.

With regard to the components of the moments affecting the rigid-body rotations, we must again use components in the selected fuselage-reference axes. That is, if the components of the moments in Equations (8.19) were developed in stability axes, then Equations (7.118) should be applied to determine the components of moments in the fuselage-reference axes. Or

$$
\begin{aligned}
l_{A_{R_F}} &= l_{A_R}\cos\alpha_0 - n_{A_R}\sin\alpha_0 \\
m_{A_{R_F}} &= m_{A_R} \\
n_{A_{R_F}} &= l_{A_R}\sin\alpha_0 + n_{A_R}\cos\alpha_0
\end{aligned}
\tag{8.48}
$$

Plus Equations (7.117), the elastic effects on these moments, must be added to the above results, where Equations (7.117) are

$$
\begin{aligned}
l_{A_E} &= \sum_{i=1}^{n}\left(\frac{\partial L_A}{\partial \eta_i}\Big|_0 \eta_i + \frac{\partial L_A}{\partial \dot{\eta}_i}\Big|_0 \dot{\eta}_i\right) = q_\infty S_W b_W \sum_{i=1}^{n}\left(C_{\text{Lroll}_{\eta_i}}\eta_i + C_{\text{Lroll}_{\dot{\eta}_i}}\dot{\eta}_i\right) \\
m_{A_E} &= \sum_{i=1}^{n}\left(\frac{\partial M_A}{\partial \eta_i}\Big|_0 \eta_i + \frac{\partial M_A}{\partial \dot{\eta}_i}\Big|_0 \dot{\eta}_i\right) = q_\infty S_W \bar{c}_W \sum_{i=1}^{n}\left(C_{M_{\eta_i}}\eta_i + C_{M_{\dot{\eta}_i}}\dot{\eta}_i\right) \\
n_{A_E} &= \sum_{i=1}^{n}\left(\frac{\partial N_A}{\partial \eta_i}\Big|_0 \eta_i + \frac{\partial N_A}{\partial \dot{\eta}_i}\Big|_0 \dot{\eta}_i\right) = q_\infty S_W b_W \sum_{i=1}^{n}\left(C_{N_{\eta_i}}\eta_i + C_{N_{\dot{\eta}_i}}\dot{\eta}_i\right)
\end{aligned}
$$

Consequently, after dividing the moments by the appropriate moment of inertia, Equations (8.21) now become

$$
\begin{aligned}
(l_A + l_P)/I_{xx} &= L_{\beta_F}\beta + L_{p_F}p + L_{r_F}r + L_{\delta_{A_F}}\delta_A + L_{\delta_{R_F}}\delta_R + \sum_{i=1}^{n}\left(L_{A_{\eta_i}}\eta_i + L_{A_{\dot{\eta}_i}}\dot{\eta}_i\right) \\
(m_A + m_P)/I_{yy} &= M_{u_F}u + M_{P_{u_F}}u + M_{\alpha_F}\alpha + M_{P_{\alpha_F}}\alpha + M_{\dot{\alpha}_F}\dot{\alpha} + M_{q_F}q + M_{\delta_{E_F}}\delta_E + M_{T_F}\delta T \\
&\quad + \sum_{i=1}^{n}\left(M_{A_{\eta_i}}\eta_i + M_{A_{\dot{\eta}_i}}\dot{\eta}_i\right) \\
(n_A + n_P)/I_{zz} &= N_{\beta_F}\beta + N_{p_F}p + N_{r_F}r + N_{\delta_{A_F}}\delta_A + N_{\delta_{R_F}}\delta_R + \sum_{i=1}^{n}\left(N_{A_{\eta_i}}\eta_i + N_{A_{\dot{\eta}_i}}\dot{\eta}_i\right)
\end{aligned}
\tag{8.49}
$$

Again note the subscript F has been added, to remind us to use dimensional derivatives developed in the fuselage-referenced axes. (See Appendix B.) The three expressions in Equations (8.49) may now be incorporated into Equations (8.6) to complete the equations governing perturbations in rigid-body angular rates.

The final set of equations to be added to complete the dynamic model is given in Equations (8.43). And from Equations (7.120) we have the expressions for the perturbations in the generalized forces given by

$$q_i \triangleq \frac{\partial Q_i}{\partial \delta \mathbf{p}}\Big|_0 \delta \mathbf{p} = q_{i_R} + q_{i_E}$$

where

$$q_{i_R} = \frac{\partial Q_i}{\partial \delta \mathbf{p}_{\text{Rigid}}}\Big|_0 \delta \mathbf{p}_{\text{Rigid}}$$

and

$$q_{i_E} = q_\infty S_W \bar{c}_W \sum_{j=1}^{n} \left(C_{Q_{i_{\eta_j}}} \eta_j + C_{Q_{i_{\dot{\eta}_j}}} \dot{\eta}_j \right)$$

In addition, we have the rigid-body perturbation vector defined as

$$\delta \mathbf{p}^T_{\text{Rigid}} = \left[u \quad \beta\left(\approx \frac{v}{U_0}\right) \quad \alpha\left(\approx \frac{w}{U_0}\right) \quad \dot{\alpha} \quad p \quad q \quad r \quad \delta_E \quad \delta_A \quad \delta_R \right],$$

the partial derivatives $\dfrac{\partial Q_i}{\partial \delta \mathbf{p}_{\text{Rigid}}}\Big|_0$ given in Equations (7.94), and the coefficients $C_{Q_{i_{\eta_j}}}$ and $C_{Q_{i_{\dot{\eta}_j}}}$ given in Equations (7.95). In terms of dimensional stability derivatives, we may now write the right-hand sides of Equations (8.43) as

$$\boxed{\begin{aligned} \frac{q_i}{\mathcal{M}_i} &= \Xi_{i_u} u + \Xi_{i_\beta} \beta + \Xi_{i_\alpha} \alpha + \Xi_{i_{\dot{\alpha}}} \dot{\alpha} + \Xi_{i_p} p + \Xi_{i_q} q + \Xi_{i_r} r \\ &\quad + \Xi_{i_{\delta_E}} \delta_E + \Xi_{i_{\delta_A}} \delta_A + \Xi_{i_{\delta_R}} \delta_R + \sum_{j=1}^{n} \left(\Xi_{i_{\eta_j}} \eta_j + \Xi_{i_{\dot{\eta}_j}} \dot{\eta}_j \right) \end{aligned}} \tag{8.50}$$

The dimensional stability derivatives $\left(\Xi_{i.}\right)$ in the above expression are defined as

$$\boxed{\Xi_{i_\bullet} = C_{Q_{i_\bullet}} q_\infty S_W \bar{c}_W / \mathcal{M}_i, \ \bullet = u, \beta, \alpha, \dot{\alpha}, p, q, r, \delta_E, \delta_A, \delta_R, \eta_j, \text{ or } \dot{\eta}_j} \tag{8.51}$$

where the effectiveness coefficients $C_{Q_{i_\bullet}}$ are all discussed in Chapter 7.

The derivative associated with propulsive thrust, or Ξ_{i_T}, has been assumed to be negligible in Equation (8.50). This assumption is justified on the following basis. First, the mode-shape components of displacement in the direction of the thrust (e.g., ν_{X_i}) are usually very small, so the virtual work associated with deforming the structure performed by the thrust will be small. Second, engine response is much slower than the responses of aerodynamic surfaces, so it is difficult to excite higher-frequency vibration modes with thrust.

If atmospheric density is not assumed constant, an additional altitude-dependent term must be added to each of Equations (8.50). Each additional term will be of the form

$$\boxed{\Xi_{i_h} = C_{Q_{i_0}} \frac{q_\infty S_w \bar{c}_W}{\mathcal{M}_i \rho_\infty} \frac{\partial \rho_\infty}{\partial h}} \tag{8.52}$$

where $C_{Q_{i_0}}$ is the value of the i'th aeroelastic coefficient C_{Q_i} evaluated at the reference flight condition, and the density-altitude gradient $\partial\rho_\infty/\partial h$ is obtained from an atmospheric model such as that given in Appendix A.

After incorporating Equations (8.50) into Equations (8.43), the linear model of the dynamics of the flexible vehicle is now almost complete. The final addition to be made is the incorporation of the effects of elastic deformation into the dynamic responses of the vehicle (i.e., the response vector $\mathbf{y}$).

For example, the responses of a rigid vehicle included the (rigid-body) pitch rate q. And for a rigid vehicle this pitch rate would be the response sensed by a rate gyro located anywhere along the centerline of the fuselage of the vehicle. But for a flexible vehicle, the rate gyro would sense the total local pitch rate, which would include the effects of local elastic deformation (e.g., bending) at that particular location on the fuselage. This local effect due to elastic deformation is depicted in Figure 8.2, in terms of the local slope of the Z component of the i'th mode shape $\nu'_{Z_i}(x)$, where

$$\nu'_{Z_i}(x) \triangleq \frac{\partial \nu_{Z_i}(x)}{\partial x}$$

Or the figure depicts the effect of the i'th vibration mode on the local pitch attitude. Consequently, the local pitch rate $q_{\text{local}}(x,t)$ at location x along the fuselage may be expressed as

$$q_{\text{local}}(x,t) = q(t) + \sum_{i=1}^{n} \nu'_{Z_i}(x)\dot{\eta}_i(t) \tag{8.53}$$

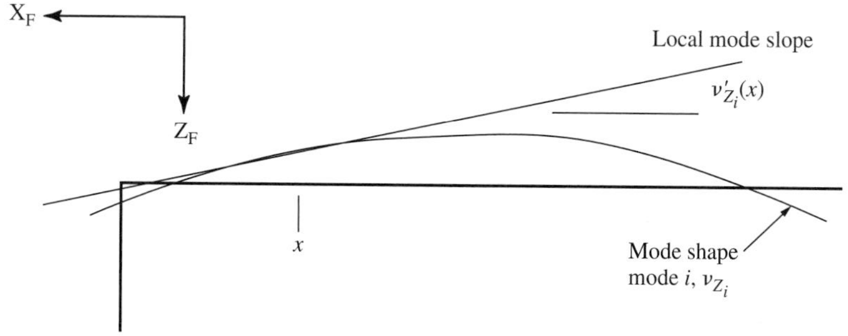

Figure 8.2 Effect of vibration mode on local pitch attitude.

or the rigid-body pitch rate plus the sum of the contributions of the elastic modes. Likewise, the local yaw rate, or $r_{\text{local}}(x,t)$, may be expressed as

$$r_{\text{local}}(x,t) = r(t) + \sum_{i=1}^{n} \nu'_{Y_i}(x)\eta_i(t) \tag{8.54}$$

In addition, the local accelerations will be affected by the elastic deformations. Let the location of interest be located at coordinates (x,y,z) on the undeformed vehicle, with this location relative to the vehicle's center of mass given as $\mathbf{p}$. Denoting the inertial acceleration (vector) at this location due to the rigid-body degrees of freedom as $\mathbf{a}_R(x,y,z,t)$, and under some mild assumptions, the total (not perturbation) local acceleration vector may be expressed as

$$\mathbf{a}_{\text{local}}(x,y,z,t) = \mathbf{a}_R(x,y,z,t) + \sum_{i=1}^{n} \boldsymbol{\nu}_i(x,y,z)\ddot{\eta}_i(t) \tag{8.55}$$

where the rigid-body acceleration vector at the sensor location $\mathbf{p}$ was given by Equation (8.25), or

$$\mathbf{a}_R(x,y,z,t) = \mathbf{a}_R(\mathbf{p},t) = \frac{d\mathbf{V}_V}{dt}\Big|_V + (\boldsymbol{\omega}_{V,I} \times \mathbf{V}_V) + (\boldsymbol{\omega}_{V,I} \times (\boldsymbol{\omega}_{V,I} \times \mathbf{p})) + \left(\frac{d\boldsymbol{\omega}_{V,I}}{dt}\Big|_V \times \mathbf{p}\right)$$

After linearizing the expressions for the components of the above acceleration under the small-perturbation assumption (see Section 8.2.3 and Problem 8.4), the three (perturbation) components of $\mathbf{a}_{\text{local}}$ may be written as

$$\begin{aligned}
a_{X_{\text{local}}}(x,y,z,t) &= a_{X_R}(x,y,z,t) + \sum_{i=1}^{n} \nu_{X_i}(x,y,z)\ddot{\eta}_i(t) \\
a_{Y_{\text{local}}}(x,y,z,t) &= a_{Y_R}(x,y,z,t) + \sum_{i=1}^{n} \nu_{Y_i}(x,y,z)\ddot{\eta}_i(t) \\
a_{Z_{\text{local}}}(x,y,z,t) &= a_{Z_R}(x,y,z,t) + \sum_{i=1}^{n} \nu_{Z_i}(x,y,z)\ddot{\eta}_i(t)
\end{aligned} \tag{8.56}$$

(Note that these equations are now all expressed in terms of perturbation variables.) Although in the general case the above three equations are rather involved, for a reference flight condition involving steady, level flight, Equations (8.56) simply become

$$\boxed{\begin{aligned}
a_{X_{\text{local}}} &= \dot{u} + W_0 q + z\dot{q} - y\dot{r} + \sum_{i=1}^{n} \nu_{X_i}(x,y,z)\ddot{\eta}_i(t) \\
a_{Y_{\text{local}}} &= \dot{v} + U_0 r - W_0 p + x\dot{r} - z\dot{p} + \sum_{i=1}^{n} \nu_{Y_i}(x,y,z)\ddot{\eta}_i(t) \\
a_{Z_{\text{local}}} &= \dot{w} - U_0 q + y\dot{p} - x\dot{q} + \sum_{i=1}^{n} \nu_{Z_i}(x,y,z)\ddot{\eta}_i(t)
\end{aligned}} \tag{8.57}$$

These three components of the local-perturbation-acceleration vector, plus the local pitch and yaw rates, are additional responses frequently included in models of flexible vehicles. Furthermore, these additional responses may be expressed in terms of linear combinations of rigid-body plus elastic degrees of freedom, or state-variables. So they may be expressed easily in state-variable format.

Let us now outline how to assemble the state-variable model of the dynamics of a flexible vehicle. We begin with the model for the rigid-body degrees of freedom, similar to Equations (8.1) and (8.16), and Equations (8.6) and (8.22), which as shown in Section 8.1.4 we may write as

$$\begin{aligned}\dot{\mathbf{x}}_R &= \mathbf{A}_R\mathbf{x}_R + \mathbf{B}_R\mathbf{u}_R\\ \mathbf{y}_R &= \mathbf{C}_R\mathbf{x}_R + \mathbf{D}_R\mathbf{u}_R\end{aligned} \tag{8.58}$$

If the aeroelastic dimensional stability derivatives in Equations (8.46) and (8.49) are then added to the rigid-body equations of motion in the above state-variable model, the new model will now take on the following form.

$$\begin{aligned}\dot{\mathbf{x}}_R &= \mathbf{A}_R\mathbf{x}_R + \mathbf{B}_R\mathbf{u}_R + \mathbf{A}_{ER}\mathbf{x}_E\\ \mathbf{y}_R &= \mathbf{C}_R\mathbf{x}_R + \mathbf{D}_R\mathbf{u}_R\end{aligned} \tag{8.59}$$

Here $\mathbf{x}_E$ is the vector of vibration-modal coordinates and rates included in the model, or

$$\mathbf{x}_E = \begin{Bmatrix}\eta_1\\ \dot{\eta}_1\\ \vdots\\ \eta_n\\ \dot{\eta}_n\end{Bmatrix} \tag{8.60}$$

and the elements of the matrix $\mathbf{A}_{ER}$ include the aeroelastic stability derivatives in Equations (8.46) and (8.49). (Note that if we were considering only longitudinal dynamics here, only symmetric vibration modes need be included in the modeling. And if we were considering only the lateral-directional dynamics, only antisymmetric modes need be included.)

Now incorporating the equations of motion governing the elastic degrees of freedom (Equations (8.43) and (8.50)) into the above model, the form of the state-variable representation then becomes

$$\begin{aligned}\begin{Bmatrix}\dot{\mathbf{x}}_R\\ \dot{\mathbf{x}}_E\end{Bmatrix} &= \begin{bmatrix}\mathbf{A}_R & \mathbf{A}_{ER}\\ \mathbf{A}_{RE} & \mathbf{A}_E\end{bmatrix}\begin{Bmatrix}\mathbf{x}_R\\ \mathbf{x}_E\end{Bmatrix} + \begin{bmatrix}\mathbf{B}_R\\ \mathbf{B}_E\end{bmatrix}\mathbf{u}_R\\ \mathbf{y}_R &= \begin{bmatrix}\mathbf{C}_R & \mathbf{0}\end{bmatrix}\mathbf{x}_R + \mathbf{D}_R\mathbf{u}_R\end{aligned} \tag{8.61}$$

Finally, adding any total measured responses that include the effects of local elastic deformation, such as Equations (8.53–8.57), results in the final form of the state-variable model for an elastic vehicle. Or

$$\begin{Bmatrix} \dot{\mathbf{x}}_R \\ \dot{\mathbf{x}}_E \end{Bmatrix} = \begin{bmatrix} \mathbf{A}_R & \mathbf{A}_{ER} \\ \mathbf{A}_{RE} & \mathbf{A}_E \end{bmatrix} \begin{Bmatrix} \mathbf{x}_R \\ \mathbf{x}_E \end{Bmatrix} + \begin{bmatrix} \mathbf{B}_R \\ \mathbf{B}_E \end{bmatrix} \mathbf{u}_R$$
$$\begin{Bmatrix} \mathbf{y}_R \\ \mathbf{y}_{\text{local}} \end{Bmatrix} = \begin{bmatrix} \mathbf{C}_R & \mathbf{0} \\ \mathbf{C}_{RE} & \mathbf{C}_E \end{bmatrix} \begin{Bmatrix} \mathbf{x}_R \\ \mathbf{x}_E \end{Bmatrix} + \begin{bmatrix} \mathbf{D}_R \\ \mathbf{D}_{\text{local}} \end{bmatrix} \mathbf{u}_R \tag{8.62}$$

Note that in the above model, the matrix $\mathbf{A}_{ER}$ reflects the elastic-to-rigid aerodynamic coupling in the dynamics, while the matrix $\mathbf{A}_{RE}$ reflects the rigid-to-elastic aerodynamic coupling. If both of these matrices were null, the rigid-body and elastic degrees of freedom would be dynamically decoupled, as with a spacecraft.

8.1.6 Adding Feedback Control Laws to a Simulation Model

Frequently one is interested in simulating a vehicle's dynamics that includes one or more feedback-control laws embedded in those dynamics. There are several techniques for accomplishing this task, one of which is to use the Simulink tool that is part of the MATLAB software package. But a key point to be made here is that linear feedback-control laws may also be easily simulated by incorporating them directly into the state-variable model, because linear control laws are also linear dynamic systems. That is, any linear control law may be modeled in state-variable format.

For example, one type of controller, Control Law A that feeds back vehicle responses $\mathbf{y}$ to generate a feedback-control input $\mathbf{u}_c$, may be expressed as

$$\begin{aligned} \dot{\mathbf{x}}_c &= \mathbf{A}_c\mathbf{x}_c + \mathbf{B}_c\mathbf{y} \\ \mathbf{u}_c &= \mathbf{C}_c\mathbf{x}_c + \mathbf{D}_c\mathbf{y} \end{aligned} \tag{8.63}$$

where $\mathbf{x}_c$ is the vector of control-law state variables. (Note that the <u>input</u> to this system is the vehicle-response vector $\mathbf{y}$, and the <u>response</u> of this system is the vector of control inputs to the vehicle $\mathbf{u}_c$.) Another type of controller, Control Law B operating on an error vector of responses $\mathbf{e} = \mathbf{y}_c - \mathbf{y}$, where $\mathbf{y}_c$ is a vector of external commands, may be expressed as

$$\begin{aligned} \dot{\mathbf{x}}_c &= \mathbf{A}_c\mathbf{x}_c + \mathbf{B}_c\mathbf{e} \\ \mathbf{u}_c &= \mathbf{C}_c\mathbf{x}_c + \mathbf{D}_c\mathbf{e} \end{aligned} \tag{8.64}$$

So, for example, using Control Law A, let the total vector of control inputs to the vehicle be defined as

$$\mathbf{u} = \mathbf{u}_i - \mathbf{u}_c \tag{8.65}$$

and let the model of the vehicle dynamics be given as

$$\begin{aligned} \dot{\mathbf{x}}_v &= \mathbf{A}_v\mathbf{x}_v + \mathbf{B}_v\mathbf{u} \\ \mathbf{y} &= \mathbf{C}_v\mathbf{x}_v + \mathbf{D}_v\mathbf{u} \end{aligned} \tag{8.66}$$

Then the state-variable model of the vehicle's dynamics, including the actions of Control Law A, is given by

$$\begin{Bmatrix} \dot{\mathbf{x}}_v \\ \dot{\mathbf{x}}_c \end{Bmatrix} = \begin{bmatrix} (\mathbf{A}_v - \mathbf{B}_v\mathbf{M}^{-1}\mathbf{D}_c\mathbf{C}_v) & -\mathbf{B}_v\mathbf{M}^{-1}\mathbf{C}_c \\ \mathbf{B}_c(\mathbf{I} - \mathbf{D}_v\mathbf{M}^{-1}\mathbf{D}_c)\mathbf{C}_v & (\mathbf{A}_c - \mathbf{B}_c\mathbf{D}_v\mathbf{M}^{-1}\mathbf{C}_c) \end{bmatrix} \begin{Bmatrix} \mathbf{x}_v \\ \mathbf{x}_c \end{Bmatrix} + \begin{bmatrix} \mathbf{B}_v(\mathbf{I} - \mathbf{M}^{-1}\mathbf{D}_c\mathbf{D}_v) \\ \mathbf{B}_c(\mathbf{I} - \mathbf{D}_v\mathbf{M}^{-1}\mathbf{D}_c)\mathbf{D}_v \end{bmatrix} \mathbf{u}_i \tag{8.67}$$

$$\mathbf{y} = \begin{bmatrix} (\mathbf{I} - \mathbf{D}_v\mathbf{M}^{-1}\mathbf{D}_c)\mathbf{C}_v & -\mathbf{D}_v\mathbf{M}^{-1}\mathbf{C}_c \end{bmatrix} \begin{Bmatrix} \mathbf{x}_v \\ \mathbf{x}_c \end{Bmatrix} + (\mathbf{I} - \mathbf{D}_v\mathbf{M}^{-1}\mathbf{D}_c)\mathbf{D}_v\mathbf{u}_i$$

$$\mathbf{M} = \mathbf{I} + \mathbf{D}_c\mathbf{D}_v$$

But this is just another state-variable system. A similar state-variable model may be developed using Control Law B (see Problem 8.6). These composite models may be simulated using the techniques discussed in Section 8.1.8.

Frequently, control laws include dynamic elements given in terms of transfer functions. In the following example, we will show how three common dynamic elements may be converted to state-variable form. The procedure may be applied to other dynamic elements as well. In addition, MATLAB's Control Toolbox includes routines for converting transfer functions into state-variable models. We will demonstrate the use of some of these MATLAB routines in Example 8.4.

EXAMPLE 8.3

Converting Dynamic Control Elements to State-Variable Form

Consider three dynamic elements consisting of a proportion-plus-integral (PI) element, a first-order lag (or low-pass filter), and a washout (or high-pass filter). Starting with the transfer functions defining each element, find a state-variable representation in each case.

■ Solution

We will address each element individually, starting with a first-order lag defined by the transfer function

$$\boxed{\frac{u_L(s)}{y_L(s)} = \frac{K_L}{s + p_L}} \tag{8.68}$$

This transfer function corresponds to the differential equation given by

$$\dot{u}_L + p_L u_L = K_L y_L(t) \tag{8.69}$$

So the problem involves finding a state-variable representation for this differential equation, and several approaches may be used.

One such approach is to define a state variable x_L, and let the controller output be given by

$$u_L = cx_L + dy_L$$

where the constants c and d are as yet undefined. Now let the equation governing x_L be taken to be

$$\dot{x}_L = ax_L + by_L$$

where a and b are also yet to be defined. Therefore, we have

$$\dot{u}_L = c\dot{x}_L + d\dot{y}_L = c(ax_L + by_L) + d\dot{y}_L = c\left(\frac{a}{c}(u_L - dy_L) + by_L\right) + d\dot{y}_L$$
$$= au_L - (ad - cb)y_L + d\dot{y}_L$$

So we see that this equation can be made to agree with Equation (8.69) if we let

$$c = 1, d = 0, a = -p_L, \text{ and } b = K_L$$

Therefore, a state-variable representation for the given first-order lag is

$$\boxed{\begin{aligned} \dot{x}_L &= -p_L x_L + K_L y_L \\ u_L &= x_L \end{aligned}} \tag{8.70}$$

Now let the transfer function of a washout element be given as

$$\boxed{\frac{u_w(s)}{y_w(s)} = \frac{K_w s}{s + p_w}} \tag{8.71}$$

This transfer function is associated with the differential equation given by

$$\dot{u}_w + p_w u_w = K_w \dot{y}_w(t) \tag{8.72}$$

So define a state-variable x_w, and let the controller output again be given by

$$u_w = cx_w + dy_w$$

where the constants c and d are as yet undefined. As before, let the state variable x_w be governed by

$$\dot{x}_w = ax_w + by_w$$

where a and b are yet to be defined. Therefore we again have

$$\dot{u}_w = c\dot{x}_w + d\dot{y}_w = c(ax_w + by_w) + d\dot{y}_w = au_w - (ad - cb)y_w + d\dot{y}_w$$

This equation can be made to agree with Equation (8.72) if we let

$$d = K_w, c = 1, a = -p_y, \text{ and } b = -p_w K_w$$

Therefore, a state-variable representation for the given washout element is

$$\boxed{\begin{aligned} \dot{x}_w &= -p_w x_w - p_w K_w y_w \\ u_w &= x_w + K_w y_w \end{aligned}} \tag{8.73}$$

Finally, let the transfer function for a PI control element be given as

$$\boxed{\frac{u_{\text{PI}}(s)}{y_{\text{PI}}(s)} = \frac{K_{\text{PI}}(s + z_{\text{PI}})}{s}} \tag{8.74}$$

This element is associated with the differential equation given by

$$\dot{u}_{\text{PI}} = K_{\text{PI}}(\dot{y}_{\text{PI}} + z_{\text{PI}} y_{\text{PI}}) \tag{8.75}$$

Define a state-variable x_{PI}, and let the controller output again be given by

$$u_{\text{PI}} = c x_{\text{PI}} + d y_{\text{PI}}$$

where the constants c and d are as yet undefined. As before, let the state variable be governed by

$$\dot{x}_{\text{PI}} = a x_{\text{PI}} + b y_{\text{PI}}$$

where the constants a and b are yet to be defined. Therefore we now have

$$\dot{u}_{\text{PI}} = c\dot{x}_{\text{PI}} + d\dot{y}_{\text{PI}} = c(a x_{\text{PI}} + b y_{\text{PI}}) + d\dot{y}_{\text{PI}} = a u_{\text{PI}} + (cb - ad) y_{\text{PI}} + d\dot{y}_{\text{PI}}$$

This equation can be made to agree with Equation (8.75) if we let

$$d = K_{\text{PI}},\ c = 1,\ b = K_{\text{PI}} z_{\text{PI}},\ \text{and } a = 0$$

Therefore, a state-variable representation for the given PI element is

$$\boxed{\begin{aligned} \dot{x}_{\text{PI}} &= (K_{\text{PI}} z_{\text{PI}}) y_{\text{PI}} \\ u_{\text{PI}} &= x_{\text{PI}} + K_{\text{PI}} y_{\text{PI}} \end{aligned}} \tag{8.76}$$

MATLAB EXAMPLE 8.4

Numerically Converting Control Elements to State-Variable Form

Use the Control Toolbox in MATLAB to convert the following elements to state-variable form:

$$\text{(a)}\ \frac{u_L(s)}{y_L(s)} = \frac{2}{s+1}, \quad \text{(b)}\ \frac{u_w(s)}{y_w(s)} = \frac{2s}{s+1}, \ \text{and} \quad \text{(c)}\ \frac{u_{\text{PI}}(s)}{y_{\text{PI}}(s)} = \frac{2(s+0.5)}{s}$$

■ **Solution**

We will perform the conversions by using the routine zpk to specify the transfer function, and then find the state-variable representation for that element using the routine ss.

(a) »lagtf=zpk([],-1,2)

```
Zero/pole/gain:

 2
------
(s+1)

»lagss=ss(lagtf)

a =
               x1
      x1       -1
b =
               u1
      x1    1.4142
c =
               x1
      y1    1.4142
d =
               u1
      y1        0

Continuous-time system.
```

By comparing the above results with those from Example 8.3, we see that this state-variable representation differs from the earlier results. But both state-variable models yield the same transfer function. This demonstrates the earlier-stated fact that state-variable models are not unique.

(b) »washtf=zpk(0,-1,2)

```
Zero/pole/gain:

 2 s
------
(s+1)

»washss=ss(washtf)

a =
               x1
      x1       -1
b =
               u1
      x1    1.4142
c =
               x1
      y1   -1.4142
d =
               u1
      y1        2

Continuous-time system.
```

(c) »ptf=zpk(-0.5,0,2)

```
Zero/pole/gain:
2 (s+0.5)
---------
   s

»pss=ss(ptf)
a =
              x1
        x1     0
b =
              u1
        x1   1.4142
c =
              x1
        y1   0.70711
d =
              u1
        y1     2
Continuous-time system.
```

8.1.7 Adding Atmospheric Turbulence to a Simulation Model

The effects of atmospheric turbulence on the forces and moments acting on the vehicle were discussed in Section 6.8. We found that gusts affected the velocity of the vehicle relative to the air mass, and these effects could be reflected in the model of a vehicle's dynamics by making the following substitutions in the equations of motion.

$$
\begin{aligned}
u_{\text{total}} &= u + u_g \\
v_{\text{total}} &= v + v_g \left(\text{or } \beta_{\text{total}} = \beta + \beta_g\right) \\
w_{\text{total}} &= w + w_g \left(\text{or } \alpha_{\text{total}} = \alpha + \alpha_g\right) \\
\dot{\alpha}_{\text{total}} &= \dot{\alpha} + \dot{\alpha}_g
\end{aligned}
\tag{8.77}
$$

That is, the velocity of the vehicle relative to the air mass is the sum of the velocity of the vehicle with respect to the earth, plus the velocity of the air mass relative to the earth.

Let us demonstrate the procedure for incorporating gusts into the linear model by assuming a rigid vehicle with a reference flight condition consisting of straight and level flight, and with constant atmospheric density. Therefore, the decoupled state-variable models developed in Section 8.1.4 are appropriate. Consider first the longitudinal model given in Equations (8.34–8.36), and note that they were developed starting with Equations (8.29). Beginning at the same point, we may now make the substitutions indicated in Equations (8.77), resulting in

$$\begin{aligned}\dot{u} &= -g\theta + \left(X_u + X_{P_u}\right)\left(u + u_g\right) + X_\alpha\left(\alpha + \alpha_g\right) + X_{\dot{\alpha}}\left(\dot{\alpha} + \dot{\alpha}_g\right) \\ &\quad + X_q q + X_{\delta_E}\delta_E + X_T\delta T \\ \dot{w} &= \left(Z_u + Z_{P_u}\right)\left(u + u_g\right) + Z_\alpha\left(\alpha + \alpha_g\right) + Z_{\dot{\alpha}}\left(\dot{\alpha} + \dot{\alpha}_g\right) + \left(Z_q + \mathrm{U}_0\right)q \\ &\quad + Z_{\delta_E}\delta_E + Z_T\delta T \\ \dot{q} &= \left(M_u + M_{P_u}\right)\left(u + u_g\right) + \left(M_\alpha + M_{P\alpha}\right)\left(\alpha + \alpha_g\right) + M_{\dot{\alpha}}\left(\dot{\alpha} + \dot{\alpha}_g\right) + M_q q \\ &\quad + M_{\delta_E}\delta_E + M_T\delta T\end{aligned} \tag{8.78}$$

Noting that $\dot{\alpha} = \dot{w}/\mathrm{U}_0$, the second of the above equations becomes

$$\begin{aligned}\dot{\alpha} &= \left(\frac{1}{\mathrm{U}_0 - Z_{\dot{\alpha}}}\right)\Big(\left(Z_u + Z_{P_u}\right)\left(u + u_g\right) + Z_\alpha\left(\alpha + \alpha_g\right) + Z_{\dot{\alpha}}\left(\dot{\alpha}_g\right) \\ &\quad + \left(Z_q + \mathrm{U}_0\right)q + Z_{\delta_E}\delta_E + Z_T\delta T\Big)\end{aligned} \tag{8.79}$$

instead of Equation (8.31). Substituting this expression into the other two equations in Equations (8.78), and assuming $Z_{P_u} = 0$, then yields

$$\begin{aligned}\dot{u} &= \left(X_u + X_{P_u} + \frac{X_{\dot{\alpha}}Z_u}{\mathrm{U}_0 - Z_{\dot{\alpha}}}\right)\left(u + u_g\right) + \left(X_\alpha + \frac{X_{\dot{\alpha}}Z_\alpha}{\mathrm{U}_0 - Z_{\dot{\alpha}}}\right)\left(\alpha + \alpha_g\right) + X_{\dot{\alpha}}\left(1 + \frac{Z_{\dot{\alpha}}}{\mathrm{U}_0 - Z_{\dot{\alpha}}}\right)\dot{\alpha}_g - g\theta \\ &\quad + \left(X_q + X_{\dot{\alpha}}\left(\frac{\mathrm{U}_0 + Z_q}{\mathrm{U}_0 - Z_{\dot{\alpha}}}\right)\right)q + \left(X_{\delta_E} + \frac{X_{\dot{\alpha}}Z_{\delta_E}}{\mathrm{U}_0 - Z_{\dot{\alpha}}}\right)\delta_E + \left(X_T + \frac{X_{\dot{\alpha}}Z_T}{\mathrm{U}_0 - Z_{\dot{\alpha}}}\right)\delta T \\ \dot{\alpha} &= \left(\frac{1}{\mathrm{U}_0 - Z_{\dot{\alpha}}}\right)\left(Z_u\left(u + u_g\right) + Z_\alpha\left(\alpha + \alpha_g\right) + Z_{\dot{\alpha}}\left(\dot{\alpha}_g\right) + \left(Z_q + \mathrm{U}_0\right)q + Z_{\delta_E}\delta_E + Z_T\delta T\right) \\ \dot{q} &= \left(M_u + M_{P_u} + \frac{M_{\dot{\alpha}}Z_u}{\mathrm{U}_0 - Z_{\dot{\alpha}}}\right)\left(u + u_g\right) + \left(M_\alpha + M_{P\alpha} + \frac{M_{\dot{\alpha}}Z_\alpha}{\mathrm{U}_0 - Z_{\dot{\alpha}}}\right)\left(\alpha + \alpha_g\right) + M_{\dot{\alpha}}\left(1 + \frac{Z_{\dot{\alpha}}}{\mathrm{U}_0 - Z_{\dot{\alpha}}}\right)\dot{\alpha}_g \\ &\quad + \left(M_q + M_{\dot{\alpha}}\left(\frac{\mathrm{U}_0 + Z_q}{\mathrm{U}_0 - Z_{\dot{\alpha}}}\right)\right)q + \left(M_{\delta_E} + \frac{M_{\dot{\alpha}}Z_{\delta_E}}{\mathrm{U}_0 - Z_{\dot{\alpha}}}\right)\delta_E + \left(M_T + \frac{M_{\dot{\alpha}}Z_T}{\mathrm{U}_0 - Z_{\dot{\alpha}}}\right)\delta T\end{aligned} \tag{8.80}$$

The above three equations, plus the kinematics equations

$$\dot{\theta} = q$$

and

$$\dot{h} = -w + \mathrm{U}_0\theta$$

constitute the model for the longitudinal dynamics, including atmospheric turbulence. These five equations may be written in the following state-variable form.

$$\begin{aligned}\dot{\mathbf{x}}_v &= \mathbf{A}_v\mathbf{x}_v + \mathbf{B}_v\mathbf{u}_v + \mathbf{G}_v\mathbf{y}_g \\ \mathbf{y}_v &= \mathbf{C}_v\mathbf{x}_v + \mathbf{D}_v\mathbf{u}_v\end{aligned} \tag{8.81}$$

with the response, state-variable, and input vectors, and the matrices as given in Equations (8.34–8.36).

Also we have

$$\mathbf{y}_g = \begin{Bmatrix} u_g \\ \beta_g \\ \alpha_g \\ \dot{\alpha}_g \end{Bmatrix} \tag{8.82}$$

and

$$\mathbf{G}_v = \begin{bmatrix} \left(X_u + X_{P_u} + \dfrac{X_{\dot{\alpha}} Z_u}{U_0 - Z_{\dot{\alpha}}}\right) & 0 & \left(X_\alpha + \dfrac{X_{\dot{\alpha}} Z_\alpha}{U_0 - Z_{\dot{\alpha}}}\right) & X_{\dot{\alpha}}\left(1 + \dfrac{Z_{\dot{\alpha}}}{U_0 - Z_{\dot{\alpha}}}\right) \\ \left(\dfrac{Z_u}{U_0 - Z_{\dot{\alpha}}}\right) & 0 & \left(\dfrac{Z_\alpha}{U_0 - Z_{\dot{\alpha}}}\right) & \left(\dfrac{Z_{\dot{\alpha}}}{U_0 - Z_{\dot{\alpha}}}\right) \\ 0 & 0 & 0 & 0 \\ \left(M_u + M_{P_u} + \dfrac{M_{\dot{\alpha}} Z_u}{U_0 - Z_{\dot{\alpha}}}\right) & 0 & \left(M_\alpha + M_{P_\alpha} + \dfrac{M_{\dot{\alpha}} Z_\alpha}{U_0 - Z_{\dot{\alpha}}}\right) & M_{\dot{\alpha}}\left(1 + \dfrac{Z_{\dot{\alpha}}}{U_0 - Z_{\dot{\alpha}}}\right) \\ 0 & 0 & 0 & 0 \end{bmatrix} \tag{8.83}$$

Now the atmospheric turbulence must be incorporated into the model. From Appendix C, the state-variable model for the turbulence may be expressed in the following form.[1]

$$\begin{aligned} \dot{\mathbf{x}}_g &= \mathbf{A}_g \mathbf{x}_g + \mathbf{B}_g n \\ \mathbf{y}_g &= \mathbf{C}_g \mathbf{x}_g + \mathbf{D}_g n \end{aligned} \tag{8.84}$$

where $\mathbf{y}_g$ is as given in Equation (8.82), $\mathbf{x}_g$ is the vector of gust state-variables, and n is the pseudo-random number sequence used to generate the stochastic gusts. Combining Equations (8.81) with Equations (8.84) then yields the complete state-variable model for the vehicle's longitudinal dynamics, excited by atmospheric turbulence. Or

$$\boxed{\begin{aligned} \begin{Bmatrix} \dot{\mathbf{x}}_g \\ \dot{\mathbf{x}}_v \end{Bmatrix} &= \begin{bmatrix} \mathbf{A}_g & \mathbf{0} \\ \mathbf{G}_v \mathbf{C}_g & \mathbf{A}_v \end{bmatrix} \begin{Bmatrix} \mathbf{x}_g \\ \mathbf{x}_v \end{Bmatrix} + \begin{Bmatrix} \mathbf{0} \\ \mathbf{B}_v \end{Bmatrix} \mathbf{u}_v + \begin{Bmatrix} \mathbf{B}_g \\ \mathbf{G}_v \mathbf{D}_g \end{Bmatrix} n \\ \mathbf{y}_v &= \begin{bmatrix} \mathbf{0} & \mathbf{C}_v \end{bmatrix} \begin{Bmatrix} \mathbf{x}_g \\ \mathbf{x}_v \end{Bmatrix} + \mathbf{D}_v \mathbf{u}_v \end{aligned}} \tag{8.85}$$

This state-variable model is now ready for numerical simulation. Note that the random-number sequence n must be generated in the simulation as discussed in Appendix C.

---

[1] This is the generic format for generating a stochastic process (Ref. 4).

Regarding simulation of a vehicle's lateral-directional dynamics, one may begin with Equations (8.30) and follow the procedure just presented. The substitutions reflected in Equations (8.77) are made into Equations (8.30), and a state-variable model of the dynamics (including gust velocities) is developed in the form given in Equations (8.81). In the resulting state-variable model for the lateral-directional dynamics, $\mathbf{y}_v$, $\mathbf{x}_v$, $\mathbf{u}_v$, and the four matrices are all as given in Equations (8.39–8.41), plus now

$$\mathbf{G}_v = \begin{bmatrix} 0 & Y_\beta/U_0 & 0 & 0 \\ 0 & 0 & 0 & 0 \\ 0 & L'_\beta & 0 & 0 \\ 0 & N'_\beta & 0 & 0 \\ 0 & 0 & 0 & 0 \end{bmatrix} \tag{8.86}$$

Then the gust model, Equations (8.84), is added to yield the final state-variable model of the form given in Equations (8.85). This is the complete simulation model for the linear lateral-directional dynamics excited by gusts.

Finally, note that the gust model given by Equations (8.84) may also be incorporated into a state-variable model for the coupled longitudinal and lateral-directional dynamics, and the $\mathbf{G}_v$ matrix, in this case, would be the sum of the matrices given in Equations (8.83) and (8.86).

8.1.8 Numerical Simulation Methods for Linear Models—A JITT*

We will first assume that the vehicle dynamics to be simulated are modeled in state-variable format. Or the system is given in generic form as

$$\begin{aligned} \dot{\mathbf{x}} &= \mathbf{A}\mathbf{x} + \mathbf{B}\mathbf{u} \\ \mathbf{y} &= \mathbf{C}\mathbf{x} + \mathbf{D}\mathbf{u} \end{aligned} \tag{8.87}$$

where the four matrices are constant matrices. This implies a constant-mass vehicle. This model may consist of just vehicle dynamics, as discussed in Section 8.1.4 or 8.1.5, or vehicle dynamics plus a feedback-control law, as discussed in Section 8.1.6. The model may also include the effects of atmospheric turbulance.

There are many techniques for numerically simulating linear dynamic systems. But the techniques discussed here include state-transition methods plus several methods in MATLAB, including Simulink. Although less numerically efficient than state-transition methods, nonlinear numerical integration techniques may also be used, and these will be discussed in Section 8.2.7. Such integration methods are frequently used for simulating linear state-variable systems with time-dependent matrices.

---

* A Just-in-Time Tutorial.

State-transition methods are based on the analytical solution to the vector differential equation in the state-variable model, or

$$\dot{\mathbf{x}} = \mathbf{A}\mathbf{x} + \mathbf{B}\mathbf{u} \tag{8.88}$$

(Ref. 3). To develop this analytical solution, we will first find the solution to the homogeneous vector equation

$$\dot{\mathbf{x}} = \mathbf{A}\mathbf{x} \tag{8.89}$$

If **x** and **A** were scalar quantities, we know that the solution to the equation

$$\dot{x} = ax$$

given in terms of an initial condition at t_0 is

$$x(t) = e^{a(t-t_0)}x(t_0)$$

where the exponential function, defined in terms of its infinite series, is

$$e^{a(t-t_0)} = \sum_{i=0}^{\infty} \frac{a^i(t-t_0)^i}{i!}$$

By analogy, then, the solution to the set of homogeneous equations, Equations (8.89), is

$$\mathbf{x}_H(t) = e^{\mathbf{A}(t-t_0)}\mathbf{x}(t_0) \tag{8.90}$$

where the matrix exponential may also be defined in terms of an infinite series, or

$$e^{\mathbf{A}(t-t_0)} = \sum_{i=0}^{\infty} \frac{\mathbf{A}^i(t-t_0)^i}{i!} \tag{8.91}$$

We can confirm that Equation (8.90) is indeed the solution to Equation (8.89) by differentiation and direct substitution, using the above infinite series.

Note that $e^{\mathbf{A}(t-t_0)}$ is a matrix quantity, and is called the *state-transition matrix*, also denoted as $\mathbf{\Phi}(t-t_0)$. So

$$\mathbf{\Phi}(t-t_0) \triangleq e^{\mathbf{A}(t-t_0)} \tag{8.92}$$

Some useful properties of the state-transition matrix, which can be proved using the definition Equation (8.92), or the series definition Equation (8.91), are:

1. $\mathbf{\Phi}(0) = \mathbf{I}$ the identity matrix
2. $\mathbf{\Phi}(t_1 + t_2) = \mathbf{\Phi}(t_1)\,\mathbf{\Phi}(t_2) = \mathbf{\Phi}(t_2)\,\mathbf{\Phi}(t_1)$
3. $\mathbf{\Phi}^{-1}(t - t_0) = \mathbf{\Phi}(-(t - t_0))$
4. $[\mathbf{\Phi}(t)]^k = \mathbf{\Phi}(kt)$, k an integer

The complete solution to the nonhomogeneous vector differential equation, Equation (8.88), is the sum of the homogeneous solution plus the particular solution, or

$$\mathbf{x}(t) = \mathbf{x}_H(t) + \mathbf{x}_P(t) \tag{8.93}$$

The method of variation of parameters may be used to find the particular solution (Ref. 3). This method attempts to find the particular solution in terms of the homogeneous solution, which here is given by Equation (8.90), or $\mathbf{\Phi}(t - t_0)\mathbf{x}(t_0)$.

So let the assumed particular solution be chosen to be

$$\mathbf{x}_P(t) = \mathbf{\Phi}(t - t_0)\mathbf{P}(t)\mathbf{x}(t_0) \tag{8.94}$$

where $\mathbf{P}(t)$ is an n $\times$ n matrix of parameters to be determined. The complete solution is now

$$\begin{aligned} \mathbf{x}(t) &= \mathbf{x}_H(t) + \mathbf{x}_P(t) = \mathbf{\Phi}(t - t_0)\mathbf{x}(t_0) + \mathbf{\Phi}(t - t_0)\mathbf{P}(t)\mathbf{x}(t_0) \\ &= \mathbf{\Phi}(t - t_0)[\mathbf{I} + \mathbf{P}(t)]\mathbf{x}(t_0) \triangleq \mathbf{\Phi}(t - t_0)\mathbf{z}(t) \end{aligned} \tag{8.95}$$

and note that $\mathbf{z}(t_0) = \mathbf{x}(t_0)$ since $\mathbf{\Phi}(t_0 - t_0) = \mathbf{I}$. Substituting Equation (8.95) back into Equation (8.88) and rearranging yields

$$\left[\dot{\mathbf{\Phi}}(t - t_0) - \mathbf{A\Phi}(t - t_0)\right]\mathbf{z}(t) + \mathbf{\Phi}(t - t_0)\dot{\mathbf{z}}(t) = \mathbf{Bu}(t)$$

Now using the series definition for the state-transition matrix, or Equation (8.91), we can show that

$$\dot{\mathbf{\Phi}}(t - t_0) - \mathbf{A\Phi}(t - t_0) = \mathbf{0} \tag{8.96}$$

or the state-transition matrix satisfies the original homogeneous differential equation. Therefore,

$$\mathbf{\Phi}(t - t_0)\dot{\mathbf{z}}(t) = \mathbf{Bu}(t)$$

or

$$\dot{\mathbf{z}}(t) = \mathbf{\Phi}^{-1}(t - t_0)\mathbf{Bu}(t). \tag{8.97}$$

But Equation (8.97) may be integrated directly to find that

$$\mathbf{z}(t) - \mathbf{z}(t_0) = \int_{t_0}^{t} \mathbf{\Phi}^{-1}(\tau - t_0)\mathbf{Bu}(\tau)d\tau$$

or

$$\mathbf{z}(t) = \mathbf{x}(t_0) + \int_{t_0}^{t} \mathbf{\Phi}^{-1}(\tau - t_0)\mathbf{Bu}(\tau)d\tau \tag{8.98}$$

So from Equation (8.95) we can write the complete solution (homogeneous plus particular) as

$$\mathbf{x}(t) = \mathbf{\Phi}(t - t_0)\mathbf{z}(t) = \mathbf{\Phi}(t - t_0)\mathbf{x}(t_0) + \mathbf{\Phi}(t - t_0)\int_{t_0}^{t} \mathbf{\Phi}^{-1}(\tau - t_0)\mathbf{B}\mathbf{u}(\tau)d\tau \quad (8.99)$$

Or, noting that

$$\mathbf{\Phi}(t - t_0)\mathbf{\Phi}^{-1}(\tau - t_0) = e^{\mathbf{A}(t-t_0)}e^{-\mathbf{A}(\tau-t_0)} = e^{\mathbf{A}(t-t_0)}e^{\mathbf{A}(t_0-\tau)} = e^{\mathbf{A}(t-\tau)} = \mathbf{\Phi}(t-\tau) \quad (8.100)$$

another form of the complete solution to the nonhomogeneous system of differential equations (Equation (8.88)) is

$$\mathbf{x}(t) = \mathbf{\Phi}(t - t_0)\mathbf{x}(t_0) + \int_{t_0}^{t} \mathbf{\Phi}(t - \tau)\mathbf{B}\mathbf{u}(\tau)d\tau \quad (8.101)$$

Although Equation (8.99) or (8.101) is the key to the state-transition method of simulation, we do not use either of them directly. Rather, we use them to derive a *discrete equivalent* to the continuous state-variable system in Equations (8.87) (Ref. 4). Consider finding the solution to the state-variable differential equations, or $\mathbf{x}(t_{k+1})$, in terms of the solution at an earlier time, or $\mathbf{x}(t_k)$, where

$$t_{k+1} - t_k = \Delta t$$

is a small time increment. Now assume that the input vector $\mathbf{u}(t)$ is constant over the interval Δt, and define the dummy variable p to equal $(t_{k+1} - \tau)$. We may now rewrite the integral in Equation (8.101) as

$$\int_{t_k}^{t_{k+1}} e^{\mathbf{A}(t_{k+1}-\tau)}\mathbf{B}\mathbf{u}(\tau)d\tau = -\left(\int_{\Delta t}^{0} e^{\mathbf{A}p}dp\right)\mathbf{B}\mathbf{u}(t_k) = \left(\int_{0}^{\Delta t} e^{\mathbf{A}p}dp\right)\mathbf{B}\mathbf{u}(t_k) \quad (8.102)$$

But

$$\int_{0}^{\Delta t} e^{\mathbf{A}p}dp = \left(e^{\mathbf{A}\Delta t} - \mathbf{I}\right)\mathbf{A}^{-1} \quad (8.103)$$

And so the solution for $\mathbf{x}(t_{k+1})$ is

$$\boxed{\begin{aligned} \mathbf{x}(t_{k+1}) &= \left[e^{\mathbf{A}\Delta t}\right]\mathbf{x}(t_k) + \left[\left(e^{\mathbf{A}\Delta t} - \mathbf{I}\right)\mathbf{A}^{-1}\mathbf{B}\right]\mathbf{u}(t_k) \\ &\triangleq \left[\mathbf{M}(\Delta t)\right]\mathbf{x}(t_k) + \left[\mathbf{N}(\Delta t)\right]\mathbf{u}(t_k) \end{aligned}} \quad (8.104)$$

Of course, knowing $\mathbf{x}(t_k)$ and $\mathbf{u}(t_k)$, we have the responses $\mathbf{y}(t_k)$ given by

$$\boxed{\mathbf{y}(t_k) = \mathbf{C}_v x(t_k) + \mathbf{D}_v u(t_k)} \quad (8.105)$$

From Equation (8.104), we see that $\mathbf{N}(\Delta t)$ is a function of $\mathbf{A}^{-1}$. And we might think this a problem if $\mathbf{A}$ is not invertible. But there are several methods available to find the discrete-equivalent system even if $\mathbf{A}$ is not invertible, including the use of MATLAB routines demonstrated later in this section. Plus, it is instructive to demonstrate using a series expansion to numerically find both $\mathbf{M}(\Delta t)$ and $\mathbf{N}(\Delta t)$ directly.

From the series definition of the state-transition matrix, Equation (8.91), we find that

$$\mathbf{M}(\Delta t) = \mathbf{I} + \mathbf{A}\Delta t + \frac{1}{2!}\mathbf{A}^2\Delta t^2 + \ldots$$

and that (8.106)

$$\mathbf{N}(\Delta t) = \Delta t\left(\mathbf{I} + \frac{1}{2!}\mathbf{A}\Delta t + \frac{1}{3!}\mathbf{A}^2\Delta t^2 + \ldots\right)\mathbf{B}$$

Note that for a sufficiently small Δt, the two series in the above expressions will converge after a finite number of terms. For example, Δt might be selected to be

$$\Delta t = \frac{1}{\omega_{n_{\max}}} \text{ sec} \tag{8.107}$$

where

$$\omega_{n_{\max}} \triangleq |\lambda_{\max}|$$

and $\lambda_{\max}$ is the eigenvalue of $\mathbf{A}$ with the largest magnitude. In addition, we also need to keep Δt small since the inputs $\mathbf{u}(t)$ were assumed constant over that time interval. Consequently, $\mathbf{M}(\Delta t)$ and $\mathbf{N}(\Delta t)$ may be found numerically by simply summing the first several terms in the series given in Equations (8.106).

After $\mathbf{M}(\Delta t)$ and $\mathbf{N}(\Delta t)$ are found by whatever method chosen, Equations (8.104) and (8.105) may be used recursively to generate the solution for the time histories of $\mathbf{y}(t_k)$, given an input time history $\mathbf{u}(t_k)$. This process constitutes the simulation of the linear dynamical system.

8.1.9 Linear-Simulation Examples

We conclude Section 8.1 with several examples involving linear simulations of a vehicle's flight dynamics.

MATLAB EXAMPLE 8.5

Simulation of the Navion's Longitudinal Dynamics

Using the state-transition method, simulate the longitudinal dynamics of the Navion aircraft developed in Example 8.2. Use a one-degree, two-second doublet as the elevator input, and determine the first 10 seconds of the responses.

■ Solution

The perturbation elevator deflection is specified to be a one-degree, two-second doublet. So the elevator input will be -1 deg for two seconds, $+1$ deg for two seconds, and then remain zero for the remainder of the simulation. We will select $\Delta t = 0.05$ sec, which is a small time interval when dealing with the dynamic response of a rigid aircraft. Therefore, there will be 20 time intervals per second, for a total of 200 intervals over the 10 second simulated time.

The state-variable model for the system of interest is given in Equations (8.37), repeated here for convenience. (The thrust input is not used here, so **B** and **D** only reflect the single control input.)

$$\mathbf{A} = \begin{bmatrix} -0.0451 & 6.348 & -32.2 & 0 & 0 \\ -0.0021 & -2.0244 & 0 & 1 & 0 \\ 0 & 0 & 0 & 1 & 0 \\ 0.0021 & -6.958 & 0 & -3.0757 & 0 \\ 0 & -176 & 176 & 0 & 0 \end{bmatrix}, \mathbf{B} = \begin{bmatrix} 0 \\ -0.160 \\ 0 \\ -11.029 \\ 0 \end{bmatrix}$$

$$\mathbf{C} = \begin{bmatrix} 1 & 0 & 0 & 0 & 0 \\ 0 & 1 & 0 & 0 & 0 \\ 0 & 0 & 1 & 0 & 0 \\ 0 & 0 & 0 & 1 & 0 \\ 0 & 0 & 0 & 0 & 1 \end{bmatrix}, \mathbf{D} = \begin{bmatrix} 0 \\ 0 \\ 0 \\ 0 \\ 0 \end{bmatrix}$$

The input vector **u** is the perturbation elevator deflection δ_E in radians, and the response and state vectors are

$$\mathbf{y} = \mathbf{x} = \begin{Bmatrix} u\,(\text{fps}) \\ \alpha\,(\text{rad}) \\ \theta\,(\text{rad}) \\ q\,(\text{rad/sec}) \\ h\,(\text{ft}) \end{Bmatrix}$$

We will next find the matrices **M** and **N** using Equations (8.106) with a $\Delta t = 0.05$ sec. The terms in the series expansions for **M** and **N** are found as follows from MATLAB.

```
A =

  4.5100e-02   6.3480e+00  -3.2200e+01        0         0
 -2.1000e-03  -2.0244e+00        0       1.0000e+00     0
       0            0            0       1.0000e+00     0
  2.1000e-03  -6.9580e+00        0      -3.0757e+00     0
       0      -1.7600e+02   1.7600e+02        0         0

»term1=.05*A

term1 =

  2.2550e-03   3.1740e-01  -1.6100e+00        0         0
 -1.0500e-04  -1.0122e-01        0       5.0000e-02     0
       0            0            0       5.0000e-02     0
  1.0490e-04  -3.4790e-01        0      -1.5379e-01     0
       0      -8.8000e+00   8.8000e+00        0         0
```

```
»M=eye(5)+term1
```
Only the first two terms in the series

M =

1.0023e+00	3.1740e-01	-1.6100e+00	0	0
-1.0500e-04	8.9878e-01	0	5.0000e-02	0
0	0	1.0000e+00	5.0000e-02	0
1.0490e-04	-3.4790e-01	0	8.4621e-01	0
0	-8.8000e+00	8.8000e+00	0	1.0000e+00

Using the notation termk $= \mathbf{A}^k(\Delta t)^k$ below, we iterated until $k = 5$, at which time the largest change in the elements of the **M** matrix is on the order of 1.0×10^{-6}.

```
»term5=term4*term1
```

term5 =

5.7032e-07	5.4134e-04	2.9674e-06	-2.7602e-04	0
1.1391e-07	1.0623e-04	6.0368e-07	-5.5591e-05	0
7.5765e-08	-4.0037e-06	6.9359e-07	-5.4625e-05	0
-1.0278e-07	3.8866e-04	-2.4396e-06	1.6470e-04	0
-4.9143e-07	4.6269e-03	-2.2382e-05	1.4417e-03	0

```
»M=M+(1/120)*term5
```

M =

1.0022e+00	3.0601e-01	-1.6118e+00	-3.0945e-02	0
-9.7288e-05	8.9601e-01	8.0371e-05	4.3893e-02	0
2.7771e-06	-7.9812e-03	1.0000e+00	4.6214e-02	0
1.1384e-04	-3.0538e-01	-8.9424e-05	8.4988e-01	0
4.4733e-04	-8.3698e+00	8.7998e+00	6.9591e-03	1.0000e+00

Likewise for the **N** matrix we have

B =

0
-1.6000e-01
0
-1.1029e+01
0

```
»N=0.05*(eye(5)+0.5*term1+(1/6)*term2+(1/24)*term3+(1/120)*term4+(1/720)*term5)*B
```

N =

4.5138e-03
-2.0239e-02
-1.3066e-02
-5.0842e-01
3.3071e-02

And again, the largest change in the elements of **N** after five terms in the series is on the order 1.0×10^{-6}.

Turning again to the MATLAB command line, the input and time sequences and the desired number of simulation time steps are defined as

```
»t=0:.05:10;
»u(1:20)= -1/57.3;
»u(21:40)=1/57.3;
»u(41:201)=0;
»k=200;
```

The recursive simulation was performed using the MATLAB function or m file named NavionSim, listed below.

```
% Navion Simulation
function [y]=NavionSim(M,N,C,D,u,k,x0)
x=x0;
y(:,1)=C*x+D*u(1);
for kk=2:k+1
   x=M*x+N*u(kk-1);
   y(:,kk)=C*x+D*u(kk);
end
```

The surge velocity and control input are plotted in Figure 8.3 as follows:

```
»x0=[0;0;0;0;0];
»y=NavionSim(M,N,C,D,u,k,x0)
»plot(t,y(1,:))
»grid
»xlabel('Time, t (sec)')
»ylabel('Surge Velocity, u (fps)')
»title('Surge Velocity Time History')
»plot(t,57.3*u)
»grid
»xlabel('Time, t (sec)')
»ylabel('Elevator Angle, deltaE (deg)')
»title('Input Elevator Deflection Time History')
```

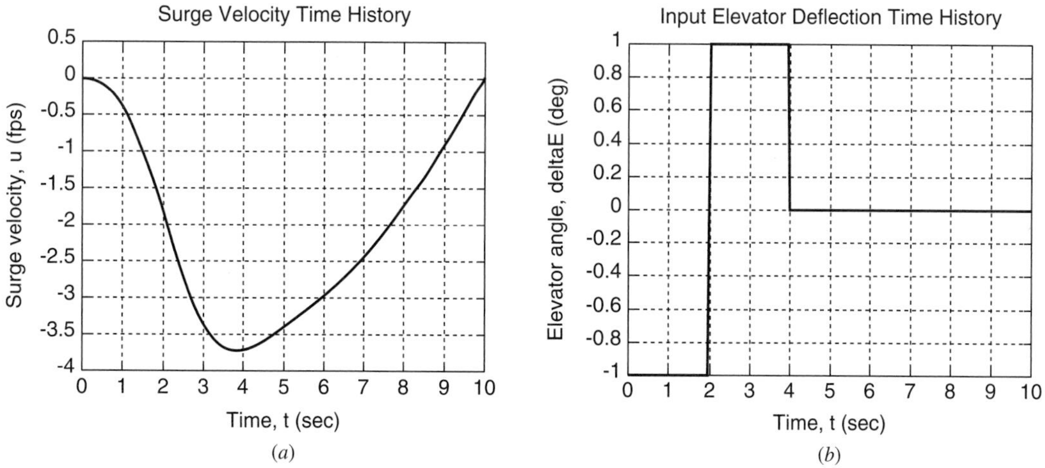

Figure 8.3 Time histories of surge velocity and elevator input.

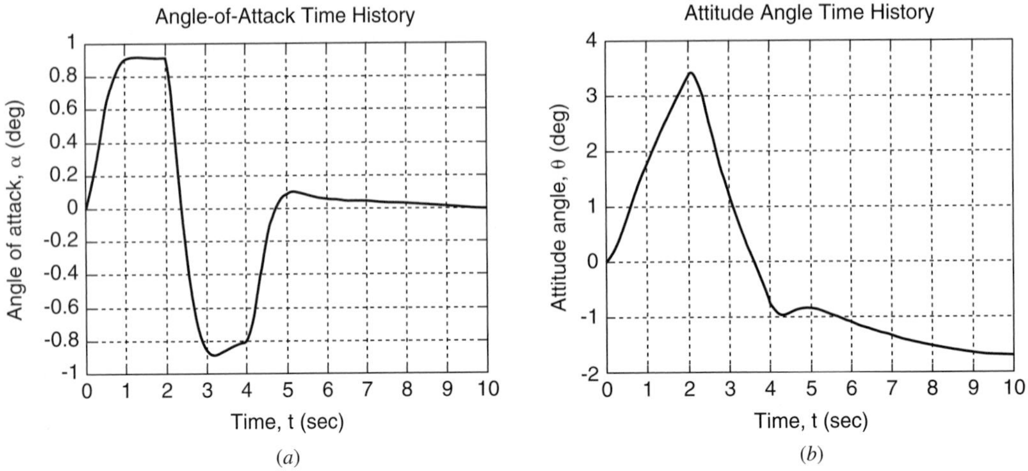

Figure 8.4 Time histories of angle of attack and pitch attitude.

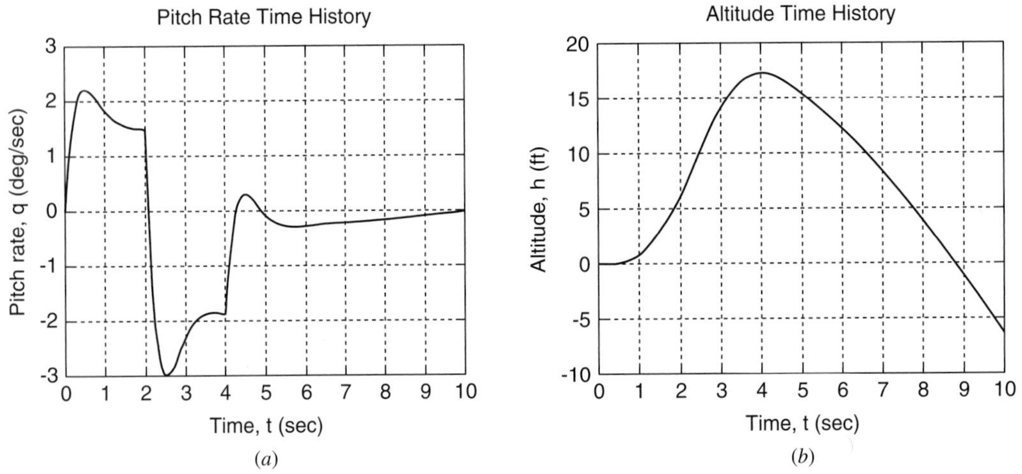

Figure 8.5 Time histories of pitch rate and altitude.

The other four responses were plotted in the same fashion, and are shown in Figures 8.4 and 8.5. Note that elevator deflection, angle of attack, and attitude angle were converted to degrees, and pitch rate to degrees per second, for plotting.

Also note at this point that the elevator input and the total simulated time of 10 seconds was intentionally kept small to assure that the responses would also remain small, and thus not violate the small-perturbation assumptions made in the derivations of the linear models. Students sometimes forget this concept and allow simulated responses to become too large. Such results are frequently not valid and must be used with caution.

A routine in MATLAB may also be used to directly determine the matrices $\mathbf{M}(\Delta t)$ and $\mathbf{N}(\Delta t)$. MATLAB's Control Toolbox contains the routine c2d, which is used to convert a linear continuous system into a discrete-equivalent system. The matrices $\mathbf{M}(\Delta t)$, $\mathbf{N}(\Delta t)$, $\mathbf{C}$, and $\mathbf{D}$ in Example 8.5 constitute the state-variable description of the discrete-equivalent system. We will demonstrate the use of c2d in Example 8.6.

MATLAB EXAMPLE 8.6

Determining the Simulation Matrices M and N Using c2d

Using MATLAB's routine c2d, find the matrices **M** and **N** found in Example 8.5, which may be used for simulating the longitudinal dynamics of the Navion.

■ Solution

The simulation time interval is again $\Delta t = 0.05$ sec. Using the **A** and **B** matrices for the Navion model, as given in Example 8.5, we turn to the MATLAB command line and directly find **M** and **N**.

```
»A=[0.0451 6.348 -32.2 0 0;-2.1e-3 -2.0244 0 1 0;
0 0 0 1 0;2.1e-3 -6.958 0 -3.0757 0;0 -176 176 0 0]

A =

  4.5100e-02   6.3480e+00  -3.2200e+01            0   0
 -2.1000e-03  -2.0244e+00            0   1.0000e+00   0
           0            0            0   1.0000e+00   0
  2.1000e-03  -6.9580e+00            0  -3.0757e+00   0
           0  -1.7600e+02   1.7600e+02            0   0

»B=[0;-.16;0;-11.029;0]

B =

           0
 -1.6000e-01
           0
 -1.1029e+01
           0

»C=eye(5)

C =

  1  0  0  0  0
  0  1  0  0  0
  0  0  1  0  0
  0  0  0  1  0
  0  0  0  0  1
```

```
»D(1:5)=0
D =
  0  0  0  0  0
»D=D'
D =
  0
  0
  0
  0
  0
»sys=ss(A,B,C,D)
a =
           x1         x2       x3       x4      x5
  x1   0.0451     6.3480    -32.2        0       0
  x2  -0.0021    -2.0244        0        1       0
  x3        0          0        0        1       0
  x4   0.0021     -6.958        0  -3.0757       0
  x5        0       -176      176        0       0
b =
          u1
  x1       0
  x2   -0.16
  x3       0
  x4 -11.029
  x5       0
c =
      x1  x2  x3  x4  x5
  y1   1   0   0   0   0
  y2   0   1   0   0   0
  y3   0   0   1   0   0
  y4   0   0   0   1   0
  y5   0   0   0   0   1
d =
      u1
  y1   0
  y2   0
  y3   0
  y4   0
  y5   0
```

Continuous-time system.	The above system is a continuous-time system.
»sysd=c2d(sys,0.05)	This command performs the conversion from continuous time to the discrete equivalent.

The "**M**" matrix

a =

	x1	x2	x3	x4	x5
x1	1.0022	0.30601	-1.6118	-0.030945	0
x2	-9.7288e-05	0.89601	8.0371e-05	0.043893	0
x3	2.7771e-06	-0.0079812	1	0.046214	0
x4	0.00011384	-0.30538	-8.9423e-05	0.84988	0
x5	0.00044733	-8.3698	8.7998	0.0069591	1

The "**N**" matrix

b =

	u1
x1	0.0045138
x2	-0.020239
x3	-0.013066
x4	-0.50842
x5	0.033071

c =

	x1	x2	x3	x4	x5
y1	1	0	0	0	0
y2	0	1	0	0	0
y3	0	0	1	0	0
y4	0	0	0	1	0
y5	0	0	0	0	1

d =

	u1
y1	0
y2	0
y3	0
y4	0
y5	0

Sampling time: 0.05
Discrete-time system.

Note that MATLAB denotes the matrices defining the discrete-equivalent system as **a**, **b**, **c**, and **d**, while we had denoted them as **M**, **N**, **C**, and **D**. We can see that for the above discrete-equivalent system found using c2d, the matrices denoted **a** and **b** are the same as the **M** and **N** matrices found in Example 8.5.

As noted previously, the Control Toolbox in MATLAB also has several routines for directly simulating linear dynamic models. Two that are particularly useful are step and lsim. The step routine generates the step responses of a linear dynamic model, while lsim plots the responses of the linear model to user-defined time histories.

MATLAB EXAMPLE 8.7

Simulation of the Navion's Longitudinal Dynamics with lsim

Using the MATLAB routine lsim, plot the Navion's longitudinal responses to a two-second, one-degree elevator doublet, as in Example 8.5.

■ Solution

The **A**, **B**, **C**, and **D** matrices, and the time and elevator-input sequences, will be the same as in Example 8.5. So turning directly to the MATLAB command line we have

```
»sys=ss(A,B,C,D)

a =
           x1        x2      x3       x4     x5
   x1   0.0451    6.3480  -32.2       0      0
   x2  -0.0021   -2.0244    0         1      0
   x3       0         0     0         1      0
   x4   0.0021   -6.958     0   -3.0757      0
   x5       0      -176   176         0      0

b =
          u1
   x1      0
   x2  -0.16
   x3      0
   x4 -11.029
   x5      0

c =
        x1  x2  x3  x4  x5
   y1    1   0   0   0   0
   y2    0   1   0   0   0
   y3    0   0   1   0   0
   y4    0   0   0   1   0
   y5    0   0   0   0   1

d =
        u1
   y1    0
   y2    0
   y3    0
   y4    0
   y5    0

Continuous-time system.

»lsim(sys,u,t,x0)
```

Note that the labels in Figure 8.6 were added to the plots later in Microsoft Powerpoint

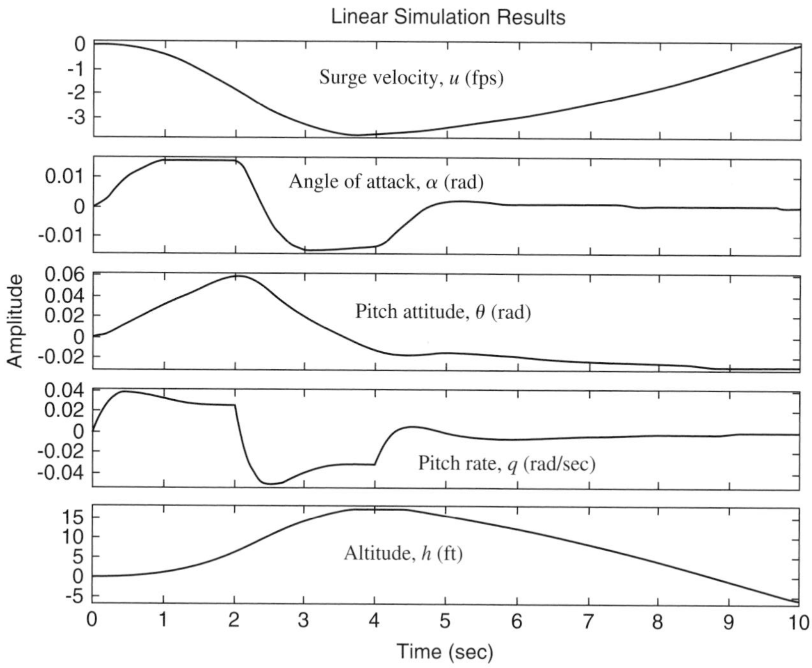

Figure 8.6 Time histories of the Navion's longitudinal responses.

Noting that the angles and pitch rate are plotted in Figure 8.6 in radians, we can see that the results agree with those given in Example 8.5. Although lsim is quick and easy, we do not have the ability to adjust the format of the plots, and must still plot the control-input sequence separately, if it is desired, as in the previous example.

Finally, the graphics-based simulation tool Simulink, which is included in the MATLAB package, is also effective for building both linear and nonlinear simulations. The basic idea in using Simulink is to draw a simulation diagram, which is simply a flow chart describing the simulation. The elements in the Simulink diagram include blocks such as those containing the dynamic models (either state-variable models or transfer functions), blocks describing the inputs to the simulation, and blocks describing how the outputs from the simulation will be recorded and/or plotted. Then all the blocks are connected graphically to describe how the various blocks interact.

SIMULINK EXAMPLE 8.8

Simulating the Longitudinal Dynamics of the Navion

Simulate the longitudinal dynamics of the Navion aircraft, as presented in Example 8.5, except using Simulink.

■ Solution

We will first build the Simulink blocks that define the dynamic elements of the simulation. The blocks are selected from the Simulink Library, which appears when the command

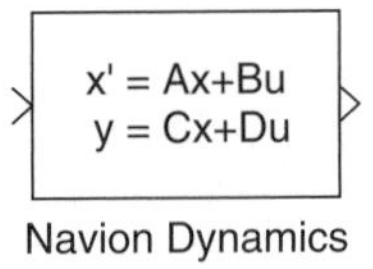

Figure 8.7 Dynamics Simulink block.

Simulink is typed in the MATLAB command line, and drug onto the Simulink file window that also opens along with the Library.

The first block, titled Navion Dynamics, will describe the vehicle's dynamics in state-variable format. It appears as shown in Figure 8.7, with the label we've provided. Double-clicking on the block opens a dialog box in which the dynamics can be defined, along with the initial conditions on the state variables. In this case we have provided the names of the matrices describing the state-variable model, that is, **A**, **B**, **C**, and **D**, which we define in the MATLAB workspace.

Next we introduce the group of blocks shown in Figure 8.8, including three input blocks and one summation block, with which we define the control-input. The input here is the perturbation elevator deflection in degrees. With this block group we create the one-degree, two-second elevator doublet. Note that we have connected the blocks with arrows defining how the blocks interact. The arrows are added using MATLAB's graphical user interface in the Simulink window.

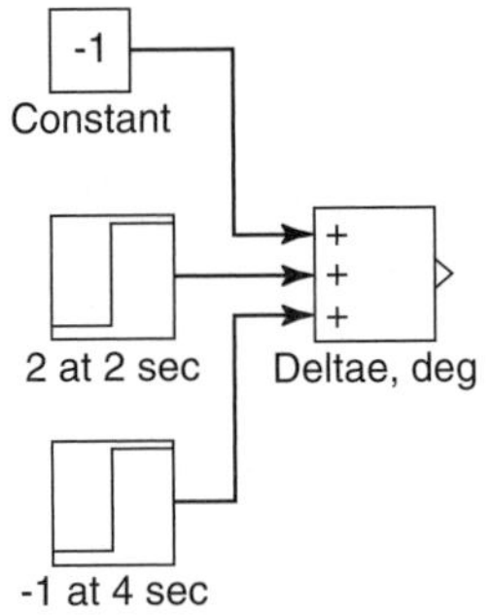

Figure 8.8 Interconnected constant, step, and summation Simulink blocks.

The third block, shown in Figure 8.9, is a demultiplexer box that splits the output vector **y** into the scalar responses (five in this case).

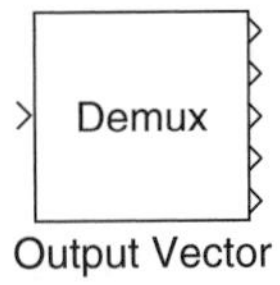

Figure 8.9 Demultiplexer Simulink block.

Four additional blocks are now introduced. As shown in Figure 8.10, these blocks simply multiply the input to the block by a user-specified constant. In our case, we want to convert the angular units between radians and degrees for plotting.

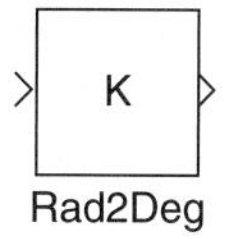

Figure 8.10 Multiplier, or Gain, Simulink block.

And finally, six Scope blocks are introduced, which generate plots of the five system responses plus the elevator input. Each Scope block is labeled to define the output being plotted. The block for plotting the elevator input is shown in Figure 8.11.

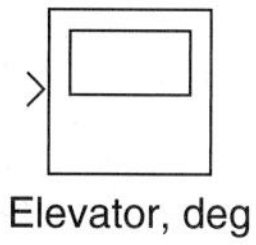

Figure 8.11 Scope Simulink block for plotting.

Instead of using Scope blocks, the responses may also be written to files and then plotted normally in MATLAB.

Now all the blocks above are connected using the graphical user interface to obtain the Simulink block diagram shown in Figure 8.12.

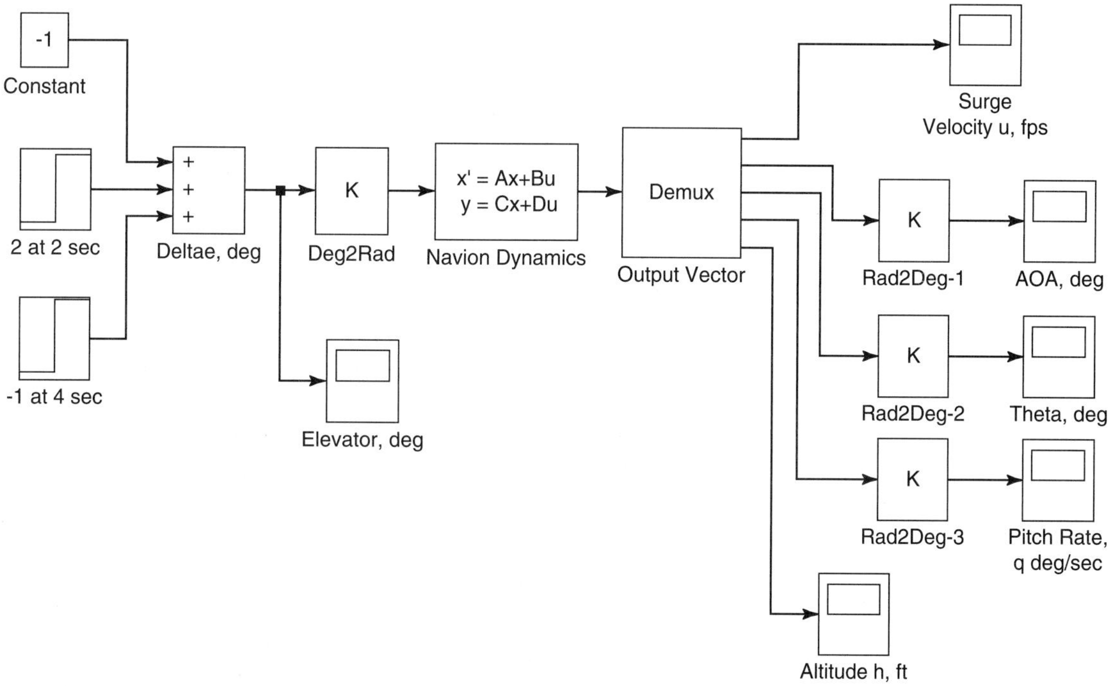

Figure 8.12 Simulink block diagram for simulating the Navion dynamics.

The simulation time of 10 seconds is set in the pull-down menu under Simulation, and the simulation is executed by clicking on Start in the same pull-down menu.

The time histories of the five responses plus elevator input are plotted in Figures 8.13, 8.14, and 8.15. Since we used the Scope blocks to obtain the plots we are not able to

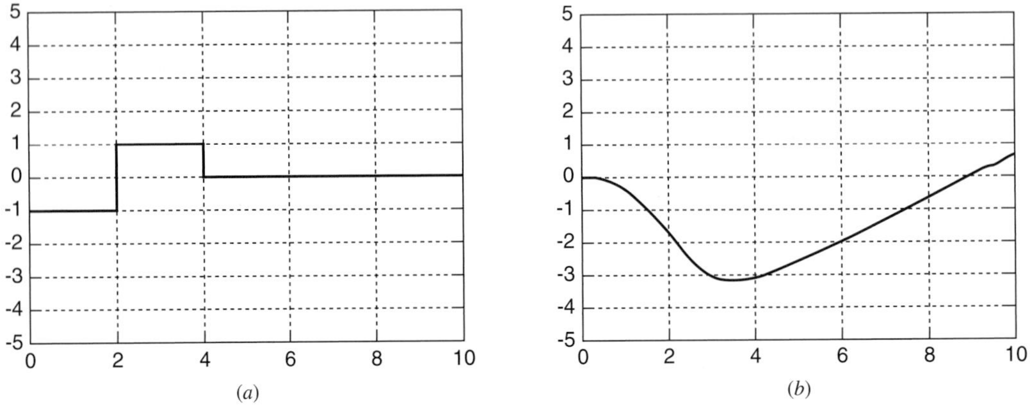

Figure 8.13 Time histories of elevator input (deg) and surge velocity (fps)—time in seconds.

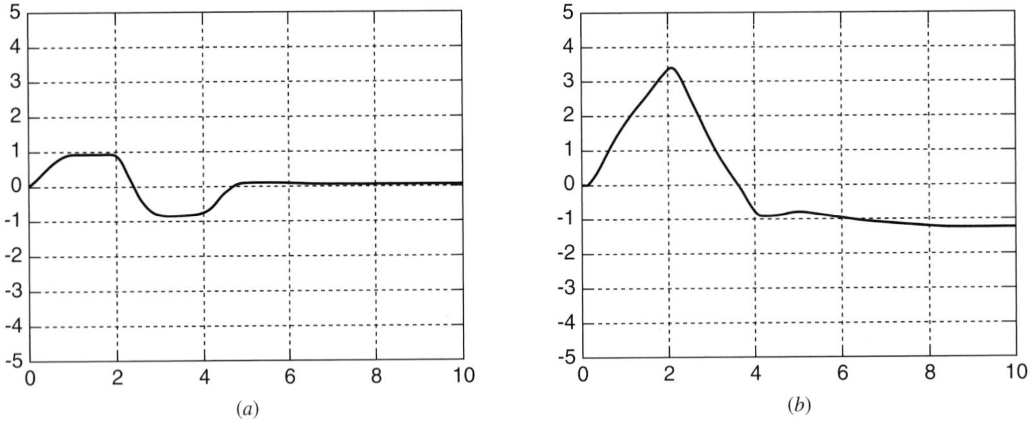

Figure 8.14 Time histories of angle of attack and pitch attitude (deg)—time in seconds.

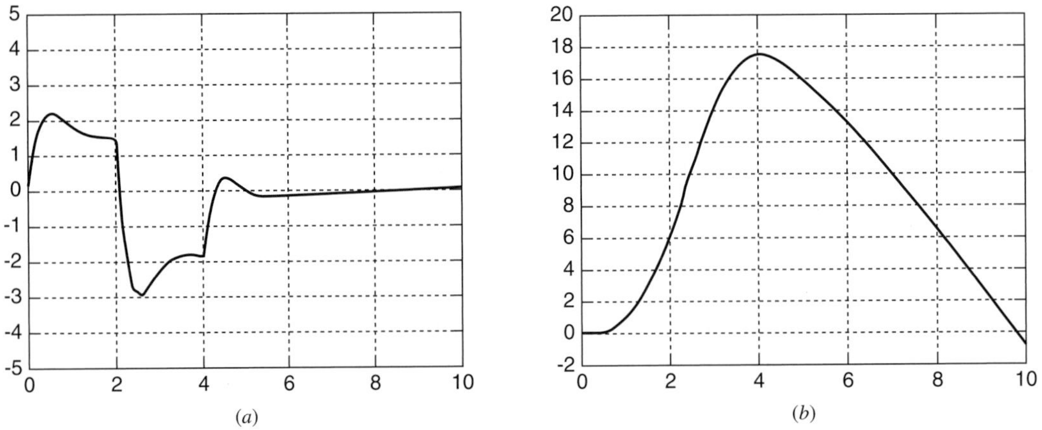

Figure 8.15 Time histories of pitch rate (deg/sec) and altitude (ft)—time in seconds.

label the axes. These blocks are for quickly obtaining plots of the outputs. Comparing these plots with those in Example 8.5 we see that they are in close agreement.

8.2 NONLINEAR MODEL ASSEMBLY AND SIMULATION

We will now turn our attention to nonlinear simulations. As noted in the introduction to this chapter, nonlinear simulations use the nonlinear equations of motion in the mathematical model. We will assemble these nonlinear models of the vehicle's dynamics, as well as discuss various nonlinear-simulation techniques and tools.

8.2.1 Nonlinear Equations of Motion

As in the linear case discussed in Section 8.1, the major pieces of the mathematical model of the vehicle's dynamics have been discussed in earlier chapters. Now we will assemble these pieces. The core of the models consists of the nonlinear equations of motion developed in Chapter 2 for rigid vehicles, or in Chapter 4 for flexible vehicles.

Let us first consider the equations developed in Chapter 2. We will focus on the flat-earth case, and assume that the vehicle is symmetric with respect to its XZ plane (thus several cross-products of inertia are zero). We will also ignore the effects of rotating machinery on the vehicle. The procedures followed here may be applied directly to handle these other cases if required, by starting with the appropriate set of equations from Chapter 2.

From Equations (2.22), the nonlinear equations governing the rigid-body translation of the vehicle are

$$
\begin{aligned}
m(\dot{U} + QW - VR) &= -mg\sin\theta + F_{A_X} + F_{P_X} \\
m(\dot{V} + RU - PW) &= mg\cos\theta\sin\phi + F_{A_Y} + F_{P_Y} \\
m(\dot{W} + PV - QU) &= mg\cos\theta\cos\phi + F_{A_Z} + F_{P_Z}
\end{aligned}
$$

Rewriting these in a more convenient form we have

$$
\boxed{
\begin{aligned}
\dot{U} &= -QW + VR - g\sin\theta + (F_{A_X} + F_{P_X})/m \\
\dot{V} &= -RU + PW + g\cos\theta\sin\phi + (F_{A_Y} + F_{P_Y})/m \\
\dot{W} &= -PV + QU + g\cos\theta\cos\phi + (F_{A_Z} + F_{P_Z})/m
\end{aligned}
}
\tag{8.108}
$$

Recall that U, V, and W are the components (in the vehicle-fixed frame) of the velocity of the vehicle's center of mass with respect to the earth-fixed inertial frame, and m is the vehicle mass.

And from Equations (2.27), the equations governing the rigid-body rotation of a symmetric vehicle are given by

$$I_{xx}\dot{P} - I_{xz}(\dot{R} + PQ) + (I_{zz} - I_{yy})RQ = L_A + L_P$$
$$I_{yy}\dot{Q} + (I_{xx} - I_{zz})PR + I_{xz}(P^2 - R^2) = M_A + M_P$$
$$I_{zz}\dot{R} - I_{xz}(\dot{P} - QR) + (I_{yy} - I_{xx})PQ = N_A + N_P$$

The $I_{..}$ terms are the various moments and products of inertia, and P, Q, and R are the components (in the vehicle-fixed frame) of the rate of rotation of the vehicle-fixed frame with respect to the earth-fixed inertial frame.

But Equations (2.27) may be converted to a more usable form for simulation. Rewriting the first and third of these equations in a convenient matrix form we have

$$\begin{aligned} &I_{yy}\dot{Q} = (I_{zz} - I_{xx})PR + I_{xz}(R^2 - P^2) + M_A + M_P \\ &\begin{bmatrix} I_{xx} & -I_{xz} \\ -I_{xz} & I_{zz} \end{bmatrix} \begin{Bmatrix} \dot{P} \\ \dot{R} \end{Bmatrix} = \begin{Bmatrix} I_{xz}PQ + (I_{yy} - I_{zz})RQ + L_A + L_P \\ -I_{xz}QR + (I_{xx} - I_{yy})PQ + N_A + N_P \end{Bmatrix} \end{aligned} \tag{8.109}$$

And so we have the desired equations given by

$$\begin{aligned} &\dot{Q} = \frac{1}{I_{yy}}\left((I_{zz} - I_{xx})PR + I_{xz}(R^2 - P^2) + M_A + M_P\right) \\ &\begin{Bmatrix} \dot{P} \\ \dot{R} \end{Bmatrix} = \frac{1}{I_{xx}I_{zz} - I_{xz}^2} \begin{bmatrix} I_{zz} & I_{xz} \\ I_{xz} & I_{xx} \end{bmatrix} \begin{Bmatrix} I_{xz}PQ + (I_{yy} - I_{zz})RQ + L_A + L_P \\ -I_{xz}QR + (I_{xx} - I_{yy})PQ + N_A + N_P \end{Bmatrix} \end{aligned} \tag{8.110}$$

The kinematic equations relating the vehicle's inertial rate of rotation to the Euler-angle rates are given in Equations (2.37), or

$$\begin{aligned} &\dot{\phi} = P + Q\sin\phi\tan\theta + R\cos\phi\tan\theta \\ &\dot{\theta} = Q\cos\phi - R\sin\phi \\ &\dot{\psi} = (Q\sin\phi + R\cos\phi)\sec\theta \end{aligned} \tag{8.111}$$

where ψ, θ, and ϕ are the 3-2-1 Euler angles, respectively, defining the orientation of the vehicle-fixed frame with respect to the inertial earth-fixed frame. Finally, the three kinematic equations relating the inertial velocity to the inertial position of the vehicle are given by Equations (2.40), or

$$\begin{aligned} &\dot{X}_E = U\cos\theta\cos\psi + V(\sin\phi\sin\theta\cos\psi - \cos\phi\sin\psi) + W(\cos\phi\sin\theta\cos\psi + \sin\phi\sin\psi) \\ &\dot{Y}_E = U\cos\theta\sin\psi + V(\sin\phi\sin\theta\sin\psi + \cos\phi\cos\psi) + W(\cos\phi\sin\theta\sin\psi - \sin\phi\cos\psi) \\ &\dot{h} = U\sin\theta - V\sin\phi\cos\theta - W\cos\phi\cos\theta \end{aligned} \tag{8.112}$$

Table 8.3 Nonlinear Translation and Rotational Equations of Motion

Case	Translational Equations	Rotational Equations
Rotating machinery	Equations (8.108) (unchanged)	Equations (8.110) plus underlined terms in Equations (2.75)
Variable mass	Equations (8.108) plus underlined terms in Equations (2.89) (see Example 2.4)	Equations (8.110) plus underlined terms in Equations (2.100) (see Example 2.4)
Rotating, spherical earth	Equations (2.123)	Equations (8.110) (*Note*: also need Equations (2.114), (2.115), and (2.125) for new kinematic equations)
Performance (wind axes)	Equations (2.136)	Not applicable

where X_E, and Y_E are the "north" and "east" coordinates of the vehicle, respectively, and h is its altitude. We use the above 12 equations of motion in the nonlinear simulation.

The final additions to the simulation model are auxiliary responses such as the inertial acceleration of a particular location on the vehicle. This acceleration (vector) was given by Equation (8.25).

In the presentation in this section (8.2.1), we considered only the case involving a flat, non-rotating earth, ignoring the effects of rotating machinery or variable mass. If one is considering these other effects, then one must begin with the appropriate set of equations governing the vehicles translation and rotation. In Table 8.3, these other cases are listed along with the corresponding equation set governing translation and rotation.

8.2.2 Models for the Aerodynamic and Propulsive Forces and Moments

There are a variety of ways to incorporate models for the forces and moments into nonlinear simulations, because there are a variety of such models. As discussed in Chapter 6, for example, models for the aerodynamic forces and moments may be based on wind-tunnel data, computational results, semi-empirical methods, or combinations of these. The models developed in Chapter 6, based on semi-empirical methods, were basically locally linear models. That is, except for the models for drag, the models for the aerodynamic forces and moments were linear in the motion variables such as angle of attack and control-surface deflections. But aerodynamic models based on wind-tunnel data or computational techniques typically capture more fully the nonlinear characteristics of the forces and moments. Furthermore, such nonlinear models are frequently desired for nonlinear simulations.

Perhaps the most general and most common form of nonlinear force and moment model is the table-lookup model. In such a model the data is summarized in tables, or databases, in which the entries depend on the independent variables (e.g., Mach number, angles of attack and sideslip, control-surface deflections, and configuration such as landing-gear or flap position). Then whenever a force or moment is needed during the computation, it is extracted from the tabulated databases using numerical interpolation schemes. Although somewhat computation-intensive, this approach is quite straightforward, and the databases are easily modified as additional data become available.

Alternatives to table-lookup models are analytical models based on polynomial or spline fits to the databases. Such approaches speed up the computations, but are more difficult to set up or to change as new data becomes available. The simplest analytical models for the aerodynamic and propulsive forces are similar to those developed in Chapter 6. To demonstrate the use of the force and moment models in nonlinear simulation, we will use these models in the developments to follow.

Consistent with Equations (6.17), (6.24), and (6.29), which were developed assuming a conventional vehicle geometry, we may write the coefficients for aerodynamic lift, side force and drag as

$$\boxed{\begin{aligned} C_L &= C_{L_0} + C_{L_\alpha}\alpha + C_{L_q}Q + C_{L_{\dot\alpha}}\dot\alpha + C_{L_{i_H}}i_H + C_{L_{\delta_E}}\delta_E \\ C_S &= C_{S_\beta}\beta + C_{S_p}P + C_{S_r}R + C_{S_{\delta_A}}\delta_A + C_{S_{\delta_R}}\delta_R \\ C_D &= C_{D_0} + \left(\frac{C_{L_W}^2}{\pi A_W e_W} + \frac{C_{L_H}^2}{\pi A_H e_H}\frac{q_H}{q_\infty}\frac{S_H}{S_W} + \frac{C_{S_V}^2}{\pi A_V e_V}\frac{q_H}{q_\infty}\frac{S_V}{S_W}\right) \end{aligned}} \tag{8.113}$$

To use the above equation for drag coefficient, we may, for example, apply the following models for the lift on the wing and tail surfaces.

$$\begin{aligned} C_{L_W} &= C_{L_{\alpha_W}}(\alpha + i_W - \alpha_{0_W}) \\ C_{L_H} &= C_{L_{\alpha_H}}\left(\left(1 - \frac{d\varepsilon}{d\alpha}\right)(\alpha + i_W) + \frac{d\varepsilon}{d\alpha}\alpha_{0_W} + i_H - \alpha_{0_H} + \alpha_\delta\delta_E\right) \\ C_{S_V} &= C_{S_{\beta_V}}(\beta + \beta_\delta\delta_R) \end{aligned}$$

These expressions were developed during the derivations of Equations (6.17), (6.24), and (6.29). The aerodynamic forces then are given by

$$\boxed{L = C_L q_\infty S_W,\ \mathsf{S} = C_S q_\infty S_W,\ \text{and}\ D = C_D q_\infty S_W} \tag{8.114}$$

with

$$q_\infty = \frac{1}{2}\rho_\infty\left(U^2 + V^2 + W^2\right) \tag{8.115}$$

Consistent with Equations (6.43), (6.55), and (6.67) we may write the expressions for the rolling-, pitching-, and yawing-moment coefficients as

$$\boxed{\begin{aligned} C_{L_{\text{Roll}}} &= C_{L_\beta}\beta + C_{L_p}P + C_{L_r}R + C_{L_{\delta_A}}\delta_A + C_{L_{\delta_R}}\delta_R \\ C_M &= C_{M_{\alpha=\delta_E=i_H=0}} + C_{M_\alpha}\alpha + C_{M_{\dot{\alpha}}}\dot{\alpha} + C_{M_q}Q + C_{M_{i_H}}i_H + C_{M_{\delta_E}}\delta_E \\ C_N &= C_{N_\beta}\beta + C_{N_p}P + C_{N_r}R + C_{N_{\delta_A}}\delta_A + C_{N_{\delta_R}}\delta_R \end{aligned}} \tag{8.116}$$

And the aerodynamic rolling, pitching, and yawing moments, respectively, are given by

$$\boxed{L_A = C_{L_{\text{Roll}}}q_\infty S_W b_W,\ M_A = C_M q_\infty S_W \bar{c}_W,\ \text{and}\ N_A = C_N q_\infty S_W b_W} \tag{8.117}$$

Consistent with Section 6.3, the propulsive thrust T may be written as

$$\boxed{T = C_T(V_{\text{inlet}}, \text{h}, \pi)q_\infty S_W} \tag{8.118}$$

where the thrust coefficient C_T, as well as any propulsive normal forces F_N, are assumed available from engine data. If a lag in the engine response is to be included, Equation (8.118) may be modified to include the transfer function

$$C_T(s) = \left(\frac{p_T}{s + p_T}\right)C_{T_C}(V_{\text{inlet}}, \text{h}, \pi) \tag{8.119}$$

with p_T selected to obtain the desired engine lag.

Also, from Equations (6.268), the propulsive pitching moment, for example, may be written as

$$\boxed{M_P = C_{P_M}q_\infty S_W \bar{c}_W} \tag{8.120}$$

where the propulsive pitching-moment coefficient is

$$C_{P_M} = \frac{1}{q_\infty S_W \bar{c}_W}\sum_{i=1}^{n_P}\left(T_i(V_{\text{inlet}}, \text{h}, \pi)\left(d_{T_i}\cos\phi_{T_i} - x_{T_i}\sin\phi_{T_i}\right) - F_{N_{Z_i}}\left(x_{T_i}\cos\phi_{T_i} + d_{T_i}\sin\phi_{T_i}\right)\right) \tag{8.121}$$

with d_T, x_T, and ϕ_T as defined in Figure 6.13. The engine data used to obtain the thrust coefficient C_T and normal forces may either be tabulated, leading to a table lookup, or fit to some function of altitude, inlet Mach number, and throttle setting, for example.

8.2.3 Assembling the Nonlinear Mathematical Model

To assemble the complete nonlinear mathematical model, we must first develop the expressions for the components of the forces and moments given in the equations of motion, or Equations (8.108) and (8.110). To do so, we must select the vehicle-fixed axes to be used. For nonlinear simulations, fuselage-referenced axes are usually the appropriate choice. In these axes the moments and products of inertia will be constant for a constant mass vehicle of fixed geometry. Also, we will not be dealing with a fixed reference flight condition, as in a linear perturbation analysis.

We may now develop the components of the aerodynamic and propulsive forces in the vehicle-fixed axes. With reference to Figure 6.4, the components of the aerodynamic forces were given by Equations (6.264), or

$$\boxed{\begin{aligned} F_{A_X} &= C_X q_\infty S_W = -D\cos\alpha\cos\beta - \mathrm{S}\cos\alpha\sin\beta + L\sin\alpha \\ F_{A_Y} &= C_Y q_\infty S_W = -D\sin\beta + \mathrm{S}\cos\beta \\ F_{A_Z} &= C_Z q_\infty S_W = -D\sin\alpha\cos\beta - \mathrm{S}\sin\alpha\sin\beta - L\cos\alpha \end{aligned}} \tag{8.122}$$

or, in terms of coefficients, we have

$$\begin{aligned} C_X &= -C_D\cos\alpha\cos\beta - C_S\cos\alpha\sin\beta + C_L\sin\alpha \\ C_Y &= -C_D\sin\beta + C_S\cos\beta \\ C_Z &= -C_D\sin\alpha\cos\beta - C_S\sin\alpha\sin\beta - C_L\cos\alpha \end{aligned} \tag{8.123}$$

The components of the propulsive forces, assuming the thrust is symmetric with respect to the vehicle XZ plane, are given by Equations (6.265), or

$$\boxed{\begin{aligned} F_{P_X} &= C_{P_X} q_\infty S_W = T(V_{\text{inlet}}, \mathrm{h}, \pi)\cos\phi_T \\ F_{P_Y} &= C_{P_Y} q_\infty S_W = F_{N_Y} \\ F_{P_Z} &= C_{P_Z} q_\infty S_W = -T(V_{\text{inlet}}, \mathrm{h}, \pi)\sin\phi_T + F_{N_Z} \end{aligned}} \tag{8.124}$$

The aerodynamic moments are as given in Equations (6.267), which are

$$\boxed{\begin{aligned} L_A &= C_{L_{\text{Roll}}} q_\infty S_W b_W \\ M_A &= C_M q_\infty S_W \bar{c}_W \\ N_A &= C_N q_\infty S_W b_W \end{aligned}} \tag{8.125}$$

while the components of the propulsive moments are as given in Equations (6.268), or

$$\boxed{\begin{aligned} L_P &= \sum_{i=1}^{n_P}\left(y_{T_i}\left(F_{N_{Z_i}}\cos\phi_{T_i} - T_i(V_{\text{inlet}}, \mathrm{h}, \pi)\sin\phi_{T_i}\right) - d_{T_i}F_{N_{Y_i}}\right) \\ M_P &= \sum_{i=1}^{n_P}\left(T_i(V_{\text{inlet}}, \mathrm{h}, \pi)\left(d_{T_i}\cos\phi_{T_i} - x_{T_i}\sin\phi_{T_i}\right) - F_{N_{Z_i}}\left(x_{T_i}\cos\phi_{T_i} + d_{T_i}\sin\phi_{T_i}\right)\right) \\ N_P &= -\sum_{i=1}^{n_P}\left(y_{T_i}T_i(V_{\text{inlet}}, \mathrm{h}, \pi)\cos\phi_{T_i} + x_{T_i}F_{N_{Y_i}}\right) \end{aligned}} \tag{8.126}$$

The complete model for the dynamics of a rigid vehicle then consists of the six equations of motion, Equations (8.108) and (8.110), the six kinematic equations, Equations (8.111) and (8.112), and the models for the forces and moments. The force and moment models consist of Equations (8.122–8.126). If the model presented in Section 8.2.2 is used for lift, drag, side force, and so on, then Equations (8.113–8.120) provide the necessary force, moment, and thrust coefficients. A model for the atmospheric density plus additional auxiliary equations consisting of

$$\boxed{\begin{aligned} \tan\alpha &= W/U \left(\text{or } \alpha \approx W/U\right) \\ \tan\beta &= V/U \left(\text{or } \beta \approx V/U\right) \\ V_\infty &= \sqrt{U^2 + V^2 + W^2} \end{aligned}} \tag{8.127}$$

must also be included. A model for the atmospheric density is given in Appendix A.

Finally, additional responses consisting of the local acceleration at a point $\mathbf{p}(x,y,z)$ relative to the vehicle center of mass are frequently included in the model. This acceleration (vector) was given for a rigid vehicle in Equation (8.25), or

$$\mathbf{a}_R(x,y,z,t) = \mathbf{a}_R(\mathbf{p},t) = \frac{d\mathbf{V}_V}{dt}\Big|_V + \left(\boldsymbol{\omega}_{V,I} \times \mathbf{V}_V\right) + \left(\boldsymbol{\omega}_{V,I} \times \left(\boldsymbol{\omega}_{V,I} \times \mathbf{p}\right)\right) + \left(\frac{d\boldsymbol{\omega}_{V,I}}{dt}\Big|_V \times \mathbf{p}\right)$$

The components of $\mathbf{a}_R$ in the vehicle-fixed frame are

$$\boxed{\begin{aligned} a_{X_R} &= \dot{U} + Q\left(W + (Py - Qx)\right) - R\left(V + (Rx - Pz)\right) + \left(\dot{Q}z - \dot{R}y\right) \\ a_{Y_R} &= \dot{V} + R\left(U + (Qz - Ry)\right) - P\left(W + (Py - Qx)\right) + \left(\dot{R}x - \dot{P}z\right) \\ a_{Z_R} &= \dot{W} + P\left(V + (Rx - Pz)\right) - Q\left(U + (Qz - Ry)\right) + \left(\dot{P}y - \dot{Q}x\right) \end{aligned}} \tag{8.128}$$

So the three components of Equation (8.25) are now known.

In closing this section it is important to note that if a nonlinear model of only the longitudinal equations is desired, and no effects of rotating machinery are included, one may use the equations for $\dot{U}$, $\dot{W}$, and $\dot{Q}$ from Equations (8.108) and (8.110), and hold V, P, and R equal to zero. A similar approach may be used if a nonlinear simulation of only the lateral-directional dynamics is desired.

8.2.4 Models for Flexible Vehicles

As discussed in Section 8.1.5, when developing a simulation of a flexible vehicle we must include the equations of motion governing the elastic degrees of freedom. These equations were developed in Chapter 4. The equations governing rigid-body translations and rotations remain unchanged except for necessary adjustments in the forces and moments acting on the vehicle. Adjustments to the models for these forces and moments were developed in Chapter 7. So for a flexible vehicle we use the nonlinear equations of motion, Equations (8.108), (8.110), (8.111), and (8.112), all of which remain unchanged in form.

From Equations (4.84), repeated here, the equations governing the dynamics of the elastic degrees of freedom were given by

$$\ddot{\eta}_i + \omega_i^2 \eta_i = \frac{Q_i}{\mathcal{M}_i}, \; i = 1, \ldots, n$$

Recall that η_i is the total (not perturbation) modal coordinate associated with the i'th vibration mode included in the model, Q_i is the generalized force associated with this degree of freedom, ω_i is the in-vacuo vibration frequency of the mode, and $\mathcal{M}_i$ is the modal generalized mass. The latter two quantities are assumed available from a prior vibration analysis of the vehicle's structure.

Note that frequently a small modal damping ζ_i of approximately 0.02 is added to each of Equations (4.84) above, to improve agreement between the analytical and experimental results from the vibration analysis. With damping included, Equations (4.84) become

$$\boxed{\ddot{\eta}_i + 2\zeta_i\omega_i\dot{\eta}_i + \omega_i^2 \eta_i = \frac{Q_i}{\mathcal{M}_i}, \; i = 1, \ldots, n} \tag{8.129}$$

The forces and moments acting on a flexible vehicle were modeled in Chapter 7, and since vibration mode shapes are defined in terms of fuselage-referenced axes, those axes are usually used in the development of the models for the forces and moments. The components of the forces acting on the vehicle were given in Equations (7.3), or

$$\begin{aligned} F_{A_X} &= C_X q_\infty S_W = -D\cos\alpha\cos\beta - \mathsf{S}\cos\alpha\sin\beta + L\sin\alpha \\ F_{A_Y} &= C_Y q_\infty S_W = -D\sin\beta + \mathsf{S}\cos\beta \\ F_{A_Z} &= C_Z q_\infty S_W = -D\sin\alpha\cos\beta - \mathsf{S}\sin\alpha\sin\beta - L\cos\alpha \end{aligned} \tag{8.130}$$

and these expressions were developed assuming fuselage-referenced axes.

Equations (7.99), or

$$\begin{aligned} F_{A_{X_E}} &= \sum_{i=1}^{n}\left(\frac{\partial F_{A_X}}{\partial \eta_i}\eta_i + \frac{\partial F_{A_X}}{\partial \dot{\eta}_i}\dot{\eta}_i\right) \\ F_{A_{Y_E}} &= \sum_{i=1}^{n}\left(\frac{\partial F_{A_Y}}{\partial \eta_i}\eta_i + \frac{\partial F_{A_Y}}{\partial \dot{\eta}_i}\dot{\eta}_i\right) \\ F_{A_{Z_E}} &= \sum_{i=1}^{n}\left(\frac{\partial F_{A_Z}}{\partial \eta_i}\eta_i + \frac{\partial F_{A_Z}}{\partial \dot{\eta}_i}\dot{\eta}_i\right) \end{aligned} \tag{8.131}$$

were developed to model the effects of elastic deformation on the three aerodynamic force components in Equations (8.130). Therefore, the three force components to be incorporated into the equations of motion governing rigid-body translations (Equations (8.108)) are the sums of the "rigid-body" contributions to Equations (8.130) (e.g., Equations (8.113)) plus the elastic contributions to these force components (Equations (8.131)).

Regarding the moments on the vehicle, the components of the aerodynamic and propulsive moments expressed in the fuselage-referenced axes for a rigid vehicle were developed in Section 6.10.1. These components were denoted

$$\begin{aligned} &L_A + L_P \ \text{(rolling moment)} \\ &M_A + M_P \ \text{(pitching moment)} \\ &N_A + N_P \ \text{(yawing moment)} \end{aligned} \tag{8.132}$$

And the effects of elastic deformation on these moments were given in Equations (7.101), or

$$\begin{aligned} L_{A_E} &= \sum_{i=1}^{n} \left(\frac{\partial L_A}{\partial \eta_i} \eta_i + \frac{\partial L_A}{\partial \dot{\eta}_i} \dot{\eta}_i \right) \\ M_{A_E} &= \sum_{i=1}^{n} \left(\frac{\partial M_A}{\partial \eta_i} \eta_i + \frac{\partial M_A}{\partial \dot{\eta}_i} \dot{\eta}_i \right) \\ N_{A_E} &= \sum_{i=1}^{n} \left(\frac{\partial N_A}{\partial \eta_i} \eta_i + \frac{\partial N_A}{\partial \dot{\eta}_i} \dot{\eta}_i \right) \end{aligned} \tag{8.133}$$

These three expressions may then simply be added to the components of moments for the rigid vehicle, Equations (8.132), and the results inserted into the right-hand sides of the three equations of motion governing the rigid-body rotations (Equations (8.110)).

To complete the model of the elastic effects, the generalized forces acting on the elastic degrees of freedom themselves were given in Equations (7.103), or

$$Q_i = \left(Q_{i_{\mathbf{p}=0}} + \frac{\partial Q_i}{\partial \mathbf{p}_{\text{Rigid}}} \mathbf{p}_{\text{Rigid}} \right) + \sum_{i=1}^{n} \left(\frac{\partial Q_i}{\partial \eta_j} \eta_j + \frac{\partial Q_i}{\partial \dot{\eta}_j} \dot{\eta}_j \right) \tag{8.134}$$

with

$$\mathbf{p}_{\text{Rigid}}^T = \left[U \ \ \beta\left(\approx \frac{V}{U}\right) \ \ \alpha\left(\approx \frac{W}{U}\right) \ \ P \ \ Q \ \ R \ \ i_H \ \ \delta_E \ \ \delta_A \ \ \delta_R \right]$$

and n the number of elastic modes included in the model. But Equations (7.94) correspond to the terms in the first parentheses on the right of Equation (8.134) above, and Equations (7.95) correspond to the partial derivatives in the second parentheses. Therefore, the n generalized forces Q_i given by Equations (8.134) are known, and may be inserted into the right-hand sides of the corresponding n equations of motion governing the elastic degrees of freedom (Equations (8.129)).

After incorporating Equations (8.134) into Equations (8.129), the model of the dynamics of the flexible vehicle is now almost complete. The final addition is to incorporate the effects of elastic deformation into the dynamic responses of the vehicle (i.e., the response vector **y**).

As discussed in Section 8.1.5, the additional dynamic responses typically included in linear models are those given in Equations (8.53–8.57). These linearized equations were expressed in terms of perturbation variables. But for the nonlinear model of the elastic vehicle, we are interested in the responses of <u>total</u> motion variables. So the total local pitch and yaw rates are now given by

$$\begin{aligned} Q_{\text{local}}(x,t) &= Q(t) + \sum_{i=1}^{n} \nu'_{Z_i}(x)\dot{\eta}_i(t) \\ R_{\text{local}}(x,t) &= R(t) + \sum_{i=1}^{n} \nu'_{Y_i}(x)\dot{\eta}_i(t) \end{aligned} \tag{8.135}$$

Regarding the local accelerations, Equation (8.55) gave the total acceleration, including elastic deformation, while Equation (8.25) is the expression for the contribution of rigid-body motion to this local inertial acceleration at a point **p**, or (x,y,z), referenced relative to the vehicle's center of mass. These equations are repeated here for convenience.

$$\mathbf{a}_{\text{local}}(x,y,z,t) = \mathbf{a}_R(x,y,z,t) + \sum_{i=1}^{n} \boldsymbol{\nu}_i(x,y,z)\ddot{\eta}_i(t)$$

$$\mathbf{a}_R(x,y,z,t) = \mathbf{a}_R(\mathbf{p},t) = \frac{d\mathbf{V}_V}{dt}\Big|_V + (\boldsymbol{\omega}_{V,I} \times \mathbf{V}_V) + (\boldsymbol{\omega}_{V,I} \times (\boldsymbol{\omega}_{V,I} \times \mathbf{p})) + \left(\frac{d\boldsymbol{\omega}_{V,I}}{dt}\Big|_V \times \mathbf{p}\right)$$

After expanding Equation (8.55) and equating $\mathbf{i}_V$, $\mathbf{j}_V$, and $\mathbf{k}_V$ components in the two above vector equations, again under mild assumptions, the results may be expressed as

$$\boxed{\begin{aligned} a_{X_{\text{local}}}(x,y,z,t) &= a_{X_R}(x,y,z,t) + \sum_{i=1}^{n} \nu_{X_i}(x,y,z)\ddot{\eta}_i(t) \\ a_{Y_{\text{local}}}(x,y,z,t) &= a_{Y_R}(x,y,z,t) + \sum_{i=1}^{n} \nu_{Y_i}(x,y,z)\ddot{\eta}_i(t) \\ a_{Z_{\text{local}}}(x,y,z,t) &= a_{Z_R}(x,y,z,t) + \sum_{i=1}^{n} \nu_{Z_i}(x,y,z)\ddot{\eta}_i(t) \end{aligned}} \tag{8.136}$$

where a_{X_R}, a_{Y_R}, and a_{Z_R} are given by Equations (8.128), or

$$\begin{aligned} a_{X_R} &= \dot{U} + Q(W + (Py - Qx)) - R(V + (Rx - Pz)) + (\dot{Q}z - \dot{R}y) \\ a_{Y_R} &= \dot{V} + R(U + (Qz - Ry)) - P(W + (Py - Qx)) + (\dot{R}x - \dot{P}z) \\ a_{Z_R} &= \dot{W} + P(V + (Rx - Pz)) - Q(U + (Qz - Ry)) + (\dot{P}y - \dot{Q}x) \end{aligned}$$

With all the information now given, the three expressions in Equations (8.136) complete the set of additional responses frequently included in simulations of flexible vehicles. The nonlinear model of the dynamics of a flexible vehicle has now been assembled.

8.2.5 Adding Feedback Control Laws to a Simulation Model

The addition of linear feedback control laws to linear models of the vehicle dynamics was addressed in Section 8.1.6. The same approach is used when linear control laws are to be incorporated into nonlinear dynamic models. The state-variable models for the control laws are simply appended to the set of differential equations in the dynamic model. The key exception involves control laws with nonlinear elements, and many times the behavior of the system including the nonlinear control elements is a key aspect of the nonlinear simulation. Examples of such nonlinear elements are limiters, breakouts, or nonlinear gains, and only such nonlinear control elements will be discussed here.

Examples of three nonlinear elements are shown in Figure 8.16, in which the input-output (IO) relationships for a limiter, a breakout, and a nonlinear gain are depicted from left to right, respectively. A linear analytical model can capture none of these IO relationships, but such relationships can be captured using a table-lookup scheme or a nonlinear analytical model. For example, the following model may be used for a linear dynamic control element followed by a limiter.

$$\begin{aligned}\dot{\mathbf{x}}_c &= \mathbf{A}_c\mathbf{x}_c + \mathbf{B}_c y(t)\\ u(t) &= \mathbf{C}_c\mathbf{x}_c + D_c y,\ \ \text{if } |u(t)| > U_{\max}\ \ \text{then } u(t) = U_{\max}\left(\frac{u(t)}{|u(t)|}\right)\end{aligned} \tag{8.137}$$

Note that this model consists of a state-variable model for the linear portion of the dynamic control element, plus an algorithm for limiting the output of the control element. This model can be easily incorporated into the numerical integration scheme used in the simulation. Models of breakout elements may also be developed in a similar fashion.

Models of nonlinear gains, or gain schedules, frequently involve table look-ups. For example, consider a dynamic control law of the form

$$\begin{aligned}\dot{\mathbf{x}}_c &= \mathbf{A}_c\mathbf{x}_c + \mathbf{B}_c y(t)\\ u(t) &= \mathbf{C}_c\mathbf{x}_c + D_{\text{Var}} y,\ \ D_{\text{Var}} = K_{\text{Var}}(\text{h}, M_\infty)\end{aligned} \tag{8.138}$$

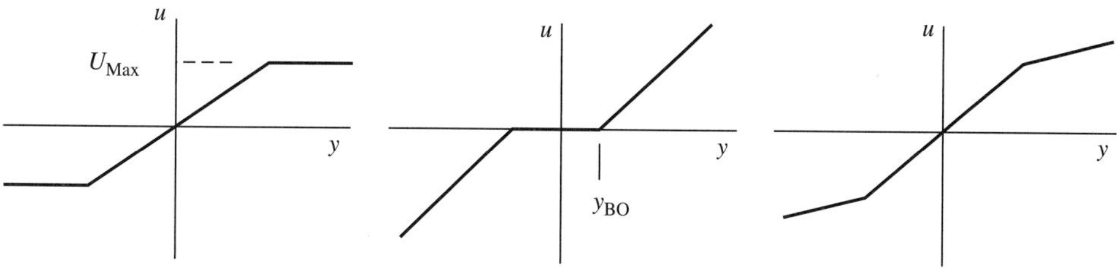

Figure 8.16 Three nonlinear control elements—limiter, breakout, and nonlinear gain.

This control law is linear but with a variable gain that is a function of flight altitude and Mach number. Since during the numerical integration of the dynamic model the vehicle altitude and Mach number at a particular simulated time are available, they may be used to look up the variable gain K_{Var} at that particular point in the simulation.

So we see that by using a numerical scheme that can integrate nonlinear equations, the addition of logic or table lookups in the models of nonlinear control elements poses little difficulty.

8.2.6 Adding Atmospheric Turbulence to a Simulation Model

Incorporating models of atmospheric turbulence into linear simulations was addressed in Section 8.1.7, and a similar approach is used when incorporating turbulence into nonlinear simulations. The same analytical models for the turbulence, such as those discussed in Appendix C, are used in the nonlinear case. The dynamic model for the turbulence is simply appended to the set of differential equations to be integrated.

Section 6.8 addressed the effects of turbulence on the aerodynamic forces and moments. In Sections 6.8 and 8.1.7 the gust velocities were treated as perturbation variables. However, since the reference values of the gust velocities may be taken to be zero, the perturbation and total gust velocities are the same. That is, we can set $w_g = W_g$, for example. Therefore, consistent with Equations (8.77), the effects of turbulence are incorporated into the models for the aerodynamic forces and moments by making the following substitutions in those models (i.e., Equations (8.113–8.117)).

$$\begin{aligned} U_{\text{total}} &= U + U_g \\ V_{\text{total}} &= V + V_g \left(\text{or } \beta_{\text{total}} = \beta + \beta_g\right) \\ W_{\text{total}} &= W + W_g \left(\text{or } \alpha_{\text{total}} = \alpha + \alpha_g\right) \\ \dot{\alpha}_{\text{total}} &= \dot{\alpha} + \dot{\alpha}_g \end{aligned} \tag{8.139}$$

In the equations of motion, U, V (or β), W (or α), and $\dot{\alpha}$ are each governed by their respective nonlinear differential equation, while U_g, V_g (or β_g), W_g (or α_g) and $\dot{\alpha}_g$ are obtained from the turbulence model, such as that given in Appendix C.

8.2.7 Numerical Simulation Techniques—A JITT*

We will address two topics in this section: numerical techniques for integrating nonlinear differential equations (or solving what is known as the *initial-value problem*) and numerical techniques for finding an equilibrium or "trim" flight condition.

---

* A Just-in-Time Tutorial.

Numerical-Integration Techniques To introduce the basic concepts underpinning numerical integration (Ref. 4), consider the evaluation of the definite integral given by

$$A = \int_{t_0}^{t_f} f(t)\,dt$$

where t_0 and t_f are given, and $f(t)$ is a known function. The integral, of course, represents the area under the plot of $f(t)$, evaluated from t_0 to t_f, as depicted on the left side of Figure 8.17.

Numerical approaches for evaluating this integral consist of subdividing the interval $[t_0, t_f]$ into small subintervals, and then approximating the integrand $f(t)$ in each subinterval. For example, if the width of each subinterval is taken to be Δt, and we approximate $f(t)$ in each subinterval by a first-order polynomial (i.e., straight line), the area A is approximated by the following relation.

$$A = \int_{t_0}^{t_f} f(t)\,dt \approx \Delta t\left(\frac{1}{2}f(t_0) + \sum_{i=1}^{n-1} f(t_i) + \frac{1}{2}f(t_f)\right) \tag{8.140}$$

This approach is depicted on the right side of Figure 8.17 (in which $n = 2$) and is the familiar trapezoidal-integration technique.

Now consider finding the solution to the first-order nonlinear differential equation

$$\frac{dx}{dt} = f(x,t) \tag{8.141}$$

over the interval $[t_0, t_f]$, given the initial condition $x(t_0)$. It then follows that

$$x(t_f) - x(t_0) = \int_{t_0}^{t_f} f(x,t)\,dt \tag{8.142}$$

From the above discussion on numerically evaluating a definite integral, it is clear that if we knew $f(x,t)$ over the interval $[t_0, t_f]$, we could evaluate the above expression. But $f(x,t)$ is a function of x, the solution we seek, so an alternate approach is required.

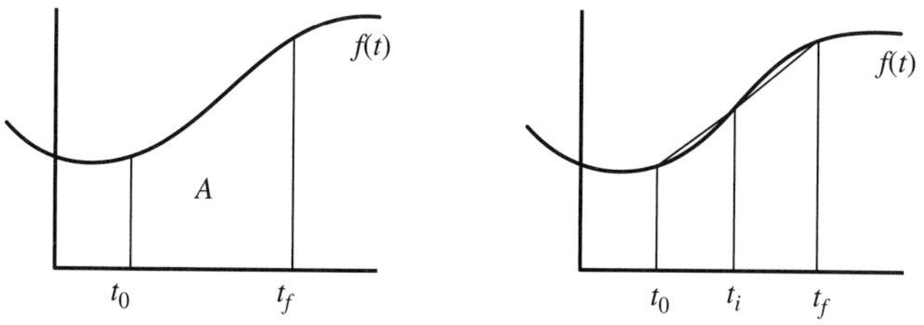

Figure 8.17 Graphical interpretation of the definite integral.

Let us instead approximate the solution for $x(t_i + 1)$ in terms of a Taylor series expanded about $x(t_i)$. Or, let

$$\begin{aligned} x(t_{i+1}) &= x(t_i) + \Delta t\frac{dx}{dt} + \frac{\Delta t^2}{2!}\frac{d^2x}{dt^2} + \frac{\Delta t^3}{3!}\frac{d^3x}{dt^3} + \cdots \\ &= x(t_i) + \Delta t\, f(x(t_i),t_i) + \frac{\Delta t^2}{2!}\frac{\partial f(x(t_i),t_i)}{\partial t} + \frac{\Delta t^3}{3!}\frac{\partial^2 f(x(t_i),t_i)}{\partial t^2} + \cdots \end{aligned} \tag{8.143}$$

If the interval step size Δt is sufficiently small, the higher-order terms in this series will be negligible. As a first approximation, let us include only the first two terms in the series, or let

$$x(t_{i+1}) \approx x(t_i) + \Delta t\, f(x(t_i),t_i) \tag{8.144}$$

So the iterative solution to the differential equation is just the above first-order expression. This technique is known as *Euler integration.*

The omission of the higher-order terms in Equation (8.143) causes an error in the integration method known as *truncation error,* and the truncation error of the Euler method is of the order of Δt^2. But if Δt is taken small enough, the Euler method can be a viable integration technique, and is frequently used to obtain a quick approximation. But when Δt is extremely small, many iteration steps are required, and another error becomes important. This *round-off error* arises due to the finite word length of a digital computer. So for each differential equation there will be an optimum step size Δt that balances truncation and round-off error, and this optimum step size depends on the characteristics of the differential equation being solved.

A frequently used technique that allows for larger step sizes and is very numerically stable is the *Runge–Kutta* method of numerical integration. Two variations of this method will be presented here. To aid in understanding the approach, we will first consider a second-order Runge–Kutta technique. Consider first the left side of Figure 8.18, where we are seeking to integrate over one time interval Δt, or from t_0 to t_1, and the exact solution to the differential equation is depicted with the heavier curved line. Let the solution at t_1 be denoted as x_1.

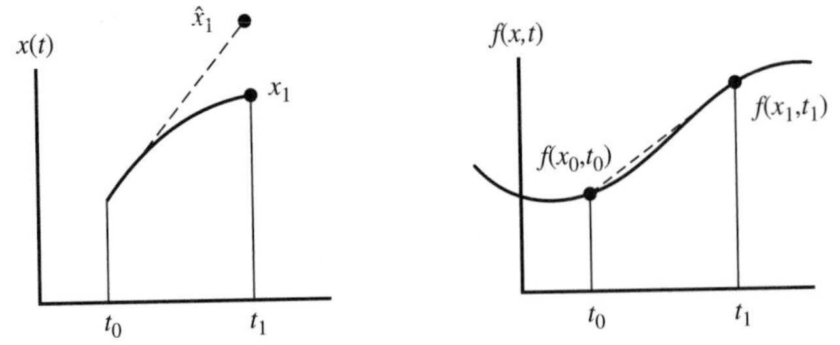

Figure 8.18 Graphical interpretation of second-order Runge-Kutta algorithm.

Using a trapezoidal approximation to the area under $f(x,t)$, shown on the right side of Figure 8.18, we have

$$x_1 \approx x_0 + \frac{\Delta t}{2}\left(f(x_0,t_0) + f(x_1,t_1)\right) \tag{8.145}$$

But since x_1 is not known we will use a first approximation for it, denoted $\hat{x}_1$ to evaluate $f(x_1,t_1)$ in the right-hand side of the above expression. This approximation will be obtained using Euler integration, or

$$\hat{x}_1 = x_0 + \Delta t\, f(x_0,t_0) \tag{8.146}$$

Therefore, from Equation (8.145) we have the numerical estimate for x_1 given by

$$x_1 \approx x_0 + \frac{\Delta t}{2}\left(f(x_0,t_0) + f(\hat{x}_1,t_1)\right) \tag{8.147}$$

The second-order Runge–Kutta method then consists of recursively using Equations (8.146) and (8.147) until reaching the desired final time t_f. Note that two evaluations of the function $f(x,t)$ are required per time step here, as opposed to only one for the Euler method. But the truncation error of this Runge–Kutta method is of the order of Δt^3.

The most common method used in practice, however, is the *fourth-order* Runge–Kutta method, which has truncation error of the order of Δt^5. The method is depicted graphically in Figure 8.19. Instead of using the trapezoidal method for approximating the area under $f(x,t)$, we use Simpson's rule based on a polynomial approximation for $f(x,t)$, which gives

$$x_1 \approx x_0 + \frac{\Delta t}{6}\left(f(x_0,t_0) + 4f\left(x_{1/2}, t_0 + \frac{\Delta t}{2}\right) + f(x_1,t_1)\right) \tag{8.148}$$

But we can see that we now need estimates for two future values of x, or $x_{1/2}$ and x_1, to evaluate the right-hand side of the above expression. We will obtain these estimates using the second-order method just discussed, with steps consisting of half intervals or $\Delta t/2$.

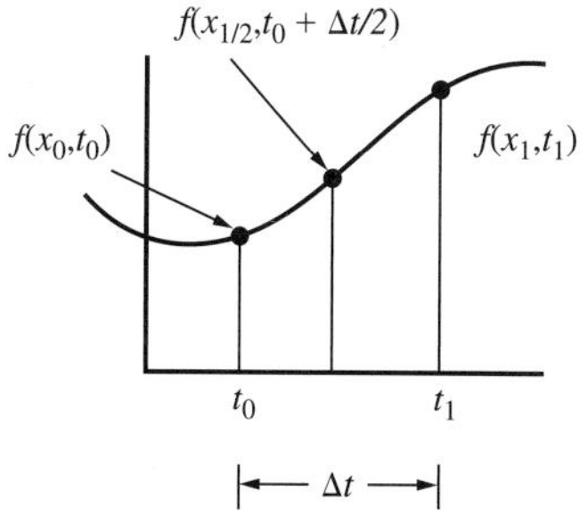

Figure 8.19 Graphical representation of the fourth order Runge–Kutta algorithm.

The approximation for $x_{1/2}$, or $\tilde{x}_{1/2}$ found from the second-order method, is given by

$$\hat{x}_{1/2} = x_0 + \frac{\Delta t}{2} f(x_0, t_0)$$
$$\tilde{x}_{1/2} = x_0 + \frac{\Delta t}{4}\left(f(x_0, t_0) + f\left(\hat{x}_{1/2}, t_0 + \frac{\Delta t}{2}\right)\right) \tag{8.149}$$

Then the procedure is repeated over a full time step Δt, using $f\left(\tilde{x}_{1/2}, t_0 + \frac{\Delta t}{2}\right)$ to initially estimate x_1. This is accomplished using the Euler step

$$\hat{x}_1 = x_0 + \Delta t f\left(\tilde{x}_{1/2}, t_0 + \frac{\Delta t}{2}\right) \tag{8.150}$$

The second estimate for x_1 is then obtained from

$$\tilde{x}_1 = x_0 + \frac{\Delta t}{2}\left(f\left(\tilde{x}_{1/2}, t_0 + \frac{\Delta t}{2}\right) + f(\hat{x}_1, t_1)\right) \tag{8.151}$$

And finally this estimate for x_1 is used to evaluate Equation (8.148). So, in summary, the fourth-order Runge–Kutta routine involves using the following recursive scheme.

$$x_{i+1} = x_i + \frac{1}{6}(a_i + 2b_i + 2c_i + d_i) \tag{8.152}$$

with

$$a_i = \Delta t f(x_i, t_i)$$
$$b_i = \Delta t f\left(x_i + \frac{a_i}{2}, t_i + \frac{\Delta t}{2}\right)$$
$$c_i = \Delta t f\left(x_i + \frac{b_i}{2}, t_i + \frac{\Delta t}{2}\right)$$
$$d_i = \Delta t f(x_i + c_i, t_{i+1})$$

The extension of this algorithm to systems of coupled first-order differential equations is straightforward. Instead of the scalar differential equation given by Equation (8.141), consider now the vector **x** governed by the n nonlinear differential equations

$$\dot{\mathbf{x}} = \mathbf{f}(\mathbf{x}, t) \tag{8.153}$$

with the n initial conditions given by $\mathbf{x}(t_0)$. The numerical integration of this set of differential equations using the Runge–Kutta method is obtained by using the same recursive routine given in Equations (8.152), except that **x**, **f**, **a**, **b**, **c**, and **d** are now vectors. Or

$$\mathbf{x}_{i+1} = \mathbf{x}_i + \frac{1}{6}\left(\mathbf{a}_i + 2\mathbf{b}_i + 2\mathbf{c}_i + \mathbf{d}_i\right) \tag{8.154}$$

with

$$\mathbf{a}_i = \Delta t\mathbf{f}\left(\mathbf{x}_i, t_i\right)$$

$$\mathbf{b}_i = \Delta t\mathbf{f}\left(\mathbf{x}_i + \frac{\mathbf{a}_i}{2}, t_i + \frac{\Delta t}{2}\right)$$

$$\mathbf{c}_i = \Delta t\mathbf{f}\left(\mathbf{x}_i + \frac{\mathbf{b}_i}{2}, t_i + \frac{\Delta t}{2}\right)$$

$$\mathbf{d}_i = \Delta t\mathbf{f}\left(\mathbf{x}_i + \mathbf{c}_i, t_{i+1}\right)$$

Finally, it is noted that MATLAB also includes numerical integration routines. Two examples are the routines ode23 and ode45. The first of these routines implements a second-order Runge–Kutta algorithm, while the second implements a fourth-order Runge–Kutta algorithm. Both are used in essentially the same manner as our routine rk4, developed in Example 8.8.

MATLAB EXAMPLE 8.8

A Fourth-Order Runge–Kutta Integration Routine

Using MATLAB, develop a routine that implements the fourth-order Runge–Kutta algorithm to integrate a set of n nonlinear differential equations over one integration step. The differential equations are given as

$$\dot{\mathbf{x}} = \mathbf{f}(\mathbf{x}, t), \quad \mathbf{x}(t_0) = \mathbf{x}_0$$

■ Solution

The solution involves developing two MATLAB m files, one called rates for calculating the rates, or $\dot{\mathbf{x}}(\mathbf{x}, t)$, and one called rk4 for implementing the Runge–Kutta algorithm.

The rates routine is presented first. Noting that the functions $f_i(\mathbf{x}, t)$, $i = 1 \ldots n$ are known, we have

```
% Function for Finding the Rates xdot
function [xdot]=rates(x,t)

xdot(1)= insert function f1(x,t) here
xdot(2)= insert function f2(x,t) here
.
.
.
xdot(n)= insert function fn(x,t) here
return
```

Note that this routine calculates the right-hand sides of the differential equations, given the current value of **x**, the vector containing the dependent variables being determined, and *t* time.

The routine rk4 then implements the algorithm given in Equations (8.154), using the rates routine to find the values of xdot. When executing rk4, the string identifying the rates routine is passed through the parameter list, and one integration step is taken.

```
% Function for Integrating Non-Linear DE's via 4th-Order Runge–Kutta
function [xnew,tnew]=rk4(rates,x0,t0,delt)
x=x0;
t=t0;
xdot=feval(rates,x,t);
a=delt*xdot;
xdot=feval(rates,x+a/2,t+delt/2);
b=delt*xdot;
xdot=feval(rates,x+b/2,t+delt/2);
c=delt*xdot;
xdot=feval(rates,x+c,t+delt);
d=delt*xdot;
xnew=x0+(a+2*b+2*c+d)/6;
tnew=t+delt;
return
```

Numerically Solving for an Equilibrium Flight Condition One difficulty encountered in nonlinear simulations is the need to find an equilibrium flight condition to use for an initial condition. For a linear simulation, the equilibrium flight condition is specified—it's the chosen reference condition. We will now develop a technique for solving for equilibrium flight conditions using a numerical-optimization method available in MATLAB. In MATLAB 5.2 the routine is called fmins, while in later versions of MATLAB it is called fminsearch. Both are in the Optimization Toolbox. The routine employs the Simplex Method of parameter optimization. The method developed here is similar to that used in MATLAB's Trim function, which may also be used instead of developing one's own code.

We must use our knowledge of flight dynamics to define an equilibrium flight condition appropriate for the model we are using for the vehicle's dynamics. Consider as a first case a flat-earth model and an equilibrium flight condition corresponding to <u>straight-and-level flight</u> at a chosen altitude h_0 and velocity V_{∞_0}. This flight condition might be defined in terms of the following constraints.

$$
\begin{array}{ll}
 & h = h_0 \\
\dot{U} = \dot{V} = \dot{W} = 0 & U^2 + W^2 = V_{\infty_0}^2 \\
\dot{P} = \dot{Q} = \dot{R} = 0 & V = \beta = P = Q = R = \phi = 0 \\
 & \theta = \alpha \\
\underline{\text{Steady State}} & \underline{\text{Straight-and-Level Flight}}
\end{array}
\qquad (8.155)
$$

Here the flight-path angle ($\gamma = \theta - \alpha$), sideslip angle β, and bank angle ϕ are all defined to be zero, as well as the translational accelerations and the rotational velocities and accelerations. So the problem is to find the angle of attack; elevator, aileron, and rudder deflection; and thrust or throttle setting to satisfy these requirements.

In another case, for example, we may define an equilibrium flight condition corresponding to a steady, level, coordinated turn, with a turn rate given as $\dot{\Psi}_0$, a flight velocity given as V_{∞_0}, and an altitude given as h_0. In a steady, level, coordinated turn, altitude and velocity are held constant, the bank angle is that required to sustain the constant turn rate, and all forces along the Y_V axis sum to zero. So the flight condition may be defined in terms of the following:

$$
\begin{array}{lll}
 & h = h_0 & \\
 & & P = -\dot{\Psi}_0 \sin\theta \\
\dot{U} = \dot{V} = \dot{W} = 0 & \dot{h} = 0 & \\
 & & Q = \dot{\Psi}_0 \cos\theta \sin\phi \\
\dot{P} = \dot{Q} = \dot{R} = 0 & U^2 + V^2 + W^2 = V_{\infty_0}^2 & \\
 & & R = \dot{\Psi}_0 \cos\theta \cos\phi \\
 & g \tan\phi = \dfrac{\dot{\Psi}_0 V_{\infty_0}}{\cos\theta} & \\
\underline{\text{Steady State}} & & \underline{\text{Level Turn}}
\end{array}
\tag{8.156}
$$

Here the altitude and the translational and rotational velocities are constant, and the bank angle and angular velocity correspond to the turning-flight condition. (These constraints for turning flight will be derived in Chapter 9.) So again the problem is to find the angles of attack and sideslip; elevator, aileron, and rudder deflection; and thrust or throttle setting to satisfy these requirements.

From the above two cases we see that the problem involves finding the control inputs and necessary flight variables such that a set of constraints are satisfied. And since these constraints are nonlinear functions of all the variables, a function-minimization approach seems appropriate. In such an approach, a "cost" function selected by the analyst is minimized, subject to certain constraints, by selecting the remaining set of free parameters.

MATLAB EXAMPLE 8.9

Numerically Finding a Steady, Level Flight Condition

Consider the first case above, involving a steady, level flight condition with altitude h_0 and velocity V_{∞_0}. Develop a numerical algorithm to determine the response and control variables that satisfy the constraints corresponding to this condition. Consider only the longitudinal equations.

■ Solution

We will assume that the sideslip and aileron and rudder deflections are all zero, and we will therefore ignore the lateral-directional equations of motion governing $\dot{V}$, $\dot{P}$, and $\dot{R}$.

Thus, we are only dealing with three degrees of freedom. (The approach taken here may be expanded quite easily to the problem involving six degrees of freedom.)

From Equations (8.155), we select the following equations as constraints that will be exactly satisfied.

$$h = h_0$$

$$U^2 + W^2 = V_{\infty_0}^2 \quad \text{or} \quad U = \frac{V_{\infty_0}}{\sqrt{1 + \tan^2\alpha}}$$

$$V = \beta = P = Q = R = \phi = 0$$

$$\theta = \alpha$$

This leaves angle of attack α, elevator deflection δ_E, and thrust T (or throttle position) still to be found. We will determine these variables by selecting them such that the following "cost" function is minimized

$$\text{Cost} = c_1\dot{U}^2 + c_2\dot{W}^2 + c_3\dot{Q}^2$$

with the constants c_1, c_2, and c_3 taken to equal unity. (These constants may be adjusted by the analyst to increase accuracy, if required. But as we will see, selecting them to be unity seems to work fine in this case.) Note that selecting the remaining parameters to minimize this cost function does not guarantee that all three accelerations in the cost function will exactly equal zero.

Turning to MATLAB we will write three m files, rates3dof, cost3dof, and constr3dof. The rates3dof routine, similar to rates given in Example 8.8, calculates $\dot{U}$, $\dot{W}$, and $\dot{Q}$ given a trial set of parameters stored in the MATLAB vector p. The cost3dof routine calculates the cost function, and the constr3dof routine finds U and θ given α. The complete algorithm may then be executed from the MATLAB command line.

```
% Function for Finding the Rates xdot

function [xdot]=rates3dof(x,u,t)
% x = [U, alpha, q, and theta]
rho=XXX;          Set the atmospheric density corresponding to the reference altitude h0
W=x(1)*tan(x(2));
Vinf=sqrt(x(1)^2+W^2);
q=0.5*rho*Vinf*Vinf;
xdot(1)= . . .  U̇ equation here
xdot(2)= . . .  Ẇ equation here
xdot(3)= . . .  Q̇ equation here
xdot(4)= . . .  θ̇ equation here
xdot(5)= . . .  ḣ equation here

% Function for Calculating the Cost for 3 DOF

function cost=cost3dof(p)
global Vinf c
```

`x(2)=p(1);`	α in rad
`x(3)=0;`	$Q = 0$
`[U,theta]=constr3dof(x(2));`	Initial U and θ
`x(1)=U;x(4)=theta;`	
`u(1)=p(2);`	δ_E (rad)
`u(2)=p(3);`	*Thrust* (lbs)

```
xdot=rates3dof(x,u,0);
cost=c(1)*xdot(1)*xdot(1)+c(2)*xdot(2)*xdot(2)+c(3)*xdot(3)*xdot(3);

% Function Using Level-Flight Constraints to Find Theta and U

function [U,theta]=constr3dof(alpha)
global Vinf
theta=alpha;
U=Vinf/sqrt(1+tan(alpha/57.3)*tan(alpha/57.3))
```

The commands, input at the command line, to execute this trimming algorithm are as given below, and will be demonstrated in the next example.

```
global Vinf c
```

`Vinf= XXX`	Set V_{∞_0}, the reference velocity
`p0=[.05;-.05;500]`	Initial α (rad), δ_E (rad), *Thrust* (lbs)
`c=[1;1;1]`	Cost weightings
`cost=cost3dof(p0);`	Initial cost
`p=fmins('cost3dof',p0)`	Find the minimizing parameter vector p
`cost=cost3dof(p);`	Final cost after minimization

8.2.8 Examples of Nonlinear Simulations

In the first example of a nonlinear simulation we will use the trim routine and the fourth-order Runge–Kutta algorithm presented in Examples 8.8 and 8.9 to simulate the nonlinear longitudinal equations of motion for the Navion aircraft with an initial velocity of 176 fps at sea level.

MATLAB EXAMPLE 8.10

Nonlinear Simulation of the Navion's Longitudinal Equations

Using the trim algorithm described in Example 8.9 and the Runge–Kutta routine presented in Example 8.8, simulate the longitudinal dynamics of the Navion for 20 seconds, using a two-second, one-degree doublet as the elevator input. Compare the results with those from Example 8.5.

■ Solution

The aerodynamic data for the Navion is given in Appendix B, and we will build the aerodynamic model using the effectiveness coefficients, not the dimensional stability derivatives. Based on the data in the appendix, the lift, drag, and aerodynamic pitching moment coefficients are taken to be

$$C_L = 4.44(\alpha + 4.7/57.3) + 0.355\delta_E$$

$$C_D = 0.03 + \frac{C_L^2}{\pi(3.51)}$$

$$C_M = -0.683\alpha - 0.071\dot{\alpha} - 0.161Q - 0.87\delta_E$$

The zero-lift angle of attack of 4.7 deg was found from the vehicle's lift effectiveness C_{L_α} and trim lift coefficient C_{L_0} at 0.6 deg, all given in the appendix.

The longitudinal equations of motion, from Equations (8.108), are

$$\dot{U} = -QW - g\sin\theta + (F_{A_X} + F_{P_X})/m$$

$$\dot{W} = QU + g\cos\theta + (F_{A_Z} + F_{P_Z})/m$$

$$\dot{Q} = (M_A + M_P)/I_{yy}$$

while the components of the aerodynamic and propulsive forces and moments are given by Equations (8.122–8.126), or

$$F_{A_X} = -D\cos\alpha + L\sin\alpha \qquad L = C_L q_\infty S_W$$

$$F_{A_Z} = -D\sin\alpha - L\cos\alpha \qquad D = C_D q_\infty S_W$$

$$F_{P_X} = T\cos\phi_T \qquad M_A = C_M q_\infty S_W \bar{c}_W$$

$$F_{P_Z} = T\sin\phi_T \qquad M_P = C_{P_M} q_\infty S_W \bar{c}_W$$

For our purposes here, we will let $\phi_T = C_{P_M} = 0$, corresponding to the thrust vector acting through the vehicle's center of mass.

Therefore the rates3dof routine is

```
% Function for Finding the Rates xdot
function [xdot]=rates3dof(x,u,t)    Note that elevator input u is passed to the routine
rho=0.002378;                       Sea level density
W=x(1)*tan(x(2));
Vinf=sqrt(x(1)^2+W^2);
q=0.5*rho*Vinf*Vinf;
CL=4.44*(x(2)+4.7/57.3)+0.355*u(1);                α0 = 4.7 deg
CD=0.03+(CL*CL)/(3.14159*3.51);
L=CL*q*184;                                         S = 184 ft^2
D=CD*q*184;
Fax=-D*cos(x(2))+L*sin(x(2));
Faz=-D*sin(x(2))-L*cos(x(2));
Fpx=u(2);
xdot(1)=-x(3)*W-32.2*sin(x(4))+(Fax+Fpx)/85.4;     U̇ (m = 85.4 sl)
xdot(2)=x(3)+32.2*cos(x(4))/x(1)+Faz/(85.4*x(1));  α̇
```

CM=-0.683*x(2)-4.36*xdot(2)-9.96*x(3)-0.87*u(1);	
Ma=CM*q*184*5.7	$\bar{c} = 5.7$ ft
xdot(3)=Ma/3000;	$\dot{Q}$ ($Iyy = 3000$ sl-ft^2)
xdot(4)=x(3);	$\dot{\theta} = Q$
xdot(5)=Vinf*sin(x(4)-x(2));	$\dot{h}$

The following are the responses from inputs from the command line executing the trim routine discussed in Example 8.9.

```
»global Vinf c
»Vinf=176
Vinf =
  176
```

»c=[1;1;1]	Cost-function weighting coefficients

```
c =
   1
   1
   1
```

»p0=[.05;-.05;500]	Initial guesses for α (rad), δ_E (rad), and *Thrust* (lbs)

```
p0 =
  5.0000e-02
 -5.0000e-02
  5.0000e+02
```

»cost=cost3dof(p0)	Initial value of the cost function

```
cost =
  2.2184e+01
```

»p=fmins('cost3dof',p0)	Minimize the cost function using fmins
p =	The minimizing parameter vector (α, δ_E, and *Thrust*)

```
  9.8783e-03
 -7.7549e-03
  3.0430e+02
```

»cost=cost3dof(p)	The value of the cost function after minimization

```
cost =
  1.6566e-15
```

So we see that the equilibrium, or trim values, of the unknown parameters are

$$\alpha_{\text{Trim}} = 0.00988 \text{ rad} = 0.566 \text{ deg } (0.6 \text{ deg given in Appendix B})$$

$$\delta_{E\text{Trim}} = -0.00775 \text{ rad} = -0.444 \text{ deg}$$

$$Thrust_{\text{Trim}} = 304.3 \text{ lbs}$$

Consistent with Example 8.5, the elevator input for the simulation is a two-second, one-degree elevator doublet, but the trim elevator deflection of −0.444 deg here is held initially for one second. The one-degree input is superimposed on the trim elevator deflection of −0.444, so the total deflection after the first second is −1.444 deg, which is held for two seconds. At three seconds the total elevator deflection is changed to 0.556 deg, which is held for two seconds. Then the elevator is changed back to the trim value of −0.444 deg. The control input is held constant over each integration step. The following commands were input at the command line to execute the nonlinear simulation.

```
»global Vinf c h
»Vinf=176
Vinf =
   176
»alpha0=0.0098783
alpha0 =
   9.8783e-03
»[U0,theta0]=constr3dof(alpha0)
U0 =
   1.7600e+02
theta0 =
   9.8783e-03
»x0=[176 alpha0 0 theta0 0]
x0 =
   1.7600e+02   9.8783e-03   0   9.8783e-03   0
»[x,t]=navNLsim(x0,0,10);
```

At this point commands similar to those given below may be executed from the command line to obtain the plots shown in Figures 8.20, 8.21, and 8.22. (Note that angle

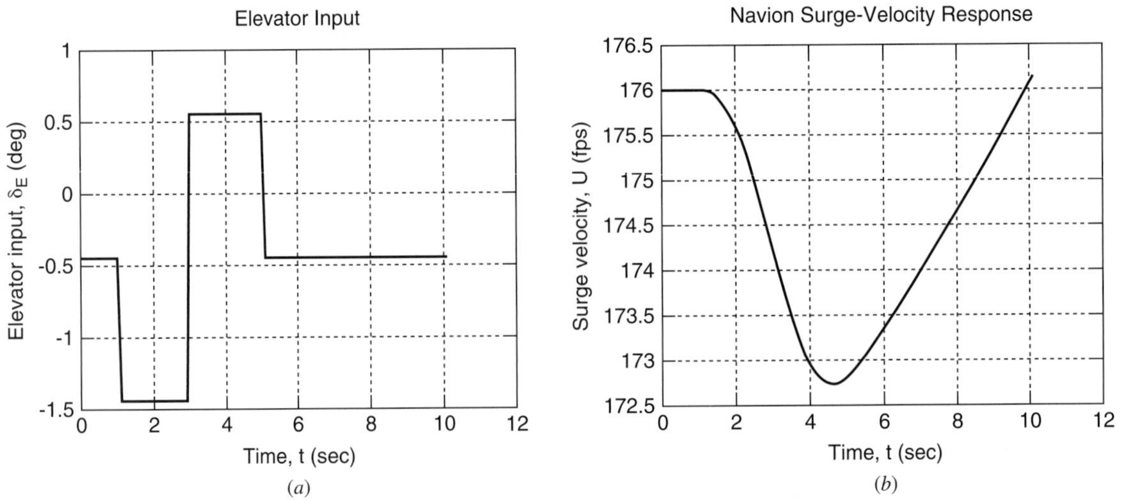

Figure 8.20 Time histories of elevator input and surge velocity.

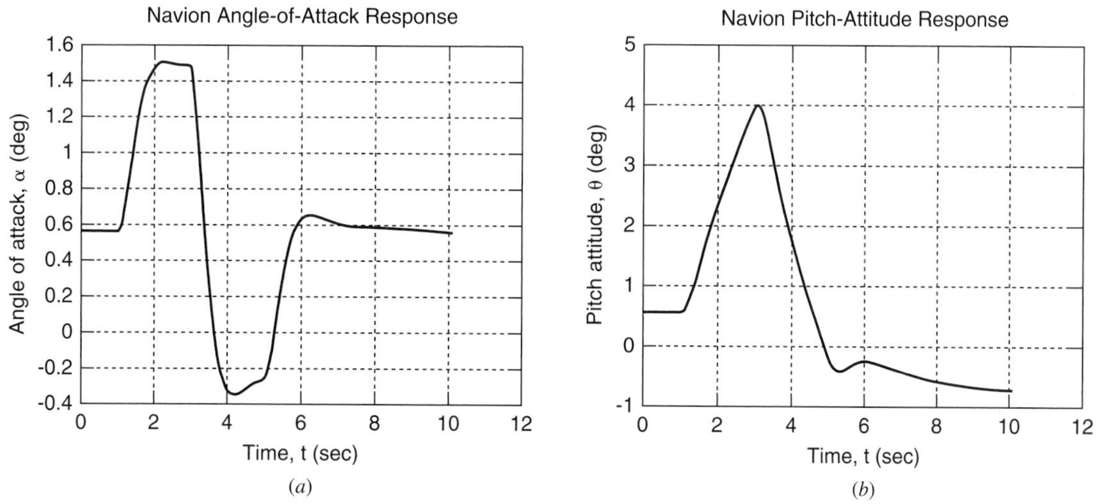

Figure 8.21 Time histories of angle of attack and pitch attitude.

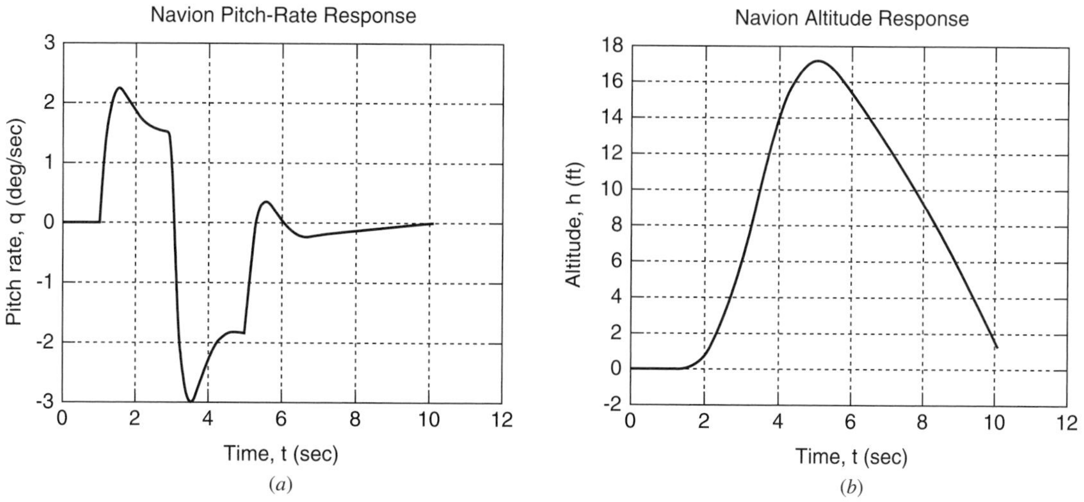

Figure 8.22 Time histories of pitch rate and altitude.

of attack, attitude angle, pitch rate, and elevator deflection have all been converted to degrees for plotting.)

```
»plot(t,x(:,1))
»grid;xlabel('Time, t (sec)')
»ylabel('Surge Velocity, U (fps)')
»title('Navion Surge-Velocity Response')
```

Comparing these results to those in Example 8.5, we see that the results agree quite well. This is as expected since the inputs and responses are small, and hence the small-perturbation assumptions made to derive the linear model remain valid. There is a slight difference noted between the final altitudes, partially due to the fact that in the nonlinear simulation, the elevator was held at the calculated trim value for one second before the doublet was initiated. This was done in part to confirm that the trim conditions were satisfied. But they are never exactly satisfied, and so after one second a small pitch rate is encountered. If the doublet was initiated at $t = 0$, the final altitude reduces to almost -2 ft, making the difference in the altitude histories even less.

These results show very good agreement between the linear and nonlinear simulations. However, if the elevator inputs were sufficiently large, such that the small-perturbation assumptions were no longer valid for the linear model, much larger differences would be observed between the two sets of results.

EXAMPLE 8.11

Case Study—A Nonlinear Aircraft-Performance Simulation[2]

In this case study we will develop a nonlinear performance simulation for a rigid aircraft operating in a steady wind (as opposed to gusts). This simulation will allow for the investigation of an aircraft's responses while following a desired flight profile. The profiles are defined in terms of commanded velocities, rates of climb, and/or headings, similar in some respects to commands given by air traffic control.

Up to this point in the chapter, we have always assumed that we wished to accurately simulate both the translational and rotational degrees of freedom of the vehicle. But now we wish to focus more on the translational performance of an aircraft being guided by a set of feedback guidance laws, and only approximate the attitude dynamics of the vehicle. This makes the simulation more numerically efficient, plus it allows us to avoid the details of an inner-loop attitude-control system at this stage of analysis. As we shall see, it also avoids the necessity of numerically solving for the initial trim flight condition.

We also take two other digressions here. The first is that the aircraft may be operating in a steady wind. If so, the air mass is assumed to be uniformly translating with respect to the earth-fixed inertial reference frame. The aerodynamic forces and moments acting on the vehicle depend on the vehicle's velocity (i.e., airspeed) and orientation relative to the air mass. And the presence of winds gives rise to differences between the vehicle's inertial velocity and its airspeed. Therefore, consistent with the material presented in Section 8.2.6, the presence of winds affects the forces on the vehicle. Finally, the simulation itself will be developed using the Simulink tools in MATLAB.

The equations of motion include a set of translational performance equations plus kinematic equations relating the vehicle's inertial position to its inertial velocity, all developed in Section 2.7. The translational equations of motion governing the magnitude and direction of the vehicle's inertial velocity (vector) were given in Equations (2.137), for a vehicle with propulsive thrust defined to be aligned with the fuselage-referenced X axis. Or

$$\begin{aligned}
m\dot{V}_V &= T\cos\alpha\cos\beta - D - mg\sin\gamma \\
mV_V(\dot{\psi}_W\cos\phi_W\cos\gamma - \dot{\gamma}\sin\phi_W) &= \mathsf{S} + T\cos\alpha\sin\beta + mg\sin\phi_W\cos\gamma \qquad (8.157)\\
mV_V(\dot{\gamma}\cos\phi_W + \dot{\psi}_W\sin\phi_W\cos\gamma) &= L + T\sin\alpha - mg\cos\phi_W\cos\gamma
\end{aligned}$$

[2] This simulation is based on one developed by Dr. John Schierman while he served as a post-doctoral research associate in the Flight Dynamics and Control Lab, University of Maryland, College Park.

Note that γ, ϕ_W, and ψ_W are the flight-path angle, and wind-axes bank and heading angles, respectively, and that V_V is the inertial velocity of the vehicle.

We now assume that the vehicle's sideslip angle β and the aerodynamic side force S are both zero (as in steady, level flight or in a steady coordinated turn), and that the vehicle's angle of attack α is sufficiently small such that

$$T\cos\alpha \approx T, \quad T\sin\alpha \ll L \tag{8.158}$$

Under these assumptions, Equations (2.137) may be rearranged to become simply

$$\boxed{\begin{aligned} \dot{V}_V &= \frac{T-D}{m} - g\sin\gamma \\ \dot{\gamma} &= \frac{1}{mV_V}\left(L\cos\phi_W - mg\cos\gamma\right) \\ \dot{\psi}_W &= \frac{L\sin\phi_W}{mV_V\cos\gamma} \end{aligned}} \tag{8.159}$$

From Equations (8.159) we can readily see how the two forces thrust T and lift L (magnitude and direction) are used to control velocity, flight-path angle, and heading angle, respectively. The angle of attack is used to adjust the <u>magnitude</u> of the lift vector, while the wind-axes bank angle ϕ_W is used to rotate the orientation of the lift vector relative to the vehicle's velocity vector.

Let us now approximate the responses of the engine and airframe with the following transfer functions or first-order differential equations.

$$\boxed{\begin{aligned} \frac{T(s)}{T_c(s)} &= \frac{p_T}{s+p_T} & & \dot{T} = -p_T T + p_T T_c \\ \frac{L(s)}{L_c(s)} &= \frac{p_L}{s+p_L} \quad \text{or} & & \dot{L} = -p_L L + p_L L_c \\ \frac{\phi_W(s)}{\phi_{W_c}(s)} &= \frac{p_\phi}{s+p_\phi} & & \dot{\phi}_W = -p_\phi \phi_W + p_\phi \phi_c \end{aligned}} \tag{8.160}$$

The parameters p_T, p_L, and p_ϕ are time constants selected to approximate the responses of the engine and the airframe attitude. Also, let the limits on these responses be taken to be

$$0 \le T \le T_{\max}, \quad L \le K_{L_{\max}} V_V^2, \quad -\phi_{W_{\max}} \le \phi_W \le \phi_{W_{\max}} \tag{8.161}$$

where the maximum values again depend on the aircraft being simulated.

The three kinematic equations relating the vehicle's velocity to the inertial position were given in Equations (2.141), repeated here.

$$\boxed{\begin{aligned}\dot{X}_I &= V_V\cos\gamma\cos\psi_W\\ \dot{Y}_I &= V\cos\gamma\sin\psi_W\\ \dot{\mathrm{h}} &= V\sin\gamma\end{aligned}}$$

Finally, let the mass of the aircraft be given by the differential equation

$$\boxed{\dot{m} = -\dot{w}_f/g = -K_{\dot{w}}T} \tag{8.162}$$

with the initial condition $m = m_0$, where $\dot{w}_f$ is the fuel flow rate and $K_{\dot{w}}$ is a constant depending on the aircraft. So Equations (8.159–8.162), along with Equations (2.141) given above, constitute the equations of motion to be used in the simulation.

The model for the aerodynamic lift and drag is taken to be

$$\boxed{\begin{aligned}&C_L = C_{L_\alpha}(\alpha - \alpha_0)\\ &C_D = C_{D_0} + \frac{C_L^2}{K_D}\\ &L = C_L q_\infty S_W,\quad D = C_D q_\infty S_W,\quad q_\infty = \frac{1}{2}\rho_\infty V_\infty^2,\quad K_D = \pi A e_{\mathrm{eff}}\end{aligned}} \tag{8.163}$$

with

and note that in the presence of wind, $V_\infty \neq V_V$. The relation between these two velocities may be described as follows. Since the inertial velocity $\mathbf{V}_V$ is the vector sum of the velocity relative to the air mass $\mathbf{V}_\infty$ plus the wind velocity $\mathbf{W}$, or

$$\mathbf{V}_V = \mathbf{V}_\infty + \mathbf{W}$$

we have

$$\mathbf{V}_\infty = \mathbf{V}_V - \mathbf{W} \tag{8.164}$$

So then

$$\boxed{V_\infty = \sqrt{\dot{X}_W^2 + \dot{Y}_W^2 + \dot{\mathrm{h}}_W^2}} \tag{8.165}$$

where

$$\begin{aligned}\dot{X}_W &= V_V\cos\gamma\cos\psi_W - W_X\\ \dot{Y}_W &= V\cos\gamma\sin\psi_W - W_Y\\ \dot{\mathrm{h}}_W &= V\sin\gamma - W_h\end{aligned} \tag{8.166}$$

with the wind velocity vector given by

$$\mathbf{W} = W_X\mathbf{i}_I + W_Y\mathbf{j}_I - W_h\mathbf{k}_I \tag{8.167}$$

Now since we are using aerodynamic lift L directly as an independent or control variable in the equations of motion, we will "invert" Equations (8.163) to find the inferred angle of attack α and drag D for a given value of lift L. That is, we have

$$
\begin{aligned}
D &= K_{D_0}V_\infty^2 + K_{D_1}\frac{L^2}{V_\infty^2} \\
\alpha &= K_L\frac{L}{V_\infty^2} + \alpha_0
\end{aligned}
\tag{8.168}
$$

with $K_{D_0} = \frac{1}{2}\rho_\infty S_W C_{D_0}, \quad K_{D_1} = \frac{2}{\rho_\infty S_W K_D}, \quad K_L = \frac{2}{\rho_\infty S_W C_{L_\alpha}}$

Of course, all these aerodynamic parameters also depend on the vehicle. This completes the mathematical model of the vehicle's dynamics.

The guidance laws given in Equations (8.169) provide the commanded thrust T_c, lift L_c, and bank angle ϕ_{Wc} by feeding back the relevant inertial-velocity, inertial-velocity-heading, and inertial rate-of-climb and comparing them to commanded values. The commanded velocity V_c, rate of climb $\dot{h}_c = V_c \sin\gamma_c$, and heading ψ_c are user-defined inputs to the simulation that describe the desired trajectory. The mathematic model is now complete.

$$
\begin{aligned}
&\frac{T_c(s)}{V_E(s)} = \frac{mK_{T_P}\left(s + \left(K_{T_I}/K_{T_P}\right)\right)}{s} \qquad && \dot{x}_T = mV_E \\
& && T_c = K_{T_I}x_T + K_{T_P}mV_E, \quad V_E \triangleq \left(V_c - V_V\right) \\
&\frac{L_c(s)}{\dot{h}_E(s)} = \frac{mK_{L_P}\left(s + \left(K_{L_I}/K_{L_P}\right)\right)}{s} \qquad \text{or} && \dot{x}_L = m\dot{h}_E \\
& && L_c = K_{L_I}x_L + K_{L_P}m\dot{h}_E, \quad \dot{h}_E \triangleq V_c\left(\sin\gamma_c - \sin\gamma\right) \\
&\frac{\phi_{Wc}}{\psi_E} = K_{\phi_P}\left(V_c/g\right) && \phi_{W_c} = K_{\phi_P}\left(V_c/g\right)\psi_E, \qquad \psi_E \triangleq \left(\psi_c - \psi_W\right)
\end{aligned}
\tag{8.169}
$$

In the case to be investigated, let the vehicle to be simulated be a large turbo-prop transport aircraft, similar to the C-130 aircraft shown in Figure 8.23. The modeling data for the aircraft is given in Table 8.4, while the values of the gains in the guidance laws for this vehicle are given in Table 8.5.

We are now ready to perform the simulation. Let the <u>initial conditions</u> be as follows:

$V_0 = 400$ mph (347 kts) (inertial velocity)

$\gamma_0 = 0$ (flight-path angle)

$\psi_{W_0} = 0$ (velocity-heading angle = north)

Figure 8.23 C-130 Aircraft. (Photo courtesy of NASA)

Table 8.4 Time Constants and Other Vehicle-Dependent Parameters

Weight, mg = 157,000–327,000 lbs		Airspeed Range, 200–600 mph	
$p_T = 2$ rad/sec	$T_{max} = 72{,}000$ lbs	$K_{\dot{w}} = 4 \times 10^{-6}$ sl/(lbs-sec)	$\alpha_0 = -0.05$ deg
$p_L = 2.5$ rad/sec	$K_{L\max} = 2.6$ lbs/fps^2	$K_{D_1} = 2.48 \times 10^{-2}$ ft^2/lb-sec^2	$K_L = 5.24$ deg-ft/sl
$p_\phi = 1$ rad/sec	$\phi_{W_{\max}} = 30$ deg	$K_{D_0} = 3.8 \times 10^{-2}$ sl/ft	

Table 8.5 Gains in the Guidance Laws

$K_{T_P} = 0.08$ sec	$K_{T_I} = 0.002/\text{sec}^2$	$K_{L_P} = 0.5/\text{sec}$	$K_{L_I} = 0.01/\text{sec}^2$	$K_{\phi_P} = 0.075$ sec

And let the commanded inertial velocity, flight-path angle (or rate of climb), and velocity heading be

$$V_c = 450 \text{ mph } (391 \text{ kts}), \quad \gamma_c = 5 \text{ deg } (\dot{h}_c = 3455 \text{ fpm}), \quad \psi_c = 15 \text{ deg}$$

So the vehicle must transition from level flight in a northerly direction at 400 mph to a heading of 15 deg at 450 mph while climbing at approximately 3500 fpm. A steady wind of about 30 mph from the southwest ($W_X = W_Y = 25$ mph, $W_h = 0$) is included. And we wish to simulate the vehicle's dynamics for two minutes and evaluate its responses.

As noted earlier, this simulation was performed using the MATLAB tool Simulink. The block diagram used in the Simulink simulation is shown in Figure 8.24. Initially, the various simulation parameters are set using a MATLAB script or m file named perfsim.m (included in the accompanying set of MATLAB files located at www.mhhe.com/schmidt).

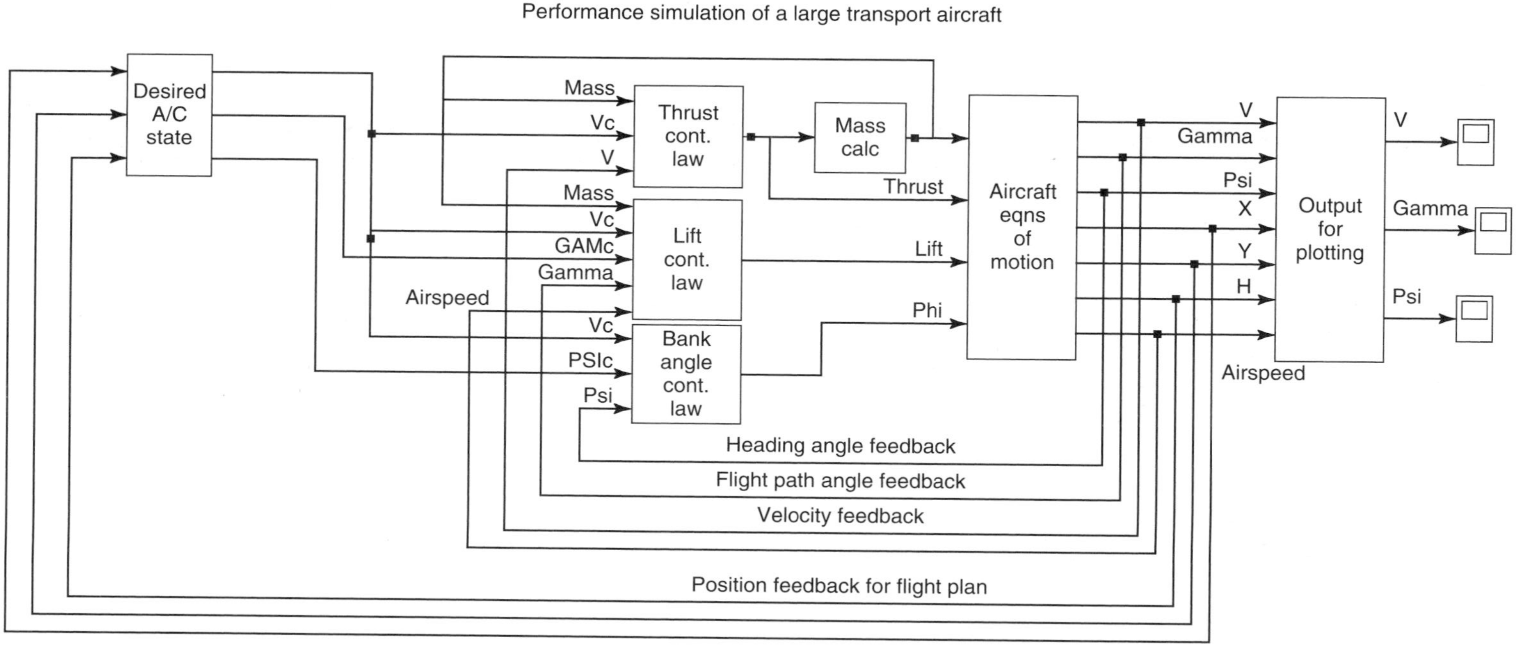

Figure 8.24 Simulink simulation diagram.

Then the Simulink model is executed for the desired simulated time of two minutes. And finally, another script file called plothist.m (also included in the accompanying set of MATLAB files) is used to plot the desired responses. The following results reveal that the vehicle responses indeed follow the desired trajectory.

The time histories of the airspeed V_∞, inertial velocity V_V, and thrust are shown first in Figure 8.25. The commanded inertial velocity is reached in approximately one minute. Since a significant increase in both airspeed and flight-path angle is commanded, a large increase in thrust is required. But this thrust does not exceed the maximum available thrust set in the model.

Next shown in Figure 8.26 are the time histories of the inertial flight-path angle and the inertial rate of climb, both compared to their commanded values, while in Figure 8.27

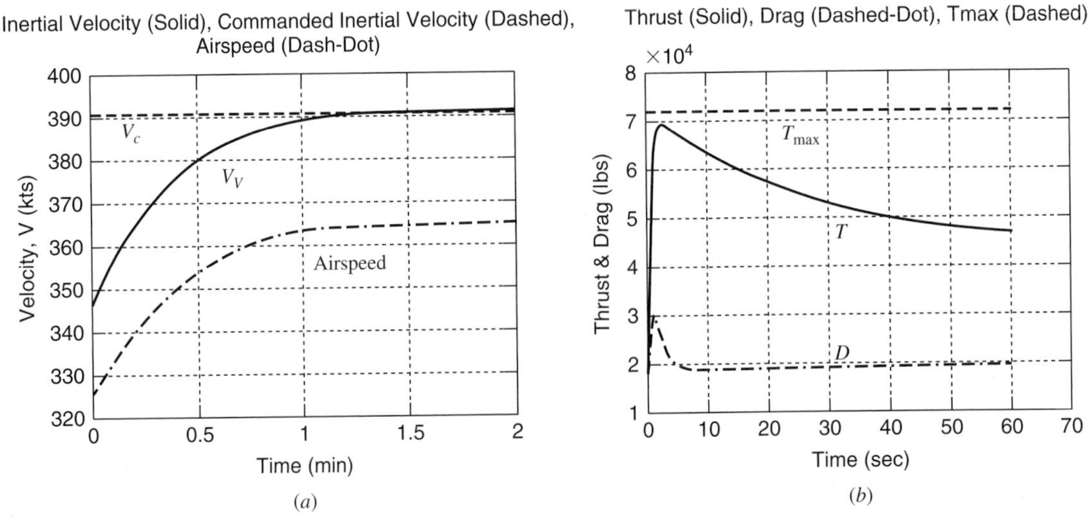

Figure 8.25 Time histories of velocities, thrust, and drag.

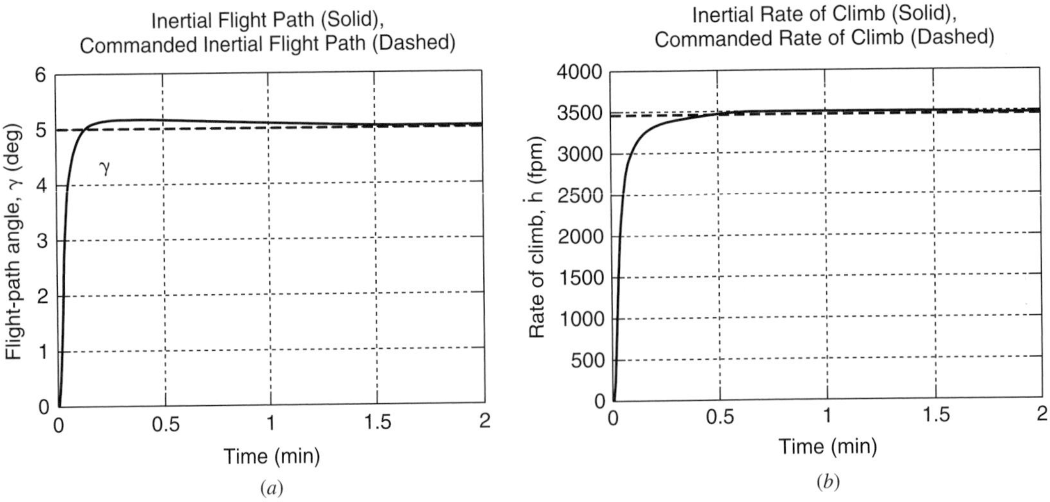

Figure 8.26 Time histories of flight-path angles and rates of climb.

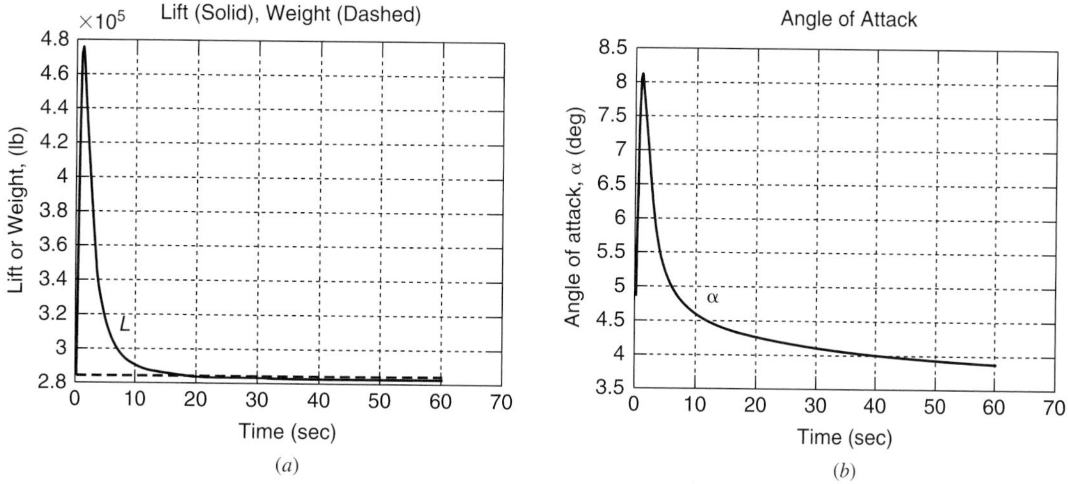

Figure 8.27 Time histories of lift, weight, and angle of attack.

the corresponding lift, weight, and angle-of-attack time histories are shown. The change in flight-path angle is rather rapid, reaching the commanded value in approximately 5 seconds. As shown in Figure 8.27*a*, a fairly aggressive increase in lift is required (due to the desired change in flight path as well as heading), which results in a peak vertical acceleration of approximately 1.7 g's. The maximum angle of attack of 8 deg, shown in Figure 8.27*b*, is achieved at approximately 1.5 seconds.

And finally we have the time history of the inertial heading compared to the commanded value, along with the corresponding bank-angle history, all shown in Figure 8.28. Note that the desired change in heading is accomplished smoothly in less than one minute. The maximum bank angle of slightly over 18 degrees occurs at about 3 sec.

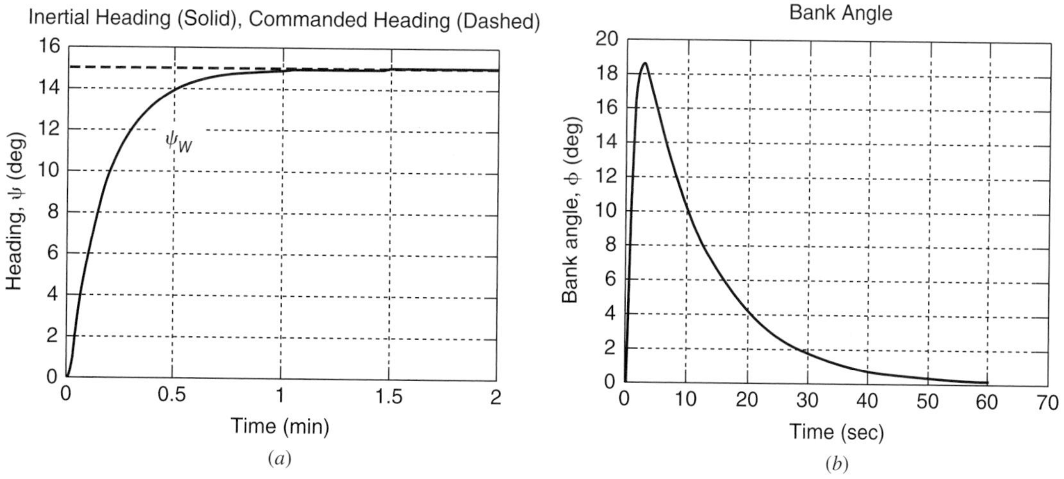

Figure 8.28 Time histories of inertial heading and bank angle.

8.3 Summary

As noted at the outset, this chapter marks a change in focus in the book—from that of building the mathematical models of the vehicle's dynamics, to analyzing those dynamics using the models we've developed. In this chapter we discussed both linear and nonlinear numerical simulation of the dynamics of both rigid and flexible vehicles. We presented numerical methods that may be used to develop these simulations. And we discussed methods for including feedback-control laws and the effects of atmospheric turbulence in the simulations. The numerical methods presented in this chapter included those developed from first principles, as well as direct applications of algorithms available in the MATLAB software suite. Finally, the concepts presented were reinforced through several examples of both linear and nonlinear flight simulations.

8.4 Problems

8.1 Consider a reference flight condition involving flight in a steady turn at constant altitude. Let the turn rate be $\dot{\Psi}_0$ and let the corresponding bank angle be Φ_0. What are the linear equations of motion for this flight condition? (You do not have to include the detailed models for the aerodynamic and propulsive forces and moments in these results. Just list the forces and moments in the relevant equations.)

8.2 For the flight condition described in Problem 8.1, do the equations of motion now decouple as they did in Section 8.1.3? Write the equations of motion in state-variable form.

8.3 In the case of a variable atmospheric density, ρ_∞ becomes a function of altitude h. Modify the linear models given in Section 8.1.2 for the aerodynamic forces and moments to include the effect of variable density. In this case, will altitude be a state variable or just a response variable in the state-variable representation of the equations of motion?

8.4 Assuming a rigid vehicle in a steady, level, reference flight condition, derive the linear equation for the local perturbation vertical acceleration a_Z at a location x along the vehicle's centerline (measured positive forward from the center of mass). How would the state-variable model for the longitudinal dynamics be modified to include this acceleration as one of the system responses? That is, what changes are necessary in the **A**, **B**, **C**, and **D** matrices?

8.5 Similar to Example 8.2, assemble the state-variable model for the lateral-directional dynamics of the Navion aircraft. The data for this aircraft is given in Appendix B. Include heading ψ in the model responses, and list the response, state, and input vectors for your model.

8.6 Assume the state-variable model for a vehicle's dynamics without any control law included is given as

$$\dot{\mathbf{x}}_v = \mathbf{A}_v\mathbf{x}_v + \mathbf{B}_v\mathbf{u}$$
$$\mathbf{y} = \mathbf{C}_v\mathbf{x}_v + \mathbf{D}_v\mathbf{u}$$

and consider the type of control law corresponding to Control Law B as given in Section 8.1.6. Develop the state-variable model for the vehicle dynamics with such a control law included.

8.7 Using a method similar to that used in Example 8.3, find a state-variable representation for a lead-lag control element that has the transfer function

$$\frac{u_{LL}(s)}{y_{LL}(s)} = \frac{K_{LL}(s + z_{LL})}{(s + p_{LL})}$$

8.8 Consider the state-variable model of the Navion developed in Problem 8.5. Using a one-degree, two-second aileron input only (the aileron is returned to zero after two seconds), use the state-transition method to simulate the linear lateral-directional responses, and plot the results. Compare these results to those obtained using MATLAB's `lsim` routine.

8.9 Consider the large, flexible aircraft modeled in Appendix B. Ignoring any control laws, and using MATLAB's `lsim` routine, perform a linear simulation of the vehicle's longitudinal perturbation dynamics at Flight Condition 1 including only the first elastic mode, and plot the rigid-body and local pitch rate (which includes elastic deformation) at the cockpit. Use a two-second, one-degree elevator doublet as used in Example 8.5.

8.10 Using the data for the Navion in Appendix B, develop a nonlinear simulation of the Navion's lateral-directional dynamics based on the routine `rk4`, and simulate the vehicle's response to a two-second, one-degree aileron input. Compare your results to those from Problem 8.8.

References

1. Waszak, M. R., J. B. Davidson, and D. K. Schmidt: "A Simulation Study of the Flight Dynamics of Elastic Aircraft," NASA Contractor Report 4102, prepared by Purdue University for NASA Langley Research Center, vols. I and II, Dec. 1987.
2. Stevens, B. L. and F. L. Lewis: *Aircraft Control and Simulation,* 2nd ed., Wiley, New York, 2003.
3. DeRusso, P. M., R. J. Roy, and C. M. Close: *State Variables for Engineers,* Wiley, New York, 1965.
4. Schmidt, D. K., and D. Andrisani: *Flight Dynamics and Control Lab Manual,* School of Aeronautics and Astronautics, Purdue University, West Lafayette, IN, 1986.
5. Papoulis, A.: *Probability, Random Variables, and Stochastic Processes,* 3rd ed., McGraw-Hill, New York, 1991.

9 CHAPTER

Analysis of Steady and Quasi-Steady Flight

Chapter Roadmap: *In this chapter, which is fundamental to any first course in flight dynamics, we focus on the definition and analysis of equilibrium or reference flight conditions. Critical concepts such a static aerodynamic stability, static margin, and forces on the cockpit-control manipulators are treated.*

In this chapter we continue the analysis of the vehicle's flight characteristics and focus on equilibrium or "steady" flight conditions. We will address such topics as trim analysis, aerodynamic static stability, and control power for maneuverability. Most of the topics we discuss are classically referred to as *static stability and control.*

Think back now to Chapter 1 and the discussion of perturbation analysis of nonlinear systems. After the development of the nonlinear equations of motion and the derivations of the reference and perturbation equations, the next step is to analyze the reference conditions for the system. Recall that in the simple-pendulum example in that chapter, we found that the system had two distinct equilibrium reference conditions, one at $\theta = 0$ and one at $\theta = \pi$.

In this chapter, we will analyze the reference or equilibrium conditions for flight vehicles, and we will find that multiple reference conditions are also possible. This should not be surprising since we know that aircraft can fly at different equilibrium, or "trim" conditions, for example, straight-and-level flight and steady level turns. But here we will more formally analyze these reference conditions, and explore criteria for finding such conditions.

The results obtained are critical in the design of flight vehicles, and many are frequently included in aircraft design requirements. Meeting these requirements has profound effects on the resulting vehicle geometry. For example, most aircraft have tail surfaces to meet the flight dynamics requirements discussed in this chapter. Designers would be delighted not to have to include tail surfaces on aircraft, because they add weight and drag and thus affect performance. As an example, consider the Beechcraft Bonanza general aviation aircraft, which has a "V" tail. Designers used only two tail surfaces instead of three to increase performance. (But feedback stability augmentation was also added to meet flight dynamics requirements.)

We will perform our analysis by considering only the rigid-body degrees of freedom. So we will assume either a rigid vehicle or that the aerodynamic coefficients used in the analyses include static-elastic corrections, as discussed in Section 7.11. We will also assume constant vehicle mass, a flat nonrotating earth, constant atmospheric density, and ignore the effects of any rotating masses on the vehicle. The vehicle will be conventional, implying that the XZ plane is a plane of symmetry, aft tail surfaces are present, and the horizontal tail has a variable incidence. If other vehicle configurations need to be considered, the techniques discussed here may be applied to them as well, and such cases will be addressed in exercises listed at the end of the chapter. Much of this chapter draws from material developed in Chapter 6, dealing with the forces and moments acting on the vehicle. We recommend reviewing Chapter 6, especially Sections 6.1-6.5, prior to beginning the study of this chapter.

9.1 EQUILIBRIUM REFERENCE CONDITIONS

Consider an aircraft shown schematically in Figure 9.1. Under the assumptions stated above, the Reference Set of equations derived in Chapter 2 were given by Equations (2.44), (2.48), and (2.53). These equations are repeated below for convenience. As noted in Chapter 2, the first three equations govern the translational velocity, the next three, the rotational velocity, and the final three provide kinematic relationships between angular velocities.

$$\begin{aligned} m(\dot{\mathrm{U}}_0 + \mathrm{Q}_0\mathrm{W}_0 - \mathrm{V}_0\mathrm{R}_0) &= -mg\sin\Theta_0 + \mathrm{F}_{A_{X_0}} + \mathrm{F}_{P_{X_0}} \\ m(\dot{\mathrm{V}}_0 + \mathrm{R}_0\mathrm{U}_0 - \mathrm{P}_0\mathrm{W}_0) &= mg\cos\Theta_0\sin\Phi_0 + \mathrm{F}_{A_{Y_0}} + \mathrm{F}_{P_{Y_0}} \\ m(\dot{\mathrm{W}}_0 + \mathrm{P}_0\mathrm{V}_0 - \mathrm{Q}_0\mathrm{U}_0) &= mg\cos\Theta_0\cos\Phi_0 + \mathrm{F}_{A_{Z_0}} + \mathrm{F}_{P_{Z_0}} \end{aligned} \tag{9.1}$$

$$\begin{aligned} I_{xx}\dot{\mathrm{P}}_0 - I_{xz}(\dot{\mathrm{R}}_0 + \mathrm{P}_0\mathrm{Q}_0) + (I_{zz} - I_{yy})\mathrm{Q}_0\mathrm{R}_0 &= \mathrm{L}_{A_0} + \mathrm{L}_{P_0} \\ I_{yy}\dot{\mathrm{Q}}_0 + (I_{xx} - I_{zz})\mathrm{P}_0\mathrm{R}_0 + \mathrm{I}_{xz}(\mathrm{P}_0^2 - \mathrm{R}_0^2) &= \mathrm{M}_{A_0} + \mathrm{M}_{P_0} \\ I_{zz}\dot{\mathrm{R}}_0 - I_{xz}(\dot{\mathrm{P}}_0 - \mathrm{Q}_0\mathrm{R}_0) + (I_{yy} - I_{xx})\mathrm{P}_0\mathrm{Q}_0 &= \mathrm{N}_{A_0} + \mathrm{N}_{P_0} \end{aligned} \tag{9.2}$$

$$\begin{aligned} \dot{\Phi}_0 &= \mathrm{P}_0 + \mathrm{Q}_0\sin\Phi_0\tan\Theta_0 + \mathrm{R}_0\cos\Phi_0\tan\Theta_0 \\ \dot{\Theta}_0 &= \mathrm{Q}_0\cos\Phi_0 - \mathrm{R}_0\sin\Phi_0 \\ \dot{\Psi}_0 &= (\mathrm{Q}_0\sin\Phi_0 + \mathrm{R}_0\cos\Phi_0)\sec\Theta_0 \end{aligned} \tag{9.3}$$

Recall from Chapter 2 that the components of the vehicle's (translational) velocity vector in an equilibrium or steady flight condition are defined by

$$\mathbf{V}_{V_0} \triangleq \mathrm{U}_0\mathbf{i}_V + \mathrm{V}_0\mathbf{j}_V + \mathrm{W}_0\mathbf{k}_V \tag{9.4}$$

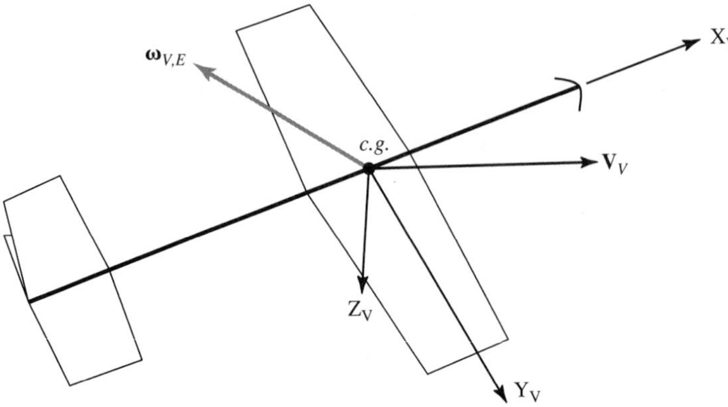

Figure 9.1 Aircraft schematic, with translational and rotational velocity vectors.

in which $\mathbf{i}_V$, $\mathbf{j}_V$, and $\mathbf{k}_V$ are unit vectors defining the chosen vehicle-fixed coordinate frame. Likewise, the components of the vehicle's angular-velocity vector in an equilibrium or steady flight condition are defined by

$$\boldsymbol{\omega}_{V,E_0} \triangleq \mathrm{P}_0\mathbf{i}_V + \mathrm{Q}_0\mathbf{j}_V + \mathrm{R}_0\mathbf{k}_V \tag{9.5}$$

Also, Ψ_0 (heading), Θ_0 (pitch attitude), and Φ_0 (bank angle) are the 3-2-1 Euler angles defining the orientation of the vehicle-fixed frame with respect to the inertial earth-fixed frame. Finally, the aerodynamic-plus-propulsive moment (vector) acting on the vehicle has components defined by

$$(\mathbf{M}_A + \mathbf{M}_P) \triangleq (L_A + L_P)\mathbf{i}_V + (M_A + M_P)\mathbf{j}_V + (N_A + N_P)\mathbf{k}_V \tag{9.6}$$

while the components of the forces are listed on the right-hand sides of Equations (9.1).

We will now use Equations (9.1–9.3) to define and analyze reference flight conditions for a flight vehicle. The formal criterion to be used for an equilibrium reference condition is given by the following definition.

Definition: Given a nonlinear dynamic system defined by the vector differential equation (Ref. 1)

$$\dot{\mathbf{x}} = \mathbf{f}(\mathbf{x},\mathbf{u}) \tag{9.7}$$

an *equilibrium solution* $\mathbf{x} = \mathbf{x}_0$ is a solution to Equation (9.7), with

$$\dot{\mathbf{x}} = \mathbf{0}, \text{ and } \mathbf{u} = \mathbf{0} \text{ or constant}$$

Equations (9.1–9.3) correspond to the equations implied by Equation (9.7). However, care must be taken in applying the above definition. In Chapter 8 we indicated that a state variable is necessary to describe the behavior of the system. In looking at Equations (9.1–9.3) we might be tempted to say that all nine variables governed by these equations are necessary to describe the system.

However, looking more closely, none of these equations depend explicitly on heading Ψ_0. Also, if we reviewed the models for the forces and moments developed in Chapter 6, and appearing in Equations (9.1) and (9.2), we would find that these forces and moments are also independent of heading Ψ_0. Therefore, heading Ψ_0 is not a state variable in the sense used in Ref. 1, but just an auxiliary variable that depends on the other states. For example, if we were to numerically integrate the nine differential equations, Equations (9.1–9.3), we could integrate the first eight without having to include the ninth governing $\dot{\Psi}_0$! So for our nonlinear system consisting of the dynamics of the flight vehicle, the state vector does not include Ψ_0. It is

$$\mathbf{x}_0 = \begin{bmatrix} U_0 & V_0 & W_0 & P_0 & Q_0 & R_0 & \Phi_0 & \Theta_0 \end{bmatrix}^T \tag{9.8}$$

Note that the above argument also applies to the other three kinematic equations (not listed above) that govern the inertial position, or X_0, Y_0, and h_0 (see Equations (2.40)). If atmospheric density is constant, as we have assumed, then none of the previous nine differential equations of motion depend on inertial position. And so X_0, Y_0, and h_0 are also just auxiliary variables.

Finally, for a conventional vehicle with propulsive thrust T, elevator δ_E, aileron δ_A, and rudder δ_R, the reference input or control vector $\mathbf{u}_0$ would be

$$\mathbf{u}_0 = \begin{bmatrix} T_0 & \delta_{E_0} & \delta_{A_0} & \delta_{R_0} \end{bmatrix}^T \tag{9.9}$$

Therefore, assuming the above set of control variables, an *equilibrium or reference flight condition* will satisfy the following criteria:

$$\boxed{\begin{aligned} &\dot{U}_0 = \dot{V}_0(\text{or } \dot{\beta}_0) = \dot{W}_0(\text{or } \dot{\alpha}_0) = 0 \\ &\dot{P}_0 = \dot{Q}_0 = \dot{R}_0 = 0 \\ &\dot{\Phi}_0 = \dot{\Theta}_0 = 0 \\ &\begin{bmatrix} T_0 & \delta_{E_0} & \delta_{A_0} & \delta_{R_0} \end{bmatrix} \text{ all constant} \end{aligned}} \tag{9.10}$$

The first six of these constraints are intuitively pleasing because they correspond to a condition involving no translational or rotational accelerations. The next two ($\dot{\Phi}_0 = \dot{\Theta}_0 = 0$) require the X_V, Y_V, and Z_V components of gravity appearing in the three translational equations to be constant. Also, the requirements that $\dot{P}_0 = \dot{Q}_0 = \dot{R}_0 = 0$ imply that the angular rates are either zero or constant (as in a steady turn), and therefore the aerodynamic plus propulsive moments about each axis must sum either to zero or to constants. Furthermore, the first eight requirements taken collectively imply that the aerodynamic plus propulsive forces along each vehicle axis also sum to zero or to constants.

Applying the criteria given in Equations (9.10) allows us to now list the criteria defining the three most common reference flight conditions to be considered. The reference equations associated with these flight conditions will also be developed. As a final note, equilibrium conditions may exist other than those to be discussed here, depending on the vehicle's aerodynamic and mass properties, and they are usually found numerically by using Equations (9.1–9.3) and Constraints 9.10. An example of such an equilibrium condition is a flat spin!

1. *Steady Rectilinear Flight:* The criteria defining this flight condition include Equations (9.10) plus $\dot{\Psi}_0 = 0$, and therefore $P_0 = Q_0 = R_0 = 0$. This condition includes flight in a straight line at constant altitude, or in shallow climbs or descents, since atmospheric density has been assumed constant. If density is variable, only flight at a constant altitude strictly satisfies the criteria for an equilibrium condition. From the complete set of nine reference equations (Equations (9.1–9.3)), we find that the reference equations corresponding to steady rectilinear flight reduce to those given in Equations (9.11) below. Assumed here is that the vehicle-fixed axes used are the stability axes, so W_0 is zero, and for flight at constant altitude, Θ_0 is zero. (Recall that for the stability axes, $\Theta_0 = \gamma_0$ flight-path angle.) So the Reference Set of equations for steady rectilinear flight reduce to simply

$$
\begin{aligned}
mg\sin\gamma_0 &= F_{A_{X_0}} + F_{P_{X_0}} & 0 &= L_{A_0} + L_{P_0}\\
(0 \text{ if } \Phi_0 = 0) = -mg\cos\gamma_0\sin\Phi_0 &= F_{A_{Y_0}} + F_{P_{Y_0}} & 0 &= M_{A_0} + M_{P_0}\\
(-mg\cos\gamma_0 \text{ if } \Phi_0 = 0) = -mg\cos\gamma_0\cos\Phi_0 &= F_{A_{Z_0}} + F_{P_{Z_0}} & 0 &= N_{A_0} + N_{P_0}
\end{aligned}
\tag{9.11}
$$

For flight at constant altitude $\gamma_0 = 0$, and the equations may be simplified accordingly. The three kinematic equations (Equations (9.3)) are trivially satisfied.

2. *Steady Turning Flight:* The criteria defining this flight condition include Equations (9.10) plus $\dot{\Psi}_0 = \text{constant} \neq 0$. Under these conditions, the vehicle's angular-velocity vector $\boldsymbol{\omega}_{V,E}$ is vertical with respect to the earth-fixed coordinate system. And as with rectilinear flight, since atmospheric density is assumed constant, shallow climbing and descending turns are possible, in addition to constant-altitude turns ($\gamma_0 = 0$). Again assuming the use of the stability axes, the Reference Set of equations corresponding to a steady turn reduce to

$$
\begin{aligned}
-m(V_0R_0) + mg\sin\gamma_0 &= F_{A_{X_0}} + F_{P_{X_0}} & (I_{zz} - I_{yy})Q_0R_0 &= L_{A_0} + L_{P_0}\\
m(R_0U_0) - mg\cos\gamma_0\sin\Phi_0 &= F_{A_{Y_0}} + F_{P_{Y_0}} & -(I_{xz}(R_0^2)) &= M_{A_0} + M_{P_0}\\
m(P_0V_0 - Q_0U_0) - mg\cos\gamma_0\cos\Phi_0 &= F_{A_{Z_0}} + F_{P_{Z_0}} & I_{xz}(Q_0R_0) &= N_{A_0} + N_{P_0}
\end{aligned}
\tag{9.12}
$$

$$
\begin{aligned}
\dot{\Phi}_0 &= P_0 = 0 & & P_0 = 0\\
\dot{\Theta}_0 &= Q_0\cos\Phi_0 - R_0\sin\Phi_0 = 0 & \text{or}\quad & Q_0 = \dot{\Psi}_0\sin\Phi_0\\
\dot{\Psi}_0 &= (Q_0\sin\Phi_0 + R_0\cos\Phi_0)\sec\gamma_0 = \text{constant} & & R_0 = \dot{\Psi}_0\cos\Phi_0
\end{aligned}
$$

Note that if one is considering a turn at constant altitude, then $\Theta_0 = \gamma_0 = 0$, and Equations (9.12) may be further simplified.

Additional flight conditions are sometimes of interest, but are not strictly equilibrium reference conditions. An example is a quasi-steady pull up, which is defined by the following criteria.

3. *Quasi-Steady Pull Up:* The criteria are the same as for steady rectilinear flight, except that $\dot{\Theta}_0 \neq 0$. Note that this condition can only be analyzed in a quasi-steady fashion because the components of gravity along the three axes of the vehicle are not constant during the maneuver. Hence, it is not strictly an equilibrium flight condition. In any event, this flight condition is frequently of interest, and the reference equations are given in Equations (9.13) below.

   The pull-up maneuver is performed in a vertical plane, and hence the vehicle's angular velocity vector $\boldsymbol{\omega}_{V,E}$ is parallel to the earth with only its $\mathbf{j}_V$ component unequal to zero. So the roll P_0 and yaw R_0 rates are both zero. And usually, such maneuvers involve zero bank Φ_0 and sideslip β_0 angles. So we have the Reference Set of equations for the pull-up maneuver given by

$$
\begin{array}{lll}
mg\sin\gamma_0 = F_{A_{X_0}} + F_{P_{X_0}} & 0 = L_{A_0} + L_{P_0} & \dot{\Phi}_0 = 0 \\
0 = F_{A_{Y_0}} + F_{P_{Y_0}} & 0 = M_{A_0} + M_{P_0} & \dot{\Theta}_0 = Q_0 \neq 0 \\
-m(Q_0U_0) - mg\cos\gamma_0 = F_{A_{Z_0}} + F_{P_{Z_0}} & 0 = N_{A_0} + N_{P_0} & \dot{\Psi}_0 = 0
\end{array}
\tag{9.13}
$$

Summarizing the key results in this section, we presented a precise definition of an equilibrium condition for a nonlinear system, and we used that definition to determine the Reference Set of equations corresponding to three specific equilibrium flight conditions. These conditions included steady rectilinear flight, steady turning flight, and quasi-steady pull-up maneuvers. We will utilize the reference equations developed in this section as we further analyze these reference conditions. Next, we need to introduce the concept of aerodynamic static stability.

9.2 CONCEPT OF AERODYNAMIC STATIC STABILITY—AND CRITERIA

The phrase "static stability" frequently confuses students just introduced to the concept. To some, the phrase is an oxymoron, since stability is strictly a mathematical property of dynamic systems. For this reason, we have chosen to call the concept *aerodynamic static stability* to distinguish it from dynamic stability, and because the concept of static stability was more common within the aerodynamics community.

Aerodynamic static stability is a property of a dynamic system in a static condition—the system's dynamics are not addressed at all. Plus aerodynamic static stability depends entirely on the sense (or direction) of the instantaneous static forces or moments acting on the vehicle due to a slight displacement from an equilibrium condition. We will define static stability of a mechanical system as follows:

Definition: A mechanical system is *statically stable* with respect to a given equilibrium condition if the force and/or moment acting on the system due to a small static displacement from that equilibrium condition is in such a direction that it would tend to return the system to the given equilibrium condition.

Therefore, the concept of static stability depends on the direction of the force or moment acting on the system that is in a fixed or static position slightly displaced from the given equilibrium condition. A clarifying example follows.

EXAMPLE 9.1

Static Stability of the Simple Pendulum

Let us consider the mechanical system consisting of the simple pendulum introduced in Chapter 1, and depicted in Figure 9.2. In analyzing this system previously, we determined that it possessed two distinct equilibrium conditions: one corresponding to $\Theta_0 = 0$ and one corresponding to $\Theta_0 = \pi$. Now if we simply displacc the pendulum a small distance from the condition $\Theta_0 = 0$, indicated as Position 1 in the figure, we note that the moment due to gravity acting on the pendulum fixed at that position, or $-mgl \sin\theta$, is clockwise, which will tend to move the pendulum toward the equilibrium condition at $\Theta_0 = 0$. Hence, the pendulum is statically stable with respect to the equilibrium condition at $\Theta_0 = 0$.

Note further that the sign of the gradient $\frac{\partial M}{\partial \theta}|_{\Theta_0=0}$ can also indicate static stability. At the selected equilibrium condition this gradient is negative, or

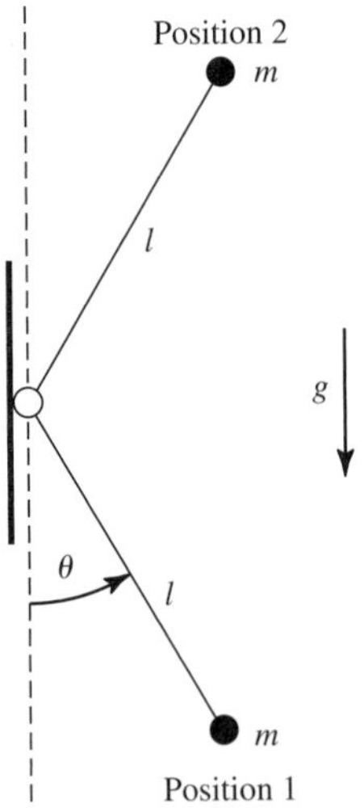

Figure 9.2 Example pendulum with two equilibrium positions.

$$\frac{\partial M}{\partial \theta}\Big|_{\Theta_0=0} = -mgl < 0$$

and so a negative gradient of the moment with respect to θ, evaluated at the equilibrium condition, may be used as a criterion for static stability of the pendulum.

Now consider the equilibrium condition at $\Theta_0 = \pi$. And let us displace the pendulum a small distance from that equilibrium condition, indicated as Position 2 in the figure. Now we note that the moment acting on the pendulum fixed at this position is clockwise, which will tend to move the pendulum away from the equilibrium condition at $\Theta_0 = \pi$. Hence, the pendulum is statically unstable with respect to the equilibrium condition at $\Theta_0 = \pi$. This is also indicated by the fact the moment gradient evaluated at this equilibrium condition is nonnegative, or

$$\frac{\partial M}{\partial \theta}\Big|_{\Theta_0=\pi} = mgl > 0$$

Let us now consider the aerodynamic static stability of a flight vehicle. First let us address in some detail the case of a vehicle perturbed only in pitch attitude θ from steady straight-and-level flight. The situation is depicted in Figure 9.3, in which the vehicle's change in pitch attitude is indicated by the change in attitude of the vehicle-fixed axes.

Note that a static displacement, or perturbation, in pitch $\Delta\theta$ produces a corresponding static displacement, or perturbation, in angle of attack $\Delta\alpha$ relative to the equilibrium or reference angle of attack α_0. So for the vehicle to possess aerodynamic static stability in pitch, there must be a nose-down restoring moment about its center of mass, or *c.g.*, created due to the perturbation in pitch attitude $\Delta\theta$ or in angle of attack $\Delta\alpha$. Since the aerodynamic and propulsive pitching moments are independent of θ, the nose-down moment must arise due to the change in angle of attack. So pitch stability will be present if the aerodynamic plus propulsive angle-of-attack pitching-moment effectiveness $\left(\frac{\partial M_A}{\partial \alpha}\Big|_0 + \frac{\partial M_P}{\partial \alpha}\Big|_0\right)$ is negative. Furthermore, since

$$M_A + M_P = (C_M + C_{P_M})q_\infty S_W \bar{c}_W \tag{9.14}$$

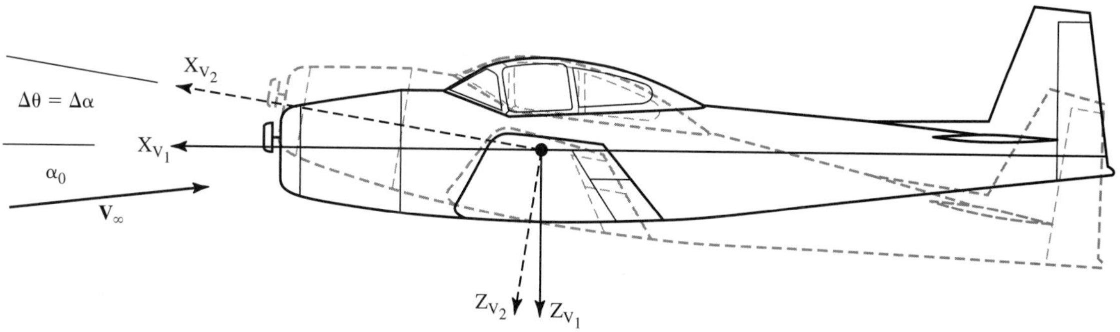

Figure 9.3 Change in angle of attack due to change in pitch attitude.

the required negativity of the moment effectiveness will be satisfied if the sum of the corresponding effectiveness coefficients is negative, or $\left(C_{M_\alpha} + C_{P_{M_\alpha}}\right) < 0$. This inequality therefore constitutes the criterion for aerodynamic static stability in pitch for an aircraft or missile.

Now note that for the pendulum, the displacement from equilibrium was defined only in terms of a single displacement, or perturbation, in the angle θ. And the restoring action was a moment arising due to that angular displacement. But when dealing with flight vehicles, more than one displacement variable is involved. And both forces and moments must be considered. These displacement variables will be grouped into the longitudinal set and the lateral-directional set.

9.2.1 Longitudinal Static Stability

The longitudinal set of displacement variables are listed in Table 9.1, along with the corresponding restoring actions. In addition, for each displacement variable the criterion for aerodynamic static stability with respect to that particular displacement is also given. The development of the criteria follows in the same fashion as in longitudinal pitch stability discussed above. Note that in developing the criteria involving forces, we assume that the reference or equilibrium angle of attack α_0 and sideslip angle β_0 are sufficiently small such that $\cos\alpha_0 \approx \cos\beta_0 \approx 1$, plus the lift L, drag D, and side force S are such that

$$|L\sin\alpha - \mathsf{S}\sin\beta| \ll |D|$$
$$|D\sin\beta| \ll |\mathsf{S}|$$
$$|D\sin\alpha + \mathsf{S}\sin\alpha\sin\beta| \ll |L|$$

If this is not the case, the criteria are slightly more complex algebraically, but are developed in the same manner.

Table 9.1 Longitudinal Static-Stability Criteria

Displacement Variable	Restoring Action	Aerodynamic Static Stability Criterion
Surge velocity, u	Axial force, $\left(F_{A_X} + F_{P_X}\right)$	$\left(-C_{D_u} + C_{P_{X_u}}\right) < 0$ "speed stability"
	Pitching moment, $\left(M_A + M_P\right)$	$\left(C_{M_u} + C_{P_{M_u}}\right) > 0$
Plunge velocity, w	Vertical force, $\left(F_{A_Z} + F_{P_Z}\right)$	$\left(-C_{L_\alpha} + C_{P_{Z_\alpha}}\right) < 0$
Angle of attack, α $\left(w/U_0\right)$	Pitching moment, $\left(M_A + M_P\right)$	$\left(C_{M_\alpha} + C_{P_{M_\alpha}}\right) < 0$ "pitch stability"
Pitch rate, q	Pitching moment, $\left(M_A + M_P\right)$	$\left(C_{M_q} + C_{P_{M_q}}\right) < 0$ "pitch-damping"

With regard to the criteria associated with surge velocity u, it should be clear that if $\left(-C_{D_u} + C_{P_{X_u}}\right) < 0$, the force gradient is such that the vehicle will tend to hold constant velocity. That is, a slight increase in velocity leads to an increase in drag, which tends to reduce the velocity. But the second criterion associated with surge velocity is less obvious. Basically, if $\left(C_{M_u} + C_{P_{M_u}}\right) > 0$, the vehicle will tend to pitch up if the velocity is increased. As a result of the vehicle pitching up, the velocity will tend to decrease due to the effect of gravity. This criterion does not stem strictly from the definition of static stability, but is frequently included in the list of static-stability criteria anyway.

Frequently, the effectiveness coefficients associated with the propulsion system $C_{P.}$ are either small or zero. In this case, the criteria are only functions of the signs of the aerodynamic effectiveness coefficients. This would be especially true for gliders, or in flight conditions in which the engines were throttled back.

Note that some of the stability criteria listed in Table 9.1 are more difficult to satisfy than others, and some are more critical than others. For example, the vehicle's angle-of-attack lift effectiveness C_{L_α} (per rad) is always larger than unity, and the propulsive angle-of-attack force effectiveness $C_{P_{Z_\alpha}}$ (see Equation (6.155)) will never be positive. So it is easy to assure that $\left(-C_{L_\alpha} + C_{P_{Z_\alpha}}\right) < 0$ is negative.

Likewise, pitch damping is always present unless significant flow separation is present. This assertion may be corroborated by recalling that from Equation (6.188) the aerodynamic pitch-damping coefficient for a vehicle with an aft horizontal tail is

$$C_{M_q} \approx -C_{L_{\alpha_H}} \frac{\left(X_{\text{Ref}} - X_{\text{AC}_H}\right)^2}{U_0 \bar{c}_W} \frac{q_H}{q_\infty} \frac{S_H}{S_W} \tag{9.15}$$

And the pitch-damping coefficient associated with the propulsion system is given by Equation (6.196), or

$$C_{P_{M_q}} \approx -\frac{x_T^2}{q_\infty S_W \bar{c}_W} \frac{\partial F_N}{\partial v_{\text{Transverse}}}\Big|_0 \tag{9.16}$$

Recall that in all expressions for pitching moments developed in Chapter 6, axial (X) locations are taken to be positive forward. Also recall that the moments acting on the vehicle are taken about a particular reference point, the vehicle's center of mass, or $c.g.$ Hence, in Equation (9.15) we must set $X_{\text{Ref}} = X_{\text{cg}}$.

Then, since $C_{L_{\alpha_H}}$ is always positive, we see that the aerodynamic pitch-damping coefficient given by Equation (9.15) will always be negative. In addition, since the propulsive normal force gradient to inlet cross flow, or $\delta F_N/\delta v_{\text{Transverse}}$, will never be negative, the propulsive pitch damping given by Equation (9.16) will never be positive. So this proves the assertion that pitch damping will always be present.

Speed Stability Speed stability, however, and, more importantly, aerodynamic pitch stability must be evaluated more carefully than the cases just discussed. The effectiveness coefficients in the speed-stability criteria, $-C_{D_u} + C_{P_{X_u}}$ and $C_{M_u} + C_{P_{M_u}}$, were discussed in Section 6.6.1. First, note that we found that the partial derivatives of the aerodynamic and propulsive pitching moments with respect to flight velocity were

$$\frac{\partial M_A}{\partial u}\Big|_0 = \left(C_{M_u} + \frac{2C_{M_0}}{U_0}\right) q_\infty S_W \bar{c}_W \tag{9.17}$$

and

$$\frac{\partial M_P}{\partial u}\Big|_0 = \left(C_{P_{X_u}} + \frac{2C_{P_{M_0}}}{U_0}\right) q_\infty S_W \bar{c}_W \tag{9.18}$$

with

$$C_{P_{X_u}} = \frac{1}{q_\infty S_W \bar{c}_W}\left((d_T - x_T \phi_T)\frac{\partial T}{\partial v_{\text{Axial}}}\Big|_0 - x_T \frac{\partial F_N}{\partial v_{\text{Transverse}}}\Big|_0 (\phi_T + \alpha_0)\right) \tag{9.19}$$

Next, note that the pitch-speed-stability criterion may be stated more fundamentally as

$$\left(\frac{\partial M_A}{\partial u}\Big|_0 + \frac{\partial M_P}{\partial u}\Big|_0\right) > 0 \tag{9.20}$$

And the quantity on the left in the above inequality is the sum of the terms given in Equations (9.17) and (9.18). Now if the vehicle is in steady, level flight, for example, the total pitching moment will be zero, and so

$$\left(C_{M_0} + C_{P_{M_0}}\right) = 0 \tag{9.21}$$

Therefore, the criterion stated in Equation (9.20) reduces to that given in Table 9.1.

We noted in Chapter 6 that C_{M_u} is typically very small except in the transonic speed range, where it can be either positive or negative. This is due to the shift in the locations of the aerodynamic centers of the lifting surfaces as the flight velocity transitions from subsonic to supersonic. We also noted that in the transonic speed range, C_{M_u} was difficult to predict. Likewise, C_{D_u} was also quite small, except in the transonic speed range. In that speed range this effectiveness coefficient may also be either positive or negative. So the transonic speed range can be problematic, which led to many crashes during early flight tests when the aeronautics community was trying to "break the sound barrier."

The effect of the propulsive forces on speed stability, or $C_{P_{X_u}}$, is indicated by Equation (6.139), which states that

$$C_{P_{X_u}} \approx \frac{1}{q_\infty S_W} \frac{\partial T}{\partial v_{\text{Axial}}}\Big|_0 \tag{9.22}$$

Hence, this term will take on the sign of the thrust gradient with respect to velocity.

Additionally, the effect of propulsive forces on pitch speed stability may be investigated by considering Equation (6.143), which states that

$$C_{P_{M_u}} \approx \frac{1}{q_\infty S_W \bar{c}_W}\left(\left(d_T - x_T\phi_T\right)\frac{\partial T}{\partial v_{\text{Axial}}}\bigg|_0 - x_T\frac{\partial F_N}{\partial v_{\text{Transverse}}}\bigg|_0\left(\phi_T + \alpha_0\right)\right) \qquad (9.23)$$

The location of the engine(s) relative to $X_{\text{Ref}} = X_{\text{cg}}$, given by the parameters x_T and d_T, and the pitch orientation of the engine(s) relative to the fuselage-referenced axis, given by ϕ_T, were all defined in Figure 6.13. We see that even though the normal-force gradient in the above expression is typically positive, $C_{P_{M_u}}$ can be positive or negative depending on the location of the engines on the vehicle and on the sign of the thrust gradient with respect to axial velocity.

So the signs on the two propulsive effectiveness coefficients given in Equations (9.22) and (9.23) may be either positive or negative. Thus analysis of speed stability requires some care.

Static Pitch Stability Revisited We will now address static pitch stability in more detail, first by looking at the contributions of the various parts of the vehicle. Recall that static pitch stability is determined by the sign of $\left(C_{M_\alpha} + C_{P_{M_\alpha}}\right)$. In Equation (6.56) we had the angle-of-attack pitching-moment effectiveness C_{M_α} given by

$$\begin{aligned}
C_{M_\alpha} &= C_{L_{\alpha_W}}\left(\frac{X_{\text{AC}_{W\,\&\,F}} - X_{\text{Ref}}}{\bar{c}_W}\right) - C_{L_{\alpha_H}}\left(1 - \frac{d\varepsilon}{d\alpha}\right)\left(\frac{X_{\text{Ref}} - X_{\text{AC}_H}}{\bar{c}_W}\right)\frac{q_H}{q_\infty}\frac{S_H}{S_W} \\
&= C_{L_{\alpha_W}}\Delta\bar{X}_{\text{AC}_{W\,\&\,F}} - C_{L_{\alpha_H}}\left(1 - \frac{d\varepsilon}{d\alpha}\right)\Delta\bar{X}_{\text{AC}_H}\frac{q_H}{q_\infty}\frac{S_H}{S_W} \qquad (9.24) \\
&= \text{Wing Contribution} - \text{Tail Contribution}
\end{aligned}$$

Also recall that in the derivation of the above expression, X locations were taken to be positive <u>forward</u> on the vehicle, and that we must set $X_{\text{Ref}} = X_{\text{cg}}$. Now looking at the wing contribution, if the aerodynamic center of the wing-fuselage combination is located forward of the *c.g.*, then $\Delta\bar{X}_{\text{AC}_{W\,\&\,F}} > 0$, and increasing wing lift (by making the wing larger, for example) increases C_{M_α}, which tends to <u>reduce</u> static pitch stability. Conversely, since an aft horizontal tail is behind the *c.g.*, then $\Delta\bar{X}_{\text{AC}_H} > 0$, and increasing the lift of the tail (by making the tail larger, for example) decreases C_{M_α}, or <u>increases</u> static pitch stability. So the tail provides for positive static pitch stability, which is the primary reason for use of an aft tail. If a forward horizontal surface, or canard, is used, then the location of the wing's aerodynamic center must be aft of the *c.g.* to obtain static pitch stability (See Problem 9.2). Finally, from Equation (9.24) also note the effect of relative tail size S_H/S_W and downwash gradient $d\varepsilon/d\alpha$ on static pitch stability.

From Equation (6.159), we may write the contribution of the propulsive forces to the angle-of-attack pitching-moment effectiveness as

$$C_{P_{M_\alpha}} = -\frac{U_0}{q_\infty S_W \bar{c}_W}\left(\left(d_T - x_T\phi_T\right)\frac{\partial T}{\partial v_{\text{Axial}}}\bigg|_0 (\phi_T + \alpha_0) + \left(x_T + d_T\phi_T\right)\frac{\partial F_N}{\partial v_{\text{Transverse}}}\bigg|_0\right)\left(1 \pm \frac{d\varepsilon_{\text{inlet}}}{d\alpha}\bigg|_0\right) \tag{9.25}$$

This parameter depends on the location and orientation of the engine(s), and the gradients of the thrust T and normal force F_N with respect to inlet-flow velocities. The location of the engine(s) relative to $X_{\text{Ref}} = X_{\text{cg}}$, given by the parameters x_T and d_T, and the pitch orientation of the engine(s) relative to the fuselage-referenced axes, given by ϕ_T, were all defined in Figure 6.13.

The gradients of the thrust and normal force are properties of the engine(s), and the gradient of the normal force is typically positive. So the angle-of-attack pitching-moment effectiveness of the propulsion system may be either positive or negative, depending on the characteristics of the propulsion systems and their locations on the vehicle. Locating the engine(s) aft and/or below the *c.g.* usually tends to increase static pitch stability. However, if the line of action of the thrust T passes near the *c.g.*, the term $\left(d_T \cos\phi_T - x_T \sin\phi_T\right) \approx \left(d_T - x_T\phi_T\right)$ will be approximately zero, and then $C_{P_{M_\alpha}} \approx 0$. But in any event, typically the most important factor is the aerodynamic static pitch stability discussed above.

The Static Margin One very important parameter is an aircraft's or missile's *static margin.* As fuel is expended or payload is added or off loaded, the axial location of the *c.g.* will move fore and aft on the vehicle, which will affect the static margin. In several sections of this chapter, including this one, we will develop criteria that will limit the allowable variation in *c.g.* location.

Referring to Figure 9.4, for now let the two locations labeled X_{cg_1} and X_{cg_2} be any two *c.g.* locations. From simple statics, and assuming the vehicle's angle of attack is small, the (aerodynamic plus propulsive) pitching moment taken about

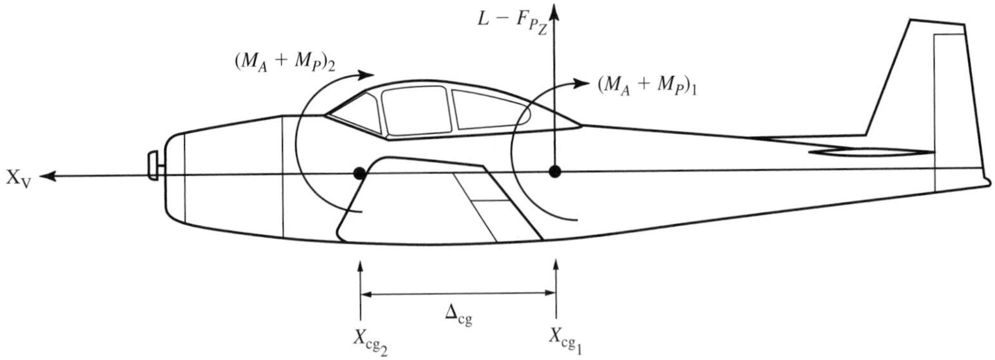

Figure 9.4 Change in pitching-moment reference location.

one *c.g.* location X_{cg_2} expressed in terms of the pitching moment taken about the other *c.g.* location X_{cg_1} is

$$(M_A + M_P)_2 = (M_A + M_P)_1 - (L - F_{P_Z})\Delta_{cg} \tag{9.26}$$

where

$$\Delta_{cg} \triangleq X_{cg_2} - X_{cg_1}$$

Here X_{cg} is measured positive forward on the vehicle, F_{P_Z} is the Z component of propulsive thrust, and L is the aerodynamic lift acting on the vehicle.

Dividing Equation (9.26) by $q_\infty S_W \bar{c}_W$ yields a similar expression given in terms of force and moment coefficients. Therefore, the sum of pitching-moment coefficients referenced to the new *c.g.* location is given by

$$(C_M + C_{P_M})_2 = (C_M + C_{P_M})_1 - (C_L - C_{P_Z})\left(\frac{\Delta_{cg}}{\bar{c}_W}\right) = (C_M + C_{P_M})_1 - (C_L - C_{P_Z})\bar{\Delta}_{cg} \tag{9.27}$$

(Note the use of the overbar notation in the above expression.) Taking the partial derivative of Equation (9.27) with respect to vehicle angle of attack yields

$$\boxed{(C_{M_\alpha} + C_{P_{M_\alpha}})_2 = (C_{M_\alpha} + C_{P_{M_\alpha}})_1 - (C_{L_\alpha} - C_{P_{Z_\alpha}})\bar{\Delta}_{cg}} \tag{9.28}$$

We may now use this expression in defining the *static margin* for a flight vehicle.

Definition: The static margin (SM) of a flight vehicle is the normalized distance its *c.g.* may be shifted aft, or $-\bar{\Delta}_{cg}$, before $(C_{M_\alpha} + C_{P_{M_\alpha}})_{cg} \rightarrow 0$.

Therefore, by setting $(C_{M_\alpha} + C_{P_{M_\alpha}})_2 = 0$ in Equation (9.28) and solving, the vehicle's static margin is found to be

$$\boxed{SM \triangleq -\bar{\Delta}_{cg}\big|_{(C_{M_\alpha} + C_{P_{M_\alpha}})_{new} \rightarrow 0} = -\frac{(C_{M_\alpha} + C_{P_{M_\alpha}})_{\text{Current cg}}}{(C_{L_\alpha} - C_{P_{Z_\alpha}})}} \tag{9.29}$$

and if the propulsive thrust does not contribute significantly to the pitching moment or vertical force, a common situation, we have the static margin given by

$$\boxed{SM \triangleq -\frac{C_{M_\alpha}}{C_{L_\alpha}}} \tag{9.30}$$

Note that if the vehicle is statically stable in pitch, or $(C_{M_\alpha} < 0)$, the static margin will be positive.

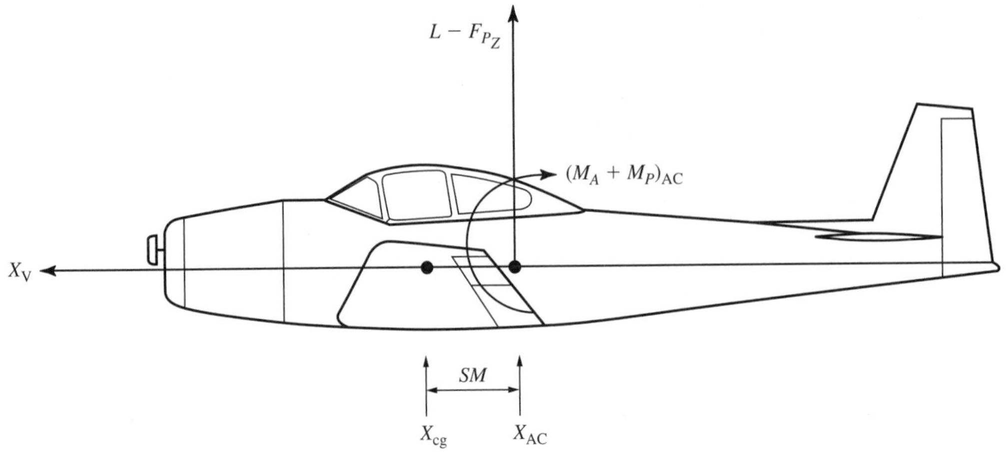

Figure 9.5 The static margin of an aircraft.

Also note that the criterion for static pitch stability, or that $(C_{M_\alpha} + C_{P_{M_\alpha}}) < 0$, requires the vehicle's *c.g.* to be forward of the vehicle's aerodynamic center. This follows from the definition of the aerodynamic center, which is the axial (*X*) location of the vehicle about which the pitching moment is invariant with respect to angle of attack, or

$$(C_{M_\alpha} + C_{P_{M_\alpha}})_{AC} = 0 \tag{9.31}$$

To prove this assertion, let the vehicle's aerodynamic center be located at the point labeled X_{cg_1} in Figure 9.4. Applying Equation (9.28) we have

$$(C_{M_\alpha} + C_{P_{M_\alpha}})_2 = (C_{M_\alpha} + C_{P_{M_\alpha}})_1 - (C_{L_\alpha} - C_{P_{Z_\alpha}})\overline{\Delta}_{cg} \tag{9.32}$$

Therefore, requiring $(C_{M_\alpha} + C_{P_{M_\alpha}}) < 0$ implies that $\overline{\Delta}_{cg} > 0$, or the *c.g.* must be forward of the aerodynamic center. Consequently, as indicated in Figure 9.5, the static margin is also the normalized distance between the vehicle's *c.g.* and its aerodynamic center, and is positive when X_{cg} is forward of X_{AC}. All this indicates that the location of the *c.g.* relative to the location of the *ac* is obviously an important design parameter.

EXAMPLE 9.2

Static Margin, Tail Size, and *c.g.* Location

Consider an aircraft with lifting surfaces as described in Examples 6.1–6.2, and for which the following data is provided.

$$C_{L_\alpha} = 4.59 \text{ /rad}, \quad C_{M_\alpha} = -1.59 \text{ /rad}, \quad \overline{X}_{Ref} = \overline{X}_{cg} = 0.5 \text{ aft of wing apex}$$

For the wing and fuselage combination, we have

$$S_W = 169\ \text{ft}^2,\ \bar{c}_W = 5.825\ \text{ft},\ C_{L_{\alpha_W}} = 4.19\ /\text{rad},\ \overline{X}_{\text{AC}_{W\,\&\,F}} = \frac{4.79 - 1.19}{5.825} = 0.618\ \text{aft of wing apex}$$

And for the horizontal tail we have

$$S_H = 42.2\ \text{ft}^2,\ S_H/S_W = 0.250,\ C_{L_{\alpha_H}} = 4.19\ /\text{rad},\ \frac{d\varepsilon_H}{d\alpha_W} = 0.57$$

$$\overline{X}_{\text{AC}_H} = \frac{18.75}{5.825} = 3.219\ \text{aft of wing apex}$$

Assuming the propulsive forces do not contribute significantly to the vertical (Z) force or pitching moment on the vehicle, determine an expression for the static margin *SM* for the vehicle as a function of the size (planform area) of the horizontal tail. How much must the tail planform area be scaled up or down to yield a static margin of 20 percent? Conversely, how much farther forward or aft can the *c.g.* be located to achieve a static margin of 20 percent with the original tail size?

■ Solution

From Equation (9.30), the static margin for this vehicle is

$$SM \triangleq -\frac{C_{M_\alpha}}{C_{L_\alpha}} = -\frac{-1.59}{4.59} = 0.346$$

which is positive since the vehicle is statically stable in pitch. From Equation (9.24), the expression for C_{M_α} of the vehicle is given by

$$C_{M_\alpha} = C_{L_{\alpha_W}}\left(\frac{X_{\text{AC}_{W\,\&\,F}} - X_{\text{Ref}}}{\bar{c}_W}\right) - C_{L_{\alpha_H}}\left(1 - \frac{d\varepsilon}{d\alpha}\right)\left(\frac{X_{\text{Ref}} - X_{\text{AC}_H}}{\bar{c}_W}\right)\frac{q_H}{q_\infty}\frac{S_H}{S_W}$$

$$= C_{L_{\alpha_W}}\Delta\overline{X}_{\text{AC}_{W\,\&\,F}} - C_{L_{\alpha_H}}\left(1 - \frac{d\varepsilon}{d\alpha}\right)\Delta\overline{X}_{\text{AC}_H}\frac{q_H}{q_\infty}\frac{S_H}{S_W}$$

in which axial (*X*) locations are taken to be positive forward on the vehicle. Now taking the partial derivative of this expression with respect to tail-to-wing area ratio yields

$$\frac{\partial C_{M_\alpha}}{\partial(S_H/S_W)} = -C_{L_{\alpha_H}}\Delta\overline{X}_{\text{AC}_H}\frac{q_H}{q_\infty}\left(1 - \frac{d\varepsilon}{d\alpha}\right)$$

which, for this vehicle, becomes

$$\frac{\partial C_{M_\alpha}}{\partial(S_H/S_W)} = -4.19(-0.5 + 3.219)(0.9)(1 - 0.57) = -4.409$$

(Note that the signs on X_{Ref} and X_{AC_H} have been reversed because they were given originally as measured positive <u>aft</u> of the wing apex.) So, as expected, the static stability of the vehicle will be <u>increased</u>, or C_{M_α} will become more negative, with increased tail area.

Next, from Equation (9.24) we write the angle-of-attack pitching-moment effectiveness in terms of the tail-to-wing area ratio as

$$C_{M_\alpha} = C_{L_{\alpha_W}} \Delta\overline{X}_{\mathrm{AC}_{W\,\&\,F}} - C_{L_{\alpha_H}} \Delta\overline{X}_{\mathrm{AC}_H} \frac{q_H}{q_\infty} \frac{S_H}{S_W}\left(1 - \frac{d\varepsilon}{d\alpha}\right)$$

$$= 4.19(-0.618 + 0.5) - 4.19(-0.5 + 3.219)(0.9)\frac{S_H}{S_W}(1 - 0.57)$$

So the static margin is linear in terms of the tail area ratio, or

$$SM = -\frac{-4.409(S_H/S_W) - 0.494}{4.59} = 0.961(S_H/S_W) + 0.108$$

Note that in this example, the *SM* is positive even if $S_H = 0$ since the *ac* of the wing and fuselage is aft of the *c.g.*

To obtain a static margin of 20 percent requires that

$$0.20 = 0.961(S_H/S_W) + 0.108$$

and so the required tail-area ratio is

$$\frac{S_H}{S_W} = 0.096$$

Given that the original tail-area ratio is 0.250, we see that the tail area must be decreased by the following amount to obtain a 20 percent static margin.

$$\Delta S_H = S_W \Delta\left(\frac{S_H}{S_W}\right) = 169(0.096 - 0.250) = -26.03 \text{ ft}^2$$

Or the tail planform area must be reduced by more than 50 percent!

Finally, for the original tail size the static margin is 0.346 (a rather large margin). And by definition, the static margin is the normalized distance between the *c.g.* and the aerodynamic center of the vehicle, or $SM = \overline{X}_{\mathrm{cg}} - \overline{X}_{\mathrm{AC}}$, if these locations are measured positive forward on the vehicle. To obtain a static margin of 20 percent implies a change in static margin of $\Delta SM = -0.146$. This implies an allowable aft shift in *c.g.* of

$$\Delta\overline{X}_{\mathrm{cg}} = -0.146.$$

Therefore, the *c.g.* can be moved aft $0.146 \times 5.825 = 0.85$ ft and still maintain a static margin of 20 percent. This would place the *c.g.* slightly aft of the aerodynamic center of the wing-fuselage combination $X_{\mathrm{AC}_{W\,\&\,F}}$.

9.2.2 Lateral-Directional Static Stability

Continuing with the discussion of static stability, now consider the lateral-direction set of displacement variables listed in Table 9.2. The corresponding restoring actions are also listed, and in each case the criterion for aerodynamic static stability with respect to the particular displacement is given.

Table 9.2 Lateral-Directional Static Stability Criteria

Displacement Variable	Restoring Action	Aerodynamic Static Stability Criteria
Lateral velocity, v	Side force, $\left(F_{A_Y} + F_{P_{A_Y}}\right)$	$\left(C_{S_\beta} + C_{P_{Y_\beta}}\right) < 0$
Side-slip angle, β $\left(= v/U_0\right)$	Yawing moment, $\left(N_A + N_P\right)$	$\left(C_{N_\beta} + C_{P_{N_\beta}}\right) > 0$ "directional stability"
	Rolling moment, $\left(L_A + L_P\right)$	$\left(C_{L_\beta} + C_{P_{L_\beta}}\right) < 0$ "dihedral effect"
Roll rate, p	Rolling moment, $\left(L_A + L_P\right)$	$\left(C_{L_p} + C_{P_{L_p}}\right) < 0$ "roll damping"
Yaw rate, r	Yawing moment, $\left(N_A + N_P\right)$	$\left(C_{N_r} + C_{P_{N_r}}\right) < 0$ "yaw damping"

NOTE TO STUDENT

A reminder regarding notation—since rolling moment and lift are by convention both denoted as L, we have sometimes labeled rolling moment L_{Roll}. However, we have not used this more complex notation for effectiveness coefficients such as $C_{L_\beta}\left(= \partial C_{L\text{roll}}/\partial\beta\right)$ because sideslip lift effectiveness $\partial C_{\text{Lift}}/\partial\beta$ is not encountered often. Hence, dropping the "roll" subscript in effectiveness coefficients should hopefully cause little confusion.

As noted in Section 9.2.1, the effectiveness coefficients associated with propulsive forces $C_{P.}$ are frequently small or zero. In such cases, the criteria are only functions of the signs of the aerodynamic effectiveness coefficients. And, as in the longitudinal case, some of the lateral-directional criteria in Table 9.2 are more difficult to satisfy, and some are more important than others. Also, with the exception of the sign on the directional-stability criterion and the dihedral effect, the criteria given in Table 9.2 are quite straightforward and intuitive. Therefore, these latter two criteria will be discussed in greater detail.

With respect to lateral velocity, C_{S_β} and $C_{P_{Y_\beta}}$ are always negative, so that criterion will always be met. (See, for example, Equations (6.25) and (6.170), respectively.) Also, in the absence of significant flow separation, roll and yaw damping will always be present. (See, for example, Equations (6.207), (6.214), (6.219), and (6.232).) Therefore, these three stability criteria are easily satisfied.

Regarding directional stability, the sum of the aerodynamic plus propulsive sideslip yaw effectiveness must be positive due to the sign convention on yawing moment and sideslip angle. Under the convention defined, a positive sideslip angle, depicted in Figure 9.6, must induce a positive yawing moment (nose right) for stability. That is, the nose of the vehicle must tend to turn into the oncoming flow $\mathbf{V}_\infty$, a characteristic also called *weathercock stability*.

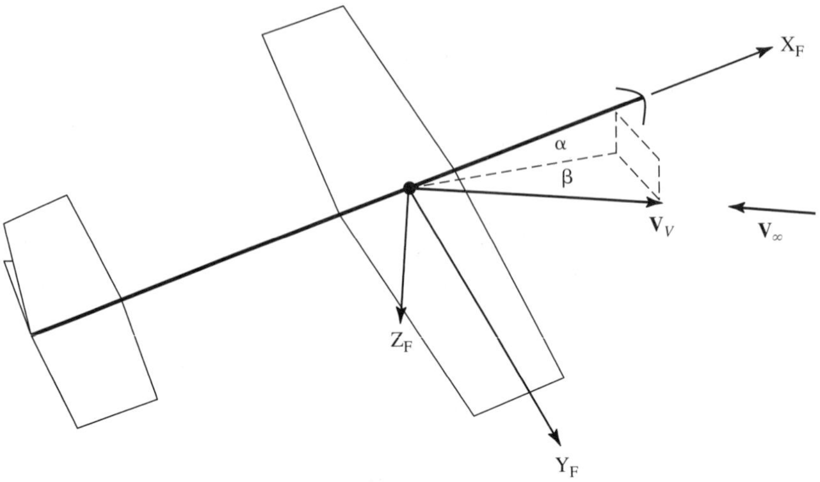

Figure 9.6 Vehicle angle of attack, sideslip angle, and fuselage-referenced axes.

The contributors to aerodynamic static directional stability may be identified by considering Equation (6.68), which states that

$$C_{N_\beta} = -C_{S_{\beta_V}}\left(\frac{X_{\text{Ref}} - X_{\text{AC}_V}}{b_W}\right)\frac{q_H}{q_\infty}\frac{S_V}{S_W} = -C_{S_{\beta_V}}\Delta\overline{X}_{\text{AC}_V}\frac{q_H}{q_\infty}\frac{S_V}{S_W} \tag{9.33}$$

Noting that this equation assumes X locations are measured positive forward, for an aft vertical tail $\Delta\overline{X}_{\text{AC}_V}$ will be positive. However, a positive sideslip angle β generates a force on the vertical tail that is in the negative Y direction. Hence $C_{S_{\beta_V}}$ will always be negative. This guarantees that an aft vertical tail will tend to yield a positive C_{N_β}, leading to positive static directional stability.

The contribution of the propulsive forces to static directional stability may be investigated by considering Equation (6.175), which states that for a vehicle in yaw trim (i.e., $C_{P_{N_0}} = C_{N_0} = 0$),

$$C_{P_{N_\beta}} = \frac{U_0}{q_\infty S_W b_W}(x_T - d_T\alpha_0)\frac{\partial F_N}{\partial v_{\text{Transverse}}} \tag{9.34}$$

If the engine(s) is (are) located aft of the *c.g.*, x_T will be positive. So assuming $(x_T\cos\alpha_0 - d_T\sin\alpha_0) \approx (x_T - d_T\alpha_0) > 0$, and given that the propulsive normal-force gradient $\partial F_N/\partial v_{\text{Transverse}}$ is positive, engines located aft of the *c.g.* will tend to increase static directional stability. The opposite will hold for engines forward of the *c.g.*, and static stability must be provided strictly through aerodynamic means.

The criterion in Table 9.2 defined as "dihedral effect" implies that a positive sideslip angle β must induce a negative rolling moment (right wing tip up). This is sometimes confusing to students. But imagine a vehicle initially flying with wings level at constant altitude. If the vehicle was then banked to the right to some positive bank angle, such a vehicle would experience a positive sideslip angle as it begins to lose altitude. Now if the rolling moment induced due to this sideslip is negative, the vehicle will tend to roll left—or the bank angle would tend back towards zero. This is the desired dihedral effect, though not strictly meeting the definition of static stability.

The contributions to dihedral effect may be investigated by considering Equation (6.44), which states that

$$C_{L_\beta} = C_{L_{\beta_W}} + C_{L_{\beta_H}} \frac{q_H}{q_\infty} \frac{S_H}{S_W} \frac{b_H}{b_W} + C_{L_{\beta_V}} \frac{q_H}{q_\infty} \frac{S_V}{S_W} \frac{b_V}{b_W} \tag{9.35}$$

$$= \text{Wing Contribution} + \text{Horizontal} + \text{Vertical Tail Contributions}$$

So the dihedral effect of the vehicle depends on the dihedral effects of the wing and the horizontal tail, and the rolling moment generated by the vertical tail. The wing- and horizontal-tail-dihedral effects were discussed in Section 5.3.4. In that section we showed that positive leading-edge wing sweep and positive dihedral angle both increased dihedral effect, or makes C_{L_β} of the wing or tail more negative.

The rolling moment due to the vertical tail was given in Equation (6.41), which states that

$$C_{L_{\beta_V}} = C_{S_{\beta_V}} \frac{Z_{AC_V}}{b_V} \tag{9.36}$$

As noted when discussing Equation (9.33), $C_{S_{\beta_V}}$ will always be negative. Hence, if the aerodynamic center of the vertical tail is located above the X_V axis, or $Z_{AC_V} > 0$, the vertical tail will tend to increase the dihedral effect of the vehicle. Note, however, that when using stability axes with a very high trim angle of attack α_0, Z_{AC_V} may be small or even less than zero (see Figure 6.8).

The effect of the propulsive forces on the dihedral effect is indicated in Equation (6.175), which states that

$$C_{P_{L_\beta}} = \frac{U_0}{q_\infty S_W b_W}\left(d_T + x_T \alpha_0\right) \frac{\partial F_N}{\partial v_{\text{Transverse}}} \tag{9.37}$$

Given that the propulsive gradient $\partial F_N / \partial v_{\text{Transverse}}$ is typically positive, we can see that engines located <u>above</u> the fuselage-referenced X axis ($d_T < 0$) and/or <u>forward</u> of the *c.g.* ($x_T < 0$) will tend to increase dihedral effect.

The astute reader will note that the static-stability criteria presented in this section can lead to conflicting requirements on the vehicle geometry. For example, one of the stability criteria might suggest that the engine(s) should be located forward of the *c.g.*, while another criterion suggests the engine(s) should be located aft. This emphasizes the fact that the design of aerospace vehicles, and design in general, always requires the designer to perform a series of tradeoffs to find the best overall solution. But this is the essence of engineering.

Finally, note that it is quite possible that some of the criteria stated in Sections 9.2.1 and 9.2.2 for aerodynamic static stability may be relaxed in terms of design criteria for a particular vehicle, due to the inherent tradeoffs that exist. The criteria given here are criteria for aerodynamic static stability, and some or all may be used as vehicle-design criteria. In contrast, some high-performance aircraft are either neutrally statically stabile or even statically unstable in pitch in many flight conditions. Reasons for this will be discussed more in Section 9.3 and later sections.

Summarizing the key results in this section, we introduced the concept of aerodynamic static stability, along with several criteria for static stability of a flight vehicle. We also investigated the effects of different aspects of the vehicle, or the vehicle's design, that contributed to meeting the static-stability criteria. In addition, we introduced an important parameter known as the *static margin* of the aircraft, and showed that to assure static pitch stability the vehicle's center of mass, or center of gravity, must be located forward of the vehicle's aerodynamic center.

9.3 ANALYSIS OF STEADY RECTILINEAR FLIGHT

We will now turn our attention to the analysis of the vehicle while in steady rectilinear flight, the first of the equilibrium flight conditions defined in Section 9.1. Of particular interest will be trim analysis, control power, engine-out effects, control (or stick) forces, and the effects of aerodynamic static stability.

Recall that we are assuming conventional vehicle geometry, with the vehicle's XZ plane a plane of symmetry. We will include a variable incidence angle i_H on the aft horizontal tail, and we will also assume that the propulsive thrust is symmetric with respect to the XZ plane of the vehicle, which implies no side force or yawing moment due to thrust. A special case dealing with a yawing moment due to asymmetric thrust will be considered in Section 9.3.3. For other types of vehicles, the analysis below must be modified accordingly.

The reference equations for this flight condition, assuming a stability-axis system, were given in Equations (9.11). If we now substitute the expressions for the aerodynamic and propulsive forces and moments into these equations, the reference equations become

$$
\begin{aligned}
&mg\sin\gamma_0 = -D_0\cos\beta_0 - \mathrm{S}_0\sin\beta_0 + \mathrm{T}_0\cos(\phi_T + \alpha_0) && \mathrm{L}_{A_0} = 0\\
&-mg\cos\gamma_0\sin\Phi_0 = \mathrm{S}_0\cos\beta_0 - D_0\sin\beta_0 && \mathrm{M}_{A_0} + \mathrm{T}_0(d_T - x_T\phi_T) = 0\\
&mg\cos\gamma_0\cos\Phi_0 = L_0 + \mathrm{T}_0\sin(\phi_T + \alpha_0) && \mathrm{N}_{A_0} = 0
\end{aligned}
\tag{9.38}
$$

For a given flight condition (i.e., altitude and velocity), the six equations above may, in principle, be solved for the six unknowns. These unknowns consist of the trim values of angle of attack α_0 and sideslip β_0, the required thrust T_0, and the three control deflections—elevator δ_{E_0}, aileron δ_{A_0}, and rudder δ_{R_0}. One approach would be to use the numerical search technique introduced in Chapter 8 for determining the initial trim condition for a nonlinear simulation. But both analytical and graphical techniques will be presented below that will give us greater insight into the results.

Assume for now that in rectilinear flight the sideslip β_0 and bank angle Φ_0 are both zero. (In this situation, the aileron and rudder deflections would typically remain zero as well.) Writing the aerodynamic forces and moment in terms of the corresponding coefficients, and assuming that ϕ_T is small, the longitudinal set of the reference equations becomes

$$\boxed{\begin{aligned} &mg\sin\gamma_0 = -C_{D_0}q_\infty S_W + \mathrm{T}_0\cos(\phi_T + \alpha_0) \\ &mg\cos\gamma_0 = C_{L_0}q_\infty S_W + \mathrm{T}_0\sin(\phi_T + \alpha_0) \\ &C_{M_0}q_\infty S_W\bar{c}_W + \mathrm{T}_0(d_T - x_T\phi_T) = 0 \end{aligned}} \tag{9.39}$$

We will now employ these equations to perform the equilibrium, or trim, analysis. The lateral-directional set will be addressed in Section 9.3.3.

9.3.1 Longitudinal Trim Analysis

To allow for increased clarity of presentation, let us initially insert linear expressions for the lift, drag, and pitching-moment coefficients, such as those derived in Chapter 6, into Equations (9.39). A graphical procedure will be discussed later in this section that allows for nonlinear aerodynamic effects. With linear expressions used for the coefficients, and ignoring any effects of aileron and rudder deflection on the drag, Equations (9.39) now become

$$\begin{aligned} &mg\sin\gamma_0 = -\left(C_{D_{\alpha=\delta=i_H=0}} + C_{D_\alpha}\alpha_0 + C_{D_{i_H}}i_H + C_{D_{\delta_E}}\delta_{E_0}\right)q_\infty S_W + \mathrm{T}_0\cos(\phi_T + \alpha_0) \\ &mg\cos\gamma_0 = \left(C_{L_{\alpha=\delta=i_H=0}} + C_{L_\alpha}\alpha_0 + C_{L_{i_H}}i_H + C_{L_{\delta_E}}\delta_{E_0}\right)q_\infty S_W + \mathrm{T}_0\sin(\phi_T + \alpha_0) \\ &\left(C_{M_{\alpha=\delta=i_H=0}} + C_{M_\alpha}\alpha_0 + C_{M_{i_H}}i_H + C_{M_{\delta_E}}\delta_{E_0}\right)q_\infty S_W\bar{c}_W + \mathrm{T}_0(d_T - x_T\phi_T) = 0 \end{aligned} \tag{9.40}$$

Observe now that for a given velocity and altitude (or q_∞), flight-path angle γ_0, and tail incidence angle i_H, the three equations above govern the three unknowns α_0, δ_{E_0}, and T_0.

To determine the reference or trim angle of attack α_0, elevator deflection δ_{E_0}, and thrust T_0, let the trim lift and thrust coefficients, respectively, be defined as

$$C_{L_{\mathrm{Trim}}} \triangleq \frac{mg}{q_\infty S_W} \qquad C_{T_{\mathrm{Trim}}} \triangleq \frac{\mathrm{T}_0}{q_\infty S_W}$$

Also assume that

$$\mathrm{T}_0\sin(\phi_T + \alpha_0) \ll L_0 \quad \text{and} \quad \cos(\phi_T + \alpha_0) \approx 1$$

Then writing Equations (9.40) in matrix format we have

$$\begin{bmatrix} -C_{D_\alpha} & -C_{D_{\delta_E}} & 1 \\ C_{L_\alpha} & C_{L_{\delta_E}} & 0 \\ C_{M_\alpha} & C_{M_{\delta_E}} & \dfrac{(d_T - x_T\phi_T)}{\bar{c}_W} \end{bmatrix} \begin{Bmatrix} \alpha_0 \\ \delta_{E_0} \\ C_{T_{\text{Trim}}} \end{Bmatrix} = \begin{bmatrix} C_{L_{\text{Trim}}} \sin\gamma_0 + C_{D_{\alpha=\delta=i_H=0}} + C_{D_{i_H}} i_H \\ C_{L_{\text{Trim}}} \cos\gamma_0 - C_{L_{\alpha=\delta=i_H=0}} - C_{L_{i_H}} i_H \\ -C_{M_{\alpha=\delta=i_H=0}} - C_{M_{i_H}} i_H \end{bmatrix} \tag{9.41}$$

The three unknowns may now be determined using MATLAB or Cramer's rule for solving linear simultaneous equations. (See Appendix D for a discussion of Cramer's rule.)

If the thrust contribution to pitching moment is negligible, that is, if

$$\mathrm{T}_0\big(d_T\cos\phi_T - x_T\sin\phi_T\big) \approx \mathrm{T}_0\big(d_T - x_T\phi_T\big) \approx 0$$

then the last two equations in Equations (9.40) decouple from the first, and may be used to solve for angle of attack α_0 and elevator deflection δ_{E_0}. Writing these two equations in matrix format we have

$$\begin{bmatrix} C_{L_\alpha} & C_{L_{\delta_E}} \\ C_{M_\alpha} & C_{M_{\delta_E}} \end{bmatrix} \begin{Bmatrix} \alpha_0 \\ \delta_{E_0} \end{Bmatrix} = \begin{bmatrix} C_{L_{\text{Trim}}} \cos\gamma_0 - C_{L_{\alpha=\delta=i_H=0}} - C_{L_{i_H}} i_H \\ -C_{M_{\alpha=\delta=i_H=0}} - C_{M_{i_H}} i_H \end{bmatrix} \tag{9.42}$$

And from Cramer's rule, the equilibrium (trim) angle of attack and elevator deflection for the conditions stated are expressed as

$$\begin{aligned} \alpha_{\text{Trim}} &= \Big(\big(C_{L_{\text{Trim}}}\cos\gamma_0 - C_{L_{\alpha=\delta=i_H=0}} - C_{L_{i_H}} i_H\big)C_{M_{\delta_E}} + \big(C_{M_{\alpha=\delta=i_H=0}} + C_{M_{i_H}} i_H\big)C_{L_{\delta_E}}\Big)/\Delta \\ \delta_{E_{\text{Trim}}} &= -\Big(\big(C_{M_{\alpha=\delta=i_H=0}} + C_{M_{i_H}} i_H\big)C_{L_\alpha} + \big(C_{L_{\text{Trim}}}\cos\gamma_0 - C_{L_{\alpha=\delta=i_H=0}} - C_{L_{i_H}} i_H\big)C_{M_\alpha}\Big)/\Delta \\ \Delta &= C_{L_\alpha}C_{M_{\delta_E}} - C_{M_\alpha}C_{L_{\delta_E}} \end{aligned} \tag{9.43}$$

Finally, after solving for the trim angle of attack and elevator deflection from the above expressions, the first of Equations (9.40) may then be used to solve for the required thrust T_0. Or T_0 may be found from

$$\begin{aligned} \mathrm{T}_0\cos\big(\phi_T + \alpha_{\text{Trim}}\big) &= \big(C_{D_{\alpha=\delta=i_H=0}} + C_{D_\alpha}\alpha_{\text{Trim}} + C_{D_{i_H}} i_H + C_{D_{\delta_E}}\delta_{E_{\text{Trim}}}\big)q_\infty S_W + mg\sin\gamma_0 \\ &= C_{D_{\text{Trim}}} q_\infty S_W + mg\sin\gamma_0 \end{aligned} \tag{9.44}$$

The forgoing analysis allows for the determination of the required angle of attack, elevator deflection, and thrust for a given trim lift coefficient, or for a given weight, flight velocity, and altitude. If the thrust contributes significantly to vehicle pitching moment, Equations (9.41) may be used; if not, Equations (9.43) and (9.44) may be used. In either case, clearly the three dependent variables α_0, δ_{E_0}, and T_0 are all functions of the specified flight condition.

Although finding the three dependent variables for a particular flight condition is important, there are several other important issues as well. These issues involve limits on, and trends in, the solutions. For example, given that there are limits on the maximum elevator deflection and thrust available for a given vehicle, these limits will in turn limit the achievable lift coefficient and the maximum velocity at a given weight and altitude.

These achievable limits, along with the trends in the solutions, may be depicted schematically as in Figure 9.7. Figure 9.7*a* depicts trim angle of attack and elevator deflection for a statically stable vehicle in pitch, as functions of both the trim lift coefficient $C_{L_{\text{Trim}}} = mg/q_\infty S_W$ and of the corresponding velocity at a given altitude. Figure 9.7*b* depicts trim thrust, also as a function of the same two quantities. Plots like these may be obtained by iteratively solving Equations (9.41) or Equations (9.43) and (9.44) while varying trim lift coefficient $C_{L_{\text{Trim}}}$.

Also indicated by dashed lines in Figure 9.7 are the maximum angle of attack (to avoid stall, for example), minimum elevator deflection (due to mechanical limits), and maximum achievable thrust. Clearly, one can see how all three limits may constrain the achievable lift coefficient and flight velocity. And, in particular, the limit on elevator deflection may lead to a lower achievable lift coefficient than does stall. In such cases, looking only at the maximum lift coefficient for the aircraft, shown on a classical plot of lift coefficient versus angle of attack, for example, would be misleading because that maximum lift coefficient may not be achievable due to limited elevator deflection. So this achievable lift is constrained due to limited elevator *control-power.* A similar analysis may be

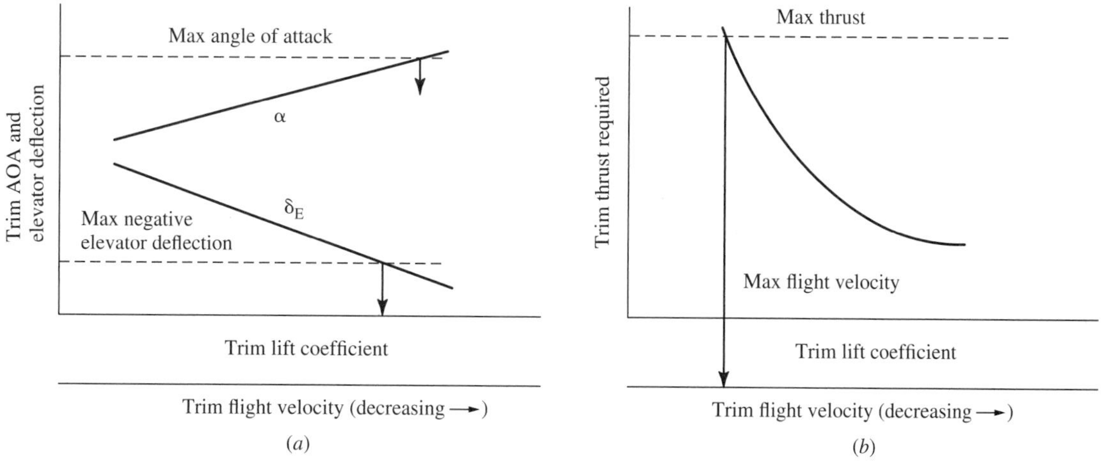

Figure 9.7 Trim solution trends and limits on achievable lift coefficient.

performed for vehicles with other pitch control devices, such as vehicles with forward control surfaces (canards).

We also see in Figure 9.7*a* that, as expected, the trim angle of attack increases with required lift coefficient. And for a statically stable aircraft, the trim elevator deflection decreases with increasing trim lift coefficient. This is due to the fact that for a statically stable vehicle, as the trim angle of attack increases, the nose-down aerodynamic pitching moment also increases. This increase in nose-down moment requires a greater downward force on the tail, or more negative elevator deflection, to trim the pitching moment on the vehicle. This trend in required elevator deflection, or the *trim elevator gradient* $\Delta\delta_{E_{\text{Trim}}}/\Delta C_{L_{\text{Trim}}}$, is an important parameter with regard to the aircraft's handling qualities. Furthermore, it can be measured in flight test, which provides a means to estimate the location of the vehicle's aerodynamic center, as discussed below.

As shown in Figure 9.7*a,* for the assumed statically stable aircraft the trim elevator gradient $\Delta\delta_{E_{\text{Trim}}}/\Delta C_{L_{\text{Trim}}}$ (slope) is negative. If the vehicle becomes more statically stable (due to a *c.g.* shift, for example), the required elevator deflection will become greater (more negative) at a given angle of attack and lift coefficient. Therefore, the trim elevator gradient $\Delta\delta_{E_{\text{Trim}}}/\Delta C_{L_{\text{Trim}}}$ (slope) will become more negative. As a result, the maximum achievable lift coefficient due to the elevator limit will be reduced. That is, increased static stability will reduce the maximum achievable lift coefficient, and the vehicle will become even more control-power limited.

Conversely, as the static stability is decreased, $\Delta\delta_{E_{\text{Trim}}}/\Delta C_{L_{\text{Trim}}}$ will become less negative, and will eventually become zero when the aircraft becomes neutrally statically stable. For example, from Equations (9.43) we can see that for level flight

$$\boxed{\frac{\Delta\delta_{E_{\text{Trim}}}}{\Delta C_{L_{\text{Trim}}}} \approx \frac{\partial\delta_{E_{\text{Trim}}}}{\partial C_{L_{\text{Trim}}}} = \frac{-C_{M_\alpha}}{C_{L_\alpha}C_{M_{\delta_E}} - C_{M_\alpha}C_{L_{\delta_E}}}} \tag{9.45}$$

which goes to zero as $C_{M_\alpha} \to 0$. Neutral static stability occurs when the *c.g.* is located at the vehicle's aerodynamic center, or the static margin is zero. Because such a situation causes the elevator trim gradient to equal zero, the axial location of the aerodynamic center is also called the *neutral point.* Due to this relationship between trim elevator gradient and static margin, a plot of the elevator gradient versus trim lift coefficient, which may be obtained during flight test, can be used to locate the vehicle's neutral point or aerodynamic center.

A related parameter of interest is the *trim-elevator speed gradient* $\Delta\delta_{E_{\text{Trim}}}/\Delta V_{\infty_{\text{Trim}}}$, which can be expressed as

$$\boxed{\frac{\Delta\delta_{E_{\text{Trim}}}}{\Delta V_{\infty_{\text{Trim}}}} = \frac{\Delta\delta_{E_{\text{Trim}}}}{\Delta C_{L_{\text{Trim}}}}\frac{\Delta C_{L_{\text{Trim}}}}{\Delta V_{\infty_{\text{Trim}}}}} \tag{9.46}$$

This gradient can also be measured in flight test, and note that

$$\frac{\Delta\delta_{E_{\text{Trim}}}}{\Delta V_{\infty_{\text{Trim}}}} \to 0 \text{ when } \frac{\Delta\delta_{E_{\text{Trim}}}}{\Delta C_{L_{\text{Trim}}}} \to 0. \tag{9.47}$$

So the elevator speed gradient may also be used to determine the location of the vehicle's aerodynamic center. Furthermore, since the change in trim lift coefficient due to a change in trim flight velocity $\Delta C_{L_{\text{Trim}}}/\Delta V_{\infty_{\text{Trim}}}$ will always be negative, the trim-elevator speed gradient $\Delta\delta_{E_{\text{Trim}}}/\Delta V_{\infty_{\text{Trim}}}$ will be <u>positive</u> for a statically stable aircraft. This is another important parameter with regards to a vehicle's handling qualities.

EXAMPLE 9.3

Trim Solutions and Limits on Maximum Achievable Lift Coefficient

Consider an aircraft similar to an F-5 or T-38, for which a sketch is shown in Figure 9.8. The vehicle is in level flight at Mach = 0.8 at 30,000 ft altitude ($q_\infty = 282$ psf), and the aerodynamic and other data for the vehicle are given below.

$C_{L_\alpha} = 4.58$ /rad	$C_{M_\alpha} = -1.40$ /rad	$C_{M_0} = 0.0017$	$\text{h}_0 = 30{,}000$ ft	$S_W = 170\ \text{ft}^2$
$C_{L_{\delta_E}} = 0.444$ /rad	$C_{M_{\delta_E}} = -0.7$ /rad	$C_{L_{\alpha=\delta=i_H=0}} = 0$	$mg = 17{,}000$ lbs	$V_{\infty_0} = 796$ fps

Letting the tail incidence angle $i_H = 0$, and assuming that the propulsive thrust does not contribute significantly to the vertical force or pitching moment on the vehicle, find the trim angle of attack and elevator deflection at this flight condition. Then calculate

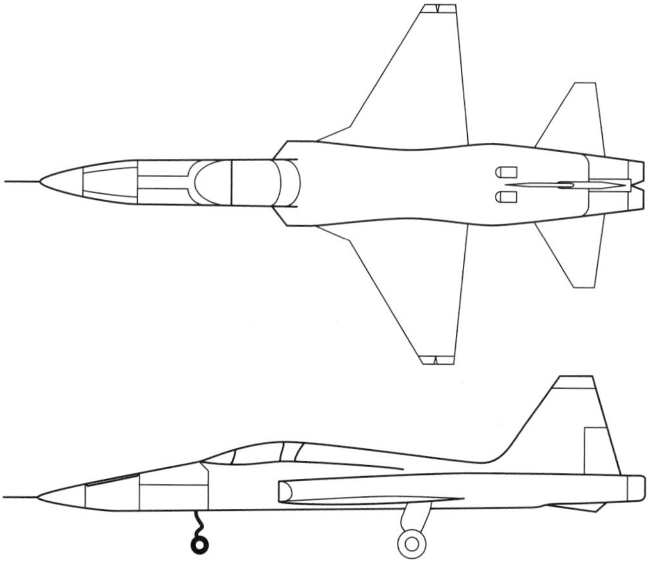

Figure 9.8 Sketch of F-5A aircraft.

and plot the trim solutions as a function of both trim lift coefficient and flight velocity at the given altitude. Do these results agree with the fact that the vehicle is statically stable in pitch? Finally, if the mechanical limits on elevator deflection are at ± 30 deg, and the maximum angle of attack is 16 deg, determine the limits these constraints place on achievable lift coefficient.

■ Solution

For the given flight condition the flight-path angle is zero and the trim lift coefficient is

$$C_{L_{\text{Trim}}} = \frac{mg}{q_\infty S_W} = 0.355$$

For the situation described, the solution given by Equations (9.43) is valid, and from these equations we find that

$$\Delta = C_{L_\alpha} C_{M_{\delta_E}} - C_{M_\alpha} C_{L_{\delta_E}} = (4.58)(-0.7) - (-1.4)(0.444) = -2.58 \text{ /rad}^2$$

The trim angle of attack and elevator deflection at the given flight condition are therefore

$$\alpha_{\text{Trim}} = \big((0.355 - 0 - 0)(-0.7) + (0.0017 + 0)(0.444)\big)/(-2.58) = 0.096 \text{ rad} = 5.50 \text{ deg}$$

$$\delta_{E_{\text{Trim}}} = -\big((0.0017 + 0)(4.58) + (0.355 - 0 - 0)(-1.4)\big)/(-2.58) = -0.190 \text{ rad} = -10.87 \text{ deg}$$

As an aside, if for this angle of attack and elevator deflection the trim drag coefficient is known to be

$$C_{D_{\text{Trim}}} = 0.042$$

the required thrust may then be determined from Equation (9.44). Or

$$\text{T}_0 \cos(\phi_T + \alpha_{\text{Trim}}) = C_{D_{\text{Trim}}} q_\infty S_W + mg \sin\gamma_0 = (0.042)(282)(170) + 0 = 2013 \text{ lbs}$$

So the required thrust T_0 at this flight condition is approximately 2015 lbs.

Next, by parametrically varying trim lift coefficient $C_{L_{\text{Trim}}}$ in Equations (9.43), one may find corresponding pairs of trim angle of attack and elevator deflection, and these results are shown in Figure 9.9. This plot should be compared to Figure 9.7*a*. Note that the maximum negative elevator deflection of -30 deg (trailing-edge up) is the limiting factor, leading to a maximum achievable lift coefficient of 0.96. Two data points corresponding to a trim lift coefficient of 0.355, the earlier result, are also shown.

Finally, given that the trim lift coefficient is

$$C_{L_{\text{Trim}}} = \frac{mg}{q_\infty S_W}$$

the trim angle of attack and elevator deflection may also be plotted against trim flight velocity. Taking the density at 30,000 ft to be $\rho_\infty = 8.9 \times 10^{-4}$ sl/ft^3, the velocity axis is as shown below the figure. Note from the figure that the trim

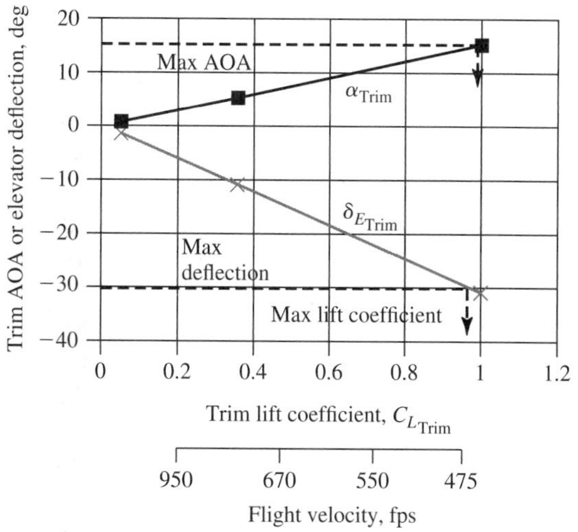

Figure 9.9 Effect of elevator limit on maximum achievable lift coefficient.

elevator gradient $\Delta\delta_{E_{\text{Trim}}}/\Delta C_{L_{\text{Trim}}}$ is negative, while the elevator speed gradient $\Delta\delta_{E_{\text{Trim}}}/\Delta V_{\infty_{\text{Trim}}}$ is positive. Both these results are as expected since the vehicle is statically stable in pitch.

Example 9.3 involved a case in which the maximum achievable lift coefficient, which determines the minimum sustainable flight velocity in level flight, was constrained by elevator control power. To avoid this condition, either the size of the elevator may be increased, or the static stability of the vehicle decreased. And again we see that too much static pitch stability is not desirable!

Next we will present a graphical technique for performing the trim analysis that does not rely on linear approximations for the aerodynamic coefficients. In fact, the analysis can be performed using wind-tunnel data directly. The advantages of this graphical approach include that fact that nonlinear aerodynamic data may be used, all the results are contained in one plot, and as we shall show, the effects of varying *c.g.* locations are easily revealed.

Shown in Figure 9.10 are two schematic diagrams, one of lift coefficient (Figure 9.10*a*) and one of pitching moment coefficient (Figure 9.10*b*) taken about a particular *c.g.* location, both plotted against vehicle angle of attack for several elevator deflections. Note that the given plot of pitching-moment coefficient reflects a vehicle that is statically stable in pitch at all angles of attack.

If the data in these two plots are now cross-plotted onto one plot, the resulting graph will appear as shown in Figure 9.11, referred to as a *stability map* for reasons that will become apparent shortly. First note that at each point along the horizontal axis, or the lift-coefficient axis, the pitching moment is zero. Hence, each point along this horizontal axis is a possible trim point, corresponding to a set of trim angle of attack α_{Trim} and elevator deflection $\delta_{E_{\text{Trim}}}$. This is analogous

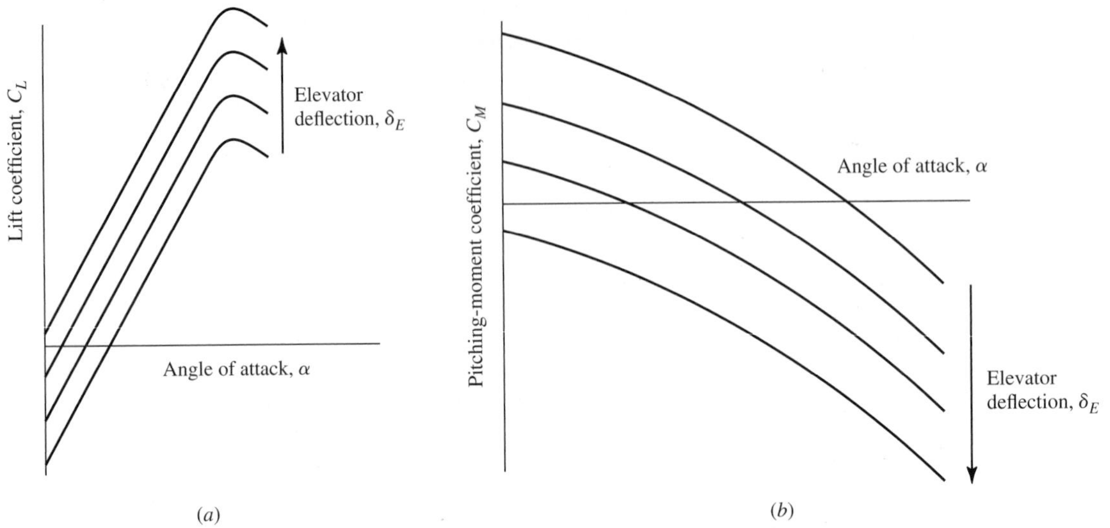

Figure 9.10 Plots of vehicle lift and pitching-moment coefficients.

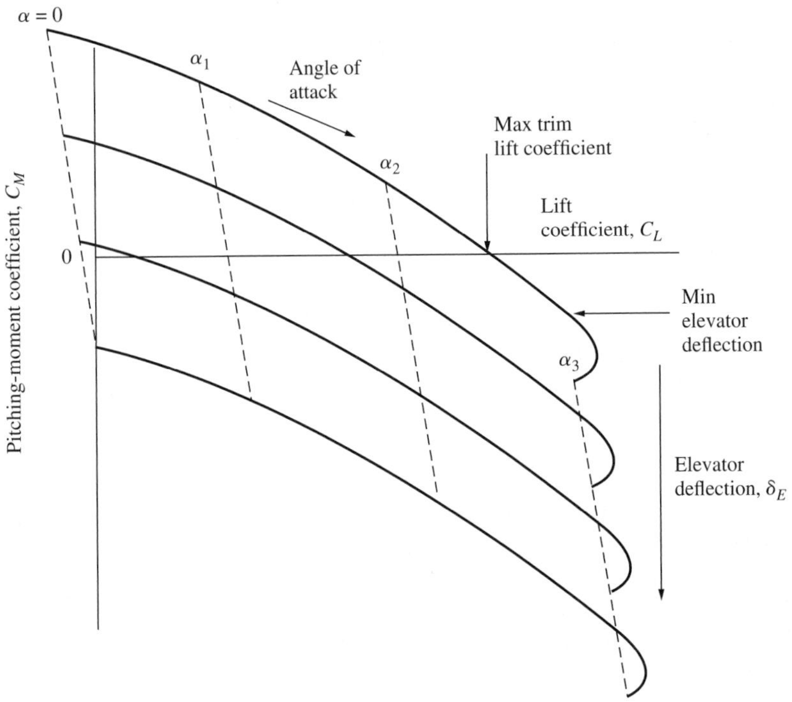

Figure 9.11 Cross plot of pitching-moment vs. lift coefficients.

to a graphical solution of Equations (9.43). (If the propulsive thrust contributes significantly to the vertical force and/or pitching moment on the vehicle, the data must be corrected for this fact.)

Also note that the trim elevator gradient $\Delta\delta_{E_{\text{Trim}}}/\Delta C_{L_{\text{Trim}}}$ may be determined graphically from the stability map by using data points along the locus of trim points on the horizontal axis. In addition, the maximum achievable trim lift coefficient $C_{L_{\text{Max}}}$ may be directly determined. Consider the four solid curves in Figure 9.11, each corresponding to a particular elevator deflection. As shown, the top curve corresponds to the minimum possible elevator deflection for the vehicle. Then the maximum achievable trim lift coefficient is that indicated on the plot. Therefore, this plot reveals all the information presented schematically in Figure 9.7*a,* but in a different format.

But there is another advantage of using the data plotted in the format shown in Figure 9.11. Let us now consider a vehicle *c.g.* location other than that for which the data in Figures 9.10*b* and 9.11 were plotted. And let the difference between these two *c.g.* locations be defined as Δ_{cg}, where

$$\Delta_{\text{cg}} \triangleq X_{\text{cg}_2} - X_{\text{cg}_1} \tag{9.48}$$

Here X_{cg_1} is the location of the *c.g.* for the data originally plotted, and X_{cg_2} is some new *c.g.* location of interest. If axial locations on the vehicle are taken to be positive forward, and if the new *c.g.* is forward of the original, then Δ_{cg} will be positive. The opposite will be true for an aft shift in *c.g.*

Consistent with Equation (9.26), the pitching moment about the new *c.g.* location X_{cg_2} is

$$M_2 = M_1 - L\Delta_{\text{cg}} \tag{9.49}$$

where L is the total lift acting on the vehicle, and the angle of attack is assumed to be small. Therefore, the pitching-moment coefficient about X_{cg_2}, or C_{M_2}, is

$$C_{M_2} = C_{M_1} - C_L\left(\frac{\Delta_{\text{cg}}}{\bar{c}_W}\right) = C_{M_1} - C_L\bar{\Delta}_{\text{cg}} \tag{9.50}$$

Here again, $\bar{\Delta}_{\text{cg}}$ is the change in *c.g.* location normalized by the wing mean aerodynamic chord, and C_L is the vehicle's total lift coefficient.

Consider now the horizontal axis in Figure 9.11, along which the moment coefficient about X_{cg_1} is zero, or $C_{M_1} = 0$. If the new *c.g.* location is located $\bar{\Delta}_{\text{cg}}$ from the old, what value does C_{M_1} need to take on, for a given value of lift coefficient, to make the new moment coefficient $C_{M_2} = 0$? This question is answered by simply setting $C_{M_2} = 0$ and solving Equation (9.50) for C_{M_1}, which becomes

$$\boxed{C_{M_2} = 0 \Rightarrow C_{M_1} = C_L\bar{\Delta}_{\text{cg}}} \tag{9.51}$$

Note that this equation is simply a straight line on a stability map (e.g., see Figure 9.12)—a plot of moment coefficient versus lift coefficient—with a slope equal to $\bar{\Delta}_{\text{cg}}$ and an intercept of zero. This line will have a positive slope for

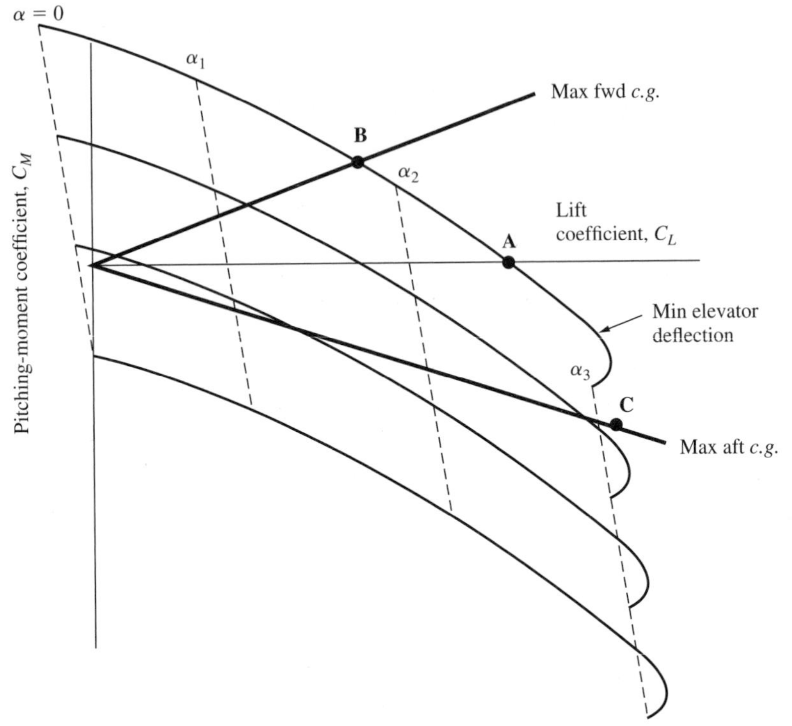

Figure 9.12 Region of possible trim points for a given $c.g.$ range.

a forward shift in $c.g.$, and a negative slope for an aft shift. Furthermore, since we defined $C_{M_2} = 0$ everywhere along this line, the line becomes a new locus of trim points for the new $c.g.$ location!

Let us now consider two new $c.g.$ locations of interest, one located forward (at $X_{cg_{fwd}}$) and one located aft (at $X_{cg_{aft}}$) of the $c.g.$ location for the original data plotted. So we now have $c.g.$ shifts of $\overline{\Delta}_{cg_{fwd}}$ and $\overline{\Delta}_{cg_{aft}}$, respectively, with $\overline{\Delta}_{cg_{fwd}} > 0$ and $\overline{\Delta}_{cg_{aft}} < 0$. Recall that the slopes of the new trim lines will be given by these two $\overline{\Delta}_{cg}$'s. The two new loci of trim solutions ($C_{M_2} = 0$) associated with these two new $c.g.$ locations are now shown in Figure 9.12, which includes the plot originally given in Figure 9.11. Note that this new figure now reveals a region of possible trim solutions enclosed between the two trim lines just added. This region contains all the trim points possible as the location of the $c.g.$ varies between $X_{cg_{fwd}}$ and $X_{cg_{aft}}$. Furthermore, if these two $c.g.$ locations represent the most forward and the most aft that the vehicle may experience, then the region between the two added trim lines contains all the possible trim points for the vehicle.

Now note that as the $c.g.$ location moves aft from its original location, the vehicle becomes less statically stable in pitch. This is due to the facts that the trim elevator gradient $\Delta\delta_{E_{Trim}}/\Delta C_{L_{Trim}}$ becomes less negative, and that the slopes of the solid curved lines relative to the trim lines also become less negative. (Confirm these results by simply considering Figures 9.11 and 9.12.) Also note that as the *c.g.* moves aft, the maximum achievable lift coefficient increases, as indicated by the points labeled A, B, and C. Point A corresponds to the maximum achievable lift coefficient for the original *c.g.* location for the data as plotted. Point B corresponds to the maximum achievable lift coefficient for the maximum forward (most stable) *c.g.* location. Finally, Point C corresponds to the maximum achievable lift coefficient for the most aft (least stable) *c.g.* location. Therefore, the maximum achievable lift increases as static pitch stability decreases, which again demonstrates that it is possible for a vehicle to be too statically stable in pitch.

EXAMPLE 9.4

Trim Analysis with a Stability Map

A stability map for an aircraft is shown in Figure 9.13. Assume the propulsive thrust does not contribute significantly to the vertical force or pitching moment on the vehicle. If the vehicle weight is 10,000 lb, the wing planform area is 170 ft^2, and the flight velocity is 330 fps, estimate the trim angle of attack and elevator deflection required for steady, level flight at sea level, along with the elevator trim gradient $\Delta\delta_{E_{\text{Trim}}}/\Delta C_{L_{\text{Trim}}}$ at low angles of attack. Also, what is the maximum achievable lift coefficient if the minimum elevator deflection due to mechanical limits is -30 deg, and what is the corresponding angle of attack? If the *c.g.* moves aft from the location for which the data is plotted such that $\Delta\overline{X}_{\text{cg}} = -0.2$, find the new elevator trim gradients at both low and high angles of attack. Finally, what is the maximum achievable lift coefficient and the corresponding angle of attack for the new *c.g.* location?

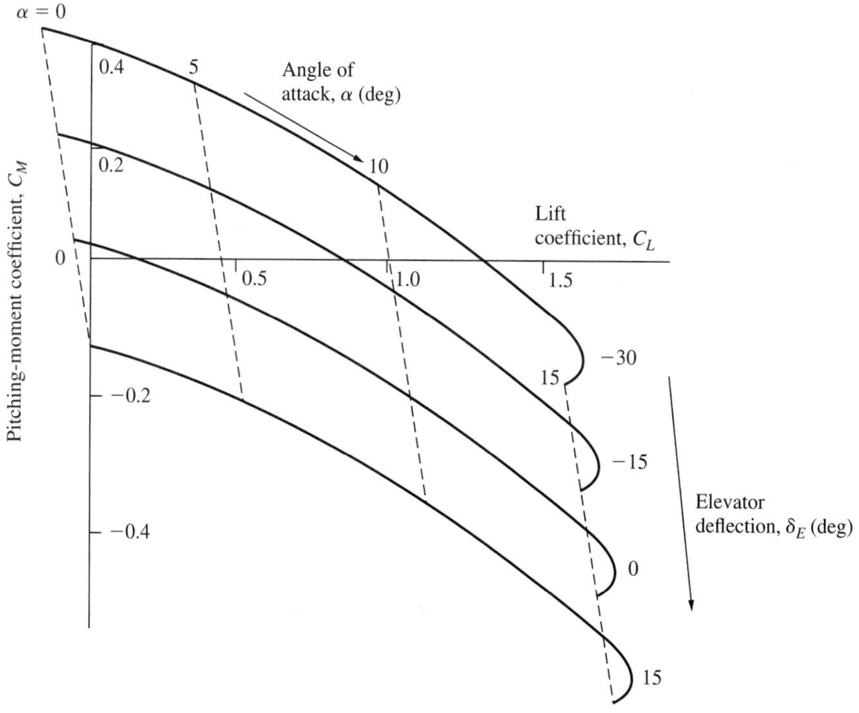

Figure 9.13 Sample stability map.

■ Solution

For the given flight condition, taking sea-level density to be 2.377×10^{-3} sl/ft^3, the dynamic pressure is

$$q_\infty = \frac{1}{2}(2.377 \times 10^{-3})(330)^2 = 129 \text{ psf}$$

and the trim lift coefficient is

$$C_{L_{\text{Trim}}} = \frac{mg}{q_\infty S_W} = \frac{10{,}000}{(129)(170)} = 0.456$$

From the stability map shown in Figure 9.14, this corresponds to Point A, or a trim angle of attack α_0 of approximately 5 deg, and a trim elevator deflection δ_{E_0} of approximately -5 deg.

A lift coefficient of zero (Point B) corresponds to a trim elevator deflection of approximately 2 deg. Thus, the trim-elevator gradient is approximately

$$\frac{\Delta\delta_{E_{\text{Trim}}}}{\Delta C_{L_{\text{Trim}}}} \approx \frac{-5-2}{0.456-0} = -15.4 \text{ deg}$$

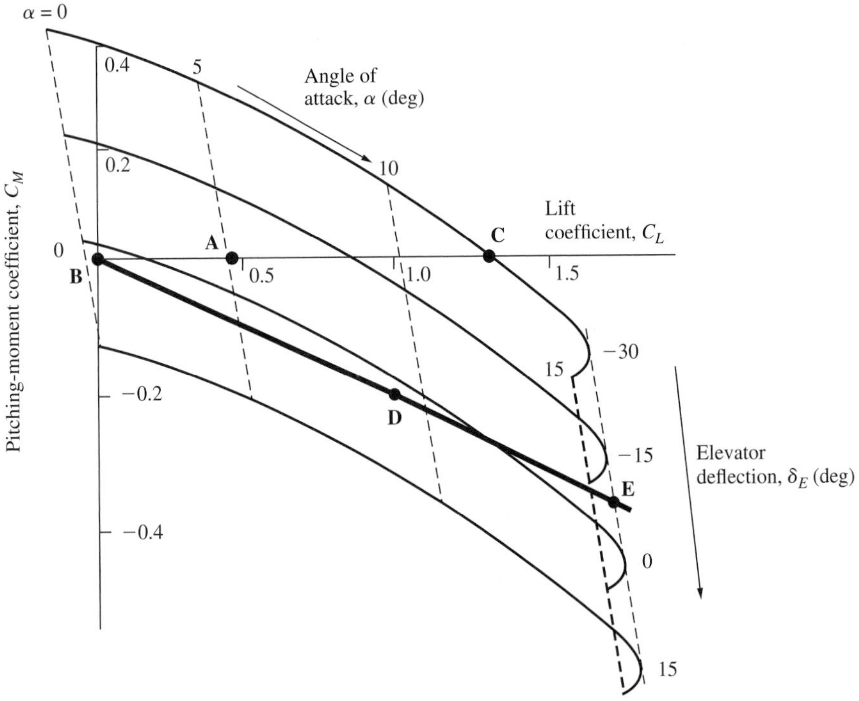

Figure 9.14 Stability map showing points of interest.

which is negative. For the *c.g.* location for the data as plotted, the maximum achievable lift coefficient as limited by the elevator control power (min $\delta_E = -30$ deg), and is approximately 1.3 (Point C). This corresponds to a trim angle of attack of approximately 12 deg.

If the *c.g.* is now moved aft such that the normalized *c.g.* shift is $\Delta\overline{X}_{\text{cg}} = -0.2$, the new trim line ($C_M = 0$) is the heavy line also shown in Figure 9.14. Note that this line passes through the point $C_L = 1.0$, $C_M = -0.2$ (Point D), such that the slope is -0.2. For this new *c.g.* location, the maximum achievable lift coefficient is now approximately 1.7 at a trim angle of attack of approximately 14 deg (Point E). (Note that if the angle of attack is increased to 15 deg, well into a stall, the maximum lift coefficient reduces to about 1.6.)

Finally using Points D and E, corresponding to 14 deg and 9 deg trim angle of attack, respectively, the elevator-trim gradient is now approximately

$$\frac{\Delta\delta_{E_{\text{Trim}}}}{\Delta C_{L_{\text{Trim}}}} \approx \frac{-8-4}{1.7-1} = -17.1 \text{ deg}$$

This gradient does not differ much from that found previously. However, if the gradient is calculated at smaller angles of attack, using Points B and D, for example, it becomes

$$\frac{\Delta\delta_{E_{\text{Trim}}}}{\Delta C_{L_{\text{Trim}}}} \approx \frac{4-2}{1-0} = 2.0$$

indicating an almost neutral to slightly statically unstable vehicle. So we see that elevator trim gradient is not only a function of *c.g.* location, but may also be a function of trim angle of attack. Finally, due to the nonlinear aerodynamic characteristics, the vehicle is more statically stable at high angles of attack.

A vehicle's *flight envelope* is a plot that reveals the region, in terms of altitude and velocity (or Mach number), in which the vehicle can sustain level flight. Such a plot is shown schematically in Figure 9.15, which includes three curves. The curve labeled "Max lift coefficient" is a locus of altitude-velocity pairs at

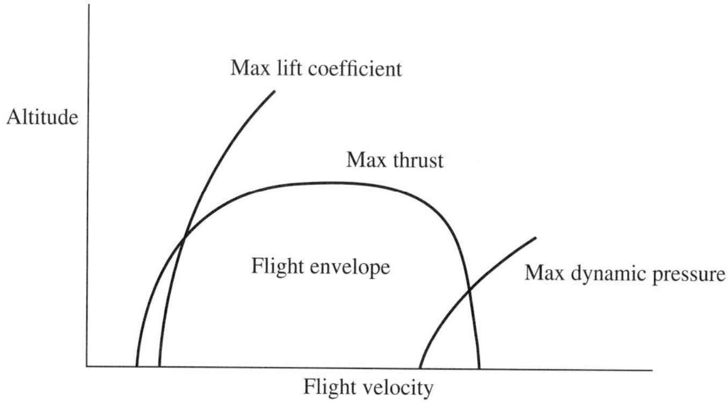

Figure 9.15 Schematic flight envelope.

which the vehicle is flying in steady, level flight at the maximum achievable lift coefficient. This maximum lift may be limited by elevator control power, as discussed in this section, or by stall, whichever yields the smaller lift coefficient. The second curve in the figure, labeled "Max thrust," reflects a locus of altitude-velocity pairs at which the vehicle is operated at maximum thrust. This flight condition corresponds to a high-drag flight condition, due to either a high flight velocity or a high lift coefficient. The third line labeled "Max dynamic pressure" reflects a locus of altitude-velocity pairs at which the dynamic pressure is such that the structural loads or flutter are the limiting factors. Inside the region defined by these three curves, the vehicle can sustain steady, level flight. So based on this discussion, limited control power can limit an aircraft's flight envelope by limiting the maximum achievable lift coefficient.

9.3.2 Control Forces

If the aircraft is to be controlled by a human pilot, the forces on the cockpit control manipulators, such as the stick or wheel, are critical factors affecting the aircraft's handling qualities. They are so important that even if the control surfaces are deflected with power systems, such as hydraulic, the force feedback to the pilot from the cockpit manipulators must be artificially generated to achieve acceptable handling qualities. This is especially true for aerodynamic statically unstable vehicles, which are stabilized with feedback augmentation systems. In this section we will highlight the key aspects of control forces that are strongly influenced by the aerodynamic characteristics of the vehicle. These aspects include control-force gradients with respect to airspeed, lift coefficient, and load factor.

Shown in Figure 9.16 is a schematic of the mechanisms used to deflect the elevator on a horizontal tail, a primary example of one aircraft control surface. The elevator is hinged to allow for deflections, and the cockpit manipulator moves the elevator either through direct mechanical connections or through commands to a powered system affecting the deflection. To keep the discussion simple, we will assume a direct mechanical connection, and the results developed in this section may be extrapolated if considering a powered system.

A moment acting on the elevator about the hinge will arise due to the pressure distribution acting on the elevator. This *hinge moment* obviously depends on

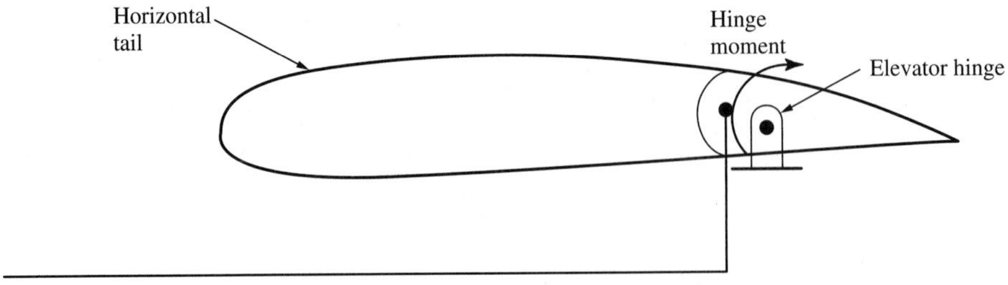

Figure 9.16 Elevator schematic.

the location of the hinge relative to the center of pressure, the dynamic pressure, the angle of attack of the tail surface, and the elevator deflection.

Let us express this elevator hinge moment HM_E as

$$HM_E = C_h q_H S_E \bar{c}_E \tag{9.52}$$

where C_h is a nondimensional aerodynamic hinge-moment coefficient, and S_E and $\bar{c}_E$ are the planform area and mean aerodynamic chord, respectively, of the elevator. Furthermore, let the hinge-moment coefficient be expressed in terms of the local angle of attack and elevator deflection as

$$C_h = C_{h_{\alpha_H = \delta_E = 0}} + C_{h_\alpha} \alpha_H + C_{h_{\delta_E}} \delta_E \tag{9.53}$$

(Note that if the horizontal tail and elevator have symmetric 2-D cross sections, $C_{h_{\alpha_H = \delta_E = 0}} = 0$.)

Now assume that the mechanical connections between the stick or wheel and the elevator produce a gearing ratio G_E such that the *stick force* due to elevator deflection F_{S_E} can be written in terms of the hinge moment as

$$F_{S_E} = G_E \times HM_E \tag{9.54}$$

And let us define a positive stick force to be one requiring the pilot to pull back on the stick to balance that stick force. Now, recall from Chapter 6 that the angle of attack of the horizontal tail may be expressed in terms of the vehicle angle of attack, tail incidence angle i_H, and downwash gradient, or

$$\alpha_H = \alpha\left(1 - \frac{d\varepsilon_H}{d\alpha_W}\right) + i_H - \varepsilon_{\alpha_W = 0} \tag{9.55}$$

So the stick force at a given trim angle of attack and elevator deflection becomes

$$F_{S_E} = q_H S_E \bar{c}_E G_E \left(C_{h_{\alpha_H = \delta_E = 0}} + C_{h_\alpha}\left(\alpha\left(1 - \frac{d\varepsilon_H}{d\alpha_W}\right) + i_H - \varepsilon_{\alpha_W = 0}\right) + C_{h_{\delta_E}} \delta_E\right) \tag{9.56}$$

which is also a function of tail incidence angle. Note that for acceptable handling qualities, the maximum stick force must be less than about 60 lbs, and the maximum force over a prolonged period should be less than about 10 lbs.

The first thing to observe when considering Equation (9.56) is that, given some tail incidence angle i_H, this expression yields the trim stick force corresponding to a given set of trim angle of attack and elevator deflection, as found from the analyses presented in the previous section. And since this angle of attack and elevator deflection is a function of trim lift coefficient, which depends on flight velocity, Equation (9.56) also yields the stick force as a function of either trim lift coefficient or flight velocity.

The gradient of this trim stick force to trim flight velocity, or $\partial F_{S_E}/\partial V_{\infty_{\text{Trim}}}$, sometimes referred to as *stick force per knot,* is another parameter important to the aircraft's handling qualities. For acceptable handling qualities, this gradient must be negative. That is, as the trim flight velocity increases, the stick force must decrease, providing an additional cue to the pilot. By differentiating Equation (9.56) with respect to trim flight velocity, one can develop criteria on the hinge-moment coefficients C_{h_α} and $C_{h_{\delta_E}}$ such that this requirement is met.

Also note that

$$\boxed{\frac{\partial F_{S_E}}{\partial V_{\infty_{\text{Trim}}}} = \frac{\partial F_{S_E}}{\partial \delta_{E_{\text{Trim}}}} \frac{\Delta \delta_{E_{\text{Trim}}}}{\Delta V_{\infty_{\text{Trim}}}}} \tag{9.57}$$

So the stick force per knot depends on the trim-elevator speed gradient $\Delta\delta_{E_{\text{Trim}}}/\Delta V_{\infty_{\text{Trim}}}$ discussed in the previous section, and appearing in Equation (9.46). As indicated by that equation, this gradient is in turn a function of the trim elevator gradient $\Delta\delta_{E_{\text{Trim}}}/\Delta C_{L_{\text{Trim}}}$. And recall for a statically stable vehicle, the elevator speed gradient is positive while the elevator gradient $\Delta\delta_{E_{\text{Trim}}}/\Delta C_{L_{\text{Trim}}}$ is negative.

However, if the vehicle is aerodynamically statically unstable (and the vehicle is stabilized through feedback augmentation), the elevator speed gradient will be negative because $\Delta\delta_{E_{\text{Trim}}}/\Delta C_{L_{\text{Trim}}}$ will be positive. In such cases the stick force per knot given by Equation (9.57) can still provide the proper cues (i.e., remain negative) by properly tailoring the gradient $\partial F_{S_E}/\partial\delta_{E_{\text{Trim}}}$ using an artificial force-feel system, or perhaps by providing the required hinge-moment coefficients C_{h_α} and $C_{h_{\delta_E}}$.

The second thing to observe from Equation (9.56) is that if the tail incidence angle is variable, as it is in many cases, it may be varied to adjust the stick force. In fact, this is the reason for including a variable incidence in the design. Pursuing this point further, let us expand the trim angle of attack and elevator deflection in Taylor series to first order about a given trim solution of interest α_{Trim_0} and $\delta_{E_{\text{Trim}_0}}$. That is, let

$$\begin{aligned} \alpha_{\text{Trim}} &= \alpha_{\text{Trim}_0} + \frac{\Delta\alpha_{\text{Trim}}}{\Delta C_{L_{\text{Trim}}}} C_{L_{\text{Trim}}} + \frac{\Delta\alpha_{\text{Trim}}}{\Delta i_H} i_H \\ \delta_{E_{\text{Trim}}} &= \delta_{E_{\text{Trim}_0}} + \frac{\Delta\delta_{E_{\text{Trim}}}}{\Delta C_{L_{\text{Trim}}}} C_{L_{\text{Trim}}} + \frac{\Delta\delta_{E_{\text{Trim}}}}{\Delta i_H} i_H \end{aligned} \tag{9.58}$$

Note that the trim elevator gradient $\Delta\delta_{E_{\text{Trim}}}/\Delta C_{L_{\text{Trim}}}$ has already been discussed in the previous section, and the remaining three gradients in the above two expressions may be determined analytically from Equations (9.43).

Now substituting Equations (9.58) for α and δ_E into Equation (9.56) and setting F_{S_E} equal to zero, we find that the tail incidence angle required to null the stick force for a given trim solution $\left(C_{L_{\text{Trim}}}, \alpha_{\text{Trim}_0}, \text{ and } \delta_{E_{\text{Trim}_0}}\right)$ in steady, level flight is

$$i_H\big|_{F_S=0} = \frac{-\left(C_{h_{\alpha_H=\delta_E=0}} + C_{h_\alpha}\left(\left(\alpha_{\text{Trim}_0} + \frac{\Delta\alpha_{\text{Trim}}}{\Delta C_{L_{\text{Trim}}}} C_{L_{\text{Trim}}}\right)\left(1 - \frac{d\varepsilon_H}{d\alpha_W}\right) - \varepsilon_{\alpha_W=0}\right) + C_{h_{\delta_E}}\left(\delta_{E_{\text{Trim}_0}} + \frac{\Delta\delta_{E_{\text{Trim}}}}{\Delta C_{L_{\text{Trim}}}} C_{L_{\text{Trim}}}\right)\right)}{\left(C_{h_\alpha}\left(\left(\frac{\Delta\alpha_{\text{Trim}}}{\Delta i_H}\right)\left(1 - \frac{d\varepsilon_H}{d\alpha_W}\right) + 1\right) + C_{h_{\delta_E}}\left(\frac{\Delta\delta_{E_{\text{Trim}}}}{\Delta i_H}\right)\right)} \tag{9.59}$$

Instead of using a variable incidence angle, another design approach for reducing stick force is the use of trim tabs. A trim tab is a deflectable surface located along the trailing edge of the elevator. This surface, when deflected relative to the elevator, changes the elevator hinge moment since the tab surface is located aft of the elevator hinge. By following an analysis paralleling the above, one may develop similar expressions for the stick force in terms of tab deflection, and tab deflection to null the stick force. (See Problem 9.9).

If the elevator is not held in place by an auxiliary power system, that is, the control mechanism is *reversible,* then the elevator will float if the pilot releases the stick. In such a case, the aerodynamic hinge moment HM_E acting on the elevator must be zero, or

$$C_{h_{\alpha_H=\delta_E=0}} + C_{h_\alpha}\alpha_H + C_{h_{\delta_E}}\delta_E = 0 \tag{9.60}$$

This *stick-free* condition establishes a relationship between the vehicle angle of attack and the elevator deflection. Using the above expression, along with the downwash gradient, we now find that in the stick-free condition,

$$\frac{\partial\delta_E}{\partial\alpha} = \frac{\partial\delta_E}{\partial\alpha_H}\frac{\partial\alpha_H}{\partial\alpha} = -\frac{C_{h_\alpha}}{C_{h_{\delta_E}}}\left(1 - \frac{d\varepsilon_H}{d\alpha}\right) \tag{9.61}$$

In Equation 6.55 we had the expression for the pitching moment on the vehicle expressed in terms of the vehicle angle of attack and elevator deflection. Differentiating that equation with respect to angle of attack, and noting that a relationship now exists between elevator and angle of attack (Equation (9.61)), yields the angle-of-attack pitching-moment effectiveness when the elevator is free to float, or

$$C_{M_\alpha}\big|_{\text{Stick Free}} = C_{L_{\alpha_W}}\Delta\bar{X}_{\text{AC}_{W\,\&\,F}} - C_{L_{\alpha_H}}\Delta\bar{X}_{\text{AC}_H}\frac{q_H}{q_\infty}\frac{S_H}{S_W}\left(\left(1 - \frac{d\varepsilon}{d\alpha}\right) + \alpha_\delta\frac{\partial\delta_E}{\partial\alpha}\right) \tag{9.62}$$

Here recall that the elevator effectiveness was defined in Chapter 5 to be

$$\alpha_\delta \triangleq \frac{C_{L_{\delta_E}}}{C_{L_{\alpha_H}}} \tag{9.63}$$

and that X locations on the vehicle were defined to be positive forward.

When previously discussing the trim elevator gradient and Equation (9.45), we noted that the location of the vehicle's aerodynamic center was also called the *neutral point.* More precisely now, it is also called the *stick-fixed neutral point* since the elevator was held fixed in the determination of C_{M_α} at that time.

Let us now find that axial (X) location on the vehicle about which the stick-free $C_{M_\alpha} = 0$ (Equation (9.62)). This point will be referred to as the *stick-free neutral point* since the elevator is now free to float. Setting Equation (9.62) to zero and solving for X_{Ref}, we find that the stick-free neutral point is located at

$$\overline{X}_{\text{AC}}\Big|_{\text{Stick Free}} = \frac{C_{L_{\alpha_W}}\overline{X}_{\text{AC}_{W\&F}} + C_{L_{\alpha_H}}\overline{X}_{\text{AC}_H}\dfrac{q_H}{q_\infty}\dfrac{S_H}{S_W}\left(1-\dfrac{d\varepsilon}{d\alpha}\right)\left(1-\dfrac{C_{h_\alpha}}{C_{h_{\delta_E}}}\alpha_\delta\right)}{C_{L_{\alpha_W}} + C_{L_{\alpha_H}}\dfrac{q_H}{q_\infty}\dfrac{S_H}{S_W}\left(1-\dfrac{d\varepsilon}{d\alpha}\right)\left(1-\dfrac{C_{h_\alpha}}{C_{h_{\delta_E}}}\alpha_\delta\right)} \tag{9.64}$$

The stick-free neutral point is generally forward of the vehicle's aerodynamic center, since with the floating elevator the horizontal tail is less effective in generating a restoring pitching moment. Noting that the *c.g.* must be located forward of the most forward neutral point to assure static pitch stability in both the stick-fixed and stick-free conditions, the location of the stick-free neutral point frequently limits the allowable *c.g.* locations.

9.3.3 Engine-Out Effects

We will now turn our attention to the yawing moment acting on the vehicle while still addressing steady rectilinear flight. Of special interest is the situation in which the propulsive thrust is not symmetric with respect to the XY plane of the vehicle. For example, consider the case of a vehicle with multiple engines mounted along the wings, and one engine fails. This, of course, generates a propulsive yawing-moment imbalance (N_P) that must be controlled using the aerodynamic control surfaces. But can these surfaces generate sufficient yawing moment to overcome the imbalance?

Let us extract the lateral-directional reference equations from Equations (9.11), yielding

$$\begin{aligned} -mg\cos\gamma_0\sin\Phi_0 &= \mathrm{S}_0\cos\beta_0 - D_0\sin\beta_0 + \mathrm{F}_{P_{Y_0}} \\ \mathrm{L}_{A_0} + \mathrm{L}_{P_0} &= 0 \\ \mathrm{N}_{A_0} + \mathrm{N}_{P_0} &= 0 \end{aligned} \tag{9.65}$$

Similar to the approach taken for lift, drag, and pitching moment in Section 9.3.1, let us write the aerodynamic side force, rolling, and yawing moment as

$$
\begin{aligned}
\mathrm{S}_0 &= (C_{S_{\beta=\delta_A=\delta_R=0}} + C_{S_\beta}\beta_0 + C_{S_{\delta_A}}\delta_{A_0} + C_{S_{\delta_R}}\delta_{R_0})q_\infty S_W \\
\mathrm{L}_{A_0} &= (C_{L_{\beta=\delta_A=\delta_R=0}} + C_{L_\beta}\beta_0 + C_{L_{\delta_A}}\delta_{A_0} + C_{L_{\delta_R}}\delta_{R_0})q_\infty S_W b_W \qquad (9.66)\\
\mathrm{N}_{A_0} &= (C_{N_{\beta=\delta_A=\delta_R=0}} + C_{N_\beta}\beta_0 + C_{N_{\delta_A}}\delta_{A_0} + C_{N_{\delta_R}}\delta_{R_0})q_\infty S_W b_W
\end{aligned}
$$

Note that typically the leading coefficients in the above three expressions are all zero. Now following the lead from Section 9.3.1, we can substitute the above force and moments into Equations (9.65) and write them in matrix format. Assuming that β_0 is sufficiently small such that

$$\sin\beta_0 = \beta_0 \text{ and } \cos\beta_0 = 1$$

we have

$$
\begin{bmatrix} C_{S_\beta} - C_{D_{\text{Trim}}} & C_{S_{\delta_A}} & C_{S_{\delta_R}} \\ C_{L_\beta} & C_{L_{\delta_A}} & C_{L_{\delta_R}} \\ C_{N_\beta} & C_{N_{\delta_A}} & C_{N_{\delta_R}} \end{bmatrix}
\begin{Bmatrix} \beta_0 \\ \delta_{A_0} \\ \delta_{R_0} \end{Bmatrix}
=
\begin{bmatrix} \dfrac{-(mg\cos\gamma_0 \sin\Phi_0 + \mathrm{F}_{P_{Y_0}})}{q_\infty S_W} \\ \dfrac{-\mathrm{L}_{P_0}}{q_\infty S_W b_W} \\ \dfrac{-\mathrm{N}_{P_0}}{q_\infty S_W b_W} \end{bmatrix} \qquad (9.67)
$$

Hence, for a given weight, trim drag, flight condition, and thrust configuration, the equilibrium, or trim sideslip angle and aileron and rudder deflections may be found by using Cramer's rule or MATLAB. The trim drag coefficient $C_{D_{\text{Trim}}} = D_0/q_\infty S_W$ is usually determined during the longitudinal analysis, and its magnitude is usually small compared to that of C_{S_β}. Note as a check, if the thrust is symmetric with respect to the XZ plane (such that the reference propulsive side force and rolling and yawing moments are zero) and the bank angle Φ_0 is zero, the sideslip angle and two control deflections all equal zero.

Noting that the terms on the right-hand side of Equation (9.67) are inversely proportional to flight dynamic pressure q_∞, there will be some velocity below which the required control deflections will exceed their mechanical limits. We will address this question further under some simplifying assumptions.

For the engine-out condition cited previously, assume that the only propulsive moment generated is the yawing moment N_{P_0}, which includes the effect of any increased drag on the failed engine. Also assume that no propulsive side force is created, the flight condition is that of level flight (i.e., $\Theta_0 = \gamma_0 = 0$), and that the bank angle is set to some small maximum allowable value $\Phi_{\max}$. In this case we have

$$\begin{bmatrix} C_{S_\beta} - C_{D_{\text{Trim}}} & C_{S_{\delta_A}} & C_{S_{\delta_R}} \\ C_{L_\beta} & C_{L_{\delta_A}} & C_{L_{\delta_R}} \\ C_{N_\beta} & C_{N_{\delta_A}} & C_{N_{\delta_R}} \end{bmatrix} \begin{Bmatrix} \beta_0 \\ \delta_{A_0} \\ \delta_{R_0} \end{Bmatrix} = \begin{bmatrix} -C_{L_{\text{Trim}}} \sin\Phi_{\max} \\ 0 \\ \dfrac{-\mathrm{N}_{P_0}}{q_\infty S_W b_W} \end{bmatrix} \tag{9.68}$$

This equation may, of course, be solved using MATLAB or Cramer's rule. But to gain further insight, let us look at only the yawing-moment equation and assume no aileron deflection. Solving this yawing-moment equation for rudder deflection yields the required rudder deflection for a given sideslip angle, or

$$\delta_{R_0} = -\frac{C_{N_\beta}\beta_0 + \dfrac{\mathrm{N}_{P_0}}{q_\infty S_W b_W}}{C_{N_{\delta_R}}} \tag{9.69}$$

If one is required to balance the engine-out moment using only rudder with no sideslip, then the required rudder deflection is

$$\delta_{R_0} = -\frac{\left(\dfrac{\mathrm{N}_{P_0}}{q_\infty S_W b_W}\right)}{C_{N_{\delta_R}}} \tag{9.70}$$

Now setting the rudder deflection equal to its maximum mechanical limit and solving the above equation for the corresponding flight velocity would yield a simple estimate for the minimum velocity at which the engine-out moment can be balanced with the rudder. This velocity is called the *minimum control speed* V_{MC}, and it also limits the vehicle's sustainable flight conditions. Minimum control speed is frequently an important consideration in sizing the vertical tail and rudder.

The above procedure only yields an estimate for V_{MC} because it was based on several assumptions. But to meet certain requirements on V_{MC}, a small bank angle (≤ 5 deg) may be allowed. Therefore, a more accurate estimate is obtained by using Equation (9.68) rather than just Equation (9.70), although the latter equation frequently yields reasonable results in preliminary design. We will demonstrate the determination of V_{MC} through the following example.

EXAMPLE 9.5

Control Deflections for Engine Out and Minimum Control Speed

Consider a two-engine turboprop aircraft, for which the relevant data is given in Table 9.3. For a right-engine-out condition, if the maximum rudder deflection is 30 deg, use Equation (9.70) to find the minimum-control speed V_{MC} at sea level. Then letting $\Phi_{\max} \pm 5$ deg, use

Table 9.3 Data for Two-Engine Turboprop Aircraft

Mass Properties	Geometry	Effectiveness Coefficients			
Weight 40,000 lb	Planform area $S_W = 954$ ft^2	$C_{S_\beta} = -0.362$	$C_{L_p} = -1.69$ sec	$C_{N_r} = -0.613$ sec	$C_{L_{\delta_A}} = 0.20$
Inertia $I_{zz} = 4.47 \times 10^5$ sl-ft^2	Wing span $b_W = 96$ ft	$C_{L_\beta} = -0.125$	$C_{N_p} = -0.922$ sec	$C_{L_{\delta_R}} = 0.024$	$C_{N_{\delta_A}} = 0$
Drag model $C_D = 0.032 + 0.042C_L^2$	Engine location $y_P = 32$ ft	$C_{N_\beta} = 0.101$	$C_{L_r} = 1.34$ sec	$C_{N_{\delta_R}} = -0.107$	$C_{S_{\delta_R}} = 0.233$

Equation (9.68) to find V_{MC} and compare the results. Assume there is no propulsive side force, and that the propeller on the failed engine is feathered such that the increased drag on the engine is negligible.

■ Solution

The most straightforward way to determine minimum control speed is to calculate and plot the required rudder deflection for several assumed flight velocities, and then determine the V_{MC} graphically. An example calculation using Equation (9.70) at a flight velocity V_∞ of 150 fps is presented first.

At sea level at a velocity of 150 fps, the dynamic pressure is

$$q_\infty = \frac{1}{2}\rho_\infty V_\infty^2 = \frac{1}{2}(0.002377)(150)^2 = 26.74 \text{ psf}$$

So the trim lift and drag coefficients are found to be

$$C_{L_{\text{Trim}}} = \frac{mg}{q_\infty S_W} = \frac{40{,}000}{(26.74)(945)} = 1.583$$

$$C_{D_{\text{Trim}}} = 0.032 + 0.042C_{L_{\text{Trim}}}^2 = 0.137$$

The required engine thrust is then

$$\text{T}_0 = D_0 = C_{D_{\text{Trim}}} q_\infty S_W = 3462 \text{ lbs}$$

With the right engine out this generates a positive propulsive yawing moment of

$$\text{N}_{P_0} = \text{T}_0 y_P = (3462)(32) = 110{,}784 \text{ ft-lb}$$

From Equation (9.70) we have the required positive rudder deflection given by

$$\delta_{R_0} = -\frac{\left(\dfrac{\text{N}_{P_0}}{q_\infty S_W b_W}\right)}{C_{N_{\delta_R}}} = -\frac{(110{,}784/(26.74)(945)(96))}{-0.107} = 0.43 \text{ rad}$$

Repeating this process for several other flight velocities and plotting the results yields Figure 9.17. From the figure we see that V_{MC} using Equation (9.70) is 140 fps.

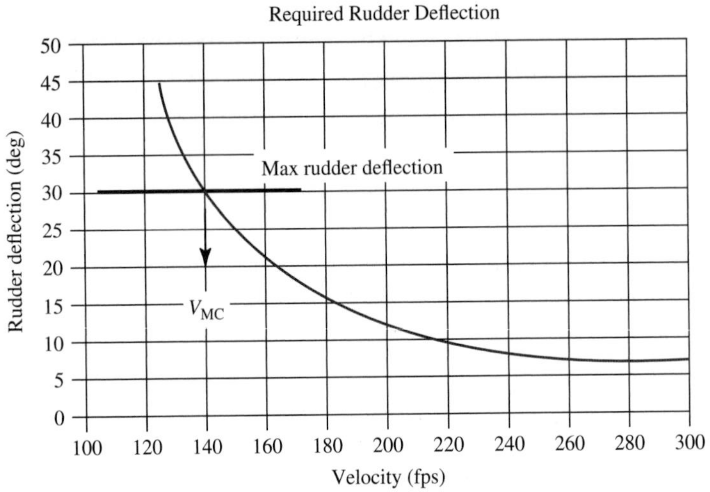

Figure 9.17 Graphical determination of minimum control speed (Equation (9.70)).

Now attention turns to finding V_{MC} using Equation (9.68), which is

$$\begin{bmatrix} C_{S_\beta} - C_{D_{\text{Trim}}} & C_{S_{\delta_A}} & C_{S_{\delta_R}} \\ C_{L_\beta} & C_{L_{\delta_A}} & C_{L_{\delta_R}} \\ C_{N_\beta} & C_{N_{\delta_A}} & C_{N_{\delta_R}} \end{bmatrix} \begin{Bmatrix} \beta_0 \\ \delta_{A_0} \\ \delta_{R_0} \end{Bmatrix} = \begin{bmatrix} -C_{L_{\text{Trim}}} \sin \Phi_{\max} \\ 0 \\ \dfrac{-N_{P_0}}{q_\infty S_W b_W} \end{bmatrix}$$

Again using a velocity of 150 fps for the sample calculation and inserting the given data, the matrix equation becomes

$$\begin{bmatrix} -0.362 - 0.137 & 0 & 0.233 \\ -0.125 & 0.200 & 0.024 \\ 0.101 & 0 & -0.107 \end{bmatrix} \begin{Bmatrix} \beta_0 \\ \delta_{A_0} \\ \delta_{R_0} \end{Bmatrix} = \begin{bmatrix} -1.583 \sin \Phi_{\max} \\ 0 \\ \dfrac{-110{,}784}{(26.74)(945)(96)} \end{bmatrix}$$

Solving the above equation for rudder deflection as a function of $\Phi_{\max}$ leads to the results shown in Figure 9.18. It is clear that the allowed bank angle has a significant effect on required rudder deflection.

Setting the bank angle to -5 deg, which from Figure 9.18 yields the smallest required rudder deflection, and iteratively solving for the rudder deflection as a function of flight velocity yields the results shown in Figure 9.19. From these results we see that the minimum control speed is now 135 fps. Although this V_{MC} is slightly less than that found using Equation (9.70), we had to allow a bank angle of -5 deg. Had we required the bank angle to be zero, the V_{MC} from Equation (9.68) would actually have been higher than 150 fps, as reflected in Figure 9.18. The main cause for the differences in required rudder is the yawing moment associated with the nonzero sideslip angle.

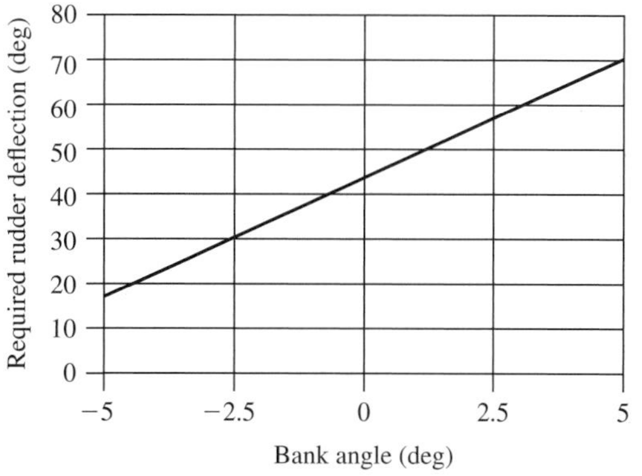

Figure 9.18 Effect of bank angle on required rudder deflection ($V_\infty = 150$ fps).

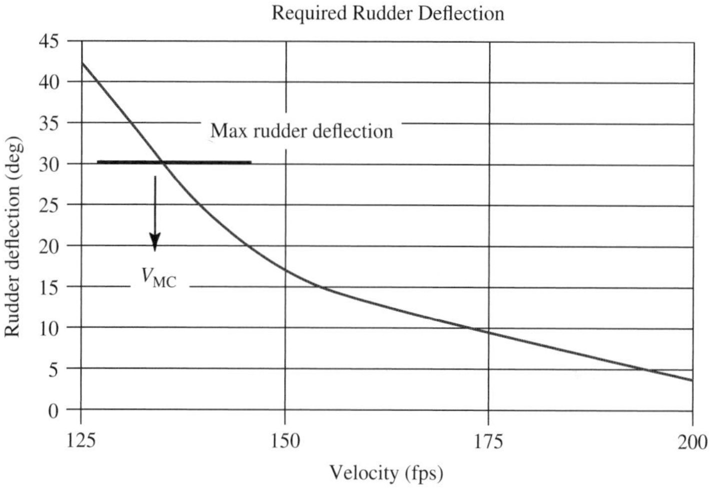

Figure 9.19 Graphical determination of minimum control speed (Equation (9.68)).

Summarizing the key results in this section, we developed two techniques to determine the required angle of attack, elevator deflection, and thrust in steady rectilinear flight. We presented a method for determining the control deflections required to control an asymmetric thrust condition such as an engine-out condition for a multi-engine aircraft. We also defined and analyzed minimum control speed.

In addition, we introduced cockpit-control forces, along with methods for estimating those forces for reversible control systems. We also defined several important control-deflection and control-force gradients, along with criteria on these gradients based on handling-qualities considerations.

Finally, we defined the stick-free neutral point, along with a method for determining its location. We noted that the location of the stick-free neutral point was normally forward of the stick-fixed neutral point, which is the same as the vehicle's aerodynamic center.

9.4 ANALYSIS OF STEADY TURNING FLIGHT

We will now direct our attention to the analysis of the second equilibrium reference flight condition identified in Section 9.1, that of steady turning flight. For this analysis, we will use both the longitudinal and the lateral-directional sets of reference equations. Of interest will be the required control deflections as a function of flight condition, and again, the limits on achievable lift coefficient.

Recall from Section 9.1 that for an equilibrium turning flight condition we require that the criteria stated in Equations (9.10) be satisfied, plus the turning rate must be $\dot{\Psi}_0 = \text{constant} \neq 0$. So, assuming a stability axis is employed, and the vehicle geometry is symmetric with respect to its XZ plane, Equations (9.12) provide the appropriate reference equations. Grouping these equations into the longitudinal and lateral-directional sets we have

Longitudinal Set

$$
\begin{aligned}
(0 \text{ if } \beta_0 = 0) &= -m(\mathrm{V}_0\mathrm{R}_0) = -mg\sin\gamma_0 + \mathrm{F}_{A_{X_0}} + \mathrm{F}_{P_{X_0}} \\
(-m\mathrm{Q}_0\mathrm{U}_0 \text{ if } \beta_0 = 0) &= m(\mathrm{P}_0\mathrm{V}_0 - \mathrm{Q}_0\mathrm{U}_0) \\
&= mg\cos\gamma_0\cos\Phi_0 + \mathrm{F}_{A_{Z_0}} + \mathrm{F}_{P_{Z_0}} \\
-I_{xz}(\mathrm{R}_0^2) &= \mathrm{M}_{A_0} + \mathrm{M}_{P_0}
\end{aligned}
$$

Lateral-Directional Set

$$
\begin{aligned}
m(\mathrm{R}_0\mathrm{U}_0) &= mg\cos\gamma_0\sin\Phi_0 + \mathrm{F}_{A_{Y_0}} + \mathrm{F}_{P_{Y_0}} \\
(I_{zz} - I_{yy})\mathrm{Q}_0\mathrm{R}_0 &= \mathrm{L}_{A_0} + \mathrm{L}_{P_0} \\
I_{xz}(\mathrm{Q}_0\mathrm{R}_0) &= \mathrm{N}_{A_0} + \mathrm{N}_{P_0}
\end{aligned}
\tag{9.71}
$$

with the kinematic equations given by

$$
\begin{aligned}
\mathrm{P}_0 &= 0 \\
\mathrm{Q}_0 &= \dot{\Psi}_0\sin\Phi_0 \\
\mathrm{R}_0 &= \dot{\Psi}_0\cos\Phi_0
\end{aligned}
\tag{9.72}
$$

9.4.1 Kinematic Analysis of the Turn

We shall now show that for a turning flight condition for which the turn rate $\dot{\Psi}_0$ and radius of turn R_{Turn} are specified, the above reference equations, Equations (9.71) and (9.72), may be used to determine the required angles of attack and sideslip, thrust, and control deflections. In this analysis, let us assume that the propulsive thrust is symmetric with respect to the vehicle's XZ plane, and hence

$$
\mathrm{F}_{P_{Y_0}} = \mathrm{L}_{P_0} = \mathrm{N}_{P_0} = 0 \tag{9.73}
$$

Consider the schematic in Figure 9.20 of an aircraft in a steady, level ($\gamma_0 = 0$) right turn. Included in the figure are the vehicle's angular-velocity vector relative to the earth-fixed frame $\boldsymbol{\omega}_{V,E} = \dot{\Psi}_0\mathbf{k}_E$, the radius of turn R_{Turn}, and the bank angle Φ_0. Also shown is the lift force L_0, the centrifugal "force" *CF,* and the weight mg. Not shown is the thrust T_0 directed out of the page.

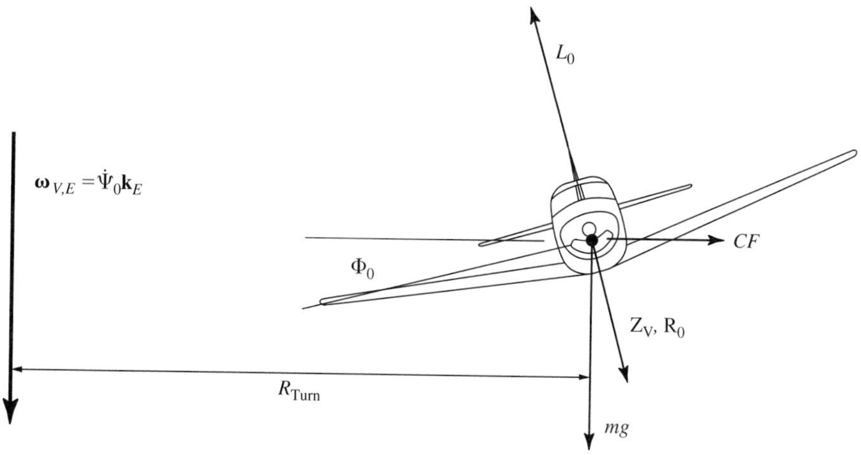

Figure 9.20 Aircraft in steady, level turn.

Now letting the vehicle's inertial velocity be denoted as $\mathbf{V}_V (= R_{Turn} \dot{\Psi}_0 \mathbf{i}_E)$, note that the centrifugal force can be determined from Newton's second law, which states that

$$\mathbf{F} = m \frac{d\mathbf{V}_V}{dt}\Big|_I = m\left(\frac{d\mathbf{V}_V}{dt}\Big|_V + \boldsymbol{\omega}_{V,E} \times \mathbf{V}_V\right) = m\left(\boldsymbol{\omega}_{V,E} \times \mathbf{V}_V\right) \tag{9.74}$$

since in equilibrium (see Equations (9.10))

$$\frac{d\mathbf{V}_V}{dt}\Big|_V = 0 \tag{9.75}$$

Therefore we see that the centrifugal force CF must be

$$CF = mR_{Turn} \dot{\Psi}_0^2 \tag{9.76}$$

Let us <u>assume</u> that this centrifugal force is balanced by the lateral component of lift plus any vertical (Z_V) component of thrust. That is, assume that

$$F_{A_{Y_0}} = 0 \tag{9.77}$$

so that

$$CF = \left(L_0 + T_0 \sin\left(\phi_T + \alpha_0\right)\right) |\sin \Phi_0| \tag{9.78}$$

(The absolute value allows for left turns with negative bank angles.) And since from Figure 9.20

$$mg = \left(L_0 + T_0 \sin\left(\phi_T + \alpha_0\right)\right) \cos \Phi_0 \tag{9.79}$$

the following must hold.

$$CF = mg\,|\tan \Phi_0| \tag{9.80}$$

Now from the side-force equation in Equations (9.71) and the kinematic Equations (9.72), and assuming Equation (9.77) is true, we have

$$m(\mathrm{R}_0\mathrm{U}_0) = mg\sin\Phi_0$$

$$\mathrm{R}_0 = \dot{\Psi}_0\cos\Phi_0$$

Therefore,

$$\boxed{\dot{\Psi}_0 = \frac{g}{\mathrm{U}_0}\tan\Phi_0} \tag{9.81}$$

which gives the steady turn rate in terms of the steady bank angle and velocity. Note that for a left turn with a negative bank angle, $\dot{\Psi}_0 < 0$. Plus, from Equations (9.76), (9.80), and (9.81) we have the radius of the turn given as a function of velocity and bank angle, or

$$\boxed{R_{\text{Turn}} = \frac{\mathrm{U}_0^2}{g|\tan\Phi_0|}} \tag{9.82}$$

Now let's define *normal load factor n* by the relation

$$\boxed{n(mg) = \left(L + T\sin(\phi_T + \alpha)\right)} \tag{9.83}$$

(Note that n is dimensionless, but by convention is typically referred to in terms of "g's." For example, in steady, level flight one frequently says that $n = 1$ g.) From Equation (9.79) we now see that steady normal load factor is a function of only steady bank angle, or

$$\boxed{n_0 = \frac{1}{\cos\Phi_0}} \tag{9.84}$$

Therefore, we can define a steady turning flight condition in terms of either bank angle or load factor. But if we define the turn in terms of load factor, we must also specify the direction of the turn—right or left.

Finally we can now determine the steady pitch and yaw rates Q_0 and R_0 in terms of the turn rate and bank angle (or load factor). From the kinematic Equations (9.72) and from Equation (9.81) we have

$$\boxed{\begin{aligned} \mathrm{Q}_0 &= \dot{\Psi}_0\sin\Phi_0 = \frac{g}{\mathrm{U}_0}\tan\Phi_0\sin\Phi_0 = \frac{g}{\mathrm{U}_0}\left(n_0 - \frac{1}{n_0}\right) \\ \mathrm{R}_0 &= \dot{\Psi}_0\cos\Phi_0 = \frac{g}{\mathrm{U}_0}\sin\Phi_0 = \pm\frac{g}{\mathrm{U}_0 n_0}\sqrt{n_0^2 - 1} \end{aligned}} \tag{9.85}$$

Note that Q_0 will always be positive, but that R_0 can be positive or negative, depending on the direction of turn. And if the turn is defined in terms of load factor, we must select the correct sign for R_0 based on the direction of the turn.

Also note that in a steady turn, the steady pitch rate Q_0 is not zero. This is somewhat counter-intuitive at first, since from the cockpit it might appear that in a steady turn the pitch rate would zero. But recall that we defined pitch rate as the component of the angular velocity vector $\boldsymbol{\omega}_{V,E}$ along the Y_V axis of the vehicle. So in a steady turn with the aircraft banked, this component of $\boldsymbol{\omega}_{V,E}$ is clearly not zero.

9.4.2 Lateral-Directional Trim Analysis

With the steady pitch and yaw rates now given in terms of the velocity and bank angle (or load factor) in Equations (9.85), we may now proceed to determine the required sideslip angle and aileron and rudder deflection in a steady level ($\gamma_o = 0$) turn. Our approach will follow that taken in Section 9.3, in which we were analyzing steady rectilinear flight.

Specifically, let us insert the linear expressions for the side-force and rolling- and yawing-moment coefficients, derived in Chapter 6, into the lateral-directional equations in Equations (9.71). Note that we have previously assumed that the thrust is symmetric with respect to the XZ plane of the vehicle such that

$$F_{P_{Y_0}} = L_{P_0} = N_{P_0} = 0 \tag{9.86}$$

Plus, let us invoke Equation (9.77), which states that

$$F_{A_{Y_0}} = S_0 \cos\beta_0 - D_0 \sin\beta_0 = 0$$

and let β_0 be assumed sufficiently small such that $\sin\beta_0 = \beta_0$, and $\cos\beta_0 = 1$. Under these conditions, the lateral-directional set in Equations (9.71) may be written as

$$\begin{aligned}
0 &= \left(\left(C_{S_\beta} - C_{D_{\text{Trim}}}\right)\beta_0 + C_{S_r}R_0 + C_{S_{\delta_A}}\delta_{A_0} + C_{S_{\delta_R}}\delta_{R_0}\right)q_\infty S_W \\
(I_{zz} - I_{yy})Q_0R_0 &= \left(C_{L_\beta}\beta_0 + C_{L_r}R_0 + C_{L_{\delta_A}}\delta_{A_0} + C_{L_{\delta_R}}\delta_{R_0}\right)q_\infty S_W b_W \\
I_{xz}(Q_0R_0) &= \left(C_{N_\beta}\beta_0 + C_{N_r}R_0 + C_{N_{\delta_A}}\delta_{A_0} + C_{N_{\delta_R}}\delta_{R_0}\right)q_\infty S_W b_W
\end{aligned} \tag{9.87}$$

Substituting Equations (9.85) for Q_0 and R_0 (in terms of bank angle) into the above three equations, rearranging, and writing the result in matrix format we have

$$\begin{bmatrix} C_{S_\beta} - C_{D_{\text{Trim}}} & C_{S_{\delta_A}} & C_{S_{\delta_R}} \\ C_{L_\beta} & C_{L_{\delta_A}} & C_{L_{\delta_R}} \\ C_{N_\beta} & C_{N_{\delta_A}} & C_{N_{\delta_R}} \end{bmatrix} \begin{Bmatrix} \beta_0 \\ \delta_{A_0} \\ \delta_{R_0} \end{Bmatrix} = \begin{bmatrix} -C_{S_r}\dfrac{g}{U_0}\sin\Phi_0 \\ \dfrac{(I_{zz} - I_{yy})}{q_\infty S_W b_W}\left(\left(\dfrac{g}{U_0}\right)^2 \dfrac{\sin^3\Phi_0}{\cos\Phi_0}\right) - C_{L_r}\dfrac{g}{U_0}\sin\Phi_0 \\ \dfrac{I_{xz}}{q_\infty S_W b_W}\left(\left(\dfrac{g}{U_0}\right)^2 \dfrac{\sin^3\Phi_0}{\cos\Phi_0}\right) - C_{N_r}\dfrac{g}{U_0}\sin\Phi_0 \end{bmatrix} \tag{9.88}$$

Or, writing Q_0 and R_0 in terms of load factor, the same equation becomes

$$\begin{bmatrix} C_{S_\beta} - C_{D_{\text{Trim}}} & C_{S_{\delta_A}} & C_{S_{\delta_R}} \\ C_{L_\beta} & C_{L_{\delta_A}} & C_{L_{\delta_R}} \\ C_{N_\beta} & C_{N_{\delta_A}} & C_{N_{\delta_R}} \end{bmatrix} \begin{Bmatrix} \beta_0 \\ \delta_{A_0} \\ \delta_{R_0} \end{Bmatrix} = \begin{bmatrix} \mp C_{S_r} \dfrac{g}{U_0 n_0} \sqrt{n_0^2 - 1} \\ \pm \dfrac{(I_{zz} - I_{yy})}{q_\infty S_W b_W} \left(\dfrac{g}{U_0}\right)^2 \left(1 - \dfrac{1}{n_0^2}\right) \sqrt{n_0^2 - 1} \mp C_{L_r} \dfrac{g}{U_0 n_0} \sqrt{n_0^2 - 1} \\ \pm \dfrac{I_{xz}}{q_\infty S_W b_W} \left(\dfrac{g}{U_0}\right)^2 \left(1 - \dfrac{1}{n_0^2}\right) \sqrt{n_0^2 - 1} \mp C_{N_r} \dfrac{g}{U_0 n_0} \sqrt{n_0^2 - 1} \end{bmatrix} \tag{9.89}$$

where again, the correct signs must be selected, based on the direction of turn. For a given trim drag coefficient, Equation (9.88) or (9.89) may be solved using MATLAB or Cramer's rule (see Appendix D) to find the required bank angle and aileron and rudder deflections for the specified turn. Noting that the right-hand sides of both of the above equations are normally small numerically, we see that typically the required sideslip angle and control deflections will be small.

The astute reader will note that a trim drag coefficient may not be available until the longitudinal trim analysis has been performed. This dilemma is resolved by initially assuming that this drag coefficient is negligible compared to C_{S_β}, and solving either Equation (9.88) or (9.89) under this assumption. Later, after the longitudinal trim analysis has been performed and a better estimate for the trim drag coefficient is available, Equation (9.88) or (9.89) may be resolved if needed. Frequently, however, sufficiently accurate results are obtained without this iteration.

9.4.3 Longitudinal Trim Analysis

With the steady pitch and yaw rates given in terms of the velocity and bank angle (or load factor) in Equations (9.85), and with the sideslip angle, aileron, and rudder deflections now available from the previous lateral-directional analysis, we may next proceed to determine the required angle of attack, thrust, and elevator deflection in a steady, level ($\gamma_0 = 0$) turn.

Expressed in terms of lift, drag, side force, and thrust, the longitudinal equations in Equations (9.71) are

$$\begin{aligned} -m(\beta_0 U_0 R_0) &= -D_0 \cos\beta_0 - S_0 \sin\beta_0 + T_0 \cos(\phi_T + \alpha_0) \\ mU_0 Q_0 + mg\cos\Phi_0 &= L_0 + T_0 \sin(\phi_T + \alpha_0) \\ -I_{xz}(R_0^2) &= M_{A_0} + T_0(d_T \cos\phi_T - x_T \sin\phi_T) \end{aligned} \tag{9.90}$$

So with linear expressions used for the aerodynamic coefficients, and assuming $\cos \beta_0 = 1$, $\sin \beta_0 = \beta_0$, and $C_{D_q} \approx 0$, Equations (9.90) become

$$\begin{aligned}
-m(\beta_0 U_0 R_0) &= -(C_{D_{\alpha=\delta=i_H=0}} + C_{D_\alpha}\alpha_0 + C_{D_{i_H}} i_H + C_{D_{\delta_E}}\delta_{E_0} + C_{D_{\delta_A}}\delta_{A_0} + C_{D_{\delta_R}}\delta_{R_0}) q_\infty S_W \\
&\quad - S_0\beta_0 + T_0\cos(\phi_T + \alpha_0) \\
mU_0Q_0 + mg\cos\Phi_0 &= (C_{L_{\alpha=\delta=i_H=0}} + C_{L_\alpha}\alpha_0 + C_{L_q}Q_0 + C_{L_{i_H}} i_H + C_{L_{\delta_E}}\delta_{E_0}) q_\infty S_W \\
&\quad + T_0\sin(\phi_T + \alpha_0) \\
-I_{xz}(R_0^2) &= (C_{M_{\alpha=\delta=i_H=0}} + C_{M_\alpha}\alpha_0 + C_{M_q}Q_0 + C_{M_{i_H}} i_H + C_{M_{\delta_E}}\delta_{E_0}) q_\infty S_W \bar{c}_W \\
&\quad + T_0(d_T\cos\phi_T - x_T\sin\phi_T)
\end{aligned} \tag{9.91}$$

Substituting Equations (9.85) for pitch rate Q_0 and yaw rate R_0, again introducing the thrust coefficient $C_{T_0} = T_0/(q_\infty S_W)$, and letting ϕ_T be small along with

$$\cos(\phi_T + \alpha_0) = 1$$

$$T_0\sin(\phi_T + \alpha_0) \ll L_0$$

we may write Equations (9.91) in the following matrix format

$$\begin{bmatrix} -C_{D_\alpha} & -C_{D_{\delta_E}} & 1 \\ C_{L_\alpha} & C_{L_{\delta_E}} & 0 \\ C_{M_\alpha} & C_{M_{\delta_E}} & \dfrac{(d_T - x_T\phi_T)}{\bar{c}_W} \end{bmatrix} \begin{Bmatrix} \alpha_0 \\ \delta_{E_0} \\ C_{T_0} \end{Bmatrix}$$

$$= \begin{bmatrix} \left(C_{S_{\text{Trim}}} - \dfrac{mg}{q_\infty S_W}\sin\Phi_0\right)\beta_0 + C_{D_{\alpha=\delta=i_H=0}} + C_{D_{i_H}} i_H + C_{D_{\delta_A}}\delta_{A_0} + C_{D_{\delta_R}}\delta_{R_0} \\ \dfrac{mg}{q_\infty S_W \cos\Phi_0} - C_{L_q}\dfrac{g}{U_0}\dfrac{\sin^2\Phi_0}{\cos\Phi_0} - C_{L_{\alpha=\delta=i_H=0}} - C_{L_{i_H}} i_H \\ -I_{xz}\left(\dfrac{g}{U_0}\right)^2 \sin^2\Phi_0 - C_{M_q}\dfrac{g}{U_0}\dfrac{\sin^2\Phi_0}{\cos\Phi_0} - C_{M_{\alpha=\delta=i_H=0}} - C_{M_{i_H}} i_H \end{bmatrix} \tag{9.92}$$

Or these same equations given in terms of load factor become

$$\begin{bmatrix} -C_{D_\alpha} & -C_{D_{\delta_E}} & 1 \\ C_{L_\alpha} & C_{L_{\delta_E}} & 0 \\ C_{M_\alpha} & C_{M_{\delta_E}} & \dfrac{(d_T - x_T\phi_T)}{\bar{c}_W} \end{bmatrix} \begin{Bmatrix} \alpha_0 \\ \delta_{E_0} \\ C_{T_0} \end{Bmatrix} = \begin{bmatrix} \left(C_{S_{\text{Trim}}} - \left(\dfrac{mg}{q_\infty S_W}\right)\sqrt{1 - \dfrac{1}{n_0^2}}\right)\beta_0 + C_{D_{\alpha=\delta=i_H=0}} + C_{D_{i_H}} i_H + C_{D_{\delta_A}}\delta_{A_0} + C_{D_{\delta_R}}\delta_{R_0} \\ n_0\left(\dfrac{mg}{q_\infty S_W}\right) - C_{L_q}\dfrac{g}{U_0}\left(n_0 - \dfrac{1}{n_0}\right) - C_{L_{\alpha=\delta=i_H=0}} - C_{L_{i_H}} i_H \\ -I_{xz}\left(\dfrac{g}{U_0}\right)^2\left(1 - \dfrac{1}{n_0^2}\right) - C_{M_q}\dfrac{g}{U_0}\left(n_0 - \dfrac{1}{n_0}\right) - C_{M_{\alpha=\delta=i_H=0}} - C_{M_{i_H}} i_H \end{bmatrix} \tag{9.93}$$

The three unknowns may now be determined using MATLAB or Cramer's rule.

Now let the effects of thrust on the required pitching moment be assumed negligible, that is, let

$$T_0\left(d_T \cos\phi_T - x_T \sin\phi_T\right) \approx T_0\left(d_T - x_T\phi_T\right) \approx 0 \tag{9.94}$$

Then the first equation in Equations (9.92) or (9.93) decouples from the other two, and the last two equations may be used to solve for angle of attack α_0 and elevator deflection δ_{E_0}. The first equation is used finally to obtain the required thrust.

That is, if Equation (9.94) is satisfied, we may find the required angle of attack and elevator deflection from

$$\begin{bmatrix} C_{L_\alpha} & C_{L_{\delta_E}} \\ C_{M_\alpha} & C_{M_{\delta_E}} \end{bmatrix} \begin{Bmatrix} \alpha_0 \\ \delta_{E_0} \end{Bmatrix} = \begin{bmatrix} n_0\left(\dfrac{mg}{q_\infty S_W}\right) - C_{L_q}\dfrac{g}{U_0}\left(n_0 - \dfrac{1}{n_0}\right) - C_{L_{\alpha=\delta=i_H=0}} - C_{L_{i_H}} i_H \\ -I_{xz}\left(\dfrac{g}{U_0}\right)^2\left(1 - \dfrac{1}{n_0^2}\right) - C_{M_q}\dfrac{g}{U_0}\left(n_0 - \dfrac{1}{n_0}\right) - C_{M_{\alpha=\delta=i_H=0}} - C_{M_{i_H}} i_H \end{bmatrix} \tag{9.95}$$

Solving for the two unknowns here, we have the trim angle of attack and elevator deflection in the steady, level turn given by

$$\alpha_{\text{Trim}} = \left(C_1(n)C_{M_{\delta_E}} - C_2(n)C_{L_{\delta_E}}\right)/\Delta$$
$$\delta_{E_{\text{Trim}}} = \left(C_2(n)C_{L_\alpha} - C_1(n)C_{M_\alpha}\right)/\Delta \qquad (9.96)$$
$$\Delta = C_{L_\alpha}C_{M_{\delta_E}} - C_{M_\alpha}C_{L_{\delta_E}}$$

with

$$C_1(n) \triangleq n_0\left(\frac{mg}{q_\infty S_W}\right) - C_{L_q}\frac{g}{U_0}\left(n_0 - \frac{1}{n_0}\right) - C_{L_{\alpha=\delta=i_H=0}} - C_{L_{i_H}} i_H$$

$$C_2(n) \triangleq -I_{xz}\left(\frac{g}{U_0}\right)^2\left(1 - \frac{1}{n_0^2}\right) - C_{M_q}\frac{g}{U_0}\left(n_0 - \frac{1}{n_0}\right) - C_{M_{\alpha=\delta=i_H=0}} - C_{M_{i_H}} i_H$$

So by using either Equations (9.93) or (9.96), we may determine the trim angle of attack α_{Trim} and elevator deflection $\delta_{E_{\text{Trim}}}$ in a steady, level turn as a function of the normal load factor n_0 in the turn. Of course, similar expressions may also be derived that define the turn in terms of bank angle Φ_0, starting from Equations (9.92).

With regard to using a graphical procedure to determine the trim angle of attack and elevator deflection, as discussed in Section 9.3.1, the procedure is complicated by the presence of a nonzero pitch rate in a turn. And the pitch rate has some effect on the pitching moment and perhaps the lift. However, the graphical procedure presented in the discussion of rectilinear flight may still be used to solve approximately for the trim angle of attack and elevator deflection in a turn. If one ignores the effect of the pitch rate on the pitching moment and lift, and assumes the term $I_{xz}(g/U_0)^2$ is approximately zero, the same plot presented in Section 9.3.1 may be used for analysis of turning flight if used with care. In the analysis of a turn, the value of trim lift coefficient $C_{L_{\text{Trim}}}$ to be used in the graphical analysis is increased by the load factor n. That is, in a turn the value of the trim lift coefficient equals $n(mg/q_\infty S_W)$.

As we also discussed in Section 9.3.1, one can again see that the limits on elevator deflection may limit the maximum achievable lift coefficient or normal load factor n in a turn. This fact may be verified either by the graphical trim analysis just discussed, or by looking at the equation for the required elevator deflection in Equations (9.96). For example, setting the elevator deflection to its maximum value in the cited equation, the equation may then be solved for the corresponding load factor n. This load factor becomes the maximum achievable at the altitude and velocity for which the analysis was performed.

The limit on maximum achievable load factor is another important aircraft design issue, as it limits a different type of flight envelope for the vehicle known as the *V-n diagram.* Such a diagram is shown schematically in Figure 9.21. The interior region defines the velocity and load-factor combinations that the aircraft can sustain, and hence defines the vehicle's maneuverability. The top and bottom

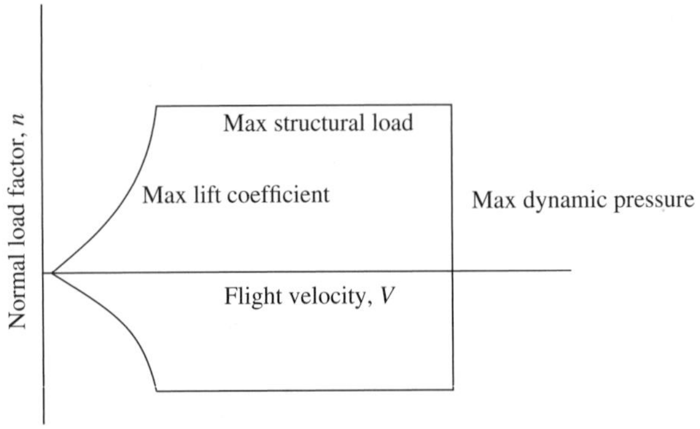

Figure 9.21 Schematic V-n diagram.

boundaries of the sustainable region are defined by the maximum structural loads allowed. And the right boundary of the region is defined by the maximum allowable dynamic pressure, due to a flutter limit, for example.

The curved line labeled "Max lift coefficient" is defined by the maximum achievable lift coefficient. And as we have just discussed, plus recalling Example 9.3, for a stable vehicle this maximum lift coefficient is frequently constrained by elevator control power, or maximum elevator deflection. The greater the static stability, the more severely the control-deflection limits constrain the maximum achievable lift coefficient. Thus again, sufficient control power is required for maneuverability, and too much static pitch stability can limit this maneuverability.

9.4.4 Control Forces and Gradients

Turning now to the subject of elevator and stick-force gradients, these gradients, introduced in Section 9.3 when discussing rectilinear flight, are still of interest when addressing turning flight. In addition, the trim *elevator-per-g* gradient, or $\partial \delta_{E_{\text{Trim}}} / \partial n_0$, is another parameter affecting the aircraft's handling qualities. From Equations (9.96) we find that in steady turning flight this gradient is expressed as

$$\boxed{\frac{\partial \delta_{E_{\text{Trim}}}}{\partial n_0} = \left(C_{L_\alpha} \frac{\partial C_2(n)}{\partial n_0} - C_{M_\alpha} \frac{\partial C_1(n)}{\partial n_0} \right) / \Delta} \tag{9.97}$$

where

$$\boxed{\begin{aligned} \frac{\partial C_1(n)}{\partial n_0} &= \left(\frac{mg}{q_\infty S_W} \right) - C_{L_q} \frac{g}{U_0} \left(1 + \frac{1}{n_0^2} \right) \\ \frac{\partial C_2(n)}{\partial n_0} &= -2I_{xz} \left(\frac{g}{U_0} \right)^2 \left(\frac{1}{n_0^3} \right) - C_{M_q} \frac{g}{U_0} \left(1 + \frac{1}{n_0^2} \right) \end{aligned}}$$

For acceptable handling qualities, this elevator-per-g gradient should be negative. Given that Δ in Equations (9.96) is typically negative and that the two derivatives

$\partial C_1(n)/\partial n_0$ and $\partial C_2(n)/\partial n_0$ are typically positive (since $\partial C_1(n)/\partial n_0 \approx (mg/q_\infty S_W)$ and usually $\partial C_2(n)/\partial n_0 \approx -C_{M_q}(g/U_0)(1 + 1/n_0^2)$), this requirement on elevator-per-g will typically be met for a vehicle statically stable in pitch.

Of interest at this point is the *c.g.* location for which the elevator-per-g gradient becomes zero. This location can be determined by first setting Equation (9.97) to zero, which yields

$$\left(C_{L_\alpha}\frac{\partial C_2(n)}{\partial n_0} - C_{M_\alpha}\frac{\partial C_1(n)}{\partial n_0}\right) = 0$$

or

$$C_{L_\alpha}\left(C_{M_q}\frac{g}{U_0}\left(1+\frac{1}{n_0^2}\right) + 2I_{xz}\left(\frac{g}{U_0}\right)^2\left(\frac{1}{n_0^3}\right)\right) + C_{M_\alpha}\left(\left(\frac{mg}{q_\infty S_W}\right) - C_{L_q}\frac{g}{U_0}\left(1+\frac{1}{n_0^2}\right)\right) = 0 \quad (9.98)$$

Solving the above expression for the ratio $C_{M_\alpha}/C_{L_\alpha}$ yields

$$\left.\frac{C_{M_\alpha}}{C_{L_\alpha}}\right|_{\frac{\partial\delta_{E_0}}{\partial n_0}=0} = \frac{-\left(C_{M_q}\frac{g}{U_0}\left(1+\frac{1}{n_0^2}\right) + 2I_{xz}\left(\frac{g}{U_0}\right)^2\left(\frac{1}{n_0^3}\right)\right)}{\left(\left(\frac{mg}{q_\infty S_W}\right) - C_{L_q}\frac{g}{U_0}\left(1+\frac{1}{n_0^2}\right)\right)} \quad (9.99)$$

Now note from Equation (9.30) and Figure 9.5, assuming that the propulsive thrust does not contribute significantly to the pitching moment, that the ratio $C_{M_\alpha}/C_{L_\alpha}$ is related to the static margin, or

$$\frac{C_{M_\alpha}}{C_{L_\alpha}} = -SM = \begin{cases}(\overline{X}_{AC} - \overline{X}_{cg}), \overline{X} \text{ positive forward}\\ (\overline{X}_{cg} - \overline{X}_{AC}), \overline{X} \text{ positive aft}\end{cases}$$

Recalling that $\overline{X}_{AC}$ is the normalized location of the vehicle's aerodynamic center, or stick-fixed neutral point, we can equate the above expression to Equation (9.99) and solve for the *c.g.* location for which the elevator-per-g gradient goes to zero. Or, assuming the *X* locations are measured <u>positive aft</u>, we have

$$\boxed{(\overline{X}_{cg})_{\frac{\partial\delta_{E_0}}{\partial n_0}=0} = \overline{X}_{AC} - \frac{\left(C_{M_q}\frac{g}{U_0}\left(1+\frac{1}{n_0^2}\right) + 2I_{xz}\left(\frac{g}{U_0}\right)^2\left(\frac{1}{n_0^3}\right)\right)}{\left(\left(\frac{mg}{q_\infty S_W}\right) - C_{L_q}\frac{g}{U_0}\left(1+\frac{1}{n_0^2}\right)\right)}} \quad (9.100)$$

The *c.g.* location given by Equation (9.100) is called the *maneuver point* in turning flight, and is typically located aft of the vehicle's aerodynamic center since C_{M_q} is negative. Hence, the limiting condition remains that the *c.g.* must be forward of the aerodynamic center for the proper elevator gradients to be achieved.

Knowing the trim angle of attack and elevator deflection from Equations (9.93) or (9.96), one may also find the trim stick force F_{S_E} at a given tail incidence angle i_H from Equation (9.56). Note that proceeding in this manner yields the trim stick force F_{S_E} as a function of trim load factor n_0. And from this relationship, another important stick gradient called *stick-force-per-g* may be found. For example, from Equation (9.56) we find that the stick-force-per-g gradient is

$$\frac{\partial F_{S_E}}{\partial n_0} = q_H S_E \bar{c}_E G_E \left(C_{h_\alpha} \frac{\partial \alpha_{\text{Trim}}}{\partial n_0} \left(1 - \frac{d\varepsilon_H}{d\alpha_W} \right) + C_{h_{\delta_E}} \frac{\partial \delta_{E_{\text{Trim}}}}{\partial n_0} \right) \tag{9.101}$$

where the elevator-per-g gradient $\partial \delta_{E_{\text{Trim}}}/\partial n_0$ is given by Equation (9.97), for example, and the angle-of-attack gradient $\partial \alpha_{\text{Trim}}/\partial n_0$ may be found by differentiating the expression for trim angle of attack in Equation (9.96). Doing so yields

$$\frac{\partial \alpha_{\text{Trim}}}{\partial n_0} = \left(\frac{\partial C_1(n)}{\partial n_0} C_{M_{\delta_E}} - \frac{\partial C_2(n)}{\partial n_0} C_{L_{\delta_E}} \right) / \Delta \tag{9.102}$$

where the two partial derivatives in the parentheses were given in Equations (9.97). For acceptable handling qualities, the stick-force-per-g gradient must be positive, or the pilot must pull back harder in a tighter turn (higher load factor). Note that this requirement on stick-force-per-g places requirements on the signs of the hinge-moment coefficients C_{h_α} and $C_{h_{\delta_E}}$.

As discussed when treating stick force per knot, note that stick-force-per-g may also be written as

$$\frac{\partial F_{S_E}}{\partial n_0} = \frac{\partial F_{S_E}}{\partial \delta_{E_{\text{Trim}}}} \frac{\partial \delta_{E_{\text{Trim}}}}{\partial n_0} \tag{9.103}$$

And if the vehicle is aerodynamically statically unstable (with stability restored with feedback augmentation), the sign of its elevator-per-g radiant will become positive. Consequently, the sign of the stick-force-per-elevator gradient $\partial F_{S_E}/\partial \delta_{E_{\text{Trim}}}$ in Equation (9.103) must be tailored by using a force-feel system such that the vehicle's stick-force-per-g remains positive.

EXAMPLE 9.6

Lateral-Directional Trim and Elevator Gradients in a Steady Turn

Consider the Navion aircraft in a steady, level 1.2 g right turn at sea level. The aerodynamic coefficients and other relevant data for the vehicle are given in Appendix B. First, does the vehicle satisfy all the requirements for lateral-directional static stability? Next, if the flight velocity is 176 fps, find the trim values of sideslip angle, aileron, and rudder deflections. Also estimate the elevator-speed $\Delta\delta_{E_{\text{Trim}}}/\Delta V_{\infty_0}$ and the elevator-per-g $\Delta\delta_{E_{\text{Trim}}}/\Delta n_0$ gradients, and the location of the maneuver point in turning flight.

Solution

We first list data from Appendix B, along with the static-stability requirements from Table 9.2. Note that all the requirements are met.

$$
\begin{aligned}
C_{S_\beta} &= -0.564 & C_{S_\beta} &< 0\\
C_{N_\beta} &= 0.0701 & C_{N_\beta} &> 0\\
C_{L_\beta} &= -0.074 & C_{L_\beta} &< 0\\
C_{L_p} &= -0.0389 \text{ sec} & C_{L_p} &< 0\\
C_{N_r} &= -0.0119 \text{ sec} & C_{N_r} &< 0
\end{aligned}
$$

Next, for the flight condition given,

$$q_\infty = 36.8 \text{ psf}$$

$$\frac{mg}{q_\infty S_W} = \frac{2750}{(36.8)(184)} = 0.41$$

$$\cos\Phi_0 = \frac{1}{1.2} = 0.8333,\ \Phi_0 = \pm 33.6 \text{ deg (positive for a right turn)}$$

and from Appendix B

$$
\begin{aligned}
C_{L_r} &= 0.0102 \text{ sec}\\
C_{N_r} &= -0.0119 \text{ sec}\\
C_{M_q} &= -0.161 \text{ sec}
\end{aligned}
$$

From Equation (9.89) we have the trim solution in a right turn given by

$$
\begin{bmatrix}
C_{S_\beta} - C_{D_{\text{Trim}}} & C_{S_{\delta_A}} & C_{S_{\delta_R}}\\
C_{L_\beta} & C_{L_{\delta_A}} & C_{L_{\delta_R}}\\
C_{N_\beta} & C_{N_{\delta_A}} & C_{N_{\delta_R}}
\end{bmatrix}
\begin{Bmatrix}
\beta_0\\
\delta_{A_0}\\
\delta_{R_0}
\end{Bmatrix}
=
\begin{bmatrix}
-C_{S_r}\dfrac{g}{U_0 n_0}\sqrt{n_0^2-1}\\
\dfrac{(I_{zz}-I_{yy})}{q_\infty S_W b_W}\left(\dfrac{g}{U_0}\right)^2\left(1-\dfrac{1}{n_0^2}\right)\sqrt{n_0^2-1} - C_{L_r}\dfrac{g}{U_0 n_0}\sqrt{n_0^2-1}\\
\dfrac{I_{xz}}{q_\infty S_W b_W}\left(\dfrac{g}{U_0}\right)^2\left(1-\dfrac{1}{n_0^2}\right)\sqrt{n_0^2-1} - C_{N_r}\dfrac{g}{U_0 n_0}\sqrt{n_0^2-1}
\end{bmatrix}
$$

Inserting the data from the appendix, including $C_{D_{\text{Trim}}} \approx 0.05$, we have

$$\begin{bmatrix} (-0.564 - 0.05) & 0 & 0.157 \\ -0.074 & 0.134 & 0.012 \\ 0.0701 & -0.0035 & -0.0717 \end{bmatrix} \begin{Bmatrix} \beta_0 \\ \delta_{A_0} \\ \delta_{R_0} \end{Bmatrix} = \begin{bmatrix} 0 \\ \dfrac{(530)}{226{,}158}(0.183)^2\left(1 - \dfrac{1}{1.44}\right)\sqrt{0.44} - 0.0101\dfrac{(0.183)}{1.2}\sqrt{0.44} \\ 0.0119\dfrac{(0.183)}{1.2}\sqrt{0.44} \end{bmatrix}$$

$$= \begin{bmatrix} 0 \\ 1.6 \times 10^{-5} - 0.001021 \\ 0.001203 \end{bmatrix}$$

From MATLAB or Cramer's Rule we find that the trim solution is

$$\begin{Bmatrix} \beta_{\text{Trim}} \\ \delta_{A_{\text{Trim}}} \\ \delta_{R_{\text{Trim}}} \end{Bmatrix} = \begin{Bmatrix} -0.005824 \\ 0.006280 \\ -0.02278 \end{Bmatrix} \text{rad} = \begin{Bmatrix} -0.334 \\ 0.360 \\ -1.305 \end{Bmatrix} \text{deg}$$

The negative rudder deflection (trailing-edge right) is consistent with a positive (right) yaw rate in a right-hand turn, but will produce a negative side force. This side force is balanced by a small negative sideslip angle, or the vehicle is pointed slightly to the right of the free-stream velocity vector towards the direction of turn. Finally the positive aileron deflection is consistent with a positive bank angle.

The static margin, assuming that the thrust doesn't contribute to the pitching moment, is found from Equation (9.30). For this vehicle it is

$$SM \triangleq -\frac{C_{M_\alpha}}{C_{L_\alpha}} = -\frac{-0.683}{4.44} = 0.154 = 15.4\% \text{ MAC}$$

Therefore, since the *c.g.* is located at 29.5 percent MAC, we have the vehicle's aerodynamic center located at

$$\overline{X}_{\text{AC}} = \overline{X}_{\text{cg}} + SM = 0.295 + 0.154 = 0.449 = 44.9\% \text{ MAC}$$

The location of the maneuver point in the turn may now be found from Equation (9.100), or

$$\left(\overline{X}_{\text{cg}}\right)_{\frac{\partial \delta_{E_0}}{\partial n_0} = 0} = \overline{X}_{\text{AC}} - \frac{\left(C_{M_q}\dfrac{g}{U_0}\left(1 + \dfrac{1}{n_0^2}\right) + 2I_{xz}\left(\dfrac{g}{U_0}\right)^2\left(\dfrac{1}{n_0^3}\right)\right)}{\left(\left(\dfrac{mg}{q_\infty S_W}\right) - C_{L_q}\dfrac{g}{U_0}\left(1 + \dfrac{1}{n_0^2}\right)\right)}$$

for $\overline{X}$ locations taken to be positive aft. For the data given, the last term in the above expression becomes

$$\frac{\left(-0.161(0.183)\left(1 + \dfrac{1}{1.44}\right) + 0\right)}{(0.41 - 0)} = -0.122$$

As expected, the location of the maneuver point is aft of the neutral point, and is at

$$\bar{X}_{\mathrm{MP}} = 0.449 + 0.122 = 0.571 = 57.1\%\ \mathrm{MAC}$$

The elevator speed gradient is given in Equation (9.46), or

$$\frac{\Delta\delta_{E_{\mathrm{Trim}}}}{\Delta V_{\infty_0}} = \frac{\Delta\delta_{E_{\mathrm{Trim}}}}{\Delta C_{L_{\mathrm{Trim}}}}\frac{\Delta C_{L_{\mathrm{Trim}}}}{\Delta V_{\infty_0}}$$

where the elevator trim gradient is given by Equation (9.45), or

$$\frac{\Delta\delta_{E_{\mathrm{Trim}}}}{\Delta C_{L_{\mathrm{Trim}}}} \approx \frac{\partial\delta_{E_{\mathrm{Trim}}}}{\partial C_{L_{\mathrm{Trim}}}} = \frac{-C_{M_\alpha}}{C_{L_\alpha}C_{M_{\delta_E}} - C_{M_\alpha}C_{L_{\delta_E}}} = \frac{0.683}{(4.44)(-0.87) - (-0.683)(0.355)} = -0.189\ \mathrm{rad}$$

And since

$$nmg = C_L q_\infty S_W$$

we have

$$\frac{\Delta C_{L_{\mathrm{Trim}}}}{\Delta V_{\infty_0}} = \frac{-4n_0 mg}{\rho_\infty S_W V_{\infty_0}^3} = \frac{-2n_0 mg}{\left(\frac{1}{2}\rho_\infty V_{\infty_0}^2\right)S_W V_{\infty_0}} = \frac{-2}{V_{\infty_0}}\left(\frac{n_0 mg}{q_\infty S_W}\right) = -0.01136(0.492) = -0.00559\ /\mathrm{fps}$$

So we have the elevator speed gradient given by

$$\frac{\Delta\delta_{E_{\mathrm{Trim}}}}{\Delta V_{\infty_0}} = (-0.189)(-0.00559) = 0.00105\ \mathrm{rad/fps} = 0.102\ \mathrm{deg/kt}$$

which is positive for the statically stable vehicle.

Finally, the elevator-per-g gradient is given by Equation (9.97). For this vehicle we have

$$\frac{\partial C_1(n)}{\partial n_0} = \left(\frac{mg}{q_\infty S_W}\right) - C_{L_q}\frac{g}{\mathrm{U}_0}\left(1 + \frac{1}{n_0^2}\right) = 0.41 - 0 = 0.41$$

and

$$\frac{\partial C_2(n)}{\partial n_0} = -2I_{xz}\left(\frac{g}{\mathrm{U}_0}\right)^2\left(\frac{1}{n_0^3}\right) - C_{M_q}\frac{g}{\mathrm{U}_0}\left(1 + \frac{1}{n_0^2}\right) = 0 - (-0.05) = 0.05$$

And so the elevator-per-g gradient is

$$\frac{\partial\delta_{E_{\mathrm{Trim}}}}{\partial n_0} = \left(C_{L_\alpha}\frac{\partial C_2(n)}{\partial n_0} - C_{M_\alpha}\frac{\partial C_1(n)}{\partial n_0}\right)/\Delta = \frac{(4.44(0.05) - (-0.683)(0.41))}{(4.44)(-0.87) - (-0.683)(0.355)}$$

$$= -0.139\ \mathrm{rad/g} = -7.94\ \mathrm{deg/g}$$

which is negative for the statically stable vehicle.

Summarizing the key results in this section, we developed techniques to determine the required angle of attack, sideslip, control-surface deflections, and thrust in steady turning flight. We defined and used normal load factor to specify the flight condition in the turn. In addition, we introduced cockpit-control forces, along with methods for estimating those forces for reversible control systems. We defined several important control-deflection and control-force gradients, along with criteria on these gradients based on handling-qualities considerations. Finally, we defined the vehicle's maneuver point for turning flight, along with a method for determining its location.

9.5 ANALYSIS OF QUASI-STEADY PULL-UP MANEUVERS

We will now turn to the analysis of the third reference flight condition defined in Section 9.1, a quasi-steady pull-up maneuver. As discussed in Section 9.1, the pull-up maneuver does not strictly satisfy the equilibrium criteria set forth in Equations (9.10) in that neither pitch rate Q_0 nor pitch attitude Θ_0 are constant during the maneuver. In spite of this fact, the maneuver is still of interest, and we will perform a quasi-steady analysis of this flight condition. In this analysis we will use only the longitudinal set of reference equations, and of particular interest will be the elevator-per-g gradient and the required control deflections as a function of flight condition.

Assuming stability axes are employed, the vehicle geometry is symmetric with respect to its XZ plane, and a wings-level ($\Phi_0 = 0$) maneuver with no sideslip ($\beta_0 = 0$), Equations (9.13) provide the appropriate reference equations. Extracting the longitudinal set of reference equations, and under the conditions stated, we have

$$
\boxed{\begin{aligned}
& mg\sin\gamma_0 = -D_0 + T_0\cos(\phi_T + \alpha_0) \\
& m(Q_0 U_0) + mg\cos\gamma_0 = L_0 + T_0\sin(\phi_T + \alpha_0) \\
& 0 = M_{A_0} + T_0(d_T\cos\phi_T - x_T\sin\phi_T)
\end{aligned}} \tag{9.104}
$$

9.5.1 Kinematic Analysis of the Pull-Up Maneuver

In order to apply Equations (9.104) to find the required thrust, angle of attack and elevator deflection, we must first determine the pitch rate Q_0 and select a flight-path angle γ_0 for analysis. Consider the schematic of the maneuver as depicted in Figure 9.22, and assume the maneuver is defined in terms of a circular trajectory corresponding to some constant pull-up radius $R_{\text{Pull Up}}$. Using stability axes, the velocity of the vehicle along the circular trajectory is then

$$U_0 = Q_0 R_{\text{Pull Up}}$$

and so

$$R_{\text{Pull Up}} = U_0/Q_0 = \text{constant} \tag{9.105}$$

(Note that this does not require both U_0 and Q_0 to be constant.)

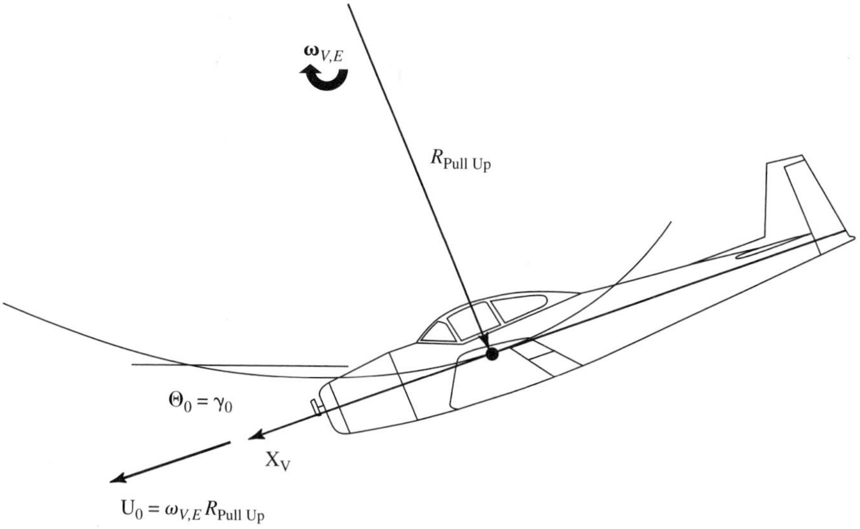

Figure 9.22 Schematic of pull-up maneuver.

Looking now at the second of Equations (9.104) we have

$$m(Q_0 U_0) + mg\cos\gamma_0 = L_0 + T_0 \sin(\phi_T + \alpha_0) \tag{9.106}$$

Expressing the right-hand side of this equation as $n_0(mg)$, where n_0 is the reference normal load factor introduced in Section 9.4.1, and solving for the pitch rate Q_0 we have

$$\boxed{Q_0 = \frac{g(n_0 - \cos\gamma_0)}{U_0}} \tag{9.107}$$

While from Equation (9.105) we find that the radius of the pull-up maneuver is

$$\boxed{R_{\text{Pull Up}} = \frac{U_0^2}{g(n_0 - \cos\gamma_0)}} \tag{9.108}$$

Finally, from Equations (9.106) and (9.108) we have the expression for the reference normal load factor given by

$$\boxed{n_0 = \cos\gamma_0 + \frac{U_0^2}{gR_{\text{Pull Up}}}} \tag{9.109}$$

Therefore, the pull-up maneuver may be defined in terms of either reference load factor n_0 or pull-up radius $R_{\text{Pull Up}}$.

For our purposes we will select the reference flight-path angle to be $\gamma_0 = 0$, or we will focus on the flight condition at the bottom of the pull up. At this point in the maneuver we have

$$\boxed{\begin{aligned} Q_0 &= \frac{g(n_0 - 1)}{U_0} \\ R_{\text{Pull Up}} &= \frac{U_0^2}{g(n_0 - 1)} \\ n_0 &= 1 + \frac{U_0^2}{g R_{\text{Pull Up}}} \end{aligned}} \tag{9.110}$$

9.5.2 Longitudinal Trim Analysis

With the quasi-steady pitch rate Q_0 in the pull up now given in terms of the velocity and load factor in Equations (9.110), we may now determine the required angle of attack, thrust, and elevator deflection for the maneuver. Our approach will follow that taken in Section 9.4.3, in which we were analyzing steady turning flight.

Initially, let us insert linear expressions for the lift, drag, and pitching-moment coefficients into the longitudinal equations given by Equations (9.104), repeated here.

$$\begin{aligned} mg\sin\gamma_0 &= -D_0 + T_0\cos(\phi_T + \alpha_0) \\ m(Q_0U_0) + mg\cos\gamma_0 &= L_0 + T_0\sin(\phi_T + \alpha_0) \\ 0 &= M_{A_0} + T_0(d_T\cos\phi_T - x_T\sin\phi_T) \end{aligned}$$

After this substitution, noting that $Q_0 = g(n_0 - 1)/U_0$ and $\gamma_0 = 0$, and assuming $C_{D_q} \approx 0$, Equations (9.104) now become

$$\begin{aligned} 0 &= -(C_{D_{\alpha=\delta=i_H=0}} + C_{D_\alpha}\alpha_0 + C_{D_{i_H}} i_H + C_{D_{\delta_E}}\delta_{E_0})q_\infty S_W + T_0\cos(\phi_T + \alpha_0) \\ n_0(mg) &= \left(C_{L_{\alpha=\delta=i_H=0}} + C_{L_q}\left(\frac{g(n_0 - 1)}{U_0}\right) + C_{L_\alpha}\alpha_0 + C_{L_{i_H}} i_H + C_{L_{\delta_E}}\delta_{E_0}\right)q_\infty S_W + T_0\sin(\phi_T + \alpha_0) \\ 0 &= \left(C_{M_{\alpha=\delta=i_H=0}} + C_{M_q}\left(\frac{g(n_0 - 1)}{U_0}\right) + C_{M_\alpha}\alpha_0 + C_{M_{i_H}} i_H + C_{M_{\delta_E}}\delta_{E_0}\right)q_\infty S_W\bar{c}_W + T_0(d_T\cos\phi_T - x_T\sin\phi_T) \end{aligned} \tag{9.111}$$

Recalling the trim thrust coefficient given by

$$C_{T_0} \triangleq \frac{T_0}{q_\infty S_W} \tag{9.112}$$

assuming that ϕ_T is small, and that

$$\cos(\phi_T + \alpha_0) = 1$$
$$T_0 \sin(\phi_T + \alpha_0) \ll L_0$$

we may write Equations (9.111) in the following matrix format

$$\begin{bmatrix} -C_{D_\alpha} & -C_{D_{\delta_E}} & 1 \\ C_{L_\alpha} & C_{L_{\delta_E}} & 0 \\ C_{M_\alpha} & C_{M_{\delta_E}} & \dfrac{(d_T - x_T\phi_T)}{\bar{c}_W} \end{bmatrix} \begin{Bmatrix} \alpha_0 \\ \delta_{E_0} \\ C_{T_0} \end{Bmatrix} = \begin{bmatrix} C_{D_{\alpha=\delta=i_H=0}} + C_{D_{i_H}} i_H \\ n_0\left(\dfrac{mg}{q_\infty S_W}\right) - C_{L_q}\left(\dfrac{g(n_0-1)}{U_0}\right) - C_{L_{\alpha=\delta=i_H=0}} - C_{L_{i_H}} i_H \\ -C_{M_q}\left(\dfrac{g(n_0-1)}{U_0}\right) - C_{M_{\alpha=\delta=i_H=0}} - C_{M_{i_H}} i_H \end{bmatrix} \quad (9.113)$$

This matrix equation may be solved for the three unknowns by using MATLAB or by Cramer's rule.

Now if the thrust vector acts through the *c.g.* of the vehicle, then

$$d_T \cos\phi_T - x_T \sin\phi_T \approx d_T - x_T\phi_T = 0$$

and the second and third equations in Equation (9.113) are independent of the first. In this case, the first of the three above equations may be used to determine the required trim thrust coefficient C_{T_0} or the thrust T_0, and the second and third may now be more simply written as

$$\begin{bmatrix} C_{L_\alpha} & C_{L_{\delta_E}} \\ C_{M_\alpha} & C_{M_{\delta_E}} \end{bmatrix} \begin{Bmatrix} \alpha_0 \\ \delta_{E_0} \end{Bmatrix} = \begin{bmatrix} n_0\left(\dfrac{mg}{q_\infty S_W}\right) - C_{L_q}\left(\dfrac{g(n_0-1)}{U_0}\right) - C_{L_{\alpha=\delta=i_H=0}} - C_{L_{i_H}} i_H \\ -C_{M_q}\left(\dfrac{g(n_0-1)}{U_0}\right) - C_{M_{\alpha=\delta=i_H=0}} - C_{M_{i_H}} i_H \end{bmatrix} \quad (9.114)$$

Solving this equation for the <u>trim angle of attack and elevator deflection in the pull up</u> yields

$$\begin{aligned} \alpha_{\text{Trim}} &= \left(C_{M_{\delta_E}} C_1(n) - C_{L_{\delta_E}} C_2(n)\right)/\Delta \\ \delta_{E_{\text{Trim}}} &= \left(C_{L_\alpha} C_2(n) - C_{M_\alpha} C_1(n)\right)/\Delta \\ \Delta &= C_{L_\alpha} C_{M_{\delta_E}} - C_{M_\alpha} C_{L_{\delta_E}} \end{aligned} \quad (9.115)$$

with

$$
\begin{aligned}
C_1(n) &\triangleq n_0\left(\frac{mg}{q_\infty S_W}\right) - C_{L_q}\left(\frac{g(n_0-1)}{U_0}\right) - C_{L_{\alpha=\delta=i_H=0}} - C_{L_{i_H}} i_H \\
C_2(n) &\triangleq -C_{M_q}\left(\frac{g(n_0-1)}{U_0}\right) - C_{M_{\alpha=\delta=i_H=0}} - C_{M_{i_H}} i_H
\end{aligned}
$$

These results will be used to find the stick forces and control gradients to be discussed below.

With regard to using a graphical procedure to determine the trim angle of attack and elevator deflection, as discussed in Section 9.3.1, we find the procedure is again complicated by the presence of a nonzero pitch rate in a pull up. However, the graphical procedure presented in the discussion of rectilinear flight may still be used to approximately solve for the trim angle of attack and elevator deflection in a pull up. If one ignores the effect of the pitch rate on the pitching moment and lift, the stability maps presented in Section 9.3.1 may be used for analysis of the pull-up maneuver if used with care. In the case of the pull up, the value of trim lift coefficient $C_{L_{\text{Trim}}}$ used in the graphical analysis is increased by the factor n_0. That is, in a pull up the trim lift coefficient equals $n_0(mg/q_\infty S_W)$.

9.5.3 Control Forces and Gradients

Turning to the subject of elevator and stick-force gradients, from Equations (9.115), or from a graphical analysis, we may now determine the elevator-per-g gradient $\partial\delta_{E_{\text{Trim}}}/\partial n_0$ for the pull-up maneuver. Differentiating the expression for trim elevator deflection in Equations (9.115), for example, yields

$$
\begin{aligned}
&\frac{\partial \delta_{E_{\text{Trim}}}}{\partial n_0} = \left(C_{L_\alpha}\frac{\partial C_2(n)}{\partial n_0} - C_{M_\alpha}\frac{\partial C_1(n)}{\partial n_0}\right)\Big/\Delta \\
&\text{with} \\
&\Delta = C_{L_\alpha}C_{M_{\delta_E}} - C_{M_\alpha}C_{L_{\delta_E}} \\
&\frac{\partial C_1(n)}{\partial n_0} = \left(\frac{mg}{q_\infty S_W}\right) - C_{L_q}\frac{g}{U_0} \\
&\frac{\partial C_2(n)}{\partial n_0} = -C_{M_q}\frac{g}{U_0}
\end{aligned}
\tag{9.116}
$$

Given that Δ in Equations (9.116) is typically negative, while the derivatives $\partial C_1(n)/\partial n_0$ and $\partial C_2(n)/\partial n_0$ are typically positive, the requirement that the elevator-per-g gradient be negative will normally be satisfied if the vehicle has positive static longitudinal stability. That is, if $C_{M_\alpha} < 0$.

As we did with turning flight, we will now determine the *c.g.* location for which the elevator-per-g gradient becomes zero. We can determine this location as we did in Section 9.4.3 by first setting Equation (9.116) to zero, which yields

$$\left(C_{L\alpha}\frac{\partial C_2(n)}{\partial n_0} - C_{M\alpha}\frac{\partial C_1(n)}{\partial n_0}\right) = 0$$

or

$$C_{L\alpha}\left(C_{Mq}\frac{g}{U_0}\right) + C_{M\alpha}\left(\left(\frac{mg}{q_\infty S_W}\right) - C_{Lq}\frac{g}{U_0}\right) = 0 \tag{9.117}$$

Solving the above expression for the ratio $C_{M_\alpha}/C_{L_\alpha}$ yields

$$\left.\frac{C_{M\alpha}}{C_{L\alpha}}\right|_{\frac{\partial\delta_{E_0}}{\partial n_0}=0} = -\frac{\left(C_{Mq}\frac{g}{U_0}\right)}{\left(\left(\frac{mg}{q_\infty S_W}\right) - C_{Lq}\frac{g}{U_0}\right)} \tag{9.118}$$

Now again note from Equation (9.30) and Figure 9.5 that, assuming that the propulsive thrust does not contribute significantly to the pitching moment, the ratio $C_{M_\alpha}/C_{L_\alpha}$ is related to the static margin by

$$\frac{C_{M_\alpha}}{C_{L_\alpha}} = -SM = \begin{cases}(\overline{X}_{AC} - \overline{X}_{cg}), \overline{X} \text{ positive forward}\\ (\overline{X}_{cg} - \overline{X}_{AC}), \overline{X} \text{ positive aft}\end{cases}$$

Recalling that $\overline{X}_{AC}$ is the normalized location of the vehicle's aerodynamic center, or stick-fixed neutral point, we can equate the above expression for *SM* to Equation (9.118) and solve for the *c.g.* location for which the elevator-per-g gradient goes to zero. Or, assuming the *X* locations are measure positive aft, we have

$$\boxed{(\overline{X}_{cg})_{\frac{\partial\delta_{E_0}}{\partial n_0}=0} = \overline{X}_{AC} - \frac{\left(C_{Mq}\frac{g}{U_0}\right)}{\left(\left(\frac{mg}{q_\infty S_W}\right) - C_{Lq}\frac{g}{U_0}\right)}} \tag{9.119}$$

The *c.g.* location given by Equation (9.119) is called the *maneuver point* in the pull-up maneuver, and is normally located aft of the vehicle's aerodynamic center since C_{M_q} is always negative. Hence again, the limiting condition remains that the *c.g.* must be forward of the aerodynamic center for the proper elevator gradients to be achieved. Note that the maneuver point in a pull up is usually not as far aft as the maneuver point in a level turn (Equation (9.100)).

From the results given in Equations (9.115) we may also determine the required stick force for a given tail incidence angle i_H, using Equation (9.56), for example.

Plus, we may also determine the stick-force-per-g gradient $\partial F_{S_E}/\partial n_0$. Differentiating Equation (9.56) with respect to normal load factor yields

$$\frac{\partial F_{S_E}}{\partial n_0} = q_H S_E \bar{c}_E G_E \left(C_{h_\alpha} \frac{\partial \alpha_{\text{Trim}}}{\partial n_0} \left(1 - \frac{d\varepsilon_H}{d\alpha_W} \right) + C_{h_{\delta_E}} \frac{\partial \delta_{E_{\text{Trim}}}}{\partial n_0} \right) \tag{9.120}$$

where the trim elevator-per-g gradient $\partial \delta_{E_{\text{Trim}}}/\partial n_0$ is as given in Equation (9.116). The trim angle-of-attack-per-g gradient $\partial \alpha_{\text{Trim}}/\partial n_0$ may be obtained by differentiating the expression for trim angle of attack given in Equations (9.115), which yields

$$\frac{\partial \alpha_{\text{Trim}}}{\partial n_0} = \left(C_{M_{\delta_E}} \frac{\partial C_1(n)}{\partial n_0} - C_{L_{\delta_E}} \frac{\partial C_2(n)}{\partial n_0} \right) / \Delta \tag{9.121}$$

The two partial derivatives in the parentheses are as given in Equations (9.116). For acceptable handling qualities, the stick-force-per-g gradient must be positive. This requirement again places constraints on the signs of the hinge-moment coefficients.

As discussed when addressing stick-force-per-g in turning flight and Equation (9.103), the stick-force-per-g may be written as

$$\frac{\partial F_{S_E}}{\partial n_0} = \frac{\partial F_{S_E}}{\partial \delta_{E_{\text{Trim}}}} \frac{\partial \delta_{E_{\text{Trim}}}}{\partial n_0}$$

And if the vehicle is aerodynamically statically unstable, with stability restored with feedback augmentation, the sign of the elevator per g $\partial \delta_{E_{\text{Trim}}}/\partial n_0$ in the pull up will be positive instead of negative. So the sign of the stick-force gradient $\partial F_{S_E}/\partial \delta_{E_{\text{Trim}}}$ must again be tailored using a force-feel system, or perhaps by adjusting the hinge-moment coefficients, to maintain a positive stick-force-per-g as required for acceptable handling qualities.

Summarizing the key results in this section, we developed techniques to determine the required angle of attack, control-surface deflection, and thrust in a quasi-steady pull-up maneuver. We considered cockpit-control forces, along with methods for estimating those forces for reversible control systems. We also defined several important control-deflection and control-force gradients, along with criteria on these gradients based on handling-qualities considerations. Finally, we defined a vehicle's maneuver point for a pull up, along with a method for determining its location.

9.6 Summary

To summarize the key results in this chapter, we presented a precise definition of an equilibrium condition for a nonlinear system, and we used that definition to determine the reference equations corresponding to three specific flight conditions. These conditions included steady rectilinear flight, steady turning flight, and quasi-steady pull-up maneuvers. We utilized the reference equations listed in this chapter to further analyze these reference flight conditions.

We introduced the concept of aerodynamic static stability, along with several criteria for longitudinal as well as lateral-directional static stability of a flight vehicle. We also investigated the effects of different aspects of the vehicle, or the vehicle's design, that contribute to meeting the static-stability criteria, and we introduced an important parameter known as the static margin of the aircraft. We showed that to assure static pitch stability, the vehicle's center of mass, or center of gravity, must be located forward of the vehicle's aerodynamic center.

We developed techniques to determine the required angles of attack and sideslip, control deflections, and thrust in the three reference flight conditions addressed. Normal load factor was defined, and used to specify the flight condition in turning flight and in pull-up maneuvers. We also presented a method for determining the control deflections required to control an asymmetric thrust condition such as an engine-out condition for a multi-engine aircraft. We then defined and analyzed the minimum control speed.

We introduced cockpit-control forces, along with methods for estimating those forces. We defined several important control-deflection and control-force gradients, along with criteria on these gradients based on handling-qualities considerations. We also defined the stick-free neutral point, along with a method for determining its location. We demonstrated that the location of the stick-free neutral point was always forward of the stick-fixed neutral point, which is the same as the vehicle's aerodynamic center. In addition, we defined a vehicle's maneuver points for turning flight and for a pull-up maneuver, along with methods for determining their locations.

9.7 Problems

9.1 For the vehicle analyzed in Example 9.2, find the *c.g.* location that would yield a positive static margin of 10 percent.

9.2 Consider a vehicle with all movable canards, such as the forward-swept X-29 aircraft shown in Figure 9.23. For such a vehicle, and using the schematic as shown, prove that the aerodynamic center of the wing must be aft of the *c.g.* to obtain a positive static margin. Show that under some reasonable assumptions, the location of the vehicle's stick-fixed neutral point may be expressed as

$$\overline{X}_{NP} \approx \frac{\overline{X}_{AC_W} + \overline{X}_{AC_C}\dfrac{S_C}{S_W}}{1 + \dfrac{S_C}{S_W}}$$

9.3 For an aircraft with an aft horizontal tail, show that an expression limiting the distance the aerodynamic center of the wing-fuselage combination may be forward of the vehicle's *c.g.* while maintaining static pitch stability is

$$\Delta\overline{X}_{AC_{W\,\&\,F}} < \frac{C_{L_{\alpha_H}}}{C_{L_{\alpha_W}}}\left(1 - \frac{d\varepsilon_H}{d\alpha_W}\right)\frac{q_H}{q_\infty}\frac{S_H}{S_W}\Delta\overline{X}_{AC_H}$$

(Here $\Delta\overline{X}_{AC_{W\,\&\,F}} > 0$ indicates that the *ac* of the wing and fuselage is forward of the *c.g.*, and $\Delta\overline{X}_{AC_H} > 0$ for an aft tail.) Note that the above inequality does not state

Figure 9.23 NASA X-29 aircraft, plus schematic. *(Photo courtesy of NASA.)*

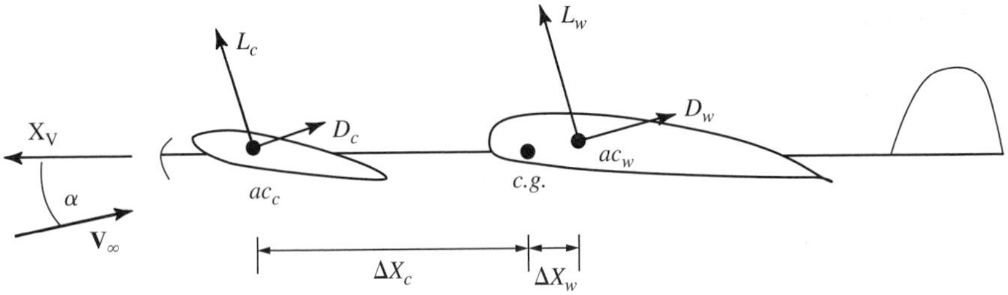

that the *ac* of the wing and fuselage must be aft of the *c.g.* Can the wing-fuselage aerodynamic center be aft of the *c.g.* while maintaining static pitch stability?

9.4 Show that for an aircraft with an aft horizontal tail in trimmed level flight, the tail will be generating a download (L_H down) if the normalized distance between the vehicle's *c.g.* and the aerodynamic center of the wing-fuselage combination satisfies the inequality $\Delta\overline{X}_{\mathrm{AC}_{W\,\&\,F}} < -\dfrac{C_{M_{\mathrm{AC}_{W\,\&\,F}}}}{C_{L_W}}$, where $\Delta\overline{X}_{\mathrm{AC}_{W\,\&\,F}} > 0$ if the wing/fuselage *ac* is forward of the *c.g.* Note that frequently this inequality is satisfied, and the horizontal tail is generating a download in trim, which increases the lift (and drag) generated by the wing in cruise. For example, for the vehicle considered in Examples 6.2 and 9.2, the tail will be generating a download whenever

$$\left(\frac{mg}{q_\infty S_W}\right) = C_{L_{\mathrm{Trim}}} > 0.7$$

9.5 Consider the Navion aircraft in steady, level flight at a velocity of 176 fps at sea level. Using the aerodynamic coefficients in Appendix B, find the trim angle of attack α_{Trim} and elevator deflection $\delta_{E_{\text{Trim}}}$, the static margin, and the elevator trim gradient $\partial\delta_{E_{\text{Trim}}}/\partial C_{L_{\text{Trim}}}$. Assume that $C_{L_{\alpha=\delta=i_H=0}} = 0.364$, $C_{M_{\alpha=\delta=i_H=0}} = 0$, and $i_H = 0$. Is the aircraft statically stable in pitch, and is the sign of the elevator trim gradient consistent with the static stability of the vehicle?

9.6 For the same vehicle in Problem 9.5, plot the stability map at least out to a lift coefficient of 1.0. Plot the data at 5 degree intervals of elevator deflection, and assume the maximum and minumum elevator deflections are ± 30 deg. For the flight condition given in Problem 9.5, find the trim angle of attack and elevator deflection. Also find the trim-elevator gradient $\Delta\delta_{E_{\text{Trim}}}/\Delta C_{L_{\text{Trim}}}$ and the maximum achievable C_L for the *c.g.* location given. Compare your results with those for Problem 9.5. Should they agree? Why?

9.7 For the Navion aircraft, use the stability map developed in Problem 9.6 to estimate the location of the stick-fixed neutral point. Does your result agree with the static margin found in Problem 9.5?

9.8 Consider the Navion aircraft at the same flight condition given in Problem 9.5. Instead of level flight, find the angle of attack and elevator deflection at the bottom of a pull-up maneuver if the load factor n_0 is 1.2 g. What is the elevator-per-g gradient and the pull-up radius $R_{\text{Pull Up}}$?

9.9 If an elevator is equipped with a trim tab rather than having variable incidence, express the lift and moment coefficients for the vehicle and the elevator-hinge-moment coefficient in terms of tab deflection δ_{Tab}. Then find the tab deflection that yields zero stick force for a reversible control system. (Recall that a trim tab is a hinged surface on the trailing edge of the elevator.)

References

1. Vidyasagar, M.: *Nonlinear Systems Analysis,* Prentice Hall, Upper Saddle River, NJ, 1978.
2. Perkins, C. D. and R. E. Hage: *Airplane Performance Stability and Control,* Wiley, New York, 1949.
3. Etkin, B.: *Dynamics of Flight—Stability and Control,* 2nd ed., Wiley, New York, 1982.
4. Cook, M. V.: *Flight Dynamics Principles,* 2nd ed., Elsevier, New York, 2007.
5. Roskam, J.: *Airplane Flight Dynamics and Automatic Flight Controls,* Roskam Aviation and Engineering Corp., Lawrence, KS, 1979.

10 CHAPTER

Linear Flight-Dynamics Analysis

Chapter Roadmap: *The material in this chapter is fundamental to any first course in flight dynamics. We introduce modal analysis of linear dynamic systems and describe the natural modes of atmospheric flight vehicles in terms of the modal eigenvalues and eigenvectors. Also, useful modal approximations are developed, based on the results from the modal analyses. These approximations will be used extensively in Chapter 11.*

We continue the analysis of the vehicle's dynamics in this chapter, now focusing on the linear analysis using the small-perturbation equations of motion. Recall from Chapter 1 that the perturbation quantities are displacements relative to a particular reference flight condition, and hence the results from the linear analysis aid in understanding the dynamics, including the stability, of the nonlinear system in the neighborhood of that reference flight condition. Of particular interest will be the determination and analysis of the natural modes of motion of the vehicle.

We will also investigate the transfer functions and time responses of the vehicle, and relate them to the vehicle's natural modes. We will develop low-order approximations for the models of the vehicle's dynamics, and find that these simpler models provide us with important insight into the dominant parameters (e.g., dimensional stability derivatives) affecting the dynamics. Since these parameters are determined by the vehicle's aerodynamic characteristics, mass properties, and ultimately the geometry, we then obtain information critical in vehicle design.

10.1 LINEAR SYSTEMS ANALYSIS—A JITT*

We will now introduce (or review) techniques for the dynamic analysis of linear, time-invariant dynamic systems. (Students who are already very familiar with this subject may proceed directly to Section 10.2, if desired.) We will consider

---

* A Just-In-Time Tutorial.

systems expressed in both state-variable and transfer-function format. We will assume the vehicle's linear dynamic model is given in state-variable format, as discussed in Section 8.1.4, and it may be helpful to review that section prior to continuing here.

10.1.1 State-Variable Descriptions and Modal Analysis

An n'th order linear dynamic system may be expressed in state-variable format as

$$\begin{aligned}\dot{\mathbf{x}}(t) &= \mathbf{A}\mathbf{x}(t) + \mathbf{B}\mathbf{u}(t)\\ \mathbf{y}(t) &= \mathbf{C}\mathbf{x}(t) + \mathbf{D}\mathbf{u}(t)\end{aligned} \tag{10.1}$$

where $\mathbf{x}$ is the vector of n state variables defining the system, $\mathbf{u}$ is the vector of m inputs, or control variables, and $\mathbf{y}$ is the vector of p response variables. The response and control-input vectors will always consist of physical variables, but the state vector may or may not, as we will see below. The coefficient matrices $\mathbf{A}$ $(n \times n)$, $\mathbf{B}$ $(n \times m)$, $\mathbf{C}$ $(p \times n)$, and $\mathbf{D}$ $(p \times m)$ are assumed to be constant.

The solution to Equation (10.1) was developed in Section 8.1.8, and from Equation (8.101) we have

$$\begin{aligned}\mathbf{x}(t) &= \mathbf{\Phi}(t - t_0)\mathbf{x}(t_0) + \int_{t_0}^{t} \mathbf{\Phi}(t - \tau)\mathbf{B}\mathbf{u}(\tau)\,d\tau\\ \mathbf{y}(t) &= \mathbf{C}\mathbf{x}(t) + \mathbf{D}\mathbf{u}(t)\end{aligned} \tag{10.2}$$

where $\mathbf{\Phi}(t)$ is the $n \times n$ state-transition matrix (see Equation (8.92)) given by

$$\mathbf{\Phi}(t) \triangleq e^{\mathbf{A}t} = \mathbf{I} + \mathbf{A}t + \frac{1}{2!}\mathbf{A}^2 t^2 + \ldots$$

The solution given in Equations (10.2) includes the sum of homogeneous plus particular solutions, where $\mathbf{\Phi}(t - t_0)\mathbf{x}(t_0)$ is the homogeneous solution. This solution gives the component of the system responses from initial conditions, while the particular solution gives the component of the responses due to control inputs.

It is particularly useful at this point to transform Equations (10.1) into *modal coordinates,* η_i, using the *modal matrix* $\mathbf{M}$ constructed from the eigenvectors of the matrix $\mathbf{A}$. That is, assume $\mathbf{A}$ has distinct eigenvalues λ_i, $i = 1 \ldots n$, where the eigenvalues are the roots of the characteristic polynomial given by

$$\det\left[\lambda\mathbf{I} - \mathbf{A}\right] = 0 \tag{10.3}$$

Each eigenvalue has an associated (right) eigenvector $\boldsymbol{\nu}_i$, and since the eigenvalues are assumed distinct, these n eigenvectors will be linearly independent. The columns of the modal matrix of $\mathbf{A}$ consist of the eigenvectors of $\mathbf{A}$, or

$$\mathbf{M} = \left[\boldsymbol{\nu}_1 \quad \cdots \quad \boldsymbol{\nu}_n\right] \tag{10.4}$$

By definition, each eigenvalue and right-eigenvector pair satisfies the relation

$$\mathbf{A}\boldsymbol{\nu}_i = \lambda_i \boldsymbol{\nu}_i,\ i = 1 \ldots n \tag{10.5}$$

So we may write

$$\mathbf{AM} = \mathbf{A}\big[\boldsymbol{\nu}_1 \ \cdots \ \boldsymbol{\nu}_n\big] = \big[\lambda_1\boldsymbol{\nu}_1 \ \cdots \ \lambda_n\boldsymbol{\nu}_n\big] = \big[\boldsymbol{\nu}_1 \ \cdots \ \boldsymbol{\nu}_n\big]\begin{bmatrix} \lambda_1 & 0 & \cdots & 0 \\ 0 & \ddots & \ddots & \vdots \\ \vdots & \ddots & \ddots & 0 \\ 0 & \cdots & 0 & \lambda_n \end{bmatrix} \triangleq \mathbf{M\Lambda} \quad (10.6)$$

where $\mathbf{\Lambda}$ is a diagonal matrix with the eigenvalues of $\mathbf{A}$ along its diagonal. Note that by premultiplying Equation (10.6) by the inverse of $\mathbf{M}$ we have

$$\mathbf{M}^{-1}\mathbf{AM} = \mathbf{\Lambda} \quad (10.7)$$

By defining the *modal state vector* $\boldsymbol{\eta}$ such that $\mathbf{x} = \mathbf{M}\boldsymbol{\eta}$, the vector differential equation in Equations (10.1) now becomes

$$\mathbf{M}\dot{\boldsymbol{\eta}}(t) = \mathbf{AM}\boldsymbol{\eta}(t) + \mathbf{Bu}(t)$$

or

$$\dot{\boldsymbol{\eta}}(t) = \mathbf{M}^{-1}\mathbf{AM}\boldsymbol{\eta}(t) + \mathbf{M}^{-1}\mathbf{Bu}(t) = \mathbf{\Lambda}\boldsymbol{\eta}(t) + \mathbf{M}^{-1}\mathbf{Bu}(t)$$

with (10.8)

$$\mathbf{y}(t) = \mathbf{CM}\boldsymbol{\eta}(t) + \mathbf{Du}(t)$$

This transformation decouples the original vector differential equation into n scalar equations governing the modal coordinates, or

$$\dot{\eta}_i(t) = \lambda_i\eta_i(t) + \boldsymbol{\mu}_i\mathbf{Bu}(t), \ i = 1 \ldots n \quad (10.9)$$

where $\boldsymbol{\mu}_i$ is the i'th row of $\mathbf{M}^{-1}$, or the i'th left eigenvector of $\mathbf{A}$. We also see that the transition matrix corresponding to the system given in Equation (10.8) is now diagonal, or

$$\mathbf{\Phi}(t - t_0) = e^{\mathbf{\Lambda}(t-t_0)} = \begin{bmatrix} e^{\lambda_1(t-t_0)} & 0 & \cdots & 0 \\ 0 & \ddots & \ddots & \vdots \\ \vdots & \ddots & \ddots & 0 \\ 0 & \cdots & 0 & e^{\lambda_n(t-t_0)} \end{bmatrix} = diag\big(e^{\lambda_i(t-t_0)}\big) \quad (10.10)$$

Note from Equation (10.9) that the left eigenvectors determine how the control inputs excite the modal responses. In the case with nonzero initial conditions, the left eigenvectors also determine how the initial conditions excite the corresponding modes.

In the absence of control inputs, the character of the solutions to the n equations given in Equation (10.9) is completely determined by the eigenvalues λ_i since each solution will be of the form $e^{\lambda_i(t-t_0)}$. If λ_i is real and negative, $\eta_i(t)$ will decay exponentially, and if λ_i is real and positive, $\eta_i(t)$ will grow exponentially. If any of the eigenvalues are complex, they will always occur in complex–conjugate pairs. Letting such a complex pair be denoted as $\lambda_i = \sigma_i \pm j\omega_i$, where $j = \sqrt{-1}$, then $\eta_i(t)$ will decay or grow exponentially, depending on the sign of σ_i, and will oscillate at the frequency ω_i. For this reason, the complex part of an eigenvalue is sometimes called the *damped frequency* ω_d. Finally, dynamic stability of the system requires that all n eigenvalues have negative real parts.

Each complex pair of eigenvalues are the roots of a quadratic factor in the characteristic polynomial. The standard form for such quadratic factors is defined to be

$$\lambda^2 + 2\zeta\omega_n\lambda + \omega_n^2 \tag{10.11}$$

The parameters in the two coefficients are the

Damping ratio—ζ

Undamped natural frequency—ω_n

And the two roots of the quadratic given in the standard form are

$$\lambda_{1,2} = -\zeta\omega_n \pm j\omega_n\sqrt{1-\zeta^2} = \sigma \pm j\omega_d \tag{10.12}$$

So we see that the damped natural frequency, written in terms of the damping ratio and undamped natural frequency, is

$$\omega_d = \omega_n\sqrt{1-\zeta^2} \tag{10.13}$$

The right eigenvectors define the relationship between the modal coordinates and the original states. Since by definition $\mathbf{x} = \mathbf{M}\boldsymbol{\eta}$, then

$$\mathbf{x}(t) = \mathbf{M}\boldsymbol{\eta}(t) = \boldsymbol{\nu}_1\eta_1(t) + \boldsymbol{\nu}_2\eta_2(t) + \ldots + \boldsymbol{\nu}_n\eta_n(t)$$

and if $\mathbf{D} = \mathbf{0}$ (10.14)

$$\mathbf{y} = \mathbf{Cx} = \mathbf{C}\boldsymbol{\nu}_1\eta_1(t) + \mathbf{C}\boldsymbol{\nu}_2\eta_2(t) + \ldots + \mathbf{C}\boldsymbol{\nu}_n\eta_n(t)$$

This latter expression is especially useful when the original state vector consists of physical variables (e.g., $\mathbf{C} = \mathbf{I}$), or at least when there are n responses. The eigenvectors are now seen to also be the *mode shapes,* defining how each modal response $\eta_i(t)$ contributes to the physical responses of the system. That is, some physical responses may be dominated by certain modes, while other physical responses may be dominated by other modes.

To expand on this last point further, assume that $\mathbf{C} = \mathbf{I}$, in which case

$$\mathbf{y}(t) = \sum_{i=1}^{n} \boldsymbol{\nu}_i\eta_i(t) \tag{10.15}$$

and we can see that each element of an eigenvector is associated with a physical variable and therefore has the units of that variable associated with it (since the modal response $\eta(t)$ is dimensionless)! This fact allows for a further *mode-shape* interpretation of the eigenvectors.

Given the elements of each eigenvector are in general complex, one can plot each eigenvector in a phasor diagram. For example, let the i'th eigenvector be given as

$$\boldsymbol{v}_i = \begin{Bmatrix} m_1 e^{j\phi_1} \\ m_2 e^{j\phi_2} \\ \vdots \\ m_n e^{j\phi_n} \end{Bmatrix} \tag{10.16}$$

where each complex element has been expressed in polar form consisting of a magnitude and phase angle. Recall that only the relative magnitudes of each element in an eigenvector are uniquely determined. So some form of normalization is always involved. For example, MATLAB normalizes each eigenvector to unity total magnitude. That is,

$$\sqrt{m_1^2 + \cdots + m_n^2} = 1$$

Now note that each element magnitude m_i has the units of its corresponding physical response y_i. So by plotting each element in a phasor diagram, we may construct a diagram that graphically describes the eigenvector, as shown schematically in Figure 10.1. This figure shows the mode shape of the eigenvector, with the relative magnitudes of each element indicating the extent to which the modal response contributes to the corresponding physical response.

In fact, one can visualize the motion of the vehicle undergoing only this mode's response by first visualizing the phasor diagram rotating at a rate equal to the frequency of the corresponding eigenvalue (the imaginary part). Since positive phase angle is indicated counterclockwise in the diagram, a positive modal frequency rotates the diagram counterclockwise at the rate indicated by that frequency. Next, visualize each vector element of the diagram shrinking (or growing) exponentially, as determined by the real part of the eigenvalue. Finally, imagine the page sliding to the left under the rotating, shrinking (or growing)

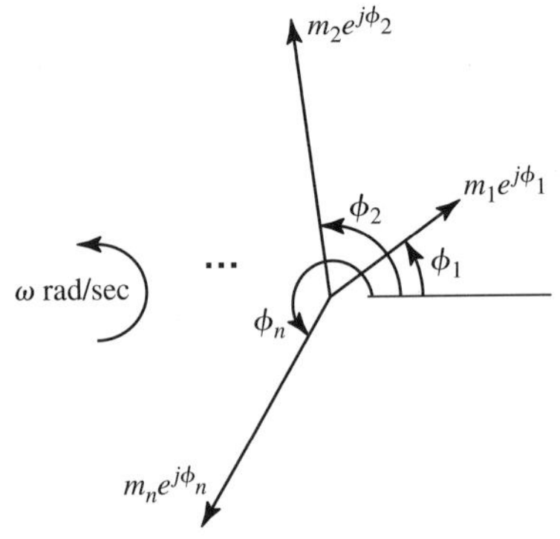

Figure 10.1 Phasor diagram of (right) eigenvector.

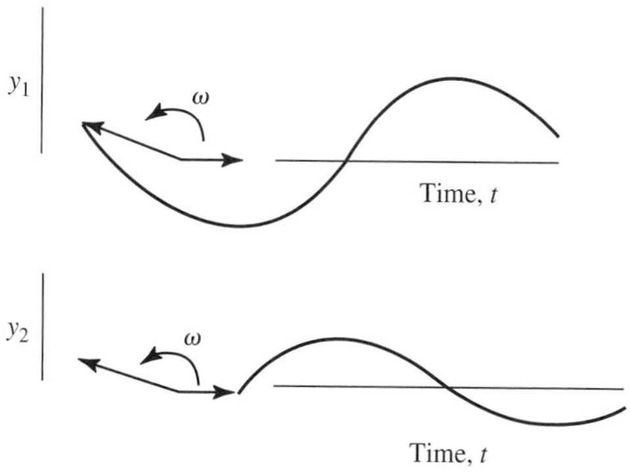

Figure 10.2 Rotating phasor diagram generating the physical responses.

phasor diagram. Now, as shown schematically in Figure 10.2, the tips of each eigenvector element trace out the time histories of the physical responses onto the page, as the page moves to the left.

This modal-visualization scheme is especially useful when each physical response is scaled to roughly equal "engineering equivalence." For example, one might scale each element of the response vector to the minimum (or maximum) magnitude that could be measured by a typical transducer (sensor). Such scaling will be demonstrated in Example 10.2.

10.1.2 Transfer Functions, Bode Plots, and Residues

The system defined by Equations (10.1) may also be expressed in terms of a matrix of *transfer functions*. Taking the Laplace transform of Equations (10.1), assuming all initial conditions are zero, and rearranging, we find that the transforms of the responses of the system are related to the transforms of the control inputs by the following relationship

$$\mathbf{y}(s) = \left[\mathbf{C}\left[s\mathbf{I} - \mathbf{A}\right]^{-1}\mathbf{B} + \mathbf{D}\right]\mathbf{u}(s) = \mathbf{TF}(s)\mathbf{u}(s) = \begin{bmatrix} g_{1,1}(s) & \cdots & g_{1,n}(s) \\ \vdots & g_{i,j}(s) & \vdots \\ g_{n,1}(s) & \cdots & g_{n,n}(s) \end{bmatrix}\mathbf{u}(s) \quad (10.17)$$

Here each element of the matrix $\mathbf{TF}(s)$, or $g_{i,j}(s)$, is a transfer function, and $\mathbf{TF}(s)$ is called a *transfer-function matrix*. Each transfer function element consists of a ratio of polynomials in the complex variable s, and the denominator of each transfer function will be the characteristic polynomial of $\mathbf{A}$, or

$$\det\left[s\mathbf{I} - \mathbf{A}\right] \quad (10.18)$$

Roots of this polynomial are called *system poles,* and note that they are also the eigenvalues of $\mathbf{A}$. Roots of the numerator polynomials in each transfer function are called *zeros.* All transfer functions will have the same poles, but each will have a unique set of zeros.

A useful graphical representation of a transfer function may be obtained by setting $s = j\omega$, and plotting the magnitude and phase of the resulting complex number versus frequency ω. Such a graphical representation is called a *Bode plot,* and is frequently used for closed-loop analysis of flight vehicles (i.e, when the vehicle is controlled by a pilot or autopilot). Closed-loop analysis will be discussed further in Chapters 11–13.

Specifically, now, note that $g(s)|_{s=j\omega}$ is a complex number having magnitude M and phase ϕ. Furthermore, this magnitude and phase are both functions of the frequency ω. Or this complex number is

$$g(s)|_{s=j\omega} = M(\omega)e^{j\phi(\omega)} \tag{10.19}$$

A Bode plot is a two-part plot of M and ϕ versus ω.

Such a plot is easily obtained using MATLAB, for example, and the shapes of the plots are closely tied to the poles and zeros of the transfer function. For example, consider a transfer function of the following form.

$$\begin{aligned} g(s)|_{s=j\omega} &= \frac{K(s+a)}{(s^2 + 2\zeta\omega_n s + \omega_n^2)}\Big|_{s=j\omega} = \frac{(Ka/\omega_n^2)(s/a+1)}{(s^2/\omega_n^2 + 2\zeta\omega_n s/\omega_n^2 + 1)}\Big|_{s=j\omega} \\ &= \frac{K_B(j\omega/a+1)}{((j\omega)^2/\omega_n^2 + 2\zeta\omega_n j\omega/\omega_n^2 + 1)} = M_B e^{j\phi_B}\frac{M_N e^{j\phi_N}}{M_D e^{j\phi_D}} \end{aligned} \tag{10.20}$$

where the numerator constant K_B is called the *Bode gain.* Also note that the subscript N identifies the complex number in the numerator given in polar form, while the subscript D identifies the complex number in the denominator.

We will now analyze the asymptotic behavior of this numerator and denominator. When ω is small compared to $|a|$, $M_N \sim 1$ and $\phi_N \sim 0$. And when ω is large compared to $|a|$, $M_N \sim \omega$ and $\phi_N \sim 90$ deg (-90 deg if a is negative). Since the low- and high-frequency behavior hinges on the value of a, it is called the *corner frequency* of the numerator. A similar analysis of the second-order denominator reveals that when ω is small compared to its corner frequency $|\omega_n|$, $M_D \sim 1$ and $\phi_D \sim 0$. And when ω is large compared to $|\omega_n|$, $M_D \sim \omega^2$ and $\phi_D \sim \pm 180$ deg. (0 deg if ω^2 is negative.) The damping ratio ζ does not affect this asymptotic behavior, but does affect both the magnitude and the phase in the frequency range near the corner frequency ω_n. If the damping is low, there will be a spike in the magnitude M_D at ω_n, and the phase ϕ_D will change rapidly between the asymptotic values. If the damping is high, there will be no peak in the magnitude, and the phase will change more gradually.

By plotting the magnitude of $g(s)|_{s=j\omega}$ on a logarithmic scale, the logarithm of the total magnitude is the appropriate sum of the logarithms of the magnitudes of the three terms at a given frequency (i.e., $\log(M_B) + \log(M_N) - \log(M_D)$). Likewise, the phase of $g(s)|_{s=j\omega}$ may similarly be obtained by appropriately summing the three phase angles. Consequently, by evaluating each factor in the numerator and denominator in like fashion, the asymptotic behavior of the Bode

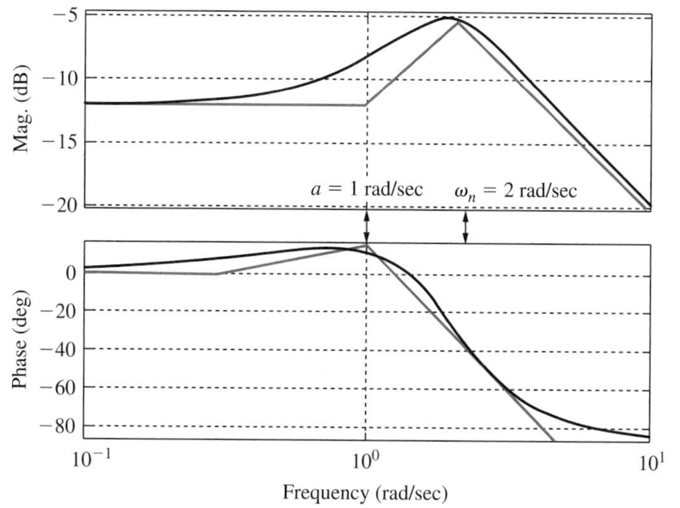

Figure 10.3 Bode plot of example transfer function.

plot may be easily sketched. Conversely, given a transfer function's Bode plot, the corresponding factors in the transfer function may be approximately identified. As a final note, magnitudes are typically plotted in units of decibels, or dB, where X in dB = 20 log (X).

A Bode plot of the previous example transfer function is shown in Figure 10.3 for $K = 1$, $a = 1$ /sec, $\omega_n = 2$ rad/sec, and $\zeta = 0.5$. For this set of parameters the Bode gain $K_B = ¼ = -12.04$ dB. Also shown in the plot are several straight lines indicating the low- and high-frequency asymptotic approximations for the factors in the transfer function. There is a slight peak in the magnitude plot, primarily due to the factor in the numerator of $g(s)$. The phase changes somewhat rapidly near 2 rad/sec, due to the modest damping ratio.

Turning now to *residues,* first recall that they are the constant numerators in each factor of a partial-fraction expansion of the transform of a system's response. Or the n residues R_i in such an expansion appear as

$$y(s) = \frac{K(s - z_1)\cdots(s - z_m)}{(s - \lambda_1)\cdots(s - \lambda_n)} = \frac{R_1}{(s - \lambda_1)} + \cdots + \frac{R_n}{(s - \lambda_n)} \tag{10.21}$$

Hence residues are important because they determine the magnitude of the contribution of the corresponding eigenvalue (or mode) to the system's time response.

Let the linear dynamic system be described here in terms of modal coordinates, as in Equations (10.8), with $\mathbf{D} = \mathbf{0}$. In this case, the transfer-function matrix in Equation (10.17) becomes

$$\mathbf{TF}(s) = \left[\mathbf{CM}\left[diag\left(\frac{1}{s - \lambda_i}\right)\right]\mathbf{M}^{-1}\mathbf{B}\right] \tag{10.22}$$

Since the modal matrix was given as

$$\mathbf{M} = \begin{bmatrix} \boldsymbol{\nu}_1 & \cdots & \boldsymbol{\nu}_n \end{bmatrix}$$

and the inverse modal matrix is

$$\mathbf{M}^{-1} = \begin{bmatrix} \boldsymbol{\mu}_1 \\ \vdots \\ \boldsymbol{\mu}_n \end{bmatrix} \tag{10.23}$$

the transfer-function matrix may be written in terms of the following partial-fraction expansion.

$$\mathbf{TF}(s) = \mathbf{C} \sum_{k=1}^{n} \frac{\left[\boldsymbol{\nu}_k \boldsymbol{\mu}_k\right]}{\left(s - \lambda_k\right)} \mathbf{B} \tag{10.24}$$

Therefore we can see that the residues in this partial-fraction expansion (which corresponds to the system's impulse response) are explicit functions of the left and right eigenvectors. Furthermore, looking at specific elements (transfer functions) in **TF**, or $g_{i,j}(s)$, where i indicates the specific response and j the specific control input, we have the partial-fraction expansion of transfer function $g_{i,j}(s)$ given by

$$g_{i,j}(s) = \sum_{k=1}^{n} \frac{\left(\mathbf{c}_i \boldsymbol{\nu}_k\right)\left(\boldsymbol{\mu}_k \mathbf{b}_j\right)}{\left(s - \lambda_k\right)} \tag{10.25}$$

Here, $\mathbf{c}_i$ is the i'th row of **C**, while $\mathbf{b}_j$ is the j'th column of **B**. Further, note that the two terms in the parentheses in the numerators (residues) are both scalar quantities. From Equation (10.25) we can now observe that the right eigenvectors $\boldsymbol{\nu}_k$ determine how the k'th modal response affects the i'th physical response (y_i), while the left eigenvectors $\boldsymbol{\mu}_j$ determine how the j'th control input affects that same k'th modal response. If any residue is zero, for example, either that mode is not excited by the j'th control input (the mode is called *uncontrollable*), or that mode does not contribute to the i'th physical response (the mode is called *unobservable*).

These residues may also be obtained by expanding the transfer function in a Taylor series about a given pole at $s = \lambda_k$. If the eigenvalues are unique, as assumed, then the k'th residue R_k is found from the following expression.

$$R_k = \left(\left(s - \lambda_k\right) g_{i,j}(s)\right)\big|_{s=\lambda_k} \tag{10.26}$$

From this expression we see that the residues depend on both the numerator polynomial of the transfer function as well as its characteristic polynomial. Hence eigenvectors and transfer-function numerators of a system must be related. This relationship is left as an exercise to the reader.

It is important to note here that if the transfer function $g_{i,j}(s)$ includes a *pole-zero cancellation* at $s = \lambda_k$, that is, it contains a zero with value equal to that of the pole in question, or λ_k, the residue $R_k = 0$! Therefore, the k'th mode will not contribute to the impulse response. Such a pole-zero cancellation indicates that the corresponding mode is either unobservable or uncontrollable. Likewise, if there is an approximate pole-zero cancellation, the residue will be small, and the contribution of the k'th mode to the impulse response will be small.

Complex eigenvalues or poles will always occur in complex pairs, as noted earlier, and the associated residues will also be complex pairs. This is necessary for the system's responses to be real. For example, consider a second-order transfer function with partial-fraction expansion given by

$$g(s) = \frac{R}{s-\lambda} + \frac{\overline{R}}{s-\overline{\lambda}} \tag{10.27}$$

Here, the two poles are complex pairs, with $\lambda = \sigma + j\omega$, and the overbar notation indicates a complex conjugate. Inverse Laplace transforming Equation (10.27) the system's impulse response will be

$$y(t) = Re^{\lambda t} + \overline{R}e^{\overline{\lambda}t} = Re^{(\sigma + j\omega)t} + \overline{R}e^{(\sigma - j\omega)t} = e^{\sigma t}\left(Re^{j\omega t} + \overline{R}e^{-j\omega t}\right) \tag{10.28}$$

Now let the residue $R = re^{j\phi}$, so we have

$$y(t) = re^{\sigma t}\left(e^{j\phi}e^{j\omega t} + e^{-j\phi}e^{-j\omega t}\right) = re^{\sigma t}\left(e^{j(\omega t+\phi)} + e^{-j(\omega t+\phi)}\right) \tag{10.29}$$

But from the following identity

$$e^{j\theta} \triangleq \cos\theta + j\sin\theta \tag{10.30}$$

we see that the response may be written as

$$y(t) = 2re^{\sigma t}\cos\left(\omega t + \phi\right) \tag{10.31}$$

And this response is real, as required.

10.1.3 Polynomial-Matrix System Descriptions

Based on what we've learned in Section 10.1.2, we see that the transfer-function matrix contains a complete description of the system, just as with the state-variable description. A third system description also containing all the information about the system is the *polynomial-matrix description.* With the system given in polynomial-matrix format, it is rather straightforward to analytically find the system's transfer functions, or the elements of the transfer-function matrix. This approach is frequently preferred instead of using Equation (10.17), which involves a matrix inversion.

A linear dynamic system in polynomial-matrix format is given as

$$\mathbf{P}(s)\mathbf{y}(s) = \mathbf{Q}(s)\mathbf{u}(s) \tag{10.32}$$

where $\mathbf{P}(s)$ $(p \times p)$ and $\mathbf{Q}(s)$ $(p \times m)$ are matrices of polynomials in s, and $\mathbf{y}(s)$ and $\mathbf{u}(s)$ are again the transforms of the response and control-input vectors, respectively. Note that the form of Equation (10.32) is that of a system of linear simultaneous equations, and we will find that Cramer's rule (see Appendix D) is useful here. Clearly, the system's transfer-function matrix $\mathbf{TF}(s)$ is

$$\mathbf{TF}(s) = \mathbf{P}^{-1}(s)\mathbf{Q}(s)$$

and the characteristic polynomial of the system is det $\mathbf{P}(s)$.

The *Routh-Hurwitz stability criterion* may be applied directly to the coefficients of this characteristic polynomial, avoiding the need to find the roots to assess stability. This criterion will be stated here for a second-, third-, and fourth-order characteristic polynomial, as these are the most frequently encountered in flight dynamics. First, for any characteristic polynomial, a necessary condition for stability is that all the coefficients must be positive. If this condition is not satisfied, the system is unstable. Second, if this condition is satisfied, then necessary and sufficient conditions for stability may be applied, as listed in Table 10.1.

Finally, Cramer's Rule (Appendix D) is especially useful in finding the individual transfer functions. This approach involves finding the determinants of polynomial matrices. For example, consider a system with two responses and two inputs with polynomial-matrix description written as

$$\begin{bmatrix} p_{1,1}(s) & p_{1,2}(s) \\ p_{2,1}(s) & p_{2,2}(s) \end{bmatrix} \begin{Bmatrix} y_1(s) \\ y_2(s) \end{Bmatrix} = \begin{bmatrix} q_{1,1}(s) & q_{1,2}(s) \\ q_{2,1}(s) & q_{2,2}(s) \end{bmatrix} \begin{Bmatrix} u_1(s) \\ u_2(s) \end{Bmatrix}$$
$$\mathbf{P}(s)\mathbf{y}(s) = \mathbf{Q}(s)\mathbf{u}(s) \tag{10.33}$$

And recall that $p_{i,j}$ and $q_{i,j}$ are each polynomials in s. Then the transfer-function matrix of the system is given by

$$\mathbf{TF}(s) = \begin{bmatrix} g_{1,1}(s) & g_{1,2}(s) \\ g_{2,1}(s) & g_{2,2}(s) \end{bmatrix} \tag{10.34}$$

where

$$g_{i,j}(s) = \frac{\det \mathbf{N}_{i,j}(s)}{\det \mathbf{P}(s)} \tag{10.35}$$

Table 10.1 Routh-Hurwitz Stability Criteria

Characteristic Polynomial	Stability Criteria
$As^2 + Bs + C$	A, B, C all > 0
$As^3 + Bs^2 + Cs + D$	A, B, C, D all > 0 plus $(BC - AD) > 0$
$As^4 + Bs^3 + Cs^2 + Ds + E$	A, B, C, D, E all > 0 plus $(BC - AD) > 0$ and $\left(\left(\frac{BC - AD}{B}\right)D - BE\right) > 0$

and $\mathbf{N}_{i,j}$ is a polynomial matrix constructed from $\mathbf{P}$ by replacing its i'th column with the j'th column of $\mathbf{Q}$. So we see that rather than taking a matrix inverse, we only need to calculate matrix determinants. We also see that the transfer-function zeros depend on elements of both $\mathbf{P}$ and $\mathbf{Q}$, while the system poles depend on the elements of $\mathbf{P}$ only.

EXAMPLE 10.1

Analysis of the Simple Pendulum

Consider again the simple pendulum introduced in Chapter 1, and assume that a torque T about the pivot may be applied to the pendulum. Let the equation of motion for such a pendulum be given as

$$\ddot{\theta} + 4\theta = T$$

And let the state-variable description for the pendulum be given by

$$\begin{Bmatrix} \dot{\theta} \\ \ddot{\theta} \end{Bmatrix} = \begin{bmatrix} 0 & 1 \\ -4 & 0 \end{bmatrix} \begin{Bmatrix} \theta \\ \dot{\theta} \end{Bmatrix} + \begin{bmatrix} 0 \\ 1 \end{bmatrix} T$$

Here, $\mathbf{D} = \mathbf{0}$ and let

$$\mathbf{C} = \mathbf{I}_2$$

Find the system's transfer function from the equation of motion, plot its Bode plot, perform a modal analysis, and find the partial-fraction expansion of the system's transfer function using two methods.

■ Solution

Taking the Laplace transform of the equation of motion, and setting the initial conditions to zero, we have

$$\left(s^2 + 4\right)\theta(s) = T(s)$$

(Note that this is the system's polynomial-matrix description.) So the single transfer function is

$$\frac{\theta(s)}{T(s)} = g(s) = \frac{1}{\left(s^2 + 4\right)}$$

The two system poles are

$$s = \pm j2 \text{ /sec}$$

indicating the pendulum's unforced ($T = 0$) response will consist of undamped oscillation at 2 rad/sec.

Using MATLAB, and including a finite damping ratio ζ of 0.001 for purposes of plotting the Bode plot, we have

```
» a=[0 1;-4 -0.004];b=[0;1];c=[1 0];d=0;
» sys=ss(a,b,c,d);
» bode(sys)
```

This Bode plot is shown in Figure 10.4.

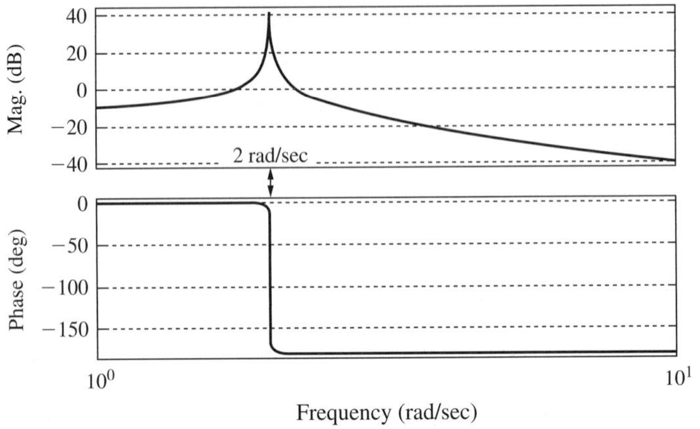

Figure 10.4 Bode plot for the simple pendulum.

Note that since the damping ratio is (almost) zero here there is a high narrow spike in the magnitude plot and a sharp change in phase of 180 deg, both at the corner frequency of 2 rad/sec.

The partial-fraction expansion of the system's transfer function (or impulse response) is

$$g(s) = \frac{1}{(s^2+4)} = \frac{R}{s+j2} + \frac{\bar{R}}{s-j2}$$

Using the residue rule (Equation (10.26)) we have

$$R = \left((s+j2)g(s)\right)|_{s=-j2} = \frac{(s+j2)}{(s+j2)(s-j2)}|_{s=-j2} = \frac{1}{-j4} = 1/4j$$

$$\bar{R} = \left((s-j2)g(s)\right)|_{s=j2} = \frac{(s-j2)}{(s+j2)(s-j2)}|_{s=j2} = \frac{1}{j4} = -1/4j$$

So this is the first method of finding the residues.

Given that for this system

$$\mathbf{A} = \begin{bmatrix} 0 & 1 \\ -4 & 0 \end{bmatrix}$$

the eigenvalues and eigenvectors of **A** are (from MATLAB)

```
»[M,D] = eig(a)

M =                                   The two right eigenvectors are the columns of M

    4.4721e-01          4.4721e-01
    0 + 8.9443e-01i     0 - 8.9443e-01i

D =                                   The two eigenvalues are on the diagonal of D

    0 + 2.0000e+00i     0
    0                   0 - 2.0000e+00i
```

```
»Minv=inv(M)

Minv =
```

The two left eigenvectors are the rows of M^{-1}

1.1180e+00	0 - 5.5902e-01i
1.1180e+00	0 + 5.5902e-01i

So from MATLAB, the eigenvalues are $\lambda_1 = j2$ and $\lambda_2 = -j2$, the right eigenvectors of **A** are

$$\boldsymbol{\nu}_1 = \begin{Bmatrix} 0.4472 \\ 0 + j0.8944 \end{Bmatrix} = \begin{Bmatrix} 0.4472e^{j0} \\ 0.8944e^{j\pi/2} \end{Bmatrix} \text{ and } \boldsymbol{\nu}_2 = \begin{Bmatrix} 0.4472 \\ 0 - j0.8944 \end{Bmatrix} = \begin{Bmatrix} 0.4472e^{j0} \\ 0.8944e^{-j\pi/2} \end{Bmatrix}$$

and the left eigenvectors are

$$\boldsymbol{\mu}_1 = \begin{bmatrix} 1.118 & 0 - j0.559 \end{bmatrix} = \begin{bmatrix} 1.118e^{j0} & 0.559e^{-j\pi/2} \end{bmatrix}$$

$$\boldsymbol{\mu}_2 = \begin{bmatrix} 1.118 & 0 + j0.559 \end{bmatrix} = \begin{bmatrix} 1.118e^{j0} & 0.559e^{j\pi/2} \end{bmatrix}$$

We have here converted the complex numbers to polar form for ease of interpretation and plotting.

Note from the right eigenvectors and the **C** (= **I**) matrix that the modal response will contribute significantly to the responses of both the angular displacement and rate, not surprisingly. And from the left eigenvectors and **B**, we see that the external torque will excite the natural mode, again as expected. The phasor diagram of $\boldsymbol{\nu}_1$ is shown in Figure 10.5. Since both of the states (or responses) are angles, no relative scaling of the units of the states (or responses) is needed.

From the above modal analysis, and from Equation (10.25), we find that the residue for $\lambda_1 = j2$ is

$$\overline{R} = (\mathbf{c}_1\boldsymbol{\nu}_1)(\boldsymbol{\mu}_1\mathbf{B}) = \left(\begin{bmatrix} 1 & 0 \end{bmatrix}\begin{Bmatrix} 0.4472e^{j0} \\ 0.8944e^{j\pi/2} \end{Bmatrix}\right)\left(\begin{bmatrix} 1.118e^{j0} & 0.559e^{-j\pi/2} \end{bmatrix}\begin{Bmatrix} 0 \\ 1 \end{Bmatrix}\right)$$

$$= \left(0.4472e^{j0}\right)\left(0.559e^{-j\pi/2}\right) = 0.25e^{-j\pi/2} = -j0.25$$

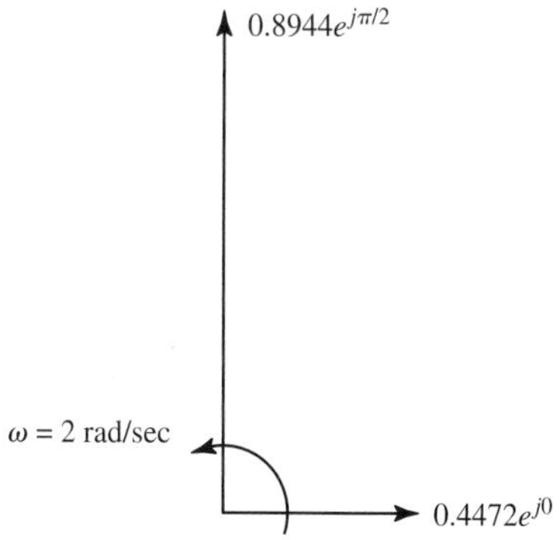

Figure 10.5 Phasor diagram for simple pendulum.

and the residue for $\lambda_2 = -j2$ is

$$R = (\mathbf{c}_1\mathbf{v}_2)(\boldsymbol{\mu}_2\mathbf{B}) = \left(\begin{bmatrix}1 & 0\end{bmatrix}\begin{Bmatrix}0.4472e^{j0}\\0.8944e^{-j\pi/2}\end{Bmatrix}\right)\left(\begin{bmatrix}1.118e^{j0} & 0.559e^{j\pi/2}\end{bmatrix}\begin{Bmatrix}0\\1\end{Bmatrix}\right)$$

$$= (0.4472e^{j0})(0.559e^{j\pi/2}) = 0.25e^{j\pi/2} = j0.25$$

Note that these residues agree with those found previously.

10.2 LINEAR FLIGHT-DYNAMICS PERTURBATION EQUATIONS

To begin our analyses of a vehicle's flight dynamics, we will start by assembling the linear equations governing the perturbation motion variables. These equations will be the same as those developed in Section 8.1.1 when we were preparing to discuss linear simulation of the vehicle's dynamics.

From Equations (8.1), the equations governing perturbations in the translational velocities are

$$\begin{aligned}
\dot{u} &= (\mathrm{V}_0 r + \mathrm{R}_0 v) - (\mathrm{Q}_0 w + \mathrm{W}_0 q) - g\cos\Theta_0\theta + (f_{A_X} + f_{P_X})/m\\
\mathrm{U}_0\dot{\beta} &= (\mathrm{P}_0 w + \mathrm{W}_0 p) - (\mathrm{R}_0 u + \mathrm{U}_0 r) + g(\cos\Theta_0\cos\Phi_0\phi - \sin\Theta_0\sin\Phi_0\theta) + (f_{A_Y} + f_{P_Y})/m\\
\mathrm{U}_0\dot{\alpha} &= (\mathrm{Q}_0 u + \mathrm{U}_0 q) - (\mathrm{P}_0 v + \mathrm{V}_0 p) - g(\cos\Theta_0\sin\Phi_0\phi + \sin\Theta_0\cos\Phi_0\theta) + (f_{A_Z} + f_{P_Z})/m
\end{aligned} \tag{10.36}$$

where we may use the approximations

$$\alpha \approx \frac{w}{\mathrm{U}_0} \text{ and } \beta \approx \frac{v}{\mathrm{U}_0} \tag{10.37}$$

in the above expressions. Recall that u, v, and w are the three components of the vehicle's perturbation translational-velocity vector, expressed in the selected vehicle-fixed axes. Or

$$\delta\mathbf{V}_V = u\mathbf{i}_V + v\mathbf{j}_V + w\mathbf{k}_V = \begin{bmatrix}u & v & w\end{bmatrix}\begin{Bmatrix}\mathbf{i}_V\\\mathbf{j}_V\\\mathbf{k}_V\end{Bmatrix} \tag{10.38}$$

From Equations (8.6), we have the equations governing perturbations in the rotational velocities, or

$$\dot{q} = \frac{1}{I_{yy}}\left(\left(I_{zz} - I_{xx}\right)\left(\mathrm{R}_0 p + \mathrm{P}_0 r\right) + 2I_{xz}\left(\mathrm{R}_0 r - \mathrm{P}_0 p\right) + \left(m_A + m_p\right)\right)$$

$$\begin{Bmatrix} \dot{p} \\ \dot{r} \end{Bmatrix} = \frac{1}{1 - \left(\dfrac{I_{xz}^2}{I_{xx}I_{zz}}\right)} \begin{bmatrix} 1 & \dfrac{I_{xz}}{I_{xx}} \\ \dfrac{I_{xz}}{I_{zz}} & 1 \end{bmatrix} \begin{Bmatrix} \dfrac{1}{I_{xx}}\left(I_{xz}\left(\mathrm{Q}_0 p + \mathrm{P}_0 q\right) + \left(I_{yy} - I_{zz}\right)\left(\mathrm{R}_0 q + \mathrm{Q}_0 r\right) + \left(l_A + l_P\right)\right) \\ \dfrac{1}{I_{zz}}\left(-I_{xz}\left(\mathrm{R}_0 q + \mathrm{Q}_0 r\right) + \left(I_{xx} - I_{yy}\right)\left(\mathrm{Q}_0 p + \mathrm{P}_0 q\right) + \left(n_A + n_P\right)\right) \end{Bmatrix} \tag{10.39}$$

Here p, q, and r are the three components of the vehicle's perturbation rotational-velocity vector expressed in the selected vehicle-fixed axes. Or

$$\delta\boldsymbol{\omega}_{V,I} = p\mathbf{i}_V + q\mathbf{j}_V + r\mathbf{k}_V = \begin{bmatrix} p & q & r \end{bmatrix} \begin{Bmatrix} \mathbf{i}_V \\ \mathbf{j}_V \\ \mathbf{k}_V \end{Bmatrix} \tag{10.40}$$

Finally, from Equations (8.7) and (2.53), the three kinematic equations relating perturbations in angular displacement to angular velocity are

$$\begin{aligned} \dot{\phi} &= p + \tan\Theta_0\left(\sin\Phi_0 q + \cos\Phi_0 r + \left(\mathrm{Q}_0\cos\Phi_0 - \mathrm{R}_0\sin\Phi_0\right)\phi\right) \\ &\quad + \left(\mathrm{Q}_0\sin\Phi_0 + \mathrm{R}_0\cos\Phi_0 + \dot{\Psi}_0\sin\Theta_0\tan\Theta_0\right)\theta \\ \dot{\theta} &= \cos\Phi_0 q - \sin\Phi_0 r - \dot{\Psi}_0\cos\Theta_0\phi \\ \dot{\psi} &= \dot{\Psi}_0\tan\Theta_0\theta + \left(\sin\Phi_0 q + \cos\Phi_0 r - \left(\mathrm{R}_0\sin\Phi_0 - \mathrm{Q}_0\cos\Phi_0\right)\phi\right)/\cos\Theta_0 \end{aligned} \tag{10.41}$$

Here ψ, θ, and ϕ, are perturbations in the 3-2-1 Euler angles defining the orientation of the vehicle-fixed frame with respect to the inertial frame. In our linear analysis in this chapter, the remaining three kinematic equations relating inertial position to translational velocity are of little interest if constant atmospheric density is assumed. If perturbations in altitude will be of interest, then from Equations (8.9) the following equation must be included in the model.

$$\begin{aligned} \dot{h} &= S_{\Theta_0}u - C_{\Theta_0}\left(S_{\Phi_0}v + C_{\Phi_0}w\right) - \left(\mathrm{V}_0 C_{\Phi_0} - \mathrm{W}_0 S_{\Phi_0}\right)C_{\Theta_0}\phi \\ &\quad + \left(\mathrm{U}_0 C_{\Theta_0} + \mathrm{V}_0 S_{\Phi_0}S_{\Theta_0} + \mathrm{W}_0 C_{\Phi_0}S_{\Theta_0}\right)\theta \end{aligned} \tag{10.42}$$

Note that in this expression, the following shorthand notation has been used: $S_{\bullet}$ = sine • and $C_{\bullet}$ = cosine •.

Now from Equations (8.16), the perturbation aerodynamic and propulsive forces appearing in Equations (10.36) may be expressed in terms of dimensional stability derivatives, or

$$
\begin{aligned}
(f_{A_X} + f_{P_X})/m &= (X_u + X_{P_u})(u + u_g) + X_\alpha(\alpha + \alpha_g) + X_{\dot{\alpha}}\dot{\alpha} + X_q q + X_{\delta_E}\delta_E + X_T\delta T \\
(f_{A_Y} + f_{P_Y})/m &= Y_\beta(\beta + \beta_g) + Y_p p + Y_r r + Y_{\delta_A}\delta_A + Y_{\delta_R}\delta_R \\
(f_{A_Z} + f_{P_Z})/m &= (Z_u + Z_{P_u})(u + u_g) + Z_\alpha(\alpha + \alpha_g) + Z_{\dot{\alpha}}\dot{\alpha} + Z_q q + Z_{\delta_E}\delta_E + Z_T\delta T
\end{aligned} \tag{10.43}
$$

The definitions of these dimensional derivatives, assuming stability axes are the selected vehicle-fixed axes, are made apparent by comparing Equations (8.16) with (8.15). Note also that u_g, α_g, and β_g, arising due to atmospheric turbulence, have been included in the expressions for these forces, consistent with the discussion in Section 8.1.7.

Likewise, from Equations (8.21) the expressions for the perturbations in the aerodynamic and propulsive moments appearing in Equations (10.39), expressed in terms of dimensional stability derivatives, are

$$
\begin{aligned}
(l_A + l_P)/I_{xx} &= L_\beta(\beta + \beta_g) + L_p p + L_r r + L_{\delta_A}\delta_A + L_{\delta_R}\delta_R \\
(m_A + m_P)/I_{yy} &= (M_u + M_{P_u})(u + u_g) + (M_\alpha + M_{P_\alpha})(\alpha + \alpha_g) \\
&\quad + M_{\dot{\alpha}}\dot{\alpha} + M_q q + M_{\delta_E}\delta_E + M_T\delta T \\
(n_A + n_P)/I_{zz} &= N_\beta(\beta + \beta_g) + N_p p + N_r r + N_{\delta_A}\delta_A + N_{\delta_R}\delta_R
\end{aligned} \tag{10.44}
$$

Again, the definitions of the dimensional derivatives, assuming stability axes are the selected vehicle-fixed axes, are made apparent by comparing Equations (8.21) with Equations (8.20). Plus, the gust terms u_g, α_g, and β_g have again been included.

Furthermore, from Equations (8.22), we may also write the expressions for the perturbations in rolling and yawing moments appearing in Equations (10.39) as

$$
\begin{aligned}
&\left(\frac{1}{1 - I_{xz}^2/(I_{xx}I_{zz})}\right)\begin{bmatrix} 1 & I_{xz}/I_{xx} \\ I_{xz}/I_{zz} & 1 \end{bmatrix}\begin{Bmatrix} (l_A + l_P)/I_{xx} \\ (n_A + n_P)/I_{zz} \end{Bmatrix} \\
&= \begin{Bmatrix} L'_\beta(\beta + \beta_g) + L'_p p + L'_r r + L'_{\delta_A}\delta_A + L'_{\delta_R}\delta_R \\ N'_\beta(\beta + \beta_g) + N'_p p + N'_r r + N'_{\delta_A}\delta_A + N'_{\delta_R}\delta_R \end{Bmatrix}
\end{aligned} \tag{10.45}
$$

where the primed dimensional derivatives are given by

$$
\begin{aligned}
L'_\beta &= \left(L_\beta + N_\beta I_{xz}/I_{xx}\right)D & N'_\beta &= \left(N_\beta + L_\beta I_{xz}/I_{zz}\right)D \\
L'_p &= \left(L_p + N_p I_{xz}/I_{xx}\right)D & N'_p &= \left(N_p + L_p I_{xz}/I_{zz}\right)D \\
L'_r &= \left(L_r + N_r I_{xz}/I_{xx}\right)D & N'_r &= \left(N_r + L_r I_{xz}/I_{zz}\right)D \\
L'_{\delta_A} &= \left(L_{\delta_A} + N_{\delta_A} I_{xz}/I_{xx}\right)D & N'_{\delta_A} &= \left(N_{\delta_A} + L_{\delta_A} I_{xz}/I_{zz}\right)D \\
L'_{\delta_R} &= \left(L_{\delta_R} + N_{\delta_R} I_{xz}/I_{xx}\right)D & N'_{\delta_R} &= \left(N_{\delta_R} + L_{\delta_R} I_{xz}/I_{zz}\right)D
\end{aligned}
\qquad (10.46)
$$

with

$$D = \frac{1}{1 - I_{xz}^2/\left(I_{xx}I_{zz}\right)}$$

NOTE TO STUDENT

In the development of Equations (10.36–10.46) (or Equations (8.1–8.22) in Chapter 8), the thrust was assumed to be symmetric with respect to the XZ plane of the vehicle, and thus to contribute no side force or rolling or yawing moments. In addition, a vehicle with control inputs consisting of elevator, aileron, and rudder deflections was assumed. If any of these assumptions are not valid for the vehicle being considered, the development in Chapter 8 should be reconsidered, and appropriate adjustments to the above linear dynamic model must be made.

Equations (10.36–10.46) then constitute the model for the dynamics of the perturbation variables, or the linear equations of motion, to be analyzed. Clearly, however, these linear equations depend on a selected reference flight condition, and frequently, straight and level flight is the only reference condition considered for linear analysis. But this can be misleading and sometimes dangerous, as we will see in Section 10.9 when discussing cases involving other reference flight conditions.

10.3 DECOUPLED LONGITUDINAL AND LATERAL-DIRECTIONAL LINEAR MODELS

For a reference flight condition involving rectilinear flight (which may in general include shallow climbs and dives) with bank angle $\Phi_0 = 0$, angular rates $P_0 = Q_0 = R_0 = 0$, and (assuming stability axes) $V_0 = W_0 = 0$, we showed in Chapter 8 that the equations of motion decouple into the longitudinal and the lateral directional sets. In such a case, from Equations (10.36), (10.39), (10.43), and (10.44), the longitudinal set of equations is

$$\dot{u} = -g\cos\Theta_0\theta + \left(X_u + X_{P_u}\right)\left(u + u_g\right) + X_\alpha\left(\alpha + \alpha_g\right) + X_{\dot{\alpha}}\dot{\alpha} + X_q q + X_{\delta_E}\delta_E + X_T\delta T$$
$$\mathrm{U}_0\dot{\alpha} = -g\sin\Theta_0\theta + \left(Z_u + Z_{P_u}\right)\left(u + u_g\right) + Z_\alpha\left(\alpha + \alpha_g\right) + Z_{\dot{\alpha}}\dot{\alpha} + \left(Z_q + \mathrm{U}_0\right)q + Z_{\delta_E}\delta_E + Z_T\delta T$$
$$\dot{q} = \left(M_u + M_{P_u}\right)\left(u + u_g\right) + \left(M_\alpha + M_{P_\alpha}\right)\left(\alpha + \alpha_g\right) + M_{\dot{\alpha}}\dot{\alpha} + M_q q + M_{\delta_E}\delta_E + M_T\delta T \quad (10.47)$$

plus, from Equations (10.41) and (10.42) we have

$$\dot{\theta} = q$$
$$\dot{h} = \sin\Theta_0 u - \cos\Theta_0 w + \mathrm{U}_0\cos\Theta_0\theta \quad (10.48)$$

From the same earlier equations plus Equations (10.45), the lateral-directional set of linear equations is given as

$$\mathrm{U}_0\dot{\beta} = g\cos\Theta_0\phi + Y_\beta\left(\beta + \beta_g\right) + Y_p p + \left(Y_r - \mathrm{U}_0\right)r + Y_{\delta_A}\delta_A + Y_{\delta_r}\delta_R$$
$$\dot{p} = L'_\beta\left(\beta + \beta_g\right) + L'_p p + L'_r r + L'_{\delta_A}\delta_A + L'_{\delta_R}\delta_R \quad (10.49)$$
$$\dot{r} = N'_\beta\left(\beta + \beta_g\right) + N'_p p + N'_r r + N'_{\delta_A}\delta_A + N'_{\delta_R}\delta_R$$

plus, from Equations (10.41)

$$\dot{\phi} = p + \tan\Theta_0 r$$
$$\dot{\psi} = r/\cos\Theta_0 \quad (10.50)$$

The perturbation accelerations at a particular (x,y,z) location on the vehicle are three auxiliary responses of frequent interest. From Equations (8.26), these local acceleration components for an equilibrium condition consisting of steady, rectilinear flight are given by

$$a_X(x,y,z) = \dot{u} + \mathrm{W}_0 q + z\dot{q} - y\dot{r}$$
$$a_Y(x,y,z) = \dot{v} + \mathrm{U}_0 r - W_0 p + x\dot{r} - z\dot{p} \quad (10.51)$$
$$a_Z(x,y,z) = \dot{w} - \mathrm{U}_0 q + y\dot{p} - x\dot{q}$$

Assuming the location of interest lies in the XZ plane of the vehicle, these components of acceleration may also be grouped with the longitudinal and lateral-directional sets of linear equations. In this case, assuming stability axes, the local accelerations grouped with the longitudinal set are

$$\boxed{\begin{aligned} a_X &= \dot{u} + z\dot{q} \\ a_Z &= \dot{w} - \mathrm{U}_0 q - x\dot{q} \end{aligned}} \tag{10.52}$$

while the local acceleration grouped with the lateral-directional set is

$$\boxed{a_Y = \dot{v} + \mathrm{U}_0 r + x\dot{r} - z\dot{p}} \tag{10.53}$$

Laplace transforming Equations (10.47), (10.48), and (10.52), ignoring the gust inputs for now, and assuming all initial conditions are zero, we have the longitudinal dynamics given by

$$\begin{aligned} su(s) &= -g\cos\Theta_0\theta(s) + \left(X_u + X_{P_u}\right)u(s) + X_\alpha\alpha(s) + X_{\dot{\alpha}}s\alpha(s) \\ &\quad + X_q q(s) + X_{\delta_E}\delta_E(s) + X_T\delta T(s) \\ \mathrm{U}_0 s\alpha(s) &= -g\sin\Theta_0\theta(s) + \left(Z_u + Z_{P_u}\right)u(s) + Z_\alpha\alpha(s) + Z_{\dot{\alpha}}s\alpha(s) \\ &\quad + \left(Z_q + \mathrm{U}_0\right)q(s) + Z_{\delta_E}\delta_E(s) + Z_T\delta T(s) \\ sq(s) &= \left(M_u + M_{P_u}\right)u(s) + \left(M_\alpha + M_{P_\alpha}\right)\alpha(s) + M_{\dot{\alpha}}s\alpha(s) + M_q q(s) \\ &\quad + M_{\delta_E}\delta_E(s) + M_T\delta T(s) \\ s\theta(s) &= q(s) \end{aligned} \tag{10.54}$$

plus, the three auxiliary equations given by

$$\begin{aligned} sh(s) &= \sin\Theta_0 u(s) - \cos\Theta_0 w(s) + \mathrm{U}_0\cos\Theta_0\theta(s) \\ a_X(s) &= su(s) + zsq(s) \\ a_Z(s) &= sw(s) - \mathrm{U}_0 q(s) - xsq(s) \end{aligned} \tag{10.55}$$

Using the last of Equations (10.54) to write pitch rate $q(s)$ in terms of $\theta(s)$, we may write Equations (10.54) in polynomial-matrix format as

$$\boxed{\begin{bmatrix} s-\left(X_u + X_{P_u}\right) & -\left(X_{\dot{\alpha}}s + X_\alpha\right) & -X_q s + g\cos\Theta_0 \\ -\left(Z_u + Z_{P_u}\right) & \left(\mathrm{U}_0 - Z_{\dot{\alpha}}\right)s - Z_\alpha & -\left(Z_q + \mathrm{U}_0\right)s + g\sin\Theta_0 \\ -\left(M_u + M_{P_u}\right) & -\left(M_{\dot{\alpha}}s + \left(M_\alpha + M_{P_\alpha}\right)\right) & s^2 - M_q s \end{bmatrix} \begin{Bmatrix} u(s) \\ \alpha(s) \\ \theta(s) \end{Bmatrix} = \begin{bmatrix} X_{\delta_E} & X_T \\ Z_{\delta_E} & Z_T \\ M_{\delta_E} & M_T \end{bmatrix} \begin{Bmatrix} \delta_E(s) \\ \delta T(s) \end{Bmatrix}} \tag{10.56}$$

which we see is in the required form of

$$\mathbf{P}(s)\mathbf{y}(s) = \mathbf{Q}(s)\mathbf{u}(s) \tag{10.57}$$

Note how the elements of $\mathbf{P}(s)$ and $\mathbf{Q}(s)$ each depend explicitly on the dimensional stability derivatives, which are determined by the vehicle aerodynamic characteristics, geometry, and mass properties. Since the system's characteristic polynomial is det $\mathbf{P}(s)$, we also see that this polynomial will be fourth-order in s written as

$$As^4 + Bs^3 + Cs^2 + Ds + E \tag{10.58}$$

where the five coefficients obviously depend on the dimensional stability derivatives in $\mathbf{P}(s)$. So the system has four characteristic roots. For the $\mathbf{P}(s)$ given above, the five coefficients of the characteristic polynomial are

$$\begin{aligned}
A &= U_0 - Z_{\dot{\alpha}} \\
B &= -(U_0 - Z_{\dot{\alpha}})\left((X_u + X_{P_u}) + M_q\right) - M_{\dot{\alpha}}(U_0 + Z_q) - Z_\alpha \\
C &= (X_u + X_{P_u})\left(M_q(U_0 - Z_{\dot{\alpha}}) + M_{\dot{\alpha}}(U_0 + Z_q) + Z_\alpha\right) + Z_\alpha M_q - (M_\alpha + M_{P_\alpha})(U_0 + Z_q) \\
&\quad - X_\alpha(Z_u + Z_{P_u}) + M_{\dot{\alpha}}\, g \sin\Theta_0 \\
D &= (X_u + X_{P_u})\left((M_\alpha + M_{P_\alpha})(U_0 + Z_q) - Z_\alpha M_q\right) + (Z_u + Z_{P_u})X_\alpha M_q - (M_u + M_{P_u})X_\alpha(U_0 + Z_q) \\
&\quad + g\cos\Theta_0\left((M_u + M_{P_u})(U_0 - Z_{\dot{\alpha}}) + M_{\dot{\alpha}}(Z_u + Z_{P_u})\right) + g\sin\Theta_0\left((M_\alpha + M_{P_\alpha}) - M_{\dot{\alpha}}(X_u + X_{P_u})\right) \\
E &= g\cos\Theta_0\left((Z_u + Z_{P_u})(M_\alpha + M_{P_\alpha}) - (M_u + M_{P_u})Z_\alpha\right) \\
&\quad + g\sin\Theta_0\left((M_u + M_{P_u})X_\alpha - (X_u + X_{P_u})(M_\alpha + M_{P_\alpha})\right)
\end{aligned} \tag{10.59}$$

In developing the coefficients A–E, it has been assumed that

$$X_{\dot{\alpha}} = X_q = 0 \tag{10.60}$$

which is frequently a good assumption. If this assumption is not valid, the coefficients must be adjusted accordingly.

We will now express the longitudinal dynamics in state-variable format. Following the same procedure used in Section 8.1.4, and again ignoring the gust inputs for now, we first rewrite the $\dot{\alpha}$ equation in Equations (10.47) as

$$\dot{\alpha} = \frac{1}{(U_0 - Z_{\dot{\alpha}})}\left(-g\sin\Theta_0\theta + (Z_u + Z_{P_u})u + Z_\alpha\alpha + (Z_q + U_0)q + Z_{\delta_E}\delta_E + Z_T\delta T\right) \tag{10.61}$$

and then use this result to eliminate the $\dot{\alpha}$ terms in the $\dot{u}$ and $\dot{q}$ equations in Equations (10.47). Finally, the equation governing altitude is

$$\dot{h} = \sin\Theta_0 u - U_0\cos\Theta_0\alpha + U_0\cos\Theta_0\theta \tag{10.62}$$

Now let the response, state, and control-input vectors be defined as

$$\begin{aligned} \mathbf{x}^T &= \begin{bmatrix} u & \alpha & \theta & q & h \end{bmatrix} \\ \mathbf{y}^T &= \begin{bmatrix} \mathbf{x}^T & a_X & a_Z \end{bmatrix} \\ \mathbf{u}^T &= \begin{bmatrix} \delta_E & \delta T \end{bmatrix} \end{aligned} \tag{10.63}$$

Then from Equations (10.47), (10.61), and (10.62) we have the four matrices in the state-variable description of the longitudinal dynamics, consistent with Equations (8.36), given by

$$\mathbf{A} = \begin{bmatrix} \left(X_u + X_{P_u} + \dfrac{X_{\dot{\alpha}}(Z_u + Z_{P_u})}{U_0 - Z_{\dot{\alpha}}}\right) & \left(X_\alpha + \dfrac{X_{\dot{\alpha}} Z_\alpha}{U_0 - Z_{\dot{\alpha}}}\right) & -g\cos\Theta_0 & \left(X_q + X_{\dot{\alpha}}\left(\dfrac{U_0 + Z_q}{U_0 - Z_{\dot{\alpha}}}\right)\right) & 0 \\ \left(\dfrac{Z_u + Z_{P_u}}{U_0 - Z_{\dot{\alpha}}}\right) & \left(\dfrac{Z_\alpha}{U_0 - Z_{\dot{\alpha}}}\right) & \left(\dfrac{-g\sin\Theta_0}{U_0 - Z_{\dot{\alpha}}}\right) & \left(\dfrac{U_0 + Z_q}{U_0 - Z_{\dot{\alpha}}}\right) & 0 \\ 0 & 0 & 0 & 1 & 0 \\ \left(M_u + M_{P_u} + \dfrac{M_{\dot{\alpha}}(Z_u + Z_{P_u})}{U_0 - Z_{\dot{\alpha}}}\right) & \left(M_\alpha + M_{P_\alpha} + \dfrac{M_{\dot{\alpha}} Z_\alpha}{U_0 - Z_{\dot{\alpha}}}\right) & 0 & \left(M_q + M_{\dot{\alpha}}\left(\dfrac{U_0 + Z_q}{U_0 - Z_{\dot{\alpha}}}\right)\right) & 0 \\ \sin\Theta_0 & -U_0\cos\Theta_0 & U_0\cos\Theta_0 & 0 & 0 \end{bmatrix}$$

(10.64)

$$\mathbf{B} = \begin{bmatrix} \left(X_{\delta_E} + \dfrac{X_{\dot{\alpha}} Z_{\delta_E}}{U_0 - Z_{\dot{\alpha}}}\right) & \left(X_T + \dfrac{X_{\dot{\alpha}} Z_T}{U_0 - Z_{\dot{\alpha}}}\right) \\ \left(\dfrac{Z_{\delta_E}}{U_0 - Z_{\dot{\alpha}}}\right) & \left(\dfrac{Z_T}{U_0 - Z_{\dot{\alpha}}}\right) \\ 0 & 0 \\ \left(M_{\delta_E} + \dfrac{M_{\dot{\alpha}} Z_{\delta_E}}{U_0 - Z_{\dot{\alpha}}}\right) & \left(M_T + \dfrac{M_{\dot{\alpha}} Z_T}{U_0 - Z_{\dot{\alpha}}}\right) \\ 0 & 0 \end{bmatrix}$$

$$\mathbf{C} = \begin{bmatrix} \mathbf{I}_5 \\ \mathbf{C}_a \end{bmatrix}, \quad \mathbf{D} = \begin{bmatrix} \mathbf{0}_{5\times 2} \\ \mathbf{D}_a \end{bmatrix}$$

Here $\mathbf{I}_5$ is a 5×5 identity matrix, $\mathbf{0}_{5\times 2}$ is a null 5×2 matrix, and $\mathbf{C}_a$ (2×5), and $\mathbf{D}_a$ (2×2) are matrices with elements corresponding to the coefficients in the acceleration equations, Equations (10.52). These accelerations are now given by

$$
\begin{aligned}
a_X = &\left(\left(X_u + X_{P_u} + \frac{X_{\dot\alpha}(Z_u + Z_{P_u})}{U_0 - Z_{\dot\alpha}}\right) + z\left(M_u + M_{P_u} + \frac{M_{\dot\alpha}(Z_u + Z_{P_u})}{U_0 - Z_{\dot\alpha}}\right)\right)u \\
&+ \left(\left(X_\alpha + \frac{X_{\dot\alpha} Z_\alpha}{U_0 - Z_{\dot\alpha}}\right) + z\left(M_\alpha + M_{P_\alpha} + \frac{M_{\dot\alpha} Z_\alpha}{U_0 - Z_{\dot\alpha}}\right)\right)\alpha \\
&+ \left(\left(X_q + X_{\dot\alpha}\left(\frac{U_0 + Z_q}{U_0 - Z_{\dot\alpha}}\right)\right) + z\left(M_q + M_{\dot\alpha}\left(\frac{U_0 + Z_q}{U_0 - Z_{\dot\alpha}}\right)\right)\right)q - g\theta \\
a_Z = &\left((Z_u + Z_{P_u}) - x\left(M_u + M_{P_u} + \frac{M_{\dot\alpha}(Z_u + Z_{P_u})}{U_0 - Z_{\dot\alpha}}\right)\right)u + \left(Z_\alpha - x\left(M_\alpha + M_{P_\alpha} + \frac{M_{\dot\alpha} Z_\alpha}{U_0 - Z_{\dot\alpha}}\right)\right)\alpha \\
&+ \left(Z_q - x\left(M_q + M_{\dot\alpha}\left(\frac{U_0 + Z_q}{U_0 - Z_{\dot\alpha}}\right)\right)\right)q + \left(Z_{\delta_E} - x\left(M_{\delta_E} + \frac{M_{\dot\alpha} Z_{\delta_E}}{U_0 - Z_{\dot\alpha}}\right)\right)\delta_E \\
&+ \left(Z_T - x\left(M_T + \frac{M_{\dot\alpha} Z_T}{U_0 - Z_{\dot\alpha}}\right)\right)\delta T
\end{aligned}
\tag{10.65}
$$

Note that the equation for a_Z assumes that

$$\frac{U_0}{U_0 - Z_{\dot\alpha}} = 1$$

The system matrix **A** in Equations (10.64) will have five eigenvalues, one at zero due to the inclusion of altitude h in the state vector. Under the constant-density assumption, the dynamics are independent of altitude (the last column of **A** is zero). So if altitude h is not a response of interest, it need not be included in the response or state vectors **y** or **x**. In which case the last row and column of **A** and the last row of **B** may be removed, **C** and **D** adjusted accordingly, and the order of the system reduced to four.

Turning now to the lateral-directional equations of motion, Laplace transforming Equations (10.49), setting $\phi(s) = \frac{p(s)}{s} + \tan\Theta_0 \frac{r(s)}{s}$, ignoring gust inputs for now, and writing the results in polynomial-matrix format yields

$$
\begin{bmatrix} U_0 s - Y_\beta & -Y_p s - g\cos\Theta_0 & (U_0 - Y_r)s - g\sin\Theta_0 \\ -L'_\beta & s^2 - L'_p s & -L'_r s \\ -N'_\beta & -N'_p s & s^2 - N'_r s \end{bmatrix}
\begin{Bmatrix} \beta(s) \\ p(s)/s \\ r(s)/s \end{Bmatrix}
=
\begin{bmatrix} Y_{\delta_A} & Y_{\delta_R} \\ L'_{\delta_A} & L'_{\delta_R} \\ N'_{\delta_A} & N'_{\delta_R} \end{bmatrix}
\begin{Bmatrix} \delta_A(s) \\ \delta_R(s) \end{Bmatrix}
\tag{10.66a}
$$

Or if $\Theta_0 = 0$, then $\phi(s) = \dfrac{p(s)}{s}$, and we can also write this polynomial-matrix description as

$$\begin{bmatrix} U_0 s - Y_\beta & -Y_p s - g & (U_0 - Y_r) \\ -L'_\beta & s^2 - L'_p s & -L'_r \\ -N'_\beta & -N'_p s & s - N'_r \end{bmatrix} \begin{Bmatrix} \beta(s) \\ \phi(s) \\ r(s) \end{Bmatrix} = \begin{bmatrix} Y_{\delta_A} & Y_{\delta_R} \\ L'_{\delta_A} & L'_{\delta_R} \\ N'_{\delta_A} & N'_{\delta_R} \end{bmatrix} \begin{Bmatrix} \delta_A(s) \\ \delta_R(s) \end{Bmatrix} \tag{10.66b}$$

Both these descriptions are again of the form $\mathbf{P}(s)\mathbf{y}(s) = \mathbf{Q}(s)\mathbf{u}(s)$. Also, from Equations (10.49) and (10.53), the kinematic and auxiliary equations are

$$\begin{aligned} s\phi(s) &= p(s) + \tan\Theta_0 r(s) \\ s\psi(s) &= r(s)/\cos\Theta_0 \\ a_Y(s) &= U_0 s\beta(s) + U_0 r(s) + xsr(s) - zsp(s) \end{aligned} \tag{10.67}$$

The above two sets of equations (Equations (10.66) and (10.67)) complete the polynomial-matrix description of the lateral-directional dynamics.

We see from Equation (10.66*a*) that the characteristic polynomial for this system (det $\mathbf{P}(s)$) is fifth order, so there will be five system poles or eigenvalues. But for level flight (i.e., $\Theta_0 = 0$), Equation (10.66*b*) is valid, and the characteristic polynomial in this case is fourth order. In other words, in level flight ($\Theta_0 = 0$) one pole and one zero in each transfer function will always be at the origin. So in level flight the characteristic polynomial may be written as

$$As^4 + Bs^3 + Cs^2 + Ds + E \tag{10.68}$$

The five A–E coefficients obviously depend on the dimensional stability derivatives in $\mathbf{P}(s)$. And these five coefficients in the characteristic polynomial are

$$\begin{aligned} A &= U_0 \\ B &= -Y_\beta - U_0(L'_p + N'_r) \\ C &= (U_0 - Y_r)N'_\beta - Y_p L'_\beta + U_0(L'_p N'_r - N'_p L'_r) + Y_\beta(N'_r + L'_p) \\ D &= (U_0 - Y_r)(L'_\beta N'_p - N'_\beta L'_p) + Y_\beta(N'_p L'_r - L'_p N'_r) + Y_p(L'_\beta N'_r - N'_\beta L'_r) - gL'_\beta \\ E &= g(L'_\beta N'_r - N'_\beta L'_r) \end{aligned} \tag{10.69}$$

The assembly of the state-variable description of the lateral-directional dynamics simply involves writing Equations (10.66) and (10.67) in state-variable format. A similar system description was given in Equations (8.41). Using the same approach here, we can take the response vector **y**, state vector **x**, and input vector **u** to be

$$\mathbf{x}^T = \begin{bmatrix}\beta & \phi & p & r & \psi\end{bmatrix}$$
$$\mathbf{y}^T = \begin{bmatrix}\mathbf{x}^T & a_Y\end{bmatrix} \tag{10.70}$$
$$\mathbf{u}^T = \begin{bmatrix}\delta_A & \delta_R\end{bmatrix}$$

Then from Equations (10.66) and (10.67), the state-variable description for the lateral-directional dynamics in level flight may be written as

$$\mathbf{A} = \begin{bmatrix} \frac{Y_\beta}{U_0} & \frac{g}{U_0} & \frac{Y_p}{U_0} & \left(\frac{Y_r}{U_0} - 1\right) & 0 \\ 0 & 0 & 1 & 0 & 0 \\ L'_\beta & 0 & L'_p & L'_r & 0 \\ N'_\beta & 0 & N'_p & N'_r & 0 \\ 0 & 0 & 0 & 1 & 0 \end{bmatrix}, \mathbf{B} = \begin{bmatrix} \frac{Y_{\delta_A}}{U_0} & \frac{Y_{\delta_R}}{U_0} \\ 0 & 0 \\ L'_{\delta_A} & L'_{\delta_R} \\ N'_{\delta_A} & N'_{\delta_R} \\ 0 & 0 \end{bmatrix}$$
$$\mathbf{C} = \begin{bmatrix} \mathbf{I}_5 \\ \mathbf{C}_a \end{bmatrix}, \text{ and } \mathbf{D} = \begin{bmatrix} \mathbf{0}_{5\times 2} \\ \mathbf{D}_a \end{bmatrix} \tag{10.71}$$

Here $\mathbf{I}_5$ is a 5 × 5 identity matrix, $\mathbf{0}_{5\times 2}$ is a null 5 × 2 matrix, and $\mathbf{C}_a(1 \times 5)$, and $\mathbf{D}_a(1 \times 2)$ are row matrices with elements corresponding to the coefficients in the acceleration equation, from Equations (10.53).

Note that as with the longitudinal equations, **A** here will have five eigenvalues, one at zero due to the inclusion of heading ψ in the state vector. But as with altitude, the dynamics are independent of heading (the last column of **A** above is zero). So if heading ψ is not a response of interest, it need not be included in the response or state vectors **y** or **x**. In such a case the last row and column of **A** and the last row of **B** may be removed, **C** and **D** adjusted accordingly, and the order of the system reduced to four.

10.4 LONGITUDINAL TRANSFER FUNCTIONS AND MODAL ANALYSIS

We will now develop the vehicle's longitudinal transfer functions with control inputs, and investigate the longitudinal natural modes of motion. (The treatment of the dynamic response from gust inputs is addressed in Problems 10.7

and 10.8.) Applying Cramer's rule to Equation (10.56), and letting $\bullet = \delta_E$ or δT, we have the six longitudinal transfer functions given by

$$\frac{u(s)}{\bullet(s)} = \det\begin{bmatrix} X_\bullet & -(X_{\dot{\alpha}}s + X_\alpha) & -X_q s + g\cos\Theta_0 \\ Z_\bullet & (U_0 - Z_{\dot{\alpha}})s - Z_\alpha & -(Z_q + U_0)s + g\sin\Theta_0 \\ M_\bullet & -\left(M_{\dot{\alpha}}s + \left(M_\alpha + M_{P_\alpha}\right)\right) & s^2 - M_q s \end{bmatrix} / \det\mathbf{P}(s),\ \bullet = \delta_E \text{ or } \delta T \tag{10.72}$$

$$\frac{\alpha(s)}{\bullet(s)} = \det\begin{bmatrix} s - \left(X_u + X_{P_u}\right) & X_\bullet & -X_q s + g\cos\Theta_0 \\ -\left(Z_u + Z_{P_u}\right) & Z_\bullet & -(Z_q + U_0)s + g\sin\Theta_0 \\ -\left(M_u + M_{P_u}\right) & M_\bullet & s^2 - M_q s \end{bmatrix} / \det\mathbf{P}(s),\ \bullet = \delta_E \text{ or } \delta T \tag{10.73}$$

and

$$\frac{\theta(s)}{\bullet(s)} = \det\begin{bmatrix} s - \left(X_u + X_{P_u}\right) & -(X_{\dot{\alpha}}s + X_\alpha) & X_\bullet \\ -\left(Z_u + Z_{P_u}\right) & (U_0 - Z_{\dot{\alpha}})s - Z_\alpha & Z_\bullet \\ -\left(M_u + M_{P_u}\right) & -\left(M_{\dot{\alpha}}s + \left(M_\alpha + M_{P_\alpha}\right)\right) & M_\bullet \end{bmatrix} / \det\mathbf{P}(s),\ \bullet = \delta_E \text{ or } \delta T$$

$$\frac{q(s)}{\bullet(s)} = \frac{s\theta(s)}{\bullet(s)} \tag{10.74}$$

We noted in the previous section that the characteristic polynomial det $\mathbf{P}(s)$ is fourth order, or

$$\det\mathbf{P}(s) = As^4 + Bs^3 + Cs^2 + Ds + E \tag{10.75}$$

with the five coefficients given in Equations (10.59). Hence there will be four system poles.

Note that the surge-velocity u and angle-of-attack α transfer functions given by Equations (10.72) and (10.73) have third-order numerator polynomials; hence they will each have three zeros. The pitch-attitude θ transfer functions given by Equation (10.74) have second-order numerator polynomials, and hence will have two zeros.

The longitudinal transfer functions for elevator input typically take on standard forms, which are given by

$$
\begin{aligned}
\frac{u(s)}{\delta_E(s)} &= \frac{K_\delta^u\left(s + 1/T_u\right)\left(s^2 + 2\zeta_u\omega_u + \omega_u^2\right)}{\left(s^2 + 2\zeta_P\omega_P + \omega_P^2\right)\left(s^2 + 2\zeta_{SP}\omega_{SP} + \omega_{SP}^2\right)} \\
\frac{\alpha(s)}{\delta_E(s)} &= \frac{K_\delta^\alpha\left(s + 1/T_\alpha\right)\left(s^2 + 2\zeta_\alpha\omega_\alpha + \omega_\alpha^2\right)}{\left(s^2 + 2\zeta_P\omega_P + \omega_P^2\right)\left(s^2 + 2\zeta_{SP}\omega_{SP} + \omega_{SP}^2\right)} \\
\frac{\theta(s)}{\delta_E(s)} &= \frac{K_\delta^\theta\left(s + 1/T_{\theta_1}\right)\left(s + 1/T_{\theta_2}\right)}{\left(s^2 + 2\zeta_P\omega_P + \omega_P^2\right)\left(s^2 + 2\zeta_{SP}\omega_{SP} + \omega_{SP}^2\right)} \\
\frac{\gamma(s)}{\delta_E(s)} &= \frac{K_\delta^\gamma\left(s + 1/T_{\gamma_1}\right)\left(s + 1/T_{\gamma_2}\right)\left(s + 1/T_{\gamma_3}\right)}{\left(s^2 + 2\zeta_P\omega_P + \omega_P^2\right)\left(s^2 + 2\zeta_{SP}\omega_{SP} + \omega_{SP}^2\right)}
\end{aligned}
\qquad (10.76)
$$

That is, the fourth-order characteristic polynomial usually factors into two quadratic terms, the surge-velocity and angle-of-attack numerators typically factor into a quadratic and a first-order term, while the pitch-attitude numerator typically factors into two first-order terms. The common notation used for all the roots of these factors is defined in Equations (10.76).

Note, however, that these transfer functions do not always take on the above forms. For example, the phugoid quadratic in the characteristic polynomial sometimes becomes two first-order factors. Also, for a statically unstable aircraft the short-period quadratic frequently becomes two first-order factors with one root being positive.

Finally, the transfer functions governing the altitude and accelerations may be found from the transfer functions given in Equations (10.72)–(10.74) and the following relationships, obtained using Equations (10.55).

$$
\begin{aligned}
\frac{h(s)}{\bullet(s)} &= \frac{\sin\Theta_0}{s}\frac{u(s)}{\bullet(s)} + \frac{U_0\cos\Theta_0}{s}\left(\frac{\theta(s)}{\bullet(s)} - \frac{\alpha(s)}{\bullet(s)}\right) \\
\frac{a_X(s)}{\bullet(s)} &= \frac{su(s)}{\bullet(s)} + zs^2\frac{\theta(s)}{\bullet(s)} \qquad \bullet = \delta_E \text{ or } \delta T \\
\frac{a_Z(s)}{\bullet(s)} &= U_0 s\frac{\alpha(s)}{\bullet(s)} - xs^2 + U_0 s\frac{\theta(s)}{\bullet(s)}
\end{aligned}
\qquad (10.77)
$$

At this point, it is difficult to say much more in general about the vehicle's longitudinal dynamics. For example, the characteristic polynomial and all the numerator polynomials are rather involved functions of the dimensional stability derivatives. So finding the poles and zeros usually requires numerical techniques. However, after we perform some additional analysis in the case study in Example 10.2, we will be able to make some simplifying approximations that will give us additional insight into the characteristics of the longitudinal dynamics of flight vehicles.

EXAMPLE 10.2

A Case Study—Longitudinal Dynamics of a Conventional Aircraft

Let us now determine the transfer functions and relevant Bode plots for a conventional aircraft's longitudinal dynamics, and perform a modal analysis. We will consider the Navion aircraft, shown in the sketch in Figure 10.6, in steady, level ($\Theta_0 = \gamma_0 = 0$) flight at sea level, as discussed in Example 8.2 and simulated in Example 8.5. The reference flight velocity U_0 is 176 fps, the aircraft weighs 2750 lbs, and has a wing span of 33.4 ft.

The state, response, and control-input vectors are taken to be

$$\mathbf{x}^T = \mathbf{y}^T = \left[u(\text{fps}) \quad \alpha(\text{rad}) \quad \theta(\text{rad}) \quad q(\text{rad/sec}) \right]$$

$$\mathbf{u}^T = \left[\delta_E(\text{rad}) \quad \delta T(\text{lb}) \right]$$

So from Equations (10.64) and Appendix B we have the state-variable description of the dynamics given by

$$\mathbf{A} = \begin{bmatrix} -0.0451 & 6.348 & -32.2 & 0 \\ -0.0021 & -2.0244 & 0 & 1 \\ 0 & 0 & 0 & 1 \\ 0.0021 & -6.958 & 0 & -3.0757 \end{bmatrix}, \quad \mathbf{B} = \begin{bmatrix} 0 & 0.0117 \\ -0.160 & 0 \\ 0 & 0 \\ -11.029 & 0 \end{bmatrix} \tag{10.78}$$

$$\mathbf{C} = \mathbf{I}_4, \text{ and } \mathbf{D} = \mathbf{0}$$

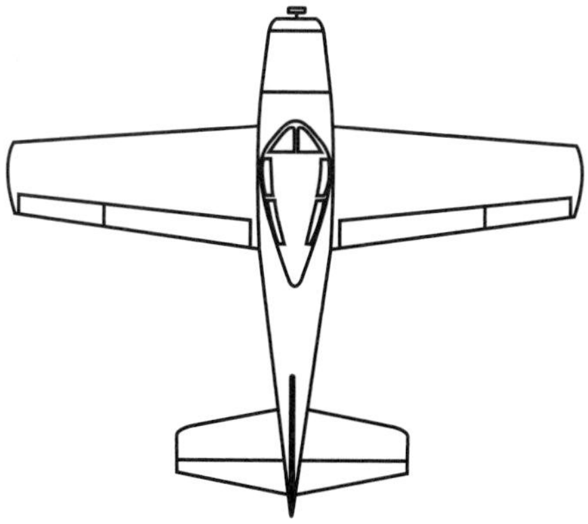

Figure 10.6 Sketch of the Navion general-aviation aircraft.

Before proceeding further, we will perform a unit conversion that converts the units of all angular quantities from radians to degrees. This conversion will make all four physical responses roughly equivalent in terms of engineering significance. For example, at the given flight condition one degree of angle of attack corresponds to approximately three feet per second plunge velocity *w*. So the plunge velocity and the corresponding angle of attack are of the same order of magnitude. On the other hand, one radian of angle of attack would correspond to approximately 150 fps plunge velocity, so with the original units, velocity and the angles would not be roughly equivalent. This balancing is necessary for the proper interpretation of the phasor diagrams of the eigenvectors.[1]

So now define the diagonal unit-conversion matrix **U** such that

$$\mathbf{y}_{\text{New}} = \mathbf{x}_{\text{New}} = \mathbf{U}\mathbf{y}_{\text{Old}} = \mathbf{U}\mathbf{x}_{\text{Old}} \tag{10.79}$$

where the subscripts "Old' and "New" refer to the units used for the responses and states. Since we are converting radians to degrees we have

$$\mathbf{U} = \begin{bmatrix} 1 & 0 & 0 & 0 \\ 0 & 57.3 & 0 & 0 \\ 0 & 0 & 57.3 & 0 \\ 0 & 0 & 0 & 57.3 \end{bmatrix} \tag{10.80}$$

The new state-variable system description now becomes

$$\begin{aligned} \dot{\mathbf{x}}_{\text{New}} &= \mathbf{U}\mathbf{A}\mathbf{U}^{-1}\mathbf{x}_{\text{New}} + \mathbf{U}\mathbf{B}\mathbf{u} = \mathbf{A}_{\text{New}}\mathbf{x}_{\text{New}} + \mathbf{B}_{\text{New}}\mathbf{u} \\ \mathbf{y}_{\text{New}} &= \mathbf{U}\mathbf{C}\mathbf{U}^{-1}\mathbf{x}_{\text{New}} + \mathbf{U}\mathbf{D}\mathbf{u} = \mathbf{I}_4\mathbf{x}_{\text{New}} \end{aligned} \tag{10.81}$$

From MATLAB, $\mathbf{A}_{\text{New}}$ is

$$\mathbf{A}_{\text{New}} = \mathbf{U}\mathbf{A}\mathbf{U}^{-1} = \begin{bmatrix} -0.0451 & 0.1109 & -0.5620 & 0 \\ -0.1203 & -2.0244 & 0 & 1 \\ 0 & 0 & 0 & 1 \\ 0.1203 & -6.958 & 0 & -3.0757 \end{bmatrix} \tag{10.82}$$

while $\mathbf{B}_{\text{New}}$ is

$$\mathbf{B}_{\text{New}} = \mathbf{B}$$

if the elevator deflection is also converted to degrees (verify this for yourself).

The eigenvalues of **A** (and of $\mathbf{A}_{\text{New}}$) are

$$\begin{aligned} \lambda_{1,2} &= -2.5554 \pm j2.5838 \text{ /sec} \\ \lambda_{3,4} &= -0.01722 \pm j0.2138 \text{ /sec} \end{aligned} \tag{10.83}$$

So the modal responses consists of two stable oscillatory modes, one lightly damped at low frequency (0.2138 rad/sec), and one reasonably damped with a higher frequency (2.5838 rad/sec). The corresponding right eigenvectors of $\mathbf{A}_{\text{New}}$ are

---

[1] Thanks to Professor Art Bryson of Stanford University for providing insight regarding this point.

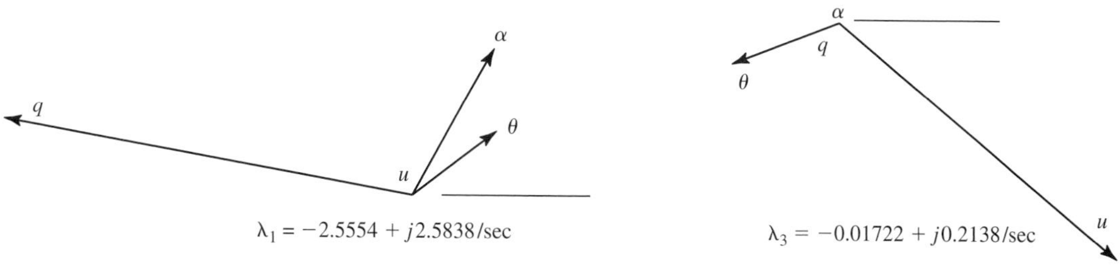

Figure 10.7 Phasor diagrams of the Navion's longitudinal mode shapes.

$$\boldsymbol{\nu}_{1,2} = \begin{Bmatrix} 0.00839 \pm j0.02940 \\ 0.10328 \pm j0.32746 \\ 0.19302 \pm j0.15739 \\ -0.89990 \pm j0.09652 \end{Bmatrix}, \text{ or } \boldsymbol{\nu}_1 = \begin{Bmatrix} 0.0306e^{j74.1^\circ} \\ 0.3436e^{j72.5^\circ} \\ 0.2491e^{j39.2^\circ} \\ 0.9051e^{j173.9^\circ} \end{Bmatrix}$$
$$\boldsymbol{\nu}_{3,4} = \begin{Bmatrix} 0.59340 \mp j0.71685 \\ -0.01157 \pm j0.01340 \\ -0.30439 \mp j0.18750 \\ 0.04532 \mp j0.06183 \end{Bmatrix}, \text{ or } \boldsymbol{\nu}_3 = \begin{Bmatrix} 0.9306e^{-j50.4^\circ} \\ 0.0177e^{j130.8^\circ} \\ 0.3575e^{-j148.4^\circ} \\ 0.0767e^{-j53.8^\circ} \end{Bmatrix} \tag{10.84}$$

And sketches of the phasor diagrams of $\boldsymbol{\nu}_1$ and $\boldsymbol{\nu}_3$ are shown in Figure 10.7.

We see from the first eigenvector that the corresponding higher-frequency mode contributes little to the surge-velocity response $u(t)$ (with a component too small to plot), indicating that this modal response occurs at almost constant velocity. So this mode might be described as a higher-frequency, well-damped mode involving oscillations of pitch attitude and rate and angle of attack. Or it primarily involves pitching oscillations at almost constant velocity. Such a mode is called a *short-period mode* of motion.

Conversely, from the second eigenvector we see that the corresponding lower-frequency mode contributes primarily to the surge-velocity and pitch-attitude responses, $u(t)$ and $\theta(t)$. It contributes negligibly to the angle-of-attack response (with a component too small to plot). That is, this mode's response occurs at almost constant angle of attack. (Note that there is little pitch rate in the modal response as well. But since there is significant pitch attitude in the response, the small pitch-rate is primarily due to the fact that the modal frequency is low.)

Therefore, this mode might be described as a lower-frequency, lightly damped mode involving oscillations of speed and attitude at an approximately constant angle of attack. Such a model is called a *phugoid mode* of motion. If the vehicle's response consisted of purely phugoid motion, we see from the eigenvector that the oscillation in speed would lead the oscillations in attitude by approximately 100 degrees. So as the aircraft speeds up, it begins to pitch up, which in turn slows the vehicle, and the vehicle will then pitch down, and so on, as on a roller coaster. Basically, the vehicle is exchanging kinetic for potential energy, and vice versa, at almost constant angle of attack.

An examination of the left eigenvectors of the Navion's longitudinal dynamics would reveal that the elevator excites both the phugoid and the short-period modes, while thrust primarily excites only the phugoid mode.

To complete the analysis, we will determine the six longitudinal transfer functions for the Navion. Using the original state-variable description and the zpk command in MATLAB, the transfer functions for elevator input are

$$
\begin{aligned}
\frac{u(s)}{\delta_E(s)} &= \frac{-1.0166(s+2.402)(s-279.7)}{(s^2+0.03444s+0.04598)(s^s+5.111s+13.21)}\ \text{fps/rad}\\
\frac{\alpha(s)}{\delta_E(s)} &= \frac{-0.16(s+72.01)(s^2+0.04419s+0.06567)}{(s^2+0.03444s+0.04598)(s^s+5.111s+13.21)}\ \text{rad/rad or deg/deg}\\
\frac{\theta(s)}{\delta_E(s)} &= \frac{-11.029(s+0.05233)(s+1.916)}{(s^2+0.03444s+0.04598)(s^s+5.111s+13.21)}\ \text{rad/rad or deg/deg}
\end{aligned}
\tag{10.85}
$$

Comparing these to the standard forms given in Equations (10.76), the two quadratic terms in the characteristic polynomial are evident, with the first associated with the phugoid mode ($\omega_n = 0.21$ rad/sec) and the second associated with the short-period mode ($\omega_n = 3.63$ rad/sec). (We know this based on the eigen-analysis just performed.) Also note the values of the factors in the numerators when compared to the standard forms. Here, for example,

$$1/T_{\theta_1} = 0.05233 \text{ /sec},$$

corresponding to the smaller of the two pitch-attitude transfer-function zeros, while

$$1/T_{\theta_2} = 1.916 \text{ /sec}$$

Finally, note that the phugoid quadratic is nearly cancelled by a similar quadratic in the numerator of the angle-of-attack transfer function. This is consistent with the results from the modal analysis that indicated that the phugoid mode contains little angle-of-attack response, or is almost unobservable in this response.

A plot of the poles and zeros of the $\theta(s)/\delta_E(s)$ transfer function is shown in Figure 10.8. Such a plot helps in the interpretation of the transfer functions. Note that all six transfer functions will have the same pole locations, but each transfer function has its unique set of zeros. A pole-zero plot of the $\alpha(s)\delta_E(s)$ transfer function, for example, would reveal the near pole-zero cancellation of the phugoid mode mentioned above.

The phugoid poles appear close to the origin in the complex plane, while the short-period poles are much farther from the origin. One can prove that the distance from the origin to the pole equals the undamped natural frequency ω_n of the corresponding mode. Also, the cosine of the angle between the negative real axis and the line from the origin to the pole, as shown in Figure 10.8, equals the damping ratio for that mode.

The Bode plots of the angle-of-attack and pitch-attitude transfer functions for negative elevator deflection are shown in Figure 10.9. Negative deflection simply changes the phase of both Bode plots by 180 deg, for ease of plotting. The peak in the magnitude for pitch attitude is due to the low phugoid modal damping, and both magnitudes become smaller ("roll off") above the short-period undamped natural frequency (corner frequency). The fact that the phugoid quadratic factor is nearly cancelled by a similar numerator factor in the angle-of-attack transfer function is evident in that Bode plot. Note also that consistent

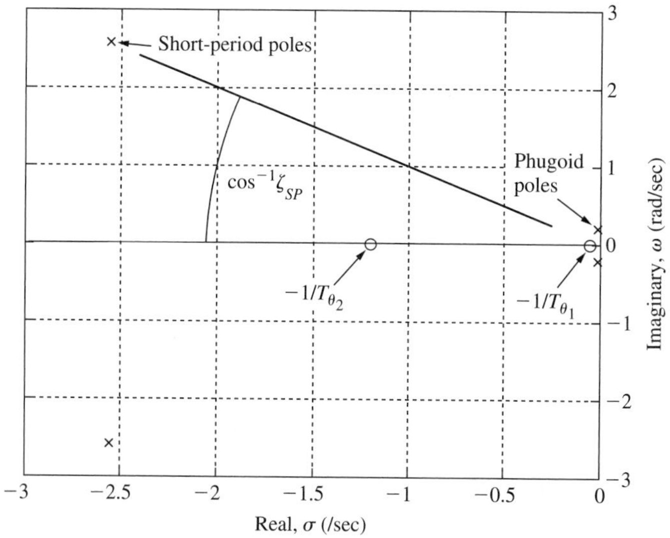

Figure 10.8 Pole-zero plot of Navion θ/δ_E transfer function.

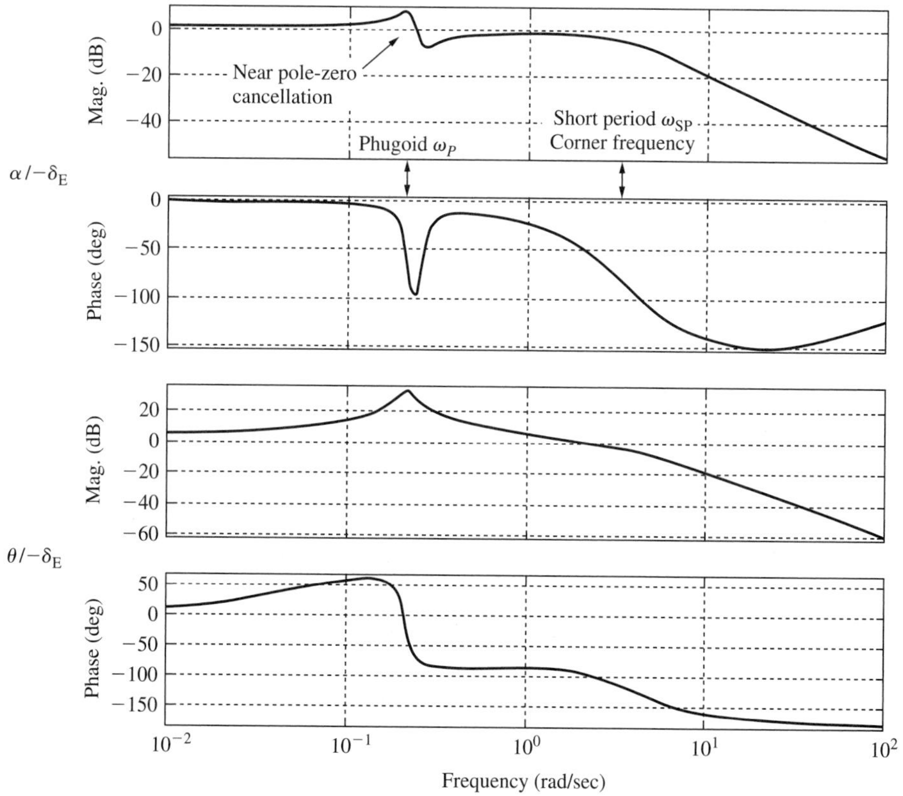

Figure 10.9 Bode plots for angle of attack and pitch attitude with elevator input—Navion aircraft.

with the previous eigen-analysis, the elevator excites both the phugoid and short-period modes, as demonstrated by the facts that the phugoid mode is quite evident in the pitch-attitude Bode plot, and the short-period mode is quite evident in both Bode plots.

Also from MATLAB, the Navion's surge-velocity transfer function for thrust input is

$$\frac{u(s)}{\delta T(s)} = \frac{0.0117s\left(s^2 + 5.1s + 13.18\right)}{\left(s^2 + 0.03444s + 0.04598\right)\left(s^s + 5.111s + 13.21\right)}\ \text{fps/lb} \tag{10.86}$$

This is the only transfer function of significance with thrust input for this aircraft. Since the thrust passes through the *c.g.*, $Z_{\delta T}$ and $M_{\delta T}$ are both zero, and the angle-of-attack and pitch-attitude transfer functions have leading numerator constants equal to zero. Hence these two transfer functions are zero.

Looking closely at the above transfer function we see that the short-period quadratic factor in the characteristic polynomial is almost exactly cancelled by a similar quadratic in the numerator. This cancellation is consistent with the findings in the eigen-analysis that indicated that the thrust primarily excites only the phugoid mode, and that this mode contributes significantly to the surge-velocity response. All these observations are confirmed as well by considering the transfer function's Bode plot shown in Figure 10.10. Only the presence of the lightly damped phugoid quadratic in the characteristic polynomial and the zero at the origin of the complex plane are evident in this plot.

Finally, we found the vehicle's time responses to an elevator doublet in Example 8.5. But from MATLAB's step command, for example, we can also easily determine the vehicle's step responses to elevator and thrust inputs. The step responses are shown in Figure 10.11 for a step elevator input of negative one degree. The response during the first one to two seconds is almost entirely associated with the higher-frequency, well-damped short-period mode, while the low-frequency responses most evident in the surge-velocity and attitude responses are associated with the phugoid mode. The short-period mode is also seen to contribute little to the surge-velocity response. (The dashed lines indicate the steady-state responses.)

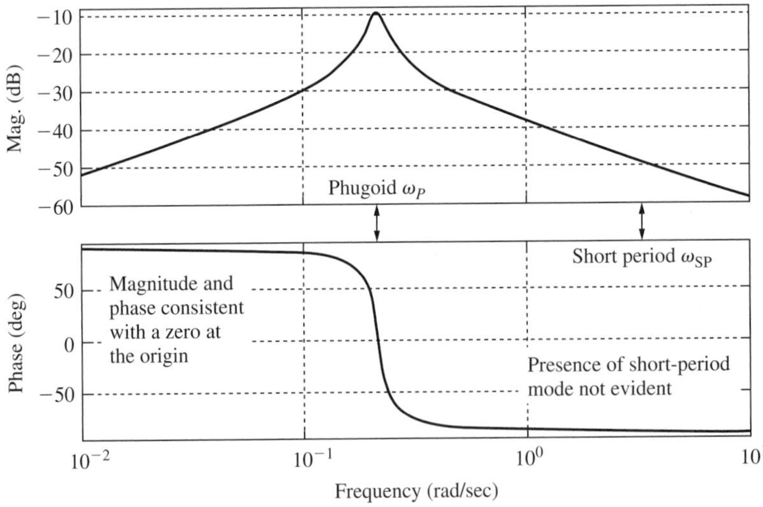

Figure 10.10 Bode plot for surge velocity with thrust input—Navion aircraft.

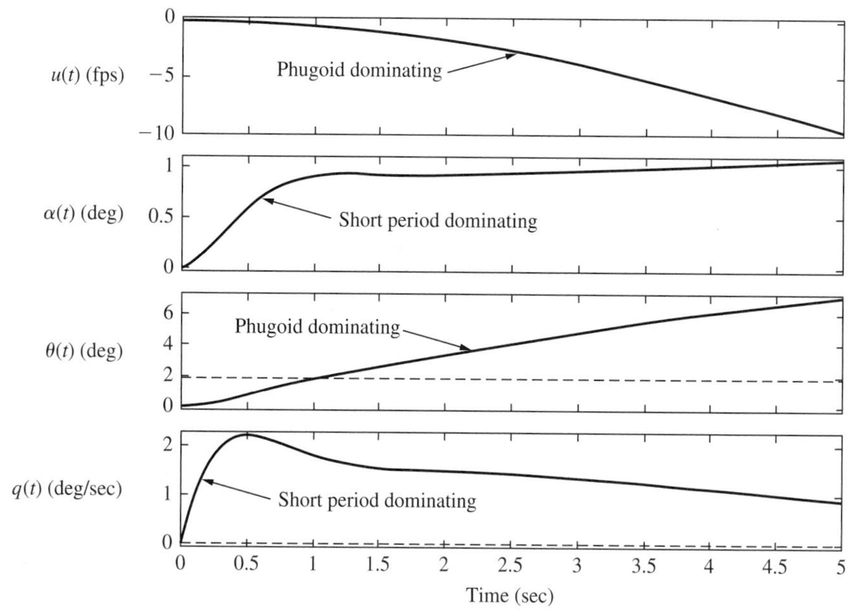

Figure 10.11 Longitudinal elevator (−1 deg) step responses of the Navion aircraft.

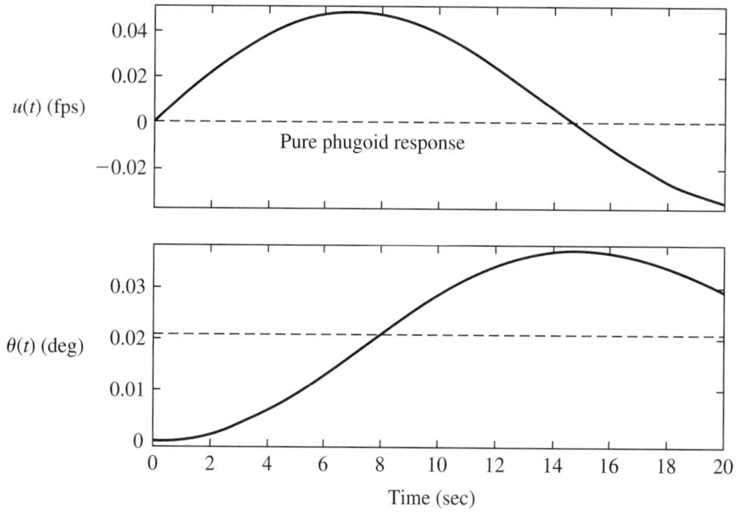

Figure 10.12 Longitudinal (1 lb) step thrust responses of the Navion aircraft.

The step responses of surge velocity and pitch attitude are shown in Figure 10.12 for a one pound thrust input. The other responses (i.e, angle of attack and pitch rate) are essentially zero. We see that the two responses shown almost entirely correspond to the lower-frequency, lightly damped phugoid mode. Also note that the surge-velocity response leads the attitude response by approximately 90 degrees, consistent with the phasor diagram for the phugoid mode. The dolphin-like or "roller-coaster" type of response is evident.

Virtually all conventional, statically stable aircraft have longitudinal modes of motion similar to those just described for the Navion. (We will verify this fact in Problems 10.3 and 10.5.) Because the oscillatory frequency is low for the phugoid mode, it is easily controlled by the pilot's control of attitude and/or altitude. However, an unstable phugoid would definitely be a nuisance, causing the speed and/or altitude to drift off the desired values if the vehicle is unattended for a short while. The short-period mode, however, is a different matter. Since it is a much higher frequency mode, it is difficult for the pilot to stabilize if it were unstable. And handling-qualities specifications (e.g., Ref. 1) have been developed to provide guidance in the vehicle design. Such specifications define the required dynamic characteristics of the vehicle, including modal frequencies and damping, for example, for different classes of vehicle and different phases of flight.

But all aircraft do not naturally possess the classical longitudinal modal characteristics described above. This is very important to remember. For example, sometimes the two phugoid eigenvalues are real and one may be slightly unstable. And when the airframe does not exhibit modal characteristics that provide acceptable handling characteristics, either the vehicle configuration must be modified or feedback augmentation must be added to achieve the desired characteristics. In Example 10.3 we will perform a modal analysis on a very different type of flight vehicle and examine its longitudinal modes of motion. Plus in Chapter 11 we will discuss the topic of feedback augmentation.

EXAMPLE 10.3

A Case Study—Modal Analysis of the Longitudinal Dynamics of a Hypersonic Vehicle

We will consider here the longitudinal dynamics of a vehicle very different from the Navion aircraft. This vehicle, first considered in Problem 6.3, is similar in concept to NASA's hypersonic X-43 HyperX research aircraft, shown in Figure 10.13. An airbreathing SCRAMjet engine is mounted under the vehicle, aerodynamic lift is generated by the fore-body lifting surface, and pitch control is provided via aft horizontal aerodynamic surfaces. The notional study vehicle analyzed here is 150 ft in length, weighs 80,000 lbs, and has a pitch inertia $I_{yy} = 5.0 \times 10^6$ sl-ft^2. The flight condition is Mach = 8 at 80,000 ft altitude.

The state-variable model, using data from Refs. 2 or 3 or Appendix B, is defined for the following state, response, and control-input vectors, respectively.

$$\mathbf{x}^T = \mathbf{y}^T = \left[u(\text{fps}), \alpha(\text{rad}), \theta(\text{rad}), q(\text{rad/sec})\right], \quad \mathbf{u} = \delta_H(\text{rad})$$

At Mach 8, the units of Mach and radians are of roughly equal engineering significance, so unit scaling was performed. After scaling, the state-variable model becomes

$$\mathbf{A} = \begin{bmatrix} -0.001936 & 0.02502 & -0.03317 & 0.000635 \\ -0.002028 & -0.06303 & 0 & 1 \\ 0 & 0 & 0 & 1 \\ 0.3287 & 11.023 & 0 & -0.0816 \end{bmatrix}, \ \mathbf{B} = \begin{bmatrix} -0.00058 \\ -0.00276 \\ 0 \\ -0.47936 \end{bmatrix}$$

$$\mathbf{C} = \mathbf{I}_4, \ \mathbf{D} = \mathbf{0}_{4\times 1}$$

Figure 10.13 NASA X-43A hypersonic research vehicle. *(Photo courtesy of NASA Dryden Flight Research Center, 1999).*

Using the MATLAB command

```
[m,d]=eig(A)
```

the four eigenvalues of **A** are found to be

$$\lambda_{1,2} = -0.000848 \pm j0.002048 \text{ /sec}$$
$$\lambda_3 = 3.253 \text{ /sec}$$
$$\lambda_4 = -3.398 \text{ /sec}$$

and the associated right eigenvectors are

$$\boldsymbol{\nu}_{1,2} = \begin{Bmatrix} 0.93504 \mp j0.34353 \\ -0.027877 \pm j0.010241 \\ -0.072908 \mp j0.038729 \\ 0.000141 \mp j0.000116 \end{Bmatrix} = \begin{Bmatrix} 0.99615e^{\mp j20.2^\circ} \\ 0.02970e^{\pm j159.8^\circ} \\ 0.08256e^{\mp j152.0^\circ} \\ 0.00018e^{\mp j39.5^\circ} \end{Bmatrix}$$

$$\boldsymbol{\nu}_3 = \begin{Bmatrix} -0.00056 \\ 0.27777 \\ 0.28230 \\ 0.91823 \end{Bmatrix}, \text{ and } \boldsymbol{\nu}_4 = \begin{Bmatrix} 0.00078 \\ 0.27724 \\ 0.27129 \\ -0.92170 \end{Bmatrix}$$

The phasor diagram for the first eigenvector $\boldsymbol{\nu}_1$ is shown in Figure 10.14. Based on this phasor diagram and the associated eigenvalue, the first mode would appear to be

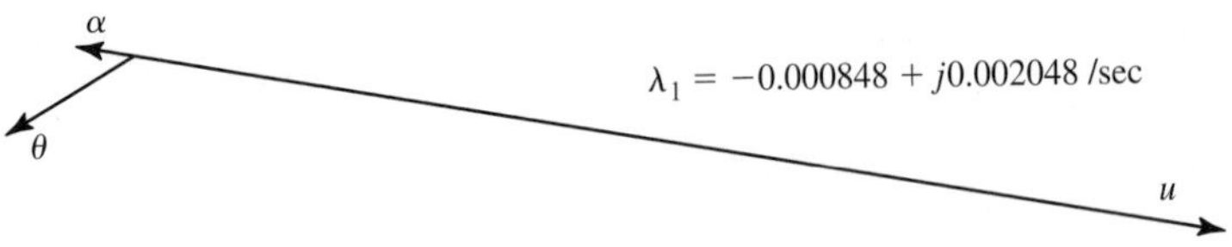

Figure 10.14 First right eigenvector—hypersonic vehicle.

a rather conventional phugoid mode. It has a low oscillatory frequency, is lightly damped, and consists almost entirely of surge velocity and pitch-attitude oscillations. Furthermore, the surge-velocity leads the attitude by approximately 130 degrees. Therefore, this mode may be characterized in terms of the lightly damped oscillatory "roller-coaster" motion associated with a phugoid mode. One surprising thing about this mode, however, is its extremely low frequency (0.002 rad/sec compared to 0.2 rad/sec for the Navion). We will provide an explanation for this result when discussing the phugoid approximation in Section 10.5.2.

Although this fairly conventional phugoid mode is present, there is no oscillatory short-period mode. It has been replaced by two nonoscillatory modes, and one of them is highly unstable. From the eigenvectors we see that these two latter modes consist of almost entirely pitch rate, with virtually no surge velocity component. So these two latter modes are similar in mode shape to a conventional short-period mode, except they are non-oscillatory and one of them is unstable. We will address this result further in Section 10.5.1.

10.5 APPROXIMATE MODELS FOR AIRCRAFT LONGITUDINAL DYNAMICS

The knowledge of the characteristics of an aircraft's natural longitudinal modes based on an eigen-analysis suggests some reasonable approximations that will allow us to make some modeling simplifications. But for a particular aircraft, an eigen-analysis is necessary to confirm that assumptions made in the approximations are valid.

NOTE TO STUDENT

The astute reader might wonder about the benefit of such modeling simplifications, given the fact that using numerical-analysis routines like MATLAB allows us to analyze the full-order model. This is a good question. It is certainly true that given a numerical model for the perturbation dynamics of a vehicle, MATLAB can be used to perform the linear analysis. However, remember that in addition to analyzing a given vehicle, we also seek to gain an understanding of the design factors (geometry, aerodynamic characteristics, etc.) that determine a vehicle's dynamic behavior. In fact, this may be the most important insight gained through the analysis of the perturbation dynamics. We will gain this insight from the approximate models developed in this section and in Section 10.6. These approximate models are widely used in practice.

10.5.1 The Short-Period Approximation

Let's first consider the short-period mode, which for conventional aircraft involves pitch oscillations at almost constant speed, or $u(t) \approx 0$. This suggests that the surge-velocity degree of freedom participates little in the short-period modal response, and hence might be eliminated from the model.

If we return now to the polynomial-matrix description of the longitudinal dynamics, or Equations (10.56), and eliminate the surge-velocity degree of freedom (the *short-period approximation*), we have a two-degree-of-freedom system given by

$$\begin{bmatrix} (\mathrm{U}_0 - Z_{\dot{\alpha}})s - Z_\alpha & -(Z_q + \mathrm{U}_0)s + g\sin\Theta_0 \\ -\left(M_{\dot{\alpha}}s + \left(M_\alpha + M_{P_\alpha}\right)\right) & s^2 - M_q s \end{bmatrix} \begin{Bmatrix} \alpha(s) \\ \theta(s) \end{Bmatrix} = \begin{bmatrix} Z_{\delta_E} & Z_T \\ M_{\delta_E} & M_T \end{bmatrix} \begin{Bmatrix} \delta_E(s) \\ \delta T(s) \end{Bmatrix} \quad (10.87)$$

The above expression is, of course, in the familiar form of

$$\mathbf{P}(s)\mathbf{y}(s) = \mathbf{Q}(s)\mathbf{u}(s)$$

and the system's characteristic polynomial is given by

$$\det \mathbf{P}(s) = \det \begin{bmatrix} (\mathrm{U}_0 - Z_{\dot{\alpha}})s - Z_\alpha & -(Z_q + \mathrm{U}_0)s + g\sin\Theta_0 \\ -\left(M_{\dot{\alpha}}s + \left(M_\alpha + M_{P_\alpha}\right)\right) & s^2 - M_q s \end{bmatrix}$$

$$= \left(s^2 - M_q s\right)\left((\mathrm{U}_0 - Z_{\dot{\alpha}})s - Z_\alpha\right) - \left(M_{\dot{\alpha}}s + \left(M_\alpha + M_{P_\alpha}\right)\right)\left((\mathrm{U}_0 + Z_q)s - g\sin\Theta_0\right) \quad (10.88)$$

$$= (\mathrm{U}_0 - Z_{\dot{\alpha}})s^3 - \left(Z_\alpha + M_{\dot{\alpha}}(\mathrm{U}_0 + Z_q) + M_q(\mathrm{U}_0 - Z_{\dot{\alpha}})\right)s^2$$
$$+ \left(M_q Z_\alpha + M_{\dot{\alpha}} g\sin\Theta_0 - (M_\alpha + M_{P_\alpha})(\mathrm{U}_0 + Z_q)\right)s + (M_\alpha + M_{P_\alpha})g\sin\Theta_0$$

This is a third-order characteristic equation, in which the phugoid quadratic factor does not appear.

For level flight $(\Theta_0 = \gamma_0 = 0)$, one of the poles in the above characteristic polynomial equals zero, and this pole is left over from the two absent phugoid characteristic roots that lie close to the origin in the complex plane. The remaining second-order quadratic approximating the short-period factor in the longitudinal characteristic equation is seen to be

$$s^2 - \left(\frac{Z_\alpha}{(\mathrm{U}_0 - Z_{\dot{\alpha}})} + M_{\dot{\alpha}}\frac{(\mathrm{U}_0 + Z_q)}{(\mathrm{U}_0 - Z_{\dot{\alpha}})} + M_q\right)s + \left(\frac{M_q Z_\alpha}{(\mathrm{U}_0 - Z_{\dot{\alpha}})} - (M_\alpha + M_{P_\alpha})\frac{(\mathrm{U}_0 + Z_q)}{(\mathrm{U}_0 - Z_{\dot{\alpha}})}\right) \quad (10.89)$$

which is of the form

$$s^2 + 2\zeta_{\mathrm{SP}}\omega_{\mathrm{SP}}s + \omega_{\mathrm{SP}}^2$$

Comparing these two quadratic polynomials, we see that the coefficient of s is significantly affected by the pitch-damping dimensional derivatives M_q and $M_{\dot{\alpha}}$, since

$$2\zeta_{SP}\omega_{SP} = -\left(\frac{Z_\alpha}{(U_0 - Z_{\dot{\alpha}})} + M_{\dot{\alpha}}\frac{(U_0 + Z_q)}{(U_0 - Z_{\dot{\alpha}})} + M_q\right) \approx -\left(\frac{Z_\alpha}{U_0} + M_{\dot{\alpha}} + M_q\right) \tag{10.90}$$

On the other hand, the last coefficient in the quadratic, or the square of the undamped short-period frequency, is significantly affected by the pitch-stiffness dimensional derivative. Or

$$\omega_{SP}^2 = \frac{M_q Z_\alpha}{(U_0 - Z_{\dot{\alpha}})} - (M_\alpha + M_{P_\alpha})\frac{(U_0 + Z_q)}{(U_0 - Z_{\dot{\alpha}})} \approx M_q\frac{Z_\alpha}{U_0} - (M_\alpha + M_{P_\alpha}) \tag{10.91}$$

For dynamic stability, both of these terms (Equations (10.90) and (10.91)) must be positive, and the first is always positive. Equation (10.91) will be positive if

$$(M_\alpha + M_{P_\alpha}) < M_q\frac{Z_\alpha}{U_0} \tag{10.92}$$

and note that $M_q Z_\alpha$ is always positive. So from the short-period approximation we see the first connection between static and dynamic stability. By guaranteeing that Equation (10.92) is satisfied, static pitch stability (i.e., $(M_\alpha + M_{P_\alpha}) < 0$) assures dynamic short-period stability under the short-period approximation.

Furthermore, from the definitions of the dimensional stability derivatives, we can see that dynamic stability is closely tied to C_{M_α}, C_{M_q}, and $C_{M_{\dot{\alpha}}}$, I_{yy}, and q_∞. Since the Navion has positive pitch stiffness, its short-period eigenvalues were both stable. But since the hypersonic vehicle has negative pitch stiffness (it was very statically unstable), Equation (10.91) is negative, leading to two real short-period eigenvalues with one being unstable.

Solving now for the four transfer functions under the short-period approximation, and noting that det $\mathbf{P}(s)$ is given in Equation (10.88), the angle-of-attack transfer functions are

$$\begin{aligned}\frac{\alpha(s)}{\bullet(s)} &= \left(Z_\bullet(s^2 - M_q s) + M_\bullet\left((Z_q + U_0)s + g\sin\Theta_0\right)\right)/\det\mathbf{P}(s) \qquad \bullet = \delta_E \text{ or } \delta T\\ &= \left(Z_\bullet s^2 + (M_\bullet(U_0 + Z_q) - Z_\bullet M_q)s + M_\bullet g\sin\Theta_0\right)/\det\mathbf{P}(s)\end{aligned} \tag{10.93}$$

Since $Z_{\delta T}$ and $M_{\delta T}$ are typically either small or zero, the magnitude of the numerator with thrust input will typically be very small or zero. The above numerator polynomial is second order, and for steady level ($\Theta_0 = \gamma_0 = 0$) flight it becomes

$$N_{\alpha,\bullet}(s) \approx Z_{\bullet}s^2 + \left(M_{\bullet}\left(\mathrm{U}_0 + Z_q\right) - Z_{\bullet}M_q\right)s \tag{10.94}$$

So the pole at the origin (of the complex plane) will be cancelled by a zero at the same location. By comparing the above numerator to the standard form given in Equations (10.76), we see that

$$1/T_\alpha \approx \left(-M_q + \left(\mathrm{U}_0 + Z_q\right)M_{\delta_E}/Z_{\delta_E}\right) \tag{10.95}$$

The pitch-attitude transfer functions under the short-period approximation are

$$\frac{\theta(s)}{\bullet(s)} = \left(M_{\bullet}\left(\left(\mathrm{U}_0 - Z_{\dot\alpha}\right)s - Z_\alpha\right) + Z_{\bullet}\left(M_{\dot\alpha}s + \left(M_\alpha + M_{P_\alpha}\right)\right)\right)/\det\mathbf{P}(s)$$

$$\bullet = \delta_E \text{ or } \delta T$$

$$= \left(\left(M_{\bullet}\left(\mathrm{U}_0 - Z_{\dot\alpha}\right) + Z_{\bullet}M_{\dot\alpha}\right)s + \left(Z_{\bullet}\left(M_\alpha + M_{P_\alpha}\right) - M_{\bullet}Z_\alpha\right)\right)/\det\mathbf{P}(s) \tag{10.96}$$

Of course, the pitch-rate transfer functions are just the above transfer functions multiplied by s. Again, since $M_{\delta T}$ and $Z_{\delta T}$ are frequently very small or zero, the pitch-attitude transfer function with thrust input is frequently small.

The above numerator polynomial is first order, and the single zero for elevator input yields an approximation for $1/T_{\theta_2}$, defined in Equations (10.76). The other standard numerator factor, or $1/T_{\theta_1}$, does not appear in the short-period approximation. Instead it appears in the phugoid approximation discussed in Section 10.5.2. So under this short-period approximation

$$1/T_{\theta_2} \approx \left(Z_{\delta_E}\left(M_\alpha + M_{P_\alpha}\right) - M_{\delta_E}Z_\alpha\right)/\left(M_{\delta_E}\left(\mathrm{U}_0 - Z_{\dot\alpha}\right) + Z_{\delta_E}M_{\dot\alpha}\right) \approx -Z_\alpha/\mathrm{U}_0 \tag{10.97}$$

This term is usually dominated by $-Z_\alpha/\mathrm{U}_0$, or the vehicle's lift effectiveness.

Of interest at this point is the validity of the short-period approximation. To address this issue we will perform an eigen-analysis of the Navion's longitudinal dynamics under the approximation, and compare the results to those from the complete longitudinal model. After eliminating the first row and column of the **A** matrix in the Navion's state-variable model (Equation (10.78) or (10.82)), as well at the first row of the **B** matrix (why these rows and column?), the state-variable model under the short-period approximation becomes

$$\mathbf{A}_{\mathrm{SP}} = \begin{bmatrix} -2.0244 & 0 & 1 \\ 0 & 0 & 1 \\ -6.958 & 0 & -3.0757 \end{bmatrix}, \quad \mathbf{B}_{\mathrm{SP}} = \begin{bmatrix} -0.160 & 0 \\ 0 & 0 \\ -11.029 & 0 \end{bmatrix} \tag{10.98}$$

$$\mathbf{C}_{\mathrm{SP}} = \mathbf{I}_3, \quad \mathbf{D}_{\mathrm{SP}} = \mathbf{0}_{3\times2}$$

The first thing to note is that under the short-period approximation, the Navion's longitudinal transfer functions for thrust input are all zero. That is, for the Navion at this flight condition the thrust neither contributes to the vertical (Z) force on the vehicle, nor contributes to the pitching moment on the vehicle.

From MATLAB, the three eigenvalues of $\mathbf{A}_{SP}$ are

$$\begin{aligned} \lambda_{SP} &= -2.5500 \pm j2.5849 \text{ /sec} \\ \lambda_3 &= 0 \end{aligned} \tag{10.99}$$

Comparing the above short-period eigenvalues with those from Example 10.2 ($\lambda_{1,2} = -2.554 \pm j2.5838$ /sec) shows very good agreement. The corresponding eigenvectors under the short-period approximation are

$$\boldsymbol{\nu}_{SP} = \begin{Bmatrix} 0.33640 \mp j0.06841 \\ 0.17753 \mp j0.17514 \\ 0 \pm j0.90552 \end{Bmatrix}, \text{ or } \boldsymbol{\nu}_{SP_1} = \begin{Bmatrix} 0.3433e^{-j11.5^\circ} \\ 0.2494e^{-j44.6^\circ} \\ 0.9055e^{j90^\circ} \end{Bmatrix}, \boldsymbol{\nu}_3 = \begin{Bmatrix} 0 \\ 1 \\ 0 \end{Bmatrix} \tag{10.100}$$

The phasor diagram for $\boldsymbol{\nu}_{SP_1}$ is shown in Figure 10.15. Comparing this diagram with that for the short-period mode in Example 10.2 also shows very good agreement. (The fact that the diagram below has been rotated approximately 60 degrees clockwise compared to that in Example 10.2 is not significant. An eigenvector may be multiplied by any nonzero scalar complex quantity and still be an eigenvector of the given matrix.)

We therefore conclude that at least for the Navion aircraft, the short-period approximation is quite accurate, especially for the angle-of-attack and pitch rate responses. This is really not too surprising since the original eigen-analysis indicated that the short-period modal response included virtually no surge velocity. In fact, the short-period approximation is quite valid for any aircraft's dynamics that exhibit a conventional short-period mode. But the student is cautioned to first perform an eigen-analysis on the longitudinal dynamics of any vehicle under consideration before invoking the short-period approximation.

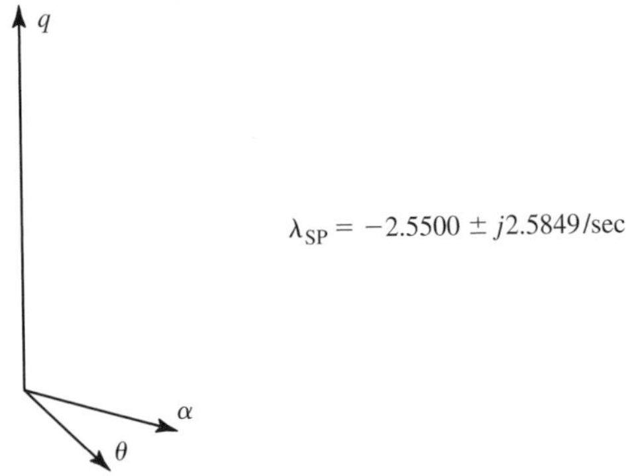

Figure 10.15 Phasor diagram under the short-period approximation for the Navion.

10.5.2 The Phugoid Approximation

We now turn our attention to the other longitudinal mode of motion, or the phugoid mode. The eigen-analysis of the Navion in Example 10.2 revealed that the phugoid's low-frequency, lightly damped modal response was characterized by an exchange of kinetic and potential energy with virtually no perturbation in angle-of-attack or pitch rate.

Consistent with these facts, let's look at a classical model of phugoid motion first developed by Lanchester in 1908 (Ref. 4). Lanchester developed his model under the following assumptions.

1. The reference flight condition consists of steady, level flight ($\Theta_0 = 0$) at a velocity of U_0.
2. The total energy (kinetic plus potential) remains constant.
3. The angle of attack remains constant at the reference value α_0.
4. Thrust balances the drag of the vehicle.
5. The motion is sufficiently slow such that the effects of pitch rate may be ignored.

From Assumptions 1 and 2 we can write

$$\frac{1}{2}mU_0^2 + mgh_0 = \frac{1}{2}m\left(U_0 + u\right)^2 + mg\left(h_0 + h\right) = \text{constant} \qquad (10.101)$$

and so

$$\left(U_0 + u\right)^2 = U_0^2 - 2gh \qquad (10.102)$$

In steady, level flight

$$L_0 = \frac{1}{2}\rho_\infty U_0^2 S_W C_{L_0} = mg \qquad (10.103)$$

and in perturbed flight, with lift coefficient constant (for constant $\alpha = \alpha_0$), the lift may be written as

$$\begin{aligned} L_0 + \delta L &= \frac{1}{2}\rho_\infty(U_0 + u)^2 S_W C_{L_0} = \frac{1}{2}\rho_\infty(U_0^2 - 2gh) S_W C_{L_0} \\ &= mg - \rho_\infty g h S_W C_{L_0} = L_0 - \rho_\infty g h S_W C_{L_0} \end{aligned} \qquad (10.104)$$

Summing vertical forces and looking at the altitude degree of freedom, we may write its equation of motion as

$$m\left(\ddot{h}_0 + \ddot{h}\right) = F_Z = \left(L_0 + \delta L\right)\cos\left(\Theta_0 + \theta\right) - mg \qquad (10.105)$$

The homogeneous equation of motion for perturbation altitude h is then

$$m\ddot{h} = \delta L \cos\Theta_0 = -\rho_\infty g h S_W C_{L_0}$$

or

$$\ddot{h} + \frac{\rho_\infty g S_W C_{L_0}}{m} h = 0 \tag{10.106}$$

This equation indicates that the phugoid motion involves undamped oscillations in altitude, at a frequency of

$$\omega_P = \sqrt{\frac{\rho_\infty g S_W C_{L_0}}{m}} = \frac{g\sqrt{2}}{U_0} \tag{10.107}$$

after invoking Equation (10.103). So Lanchester's model indicates that the phugoid mode is undamped, with a frequency of oscillation that is inversely proportional to flight velocity.

Now regarding the development of more general approximations, there are several variations such as those offered in Refs. 5, 6, and 7. One is tempted initially to simply eliminate the angle-of-attack degree of freedom, similar to the approach used for the short-period approximation. Doing so yields the following polynomial-matrix description of the phugoid dynamics

$$\begin{bmatrix} s-(X_u+X_{P_u}) & -X_q s + g\cos\Theta_0 \\ -(M_u+M_{P_u}) & s^2 - M_q s \end{bmatrix} \begin{Bmatrix} u(s) \\ \theta(s) \end{Bmatrix} = \begin{bmatrix} X_{\delta_E} & X_T \\ M_{\delta_E} & M_T \end{bmatrix} \begin{Bmatrix} \delta_E(s) \\ \delta T(s) \end{Bmatrix} \tag{10.108}$$

which again is of the form

$$\mathbf{P}(s)\mathbf{y}(s) = \mathbf{Q}(s)\mathbf{u}(s)$$

This system's characteristic polynomial is

$$\begin{aligned} \det \mathbf{P}(s) &= (s^2 - M_q s)(s-(X_u+X_{P_u})) - (M_u+M_{P_u})(X_q s - g\cos\Theta_0) \\ &= s^3 - (M_q + (X_u+X_{P_u}))s^2 \\ &\quad + (M_q(X_u+X_{P_u}) - X_q(M_u+M_{P_u}))s + (M_u+M_{P_u})g\cos\Theta_0 \end{aligned} \tag{10.109}$$

a third-order polynomial. But the above result is not very satisfying. Note, for example, that when $(M_u + M_{P_u}) = 0$ (which may occur at low subsonic flight velocity, as with the Navion), one of the characteristic roots is zero, and the other two satisfy

$$s^2 - (M_q + (X_u+X_{P_u}))s + M_q(X_u+X_{P_u}) = 0 \tag{10.110}$$

For the Navion, this would yield two more real roots—not a very good approximation for its phugoid characteristic roots.

As an alternative, Refs. 5 & 6 suggest eliminating the pitching-moment equation and letting the perturbation angle of attack remain zero. The argument for this is based on the fact that the phugoid modal response involves little angle-of-attack response and does not involve high pitch rates.

Following this approach, the polynomial-matrix system description becomes

$$\begin{bmatrix} s - (X_u + X_{P_u}) & -X_q s + g\cos\Theta_0 \\ -(Z_u + Z_{P_u}) & -(Z_q + \mathrm{U}_0)s + g\sin\Theta_0 \end{bmatrix} \begin{Bmatrix} u(s) \\ \theta(s) \end{Bmatrix} = \begin{bmatrix} X_{\delta_E} & X_T \\ Z_{\delta_E} & Z_T \end{bmatrix} \begin{Bmatrix} \delta_E(s) \\ \delta T(s) \end{Bmatrix} \tag{10.111}$$

for which the characteristic polynomial (with $\Theta_0 = 0$) is

$$\begin{aligned} \det \mathbf{P}(s) &= \left(-(Z_q + \mathrm{U}_0)s\right)\left(s - (X_u + X_{P_u})\right) - (Z_u + Z_{P_u})(X_q s - g) \\ &= -(Z_q + \mathrm{U}_0)s^2 + \left((Z_q + \mathrm{U}_0)(X_u + X_{P_u}) - X_q(Z_u + Z_{P_u})\right)s \\ &\quad + (Z_u + Z_{P_u})g \end{aligned} \tag{10.112}$$

By comparing this polynomial with the standard quadratic form we see that the phugoid undamped frequency and damping are given by

$$\omega_P^2 = -\frac{g(Z_u + Z_{P_u})}{(\mathrm{U}_0 + Z_q)} \tag{10.113}$$

and

$$2\zeta_P\omega_P = -(X_u + X_{P_u}) + \frac{X_q(Z_u + Z_{P_u})}{(\mathrm{U}_0 + Z_q)} \tag{10.114}$$

For the Navion, for example, this approximation gives the characteristic roots as

$$\lambda_P = -0.02255 \pm j0.25906 \text{ /sec}$$

compared to $-0.01722 \pm j0.21375$ /sec for the full-order model. Also, the eigenvector phasor diagrams from this approximation are in fairly good agreement with the full-order results presented earlier.

But although this approximation might yield acceptable results when $(M_u + M_{P_u}) = 0$, it does not capture the case when this term is large. As discussed in Chapter 6, C_{M_u} may become large and be either positive or negative in the transonic range. And phugoid instabilities ("Mach tuck") are known to occur when $(M_u + M_{P_u})$ becomes large and negative. Hence this approximation does not adequately model an important situation, and hence is not recommended.

The *preferred phugoid approximation* is taken from Ref. 8. Unlike the others cited above, this approximation retains both the pitching-moment and angle-of-attack equations, but sets $\dot{\alpha} = \dot{q} = 0$ in all equations. This approach is based on the argument that since the phugoid is a low-frequency, long-period mode, the angle of attack and pitch rate will quickly reach steady state over the period of the phugoid. Recall also that the eigenvectors of the Navion's full-order longitudinal model indicated that its phugoid modal response contained little angle of attack or pitch rate, so both of their rates will be quite small for a low-frequency mode. Therefore, this approach is consistent with the results from the eigen-analysis, and partially consistent with that in Ref. 7, in that pitch acceleration $\dot{q}$ is eliminated from the model.

Under this preferred phugoid approximation, the polynomial-matrix description of the system is

$$\begin{bmatrix} s-(X_u+X_{P_u}) & -X_\alpha & -X_q s + g\cos\Theta_0 \\ -(Z_u+Z_{P_u}) & -Z_\alpha & -(Z_q+\mathrm{U}_0)s + g\sin\Theta_0 \\ -(M_u+M_{P_u}) & -(M_\alpha+M_{P_\alpha}) & -M_q s \end{bmatrix} \begin{Bmatrix} u(s) \\ \alpha(s) \\ \theta(s) \end{Bmatrix} = \begin{bmatrix} X_{\delta_E} & X_T \\ Z_{\delta_E} & Z_T \\ M_{\delta_E} & M_T \end{bmatrix} \begin{Bmatrix} \delta_E(s) \\ \delta T(s) \end{Bmatrix} \quad (10.115)$$

which for level flight ($\Theta_0 = 0$) has the following characteristic polynomial.

$$\begin{aligned} \det\mathbf{P}(s) &= s\big(Z_\alpha M_q - (M_\alpha+M_{P_\alpha})(Z_q+\mathrm{U}_0)\big)\big(s-(X_u+X_{P_u})\big) \\ &\quad -X_\alpha s\big((M_u+M_{P_u})(\mathrm{U}_0+Z_q) - M_q(Z_u+Z_{P_u})\big) \\ &\quad -(X_q s - g)\big((M_\alpha+M_{P_\alpha})(Z_u+Z_{P_u}) - Z_\alpha(M_u+M_{P_u})\big) \\ &= \big(Z_\alpha M_q - (M_\alpha+M_{P_\alpha})(\mathrm{U}_0+Z_q)\big)s^2 \\ &\quad + \begin{pmatrix} (M_\alpha+M_{P_\alpha})\big((X_u+X_{P_u})(\mathrm{U}_0+Z_q) - X_q(Z_u+Z_{P_u})\big) \\ + M_q\big(X_\alpha(Z_u+Z_{P_u}) - Z_\alpha(X_u+X_{P_u})\big) + (M_u+M_{P_u})\big(X_q Z_\alpha - X_\alpha(\mathrm{U}_0+Z_q)\big) \end{pmatrix} s \\ &\quad + g\big((M_\alpha+M_{P_\alpha})(Z_u+Z_{P_u}) - Z_\alpha(M_u+M_{P_u})\big) \end{aligned} \quad (10.116)$$

So the above phugoid quadratic when compared to the standard form of

$$s^2 + 2\zeta_P\omega_P s + \omega_P^2$$

yields an undamped natural frequency and a damping coefficient corresponding to

$$\boxed{\omega_P^2 = g\frac{(M_\alpha+M_{P_\alpha})(Z_u+Z_{P_u}) - Z_\alpha(M_u+M_{P_u})}{Z_\alpha M_q - (M_\alpha+M_{P_\alpha})(\mathrm{U}_0+Z_q)}} \quad (10.117)$$

$$\boxed{2\zeta_P\omega_P = -(X_u+X_{P_u}) + \frac{\big((Z_u+Z_{P_u})(M_q X_\alpha - X_q(M_\alpha+M_{P_\alpha})) + (M_u+M_{P_u})(X_q Z_\alpha - X_\alpha(\mathrm{U}_0+Z_q))\big)}{Z_\alpha M_q - (M_\alpha+M_{P_\alpha})(\mathrm{U}_0+Z_q)}} \quad (10.118)$$

The first thing to observe from the above results is the fact that the frequency and damping depend on $(M_u + M_{P_u})$ in addition to $(Z_u + Z_{P_u})$. So if compressibility effects are significant, as evidenced by the magnitudes of these two terms, the phugoid frequency and damping will be significantly affected, as desired. In fact, if $(M_u + M_{P_u})$ is large and negative we can see that Equation (10.117) can become negative, indicating phugoid instability as we would hope in a good approximation. Consequently, this approximation is superior to those discussed previously.

This phugoid approximation also admits a simple, two-degree of freedom state-variable model. From Equations (10.47) and (10.48), after setting all $\dot{\alpha}$ and $\dot{q}$ terms equal to zero we have the reduced-order model given by

$$\begin{Bmatrix} \dot{u} \\ 0 \\ 0 \\ \dot{\theta} \end{Bmatrix} = \begin{bmatrix} (X_u + X_{P_u}) & X_\alpha & X_q & -g\cos\Theta_0 \\ (Z_u + Z_{P_u}) & Z_\alpha & (U_0 + Z_q) & -g\sin\Theta_0 \\ (M_u + M_{P_u}) & (M_\alpha + M_{P_\alpha}) & M_q & 0 \\ 0 & 0 & 1 & 0 \end{bmatrix} \begin{Bmatrix} u \\ \alpha \\ q \\ \theta \end{Bmatrix} + \begin{bmatrix} X_{\delta_E} & X_T \\ Z_{\delta_E} & Z_T \\ M_{\delta_E} & M_T \\ 0 & 0 \end{bmatrix} \begin{Bmatrix} \delta_E \\ \delta T \end{Bmatrix} \quad (10.119)$$

Now the second and third equations in Equations (10.119) may be used to eliminate the α and q terms in the other two state-variable equations. Solving for α and q for level flight ($\Theta_0 = 0$) yields

$$\begin{Bmatrix} \alpha \\ q \end{Bmatrix} = \frac{1}{\det \mathbf{M}} \left(\begin{Bmatrix} M_q(Z_u + Z_{P_u}) - (M_u + M_{P_u})(U_0 + Z_q) \\ Z_\alpha(M_u + M_{P_u}) - (M_\alpha + M_{P_\alpha})(U_0 + Z_q) \end{Bmatrix} u + \begin{Bmatrix} Z_\bullet M_q - M_\bullet(U_0 + Z_q) \\ -Z_\bullet(M_\alpha + M_{P_\alpha}) + M_\bullet Z_\alpha \end{Bmatrix} \bullet \right) \quad (10.120)$$

where $\bullet = \delta_E$ and δT and

$$\det \mathbf{M} = \det \begin{bmatrix} Z_\alpha & (U_0 + Z_q) \\ (M_\alpha + M_{P_\alpha}) & M_q \end{bmatrix} = Z_\alpha M_q - (U_0 + Z_q)(M_\alpha + M_{P_\alpha}) \quad (10.121)$$

Inserting this result into the remaining state-variable equations yields the following two-degree-of-freedom model for the phugoid dynamics.

$$\begin{Bmatrix} \dot{u} \\ \dot{\theta} \end{Bmatrix} = \begin{bmatrix} a_{1,1} & -g \\ a_{2,1} & 0 \end{bmatrix} \begin{Bmatrix} u \\ \theta \end{Bmatrix} + \begin{bmatrix} b_{1,1} & b_{1,2} \\ b_{2,1} & b_{2,2} \end{bmatrix} \begin{Bmatrix} \delta_E \\ \delta T \end{Bmatrix} \quad (10.122)$$

with

$$a_{1,1} = (X_u + X_{P_u}) + \left((X_\alpha M_q - X_q(M_\alpha + M_{P_\alpha}))(Z_u + Z_{P_u}) - (M_u + M_{P_u})(X_\alpha(U_0 + Z_q) - X_q Z_\alpha)\right)/D$$

$$a_{2,1} = \left(Z_\alpha(M_u + M_{P_u}) - (Z_u + Z_{P_u})(M_\alpha + M_{P_\alpha})\right)/D$$

$$b_{1,1} = X_{\delta_E} + X_\alpha\left(Z_{\delta_e} M_q - M_{\delta_E}(U_0 + Z_q)\right)/D \qquad b_{1,2} = X_T + X_\alpha\left(Z_T M_q - M_T(U_0 + Z_q)\right)/D$$

$$+ X_q\left(-Z_{\delta_e}(M_\alpha + M_{P_\alpha}) + M_{\delta_E} Z_\alpha\right)/D \qquad + X_q\left(-Z_T(M_\alpha + M_{P_\alpha}) + M_T Z_\alpha\right)/D$$

$$b_{2,1} = \left(-Z_{\delta_e}(M_\alpha + M_{P_\alpha}) + M_{\delta_E} Z_\alpha\right)/D \qquad b_{2,2} = \left(-Z_T(M_\alpha + M_{P_\alpha}) + M_T Z_\alpha\right)/D$$

$$D = Z_\alpha M_q - (M_\alpha + M_{P_\alpha})(U_0 + Z_q)$$

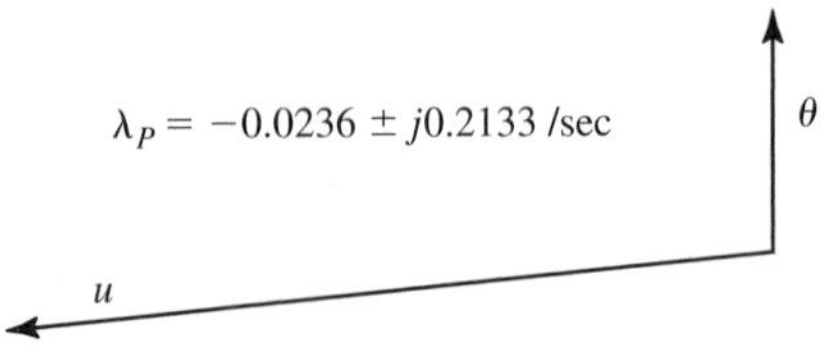

Figure 10.16 Navion phasor diagram—phugoid approximation.

As a numerical example, we will again consider the Navion. Using MATLAB we determine the $\mathbf{A}_P$ matrix for the above two-degree-of-freedom model to be

$$\mathbf{A}_P = \begin{bmatrix} -0.047202 & -32.2 \\ 0.001430 & 0 \end{bmatrix} \tag{10.123}$$

After converting the units on θ to degrees, this matrix becomes

$$\mathbf{A}_{P_{\text{New}}} = \begin{bmatrix} -0.047202 & -0.56195 \\ 0.081962 & 0 \end{bmatrix} \tag{10.124}$$

with eigenvalues given by

$$\lambda_P = -0.02360 \pm j0.21331\,/\text{sec} \tag{10.125}$$

The phasor diagram for the first eigenvector is as shown in Figure 10.16. Comparing these results to those from the full-order model in Example 10.2 indicates quite good agreement.

Although Equations (10.117–10.122) appear rather complicated, they frequently reduce to something much simpler. For example, since the denominator in all the terms will frequently be dominated by $(M_\alpha + M_{P_\alpha})\mathrm{U}_0$, we see that the phugoid undamped natural frequency ω_P^2 is proportional to $-\dfrac{g}{\mathrm{U}_0}\left((Z_u + Z_{P_u}) - Z_\alpha(M_u + M_{P_u})/(M_\alpha + M_{P_\alpha})\right)$. So the phugoid frequency is significantly affected by $(Z_u + Z_{P_u})$ and $(M_u + M_{P_u})$.

Furthermore, here we observe a second connection between static and dynamic stability. From Equation (10.117), this phugoid approximation indicates that phugoid instability will occur if

$$(M_u + M_{P_u}) < (M_\alpha + M_{P_\alpha})(Z_u + Z_{P_u})/Z_\alpha \tag{10.126}$$

And this inequality could be satisfied if the vehicle has a static speed instability indicated by $(M_u + M_{P_u}) < 0$. Consequently, under this phugoid approximation, static speed stability tends to lead to dynamic stability.

Plus when $(M_u + M_{P_u}) = 0$, as in low subsonic flight, the undamped natural frequency reduces to approximately

$$\omega_P^2 \approx \frac{-g}{\mathrm{U}_0}(Z_u + Z_{P_u}) \tag{10.127}$$

which agrees with the result from the phugoid approximation suggested in Refs. 5 and 6. Additionally, in low subsonic flight we may also assume that $C_{L_u} \approx C_{P_{Z_u}} \approx 0$. So from Equations (8.15) and (8.16), along with the fact that the vertical propulsive plus aerodynamic force equals the weight, we have

$$\left(Z_u + Z_{P_u}\right) \approx -\frac{q_\infty S_W}{m}\frac{2}{\mathrm{U}_0}\left(C_{L_0} - C_{P_{Z_0}}\right) = -\frac{2mg}{m\mathrm{U}_0} = -\frac{2g}{\mathrm{U}_0} \tag{10.128}$$

Under these conditions, the phugoid natural frequency is approximately

$$\omega_P \approx \frac{g}{\mathrm{U}_0}\sqrt{2} \tag{10.129}$$

indicating that the phugoid natural frequency is inversely proportional to flight velocity, which agrees with the classical Lanchester result.

The result in Equation (10.129), that the phugoid undamped natural frequency ω_P is roughly inversely proportional to the flight velocity, tends to explain why the phugoid frequency of the hypersonic vehicle was so much smaller that that of the Navion. The flight velocity of the Navion was 176 fps, while the velocity of the hypersonic vehicle was Mach = 8, or approximately 8000 fps. These differ by a factor of approximately 50.

With regard to the phugoid damping under this approximation, by letting $X_q \approx 0$ (which is typically a good approximation) we find that the phugoid damping may be found from

$$2\zeta_P\omega_P = -\left(X_u + X_{P_u}\right) + X_\alpha\frac{\left(M_q\left(Z_u + Z_{P_u}\right) - \left(M_u + M_{P_u}\right)\left(\mathrm{U}_0 + Z_q\right)\right)}{Z_\alpha M_q - \left(M_\alpha + M_{P_\alpha}\right)\left(\mathrm{U}_0 + Z_q\right)} \tag{10.130}$$

Now assume that

$$M_q\left(Z_u + Z_{P_u}\right) \ll \left(Z_\alpha M_q - \left(M_\alpha + M_{P_\alpha}\right)\left(\mathrm{U}_0 + Z_q\right)\right). \tag{10.131}$$

Then when $\left(M_u + M_{P_u}\right) = 0$, Equation (10.130) reduces to simply

$$2\zeta_P\omega_P \approx -\left(X_u + X_{P_u}\right) \tag{10.132}$$

But from Equations (8.15) and (8.16), if $C_{D_u} = C_{P_{X_u}} = 0$ we have

$$-\left(X_u + X_{P_u}\right) = \frac{q_\infty S_W}{m}\left(\left(C_{D_u} + \frac{2}{\mathrm{U}_0}C_{D_0}\right) - \left(C_{P_{X_u}} + \frac{2}{\mathrm{U}_0}C_{P_{X_0}}\right)\right) = \frac{q_\infty S_W}{m}\frac{2}{\mathrm{U}_0}\left(C_{D_0} - C_{P_{X_0}}\right) \tag{10.133}$$

so the phugoid damping under these approximations is found from

$$\zeta_P\omega_P \approx \frac{gq_\infty S_W}{mg}\frac{\left(C_{D_0} - C_{P_{X_0}}\right)}{\mathrm{U}_0} \approx \frac{g}{\mathrm{U}_0}\frac{\left(C_{D_0} - C_{P_{X_0}}\right)}{\left(C_{L_0} - C_{P_{Z_0}}\right)} \tag{10.134}$$

Thus again when compressibility effects are negligible, under this phugoid approximation we see that the phugoid damping ratio ζ_P is approximately zero for a powered vehicle, or inversely proportional to the trim L/D ratio for an unpowered glider. And since a high L/D ratio is usually desired for aircraft, a low phugoid damping will naturally result. This result also agrees with those from Refs. 5–7.

Using the polynomial-matrix description in Equation (10.115), the transfer functions under this phugoid approximation (letting $\bullet = \delta_E$ or δT) are

$$\frac{u(s)}{\bullet(s)} = \begin{pmatrix} X_\bullet s\big(Z_\alpha M_q - (M_\alpha + M_{P_\alpha})(U_0 + Z_q)\big) + X_\alpha s\big(M_\bullet(U_0 + Z_q) - Z_\bullet M_q\big) \\ +X_q s\big(Z_\bullet(M_\alpha + M_{P_\alpha}) - M_\bullet Z_\alpha\big) - g\big(Z_\bullet(M_\alpha + M_{P_\alpha}) - M_\bullet Z_\alpha\big) \end{pmatrix} \Big/ \det \mathbf{P}(s), \bullet = \delta_E \text{ or } \delta T$$

$$= \begin{pmatrix} \big(Z_\alpha(X_\bullet M_q - M_\bullet X_q) + (M_\alpha + M_{P_\alpha})(Z_\bullet X_q - X_\bullet(U_0 + Z_q)) + X_\alpha(M_\bullet(U_0 + Z_q) - Z_\bullet M_q)\big)s \\ -g\big(Z_\bullet(M_\alpha + M_{P_\alpha}) - M_\bullet Z_\alpha\big) \end{pmatrix} \Big/ \det \mathbf{P}(s)$$

$$\frac{\theta(s)}{\bullet(s)} = \begin{pmatrix} \big(s - (X_u + X_{P_u})\big)\big(-M_\bullet Z_\alpha + Z_\bullet(M_\alpha + M_{P_\alpha})\big) + X_\alpha\big(Z_\bullet(M_u + M_{P_u}) - M_\bullet(Z_u + Z_{P_u})\big) \\ +X_\bullet\big((M_\alpha + M_{P_\alpha})(Z_u + Z_{P_u}) - Z_\alpha(M_u + M_{P_u})\big) \end{pmatrix} \Big/ \det \mathbf{P}(s)$$

$$= \begin{pmatrix} \big(Z_\bullet(M_\alpha + M_{P_\alpha}) - M_\bullet Z_\alpha\big)s - (X_u + X_{P_u})\big(Z_\bullet(M_\alpha + M_{P_\alpha}) - M_\bullet Z_\alpha\big) \\ +(Z_u + Z_{P_u})\big(X_\bullet(M_\alpha + M_{P_\alpha}) - M_\bullet X_\alpha\big) + (M_u + M_{P_u})\big(Z_\bullet X_\alpha - X_\bullet Z_\alpha\big) \end{pmatrix} \Big/ \det \mathbf{P}(s), \bullet = \delta_E \text{ or } \delta T \tag{10.135}$$

where $\det \mathbf{P}(s)$ is the characteristic polynomial given in Equation (10.116). Note that both sets of transfer functions have one numerator zero, and note that these numerators may usually be considerably simplified. For example, we frequently have

$$X_q \approx X_{\delta_E} \approx 0 \tag{10.136}$$

and therefore the standard factor in the $\theta(s)/\delta_E(s)$ numerator becomes

$$\frac{1}{T_{\theta_1}} \approx -(X_u + X_{P_u}) + \frac{X_\alpha\big(Z_{\delta_E}(M_u + M_{P_u}) + M_{\delta_E}(Z_u + Z_{P_u})\big)}{\big(Z_{\delta_E}(M_\alpha + M_{P_\alpha}) - M_{\delta_E} Z_\alpha\big)} \tag{10.137}$$

This factor, recall, was not present in the short-period approximation.

Since the model under this phugoid approximation has only two degrees of freedom, the transfer functions given in Equations (10.135) completely define the system. However, the angle-of-attack transfer functions may also be found, either from the polynomial-matrix description (Equation (10.115)), or by using the equation for angle of attack in Equations (10.120).

10.6 LATERAL-DIRECTIONAL TRANSFER FUNCTIONS AND MODAL ANALYSIS

Attention now turns to the aircraft's lateral-directional axes. In this section we will develop the transfer functions and investigate the natural modes of motion for these axes.

Applying Cramer's rule to the polynomial-matrix description given in Equation (10.66*a*), and letting $\bullet = \delta_A$ or δ_R, we have the six lateral-directional transfer functions given by

$$\frac{\beta(s)}{\bullet(s)} = \det\begin{bmatrix} Y_\bullet & -(Y_p s + g\cos\Theta_0) & (U_0 - Y_r)s - g\sin\Theta_0 \\ L'_\bullet & s^2 - L'_p s & -L'_r s \\ N'_\bullet & -N'_p s & s^2 - N'_r s \end{bmatrix} / \det \mathbf{P}(s)$$

$$\frac{1}{s}\left(\frac{p(s)}{\bullet(s)}\right) = \det\begin{bmatrix} U_0 s - Y_\beta & Y_\bullet & (U_0 - Y_r)s - g\sin\Theta_0 \\ -L'_\beta & L'_\bullet & -L'_r s \\ -N'_\beta & N'_\bullet & s^2 - N'_r s \end{bmatrix} / \det \mathbf{P}(s)$$

$$\frac{1}{s}\left(\frac{r(s)}{\bullet(s)}\right) = \det\begin{bmatrix} U_0 s - Y_\beta & -(Y_p s + g\cos\Theta_0) & Y_\bullet \\ -L'_\beta & s^2 - L'_p s & L'_\bullet \\ -N'_\beta & -N'_p s & N'_\bullet \end{bmatrix} / \det \mathbf{P}(s), \quad \bullet = \delta_A \text{ or } \delta_R \tag{10.138}$$

We noted in Section 10.3 when discussing Equation (10.66*a*) that the characteristic polynomial det **P**(s) is fifth order, and when $\Theta_0 = 0$ one pole will be at the origin. This pole will always be cancelled by the $1/s$ appearing on the left-hand side of the above equations for the roll-rate $p(s)$ and yaw-rate $r(s)$ transfer functions. Usually for conventional aircraft, two of the four remaining poles are real, and the other two consist of a complex–conjugate pair, although there are exceptions.

The above sideslip-angle transfer function has a fourth-order numerator polynomial, with one root (zero) at the origin. This zero will always be cancelled by the pole at the origin (when $\Theta_0 = 0$), leaving three remaining zeros. The roll-rate $p(s)$ transfer function has a third-order numerator polynomial, and so will have three zeros. But when $\Theta_0 = 0$, one of these zeros will also be at the origin. Finally, the yaw-rate $r(s)$ transfer function has a third-order numerator polynomial, and hence has three zeros.

For a conventional aircraft in level flight ($\Theta_0 = 0$), the lateral-directional transfer functions typically take on standard forms, which are given by

$$\begin{aligned}
\frac{\beta(s)}{\bullet(s)} &= \frac{K_\beta\left(s + 1/T_{\beta_1}\right)\left(s + 1/T_{\beta_2}\right)\left(s + 1/T_{\beta_3}\right)}{\left(s + 1/T_S\right)\left(s + 1/T_R\right)\left(s^2 + 2\zeta_{\mathrm{DR}}\omega_{\mathrm{DR}} + \omega_{\mathrm{DR}}^2\right)} \\
\frac{\phi(s)}{\bullet(s)} &= \frac{K_\phi\left(s^2 + 2\zeta_\phi\omega_\phi + \omega_\phi^2\right)}{\left(s + 1/T_S\right)\left(s + 1/T_R\right)\left(s^2 + 2\zeta_{\mathrm{DR}}\omega_{\mathrm{DR}} + \omega_{\mathrm{DR}}^2\right)}, \quad \bullet = \delta_A \text{ or } \delta_R \\
\frac{r(s)}{\bullet(s)} &= \frac{K_r\left(s + 1/T_{r_1}\right)\left(s^2 + 2\zeta_r\omega_r + \omega_r^2\right)}{\left(s + 1/T_S\right)\left(s + 1/T_R\right)\left(s^2 + 2\zeta_{\mathrm{DR}}\omega_{\mathrm{DR}} + \omega_{\mathrm{DR}}^2\right)}
\end{aligned} \tag{10.139}$$

As noted earlier, the fourth-order characteristic polynomial usually factors into two real factors plus one quadratic factor. The sideslip-angle numerator frequently factors into three real terms, while the bank-angle numerator frequently factors into one quadratic term. The yaw-rate numerator frequently factors into one quadratic and one real term. The notation used for all the roots of these factors is defined in Equations (10.139).

Finally, the transfer functions governing the heading and bank angles and lateral acceleration may be found from Equations 10.67, or

$$\begin{aligned}
\frac{s\phi(s)}{\bullet(s)} &= \frac{p(s)}{\bullet(s)} + \tan\Theta_0\frac{r(s)}{\bullet(s)} \\
\frac{\psi(s)}{\bullet(s)} &= \frac{1}{s\cos\Theta_0}\frac{r(s)}{\bullet(s)} \qquad \bullet = \delta_A \text{ or } \delta_R \\
\frac{a_Y(s)}{\bullet(s)} &= \mathrm{U}_0\frac{s\beta(s)}{\bullet(s)} + \mathrm{U}_0\frac{r(s)}{\bullet(s)} + x\frac{sr(s)}{\bullet(s)} - z\frac{sp(s)}{\bullet(s)}
\end{aligned} \tag{10.140}$$

At this point, as with the longitudinal transfer functions, it is difficult to say much more in general about the vehicle's dynamics. For example, the characteristic polynomial and all the numerator polynomials are rather involved functions of the dimensional stability derivatives. So finding the poles and zeros would usually require numerical techniques. However, after we perform some additional analysis in the case study in Example 10.4, we will obtain some simplified approximations that will be very useful, and are frequently used in practice.

EXAMPLE 10.4

A Case Study—Lateral-Directional Dynamics of the Navion

Here we will determine the transfer functions, plot the relevant Bode plots, and perform a modal analysis of a conventional aircraft's lateral-directional dynamics. Let's again use the Navion aircraft in steady, level flight ($\Theta_0 = \gamma_0 = 0$) at sea level, as discussed in Example 10.2. The reference flight velocity U_0 is 176 fps.

The state, response, and control-input vectors are

$$\mathbf{x}^T = \mathbf{y}^T = \begin{bmatrix} \beta(\text{rad}) & \phi(\text{rad}) & p(\text{rad/sec}) & r(\text{rad/sec}) \end{bmatrix}$$

$$\mathbf{u}^T = \begin{bmatrix} \delta_A(\text{rad}) & \delta_R(\text{rad}) \end{bmatrix}$$

So from Equations (10.71) and Appendix B we have the state-variable description of the Navion's lateral-directional dynamics given by

$$\mathbf{A} = \begin{bmatrix} \dfrac{Y_\beta}{U_0} & \dfrac{g}{U_0} & \dfrac{Y_p}{U_0} & \left(\dfrac{Y_r}{U_0} - 1\right) \\ 0 & 0 & 1 & 0 \\ L'_\beta & 0 & L'_p & L'_r \\ N'_\beta & 0 & N'_p & N'_r \end{bmatrix} = \begin{bmatrix} -0.2543 & 0.1830 & 0 & -1 \\ 0 & 0 & 1 & 0 \\ -15.982 & 0 & -8.402 & 2.193 \\ 4.495 & 0 & -0.3498 & -0.7605 \end{bmatrix}$$

$$\mathbf{B} = \begin{bmatrix} \dfrac{Y_{\delta_A}}{U_0} & \dfrac{Y_{\delta_r}}{U_0} \\ 0 & 0 \\ L'_{\delta_A} & L'_{\delta_R} \\ N'_{\delta_A} & N'_{\delta_R} \end{bmatrix} = \begin{bmatrix} 0 & 0.0708 \\ 0 & 0 \\ 28.984 & 2.548 \\ -0.2218 & -4.597 \end{bmatrix} \tag{10.141}$$

$$\mathbf{C} = \mathbf{I}_4, \text{ and } \mathbf{D} = \mathbf{0}$$

Since all the responses and inputs are angles and angular rates, no unit balancing is necessary for the eigen-analysis here.

From MATLAB, the eigenvalues of **A**, or the system's characteristic roots, are

$$\begin{aligned} \lambda_1 &= -8.4346 \text{ /sec} \\ \lambda_2 &= -0.00876 \text{ /sec} \\ \lambda_{3,4} &= -0.48674 \pm j2.3349 \text{ /sec} \end{aligned} \tag{10.142}$$

So the modal responses consist of one lightly damped stable oscillatory mode with a frequency of 2.33 rad/sec, and two stable first-order modes. One first-order pole is located very near the origin of the complex plane, and so this mode has a long time constant ($T = -1/\lambda_2 = 114$ sec). The other first-order mode decays at a rapid rate ($T = 0.12$ sec).

To describe the modal responses we must examine the corresponding eigenvectors of **A**, which are

$$\boldsymbol{\nu}_1 = \begin{Bmatrix} -0.00762 \\ 0.11763 \\ -0.99219 \\ -0.04077 \end{Bmatrix}, \ \boldsymbol{\nu}_2 = \begin{Bmatrix} -0.02830 \\ -0.98444 \\ 0.00862 \\ -0.17321 \end{Bmatrix}$$

$$\boldsymbol{\nu}_{3,4} = \begin{Bmatrix} -0.00914 \pm j0.31069 \\ -0.25157 \pm j0.04090 \\ 0.02695 \mp j0.60729 \\ 0.67727 \pm j0.10103 \end{Bmatrix}, \text{ or } \boldsymbol{\nu}_3 = \begin{Bmatrix} 0.31082e^{j91.7^\circ} \\ 0.25487e^{j170.8^\circ} \\ 0.60789e^{-j87.5^\circ} \\ 0.68476e^{j8.5^\circ} \end{Bmatrix} \tag{10.143}$$

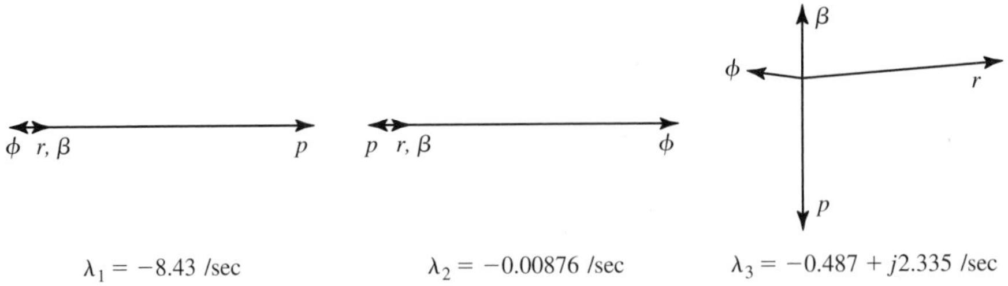

Figure 10.17 Phasor diagrams of the Navion's lateral-directional mode shapes.

The phasor diagrams of the eigenvectors are shown in Figure 10.17. Recall again that all four responses of the system in $\mathbf{y}(t)$ are angles and angular rates.

We see from the first eigenvector that this stable first-order mode consists of almost entirely roll rate p, with a small amount of bank angle ϕ. Sideslip angle and yaw rate are virtually zero. So this mode might be described as a first-order stable *roll-rate or roll-subsidence mode,* and it is an important component of an aircraft's dynamics. Conversely, from the second eigenvector we see that this slow first-order mode consists of almost entirely bank angle ϕ, with a small amount of roll rate p. Sideslip angle and yaw rate are again virtually zero. So this appears to be primarily a rolling mode as well. But it turns out that if heading ψ were included in the responses, we would see that the mode also includes a significant heading-angle response. As a consequence, this mode is called the *spiral or heading mode.* Sometimes this mode is dynamically unstable, with a real pole located slightly in the right half of the complex plane. Due to its slow response, however, it is easily stabilized by the pilot or autopilot by controlling bank angle. But if it is unstable, it is a nuisance to the pilot because the aircraft's heading will tend to drift if the pilot is distracted.

The third mode is a lightly damped oscillatory mode contributing to all four responses. Note that the yaw rate and bank angle are almost 180 degrees out of phase. That is, as the aircraft yaws right, it also banks to the left, and vice versa. Also, the sideslip angle lags the bank angle by approximately 90 degrees. So the aircraft's motion while undergoing only this modal response would consist of a lightly damped out-of-phase oscillatory rolling and yawing. This mode is called the *dutch-roll mode,* and the origins of this term are open to debate. The mode is basically a nuisance in terms of the vehicle's dynamics, and needs to be stable and reasonably damped for acceptable handling qualities.

Finally, an examination of the left eigenvectors of the Navion's lateral-directional dynamics would reveal that the aileron excites all three modes, while rudder primarily excites the roll-subsidence and dutch-roll modes.

Next we will determine the six lateral-directional transfer functions for the Navion. Using the state-variable description above and the `zpk` command in MATLAB, the transfer functions for aileron input are

$$\frac{\beta(s)}{\delta_A(s)} = \frac{0.2218\,(s+0.2286)(s+77.8)}{(s+8.435)(s+0.00876)(s^2+0.9735s+5.689)}\ \text{rad/rad or deg/deg}$$

$$\frac{\phi(s)}{\delta_A(s)} = \frac{28.984\,(s^2+0.998s+4.562)}{(s+8.435)(s+0.00876)(s^2+0.9735s+5.689)}\ \text{rad/rad or deg/deg} \qquad (10.144)$$

$$\frac{r(s)}{\delta_A(s)} = \frac{-0.2218\,(s-1.253)(s+1.543)(s+54.08)}{(s+8.435)(s+0.00876)(s^2+0.9735s+5.689)}\ \text{rad/sec/rad or deg/sec/deg}$$

Comparing the above results to the standard forms given in Equations (10.139), the quadratic dutch-roll factor is evident in the characteristic polynomial, along with the two first-order factors corresponding to the roll-subsidence and spiral modes. (We know this based on the eigen-analysis just performed.) Also note the numerator factors, including the right-half-plane zero in the yaw-rate transfer function, when compared to the standard forms.

The pole-zero plot of the ϕ/δ_A transfer function is shown in Figure 10.18. Note the locations of the complex dutch-roll poles, along with the spiral and roll-subsidence poles. Due to the low dutch-roll damping, these roots are close to the imaginary axis. In addition, in this transfer function the dutch-roll poles are almost cancelled by a pair of zeros, indicating that the dutch-roll mode is not too significant in the aileron-to-bank-angle response for this aircraft. In fact, an approximation to the roll-rate transfer function might simply include the roll-subsidence pole. That is,

$$\frac{p(s)}{\delta_A(s)} = \frac{s\phi(s)}{\delta_A(s)} = \frac{28.984\,s(s^2+0.998s+4.562)}{(s+8.435)(s+0.00876)(s^2+0.9735s+5.689)} \approx \frac{28.984}{(s+8.435)}$$

which suggests a lower-order approximation to be discussed more in Section 10.7.1. But also note that since the zeros in the other aileron transfer functions differ from those shown in Figure 10.18, this approximate pole-zero cancellation is not present in all three aileron-input transfer functions.

The Bode plots of the bank-angle and yaw-rate transfer functions with aileron input are shown in Figure 10.19. Note that consistent with the previous analysis, the Bode plot for bank angle shows only slight evidence of the dutch-roll mode. This mode is somewhat

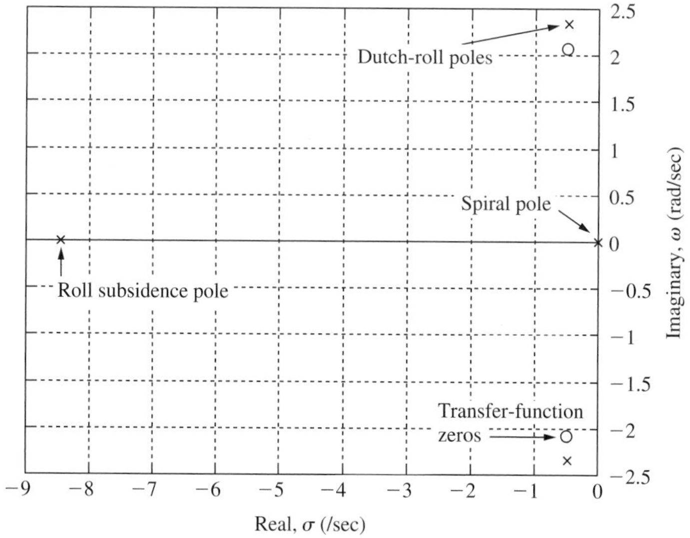

Figure 10.18 Pole-zero plot of the Navion's ϕ/δ_A transfer function.

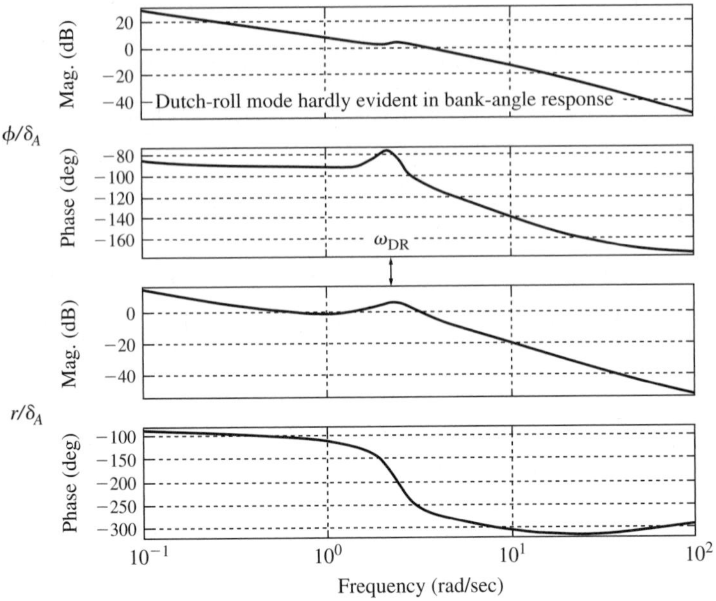

Figure 10.19 Bode plots for bank angle and yaw rate with aileron input—Navion aircraft.

more evident, however, in the Bode plot for yaw rate. From these Bode plots we can also see that the aileron excites all three modes.

The transfer functions for rudder input are

$$\frac{\beta(s)}{\delta_R(s)} = \frac{0.0708\,(s-0.03663)(s+8.795)(s+65.33)}{(s+8.435)(s+0.00876)(s^2+0.9735s+5.689)}\ \text{rad/rad or deg/deg}$$

$$\frac{\phi(s)}{\delta_R(s)} = \frac{2.548\,(s+3.606)(s-6.992)}{(s+8.435)(s+0.00876)(s^2+0.9735s+5.689)}\ \text{rad/rad or deg/deg} \tag{10.145}$$

$$\frac{r(s)}{\delta_R(s)} = \frac{-4.597\,(s+8.638)(s^2+0.1427s+0.2858)}{(s+8.435)(s+0.00876)(s^2+0.9735s+5.689)}\ \text{rad/sec/rad or deg/sec/deg}$$

Note the pole corresponding to the role-subsidence mode ($\lambda_1 = -8.435$ /sec) is almost cancelled in the sideslip-angle and yaw-rate transfer functions here. The Bode plots of the sideslip-angle and yaw-rate transfer functions with rudder input are shown in Figure 10.20. These plots all indicate the presence of the dutch-roll mode, and the quadratic term in the yaw-rate numerator is also clearly indicated. Hence the rudder clearly excites the dutch-roll mode.

Finally, from MATLAB we can easily determine the vehicle's step responses to aileron or rudder inputs using the `step` command. The step responses for step aileron or rudder inputs of one degree are shown in Figure 10.21. The first-order roll-subsidence mode clearly dominates the roll-rate response from aileron input, while the oscillatory dutch-roll mode is quite evident, particularly in the responses from the rudder. (The dashed lines indicate the steady-state responses.) All these results are consistent with the previous eigen-analysis.

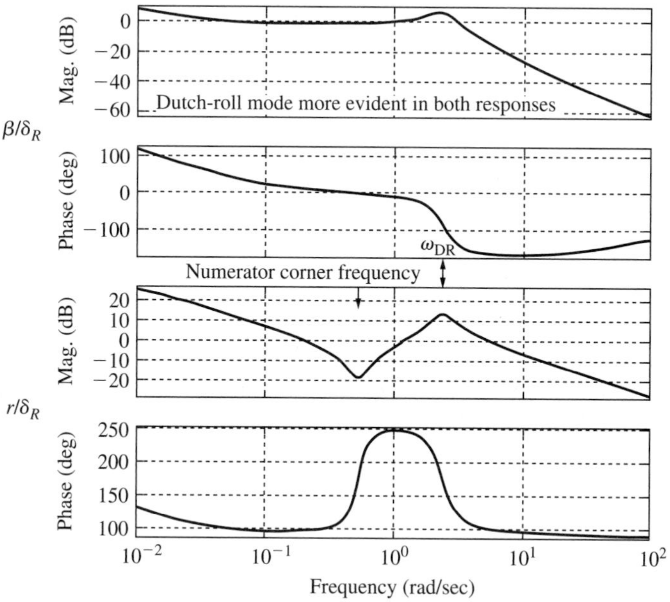

Figure 10.20 Bode plots for sideslip angle and yaw rate with rudder input—Navion aircraft.

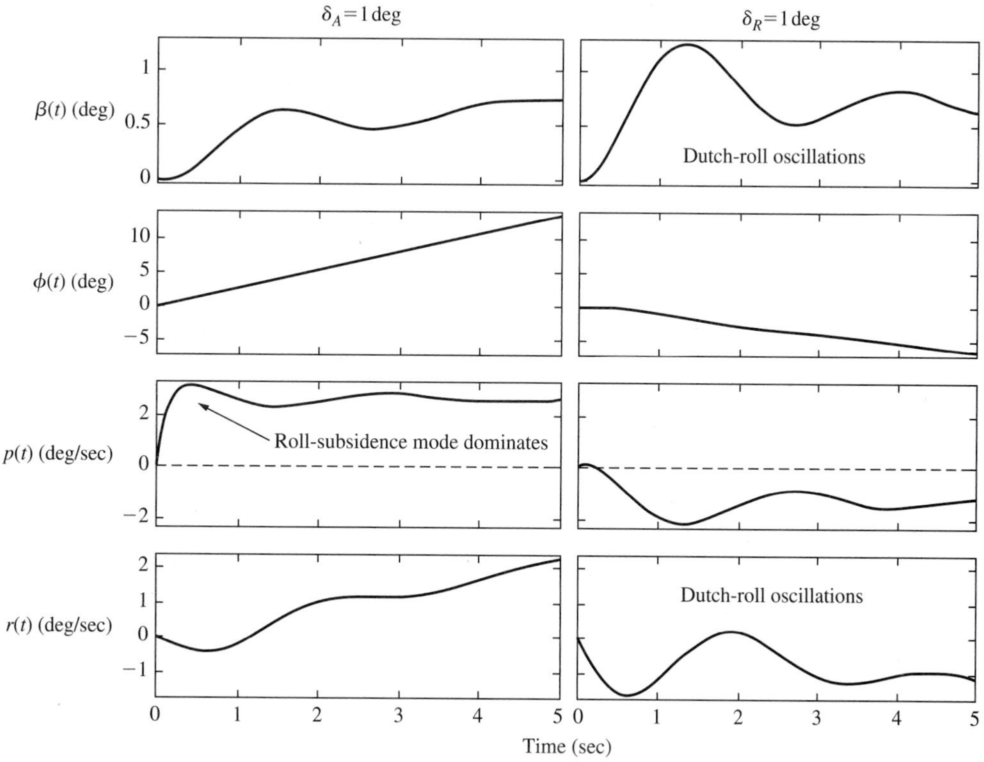

Figure 10.21 Lateral-directional step responses of the Navion aircraft.

Again, virtually all conventional aircraft exhibit lateral-directional modes of motion similar to those described for the Navion aircraft. That is, the modal responses consist of a lightly damped roll-yaw oscillatory dutch-roll mode, a real roll-subsidence mode, and a real spiral mode. However, as with the longitudinal axes, there are exceptions to this modal characterization. For example, sometimes the roll-subsidence and spiral modes coalesce into an oscillatory *roll-spiral mode.* So a complete eigen-analysis is always recommended when assessing a vehicle's dynamics.

10.7 APPROXIMATE MODELS FOR AIRCRAFT LATERAL-DIRECTIONAL DYNAMICS

Based on the knowledge of the lateral-directional natural modes gained from the above analysis, we are now ready to develop some simpler models for the lateral-directional dynamics. Recall that the main objective in developing such models is to gain insight into the major design factors (vehicle aerodynamic characteristics, geometry, etc.) that influence the dynamics.

10.7.1 The Roll-Mode Approximation

The first, and perhaps most straighforward approximation is the *roll-mode approximation.* This approximation is frequently used in practice, and the roll mode is an important component of an aircraft's dynamics. We saw in the above case study of the Navion's lateral-directional dynamics that the roll-rate or roll-subsidence modal response was first-order, containing virtually no sideslip or yaw rate. Hence a simple single-degree-of-freedom model is suggested.

Starting with the state-variable model for the dynamics given in Equations (10.70) and (10.71), eliminating the equations for sideslip angle and yaw rate, and setting these two variables to zero in the equation for roll rate yields

$$\boxed{\dot{p}(t) \approx L'_p p + L'_{\delta_A}\delta_A + L'_{\delta_R}\delta_R} \tag{10.146}$$

Or if bank angle is included we have

$$\begin{Bmatrix} \dot{\phi} \\ \dot{p} \end{Bmatrix} = \begin{bmatrix} 0 & 1 \\ 0 & L'_p \end{bmatrix} \begin{Bmatrix} \phi \\ p \end{Bmatrix} + \begin{bmatrix} 0 & 0 \\ L'_{\delta_A} & L'_{\delta_R} \end{bmatrix} \begin{Bmatrix} \delta_A \\ \delta_R \end{Bmatrix} \tag{10.147}$$

Under this approximation the roll-rate transfer function with aileron input is simply

$$\boxed{\frac{p(s)}{\delta_A(s)} \approx \frac{L'_{\delta_A}}{s - L'_p}} \tag{10.148}$$

Note that this transfer function depends on only two dimensional derivatives, and the simplicity is one of the attractive features of this approximation.

Again recalling the results from the case study in Example 10.4, we saw that Equation (10.148) is indeed a good approximation for this transfer function for the Navion. However, Equation (10.146) may not yield such accurate results for the transfer function for rudder input. But this model is most frequently used with aileron input, with the case with rudder input taking on less importance.

The two eigenvalue–eigenvector pairs for the approximation given in Equation (10.147) also agree quite well with the results obtained for the Navion in Example 10.4. For the full-order model, the roll-mode eigenvalue was $\lambda_R = -8.435$ /sec compared to -8.402 /sec under this approximation. (What would the phasor diagram look like for this eigenvector from the approximation?)

As a final point we noted that for some aircraft in certain flight conditions, there are not two separate, real roll and spiral modes. Rather, an oscillatory roll-spiral mode may exist instead. Obviously in such cases the roll-mode approximation given here is no longer appropriate.

10.7.2 The Dutch-Roll Approximation

The second lateral-directional approximation of interest is the *dutch-roll approximation.* Remember that the dutch-roll mode for the Navion was a lightly damped oscillatory mode with a response consisting of out-of-phase rolling and yawing. Since this mode's eigenvector indicated that this mode involved all the vehicle's lateral-directional responses, we may expect that a low-order approximation is not obvious. However, we still seek to determine the parameters that dominate this mode's characteristics.

Some texts (e.g., Refs. 5 and 6) eliminate the rolling-moment equation in an attempt to obtain a low-order dutch-roll approximation. But this is not supported by the eigen-analysis. From Figure 10.21, it is clear that there is a significant amount of roll rate in the Navion's dutch-roll mode. A better approach (Ref. 7) is to keep all three degrees of freedom and just ignore some of the smaller terms. Therefore, instead of eliminating the rolling-moment equation, let us instead ignore the coefficient of bank-angle and retain only the Y_β coefficient in the $\dot{\beta}$ equation. In addition, let us ignore the relatively small cross-coupling terms $(L'_r \text{ and } N'_p)$ in the $\dot{p}$ and $\dot{r}$ equations.

Under this dutch-roll approximation, and for level flight ($\Theta_0 = 0$), the polynomial-matrix description of the lateral-directional dynamics from Equation (10.66*b*) is

$$\begin{bmatrix} U_0 s - Y_\beta & 0 & U_0 \\ -L'_\beta & s^2 - L'_p s & 0 \\ -N'_\beta & 0 & s - N'_r \end{bmatrix} \begin{Bmatrix} \beta(s) \\ \phi(s) \\ r(s) \end{Bmatrix} = \begin{bmatrix} Y_{\delta_A} & Y_{\delta_r} \\ L'_{\delta_A} & L'_{\delta_R} \\ N'_{\delta_A} & N'_{\delta_R} \end{bmatrix} \begin{Bmatrix} \delta_A(s) \\ \delta_R(s) \end{Bmatrix} \tag{10.149}$$

So the transfer functions, with $\bullet = \delta_A$ or δ_R, become

$$\frac{\beta(s)}{\bullet(s)} = \left(Y_\bullet(s^2 - L_p's)(s - N_r') + N_\bullet' U_0(s^2 - L_p's)\right)/\det\mathbf{P}(s)$$
$$= Y_\bullet s(s - L_p')\left(s + (N_\bullet' U_0/Y_\bullet - N_r')\right)/\det\mathbf{P}(s)$$
$$\frac{\phi(s)}{\bullet(s)} = \left(L_\bullet'(U_0 s - Y_\beta)(s - N_r') + Y_\bullet L_\beta'(s - N_r') - U_0(N_\bullet' L_\beta' - L_\bullet' N_\beta')\right)/\det\mathbf{P}(s)$$
$$= \left(L_\bullet' U_0 s^2 - \left(L_\bullet'(U_0 N_r' + Y_\beta) - (Y_\bullet L_\beta')\right)s - \left(U_0(N_\bullet' L_\beta' - L_\bullet' N_\beta') + Y_\bullet L_\beta' N_r' - L_\bullet' N_r' Y_\beta\right)\right)/\det\mathbf{P}(s)$$
$$\frac{r(s)}{\bullet(s)} = \left(N_\bullet'(s^2 - L_p's)(U_0 s - Y_\beta) + Y_\bullet N_\beta'(s^2 - L_p's)\right)/\det\mathbf{P}(s)$$
$$= N_\bullet' s(s - L_p')\left(U_0 s + (Y_\bullet N_\beta'/N_\bullet' - Y_\beta)\right)/\det\mathbf{P}(s),\ \bullet = \delta_A \text{ or } \delta_R \tag{10.150}$$

The characteristic polynomial is given by

$$\det\mathbf{P}(s) = (s^2 - L_p's)(U_0 s - Y_\beta)(s - N_r') + U_0 N_\beta'(s^2 - L_p's)$$
$$= s(s - L_p')\left(s^2 - (N_r' + Y_\beta/U_0)s + (N_\beta' + N_r' Y_\beta/U_0)\right) \tag{10.151}$$

This polynomial is seen to contain a quadratic term approximating the dutch-roll factor, a first-order term that is the same as in the roll-mode approximation, and a root at the origin approximating the spiral root. Note that these two real poles are exactly cancelled by numerator zeros in the sideslip-angle and yaw-rate transfer functions. This leads to much simpler expressions for these transfer functions than would be assumed based on a casual evaluation of Equations (10.150) and (10.151). As a result of these pole-zero cancellations, the character of the sideslip and yaw-rate responses depends only on the dutch-roll quadratic under this approximation.

Furthermore, the dutch-roll undamped natural frequency is given by

$$\omega_{DR}^2 = \left(N_\beta' + N_r' Y_\beta/U_0\right) \tag{10.152}$$

which is typically dominated by the weathercock stability derivative N_β'. The dutch-roll damping under this approximation is determined from

$$2\zeta_{DR}\omega_{DR} = -\left(N_r' + Y_\beta/U_0\right) \tag{10.153}$$

in which the largest term is typically the yaw-damping stability derivative N_r'. Consequently we find some additional connections between static and dynamic stability. Under this dutch-roll approximation, increasing both the static directional ($N_\beta > 0$) and yaw-damping ($N_r < 0$) stability increases the likelihood of dynamic stability of the dutch-roll mode.

It is interesting to note that the dutch-roll quadratic in Equation (10.151) would appear to be independent of the vehicle's dihedral coefficient L_β, which is usually a significant term in the rolling-moment equation. However, upon closer inspection of Equation (10.152), and from Equations (10.46), we note that

$$N'_\beta = \left(N_\beta + L_\beta I_{xz}/I_{zz}\right)D \tag{10.154}$$

Hence under this approximation the dutch-roll natural frequency is in fact a function of the vehicle's dihedral effect if the cross product of inertia is not zero.

Let us now compare the results from this approximation with those from the full-order model evaluated in Example 10.4. Instead of Equations (10.141), but using the same response **y**, state **x**, and control **u** vectors, the state-variable model for the Navion under this dutch-roll approximation becomes

$$\mathbf{A}_{\text{DR}} = \begin{bmatrix} \dfrac{Y_\beta}{U_0} & 0 & 0 & -1 \\ 0 & 0 & 1 & 0 \\ L'_\beta & 0 & L'_p & 0 \\ N'_\beta & 0 & 0 & N'_r \end{bmatrix} = \begin{bmatrix} -0.2543 & 0 & 0 & -1 \\ 0 & 0 & 1 & 0 \\ -15.982 & 0 & -8.402 & 0 \\ 4.495 & 0 & 0 & -0.7605 \end{bmatrix}$$

$$\mathbf{B}_{\text{DR}} = \begin{bmatrix} \dfrac{Y_{\delta_A}}{U_0} & \dfrac{Y_{\delta_R}}{U_0} \\ 0 & 0 \\ L'_{\delta_A} & L'_{\delta_R} \\ N'_{\delta_A} & N'_{\delta_R} \end{bmatrix} = \begin{bmatrix} 0 & 0.0708 \\ 0 & 0 \\ 28.984 & 2.548 \\ -0.2218 & -4.597 \end{bmatrix} \tag{10.155}$$

$$\mathbf{C} = \mathbf{I}_4, \text{ and } \mathbf{D} = \mathbf{0}$$

Using MATLAB, the eigenvalues and phasor diagrams of the eigenvectors of $\mathbf{A}_{\text{DR}}$ are as shown in Figure 10.22.

The dutch-roll undamped natural frequencies agree rather well with those from the full-order model (ω_{DR} = 2.165 rad/sec compared to 2.385 rad/sec),

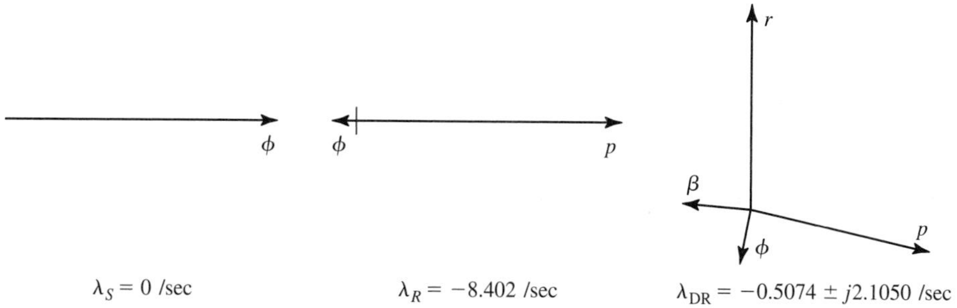

Figure 10.22 Phasor diagrams from the dutch-roll approximation for the Navion.

as do the damping ratios ($\zeta_{DR} = 0.234$ compared to 0.204). The roll and spiral eigenvalues also show good agreement, as do the phasor diagrams when compared to those from the full-order model shown in Figure 10.17.

From MATLAB, the six Navion transfer functions from this dutch-roll approximation are given below. For aileron input we have

$$\frac{\beta(s)}{\delta_A(s)} = \frac{0.2218s(s+8.402)}{s(s+8.402)(s^2+1.015s+4.688)} \text{ rad/rad or deg/deg}$$

$$\frac{\phi(s)}{\delta_A(s)} = \frac{28.984(s^2+1.015s+4.566)}{s(s+8.402)(s^2+1.015s+4.688)} \text{ rad/rad or deg/deg} \qquad (10.156)$$

$$\frac{r(s)}{\delta_A(s)} = \frac{-0.2218s(s+0.2543)(s+8.402)}{s(s+8.402)(s^2+1.015s+4.688)} \text{ rad/sec/rad or deg/sec/deg}$$

Here note that the bank-angle transfer function with aileron input is almost exactly the same as that from the roll-mode approximation. For rudder input we have

$$\frac{\beta(s)}{\delta_R(s)} = \frac{0.0708s(s+8.402)(s+65.69)}{s(s+8.402)(s^2+1.015s+4.688)} \text{ rad/rad or deg/deg}$$

$$\frac{\phi(s)}{\delta_R(s)} = \frac{2.548(s-4.671)(s+5.242)}{s(s+8.402)(s^2+1.015s+4.688)} \text{ rad/rad or deg/deg} \qquad (10.157)$$

$$\frac{r(s)}{\delta_R(s)} = \frac{-4.597s(s+0.1851)(s+8.402)}{s(s+8.402)(s^2+1.015s+4.688)} \text{ rad/sec/rad or deg/sec/deg}$$

The accuracy of these six transfer functions is perhaps best indicated by the step responses shown in Figure 10.23, obtained using this dutch-roll approximation, when compared to the responses from the full-order model, given in Figure 10.21. All the responses to rudder input compare quite well. And for aileron input, the approximation is seen to yield good results for the roll-rate and bank-angle responses. But the results for sideslip angle and yaw rate are much poorer, agreeing more with those from the roll-mode approximation. (The dashed lines indicate steady-state responses.) Hence caution is advised when using these latter two transfer functions under this dutch-roll approximation.

10.7.3 The Spiral Approximation

The final lateral-directional approximation to be considered is the *spiral approximation.* This mode has a long time constant (or time to double amplitude, if unstable), so as noted in Ref. 8 it is reasonable to assume that the degrees of freedom other than bank angle (and heading) reach their quasi-steady values quickly relative to this mode's time constant. Hence under this spiral approximation we will set the rates of change of sideslip angle and roll and yaw rates all to zero,

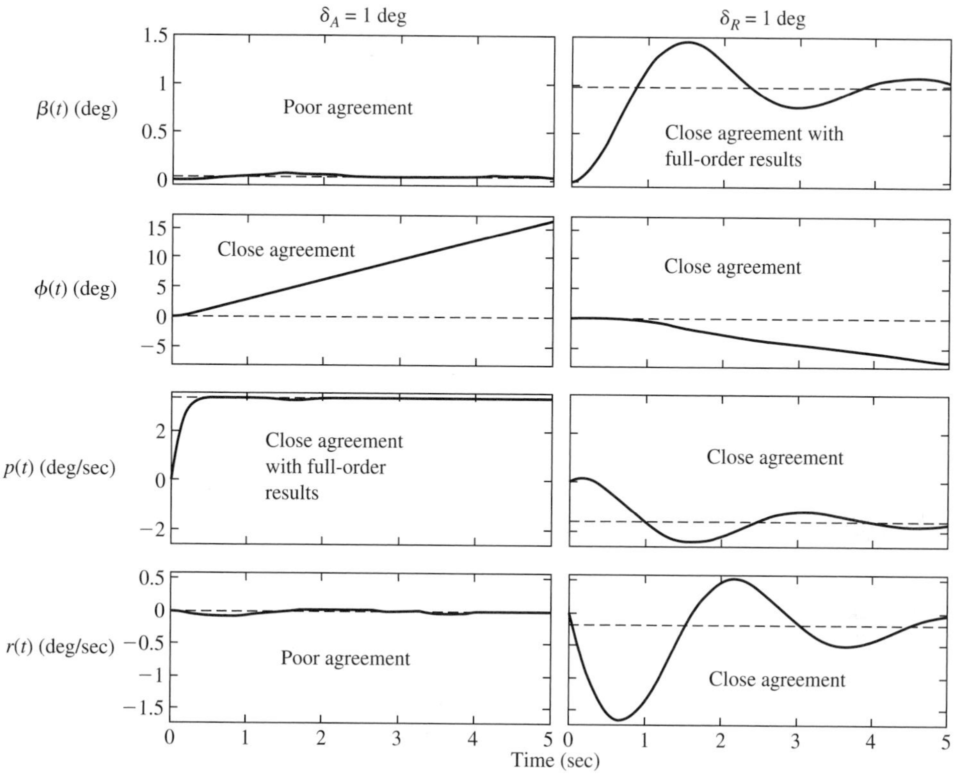

Figure 10.23 Step responses from the dutch-roll approximation for the Navion.

and use the resulting three algebraic equations to eliminate these three variables from the model. This may be accomplished using either the state-variable model or the polynomial-matrix description.

For example, under this approximation, and in level flight, the polynomial-matrix description of the system (Equation (10.66*b*)) becomes

$$\begin{bmatrix} -Y_\beta & -(Y_p s + g) & (U_0 - Y_r) \\ -L'_\beta & -L'_p s & -L'_r \\ -N'_\beta & -N'_p s & -N'_r \end{bmatrix} \begin{Bmatrix} \beta(s) \\ \phi(s) \\ r(s) \end{Bmatrix} = \begin{bmatrix} Y_{\delta_A} & Y_{\delta_R} \\ L'_{\delta_A} & L'_{\delta_R} \\ N'_{\delta_A} & N'_{\delta_R} \end{bmatrix} \begin{Bmatrix} \delta_A(s) \\ \delta_R(s) \end{Bmatrix} \tag{10.158}$$

for which the first-order characteristic equation is

$$\begin{aligned} \det \mathbf{P}(s) &= Y_\beta(L'_r N'_p - N'_r L'_p)s - (Y_p s + g)(L'_r N'_\beta - N'_r L'_\beta) + (U_0 - Y_r)(L'_\beta N'_p - N'_\beta L'_p)s \\ &= \left(Y_\beta(L'_r N'_p - N'_r L'_p) + Y_p(N'_r L'_\beta - L'_r N'_\beta) + (U_0 - Y_r)(L'_\beta N'_p - N'_\beta L'_p)\right)s \\ &\quad + g(N'_r L'_\beta - L'_r N'_\beta) \end{aligned} \tag{10.159}$$

This characteristic polynomial has a single spiral root given by

$$\lambda_S = \frac{g\left(L'_r N'_\beta - N'_r L'_\beta\right)}{Y_\beta\left(L'_r N'_p - N'_r L'_p\right) + Y_p\left(N'_r L'_\beta - L'_r N'_\beta\right) + \left(U_0 - Y_r\right)\left(L'_\beta N'_p - N'_\beta L'_p\right)} \approx \frac{g\left(L'_r N'_\beta - N'_r L'_\beta\right)}{\left(Y_\beta\left(L'_r N'_p - N'_r L'_p\right) + U_0\left(L'_\beta N'_p - N'_\beta L'_p\right)\right)} \tag{10.160}$$

where the last approximation assumes that $Y_p = Y_r = 0$. For the Navion, this approximation yields a spiral pole at -0.00930 /sec compared to -0.00896 /sec from the full-order model.

Typically, the denominator in Equation (10.160) is positive. Hence, stability of this spiral eigenvalue depends on the sign of $\left(L'_r N'_\beta - N'_r L'_\beta\right)$. This term will be negative (stable) when the vehicle exhibits sufficient yaw damping N'_r and dihedral effect L'_β, with sufficiently small static directional stability (N'_β not too negative) and rolling moment due to yaw rate L'_r (which is usually positive). So once again we see connections between static and dynamic stability. Under this spiral approximation, increased static yaw-damping and dihedral-effect stability increases the likelihood of dynamic stability. However, too much static directional stability may have a negative effect on dynamic stability of the spiral mode.

Finally, the bank-angle transfer functions for this single-degree-of-freedom system are simply given by

$$\frac{\phi(s)}{\bullet(s)} \approx \left(Y_\beta\left(L'_\bullet N'_r - N'_\bullet L'_r\right) + Y_\bullet\left(L'_r N'_\beta - N'_r L'_\beta\right) + \left(U_0 - Y_r\right)\left(L'_\bullet N'_\beta - N'_\bullet L'_\beta\right)\right)/\det \mathbf{P}(s)$$

$$= \frac{K}{s - \lambda_S}, \quad \bullet = \delta_A \text{ or } \delta_R \tag{10.161}$$

The other transfer functions may be found as needed from Equations (10.158) using Cramer's rule.

Another spiral approximation, well known for a long time, is to simply use the last two coefficients in the lateral-directional characteristic polynomial as an approximation for the spiral pole ($-E/D$ in Equations (10.69)). This is justified because the magnitude of the spiral root is known to be small, and hence higher-order terms in s in the characteristic polynomial may reasonably be neglected.

By comparing the result from Equations (10.69) with that in Equation (10.160) we see that $-E$ appears in the numerator of Equation (10.160). Hence both approximations indicate the same potential causes for spiral instability. For the Navion aircraft, setting the spiral pole equal to $-E/D$ yields a pole at -0.00873 /sec—very close to that from the full-order model. However, the spiral approximation presented earlier in this section provides a more complete system description.

10.8 CONFIGURATION DESIGN TO ACHIEVE DESIRABLE DYNAMIC CHARACTERISTICS

We will now turn our attention to demonstrating by examples how the approximate models presented in Sections 10.6 and 10.7 may be used in aircraft-configuration design. Historically, vehicle designers only considered static stability criteria in the preliminary design of aircraft. These criteria were used to size the horizontal and vertical tails and control surfaces, for example. But today it is generally recognized that this approach is not acceptable, and that the dynamic characteristics of the vehicle should be considered early in the design cycle.

We have noted frequently in this chapter that one key objective of our analysis is to understand not only the vehicle's dynamic characteristics, but also the causes which lead to those characteristics. For example, using the knowledge of critical dimensional derivatives gained from the simplified models, and the expressions derived in Chapter 6 relating the vehicle's aerodynamic characteristics to the vehicle's geometry, we may determine how to adjust the geometry to tailor the dynamic characteristics. Sometimes, however, such an approach leads to stabilizing surfaces that are too large, leading to excessive drag and performance penalties. In such cases the designer might choose to use slightly smaller surfaces, and further tailor the dynamics using feedback-control devices. Such feedback stability augmentation is the topic of Chapter 11.

10.8.1 Effects of Static Margin and Tail Size on the Longitudinal Eigenvalues

To begin our discussion, it is instructive to investigate the sensitivity of the longitudinal eigenvalues to variations in two key design parameters—*c.g.* location and the size of the horizontal tail. Limits on *c.g.* location establish allowable operating conditions for the vehicle, while tail size is a critical aspect of configuration design. To perform this study, data for the Navion aircraft from Appendix B will again be used.

We will first perform a parametric analysis using the *c.g.* location. For the Navion data in Appendix B, the *c.g.* location (in percent MAC) is $\overline{X}_{\text{cg}} = 29.5\%$ and the static margin is

$$SM \triangleq \overline{X}_{\text{AC}} - \overline{X}_{\text{cg}} = -\frac{C_{M_\alpha}}{C_{L_\alpha}} = -\frac{-0.683}{4.44} = 15.4\% \qquad (10.162)$$

The two aerodynamic coefficients most influenced by *c.g.* location are C_{M_α} and C_{M_q} (or M_α and M_q), and we will assume the effects on the remaining aerodynamic coefficients are negligible. Since C_{M_α} is directly proportional to static margin, we see that M_α is proportional to *SM* as well. Or since

$$C_{M_\alpha} = -SMC_{L_\alpha} = -4.44SM$$

then at sea level with $U_0 = 176$ fps

$$M_\alpha = -4.44SM\frac{q_\infty S_W\bar{c}_W}{I_{yy}} = -57.12SM \tag{10.163}$$

We will now take advantage of the closed-form expressions derived in Chapter 6 for the aerodynamic coefficients. Namely, from Equation (6.188) we see that C_{M_q} can also be expressed in terms of *SM*, or

$$C_{M_q} \approx -C_{L_{\alpha_H}}\frac{(X_{AC_H}-X_{cg})^2}{U_0\bar{c}_W}\frac{q_H}{q_\infty}\frac{S_H}{S_W} = -C_{L_{\alpha_H}}\frac{\bar{c}_W(\bar{X}_{AC_H}-\bar{X}_{cg})^2}{U_0}\frac{q_H}{q_\infty}\frac{S_H}{S_W}$$
$$= -C_{L_{\alpha_H}}\frac{\bar{c}_W((\bar{X}_{AC_H}-\bar{X}_{AC})+SM)^2}{U_0}\frac{q_H}{q_\infty}\frac{S_H}{S_W} = K((\bar{X}_{AC_H}-\bar{X}_{AC})+SM)^2 \tag{10.164}$$

Now assuming that for the Navion[2]

$$\bar{X}_{AC_H}-\bar{X}_{cg} = 15.5/5.7 = 2.72 \tag{10.165}$$

then

$$(\bar{X}_{AC_H}-\bar{X}_{AC}) = (\bar{X}_{AC_H}-\bar{X}_{cg}) - (\bar{X}_{AC}-\bar{X}_{cg}) = 2.72 - 0.154 = 2.566 \tag{10.166}$$

Noting that for the Navion (Appendix B)

$$C_{M_q} = -0.161 \text{ sec}$$

we then have, using Equation (10.164)

$$C_{M_q} = -0.0218(2.566 + SM)^2$$

and at the given flight condition

$$M_q = -0.0218(2.57 + SM)^2\frac{q_\infty S_W\bar{c}_W}{I_{yy}} = -0.280(2.57+SM)^2 \tag{10.167}$$

Substituting these expressions for M_α and M_q into the state-variable description of the Navion's longitudinal dynamics (Equations (10.64)), we may parametrically solve for the longitudinal eigenvalues as a function of *SM* or *c.g.* location.

A locus of the resulting longitudinal eigenvalues is shown in Figure 10.24, as the static margin is varied from its original value (0.154) to −0.044 in increments of 0.01. Four branches of the root locus begin at the locations of the original roots. As static margin is decreased, the roots move along these branches and eventually converge on the real axis. Then after all four roots move along the real axis, two of the roots again coalesce, branch off the real axis, and move

[2] http://www.eaa62.org/technotes/tail.htm accessed November 2009.

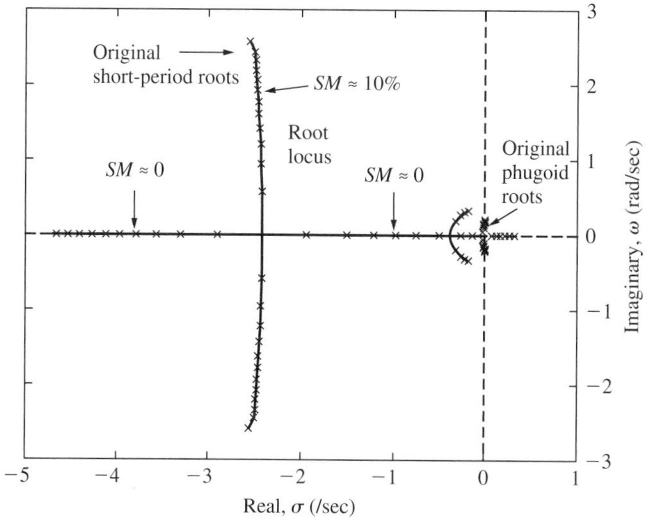

Figure 10.24 Locus of longitudinal eigenvalues as a function of static margin.

along curved paths in the direction of the original phugoid roots. The remaining two roots remain on the real axis, and one of these eventually becomes unstable, moving into the right half of the complex plane (i.e., positive real part).

It is important to note that the associated eigenvectors reveal that conventional short-period and phugoid modal responses disappear when the *SM* approaches zero, and the modes remain very unconventional as the *SM* continues to decrease. For example, one mode sometimes becomes dominated by angle-of-attack, or the largest modal response is angle of attack instead of pitch rate. While other times there are two modes—one first order and the other oscillatory—that have large surge-velocity responses similar to a phugoid mode. Therefore, to preserve traditional modal characteristics in this case it would be important to maintain at least a positive static margin, and it might be desirable to maintain a minimum static margin of about 10 percent. As the static margin increases above 10 percent, both the undamped natural frequency and the damping ratio of the short-period mode increase.

We will next address the parametric behavior of the longitudinal eigenvalues as the size of the horizontal tail is varied. We will again assume that the only vehicle aerodynamic coefficients affected are C_{M_α} and C_{M_q}, since the other coefficients like C_{L_α} should not vary significantly over the tail sizes to be considered.

From Equation (10.164) we see that C_{M_q} is proportional to the tail-area ratio S_H/S_W. Plus, from Equation (6.56) (for X measure positive aft) we have

$$C_{M_\alpha} = C_{L_{\alpha_W}}(\overline{X}_{cg} - \overline{X}_{AC_{W\&F}}) - C_{L_{\alpha_H}}\left(1 - \frac{d\varepsilon}{d\alpha}\right)(\overline{X}_{AC_H} - \overline{X}_{cg})\frac{q_H}{q_\infty}\frac{S_H}{S_W} \quad (10.168)$$

So this coefficient is linear in tail-area ratio. Again for the Navion we have

$$\overline{X}_{AC_H} - \overline{X}_{cg} = 2.72$$

$$\frac{S_H}{S_W} = 0.233$$

and we will assume that

$$C_{L_\alpha} \approx C_{L_{\alpha_W}} \approx C_{L_{\alpha_H}} = 4.4 \text{ /rad}$$

$$1 - \frac{d\varepsilon}{d\alpha} \approx 0.5 \tag{10.169}$$

$$\frac{q_H}{q_\infty} \approx 0.9$$

Now given that for the Navion $C_{M_\alpha} = -0.683$, using the above data we find that

$$\left(\overline{X}_{cg} - \overline{X}_{AC_{W\,\&\,F}}\right) \approx 0.129 \tag{10.170}$$

So given that $C_{M_q} = -0.161$ sec for the Navion, the expressions for the coefficients are

$$\begin{aligned} C_{M_q} &= -0.691\frac{S_H}{S_W} & M_q &= C_{M_q}\frac{q_\infty S_W \bar{c}_W}{I_{yy}} = -8.89\frac{S_H}{S_W} \\ C_{M_\alpha} &= 0.568 - 5.386\frac{S_H}{S_W} & M_\alpha &= C_{M_\alpha}\frac{q_\infty S_W \bar{c}_W}{I_{yy}} = \left(7.31 - 69.29\frac{S_H}{S_W}\right) \end{aligned} \tag{10.171}$$

We can now substitute these expressions into the state-variable model for the Navion's longitudinal dynamics and parametrically solve for the longitudinal eigenvalues.

The plot in Figure 10.25 shows the locus of the longitudinal eigenvalues as the tail-area ratio is reduced from its original value (0.233) to 0.093 in increments of 0.01. This root locus looks only slightly different from that shown previously for parametric *c.g.* variations. The same number of branches exist, four branches begin at the locations of the original four eigenvalues, the four roots coalesce onto the real axis as tail size is decreased, and then two eventually coalesce again and split off to form two additional branches. One root eventually becomes unstable moving into the right half of the complex plane.

Again the associated eigenvectors reveal that classical short-period and phugoid modal behavior disappears after the tail area ratio has been reduced approximately 40 percent (tail-area ratio = 0.133). After this point one mode sometimes becomes dominated by angle-of-attack, or the largest modal response is angle of attack instead of pitch rate. And other times there are two

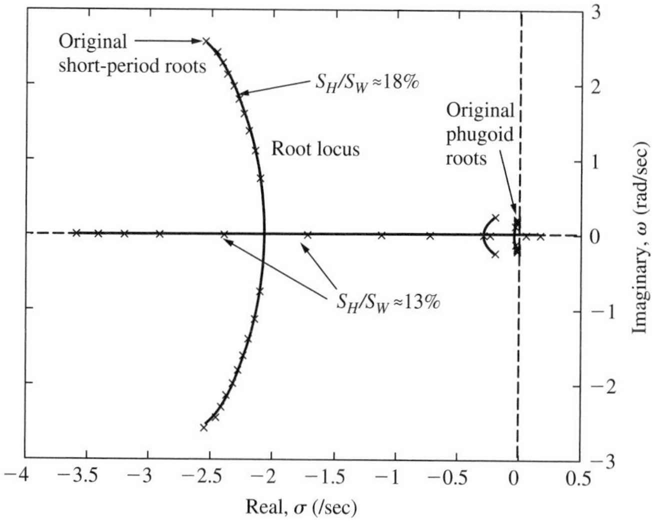

Figure 10.25 Locus of longitudinal eigenvalues as a function of horizontal tail size.

modes—one first order and the other oscillatory—that have large surge-velocity responses similar to a phugoid mode. Thus to preserve conventional modal characteristics in this case, the tail-area ratio must be larger than 13 percent, and preferably larger than 18 percent to achieve sufficient short-period damping.

10.8.2 Improving Spiral and Dutch-Roll Stability

This case, taken from Ref. 9, involves another general-aviation aircraft. The vehicle is a light, single-engine aircraft, with a high unswept wing and tricycle landing gear. The basic data for this vehicle is given in Table 10.2. The vehicle's spiral mode is slightly unstable (time to double amplitude = 200 sec), and the dutch-roll damping is low with $\zeta_{DR} = 0.065$. We want to stabilize the spiral mode with a time to half amplitude of 50 sec, and to increase the dutch-roll damping by approximately 50 percent.

Table 10.2 Characteristics of a Single-Engine, High-Wing, General Aviation Aircraft

W = 2600 lbs	U_0 = 107 fps	$I_{xz} = 0$
$S_W = 180\ \text{ft}^2$	$b_W = 36.9$ ft	$\Gamma_W = 3$ deg
$S_H = 41.4\ \text{ft}^2$	$b_H = 11.7$ ft	$S_V = 18\ \text{ft}^2$
$C_{Y_\beta} = -0.303$	$C_{Y_p} = -0.0376$ sec	$C_{Y_r} = 0.0347$ sec
$C_{L_\beta} = -0.122$	$C_{L_p} = -0.0834$ sec	$C_{L_r} = 0.0352$ sec
$C_{N_\beta} = 0.0701$	$C_{N_p} = -0.0166$ sec	$C_{N_r} = -0.0198$ sec

From the spiral approximation presented in Section 10.7.3, we noted when discussing Equation (10.160) that the spiral mode will be stable when the vehicle exhibits sufficient yaw damping N_r' and dihedral effect L_β', with sufficiently small yaw stiffness N_β'. This suggests increasing the vehicle's dihedral effect to improve spiral stability. Noting that for this vehicle $I_{xz} = 0$, we have the dihedral effect given by

$$L_\beta' \triangleq \frac{\left(L_\beta + N_\beta I_{xz}/I_{xx}\right)}{1 - I_{xz}^2/\left(I_{xx} I_{zz}\right)} = L_\beta = C_{L_\beta} \frac{q_\infty S_W b_W}{I_{xx}} \tag{10.172}$$

Plus, from Equation (6.44) we see that the wing and both tail surfaces contribute to C_{L_β}, with the wing providing the greatest contribution. Then we saw in Chapter 5 that for an unswept wing, increasing dihedral angle Γ_W increases the wing's dihedral effect. And, in fact, for the vehicle in question, it was determined that the desired spiral stability of the vehicle can be achieved by increasing the wing's dihedral angle from 3 deg to 6.5 deg.

Now with respect to increasing the dutch-roll damping, we note from the dutch-roll approximation and Equation (10.153) that the dutch-roll damping is significantly affected by the yaw damping N_r'. And for this vehicle, since $I_{xz} = 0$, we have the yaw damping given by

$$N_r' = N_r = C_{N_r} \frac{q_\infty S_W b_W}{I_{zz}} \tag{10.173}$$

Further, from Equation (6.219) we note that

$$C_{N_r} \approx -C_{L_{\alpha_V}} \frac{\left(X_{\text{Ref}} - X_{\text{AC}_V}\right)^2}{U_0 b_W} \frac{q_H}{q_\infty} \frac{S_V}{S_W} \tag{10.174}$$

So the dutch-roll damping is in turn significantly affected by the size of the vertical tail S_V. And, in fact, it was later determined that sufficient dutch-roll damping could be achieved for this vehicle by increasing the size of the vertical tail from 18 ft^2 to 30 ft^2.

10.9 CROSS-AXIS COUPLING

Up to this point in the chapter we have been dealing exclusively with the flight dynamics corresponding to flight conditions that admit a decoupling between the longitudinal and the lateral-directional axes. That is, strictly speaking we have restricted our analysis to steady rectilinear flight with no bank angle or sideslip. But for other flight conditions, such axes decoupling does not strictly exist. For flight conditions with modest bank angle and/or sideslip, the coupling may not be strong, so the decoupled longitudinal and lateral-direction models may still yield adequate results. But for flight conditions with large bank angle, roll or turning rates, or sideslip, the effects of coupling may be significant. In such cases, an analysis using a full six-degree-of-freedom linear model, built from Equations (10.36–10.46), is required to assess the significance of the coupling.

To begin the discussion of the coupled dynamic model, let's first consider the structure of the polynomial-matrix description of the complete system. Let's write this system model in the following form.

$$\begin{bmatrix} \mathbf{P}_{\text{Long}}(s) & \mathbf{P}_{\text{Long-Lat}}(s) \\ \mathbf{P}_{\text{Lat-Long}}(s) & \mathbf{P}_{\text{Lat-Dir}}(s) \end{bmatrix} \begin{Bmatrix} \mathbf{y}_{\text{Long}}(s) \\ \mathbf{y}_{\text{Lat-Dir}}(s) \end{Bmatrix} = \begin{bmatrix} \mathbf{Q}_{\text{Long}}(s) & \mathbf{Q}_{\text{Long-Lat}}(s) \\ \mathbf{Q}_{\text{Lat-Long}}(s) & \mathbf{Q}_{\text{Lat-Dir}}(s) \end{bmatrix} \begin{Bmatrix} \mathbf{u}_{\text{Long}}(s) \\ \mathbf{u}_{\text{Lat-Dir}}(s) \end{Bmatrix} \tag{10.175}$$

Here $\mathbf{P}_{\text{Long}}$ and $\mathbf{Q}_{\text{Long}}$ are the polynomial matrices and $\mathbf{y}_{\text{Long}}$ and $\mathbf{u}_{\text{Long}}$ are the vectors of responses and control inputs, respectively, associated with the longitudinal axes, analogous to those given in Equations (10.56). For example, let the response and control-input vectors here be given by

$$\begin{aligned} \mathbf{y}_{\text{Long}}^T(s) &= \begin{bmatrix} u(s) & \alpha(s) & q(s) & \theta(s) \end{bmatrix} \\ \mathbf{u}_{\text{Long}}^T(s) &= \begin{bmatrix} \delta_E(s) & \delta T(s) \end{bmatrix} \end{aligned} \tag{10.176}$$

Note that both pitch rate $q(s)$ and attitude $\theta(s)$ must be included in the response vector now because the simple kinematic relationships valid in rectilinear flight (e.g., $s\theta(s) = q(s)$) are no longer valid. The more general kinematic equation for $\dot{\theta}$ given in Equations (10.41) must now be used. Finally, we now have the two polynomial matrices associated with the longitudinal axes given by

$$\begin{aligned} \mathbf{P}_{\text{Long}}(s) &= \begin{bmatrix} s-(X_u+X_{P_u}) & -(X_{\dot{\alpha}}s+X_\alpha)+\mathrm{Q}_0\mathrm{U}_0 & -(X_q-\mathrm{W}_0) & g\cos\Theta_0 \\ -(Z_u+Z_{P_u})-\mathrm{Q}_0 & (\mathrm{U}_0-Z_{\dot{\alpha}})s-Z_\alpha & -(Z_q+\mathrm{U}_0) & g\sin\Theta_0\cos\Phi_0 \\ -(M_u+M_{P_u}) & -\left(M_{\dot{\alpha}}s+(M_\alpha+M_{P_\alpha})\right) & s-M_q & 0 \\ 0 & 0 & -1 & s\cos\Phi_0-\dot{\Psi}_0\sin\Theta_0\sin\Phi_0 \end{bmatrix} \\ \mathbf{Q}_{\text{Long}}(s) &= \begin{bmatrix} X_{\delta_E} & X_T \\ Z_{\delta_E} & Z_T \\ M_{\delta_E} & M_T \\ 0 & 0 \end{bmatrix} \end{aligned} \tag{10.177}$$

Likewise, $\mathbf{P}_{\text{Lat-Dir}}$ and $\mathbf{Q}_{\text{Lat-Dir}}$ appearing in Equations (10.175) are the polynomial matrices, and $\mathbf{y}_{\text{Lat-Dir}}$ and $\mathbf{u}_{\text{Lat-Dir}}$ are the vectors of responses and control inputs, respectively, associated with the lateral-directional axes, analogous to those given in Equations (10.66). For example, we have the lateral-directional responses and control inputs given by

$$\begin{aligned} \mathbf{y}^T_{\text{Lat-Dir}}(s) &= \begin{bmatrix} \beta(s) & p(s) & r(s) & \phi(s) & \psi(s) \end{bmatrix} \\ \mathbf{u}^T_{\text{Lat-Dir}}(s) &= \begin{bmatrix} \delta_A(s) & \delta_R(s) \end{bmatrix} \end{aligned} \tag{10.178}$$

Again note that $\phi(s)$ and $\psi(s)$ must be included in the response vector because the simple kinematic relationships valid in rectilinear flight (e.g., $s\phi(s) = p(s)$) are also no longer valid. The complete kinematic equations for $\dot{\phi}$ and $\dot{\psi}$ given in Equations (10.41) must now be used. Finally, we have the polynomial matrices associated with the lateral-directional axes given by

$$\mathbf{P}_{\text{Lat-Dir}}(s) = \begin{bmatrix} \mathrm{U}_0 s - Y_\beta & -Y_p - \mathrm{W}_0 & (\mathrm{U}_0 - Y_r) & -g\cos\Theta_0\cos\Phi_0 & 0 \\ -L'_\beta & s - L'_p - C_1\mathrm{Q}_0 & -L'_r - C_2\mathrm{Q}_0 & 0 & 0 \\ -N'_\beta & -N'_p - C_3\mathrm{Q}_0 & s - N'_r - C_4\mathrm{Q}_0 & 0 & 0 \\ 0 & -1 & 0 & s & -s\sin\Theta_0 \\ 0 & 0 & -1 & C_5 & s\cos\Theta_0\cos\Phi_0 \end{bmatrix}$$

with

$$C_1 = \left(I_{xx} - I_{yy} + I_{zz}\right)\left(I_{xz}/(I_{xx}I_{zz})\right)D \qquad C_2 = \left(\left(I_{yy} - I_{zz}\right) + I_{xz}^2/I_{zz}\right)\left(1/I_{xx}\right)D$$

$$C_3 = \left(\left(I_{xx} - I_{yy}\right) + I_{xz}^2/I_{xx}\right)\left(1/I_{zz}\right)D \qquad C_4 = -C_1$$

$$C_5 = -\dot{\Psi}_0\cos\Theta_0\sin\Phi_0 - \dot{\Theta}_0\cos\Phi_0 \qquad D = \frac{1}{1 - I_{xz}^2/(I_{xx}I_{zz})}$$

and

$$\mathbf{Q}_{\text{Lat-Dir}}(s) = \begin{bmatrix} Y_{\delta_A} & Y_{\delta_R} \\ L'_{\delta_A} & L'_{\delta_R} \\ N'_{\delta_A} & N'_{\delta_R} \\ 0 & 0 \\ 0 & 0 \end{bmatrix} \tag{10.179}$$

The four new submatrices appearing in Equation (10.175) are those on the off diagonal in the full **P** and **Q** matrices. These submatrices contain the cross-coupling terms, and the first is given as

$$\mathbf{P}_{\text{Long-Lat}}(s) = \begin{bmatrix} -\mathrm{U}_0\mathrm{R}_0 - \boxed{X_\beta} & 0 & -\mathrm{U}_0\beta_0 & 0 & 0 \\ \mathrm{U}_0\mathrm{P}_0 & \mathrm{U}_0\beta_0 & 0 & g\cos\Theta_0\sin\Phi_0 & 0 \\ 0 & C_1 & C_2 & 0 & 0 \\ 0 & 0 & 0 & C_3 & s\cos\Theta_0\sin\Phi_0 \end{bmatrix}$$

$$C_1 = \left((I_{xx} - I_{zz})\mathrm{R}_0 + 2I_{xz}\mathrm{P}_0\right)/I_{yy}$$

$$C_2 = \left((I_{xx} - I_{zz})\mathrm{P}_0 - 2I_{xz}\mathrm{R}_0\right)/I_{yy}$$

$$C_3 = \dot{\Psi}_0\cos\Theta_0\cos\Phi_0 - \dot{\Theta}_0\sin\Phi_0 \qquad (10.180)$$

The boxed matrix element, or X_β, is a new dimensional stability derivative that was not included in the original expansion of the perturbation force f_{A_X}/m (in Equations (10.43)). But for reference flight conditions involving large side-slip angles β_0 this term may be significant.

The three remaining cross-coupling matrices are given by

$$\mathbf{P}_{\text{Lat-Long}}(s) = \begin{bmatrix} \mathrm{R}_0 & -\mathrm{P}_0\mathrm{U}_0 - \boxed{Y_\alpha} & 0 & g\sin\Theta_0\sin\Phi_0 \\ 0 & -\boxed{L'_\alpha} & \left(C_1/I_{xx} + \left(I_{xz}/(I_{xx}I_{zz})\right)C_2\right)D & 0 \\ 0 & -\boxed{N'_\alpha} & \left(C_2/I_{zz} + \left(I_{xz}/(I_{xx}I_{zz})\right)C_1\right)D & 0 \\ 0 & 0 & 0 & -\dot{\Psi}_0\cos\Theta_0 \\ 0 & 0 & 0 & -s\sin\Phi_0 - \dot{\Psi}_0\sin\Theta_0\cos\Phi_0 \end{bmatrix}$$

with

$$C_1 = (I_{zz} - I_{yy})\mathrm{R}_0 - I_{xz}\mathrm{P}_0$$

$$C_2 = (I_{yy} - I_{xx})\mathrm{P}_0 + I_{xz}\mathrm{R}_0$$

$$D = \frac{1}{1 - I_{xz}^2/(I_{xx}I_{zz})}$$

and

$$\mathbf{Q}_{\text{Long-Lat}}(s) = \begin{bmatrix} 0 & \boxed{X_{\delta_R}} \\ 0 & 0 \\ 0 & 0 \\ 0 & 0 \end{bmatrix}$$

$$\mathbf{Q}_{\text{Lat-Long}}(s) = \mathbf{0}_{5\times 2} \qquad (10.181)$$

Again, the boxed matrix elements Y_α, L'_α, N'_α, and X_{δ_R} are also new dimensional stability derivatives that were not included in the original expansion of the perturbation forces and moments (in Equations (10.43) and (10.44)). But sometimes, especially at high reference sideslip angles or for vehicles with large rudders, these terms may be significant.

By reviewing the cross-coupling matrices in Equations (10.180) and (10.181), one should note that for reference flight conditions involving steady rectilinear flight (i.e., $\dot{\Psi}_0 = P_0 = Q_0 = R_0 = 0$) with zero bank and sideslip angles (i.e., $\Phi_0 = \beta_0 = 0$), all the above four cross-coupling matrices are null. (Note this also assumes the newly introduced cross-coupling stability derivatives are also negligible.) And with null cross-coupling matrices, the polynomial-matrix description for the complete dynamic system given in Equations (10.175) reduces simply to the two decoupled equations

$$\begin{aligned} \mathbf{P}_{\text{Long}}(s)\mathbf{y}_{\text{Long}}(s) &= \mathbf{Q}_{\text{Long}}(s)\mathbf{u}_{\text{Long}}(s) \\ \mathbf{P}_{\text{Lat-Dir}}(s)\mathbf{y}_{\text{Lat-Dir}}(s) &= \mathbf{Q}_{\text{Lat-Dir}}(s)\mathbf{u}_{\text{Lat-Dir}}(s) \end{aligned} \tag{10.182}$$

But this is just a reversion to the two cases with decoupled dynamics addressed in Sections 10.3–10.7, including Equations (10.56) and (10.66).

Now to demonstrate the possible effects of axis cross coupling, consider the two pole-zero plots shown in Figure 10.26, revealing the pole and zero locations of the pitch-attitude-to-elevator transfer function $\theta(s)/\delta_E(s)$ for a high-performance aircraft (Ref. 10). Aerodynamic data for the aircraft that is statically stable in both pitch and yaw was used in both cases, and the aerodynamic dimensional derivatives were functions of both α and β. The reference flight condition involves high angle of attack $\alpha_0 = 19$ deg and a nonzero sideslip angle $\beta_0 = 6$ deg.

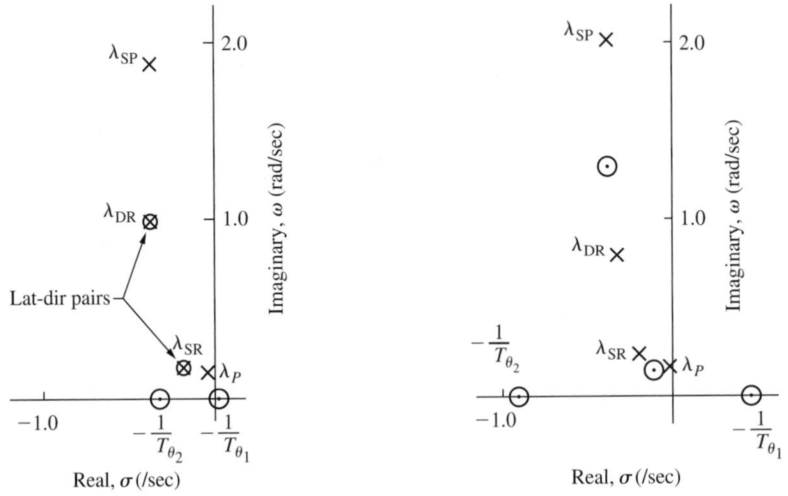

Figure 10.26 Pole-zero plots of $\theta(s)/\delta_E(s)$ indicating effects of cross-axis coupling.

For the plot on the left, a decoupled model corresponding to Equations (10.56) and (10.66) was used. That is, the four cross-coupling matrices in Equations (10.180) and (10.181) were forced to be null. In this plot, note that both the longitudinal and lateral-directional sets of eigenvalues or poles are shown, including the short period, phugoid, dutch roll, and a complex spiral-role pair. Here the roll-subsidence and spiral roots are not real, and these two modes have coalesced into a complex coupled mode.

Six zeros are present in this transfer function, with their locations also indicated. Two of the zeros, those labeled $1/T_{\theta_1}$ and $1/T_{\theta_2}$, are the normal two zeros in the pitch-attitude transfer function. But the remaining four zeros exactly cancel the four lateral-directional poles, or the dutch roll and the spiral-roll pairs. Such pole-zero cancellations will occur whenever the longitudinal and lateral-directional axes are decoupled (which was forced to be true here), and would be present in all the longitudinal transfer functions for the complete system. Likewise, all four longitudinal poles would be cancelled in all the lateral-directional transfer functions when the axes are decoupled. Therefore, with decoupled axes, the modes associated with the lateral-directional axes do not participate in any longitudinal responses, and vice versa.

However, now consider the pole-zero plot on the right. A coupled dynamic model, or the model presented in this section, was used to generate this plot. The difference between these two plots is therefore entirely due to the effects of cross-axis coupling. First note that the locations of the poles are modified. For example, the short-period and dutch-roll frequencies have clearly been affected. But more important is the fact that the locations of the six zeros are significantly different. First, the four lateral-directional eigenvalues are <u>not</u> cancelled by zeros in this case. So these modes actually will participate in the pitch-attitude response to elevator input.

Furthermore, one of the zeros located on the real axis has moved well into the right half of the complex plane. Readers familiar with the basic concepts of feedback control know that when a dynamic system is controlled via feedback, the eigenvalues or poles of the closed-loop system migrate towards the locations of the system zeros (or towards infinity) as the loop gain is increased. Therefore, when the pilot closes the feedback loop by controlling pitch attitude here, with one system zero well into the right half of the complex plane, it is quite possible that the resulting pilot-vehicle system would be unstable!

Therefore, for the case considered above it would appear that cross-coupling effects would be significant. The high angle of attack for the reference flight condition ($\alpha_0 = 19$ deg) increases the vehicle's cross-product of inertia I_{xz} in the stability axes. And the rather large sideslip angle ($\beta_0 = 6$ deg) increases the likelihood of significant kinematic and aerodynamic cross coupling.

10.10 ON THE FLIGHT DYNAMICS OF FLEXIBLE VEHICLES

If the flight vehicle is quite flexible, its dynamic responses and modal characteristics may be significantly affected by the elastic deformation. Wing deformations, for example, may change the vehicle's aerodynamic characteristics, while fuselage deformations may influence the measured responses (e.g., accelerations and/or angular rates). And such measured vehicle responses are especially important in the control of the vehicle, whether the control is performed manually or by an autopilot.

The focus in this book, with regard to elastic vehicles, is on the dynamics of the rigid-body degrees of freedom, and their possible coupling with the lower-frequency elastic modes. We are not, for example, addressing high-frequency aeroelastic flutter, which involves coupling between two elastic modes. In a typical flutter analysis the dynamics of the elastic degrees of freedom are frequently analyzed after the rigid-body degrees of freedom have been eliminated from the model. But such an approach is justified only when there is wide frequency separation between the predominately rigid-body modes and elastic modes. When this separation is present, the rigid-body and elastic degrees of freedom are essentially decoupled.

But aircraft with relatively low-frequency elastic modes may not enjoy a large frequency separation. And so there is a significant possibility for coupling between the rigid-body and elastic degrees of freedom. For example, the X-29 aircraft with forward-swept composite wings (considered in Problem 9.2) had strongly coupled rigid-body and elastic degrees of freedom. The modeling approaches discussed in this book, including both rigid-body and elastic degrees of freedom, would be appropriate for investigating such a vehicle.

To address the dynamics of such elastic vehicles, we will assume that a dynamic model is available, including both rigid-body and elastic degrees of freedom. And in previous chapters, especially Chapters 4 and 7, we have presented techniques for developing such models. We also assume that we have knowledge of the natural frequencies and mode shapes of the free vibration modes of the airframe. Such information would be obtained from a previous analysis of the free vibrations of the vehicle's structure, such as a finite-element analysis, a topic not addressed in this book.

The techniques for analyzing the linear dynamics of flexible aircraft are essentially the same as those previously discussed in Section 10.1. But special care must be used when analyzing models that include elastic degrees of freedom. For example, if the generalized modal coordinates associated with the free-vibration modes of the structure are included as state variables in the model of the vehicle's dynamics, these states should be converted to physical quantities for proper interpretation of the eigenvectors. Physical states must be used so that unit scaling may be performed. Therefore, one must use data from the mode shapes to convert generalized modal coordinates to physical coordinates, and then perform unit scaling as introduced in Example 10.2. All this will be demonstrated in Example 10.5.

The analyst must also consider which elastic deformations are expected to be critical, and then use the appropriate modal data. For example, if vertical accelerations at some fuselage location are of interest, then the modal data for Z displacement at that fuselage location would be required for each mode included in the dynamic model. On the other hand, if pitch rate measured at some fuselage location is of interest, then modal data on the slope of the Z displacements (dZ/dX) measured at that fuselage location must be obtained. Again, these concepts may best be made clear through an illustrative example.

EXAMPLE 10.5

Longitudinal Modal Analysis of an Elastic Hypersonic Vehicle

Let's again consider the notional hypersonic vehicle addressed in Example 10.3. The reference flight condition is Mach = 8 and 80,000 ft altitude. Assume a state-variable model of the longitudinal dynamics of the elastic vehicle is available (e.g., from Refs. 2 or 3 or the data in Appendix B). Also assumed available is the modal data associated with any elastic degrees of freedom included in the model. Perform a modal analysis and find the time responses to assess the significance of elastic deformation, with regard to the vehicle's dynamic characteristics.

■ Solution

The state-variable model to be used includes one elastic degree of freedom, corresponding to the lowest-frequency body-bending mode with a vibration frequency of 18 rad/sec and a generalized mass of 40 slugs. The model in Appendix B is defined for the following state, response, and control-input vectors.

$$\begin{aligned} \mathbf{x}^T = \mathbf{y}^T &= \left[u(\text{fps}) \quad \alpha(\text{rad}) \quad \theta(\text{rad}) \quad q(\text{rad/sec}) \quad \eta(-) \quad \dot{\eta}(/\text{sec})\right] \\ u &= \delta_H(\text{rad}) \end{aligned} \tag{10.183}$$

Letting $\Theta_0 = 0$ and $X_{\dot{\alpha}} = Z_{\dot{\alpha}} = Z_q = 0$, the state-variable matrices are

$$\mathbf{A} = \begin{bmatrix} X_u & X_\alpha & -g & 0 & 0 & 0 \\ \dfrac{Z_u}{U_0} & \dfrac{Z_\alpha}{U_0} & 0 & 1 & \dfrac{Z_\eta}{U_0} & \dfrac{Z_{\dot{\eta}}}{U_0} \\ 0 & 0 & 0 & 1 & 0 & 0 \\ M_u + M_{\dot{\alpha}}\dfrac{Z_u}{U_0} & M_\alpha + M_{\dot{\alpha}}\dfrac{Z_\alpha}{U_0} & 0 & M_q + M_{\dot{\alpha}} & M_\eta + M_{\dot{\alpha}}\dfrac{Z_\eta}{U_0} & M_{\dot{\eta}} + M_{\dot{\alpha}}\dfrac{Z_{\dot{\eta}}}{U_0} \\ 0 & 0 & 0 & 0 & 0 & 1 \\ 0 & \Xi_\alpha + \Xi_{\dot{\alpha}}\dfrac{Z_\alpha}{U_0} & 0 & \Xi_q + \Xi_{\dot{\alpha}} & \left(\Xi_\eta + \Xi_{\dot{\alpha}}\dfrac{Z_\eta}{U_0} - \omega^2\right) & \left(\Xi_{\dot{\eta}} + \Xi_{\dot{\alpha}}\dfrac{Z_{\dot{\eta}}}{U_0} - 2\zeta\omega\right) \end{bmatrix}$$

$$\mathbf{B} = \begin{bmatrix} X_{\delta_E} \\ \dfrac{Z_{\delta_E}}{U_0} \\ 0 \\ M_{\delta_E} + \dfrac{M_{\dot{\alpha}} Z_{\delta_E}}{U_0} \\ 0 \\ \Xi_{\delta_E} + \Xi_{\dot{\alpha}}\dfrac{Z_{\delta_E}}{U_0} \end{bmatrix}, \quad \mathbf{C} = \mathbf{I}_6, \quad \mathbf{D} = \mathbf{0}_{6\times 1} \tag{10.184}$$

The first four elements of the state vector are rigid-body degrees of freedom corresponding to the motion of the vehicle's fuselage-referenced axis, and η is the generalized coordinate associated with the first free-vibration mode of the structure. The control input is the angular deflection of the aft horizontal pitch-control surface.

First, the state-variable model is of the following form

$$\begin{Bmatrix} \dot{\mathbf{x}}_R \\ \dot{\mathbf{x}}_E \end{Bmatrix} = \left[\begin{array}{c:c} \mathbf{A}_R & \mathbf{A}_{\mathrm{RE}} \\ \hdashline \mathbf{A}_{\mathrm{ER}} & \mathbf{A}_E \end{array}\right] \begin{Bmatrix} \mathbf{x}_R \\ \mathbf{x}_E \end{Bmatrix} + \begin{bmatrix} \mathbf{B}_R \\ \mathbf{B}_E \end{bmatrix} \delta_H$$
$$\mathbf{y} = [\mathbf{I}_6] \begin{Bmatrix} \mathbf{x}_R \\ \mathbf{x}_E \end{Bmatrix} + [\mathbf{0}_{6\times 1}]\delta_H \tag{10.185}$$

which is partitioned here to indicate the submatrices corresponding to the rigid-body and elastic degrees of freedom and the cross coupling. Also, the last two states, associated with the elastic degree of freedom, are not physical quantities as desired. The desired states include pitch deflection θ_E and pitch rate $\dot{\theta}_E$ measured at the nose of the vehicle, due to elastic deformation alone. These states are given by

$$\theta_E(t) \triangleq \nu_Z'(0)\eta(t)$$
$$\dot{\theta}_E(t) \triangleq \nu_Z'(0)\dot{\eta}(t) \tag{10.186}$$

Here, $\nu_Z'(0)(= 1\,\mathrm{deg})$ is the *dZ/dX* slope, measured at fuselage station $x = 0$ (nose), of the mode shape associated with the vibration mode.

For this hypersonic vehicle at the chosen flight condition, let the new response, state, and control vectors be

$$\mathbf{y}^T = \mathbf{x}^T = \begin{bmatrix} u(\mathrm{Mach}) & \alpha(\mathrm{rad}) & \theta(\mathrm{rad}) & \dot{\theta}(\mathrm{rad/sec}) & \theta_E(\mathrm{rad}) & \dot{\theta}_E(\mathrm{rad/sec}) \end{bmatrix}$$
$$u = \delta_H\,(\mathrm{rad})$$

where the units on surge velocity have been converted to Mach number, and the last two states are now as given by Equations (10.186). The **A** and **B** matrices in this new state-variable model, given numerically, are

$$\mathbf{A} = \left[\begin{array}{cccc:cc} -0.00194 & 0.02502 & -0.03317 & 0.00064 & -0.01490 & 0.00070 \\ -0.00203 & -0.06303 & 0 & 1.0030 & -0.05904 & 0.00034 \\ 0 & 0 & 0 & 1 & 0 & 0 \\ 0.32865 & 11.023 & 0 & -0.08161 & 10.894 & -0.08021 \\ \hdashline 0 & 0 & 0 & 0 & 0 & 1 \\ 2.5807 & 82.567 & 0 & -0.64673 & -241.43 & -0.98821 \end{array}\right]$$

$$\mathbf{B} = \left[\begin{array}{c} -0.00058 \\ -0.00276 \\ 0 \\ -0.47936 \\ \hdashline 0 \\ 4.2863 \end{array}\right] \tag{10.187}$$

From MATLAB, the six eigenvalues of **A** are

$$\begin{aligned}\lambda_{1,2} &= -0.45248 \pm j15.640 \text{ /sec}\\ \lambda_3 &= -3.9344 \text{ /sec}\\ \lambda_4 &= 3.7062 \text{ /sec}\\ \lambda_{5,6} &= -8.4893 \times 10^{-4} \pm j2.0374 \times 10^{-3} \text{ /sec}\end{aligned} \tag{10.188}$$

and the corresponding right eigenvectors are

$$\boldsymbol{\nu}_1 = \begin{Bmatrix} 0.00007e^{-j112.5^\circ} \\ 0.00277e^{j0.2^\circ} \\ 0.00274e^{j5.0^\circ} \\ 0.04288e^{j96.7^\circ} \\ 0.06372e^{-j165.8^\circ} \\ 0.99704e^{-j74.1^\circ} \end{Bmatrix}, \boldsymbol{\nu}_3 = \begin{Bmatrix} 0.00093 \\ 0.23214 \\ 0.22654 \\ -0.89130 \\ 0.07804 \\ -0.30704 \end{Bmatrix}, \boldsymbol{\nu}_4 = \begin{Bmatrix} 0.00065 \\ -0.23765 \\ -0.24211 \\ -0.89731 \\ -0.07356 \\ -0.27264 \end{Bmatrix}, \boldsymbol{\nu}_5 = \begin{Bmatrix} 0.99613e^{-j1.9^\circ} \\ 0.03006e^{j178.2^\circ} \\ 0.08260e^{-j134.1^\circ} \\ 0.00018e^{-j21.4^\circ} \\ 0.00037e^{-j1.8^\circ} \\ \sim 0 \end{Bmatrix} \tag{10.189}$$

The phasor diagrams of these eigenvectors are shown in Figure 10.27, in which components too small to plot are omitted. We see that the first mode is a lightly damped oscillatory mode with modal response dominated by the rate of elastic pitch deformation at the nose. So this is primarily an elastic mode. But unlike a free vibration mode this mode now contains both elastic and rigid-body degrees of freedom in its response. This occurs because the vehicle's aerodynamics provides coupling between all the degrees of freedom. Note also that the damped frequency of 15.64 rad/sec for this mode is considerably less than the free-vibration frequency of 18 rad/sec. Plus, although the damping is low ($\zeta \approx 0.026$), it is larger than the elastic modal damping of 0.02 included in the free vibration model. This reduction in frequency and increased damping also arise due to aerodynamic effects.

The remaining three modes have characteristics somewhat similar to those obtained in Example 10.3. A low-frequency oscillatory phugoid mode (the fourth mode) is present, plus there are two real eigenvalues associated with modes dominated by rigid-body pitch rate, one of which is highly unstable. So these three modes are primarily rigid-body modes, but two also contain some elastic deformation in their responses. But perhaps the most

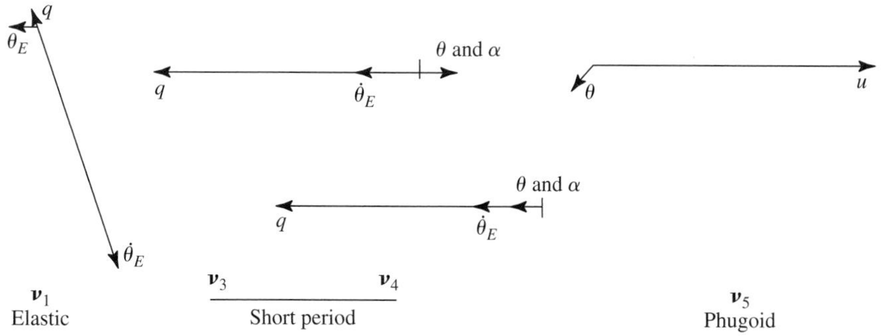

Figure 10.27 Phasor diagrams for elastic hypersonic vehicle.

significant change in the modal characteristics is the fact that the magnitudes of the two real eigenvalues have increased by approximately 15 percent, making the vehicle even more dynamically unstable in pitch.

All these changes in modal characteristics arise due to the aeroelastic effects that were not considered in Example 10.3. (Although in that analysis the increased rigid-body instability could have been more accurately predicted if static-elastic effects were included, as discussed in Section 7.11.) Although some frequency separation exists between the rigid-body and vibration modes (3.4 rad/sec versus 18 rad/sec), the inclusion of the first vibration mode in this analysis has lead to changes in the vehicle's dynamics that may be significant.

The elastic effects on the dynamics are further evaluated by considering the time responses to a control input. The vehicle responses (not including surge velocity u) for a horizontal control-surface deflection δ_H of -1 deg for 0.5 sec changing to $+5$ deg thereafter are shown in Figure 10.28. Note that all the responses are plotted in degrees or degrees per second. Since the vehicle is very unstable and the responses diverge rapidly, only the first two seconds of each response is shown. The finer aspects of the responses could not be viewed accurately, if at all, in plots of responses over longer time periods.

The rigid-body pitch attitude θ and angle of attack are diverging rapidly. But also note that the elastic pitch deflection of the nose θ_E also diverges, reaching approximately 0.5 deg after two seconds. And recall that the total pitch attitude at the nose is the sum of the rigid-body plus elastic contributions, or $\theta_{\text{Total}}(t) = \theta(t) + \theta_E(t)$. As the vehicle's angle of attack diverges, the aerodynamic load on the fore body of the vehicle increases. And the elastic deformation increases as well, thus increasing the lift on the nose leading to the increased pitch instability of the vehicle.

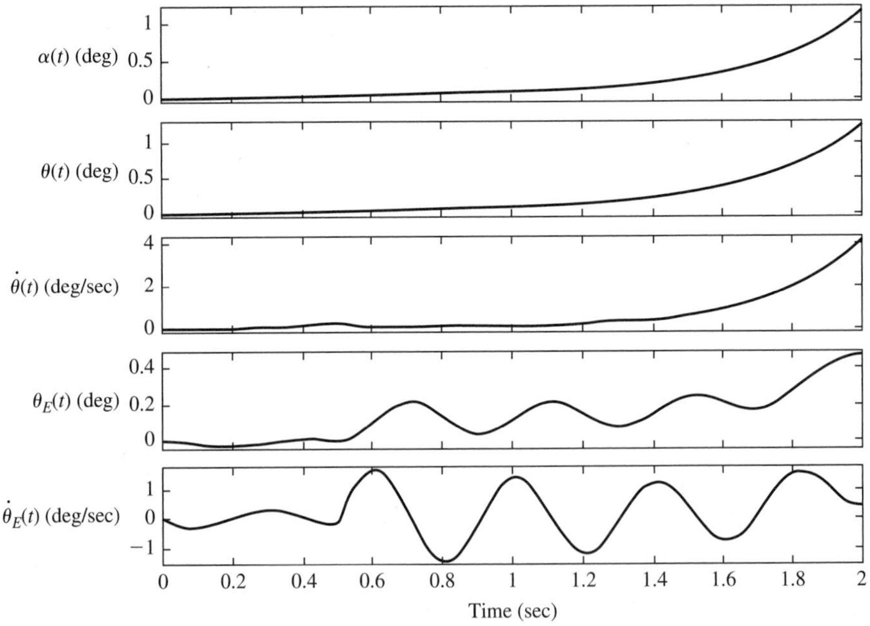

Figure 10.28 Time responses for elastic hypersonic vehicle.

During this two-second simulation, the elastic contribution to pitch rate at the nose reaches approximately 25 percent of the total pitch-rate deflection. And the oscillations of the 15 rad/sec elastic mode are also evident in the elastic-pitch-deflection and elastic-pitch-rate responses. These fore body oscillations could cause difficulties because they create pressure oscillations at the engine inlet.

10.11 Summary

In this chapter we have introduced methods for the analysis of linear dynamic systems, and applied those methods to investigate the perturbation dynamics of atmospheric flight vehicles. We determined that for conventional aircraft, the longitudinal dynamics were typically characterized in terms of two natural modes of motion—the short-period and the phugoid. For statically stable vehicles, both of these modes are typically oscillatory in nature, with the short-period mode oscillating at a higher frequency and with moderate damping, while the oscillations of the phugoid mode are of lower frequency and are lightly damped. The eigenvectors associated with these modes revealed that the short-period modal response consists of pitch oscillations occurring at almost constant surge velocity, while the phugoid modal response consists of oscillations in surge velocity and pitch attitude at almost constant angle of attack. We also demonstrated that the modal characteristics for unconventional, statically unstable aircraft could differ significantly from that just described.

We further determined that for conventional aircraft, the lateral-directional dynamics were typically characterized in terms of three natural modes of motion—the dutch roll, roll subsidence, and the spiral. The dutch-roll mode is oscillatory in nature and lightly damped, while the roll-subsidence and spiral modes are typically nonoscillatory. The eigenvectors associated with these modes revealed that the dutch-roll modal response consists of out-of-phase rolling, yawing, and sideslip oscillations. The response of the roll-subsidence mode consists of almost entirely roll rate, while the spiral mode's response consists primarily of bank and heading angles. We also noted that for some vehicles in certain flight conditions the roll-subsidence and spiral modes coalesce and form a single oscillatory roll-spiral mode.

Based on our understanding of these modal characteristics, we developed simplified models of the dynamics of flight vehicles. We then demonstrated by examples that these simplified models, along with the closed-form expressions for the vehicle's aerodynamic characteristics derived in Chapter 6, may be used effectively to influence the aircraft's configuration design and thus tailor the vehicle's dynamic characteristics.

We emphasized that the results just cited assumed that the longitudinal and lateral-directional dynamics were decoupled, which is not strictly true for reference flight conditions with nonzero bank and/or sideslip angle. In such cases, a dynamic analysis using the coupled six-degree-of-freedom linear model may be recommended. We also noted that the natural modal characteristics of an aircraft may be significantly modified due to elastic effects. And in such cases it may be necessary to use a model that includes such elastic effects.

Finally, in this chapter we have emphasized the responses of the vehicle to <u>control</u> inputs. But the responses to <u>gust</u> inputs are also important. Using the method presented in Chapter 8 for modeling the effects of atmospheric gusts, along with the analysis techniques presented in this chapter, we may also assess the response of the vehicle to gusts.

10.12 Problems

10.1 Prove that in a pole-zero plot of a transfer function, the undamped natural frequency of an oscillatory mode is equal to the distance between one of the mode's pole locations and the origin of the complex plane. Also prove that the angle between the negative real axis and the line connecting the pole's location with the origin equals the arccosine of the mode's damping ratio.

10.2 Consider a wingless missile, as in Problem 6.2, with geometry as shown below. Noting that the nose section and tail generate most of the lift, and the effects of downwash are negligible, assume that $Z_{\dot{\alpha}} = M_{\dot{\alpha}} = Z_q = M_q = 0$. Under the short-period approximation, write the polynomial-matrix description of the longitudinal dynamics of the missile, letting the control input be aft fin deflection denoted as δ_{Fin}. Next, letting $\Theta_0 = 0$ find the two transfer functions for the angle-of-attack α and pitch-rate q responses and show that the characteristic polynomial is simply $s^2 - \dfrac{Z_\alpha}{U_0}s - M_\alpha$.

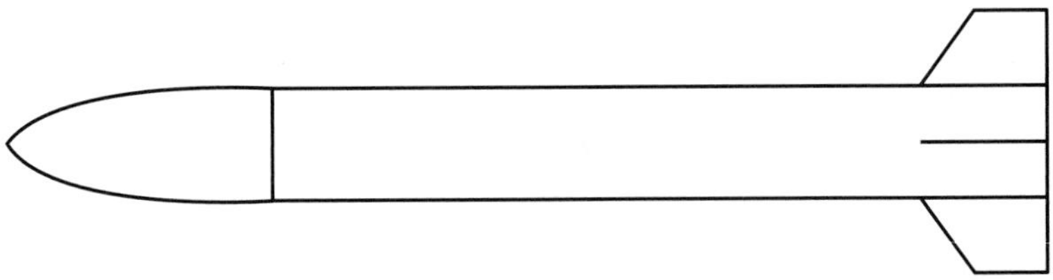

10.3 Using data from Appendix B, assemble the state-variable model for the longitudinal dynamics of the DC-8 aircraft at Flight Condition 3 and perform an eigenanalysis on these dynamics. Include phasor diagrams of the right eigenvectors after unit scaling (i.e., convert all angles to degrees). Do these dynamics exhibit traditional phugoid and short-period modal characteristics? Explain.

10.4 Using the short-period and preferred phugoid approximations presented in this chapter, compare the eigenvalues obtained using these approximations with those from the full-order model developed in Problem 10.3.

10.5 Using data from Appendix B, assemble the state-variable model for the longitudinal dynamics of the F-5A aircraft at Flight Condition 1 and perform an eigenanalysis on these dynamics. Include phasor diagrams of the right eigenvectors after unit scaling (i.e., convert all angles to degrees). Do these dynamics exhibit traditional phugoid and short-period modes? Explain.

10.6 Consider the F-5A aircraft at Flight Conditions 1 and 2, and note that while its phugoid mode is stable at Flight Condition 1 ($M_\infty = 0.875$), it is unstable at Flight Condition 2 ($M_\infty = 1.25$). Using the phugoid approximations presented in this chapter (not including the Lanchester model) determine which (if any) approximation correctly captures this phugoid instability, and suggest a cause for the instability.

10.7 Develop a polynomial-matrix description for the longitudinal dynamics of a flight vehicle and express the vehicle's transfer functions with gust input (α_g) in terms of the determinants of the relevant matrices.

10.8 Using the vehicle data developed for Problem 10.3, and the results from Problem 10.7, find the three transfer functions $u(s)/\alpha_g(s)$, $\alpha(s)/\alpha_g(s)$, and $\theta(s)/\alpha_g(s)$.

10.9 Consider the DC-8 aircraft at Flight Conditions 1 and 2, and note that while its spiral mode is stable at Flight Condition 2, it is unstable at Flight Condition 1. Determine whether the spiral approximation presented in this chapter correctly captures this spiral instability, and suggest a possible cause for the instability.

10.10 Using data from Appendix B, assemble the state-variable model for the lateral-directional dynamics of the DC-8 aircraft at Flight Condition 1 and perform an eigen-analysis on these dynamics. Include phasor diagrams of the right eigenvectors. Do these dynamics exhibit traditional modal characteristics? Explain.

10.11 Develop a state-variable model for the fully coupled dynamics developed in Section 10.9. Use the aerodynamic data from Appendix B for the DC-8 aircraft in Flight Condition 2. However, assume the flight condition involves a 20-degree banked turn at the given altitude instead of level rectilinear flight. Also assume the reference sideslip angle $\beta_0 = 0$. Using this model perform an eigen-analysis assuming the dynamics are decoupled, and compare with the results from an eigen-analysis of the fully coupled model. Are the coupling effects significant? Why?

10.12 Consider the flexible aircraft modeled in the case study in Section 7.9, and for which data is presented in Appendix B. Using the data for Flight Condition 1, perform a modal analysis of the linear model of the longitudinal dynamics including only the first and third elastic modes, and assess the significance of elastic deformation on the vehicle's dynamics. (Remember to use physical state variables, and to balance the units of the responses and states, as required. Let the physical states associated with the elastic degrees of freedom consist of pitch attitude deflections measured at the cockpit.)

References

1. MIL-F-8785B Military Specifications, "Flying Qualities of Piloted Airplanes," August 1969.
2. Chavez, F. R. and D. K. Schmidt: *An Integrated Analytical Aeropropulsive/Aeroelastic Model for the Dynamic Analysis of Hypersonic Vehicles,* prepared for NASA under Grant NAG-1-1341, Aerospace Research Center, Arizona State University, June 1992.
3. Chavez, F. R. and D. K. Schmidt: "An Analytical Model and Dynamic Analysis of Aeropropulsive/Aeroelastic Hypersonic Vehicles," *Journal of Guidance, Control, and Dynamics,* vol. 17, no. 6, Nov.–Dec. 1994.
4. Lanchester, F. W.: *Aerodonetics,* Macmillan, New York, 1908.
5. Etkin, B.: *Dynamics of Flight, Stability and Control,* 2nd ed., Wiley, New York, 1982.

6. Nelson, R.C.: *Flight Stability and Automatic Control,* 2nd ed., McGraw-Hill, New York, 1998.

7. McRuer, D., I. Ashkenas, and D. Graham: *Aircraft Dynamics and Automatic Flight Control,* Princeton Press, Princeton, NJ, 1973.

8. Cook, M.V.: *Flight Dynamics Principles,* 2nd ed., Elsevier, New Hork, 2007.

9. Roskam, J.: *Airplane Flight Dynamics and Automatic Flight Controls,* Roskam Aviation and Engineering Corp., Lawrence, KS, 1979.

10. McRuer, D. T. and D. E. Johnston: *Flight Control Systems Properties and Problems,* NASA Contractor Report CR-2500, prepared by Systems Technology, Inc., Feb. 1975.

CHAPTER 11

Feedback Stability Augmentation

Chapter Roadmap: *The subjects addressed in this chapter would typically be included in a first-course in flight dynamics. We focus on stability augmentation using feedback, and the approach builds directly from the modal analyses of the vehicle's dynamics presented in Chapter 10.*

In Chapter 10 we focused on gaining an understanding of the flight dynamics of flight vehicles. In this chapter we adjust our focus to that of using our knowledge of the vehicle's dynamics to synthesize feedback control laws that will improve that vehicle's dynamic characteristics. In Chapter 10 we demonstrated by example how an aircraft's configuration might be modified to improve its dynamics, and we noted that sometimes such modifications could lead to excessively large stabilizing surfaces, for example. So by introducing a feedback system to further adjust the dynamics, we provide the designer with an additional tool to optimize the design. The overall feedback-design objectives are to preserve conventional aircraft responses by using simple, reliable feedback systems. We will not be addressing the electronics associated with the design of the feedback system, although that is important. Rather, we will be interested in the dynamics of the airframe acted upon by the feedback interconnections, which define the *feedback architecture.*

Now a comment on the implementation of a feedback control law is in order. In this chapter, and in those to follow, we will make copious use of linear systems-analysis techniques. And yet we must always remember that we are dealing with a nonlinear system. Through perturbation theory we obtain a linearized model of the perturbation dynamics, and that allows for the use of the linear analysis techniques. But when the control law is actually implemented on the vehicle, we do not deal with perturbation quantities such as perturbation pitch rate, for example. Students sometimes are confused by this fact. The total pitch rate would be measured by a transducer and fed back through some pitch-rate feedback control law. And the control law would drive the total elevator deflection, not perturbation deflection. Finally, as different reference flight conditions are considered, the parameters in

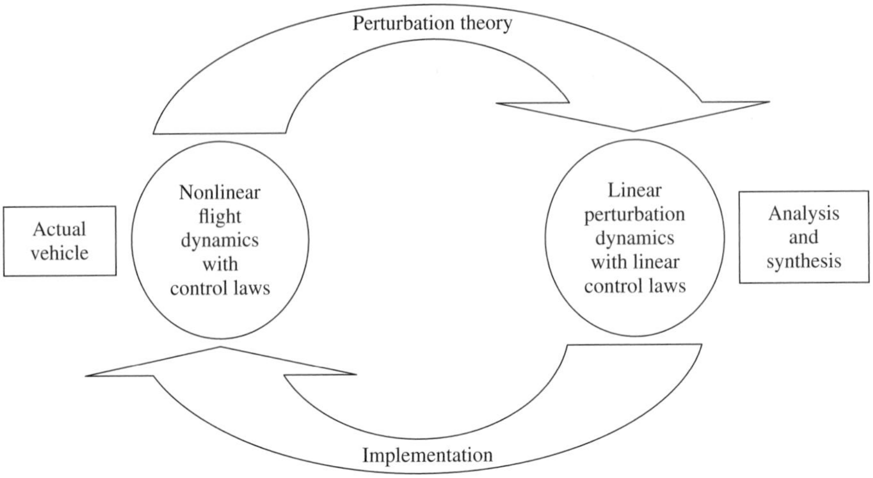

Figure 11.1 Overall analysis and synthesis concept.

the control law may be varied over the flight envelope—a technique known as *gain scheduling.* These concepts are depicted conceptually in Figure 11.1

We will begin in Section 11.1 by reviewing the concept of block diagrams to depict a dynamic system, as such block diagrams are especially useful in describing a feedback system. In the same section we will briefly review root-locus plots, which are also useful in the analysis and synthesis of feedback systems. Those readers already familiar with block diagrams and root-locus plots may choose to skip Section 11.1 and move directly to Section 11.2. That section addresses important characteristics of multi-input/multi-output systems that naturally arise in flight dynamics. Even if readers are familiar with single-input/single-output feedback systems, we still strongly encourage them to read Section 11.2.

The remainder of the chapter is devoted to augmenting the dynamics of flight vehicles. Such augmentation systems are important regardless of whether the vehicle will ultimately be controlled by a pilot or an autopilot. We will consider which vehicle responses are most appropriate to use for feedback and which control input is most effective to use, and we will determine the effects of the particular feedback architecture on the vehicle's dynamics. And, thus, we will indentify effective feedback stability augmentation systems.

11.1 BLOCK DIAGRAMS, FEEDBACK, AND ROOT-LOCUS PLOTS—A JITT*

A block diagram is simply a graphical depiction of an equation or set of equations that describe an interactive dynamic system. Typically, block diagrams are built using the transfer functions of the components of the dynamic system. Each block may represent one dynamic component or a group of components,

* A Just-in-Time Tutorial.

and by connecting the blocks one describes the interactions between the various components.

The simplest block diagram is that depicting the input-output nature of a single dynamic element, for which the transfer function is denoted as $g(s)$. By definition, the transfer function is the ratio of the Laplace transform of the response $y(s)$ of the dynamic element to the transform of the input $u(s)$ that is disturbing or forcing the dynamic element, assuming all initial conditions are zero. Or the system may be described by the equation

$$g(s) \triangleq \frac{y(s)}{u(s)}$$

The block-diagram depiction of this dynamic element is simply

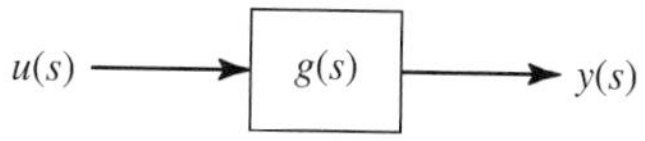

The most useful application of block diagrams, however, is in the description of interacting dynamic elements. For example, consider two dynamic elements with transfer functions $g_1(s)$ and $g_2(s)$. If the response of the first element, $y_1(s)$, is the input to the second element, $u_2(s)$, or the components are connected in *series,* then the block diagram for this system is shown in Figure 11.2.

Writing the equations indicated by the block diagram as

$$\begin{aligned} y_1(s) &= g_1(s)u_1(s) \\ y_2(s) &= g_2(s)u_2(s) \\ u_2(s) &= y_1(s) \end{aligned} \tag{11.1}$$

we find that the transfer function for the complete system in series is

$$\frac{y_2(s)}{u_1(s)} = g_2(s)g_1(s) = g_1(s)g_2(s)$$

since (11.2)

$$y_2(s) = g_2(s)u_2(s) = g_2(s)y_1(s) = g_2(s)g_1(s)u_1(s)$$

If, on the other hand, the two elements are connected in *parallel,* the inputs to the two elements are the same, and the response is the sum of the two responses. The block diagram for this system is shown in Figure 11.3.

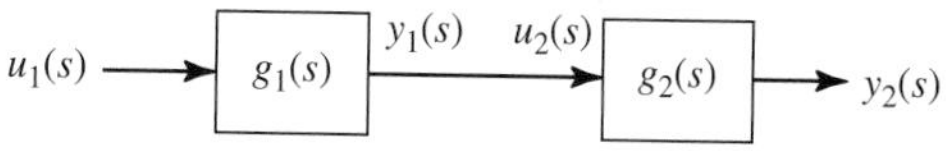

Figure 11.2 Two dynamic elements connected in series.

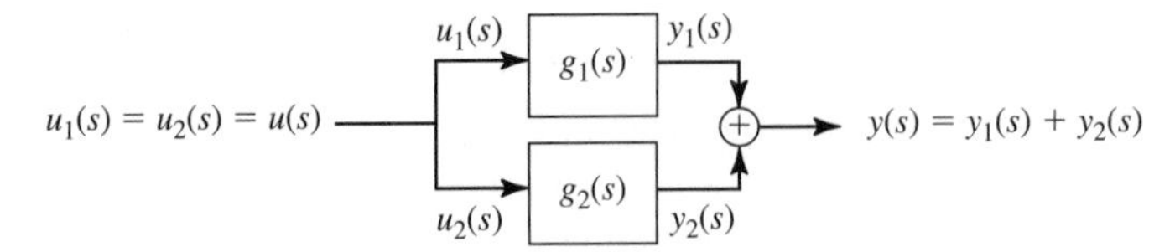

Figure 11.3 Two dynamic elements connected in parallel.

The transfer function of this complete system in parallel is now

$$\frac{y(s)}{u(s)} = \left(g_1(s) + g_2(s)\right) \tag{11.3}$$

(Students should verify this result for themselves.) Therefore, based on these two examples it should be clear that the block diagram is just a depiction of the equations defining the system. And given either the block diagram or the set of equations, the other can be found.

EXAMPLE 11.1

Block Diagram of a Flight Control System

Consider a feedback system on an aircraft, consisting of a motion sensor, or transducer, with transfer function $s(s)$, that senses the aircraft pitch rate $q(s)$; an amplifier, with transfer function $k(s)$, that amplifies its input signal by some given factor; and a servo-hydraulic actuator, with transfer function $a(s)$, that deflects the elevator on the vehicle in response to a command. Let this feedback system be defined by the *control law* given by

$$\delta_c(s) = \delta_{\text{Stick}} - k(s)y_s(s) \tag{11.4}$$

where

δ_c = commanded elevator deflection to actuator

δ_{Stick} = signal representing pilot's control-stick deflection

y_s = signal from the sensor, corresponding to measured pitch rate

Find the block-diagram representation of this feedback system.

■ Solution

Consider the aircraft pitch-rate dynamics represented by the transfer function $q(s)/-\delta_E(s)$. Without any feedback system in place, this transfer function is the *open-loop* transfer function. (Note that by introducing the negative sign in this transfer function, we have simply changed the sign convention on the elevator such that positive $\delta_E(s)$ produces a positive pitching moment. This makes all our feedback block diagrams consistent with standard notation.)

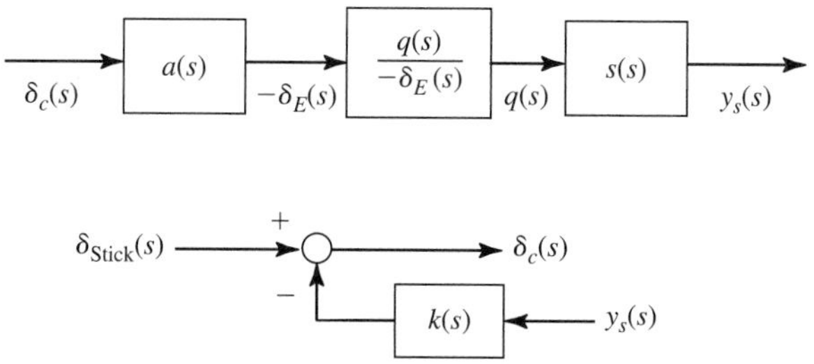

Figure 11.4 Vehicle and control-law block diagrams.

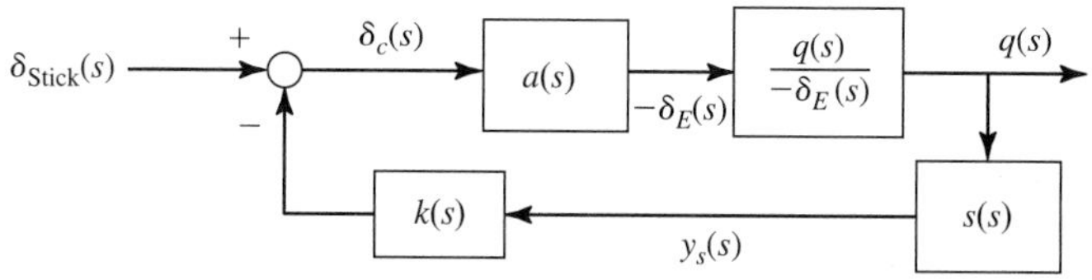

Figure 11.5 Block diagram of example flight-control system.

Now the block diagram of the actuator, sensor, and aircraft dynamics may be depicted as shown in the top of Figure 11.4. And the block diagram for the control law may be depicted as shown in the bottom of the same figure. Finally, the block diagram for the complete system with feedback, or the *closed-loop system,* is shown in Figure 11.5.

One can show by algebraic manipulations of the equations describing the feedback system depicted in Figure 11.5 that the transfer function for this closed-loop system is given by

$$\frac{q(s)}{\delta_{\text{Stick}}(s)} = \frac{a(s)g(s)}{1 + k(s)a(s)g(s)s(s)} \tag{11.5}$$

where the open-loop pitch-rate transfer function is $g(s) = q(s)/-\delta_E(s)$. To simplify the discussion at this point, let us assume that the transfer functions for the sensor and actuator are both simply unity, and that the transfer function for the amplifier is simply a constant K. (Or more precisely, the Bode plots of these three transfer functions would have constant magnitude and zero phase over a wide frequency range. In fact, this is consistent with the objectives in designing an amplifier, sensor, and actuator.)

Under these simplifying assumptions, the above transfer function becomes

$$\frac{q(s)}{\delta_{\text{Stick}}(s)} = \frac{g(s)}{1 + Kg(s)} = \frac{N_g(s)/\Delta_g(s)}{1 + KN_g(s)/\Delta_g(s)} = \frac{N_g(s)}{\Delta_g(s) + KN_g(s)} \tag{11.6}$$

where the transfer function representing the vehicle's open-loop pitch-rate dynamics, or $g(s)$, has been written in terms of its numerator polynomial $N_g(s)$ and its denominator polynomial $\Delta_g(s)$, which is also the open-loop characteristic polynomial. The roots of the characteristic polynomial $\Delta_g(s)$ are of course the system's open-loop eigenvalues, or poles, while the roots of the numerator polynomial $N_g(s)$ are the zeros of the transfer function $g(s)$.

Note by observing Equation (11.6) that although the introduction of this particular feedback control law has not affected the zeros of the closed-loop transfer function (they are roots of $N_g(s)$ as well), the characteristic polynomial of the complete system with feedback, or the closed-loop system, is now equal to

$$\Delta_{\mathrm{CL}}(s) \triangleq \Delta_g(s) + KN_g(s) \tag{11.7}$$

So the closed-loop system's characteristic polynomial is clearly modified from that of the open-loop system ($g(s)$ here) as long as $K \neq 0$, and this is one of the primary advantages of introducing feedback into a dynamic system. Not only can the system's dynamics be modified through the adjustment of the system's eigenvalues, but an unstable system may even be stabilized. Furthermore, a plot of the eigenvalues of the closed-loop system, plotted as a function of the feedback constant (*gain*) *K*, is known as a *root-locus plot.*

NOTE TO STUDENT

In Section 10.8, we introduced plots in which we labeled a *root locus,* and those plots were loci of eigenvalues plotted as a function of some design variable such as c.g. location or size of the horizontal tail. However, when one encounters the term *root-locus plot* in general, it usually means a locus of eigenvalues of some closed-loop system plotted as a function of feedback gain. And that is how we have defined the term here.

Note from Equation (11.7) that the closed-loop characteristic polynomial $\Delta_{\mathrm{CL}}(s)$ depends on the numerator and the denominator of $g(s)$, plus the feedback gain K. Hence the closed-loop eigenvalues depend on the open-loop poles and zeros and the feedback gain.

So the construction of a root-locus plot begins with a pole-zero plot of the open-loop system, and the root-locus plot may then be sketched using the rules listed in Table 11.1. These rules are developed in any textbook on feedback systems such as Ref. 1. If the root locus is to be useful in the synthesis of feedback control laws, one must know, for example, how introducing additional poles or zeros in the control law will change the root locus. Finally, numerical methods such as those in MATLAB are very useful for plotting the root locus.

Table 11.1 Rules for Sketching a Root-Locus Plot (Assuming Negative Feedback and Positive Gain K)

1. The root locus is symmetric about the real axis, and the branches of the root locus start at the location of the open-loop poles, with gain $K = 0$, and terminate at either the locations of the open-loop zeros or at infinity, with $K = \infty$.
2. If K is positive, the locus includes all points on the real axis to the left of an odd number of open-loop poles plus open-loop zeros.
3. As $K \to \infty$, the branches of the root locus asymptotically approach straight lines (asymptotes), which lie at angles relative to the real axis equal to $\frac{(2k+1)180}{n_p - n_z}$ deg, $k = 0, \pm 1, \pm 2, \ldots$, where n_p = number of open-loop poles, and n_z = number of open-loop zeros.
4. The asymptotes radiate from a point on the real axis located at $x = \frac{\Sigma poles - \Sigma zeros}{n_p - n_z}$.
5. The loci coalesce onto, or break away from, the real axis between pairs of open-loop zeros or pairs of open-loop poles, respectively.
6. Two loci coalescing onto or breaking away from the real axis do so at 90 deg angles from the real axis.
7. The angle of departure from a complex open-loop pole, or the angle of arrival at an open-loop zero (measured from a line parallel to the positive real axis) is found from ΣAngles from all zeros $-$ ΣAngles from all poles $= 180(2q+1)$, $q = 0, 1, 2, \ldots$

EXAMPLE 11.2

Demonstration of the Rules for the Root Locus

Consider a dynamic element with transfer function $g(s)$ given by

$$\frac{y(s)}{u(s)} = g(s) = \frac{s+1}{s(s^2+2s+2)}$$

and let the feedback control law here be given by

$$u = K(y_c - y) \tag{11.8}$$

(which differs from that given in Equation (11.4)). Draw the block diagram for the closed-loop system, find the closed-loop transfer function expressed in terms of K, use MATLAB to plot the root locus of closed-loop eigenvalues, and comment on how the rules in Table 11.1 apply in this case.

■ Solution

The block diagram for the closed-loop system is shown in Figure 11.6, and from algebra the closed-loop transfer function is given by

$$\frac{y}{y_c} = \frac{Kg(s)}{1+Kg(s)} \tag{11.9}$$

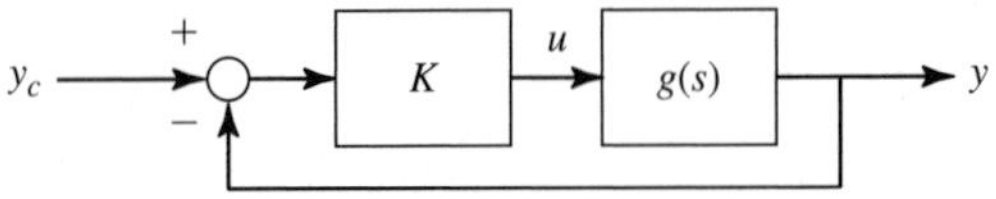

Figure 11.6 Closed-loop system.

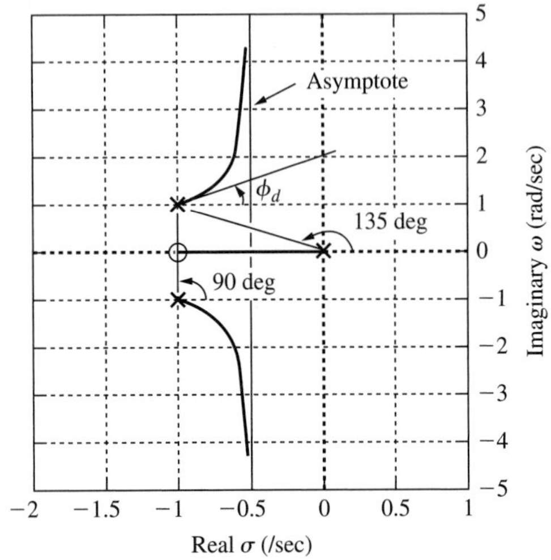

Figure 11.7 Example root-locus plot.

To plot the root locus, from MATLAB we have

```
»num=[1 1];den=[1 2 2 0];
»g=tf(num,den)
»rlocus(g)
»grid
```

And the root-locus plot from MATLAB, using rlocus, is shown in Figure 11.7.

The root locus is seen to be symmetric about the real axis, and has three branches. Each branch begins at an open-loop pole (labeled as X) and ends at an open-loop zero (labeled as O) or at infinity (Rule 1). The branch on the real axis lies to the left of the single pole at the origin—or left of an odd number of poles and zeros (Rule 2). There are two asymptotes, with angles relative to the real axis equal to

$$\frac{(2k+1)180}{n_p - n_z} = \frac{(2k+1)180}{3-1}, \quad k = 0, \pm 1, \ldots$$

or ±90 deg (Rule 3). The asymptotes intersect the real axis at

$$x = \frac{\Sigma poles - \Sigma zeros}{n_p - n_z} = \frac{(-1-1-0)-(-1)}{3-1} = -0.5 \text{ (Rule 4)}$$

There are no pairs of poles or zeros on the real axis, so there are no branches breaking away from or coalescing onto the real axis (Rule 5). Finally, the angle of departure from the complex pole at $s = -1 + j$ is denoted as ϕ_d in the plot. From Rule 7 we have

$$\Sigma\text{Angles from all zeros} - \Sigma\text{Angles from all poles} = 180(2q+1),\ q = 0, 1, 2, \ldots$$

and so

$$(90) - (135 + 90 + \phi_d) = 180, 540, \ldots$$

or

$$\phi_d = -135 - 180 = -315 = 45 \text{ deg}$$

(Note that since the axes used in the plot do not have the same scale, the departure angle may not appear to equal 45 deg.)

11.2 ON MULTI-INPUT/MULTI-OUTPUT SYSTEMS AND COUPLING NUMERATORS

Consider the conceptual block diagram in Figure 11.8, which shows three blocks: one representing the complete longitudinal dynamics of an aircraft, one a feedback stability-augmentation system, and one the pilot. Elevator and throttle, or thrust, are the two control inputs acting on the vehicle, and note that several aircraft responses are shown. The pilot is seen to be closing control loops around the complete system consisting of the aircraft and the stability-augmentation systems, using motion and visual cues as feedback quantities.

If students have been exposed to feedback control systems in other course work, they most likely will have dealt with only one input and one output, or response. Yet with reference to Figure 11.8, we see that even if we consider only the longitudinal axes, aircraft dynamics involve more than one input and one response.

Of course, state-variable system descriptions allow for multiple inputs and multiple outputs. But we will show in this section that by careful analysis using polynomial-matrix descriptions of the aircraft dynamics and the

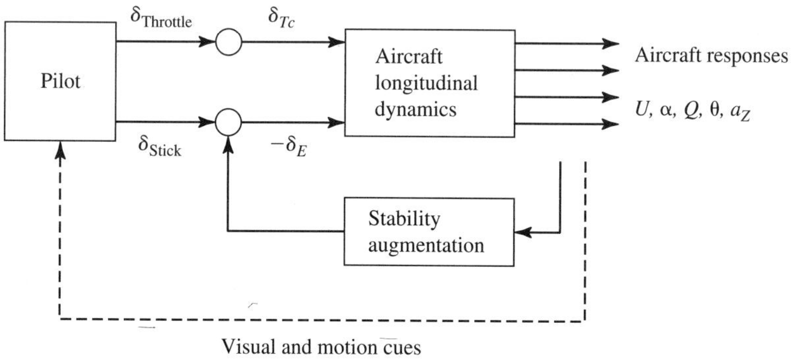

Figure 11.8 Block diagram for the longitudinal control of an aircraft.

feedback-control laws, we can deal with our multi-input/multi-output (MIMO) systems in a rigorous yet intuitive and simple fashion. All we will use is algebra and matrix determinants, and the methodology developed applies generally to any MIMO system. Of particular interest, for example, is the effect of introducing feedback on the poles and zeros of the aircraft's transfer functions, as flight-dynamics design requirements are frequently written in terms of these transfer functions.

As we have seen in Chapter 10, the vehicle's dynamics may be represented in a polynomial-matrix description of the following form (c.f., Equation (10.32)).

$$\begin{bmatrix} p_{1,1}(s) & p_{1,2}(s) & p_{1,3}(s) \\ p_{2,1}(s) & p_{2,2}(s) & p_{2,3}(s) \\ p_{3,1}(s) & p_{3,2}(s) & p_{3,3}(s) \end{bmatrix} \begin{Bmatrix} y_1(s) \\ y_2(s) \\ y_3(s) \end{Bmatrix} = \begin{bmatrix} q_{1,1}(s) & q_{1,2}(s) \\ q_{2,1}(s) & q_{2,2}(s) \\ q_{3,1}(s) & q_{3,2}(s) \end{bmatrix} \begin{Bmatrix} u_1(s) \\ u_2(s) \end{Bmatrix} \tag{11.10}$$

or

$$\mathbf{P}(s)\mathbf{y}(s) = \mathbf{Q}(s)\mathbf{u}(s)$$

The elements of the **P** and **Q** matrices are polynomials in s. For the longitudinal axes, the first response $y_1(s)$ might be surge velocity $u(s)$, and the first control input $u_1(s)$ might be elevator deflection δ_E, for example. We have already discussed how to use Cramer's Rule (Appendix D) to find all the system transfer functions, and we know that each transfer function contains the system's characteristic polynomial $\Delta(s)$ in the denominator. For example, the transfer function $y_3(s)/u_1(s)$ is

$$\frac{y_3(s)}{u_1(s)} \triangleq \frac{N_{u_1}^{y_3}(s)}{\Delta(s)} = \det \begin{bmatrix} p_{1,1}(s) & p_{1,2}(s) & q_{1,1}(s) \\ p_{2,1}(s) & p_{2,2}(s) & q_{2,1}(s) \\ p_{3,1}(s) & p_{3,2}(s) & q_{3,1}(s) \end{bmatrix} / \det \mathbf{P}(s) \tag{11.11}$$

Note how the numerator polynomial is found, and also note that we have introduced the notation we will use for transfer-function numerators.

Now consider a stability-augmentation control law of the following form, which is consistent with the stability-augmentation control laws to be considered in this chapter.

$$\boxed{u_i(s) = \delta_{\text{Pilot}_k} - K_{y_j} y_j(s)} \tag{11.12}$$

Here δ_{Pilot_k} is a cockpit-control input from the pilot. Of interest now is the effect of this control law on the system's characteristic equation (or system poles), and on the numerators of the system's transfer functions (or transfer-function zeros). As an example, and we can generalize later, let the indices in Equation (11.12) be $i = 1, j = 3, k = 1$. Substituting this control law into Equation (11.10) yields

$$\begin{bmatrix} p_{1,1}(s) & p_{1,2}(s) & p_{1,3}(s) + \boxed{K_{y_3}q_{1,1}(s)} \\ p_{2,1}(s) & p_{2,2}(s) & p_{2,3}(s) + \boxed{K_{y_3}q_{2,1}(s)} \\ p_{3,1}(s) & p_{3,2}(s) & p_{3,3}(s) + \boxed{K_{y_3}q_{3,1}(s)} \end{bmatrix} \begin{Bmatrix} y_1(s) \\ y_2(s) \\ y_3(s) \end{Bmatrix} = \begin{bmatrix} q_{1,1}(s) & q_{1,2}(s) \\ q_{2,1}(s) & q_{2,2}(s) \\ q_{3,1}(s) & q_{3,2}(s) \end{bmatrix} \begin{Bmatrix} \delta_{\text{Pilot}_1}(s) \\ u_2(s) \end{Bmatrix} \quad (11.13)$$

or

$$\mathbf{P}_{u_1}^{y_3}(s)\mathbf{y}(s) = \mathbf{Q}(s)\mathbf{u}(s)$$

where the new terms introduced due to feedback are shown in boxes, and the subscript and superscript on **P** indicate that it now corresponds to the system with a feedback loop closed between u_1 and y_3.

The characteristic polynomial for this *closed-loop system,* the system with feedback in place, is of course equal to det $\mathbf{P}_{u_1}^{y_3}(s)$, which from the properties of matrix determinants may be written as

$$\begin{aligned} \Delta_{CL}(s) &= \det \mathbf{P}_{u_1}^{y_3}(s) = \det \begin{bmatrix} p_{1,1}(s) & p_{1,2}(s) & p_{1,3}(s) + K_{y_3}q_{1,1}(s) \\ p_{2,1}(s) & p_{2,2}(s) & p_{2,3}(s) + K_{y_3}q_{2,1}(s) \\ p_{3,1}(s) & p_{3,2}(s) & p_{3,3}(s) + K_{y_3}q_{3,1}(s) \end{bmatrix} \\ &= \det \left[\begin{bmatrix} p_{1,1}(s) & p_{1,2}(s) & p_{1,3}(s) \\ p_{2,1}(s) & p_{2,2}(s) & p_{2,3}(s) \\ p_{3,1}(s) & p_{3,2}(s) & p_{3,3}(s) \end{bmatrix} + \begin{bmatrix} p_{1,1}(s) & p_{1,2}(s) & K_{y_3}q_{1,1}(s) \\ p_{2,1}(s) & p_{2,2}(s) & K_{y_3}q_{2,1}(s) \\ p_{3,1}(s) & p_{3,2}(s) & K_{y_3}q_{3,1}(s) \end{bmatrix} \right] \\ &= \det \begin{bmatrix} p_{1,1}(s) & p_{1,2}(s) & p_{1,3}(s) \\ p_{2,1}(s) & p_{2,2}(s) & p_{2,3}(s) \\ p_{3,1}(s) & p_{3,2}(s) & p_{3,3}(s) \end{bmatrix} + K_{y_3} \det \begin{bmatrix} p_{1,1}(s) & p_{1,2}(s) & q_{1,1}(s) \\ p_{2,1}(s) & p_{2,2}(s) & q_{2,1}(s) \\ p_{3,1}(s) & p_{3,2}(s) & q_{3,1}(s) \end{bmatrix} \\ &= \Delta(s) + K_{y_3}N_{u_1}^{y_3}(s) \end{aligned} \quad (11.14)$$

That is, the closed-loop characteristic polynomial can be written in terms of the original open-loop characteristic polynomial $\Delta(s)$ and the numerator of the transfer-function associated with the response and the input being used in the control law, or $N_{u_1}^{y_3}(s)$. Furthermore, this closed-loop characteristic polynomial will appear in all six closed-loop transfer functions. Finally, this result can be generalized easily for any linear feedback control law involving any one of the three responses, either of the two control inputs, and either of the two cockpit-control inputs.

So far, all this is consistent with the discussion in Section 11.1 regarding root-locus plots, except we have now developed the closed-loop characteristic polynomial in the context of the polynomial-matrix description. Among other things, this will allow us to easily address the effects of adding feedback on the numerators (or the zeros) of the closed-loop transfer functions.

Let's first consider the numerators of the closed-loop transfer functions associated with the response used in the feedback control law (or y_3 in Equation (11.12)). We will see that the addition of y_3 feedback has no effect on the zeros of the closed-loop y_3 transfer functions. From Cramer's rule and Equation (11.13) we know that the zeros of all these y_3 transfer functions are the roots of the polynomial det $\mathbf{P}_{u_1}^{y_3}(s)$, but with its third column replaced by the appropriate column of $\mathbf{Q}(s)$. However, by making this substitution all the terms introduced by adding feedback that appear in the third column of $\mathbf{P}_{u_1}^{y_3}(s)$ will be eliminated. Thus, the zeros of all transfer functions involving the response fed back, or y_3 here, will be unaffected by the introduction of that feedback.

Now let's next look at the zeros of the numerators of the closed-loop transfer functions associated with responses other than that being fed back, but with the same input used in the control law. Consider, for example, the closed-loop transfer function $y_1(s)/u_1(s)$. Again from Equation (11.13), the numerator of this transfer function will be obtained from det $\mathbf{P}_{u_1}^{y_3}(s)$, but with the first column replaced by the first column of $\mathbf{Q}(s)$. Or

$$
\begin{aligned}
N_{u_1\,\text{aug}}^{y_1}(s) &= \det\begin{bmatrix} q_{1,1}(s) & p_{1,2}(s) & p_{1,3}(s) + \boxed{K_{y_3}q_{1,1}(s)} \\ q_{2,1}(s) & p_{2,2}(s) & p_{2,3}(s) + \boxed{K_{y_3}q_{2,1}(s)} \\ q_{3,1}(s) & p_{3,2}(s) & p_{3,3}(s) + \boxed{K_{y_3}q_{3,1}(s)} \end{bmatrix} \\
&= \det\begin{bmatrix} q_{1,1}(s) & p_{1,2}(s) & p_{1,3}(s) \\ q_{2,1}(s) & p_{2,2}(s) & p_{2,3}(s) \\ q_{3,1}(s) & p_{3,2}(s) & p_{3,3}(s) \end{bmatrix} + K_{y_3}\det\begin{bmatrix} q_{1,1}(s) & p_{1,2}(s) & q_{1,1}(s) \\ q_{2,1}(s) & p_{2,2}(s) & q_{2,1}(s) \\ q_{3,1}(s) & p_{3,2}(s) & q_{3,1}(s) \end{bmatrix} \\
&= N_{u_1}^{y_1}(s)
\end{aligned}
\tag{11.15}
$$

where

$$
\det\begin{bmatrix} q_{1,1}(s) & p_{1,2}(s) & q_{1,1}(s) \\ q_{2,1}(s) & p_{2,2}(s) & q_{2,1}(s) \\ q_{3,1}(s) & p_{3,2}(s) & q_{3,1}(s) \end{bmatrix} = 0 \tag{11.16}
$$

due to the fact that the first and third columns are identical. Therefore, we can see that the zeros of all closed-loop transfer functions associated with the control input used in the control law are also unaffected by the addition of feedback using that control input.

But this is not true for the transfer-function zeros associated with other control inputs. And this is a key property of MIMO systems. Still with regard to

Equation (11.13), consider again the closed-loop transfer function for response y_1 but now using input u_2. The zeros of this transfer function will be obtained from det $\mathbf{P}_{u_1}^{y_3}(s)$, but with the <u>first</u> column replaced by the <u>second</u> column of $\mathbf{Q}(s)$. That is, this numerator is now

$$
\begin{aligned}
N_{u_2\,\text{aug}}^{y_1}(s) &= \det\begin{bmatrix} q_{1,2}(s) & p_{1,2}(s) & p_{1,3}(s) + \boxed{K_{y_3}q_{1,1}(s)} \\ q_{2,2}(s) & p_{2,2}(s) & p_{2,3}(s) + \boxed{K_{y_3}q_{2,1}(s)} \\ q_{3,2}(s) & p_{3,2}(s) & p_{3,3}(s) + \boxed{K_{y_3}q_{3,1}(s)} \end{bmatrix} \\
&= \det\begin{bmatrix} q_{1,2}(s) & p_{1,2}(s) & p_{1,3}(s) \\ q_{2,2}(s) & p_{2,2}(s) & p_{2,3}(s) \\ q_{3,2}(s) & p_{3,2}(s) & p_{3,3}(s) \end{bmatrix} + K_{y_3}\det\begin{bmatrix} q_{1,2}(s) & p_{1,2}(s) & q_{1,1}(s) \\ q_{2,2}(s) & p_{2,2}(s) & q_{2,1}(s) \\ q_{3,2}(s) & p_{3,2}(s) & q_{3,1}(s) \end{bmatrix} \\
&= N_{u_2}^{y_1}(s) + K_{y_3}N_{u_1\,u_2}^{y_3\,y_1}(s)
\end{aligned}
\tag{11.17}
$$

So this numerator polynomial is <u>no longer equal to that of the open-loop system</u>, or $N_{u_2}^{y_1}(s)$. We see that this closed-loop numerator polynomial is a function of the feedback gain K_{y_3}, plus a new polynomial $N_{u_1\,u_2}^{y_3\,y_1}(s)$ called a *coupling numerator* (Ref 2). Furthermore, note from Equation (11.17) that the coupling numerator is found from det $\mathbf{P}(s)$, except with <u>two</u> columns replaced by the appropriate columns of $\mathbf{Q}(s)$.

We actually encountered a coupling numerator previously when considering the numerators of the closed-loop transfer functions associated with the input used in the control law, or $u_1(s)$. But that coupling numerator, given in Equation (11.16), was equal to zero. From the cases considered above, and from the properties of determinants, we can see that some of the useful <u>general properties of coupling numerators</u> are

$$
\boxed{\begin{aligned}
&N_{u_i\,u_j}^{y_k\,y_l}(s) = N_{u_j\,u_i}^{y_l\,y_k}(s) \\
&N_{u_i\,u_i}^{y_k\,y_l}(s) = 0 \\
&N_{u_i\,u_j}^{y_k\,y_k}(s) = 0
\end{aligned}}
\tag{11.18}
$$

It is now relevant to make an observation regarding the form of the closed-loop <u>numerator polynomial</u> such as given in Equation (11.17). This form is

$$
\begin{aligned}
\big(\text{Closed-Loop Numerator Polynomial}\big) &= \big(\text{Open-Loop Numerator Polynomial}\big) \\
&\quad + K\big(\text{Coupling Numerator Polynomial}\big)
\end{aligned}
\tag{11.19}
$$

But this <u>form</u> is exactly the same as that for the closed-loop <u>characteristic polynomial</u> given in Equation (11.7) in the discussion of the root-locus plot.

This suggests, therefore, that the root-locus technique may also be used for determining the locations of closed-loop zeros as a function of the control gain K. And that is indeed the case, as we shall see in an example in Section 11.3.

But first we need to further discuss the subject of the closed-loop characteristic polynomial, but with more than one feedback loop closed. Equation (11.14) gave the characteristic polynomial when one feedback loop is closed using the control law given in Equation (11.12), and with $i = 1, j = 3$, and $k = 1$. Let us now add a second feedback-control law, of the same form as Equation (11.12), except now let $i = 2, j = 1$, and $k = 2$, for example. The polynomial-matrix description for the system with both control loops closed is written as

$$\begin{bmatrix} p_{1,1}(s) + \boxed{K_{y_1}q_{1,2}(s)} & p_{1,2}(s) & p_{1,3}(s) + \boxed{K_{y_3}q_{1,1}(s)} \\ p_{2,1}(s) + \boxed{K_{y_1}q_{2,2}(s)} & p_{2,2}(s) & p_{2,3}(s) + \boxed{K_{y_3}q_{2,1}(s)} \\ p_{3,1}(s) + \boxed{K_{y_1}q_{3,2}(s)} & p_{3,2}(s) & p_{3,3}(s) + \boxed{K_{y_3}q_{3,1}(s)} \end{bmatrix} \begin{Bmatrix} y_1(s) \\ y_2(s) \\ y_3(s) \end{Bmatrix} = \begin{bmatrix} q_{1,1}(s) & q_{1,2}(s) \\ q_{2,1}(s) & q_{2,2}(s) \\ q_{3,1}(s) & q_{3,2}(s) \end{bmatrix} \begin{Bmatrix} \delta_{\text{Pilot}_1}(s) \\ \delta_{\text{Pilot}_2}(s) \end{Bmatrix} \tag{11.20}$$

or

$$\mathbf{P}_{u_1\,u_2}^{y_3\,y_1}(s)\mathbf{y}(s) = \mathbf{Q}(s)\mathbf{u}(s)$$

The terms introduced due to the feedback loops are again highlighted in boxes, and the characteristic polynomial is, of course, $\det \mathbf{P}_{u_1\,u_2}^{y_3\,y_1}(s)$, which is found to be

$$\begin{aligned} \det \mathbf{P}_{u_1\,u_2}^{y_3\,y_1}(s) &= \det \begin{bmatrix} p_{1,1}(s) + \boxed{K_{y_1}q_{1,2}(s)} & p_{1,2}(s) & p_{1,3}(s) + \boxed{K_{y_3}q_{1,1}(s)} \\ p_{2,1}(s) + \boxed{K_{y_1}q_{2,2}(s)} & p_{2,2}(s) & p_{2,3}(s) + \boxed{K_{y_3}q_{2,1}(s)} \\ p_{3,1}(s) + \boxed{K_{y_1}q_{3,2}(s)} & p_{3,2}(s) & p_{3,3}(s) + \boxed{K_{y_3}q_{3,1}(s)} \end{bmatrix} \\ &= \det \begin{bmatrix} p_{1,1}(s) & p_{1,2}(s) & p_{1,3}(s) \\ p_{2,1}(s) & p_{2,2}(s) & p_{2,3}(s) \\ p_{3,1}(s) & p_{3,2}(s) & p_{3,3}(s) \end{bmatrix} + K_{y_1} \det \begin{bmatrix} q_{1,2}(s) & p_{1,2}(s) & p_{1,3}(s) \\ q_{2,2}(s) & p_{2,2}(s) & p_{2,3}(s) \\ q_{3,2}(s) & p_{3,2}(s) & p_{3,3}(s) \end{bmatrix} \\ &\quad + K_{y_3} \det \begin{bmatrix} p_{1,1}(s) & p_{1,2}(s) & q_{1,1}(s) \\ p_{2,1}(s) & p_{2,2}(s) & q_{2,1}(s) \\ p_{3,1}(s) & p_{3,2}(s) & q_{3,1}(s) \end{bmatrix} + K_{y_1}K_{y_3} \det \begin{bmatrix} q_{1,2}(s) & p_{1,2}(s) & q_{1,1}(s) \\ q_{2,2}(s) & p_{2,2}(s) & q_{2,1}(s) \\ q_{3,2}(s) & p_{3,2}(s) & q_{3,1}(s) \end{bmatrix} \\ &= \Delta(s) + K_{y_1}N_{u_2}^{y_1}(s) + K_{y_3}N_{u_1}^{y_3}(s) + K_{y_1}K_{y_3}N_{u_1\,u_2}^{y_3\,y_1}(s) \end{aligned} \tag{11.21}$$

So with two feedback loops closed, the closed-loop characteristic polynomial may now be expressed in terms of the open-loop characteristic polynomial, the two feedback gains, the numerators of the two appropriate open-loop transfer functions, plus the coupling numerator corresponding to the two inputs and two outputs used in the two control laws.

The key points in this section may be summarized as follows:

1. An aircraft's flight dynamics constitutes a dynamic system with multiple inputs and multiple responses.
2. The introduction of feedback into the system modifies the closed-loop system's characteristic polynomial as well as some of the closed-loop-transfer-function numerators.
3. The effect of feedback on the characteristic polynomial and closed-loop-transfer-function numerators may be evaluated systematically by using the system's polynomial-matrix description.
4. Coupling numerators play important roles in terms of both the closed-loop characteristic polynomial as well as the transfer-function numerators of MIMO systems.

11.3 AUGMENTING THE LONGITUDINAL DYNAMICS

Consider the conceptual block diagram shown in Figure 11.9, which is similar to Figure 11.8. Assuming this is a conventional aircraft, elevator and thrust are the two control inputs, and several aircraft responses are shown. The pilot is again seen to be closing a control loop (actually two control loops) around the complete aircraft and stability-augmentation systems, using motion and visual cues as the feedback quantities. In this architecture, the stability-augmentation systems act as *inner loops,* and the feedback-control actions of the pilot constitute *outer loops.*

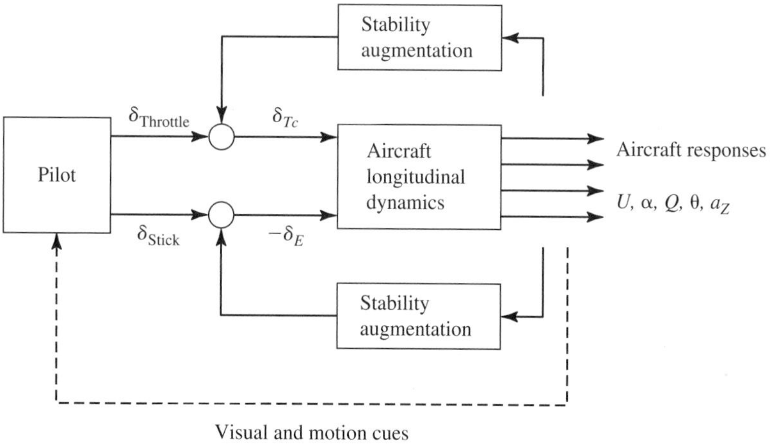

Figure 11.9 Block diagram for the longitudinal control of an aircraft.

Several transducers, or sensors, are available for use in the stability-augmentation systems. These include accelerometers, angle-of-attack and side-slip vanes, angular-rate and angular-displacement gyroscopes, altimeters, and GPS navigation devices, for example. In the longitudinal axes, angular-rate gyros could sense pitch rate, and an angular-displacement gyro could sense pitch-attitude angle. The stability-augmentation system would typically provide commands, or signals, to the existing control effectors on the vehicle, such as the elevator actuator and engine.

The topic in this section is the synthesis of stability-augmentation system(s) to tailor the vehicle's dynamics such that either its handling qualities are enhanced, or it is easier to design an outer-loop autopilot (as discussed in Chapter 12). Central to this issue is the selection of the appropriate responses to feed back, and the selection of the most effective control input to use.

To demonstrate an important fact contributing to the need for longitudinal stability augmentation, consider the root locus shown in Figures 11.10*a*–*b*. (The area around the origin is shown enlarged in Figure 11.10*b*, to see the migration of the phugoid poles.) The vehicle dynamics consist of the pitch-attitude response to elevator input for a vehicle similar to the F-5A, with no stability augmentation and with the attitude-feedback-control loop being closed by the pilot. So with reference to the system in Figure 11.9, there is no stability augmentation present, and the pilot is closing the (outer) feedback loop based on sensed pitch attitude.

Note that as the pilot's feedback gain increases, the closed-loop poles move along the root locus away from the open-loop poles. As a result, the damping of the closed-loop phugoid mode is increased, which is a desirable result of the pilot's control of pitch attitude. However, the damping of the closed-loop short-period eigenvalues is reduced below that of the open-loop aircraft, which can lead to a lightly damped closed-loop pilot-vehicle system and therefore poor handling qualities. This situation is exacerbated when the short-period damping

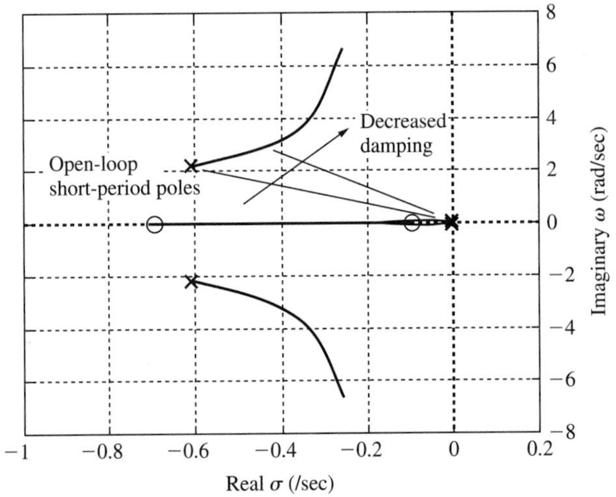

Figure 11.10*a* Root locus for piloted control of pitch attitude.

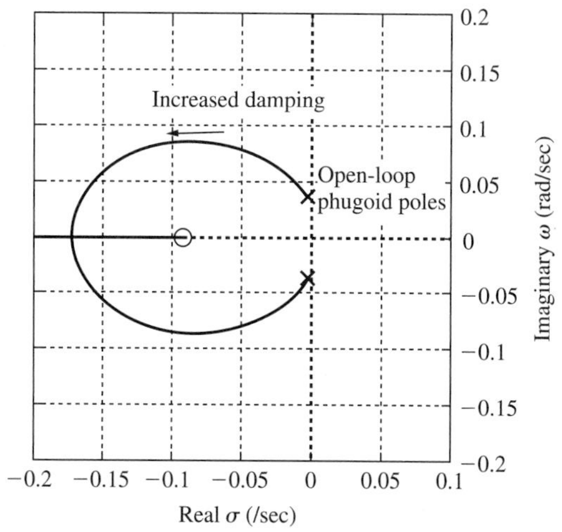

Figure 11.10*b* Root locus—area near origin enlarged.

of the open-loop aircraft is low to begin with. So increasing short-period damping is a common requirement for stability augmentation.

11.3.1 Increasing Short-Period Damping

To increase the short-period damping, let's recall that the short-period approximation (Equation (10.90)) indicated that this damping is determined from

$$2\zeta_{SP}\omega_{SP} = -\left(\frac{Z_\alpha}{(U_0 - Z_{\dot{\alpha}})} + M_{\dot{\alpha}}\frac{(U_0 + Z_q)}{(U_0 - Z_{\dot{\alpha}})} + M_q\right) \approx -\left(\frac{Z_\alpha}{U_0} + M_{\dot{\alpha}} + M_q\right) \tag{11.22}$$

This suggests that the short-period damping may be increased by increasing the effective pitch damping of the vehicle M_q. But M_q is simply the pitching moment acting on the vehicle due to its pitch rate. Therefore, we may increase effective pitch damping by providing an additional pitching moment that is proportional to pitch rate. This may be accomplished by simply feeding back measured pitch rate to the pitch-control surface (e.g., elevator), as indicated in Figure 11.11. Such a stability-augmentation system is called a *pitch damper.*

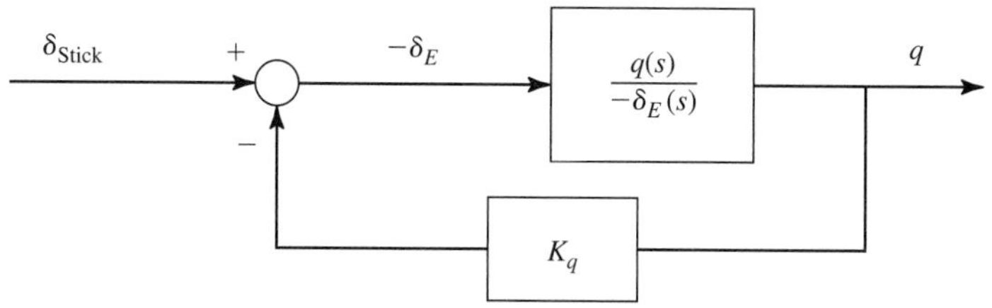

Figure 11.11 Block diagram for pitch damper.

NOTE TO STUDENT

We have applied the insight gained through the short-period approximation to select a vehicle response for feedback, as well as an appropriate control effector—namely pitch rate and the pitch-control surface. This feedback-control architecture was synthesized through our knowledge of the vehicle's dynamics. We will use this approach to develop all the stability-augmentation systems in this chapter.

To confirm our selection of this feedback architecture, consider the root-locus shown in Figures 11.12*a–b*. (The region near the origin is enlarged in Figure 11.12*b*.) The vehicle dynamics are the same as that used to generate Figures 11.10*a–b*, or a vehicle similar to the F-5A, except here we are using pitch-rate feedback rather than pitch attitude feedback. The root-locus reveals the effect of our stability-augmentation system, or feedback of pitch rate to the elevator, on the closed-loop eigenvalues. Note that the phugoid poles are not significantly modified, except that modal damping will be increased, which is usually desirable. But the short-period damping may be increased significantly as feedback gain is increased. This confirms that, for this vehicle, pitch-rate feedback is indeed effective for increasing short-period damping. This result is typical for any vehicle that exhibits conventional short-period characteristics.

To gain additional insight into the effect of pitch-rate feedback in general, consider a vehicle's polynomial-matrix description given below for level flight ($\Theta_0 = 0$).

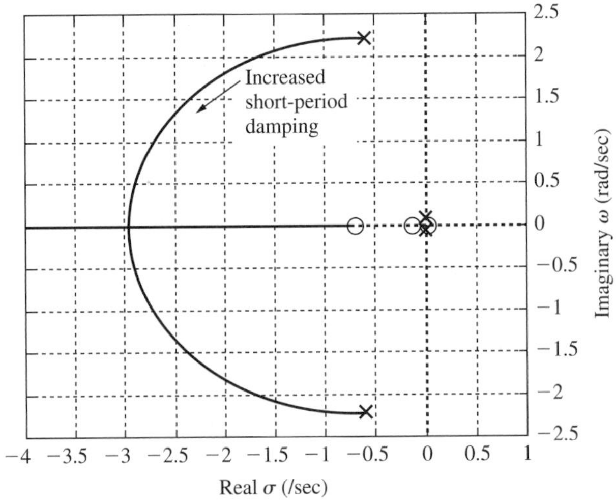

Figure 11.12*a* Root locus indicating effectiveness of pitch-rate feedback.

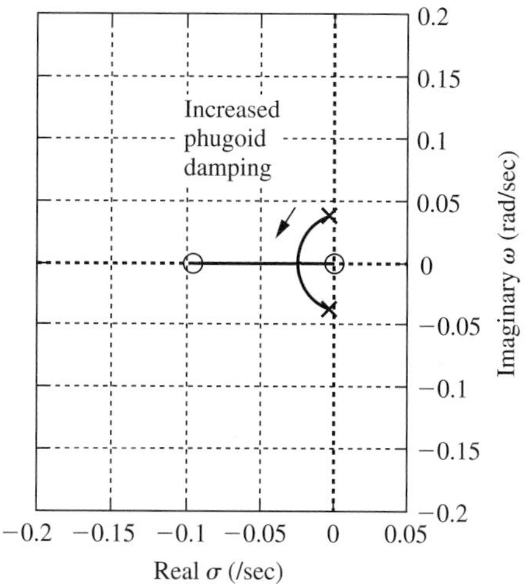

Figure 11.12*b* Region near origin of root-locus enlarged.

$$\begin{bmatrix} s-(X_u+X_{P_u}) & -(X_{\dot\alpha}s+X_\alpha) & -X_q s+g \\ -(Z_u+Z_{P_u}) & (\mathrm{U}_0-Z_{\dot\alpha})s-Z_\alpha & -(Z_q+\mathrm{U}_0)s \\ -(M_u+M_{P_u}) & -(M_{\dot\alpha}s+(M_\alpha+M_{P_\alpha})) & s^2-M_q s \end{bmatrix} \begin{Bmatrix} u(s) \\ \alpha(s) \\ \theta(s) \end{Bmatrix} = \begin{bmatrix} -X_{\delta_E} & X_T \\ -Z_{\delta_E} & Z_T \\ -M_{\delta_E} & M_T \end{bmatrix} \begin{Bmatrix} -\delta_E(s) \\ \delta T(s) \end{Bmatrix} \tag{11.23}$$

(Note that the sign convention on elevator deflection has again been adjusted.) With pitch-rate feedback, consisting of the control law

$$-\delta_E(s) = \delta_{\text{Stick}}(s) - K_q\dot\theta(s)$$

the above polynomial-matrix description becomes

$$\begin{bmatrix} s-(X_u+X_{P_u}) & -(X_{\dot\alpha}s+X_\alpha) & -(X_q+\boxed{K_qX_{\delta_E}})s+g \\ -(Z_u+Z_{P_u}) & (\mathrm{U}_0-Z_{\dot\alpha})s-Z_\alpha & -((Z_q+\boxed{K_qZ_{\delta_E}})+\mathrm{U}_0)s \\ -(M_u+M_{P_u}) & -(M_{\dot\alpha}s+(M_\alpha+M_{P_\alpha})) & s^2-(M_q+\boxed{K_qM_{\delta_E}})s \end{bmatrix} \begin{Bmatrix} u(s) \\ \alpha(s) \\ \theta(s) \end{Bmatrix} = \begin{bmatrix} X_{\delta_E} & X_T \\ Z_{\delta_E} & Z_T \\ M_{\delta_E} & M_T \end{bmatrix} \begin{Bmatrix} -\delta_{\text{Stick}}(s) \\ \delta T(s) \end{Bmatrix} \tag{11.24}$$

or

$$\mathbf{P}^q_{-\delta_E}(s)\mathbf{y}(s) = \mathbf{Q}(s)\mathbf{u}(s)$$

Two things should be observed at the outset. First, the terms introduced due to the feedback control law are highlighted in boxes; the terms appear in the third column of $\mathbf{P}^q_{-\delta_E}(s)$. Second, the effective values of the dimensional stability derivatives X_q, Z_q, and M_q will clearly be augmented by introducing pitch-rate feedback. This second feature is why we call such feedback systems *stability-augmentation systems.*

Now we need to explore the effect of adding pitch-rate feedback on the zeros of the vehicle transfer functions. We know the effect on the longitudinal eigenvalues, as that was indicated in the root loci in Figures 11.12*a*–*b*. But what are the possible effects on the transfer-function zeros?

First, we know there is no effect on the zeros of the closed-loop pitch-attitude and pitch rate transfer functions with elevator input. This has been shown previously in Section 11.2. However, there may be an effect on some of the other transfer function zeros.

To explore this possibility, again consider the vehicle's polynomial-matrix description given in Equation (11.24). In Section 11.2 we found that the zeros of any transfer function associated with elevator input would also be unaffected by the introduction of this feedback system. For example, the numerator of the $\alpha(s)/-\delta_{\text{Stick}}(s)$ transfer function is

$$N^{\alpha}_{-\delta_{\text{Stick}}\,\text{aug}}(s) = \det\begin{bmatrix} s-(X_u+X_{P_u}) & X_{\delta_E} & -(X_q+\boxed{K_qX_{\delta_E}})s+g \\ -(Z_u+Z_{P_u}) & Z_{\delta_E} & -\left((Z_q+\boxed{K_qZ_{\delta_E}})+U_0\right)s \\ -(M_u+M_{P_u}) & M_{\delta_E} & s^2-(M_q+\boxed{K_qM_{\delta_E}})s \end{bmatrix}$$

$$= \det\begin{bmatrix} s-(X_u+X_{P_u}) & X_{\delta_E} & -X_qs+g \\ -(Z_u+Z_{P_u}) & Z_{\delta_E} & -(Z_q+U_0)s \\ -(M_u+M_{P_u}) & M_{\delta_E} & s^2-M_qs \end{bmatrix} - K_qs\det\begin{bmatrix} s-(X_u+X_{P_u}) & X_{\delta_E} & X_{\delta_E} \\ -(Z_u+Z_{P_u}) & Z_{\delta_E} & Z_{\delta_E} \\ -(M_u+M_{P_u}) & M_{\delta_E} & M_{\delta_E} \end{bmatrix}$$

$$= N^{\alpha}_{\delta_E}(s) - K_qsN^{\theta\alpha}_{\delta_E\delta_E}(s) = N^{\alpha}_{\delta_E}(s) - 0 \tag{11.25}$$

So this transfer-function's zeros are unaffected by the addition of pitch-rate feedback to the elevator.

But again from Section 11.2, we also know that the same cannot be said for the transfer functions associated with thrust input. Consider the transfer function $u(s)/\delta T(s)$, for example, which will usually be the most affected. The numerator of this transfer function involves a coupling numerator, or

$$N^{u}_{\delta T\text{aug}}(s) = \det\begin{bmatrix} X_T & -(X_{\dot\alpha}s+X_\alpha) & -(X_q+\boxed{K_qX_{\delta_E}})s+g \\ Z_T & (U_0-Z_{\dot\alpha})s-Z_\alpha & -\left((Z_q+\boxed{K_qZ_{\delta_E}})+U_0\right)s \\ M_T & -\left(M_{\dot\alpha}s+(M_\alpha+M_{P_\alpha})\right) & s^2-(M_q+\boxed{K_qM_{\delta_E}})s \end{bmatrix}$$

$$= \det\begin{bmatrix} X_T & -(X_{\dot\alpha}s+X_\alpha) & -X_qs+g \\ Z_T & (U_0-Z_{\dot\alpha})s-Z_\alpha & -(Z_q+U_0)s \\ M_T & -\left(M_{\dot\alpha}s+(M_\alpha+M_{P_\alpha})\right) & s^2-M_qs \end{bmatrix} - K_qs\det\begin{bmatrix} X_T & -(X_{\dot\alpha}s+X_\alpha) & X_{\delta_E} \\ Z_T & (U_0-Z_{\dot\alpha})s-Z_\alpha & Z_{\delta_E} \\ M_T & -\left(M_{\dot\alpha}s+(M_\alpha+M_{P_\alpha})\right) & M_{\delta_E} \end{bmatrix}$$

$$= N^{u}_{\delta T}(s) - K_qsN^{\theta u}_{\delta_E\delta T}(s) \tag{11.26}$$

Assuming $Z_T \approx M_T \approx 0$, the coupling numerator in Equation (11.26) is

$$N^{\theta u}_{\delta_E \delta T}(s) = \det\begin{bmatrix} X_T & -\left(X_{\dot{\alpha}}s + X_\alpha\right) & X_{\delta_E} \\ Z_T = 0 & \left(\mathrm{U}_0 - Z_{\dot{\alpha}}\right)s - Z_\alpha & Z_{\delta_E} \\ M_T = 0 & -\left(M_{\dot{\alpha}}s + \left(M_\alpha + M_{P_\alpha}\right)\right) & M_{\delta_E} \end{bmatrix}$$
$$= X_T\left(M_{\delta_E}\left(\left(\mathrm{U}_0 - Z_{\dot{\alpha}}\right)s - Z_\alpha\right) + Z_{\delta_E}\left(M_{\dot{\alpha}}s + \left(M_\alpha + M_{P_\alpha}\right)\right)\right)$$
$$= X_T\left(\left(M_{\delta_E}\left(\mathrm{U}_0 - Z_{\dot{\alpha}}\right) + Z_{\delta_E}M_{\dot{\alpha}}\right)s + \left(Z_{\delta_E}\left(M_\alpha + M_{P_\alpha}\right) - M_{\delta_E}Z_\alpha\right)\right) \quad (11.27)$$

For the F-5A aircraft, for example, and letting $X_T = 1/m$ /slug, this coupling numerator becomes

$$N^{\theta u}_{\delta_E \delta T}(s) = -39.1\left(s + 0.701\right) \quad (11.28)$$

while the numerator for the open-loop transfer function is

$$N^{u}_{\delta T}(s) = 2.73s\left(s^2 + 1.214s + 3.707\right) \quad (11.29)$$

To determine the behavior of the $u(s)/\delta T(s)$ transfer-function zeros as pitch-rate feedback gain K_q is varied, we may use the root-locus technique. From Equation (11.26), the zeros in question are the roots of the polynomial given by

$$N^{u}_{\delta T}(s) - K_q s N^{\theta u}_{\delta_E \delta T}(s) = 0$$

or (11.30)

$$1 - K_q \frac{sN^{\theta u}_{\delta_E \delta T}(s)}{N^{u}_{\delta T}(s)} = 0$$

Using the second of the above two expressions, along with Equations (11.28) and (11.29), we obtain the *numerator root locus* shown in Figure 11.13. This figure indicates that the zeros of the (closed-loop) $u(s)/\delta T(s)$ transfer-function migrate from their open-loop location (indicated by the X's) along the branches shown, towards the roots of the coupling numerator (indicated by the O's) or towards infinity. Since the short-period damping will be increased as pitch-rate feedback gain is increased, and since we typically want the short-period quadratic in the $u(s)/\delta T(s)$ transfer function to be approximately cancelled by a similar term in the numerator (such that the thrust-to-speed response is dominated by the phugoid mode), it is actually desirable to have the zeros in the $u(s)/\delta T(s)$ transfer function move in the direction indicated in the numerator-root locus.

From the above analysis, we can see that pitch dampers are effective stability-augmentation systems for increasing the short-period damping. Pitch dampers are used on many aircraft, including the F-5A.

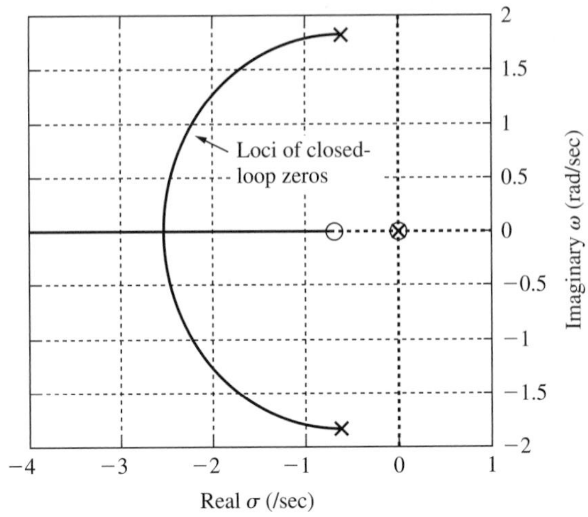

Figure 11.13 Numerator root locus—zeros of $u(s)/\,\delta T(s)$.

EXAMPLE 11.3

Stability Augmentation to Improve Short-Period Damping

Using the data for an F-5A aircraft at Flight Condition 1 (Appendix B), find the pitch-rate-feedback gain necessary to increase the short-period damping to 0.6. Also, compare the angle-of-attack and surge-velocity transfer functions that include the pitch damper to those for the unaugmented airframe.

■ Solution

Using the data from Appendix B, the state-variable model for the longitudinal dynamics may be assembled. The appendix specifies a body-fixed stability axis corresponding to the following reference flight condition.

$$U_0 = 850 \text{ fps}, \Theta_0 = \gamma_0 = 0, h_0 = 40{,}000 \text{ ft, and } \Phi_0 = 0 \left(\text{implied}\right)$$

The longitudinal dimensional stability derivatives for this flight condition are given in Table 11.2. All angles are in radians, forces in pounds, and velocities in feet per second.

Table 11.2 Dimensional Stability Derivatives for the F-5A

Stability Derivative	Value	Stability Derivative	Value
X_α	-6.826 ft/sec^2	M_α	-3.392 sec^{-2}
$X_u + X_{T_u}$	-0.011 sec^{-1}	$M_{\dot\alpha}$	-0.051 sec^{-1}
Z_α	-623.9 ft/sec^2	M_q	-0.429 sec^{-1}
$Z_u + Z_{T_u}$	-0.124 sec^{-1}	$M_u + M_{T_u}$	-0.00046 (sec-ft)$^{-1}$
Z_{δ_E}	-119 ft/sec^2	$M_{\delta E}$	-14.31/sec^2

Since we will be finding several transfer functions, it is helpful to use a state-variable model and MATLAB.

Using Equations (10.64), along with the stability derivatives in Table 11.2, we have the state-variable model given by

$$\mathbf{y}^T = \mathbf{x}^T = \left[u(\text{fps}) \quad \alpha(\text{rad}) \quad \theta(\text{rad}) \quad q(\text{rad/sec})\right], \ \mathbf{u}^T = \left[-\delta_E(\text{rad}) \quad \delta T(\text{lbs})\right]$$

$$\mathbf{A} = \begin{bmatrix} -0.011 & -6.826 & -32.2 & 0 \\ -0.000146 & -0.734 & 0 & 1 \\ 0 & 0 & 0 & 1 \\ -0.000453 & -3.355 & 0 & -0.480 \end{bmatrix}, \mathbf{B} = \begin{bmatrix} 0 & 1/311 \\ 0.14 & 0 \\ 0 & 0 \\ 14.3 & 0 \end{bmatrix} \tag{11.31}$$

$$\mathbf{C} = \mathbf{I}_4, \ \mathbf{D}_{4\times 2} = 0$$

(Note the control-input vector **u** and the signs on the elements in the first column of the **B** matrix.) The transfer functions without stability augmentation, found from MATLAB, are

$$\frac{u(s)}{-\delta_E(s)} = \frac{-0.956(s+0.579)(s+564)}{(s^2+0.00555s+0.00136)(s^2+1.219s+3.711)} \text{ fps/rad}$$

$$\frac{\alpha(s)}{-\delta_E(s)} = \frac{0.140(s+102.6)(s^2+0.01086s+0.00453)}{(s^2+0.00555s+0.00136)(s^2+1.219s+3.711)} \text{ rad/rad}$$

$$\frac{q(s)}{-\delta_E(s)} = \frac{14.29s(s+0.0095)(s+0.703)}{(s^2+0.00555s+0.00136)(s^2+1.219s+3.711)} \text{ rad/sec/rad} \tag{11.32}$$

$$\frac{u(s)}{\delta T(s)} = \frac{0.0032154\, s\,(s^2+1.214s+3.707)}{(s^2+0.00555s+0.00136)(s^2+1.219s+3.711)} \text{ fps/lb}$$

with the remaining two transfer functions for thrust input equal to zero.

The root locus for the system with the pitch damper, or

$$-\delta_E(s) = \delta_{\text{Stick}}(s) - K_q q(s)$$

is shown in Figure 11.14, with the desired short-period pole locations indicated.

The state-variable description of the system augmented with the given control law is found as follows:

$$\dot{\mathbf{x}} = \mathbf{A}\mathbf{x} + \mathbf{B}\begin{Bmatrix} -\delta_E \\ \delta T \end{Bmatrix}, \quad -\delta_E = \delta_{\text{Stick}} - K_q q$$

$$\dot{\mathbf{x}} = \mathbf{A}\mathbf{x} + \mathbf{B}\begin{Bmatrix} -\delta_{\text{Stick}} - K_q q \\ \delta T \end{Bmatrix} = (\mathbf{A} - \mathbf{b}_{-\delta_E} K_q \mathbf{c}_q)\mathbf{x} + \mathbf{B}\begin{Bmatrix} \delta_{\text{Stick}} \\ \delta T \end{Bmatrix}$$

$$\triangleq \mathbf{A}_{\text{aug}}\mathbf{x} + \mathbf{B}\begin{Bmatrix} \delta_{\text{Stick}} \\ \delta T \end{Bmatrix} \tag{11.33}$$

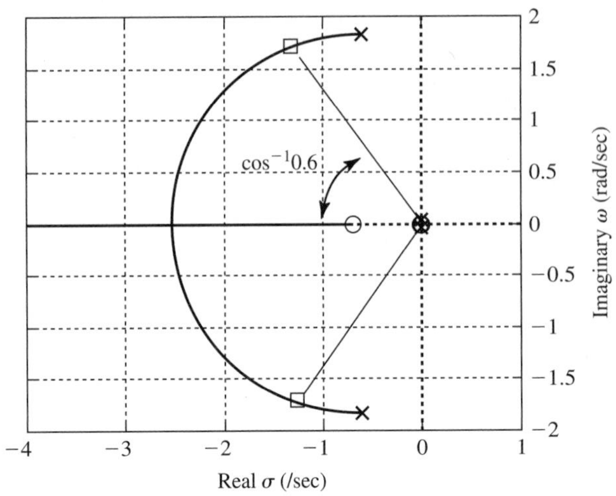

Figure 11.14 Root locus showing desired short-period pole locations.

Here $\mathbf{b}_{-\delta_E}$ is the first column of **B** corresponding to (negative) elevator input, and $\mathbf{c}_q$ is the fourth row of the **C** matrix corresponding to pitch-rate response, or

$$\mathbf{c}_q = \begin{bmatrix} 0 & 0 & 0 & 1 \end{bmatrix}$$

By selecting a rate-feedback gain K_q of 0.1 rad/rad/sec, we obtain the following set of eigenvalues for the augmented vehicle.

$$\lambda_{\mathrm{SP}} = -1.324 \pm j1.722\,/\mathrm{sec}$$

$$\lambda_P = -0.0031 \pm j0.0326\,/\mathrm{sec}$$

This yields a short-period damping and natural frequency of

$$\zeta_{\mathrm{SP}} = 0.61$$

$$\omega_{\mathrm{SP}} = 2.172\ \mathrm{rad/sec}$$

which is considered sufficiently close to the desired damping.

Now MATLAB can again be used to find all the transfer functions of the augmented vehicle. That is, we now have

$$\dot{\mathbf{x}} = \mathbf{A}_{\mathrm{aug}}\mathbf{x} + \mathbf{B}\begin{Bmatrix} \delta_{\mathrm{Stick}} \\ \delta T \end{Bmatrix} \tag{11.34}$$

$$\mathbf{C} = \mathbf{I}_4, \quad \mathbf{D} = \mathbf{0}_{4\times 2}$$

The four transfer functions become

$$
\begin{aligned}
\frac{u(s)}{\delta_{\text{Stick}}(s)} &= \frac{-0.956(s+0.579)(s+584)}{(s^2+0.00614s+0.00107)(s^2+2.648s+4.72)} \text{ fps/rad} \\
\frac{\alpha(s)}{\delta_{\text{Stick}}(s)} &= \frac{0.140(s^2+0.01086s+0.00453)(s+102.6)}{(s^2+0.00614s+0.00107)(s^2+2.648s+4.72)} \text{ rad/rad} \\
\frac{q(s)}{\delta_{\text{Stick}}(s)} &= \frac{14.29s(s+0.0095)(s+0.703)}{(s^2+0.00614s+0.00107)(s^2+2.648s+4.72)} \text{ rad/sec/rad} \\
\frac{u(s)}{\delta T(s)} &= \frac{0.00322\,s\,(s^2+2.643s+4.709)}{(s^2+0.00614s+0.00107)(s^2+2.648s+4.72)} \text{ fps/lb}
\end{aligned}
\quad (11.35)
$$

Comparing these to the transfer functions found previously for the unaugmented system, we see that the short-period frequency and damping have been increased (as expected from the root locus), and the phugoid damping has been slightly increased while the frequency has been slightly decreased (also as expected from the root locus). In addition, there have been no changes in the numerators of the transfer functions corresponding to elevator input (as expected), but the numerator of the $u(s)/\delta T(s)$ transfer function has been modified as predicted by the numerator-root locus. With the modified numerator, there is an approximate pole-zero cancellation of the augmented short-period poles in this transfer function, which is desirable.

11.3.2 Increasing Short-Period Frequency

A second dynamic deficiency that may be encountered is a low short-period frequency, which can lead to sluggish pitch response and poor handling qualities. Now recall from the short-period approximation that the short-period frequency (Equation (10.91)) is approximately

$$
\omega_{\text{SP}}^2 = \frac{M_q Z_\alpha}{(\text{U}_0 - Z_{\dot{\alpha}})} - (M_\alpha + M_{P_\alpha})\frac{(\text{U}_0 + Z_q)}{(\text{U}_0 - Z_{\dot{\alpha}})} \approx M_q \frac{Z_\alpha}{\text{U}_0} - (M_\alpha + M_{P_\alpha}) \quad (11.36)
$$

This suggests that the short-period frequency may be increased by increasing the effective pitch stiffness of the vehicle M_α, or the pitching moment acting on the vehicle due to angle of attack. We increase effective pitch stiffness by providing an additional pitching moment that is proportional to angle of attack. This may be accomplished by simply feeding back measured angle of attack to the pitch-control surface (e.g., elevator), as indicated in Figure 11.15.

To assess the effect of such an augmentation system on the system's eigenvalues, consider the root locus shown in Figures 11.16*a–b*. Here, the open-loop transfer function is that of angle of attack with (negative) elevator input, for a vehicle similar to the F-5A with conventional short-period and phugoid modal characteristics. The root locus shown in Figure 11.16*a* only shows the branches

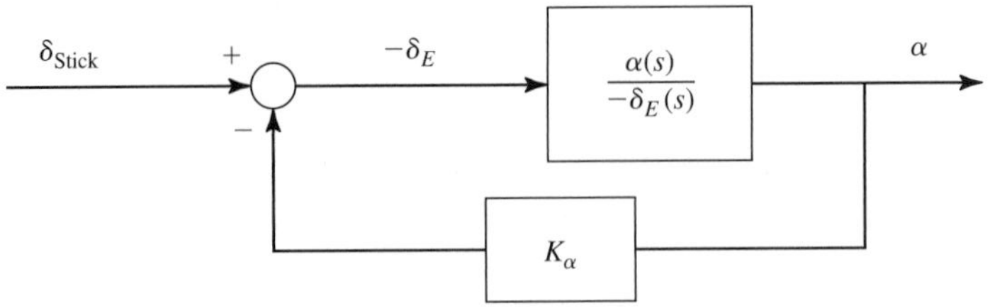

Figure 11.15 Block diagram of angle-of-attack stability augmentation system.

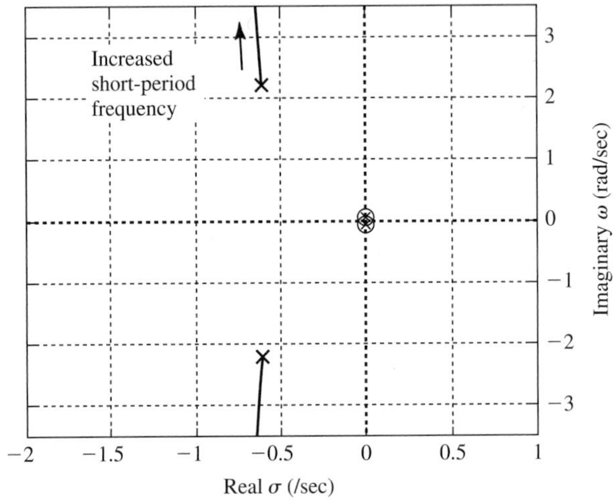

Figure 11.16*a* Root locus indicating effect of angle-of-attack feedback.

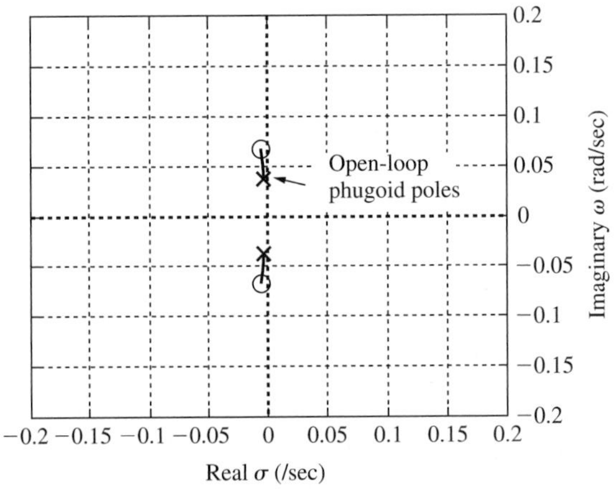

Figure 11.16*b* Root locus—region near the origin enlarged.

that start at the open-loop short-period poles for moderate values of gain. At high values of gain, the short-period eigenvalues move far to the left and eventually move onto the negative real axis, but we will only need a low gain to meet our objectives. One can see that the frequency of the short-period eigenvalues may be significantly increased through this type of feedback augmentation. And this is the desired effect. Note from Figure 11.16*b* that the effect on the phugoid eigenvalues is again negligible, as expected, since the phugoid modal response contains virtually no angle-of-attack participation.

But we must also consider the effect of adding angle-of-attack feedback on the zeros of the vehicle transfer functions. We know there is no effect on the zeros of the angle-of-attack transfer functions, just as there was no effect on the pitch-attitude transfer functions due to a pitch damper. We also know there is no effect on the zeros of the transfer functions with elevator (or stick) input. However, there is an effect on the zeros of the $u(s)/\delta T(s)$ transfer function.

To explore this issue, again consider the vehicle's polynomial-matrix description given in Equation (11.23), except here we have incorporated the effects of angle-of-attack feedback. That is, we now have

$$\begin{bmatrix} s-(X_u+X_{P_u}) & -\left(X_{\dot\alpha}s+\left(X_\alpha+\boxed{K_\alpha X_{\delta_E}}\right)\right) & -X_q s+g \\ -(Z_u+Z_{P_u}) & (\mathrm{U}_0-Z_{\dot\alpha})s-\left(Z_\alpha+\boxed{K_\alpha Z_{\delta_E}}\right) & -(Z_q+\mathrm{U}_0)s \\ -(M_u+M_{P_u}) & -\left(M_{\dot\alpha}s+\left(\left(M_\alpha+\boxed{K_\alpha M_{\delta_E}}\right)+M_{P_\alpha}\right)\right) & s^2-M_q s \end{bmatrix}\begin{Bmatrix} u(s)\\ \alpha(s)\\ \theta(s)\end{Bmatrix}=\begin{bmatrix} X_{\delta_E} & X_T\\ Z_{\delta_E} & Z_T\\ M_{\delta_E} & M_T\end{bmatrix}\begin{Bmatrix}-\delta_{\text{Stick}}(s)\\ \delta T(s)\end{Bmatrix} \tag{11.37}$$

or

$$\mathbf{P}^{\alpha}_{-\delta_E}(s)\mathbf{y}(s)=\mathbf{Q}(s)\mathbf{u}(s)$$

Here we see that the terms arising due to the angle-of-attack feedback augmentation are all in the <u>second</u> column of $\mathbf{P}^{\alpha}_{-\delta_E}(s)$. The numerator of the $u(s)/\delta T(s)$ transfer function is now

$$\begin{aligned} N^{u}_{\delta T\text{aug}}(s) &= \det\begin{bmatrix} X_T & -\left(X_{\dot\alpha}s+\left(X_\alpha+\boxed{K_\alpha X_{\delta_E}}\right)\right) & -X_q s+g \\ Z_T & (\mathrm{U}_0-Z_{\dot\alpha})s-\left(Z_\alpha+\boxed{K_\alpha Z_{\delta_E}}\right) & -(Z_q+\mathrm{U}_0)s \\ M_T & -\left(M_{\dot\alpha}s+\left(\left(M_\alpha+\boxed{K_\alpha M_{\delta_E}}\right)+M_{P_\alpha}\right)\right) & s^2-M_q s \end{bmatrix} \\ &= \det\begin{bmatrix} X_T & -(X_{\dot\alpha}s+X_\alpha) & -X_q s+g \\ Z_T & (\mathrm{U}_0-Z_{\dot\alpha})s-Z_\alpha & -(Z_q+\mathrm{U}_0)s \\ M_T & -\left(M_{\dot\alpha}s+(M_\alpha+M_{P_\alpha})\right) & s^2-M_q s \end{bmatrix} - K_\alpha\det\begin{bmatrix} X_T & X_{\delta_E} & -X_q s+g \\ Z_T & Z_{\delta_E} & -(Z_q+\mathrm{U}_0)s \\ M_T & M_{\delta_E} & s^2-M_q s \end{bmatrix} \\ &= N^{u}_{\delta T}(s)-K_\alpha N^{\alpha u}_{\delta_E \delta T}(s) \end{aligned} \tag{11.38}$$

Again assuming $Z_T = M_T = 0$, the coupling numerator in Equation (11.38) is

$$\boxed{\begin{aligned} N^{\alpha u}_{\delta_E \delta T}(s) &= X_T\left(Z_{\delta_E}\left(s^2 - M_q s\right) + M_{\delta_E}\left(U_0 + Z_q\right)s\right) \\ &= X_T s\left(Z_{\delta_E} s + \left(M_{\delta_E}\left(U_0 + Z_q\right) - Z_{\delta_E} M_q\right)\right) \end{aligned}} \tag{11.39}$$

And the open-loop numerator of the $u(s)/\delta T(s)$ transfer function, or $N^{u}_{\delta T}(s)$, was given in Equation (11.29). Again, for an aircraft similar to an F-5A, the above coupling numerator becomes

$$N^{\alpha u}_{\delta_E \delta T}(s) = -0.383s\left(s + 101.7\right)$$

And from Equation (11.29), $N^{u}_{\delta T}(s)$ is

$$N^{u}_{\delta T}(s) = 2.73\ s\left(s^2 + 1.214s + 3.707\right)$$

To find the effect of angle-of-attack feedback on the closed-loop zeros of the $u(s)/\delta T(s)$ transfer function, we may again use a numerator root locus. The relevant equation for this root locus is

$$1 - K_\alpha \frac{N^{\alpha u}_{\delta_E \delta T}(s)}{N^{u}_{\delta T}(s)} = 1 - K_\alpha \frac{-0.383s\left(s + 101.7\right)}{0.00322\ s\left(s^2 + 1.214s + 3.707\right)} = 0 \tag{11.40}$$

and the numerator root locus is shown in Figure 11.17. We see from this figure that one zero remains at the origin, while two move farther away from the origin along the branches shown. Again, this is a desirable effect, since angle-of-attack feedback is increasing the short-period frequency, and we want the numerator

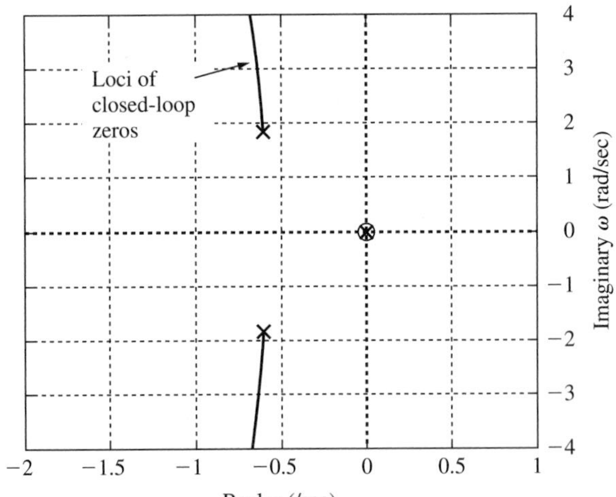

Figure 11.17 Numerator root locus—effect of AOA feedback on $u(s)/\delta T(s)$ zeros.

quadratic in the closed-loop $u(s)/\delta T(s)$ transfer function to approximately cancel the short-period quadratic in the characteristic polynomial.

Consequently, we have found that, in general, a stability-augmentation system that feeds back angle of attack is effective at increasing the short-period frequency, while also having a desirable effect on the surge-velocity transfer-function zeros. Angle-of-attack feedback is used in practice, as well as the following alternate approach.

An alternative to feeding back angle of attack is to feed back local vertical acceleration $a_{Z\,\text{local}}$, which, according to Equation (10.55), is

$$a_{Z\,\text{local}}(s) = sw(s) - U_0 q(s) - xsq(s) \tag{11.41}$$

If the accelerometer is located close to the vehicle's *c.g.* (i.e., $x \approx 0$), this reduces to

$$\boxed{a_{Z\,\text{cg}}(s) = \dot{w}(s) - U_0 q(s) \approx Z_\alpha \alpha(s) + Z_{\delta_E}\delta_E(s)} \tag{11.42}$$

if the small effect of Z_u is ignored. So we can see that vertical-acceleration feedback includes a significant angle-of-attack component. In fact, the approximate effective control law with acceleration feedback is

$$\delta_E(s) = -\delta_{\text{Stick}} - K_{a_Z} a_{Z\,\text{cg}} \approx -\delta_{\text{Stick}} - K_{a_Z}\big(Z_\alpha \alpha(s) + Z_{\delta_E}\delta_E(s)\big)$$

or

$$\delta_E(s) \approx \frac{1}{\left(1 + K_{a_Z} Z_{\delta_E}\right)}\big(-\delta_{\text{Stick}} - K_{a_Z} Z_\alpha \alpha(s)\big) \tag{11.43}$$

So we would expect acceleration feedback to have an effect very similar to that of angle-of-attack feedback.

A stability-augmentation system that consists of vertical-acceleration feedback is shown in Figure 11.18. And from Equation (11.42) we see that the transfer function indicated in the figure can be found from

$$\boxed{\frac{a_{Z\,\text{cg}}(s)}{\delta_E(s)} = U_0\left(\frac{s\alpha(s)}{\delta_E(s)} - \frac{q(s)}{\delta_E(s)}\right) = -U_0 s\frac{\gamma(s)}{\delta_E(s)}} \tag{11.44}$$

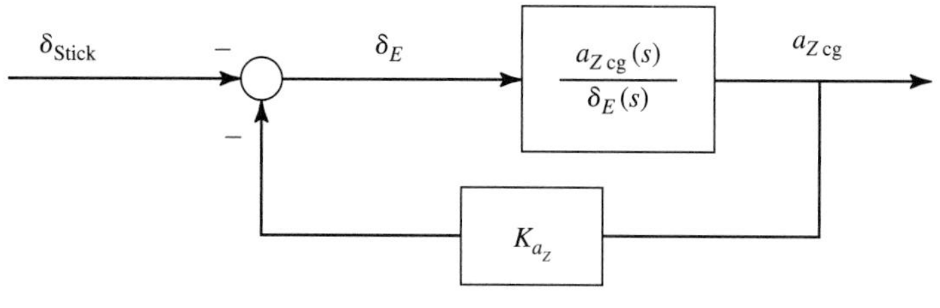

Figure 11.18 Block diagram for vertical-acceleration stability augmentation.

(Note here that a positive elevator deflection leads to a positive a_{Zcg}, so no change in the elevator sign convention is necessary.)

To assess the effect of vertical-acceleration feedback on the closed-loop eigenvalues, consider the root locus shown in Figures 11.19*a–b*. (The region near the origin is enlarged in Figure 11.19*b*.) The open-loop vehicle transfer function used for generating this root locus is

$$\frac{a_{Z\,cg}(s)}{\delta_E(s)} = \frac{-121s(s+0.0794)(s-8.806)(s+9.175)}{(s^2+0.00559s+0.00135)(s^2+1.218s+5.244)} \tag{11.45}$$

corresponding to a vehicle similar to the F-5A. If the scale used in Figure 11.19*a* was expanded, one could see the short-period eigenvalues move far to the left as feedback gain is increased, and eventually move onto the real axis. But for reasonable values of gain, this will not be an issue.

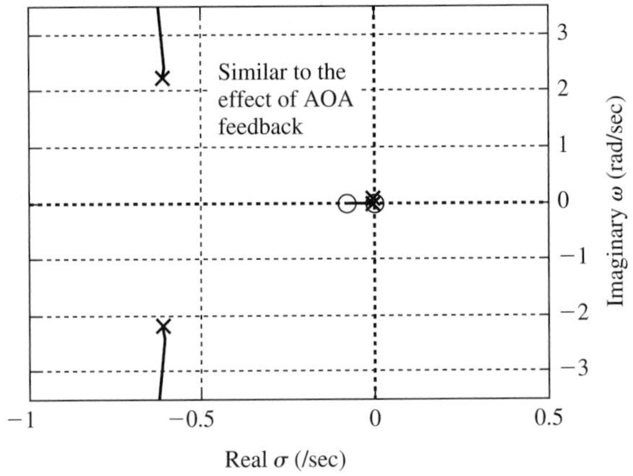

Figure 11.19*a* Root locus—effect of vertical-acceleration feedback.

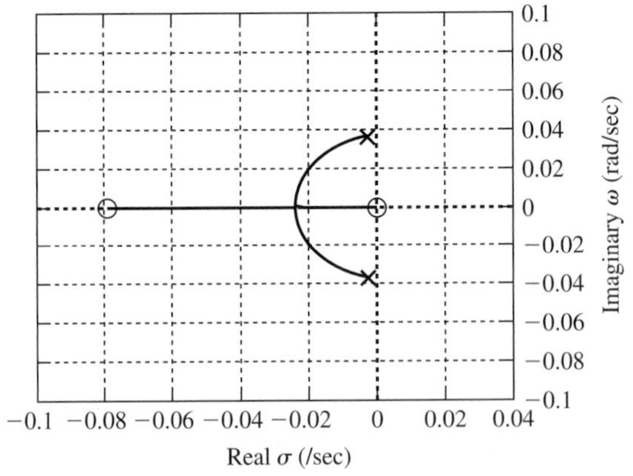

Figure 11.19*b* Enlarged region of root locus near the origin.

Note from Figure 11.19*a* that the effect of vertical-acceleration feedback on the short-period eigenvalues is almost identical to that for angle-of-attack feedback, as shown in Figure 11.16*a*. Also, the effect of vertical-acceleration feedback on the phugoid eigenvalues is similar to that for angle-of-attack feedback. Use of an accelerometer rather than an angle-of-attack vane as the sensor is the primary reason for feeding back vertical acceleration instead of angle of attack in some stability-augmentation systems, such as those for the A-4D and A-7A (see Appendix B).

Now with regard to the effect of vertical-acceleration feedback on the zeros of the $u(s)/\delta T(s)$ transfer function, recall that the effective control law here is given by

$$\delta_E(s) = -\delta_{\text{Stick}} - K_{a_Z} a_{Z\,\text{cg}} \approx \delta_{\text{Stick}} - K_{a_Z}\big(Z_\alpha \alpha(s) + Z_{\delta_E}\delta_E(s)\big)$$

or

$$\delta_E(s) \approx \frac{1}{\big(1 + K_{a_Z} Z_{\delta_E}\big)}\big(-\delta_{\text{Stick}} - K_{a_Z} Z_\alpha \alpha(s)\big) \tag{11.46}$$

But this is just angle-of-attack feedback, and the effects of angle-of-attack feedback on these zeros were already discussed earlier in this section.

11.3.3 Stabilizing an Unstable Short-Period Mode

When an aircraft has an unstable short-period mode, stability augmentation is clearly required. And, as we have seen, the instability is typically indicated by the presence of a single, real, unstable pole. With regards to the short-period quadratic in the longitudinal characteristic polynomial, or

$$s^2 + 2\zeta_{\text{SP}}\omega_{\text{SP}}s + \omega_{\text{SP}}^2$$

the case with one unstable pole corresponds to ω_{SP}^2 being negative.

Again from the short-period approximation, we found that this last coefficient, given in Equation (11.36), is a strong function of $-\big(M_\alpha + M_{P_\alpha}\big)$. So, as in Section 11.3.2, this again suggests augmenting M_α by feeding back angle of attack to the pitch-control surface. That is, the control law may again be depicted as in Figure 11.15. Although we have already discussed the effect of this control law in the context of increasing short-period natural frequency, it is instructive to assess the effectiveness of such a control law when the short-period mode is actually unstable.

For this analysis, let's consider the highly unstable notional hypersonic vehicle introduced in Example 10.3, and recall that the eigenvectors found in that example indicated that the unstable eigenvalue corresponded to a mode dominated by pitch rate and angle of attack with little surge-velocity participation. Or the unstable mode was a short-period mode. For this notional vehicle, the state, response, and control vectors are

$$\mathbf{x}^T = \mathbf{y}^T = \left[u\,(\text{Mach})\ \ \alpha\,(\text{rad})\ \ \theta\,(\text{rad})\ \ q\,(\text{rad/sec})\right],\quad \text{u} = \delta_H\,(\text{rad})$$

and the state-variable model is

$$\mathbf{A} = \begin{bmatrix} -0.001936 & 0.02502 & -0.03317 & 0.000635 \\ -0.002028 & -0.06303 & 0 & 1 \\ 0 & 0 & 0 & 1 \\ 0.3287 & 11.023 & 0 & -0.0816 \end{bmatrix},\ \mathbf{B} = \begin{bmatrix} -0.00058 \\ -0.00276 \\ 0 \\ -0.47963 \end{bmatrix},\ \mathbf{C} = \mathbf{I}_4,\ \mathbf{D} = \mathbf{0} \tag{11.47}$$

From MATLAB we find that the angle-of-attack transfer function is

$$\frac{\alpha(s)}{\delta_H(s)} = \frac{-0.003(s+174)(s^2+0.00233s+1.30\times10^{-4})}{(s-3.25)(s+3.40)(s^2+0.001695s+4.911\times10^{-6})}\ \text{rad/rad} \tag{11.48}$$

The effects of angle-of-attack feedback on the vehicle's closed-loop eigenvalues are indicated by the root-loci shown in Figures 11.20*a–c*. (The second two figures show the region near the origin enlarged for clarity.) Note from Figure 11.20*a* that the two real open-loop eigenvalues immediately move towards the origin, and with sufficiently large feedback gain two stable closed-loop eigenvalues leave the real axis and move along and to the left of the imaginary axis. As these poles move away from the origin, the natural frequency of the corresponding mode is increased, as discussed in Section 11.3.2.

Looking now at Figures 11.20*b* and *c,* we see that the phugoid eigenvalues move onto the real axis and then move apart, eventually coalescing with the

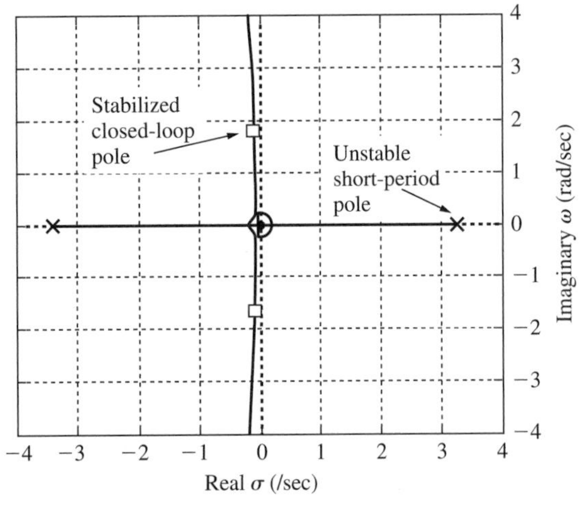

Figure 11.20*a* Effect of angle-of-attack feedback on unstable short-period mode.

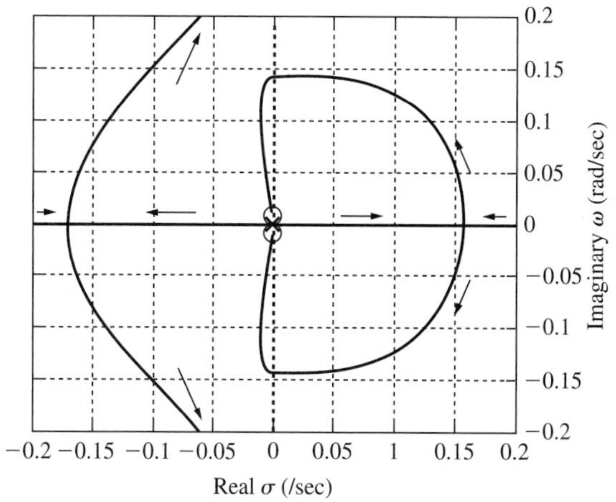

Figure 11.20*b* Region of root locus near origin expanded.

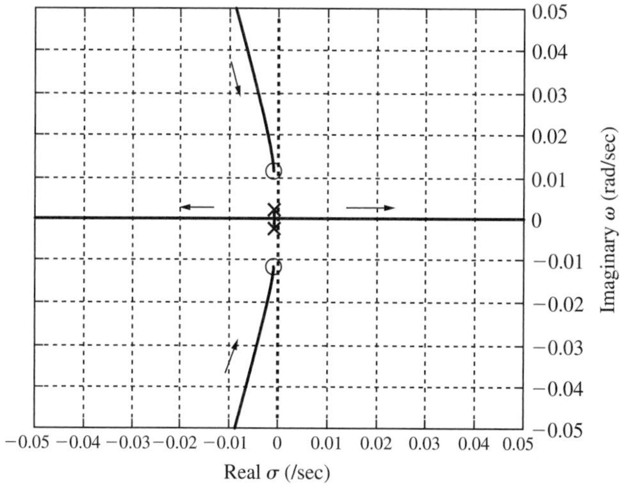

Figure 11.20*c* Region of root locus near origin further expanded.

two other closed-loop eigenvalues moving along the real axis, as just discussed. However, note that two closed-loop roots always remain near the origin, and sometimes they are unstable. So it would appear from the above discussion that the short-period mode has been stabilized, but an eigen-analysis of the closed-loop system is always recommended.

For example, at one value of feedback gain K_α the closed-loop poles take on the following values.

$$\begin{aligned}\lambda_{1,2} &= -0.1125 \pm j1.835\,/\text{sec}\\ \lambda_{3,4} &= -0.0022 \pm j0.0231\,/\text{sec}\end{aligned} \tag{11.49}$$

And the corresponding eigenvectors are

$$\boldsymbol{v}_1 = \begin{Bmatrix} 0.0060e^{-j89.6^\circ} \\ 0.4328e^{-j167.7^\circ} \\ 0.4308e^{-j172.2^\circ} \\ 0.7919e^{-j78.7^\circ} \end{Bmatrix}, \quad \boldsymbol{v}_3 = \begin{Bmatrix} 0.8179e^{j8.1^\circ} \\ 0.0798e^{j8.0^\circ} \\ 0.5697e^{-j79.5^\circ} \\ 0.0132e^{j16.0^\circ} \end{Bmatrix} \tag{11.50}$$

Note that the first eigenvalue–eigenvector pair reveals that the corresponding mode is indeed a conventional short-period mode, but with low damping. This oscillatory mode's response is dominated by pitch rate, attitude, and angle of attack, with virtually no surge-velocity variations. The third eigenvalue-eigenvector pair reveals that the corresponding mode is a conventional phugoid mode. This low-frequency, lightly damped mode's response is dominated by surge velocity and pitch attitude, with very little angle of attack or pitch-rate in the modal response.

Therefore, we may conclude that that angle-of-attack feedback has indeed stabilized the short-period mode, and has led to modal characteristics of the closed-loop system that appear to be quite conventional. To increase the short-period damping, pitch-rate feedback may be added. It is significant to note that angle-of-attack plus pitch-rate feedback were used in the stability-augmentation system for NASA's Hyper-X hypersonic research vehicle (Ref. 3), a precursor to the X-43 shown in Figure 10.13 in Example 10.3.

11.3.4 Stabilizing an Unstable Phugoid Mode

The final longitudinal dynamic deficiency to be discussed is that of phugoid instability. As noted in Chapter 6, sometimes in the transonic speed range compressibility effects give rise to a negative M_u, which is the pitching moment associated with surge-velocity perturbations. This can lead to a phugoid instability we called Mach tuck.

Recall that the phugoid quadratic is of the form

$$s^2 + 2\zeta_P\omega_P s + \omega_P^2$$

And under the phugoid approximation, the last coefficient, or ω_P^2, was given in Equation (10.117), or

$$\omega_P^2 = g\frac{(M_\alpha + M_{P_\alpha})(Z_u + Z_{P_u}) - Z_\alpha(M_u + M_{P_u})}{Z_\alpha M_q - (M_\alpha + M_{P_\alpha})(U_0 + Z_q)} \tag{11.51}$$

Since the denominator in the above expression is typically positive, phugoid instability can occur when

$$(M_\alpha + M_{P_\alpha})(Z_u + Z_{P_u}) - Z_\alpha(M_u + M_{P_u}) < 0 \tag{11.52}$$

Since the cause of such instability is the troublesome pitching moment $(M_u + M_{P_u})$, feedback of some appropriate vehicle response to the pitch-control surface is suggested. And since the phugoid modal response contains significant surge-velocity and pitch-attitude components, one of these two responses could be considered for feedback. But typically, pitch-attitude feedback leads to better overall results.

Let's consider such a control law feeding back pitch attitude, as shown in the block diagram in Figure 11.21. The effect of attitude feedback on the short-period eigenvalues was noted previously when discussing Figure 11.10*a,* but the effects of such a stability-augmentation system on unstable phugoid eigenvalues are shown in the root-locus plot in Figure 11.22. The open-loop transfer function is that of pitch attitude with negative elevator input for a vehicle with an unstable phugoid mode. Note from Figure 11.22 that indeed this unstable phugoid root can be stabilized with this control law, especially when the zero at $-1/T_{\theta_1}$ is sufficiently to the left of the origin.

We now need to assess the effects of attitude feedback on the zeros of the $u(s)/\delta T(s)$ transfer function. Again recall that the zeros of all the pitch-attitude and pitch-rate transfer functions will be unaffected, as we have discussed in Section 11.2, as well as all the zeros of the transfer functions associated with elevator input.

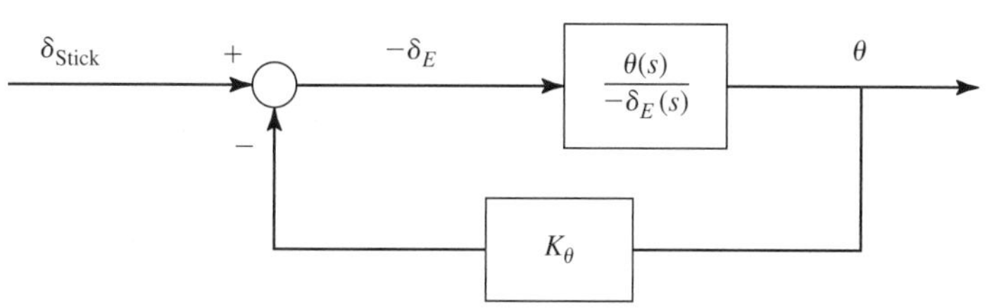

Figure 11.21 Block diagram of pitch-attitude feedback control law.

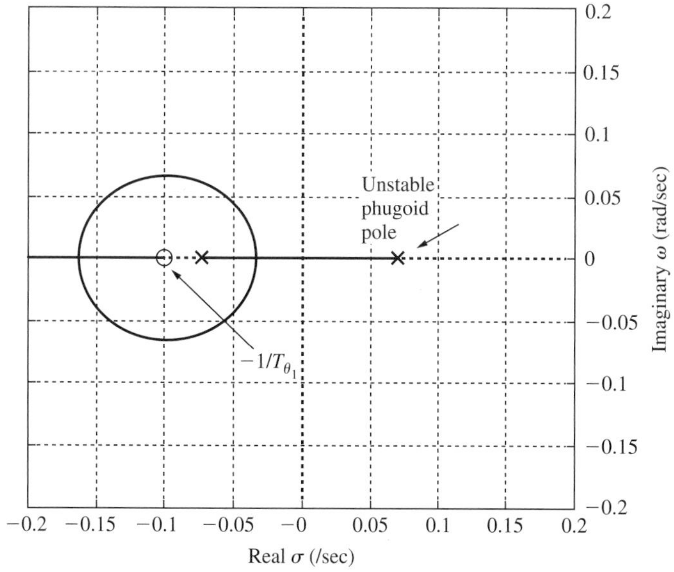

Figure 11.22 Root locus—effects of attitude feedback on unstable phugoid eigenvalues.

To evaluate the effect on the zeros of the $u(s)/\delta T(s)$, again consider the polynomial-matrix description of the dynamic system given in Equation (11.23), except we now include the effect of the pitch-attitude control law.

$$\begin{bmatrix} s-(X_u+X_{P_u}) & -(X_{\dot\alpha}s+X_\alpha) & -X_q s+\left(g-\boxed{K_\theta X_{\delta_E}}\right) \\ -(Z_u+Z_{P_u}) & (U_0-Z_{\dot\alpha})s-Z_\alpha & -(U_0+Z_q)s-\boxed{K_\theta Z_{\delta_E}} \\ -(M_u+M_{P_u}) & -\left(M_{\dot\alpha}s+(M_\alpha+M_{P_\alpha})\right) & s^2-M_q s-\boxed{K_\theta M_{\delta_E}} \end{bmatrix} \begin{Bmatrix} u(s) \\ \alpha(s) \\ \theta(s) \end{Bmatrix} = \begin{bmatrix} X_{\delta_E} & X_T \\ Z_{\delta_E} & Z_T \\ M_{\delta_E} & M_T \end{bmatrix} \begin{Bmatrix} -\delta_{\text{Stick}}(s) \\ \delta T(s) \end{Bmatrix} \tag{11.53}$$

Note that with the incorporation of this control law, we have effectively introduced three new stability derivatives, which we denote as X_θ, Z_θ, and M_θ, and where

$$\begin{aligned} X_\theta &\triangleq K_\theta X_{\delta_E} \\ Z_\theta &\triangleq K_\theta Z_{\delta_E} \\ M_\theta &\triangleq K_\theta M_{\delta_E} \end{aligned} \tag{11.54}$$

Now with regard to the numerator of the $u(s)/\delta T(s)$ transfer function, this numerator is

$$N^u_{\delta T\text{aug}} = \det \begin{bmatrix} X_T & -(X_{\dot\alpha}s+X_\alpha) & -X_q s+\left(g-\boxed{K_\theta X_{\delta_E}}\right) \\ Z_T & (U_0-Z_{\dot\alpha})s-Z_\alpha & -(U_0+Z_q)s-\boxed{K_\theta Z_{\delta_E}} \\ M_T & -\left(M_{\dot\alpha}s+(M_\alpha+M_{P_\alpha})\right) & s^2-M_q s-\boxed{K_\theta M_{\delta_E}} \end{bmatrix}$$

$$= \det \begin{bmatrix} X_T & -(X_{\dot\alpha}s+X_\alpha) & -X_q s+g \\ Z_T & (U_0-Z_{\dot\alpha})s-Z_\alpha & -(U_0+Z_q)s \\ M_T & -\left(M_{\dot\alpha}s+(M_\alpha+M_{P_\alpha})\right) & s^2-M_q s \end{bmatrix} - K_\theta \det \begin{bmatrix} X_T & -(X_{\dot\alpha}s+X_\alpha) & X_{\delta_E} \\ Z_T & (U_0-Z_{\dot\alpha})s-Z_\alpha & Z_{\delta_E} \\ M_T & -\left(M_{\dot\alpha}s+(M_\alpha+M_{P_\alpha})\right) & M_{\delta_E} \end{bmatrix}$$

$$= N^u_{\delta T}(s) - K_\theta N^{\theta u}_{\delta_E \delta T}(s) \tag{11.55}$$

Again assuming $Z_T = M_T = 0$, the coupling numerator in the above expression is

$$\begin{aligned} N^{\theta u}_{\delta_E \delta T}(s) &= X_T\left(M_{\delta_E}\left((U_0-Z_{\dot\alpha})s-Z_\alpha\right)+Z_{\delta_E}\left(M_{\dot\alpha}s+(M_\alpha+M_{P_\alpha})\right)\right) \\ &= X_T\left(\left(M_{\delta_E}(U_0-Z_{\dot\alpha})+Z_{\delta_E}M_{\dot\alpha}\right)s+\left(Z_{\delta_E}(M_\alpha+M_{P_\alpha})-M_{\delta_E}Z_\alpha\right)\right) \end{aligned} \tag{11.56}$$

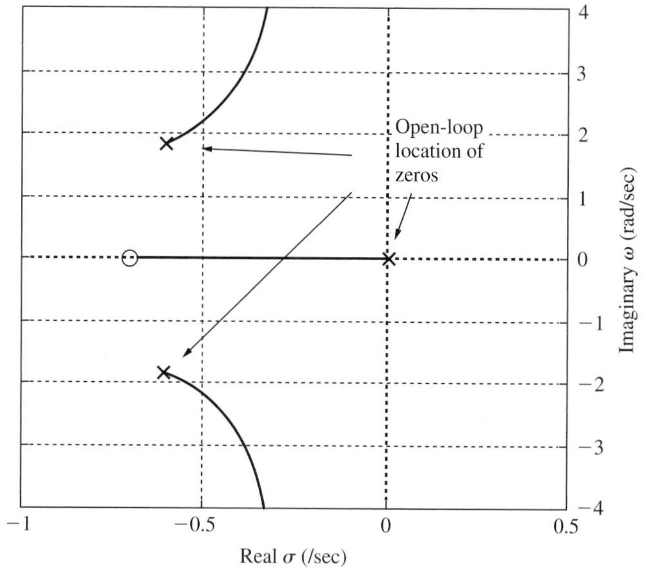

Figure 11.23 Numerator root locus—effect of attitude feedback on $u(s)/\delta T(s)$ zeros.

By comparing Equation (11.55) with Equation (11.26), we see that the only difference is the lack of a root at the origin in front of the above coupling numerator. So using Equations (11.28) and (11.29) as numerical examples, the numerator root locus for the closed-loop $u(s)/\delta T(s)$ transfer function would have the characteristics shown in Figure 11.23. That is, one zero moves to the left along the real axis, while the other two move farther away from the origin. Again, this is desirable, since with attitude feedback the short-period poles move in a similar fashion (away from the origin), and we want the short-period quadratic to be approximately cancelled by a similar numerator quadratic in the closed-loop $u(s)/\delta T(s)$ transfer function.

All the above results indicate the effectiveness of attitude feedback in terms of stabilizing an unstable phugoid mode. Such stabilization may be provided by either an automatic feedback system or by the pilot controlling the pitch-attitude.

11.4 LATERAL-DIRECTIONAL STABILITY AUGMENTATION

As we began Section 11.3, which addressed longitudinal control, we now consider the block diagram shown in Figure 11.24. The block diagram includes the complete lateral-directional dynamics of the aircraft, feedback stability-augmentation systems, and the pilot. The pilot is seen to be closing control loops around the entire system consisting of the aircraft and the stability-augmentation system(s), using motion and visual cues as the feedback quantities. Assuming

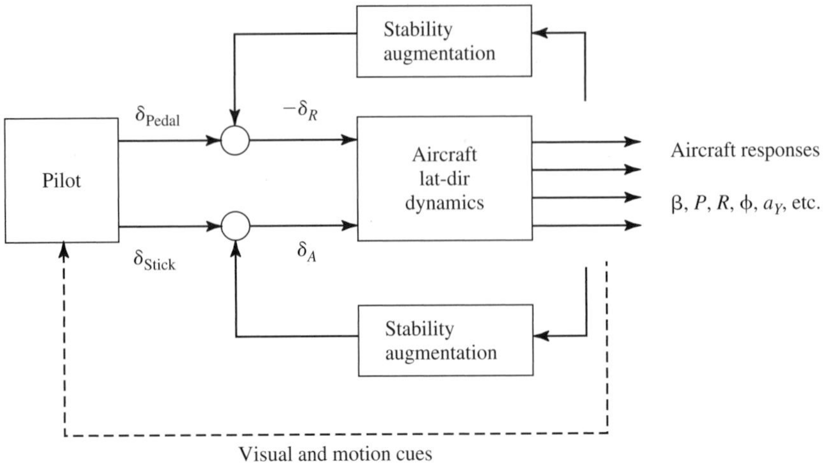

Figure 11.24 Block diagram for the lateral-directional control of an aircraft.

this is a conventional aircraft, aileron and rudder are the two control inputs acting on the vehicle. Note that several aircraft responses are shown.

Several transducers, or sensors, are available for use in lateral-directional feedback systems. For example, accelerometers could sense lateral acceleration, angular rate gyros could sense roll and/or yaw rate, and angular-displacement gyros could sense bank and/or heading angle. The feedback system would typically provide commands or signals to the existing control effectors on the vehicle, such as aileron and rudder actuators.

So as before, the topic in this section is the synthesis of stability-augmentation system(s) to tailor the dynamics such that either the handling qualities are enhanced, or it is easier to design an outer-loop autopilot. Central to this issue is the selection of the appropriate responses to feed back, and the selection of the most effective control input. And as in the longitudinal case, we note that this system is a multi-input/multi-output (MIMO) system, so the discussion of Section 11.2 is again relevant.

To demonstrate an important phenomenon that leads to the need for lateral-directional stability augmentation, consider the root locus shown in Figure 11.25. The vehicle dynamics consist of the unaugmented bank-angle response to aileron input for a conventional aircraft, with the feedback-control loop being closed by the pilot. So with reference to the system in Figure 11.24, there is no stability augmentation present and the pilot is closing the (outer) feedback loop based on sensed bank angle.

Note that, as in any feedback system, when the pilot's control gain increases, the closed-loop poles move along the root locus away from the open-loop poles of the aircraft. As a result, the closed-loop spiral mode near the origin is stabilized, which is a desirable result of the pilot's control of bank angle, while the roll-subsidence time constant[1] is increased. In addition, for the dutch-roll poles

[1] If a real eigenvalue is denoted as $\lambda = -1/T$, then T is the time constant associated with the mode.

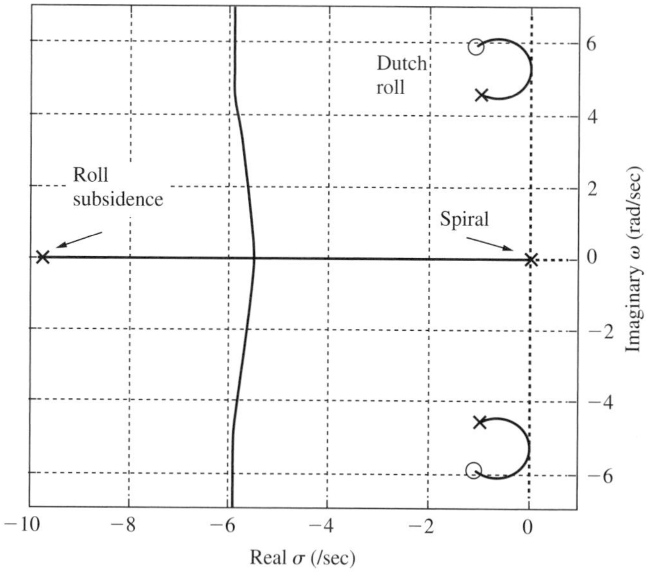

Figure 11.25 Root locus—effects of pilot control of bank angle on eigenvalues.

and neighboring zeros in their relative positions as shown, the damping of the closed-loop dutch-roll mode is reduced below that of the open-loop aircraft, which can lead to a lightly damped closed-loop pilot-vehicle system and therefore poor handling qualities. This situation is exacerbated when the dutch-roll damping of the open-loop aircraft is low to begin with, and when the dutch-roll poles and neighboring zeros are far apart. So increasing low dutch-roll damping and adjusting the relative position of the dutch-roll poles and neighboring zeros are common requirements in lateral-directional stability augmentation.

11.4.1 Increasing Dutch-Roll Damping

A common deficiency requiring stability augmentation is the need to increase the damping of the dutch-roll mode. Recall from the dutch-roll approximation that this damping may be determined from Equation (10.153), or

$$2\zeta_{DR}\omega_{DR} = -\left(N_r' + Y_\beta/U_0\right) \tag{11.57}$$

This suggests that increasing the effective yaw damping of the vehicle N_r (the major contributor to N_r') would increase the dutch-roll damping. But N_r is just the yawing moment acting on the vehicle due to yaw rate. Therefore, we can increase effective yaw damping by providing an additional yawing moment that is proportional to yaw rate.

This may be accomplished by simply feeding back measured yaw rate to the yaw-moment effector (e.g., rudder), as indicated in Figure 11.26. Note that we have again changed the sign convention on rudder deflection, such that a

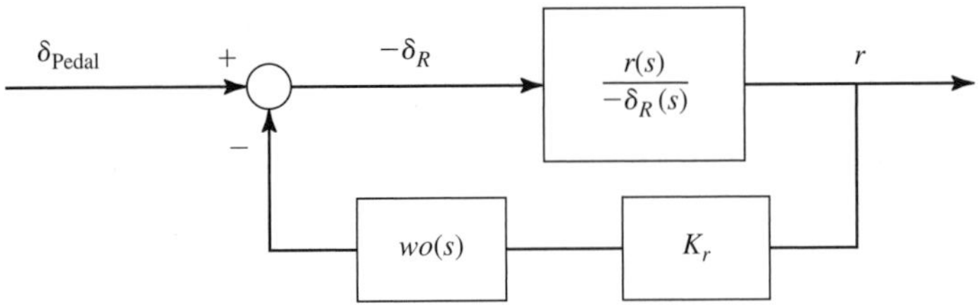

Figure 11.26 Block diagram of a yaw damper.

positive rudder deflection leads to a positive yaw rate. Note also the introduction of a new element $wo(s)$. But for now, we will just let $wo(s) = 1$. Such a stability-augmentation system is called a *yaw damper.*

NOTE TO STUDENT

We have again applied the insight gained through the dutch-roll approximation to select a vehicle response for feedback, as well as an appropriate control effector—namely yaw rate and the yaw-control surface. So this feedback-control architecture was synthesized through our knowledge of the vehicle's dynamics.

To confirm our selection of this feedback architecture, consider the root-locus shown in Figure 11.27. The open-loop transfer function here is yaw rate with rudder input (with $wo(s) = 1$) for the A-7A at Flight Condition 3 (Appendix B). The root-locus reveals the effect of the yaw damper on the closed-loop eigenvalues.

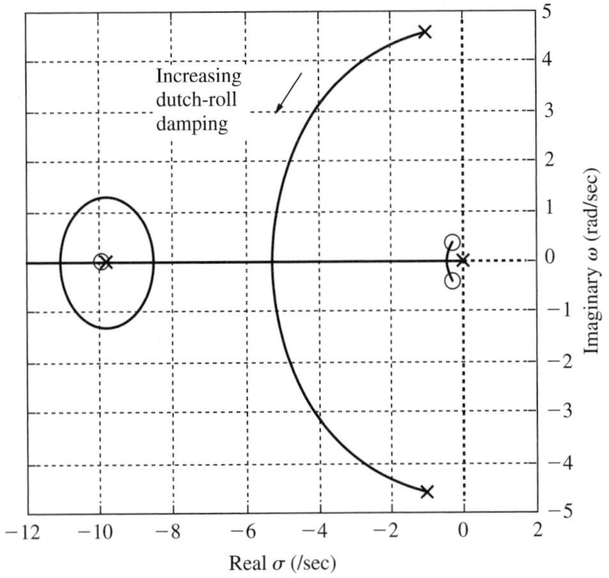

Figure 11.27 Root locus revealing the effects of a yaw damper.

Note that the spiral eigenvalue near the origin is stabilized, and that the dutch-roll damping is increased significantly as feedback gain is increased. The open-loop roll-subsidence pole is almost exactly cancelled by a neighboring zero, hence this mode's eigenvalue will be essentially unchanged by this control law. All this confirms that yaw-rate feedback is indeed effective for increasing dutch-roll damping if the two complex zeros in this transfer function are sufficiently close to the origin of the complex plane (a topic addressed further in Section 11.4.3).

Now we need to consider the effect of adding yaw-rate feedback on the zeros of the remaining closed-loop lateral-directional transfer functions. First, we know there is no effect on the zeros of the yaw-rate transfer functions, or on any transfer functions with rudder input. These facts have been shown previously in Section 11.2. However, there will be some effect on the other transfer-function zeros. To explore the effects on these zeros, consider the vehicle's polynomial-matrix description given below for level flight ($\Theta_0 = 0$), and with the effect of the yaw damper (with $wo(s) = 1$) included.

$$\begin{bmatrix} U_0 s - Y_\beta & -Y_p s - g & \left(U_0 - \left(Y_r + \boxed{K_r Y_{\delta_R}}\right)\right) \\ -L'_\beta & s^2 - L'_p s & -\left(L'_r + \boxed{K_r L'_{\delta_R}}\right) \\ -N'_\beta & -N'_p s & s - \left(N'_r + \boxed{K_r N'_{\delta_R}}\right) \end{bmatrix} \begin{Bmatrix} \beta(s) \\ \phi(s) \\ r(s) \end{Bmatrix} = \begin{bmatrix} Y_{\delta_A} & Y_{\delta_R} \\ L'_{\delta_A} & L'_{\delta_R} \\ N'_{\delta_A} & N'_{\delta_R} \end{bmatrix} \begin{Bmatrix} \delta_A(s) \\ -\delta_{\text{Pedal}}(s) \end{Bmatrix} \quad (11.58)$$

or

$$\mathbf{P}^{r}_{-\delta_R}(s)\mathbf{y}(s) = \mathbf{Q}(s)\mathbf{u}(s)$$

The new terms added due to the yaw-rate feedback are highlighted in boxes, and we see that the three stability derivatives Y_r, L'_r, and N'_r have been augmented due to this feedback action.

Of particular interest is the $\phi(s)/\delta_A(s)$ transfer function, for which the closed-loop numerator is

$$\boxed{\begin{aligned} N^{\phi}_{\delta_A \text{aug}}(s) &= \det \begin{bmatrix} U_0 s - Y_\beta & Y_{\delta_A} & \left(U_0 - \left(Y_r + \boxed{K_r Y_{\delta_R}}\right)\right) \\ -L'_\beta & L'_{\delta_A} & -\left(L'_r + \boxed{K_r L'_{\delta_R}}\right) \\ -N'_\beta & N'_{\delta_A} & s - \left(N'_r + \boxed{K_r N'_{\delta_R}}\right) \end{bmatrix} \\ &= \det \begin{bmatrix} U_0 s - Y_\beta & Y_{\delta_A} & (U_0 - Y_r) \\ -L'_\beta & L'_{\delta_A} & -L'_r \\ -N'_\beta & N'_{\delta_A} & s - N'_r \end{bmatrix} - K_r \det \begin{bmatrix} U_0 s - Y_\beta & Y_{\delta_A} & Y_{\delta_R} \\ -L'_\beta & L'_{\delta_A} & L'_{\delta_R} \\ -N'_\beta & N'_{\delta_A} & N'_{\delta_R} \end{bmatrix} \\ &= N^{\phi}_{\delta_A}(s) - K_r N^{r\phi}_{\delta_R \delta_A}(s) \end{aligned} \quad (11.59)}$$

Assuming $Y_{\delta_A} \approx 0$, the coupling numerator in the above expression is

$$N^{r\phi}_{\delta_R\delta_A}(s) = \left(U_0 s - Y_\beta\right)\left(L'_{\delta_A}N'_{\delta_R} - N'_{\delta_A}L'_{\delta_R}\right) - Y_{\delta_R}\left(N'_{\delta_A}L'_\beta - L'_{\delta_A}N'_\beta\right)$$
$$= -\left(N'_{\delta_A}L'_{\delta_R} - L'_{\delta_A}N'_{\delta_R}\right)U_0 s - Y_\beta\left(L'_{\delta_A}N'_{\delta_R} - N'_{\delta_A}L'_{\delta_R}\right) - Y_{\delta_R}\left(N'_{\delta_A}L'_\beta - L'_{\delta_A}N'_\beta\right) \tag{11.60}$$

And the open-loop numerator $N^{\phi}_{\delta_A}(s)$ may also be obtained from Equation (11.59). For the A-7A aircraft at Flight Condition 3[2], for example, the coupling numerator is

$$N^{r\phi}_{\delta_R\delta_A}(s) = -2.55 \times 10^5\left(s + 0.390\right)$$

while the open loop numerator is

$$N^{\phi}_{\delta_A}(s) = 2.14 \times 10^4\left(s^2 + 2.18s + 24.11\right)$$

To determine the effect of the yaw damper on the zeros of the closed-loop $\phi(s)/\delta_A(s)$ transfer function, we will use the above two polynomials and generate a numerator root locus. That is, we will find the roots of

$$1 - K_r\frac{N^{r\phi}_{\delta_R\delta_A}(s)}{N^{\phi}_{\delta_A}(s)} = 1 - K_r\frac{-11.9\left(s + 0.390\right)}{\left(s^2 + 2.18s + 24.11\right)} = 0 \tag{11.61}$$

This numerator root locus is shown in Figure 11.28, from which we see that the zeros of the closed-loop $\phi(s)/\delta_A(s)$ transfer function move in the same manner as the dutch-roll poles indicated in Figure 11.27. This is desirable because we would like to have the dutch-roll poles approximately cancelled by these zeros in order to have the roll-rate response to aileron dominated by the first-order, roll-subsidence mode. Based on the above discussion, we see that a yaw damper is effective at increasing the damping of the dutch-roll mode.

But in the block diagram for the yaw damper, given in Figure 11.26, we had an additional element $wo(s)$. Up to this point we had assumed $wo(s) = 1$, and we will now revisit this assumption. If $wo(s)$ remained equal to unity, it would cause problems in a sustained turn, during which the yaw rate is not zero. The yaw damper would tend to "fight" the turn by maintaining a rudder deflection opposite to that desired for the turn. To eliminate this problem, we include a *washout* circuit, or filter, in the feedback loop, which is represented by $wo(s)$ in the block diagram of the yaw damper.

A washout filter has a transfer function of the following form.

---

[2] Even though the dimensional derivatives for the A-7A were developed in fuselage-reference instead of stability axes, we will use them in this example even though the polynomial-matrix description assumes stability axes. In this flight condition the trim angle of attack is only 2 deg, so the stability and fuselage-reference axes are almost identical.

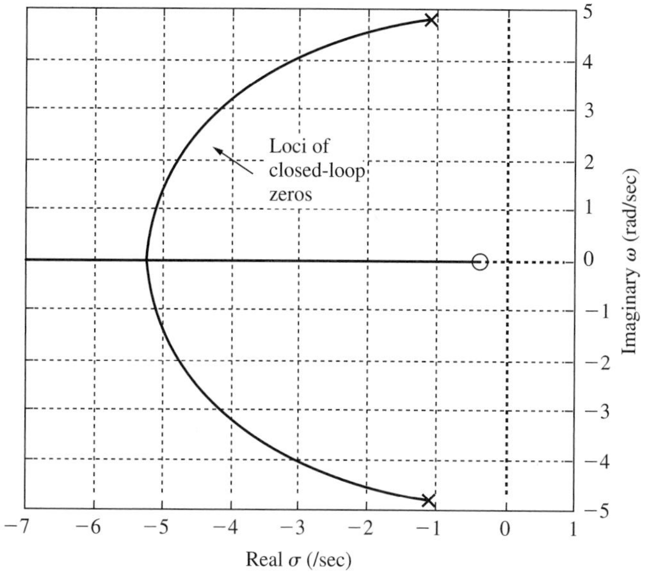

Figure 11.28 Numerator root locus—effect of yaw damper on $\phi(s)/\delta_A(s)$ zeros.

$$\boxed{wo(s) = \frac{T_W s}{T_W s + 1}} \tag{11.62}$$

The effect of the washout may be viewed from the perspective of its Bode plot, or from its influence on the root locus. The straight-line approximation for the Bode plot of a washout is shown in Figure 11.29. Note that at frequencies above the corner frequency $1/T_W$, the magnitude and phase plots indicate that $wo(s)$ is approximately unity, which is what we had originally assumed. But as the frequency approaches zero, $|wo(j\omega)|_{\omega \to 0} \to 0$, which is the same as opening up

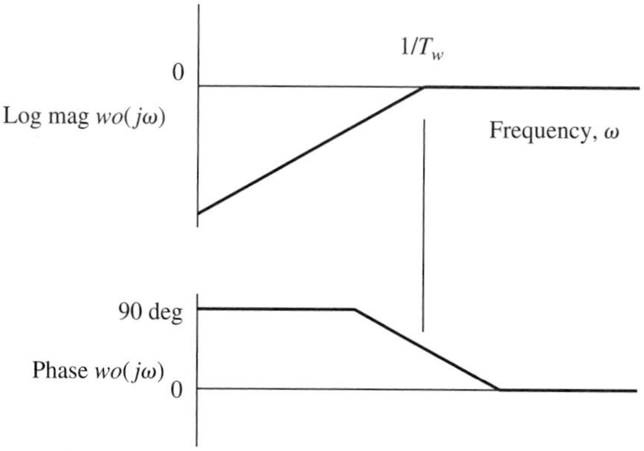

Figure 11.29 Straight-line sketch of Bode plot of washout filter.

the yaw-rate feedback loop. So we see that during a sustained turn, the washout filter effectively turns off the yaw damper!

To understand the effect of the washout on the closed-loop eigenvalues, let's revisit the root locus using the transfer function for yaw rate with rudder input, or revisit Figure 11.27. Letting $1/T_W = 1$ /sec, for example, the new root locus is shown in Figure 11.30. Note that by selecting this value of $1/T_W$, we have placed the washout pole approximately the same distance from the origin as the two complex zeros in the transfer function.

By comparing Figure 11.30 with Figure 11.27, we see that the effect of the yaw damper on the dutch-roll eigenvalues is essentially unchanged, and the damping may clearly be increased. But the effect on the spiral root is important here. The zero at the origin in the washout filter effectively cancels the spiral pole, so the spiral eigenvalue is now essentially unaffected by the yaw damper. Also note that the washout zero at the origin does not appear in the closed-loop $r(s)/\delta_A(s)$ transfer function because the washout is in the feedback path. (You should confirm this for yourself.) Leaving the spiral pole unaffected is the desired effect of the washout—we want the spiral mode to remain almost neutrally stable to allow for sustained turns.

The washout does introduce an additional eigenvalue in the closed-loop transfer functions, the position of which in indicated by the root locus in Figure 11.30. This eigenvalue corresponds to an additional system mode. But the zeros of the closed-loop $r(s)/\delta_A(s)$ transfer function will now include the term $(s + 1/T_W)$, which will approximately cancel the augmented washout pole. (Verify this for yourself as well.) The effect of the washout on the zeros of the remaining transfer functions is left as an exercise for the reader (see Problem 11.5). Finally, note that a yaw damper with washout is used on several aircraft, including the A-7A and F-5A.

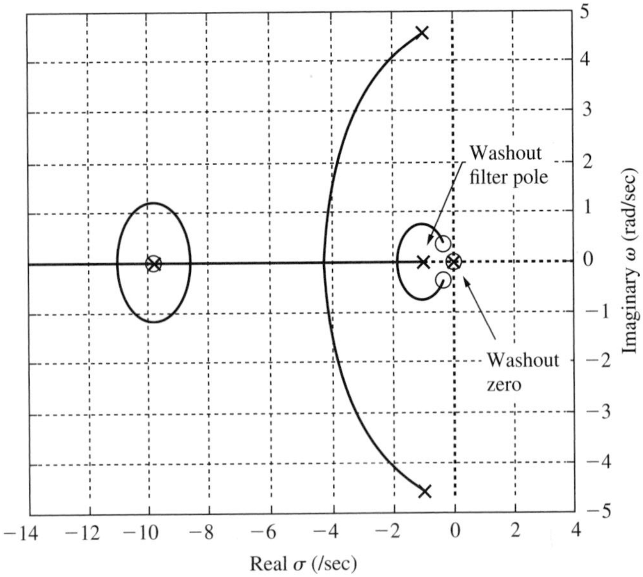

Figure 11.30 Effect of yaw damper with washout on lateral-directional eigenvalues.

11.4.2 Reducing Aileron Excitation of the Dutch-Roll

We saw in Section 11.4 that when the pilot controlled bank angle with aileron, the dutch-roll mode could cause difficulties, especially if that mode wasn't well damped or if there wasn't an approximate cancellation of the dutch-roll roots by a pair of nearby zeros. In Section 11.4.1 we addressed the issue of increasing dutch-roll damping by adding yaw-rate feedback, and we also addressed the effects of this feedback on the zeros of the $\phi(s)/\delta_A(s)$ transfer function. In this section we will again address the locations of the zeros of this transfer function, but not the effects of feedback on these locations.

Instead, we will look at modifying their location by using a *control cross feed* between the rudder and aileron. Such a cross feed is called an *aileron-rudder interconnect* (ARI), and it can reduce the level of dutch-roll excitation due to aileron deflection, or reduce any adverse yawing moment created due to aileron deflection, N_{δ_A}. The block diagram of an ARI is shown in Figure 11.31, which shows that the control cross feed is implemented by using the cross-feed "control law"

$$\delta_R(s) = -\delta_{\text{Pedal}}(s) + K_{\text{ARI}}\,\delta_A(s) \tag{11.63}$$

The effect of the ARI on the system's dynamics is captured in the following polynomial-matrix description, which includes the effects of the ARI.

$$\begin{bmatrix} \mathrm{U}_0 s - Y_\beta & -Y_p s - g & (\mathrm{U}_0 - Y_r) \\ -L'_\beta & s^2 - L'_p s & -L'_r \\ -N'_\beta & -N'_p s & s - N'_r \end{bmatrix} \begin{Bmatrix} \beta(s) \\ \phi(s) \\ r(s) \end{Bmatrix} = \begin{bmatrix} \left(Y_{\delta_A} + \boxed{K_{\text{ARI}} Y_{\delta_R}}\right) & Y_{\delta_R} \\ \left(L'_{\delta_A} + \boxed{K_{\text{ARI}} L'_{\delta_R}}\right) & L'_{\delta_R} \\ \left(N'_{\delta_A} + \boxed{K_{\text{ARI}} N'_{\delta_R}}\right) & N'_{\delta_R} \end{bmatrix} \begin{Bmatrix} \delta_A(s) \\ -\delta_{\text{Pedal}}(s) \end{Bmatrix} \tag{11.64}$$

or

$$\mathbf{P}(s)\mathbf{y}(s) = \mathbf{Q}_{\text{ARI}}(s)\mathbf{u}(s)$$

We see from this equation that the ARI will have no effect on det $\mathbf{P}(s)$, and therefore no effect on the characteristic polynomial of the system. But the cross feed will affect the numerators of all the transfer function with aileron input, since clearly the effective values of Y_{δ_A}, L'_{δ_A}, and N'_{δ_A} have all been modified.

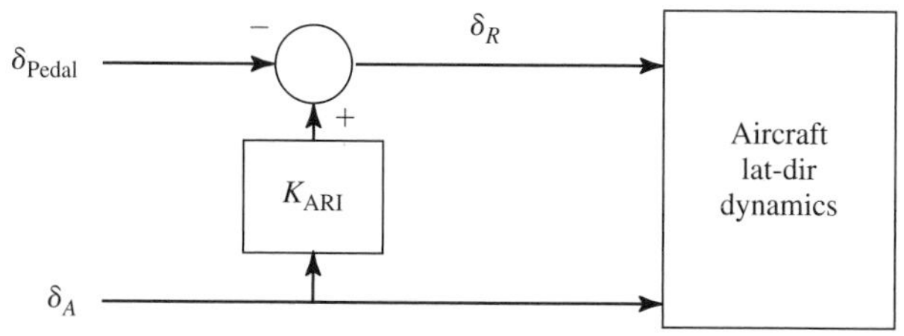

Figure 11.31 Block diagram of aileron-rudder interconnect (ARI).

To understand the motivation for including an ARI, consider the following argument. From the dutch-roll approximation, we found that the numerator of the $\phi(s)/\delta_A(s)$ transfer function (Equation (10.150)) is given by

$$\begin{aligned} N_{\delta_A}^{\phi}(s) &= L'_{\delta_A}U_0 s^2 - \left(L'_{\delta_A}(U_0 N'_r + Y_\beta) - (Y_{\delta_A}L'_\beta)\right)s \\ &\quad - \left(U_0(N'_{\delta_A}L'_\beta - L'_{\delta_A}N'_\beta) + Y_{\delta_A}L'_\beta N'_r - L'_{\delta_A}N'_r Y_\beta\right) \end{aligned} \tag{11.65}$$

and we seek to have this quadratic term approximately cancel the dutch-roll quadratic in the characteristic polynomial. Again from the dutch-roll approximation, this dutch-roll quadratic (Equation (10.151)) is given by

$$s^2 - (N'_r + Y_\beta/U_0)s + (N'_\beta + N'_r Y_\beta/U_0) \tag{11.66}$$

To maximize the likelihood of the desired cancellation, we would like the last coefficients in the above two quadratic terms to be equal. This will force the two poles and two zeros to be the same distance from the origin in the complex plane. Therefore, we want

$$-\left((N'_{\delta_A}/L'_{\delta_A})L'_\beta - N'_\beta\right) - \left((Y_{\delta_A}/L'_{\delta_A})L'_\beta N'_r - N'_r Y_\beta\right)/U_0 \approx N'_\beta + N'_r Y_\beta/U_0 \tag{11.67}$$

This expression will be satisfied when $N'_{\delta_A}/L'_{\delta_A}$ and $Y_{\delta_A}/L'_{\delta_A}$, the yawing moment and side force due to aileron, each normalized by the rolling moment due to aileron, both equal zero. Although these ratios cannot be adjusted independently with an ARI, we can usually still accomplish our goal of approximate pole-zero cancellation.

One simple way to select the ARI gain K_{ARI} is to force the effective $N'_{\delta_A} \approx 0$ by simply setting the yawing moment due to rudder deflection equal to the negative of the like moment due to the aileron deflection. That is, the adverse yaw from the aileron is cancelled with the rudder. So set

$$N'_{\delta_R}\delta_R = -N'_{\delta_A}\delta_A$$

or let (11.68)

$$\delta_R = -(N'_{\delta_A}/N'_{\delta_R})\delta_A$$

which means that the ARI gain is

$$K_{ARI} = -(N'_{\delta_A}/N'_{\delta_R}) \tag{11.69}$$

EXAMPLE 11.4

Aileron-Rudder Interconnect for an A-7A Aircraft

Consider the A-7A aircraft at Flight Condition 4 (Appendix B). Using an ARI with a gain of −0.2 deg/deg, compare the $\phi(s)/\delta_A(s)$ transfer functions with and without the ARI. Does the ARI improve the aircraft's response to aileron input? Why?

■ Solution

For this aircraft at the selected flight condition, the dimensional stability derivatives and flight condition (in the fuselage-reference axis) are given in Table 11.3. Since these data are given for a fuselage-reference axes, we must be sure to use the corresponding perturbation equations of motion. For this set of axes, $W_0 = U_0 \tan \alpha_0$ and $\Theta_0 = \alpha_0$ in level flight, and these terms must be included in the equation of motion for $\dot{\beta}$. In addition the equation for $\dot{\phi}$ is modified.

Specifically, with reference to Equations (10.36), (10.41), and (10.43), these equations are

$$
\begin{aligned}
U_0\dot{\beta} &= \underline{W_0 p} + \underline{g\cos\alpha_0}\phi + Y_\beta\beta + Y_p p + (Y_r - U_0)r + Y_{\delta_A}\delta_A + Y_{\delta_R}\delta_R \\
\dot{\phi} &= p + \underline{\tan\alpha_0 r}
\end{aligned}
\tag{11.70}
$$

where the underlined terms are included or modified appropriately for a fuselage-reference axis. Using these equations, along with Equations (10.39) and (10.45) and the above data, the state-variable representation of the lateral-directional dynamics becomes

$$
\mathbf{y}^T = \mathbf{x}^T = [\beta\,(\text{rad})\ \ \phi\,(\text{rad})\ \ p\,(\text{rad/sec})\ \ r\,(\text{rad/sec})],\quad \mathbf{u}^T = [\delta_A\,(\text{rad})\ \ \delta_R\,(\text{rad})]
$$

$$
\mathbf{A} = \begin{bmatrix} \dfrac{Y_\beta}{U_0} & \dfrac{g\cos\alpha_0}{U_0} & \dfrac{Y_p + W_0}{U_0} & \dfrac{Y_r}{U_0} - 1 \\ 0 & 0 & 1 & \tan\alpha_0 \\ L'_\beta & 0 & L'_p & L'_r \\ N'_\beta & 0 & N'_p & N'_r \end{bmatrix} = \begin{bmatrix} -0.119 & 0.096 & 0.224 & -1 \\ 0 & 0 & 1 & 0.236 \\ -8.97 & 0 & -1.38 & 0.857 \\ 0.948 & 0 & -0.031 & -0.271 \end{bmatrix}
$$

$$
\mathbf{B} = \begin{bmatrix} \dfrac{Y_{\delta_A}}{U_0} & \dfrac{Y_{\delta_R}}{U_0} \\ 0 & 0 \\ L'_{\delta_A} & L'_{\delta_R} \\ N'_{\delta_A} & N'_{\delta_R} \end{bmatrix} = \begin{bmatrix} -0.0015 & 0.030 \\ 0 & 0 \\ 3.75 & 1.82 \\ 0.280 & -1.56 \end{bmatrix},\quad \mathbf{c}_\phi = [0\ 1\ 0\ 0],\ \mathbf{d}_\phi = \mathbf{0}
\tag{11.71}
$$

Table 11.3 Dimensional Stability Derivatives for the A-7A

Stability Derivative	Value	Stability Derivative	Value	Stability Derivative	Value
$Y_\beta\ (= Y_v U_0)$	-38.67 ft/sec^2	L'_p	-1.38 /sec	N'_β	0.948 /sec^2
Y_{δ_A}	-0.476 ft/sec^2	L'_r	0.857 /sec	N'_p	-0.031 /sec
Y_{δ_R}	9.73 ft/sec^2	L'_{δ_A}	3.75 /sec^2	N'_r	-0.271 /sec
L'_β	-8.79 /sec^2	L'_{δ_R}	1.82 /sec^2	N'_{δ_A}	0.280 /sec^2
U_0	309 fps	α_0	13.3 deg	N'_{δ_R}	-1.56 /sec^2

Using MATLAB, the $\phi(s)/\delta_A(s)$ transfer function is found to be

$$\frac{\phi(s)}{\delta_A(s)} = \frac{3.81\left(s^2 + 0.4713s + 1.705\right)}{\left(s + 0.0441\right)\left(s + 0.972\right)\left(s^2 + 0.7565s + 2.760\right)} \tag{11.72}$$

With the ARI, the control-input vector becomes

$$\mathbf{u}^T = \begin{bmatrix} \delta_A & -\delta_{\text{Pedal}} \end{bmatrix}$$

and the **B** matrix is modified using Equation (11.63) to become

$$\mathbf{B}_{\text{ARI}} = \begin{bmatrix} -0.0077 & 0.030 \\ 0 & 0 \\ 3.39 & 1.82 \\ 0.592 & -1.56 \end{bmatrix} \tag{11.73}$$

This result is obtained as follows: Defining $\mathbf{B} = [\mathbf{b1}\ \mathbf{b2}]$, where **b1** and **b2** are the columns of **B**, then $\mathbf{B}_{\text{ARI}} = [(\mathbf{b1} + K_{\text{ARI}}\mathbf{b2})\ \mathbf{b2}]$. Again from MATLAB, <u>with the ARI</u> included the $\phi(s)/\delta_A(s)$ transfer function becomes

$$\left.\frac{\phi(s)}{\delta_A(s)}\right|_{\text{ARI}} = \frac{3.530\left(s^2 + 0.5923s + 2.572\right)}{\left(s + 0.0441\right)\left(s + 0.972\right)\left(s^2 + 0.7565s + 2.760\right)} \tag{11.74}$$

By comparing the two transfer functions in Equations (11.72) and (11.74), we see that the characteristic polynomial is unchanged due to the control-input cross feed, as expected. But the numerator of the transfer function is modified. With the ARI the numerator quadratic more closely approximates the dutch-roll quadratic, hence approximate pole-zero cancellation of the dutch-roll poles is achieved in the bank-angle response to aileron.

Finally, it is worth noting that an ARI us used on many aircraft, including the A-7A and the T-38.

11.4.3 Increasing Yaw-Damper Effectiveness

Recall in Section 11.4 that when the pilot controlled bank angle with aileron ($\phi(s)/\delta_A(s)$), the dutch-roll mode could cause difficulties, especially if that mode wasn't well damped or if there wasn't an approximate cancellation of the dutch-roll roots by a pair of nearby zeros. In Section 11.4.1 we addressed the issue of increasing dutch-roll damping by adding yaw-rate feedback, and in Section 11.4.2 we addressed adjusting the locations of the zeros of the $\phi(s)/\delta_A(s)$ transfer function by introducing an aileron-to-rudder interconnect (ARI).

While discussing the yaw damper and yaw-rate feedback in Section 11.4.1, we noted that the effectiveness of the yaw damper, in terms of increasing the dutch-roll damping, depended upon having the proper locations for the zeros in the $r(s)/\delta_R(s)$ transfer function (see the root locus in Figure 11.27). Specifically, these zeros must be sufficiently close to the origin in the complex plane for the yaw damper to be effective. In this section, we will look at assuring the desired locations of these zeros by using another feedback stability augmentation system. Since feedback will be used, the characteristic polynomial will be affected, which may in fact be desirable as well. But that will be discussed further in Section 11.4.4. The goal here is to properly position the zeros of $r(s)/\delta_R(s)$.

From Equations (10.139), we know that the numerator of $r(s)/\delta_R(s)$ usually factors into the following form

$$N^r_{\delta_R}(s) = K^r_{\delta_R}\left(s + 1/T_{r_1}\right)\left(s^2 + 2\zeta_r\omega_r + \omega_r^2\right) \tag{11.75}$$

where the quadratic term is our primary focus of this discussion. From Refs. 2 and 4, the last coefficient in this term and the real zero may usually be approximated by

$$\omega_r^2 \approx \frac{g}{U_0}\frac{L_\beta}{L_p} \qquad 1/T_{r_1} \approx -L_p \tag{11.76}$$

(Note the unprimed stability derivatives.) Hence, if the magnitude of L_p is increased, the real zero will move farther to the left along the negative real axis, and the two complex zeros will move closer to the origin. And recall, moving these complex zeros in this fashion is the desired effect. Moving the real zero will be discussed more in Section 11.4.4.

To increase the effective L_p, we must modify the rolling moment due to roll rate, and this can be accomplished by introducing roll-rate feedback to the aileron. The control law is typically of the following form

$$\delta_A(s) = \delta_{\text{Stick}} - K_p p(s) \tag{11.77}$$

The block-diagram representation of this control law, called a *roll damper,* is shown in Figure 11.32.

The polynomial-matrix description of the lateral-directional dynamics, including this control law, becomes (for $\Theta_0 = 0$)

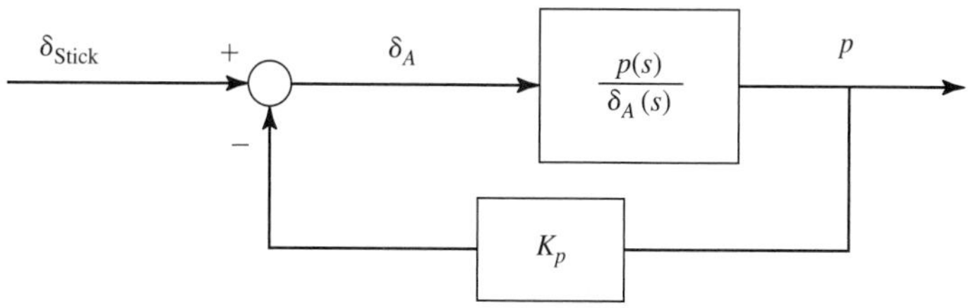

Figure 11.32 Block diagram of roll damper.

$$\begin{bmatrix} U_0 s - Y_\beta & -\left(Y_p - \boxed{K_p Y_{\delta_A}}\right)s - g & (U_0 - Y_r) \\ -L'_\beta & s^2 - \left(L'_p - \boxed{K_p L'_{\delta_A}}\right)s & -L'_r \\ -N'_\beta & -\left(N'_p - \boxed{K_p N'_{\delta_A}}\right)s & s - N'_r \end{bmatrix} \begin{Bmatrix} \beta(s) \\ \phi(s) \\ r(s) \end{Bmatrix} = \begin{bmatrix} Y_{\delta_A} & Y_{\delta_R} \\ L'_{\delta_A} & L'_{\delta_R} \\ N'_{\delta_A} & N'_{\delta_R} \end{bmatrix} \begin{Bmatrix} \delta_{\text{Stick}}(s) \\ \delta_R(s) \end{Bmatrix} \quad (11.78)$$

Note that L'_p is always negative, and since $L'_{\delta_A} > 0$, the effective L'_p becomes more negative.

As we have noted in earlier sections, the addition of feedback can affect the zeros in some of the closed-loop transfer functions. And that fact is what we are exploiting here. Specifically, the numerator of the $r(s)/\delta_R(s)$ is the focus, and this numerator now becomes

$$N^r_{\delta_R\,\text{aug}}(s) = \det\begin{bmatrix} U_0 s - Y_\beta & -\left(Y_p - \boxed{K_p Y_{\delta_A}}\right)s - g & Y_{\delta_R} \\ -L'_\beta & s^2 - \left(L'_p - \boxed{K_p L'_{\delta_A}}\right)s & L'_{\delta_R} \\ -N'_\beta & -\left(N'_p - \boxed{K_p N'_{\delta_A}}\right)s & N'_{\delta_R} \end{bmatrix}$$

$$= \det\begin{bmatrix} U_0 s - Y_\beta & -Y_p s - g & Y_{\delta_R} \\ -L'_\beta & s^2 - L'_p s & L'_{\delta_R} \\ -N'_\beta & -N'_p s & N'_{\delta_R} \end{bmatrix} + K_p s \det\begin{bmatrix} U_0 s - Y_\beta & Y_{\delta_A} & Y_{\delta_R} \\ -L'_\beta & L'_{\delta_A} & L'_{\delta_R} \\ -N'_\beta & N'_{\delta_A} & N'_{\delta_R} \end{bmatrix}$$

$$= N^r_{\delta_R}(s) + K_p s N^{\phi r}_{\delta_A \delta_R}(s) \quad (11.79)$$

or the open-loop numerator, typically of the form given in Equation (11.75), plus the roll-rate feedback gain times s times the coupling numerator.

By inspection of Equation (11.79), we can see that the coupling numerator is of the following form.

$$N^{\phi r}_{\delta_A \delta_R}(s) = A(s + B)$$

And B is typically a small positive number. The effect of the roll damper on the zeros of the $r(s)/\delta_R(s)$ transfer function may now be revealed by using a numerator root locus, or a plot of the roots of

$$1 + K_p \frac{s N^{\phi r}_{\delta_A \delta_R}(s)}{N^r_{\delta_R}(s)} = 1 + K_p \frac{As(s + B)}{K^r_{\delta_R}\left(s + 1/T_{r_1}\right)\left(s^2 + 2\zeta_r \omega_r + \omega_r^2\right)} = 0 \quad (11.80)$$

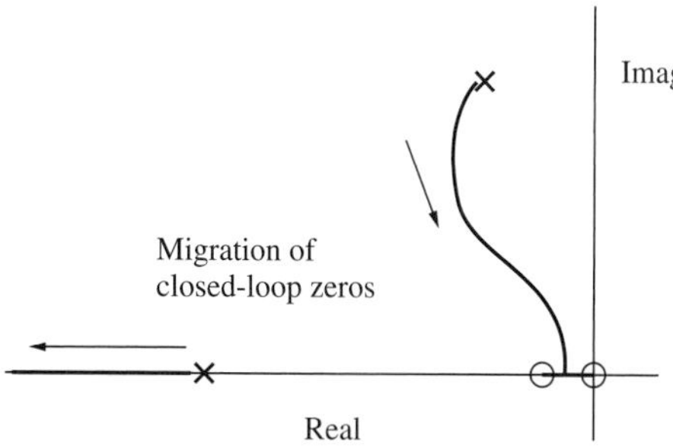

Figure 11.33 Effect of roll damper on $r(s)/\delta_R(s)$ zeros.

Given that B is typically a small positive number, the shape of this numerator root locus usually appears as in Figure 11.33. And from that figure, we see that the zeros in question are indeed moved closer to the origin of the complex plane, which was the original objective.

11.4.4 Reducing the Roll-Mode Time Constant

Sometimes the roll-mode time constant T_R must be reduced, where the eigenvalue corresponding to the roll-subsidence mode is equal to $-1/T_R$. The need is either to improve the handling qualities in roll control, or to improve the response of a roll-command autopilot (discussed in Chapter 12). From the roll-mode approximation discussed in Section 10.7.1, we know that the roll-subsidence eigenvalue is approximately given by

$$-1/T_R \approx L'_p \tag{11.81}$$

Therefore, we may reduce the roll-mode time constant by again using roll-rate feedback to the aileron, or by using a roll damper.

The use of a roll damper to reposition the zeros of the $r(s)/\delta_R(s)$ transfer function was just discussed in Section 11.4.3. This repositioning improves the effectiveness of the yaw damper with regard to increasing the dutch-roll damping. But in Section 11.4.3, we also noted that a roll damper will concomitantly reduce the roll-mode time constant, and that is the topic here. The control law for the roll damper was given in Equation (11.77), repeated here for convenience.

$$\delta_A(s) = \delta_{\text{Stick}} - K_p p(s)$$

The block diagram representing this control law was shown in Figure 11.32.

To confirm our selection of this feedback architecture to reduce the roll-mode time constant, consider the root-locus shown in Figure 11.34. The open-loop transfer function here is roll rate with aileron input for the A-7A at Flight Condition 4, and this root-locus reveals, of course, the effects of the roll damper on the closed-loop eigenvalues. Note that the effect on the spiral and dutch-roll eigenvalues is minimal. But the location of the roll-subsidence pole may be

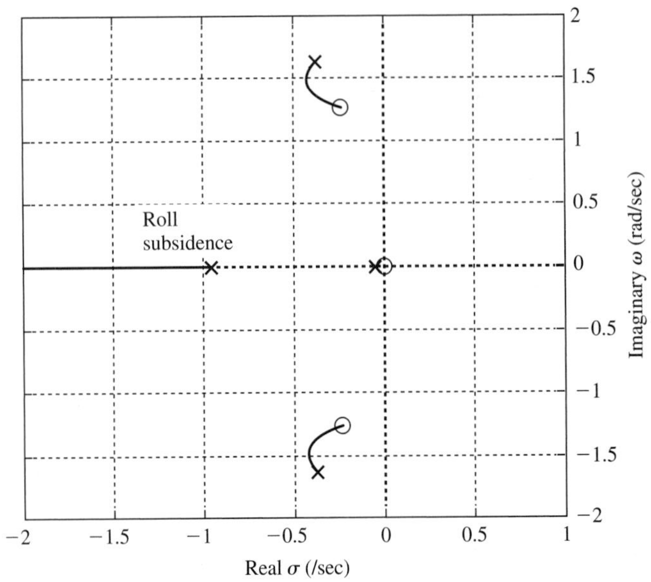

Figure 11.34 Root locus—effect of a roll damper on the closed-loop eigenvalues.

significantly altered. This confirms that roll-rate feedback to the aileron is indeed effective for reducing the roll-mode time constant.

With regard to the effect of a roll damper on zeros of the transfer-functions, we know from the discussion in Section 11.2 that roll-rate feedback to the aileron will have no effect on the zeros of any of the bank-angle or roll-rate transfer functions, and no effect on any transfer functions with aileron input. This leaves only the effects on the zeros of the $\beta(s)/\delta_R(s)$ and $r(s)/\delta_R(s)$ transfer functions. But we have already discussed the effect on the zeros of $r(s)/\delta_R(s)$ in Section 11.4.3. The effect of a roll damper on the zeros of the $\beta(s)/\delta_R(s)$ transfer function is the topic of Problem 11.7 at the end of this chapter.

Based on the discussion in this section and in Section 11.4.3, we see that a roll damper may be quite useful. The roll-mode time constant may be reduced and the critical zeros in the $r(s)/\delta_R(s)$ transfer function may be repositioned to improve yaw-damper effectiveness. For these reasons, roll dampers are used on several aircraft, including the A-7A.

11.5 COMMENTS ON ELASTIC EFFECTS

Two primary effects of elastic deformation on feedback stability augmentation are:

- Static-elastic effects on the vehicle's dimension stability derivatives.
- Effect on sensor placement.

The first of these two issues was introduced in Chapter 7, in which a method was suggested for determining the effects of static deformation of the structure on the rigid-body stability derivatives such as M_α. We will not address this issue further here, other than to say that these static-elastic effects can be significant, especially for large vehicles. So when dealing with feedback stability augmentation of such vehicles, it is important to use dimensional stability derivatives in the analysis and design that have been corrected for static-elastic effects.

The second issue arises, as we saw in Chapters 3 and 4, due to the fact that the total displacement $\mathbf{d}(t)$ of a point (x,y,z) on the vehicle's structure (e.g., a sensor location) is composed of the rigid-body plus the elastic components. That is,

$$\mathbf{d}(x,y,z,t) = \mathbf{d}_R(x,y,z,t) + \mathbf{d}_E(x,y,z,t) = \mathbf{d}_R(x,y,z,t) + \sum_{i=1}^{\infty} \boldsymbol{\nu}_i(x,y,z)\eta_i(t)$$

Here recall that $\boldsymbol{\nu}_i(x,y,z)$ is the mode shape for the i'th mode evaluated at the point $(x.y.z)$ on the structure, and $\eta_i(t)$ is the generalized coordinate associated with that vibration mode. Consequently, even though it may be desired that the sensor respond only to rigid-body motion $\mathbf{d}_R$ (e.g., rigid-body pitch rate), it will actually sense the rigid-body plus the elastic components. But by locating the sensor on a nodal point of the mode shape, the designer can minimize the contribution of that mode to the total displacement. In addition, frequency shaping of the sensor's output may be used to filter out the effects of the higher-frequency elastic modes. We will address frequency shaping in more detail when discussing the active control of elastic deformation later in Chapter 12.

11.6 Summary

In this chapter we have reviewed some linear-systems analysis techniques, and introduced important methods for dealing with the multi-input/multi-output (MIMO) systems arising in flight dynamics. Then we synthesized, analyzed, and discussed several feedback stability-augmentation systems, widely used in practice, for augmenting the longitudinal and lateral-directional dynamics of aircraft. We stressed that the architecture for these systems, including the identification of which aircraft response to feed back to which control input, was selected by direct application of our knowledge of the vehicle's dynamics. In particular, the various modal approximations and our understanding of the physics governing the forces and moments proved invaluable. We demonstrated how the control laws presented directly augmented, or modified, the dimensional stability derivatives—hence the terminology *stability augmentation*. These control laws are *dynamics based*, in that they are selected to work synergistically with the natural dynamics of the vehicle. They are also simple and relatively easy to implement.

We introduced the stability-augmentation systems in the context of correcting some deficiency in the dynamics, such as insufficient modal damping and/or frequency. And we applied the eigenvalue root locus and the numerator root locus to evaluate the effects of the stability augmentation. We also stated that a key design objective was to preserve or restore conventional modal characteristics (e.g., short period and phugoid). This is important because the handling-qualities database primarily includes design criteria on the flight dynamics that assumes the vehicle's dynamics are conventional.

Though linear-systems tools are employed, we noted that perturbation techniques are not involved when the control laws are actually implemented on the vehicle. For example, total (not perturbation) angular rates or displacements would be fed back in the control law. Perturbation theory, however, is the basis for using linear techniques in the analysis and synthesis.

11.7 Problems

11.1 Given a three-by-three matrix **P**, with

$$\mathbf{P} = \begin{bmatrix} a & b & g + kh \\ c & d & i + kj \\ e & f & l + km \end{bmatrix}$$

prove that

$$\det \mathbf{P} = \det \begin{bmatrix} a & b & g \\ c & d & i \\ e & f & l \end{bmatrix} + k \det \begin{bmatrix} a & b & h \\ c & d & j \\ e & f & m \end{bmatrix}$$

Hint: Expand the determinant in terms of co-factors of the elements of the third column.

11.2 Using MATLAB and the data for the DC-8 at Flight Condition 3 (Appendix B), generate the root-locus plot showing the effect of a pitch damper on the closed-loop longitudinal eigenvalues. Also, using a pitch damper gain of 0.1 deg/deg/sec, find the transfer functions for surge velocity u, angle of attack α, and pitch attitude θ with elevator-stick and thrust input. (Assume that $X_T = 1/\text{m}$, and that the remaining thrust-input stability derivatives Z_T and M_T equal zero.) Compare these transfer functions to those without the pitch damper. Comment on the effects of the pitch damper on the longitudinal dynamic modes and the transfer-function zeros.

11.3 Using MATLAB and the data for the DC-8 at Flight Condition 4 (Appendix B), generate the root locus plot showing the effect of pitch-attitude feedback on the closed-loop longitudinal eigenvalues. Show that the unstable phugoid mode can be stabilized with pitch-attitude feedback.

11.4 Using MATLAB and the data for the A-7A at Flight Condition 4 (Appendix B), generate the root locus plot showing the effect of vertical-acceleration feedback ($a_{Z\,\text{cg}}$) on the closed-loop longitudinal eigenvalues. (Be sure to use the system description for the dynamics assuming a fuselage-reference axis. That is, include the necessary additional terms such as W_0 in the perturbation equations.) Then using an acceleration feedback gain of 0.005 rad/ft/sec^2, find the transfer functions for surge velocity u, angle of attack α, and pitch attitude θ with elevator-stick input, and compare them to the transfer functions with no acceleration feedback. Then add a pitch damper with a gain of 0.25 rad/rad/sec and find the transfer functions with acceleration feedback and the pitch damper. Compare these to the three transfer functions found previously.

11.5 The effect of the washout filter in a yaw damper was discussed in Section 11.4.1. Using the data for the A-7A at Flight Condition 3 (Appendix B) and a washout with a time constant $T_W = 1$ /sec, determine the effect of the washout on the zeros of the $\phi(s)/\delta_A(s)$ transfer function. That is, plot the numerator root locus and compare the results to those in Figure 11.28. (Be sure to use the system descriptions for the dynamics assuming a fuselage-reference axis by including the terms indicated in Equations (11.70.)

11.6 Find all six transfer functions (not including lateral acceleration) for the case considered in Problem 11.5 using a yaw-damper gain of 0.25 rad/rad/sec. Note the effects of the washout on the zeros of the transfer functions. Do these zeros approximately cancel the augmented washout pole in all the transfer functions?

11.7 Starting with the polynomial-matrix description of the lateral-directional dynamics, including the roll damper, derive the expression for the numerator of the closed-loop $\beta(s)/\delta_R(s)$ transfer function, or $N^{\beta}_{\delta_R\,\mathrm{aug}}(s)$, in terms of the open-loop numerator of this transfer function plus a coupling numerator. Then using the data for the A-7A in Flight Condition 4 (Appendix B), use MATLAB to plot the numerator root locus showing the effects of a roll damper on the zeros of the $\beta(s)/\delta_R(s)$ transfer function. (Be sure to use the polynomial-matrix description for the dynamics assuming a fuselage-reference axis, by including the terms indicated in Equations (11.70).)

References

1. D'Azzo, J. J. and C. H. Houpis: *Linear Control System Analysis and Design,* 3rd ed., McGraw-Hill, New York, 1988.
2. McRuer, D., I. Ashkenas, and D. Graham: *Aircraft Dynamics and Automatic Control,* Princeton University Press, Princeton, NJ, 1973.
3. Davidson, J. B., et al: "Flight-Control Laws for NASA's Hyper-X Research Vehicle," AIAA Paper No. AIAA-99-4124, Proceedings of the AIAA Guidance, Navigation, and Control Conference, Portland, OR, August 1999.
4. McRuer, D. T. and D. E. Johnston: *Flight Control Systems Properties and Problems Vols. I and II,* NASA Contractor Report CR-2500, prepared by Systems Technology, Inc., Feb. 1975.

12 CHAPTER

Automatic Guidance and Control—Autopilots

Chapter Roadmap: *The material presented in this chapter may be included in a first course in flight dynamics, but frequently is not due to the time limitations of the semester or quarter. The chapter builds on the material presented in Chapter 11, and addresses the analysis and synthesis of the typical modes of an autopilot via loop shaping.*

In this chapter we continue the discussion of flight vehicles under feedback control. But unlike Chapter 11, which focused strictly on stability augmentation, this chapter will focus on control systems that can replace some of the functions involved with piloting the aircraft. As discussed in Chapter 11, stability augmentation involves using feedback to augment the vehicle's dimensional stability derivatives, that is, to tailor the natural dynamics. But the dynamics of the flight vehicle typically remain conventional, characterized in terms of the conventional natural modes. With the feedback systems discussed in this chapter, however, the system's (vehicle and control system) dynamics are typically dominated by the feedback system. This is one of the primary reasons for including this chapter in a text on flight dynamics. The flight dynamicist must not only understand the vehicle's open-loop and augmented dynamics, but also the dynamics of the vehicle under automatic feedback control. In addition, note that the architecture of the flight-control systems considered in this chapter will draw heavily from a thorough understanding of the vehicle's dynamics.

As for the tools applied in the discussion of autopilots, we could approach the problem by using polynomial-matrix descriptions of the systems and the root-locus method as done in Chapter 11. However, in this chapter, we employ an approach known as *loop shaping* that is numerically intensive and graphics based. MATLAB, state-variable descriptions, and computer graphics are tools that are very compatible with this approach.

Loop shaping is widely used in the design of feedback systems, and dates back to Bode's original work dealing with the invention of such systems. But even though it is straightforward, intuitive, and easily understood, many undergraduates have not used the technique. Therefore, this chapter begins with a "just-in-time tutorial" on loop shaping and inner and outer loops. In the remainder of the chapter, we will demonstrate the use of these techniques in discussions of the more common modes of autopilots.

12.1 FEEDBACK CONTROL-LAW SYNTHESIS VIA LOOP SHAPING—A JITT*

We will not cover all of feedback control theory, as the field is quite vast. But it is helpful to students if some key concepts are discussed, even if they have had a course on feedback control. These concepts include loop shaping in the frequency domain and the use of Bode and Nyquist plots to design simple, robust feedback control laws. Undergraduate students may have been introduced to Bode and Nyquist plots, but have had little opportunity to use them in designing a feedback system. So learning how to use these plots in a design setting will be important to us.

Several important prerequisite topics were discussed in the previous two chapters—specifically in Sections 10.1 and 11.1. The topics included state-variable system descriptions and transfer functions, Bode plots and frequency response, block diagrams, and the root-locus technique for synthesizing and analyzing a feedback system. To fully understand the material presented in this section and this chapter, students should be familiar with the material in these two previous sections.

With regard to a linear system's frequency response and its complex function $g(j\omega)$, Fourier theory tells us that the frequency response is a full and complete representation of that system. That is, no information is lost by defining the system in terms of its frequency response. Therefore using $g(j\omega)$ to represent the linear dynamic system is just as valid as using its transfer function $g(s)$ or state-variable description. We will not prove this, but refer the student to Ref. 1, for example. Throughout this chapter, we will frequently represent a linear system (or at least an input-output pair of the system, if it has multiple inputs or outputs) in terms of its complex function $g(j\omega)$, and equivalently in terms of its Bode plot.

12.1.1 Bode Plots Revisited

Although Bode plots were introduced in Section 10.1, we will do a quick review here. Recall that a complex number corresponds to a point in the complex plane, and that this complex number may be represented in polar form in terms of its

---

* A Just-In-Time Tutorial.

magnitude and argument (or phase). Specifically, we can express the complex number $g(j\omega)$ as

$$g(j\omega) = M(j\omega)e^{\phi(j\omega)} \tag{12.1}$$

where $M(j\omega)$ is the magnitude and $\phi(j\omega)$ is the argument of the complex number. Bode plots are simply plots of $M(j\omega)$ and $\phi(j\omega)$, each plotted versus frequency ω, and by convention, $M(j\omega)$ and ω are plotted on a logarithmic scale. This may best be clarified by an example.

MATLAB EXAMPLE 12.1

Plotting a Bode Plot

Let's begin with the following transfer-function representation of a system

$$g(s) = \frac{1}{s^2 + s + 1} \tag{12.2}$$

So then its alternate representation is

$$g(j\omega) = \frac{1}{-\omega^2 + j\omega + 1} = \frac{1}{(1-\omega^2) + j\omega} = M(j\omega)e^{\phi(j\omega)} \tag{12.3}$$

Using a little complex arithmetic, we can write this as

$$\frac{1}{(1-\omega^2)+j\omega} = \frac{(1-\omega^2)-j\omega}{(1-\omega^2)^2+\omega^2} = \left(\frac{(1-\omega^2)}{(1-\omega^2)^2+\omega^2}\right) + j\left(\frac{-\omega}{(1-\omega^2)^2+\omega^2}\right) = a(\omega) + jb(\omega) \tag{12.4}$$

So the complex number $g(j\omega) = a(\omega) + jb(\omega)$, and therefore the magnitude and argument are

$$\begin{aligned} M(j\omega) &= \text{sqrt}(a^2 + b^2) \\ \phi(j\omega) &= \tan^{-1}(b/a) \end{aligned} \tag{12.5}$$

The above analysis reinforces the facts that first, $g(j\omega)$ is simply a complex number, and second, that complex numbers may be represented in polar form in terms of their magnitude and argument, or *phase.*

Fortunately, we can use MATLAB to easily obtain a system's Bode plot. Again we have

$$g(s) = \frac{1}{s^2 + s + 1}$$

and we can use the following MATLAB commands to obtain the desired Bode plot shown in Figure 12.1.

```
»num=[1]; den=[1 1 1];sys=tf(num,den)
Transfer function:
    1
-----------
s^2 + s + 1
»bode(sys)
```

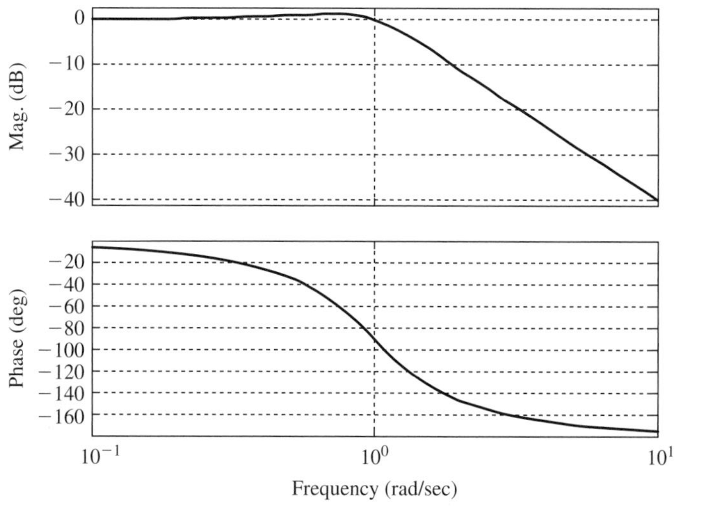

Figure 12.1 Example 12.1 Bode plot.

In Figure 12.1, the top plot is the magnitude, or $M(j\omega)$, plotted versus frequency. The bottom plot is the argument (phase), or $\phi(j\omega)$, also plotted versus frequency. Note that the scale for $M(j\omega)$ is labeled dB, for decibels. This is also by convention. Any real number N may be expressed in dB by taking the logarithm of that number. Or, by definition,

$$N(\text{dB}) = 20\log_{10}(N) \tag{12.6}$$

So even though a magnitude plot may have the dB scale, it is really just a special logarithmic scale.

12.1.2 Nyquist Stability Theory

Next we will address a key aspect of loop shaping dealing with system stability, or Nyquist stability theory. The theoretical underpinning of Nyquist's theory is a concept from the theory of complex variables known as the *Principle of the Argument.* (Ref. 2). The Principle of the Argument (P of A) states:

> Given a closed contour C in the complex plane and a complex mapping (function) $P(s)$ that is everywhere analytic on C, then if C (clockwise) encircles N zeros and M singularities of $P(s)$, the image of C under P will encircle the origin (of the complex plane) exactly $(N - M)$ times clockwise.

This principle seems a little abstract, but can be intuitively understood quite nicely with a couple simple examples. But first some notes on terminology. First, the function $P(s)$ is "analytic" on the contour C if $P(s)$ has no singularities on C. That is, for no value of s on C does $P(s) \to \infty$. Second, the "zeros" of P are simply the values of s that satisfy the equation $P(s) = 0$. For example, if $P(s)$ happened to be a polynomial, its zeros would be the roots of that polynomial. (This is consistent with our earlier terminology regarding the "zeros" of a transfer function.) For our purposes in deriving Nyquist's theory, we will always take the function

$P(s)$ to be a polynomial or a ratio of two polynomials in s (although the Principle of the Argument is not restricted to polynomials). Third, the "image of C under P" simply refers to the plot of the complex function P as s travels around the closed contour C.

EXAMPLE 12.2

Principle of the Argument—$P = (s + 1)$

Demonstrate the Principle of the Argument if $P(s) = (s + 1)$.

■ Solution

Note that the zero of the polynomial $P(s) = s + 1$ is located at the *point* -1 in the complex plane, as shown on the left in Figure 12.2. Also shown in the left side are two closed contours, C_1 and C_2, with C_1 encircling the point at -1, and C_2 not encircling that point. Now the "image of C under P" is generated graphically as follows. Let s be an arbitrary point on C (in the first case, it is a point on C_1), and graphically depict the complex number s as the vector from the origin to the point s. Using the same graphical approach, another vector in the figure represents the value $+1$. Summing these two vectors graphically yields the vector $\mathbf{P}_1$, which is the graphical representation of the function $P(s) = (s + 1)$, with s the chosen arbitrary point on C_1.

Now let that point s travel clockwise around C_1, and visualize how the magnitude and argument of the vector $\mathbf{P}_1$ change as s is so varied. In particular, note that the argument of $\mathbf{P}_1$ changes a full 360 deg as s travels around C_1 and returns to its starting location. The plot of this magnitude and argument as s moves around C_1 is shown schematically in the right side of the figure. This plot is the "image of C_1 under P." One can observe with some careful thought that since C_1 does enclose the zero of $P(s)$ at -1, the image of C_1 under P must encircle the origin (of the complex plane) one time clockwise.

Now look at the closed contour C_2 on the left, and let s be an arbitrary point on C_2. With s as defined, one can again show that the graphical representation of the function $P(s) = (s + 1)$ is the vector $\mathbf{P}_2$ shown. Now letting s traverse around C_2 clockwise will

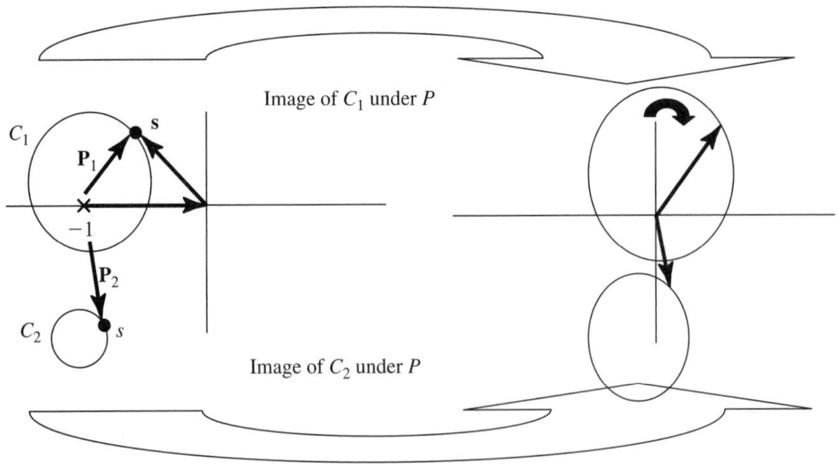

Figure 12.2 First example of the Principle of the Argument.

vary both the magnitude and argument of $\mathbf{P}_2$, and result in the plot indicated in the right side of the figure. The major difference here is that since C_2 does <u>not</u> encircle the zero of P at -1, the argument of $\mathbf{P}_2$ does <u>not</u> vary the full 360 deg. And so the image of C_2 under P does <u>not</u> encircle the origin.

By extension, now think about how the image of C under P would change if instead P had two zeros and C encircled them both. Clearly the argument of $\mathbf{P}$ in this case would have to rotate a full 720 degrees as s moved around C and returned to its starting position. As a result, the image of C under P must now encircle the origin <u>twice</u> clockwise. We will now consider another example.

EXAMPLE 12.3

The Principle of the Argument—$F = 1/P$

Demonstrate the Principle of the Argument for the function $F(s) = 1/(s + 1)$, or now consider the case with the function

$$F(s) = \frac{1}{s+1} = \frac{1}{P(s)} \tag{12.7}$$

■ Solution

Note that this function has no zeros, but rather has a singularity at $s = -1$. Also note that the contour C, shown on the left in Figure 12.3, encircles this point. As in Example 12.2, we let s lie on C and the graphical representation of the function $P(s)$ be denoted as $\mathbf{P}$, derived as in the previous example. Now let P be represented in terms of its magnitude M and argument ϕ, or

$$P(s) = M(s)e^{\phi(s)} \tag{12.8}$$

with both M and ϕ clearly functions of the variable s that lies on C. As before, ϕ will vary a full 360 degrees clockwise as s moves around C clockwise. But since

$$F(s) = \frac{1}{P(s)} = \frac{1}{M}e^{-\phi} \tag{12.9}$$

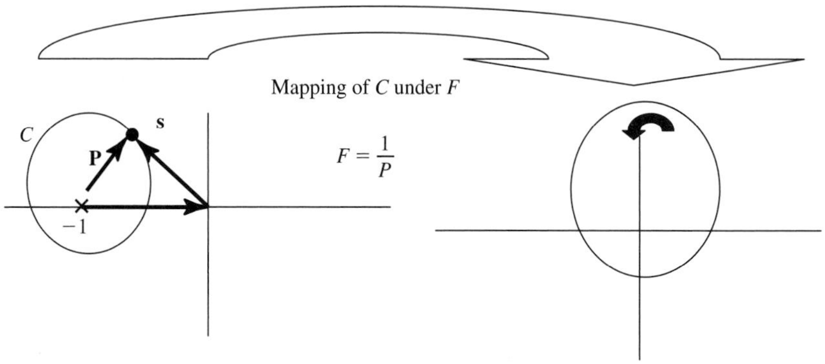

Figure 12.3 Mapping of C under F, F with one encircled singularity.

the argument of $F(s)$ is $-\phi$. Therefore, the argument of the function F will vary from zero to -360 degrees as s moves clockwise around C and returns to its initial position. So the image of C under F will encircle the origin, but will do so in a counterclockwise direction, as shown in the right of the figure. Hence, the direction of encirclement of the origin changes, depending on whether zeros or singularities are encircled by the closed contour C.

By selecting different functions for P and F, the student can develop a family of images, each depending on the number of zeros and/or the number of singularities encircled by the chosen contour C. As a result, using the intuitive concepts demonstrated in these examples we can convince ourselves that the following assertion is true. (This assertion will be useful later.)

Assertion If a polynomial $P_1(s)$ has N_1 zeros enclosed in the closed contour C, and if a polynomial $P_2(s)$ has N_2 zeros enclosed in the same closed contour, then the function

$$P(s) = P_1(s)P_2(s) \tag{12.10}$$

will have $N_1 + N_2$ zeros enclosed in C, and the image of C under P will encircle the origin $N_1 + N_2$ times clockwise.

We now have all the facts necessary to derive Nyquist's stability theory. All we need to do is to select the appropriate function $P(s)$, and the appropriate closed contour C. Since we are interested in the stability of a dynamic system, we are interested in the existence of roots of the system's characteristic polynomial in the open right half of the complex plane (where "open" refers to the real axis going to $+\infty$). The system we might consider as an example is the generic feedback system depicted in Figure 12.4.

To apply the Principle of the Argument we will select as the function $P(s)$ the system's closed-loop characteristic polynomial $\Delta_{CL}(s)$. And we will select as the closed contour the entire open right half of the complex plane, shown in Figure 12.5. This particular closed contour is called the *Nyquist D contour,* for obvious reasons. Clearly, if no zeros of $\Delta_{CL}(s)$ (which are the closed-loop poles) are enclosed in D, the system is stable.

With these selections, we can directly apply the Principle of the Argument to obtain a requirement for the stability of a (closed-loop) dynamic system:

Closed-Loop Stability Requirement A feedback system is stable if and only if the image of D under $\Delta_{CL}(s)$ does not encircle the origin of the complex plane.

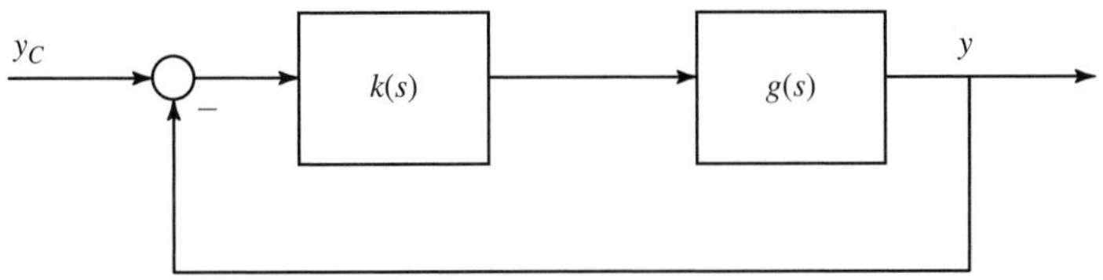

$$\frac{y}{y_C} = \frac{kg}{1 + kg}$$

Figure 12.4 Generic feedback system.

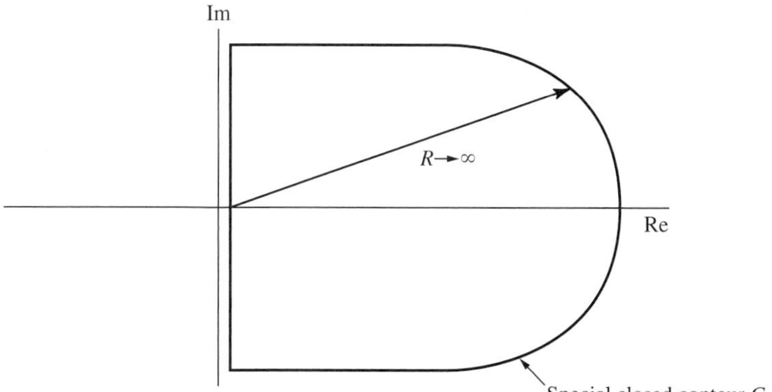

Figure 12.5 Nyquist D contour encloses entire right half of complex plane.

Now this is really Nyquist's stability requirement, but it may not look familiar. To obtain the more familiar form, we must manipulate the closed-loop system's characteristic polynomial. This polynomial $\Delta_{CL}(s)$ may be written in terms of the characteristic polynomial of the open-loop system $\Delta_{OL}(s)$, and with reference to the system depicted in Figure 12.4, note that $\Delta_{OL}(s)$ is just the denominator of the *open-loop transfer function* $k(s)g(s)$. Therefore we have

$$\Delta_{CL}(s) = \Delta_{OL}(s)\big(1 + k(s)g(s)\big) \tag{12.11}$$

Recall further that $\Delta_{OL}(s)$ is a polynomial, and that $\big(1 + k(s)g(s)\big)$ is a ratio of polynomials. Now using the Assertion presented previously, we can restate our closed-loop stability requirement.

Closed-Loop Stability Requirement 2 A feedback system is stable if and only if the image of D under $(1 + kg)$ encircles the origin $-N_{OL}$ times clockwise, where N_{OL} is the number of zeros of $\Delta_{OL}(s)$ in the right-half plane (or the number of unstable open-loop poles).

This requirement follows from the fact that for stability we must have no zeros (roots) of $\Delta_{CL}(s)$ enclosed in D. So from the Principle of the Argument and from the Assertion, for stability we must have

$$N_{OL} + N_{1+gk} = 0 \tag{12.12}$$

where N_{1+gk} is the number of clockwise encirclements of the origin made by the image of D under $(1 + gk)$ (a ratio of polynomials).

Stability Requirement 2 above is closer to, but not quite the same familiar Nyquist statement. To obtain the final form, we simply need to consider Figure 12.6. First note that the image of D under $(1 + kg)$ is a plot of the (complex) values taken on by the function $(1 + kg)$ as s moves around D. So let the value of $(1 + kg)$ for some s on D be denoted as V. Then for that same s, kg must take on the value $V - 1$. Therefore, the image of kg is just the image of $(1 + kg)$ shifted by -1. With this in mind, we can now state the familiar form of Nyquist's stability requirement.

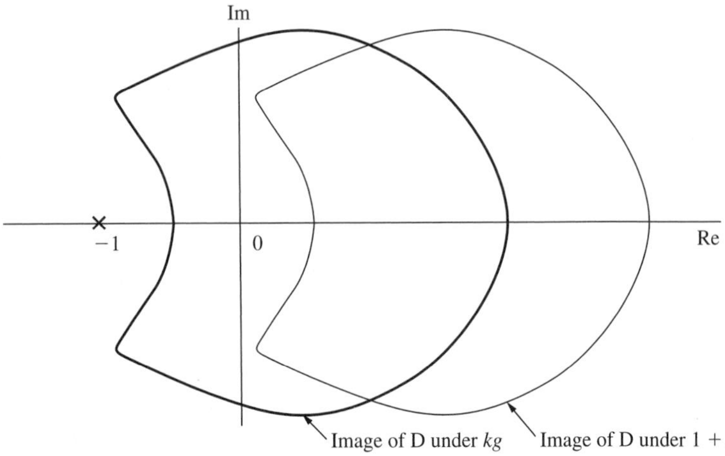

Figure 12.6 Adjustment of the mapping of $(1 + kg)$.

Nyquist's Stability Requirement A feedback system is stable if and only if the image of D under kg encircles the -1 point $-N_{\mathrm{OL}}$ times clockwise, where N_{OL} is the number of zeros of $\Delta_{\mathrm{OL}}(s)$ in the right-half plane. (Note again that a negative clockwise encirclement is a counterclockwise encirclement.)

This is the form of Nyquist's criteria usually encountered. Note that if an image, or the so-called Nyquist plot, encircles -1 clockwise, the system is unstable regardless of whether the open-loop system is unstable or not. Further, if the image is *close* to the -1 point, then it is *close* to changing its number of encirclements of that point. This *closeness,* or distance between the image and the -1 point, is typically measured in terms of *gain and phase margins.*

These margins simply quantify the distance between the mapping (the Nyquist plot) and the point at -1, measured in terms of a magnitude (along the real axis), or a phase angle (along the unit circle). The phase margin is measured at the *gain-crossover frequency,* or the frequency at which the Nyquist plot crosses the unit circle (i.e., has unity magnitude). The gain margin is measured at the *phase-crossover frequency* at which the phase of the Nyquist plot is -180 degrees (or crosses the negative real axis). For more discussion of Nyquist theory and these stability margins, see Refs. 1 and 3.

EXAMPLE 12.4

Applying the Nyquist Stability Criterion

Let $g(s)$, the system to be controlled, be the system used in Example 12.1 dealing with Bode plots, or

$$g(s) = \frac{y(s)}{u(s)} = \frac{1}{s^2 + s + 1}$$

and let the *feedback-control law* shown in Figure 12.4 be

$$u(s) = k(s)\left(y_{\mathrm{C}} - y\right) = K\left(y_{\mathrm{C}} - y\right) = 2\left(y_{\mathrm{C}} - y\right)$$

or the *feedback gain* $K = 2$ here. Plot the Nyquist plot for the system, find the gain and phase margins, and find the gain- and phase-crossover frequencies.

■ Solution

The Nyquist plot for the open-loop system $k(s)g(s)$ is obtained using MATLAB.

```
»num=[2];den=[1 1 1];olsys=tf(num,den)        Form the open-loop system description
Transfer function:
     2
-----------
s^2 + s + 1
»nyquist(olsys)                               Plot the Nyquist plot
»grid
»[Gm,Pm,Wcp,Wcg] = MARGIN(olsys)              Find the gain and phase margins and
                                              gain-crossover frequency
Gm =
  Inf
Pm =
  49.35368062792559
Wcp =
  NaN
Wcg =
  1.51748991355198
```

The Nyquist plot is shown in Figure 12.7. Note that there are no open-loop poles in the right-half plane, and the plot (image of D) does not encircle the point at -1. So the closed-loop system is stable. Furthermore, we have used the MARGIN command in MATLAB. Since the Nyquist plot passes through the origin, or crosses the real axis at the origin, the gain margin is infinite and the phase-crossover frequency is not defined (NaN). The phase margin, however, is about 49 deg, and the gain-crossover frequency is about 1.5 rad/sec.

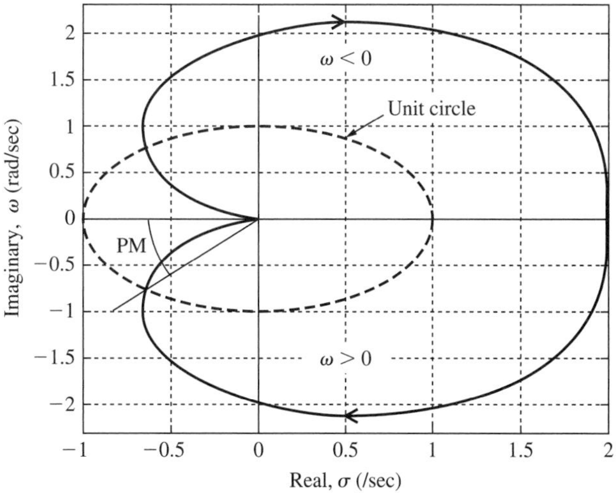

Figure 12.7 Example 12.4 Nyquist plot.

12.1.3 The Loop-Shaping Technique

We are now ready to discuss the loop-shaping approach to the design of feedback systems. In this discussion, we will still consider the generic feedback system shown in Figure 12.4, known as a *regulator* or *tracking controller*. Modifying the feedback system slightly, we now depict the system as in Figure 12.8. In this figure a generic disturbance D and a generic sensor noise N are shown, and the control objective may be stated as keeping the error e small in the presence of commands, disturbances, and sensor noise.

From the figure and some algebra, we see that the response of the closed-loop system may be expressed as

$$\boxed{y(j\omega) = \left[\frac{kg}{1+kg}\right](y_C + N) + \left[\frac{1}{1+kg}\right](D)} \tag{12.13}$$

So the response to the commands may be written as

$$\boxed{\frac{y}{y_C}(j\omega) = \frac{kg(j\omega)}{1+kg(j\omega)}} \tag{12.14}$$

The response to sensor noise is also

$$\boxed{\frac{y}{N}(j\omega) = \frac{kg(j\omega)}{1+kg(j\omega)}} \tag{12.15}$$

And the response to the external disturbance is

$$\boxed{\frac{y}{D}(j\omega) = \frac{1}{1+kg(j\omega)}} \tag{12.16}$$

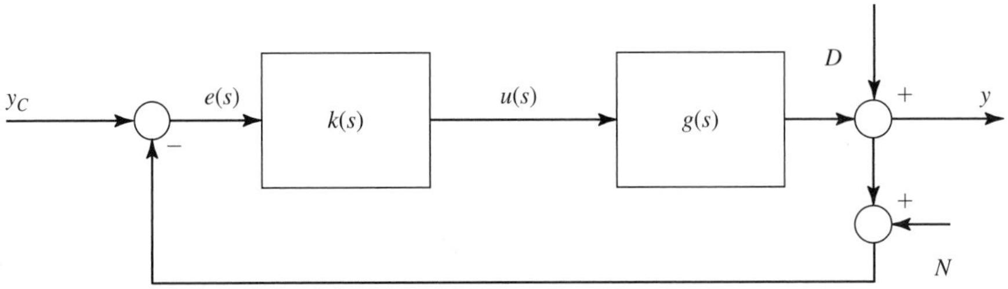

Figure 12.8 Generic feedback system with disturbance and sensor noise.

Now the control objective of the generic feedback system may be restated as one of making y track the commanded y_C, while keeping the responses due to the external disturbance and the sensor noise small. In addition, the system must be robustly stable. Let's consider responses to the commands and the disturbance first, or Equations (12.14) and (12.16).

To make y track the commanded y_C means that the complex number given by the right side of Equation (12.14) must be approximately unity. This can be accomplished by making the magnitude of $kg(j\omega)$ large compared to one. That is, when the magnitude of $kg(j\omega)$ is large, y will approximately equal y_C. Observe also that when the magnitude of $kg(j\omega)$ is large, the complex number given by Equation (12.16) will be small. So the system will respond little to the disturbance (*disturbance rejection*). This was the second of our objectives.

Unfortunately, making the magnitude of $kg(j\omega)$ large also makes the complex number given by Equation (12.15) approximately unity. So it would appear that the system would also respond to sensor noise. To satisfy the third objective, or to deal with the sensor noise, we will take into account the fact that we cannot (and we don't want to) make the magnitude of $kg(j\omega)$ large over all frequencies. We only need this magnitude to be large over the frequency range of the commands y_C and the disturbances D. If in this frequency range the sensor noise is low, then the fact that the complex number given by Equation (12.15) is approximately unity is not a problem. Usually, however, sensor noise can be large at high frequencies. So at high frequencies we will want the magnitude of $kg(j\omega)$ to be small. (And we must use sensors with low noise at low frequency.)

Taking all the above into account, we can depict the desired "loop shape" as shown in Figure 12.9. The magnitude of $kg(j\omega)$, or the *loop gain,* is large over the low-frequency range, small in the high-frequency range, and finally the loop gain must transition between large and small in the mid-frequency range. And the frequency at which the magnitude of $kg(j\omega)$ equals unity, or 0 dB, is of course the gain-crossover frequency ω_c.

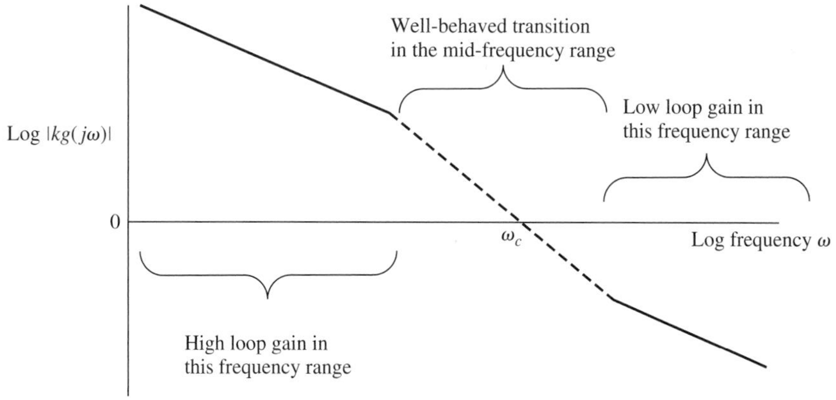

Figure 12.9 Desired "loop shape".

Though we have the desired loop shape determined qualitatively, it is important to point out the key design variables that the designer of the feedback system can manipulate. The first, and most important, is the gain-crossover frequency ω_c. By selecting this frequency, the designer is defining the mid-frequency range and, equally important, the high- and low-frequency ranges. The designer must select this crossover frequency in part based on understanding the frequency content (or *spectra*) of the commands and/or disturbances. (These frequency spectra, in the context of automatic flight control, are discussed in Section 12.3.)

The second aspect of the design the designer must determine is the actual shape of the loop, especially in the mid-frequency range. The detailed shape at high and low frequency is not as critical as the shape in the mid range, as long as the gain is "large" at low frequency and "small" at high frequency. (Look again at Equations (12.14–12.16) to see why.) But the loop shape in the mid-frequency range is determined from Nyquist's stability criteria—the system must have good closed-loop stability margins. These margins are determined by the magnitude and phase of the open-loop system (kg) in this mid-frequency range. So the control law and control compensation $k(s)$ must be chosen by the designer to yield good margins in this mid-frequency range.

Typical compensators (or *filters*) used for this shaping are summarized in Table 12.1, in which the transfer functions of the compensators are listed, and

Table 12.1 Typical Compensation Used for Loop Shaping

Compensator	Transfer Function	Contribution to Bode Magnitude Plot
Gain or proportional	K	
Integration	K/s	
Proportional plus integral (PI)	$\frac{K(s+z)}{s}$	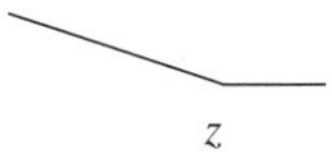
Lead-lag	$K\frac{s+z}{s+p}$, $\lvert z\rvert < \lvert p\rvert$	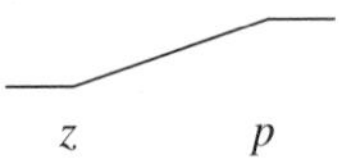
Washout or high pass	$K\frac{s}{s+p}$	
Lag-lead	$K\frac{s+z}{s+p}$, $\lvert z\rvert > \lvert p\rvert$	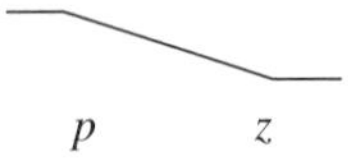
First-order lag or low pass	$K\frac{p}{s+p}$	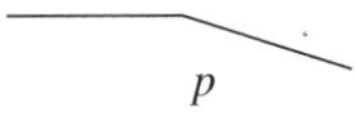

the effect of each compensator on the Bode magnitude plot is sketched. These filters are usually implemented using analog circuits or digital devices. We will not address many of the details of implementing these compensators, but note that implementation issues limit some aspects of a compensator.

For example, a lead-lag filter adds phase lead, especially over the frequency range between z and p. But the maximum amount of phase lead such a filter can generate is usually limited to approximately 70–75 degrees. Similar factors limit the maximum phase lag a lag-lead can generate to less than 90 degrees. Likewise, there are issues that relate to building a stable integrator (by definition the pole is very close to the imaginary axis). So frequently an integrator is not implemented, but a different sensor is used instead. For example, a rate gyro senses pitch rate, while an attitude gyro senses attitude angle. These two measurements are related mathematically through an integration operation, but we don't need to integrate the rate gyro output to sense attitude.

At first it seems that both open-loop (kg) magnitude and phase must be shaped in the mid-frequency range to achieve good stability margins. But it turns out that the designer may usually achieve good stability margins by only shaping the open-loop magnitude. The reason for this can be traced to Bode, who showed that the magnitude and phase are not independent. In fact, for a stable open-loop system kg they are related by the *Bode gain and phase relationship* (Ref. 3) given in Equation (12.17).

$$\arg\big(kg(j\omega)\big) = \pi \int_0^\infty \frac{d\big(20\log \mathrm{Mag}(kg)\big)}{d(\log\omega)}\big(\arctan(\omega - \omega_c)\big)d\omega \tag{12.17}$$

Though cumbersome in appearance, this equation can be interpreted quite simply. The term $d(20\log \mathrm{Mag}(kg))/d(\log\omega)$ in the integrand is just the slope of the magnitude of $kg(j\omega)$ versus frequency in dB and log scales, respectively. Or this is the slope of the magnitude portion of the Bode plot. The arc-tangent term in the integrand has the effect of magnifying the values of the integrand in the frequency range near gain crossover, ω_c. So, for example, if in the frequency range near gain crossover the slope of the Bode magnitude plot was -20 dB per decade of frequency (or -1 on the dB and log-frequency scales), the right-hand side of Equation (12.17) would equal approximately $-\pi$. Therefore the phase margin, measured at gain crossover, would be about 90 degrees.

This observation leads to the loop-shaping designer rule of thumb that states, To achieve good phase margin, make the slope of the magnitude plot near crossover approximately -20 dB per decade frequency. So assuming we have a good loop shape, the following approximation is valid near gain crossover:

$$kg(j\omega) \approx \frac{\omega_c}{(j\omega)} \tag{12.18}$$

or

$$kg(s) \approx \frac{\omega_c}{s}.$$

This is a very powerful observation because it implies that if the designer has indeed applied the rule of thumb and achieved the desired loop shape shown in Figure 12.9, then the closed-loop system's transfer function is approximately

$$\boxed{\frac{y(s)}{y_c(s)} \approx \frac{\omega_c/s}{1+\omega_c/s} = \frac{\omega_c}{s+\omega_c}} \tag{12.19}$$

Or the closed-loop system has excellent stability margins and is (approximately) a function of only one parameter—the gain-crossover frequency ω_c! Furthermore, the speed of response of the closed-loop system is seen to be tied directly to ω_c. Therefore, the most important parameter for the designer to select is the gain-crossover frequency.

If the performance requirements are met by a feedback system, it is difficult to have too low a gain-crossover frequency. But it is certainly possible to have a crossover frequency that is too high. The maximum achievable crossover frequency is limited by control power (speed and authority of the control effectors) and by the presence of unmodeled dynamics and noise in the higher-frequency range. (Recall that's why the magnitude plot must decrease or "roll off" in this frequency range.)

EXAMPLE 12.5

A Loop-Shaping Design

As the final example in this section, let the system to be controlled be

$$g(s) = \frac{1}{s^2+s+1} \tag{12.20}$$

which is the same system considered in the previous examples. Let's assume that the frequency content of the command y_C and the disturbance D is primarily below 0.5 rad/sec. Suggest a control law $k(s)$ that should lead to good performance and has good stability margins.

■ Solution

If the command y_C and the disturbance D have most of their power in frequencies below 0.5 rad/sec, the gain-crossover frequency ω_c should be above 0.5 rad/sec for good tracking and disturbance-rejection performance. Let's initially let $k_1(s) = 2$, as in the Nyquist analysis in Example 12.4, and recall that in that example the gain-crossover frequency was $\omega_c \approx 1.5$ rad/sec. The Bode plot for this open-loop system $k_1 g(j\omega)$ is again shown in Figure 12.10.

Note that this Bode plot does not have the desired loop shape. Although the phase margin is about 50 deg, which might be acceptable, the loop gain, or $|kg|$, is not very large at low frequencies (below 1 rad/sec). So the tracking and disturbance-rejection performance will be poor. To make the loop gain larger at low frequency we can introduce integral action by using a proportional-plus-integral (PI) controller, for example. So let's now select the control compensator to be

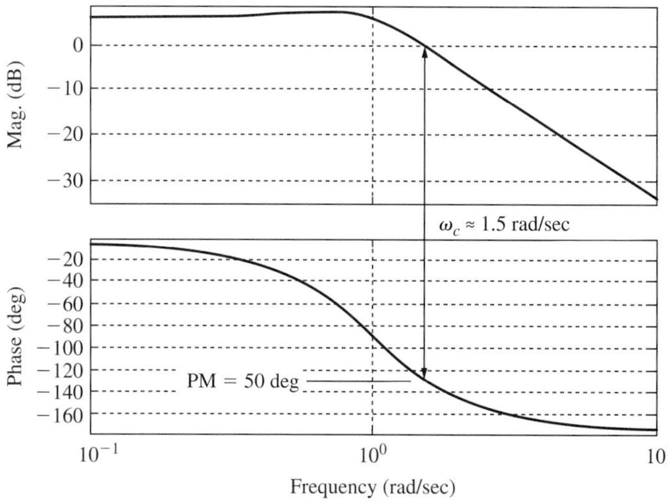

Figure 12.10 Initial loop shape.

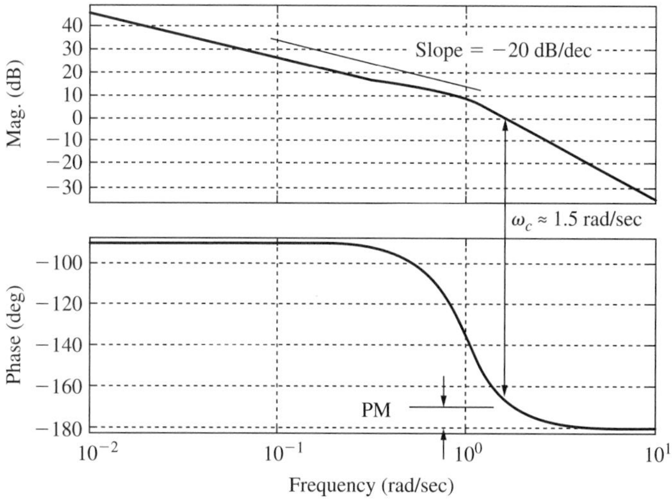

Figure 12.11 Effect of adding PI compensation.

$$k_2(s) = K\frac{(s+a)}{s} \tag{12.21}$$

If we chose the proportional gain $K = 2$, and the integral gain $Ka = 2$ (which makes $a = 1$), the Bode plot for this new open-loop system $k_2 g(j\omega)$ is shown in Figure 12.11.

The loop gain is now high at low frequency—in fact, the gain goes to infinity as $\omega \to 0$. And the gain-crossover frequency remains around 1.5 rad/sec. But the phase margin is now only about 10 deg. This is consistent with the fact that the slope of the magnitude plot near gain crossover is much more negative than −20 dB per decade.

To increase this phase margin, we could reduce the gain K so that crossover occurs where the slope of the magnitude plot is -20 dB per decade. But this would reduce the tracking and disturbance-rejection performance. If the performance achieved by using this lower feedback gain is adequate, this approach would be preferred.

If not, a second approach would be to keep gain crossover at around 1.5 rad/sec and introduce lead compensation of the form

$$k_{\text{lead}}(s) = K_{\text{lead}} \frac{(s+z)}{(s+p)} \tag{12.22}$$

with p and z both positive, and $|z| < |p|$. (Consider the Bode plot of the compensator, and think about how this compensation will change the slope of the magnitude plot of $kg(j\omega)$.) With this lead compensator added, the total compensation is now of the form

$$k_3(s) = \left(\frac{K(s+a)}{s}\right)\left(\frac{K_{\text{lead}}(s+z)}{(s+p)}\right) \tag{12.23}$$

Now select K_{lead} such that the gain-crossover frequency is not changed from its previous value, or ω_c stays at about 1.5 rad/sec. Selecting $z = 0.8$, $p = 5$, and $K_{\text{lead}} = 3.75$ yields the open-loop Bode plot shown in Figure 12.12. Note from this Bode plot that the loop shape (magnitude plot) is closer to the desired loop shape, and with the lead compensation added the phase margin has been increased to about 55 deg. Adjusting the lead compensator could increase the phase margin further, but this margin may be satisfactory. The gain margin is infinite.

The step response for this closed-loop system is shown in Figure 12.13. Note the rise time of about 1 sec is consistent with our gain-crossover frequency of about 1.5 rad/sec, and the system exhibits a little overshoot consistent with a phase margin of 55 deg.

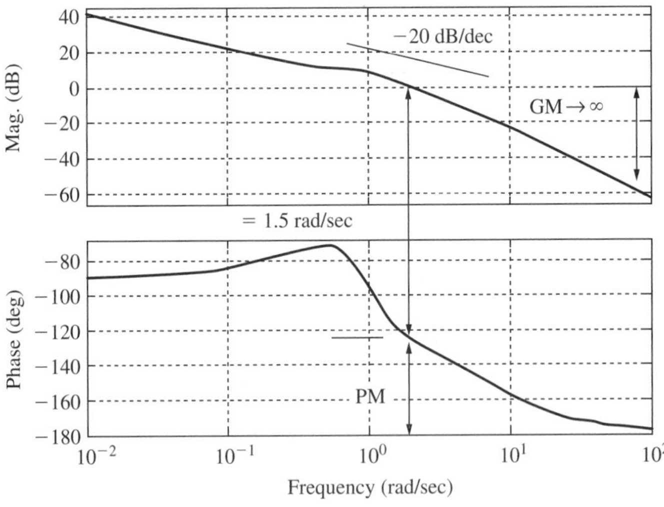

Figure 12.12 Final loop shape.

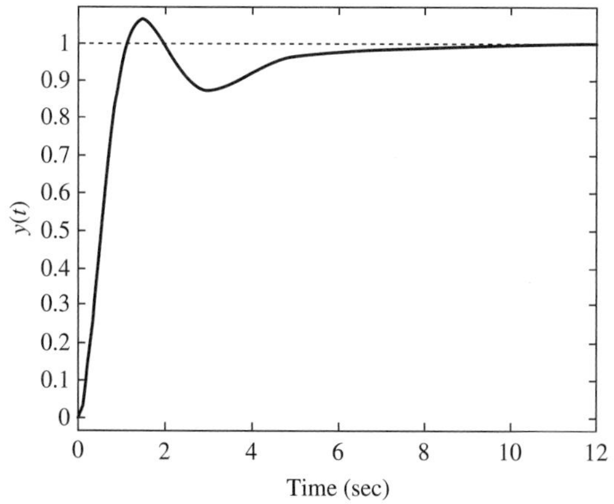

Figure 12.13 Example 12.5 step response.

Hopefully, the discussion in this section has demonstrated the utility of the loop-shaping design method. It maps closed-loop performance and stability-robustness requirements entirely onto the shape of the magnitude plot of the open-loop system, and clearly brings out the importance of the gain-crossover frequency. In the time domain, the speed of response is directly tied to crossover frequency, while overshoot is tied to gain and phase margins.

12.2 INNER AND OUTER LOOPS, AND FREQUENCY SEPARATION

Most physical systems have a natural frequency separation between their responses. The wise feedback-systems designer appreciates and takes advantage of this fact. For example, think about how an aircraft changes altitude. First the pitch attitude and angle of attack is changed so that the total lift generated by the vehicle is modified. And this change in lift corresponds to a change in the vehicle's vertical acceleration. But this vertical acceleration must be integrated twice to achieve any change in altitude. Consequently, we can see that an aircraft's pitch-attitude or angle-of-attack response occurs on a much faster time scale than its altitude response. The same can be said about bank angle and lateral position.

Speaking more abstractly, let's consider two responses of some dynamic system, y_{Fast} and y_{Slow}. Let's assume the transfer functions defining the responses to a single control input are given by

$$\begin{aligned} \frac{y_{\text{Fast}}(s)}{u(s)} &\triangleq g_{\text{Fast}}(s) \\ \frac{y_{\text{Slow}}(s)}{u(s)} &\triangleq g_{\text{Slow}}(s) \end{aligned} \tag{12.24}$$

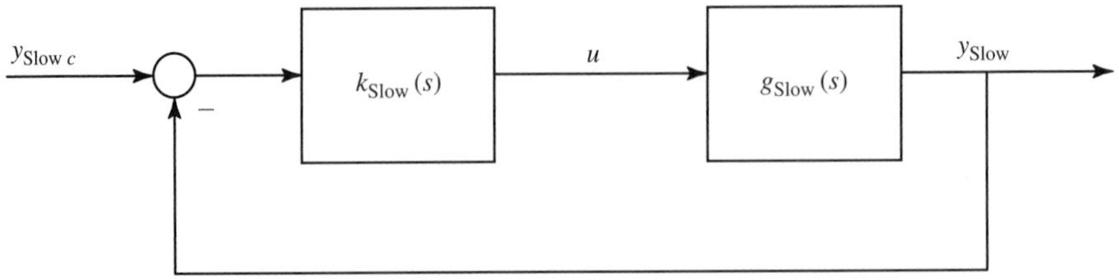

Figure 12.14 Candidate control system for y_{Slow}.

We might ask what is the best approach to control y_{Slow} using a feedback system? Of course, one approach is to proceed directly in one step to design a single control law such as

$$u(s) = k_{\text{Slow}}(s)\left(y_{\text{Slow } c}(s) - y_{\text{Slow}}(s)\right) \tag{12.25}$$

corresponding to the block diagram given in Figure 12.14.

However, let us now assume that the two responses y_{Fast} and y_{Slow} are related to one another through at least one integration, as in the case of flight-path angle and altitude, for example. And let us assume further that a change in y_{Fast} naturally causes a change in y_{Slow}, with the relationship between the two given by

$$\frac{y_{\text{Slow}}(s)}{y_{\text{Fast}}(s)} = \frac{y_{\text{Slow}}(s)/u(s)}{y_{\text{Fast}}(s)/u(s)} = \frac{g_{\text{Slow}}(s)}{g_{\text{Fast}}(s)}$$

Therefore, note that

$$\frac{y_{\text{Slow}}(s)}{u(s)} = g_{\text{Slow}}(s) = \left(\frac{g_{\text{Slow}}(s)}{g_{\text{Fast}}(s)}\right) g_{\text{Fast}}(s) \tag{12.26}$$

Under the assumption made about the integral relationship between y_{Fast} and y_{Slow}, we know that the denominator of $g_{\text{Slow}}(s)/g_{\text{Fast}}(s)$ is <u>at least of order one higher in s that the numerator</u>.

Now consider the following control-design strategy. Since a change in y_{Fast} naturally leads to a change in y_{Slow}, let's first design an *inner-loop* control system for controlling y_{Fast}, and then design a second *outer-loop* control system for controlling y_{Slow}. Or let's use the architecture depicted in Figure 12.15, with the inner loop depicted inside the dashed box. Although this seems like a more complex and convoluted approach, it is usually much simpler to design, and it usually leads to a more robust closed-loop system.

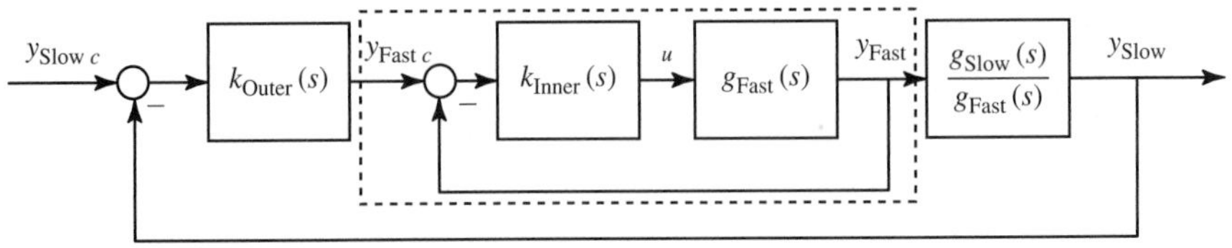

Figure 12.15 Inner- and outer-loop structure.

Following this design procedure, one synthesizes $k_{\text{Inner}}(s)$ such that y_{Fast} follow or tracks $y_{\text{Fast}\,c}$. Now note that this makes the inner-loop system in the dashed box have a transfer function of

$$\frac{y_{\text{Fast}}(s)}{y_{\text{Fast}\,c}(s)} \approx \frac{\omega_{c\ \text{inner}}}{s + \omega_{c\ \text{inner}}} \approx 1 \text{ for frequencies below } \omega_{c\ \text{inner}} \tag{12.27}$$

So this transfer function is approximately unity below the inner-loop gain-crossover frequency $\omega_{c\ \text{inner}}$.

This fact makes the preliminary design of the outer loop much easier. One simply sets the transfer function for the inner loop (in the dashed box) equal to unity, and then designs $k_{\text{Outer}}(s)$ to make y_{Slow} follow or track $y_{\text{Slow}\,c}$. And since $g_{\text{Slow}}(s)/g_{\text{Fast}}(s)$ will be simpler than $g_{\text{Slow}}(s)/u(s)$, this design step is much more straightforward than the original one-step design approach.

A word of caution. The designer must guarantee that the gain-crossover frequency of the outer loop $\omega_{c\ \text{outer}}$ is less than the inner-loop crossover $\omega_{c\ \text{inner}}$. That is, *crossover-frequency separation* must be preserved. Otherwise the assumption that the closed-loop transfer function for the inner loop in the dashed box equals unity would be violated! But keeping $\omega_{c\ \text{outer}}$ less than $\omega_{c\ \text{inner}}$ is perfectly consistent with the fact that y_{Fast} responds much more quickly than y_{Slow}. So providing this frequency separation is perfectly consistent with the physics of the system.

Providing crossover-frequency separation also allows the inner and outer loops to essentially be designed separately. In particular, let's consider the effects of adding an outer loop on the Bode/Nyquist analysis of the inner loop. When initially designing the inner loop (in the absence of the outer loop), loop shaping will result in the Bode magnitude of the open-loop transfer function of the inner loop $|k_{\text{Inner}}g_{\text{Inner}}|$ being large over the frequency range below the gain-crossover frequency of this loop $\omega_{c\ \text{inner}}$. Plus, when independently designing the outer loop (while assuming the closed-loop transfer function of the inner loop is approximately unity), loop shaping will result in the Bode magnitude of the open-loop transfer function of the outer loop $|k_{\text{Outer}}g_{\text{Outer}}|$ being small (or "rolling off") in the frequency range above the gain-crossover frequency of this loop $\omega_{c\ \text{outer}}$. Then if the designer assures frequency separation between the inner and outer loops, or guarantees that $\omega_{c\ \text{outer}} < \omega_{c\ \text{inner}}$, closing the outer loop around the inner will only slightly modify the open-loop Bode plot for the inner loop, and any modification will occur only in the frequency range below $\omega_{c\ \text{inner}}$—where the magnitude was large to begin with. Hence, with proper loop shaping and frequency separation, the addition of the outer loop will have little or no effect on the gain-crossover frequency or stability margins initially obtained in the inner loop.

Likewise, a similar analysis of the effects of the presence of the inner loop on the Bode/Nyquist analysis of the outer loop will show that with proper loop shaping and frequency separation, the presence of the inner loop will have little or no effect on the gain-crossover frequency or stability margins initially obtained in the outer loop. We will use this inner- and outer-loop approach throughout this chapter, and it is used widely in practice.

12.3 THE FLIGHT-DYNAMICS FREQUENCY SPECTRA

Recall in the discussion of loop shaping we noted that the designer must be aware of the frequency spectra of the commands and disturbances because it influences the selection of the gain-crossover frequency. To gain an appreciation for the typical frequency spectra encountered in aircraft flight control, consider Figure 12.16. The frequency ranges shown are qualitative, so no specific frequencies are shown on the plot. However, the relative magnitudes of the frequency ranges shown are appropriate.

The first row under the graph corresponds to the natural longitudinal modes of the vehicle, including surface actuators and propulsion systems. Note the frequency separation between the short-period and phugoid modes, and note the vehicle responses that are dominant in the modal responses. Also note that sometimes the surface actuators are sufficiently fast so as to excite the structural modes.

The second row under the picture contains the typical bandwidths (or gain-crossover frequencies) for the various controlled responses. For example, the bandwidth of a pitch-attitude control system is typically near the natural

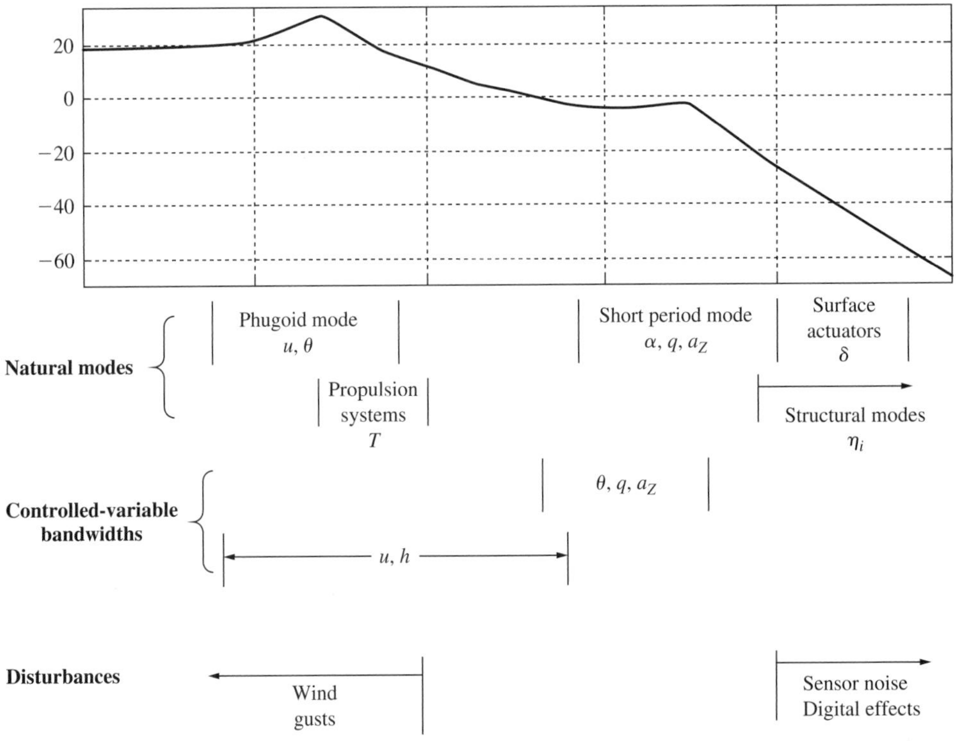

Figure 12.16 Frequency spectra for the longitudinal axes.

frequency of the short-period mode, which dominates the pitch response, while an altitude-control-system's bandwidth is naturally below the short-period natural frequency, since pitch rotation is required to generate the lift leading to the vertical acceleration.

The final row indicates the frequency ranges for important disturbances that can excite the feedback control system, and which the feedback system must reject. These disturbances include steady winds and gusts, as well as sensor noise and digital effects such as aliasing. Recall that the magnitude of the open-loop Bode plot needs to be large over the frequency range of the disturbances to be rejected (e.g., the winds), and small over the frequency range where unmodeled dynamics (e.g., digital effects) might destabilize the feedback system. We will draw from these concepts frequently as we discuss longitudinal autopilots.

A similar figure corresponding to the lateral-directional axes is shown in Figure 12.17. The same three rows of frequency ranges are again shown: the natural modes, the controlled responses, and the disturbances. One key issue to point out here is that the important roll-subsidence mode and the "nuisance" dutch-roll mode both fall in the same frequency range. We will frequently draw

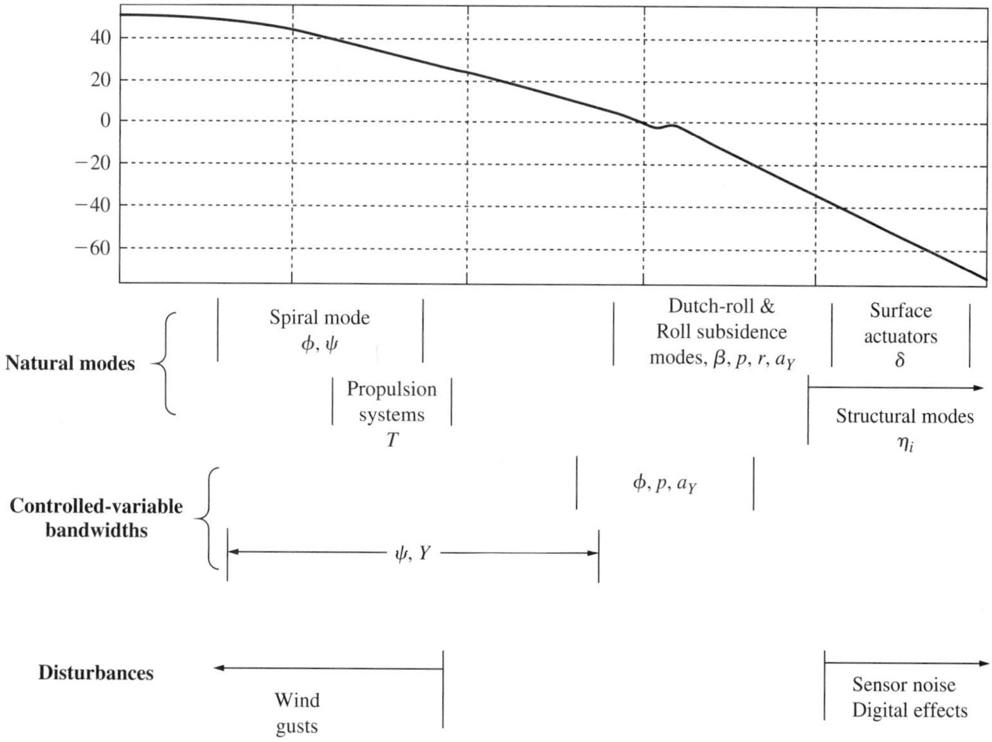

Figure 12.17 Frequency spectra for the lateral-directional axes.

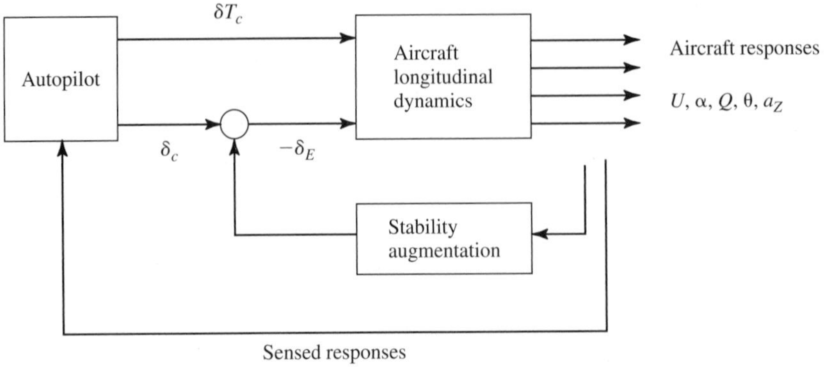

Figure 12.18 Typical control architecture for an autopilot.

from the concepts presented in this figure as we discuss lateral-directional autopilots.

Finally, before specifically dealing with the various autopilot modes, we stress again that the flight vehicle is a multi-input/multi-output system, as discussed in Chapter 11. Shown in Figure 12.18 is a schematic intended to emphasize this fact, where the figure is fashioned after Figure 11.8. In this earlier figure, the outer-loop control was performed by the pilot. Here, at least some of the outer-loop control functions will be performed by the autopilot.

We stress the multi-input/multi-output nature of the system here because, in the sections to follow, many of the autopilot modes will be described using a single-input–single-output block diagram. The use of this type of block diagram allows us to focus on the autopilot-feedback loop under discussion. But it needs to be clear that the addition of the feedback loop in question will certainly affect all of the vehicle's responses. And when a particular autopilot mode is evaluated, all of the vehicle responses will be considered.

12.4 ATTITUDE CONTROL

In flight dynamics, the primary responses to be controlled are pitch attitude and bank angle because every flight maneuver is predicated on the ability to control attitude. Recall the discussion in Section 12.2 regarding a change in altitude—the maneuver is initiated with a change in pitch attitude (or a change in angle of attack). Likewise, a change in lateral position is initiated with a change in bank angle. Simply put, angle of attack controls the magnitude of the lift vector, and bank angle controls the orientation of the lift vector. So the control of attitude is fundamental, and that's where we need to begin.

Throughout this discussion, we will assume a conventional vehicle configuration (i.e., elevator, aileron, and rudder for pitching-, rolling-, and yawing-moment effectors, respectively.) If we were considering another vehicle configuration, we would make appropriate adjustments in the notation corresponding to the moment effectors, but the basic overall approach would remain the same.

12.4.1 Pitch-Attitude Control

As with all autopilot modes, a pitch-attitude command mode is used to reduce pilot workload. For example, during a climb to a cruise altitude assigned by air traffic control, the autopilot mode may be engaged to hold pitch attitude with some fixed throttle setting.

We introduced the topic of pitch-attitude control in Chapter 11 (Section 11.3) to demonstrate the need for providing adequate short-period damping. We will continue this discussion here. Following the methodology presented in Section 12.1, if we want to synthesize a control law such that the pitch attitude θ follows, or tracks, the commanded pitch attitude θ_c, the key is to develop a control law that keeps the error e small in the presence of commands, disturbances, and sensor noise, where

$$e(s) = \theta_c(s) - \theta(s) \tag{12.28}$$

The basic control architecture is shown in the block diagram in Figure 12.19, consistent with the generic feedback system shown in Figure 12.8. (Note that, as in Chapter 11, the sign convention on elevator deflection has been reversed in this block diagram, such that a positive actuator command δ_a produces positive pitch acceleration.) The pitch dynamics include any necessary stability-augmentation systems such as a pitch damper and/or angle-of-attack feedback to enhance short-period frequency. Such feedback stability augmentation was discussed in detail in Chapter 11.

For now we will not focus on the source of the attitude command θ_c. It may come from the pilot's stick input, it may be dialed in by the pilot, or it may come from other sources to be discussed later. Also, for simplicity, the external disturbance D and the sensor noise S indicated in Figure 12.8 will no longer be shown in this and all later block diagrams. But in reality they are always present.

Finally, a pitch-surface servo-actuator is shown in Figure 12.19, and such an actuator will be included in all discussions of attitude control because the actuator is one of the key limiting factors in such systems. For our purposes, the actuators will be represented as a first-order lag with transfer function given by

$$\boxed{a(s) \triangleq \frac{-\delta_E(s)}{\delta_a(s)} = \frac{1}{T_a s + 1}} \tag{12.29}$$

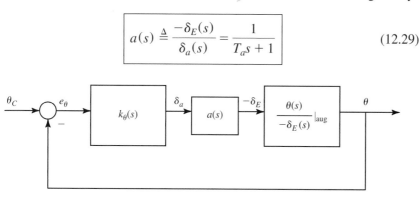

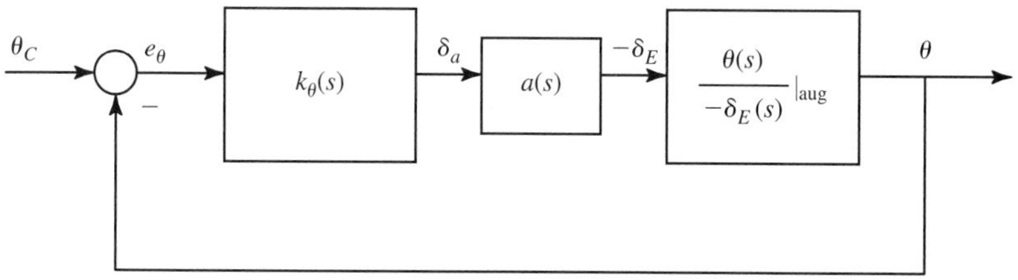

Figure 12.19 Pitch-attitude control loop.

where T_a is the actuator time constant. Actuator dynamics are actually more complex, but the use of a simple lag model captures the key characteristics.

We may now address the synthesis of the pitch-attitude-control law. Students should pay special attention to both the process, as well as how design-decisions are made. Recall from Section 11.3 that as the pitch-attitude feedback gain is increased, the damping of the short-period mode is reduced, as shown in the root locus in Figure 11.10*a*. But let's now look at the situation in terms of the Bode plot of the open-loop system, or $k_\theta(s)a(s)g(s)$ here. To make the discussion more explicit, we will use the data for the DC-8 aircraft at Flight Condition 3, a cruise condition at 33,000 ft altitude and cruise Mach number of 0.84 (U_0 = 824 fps). The vehicle weight is 230,000 lbs. (See Appendix B.) Many of the issues that will arise are generic, so they are not unique to this aircraft. The pitch-attitude transfer function in this case is

$$\frac{\theta(s)}{-\delta_E(s)} = \frac{4.57(s+0.0144)(s+0.7247)}{(s^2+0.01174s+0.0005933)(s^2+2.153s+9.896)} \text{ rad/rad} \quad (12.30)$$

Note that the frequency and damping of the short-period and phugoid modes are

$$\omega_{SP} = 3.15 \text{ rad/sec} \quad \omega_P = 0.0243 \text{ rad/sec}$$

$$\zeta_{SP} = 0.342 \qquad \zeta_P = 0.241$$

The Bode plot of $\theta(j\omega)/\delta_a(j\omega)$ ($= kag(s)$ with $k_\theta = K_\theta = 1$ rad/rad), which includes the actuator, is given in Figure 12.20. Note in the figure, reference lines indicate a -20 dB/decade slope of the magnitude plot and -90 deg phase angle.

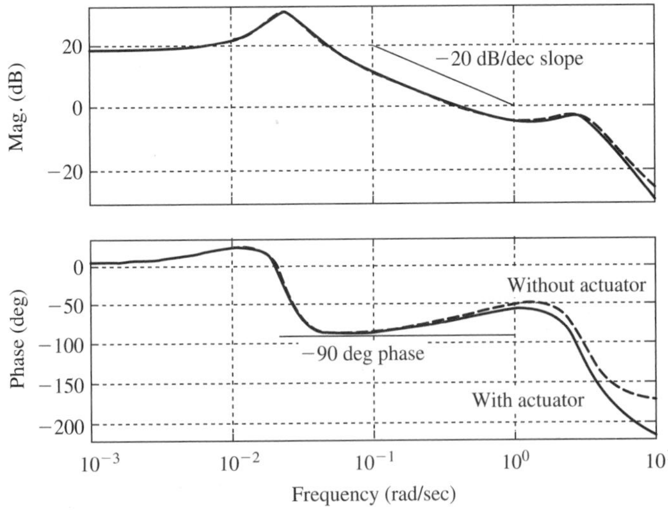

Figure 12.20 Bode plots of pitch-attitude transfer functions.

Also note that two Bode plots are shown. The dashed curves correspond to the pitch-attitude transfer function without the actuator, or Equation (12.30). Both sets of plots were included to show how the actuator affects the phase at frequencies well below the magnitude of the actuator eigenvalue, which in this case is 15 /sec.

From these Bode plots we can see that if we adjust the control gain K_θ such that the gain-crossover frequency ω_c is in the frequency range of about 0.06 to 1.0 rad/sec, the slope of the magnitude plot at ω_c would be approximately -20 dB/dec, and the feedback loop would have a robust phase margin of at least 90 deg. However, if ω_c is much below 1 rad/sec, it is unlikely that this control law would provide acceptable performance for two reasons. First, the loop gain $|kag(j\omega)|$ may not be sufficiently large at frequencies below gain crossover. Second, the gain-crossover frequency may not be sufficiently high such that the speed of response is fast enough. Recall that the time constant for the closed-loop system is approximately $T_{CL} \approx 1/\omega_c$.

We will attempt to eliminate both deficiencies by using a higher gain-crossover frequency and seeing if we can avoid adding integral action in the controller $k_\theta(s)$. But increasing the gain-crossover frequency ω_c requires care. We may certainly increase the control gain K_θ such that gain crossover ω_c occurs at about 1–2 rad/sec. But if we need ω_c to be higher, the gain margin will be degraded, leading to a system that will be quite sensitive to the gain K_θ. Note in the Bode plots how the phase drops sharply at about 3 rad/sec, and the slope of the magnitude plot at this frequency is not the required -20 dB/dec. This characteristic corresponds to the situation that was revealed in the root-locus analysis in Section 11.3, in which the short-period poles move closer to the imaginary axis as attitude-control gain is increased.

But we know how to increase the short-period damping, by adding a pitch damper. In terms of the effect of increased short-period damping on the Bode plot, recall in general that as modal damping is increased, any peaking in the magnitude plot at the corner frequency (short-period frequency here) is reduced, and the 180 deg phase transition occurs more gradually. Both are desired results.

After reviewing the discussion in Section 11.3.1, we know that in the context of the attitude-control block diagram given in Figure 12.19, a pitch damper would have the following control law.

$$\delta_a(s) = \delta_c(s) - K_q q(s) \tag{12.31}$$

Where now $\delta_c(s)$ is the output of the attitude controller $k_\theta(s)$ in Figure 12.19, and the pitch rate is sensed using a rate gyro. The pitch-attitude controller with such a pitch damper is depicted in the block diagram given in Figure 12.21. The attitude controller may still be represented as in Figure 12.19, but the pitch-attitude transfer function in that figure would become $\theta(s)/\delta_c(s)$, which includes the pitch damper and the actuator.

Following the approach taken in Example 11.3.1, and using an actuator with a time constant $T_a = 0.05$ sec, a pitch-damper gain is selected that yields a high

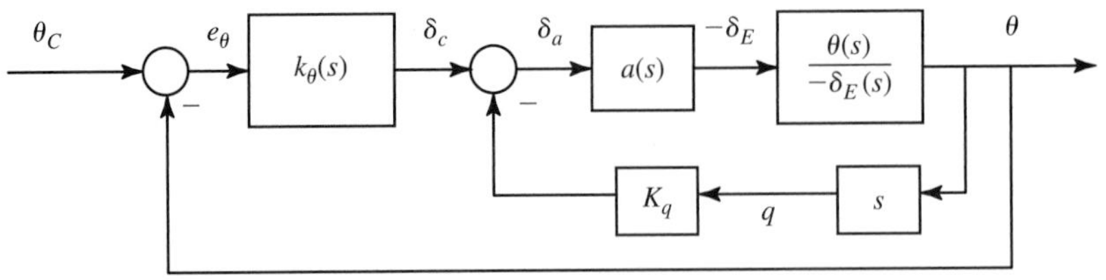

Figure 12.21 Pitch-attitude controller with pitch damping.

short-period damping, since we know that when the pitch-attitude loop is closed this damping with be reduced. A pitch-damper gain K_q of 1.0 rad/rad/sec yields the following set of eigenvalues

$$\lambda_{\text{SP 1}} = -3.0962 \text{ /sec}$$
$$\lambda_{\text{SP 2}} = -7.2041 \text{ /sec}$$
$$\lambda_P = -0.00614 \pm j0.02016 \text{ /sec}$$
$$\lambda_a = -11.85 \text{ /sec}$$

and we will proceed using this pitch-damper gain, which may be fine-tuned later if necessary. The MATLAB commands used to select this gain are:

»sys=ss(a,-b,c,d);	Forms the state-variable model of vehicle from matrices **a**, -**b**, **c**, and **d**
»dena=[1 20];aa=tf(20,dena)	Defines the actuator transfer function
»sysaa=series(aa,sys)	Forms the state-variable model for vehicle w. actuator
»[aaa,ba,ca,da]=ssdata(sysaa)	Lists the **A**, **B**, **C**, and **D** matrices for the vehicle w. actuator
»aaaug=aaa-ba*1*ca(4,:)	Finds the augmented **A** matrix for system with pitch-damper
»eig(aaaug)	Determines the eigenvalues of the augmented **A** matrix

The Bode plot of $\theta(s)/\delta_c(s)$, which includes the actuator and pitch damper, is shown in Figure 12.22. Also shown for reference is the Bode plot with the pitch damper removed. Note how the pitch damper has smoothed the 180 deg phase transition near 3.2 rad/sec, the open-loop short-period frequency. This Bode plot suggests that the gain-crossover frequency ω_c in the pitch-attitude loop may now be increased, since the phase reduction in this frequency range is now much more gradual. We could increase crossover to about 5 rad/sec, if necessary. But that would be close to the maximum gain-crossover frequency allowable without additional phase compensation, limited by a minimum allowable phase margin of approximately 60 deg. Plus, the higher the crossover frequency, the greater the elevator deflections and rates.

Shown in Figure 12.23 is the same Bode plot for the system including the pitch damper, but we have enlarged the region near the desired gain-crossover

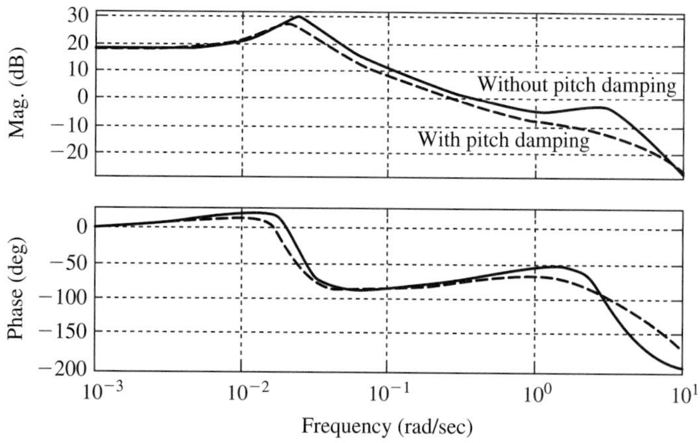

Figure 12.22 Effect of pitch damper on pitch-attitude bode plot.

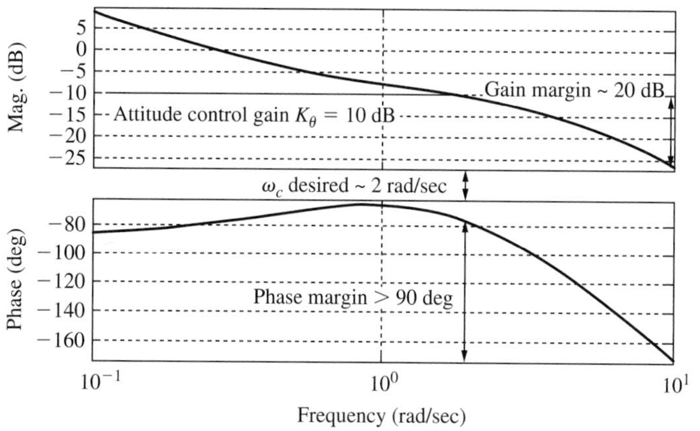

Figure 12.23 Bode plot enlarged near desired gain-crossover frequency.

frequency. Note that if the Bode magnitude plot is uniformly raised 10 dB (or 3.16 rad/rad), gain crossover will occur at about 2 rad/sec, and this sets the required value for the pitch-control gain K_θ. At this value of gain, the phase margin is over 90 deg. Also, the gain margin will be ~20 dB at the phase-crossover frequency slightly above 10 rad/sec. Both are robust stability margins.

For reference, the Nyquist plot for the same attitude-control loop with the pitch damper, or $kag_{\text{aug}}(s)$, is shown in Figure 12.24. This plot confirms the stability margins just obtained from the Bode plot. But since the Nyquist plot really doesn't offer much more additional information over the Bode plot in Figure 12.23, we will seldom show the Nyquist plots.

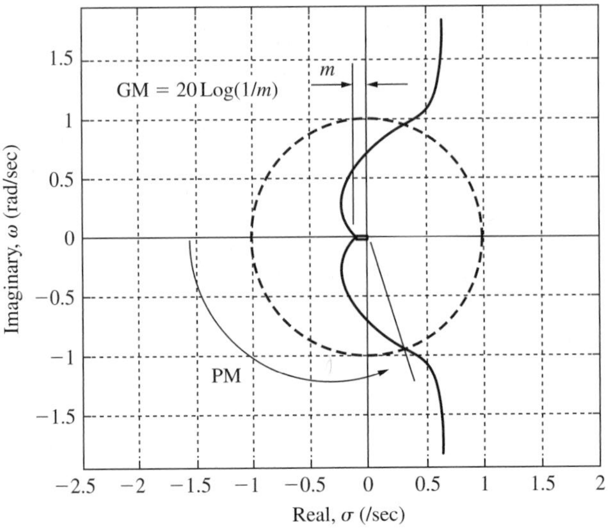

Figure 12.24 Nyquist plot of pitch-attitude control loop (w. pitch damper).

With the pitch-attitude gain selected, we may now assemble the model for the closed-loop pitch-attitude-controlled system. Assume that the state-variable representation of the aircraft, including the pitch damper, is given as

$$
\begin{aligned}
\mathbf{x}^T &= \begin{bmatrix} u & \alpha & \theta & q & \delta_E \end{bmatrix} \\
\dot{\mathbf{x}} &= \mathbf{A}_{\mathrm{PD}}\mathbf{x} + \mathbf{b}_{\mathrm{PD}}\delta_c \\
\mathbf{c}_\theta &= \begin{bmatrix} 0 & 0 & 1 & 0 & 0 \end{bmatrix} \\
d &= 0
\end{aligned}
\tag{12.32}
$$

Here $\mathbf{A}_{\mathrm{PD}}$ and $\mathbf{b}_{\mathrm{PD}}$ reflect the aircraft's longitudinal dynamics plus actuator, augmented by the pitch damper, as discussed above.

Then the state-variable representation of the closed-loop pitch-attitude-controlled system may be determined as follows:

$$
\begin{aligned}
\delta_c &= K_\theta(\theta_c - \theta), \text{ pitch-attitude control law} \\
\dot{\mathbf{x}} &= \mathbf{A}_{\mathrm{PD}}\mathbf{x} + \mathbf{b}_{\mathrm{PD}}K_\theta(\theta_c - \theta) \\
&= (\mathbf{A}_{\mathrm{PD}} - \mathbf{b}_{\mathrm{PD}}K_\theta\mathbf{c}_\theta)\mathbf{x} + (\mathbf{b}_{\mathrm{PD}}K_\theta)\theta_c
\end{aligned}
\tag{12.33}
$$

These calculations are easily performed in MATLAB using the following commands.

»sysaug=ss(aaaug,ba,ca,da)	Define state-variable model for augmented vehicle with pitch damper (discussed previously)
»acl=aaaug-ba*3.16*ca(3,:)	Closed-loop $\mathbf{A}_{CL\theta}$ matrix w. pitch controller
»bcl=ba*3.16*	Closed-loop $\mathbf{B}_{CL\theta}$ matrix w. pitch controller
»syscl=ss(acl,bcl,ca,da)	Define state-variable model for closed-loop system with pitch controller

After completing these calculations, we obtain the following closed-loop eigenvalues (for the DC-8 at Flight Condition 3) for the pitch-attitude-controlled system.

$$\lambda_{P\,1} = -0.0150\ /\text{sec}$$
$$\lambda_{P\,2} = -0.4204\ /\text{sec}$$
$$\lambda_{SP} = -3.4212 \pm j4.6499\ /\text{sec}$$
$$\lambda_a = -14.886\ /\text{sec}$$

Notice that the short-period damping has been reduced from >1 to approximately 0.6, and the two phugoid poles have become real. Both of these results were as expected, based on the discussion in Section 11.3.

The frequency response of the closed-loop attitude-controlled system is shown in Figure 12.25. Note that the system's *bandwidth,* defined here as the frequency at which the phase becomes −45 deg, is about 2 rad/sec, the same as the gain-crossover frequency ω_c. This plot indicates that, as desired, the closed-loop transfer function $\theta(s)/\theta_c(s)$ is approximately unity over the frequency range below its bandwidth. But note that there is some small phase loss below 2 rad/sec.

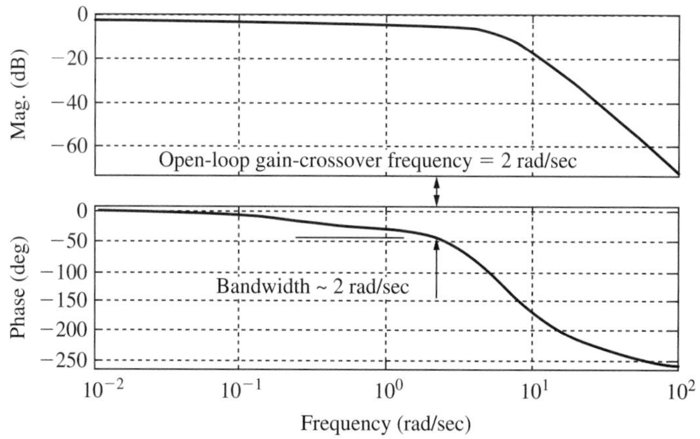

Figure 12.25 Closed-loop frequency response—pitch-attitude-controlled vehicle—$\theta(s)/\theta_c(s)$ (cruise condition).

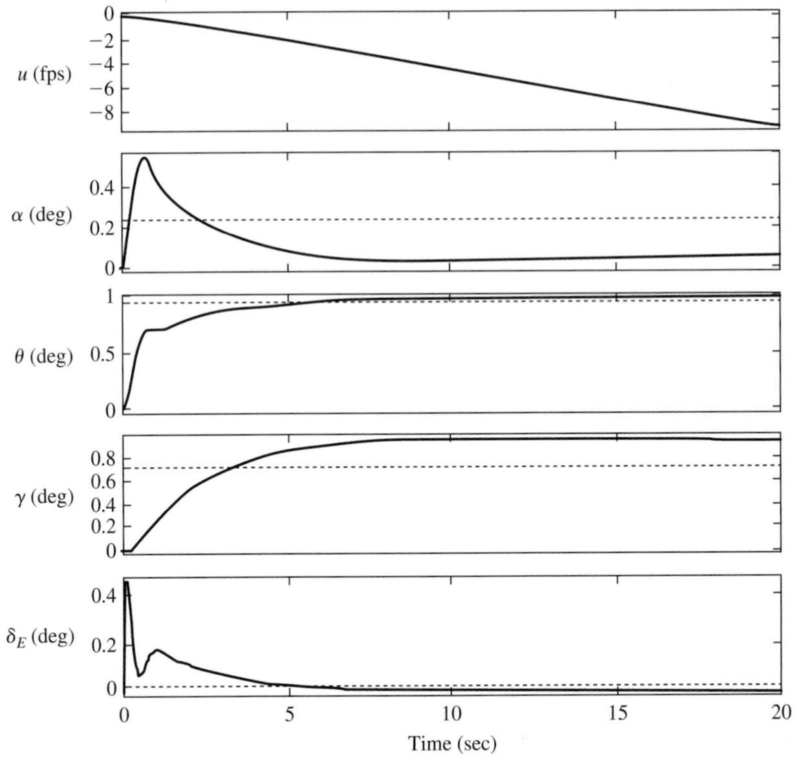

Figure 12.26 Time responses of pitch-attitude controller (cruise condition).

All these features of the closed-loop frequency response will be important to remember in our analyses later.

The time responses of the pitch-attitude-controlled vehicle for a step change in commanded attitude θ_c of one degree are shown in Figure 12.26. The pitch-attitude response indicates a rise time of about 3–4 seconds with no overshoot, with a peak elevator deflection of less than 0.5 deg, which are both probably acceptable. But note that there is a slight steady-state pitch-attitude error, since the control law did not include any integral action. (The dashed lines indicate the steady-state responses.) This may also be acceptable, however, depending on the application. Faster responses could be achieved with a higher pitch-attitude gain-crossover frequency ω_c, but that would lead to higher elevator deflection and require a higher pitch-damper gain as well. And these gains should not be set too high, due to the presence of noise and unmodeled dynamics at the higher frequencies.

Also important in these time responses is the fact that the steady-state flight-path angle γ is positive. This is important because presumably one key application of this controller would be to adjust altitude. And at least in a cruise flight condition, one would expect an increase in pitch attitude to lead to altitude gain. The steady-state flight-path angle remains positive even though the flight velocity bleeds off until reaching a steady-state loss of about 40 fps. Since the aircraft is said to be "on the front side of the power curve,"[1] a reduction in flight velocity leads to a reduction in power (thrust) required to maintain level flight, so the aircraft continues to climb. This fact will be revisited in the case below, but for now we will accept this controller.

Now consider the pitch-attitude-control problem using the same aircraft, but in an approach flight condition. We will find that the time responses of the controller will have characteristics somewhat different than those just presented. The aircraft is again the DC-8 in Flight Condition 1, an approach condition that consists of flight at sea level at a Mach number of 0.218 ($U_0 = 244$ fps) with flaps deflected. The weight is 190,000 lbs. (See Appendix B) The pitch-attitude transfer function in this case is

$$\frac{\theta(s)}{-\delta_E(s)} = \frac{1.3391(s + 0.06049)(s + 0.5352)}{(s^2 + 0.01986s + 0.02669)(s^2 + 1.69s + 2.625)} \text{ rad/rad} \qquad (12.34)$$

and the short-period and phugoid frequencies and dampings are

$$\omega_{\text{SP}} = 1.619 \text{ rad/sec} \quad \omega_P = 0.1635 \text{ rad/sec}$$

$$\zeta_{\text{SP}} = 0.522 \qquad \zeta_P = 0.0606$$

Using the same actuator time constant of 0.05 sec, a rate-feedback gain K_q of 1.5 rad/rad/sec yields a short-period frequency and damping of 2.07 rad/sec and 0.99, respectively. From the Bode plot of the open-loop system (with pitch damper) in Figure 12.27, we see that a pitch-attitude control gain K_θ of 10 dB (3.16 rad/rad) yields a gain-crossover frequency of slightly over 1 rad/sec (less than the previous case of 2 rad/sec), a phase margin of over 90 deg, and a gain margin of over 20 dB.

The time responses for the vehicle in the approach flight condition with this pitch-attitude controller are shown in Figure 12.28. We can see that the performance of this controller is significantly degraded from that for the cruise flight condition. The pitch-attitude response has a larger steady-state error of about 35 percent, the velocity bleeds off about 6 fps, and the steady-state flight-path angle is now slightly less than zero. The flight-path result is associated with the fact that the aircraft is said to now be on or near "the back side of the power curve," which means that as the aircraft slows down, more power (thrust) is required to maintain level flight. So at constant thrust, the flight-path angle cannot remain positive.

To improve performance we need to increase the loop gain $|kag_{\text{aug}}|$ at low frequencies, plus incorporate an auto-throttle to maintain constant flight velocity. We will address the auto-throttle in Section 12.6.1 when investigating longitudinal path guidance, but to improve the pitch-attitude performance of this controller, consider

[1] The "power curve" is a plot of power required (DV_V) to sustain level flight and power available (TV_V), plotted vs. flight velocity (V_V). Excess power available allows the aircraft to climb.

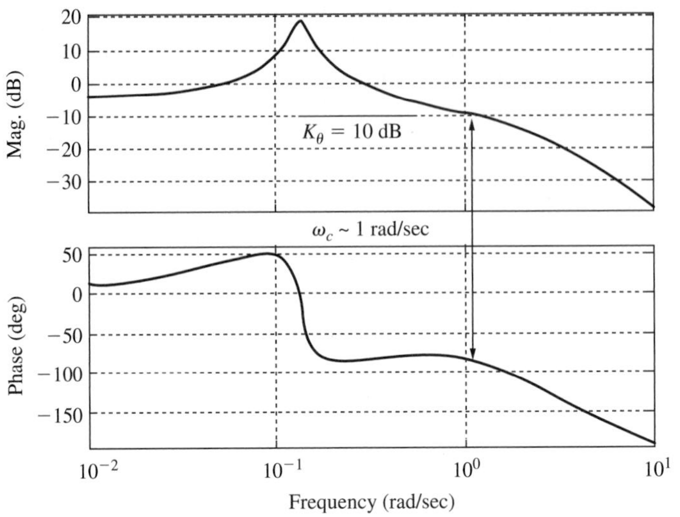

Figure 12.27 Bode plot—elevator to pitch attitude $\theta(s)/-\delta_E(s)$ w. pitch damper (approach condition).

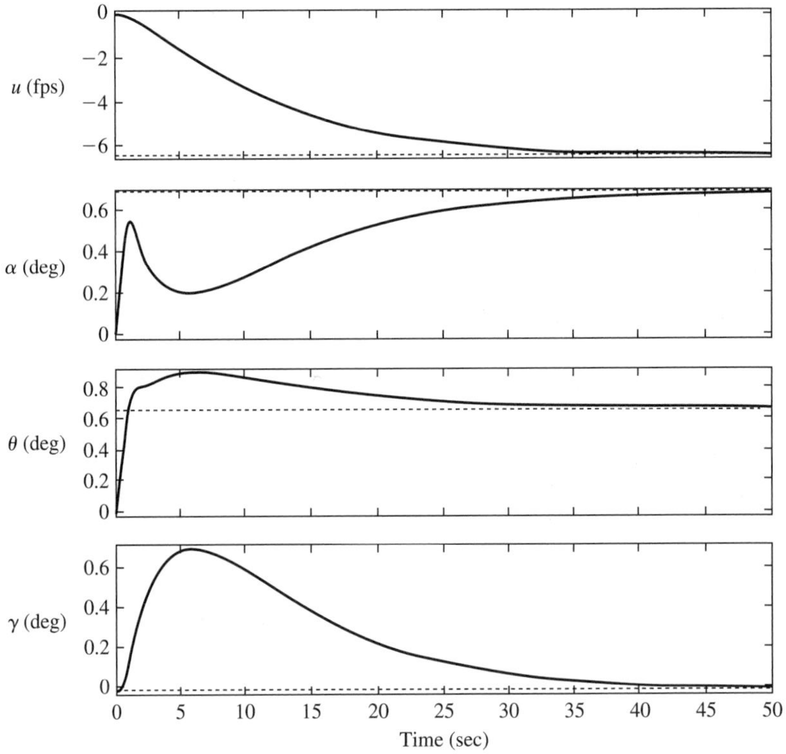

Figure 12.28 Time responses of pitch-attitude controller (gain only) (approach condition).

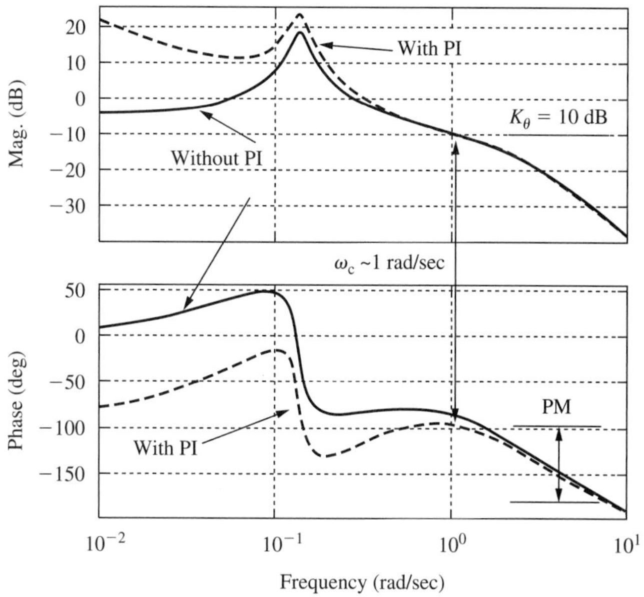

Figure 12.29 Bode plots—effect of PI compensation on open-loop pitch response (both include pitch damper, approach condition).

the Bode plots shown in Figure 12.29. One Bode plot, indicated as "Without PI," corresponds to the case just discussed (i.e., the magnitude plot does not reflect sufficiently high gain at low frequencies, which leads to the high steady-state error).

To correct this situation one could increase the gain-crossover frequency of the attitude-control loop, or add integral action to the control law. The former approach is not advisable because an increase in the crossover frequency of the attitude loop would require increasing the pitch-damper gain even further. And an inspection of the Bode plot of the pitch-rate transfer function would show that the gain-crossover frequency for the pitch-damper loop is already above 10 rad/sec. To increase it further could cause problems due to sensor noise and unmodeled dynamics, such as structural vibrations. Thus increasing the crossover frequency of this pitch-attitude loop is not recommended.

Instead, proportional-plus-integral (PI) compensation will be incorporated into the attitude control law $k_\theta(s)$. From consideration of the Bode plot of the pitch-attitude loop labeled "Without PI" in Figure 12.29, the zero in the PI compensator should be selected such that the compensator will not reduce the phase near the desired crossover frequency of around 1 rad/sec, and that the magnitude plot will remain like *K/s* near this frequency. So this zero may be set at 0.2 /sec, leading to a compensator of the following form.

$$k_\theta(s) = K_\theta \frac{(s + 0.2)}{s} \tag{12.35}$$

With the gain K_θ at unity for now, the Bode plot of the open-loop pitch-attitude dynamics (including the pitch damper) with the above PI compensator is shown as the second set of curves labeled "With PI" in Figure 12.29. Note that with PI compensation included, the low-frequency gain has been increased, while the phase loss above 1 rad/sec is minimal. Hence we have met our objectives.

Again from this same Bode plot we can see that if the loop gain is increased by 10 dB, we will obtain a gain-crossover frequency of 1 rad/sec, so the selected gain K_θ in the pitch-attitude control law is again 3.16 rad/rad. With this value of gain, the phase margin is now about 85 deg, and the gain margin remains over 20 dB. The final pitch-attitude control compensation is

$$k_\theta(s) = 3.16\frac{(s + 0.2)}{s} \text{ rad/rad} \tag{12.36}$$

The step responses for the vehicle with this pitch-attitude controller (with pitch damper) are shown in Figure 12.30. The pitch response of this controller has been significantly improved by the addition of PI compensation, and the peak elevator deflection is still only approximately 0.5 deg. With the addition of the integral action, the steady-state pitch-attitude error is now zero. However, the steady-state flight-path angle is still slightly negative, an issue that will be addressed further when we discuss the auto-throttle in Section 12.6.1.

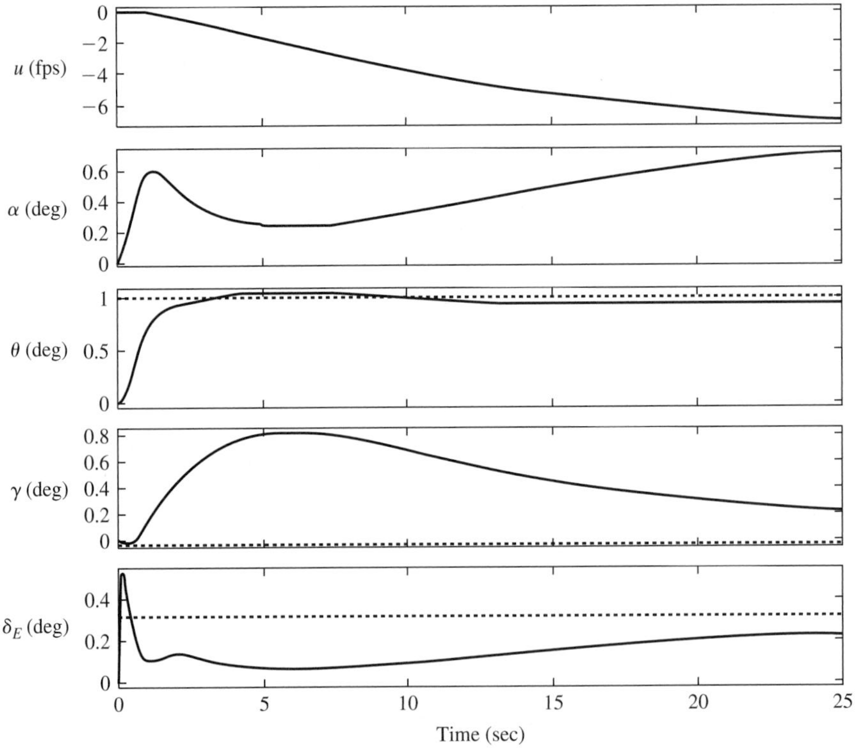

Figure 12.30 Vehicle step responses—pitch-attitude controller w. PI compensation (approach condition).

The MATLAB commands to assemble the model for the vehicle with this pitch-control law are as follows:

`»sys=ss(a,bdel,c,d)`	Vehicle model with elevator input
`»dena=[1 20];aa=tf(-20,dena);`	Elevator actuator model (with elevator sign convention reversed)
`»sysaa=series(aa,sys)`	Actuator and vehicle model in series
`» [aaa,ba,ca,da]=ssdata(sysaa)`	State-variable matrices recovered
`»aaaug=aaa-ba*1.5*ca(4,:);`	Vehicle **A** matrix augmented with pitch damper
`»sysaug=ss(aaaug,ba,ca,da)`	Define state-variable model for augmented system
`»numc=[1 .2];denc=[1 0];comp=tf(numc,denc)`	Define PI compensation
`»kg=series(comp,sysaug)`	PI and augmented vehicle in series
`»[akg,bkg,ckg,dkg]=ssdata(kg)`	Retrieve state-model matrices for open-loop *kg*
`»acl=akg-bkg*3.16*ckg(3,:)`	Closed-loop $\mathbf{A}_{CL\theta}$ matrix w. pitch controller
`»bcl=bkg*3.16`	$\mathbf{B}_{CL\theta}$ matrix for closed-loop system
`»syscl=ss(acl,bcl,ckg,dkg)`	Pitch-controlled system with θ_c input

NOTE TO STUDENT

The astute reader might notice that we have developed two different controllers for the same aircraft, using two different flight conditions. This is a fairly typical situation, but it does require switching from one controller to another as the flight condition changes. Sometimes the switching is based on altitude and Mach number, while in other cases deflecting flaps may trigger a control-law switch. This switching is referred to as *gain scheduling,* and is commonly used in flight control. But note that gain scheduling is based on a clear understanding of how the flight condition changes the dynamics and thereby affects the control law; that is, gain scheduling is physics based.

12.4.2 Other Pitch-Attitude-Control Approaches

We have focused on systems that feed back and control pitch-attitude angle directly, but other techniques control pitch attitude indirectly. For example, we could feed back and control pitch rate q, angle of attack α, or vertical acceleration a_Z. Note that we considered feedback of q, α, and a_Z in the last chapter when discussing stability augmentation, but were not attempting to control the response in terms of having it track a command as is the case here.

A conceptual block diagram of an alternate pitch control system (including a pitch damper) is shown in Figure 12.31, in which the controlled response y is either q, α, or a_Z. (The plus-and-minus signs imply that one must use the appropriate sign convention on the pitch-control surface.) This block diagram is quite similar to that shown in Figure 12.21.

The input y_c may come from an outer-loop guidance system, such as the altitude-hold system discussed in Section 12.5.2, or one could interpret the pilot's

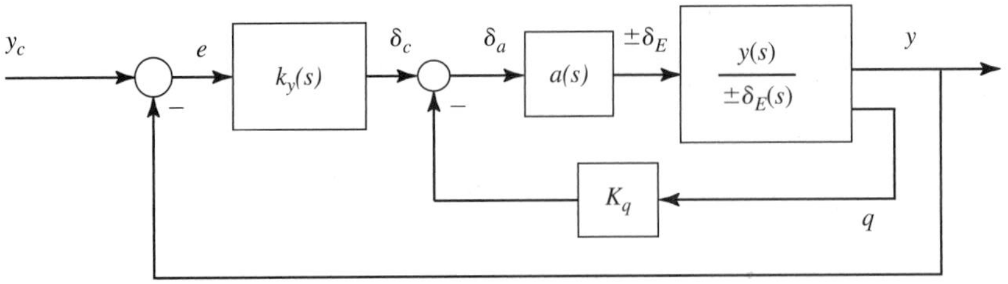

Figure 12.31 Indirect pitch-attitude control systems.

stick input as the commanded y_c. In either case, the objective of this control system is to force the response to track the command, and when the command is zero, the system attempts to hold the response at a constant value. As a result, a system designed to control pitch rate, for example, is sometimes called a *rate-command/attitude-hold system.*

As a particular example, let's consider the case of controlling angle of attack. Here we will use the data for the F-5A in Flight Condition 1 (see Appendix B), the same data set used in Example 11.3 in which a pitch-damper was synthesized. The angle-of-attack transfer function for the vehicle without an actuator or pitch damper is

$$\frac{\alpha(s)}{-\delta_E(s)} = \frac{0.14\left(s^2 + 0.01086s + 0.004533\right)\left(s + 102.6\right)}{\left(s^2 + 0.005552s + 0.001364\right)\left(s^2 + 1.219s + 3.711\right)} \text{ rad/rad} \quad (12.37)$$

As is typically the case, the phugoid quadratic is nearly cancelled by a similar quadratic in the numerator of this transfer function. Now add an actuator with a time constant $T_a = 0.05$ sec, and in the context of Figure 12.31 let the control law for the pitch damper be

$$\delta_a(s) = \delta_c(s) - K_q q(s) \quad (12.38)$$

With a pitch-damper gain $K_q = 0.25$ rad/rad/sec, for example, one obtains short-period damping of greater than unity (two real poles).

The Bode plot of the open-loop system, with the actuator and the above pitch damper, or $kag_{\text{aug}}(s)$, is shown labeled "Without PI" in Figure 12.32. Since the Bode magnitude is not large at low frequency, one may include proportional-plus-integral compensation of the form

$$k_\alpha(s) = K_P + \frac{K_I}{s} = K_P \frac{\left(s + K_I/K_P\right)}{s} \text{ rad/rad} \quad (12.39)$$

Note that if we select the compensator zero, or K_I/K_P, approximately equal to the short-period frequency, the Bode plot will be more like K/s over a wide frequency range. Setting $K_I/K_P = 2$ /sec and keeping the proportional gain K_P equal to unity for now yields the second Bode plot labeled "With PI" in Figure 12.32. This plot indicates that to maintain a 60 deg phase margin we must select a gain-crossover frequency ω_c of no greater than 2 rad/sec. Selecting 2 rad/sec for gain

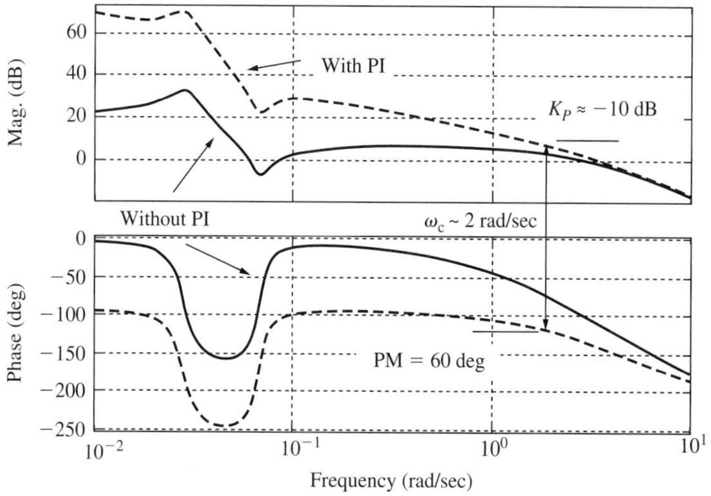

Figure 12.32 Bode plot—effect of PI on open-loop $\alpha(s)/\delta_c(s)$ (w. pitch damper).

crossover corresponds to a proportional gain in Equation (12.39) of $K_p \approx -10$ dB (0.32 rad/rad). The corresponding gain margin will be over 20 dB.

Similar to Equations (12.33), the model for the vehicle with the closed-loop angle-of-attack control law is found as follows:

$$\dot{\mathbf{x}} = \mathbf{A}'_{\text{PD+I}}\mathbf{x} + \mathbf{b}'_{\text{PD+I}}\delta_c, \text{ system } \underline{\text{with}} \text{ pitch damper + integral compensation}$$
$$\delta_c = K_P(\alpha_c - \alpha), \ \alpha\text{-control law proportional compensation} \qquad (12.40)$$
$$\dot{\mathbf{x}} = (\mathbf{A}_{\text{PD+I}} - \mathbf{b}'_{\text{PD+I}}K_P\mathbf{c}'_\alpha)\mathbf{x} + (\mathbf{b}'_{\text{PD+I}}K_P)\alpha_c, \text{ closed-loop system}$$

Using 0.32 rad/rad for K_P, the following closed-loop eigenvalues are obtained

$$\lambda_P = -0.0041562 \pm j0.067262 \text{ /sec}$$
$$\lambda_{\text{SP}} = -1.8535 \pm j1.6584 \text{ /sec}$$
$$\lambda_{\text{actuator}} = -15.7 \text{ /sec}, \quad \lambda_{\text{PI}} = -1.87 \text{ /sec}$$

which yields a closed-loop short-period damping of 0.75.

The vehicle's step responses to a one degree commanded change in angle of attack are shown in Figure 12.33. Note that the angle-of-attack response is rapid with no overshoot, and with a rise time of less than 2 sec. Plus with integral action in the control law, there is no steady-state error. The other three responses are dominated by the phugoid responses, and remain lightly damped. This is in contrast to the results for the pitch-attitude control system discussed previously, which resulted in a well-damped phugoid mode. So here the outer-loop controller, such as a pilot, will need to control the phugoid, but that should pose little problem technically. Whether this control approach leads to acceptable handling characteristics is another issue that will not be addressed here.

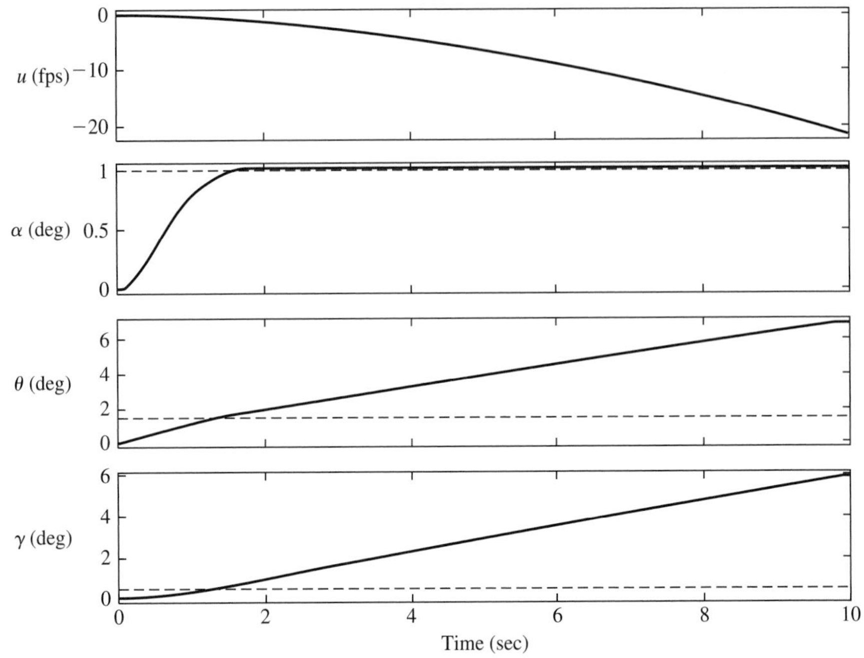

Figure 12.33 Vehicle step responses—angle-of-attack controller.

12.4.3 Bank-Angle Control

We noted that bank-angle control is a fundamental requirement for directional maneuvering, just as pitch-attitude control is for longitudinal maneuvering. (In fact a "wing leveler" was one of the first demonstrated automatic flight-control systems.) Hence a bank-angle controller will be an important building block for all lateral-directional autopilots to be discussed later in this chapter.

We introduced bank-angle control in Chapter 11 (Section 11.4) when addressing lateral-directional stability augmentation. From that discussion recall that yaw dampers, roll dampers, and aileron-rudder interconnects (ARI's) were all discussed in the context of improving the roll response of the vehicle. Also recall from the root-locus analyses in Chapter 11 that the dutch-roll mode can be troublesome, so it requires special attention.

A generic block diagram for a bank-angle controller is shown in Figure 12.34. This controller is designed to force the bank angle to track the commanded bank angle, which may be generated by the pilot or an outer-loop control law. Surface actuators are included, and the vehicle dynamics include any necessary stability augmentation such as a washed-out yaw damper and ARI. (Note that the sign on rudder deflection has been reversed.) Recall that yaw dampers are implemented to provide additional damping of the dutch-roll mode, while ARI's reduce the adverse yaw from aileron deflection that causes an undesirable excitation of that mode.

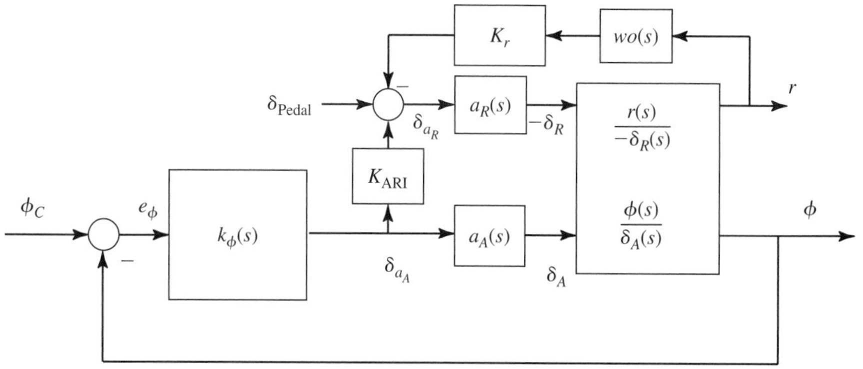

Figure 12.34 Block diagram of bank-angle control system.

To make the following discussion more concrete, let us demonstrate the synthesis of such a bank-angle controller using the data for the DC-8 aircraft in Flight Condition 3 (Appendix B). Many of the issues that will arise are generic, so they are not unique to this aircraft. The vehicle weighs 230,000 lbs, and is in a cruise flight condition at 33,000 ft altitude with Mach = 0.84. The aileron-to-bank-angle transfer function in this case is

$$\frac{\phi(s)}{\delta_A(s)} = \frac{2.11\left(s^2 + 0.3045s + 2.023\right)}{\left(s + 0.004053\right)\left(s + 1.254\right)\left(s^2 + 0.2373s + 2.235\right)} \text{ rad/rad}$$
$$\approx \frac{1.91}{s\left(s + 1.254\right)} \text{ rad/rad} \qquad (12.41)$$

The dutch-roll undamped natural frequency and damping ratio are about 1.5 rad/sec and 0.16, respectively. Note that for this flight condition the magnitude of the roll-subsidence pole (1.25 /sec) is smaller than the dutch-roll frequency, which makes the roll-control problem somewhat easier than the case with the situation reversed—a case to be addressed later in this section.

Two servo actuators are now added to deflect the aileron and rudder surfaces. These actuators are again modeled using simple first-order lags with a time constant of 0.05 sec, the same model used for pitch-attitude control discussed in Sections 12.4.1 and 12.4.2.

The root-locus and Bode plots corresponding to the dynamics represented by Equation (12.41) (but including the actuator) are shown in Figures 12.35 and 12.36, respectively. (The actuator pole is far to the left in the root locus and not shown.) The important points to observe from these two plots are the behavior of the dutch-roll mode and its effect on the bank-angle response. The root locus indicates that as the bank-angle-control gain is increased, the dutch-roll poles move away from their neighboring complex zeros. Hence the dutch-roll residues will increase, and this mode will contribute more to the bank-angle response.

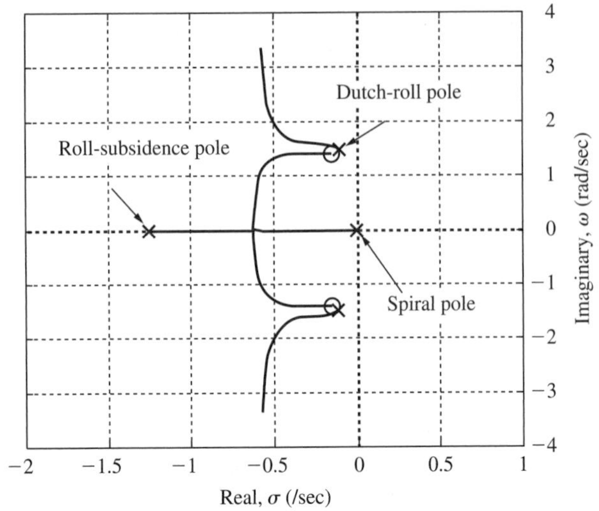

Figure 12.35 Root locus of open-loop bank-angle control dynamics $\phi(s)/\delta_{c_A}(s)$ (no augmentation).

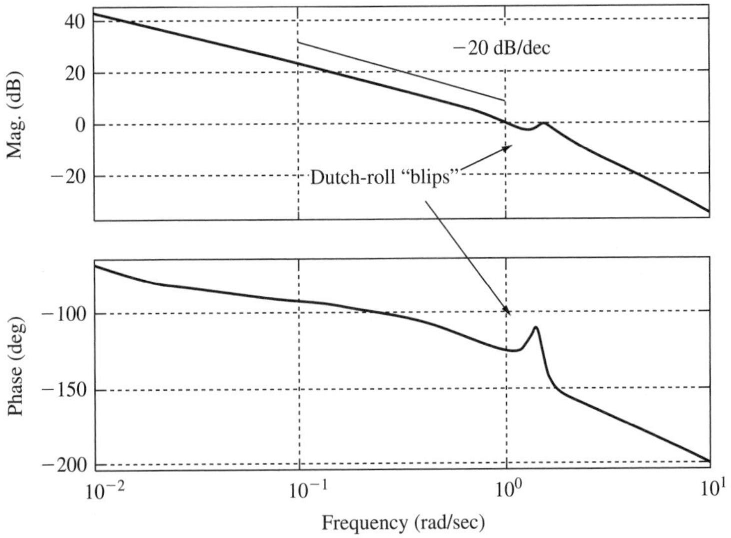

Figure 12.36 Bode plot of open-loop bank-angle-control dynamics $\phi(s)/\delta_{c_A}(s)$ (no augmentation).

Furthermore, from the Bode plots note the "blips" at the dutch-roll frequency in both the magnitude and phase plots. The size of these blips indicates the level of participation of the dutch-roll mode in the bank-angle response. Consequently, both an aileron-rudder interconnect (ARI) and a yaw damper will be added to mitigate against these undesirable characteristics.

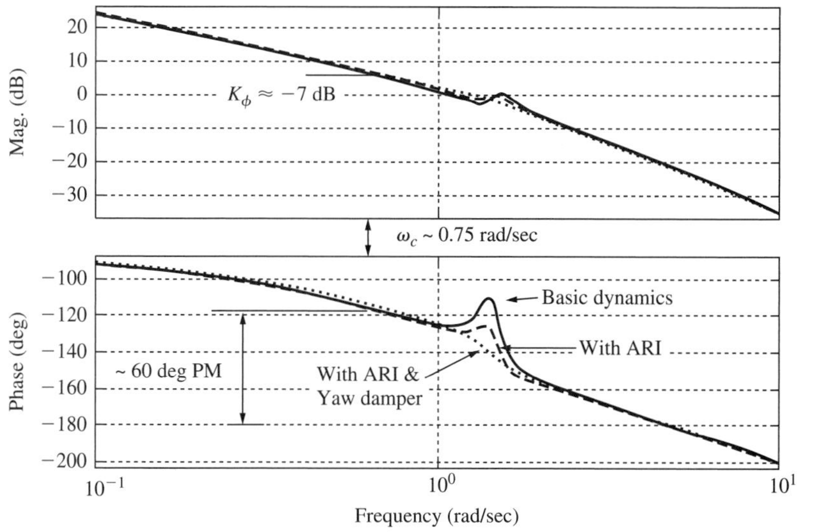

Figure 12.37 Bode plots of open-loop bank-angle controller including compensation.

The first augmentation system to be added is the ARI, which consists of a cross feed from the commanded aileron deflection to the commanded rudder deflection. The block-diagram schematic for the ARI was included in Figure 12.34, and the control law for the ARI is

$$\delta_{a_R}(s) = \delta_{\text{Pedal}}(s) + K_{\text{ARI}}\delta_{a_A}(s), \quad K_{\text{ARI}} = -\left(N'_{\delta_A}/N'_{\delta_R}\right) \tag{12.42}$$

The selection of the ARI gain K_{ARI} was discussed in Section 11.4.2, and for the vehicle in question, this gain becomes $K_{\text{ARI}} = -(-0.0652/-1.164) = -0.056$ rad/rad. The Bode plot of the aileron-to-bank-angle transfer function (including the actuator) with this ARI implemented is shown labeled "With ARI" in Figure 12.37. Comparing these plots to those for the vehicle without the ARI (also shown), we can see that the magnitudes of the "blips" have been reduced.

The next augmentation system to be incorporated is the yaw damper, which includes a washout filter, as discussed in Section 11.4.1. Recall that the washout essentially opens the yaw-rate feedback loop at low frequencies so that the yaw damper allows for the sustained yaw rate needed during a turn. Shown labeled as "Basic dynamics" in Figure 12.38 is the Bode plot of the rudder-to-yaw-rate transfer function (including the actuator), or $r(s)/\delta_{a_R}(s)$, for the aircraft in question, with the effects of the dutch-roll mode and a pair of neighboring complex zeros clearly evident.

Let the washout filter be

$$\frac{r_{\text{wo}}(s)}{r(s)} = \frac{s}{s+0.5} \tag{12.43}$$

where r_{wo} is the washed out yaw rate, or the output of the filter. The filter pole is selected to be well below the dutch-roll frequency, such that yaw-rate feedback

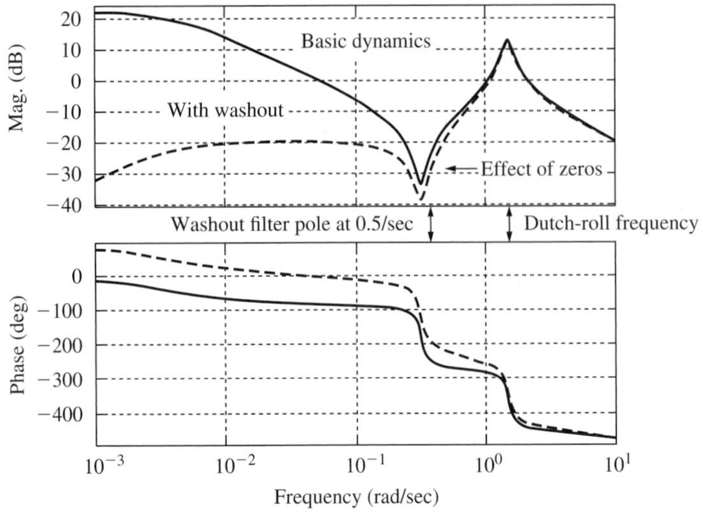

Figure 12.38 Effect of washout filter on rudder to yaw rate $r(s)/\delta_{a_R}(s)$.

will be effective at increasing the damping of that mode. The effect of introducing this filter into the yaw-rate feedback loop is also revealed in Figure 12.38, in which the second set of Bode plots, labeled "With washout," correspond to $r_{\text{wo}}(s)/\delta_{a_R}(s)$. As desired, the washout has little or no effect above 0.5 rad/sec, and the magnitude tends to zero at very low frequency. This washout filter will be used in the analysis to follow.

The yaw-damper gain K_r shown in Figure 12.34 may now be selected. A gain of 0.5 rad/rad/sec was chosen to increase the dutch-roll damping to 0.26, or an increase of over 60 percent. The eigenvalues of the system that now includes the yaw damper with washout are

$$\lambda_{\text{DR}} = -0.36025 \pm j1.3524 \text{ /sec}$$

$$\lambda_{\text{Roll}} = -1.2408 \text{ /sec}$$

$$\lambda_{\text{Spiral}} = -0.0038494 \text{ /sec}$$

$$\lambda_{\text{wo}} = -0.62575 \text{ /sec}$$

$$\lambda_{\text{Actuators}} = -20.0 \text{ /sec and } -19.4 \text{ /sec}$$

The Bode plot of the open-loop bank-angle controller (including this yaw damper) is also shown labeled "With ARI and yaw damper" in Figure 12.37. Note that the dutch-roll "blips" are now hardly evident, which was the objective of the augmentation.

Now with regard to the selection of the gain-crossover frequency ω_c for the bank-angle controller, it will be prudent to select this frequency to be below that of the dutch roll natural frequency (1.5 rad/sec here). And for effective roll control, gain crossover should be near or slightly below the magnitude of the roll-subsidence pole (or 1.24/sec here). So a desired crossover-frequency range of 0.6–1.0 rad/sec will be selected.

It will also be desirable to use a simple gain in the bank-angle control law, and avoid additional compensation. And from the Bode plots in Figure 12.37 we see that if we restrict the gain-crossover frequency to be approximately 0.6–0.8 rad/sec, using simply a gain K_ϕ in the bank-angle control law would yield a phase margin of approximately 60 deg. Selecting the gain to be approximately −7 dB (0.45 rad/rad), will result in a crossover frequency of about 0.7–0.8 rad/sec, so let the preliminary design for the bank-angle controller be

$$k_\phi(s) = K_\phi = 0.477 \text{ rad/rad} \tag{12.44}$$

To evaluate this design, the step responses to a commanded five-degree change in bank angle are shown in Figure 12.39. The rise time for the bank-angle response is about three seconds, with little oscillation or overshoot. Also, the steady-state bank-angle error is essentially zero. No dutch-roll oscillations are evident in the bank-angle response, although they are evident in the much smaller sideslip and yaw rate responses. But these responses are reasonably damped.

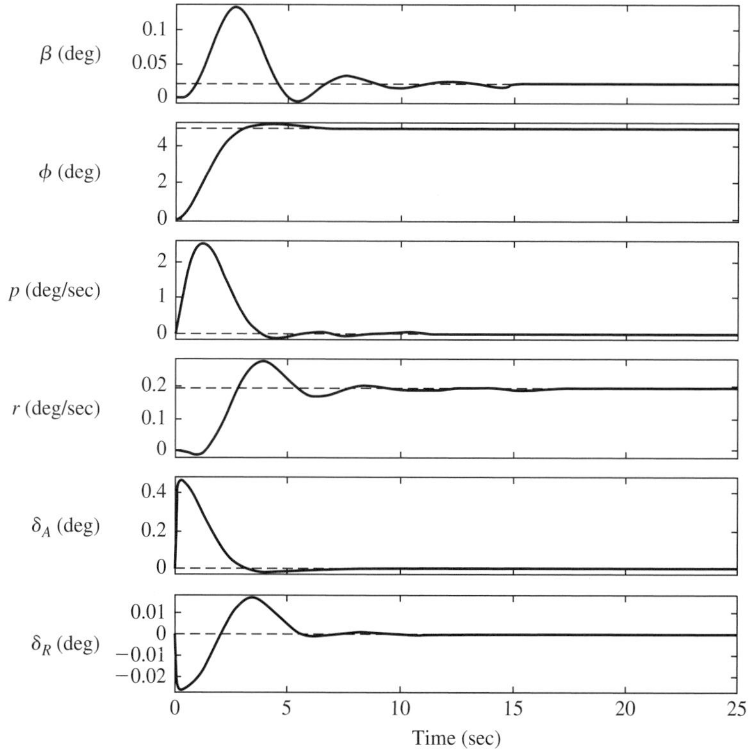

Figure 12.39 Step responses—bank-angle controller w. ARI and washed out yaw rate (cruise condition).

Also note that the scales for the sideslip and yaw rate responses are magnified. The peak-to-peak sideslip oscillations after 5 seconds are approximately 0.025 deg, and the steady-state sideslip angle is very small. The peak aileron deflection is only approximately 0.5 deg, and the small rudder deflections reflect the actions of both the ARI and yaw damper with washout. These responses are all considered quite acceptable. However, sideslip response in general will be discussed further in Section 12.4.4. Finally, the closed-loop eigenvalues are

$$\begin{aligned}
\lambda_{\text{DR}} &= -0.33435 \pm j1.3882 \text{ /sec} \\
\lambda_{\text{Roll-Spiral}} &= -0.66265 \pm j0.74523 \text{ /sec} \\
\lambda_{\text{wo}} &= -0.54449 \text{ /sec} \\
\lambda_{\text{Actuators}} &= -19.405 \text{ /sec and } -20.052 \text{ /sec}
\end{aligned}$$

The MATLAB commands to assemble the model for the bank-angle controller with ARI and yaw damper are as follows:

```
»sys=ss(a,b,c,d)                                  Basic aircraft
»numa=[20];dena=[1 20];aa=tf(numa,dena)           Actuators
»sysa=series(aa,sys);
»[aaa,ba,ca,da]=ssdata(sysa)
»bari=ba;bari(:,1)=ba(:,1)+ba(:,2)*(-.056)        Add ARI
»sysaari=ss(aaa,bari,ca,da)
»numw=[1 0];denw=[1 .5];wo=tf(numw,denw)          Add washed out yaw rate
»sysaawo=series(sysaari,wo,[4],[1]);
»[aawo,bawo,cawo,dawo]=ssdata(sysaawo)
»ca=[[0;0;0;0] ca];ca(5,:)=cawo;da=[da;0 0]
»sysaawo=ss(aawo,bawo,ca,da)
»aaaug=aawo+bawo(:,2)*.5*ca(5,:)                  Add yaw damper (w. washout)
»sysaug=ss(aaaug,bawo,ca,da)
»acl=aaaug-bawo(:,1)*.477*ca(2,:)                 Close φ control loop
»bcl=[.477*bawo(:,1) bawo(:,2)]
»syscl=ss(acl,bcl(:,1),ca,da(:,1))                Closed-loop bank-angle controller
```

The above discussion addressed the bank-angle control of a transport aircraft in a cruise flight condition. Now we will address the more challenging problem of bank-angle control in an approach flight condition. With the flight velocity significantly less than that for cruise (244 fps vs. 844 fps for the DC-8), the dutch-roll modal frequencies are lower, and therefore limit the maximum allowable gain-crossover frequency.

Consider the data for the DC-8 in Flight Condition 1 (Appendix B), an approach flight condition corresponding to flight at sea level with Mach = 0.218 (244 fps).

The vehicle weight is 190,000, and the flaps are extended. In this case the bank-angle response to aileron-actuator command (includes actuator) is given as

$$\frac{\phi(s)}{\delta_{a_A}(s)} = \frac{14.52\left(s^2 + 0.3317s + 0.6842\right)}{(s - 0.01294)(s + 1.121)\left(s^2 + 0.2188s + 0.9918\right)(s + 20)} \text{ rad/rad} \quad (12.45)$$

Three things to notice from the above transfer function are (1) the spiral mode is slightly unstable, (2) the dutch roll is even more lightly damped than before ($\zeta = 0.1$ vs. 0.16), and (3) the dutch-roll frequency (at ~1 rad/sec) is now less than the magnitude of the roll-subsidence pole (at 1.121 /sec). The unstable spiral mode will cause no problem, since it will be easily stabilized by the bank-angle controller. But consistent with our earlier comments, the low dutch-roll frequency, along with the lower damping, makes this control problem more challenging. Not only does this low frequency tend to limit the allowable gain-crossover frequency, but it also means that the dutch-roll frequency may be closer to the gain-crossover frequency than the magnitude of the roll-subsidence eigenvalue. This means that the dutch-roll mode is more likely to cause difficulty in terms of achieving acceptable bank-angle responses.

The Bode plot for the bank-angle dynamics given in Equation (12.45) is shown labeled "Basic dynamics" in Figure 12.40. Note that over the frequency range of 0.1–0.7 rad/sec, the Bode plot exhibits the desirable K/s characteristics. But the

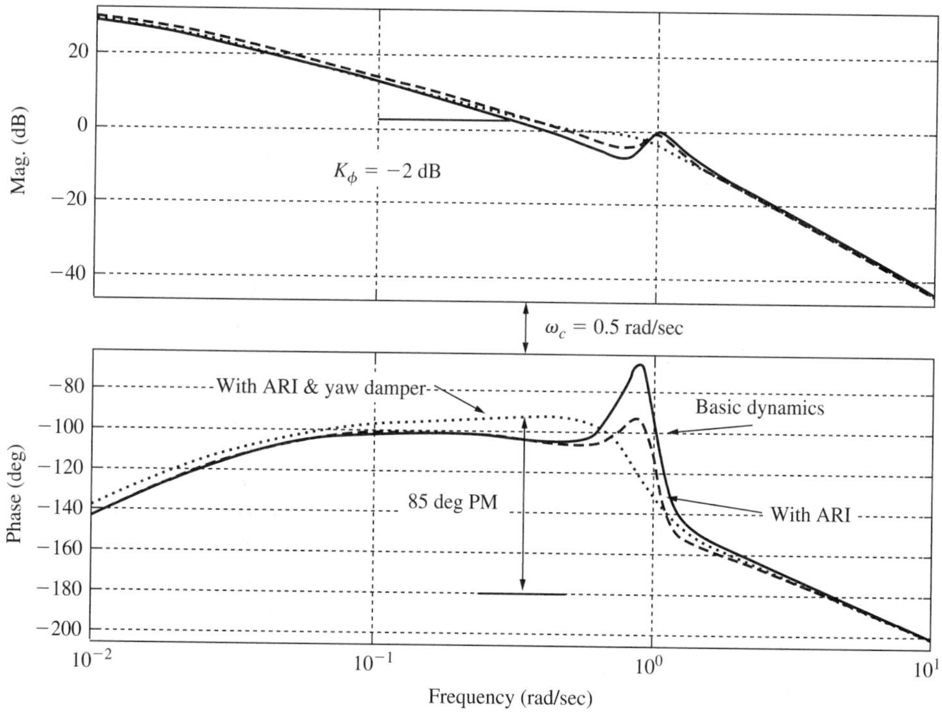

Figure 12.40 Bode plots of bank-angle responses to aileron-actuator command (approach condition).

"blips" due to the dutch-roll mode near 1 rad/sec are now quite pronounced, due to the very low dutch-roll damping and the difference between the dutch-roll frequency and the "frequency" of the quadratic term in the transfer-function numerator in Equation (12.45). The larger this difference, the larger the dutch-roll residues in the bank-angle response, and the greater the contribution of the dutch-roll mode to that response.

An aileron-rudder interconnect (ARI) will be introduced first, followed by a yaw damper. The ARI will not change the dutch-roll damping, but will reduce the distance (in the complex plane) between the dutch-roll poles and the zeros corresponding to the quadratic term in the numerator of Equation (12.45). Hence, these two zeros will more closely cancel the two dutch-roll poles.

From Section 11.4.2, the ARI control law is now given by

$$\delta_{a_R}(s) = \delta_{\text{Pedal}}(s) + K_{\text{ARI}}\delta_{a_A}(s), \quad K_{\text{ARI}} = -\left(N'_{\delta_A}/N'_{\delta_R}\right) \tag{12.46}$$

For the DC-8 in the approach flight condition, this yields a cross-feed gain of

$$K_{\text{ARI}} = -\left(-0.0532/-0.389\right) = -0.137 \text{ rad/rad} \tag{12.47}$$

However, after examining the numerators of the $\beta(s)/\delta_{a_A}(s)$ and $r(s)/\delta_{a_A}(s)$ transfer functions, this ARI cross-feed gain was increased to -0.2 rad/rad. The gain was adjusted to reduce the Bode gain in these two transfer functions—hence the responses to aileron—as well as to reduce the magnitudes of the nonminimum-phase zeros in these transfer functions. Both of these factors are indicators of the level of turn coordination present, discussed in Section 12.4.4.

With the ARI gain increased to -0.2 rad/rad, the bank-angle response to aileron-actuator command becomes

$$\frac{\phi(s)}{\delta_{a_A}(s)} = \frac{13.79\left(s^2 + 0.4072s + 0.8355\right)}{\left(s - 0.01294\right)\left(s + 1.121\right)\left(s^2 + 0.2188s + 0.9918\right)\left(s + 20\right)} \text{ rad/rad} \tag{12.48}$$

Note the desirable effect of the ARI on the two complex zeros—they are now closer to the dutch-roll poles. The Bode plot for this transfer function is shown labeled "With ARI" in Figure 12.40, and note how the prominence of the dutch-roll "blips" has been reduced.

Now the yaw damper will be addressed. First, since the dutch-roll frequency is now approximately 33 percent less than that for the cruise flight condition, the washout pole needs to be adjusted. Recall from the discussion regarding Figure 12.38, the washout should not affect the yaw-rate frequency response at and above the frequency corresponding to the location of the zeros from the quadratic term in the numerator of the aileron-to-yaw-rate transfer function. And with the lower dutch-roll frequency, these zeros now appear at a significantly lower frequency. Consequently, the washout pole is now at -0.4 /sec, instead of the previously selected -0.5 /sec. Or the washout is now

$$\frac{r_{\text{wo}}(s)}{r(s)} = \frac{s}{s + 0.4} \tag{12.49}$$

Using this washout, and selecting a yaw-rate feedback gain K_r of 1.5 rad/rad/sec, increases the dutch-roll damping by approximately a factor of three. Using this yaw-damper gain, the bank-angle response to aileron-actuator command is now

$$\frac{\phi(s)}{\delta_{a_A}(s)} = \frac{13.7948(s + 0.6107)(s^2 + 0.8318s + 0.5652)(s + 19.36)}{(s - 0.01113)(s + 0.6782)(s + 1.118)(s^2 + 0.5641s + 0.7046)(s + 19.38)(s + 20)} \text{ rad/rad} \tag{12.50}$$

Note here that the spiral and roll-subsidence poles have been essentially unaffected by the washed out yaw damper, while both the dutch-roll damping and frequency have been modified. The pole at -0.678 /sec is from the washout filter, and is approximately cancelled by the zero at -0.611 /sec. The pole at -19.38 /sec is from the rudder actuator, and is essentially cancelled by the zero at -19.36 /sec.

The Bode plot for the above transfer function, labeled "With ARI and yaw damper," is shown in Figure 12.40. Clearly the dutch-roll "blips" have been almost completely eliminated, plus the phase below the dutch-roll frequency has been increased slightly. Using this later set of Bode plots, we may now select a gain-crossover frequency for the bank-angle controller. This frequency should be as high as possible for a fast speed of response, limited by the minimum allowable phase margin in this control loop. Choosing a bank-angle-control gain $K_\phi = -2$ dB (-0.794 rad/rad) will set the crossover frequency at approximately 0.5 rad/sec, in the higher part of the frequency range where the Bode plots exhibit *K/s* characteristics. For the selected gain-crossover frequency, the phase margin is greater than 80 deg, and the gain margin is approximately 30 dB.

Using this bank-angle-control gain K_ϕ, the closed-loop eigenvalues for the bank-angle-controlled system with the ARI and yaw damper are

$$\begin{aligned}
\lambda_{\text{DR}} &= -0.27174 \pm j0.92429 \text{ /sec}\\
\lambda_{\text{RS}} &= -0.71247 \pm j0.22751 \text{ /sec}\\
\lambda_{\text{wo}} &= -0.35190 \text{ /sec}\\
\lambda_{\text{Actuators}} &= -19.378 \text{ /sec and } -20.029 \text{ /sec}
\end{aligned}$$

The step responses to a commanded change in bank angle of <u>five</u> degrees are shown in Figure 12.41. As expected, the bank-angle response includes more dutch-roll participation than was the case in the cruise flight condition. But the dutch-roll mode is reasonably damped. The bank-angle response exhibits a rise time of about 6 sec, with no overshoot and virtually no steady-state error. Due to the lower flight velocity, the steady-state yaw rate is larger than that in the previous case (cruise). This leads to the higher yaw-rate induced steady-state sideslip angle. The peak aileron deflection has increased to 0.8 deg (from 0.5 deg), and the rudder deflections again reflect the actions of the ARI and yaw damper. Although these responses may not be perfect, they are considered reasonable for this flight condition.

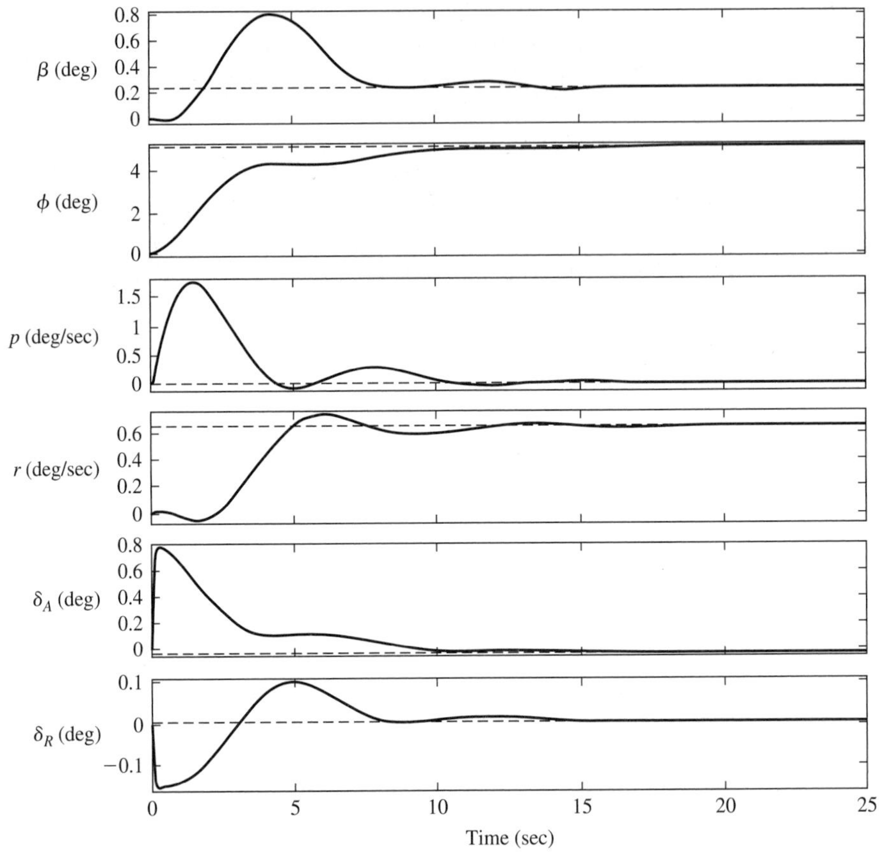

Figure 12.41 Vehicle step responses with bank-angle controller (approach condition).

The MATLAB commands for assembling the model for the vehicle with this bank-angle controller are as follows:

```
»sys=ss(a,b,c,d)                                   Vehicle dynamics
»numa=[20];dena=[1 20];aa=tf(numa,dena)            Actuators
»sysa=series(aa,sys);
»[aaa,ba,ca,da]=ssdata(sysa)
»bari=ba;bari(:,1)=ba(:,1)+ba(:,2)*(-.2)           Add ARI
»sysaari=ss(aaa,bari,ca,da)
»numw=[1 0];denw=[1 .4];wo=tf(numw,denw)           Add washed out yaw rate
»sysaawo=series(sysaari,wo,[4],[1]);
»[aawo,bawo,cawo,dawo]=ssdata(sysaawo)
»ca=[[0;0;0;0] ca];ca(5,:)=cawo;da=[da;0 0]
»sysaawo=ss(aawo,bawo,ca,da)
```

```
»aaaug=aawo+bawo(:,2)*1.5*ca(5,:)        Add yaw damper (w. washout)
»sysaug=ss(aaaug,bawo,ca,da)
»acl=aaaug-bawo(:,1)*.794*ca(2,:)        Close φ control loop
»bcl=[.794*bawo(:,1) bawo(:,2)]
»syscl=ss(acl,bcl(:,1),ca,da(:,1))       Closed-loop bank-angle controller
```

12.4.4 Turn Coordination and Turn Compensation

A coordinated turn may be defined as a turn during which any of the following criteria are met, and they are not all equivalent.

1. The sideslip angle β is zero.
2. The lateral inertial acceleration at the vehicle *c.g.* $a_{Y_{cg}}$ is zero.
3. The turn rate $\dot{\psi}$ is consistent with the bank angle and flight velocity.
4. The lateral acceleration at the cockpit $a_{Y_{CP}}$ is zero.

The first two of these criteria are self-explanatory, while the other two require addition explanation as to their precise meaning.

As for Criterion 3, recall the analysis of a steady turn $\left(\dot{U} = \dot{V} = \dot{W} = \dot{\phi} = \dot{\theta} = 0\right)$ addressed in Chapter 9 (Section 9.4.1). It was shown that during a turn at constant altitude ($\gamma = 0$), if the total external (aerodynamic plus propulsive) force along the lateral Y axis of the vehicle was zero, then the following must hold.

$$\begin{aligned}
\dot{\psi} &= \frac{g}{V_\infty}\tan\phi \\
P &= 0 \\
Q &= \dot{\psi}\sin\phi = \frac{g}{V_\infty}\tan\phi\sin\phi \\
R &= \dot{\psi}\cos\phi = \frac{g}{V_\infty}\sin\phi
\end{aligned} \tag{12.51}$$

If the first of these equations is satisfied, then Criterion 3 is satisfied.

Regarding Criterion 4, a cockpit instrument known as a *turn-and-bank indicator* includes a ball in a U-shaped tube. When during a turn the ball is in the center of the tube, or at the bottom of the "U," the turn is said to be *coordinated.* That ball location indicates that the lateral acceleration at the cockpit is zero. This criterion is more related to ride quality, and will not be considered further.

Note that to derive Equations (12.51), the assumption of zero lateral aerodynamic and propulsive forces was made. Hence, Criteria 2 and 3 are equivalent. Plus, if the side forces due to aileron, rudder, and yaw rate, $\left(Y_{\delta_A}, Y_{\delta_R} \text{ and } Y_r\right)$ are small, then Criterion 1 is approximately equivalent to Criteria 2 and 3.

But rudder side force Y_{δ_R} is frequently not negligible, in which case Criterion 1 is not equivalent to Criteria 2 and 3. But with a small adjustment to Criterion 2, the two criteria can be made almost equivalent. Instead of placing the lateral accelerometer at the *c.g.*, as indicated in Criterion 2, it may be placed at the *center of rotation* for rudder deflections.

At the center of rotation, a step change in rudder deflection will produce an instantaneous change in lateral acceleration that is exactly balanced by the acceleration due to yaw acceleration, or $x_a\dot{R}$. This location, denoted here as x_a, is located a distance

$$x_a = -\frac{Y_{\delta_R}}{N_{\delta_R}} \tag{12.52}$$

ahead of the *c.g.* for an aft rudder. If the accelerometer is located on the centerline of a rigid vehicle, and a distance x_a from the *c.g.*, then the lateral acceleration experienced by that accelerometer is

$$a_{Y_{\text{local}}} = a_{Y_{\text{cg}}} + \left(PQ + \dot{R}\right)x_a = a_{Y_{\text{cg}}} + \dot{R}x_a \text{ when } P = 0 \tag{12.53}$$

And if the effects of other motion quantities on the lateral force are small, this sensed acceleration is approximately given by

$$a_{Y_{\text{cg}}} + \dot{R}x_a \approx Y_\beta\beta + Y_{\delta_R}\delta_R + \dot{R}x_a = Y_\beta\beta \tag{12.54}$$

Therefore, zero lateral acceleration sensed at the rudder's center of rotation implies zero sideslip angle, and Criterion 1 is almost equivalent to the modified Criterion 2. Furthermore, lateral acceleration sensed at the rudder's center of rotation closely approximates a sideslip sensor.

We are now interested in methods for generating the lateral-directional control inputs (e.g., rudder and/or aileron) necessary to maintain a coordinated turn. Based on the above criteria for a coordinated turn, several methods are suggested. These methods include:

1. Feedback of sideslip angle to the rudder.
2. Feedback of lateral acceleration (measured at the rudder's center of rotation) to the rudder.
3. Feedback of computed yaw rate to the rudder.
4. Introduction of an aileron-rudder interconnect (ARI) to eliminate the initiation of sideslip due to adverse yaw from aileron deflection.

From the above discussion it should be clear that Methods 1 and 2 are basically equivalent, in terms of the dynamics. The differences are mainly due to the difficulties in sensing the required quantities for feedback. Method 3 relies on using the last of Equations (12.51) to determine the required yaw rate for a given bank angle and velocity, and feeding back this quantity to the rudder. And Method 4 has been demonstrated extensively in this chapter. Examples of Methods 1–3 are presented in Refs. 4 and 5.

Turn compensation is concerned with methods for generating the longitudinal control inputs (e.g., elevator) necessary to maintain altitude during a coordinated turn. When the vehicle is in a turning flight condition, the vertical component of lift must still support its weight if it is to maintain constant altitude. Consequently, as the vehicle enters the turn, the angle of attack must be increased, which requires an elevator deflection appropriate for the flight velocity and bank angle.

Two methods are suggested for providing this necessary elevator deflection:

1. Use a computed pitch rate to generate commanded pitch rate to a rate-command/attitude-hold control system.
2. Employ an altitude-hold control system in the longitudinal axes.

Altitude holds will be discussed in Section 12.5.2, so we will only discuss the first approach here. As noted earlier in this section, the pitch rate Q and yaw rate R corresponding to a coordinated turn are given in Equations (12.51). Combining these two expressions, one obtains a relationship for pitch rate, given in terms of the yaw rate and bank angle, or

$$Q = R\tan\phi \tag{12.55}$$

So, assuming both yaw rate and bank angle are measured, the required pitch rate is known, and can be used to generate the commanded pitch rate q_c for a pitch-rate-command system such as that cited in Section 12.4.2.

12.5 RESPONSE HOLDS

From Section 12.4 and attitude controllers, we now turn our attention to other autopilot modes that are frequently used in cruise or climbing flight conditions to further reduce pilot workload. Examples include speed or Mach holds, altitude holds, and heading holds. In each case considered, we will demonstrate the development of the control law while taking advantage of an inner-loop attitude controller like those discussed in Section 12.4. So one must be cognizant of the bandwidth and gain-crossover frequencies of the respective inner-loop controllers in selecting the gain-crossover frequencies of the outer loops discussed in this section.

12.5.1 Speed (Mach) Hold

A speed, or Mach, hold is frequently used when a vehicle is in climbing flight under air traffic control. For example, the pilot may set the commanded speed or Mach to the desired value, and then maintain a constant thrust setting as the vehicle climbs. Synthesizing a Mach-hold controller may be one of the simplest applications of the inner- and outer-loop design approach. Here the inner loop will consist of the vehicle controlled by a pitch-attitude controller.

Consider the block diagram shown in Figure 12.42, clearly showing the inner- and outer-loop structure. Note that here surge velocity u and pitch-attitude θ

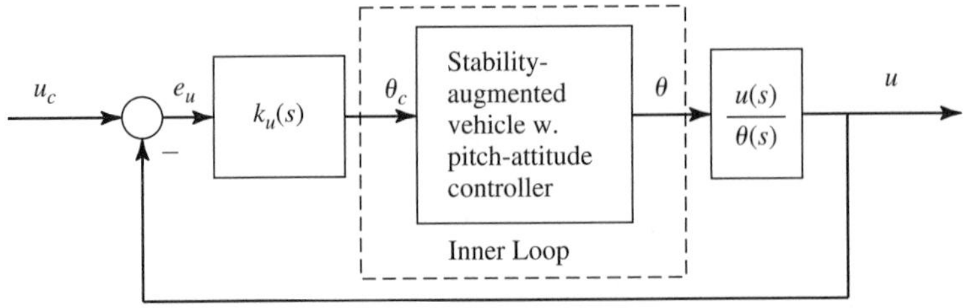

Figure 12.42 Block diagram of speed (Mach) hold.

are both responses to the pitch-control surface, such as an elevator. As such, the relationship between the two responses simply involves the numerators of their respective transfer functions. Specifically, we have

$$\frac{u(s)}{\theta(s)}=\frac{u(s)/\delta_E(s)}{\theta(s)/\delta_E(s)}=\frac{N^u_{\delta_E}(s)}{N^\theta_{\delta_E}(s)} \tag{12.56}$$

since both transfer functions $u(s)/\delta_E(s)$ and $\theta(s)/\delta_E(s)$ have the same characteristic polynomial. Care must be exercised to use the numerators for the transfer functions with the <u>same control input</u>, the input to be used in the control system.

Now the question turns to whether to use the transfer functions (or numerators) corresponding to the vehicle alone, or to the vehicle controlled by the pitch-attitude controller. In fact, either set may be used as long as one is considering only one input (here, the elevator). As discussed in Section 11.2, when considering the effect of feedback on the numerators of the transfer functions, feedback to one input will have no effect on the numerators of any of the other transfer functions with the same input.

Consider now the DC-8 in Flight Condition 3 (Appendix B), a cruise flight condition at an altitude of 33,000 ft with Mach = 0.84. Again, the issues to be addressed are generic, and not specific to this particular vehicle. For the DC-8, the relationship between the velocity u and attitude θ shown in the outer loop in Figure 12.42 is

$$\begin{aligned}\frac{u(s)}{\theta(s)}=\frac{N^u_{\delta_E}(s)}{N^\theta_{\delta_E}(s)}&=\frac{0.03254(s+0.8153)(s-880.2)}{(s+0.01441)(s+0.7247)}\\&\approx\frac{-28.64(s+0.8153)}{(s+0.01441)(s+0.7247)}\approx\frac{-28.64}{(s+0.01441)}\text{ fps/rad}\end{aligned} \tag{12.57}$$

So basically the outer-loop dynamics look like a simple first-order lag with a pole at $-1/T_{\theta_1}$. One approximation made to obtain the above result is simply an approximate pole-zero cancellation, while the other approximation involving the zero at 880 /sec will be further justified later in this section. Also, note the negative constant in the latter two terms, consistent with the fact that a pitch up will lead to a reduction in flight velocity, and this will require an appropriate

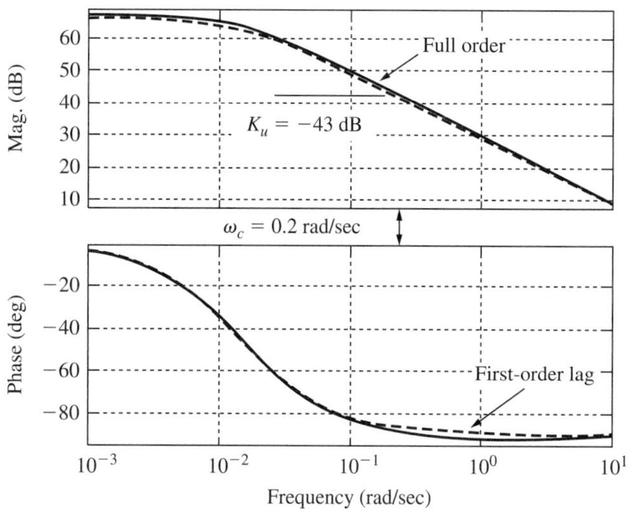

Figure 12.43 Bode plots of outer-loop dynamics $-u(s)/\theta(s)$ (note sign reversal).

sign reversal in the analysis. Finally, although we will use Equation (12.57) like a transfer function, one must remember that the two roots of the denominator at -0.0144 /sec and -0.725 /sec are not eigenvalues of the system.

The Bode plots for all three expressions in Equation (12.57) with the sign reversal are shown in Figure 12.43. Note that they are all in very close agreement in the frequency range shown. Recall that we require the gain-crossover frequency in this outer loop to be less than that of the inner-loop, which for the attitude controller in the cruise condition (discussed in Section 12.4.1) was 2 rad/sec. So we are not interested in frequencies above 10 rad/sec.

The first approximation made in Equation (12.57) may now be discussed further in the context of the above Bode plots. In particular, the only effect of the right-half-plane zero at 880 /sec seen in these Bode plots is its sign. We will set the gain-crossover frequency for this outer loop well below 10 rad/sec, so the critical frequency range is that shown in the Bode plots. With respect to this frequency range, the following approximation is seen to be valid.

$$\left(s - 880\right)\big|_{s=j\omega} = \left(-880 + j\omega\right)\big|_{\omega<10} \approx -880 \tag{12.58}$$

That is, at frequencies well below 880 rad/sec the above term is approximately -880 /sec, and this system zero is in effect at infinity in the complex plane.

Now consider the block diagram for the speed hold given in Figure 12.42, and recall that the transfer function for the inner pitch-attitude loop is approximately unity at frequencies below its bandwidth (~2 rad/sec here). This observation can be reinforced by again considering the closed-loop frequency response for the attitude controller, shown in Figure 12.25. Therefore we can use the simple model for the outer-loop dynamics given in Equation (12.57), or the Bode plots given in Figure 12.43, to preliminarily synthesize the control law for the outer loop. We will then validate the design using the full-order model of the augmented vehicle including the pitch-attitude controller.

With reference to the Bode plots in Figure 12.43, one can observe that above 0.01 rad/sec the slope of the magnitude plot is the desired −20 dB/dec. So let's initially consider an outer-loop gain-crossover frequency of 0.2 rad/sec, a decade below the inner-loop gain crossover of 2 rad/sec. With this crossover frequency, the low-frequency Bode magnitude will be almost 30 dB. One may choose to accept this magnitude as sufficiently high so as to yield acceptable tracking performance, or one may add PI compensation. But to keep the control law as simple as possible, let's just use a constant gain $k_u(s) = K_u$ instead of adding PI compensation. This decision may be revisited later if necessary.

With continuing reference to Figure 12.43, we can see that selecting an outer-loop gain of −43 dB ($K_u = -7.08 \times 10^{-3}$ rad/fps) would lead to a crossover frequency of 0.2 rad/sec, a phase margin of greater than 90 deg, and an infinite gain margin. However, we know these margins are optimistic due to the presence of the right-half-plane "zero" at −880 /sec, and because we have not yet accounted for any phase loss in the pitch-attitude inner loop. This completes the preliminary design of the outer-loop speed-hold control law.

The complete model will now be used to validate these results. Shown in Figure 12.44 is the Bode plot of the open- (outer-)loop vehicle dynamics including pitch damper and pitch-attitude controller, as shown in Figure 12.42, or the $-u(s)/\theta_c(s)$ transfer function (includes sign reversal). First, note the similarity between this Bode plot and those using the simpler models shown in Figure 12.43, further justifying the use of the simpler models in the preliminary design. For reference, the transfer function here is

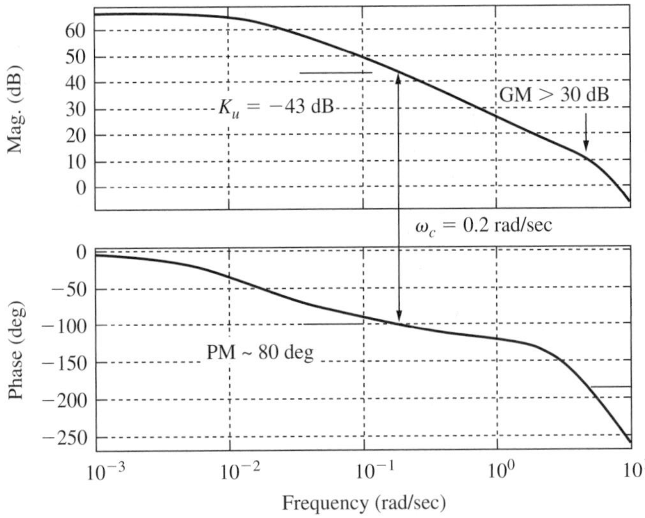

Figure 12.44 Bode plot—Outer loop of speed hold $-u(s)/\theta_c(s)$ w. pitch damper and pitch-attitude controller (complete DC-8 model, approach condition).

$$\frac{-u(s)}{\theta_c(s)} = \frac{-9.4029(s+0.8153)(s-880.2)}{(s+0.01503)(s+0.4204)(s^2+6.842s+33.33)(s+14.89)} \text{ fps/rad} \quad (12.59)$$

With reference to Figure 12.44, we confirm that the outer-loop gain of −43 dB ($K_u = 7.08 \times 10^{-3}$ rad/fps) indeed yields a crossover frequency of 0.2 rad/sec, a phase margin of approximately 80 deg, and a gain margin of over 30 dB.

The closed-loop eigenvalues of the vehicle with this speed-hold autopilot in place are

$$\lambda_{SP} = -3.3798 \pm j4.5730 \text{ /sec}$$

$$\lambda_P = -0.27196 \pm j0.17881 \text{ /sec}$$

$$\lambda_{\text{Actuator}} = -14.861 \text{ /sec}$$

Note that the short-period damping is now 0.6 and that the phugoid mode is oscillatory but well damped.

The vehicle's time responses to a commanded 10 fps step increase in flight velocity at constant thrust are shown in Figure 12.45. The rise time in the velocity response is approximately 10 sec, with little overshoot. However, since the control law does not include integral action, there is a small steady-state speed error.

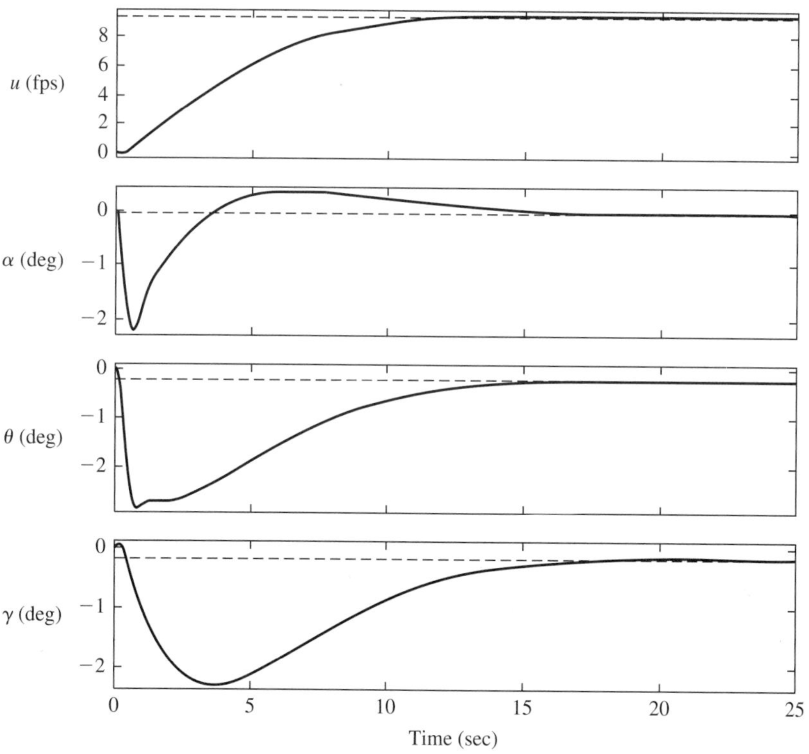

Figure 12.45 Step responses of vehicle with speed (Mach) hold, u_c = 10 fps (cruise condition).

Note that to increase the flight velocity with thrust constant, the aircraft must pitch down. So to climb, the pilot would increase the thrust setting, and the speed hold would increase the pitch attitude to maintain constant speed. All these responses are smooth and well damped, and if the steady-state error is not deemed acceptable, proportion-plus-integral compensation may be considered. (See Problem 12.3.)

The MATLAB commands for assembling the model for the vehicle with this speed hold are as follows:

»syscl=ss(acl,bcl,ca,da);	Model for vehicle with pitch-attitude controller
»aclspeed=acl+bcl*0.00708*ca(1,:);	Add speed hold control law
»bclspeed=bcl*.00708;	
»sysclspeed=ss(aclspeed,bclspeed,ca,da);	Closed-loop system with speed hold

12.5.2 Altitude Hold

The objective of an altitude-hold autopilot is, as its name implies, to hold altitude constant. Usually this autopilot is used in a cruise flight condition, with thrust or throttle setting selected based on the desired flight velocity. The block diagram of such an autopilot is shown in Figure 12.46. Again note the inner- and outer-loop structure. The inner loop consists of the longitudinal dynamics of the aircraft controlled by a pitch-attitude controller, such as that discussed in Section 12.4.1. The outer loop includes the dynamic relationship between pitch attitude and altitude. Note that an altitude hold and a speed hold using attitude cannot be engaged at the same time. Only one of these responses can be controlled through a single pitch-control input (e.g., elevator).

In the context of this block diagram, a key requirement for the inner-loop attitude-control system is to have a commanded increase in pitch attitude produce a positive steady-state flight-path angle. In Section 12.4.1 we evaluated two similar pitch-attitude-controllers but at two different flight conditions. The first controller met the flight-path angle requirement at the cruise flight condition, but the second controller for the approach flight condition did not. In this section we will discuss the development of the outer-loop altitude-hold control law in a cruise flight condition. Later in Section 12.6.1 we will introduce an auto-throttle that is needed for altitude control in the approach flight condition.

As discussed in Section 12.2, the inner-loop transfer function is approximately unity at frequencies within the bandwidth of the inner loop. The gain-crossover

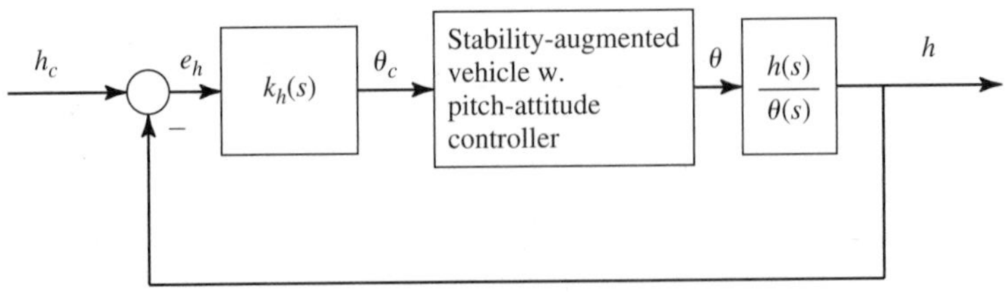

Figure 12.46 Block diagram of altitude-hold autopilot.

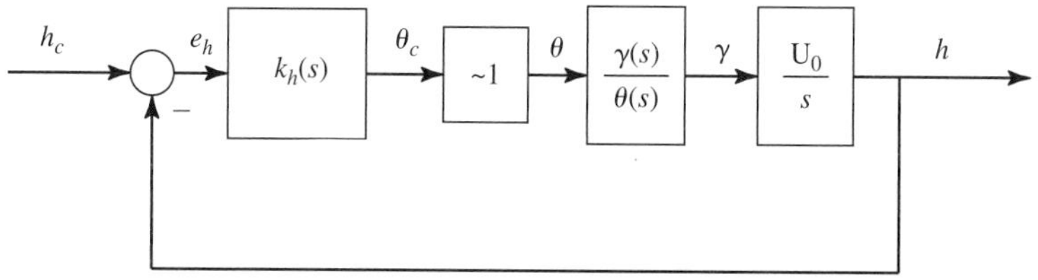

Figure 12.47 Simplified altitude-hold block diagram.

frequency in the outer loop must be well below the gain-crossover frequency of the inner loop. Therefore, we will again assume in the preliminary-design phase that $\theta(s)/\theta_c(s) \approx 1$, and use the complete model for final evaluations.

With the transfer function for the inner attitude-control loop equal to unity, the block diagram for altitude control now simplifies to that shown in Figure 12.47. The flight-path angle γ is derived from a linear combination of pitch attitude θ and angle of attack α, or at small reference bank angle Φ_0 we have

$$\gamma(t) = \theta(t) - \alpha(t) \tag{12.60}$$

Plus, the altitude is governed by

$$\dot{h} = U_0 \sin\gamma \approx U_0\gamma$$

Therefore, it must be possible to determine the relationship between pitch attitude θ and flight-path angle γ from algebraic manipulation of the numerators of the transfer-functions. That is,

$$\frac{\gamma(s)}{\theta(s)} = \frac{\gamma(s)/\delta_E(s)}{\theta(s)/\delta_E(s)} = 1 - \frac{\alpha(s)}{\theta(s)} = 1 - \frac{N^{\alpha}_{\delta_E}(s)}{N^{\theta}_{\delta_E}(s)} = \frac{N^{\gamma}_{\delta_E}(s)}{N^{\theta}_{\delta_E}(s)} \tag{12.61}$$

Recall that the transfer functions with elevator input must be used here because the controller uses only the elevator.

Again the question may be asked as to which set of transfer functions to use—those for the aircraft alone, or those for the aircraft controlled by the pitch-attitude controller. The answer again is either one because in Section 11.2, when discussing the effects of feedback on the transfer-function numerators, we saw that the numerators of all transfer functions with the same input as that used in a feedback-control law (e.g., elevator) would be unaffected by that feedback.

Using Equation (12.61) and the data for the DC-8 in the cruise flight condition (Flight Condition 3), the dynamic relationship between the flight-path angle and the pitch-attitude angle is given by

$$\begin{aligned}\frac{\gamma(s)}{\theta(s)} &= \frac{-0.009184(s+0.01075)(s+9.585)(s-8.237)}{(s+0.01441)(s+0.7247)} \\ &\approx \frac{-0.009184(s+9.585)(s-8.237)}{(s+0.7247)} \approx \frac{0.7247}{(s+0.7247)} \text{ rad/rad}\end{aligned} \tag{12.62}$$

This expression results from using the transfer functions for the vehicle including the pitch-attitude controller. So the transfer functions utilized were $\alpha(s)/\theta_c(s)$ and $\theta(s)/\theta_c(s)$. Again recall that the roots of the denominator in Equation (12.62), or -0.01441 /sec and -0.7247 /sec, are zeros of the $\theta(s)/\theta_c(s)$ transfer function, $-1/T_{\theta_1}$ and $-1/T_{\theta_2}$, not eigenvalues of the system.

The first approximation used in Equation (12.62) is the cancellation of the low-frequency "pole" and "zero." Since the value of s $(= j\omega)$ in the anticipated frequency range of interest is over a decade above the magnitudes of these two terms, both terms are approximately s. The last approximation in Equation (12.62) was obtained by setting s to zero in the two terms in the numerator, since the anticipated frequency range of interest is more than a decade below the magnitudes of these two zeros. The accuracy of these two approximations is revealed in the Bode plots in Figure 12.48. Over the frequency range of 0.1–1.0 rad/sec, the anticipated frequency range of interest, these three models are essentially indistinguishable. So we will proceed here using the simplest, or the first-order, lag model.

Regarding the selection of the gain-crossover frequency in the outer loop of the altitude hold, we know it should be above the frequency of the phugoid mode, and must be less than the crossover frequency of the pitch-attitude inner loop (2 rad/sec here). Also, it is likely that integral action will need to be added in the outer-loop controller to achieve high loop gain $|kg|$ at low frequency, and introducing integral action can reduce the phase margin. So to allow for sufficient phase margin, we will select a crossover frequency of about 0.4 rad/sec, which happens to be a little higher that that of the speed-hold.

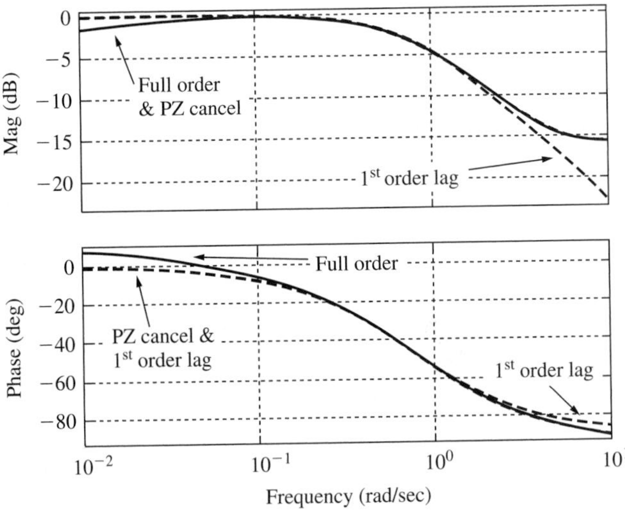

Figure 12.48 Bode plots of $\gamma(s)/\theta(s)$ (Equations (12.62)).

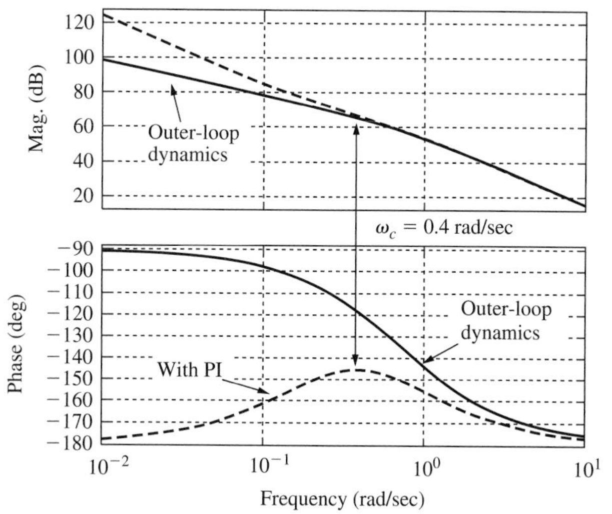

Figure 12.49 Bode plot—effect of PI on altitude-hold outer-loop dynamics (simplified model).

Now with reference to the Bode plots labeled "Outer-loop dynamics" in Figure 12.49, the simplified outer-loop altitude-hold dynamics $h(s)/\theta(s)$ without any outer-loop compensation appear to yield an acceptable loop shape. However, the desirable K/s characteristics evidenced in this Bode plot are in the dynamics (the integral relationship between flight-path angle and altitude) not in the control law. (See Figure 12.47.) One might use a pure-gain controller, or introduce PI compensation to further minimize tracking errors.

Here, proportional-plus-integral compensation will be added, with the initial compensator taken to be

$$k_h(s) = K_h \frac{(s + 0.2)}{s} \text{ rad/ft} \tag{12.63}$$

The zero in this compensator was selected to produce maximum phase at 0.4 rad/sec, the selected gain-crossover frequency in the outer loop, as shown in Figure 12.49. However, the resulting phase margin of only 35 deg will be insufficient (especially given the additional phase loss in the inner attitude loop that we haven't considered yet), so additional phase lead is required.

A lead compensator that produces almost 60 deg of phase lead near 0.4 rad/sec is

$$L(s) = \frac{s + 0.1}{s + 1} \tag{12.64}$$

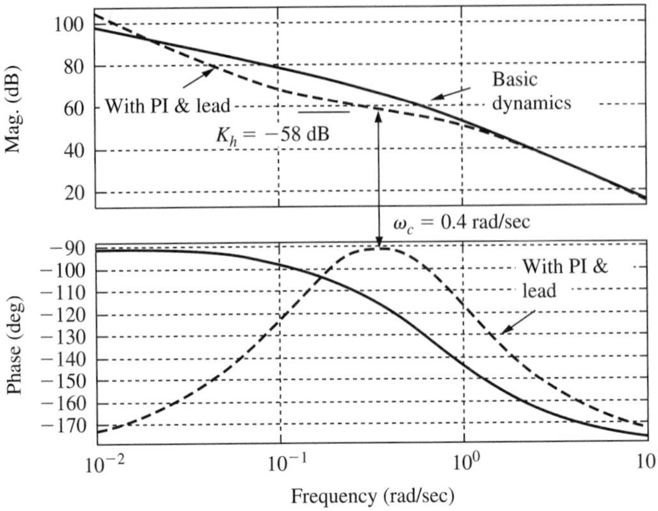

Figure 12.50 Bode plot—altitude-hold outer-loop dynamics w. and wo. PI and lead (simplified model, cruise condition).

The open-loop Bode plot of the altitude controller including the PI plus this lead compensation is now shown in Figure 12.50, and the effect of the compensation is clear. Adding this lead compensation to the control law will yield over 85 deg of phase margin. We could adjust the lead compensator to produce less lead and still have an acceptable phase margin. But accepting the given lead compensator as satisfactory, we note from this Bode plot that selecting an altitude-hold gain K_h for the controller of -58 dB (1.26×10^{-3} rad/ft) will result in the desired gain-crossover frequency of 0.4 rad/sec. The final altitude-hold compensator is

$$k_h(s) = 1.26 \times 10^{-3} \frac{(s + 0.2)}{s} \left(\frac{s + 0.1}{s + 1} \right) \text{rad/ft} \tag{12.65}$$

For reference, the MATLAB commands to produce the Bode plot in Figure 12.50, including the PI and lead compensation, are as follows:

»Z=[];P=[-.7247];K=.7247;g=zpk(Z,P,K)	Define lag model for $\gamma(s)/\theta(s)$ dynamics
»numi=[824];deni=[1 0];int=tf(numi,deni)	Define $h(s)/\gamma(s)$ dynamics
»g=series(g,int)	Model the two transfer functions in series
»numc=[1 .2];denc=[1 0];PI=tf(numc,denc)	Define PI compensator
»numl=[1 .1];denl=[1 1];lead=tf(numl,denl)	Define lead compensator
»comp=series(lead,PI)	Define total compensation (with gain = 1 rad/ft)
»kg=series(comp,g)	Model compensator and outer-loop dynamics in series
»bode(g);hold;bode(kg)	Plot Bode plots

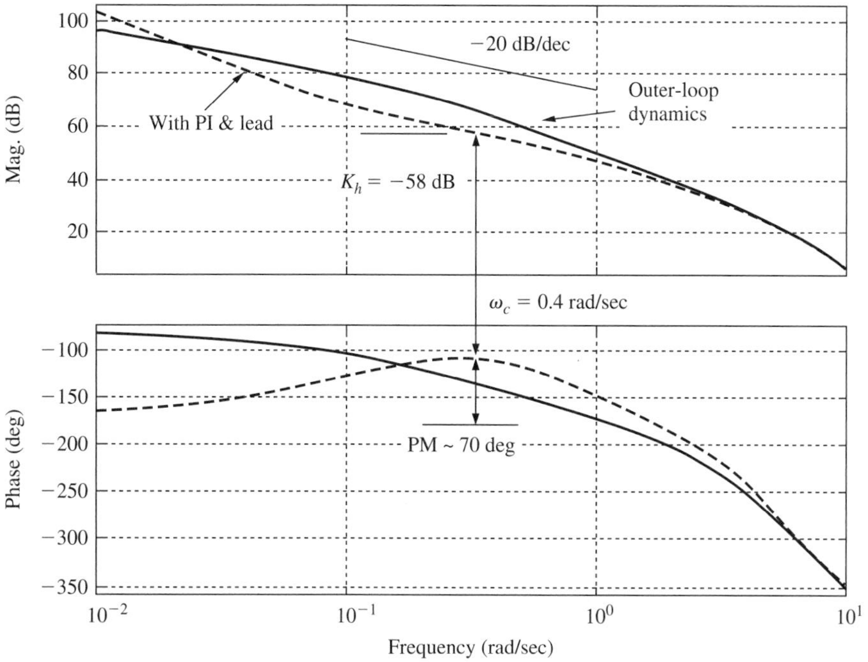

Figure 12.51 Bode plot—effect of compensation on altitude-hold outer-loop dynamics (complete model, cruise condition).

As usual, one always needs to validate the preliminary design of the controller using the full-order model of the vehicle dynamics with the pitch-attitude controller. Using this complete model yields the two Bode plots shown in Figure 12.51. One set labeled "Outer-loop dynamics" constitutes the Bode plot for the vehicle with the attitude controller (or $h(s)/\theta_c(s)$), while the second set labeled "With PI & lead" constitutes the Bode plot for the complete open-loop dynamics for the outer altitude loop ($h(s)/e_h(s)$), including the altitude-control compensator with a gain $K_h = 1$ rad/ft. By comparing this set of Bode plots with those in Figure 12.50, which were obtained using the simplified model of the outer loop dynamics, we see that the simplified model produced quite accurate results in the frequency range of interest. The main differences include the fact that the phase margin in the outer altitude loop is now about 70 degrees, compared to about 85 deg obtained using the simplified model, and that the gain margin is now about 18 dB. As a result, no adjustment in the controller will be made.

The vehicle step responses to a 10 ft commanded change in altitude are shown in Figure 12.52. The altitude response indicates a rise time of approximately 5 seconds, with a small (~10 percent) overshoot. If the rise time is too fast for passenger comfort, the gain-crossover frequency in the outer altitude loop could be reduced, and the outer-loop compensator retuned to this new crossover frequency. To increase altitude, the vehicle pitches up, and the flight-path angle quickly increases. Also, even though the flight velocity is slightly reduced (we are not directly controlling velocity here), the required altitude increase is achieved. Hence, the altitude controller appears to perform satisfactorily.

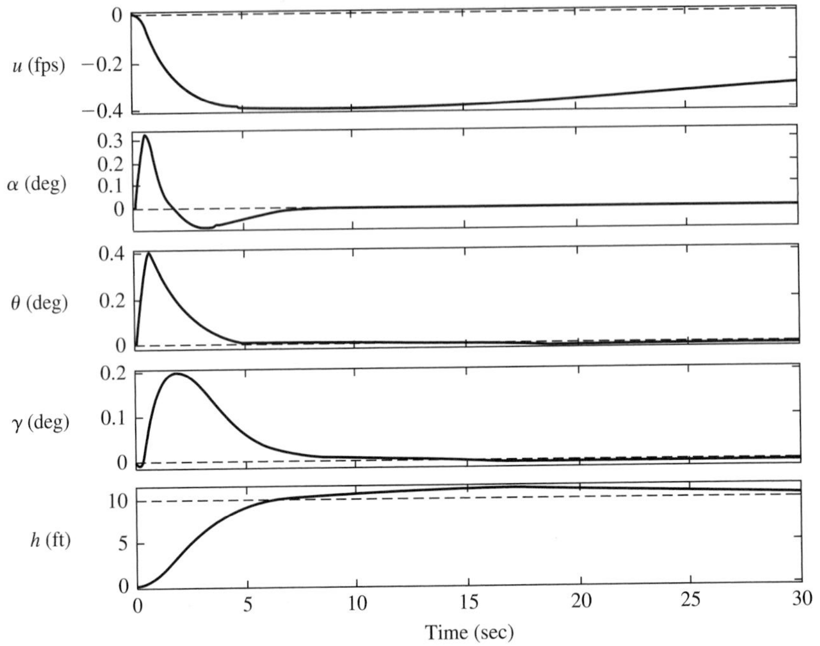

Figure 12.52 Vehicle time responses with altitude hold—h_c = 10 ft (cruise flight condition).

For reference, the MATLAB commands used to obtain the above Bode plots and step responses are as follows:

»syscl=ss(acl,bcl,ca,da)	Define state-variable model of vehicle w. attitude controller
»tfs=zpk(syscl);bode(tfs(5))	Plot Bode plot of $h(s)/\theta_c(s)$
»kgss=series(comp,syscl)	Define state-variable model of altitude compensator (defined previously) in series with vehicle including attitude controller
»tfs=zpk(syscl);bode(tfs(5))	Plot Bode plot of $h(s)/e_h(s)$ with gain K_h = 1 rad/ft
»[akg,bkg,ckg,dkg]=ssdata(kgss)	Retrieve matrices defining model kgss
»aclhhold=akg-bkg*(1.26e-3)*ckg(5,:)	Obtain closed-loop **A** matrix for altitude hold
»bclhhold=(1.26e-3)*bkg	Obtain closed-loop **B** matrix for altitude hold
»btime=10*bclhhold	Obtain **B** matrix for 10 ft commanded altitude change
»ctime=ckg;ctime(4,:)=[0 -1 1 0 0 0 0 0];	
ctime(2:4,:)=57.3*ctime(2:4,:)	Obtain **C** matrix to include flight-path angle and plot angles in degrees
»syshtime=ss(aclhhold,btime,ctime,dkg)	Form closed-loop system for plotting time responses
»step(syshtime,30)	Plot time responses

12.5.3 Heading Hold

This autopilot mode is used to maintain a specific heading ψ such as a "radar vector" commanded by air traffic control. From Section 12.4.4, in a perfectly coordinated turn at constant altitude, the bank angle is related to the rate of change of heading through the following relationship (Equation (12.51))

$$\dot{\psi} = \frac{g}{V_\infty} \tan \phi \tag{12.66}$$

Consequently, after linearization, the block diagram for a heading hold may be depicted as in Figure 12.53. As discussed in Section 12.4.4, in implementing a heading hold we must remember that since constant altitude is assumed, turn compensation must also be utilized in the longitudinal axis.

Note from the block diagram that the outer-loop dynamics $\psi(s)/\phi(s)$ are simply *K/s,* so closing this outer loop using a control law consisting only of a gain $k_\psi(s) = K_\psi$ should lead to excellent phase and gain margins. However, note again that the integration is in the outer-loop dynamics and not in the outer-loop controller $k_\psi(s)$.

Of course, the outer-loop crossover frequency must be less than that for the inner bank-angle loop. Using the data for the DC-8 in the cruise flight condition (Flight Condition 3) and the bank-angle controller discussed in Section 12.4.3, the crossover frequency in that inner-loop controller was approximately 0.75 rad/sec.

Continuing to use the data and bank-angle controller for the DC-8 in cruise, for which $U_0 = 844$ fps, the Bode plot for the heading-angle response from commanded bank angle $\psi(s)/\phi_c(s)$ is shown in Figure 12.54. From this plot, note that at frequencies less than 0.75 rad/sec, the Bode plot exhibits the desired *K/s* characteristics. Consequently, crossover must be selected to be less than 0.75 rad/sec, and we will consider a simple gain $k_\psi(s) = K_\psi$ in this controller, although PI compensation could be considered.

Selecting a crossover frequency of 0.4 rad/sec leads to a phase margin in the outer loop of 60 deg, so 0.4 rad/sec is near the maximum crossover frequency for this heading loop without adding phase compensation. A heading-hold gain K_ψ of approximately 20 dB (10 rad/rad) yields the desired crossover frequency.

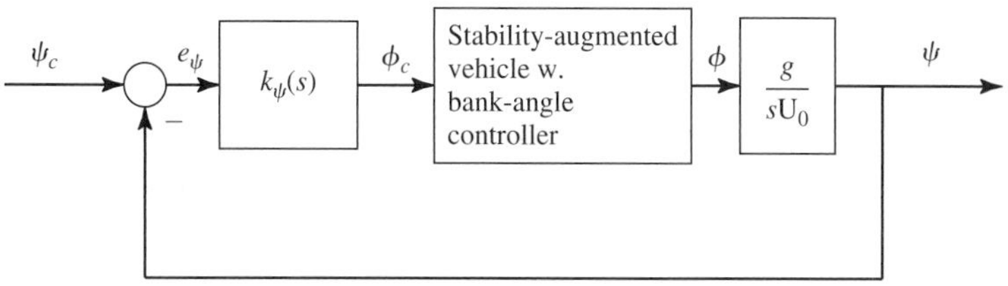

Figure 12.53 Block diagram for heading hold.

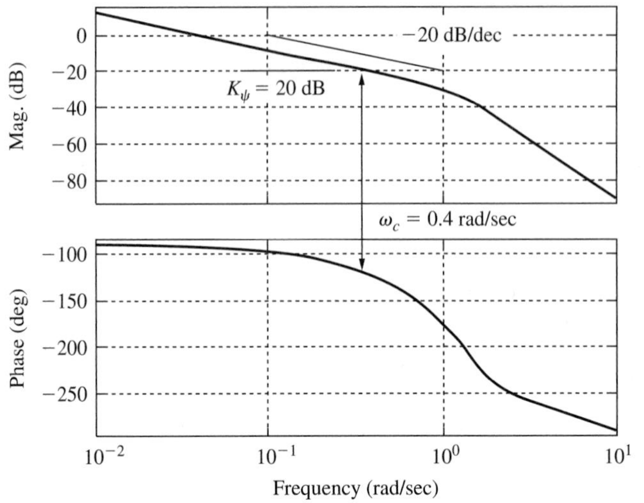

Figure 12.54 Bode plot of heading-hold outer-loop dynamics $\psi(s)/\phi_c(s)$.

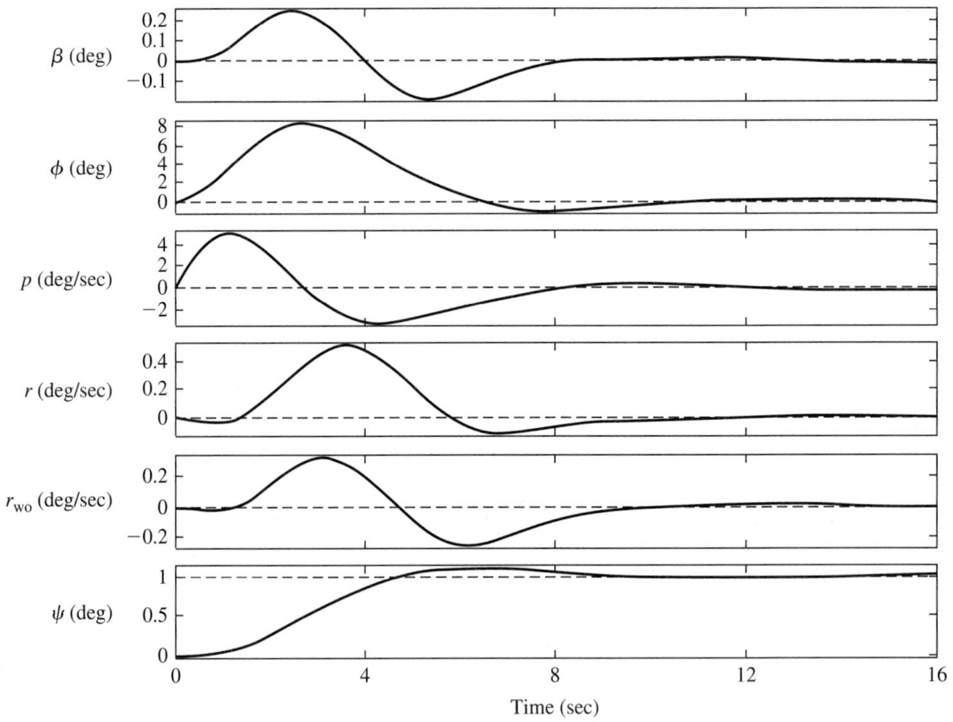

Figure 12.55 Time responses of heading hold (cruise condition).

Using this simple outer-loop gain controller, the step responses for the closed-loop heading hold are shown in Figure 12.55. The responses correspond to a one degree commanded step change in heading angle. Note that the heading response exhibits a rise time of approximately 4 seconds, with little overshoot and no steady-state error. The vehicle rolls right to accomplish the heading change. There are some sideslip excursions during the maneuver, but the dutch-roll mode is not evident in the bank-angle response. The maximum bank angle for this commanded heading change is 8 degrees. If the bank angle is deemed excessive, either the outer-loop gain-crossover frequency may be reduced, or a limiter, as discussed below, may be added. However, these time responses appear to be satisfactory.

The MATLAB commands to assemble the model for the heading hold and obtain the above time responses are as follows:

»numol=[(32.2/824)];denol=[1 0];OL=tf(numol,denol)	Outer-loop kinematics
»syspsi=series(syscl,OL,[2],[1])	syscl is ϕ/ϕ_c closed loop
»[apsi,bpsi,cpsi,dpsi]=ssdata(syspsi)	
»aclpsi=apsi-bpsi*7.94*cpsi	Close heading loop
»bclpsi=bpsi*7.94	
»cclpsi=ca;cclpsi=[[0;0;0;0;0] ca];cclpsi(6,:)=cpsi	Define responses for plotting
»dclpsi=[0;0;0;0;0;0]	
»sysclpsi=ss(aclpsi,bclpsi,cclpsi,dclpsi)	Closed-loop heading hold system
»step(sysclpsi)	

Sometimes bank angle must be limited for passenger comfort. In such a case, a limiting circuit or algorithm may be used that has characteristics similar to those shown in Figure 12.56. Such a limiter may be placed on the commanded bank angle ϕ_c, at the output of the outer-loop heading-hold controller shown in Figure 12.53. Introducing such a device will reduce the speed of response in aggressive maneuvers, and also introduces a nonlinearity in the system. Hence the performance of such systems is frequently assessed through nonlinear simulations.

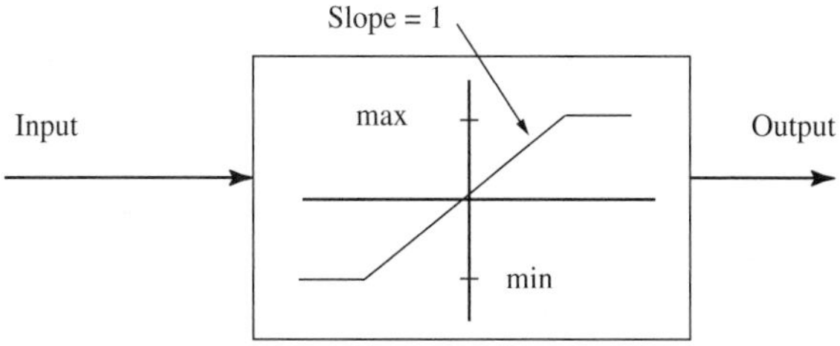

Figure 12.56 Command limiter.

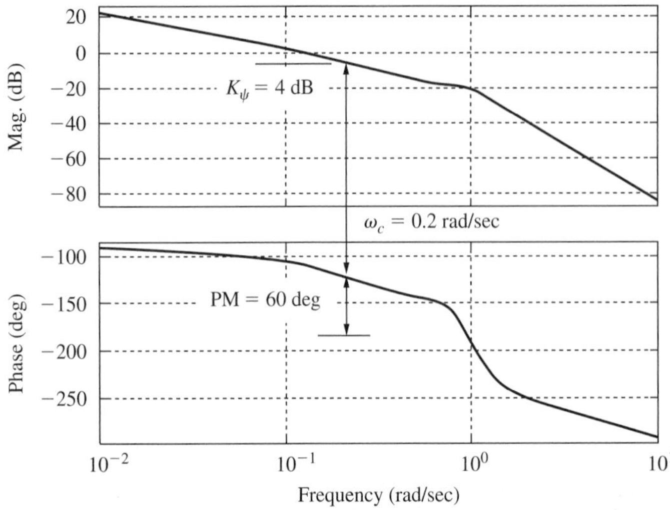

Figure 12.57 Bode plot—open-loop heading hold w. bank-angle controller (approach condition).

We will now synthesize another heading-hold controller for use in an approach flight condition. This heading hold will be used in Section 12.6.2, when dealing with lateral beam guidance. Using the data for the DC-8 in the approach condition (Flight Condition 1 in Appendix B), the Bode plot of the outer-loop heading dynamics $\psi(s)/\phi_c(s)$ is shown in Figure 12.57. These dynamics include the closed-loop bank-angle controller for the approach flight condition discussed in Section 12.4.3. Recall that the gain-crossover frequency in this heading loop must be less than the crossover-frequency of the bank-angle controller (~0.5 rad/sec in this case).

Only a gain K_ψ will be used in this heading hold. From the phase plot in Figure 12.57, we see that we may select the heading-hold crossover frequency to be 0.2 rad/sec and obtain a phase margin of 60 deg. (For reference, the crossover frequency for the heading hold in the cruise flight condition, discussed previously, was higher at 0.4 rad/sec.) The control gain K_ψ necessary to achieve the crossover frequency of 0.2 rad/sec is 4 dB (1.585 rad/rad).

The step responses for this controller to a commanded change in heading of one degree are shown in Figure 12.58. Of course, in this approach flight condition the dynamics are naturally slower, plus the crossover frequency in this heading loop is lower than that for the cruise flight condition. All this leads to the slower time responses shown. For example, the rise time for the heading response is now approximately 9 sec, about twice that for the cruise flight condition. Also, the dutch-roll mode is more evident in these responses, except for that of heading.

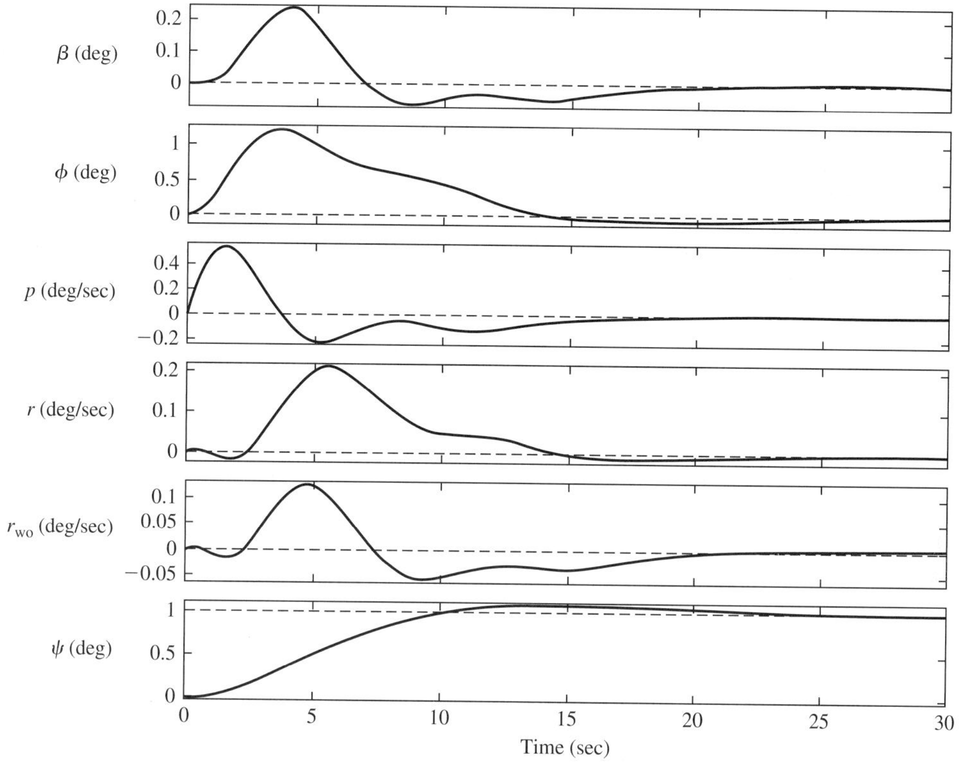

Figure 12.58 Time responses of heading hold (approach condition).

12.6 PATH GUIDANCE—ILS COUPLERS AND VOR HOMING

We now turn our attention to the subject of path-following guidance, where the objective is to control the vehicle such that it follows a prescribed path in space. Thus we wish to minimize the deviation from this path, or the position error, and we would expect this position error to define the outer loop of the controller. Again, the inner loop will consist of the vehicle under attitude control, as in all the cases considered previously in this chapter.

In this section we will consider two specific cases of path-following guidance—that of following the landing-approach path defined by an Instrument Landing System (ILS) and that of VOR homing. The ILS follower provides automatic path guidance during approach to landing, to enhance the safety of landing operations in inclement weather, and the VOR homing device is used for enroute navigation and guidance.

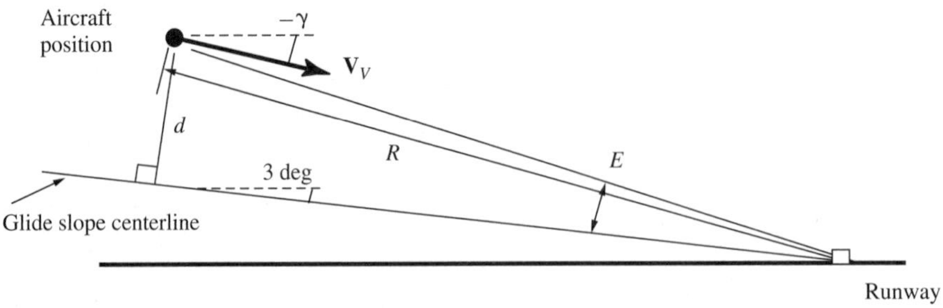

Figure 12.59 Geometry of ILS glide slope and vehicle position.

12.6.1 Longitudinal Path Guidance

Consider the schematic shown in Figure 12.59, which shows the side view of an Instrument Landing System (ILS). An antenna is located near the threshold of the runway, and the ground-based system interacts with an ILS receiver onboard the vehicle. The airborne instrument measures angular deviation E from the desired approach trajectory, or the glide-slope centerline, which is typically elevated at about 3 degrees from horizontal.

The instantaneous position of the aircraft is indicated, with its range to the antenna denoted as R. The vehicle's flight velocity vector $\mathbf{V}_V$ and instantaneous flight-path angle γ are also shown (the flight-path angle is negative). The instantaneous glide-slope offset, or position error, is indicated as d in the figure.

From the geometry of the situation one can derive the following relationships between all these quantities.

$$\sin E = d/R,\ E \approx d/R$$
$$\dot{R} = -V_V \cos\gamma \approx -V_V \tag{12.67}$$

In terms of reference and perturbation flight-path angle, the flight path may be expressed as

$$\gamma = \Gamma_0 + \gamma \tag{12.68}$$

Assume that the reference flight-path angle $\Gamma_0 = -3$ deg, a stability axis is used, and $V_{V_0} = U_0$. Then again from the geometry we have

$$\dot{d} = U_0 \sin(3° + \gamma) = U_0 \sin(3° + \Gamma_0 + \gamma) = U_0 \sin(\gamma) \approx U_0 \gamma \tag{12.69}$$

Also assuming the angular error E is the measured quantity that will be fed back, the block diagram for the glide-slope-following controller, or *glide-slope coupler*, may be depicted as shown in Figure 12.60. By comparing this figure to Figure 12.47, the block diagram for the altitude controller, we can see that the glide-slope coupler is essentially altitude control. The main difference here is that for a given offset distance d, the angular error E increases with decreasing

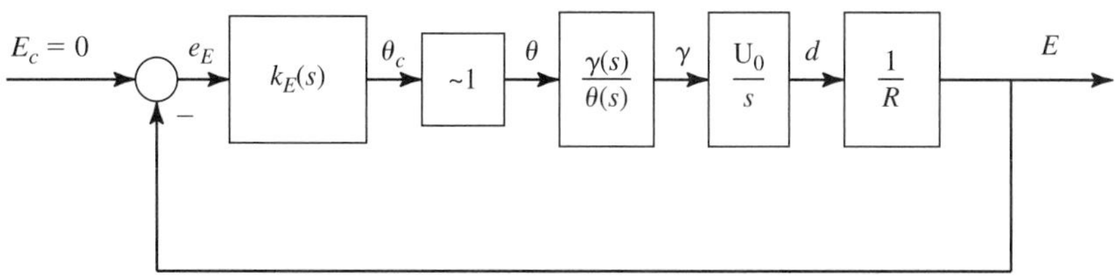

Figure 12.60 Block diagram of glide-slope coupler.

range to the antenna. But range is assumed to be measured or computed, so the gain K_E in the controller $k_E(s)$ may be adjusted to keep the effective gain K_E/R constant at the design value if necessary.

To make the discussion more concrete, we will discuss the synthesis of a glide-slope coupler using the data for the DC-8 in the approach flight condition (Flight Condition 1). Recall that we previously synthesized an attitude controller for this case in Section 12.4.1. However, also recall from the discussion of the time responses in that section that although the attitude controller was effective at controlling attitude, a positive change in steady-state attitude did not result in a positive change in steady-state flight-path angle. So it was noted that an auto-throttle was required, which will now be addressed.

Auto-Throttle For the synthesis of an auto-throttle, first consider the block diagram shown in Figure 12.61. The first question that must be addressed is whether the auto-throttle should be designed using the open-loop vehicle dynamics, or whether the design should proceed using the vehicle dynamics including a complete pitch-attitude control system with pitch damper. Either approach might work because of the inherent frequency separation between the flight-velocity and pitch-attitude responses, and the fact that the short-period mode is essentially independent of surge velocity u. But based on the rule of thumb that one should close the faster feedback loops (i.e., higher gain-crossover frequency) first, the attitude-control loops will be closed first, and then the auto-throttle control law will be synthesized.

The attitude-control block diagram was shown in Figure 12.21, and for the DC-8 in the approach flight condition with a pitch damper the attitude-control

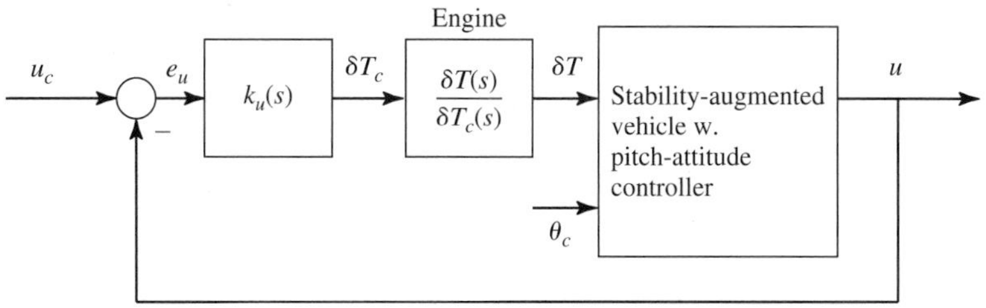

Figure 12.61 Block diagram of auto-throttle.

compensator was given in Equation (12.36). Now the thrust-to-surge-velocity transfer function (with this attitude controller and pitch damper in place) is

$$\frac{u(s)}{\delta T(s)}\Big|_{AC} = \frac{0.169s(s^2+0.4717s+0.07557)(s^2+3.309s+7.09)(s+17.9)}{s(s+0.04924)(s^2+0.4493s+0.08768)(s^2+3.312s+7.092)(s+17.9)} \text{ fps/Klbs}$$
$$\approx \frac{0.169}{(s+0.0492)} \text{ fps/Klbs} \tag{12.70}$$

The transfer functions for angle of attack and pitch rate are essentially zero. Now with the attitude-controller in place there is almost exact pole-zero cancellation of the short-period poles and the elevator-actuator pole in Equation (12.70). This is not surprising given the discussion in Section 11.3.1 on the effect of pitch-attitude or pitch-rate feedback on this transfer function. Plus, the elevator actuator should clearly not affect this transfer function. However, the attitude controller also yields approximate pole-zero cancellation of the phugoid poles, and the velocity response is now dominated by a single mode having a pole at −0.049 /sec.

The transfer functions for thrust input may be found using the following MATLAB commands.

»sys=ss(a,bdel,c,d)	Vehicle model with elevator input
»dena=[1 20];aa=tf(-20,dena);	Elevator actuator model
»sysaa=series(aa,sys)	Actuator and vehicle model in series
»[aaa,ba,ca,da]=ssdata(sysaa)	State-variable matrices obtained
»aaaug=aaa-ba*1.5*ca(4,:);	Vehicle **A** matrix augmented with pitch damper
»sysaug=ss(aaaug,ba,ca,da)	Define state-variable model for augmented system
»numc=[1 .2];denc=[1 0];comp=tf(numc,denc)	Define PI compensation
»kg=series(comp,sysaug)	PI and augmented system in series
»[akg,bkg,ckg,dkg]=ssdata(kg)	Retrieve state-model matrices for *kg*
»acl=akg-bkg*3.16*ckg(3,:)	Closed-loop $\mathbf{A}_{CL\theta}$ matrix w. pitch controller
»bclt=[1000/(230000/32.2);0;0;0;0;0]	**b** matrix for thrust input in units of Klbs
»sysclt=ss(acl,bclt,ckg,dkg)	Pitch-controlled system w. thrust input
»tfclt=zpk(sysclt)	Transfer functions

A simple first-order lag model is selected for the engine dynamics. Since the response of a propulsion system is considerably slower than that of an aerodynamic-surface actuator, an engine time constant of 0.2 sec has been selected. Adding the engine lag, the transfer function in Equation (12.70) becomes

$$\frac{u(s)}{\delta T_c(s)} = \frac{0.847\left(s^2 + 0.4717s + 0.07557\right)\left(s^2 + 3.309s + 7.09\right)(s + 17.9)}{(s + 0.04924)\left(s^2 + 0.4493s + 0.08768\right)\left(s^2 + 3.312s + 7.092\right)(s + 17.9)(s + 5)} \text{ fps/Klbs}$$
$$\approx \frac{0.847}{(s + 0.0492)(s + 5)} \text{ fps/Klbs} \qquad (12.71)$$

Using the simpler second-order model for the preliminary design, its Bode plot is shown in Figure 12.62.

Regarding the selection of the gain-crossover frequency in the auto-throttle, examination of Equation (12.71) indicates that to adequately control the velocity response, the crossover frequency should be above 0.05 rad/sec. Recall that the crossover frequency selected for the pitch-attitude controller was 1 rad/sec. Since it is not reasonable to expect the vehicle's velocity response to be as fast as its pitch-attitude response, the gain-crossover frequency should be less than this 1 rad/sec. In addition, from Figure 12.62, if a crossover frequency above 1 rad/sec were selected, phase-lead compensation would likely need to be added. Consequently, an intermediate gain-crossover frequency of approximately 0.6 rad/sec may be selected.

We will try to avoid introducing proportion-plus-integral compensation in the auto-throttle control law, which would increase the Bode gain at low frequencies, and determine later if the steady-state error can be tolerated. Therefore only a gain K_u will be used in the preliminary design of the auto-throttle control law. From the Bode plot in Figure 12.62, we see that selecting a gain K_u of 10 dB (3.16 Klbs/fps) will set the gain-crossover frequency near the desired value of 0.6 rad/sec, and the resulting phase margin is 90 deg.

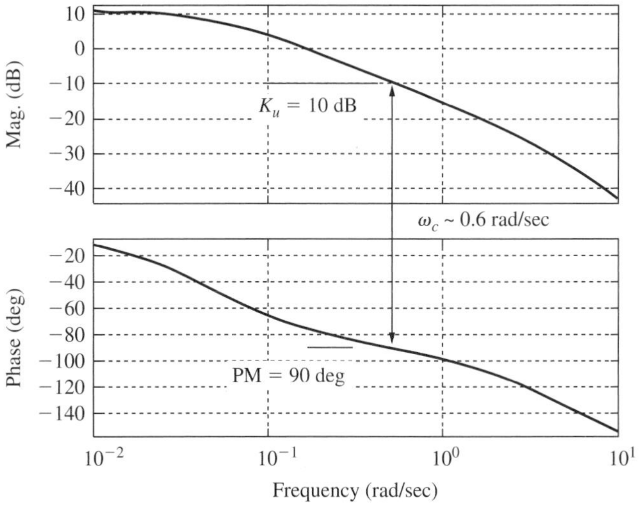

Figure 12.62 Bode plot—speed response to thrust command $u(s)/\delta T_c(s)$ (simplified model, Equation (12.71)).

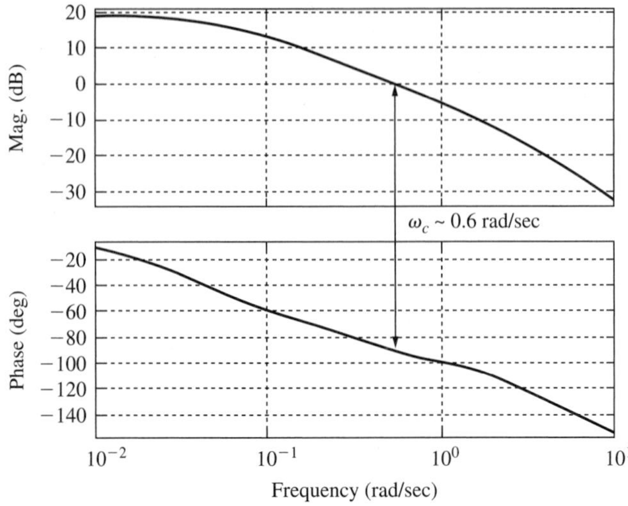

Figure 12.63 Bode plot of $K_u u(s)/\delta T_c(s)$ (w. attitude controller), $K_u = 3.16$ Klbs/fps (complete vehicle model).

This preliminary design will now be evaluated using the complete model for the vehicle's velocity response (in Equation (12.71)). Using the gain of $K_u = 3.16$ Klbs/fps, along with the complete dynamic model, yields the Bode plot for the open-loop auto-throttle controller, or $K_u u(s)/\delta T_c(s)$, shown in Figure 12.63. Note that the gain-crossover frequency is approximately 0.6 rad/sec, and the phase margin remains at ~90 deg.

Selecting this value of auto-throttle gain yields closed-loop (auto-throttle plus attitude control) eigenvalues of

$$\text{Aircraft dynamics and compensation} \begin{bmatrix} -1.6562 \pm j2.0853 \text{ /sec} \\ -0.2829 \pm j0.1160 \text{ /sec} \\ -0.5461 \text{ /sec} \end{bmatrix}$$

$$\text{Engine} \quad -4.39 \text{ /sec}$$

$$\text{Elevator Actuator} \quad -17.9 \text{ /sec}$$

The step responses to a commanded 1 fps increase in flight velocity are shown in Figure 12.64. The speed-response rise time is consistent with the 0.6 rad/sec gain-crossover frequency, and no overshoot is exhibited. But a steady-state error of approximately −10 percent is evident. Note that with the pitch-attitude held essentially constant, the angle of attack declines as a result of the speed increase, and the flight-path angle increases. So with a fixed attitude (due to the presence of the attitude controller), a commanded increase in flight velocity produces an increase in altitude, as we would expect. We will accept this auto-throttle design for now, subject to further adjustment if necessary.

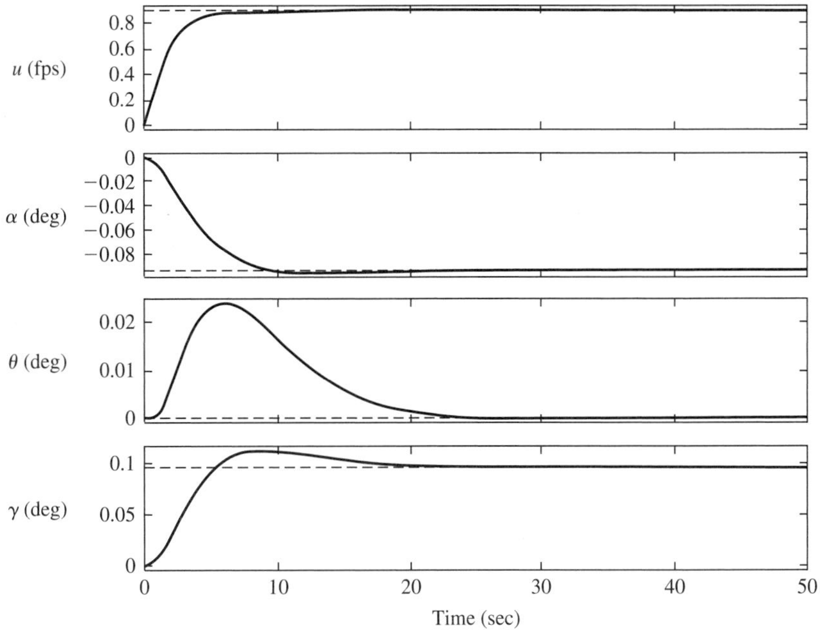

Figure 12.64 Vehicle step responses of auto-throttle and attitude controller (complete vehicle model, approach condition).

The MATLAB commands to assemble the model of the vehicle including the auto-throttle are as follows:

`»bclt=[1000/(230000/32.2);0;0;0;0;0]`	**B** matrix for thrust input in units of Klbs
`»sysclt=ss(acl,bclt,ckg,dkg)`	Pitch-controlled system w. thrust input
`»aclat=acl-bclt*3.16*ckg(1,:);`	Close auto-throttle loop
`»aclat=bclt*3.16;`	
`»sysclat=ss(aclat,bclat,ckg,dkg);`	Closed-loop system w. auto-throttle

We may now reevaluate the time response of the pitch-attitude controller <u>with</u> the auto-throttle. Recall from Figure 12.30 that without the auto-throttle, a commanded increase in pitch attitude did not result in a positive steady-state flight-path angle. But now consider the time responses shown in Figure 12.65, which shows the one-degree step responses of the attitude controller <u>with</u> auto-throttle for the DC-8 in the approach flight condition. By comparing these new time responses with those in Figure 12.30, we can see that <u>with</u> the auto-throttle the velocity loss is reduced, and an <u>increase</u> in pitch attitude now produces the desired <u>increase</u> in flight-path angle.

Glide Slope Coupler w. Auto-Throttle We now continue with the synthesis of the glide-slope coupler. With respect to the block diagram in Figure 12.60, the model for the outer-loop dynamics relating flight path and attitude must be revisited. As noted when discussing the altitude hold, one model for the outer-loop

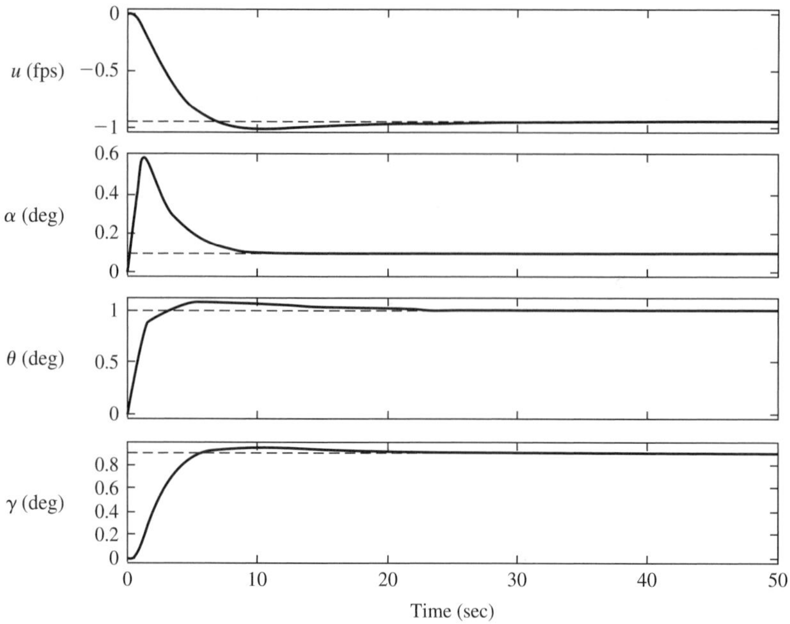

Figure 12.65 Step responses of pitch-attitude controller w. auto-throttle (approach condition).

dynamics could be developed using the model for the vehicle alone, and another using the model for the vehicle including the pitch-attitude controller. These two models are no longer equivalent because of the presence of the auto-throttle. As discussed in Section 11.2, feedback to the elevator will affect the numerators of the transfer functions with thrust input and vice versa, due to the coupling numerators.

To demonstrate this fact, for the DC-8 in the approach flight condition the dynamic relationship between flight path and attitude angle for the vehicle alone is

$$\frac{\gamma(s)}{\theta(s)} = \frac{-0.031186(s - 0.001692)(s + 4.835)(s - 3.751)}{(s + 0.06049)(s + 0.5352)} \text{ rad/rad} \quad (12.72)$$

(Since glide-slope error will be controlled through the elevator, transfer functions for elevator input were used in developing the above expression.) But for the vehicle with the pitch-attitude controller and auto-throttle, the dynamic relationship becomes

$$\frac{\gamma(s)}{\theta(s)}\Big|_{\text{AC+AT}} = \frac{-0.031186(s + 0.6085)(s + 4.815)(s + 4.411)(s - 3.752)}{(s^2 + 1.21s + 0.3828)(s + 4.386)} \text{ rad/rad}$$
$$\approx \frac{-0.031186(s + 0.6085)(s + 4.815)(s - 3.752)}{(s^2 + 1.21s + 0.3828)} \approx \frac{0.5634(s + 0.6085)}{(s^2 + 1.21s + 0.3828)} \text{ rad/rad} \quad (12.73)$$

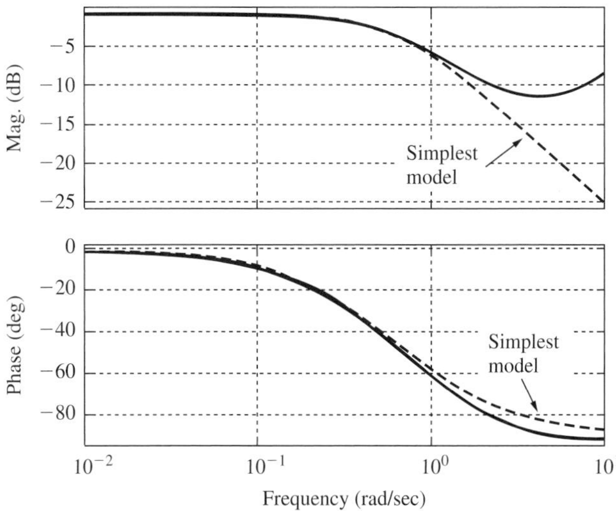

Figure 12.66 Bode plots of three models for $\gamma(s)/\theta(s)$ (Equation (12.73)).

Transfer functions for commanded-pitch-attitude input were used in developing Equation (12.73). Clearly Equations (12.72) and (12.73) are not identical, and we must use the model given in Equation (12.73).

The Bode plots for the three models given in Equation (12.73) are shown in Figure 12.66. Based on these results, using the intermediate model after the pole-zero cancellations near −4.4 /sec is suggested, since it closely matches the more complex model in the frequency range of interest (0.1–1.0 rad/sec).

To complete the model of the outer-loop dynamics, note that the reference flight velocity U_0 for this flight condition is 244 fps. We will select the range to the ILS antenna to use in the control-law design to be $R = 1200$ ft. For these values of parameters, the model for the outer-loop dynamics becomes

$$\begin{aligned}\frac{E(s)}{\theta_c(s)} &= \frac{\theta(s)}{\theta_c(s)}\frac{\gamma(s)}{\theta(s)}\frac{U_0}{s}\frac{1}{R} \approx (1)\frac{1}{R}\frac{U_0}{s}\frac{-0.031186(s+0.6085)(s+4.815)(s-3.752)}{(s^2+1.21s+0.3828)} \\ &= \frac{-0.006341(s+0.6085)(s+4.815)(s-3.752)}{s(s^2+1.21s+0.3828)}\ \text{rad/rad}\end{aligned} \quad (12.74)$$

The Bode plot for this model is shown labeled "Outer-loop dynamics" in Figure 12.67.

To select the desired gain-crossover frequency, note that the crossover frequency in the (inner-loop) pitch-attitude controller is 1 rad/sec. Letting the maximum allowable gain-crossover frequency in the (outer-loop) glide-slope coupler to be approximately one fifth of that, 0.2 rad/sec is selected here.

To minimize steady-state errors in the path-following control law, proportional-plus-integral compensation may be added, and the effects of PI compensation are shown in a second set of Bode plots labeled "With PI" in Figure 12.67.

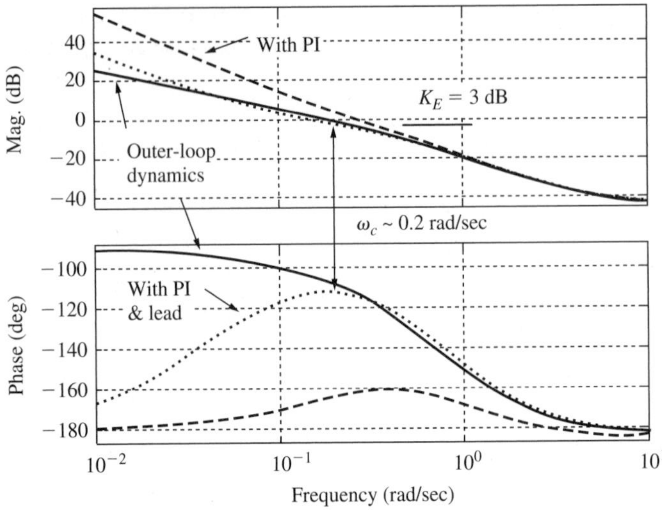

Figure 12.67 Bode plots of outer-loop dynamics of glide-slope coupler (simplified flight-path model).

From this set of Bode plots we note that lead compensation will also be required to achieve an acceptable phase margin. With lead compensation required anyway, we selected a PI zero of 0.3 /sec. Therefore, the preliminary design for the control compensation, including a lead compensator, is

$$k_E(s) = K_E \frac{(s+0.3)}{s} \frac{(s+0.04)}{(s+0.4)} \text{ rad/rad} \tag{12.75}$$

The effects of this total compensation is shown in a third set of Bode plots labeled "With PI & lead" in Figure 12.67. Finally, from this latter set of Bode plots we note that an outer-loop gain K_E of 3 dB (1.412 rad/rad) will result in the desired gain-crossover frequency of approximately 0.2 rad/sec.

The evaluation of this preliminary design is now addressed, using the model of the complete vehicle. The Bode plot for the vehicle dynamics ($E(s)/\theta_c(s)$ in Equation (12.74)), including the pitch-attitude controller and auto-throttle, is shown in Figure 12.68. This Bode plot is quite similar to that for the simplified model reflected in Figure 12.67, so no large adjustments in the control compensation are expected. However, based on further analysis using Figure 12.68, an adjustment in the lead compensation was performed such that the maximum phase lead was centered at the desired value of gain-crossover frequency, or 0.2 rad/sec. Also, the ILS controller gain K_E was increased slightly to 1.585 rad/rad (4 dB) to set the gain-crossover frequency to 0.2 rad/sec. The final compensator is given below.

$$k_E(s) = 1.585 \frac{(s+0.3)}{s} \frac{(s+0.06)}{(s+0.6)} \text{ rad/rad} \tag{12.76}$$

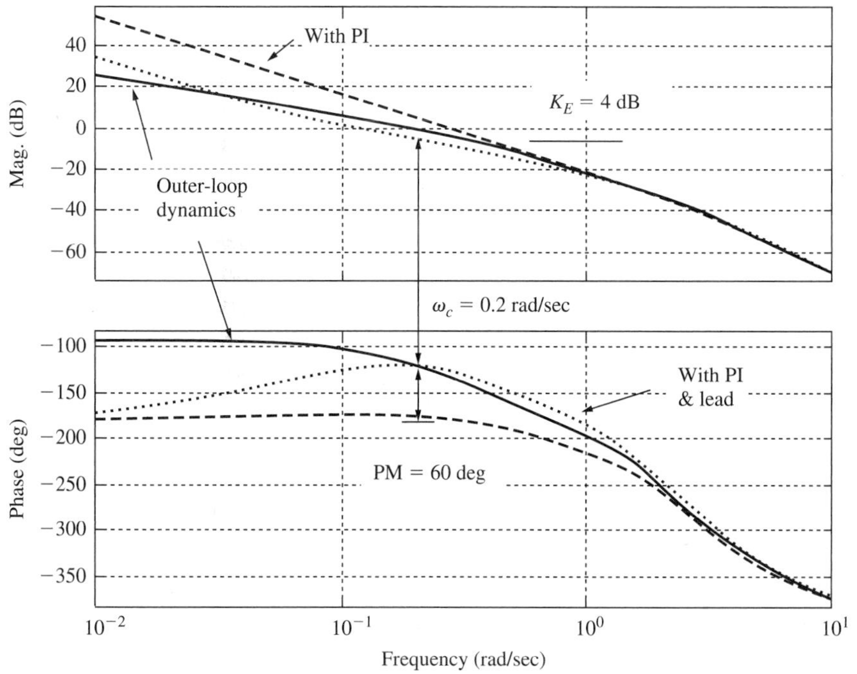

Figure 12.68 Open-loop Bode plot for glide-slope coupler $E(s)/\theta_c(s)$ (complete vehicle model).

The individual effects of the PI and lead compensators are evident in the Bode plots shown in Figure 12.68, and with this final control compensator the phase margin is 60 deg and the gain margin is approximately 20 dB.

Using the above control law, the closed-loop eigenvalues for the vehicle controlled by the glide-slope coupler, including the auto-throttle and pitch-attitude controller, are

$$\text{Aircraft plus compensators}\left|\begin{array}{l}-1.6952 \pm j2.0661\ /\text{sec}\\ -0.26152 \pm j0.29020\ /\text{sec}\\ -0.60690\ /\text{sec},\quad -0.36310\ /\text{sec}\\ -0.071764 \pm j0.052210\ /\text{sec}\end{array}\right.$$

$$\text{Actuator}\quad -17.9\ /\text{sec}$$

$$\text{Engine}\quad -4.39\ /\text{sec}$$

Also, the initial-condition responses are shown in Figure 12.69. To obtain these results all initial conditions on the states are zero, except the initial perturbation pitch attitude $\theta(0) = 3$ deg (such that the initial perturbation flight path $\gamma(0) = 3$ deg)[2] and the initial position offset was $d(0) = -40$ ft. This set of

[2] Recall for this analysis, the reference flight-path angle $\Gamma_0 = -3$ deg.

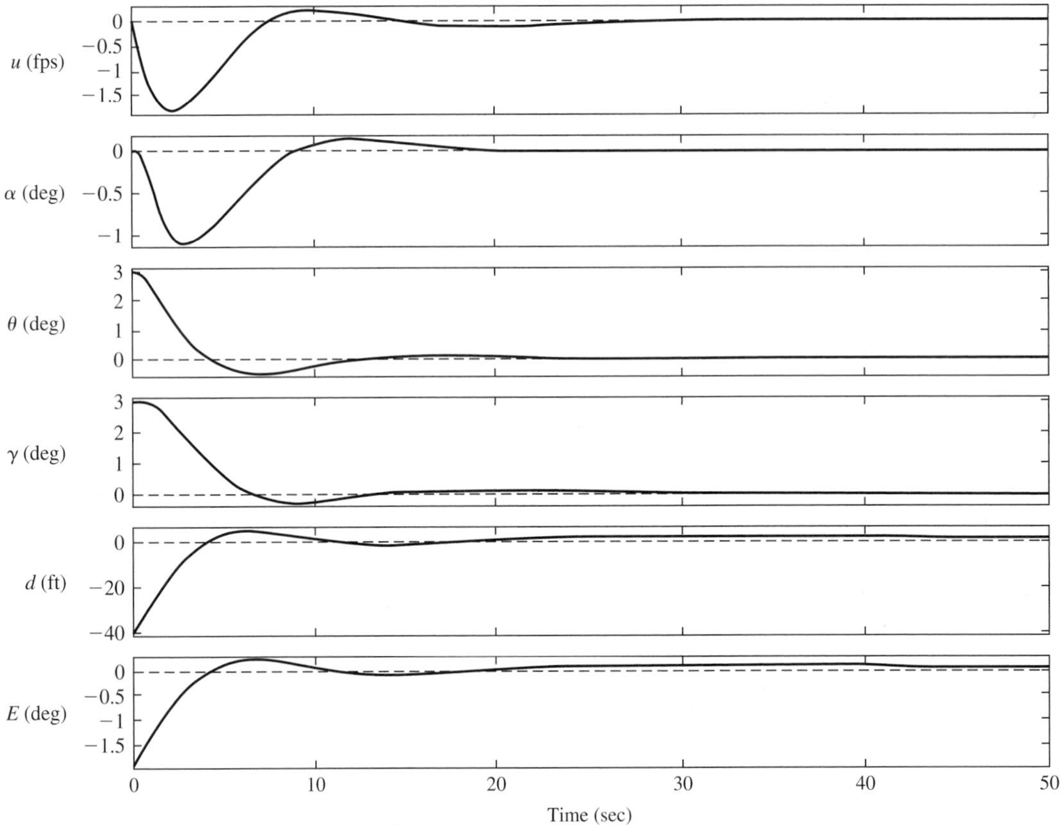

Figure 12.69 Initial condition responses of ILS glide-slope coupler.

initial conditions simulates the case in which the pilot engages the coupler while in level flight at an offset distance of 40 ft below the glide slope. Note that the steady-state errors are all zero and the position offset d error remains less than 5 ft after 4 sec. The aircraft initially pitches down and slows slightly to set up and stabilize on the glide slope, which is accomplished in approximately 10 seconds after engaging the coupler.

The MATLAB commands for assembling the model for the vehicle with the above glide-slope coupler and auto-throttle are as follows:

»sysacAT=ss(aclat,bcl,ckg,dkg);	System w. attitude control and auto throttle, θ_c input
»ckg(5,:)=[0 -1 1 0 0 0];dkg=[dkg;0];	Add γ to responses
»num2=[244/1200];den2=[1 0];OL=tf(num2,den2);	E/γ outer loop
»sysOL=series(sysacAT,OL,[5],[1]);	
»[aOL,bOL,cOL,dOL]=ssdata(sysOL);	
»cgde=[[0;0;0;0;0] ckg];cgde=[cgde;1200*cOL;cOL];dgde=[dkg;0;0];	Add d and E to **y**

```
»sysOL=ss(aOL,bOL,cgde,dgde);
»numpi=[1 .3];denpi=[1 0];pi=tf(numpi,denpi);        PI compensator
»numl=[1 .06];denl=[1 .6];lead=tf(numl,denl);        Lead compensator
»comp=pi*lead;
»syscomp=series(comp,sysOL);
»[acomp,bcomp,ccomp,dcomp]=ssdata(syscomp);
»aILS=acomp-bcomp*1.585*ccomp(7,:);                  Close glideslope loop
»bILS=bcomp*1.585;
»sysgs=ss(aILS,bILS,ccomp,dcomp);                    Closed-loop system w. coupler
```

12.6.2 Lateral-Directional Path Guidance

Lateral-directional path guidance involves controlling the vehicle in such a way so as to follow a prescribed path in the horizontal plane. One example is VOR homing guidance, which involves following a radial track that passes over an air-traffic-control navigation radio station known as a VOR, or Omni. Another is ILS localizer guidance, which involves following the ILS localizer beam aligned with the centerline of a runway.

These two types of guidance systems are basically equivalent problems in terms of control-law development. The three major differences are (1) the typical flight velocities involved, (2) the width of the beam that provides the position-error signal for the control law, and (3) range to the station. VOR homing is frequently used during cruise or during climbing and descending while entering or exiting a terminal area, and cruise flight velocities of 700 fps are common. But following the ILS localizer during landing approach typically involves flight velocities near 200 fps. Secondly, the beam width for a VOR is around 10 degrees, compared to around 3 degrees for the ILS localizer. And third, station ranges under five miles are typical for tracking an ILS localizer, while station ranges of hundreds of miles are common for VOR homing. Consequently, allowable position offsets differ widely between the two cases.

The geometry in the horizontal plane for either guidance problem is shown in Figure 12.70. The offset, or position error, is shown as d, the range to the station is shown as R, and the angular position error is given as E. From the figure, it is clear that

$$\sin E = d/R \tag{12.77}$$

and the kinematics for either guidance problem are given by

$$\dot{d} = V_V \sin\left(\psi - \Psi_{\text{Ref}}\right) \approx V_V\left(\psi - \Psi_{\text{Ref}}\right) \tag{12.78}$$

where ψ is the vehicle heading measured clockwise from north.

After linearizing the expressions in Equations (12.77) and (12.78), we may draw the block diagram for the lateral-beam tracking system as shown in Figure 12.71. Note that the innermost feedback loop is the bank-angle-controled system

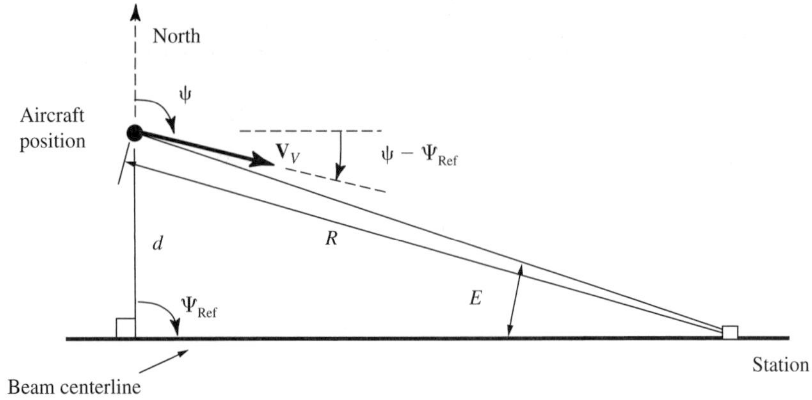

Figure 12.70 Geometry for lateral-beam guidance.

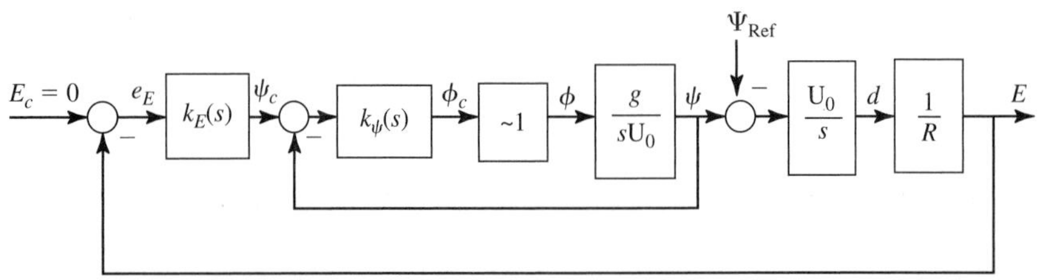

Figure 12.71 Block diagram for lateral-beam guidance.

$\phi(s)/\phi_c(s)$ (~1), the next-most inner loop is the heading-hold loop $\psi(s)/\psi_c(s)$ (~1), and the dynamics of the outer-most loop $E(s)/\psi(s)$ are simply K/s, where here K is U_0/R. So, as with ILS glide-slope tracking, once again the kinematics are range dependent, and adjusting the outer-loop gain in $k_E(s)$, or K_E, such that $K_E U_0/R$ is constant may be required.

VOR Homing First consider the case of VOR homing. We will use the data for the DC-8 in the cruise flight condition (Flight Condition 3), a range to the station R = 30,000 ft (~5 nm), the bank-angle-controller discussed in Section 12.4.3, and the heading hold discussed in Section 12.5.3. The Bode plot for the open-loop beam-tracking system, or $E(s)/\psi_c(s)$, with the two inner feedback loops closed, is shown labeled "Without PI" in Figure 12.72.

Since the outer loop is again K/s below the crossover frequency in the heading-hold loop (0.4 rad/sec here), a simple gain and an outer-loop gain-crossover frequency at or below 0.2 rad/sec would produce acceptable stability margins. But improved steady-state performance with lateral gusts, for example, is obtained with integral action in the control law. Consequently, proportional-plus-integral (PI) control will be introduced here, but this will reduce the phase at the crossover frequency. So either the magnitude of the zero in the PI compensator must be very small ($<$ 0.05 /sec) or a lead compensator will need to be introduced as well. The approach to be taken here is to avoid adding additional lead compensation. But it should be remembered that the beam-tracking performance might be improved (e.g., higher phase margin) if necessary by adding lead compensation.

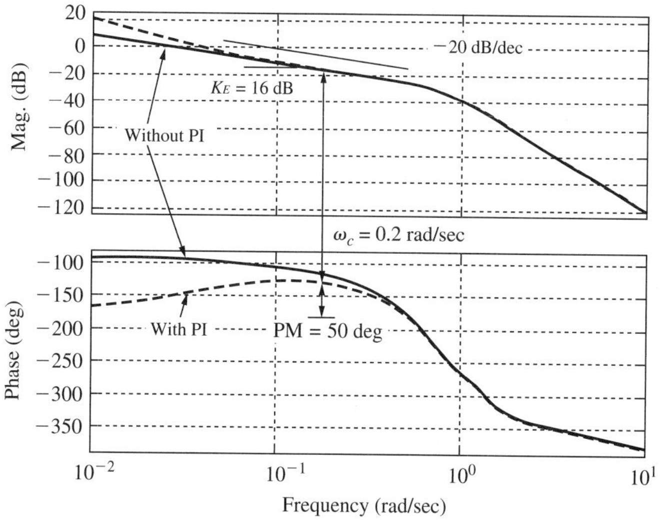

Figure 12.72 Bode plots of open-loop beam-tracking $E(s)/\psi_c(s)$ w. and wo. PI.

From Figure 12.72, note that the phase falls off rapidly above 0.2 rad/sec, so this frequency is near the upper limit in terms of gain-crossover frequency. Choosing 0.2 rad/sec as the target crossover frequency, a PI compensator is then added, with a compensator zero at 0.04 /sec. This was chosen to be well below the crossover frequency to minimize the phase loss at crossover. The form of the outer-loop control compensation is then

$$k_E(s) = K_E \frac{(s + 0.04)}{s} \text{ rad/rad} \tag{12.79}$$

and the Bode plot of the outer-loop beam-tracking dynamics including this PI compensator is shown labeled "With PI" in Figure 12.72. Note that by selecting an outer-loop gain K_E of 16 dB (6.3 rad/rad), the desired crossover frequency and a phase margin of approximately 50 degrees are obtained. Although this phase margin is not as large as desired, it may be acceptable. If not, the crossover frequency could be reduced, phase-lead compensation may be added, or pure-gain compensation may be used instead of PI.

Using the above PI control compensation and gain, the initial-condition time responses are shown in Figure 12.73. The initial conditions chosen correspond to an initial-position error E of one degree, which corresponds to an initial-position offset d of 525 ft. The initial position is left of the desired track, and the initial heading is the same as the desired final heading, so no steady-state change in heading is required. The vehicle first rolls right to acquire the beam centerline, and then rolls back left to stabilize on the beam. The response for the angular beam error E indicates a rise time of less than 10 seconds, with a slight overshoot and no steady-state error. The overshoot is due to the fact that the phase margin was a little low, as noted. Except for the rather high initial bank angle of approximately 45 deg, these responses appear to be acceptable. If the bank angle is considered excessive, one may either reduce the gain-crossover frequency in the bank-angle controller or add a bank-angle limiter, as discussed in Section 12.5.3.

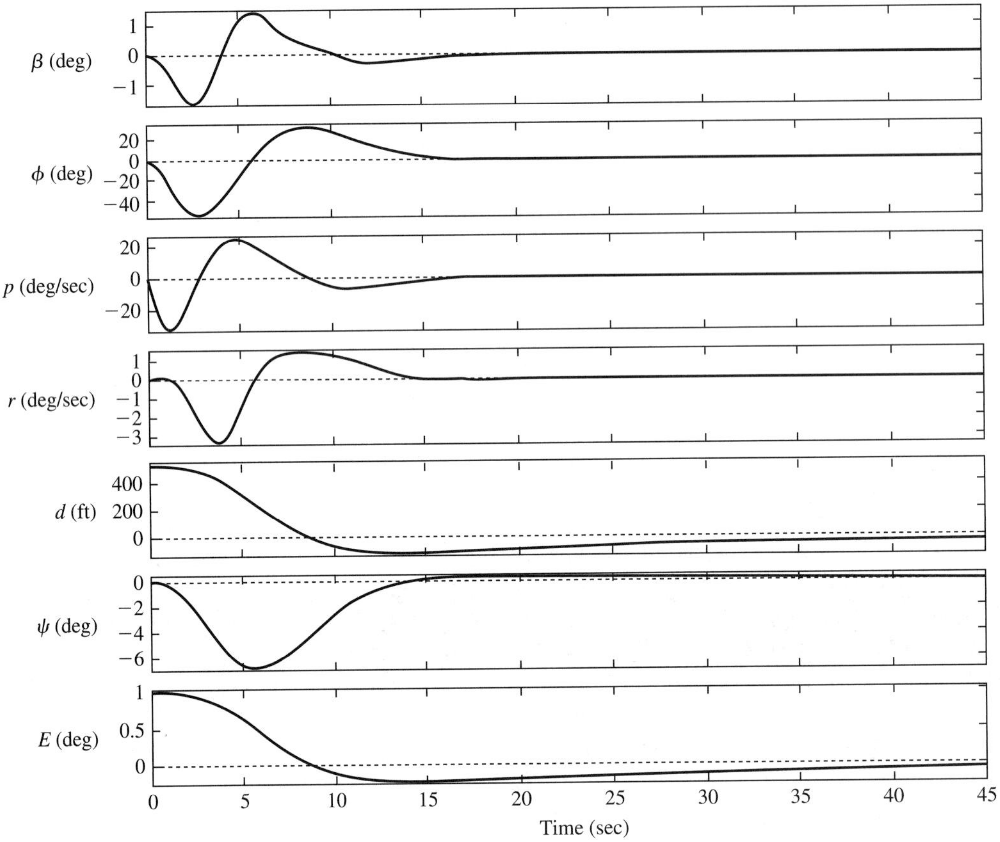

Figure 12.73 Initial-condition response for VOR homing guidance.

The MATLAB commands to assemble the model for this beam-guidance controller and obtain the above initial-condition responses are as follows:

```
»numol=[(32.2/824)];denol=[1 0];OL=tf(numol,denol)     Heading-angle control outer-loop kinematics
»syspsi=series(syscl,OL,[2],[1]);                      syscl is closed-loop bank-angle-control system
»[apsi,bpsi,cpsi,dpsi]=ssdata(syspsi)
»aclpsi=apsi-bpsi*10*cpsi
»bclpsi=bpsi*10
»cclpsi=[[0;0;0;0;0] ca];cclpsi(6,:)=cpsi
»dclpsi=[0;0;0;0;0;0]
»sysclpsi=ss(aclpsi,bclpsi,cclpsi,dclpsi);
»R=30000;numE=[824/R];denE=[1 0];OLE=tf(numE,denE)     Beam-angle control outer-loop kinematics
»sysE=series(sysclpsi,OLE,[6],[1]);
```

»numpi=[1 .04];denpi=[1 0];pi=tf(numpi,denpi)	Add PI
»sysEpi=series(pi,sysE);	
»[aEpi,bEpi,cEpi,dEpi]=ssdata(sysEpi)	
»aEpicl=aEpi-bEpi*6.3*cEpi	Gain is 16 dB
»bEpicl=bEpi*6.3	
»dEcl=[0;0;0;0;0;0];cEcl=[dEcl cclpsi dEcl]	
»cEcl=[cEcl;cEpi];dEcl=[dEcl;0]	
»cEcl(5,:)=(R/57.3)*cEcl(7,:)	
»sysEcl=ss(aEpicl,bEpicl,cEcl,dEcl)	Closed-loop VOR homing

ILS Localizer Tracking Finally, we will consider the case of ILS localizer tracking. With reference to the block diagram shown in Figure 12.71, the inner-most loop is a bank-angle controller, and the second innermost loop is a heading hold. We designed these controllers for the approach flight condition in Sections 12.4.3 and 12.5.3, respectively, and we will use them here.

Now the outer-most ILS-tracking loop may be addressed. As with the VOR homing controller considered above, the range to the station is taken to be 30,000 ft, and the approach is to introduce PI compensation and try to avoid adding additional lead compensation. But the flight velocity in the approach flight condition is significantly less than that in cruise, so the gain-crossover frequencies in all the loops now are less than those for VOR homing.

Again with reference to the block diagram given in Figure 12.71, and after closing the heading-hold loop, the Bode plot of the dynamics in the outer ILS-tracking loop $E(s)/\psi_c(s)$ is shown labeled "Outer-loop dynamics" in Figure 12.74. From the phase plot it appears that the maximum gain-crossover frequency in this ILS tracking loop is approximately 0.1 rad/sec if we want to obtain a reasonable phase margin without adding lead compensation.

Given that this maximum ILS-tracking crossover frequency is less than that for the VOR homing controller, the zero of the PI compensation was moved from -0.04 /sec to -0.03 /sec. So the form for the control compensation for the ILS localizer-tracking controller is

$$k_E(s) = K_E \frac{(s + 0.03)}{s} \text{ rad/rad} \tag{12.80}$$

and the Bode plot for this outer loop, including the PI compensator with $K_E = 1$ rad/rad, is also shown labeled "With PI" in Figure 12.74. From this plot we note that increasing the beam-tracking gain K_E to 22 dB (12.59 rad/rad) would yield a gain-crossover frequency of 0.1 rad/sec and a phase margin of approximately 50 deg.

Using the above PI compensation with a gain K_E of 12.59 rad/rad, the closed-loop ILS localizer-tracking system may now be assessed. The initial-condition responses for this system are shown in Figure 12.75. The initial conditions selected correspond to an initial angular error e_E of 1 deg, or a position error of 525 ft, with the initial heading corresponding to the desired final heading.

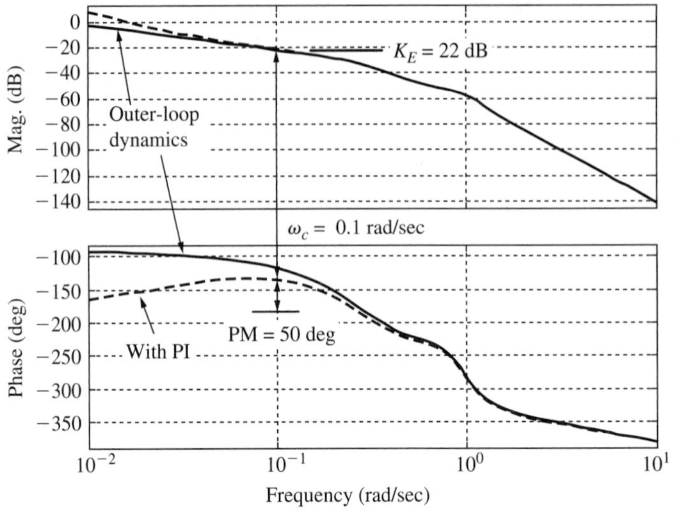

Figure 12.74 Bode plots—open-loop beam-tracking controller, including bank-angle controller and heading hold (approach condition).

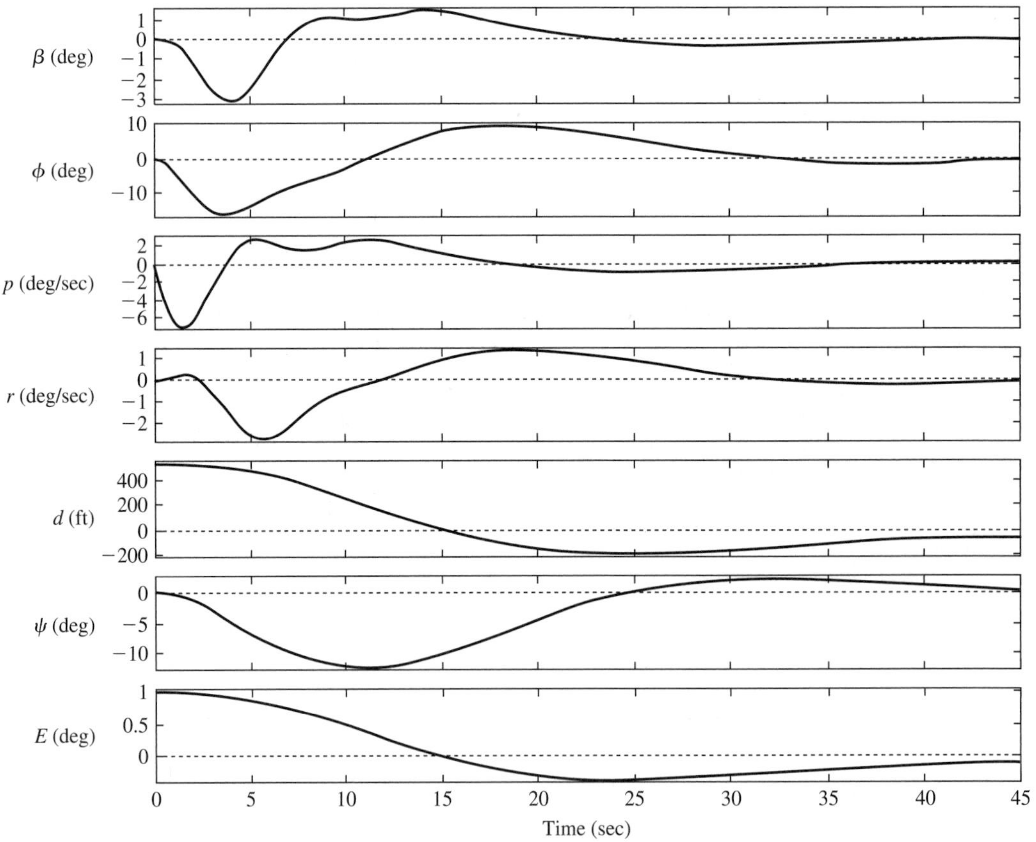

Figure 12.75 Initial-condition responses—ILS localizer-tracking controller.

So the vehicle is flying on a heading corresponding to the runway heading, but is offset 525 ft left of the runway centerline. The vehicle initially rolls right to acquire the localizer centerline, and then rolls back left to stabilize on the centerline.

Looking at the response of the position offset *d* or the angular beam error *E*, a rise time of less than 15 sec is indicated, with an overshoot of approximately 30 percent. Note that it takes approximately 40 seconds for these errors to become small. However, since the localizer-tracking controller was assumed to be engaged 30,000 ft from the runway threshold (recall that *R* was selected to be 30,000 ft), after 40–45 seconds the vehicle will be essentially on the localizer centerline and on the runway heading after traveling one third of the way to the runway, or when it is 20,000 ft from the threshold.

The MATLAB commands to assemble the model for the ILS localizer-tracking controller and obtain the above initial-condition responses are given as follows:

```
»numol=[(32.2/244)];denol=[1 0];OL=tf(numol,denol)      Outer loop of heading-
                                                         angle controller
»syspsi=series(syscl,OL,[2],[1]);
»[apsi,bpsi,cpsi,dpsi]=ssdata(syspsi)
»aclpsi=apsi-bpsi*1.585*cpsi                             Closing the heading-angle
                                                         loop
»bclpsi=bpsi*1.585
»cclpsi=[[0;0;0;0;0] ca];cclpsi(6,:)=cpsi
»dclpsi=[0;0;0;0;0;0]
»sysclpsi=ss(aclpsi,bclpsi,cclpsi,dclpsi);
»R=30000;numE=[244/R];denE=[1 0];OLE=tf(numE,denE)      Outer loop of ILS
                                                         coupler
»sysE=series(sysclpsi,OLE,[6],[1]);
»numpi=[1 .03];denpi=[1 0];pi=tf(numpi,denpi)            Add the PI
»sysEpi=series(pi,sysE);
»[aEpi,bEpi,cEpi,dEpi]=ssdata(sysEpi)
»aEpicl=aEpi-bEpi*12.589*cEpi                            Close the beam-tracking
                                                         loop
»bEpicl=bEpi*12.589
»dEcl=[0;0;0;0;0;0];cEcl=[dEcl cclpsi dEcl]              Prepare for plotting time
                                                         responses
»cEcl=[cEcl;cEpi];dEcl=[dEcl;0]
»cEcl(5,:)=(R/57.3)*cEcl(7,:)
»sysEcl=ss(aEpicl,bEpicl,cEcl,dEcl)                      Closed-loop beam-tracking
                                                         controller
»x0=[(1/.065067);0;0;0;0;0;0;0;0;0]                      Sets initial conditions
»initial(sysEcl,x0),45)                                  Solves for and plots IC
                                                         responses
```

12.7 ELASTIC EFFECTS AND STRUCTURAL-MODE CONTROL

In Section 12.3 we noted that the structural modes may be excited by aerodynamic-surface actuation. These surfaces are deflected from pilot commands as well as control inputs from the feedback systems discussed in this chapter. The elastic deformations may affect the vehicle's ride qualities, but more importantly, they can also destabilize the flight-control systems. Hence, elastic effects can be quite important in the design of flight-control laws.

The most important factor that determines whether elastic effects will be significant is the frequency separation between the higher-frequency rigid-body modes (short period, dutch roll, and roll subsidence) and the lowest-frequency vibration modes. If this separation is greater than a factor of 10–15, the elastic effects may not play an important roll. By assuring that the magnitudes of the open-loop Bode plots for the flight-control loops are small in the frequency range of the elastic modes (or sufficient roll off is included), there should be little interaction between the control system and the elastic deformations. But if the frequency separation is less than a factor of 10–15, then elastic effects may need to be considered more carefully.

When high-frequency roll off cannot sufficiently mitigate against the undesirable effects of elastic deformation, the elastic effects may be addressed by sensor placement, structural filters, and/or active structural-mode control. These three techniques are listed basically in the order of their complexity. The concepts presented here draw from the material present in Chapters 3 and 4.

Sensor placement refers to the fact that if the sensor used in the feedback-control law, such as a rate gyro or accelerometer, is properly located on the structure, the undesirable effects of elastic deformation may be minimized. Recall, for example, that two important displacements of the structure due to an elastic mode's excitation may be represented as

$$\begin{aligned} Z(t) &= \nu_Z(x,y,z)\eta_i(t) \\ \theta(t) &= \nu_Z'(x,y,z)\eta_i(t) \end{aligned} \tag{12.81}$$

where $\nu_Z(x,y,z)$ is the value of Z displacement of the mode shape evaluated at location (x,y,z) on the structure, $\nu_Z'(x,y,z)$ is the slope $\partial\nu_Z/\partial x$ of the same mode shape at the same location, and $\eta_i(t)$ is the time-dependant value of the i'th generalized coordinate associated with that elastic degree of freedom.

Now assume that this Z-displacement mode shape has the shape indicated in Figure 12.76. If an accelerometer with its axis aligned in the Z direction is located at a displacement node of the mode shape, as indicated in the figure, then that accelerometer will not sense any acceleration due to the deformation of that mode. Likewise, if a rate gyro (an angular-displacement sensor) is placed at a mode-slope node of the mode shape, as also indicated in the figure, then that sensor will not sense any angular rate due to the deformation of that mode. Consequently, by appropriate sensor placement the feedback-system designer

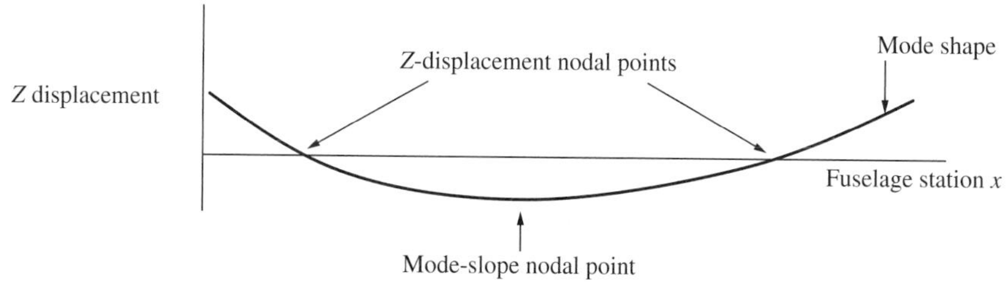

Figure 12.76 Mode-shape nodal points.

can eliminate particular vibration modes from the problem at hand, such as the lowest-frequency vibration mode. Since that mode's deformation will then not be sensed, it will not be fed back into the flight-control system.

Unfortunately, one cannot place a sensor at the nodes of all vibration modes because those nodes are located at different places on the structure. So only the most troubling modes, such as those with the lowest vibration frequencies, may be dealt with through sensor placement. Also, sensor placement as just described does not eliminate any ride-qualities issues due to a mode's displacement. If the modes are excited somehow, such as by turbulence, they will contribute to the ride quality. But they will not destabilize the feedback system.

The second technique for dealing with elastic modes is to use *structural filters,* such as notch filters on the sensor outputs. A *notch filter* has a transfer function of the following form.

$$N(s) = K_N \frac{\left(s^2 + 2\zeta_N \omega_N s + \omega_N^2\right)}{\left(s^2 + 2\zeta \omega s + \omega^2\right)} \qquad (12.82)$$

where here the subscript N indicates terms in the numerator. In a notch filter, ζ_N, ω, and ω_N are chosen to be close to the damping and frequency, respectively, of the elastic mode in question, while ζ is chosen to be much larger than ζ_N. An elastic mode manifests itself as a spike in the Bode magnitude plot, and a rapid 180 deg change in phase. If a notch filter is added at the output of the sensor and tuned to the elastic mode in question, the spike is "notched out" and the 180 deg phase change becomes more gradual. We will clarify these concepts in Example 12.6.

EXAMPLE 12.6

Effects of an Elastic Mode on Missile Attitude Control

Consider a flexible missile operating outside the atmosphere, with geometry shown in Figure 12.77. If the missile length L, mass per unit length m, and thrust T are all normalized to unity, the linearized rigid-body equation of motion is simply

$$\ddot{\theta} = \delta$$

So the pitch-rate transfer function has a pole at the origin.

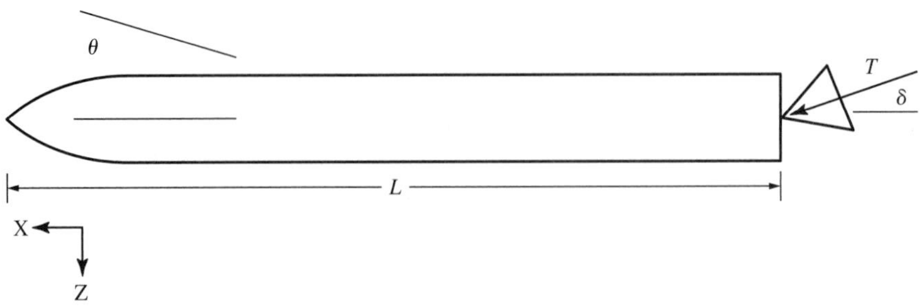

Figure 12.77 Flexible missile schematic.

Let the first-body-bending vibration mode be described by the modal equation

$$\ddot{\eta} + 2\zeta\omega\dot{\eta} + \omega^2\eta = \frac{T\nu(L)}{\mathcal{M}}\delta \tag{12.83}$$

where $\mathcal{M}$ is the generalized mass, ζ is the elastic modal damping, and ω is the vibration frequency. Also, let the mode shape be given by

$$\nu(x) = \sin\left(\pi x/L\right) - 0.5 \tag{12.84}$$

which yields a generalized mass of 0.27. Assume that a rate gyro is located at the nose of the vehicle and used as the sensor for a pitch damper. If $\zeta = 0.02$ and $\omega = 1$ rad/sec, demonstrate how high-frequency roll off and notch filters may be employed to assure that this elastic mode does not destabilize the pitch damper.

■ Solution

The state-variable model for this vehicle may be expressed as

$$\mathbf{x}^T = \begin{bmatrix}\dot{\theta} & \eta & \dot{\eta}\end{bmatrix} \quad u = \delta$$

$$\mathbf{A} = \begin{bmatrix} 0 & 0 & 0 \\ 0 & 0 & 1 \\ 0 & -1 & -0.04 \end{bmatrix} \quad \mathbf{B} = \begin{bmatrix} 1 \\ 0 \\ -0.5/0.27 \end{bmatrix} \tag{12.85}$$

Note that the Z displacement of the mode shape at $x = L$ is $\nu(L) = -0.5$, and the slope of the mode shape at the nose is $\nu'(0) = 1$. So if the responses of interest are the rigid-body pitch rate and the total pitch rate sensed by the rate gyro, or

$$y^T = \begin{bmatrix}\dot{\theta}_{\text{Rigid}}(/\text{sec}) & \dot{\theta}_{\text{Sensed}}(/\text{sec})\end{bmatrix} \tag{12.86}$$

the **C** and **D** matrices are given by

$$\mathbf{C} = \begin{bmatrix} 1 & 0 & 0 \\ 1 & 0 & 1 \end{bmatrix}, \quad \mathbf{D} = \begin{bmatrix} 0 \\ 0 \end{bmatrix} \tag{12.87}$$

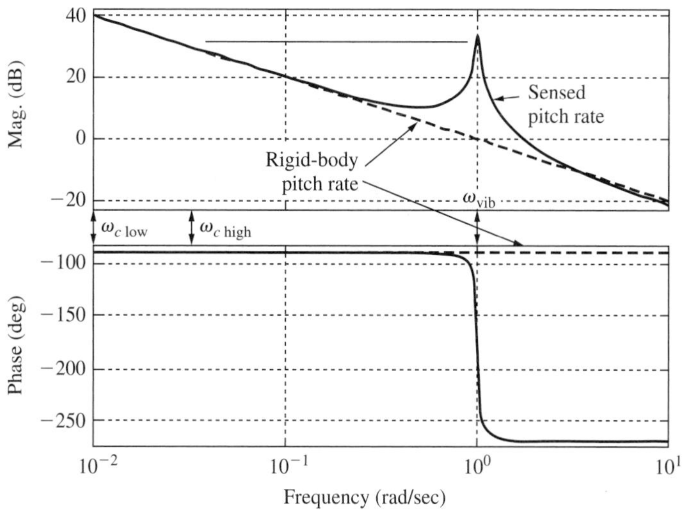

Figure 12.78 Rigid and sensed pitch-rate transfer functions.

The two transfer functions are

$$\frac{\dot{\theta}_{\text{Rigid}}(s)}{\delta(s)} = \frac{1}{s} \text{ /sec}$$

$$\frac{\dot{\theta}_{\text{Sensed}}(s)}{\delta(s)} = \frac{-0.852(s+1.06)(s-1.107)}{s(s^2+0.04s+1)} \text{ /sec} \tag{12.88}$$

and note the potentially troubling right-half-plane zero at 1.107 /sec in the second transfer function. By comparing the Bode plots for these two transfer functions in Figure 12.78, the effects of the elastic mode are clearly revealed. The spike in the magnitude plot and the rapid 180 deg phase loss, both occurring at the vibration frequency, are apparent.

A pitch damper is to be used, and the control law is given by

$$\delta = \delta_c - K_q \dot{\theta}_{\text{Sensed}} \tag{12.89}$$

If the vehicle was actually rigid, any value of K_q could be used without fear of instability, as we can see from the Bode plot in Figure 12.78 (or a root locus). The Nyquist plot for this (rigid) pitch-rate feedback loop would simply consist of a straight line along the imaginary axis of the complex plane, which could never encircle the critical negative-one point. So we might select a pitch-rate feedback gain K_q of −30 dB (0.032 sec) to set the gain-crossover frequency at approximately 0.04 rad/sec (and assumed that the vehicle was rigid).

Unfortunately, even though the selected gain-crossover frequency is well below the elastic-mode frequency, the Nyquist plot for the actual flexible vehicle with this rate-feedback gain is as shown in Figure 12.79. This feedback system is clearly unstable

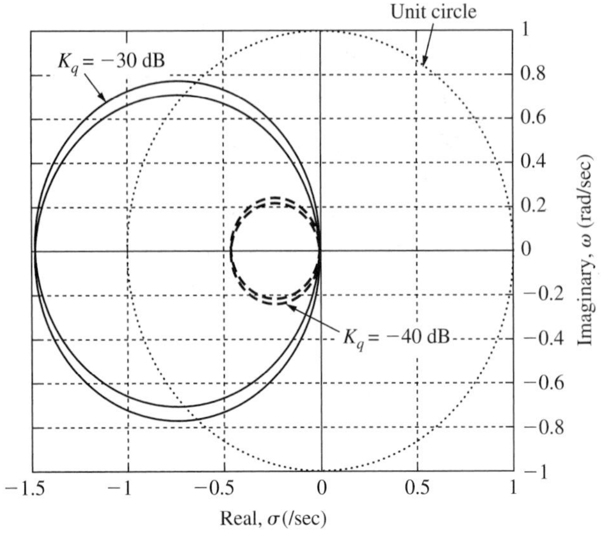

Figure 12.79 Nyquist plots for pitch damper, $K_q = -30$ and -40 dB, elastic vehicle.

(there are two clockwise encirclements of the critical point), and the instability is due to the presence of the elastic mode. With this rate-feedback gain of -30 dB, the closed-loop eigenvalues for the missile with the pitch damper are

$$-0.0319 \text{ /sec}$$

$$0.0096 \pm j1.001 \text{ rad/sec}$$

Now let the pitch-damper gain K_q be reduced to -40 dB (0.01 sec). The Nyquist plot for the pitch damper using this reduced gain is also shown in Figure 12.79, and we can see that the system is now stable. Both loops of the Nyquist plot are inside the unit circle and do not encircle the critical point at -1.

We have just demonstrated that if we assume the vehicle is rigid, the gain-crossover frequency may be set too high, and that can result in instability. Note here the important effect of the high-frequency roll off. By lowering the pitch-damper gain from -30 dB to -40 dB, Figure 12.78 shows that the gain-crossover frequency for the pitch damper was reduced from approximately 0.04 rad/sec to 0.01 rad/sec. From the same figure we see that with the gain-crossover frequency at 0.01 rad/sec, the spike due to the flexible mode is never larger than unity (0 dB). Hence, the Nyquist plot cannot encircle the critical point at -1.

Next consider the following notch-filter

$$N(s) = \frac{(s^2 + 0.04s + 1.0)}{(s^2 + 1.2s + 1.0)} = \frac{\dot{\theta}_{\text{Filtered}}(s)}{\dot{\theta}_{\text{Sensed}}(s)} \tag{12.90}$$

Note that the locations of the two zeros are identical to those of the poles of the elastic mode, as indicated in Equation (12.88). If this notch filter is inserted at the output of the

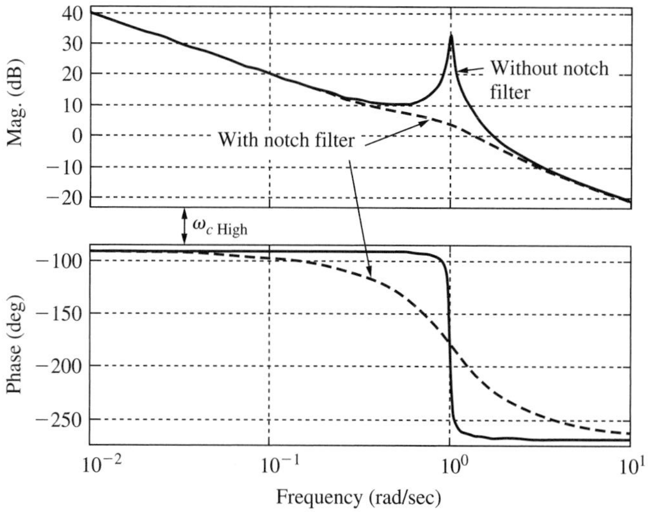

Figure 12.80 Bode plot of sensed pitch rate $\dot{\theta}_{\text{Sensed}}(s)/\delta(s)$, w. and wo. notch filter.

rate gyro, and the output of the filter is then fed back in the pitch-damper, the pitch-damper control law becomes

$$\delta = \delta_c - K_q \dot{\theta}_{\text{Filtered}}$$

The resulting open-loop Bode plot for the pitch damper with notch filter is shown labeled "With notch filter" in Figure 12.80. Now, not only is the system with the notch filter stable using the higher gain of −30 dB, but the rate-feedback gain might even be increased further and the system remain stable.

This seems like the obvious solution, were it not for the difficulty of implementing a notch filter. The filter defined above is a "perfect filter"; it was perfectly tuned to the elastic modal parameters. It placed two zeros exactly on top of the two complex poles associated with the elastic mode. In reality, one never knows the vibration damping and frequency exactly. Model-based predictions and ground vibration tests can only estimate these parameters. Furthermore, elastic-modal frequencies can change, due to fuel and/or passenger loading and external stores on fighter aircraft. And if the filter zeros are not placed correctly, instability may still occur. So using a notch filter involves some risk.

The final method for dealing with elastic effects is to actively control them, in contrast to attempting to keep them from being fed back into the control system. The disadvantage of *active structural-mode control* is its added complexity. But if the elastic modes must be modified for some reason, such as increasing their damping, then active control is one way to accomplish that goal. Active structural-mode control will be demonstrated in the following case study.

EXAMPLE 12.7

A Case Study—Structural Mode Control

To demonstrate the key concepts in structural mode control, again consider the large, high-speed aircraft discussed in the case study in Section 7.9, with additional data given in Appendix B. The vehicle is shown in the sketchs in Figure 12.81. Note the small control vanes mounted on either side of the cockpit. These vanes are on the vehicle specifically for structural-mode control because accelerations experienced by pilots significantly degraded their ability to fly the vehicle. These vanes are much smaller than the horizontal tail, and are sized to be effective at providing forces useful for controlling the structural modes but not to significantly contribute to the total lift or pitching moment on the vehicle. This provides for some decoupling between the two control inputs (vanes and elevator).

The descriptive data for the vehicle, taken from Table 7.1, is given in Table 12.2, and the flight condition used in the numerical analysis involves flight at 6000 ft altitude at a Mach number of 0.6. The first four symmetric vibration modes will be included in the model to be used. Shown in Table 12.2 are the vibration frequencies and generalized masses for these vibration modes.

Sketches of the modes shapes are shown in Figure 12.82. Note that all the mode shapes are normalized such that the displacement at the nose is unity. Before even beginning any numerical analysis, note that the second mode is basically a symmetric wing-bending mode, while the other three modes are fuselage-bending modes. Due to the nature of the second mode, and the extremely large generalized mass of the fourth mode, these two modes should not be expected to play a large role in terms of cockpit accelerations. (Why?) But they were retained in the model to demonstrate the types of vibration modes that can arise.

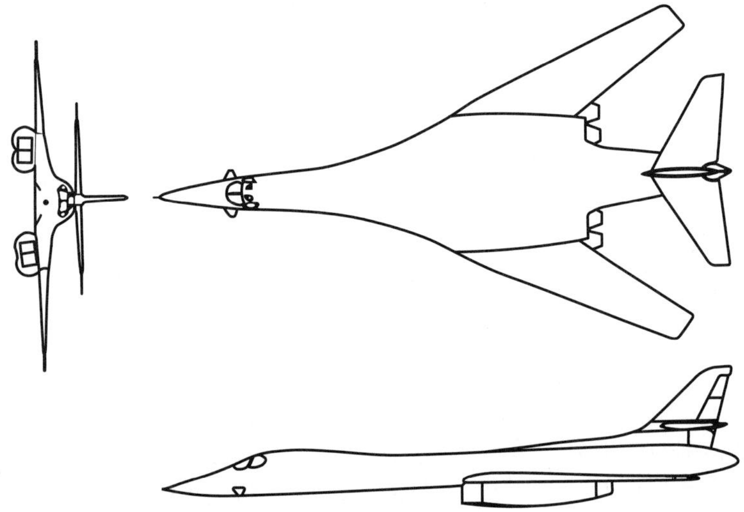

Figure 12.81 Large, high-speed aircraft.

Table 12.2 Data for the Large, High-Speed Aircraft

Wing geometry	$S_W = 1{,}950\ \text{ft}^2$	**Inertias**	$I_{xx} = 9.5 \times 10^5\ \text{sl-ft}^2$
	$\bar{c}_W = 15.3\ \text{ft}$		$I_{yy} = 6.4 \times 10^6\ \text{sl-ft}^2$
	$b_W = 70\ \text{ft}$		$I_{zz} = 7.1 \times 10^6\ \text{sl-ft}^2$
	$\Lambda_{LE} = 65\ \text{deg}$		$I_{xz} = -52{,}700\ \text{sl-ft}^2$
Weight	288,000 lbs	**Vehicle length**	143 ft
Modal generalized masses	$\mathcal{M}_1 = 184\ \text{sl-ft}^2$	**Modal frequencies**	$\omega_1 = 12.6\ \text{rad/sec}$
	$\mathcal{M}_2 = 9{,}587\ \text{sl-ft}^2$		$\omega_2 = 14.1\ \text{rad/sec}$
	$\mathcal{M}_3 = 1{,}334\ \text{sl-ft}^2$		$\omega_3 = 21.2\ \text{rad/sec}$
	$\mathcal{M}_4 = 436{,}000\ \text{sl-ft}^2$		$\omega_4 = 22.1\ \text{rad/sec}$

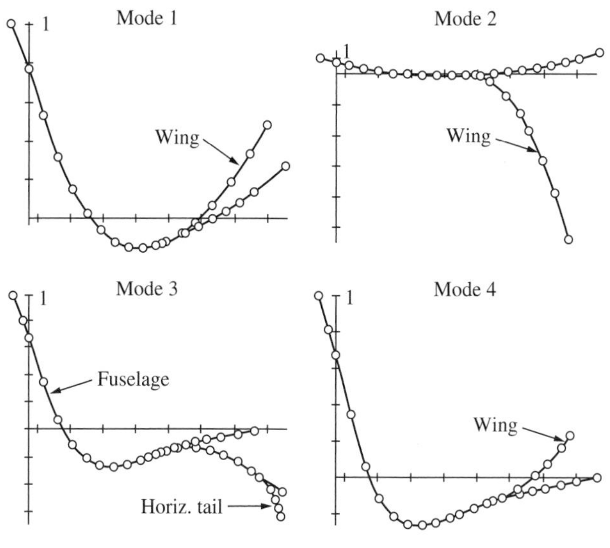

Figure 12.82 Sketches of vibration mode shapes.

There are two sources of elastic-modal excitation that contribute to the cockpit accelerations—elevator deflection and atmospheric turbulence. We will only address elevator deflections here. Shown in Figure 12.83 is the frequency response (Bode plot) of the vertical acceleration at the cockpit $a_Z(x_{cp})$ due to elevator deflection. The elevator actuator has been ignored here to emphasize the effects of elevator deflections in this discussion. But, in fact, the actual presence of the actuator, with a pole at -10 /sec, limits the ability of the elevator to excite elastic modes with frequencies much above 10 rad/sec.

The acceleration responses associated with two of the elastic modes (Mode 1 and 3) are quite pronounced, indicating that these modes can significantly contribute to the vertical

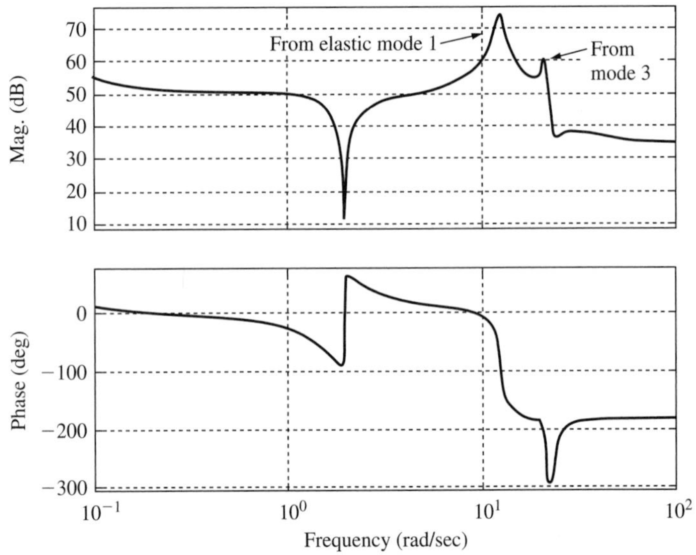

Figure 12.83 Frequency response—cockpit acceleration from elevator deflection (ft/sec^2/rad)—no elevator actuator.

accelerations at the cockpit. So a design objective of a structural-mode control (SMC) system might be to reduce these accelerations associated with these elastic modes.

The Bode plot for the accelerations at the cockpit $a_Z(x_{cp})$ due to the deflections of the control vanes is depicted in Figure 12.84. The acceleration responses at the elastic-mode frequencies are similar to those due to elevator deflection, and clearly the control

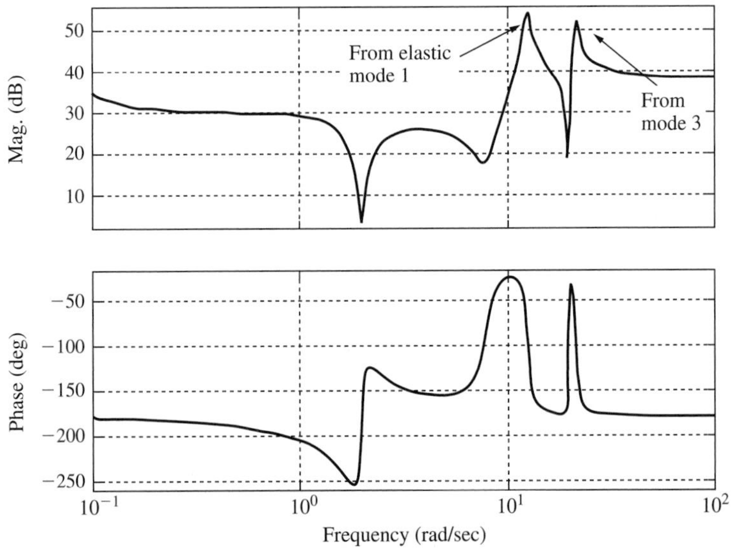

Figure 12.84 Frequency response—cockpit accelerations due to control vane deflection (ft/sec^2/rad).

vane can also excite these two elastic modes. However, the magnitude of the cockpit accelerations due to control-vane deflections is smaller than that for elevator deflection (in Figure 12.83), due to the smaller size of the control vane.

We will now address the control scheme to be considered. In 1968, John Wykes (Ref. 6) introduced what he called "Identically Located Acceleration and Force," or ILAF. In later years this concept became known as *co-located actuator and sensor*. The reasoning behind the approach is based on the following: (1) if a force applied to an object is proportion to, and in the opposite direction from the velocity of the object, the result is to increase the damping in the motion and (2) if the force and measured velocity are at the same location on a flexible structure, the proper phasing will always be present. By *proper phasing* we mean that the force will not destabilize the elastic system.

To approximate the velocity measurement one can use a low-pass filter, or lag, on the output of the accelerometer. Above the corner frequency of the lag, the filtered measurement is simply the integral of the acceleration, or velocity. A lag is used instead of an integrator for ease of implementation. In addition, since we don't want to affect the low-frequency vertical accelerations, we will insert a washout filter into the SMC control loop. Consequently, the structural-mode control law is of the following form

$$-\delta_{\text{cv}}(s) = -k_{\text{smc}}(s)a_{Z_{\text{wo}}}(x_{\text{cp}}) = -\left(K_{\text{smc}}\frac{1}{T_L s + 1}\right)\left(\frac{T_{\text{wo}}s}{T_{\text{wo}}s + 1}\right)a_Z(x_{\text{cp}}) \qquad (12.91)$$

The block diagram for this SMC system, plus a pitch damper, is shown in Figure 12.85. The pitch damper will be discussed more below.

If the lag and washout filter poles $1/T_L$ and $1/T_{\text{wo}}$ in the SMC (Equation (12.91)) are both chosen to be unity, well below the lowest elastic-mode frequency of 12.6 rad/sec, the root locus for the SMC feedback system, or $k_{\text{smc}}(s)\left(a_{Z_{\text{cp}}}(s)/-\delta_{\text{cv}}(s)\right)$, is shown in Figure 12.86. Note that as with the pitch-attitude controllers discussed in Section 12.4.1,

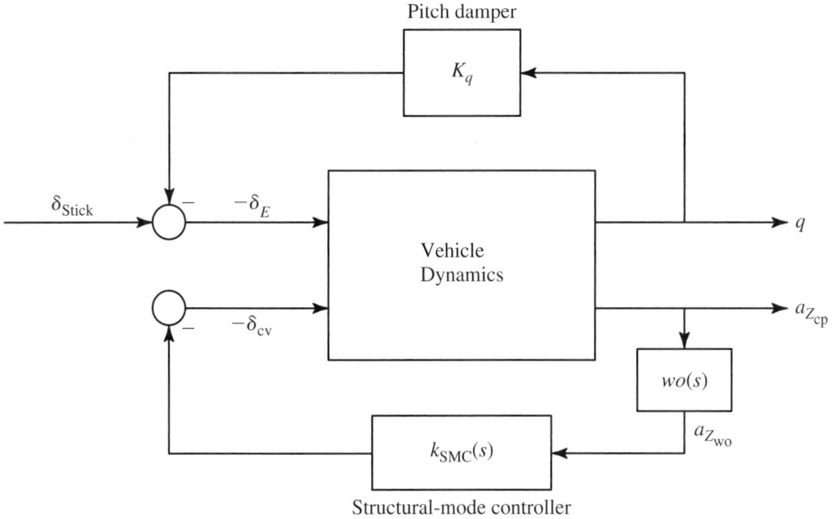

Figure 12.85 Structural-mode controller w. pitch damper.

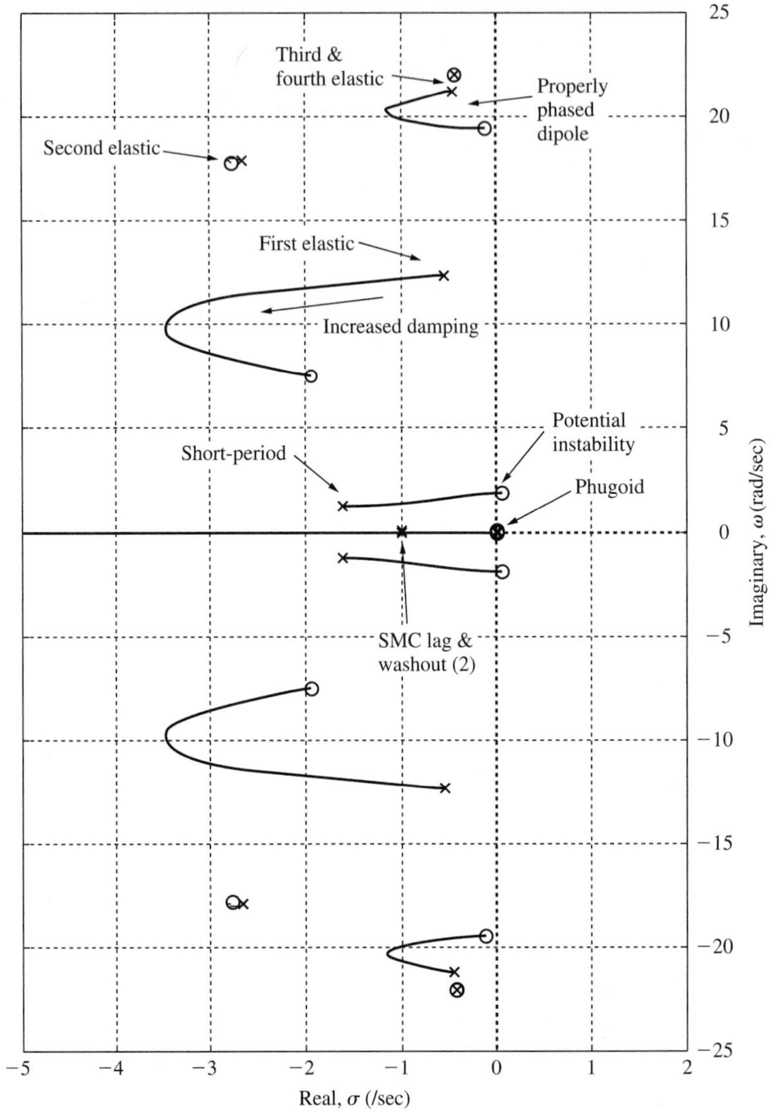

Figure 12.86 Root locus of $a_{Z_{cp}}(s)/-\delta_{cv}(s)$ w. lag and washout at 1 /sec w. pitch damper (rigid pitch rate) $K_q = 0.4$.

the sign convention for positive control-vane deflection was reversed here to obtain this and the following plots, so as to be consistent with the usual notation used in feedback systems.

There are four items to note from this root locus. First, the SMC feedback loop can significantly increase the damping of the first elastic mode, and can also improve the damping of the third elastic mode. This desirable feature is due to the attractive locations of the nearby zeros. That is, the zeros are closer to the origin than the poles, and the root-locus branches loop to the left farther into the left-half plane. This is the "proper phasing" that Wykes was referring to. Improper phasing, when the zeros in the pole-zero pair (or the *dipole*) are farther from the origin than the pole, causes the root-locus branches to loop to the right into the right-half plane.

The second observation from the root locus is the fact that the pole-zero dipoles associated with the second and fourth elastic modes are effectively pole-zero cancellations, and, as we expected from the frequency response, do not play a role here.

The third observation is the fact that the short-period mode will be destabilized if the SMC gain is too high. But the damping of the short-period mode may be increased by incorporating the pitch damper from the vehicle's longitudinal stability-augmentation system (see Appendix B). A pitch-damper feedback gain of $K_q = 0.4$ rad/rad/sec was used in generating the root locus in Figure 12.86, and the short-period damping (with no SMC) is 0.8. The rate gyro used for the pitch damper is assumed to be located close to the mode-slope nodes of the first and third elastic modes such that essentially rigid-body pitch rate may be fed back in the pitch damper.

Finally, the forth observation is that the very-low-frequency phugoid mode and two associated zeros in the above root locus are all at the origin due to the scale used in plotting. This mode is of little interest here, as it will be easily stabilized through the normal attitude-control function of the pilot or autopilot, as discussed in Section 12.4.1 when addressing pitch-attitude control.

The corresponding open-loop Bode plot of $k_{\mathrm{smc}}(s)\left(a_{Z_{\mathrm{cp}}}(s)/-\delta_{\mathrm{cv}}(s)\right)$, or loop shape, is shown in Figure 12.87 for an SMC control gain $K_{\mathrm{SMC}} = 1$ rad/ft/sec^2. Note that we

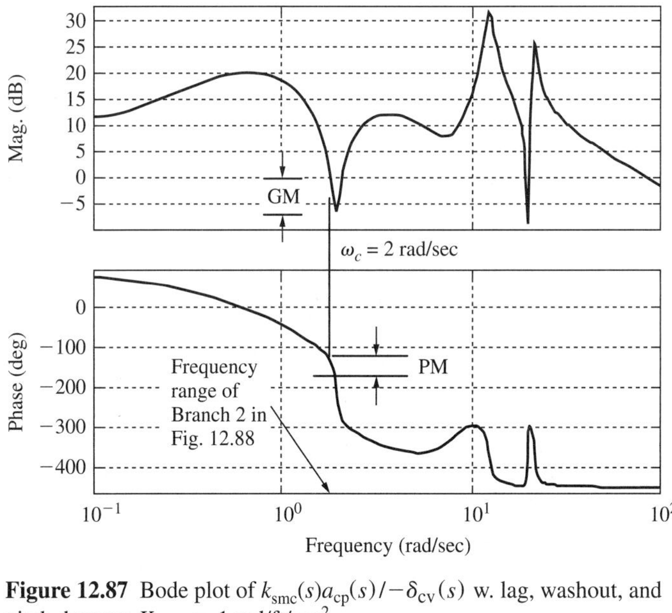

Figure 12.87 Bode plot of $k_{\mathrm{smc}}(s)a_{\mathrm{cp}}(s)/-\delta_{\mathrm{cv}}(s)$ w. lag, washout, and pitch damper, $K_{\mathrm{SMC}} = 1$ rad/ft/sec^2.

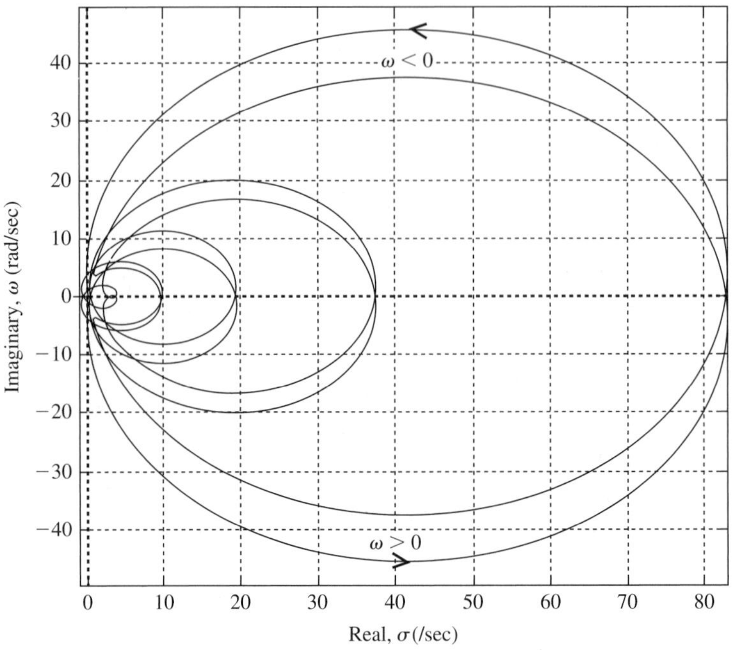

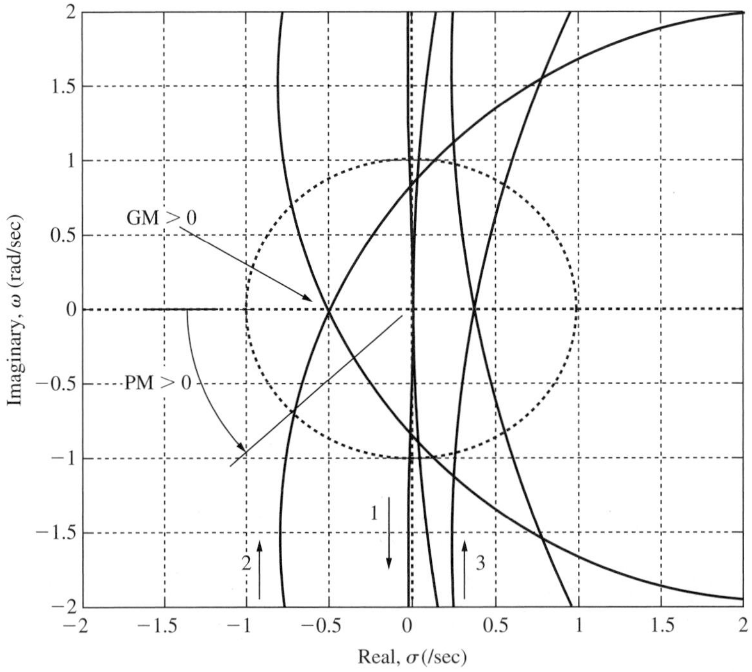

Figure 12.88 Nyquist plot for $a_{cp}(s)/-\delta_{cv}(s)$ w. lag, washout, and pitch damper.

are not seeking a conventional loop shape with this control system, such as one with *K/s* characteristics. The design objective is one of stability augmentation (increase damping) rather that command tracking, for example. However, the Bode plot (and the corresponding Nyquist plot) reveals the all important stability margins for the feedback system.

The stability margins may be best understood by considering the Nyquist plot for the SMC shown in Figure 12.88. The region near the origin has been enlarged in the bottom figure to reveal the phase and gain margins of the SMC for the gain $K_{\text{smc}} = 1$ rad/ft/sec^2. The branches labeled "1," "2," and "3" are associated with positive frequencies $\omega > 0$, which also correspond to the Bode plot in Figure 12.87. Positive gain and phase margins are indicated in both Figures 12.87 and 12.88.

There are eight large loops in the Nyquist plot, and these loops all lie to the right of the origin and do not encircle the critical point at -1. (There are eight large loops because three correspond to positive frequencies and three correspond to negative frequencies. Recall the definition of a Nyquist plot.) The Nyquist plot indicates that the system is unstable because it does not encircle the critical point (counterclockwise), and, although not mentioned previously, the open-loop system has an unstable phugoid mode. But by considering the root locus (Figure 12.86) along with the Bode and Nyquist plots (Figures 12.87 and 12.88), we can see that the gain and phase margins here are indications of how close the short-period mode is to instability. Furthermore, these margins will be increased if the pitch-damper gain is increased (moving the open-loop short-period poles in Figure 12.87 farther to the left) or if the SMC gain is decreased (moving Branch 2 in Figure 12.88 closer to the origin). With this knowledge in hand, the SMC gain can be easily adjusted to yield maximum damping on the third elastic mode and a significant increase in damping of the first elastic mode. A gain K_{smc} of -38 dB (0.15 rad/ft/sec^2) appears to yield good results, to be discussed next.

To evaluate this candidate SMC, first consider the Bode plots in Figure 12.89. These plots correspond again to frequency responses of cockpit acceleration due to elevator deflection. One set of plots is indicated as "With SMC." By comparing this set with the other in the figure, indicated as "Without SMC," it is clear that the levels of accelerations due to elevator deflection have been significantly reduced by the structural-mode controller. In particular, the peak accelerations associated with the elastic modes have been virtually eliminated, leading to a reduction in peak accelerations of approximately 15 dB, or a factor of 8.

Finally, the closed-loop eigenvalues, including the effects of the pitch damper and SMC, are

Elastic Mode 4: $-0.4267 \pm j22.045$ /sec

Elastic Mode 3: $-1.16 \pm j20.37$ rad/sec (damping increased by a factor of 3)

Elastic Mode 2: $-2.7304 \pm j17.969$ /sec

Elastic Mode 1: $-3.1248 \pm j11.214$ rad/sec (damping increased by a factor of 7)

Lag & WO: -9.879 /sec, -0.1734 /sec

Short Period: $-0.4096 \pm j1.7102$ /sec

Phugoid: $0.00616 \pm j0.05743$ /sec (unstable)

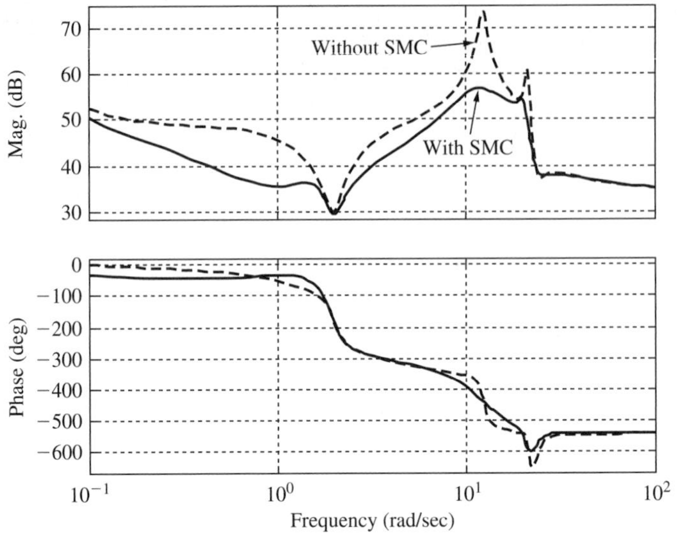

Figure 12.89 Bode plots of $a_{cp}(s)/-\delta_E(s)$, with and without SMC.

In summary then, through the proper interpretation and use of the modal data for the vibration modes, we can determine a priori which modes are expected to be significant in the analysis. Two elastic modes in particular were shown to be important, in terms of their contribution to the cockpit acceleration environment. By applying the concept of ILAF, or co-location of sensor and actuator, the damping of the problematic elastic modes was significantly increased, leading to a significant decrease in the contributions of these elastic modes to the cockpit accelerations.

12.8 Summary

In this chapter we focused primarily on autopilots, and how such feedback systems control and guide the vehicle. We introduced the various modes of autopilots and presented the architectures of these autopilot modes. These modes included attitude command/holds; heading, speed (Mach), and altitude command/holds; and longitudinal and lateral beam guidance. We demonstrated how the architectures were closely tied to the dynamics and kinematics of the vehicle, and how stability augmentation was used to improve the characteristics of the feedback-control systems.

We repeatedly used loop shaping and exploited the natural inner- and outer-loop structure of the vehicle's dynamics to develop simple, robust, practical designs for the autopilot modes considered. The discussion was enhanced through the integration of specific case studies into the presentation. Two such examples/case studies demonstrated key concepts regarding elastic effects in flight control, including active structural-mode control. Finally, it should be pointed out that the control laws synthesized in this chapter were not intended to be optimized. Rather the intent was to demonstrate the techniques and to identify the important trade offs involved.

12.9 Problems

12.1 Using the data for the F-5A at Flight Condition 1 (see Appendix B), assemble the state-variable model in MATLAB for the longitudinal dynamics, and confirm the transfer function given in Equation (12.37). Then add an elevator actuator with a time constant $T_a = 0.05$ sec, and confirm that a pitch-damper gain K_q of 0.15 rad/rad/sec yields a short-period damping of 0.76.

12.2 Using the data for the DC-8 at Flight Condition 3 (see Appendix B), assemble the state-variable model in MATLAB for the longitudinal dynamics, and confirm the pitch-attitude transfer function give in Equation (12.30). Then add an elevator actuator with a time constant $T_a = 0.05$ sec, and generate the Bode plots for the pitch-attitude transfer function with and without actuator. Compare to Figure 12.20.

12.3 Using the results from Problem 12.2, incorporate the pitch-attitude controller presented in Section 12.4.1 into the model. Then redesign the speed-hold control law discussed in Section 12.5.1 by adding proportional-plus-integral compensation. While attempting to maintain the same gain-crossover frequency of 0.2 rad/sec, use the simple lag model given in Equation (12.57) and select the zero in the PI compensator to meet the requirement for a minimum phase margin of 60 deg. Then use the full-order vehicle model and determine the gain and phase margins and plot the step responses. Compare these results to those presented in Section 12.5.1.

12.4 Using the data for the DC-8 in Flight Condition 3 (see Appendix B), assemble the state-variable model in MATLAB for the lateral-directional dynamics, and confirm the bank-angle transfer function give in Equation (12.41). Then add an aileron actuator with a time constant $T_a = 0.05$ sec, and reproduce the Bode plot for the bank-angle transfer function with actuator, shown in Figure 12.36.

12.5 Using the results from Problem 12.4, incorporate the bank-angle controller presented in Section 12.4.3 into the model. Then redesign the heading-hold controller discussed in Section 12.5.3 by adding proportional-plus-integral compensation. While attempting to maintain the same gain-crossover frequency of 0.4 rad/sec, select the zero in the PI compensator to achieve a minimum phase margin of 50 deg. Then determine the gain margin and plot the step responses. Compare these results to those presented in Section 12.5.3.

12.6 Using the model for the flexible missile in Example 12.6, use a root locus to explain the effect of the notch filter on the closed-loop eigenvalues of the vehicle with the pitch damper. Then determine how the root locus changes if (1) the numerator of the notch filter is changed to $(s^2 + 0.036s + 0.81)$ or (2) if the numerator of the notch filter is changed to $(s^2 + 0.044s + 1.21)$. What does this suggest, in terms of tuning a notch filter when the frequency of the elastic mode is not known exactly?

References

1. D'Azzo, J. J. and C. H. Houpis: *Feedback Control System Analysis and Synthesis,* 2nd ed., McGraw-Hill, New York, 1966.
2. Churchill, R. V.: *Complex Variables and Applications,* McGraw-Hill, New York, 1960.

3. Maciejowski, J. M.: *Multivariable Feedback Design,* Addison-Wesley, Reading, MA, 1989.
4. Blakelock, J. H.: *Automatic Control of Aircraft and Missiles,* 2nd ed., Wiley, New York, 1991.
5. McRuer, D., I. Ashkenas, and D. Graham: *Aircraft Dynamics and Automatic Control,* Princeton University Press, Princeton, NJ, 1973.
6. Wykes, J. H.: "Structural Dynamic Stability Augmentation and Gust Alleviation of Flexible Aircraft," AIAA Paper 68-1067, AIAA Annual Meeting, October 1968.
7. McRuer, D. T. and D. E. Johnston: *Flight Control Systems Properties and Problems Vols. I and II,* NASA Contractor Report CR-2500, prepared by Systems Technology, Inc., Feb. 1975.
8. Schmidt, D. K. and G. Hartmann: "A Short Course on Aircraft Flight Control" (bound notes), presented at the National Cheng Kung University, Tainan, Taiwan, January 1984.

CHAPTER 13

Control Characteristics of the Human Pilot

Chapter Roadmap: *Although it would be beneficial to include this chapter in a first course on flight dynamics, there may be insufficient time in a quarter or semester to do so.*

This chapter concludes our discussion of the feedback control of the aircraft, but now the focus changes to the control function performed by the human pilot. By gaining an understanding of the control characteristics of the human, we may better understand the handling-qualities requirements on the vehicle's dynamics. Before World War II Koppen stated, "Since the controlled motion of an airplane is a combination of airplane and pilot characteristics, it is necessary to know something about both airplane and pilot characteristics before a satisfactory job of airplane design can be done." (Ref. 1.)

In discussing the control characteristics of the pilot, we will draw heavily from loop-shaping concepts for feedback systems, Bode plots, and Nyquist stability criteria. Since these topics have been discussed and used at length in Chapter 12, it is assumed that the reader is sufficiently familiar with them.

13.1 BACKGROUND

The study of the characteristics of the human operator in a precision feedback-control task dates back to the early studies of Tustin in England during World War II (Refs. 2, 3). He introduced the subjects of describing functions and remnant behavior (discussed later), and used them in experimental studies of human-control behavior. During the same period, but independent of Tustin's work, research at MIT's Lincoln Laboratory and at the U.S. Army's Aberdeen Proving Grounds also developed concepts of quasi-linear models of the human operator in gun tracking (Ref. 4, 5).

Research in human control modeling has continued over the intervening years, and an extensive amount of experimental data has been collected. This research effort has produced three basic classes of human-control models.

The simplest and most widely used is the *crossover model* (Refs. 4, 5). We will discuss this model exclusively in this chapter. A second class, and one of the most elaborate descriptions of human dynamic properties as a controller, is the *structural-isomorphic model.* This model is an extension of the crossover model that attempts to account for the many subsystem aspects of the human controller, as well as its input-output characteristics (Ref. 6). A related model, also dubbed a *structural model,* lies somewhere between the basic crossover model and the structural-isomorphic model, in terms of modeling details and complexity, and this model has many varied applications (Refs. 7, 8). The third class of human-operator model is the algorithmic, or *optimal-control model* (Refs 9, 10). However, because the crossover model is the most broadly validated and best understood of the human-control descriptions, the behavior predicted by the other classes of models must "reduce" to the crossover model. Thus the more elaborate models must inevitably agree with the fundamental aspects of the crossover model as a necessary limiting case.

13.2 THE CROSSOVER MODEL

The basic paradigm, notation, and key aspects of the crossover model of the human operator in a single-axis control task are shown in Figure 13.1. The *controlled element* is denoted as $Y_C(s)$ (or $Y_C(j\omega)$), and the human-control model is indicated by the dashed box, which includes the human-dynamic model $Y_P(j\omega)$. The block diagram shown also corresponds to the experimental setup used to measure the human's control characteristics. Without going into too many details, it is important to know that the model to be discussed has been validated in the laboratory, as well as in flight. The basic measurement technique employed is to instruct the human subjects to aggressively track a random-appearing visual display driven by sinusoidal signals at several specific frequencies. The controlled element is, of course, known, and the human's control response $c(t)$ and system response $m(t)$ were recorded. Then the human's *describing function* and *remnant* were extracted from these measurements.

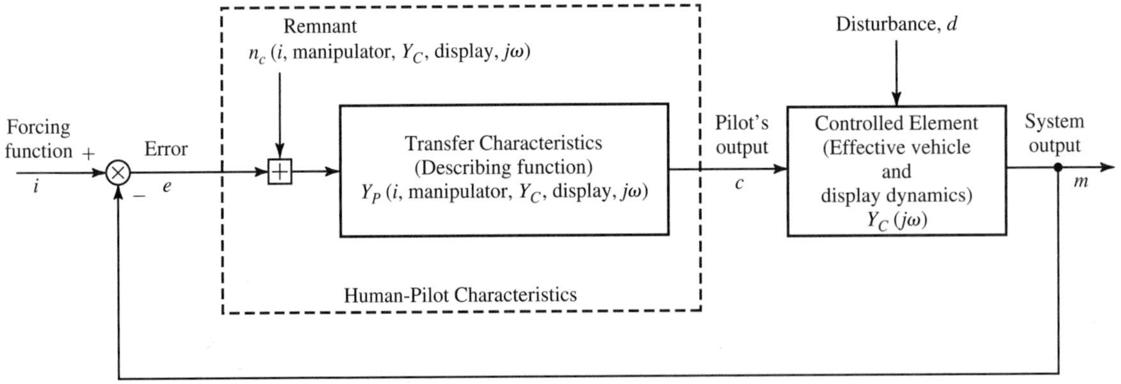

Figure 13.1 Paradigm of the human-operator control model.

When a nonlinear dynamic system is excited by an input at a given frequency ω_{in}, the system's response contains a primary component $g(j\omega_{\text{in}})$ that is only a function of that input frequency, plus additional components that are functions of frequencies other than the input frequency, called *higher harmonics* or *remnant.* This primary component $g(j\omega_{\text{in}})$, a frequency-dependent complex function, is the system's describing function or the quasi-linear description of the system. When the system is truly linear, no higher harmonics are present in the response and $g(j\omega_{\text{in}})$ is the system's frequency response. The remnant then consists of that component of the response of the nonlinear system that is not linearly correlated with the input, where linearly correlated means that it is only a function of the input frequency. If the system were linear, there would be no such remnant. So this component of remnant is a measure of the linearity of the system. (But note that part of the human operator's remnant can also be attributed to time-varying behavior and noise injection, both of which are linear phenomena.)

Examples of data obtained in this fashion are shown in Figure 13.2 (Ref. 4) for three generic controlled elements Y_C—*K, K/s,* and K/s^2. Note that these plots are in the same format as Bode plots, or magnitude and phase plotted versus frequency. And they are used in the same fashion as Bode plots. Gain and phase margins for the closed-loop system may be determined, along with gain-crossover frequency and so on. The dotted lines in the figures correspond to the Bode plots of the linear controlled-element dynamics alone, and the data points (with the fitted solid lines) are the measured describing functions for the open-loop system dynamics including the human, or $Y_P Y_C(j\omega)$ in the notation used in Figure 13.1. The gain-crossover frequency ω_c is also indicated in each case.

Focusing on only the magnitude portion of these plots at first, the most important characteristic to note is that in all three cases the magnitude of $Y_P Y_C(j\omega)$ is approximately *K/s* in the region of gain crossover! That is, the human controller in each case has performed the control function in such a fashion that the characteristics of the closed-loop system correspond to those for a well-designed feedback system (recall the loop-shaping paradigm introduced in Chapter 12). This fact has led to what has become known as *McRuer's crossover law,* after one of the pioneers in human-control modeling, Duane McRuer.

> For the tasks considered, and subject to human limitations, the human-control operator adjusts their control strategy so as to make the open-loop transfer function approximately *K*/s near the region of gain crossover.

Therefore, with reference to Figure 13.2, for the acceleration (K/s^2) controlled element, the human must introduce <u>lead</u> compensation to make $Y_P Y_C(j\omega)$ approximately *K/s,* since the slope of the magnitude plot must be increased from that of just the controlled element. For the pure gain (*K*) controlled element, the human must introduce <u>lag</u> compensation to make $Y_P Y_C(j\omega)$ approximately *K/s.* And for the velocity (*K/s*) controlled element, the human needs to introduce neither lead nor lag.

McRuer's law is the first and most important component of the crossover model of the human controller. Furthermore, experimental evidence suggests

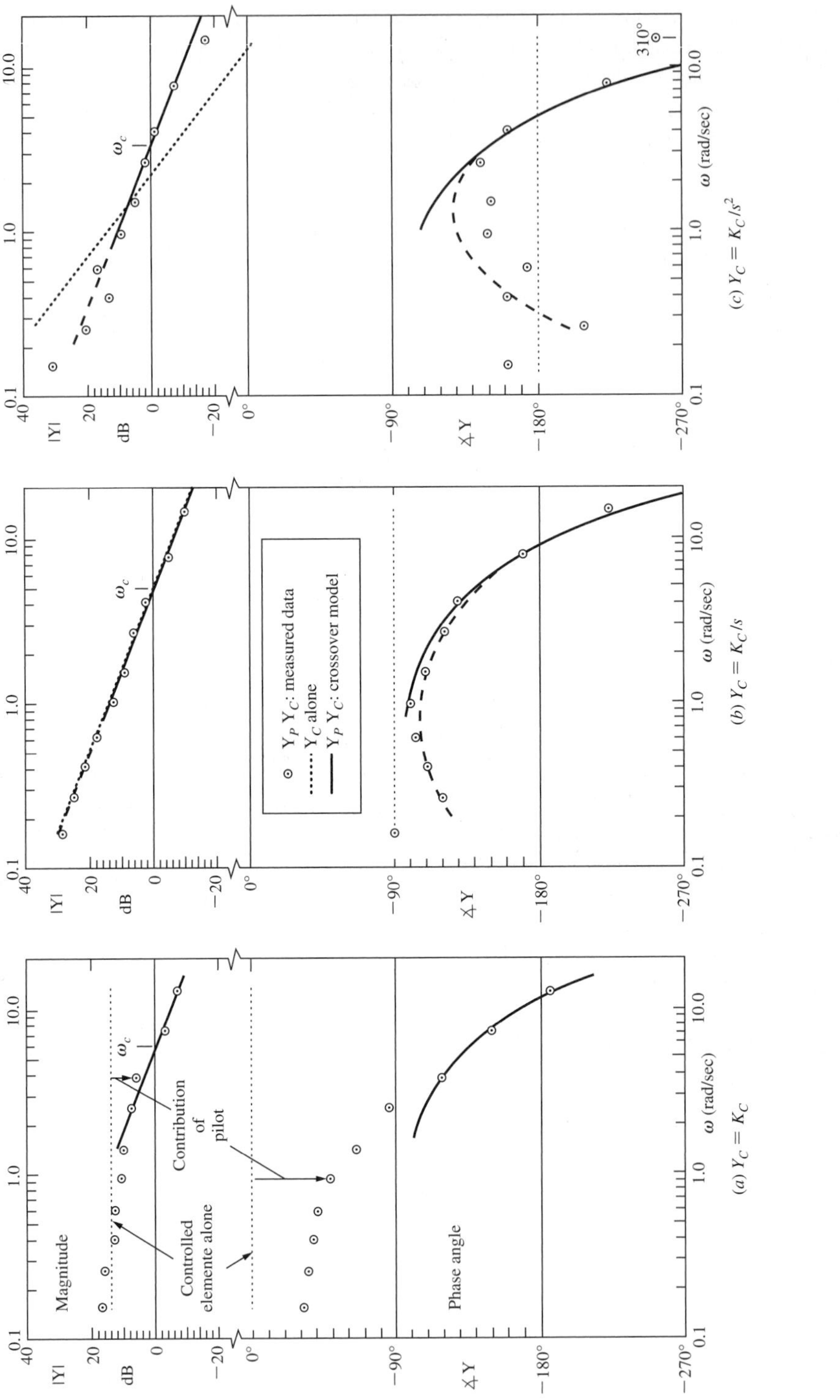

Figure 13.2 Measured human-operator characteristics in the feedback-tracking task (Ref. 4).

that this component of the model may usually be captured by using simple lead-lag or lag-lead compensation. Finally, several other aspects of the model depend on the amount of lead the human must provide in the control loop to obtain the *K/s* characteristics at gain crossover.

The remaining components of the model address the important "human limitations" noted in McRuer's law. These limitations include the delay or lag of the human's neuromuscular system, and the time delay associated with information processing. And this takes us to the phase portion of the plots in Figure 13.2. Again focusing on the critical region near gain crossover, note the data points with the fitted sold lines. Especially note these data in the center plot, corresponding to the *K/s* controlled element. Since the measured human and controlled element, or $Y_P Y_C(j\omega)$, is *K/s,* one might conclude that the compensation contributed by the human, or Y_P, is simply a pure gain K_p. However, the measured phase in this plot is much lower than the 180 deg phase associated with the controlled element. So the human must be introducing the additional phase loss in these data.

The sources of this phase loss are the two cited previously—neuromuscular delay or lag and information-processing delay. Note that the Laplace transform of a pure time delay τ is

$$L\{delay\ \tau\} = e^{-\tau s} \tag{13.1}$$

So the contribution of this delay to a Bode plot is

$$e^{-\tau s}\big|_{s=j\omega} = e^{j(-\tau\omega)} \tag{13.2}$$

which is just a phase loss of $\tau\omega$ rad. (The magnitude of the exponential function is unity.) Therefore, the human's effective time delay τ_E may be estimated by fitting the function $\tau_E\omega$ through the phase data points over the frequency region near gain crossover. (The human's phase contribution in the frequency range well below gain crossover is not of major importance in our discussion here, since the loop gain is large at these frequencies. However, this "phase droop" at low frequencies is usually attributed to low-frequency integration that is not apparent in the magnitude plots.)

Taking McRuer's law and the above human limitations into account, the parametric form for the crossover model of the human controller may now be expressed as

$$Y_P(j\omega) = K_P \frac{(T_{\text{lead}}\,j\omega + 1)}{(T_{\text{lag}}\,j\omega + 1)} e^{-j\tau_E\omega}$$

or equivalently

$$Y_P(j\omega) = K_P \frac{(T_{\text{lead}}\,j\omega + 1)}{(T_{\text{lag}}\,j\omega + 1)(T_{\text{NM}}\,j\omega + 1)} e^{-j\tau_{\text{IP}}\omega} \tag{13.3}$$

In the first of these two expressions, the information-processing and neuromuscular delays have been combined into the total effective delay τ_E. In the second expression the effect of the neuromuscular system is modeled as a first-order lag, and the pure delay includes only the delay associate with information processing. In other words, $\tau_E = \tau_{\text{IP}}$. The models are effectively equivalent, differing only in terms of how one chooses to model the neuromuscular effect. The advantage of the second of the two models above is that it properly reflects the fact that the pilot has a finite bandwidth with high-frequency roll off.

Note that the treatment of the phase loss due to time delay is straightforward using Bode plots. But to develop a model like Equations (13.3) in MATLAB, or to use the model in a root-locus analysis, one must use a rational approximation for the exponential function. One such approximation may be developed using Padé approximants (Ref. 10). For example, a first-order Padé approximation for the exponential is

$$e^{-\tau_E s} \approx \frac{1 - \frac{1}{2}\tau_E s}{1 + \frac{1}{2}\tau_E s} \tag{13.4}$$

while a second-order Padé approximation is

$$e^{-\tau_E s} \approx \frac{1 - \frac{1}{2}\tau_E s + \frac{1}{8}\left(\tau_E s\right)^2}{1 + \frac{1}{2}\tau_E s + \frac{1}{8}\left(\tau_E s\right)^2} \tag{13.5}$$

When using such approximations, note that the pole(s) and zero(s) introduced are artifacts of the approximation.

The effective time delay τ_E has been found to be a function of the characteristics of the input signal to be tracked and either the dynamics of the controlled element or equivalently the amount of phase lead the human must contribute to achieve the K/s characteristics near gain crossover. This dependence is revealed in Figure 13.3. Note that as the human is forced to introduce more lead compensation into the control loop (i.e., the controlled elements transition from K to K/s to K/s^2), the effective time delay increases. Conversely, as the bandwidth of the signal (i.e., the highest frequencies included in the signal) to be tracked increases, or the signal becomes more difficult to track, the effective delay diminishes. This reduction in time delay is associated with the human's need to increase the gain-crossover frequency to achieve adequate tracking performance, and is accomplished by the human increasing mental focus (to reduce τ_{IP}) as well as the tension in their neuromuscular system (to reduce τ_{NM}).

Finally, note from Equation (13.3) that the human-controller adjusts the gain K_p to set the gain-crossover frequency in the control loop, or ω_c. This crossover frequency is also a function of the controlled-element dynamics (or the amount of lead required on the part of the human) and the command-signal bandwidth. The relationships between these variables are depicted in Figure 13.4. As the human

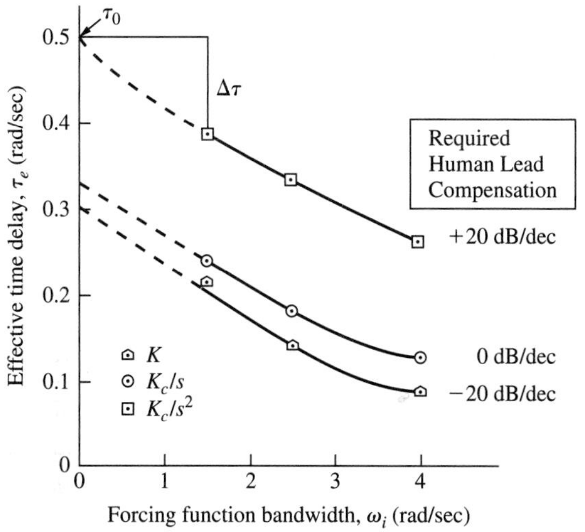

Figure 13.3 Variation of human's effective time delay with controlled element and command-signal bandwidth (Ref. 4).

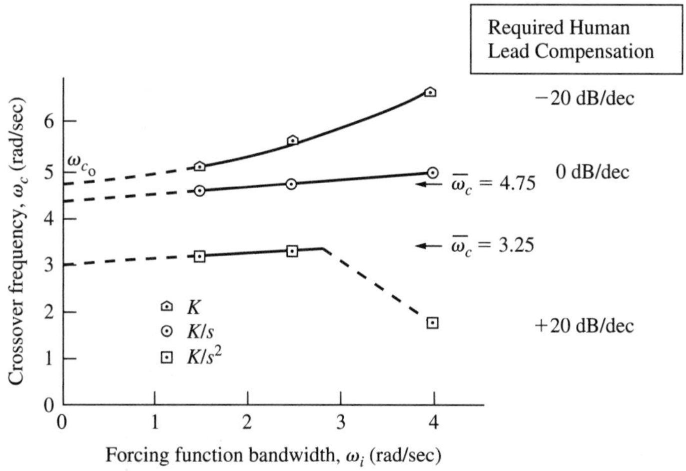

Figure 13.4 Variation of gain-crossover frequency with controlled element and command-signal bandwidth (Ref. 4).

controller is forced to introduce more lead compensation into the control loop, the human reduces the gain-crossover frequency. And as the bandwidth of the command signal increases (i.e., it becomes more difficult to track the command signal), the human-controller tends to increase the gain-crossover frequency slightly.

Due to closed-loop stability requirements, the effective time delay and gain-crossover frequency are interrelated. Without any time delay, the phase margin for the crossover model would always be 90 deg, since the open-loop dynamics are $Y_P Y_C(j\omega) \sim K/(j\omega)$. But, of course, any phase loss due to time delay and neuromuscular lag reduces this phase margin. With an effective time delay of τ_E the phase margin becomes

$$\text{PM} = \frac{\pi}{2} - \tau_E \omega_C \tag{13.6}$$

And note that for a given effective time delay τ_E, the maximum gain-crossover frequency that can be achieved and have the closed-loop system remain stable is

$$\omega_{C_{\max}} = \frac{\pi}{2\tau_E} \tag{13.7}$$

An empirical technique for selecting the effective time delay when using the crossover model is to note from Figure 13.3 that an approximate expression for the effective delay is

$$\tau_E \approx \tau_0(Y_C) - \Delta\tau(\omega_i) \tag{13.8}$$

That is, the basic time delay τ_0 is a function of only the controlled element, while the incremental time delay $\Delta\tau$ may be approximated using a straight-line fit to the data in Figure 13.3. Data for use in Equation (13.8) are given in Table 13.1.

After estimating the effective time delay for the situation at hand, the maximum gain-crossover frequency achievable may then be determined using Equation (13.7). For example, for the *K/s* controlled element with no incremental time delay, Figure 13.3 indicates that the effective time delay equals $\tau_0 = 0.33$ sec. So the maximum achievable gain-crossover frequency is

$$\omega_{C_{\max}} = \frac{\pi}{2(0.33)} = 4.76 \text{ rad/sec}$$

This is consistent with the measured gain-crossover frequency shown in the center plot in Figure 13.2. The above crossover frequency of 4.76 rad/sec is based on the basic delay τ_0 from Figure 13.3. But the subject being tested in Figure 13.2 is exhibiting a smaller delay, or there is also an incremental delay $-\Delta\tau$, which leads to a positive phase margin.

All the above results dealing with the gain-crossover frequency correspond to the situation in which the human is performing a single-axis tracking task in a laboratory. When the human is performing a multi-axis tracking task,

Table 13.1 Empirical Data for Estimating Human's Effective Time Delay (Laboratory Tracking Tasks Only)

Controlled Element	Human Compensation Required, dB/dec	Basic Time Delay, τ_0 sec	Incremental Time Delay, $\Delta\tau$ sec
K	−20	0.33	$0.070\ \omega_i$
K/s	0	0.36	$0.065\ \omega_i$
K/s^2	+20	0.50	$0.065\ \omega_i$

analogous to controlling both the longitudinal and lateral-directional axes of an aircraft in severe turbulence, for example, a phenomenon referred to as *cross-over regression* occurs. This simply means that the maximum achievable gain-crossover frequencies in the multi-axis task are lower than the gain-crossover frequency achievable if each axis was controlled separately (i.e., single-axes tasks).

The reason this occurs is because the human's effective time delay is increased in multi-axis tasks. This increase is due to the fact that the human has finite information-processing capacity, so if the rate of processing is fixed, and the amount of information increases (due to the need to control two axes, for example), the total required processing time increases accordingly (Ref. 11).

Consequently, as a rule of thumb when one is considering the task of flying a real aircraft versus performing a single-axis laboratory tracking task, at least 0.1 sec of effective time delay should be added to the delays found from Equation (13.8) and Table 13.1. For example, if one is considering a challenging piloting task in which the aircraft dynamics are like *K/s* in the region of crossover, the minimum pilot's effective time delay is approximately

$$\begin{aligned}\tau_E &\approx \tau_0(Y_C) - \Delta\tau(\omega_i) + 0.1 \\ &= 0.36 - 0.065 \times 4 + 0.1 \\ &= 0.2 \text{ sec}\end{aligned} \tag{13.9}$$

The final aspect of the crossover model to be considered is the correlation between the amount of lead compensation the pilot must contribute and the subjective assessment of difficulty, or workload, the human associates with that task. Shown in Figure 13.5 is experimental data from Ref. 4 that show that as the

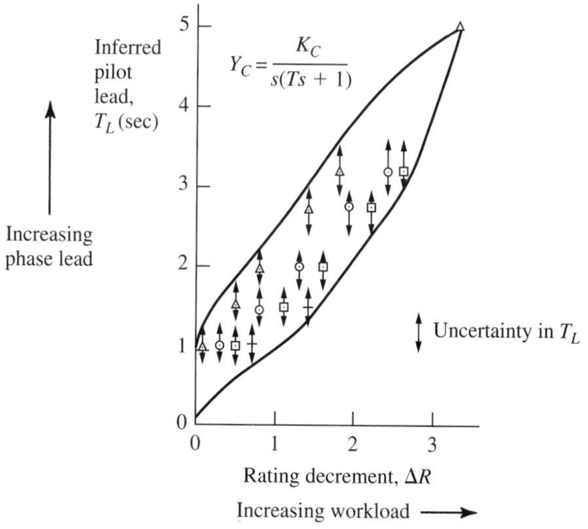

Figure 13.5 Correlation between a human's lead generation and subjective assessment of task difficulty or workload (Ref. 4).

amount of lead the human must contribute to achieve the *K/s* characteristics is increased, the subjective assessment of difficulty, or workload, also increases. Hence to make a task easier, reduce the amount of lead the human controller must contribute. This has important implications with regards to the handling qualities of aircraft.

13.3 FLIGHT-DYNAMICS IMPLICATIONS OF THE HUMAN PILOT'S CONTROL CHARACTERISTICS

The implications of the material presented in Section 13.2, with respect to the handling qualities of aircraft, may be summarized succinctly as follows. The flight dynamicist must:

1. Provide aircraft dynamics such that the pilot does not need to introduce lead compensation to stabilize and control the vehicle.
2. Recognize the bandwidth limitations and time delays inherent in the human pilot, and factor these into the requirements on the vehicle's flight dynamics. In addition, if the vehicle itself includes time delay, it is additive to the human's time delay and either the maximum achievable gain-crossover frequency will be reduced or the stability margins will be degraded.
3. Minimize the need to perform complex, multi-axis piloting tasks. (e.g., provide for proper turn coordination and/or compensation).

These implications will be reinforced through the examples to follow, which will also demonstrate how the crossover model may be applied.

Note that in the following examples a fixed effective delay $\tau_E = 0.25$ sec was selected, and the maximum gain-crossover frequency was limited by the requirement that the closed-loop pilot-vehicle system must achieve a minimum phase margin of 35–45 deg, and a minimum gain margin of 5–6 dB.[1]

EXAMPLE 13.1

The Effect of a Roll Damper on Aircraft Handling Qualities

Consider the A-7A aircraft in Flight Condition 8 (Appendix B). Show that the addition of a roll damper with a gain K_p of 0.1 rad/rad/sec improves the handling characteristics of the vehicle with regard to the pilot's ability to control the bank angle. The block diagram for the bank-angle control task is shown in Figure 13.6.

[1] Thanks to Professor Ron Hess at the University of California, Davis for his helpful suggestions regarding these examples.

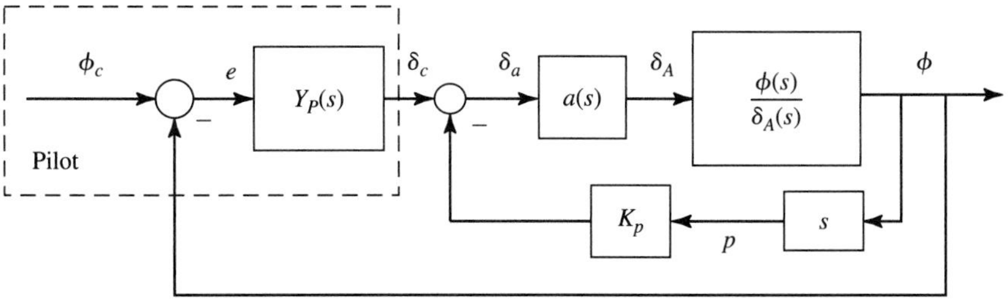

Figure 13.6 Block diagram for bank-angle control task w. roll damper.

■ Solution

Using the data from Appendix B, along with the roll-mode approximation, the aileron-to-bank-angle transfer functions are given by

$$\frac{\phi(s)}{\delta_A(s)} \approx \frac{8}{s(s+1.4)} \text{ deg/deg} \qquad \frac{\phi(s)}{\delta_A(s)} \approx \frac{8}{s(s+2.2)} \text{ deg/deg} \tag{13.10}$$

No roll damper — With roll damper

Adding an aileron actuator $a(s)$ modeled as a first-order lag with a time constant of 0.05 sec, the Bode plots for the vehicle's bank-angle responses $\phi(s)/\delta_c(s)$ are shown in Figure 13.7.

Now if the task is to be performed without excessive workload, and the pilot-vehicle system is to achieve maximum closed-loop performance, the crossover model implies that the effective controlled-element dynamics ($\phi(j\omega)/\delta_c(j\omega)$ here) should be approximately *K/s* over as large a frequency range as possible near gain crossover. In addition, as just noted, the gain-crossover frequency must be as high as possible, subject to the requirement

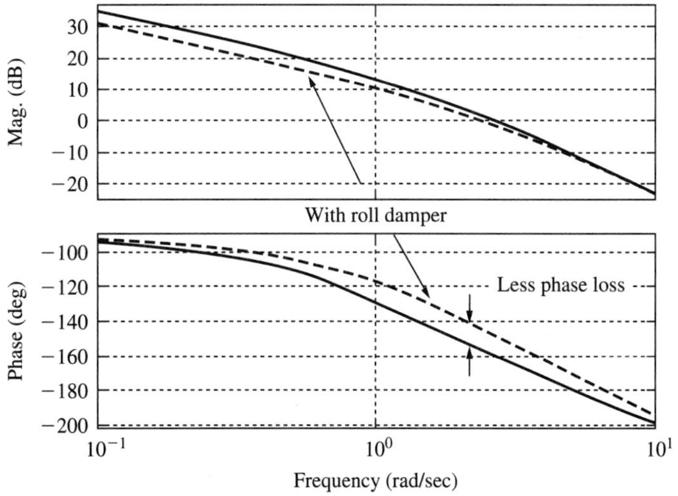

Figure 13.7 Bode plots of $\phi(s)/\delta_c(s)$ (including actuator).

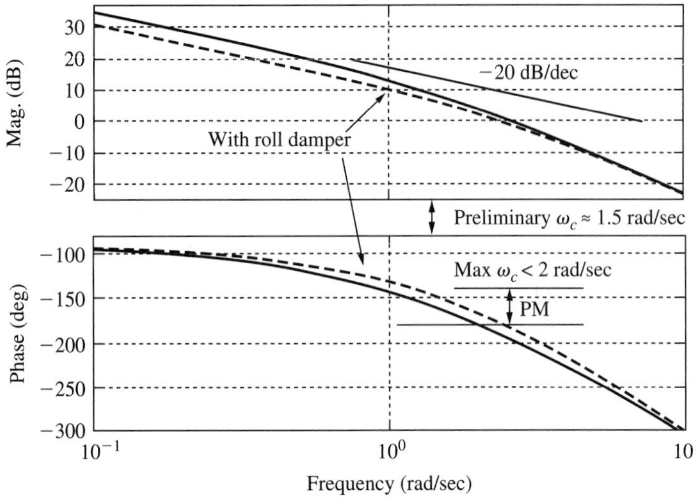

Figure 13.8 Bode plots of bank-angle responses $\phi(s)/\delta_c(s)$ w. 0.25 sec effective time delay τ_E.

that the minimum phase and gain margins are 35–45 deg and 5–6 dB, respectively, with the pilot exhibiting an effective time delay of 0.25 sec.

Incorporating this 0.25 sec time delay (or $e^{-j0.25\omega}$) into the bank-angle response to aileron $\phi(j\omega)/\delta_c(j\omega)$, the Bode plots are now as shown in Figure 13.8. (A first-order Padé approximation was actually used to model the time delay.) Note that without any pilot compensation, and due to the phase-margin requirement, the maximum gain-crossover frequency ω_c in either case is less than 2 rad/sec. So a crossover frequency of about 1.5 rad/sec will be selected for further analysis.

To achieve the required *K/s* characteristics near this frequency, the pilot must introduce some lead. First, consider the case with no roll damper. Adding the following pilot lead compensation,

$$\text{Lead}(s) = \frac{s + 0.55}{s + 2.2} \tag{13.11}$$

the Bode plot of the open-loop pilot-vehicle system, or $Y_P Y_C(j\omega)$, is as shown in Figure 13.9. The slope of the magnitude plot at gain crossover is now the desired −20 dB/dec.

For this Bode plot the pilot gain K_P is unity, and we see that selecting a pilot gain of −2.5 dB (0.75 deg/deg) would set the gain-crossover frequency at about 1.5 rad/sec. Phase and gain margins of ~50 deg and 6 dB, respectively, are obtained. The crossover pilot model in this case is therefore

$$Y_P(s) = 0.75\left(\frac{s + 0.55}{s + 2.2}\right)e^{-0.25s} \text{ deg/deg} \tag{13.12}$$

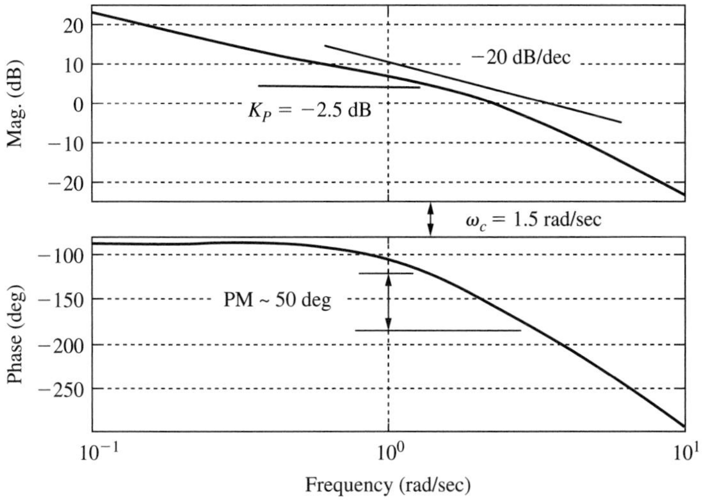

Figure 13.9 Bode plot of $Y_P Y_C$ with lead (no roll damping).

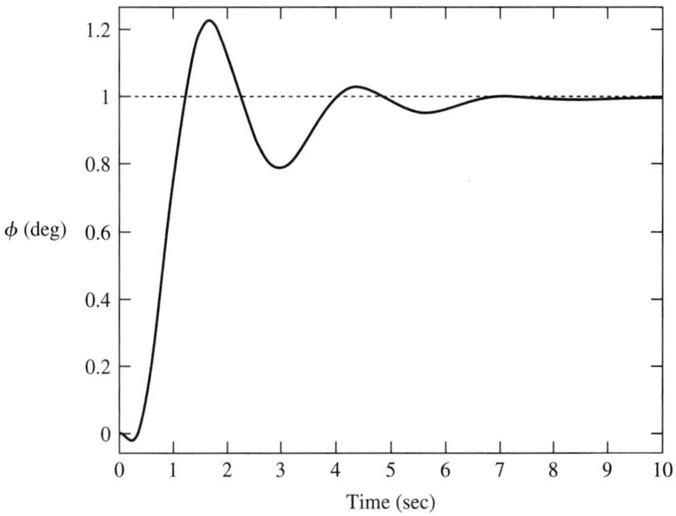

Figure 13.10 Bank-angle step response of pilot-vehicle system (no roll damper).

Using this model (with a first-order Padé approximation), the bank-angle step response for the closed-loop pilot-vehicle system is shown in Figure 13.10. The response is fairly well damped, with zero steady-state error.

Next we will consider the case with the roll damper. With reference to Figure 13.8, note that the pilot must still introduce lead near the selected gain-crossover frequency of

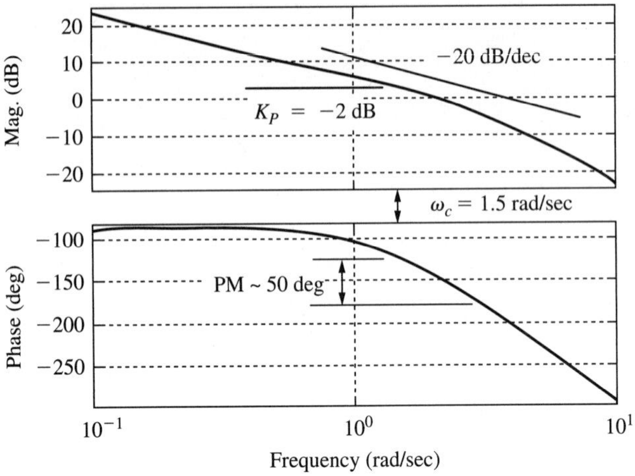

Figure 13.11 Bode plot of $Y_P Y_C$ with lead (with roll damping).

1.5 rad/sec, but not as much as in the case without the roll damper. Here, using the following pilot lead compensation,

$$\text{Lead}(s) = \frac{s + 0.7}{s + 1.75} \tag{13.13}$$

the Bode plot of the open-loop pilot-vehicle system, or $Y_P Y_C(j\omega)$, is as shown in Figure 13.11. The desired −20 dB/dec slope of the magnitude plot at gain crossover is achieved.

For this Bode plot the pilot gain K_P is unity, and we see that selecting a pilot gain K_P of −2 dB (0.79 deg/deg) again sets the gain-crossover frequency at about 1.5 rad/sec. Phase and gain margins of ~50 deg and ~6 dB, respectively, are achieved. The pilot crossover model in this case is therefore

$$Y_P(s) = 0.79\left(\frac{s + 0.7}{s + 1.75}\right)e^{-0.25s} \text{ deg/deg} \tag{13.14}$$

Using this model, the bank-angle step response for the closed-loop pilot vehicle system is essentially the same as that shown in Figure 13.10. Therefore, the performance of the pilot-vehicle systems with and without roll damping is essential identical, and the stability margins are essentially identical as well.

However, the lead that must be introduced by the pilot is not the same in each case. Without the roll damper, the pilot must introduce the lead compensation given by Equation (13.11). But with roll damping, the required pilot lead compensation is that given by Equation (13.13). The Bode plots of these two pilot compensators are shown in Figure 13.12. So without roll damping, the pilot must introduce 12 more degrees of phase lead at gain crossover, and with reference to Figure 13.5, this corresponds to a pilot's subjective assessment of more workload. Therefore, we may conclude that the roll damper allows the pilot to accomplish roll control with less workload, and therefore leads to better handling qualities.

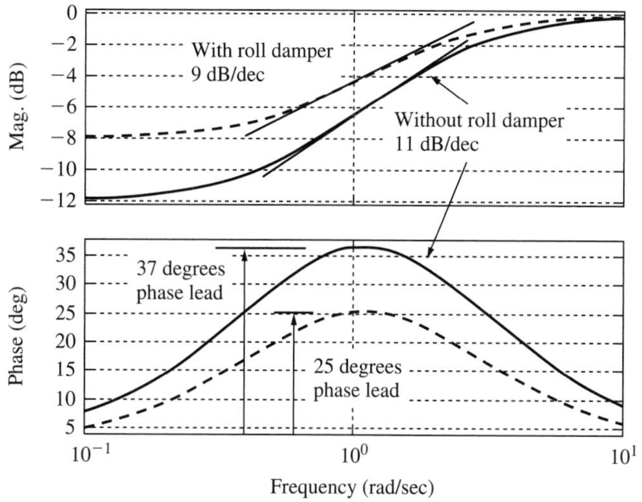

Figure 13.12 Comparison of pilot lead compensation required.

EXAMPLE 13.2

The Effect of a Pitch Damper on Aircraft Handling Qualities

Now consider the F-5A aircraft in Flight Condition 7 (Appendix B). Show that the addition of a pitch damper with a gain K_q of 0.1 rad/rad/sec improves the handling characteristics of the vehicle in the pitch-control task. The block diagram for the pitch-control task is shown in Figure 13.13.

■ Solution

Using data from Appendix B, and adding an elevator actuator modeled as a first-order lag with a time constant of 0.05 sec, the pitch-attitude transfer function is given by

$$\frac{\theta(s)}{\delta_a(s)} = \frac{290(s+0.01584)(s+0.4749)}{(s^2+0.01685s+0.003019)(s^2+1.034s+7.032)(s+20)} \text{ rad/rad} \quad (13.15)$$

The short-period natural frequency and damping ratio are 2.65 rad/sec and 0.195, respectively. With the pitch damper, the same transfer function becomes

$$\frac{\theta(s)}{\delta_c(s)} = \frac{290(s+0.01584)(s+0.4749)}{(s^2+0.01629s+0.002749)(s^2+2.625s+8.388)(s+18.4)} \text{ rad/rad} \quad (13.16)$$

and the short-period frequency and damping are now 2.9 rad/sec and 0.453, respectively.

As in the previous example, if the task is to be performed without excessive workload, and the pilot-vehicle system is to achieve maximum closed-loop performance, the crossover model implies that the effective controlled-element dynamics should be approximately *K/s* over as large a frequency range as possible near gain crossover.

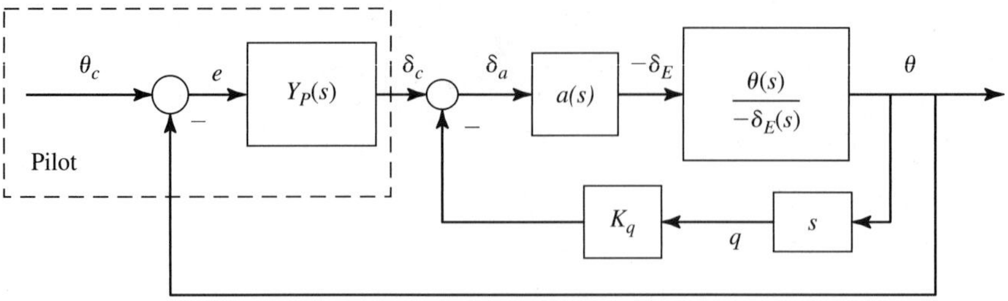

Figure 13.13 Block diagram of pitch-control task with pitch damper.

In addition, the gain-crossover frequency will be as high as possible, subject to the requirement that the minimum phase and gain margins are 35–45 deg and 5–6 dB, respectively, with the pilot exhibiting an effective time delay of 0.25 sec.

Incorporating the 0.25 sec time delay into the pitch-attitude responses to elevator command, including the actuator, the Bode plots are as shown in Figure 13.14. (A first-order Padé approximation was again used here to model this delay.) Note that without any pilot compensation, and due to the phase-margin requirement, the maximum gain-crossover frequency is approximately 3 rad/sec. However, we shall see that the pilot-vehicle gain-crossover frequency will be considerably less.

Let's first consider the case with the pitch damper. Note that at frequencies below about 0.5 rad/sec, the Bode plot reflects the desirable *K/s* characteristics. But to achieve maximum performance, the pilot will need to set the gain-crossover frequency above this value. And above 0.5 rad/sec and below the short-period frequency of 2.9 rad/sec, the Bode plot reflects control-element dynamics more like a pure gain with some time delay. Therefore, to achieve the required *K/s* characteristic in this frequency range, the pilot must introduce lag compensation.

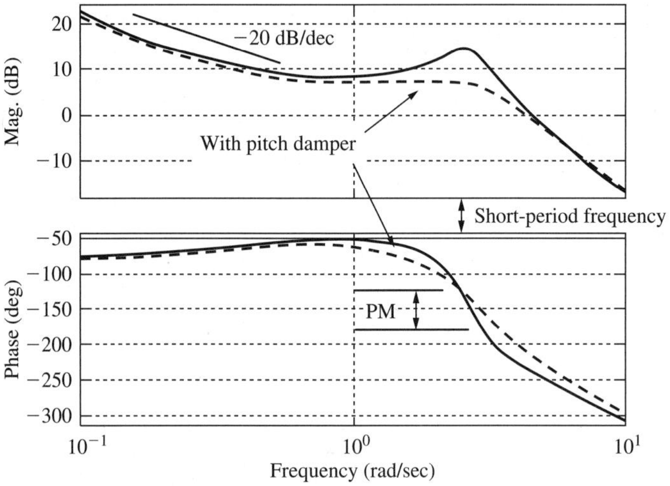

Figure 13.14 Bode plots of pitch-attitude responses $\theta(s)/\delta_c(s)$ w. 0.25 sec effective time delay τ_E.

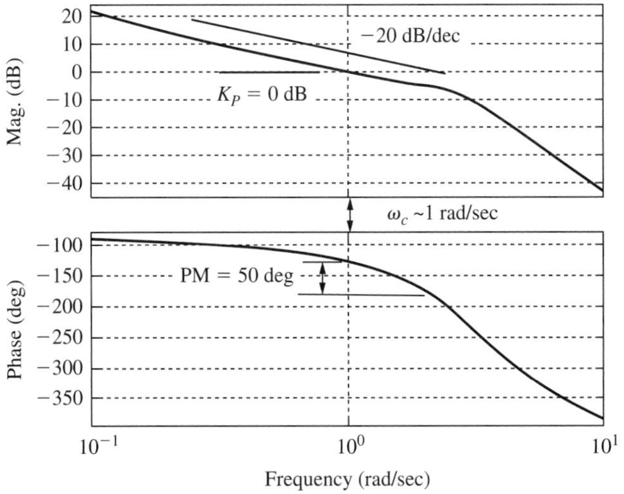

Figure 13.15 Bode plots of $Y_P Y_C(j\omega)$ (with pitch damper).

Adding the following pilot lag compensation,

$$\text{Lag}(s) = \frac{1}{2s + 1} \tag{13.17}$$

which approximately cancels the zero at $-1/T_{\theta_2}$, the Bode plots of the open-loop pilot-vehicle system, or $Y_P Y_C(j\omega)$ with a pilot gain K_P of 1 deg/deg, are as shown in Figure 13.15. From this figure we see that to achieve a phase margin of 40 deg, the crossover frequency can be no higher than about 1.2 rad/sec, which is well below the 3 rad/sec originally mentioned. The reduction in maximum crossover frequency is due to the additional lag introduced from pilot compensation, and the phase-margin requirement. However, after evaluating some pilot-vehicle step responses, it was determined that a 40 deg phase margin was probably too low, and the gain-crossover frequency was reduced to 1 rad/sec. Phase and gain margins of 50 deg and 6 dB resulted. The final pilot crossover model in this case is then

$$Y_P(s) = \left(\frac{1}{2s + 1}\right) e^{-0.25s} \text{ deg/deg} \tag{13.18}$$

Now let's look at the situation without the pitch damper. From the Bode plot for this vehicle in Figure 13.14, we can see that to achieve a gain-crossover frequency greater than 0.5 rad/sec, pilot lag must again be introduced, and the lag given in Equation (13.17) will be used. So using the crossover model given by Equation (13.18), the Bode plot of $Y_P Y_C(j\omega)$ is as now shown in Figure 13.16.

Note again that to obtain a phase margin of at least 40 deg, the gain-crossover frequency must be less than 2 rad/sec. So we will again consider a crossover frequency of 1 rad/sec. The result, however, is unsatisfactory. The gain margin for any gain-crossover frequency near 1 rad/sec will be essentially zero. The piloted step responses to a commanded change in pitch attitude of 5 deg, with and without pitch damper, are shown in

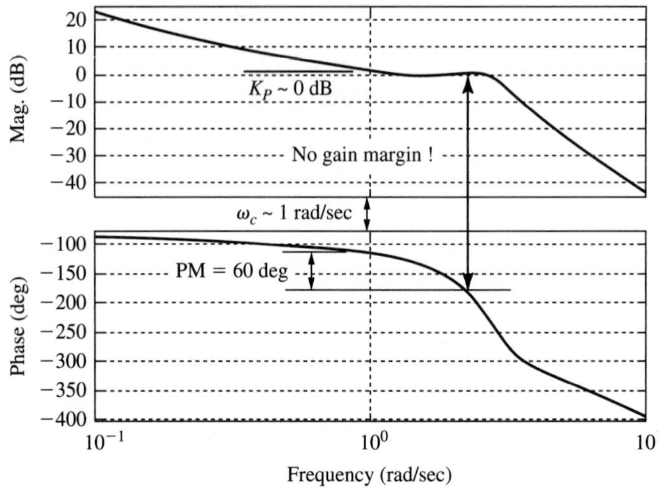

Figure 13.16 Bode Plots of $Y_PY_C(j\omega)$, wo. pitch damper.

Figure 13.17. We can see that without the pitch damper the closed-loop pilot-vehicle system is only marginally stable.

To obtain an acceptable gain margin of at least 6 dB, the gain-crossover frequency must be reduced to approximately 0.6 rad/sec. And this will lead to greatly degraded performance compared to the case with the pitch damper. Therefore, without the pitch damper, either the closed-loop pilot-vehicle performance is greatly degraded compared to the achievable performance with the pitch damper, or the stability margins are severely diminished.

In either case, the handling qualities of the vehicle without the pitch damper will be worse than those with the pitch damper. In fact, there has been at least one case involving a pilot actually turning off the pitch damper during a flight test of an early version of the

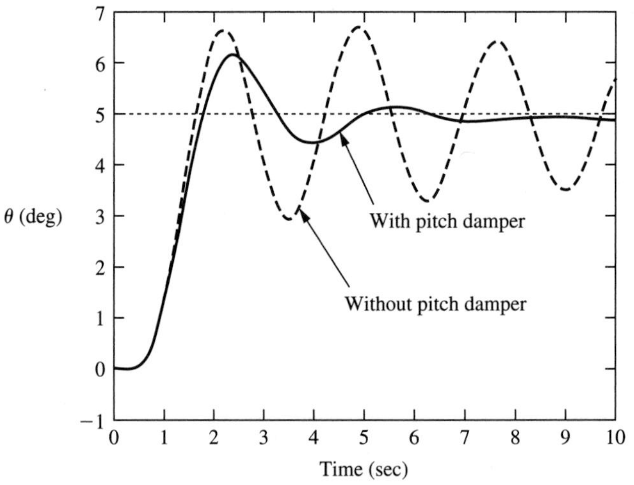

Figure 13.17 Effect of pitch damper on piloted step responses.

T-38, which is quite similar to the F-5, and experiencing neutrally-stable pitch oscillations similar to those indicated in Figure 13.17. This phenomenon is called a *pilot-induced oscillation* (PIO), and it is extremely hazardous. The occurrence of a PIO depends on the pilot-vehicle system. The oscillations would cease if the pilot released the control stick, thereby opening up the pilot's feedback loop.

As a final note, the crossover frequency for pitch-attitude control of other vehicles and/or flight conditions will frequently be higher than that obtained in this example, especially if the pilot doesn't have to introduce lag as was the case here (Ref. 7). This lag was required because $1/T_{\theta_2}$ for the aircraft was significantly smaller (~0.5/sec in the example) than the short-period frequency, and this leads to a fairly wide frequency range over which the magnitude of the elevator-to-pitch-attitude transfer function was flat (see Figure 13.14). When $1/T_{\theta_2}$ is larger, the magnitude over this frequency range will be more like K/s, and pilot lag would not be necessary. Hence the pilot's phase loss would be less, leading to higher achievable gain-crossover frequencies.

13.4 Summary

In this chapter we introduced the control characteristic of the human pilot and demonstrated how a pilot model may be used for pilot-vehicle analysis. The material presented directly follows the loop-shaping methodology presented in Chapter 12 for synthesizing feedback-control systems.

The underlying principal for human-control modeling is McRuer's crossover law, which states, "For the tasks considered, and subject to human limitations, the human-control operator adjusts their control strategy so as to make the open-loop transfer function approximately K/s near the region of gain crossover." Key human limitations include information-processing time delay and neuromuscular delay or lag. By applying the modeling concepts presented, an aircraft's dynamic characteristics may be assessed in terms of the relative ease or difficulty of the pilot's control task. Two examples demonstrated the procedure.

13.5 Problems

13.1 Again consider the single-axis pitch-control task addressed in Example 13.2. Using a short-period approximation instead of the complete pitch-attitude transfer function in the analysis, and assuming an effective pilot time delay of 0.25 sec, again estimate the maximum gain-crossover frequency, pilot compensation, and relative handling qualities with and without pitch damper when performing this task. Are your results significantly different from those obtained in the example?

13.2 Consider Example 13.1 again. If a time delay of 0.05 sec ($e^{-0.05s}$) is added to the vehicle's aileron-to-bank-angle transfer function, due to digital effects in the aircraft's flight-control system, for example, estimate the effect(s) on the maximum achievable gain-crossover frequency and the pilot compensation. Assume an effective pilot time delay of 0.25 sec, and include the roll damper used in the example. Has the addition of this time delay degraded the aircraft's roll-control handling qualities? Why?

References

1. Koppen, O. C.: "Airplane Stability and Control from a Designer's Point of View," *Journal of the Aeronautical Sciences,* vol. 7, no. 4, Feb. 1940, pp. 135–140.
2. Tustin, A.: "The Effects of Backlash and of Speed-Dependent Friction on the Stability of Closed-Cycle Control Systems," *Journal of the IEE,* vol. 94, part IIA, no. 1, May 1947.
3. Tustin, A.: "The Nature of the Operator's Response in Manual Control and Its Implications for Controller Design," *Journal of the IEE,* vol. 94, part IIA, no. 2, 1947, 190–207.
4. McRuer, D. T. and H. R. Jex: "A Review of Quasi-Linear Pilot Models." STI Paper 63, Systems Control Technology, Inc., 13766 S. Hawthorne Blvd. Hawthorne, CA, 90250. (Also published in *IEEE Transactions on Human Factors in Electronics,* vol. HFE 8, no. 3, pp. 231–249) 1967.
5. McRuer, D. T. and E. S. Krendel: "Mathematical Models of Human Pilot Behavior," STI Paper 146, Systems Control Technology, Inc., 13766 S. Hawthorne Blvd. Hawthorne, CA, 90250. (Also published as ARARDograph 188, Advisory Group on Aeronautics R&D, NATO) Jan. 1974.
6. McRuer, D. T.: "Pilot Modeling," AGARD Lecture Series 157, Advisory Group on Aeronautics R&D, NATO, 1988.
7. Hess, R.: "Structural Model of the Adaptive Human Pilot," *Journal of Guidance and Control,* vol. 3, no. 5, Sept.–Oct. 1980, pp. 416–423.
8. Hess, R.: "Unified Theory for Aircraft Handling Qualities and Adverse Aircraft-Pilot Coupling," *Journal of Guidance, Control and Dynamics,* vol. 20, no. 6, Nov.–Dec. 1997, pp. 1141–1148.
9. Kleinman, D. L., S. Barron, and W. H. Levison: "An Optimal Control Model of Human Response," Parts 1 and 2, *Automatica,* vol. 6., May 1970, pp. 357–383.
10. Davidson, J. B. and D. K. Schmidt: *Modified Optimal Control Pilot Model for Computer-Aided Design and Analysis,* NASA Technical Memorandum 4384, Oct. 1992.
11. McRuer, D. T. and D. K. Schmidt: "Pilot-Vehicle Analysis of the Multi-Axis Task," *Journal of Guidance, Control, and Dynamics,* vol. 13, no. 2, Mar.–Apr. 1990, pp. 348–355.

APPENDIX A

Properties of the Atmosphere

Under the model of the atmosphere defined as the 1962 Standard Atmosphere (see Ref. 1), the temperature of the air varies linearly with altitude above the surface of the earth. This variation, or the *temperature laps rate l,* is given in Table A.1, where it is constant from the altitude next to which it is listed to the next higher altitude given. For example, for altitudes from 0 to 36,089 ft, the laps rate is -3.5662×10^{-3} deg R/ft.

Furthermore, air at rest is assumed to satisfy the following two equations.

$$\begin{aligned} &\text{Aerostatic equation:} && dp = -\rho g dh \\ &\text{Perfect-gas equation of state:} && p = \rho R T \end{aligned} \tag{A.1}$$

Here p is the pressure, ρ the density, T the absolute temperature of the atmosphere, h the altitude, g the acceleration due to gravity (assumed constant $= 32.174$ ft/sec^2), and R the universal gas constant for air ($= 1716.5$ ft^2/sec^2-°R). Finally, the sonic velocity in air a may be expressed in terms of the absolute temperature (given in degrees R). Or

$$a = \sqrt{kRT} = 49.021\sqrt{T} \tag{A.2}$$

where k is the ratio of specific heats for air, taken to be 1.4, and a is given in fps.

Table A.1 Atmospheric Constants

Altitude h^*, ft	Atmospheric Region	Laps Rate l, deg R/ft	Temperature T^*, deg R	Pressure p^*, psf	Density ρ^*, sl/ft^3
0	Troposphere	-3.5662×10^{-3}	518.69	2,116.2	2.3769×10^{-3}
36,089	Stratosphere I	0	389.99	472.68	7.0613×10^{-4}
65,617	Stratosphere II	5.4864×10^{-4}	389.99	114.35	1.7083×10^{-4}
104,990			411.59	18.13	2.5661×10^{-5}

From Equations (A.1), plus the data in Table A.1, the following equations may be derived in the three levels of the atmosphere with fixed laps rates.

For altitudes h between 0–36,089 feet:

$$T = T^* + l(h - h^*) = 518.69 - (3.5662 \times 10^{-3})h \text{ deg R}$$
$$p = (1.1376 \times 10^{-11})T^{5.256} \text{ psf} \qquad \text{(A.3)}$$
$$\rho = (6.6277 \times 10^{-15})T^{4.256} \text{ sl/ft}^3$$

For altitudes h between 36,090–65,617 feet:

$$T = T^* + l(h - h^*) = 389.99 \text{ deg R}$$
$$p = (2678.4)e^{(-4.8063 \times 10^{-5})h} \text{ psf} \qquad \text{(A.4)}$$
$$\rho = (1.4939 \times 10^{-6})p \text{ sl/ft}^3$$

And finally, for altitudes h between 65,618–104,990 feet:

$$T = T^* + l(h - h^*) = 389.99 + 5.4864 \times 10^{-4}(h - 65{,}617) \text{ deg R}$$
$$p = (3.7930 \times 10^{90})T^{-34.164} \text{ psf} \qquad \text{(A.5)}$$
$$\rho = (2.2099 \times 10^{87})T^{-35.164} \text{ sl/ft}^3$$

The above expressions, plus Equation (A.2), were used to generate the atmospheric properties given in Table A.2. The equations themselves may be used in simulations of the vehicle dynamics, for example.

Sometimes a simple, approximate, analytical expression is needed for atmospheric density, and the following exponential expressions are frequently used.

Troposphere: $$\rho = (2.3769 \times 10^{-3})e^{(-h/29{,}730)} \text{ sl/ft}^3$$

Stratosphere I: $$\rho = (7.0613 \times 10^{-4})e^{(-(h-36{,}089)/20{,}806)} \text{ sl/ft}^3 \qquad \text{(A.6)}$$

Stratosphere II: $$\rho = (1.7083 \times 10^{-4})e^{(-(h-65{,}617)/20{,}770)} \text{ sl/ft}^3$$

Table A.2 Atmospheric Properties

h, ft	*T*, deg R	*p*, psf	*ρ*, sl/ft^3	*a*, fps
0	518.69	2,116.1	0.0023769	1,116.4
1,000	515.12	2,040.8	0.0023081	1,112.6
2,000	511.56	1,967.6	0.0022409	1,108.7
3,000	507.99	1,896.6	0.0021751	1,104.9
4,000	504.43	1,827.6	0.0021109	1,101.0
5,000	500.86	1,760.7	0.0020481	1,097.1
6,000	497.29	1,695.8	0.0019868	1,093.2
7,000	493.73	1,632.9	0.0019268	1,089.2
8,000	490.16	1,571.8	0.0018683	1,085.3
9,000	486.59	1,512.7	0.0018111	1,081.3
10,000	483.03	1,455.3	0.0017553	1,077.4
11,000	479.46	1,399.7	0.0017008	1,073.4
12,000	475.90	1,345.8	0.0016476	1,069.4
13,000	472.33	1,293.7	0.0015957	1,065.4
14,000	468.76	1,243.1	0.0015450	1,061.4
15,000	465.20	1,194.2	0.0014956	1,057.3
16,000	461.63	1,146.9	0.0014474	1,053.2
17,000	458.06	1,101.1	0.0014004	1,049.2
18,000	454.50	1,056.8	0.0013546	1,045.1
19,000	450.93	1,013.9	0.0013100	1,041.0
20,000	447.37	972.47	0.0012664	1,036.8
21,000	443.80	932.41	0.0012240	1,032.7
22,000	440.23	893.69	0.0011827	1,028.5
23,000	436.67	856.29	0.0011425	1,024.4
24,000	433.10	820.17	0.0011033	1,020.2
25,000	429.54	785.29	0.0010651	1,016.0
26,000	425.97	751.62	0.0010280	1,011.7
27,000	422.40	719.13	0.00099187	1,007.5
28,000	418.84	687.79	0.00095672	1,003.2

(continued)

h, ft	*T*, deg R	*p*, psf	ρ, sl/ft^3	*a*, fps
29,000	415.27	657.56	0.00092253	998.96
30,000	411.70	628.42	0.00088928	994.66
31,000	408.14	600.33	0.00085695	990.34
32,000	404.57	573.27	0.00082553	986.01
33,000	401.01	547.20	0.00079501	981.65
34,000	397.44	522.10	0.00076535	977.28
35,000	393.87	497.95	0.00073654	972.88
36,000	390.31	474.70	0.00070858	968.47
37,000	389.99	452.43	0.00067589	968.07
38,000	389.99	431.20	0.00064418	968.07
39,000	389.99	410.97	0.00061395	968.07
40,000	389.99	391.68	0.00058514	968.07
41,000	389.99	373.30	0.00055768	968.07
42,000	389.99	355.79	0.00053151	968.07
43,000	389.99	339.09	0.00050657	968.07
44,000	389.99	323.18	0.00048280	968.07
45,000	389.99	308.01	0.00046014	968.07
46,000	389.99	293.56	0.00043855	968.07
47,000	389.99	279.78	0.00041797	968.07
48,000	389.99	266.65	0.00039835	968.07
49,000	389.99	254.14	0.00037966	968.07
50,000	389.99	242.21	0.00036184	968.07
51,000	389.99	230.85	0.00034486	968.07
52,000	389.99	220.02	0.00032868	968.07
53,000	389.99	209.69	0.00031326	968.07
54,000	389.99	199.85	0.00029856	968.07
55,000	389.99	190.47	0.00028455	968.07
56,000	389.99	181.53	0.00027119	968.07
57,000	389.99	173.02	0.00025847	968.07
58,000	389.99	164.90	0.00024634	968.07
59,000	389.99	157.16	0.00023478	968.07
60,000	389.99	149.78	0.00022376	968.07
61,000	389.99	142.76	0.00021326	968.07
62,000	389.99	136.06	0.00020325	968.07

(continued)

h, ft	*T*, deg R	*p*, psf	*ρ*, sl/ft³	*a*, fps
63,000	389.99	129.67	0.00019372	968.07
64,000	389.99	123.59	0.00018463	968.07
65,000	389.99	117.79	0.00017596	968.07
66,000	390.20	112.26	0.00016763	968.34
67,000	390.75	107.00	0.00015955	969.02
68,000	391.30	101.99	0.00015186	969.70
69,000	391.85	97.227	0.00014456	970.38
70,000	392.39	92.689	0.00013762	971.05
71,000	392.94	88.368	0.00013103	971.73
72,000	393.49	84.255	0.00012475	972.41
73,000	394.04	80.338	0.00011879	973.09
74,000	394.59	76.608	0.00011312	973.77
75,000	395.14	73.057	0.00010772	974.44
76,000	395.69	69.675	0.00010259	975.12
77,000	396.24	66.453	9.7713e-05	975.80
78,000	396.78	63.385	9.3073e-05	976.47
79,000	397.33	60.462	8.8658e-05	977.15
80,000	397.88	57.678	8.4459e-05	977.82

REFERENCE

1. Anon.: *U.S. Standard Atmosphere, 1962,* U.S. Government Printing Office, Washington, D.C., Dec. 1962.

B APPENDIX

Data for Several Aircraft

Note: Some of the data from Reference 1 have been modified such that the data presented in this appendix is consistent with the notation and definitions of the stability derivatives given in the Nomenclature section at the front of this book.

A-7A

Nominal Cruise Configuration

Clean airplane
60% fuel
W = 21,889 lbs
CG at 30% MGC
I_x = 13,635 slug-ft^2
I_y = 58,966 slug-ft^2
I_z = 67,560 slug-ft^2
I_{xz} = 2,933 slug-ft^2
} Body ref. axes

Reference Geometry

S = 375 ft^2
c = 10.8 ft
b = 38.7 ft

Flight Envelope

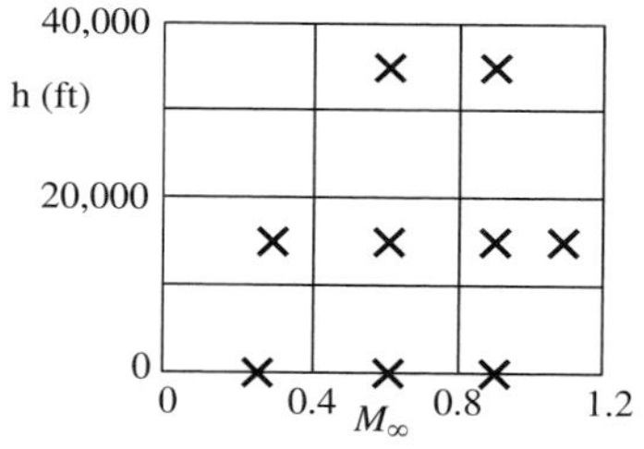

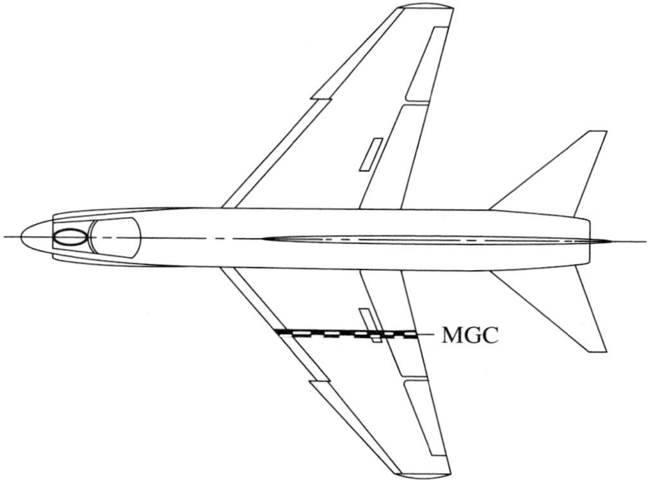

References

1. LTV Vought Aeronautics Div. Rept. No. 2-53310/5R-1981, "A-7A Aerodynamics Data Report", 21 May 1965 (U).
2. LTV Vought Aeronautics Div. Rept. No. 2-53310/5R-5121, Rev. I, "A-7A Estimated Flying Qualities", 20 August 1965(C).
3. LTV Vought Aeronautics Div., "Updated A-7A Aircraft Lateral-Directional Cruise Device Configuration Data, 25 August 1967.

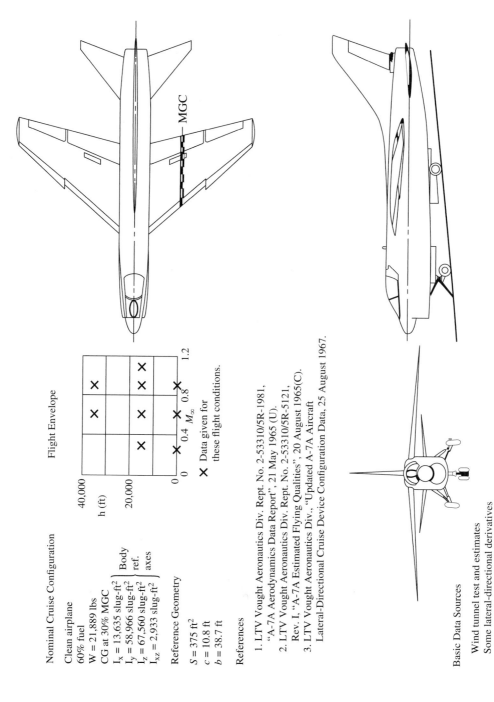

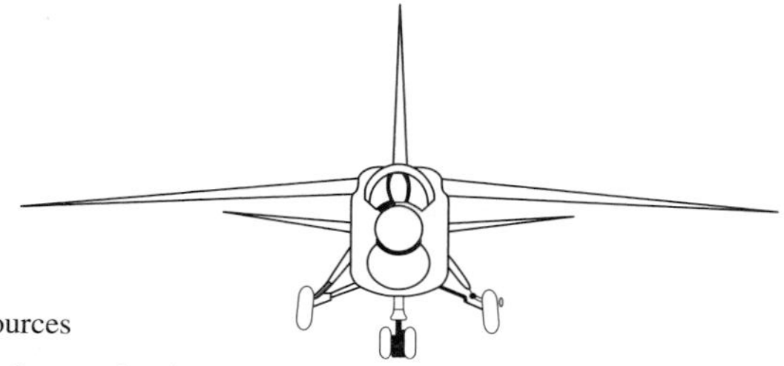

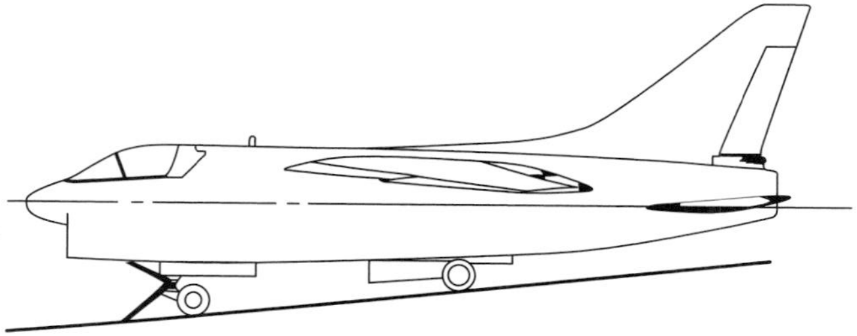

Basic Data Sources

Wind tunnel test and estimates
Some lateral-directional derivatives
adjusted after flight test

Figure B.1 A-7A aircraft configuration and data.

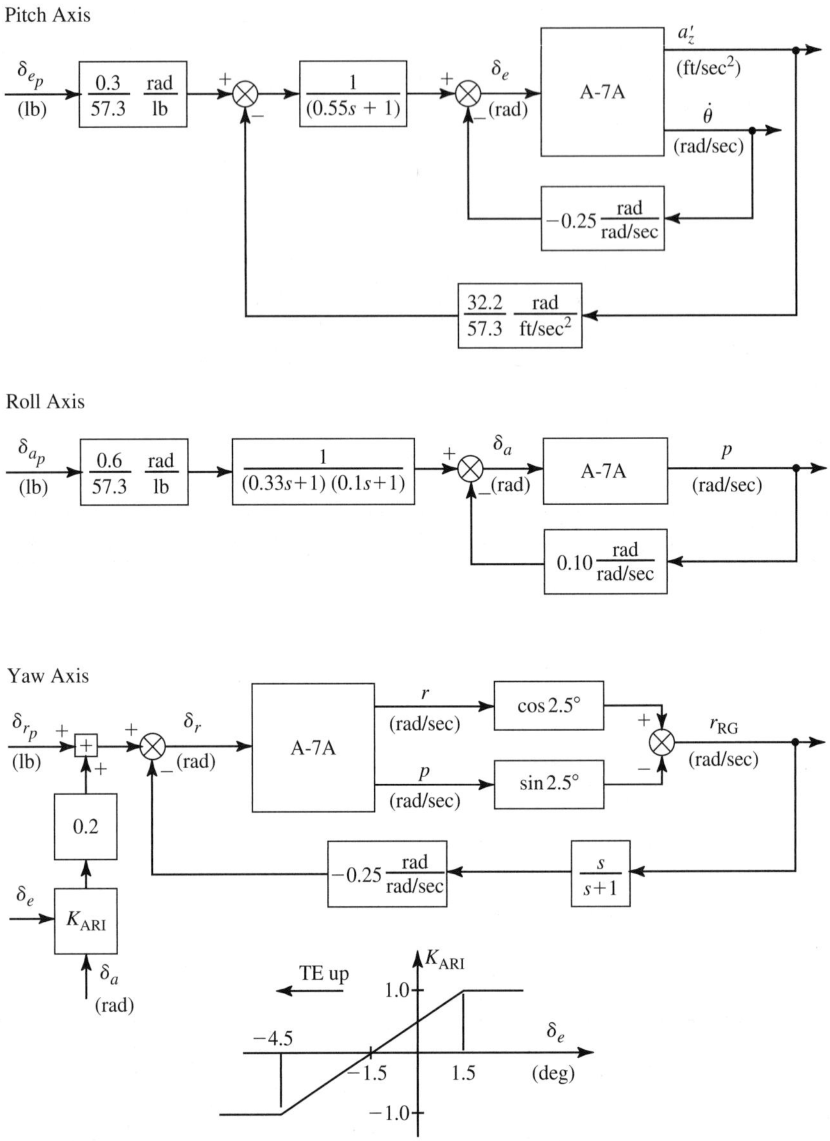

Figure B.2 A-7A stability augmentation systems.

Table B.1 A-7A Flight Conditions

	Flight Condition				Flight Condition		
	3	4	8		3	4	8
h, ft	0	15,000	35,000	α_0, deg	2.1	13.3	7.5
M_∞	0.9	0.3	0.6	U_0, fps	1,004	309	579
a, fps	1,117	1,058	973.3	W_0, fps	36.8	72.9	76.2
ρ_∞, sl/ft^3	2.378×10^{-3}	1.496×10^{-3}	7.36×10^{-4}	δ_{E_0}, deg	−3.8	−8.8	−5.4
V_V, fps	1,005	317	584	γ_0, deg	0	0	0
q_∞, psf	1,200	75.3	126				

Note: All data given in fuselage-reference axes. Clean, flexible airplane.

Table B.2 A-7A Longitudinal Dimensional Derivatives

	Flight Condition				Flight Condition		
	3	4	8		3	4	8
h, ft	0	15,000	35,000	Z_{δ_E}	−318	−23.8	−43.2
M_∞	0.9	0.3	0.6	$M_u + M_{Pu}$	0.00118	0.00183	0.000873
$X_u + X_{Pu}$	−0.0732	0.00501	0.00337	M_α	−40.401	−2.463	−4.152
X_α	−28.542	1.457	8.526	$M_{\dot{\alpha}}$	−0.372	−0.056	−0.065
X_{δ_E}	11.6	5.63	5.70	M_q	−1.57	−0.340	−0.330
$Z_u + Z_{Pu}$	0.0184	−0.0857	−0.0392	M_{δ_E}	−58.6	−4.52	−8.19
Z_α	−3,417	−172.77	−323.54				

Note: All data given in fuselage-reference axes. Clean, flexible airplane.

Table B.3 A-7A Lateral-Directional Dimensional Derivatives

	Flight Condition				Flight Condition		
	3	4	8		3	4	8
h, ft	0	15,000	35,000	L'_{δ_A}	25.2	3.75	7.96
M_∞	0.9	0.3	0.6	L'_{δ_R}	13.2	1.82	3.09
Y_β	−516.57	−38.674	−49.465	N'_β	17.2	0.948	1.38
Y_{δ_A}	−8.613	−0.4755	−1.5593	N'_p	−0.319	−0.031	−0.0799
Y_{δ_R}	62.913	9.732	15.593	N'_r	−1.54	−0.271	−0.247
L'_β	−98.0	−8.79	−14.9	N'_{δ_A}	1.56	0.280	0.652
L'_p	−9.75	−1.38	−1.40	N'_{δ_R}	−11.1	−1.56	−2.54
L'_r	1.38	0.857	0.599				

Note: All data given in fuselage-reference axes. Clean, flexible airplane.

A-4D

Nominal Cruise Configuration

Clean airplane
W = 17,578 lbs
CG at 25% MGC
$I_x = 8{,}090$ slug-ft^2
$I_y = 25{,}900$ slug-ft^2
$I_z = 29{,}200$ slug-ft^2
$I_{xz} = 1{,}300$ slug-ft^2
} Body ref. axes

Reference Geometry

$S = 260$ ft^2
$c = 10.8$ ft
$b = 27.5$ ft

Flight Envelope

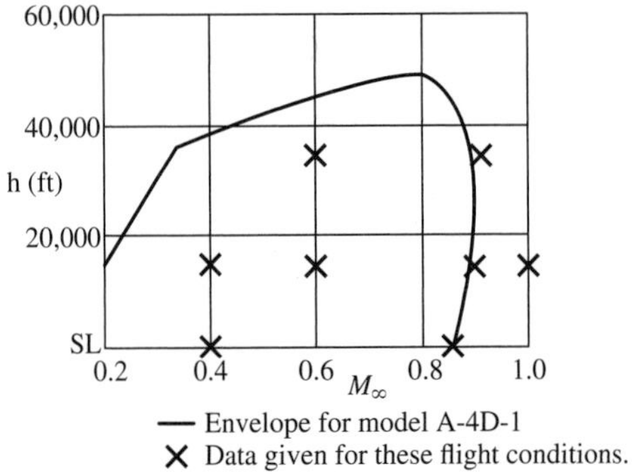

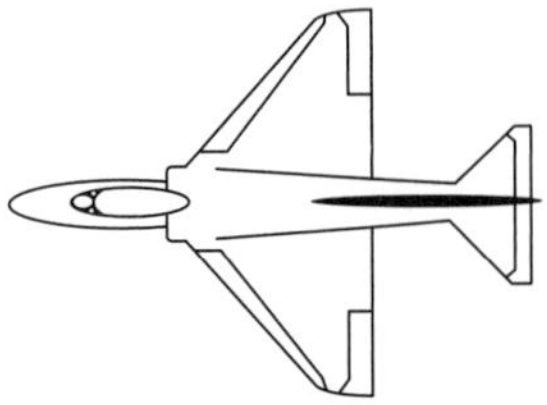

References

1. Abzug, M. J. and R. L. Faith, "Aerodynamic Data for Model A4D-1 Operational Flight Trainer", Douglas Aircraft Co. Report ES-26104, November 1, 1955.
2. Johnston, D. E. and D. H. Weir, "Study of Pilot-Vehicle-Controller Integration for a Minimum Complexity AFCS", Systems Technology, Inc. Technical Report No.127-1, July 1964.

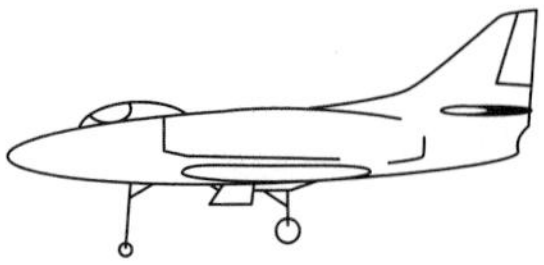

Basic Data Sources

Wind tunnel test

Figure B.3 A-4D configuration and data.

Pitch

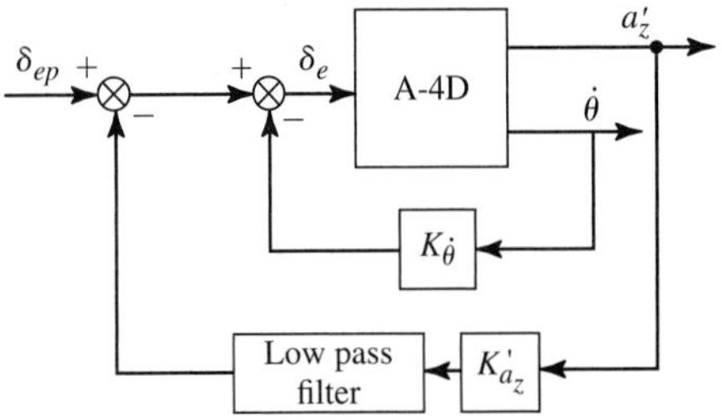

$K_{\dot\theta}$, K_{a_z}': Scheduled for indicated air speed

Note: System used on the A4D-2N model only.
Control stick steering mode shown.

Roll

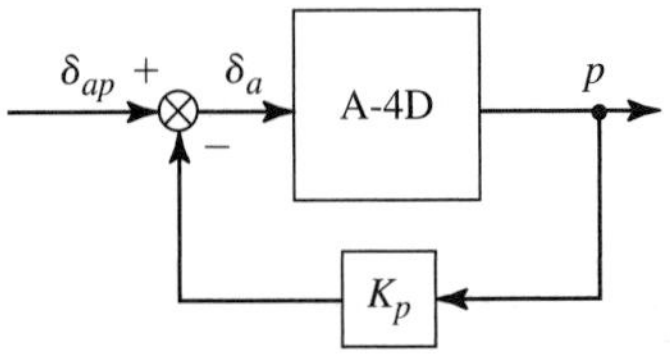

K_p: Gain in deg/deg/sec, scheduled for indicated air speed

Yaw

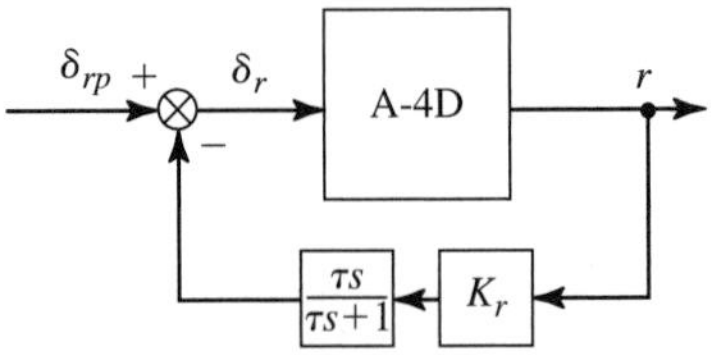

K_r: Gain in deg/deg/sec, scheduled for indicated air speed

Figure B.4 A-4D stability augmentation systems.

Table B.4 A-4D Flight Conditions

	Flight Condition			Flight Condition	
	4	7		4	7
h, ft	15,000	35,000	q_∞, psf	301	126
M_∞	0.6	0.6	α_0, deg	3.4	8.8
a, fps	1,058	973.3	U_0, fps	634	577
ρ_∞, sl/ft^3	1.496×10^{-3}	7.36×10^{-4}	W_0, fps	37.7	89.3
V_V, fps	635	584	γ_0, deg	0	0

Note: All data given in fuselage-reference axes. Clean, flexible airplane.

Table B.5 A-4D Longitudinal Dimensional Derivatives

	Flight Condition			Flight Condition	
	4	7		4	7
h, ft	15,000	35,000	Z_{δ_E}	−56.68	−23.037
M_∞	0.6	0.6	$M_u + M_{Pu}$	0.00162	0.001824
$X_u + X_{Pu}$	−0.00938	0.000806	M_α	−12.954	−5.303
X_α	26.797	13.257	$M_{\dot{\alpha}}$	−0.3524	−0.1577
X_{δ_E}	7.396	6.288	M_q	−1.071	−0.484
$Z_u + Z_{Pu}$	−0.0533	−0.0525	M_{δ_E}	−19.456	−8.096
Z_α	−521.97	−226.242			

Note: All data given in fuselage-reference axes. Clean, flexible airplane.

Table B.6 A-4D Lateral-Directional Dimensional Derivatives

	Flight Condition			Flight Condition	
	4	7		4	7
h, ft	15,000	35,000	L'_{δ_A}	21.203	8.170
M_∞	0.6	0.6	L'_{δ_R}	10.398	4.168
Y_β	−144.78	−60.386	N'_β	16.629	6.352
Y_{δ_A}	−2.413	−0.4783	N'_p	−0.02173	−0.02513
Y_{δ_R}	25.133	10.459	N'_r	−0.5144	−0.2468
L'_β	−35.95	−17.557	N'_{δ_A}	1.769	0.5703
L'_p	−1.566	−0.761	N'_{δ_R}	−7.78	−3.16
L'_r	0.812	0.475			

Note: All data given in fuselage-reference axes. Clean, flexible airplane.

F-5A

Configurations

GAR-8-GAR-8 on wing tips
- I – Center line Tank
  - 150 gal. tanks at W.S.85
  - 750 lb. stores at W.S.114.5
  - 50 gal. tip tanks
- I-A – as I with 50% fuel
- II – 2000 lb centerline store
  - 1000 lb stores at W.S. 85
  - 750 lb stores at W.S. 114.5
  - 50 gal. tip tanks

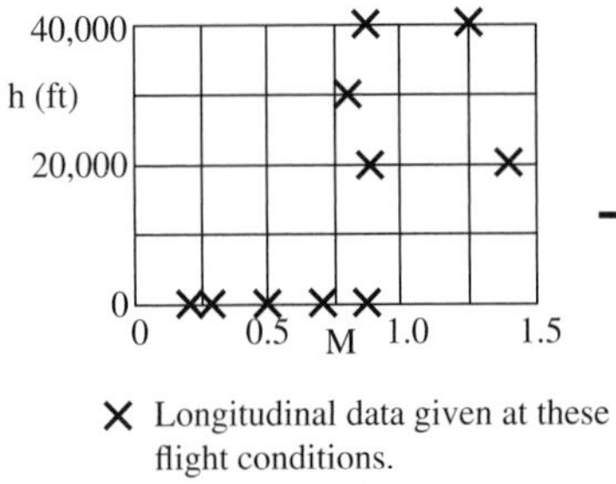

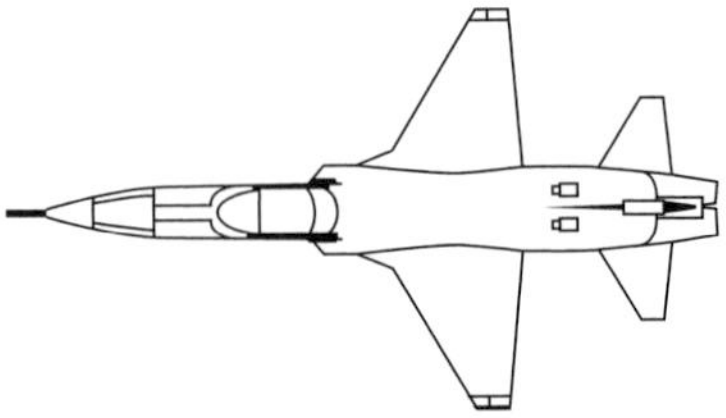

Reference Geometry

$S = 170\ \text{ft}^2$
$b = 25.25$ ft
$c = 7.75$ ft

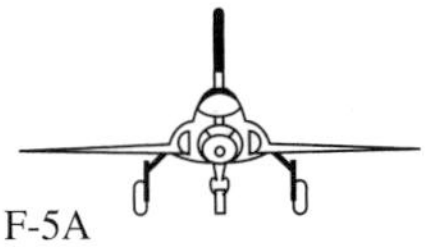

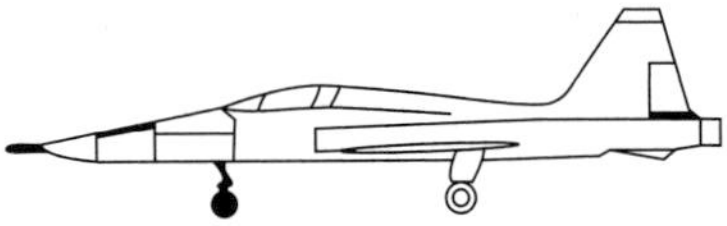

Reference

1. Jex, H. R. and J. Nakagawa, "Typical F-5A Longitudinal Aerodynamic Data and Transfer Functions for 14 Conditions", Systems Technology, Inc., Technical Memorandum No.239-4, March 1964.

Basic Data Sources

Wind tunnel tests with corrections made per flight test

Figure B.5 F-5A Configuration and data.

Table B.7 F-5A Flight Conditions

	Flight Condition				Flight Condition		
	1 (GAR-8)	2 (GAR-8)	7 (I)		1 (GAR-8)	2 (GAR-8)	7 (I)
h, ft	40,000	40,000	30,000	U_0, fps	850	1,210	796
M_∞	0.875	1.25	0.8	W_0, fps	0	0	0
a, fps	971.4	971.4	995	δ_{E_0}, deg	−1.15	−1.80	−2.57
ρ_∞, sl/ft^3	5.813×10^{-4}	5.813×10^{-4}	8.901×10^4	γ_0, deg	0	0	0
V_V, fps	850	1,210	796	$\bar{x}_{cg}$	0.22	0.22	0.12
q_∞, psf	210	428	282	I_{yy}, sl-ft^2	30,000	30,000	34,600
α_0, deg	3.2	1.0	9.0			$\phi_T = 0.5$ deg	

Note: All data given for stability axes.

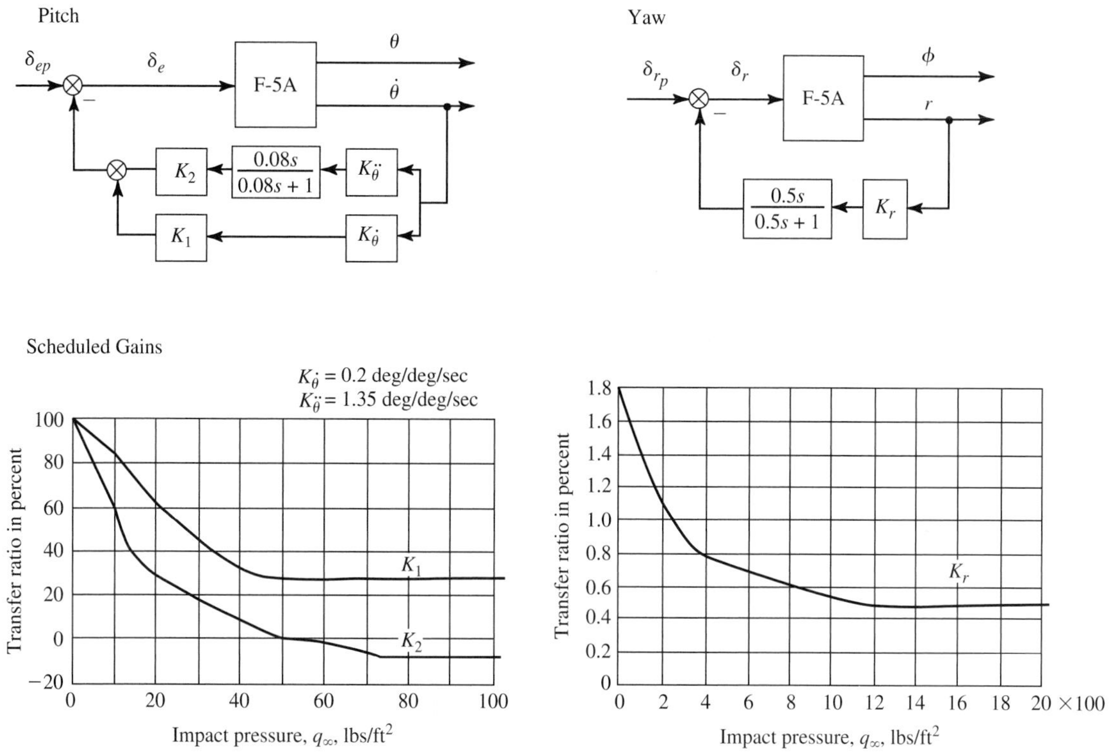

Figure B.6 F-5A stability augmentation systems.

Table B.8 F-5A Longitudinal Effectiveness Coefficients

	Flight Condition				Flight Condition		
	1	2	7		1	2	7
h (ft)	40,000	40,000	30,000	C_{D_M}	0.045	0	0.100
M_∞	0.876	1.25	0.8	$C_{D_{\delta_E}}$	0	0	0
$C_{L\,\text{Trim}}$	0.280	0.132	0.355	C_{M_0}	0.000902	0.00146	0.001695
$C_{D\,\text{Trim}}$	0.0279	0.0451	0.0422	C_{M_α} (/rad)	−0.367	−1.46	−0.691
C_{L_α} (/rad)	5.38	5.38	4.58	$C_{M_{\dot{\alpha}}}$ (sec)	−0.00546	0.0096	−0.00243
C_{L_q} (sec)	0.0355	0.0176	0.0257	C_{M_q} (sec)	−0.04638	−0.0288	−0.04613
C_{L_M}	0.40	−0.70	−0.25	C_{M_M}	−0.050	−0.100	0.020
$C_{L_{\delta_E}}$ (/rad)	1.03	0.745	0.888	$C_{M_{\delta_E}}$ (/rad)	−1.55	−1.29	−1.39
C_{D_α} (/rad)	0.339	1.97	0.352	$\frac{\partial T}{\partial M_\infty}$ (lbs)	600	3,500	1,400
C_{D_q} (sec)	0	0	0				

Note: All data given for stability axes.

Table B.9 F-5A Longitudinal Dimensional Derivatives

	Flight Condition				Flight Condition		
	1 (GAR-8)	2 (GAR-8)	7 (I)		1 (GAR-8)	2 (GAR-8)	7 (I)
h, ft	40,000	40,000	30,000	Z_{δ_E}	−119	−175	−78.6
M_∞	0.875	1.25	0.8	$M_u + M_{Pu}$	−0.000462	−0.00193	0.000325
$X_u + X_{Pu}$	−0.0109	−0.00589	−0.0158	M_α	−3.392	−27.467	−6.782
X_α	−6.826	−15.246	0.269	$M_{\dot{\alpha}}$	−0.0506	0.180	−0.0260
X_{δ_E}	0	0	0	M_q	−0.429	−0.540	−0.488
$Z_u + Z_{Pu}$	−0.124	0.118	−0.0575	M_{δ_E}	−14.3	−24.3	−14.5
Z_α	−623.9	−1,271	−414.7				

Note: All data given for stability axes.

NAVION

Nominal Flight Condition

h (ft) = 0; $M_\infty = 0.158$; $V_{T_0} = 176$ ft/sec

W = 2750 lbs
CG at 29.5% MAC
$I_x = 1048$ slug ft^2
$I_y = 3000$ slug ft^2
$I_z = 3530$ slug ft^2
$I_{xz} = 0$

Reference Geometry

$S = 184$ ft^2
$c = 5.7$ ft
$b = 33.4$ ft

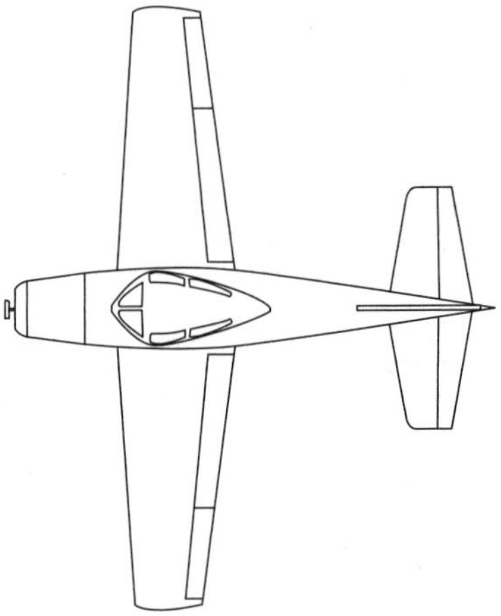

Figure B.7 Navion configuration and data.

Table B.10 Navion Flight Condition and Effectiveness Coefficients

S_W, ft^2	184	$C_{L\,\text{Trim}}$	0.41	C_{S_β} (/rad)	−0.564
b, ft	33.4	$C_{D\,\text{Trim}}$	0.05	$C_{S_{\delta_A}}$ (/rad)	0
$\bar{c}_w$, ft	5.7	C_{L_α} (/rad)	4.44	$C_{S_{\delta_R}}$ (/rad)	0.157
h, ft	0	$C_{L_{\dot{\alpha}}}$ (sec)	0	C_{L_β} (/rad)	−0.074
M_∞	0.158	C_{L_M}	0	C_{L_p} (sec)	−0.0389
a, fps	1,117	$C_{L_{\delta_E}}$ (/rad)	0.355	C_{L_r} (sec)	0.0102
ρ_∞, sl/ft^3	0.002378	C_{D_α} (/rad)	0.330	$C_{L_{\delta_A}}$ (/rad)	0.1342
V_V, fps	176	C_{D_M}	0	$C_{L_{\delta_R}}$ (/rad)	0.0118
q_∞, psf	36.8 psf	$C_{D_{\delta_E}}$	0	C_{N_β} (/rad)	0.0701
α_0, deg	0.6	C_{M_α} (/rad)	−0.683	C_{N_p} (sec)	−0.0055
U_0, fps	176	$C_{M_{\dot{\alpha}}}$ (sec)	−0.0706	C_{N_r} (sec)	−0.0119
W_0, fps	0	C_{M_q} (sec)	−0.161	$C_{N_{\delta_A}}$ (/rad)	−0.00346
γ_0, deg	0	C_{M_M}	0	$C_{N_{\delta_R}}$ (/rad)	−0.0717
Φ_0, deg	0	$C_{M_{\delta_E}}$ (/rad)	−0.87		

Note: All data for stability axes.

Table B.11 Navion Dimensional Derivatives

$X_u + X_{Pu}$	−0.0451	Y_β	−44.757
X_α	6.348	Y_{δ_A}	0
X_{δ_E}	0	Y_{δ_R}	12.461
$Z_u + Z_{Pu}$	−0.3697	L'_β	−15.982
Z_α	−356.29	L'_p	−8.402
Z_{δ_E}	−28.17	L'_r	2.193
$M_u + M_{Pu}$	0	L'_{δ_A}	28.984
M_α	−8.795	L'_{δ_R}	2.548
$M_{\dot{\alpha}}$	−0.9090	N'_β	4.495
M_q	−2.0767	N'_p	−0.3498
M_{δ_E}	−11.189	N'_r	−0.7605
		N'_{δ_A}	−0.2218
		N'_{δ_R}	−4.597

Note: All data for stability axes.

DC-8

Flight Conditions

Flight Condition	Approach	Holding	Cruise	V_{NE}
h (ft)	0	15,000	33,000	33,000
M_∞	0.219	0.443	0.84	0.88
W (lbs)	190,000	190,000	230,000	230,000
I_x (slug-ft^2)	3.09×10^6	3.11×10^6	3.77×10^6	3.77×10^6
I_y (slug-ft^2)	2.94×10^6	2.94×10^6	3.56×10^6	3.56×10^6
I_z (slug-ft^2)	5.58×10^6	5.88×10^6	7.13×10^6	7.13×10^6
I_{xz} (slug-ft^2)	28×10^3	-64.5×10^3	45×10^3	53.7×10^3
X_{cg}/c	0.15	0.15	0.15	0.15

Stability axes (I_x, I_y, I_z)

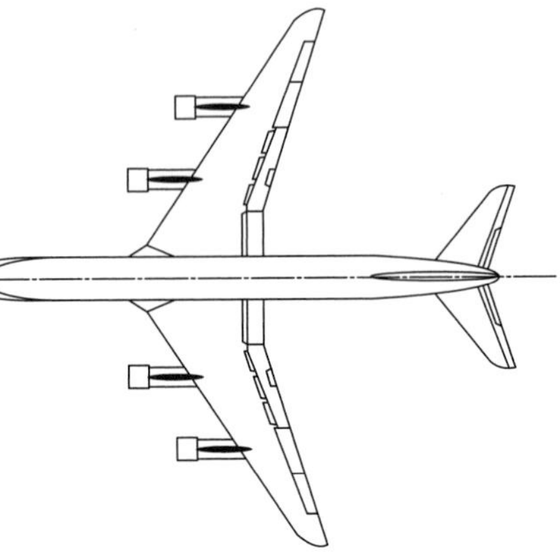

Reference Geometry

$S = 2600$ ft^2
$b = 142.3$ ft
$c = 23$ ft

References

Unpublished data

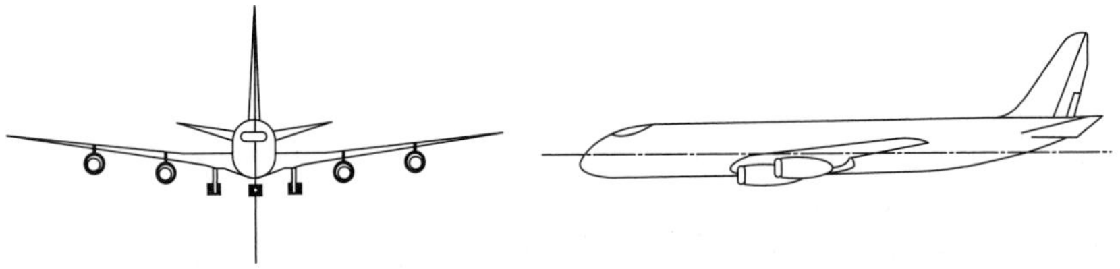

Figure B.8 DC-8 configuration and data.

Table B.12 DC-8 Flight Conditions

	Flight Condition			
	1 Approach	**2 Holding**	**3 Cruise**	**4 V_{NE}**
h, ft	0	15,000	33,000	33,000
M_∞	0.218	0.443	0.84	0.88
a, fps	1,117	1,058	982	982
ρ_∞, sl/ft^3	2.378×10^{-3}	1.496×10^{-3}	7.95×10^4	7.95×10^4
V_V, fps	243.5	468.2	824.2	863.46
q_∞, psf	71.02	163.97	270	296.36
U_0, fps	243.5	468.2	824.2	863.46
W_0, fps	0	0	0	0
γ_0, deg	0	0	0	0
δ_F, deg	35	0	0	0

Note: All data given for stability axes.

Table B.13 DC-8 Longitudinal Effectiveness Coefficients

	Flight Condition					Flight Condition			
	1	2	3	4		1	2	3	4
h, ft	0	15,000	33,000	33,000	C_{D_α} (/rad)	0.487	0.212	0.272	0.486
M_∞	0.218	0.443	0.84	0.88	C_{D_M}	0.0202	0.0021	0.1005	0.3653
$C_{L\,\text{Trim}}$	0.98	0.42	0.308	0.279	$C_{D_{\delta_E}}$	0	0[1]	0	0
$C_{D\,\text{Trim}}$	0.1095	0.0224	0.0188	0.0276	C_{M_α} (/rad)	−1.478	−1.501	−2.017	−2.413
C_{L_α} (/rad)	4.810	4.876	6.744	6.899	$C_{M_{\dot\alpha}}$ (sec)	−0.1814	−0.1007	−0.0924	−0.0910
$C_{L_{\dot\alpha}}$ (sec)	0	0	0	0	C_{M_q} (sec)[2]	−0.5485	−0.2971	−0.2037	−0.2025
C_{L_M}	0.02	0.048	0	−1.2	C_{M_M}	−0.006	−0.020	−0.170	−0.500
$C_{L_{\delta_E}}$ (/rad)	0.328	0.328	0.352	0.358	$C_{M_{\delta_E}}$ (/rad)[2]	−0.9354	−0.9715	−1.0120	−1.0285

Note: All data given for stability axes.

[1] Incorrectly listed as −0.9712 in original report.

[2] Found from M_q and M_{δ_E}, respectively.

Table B.14 DC-8 Lateral-Directional Effectiveness Coefficients

	Flight Condition					Flight Condition			
	1	2	3	4		1	2	3	4
h, ft	0	15,000	33,000	33,000	$C_{L_{\delta_A}}$ (/rad)	0.0860	0.0831	0.0797	0.0791
M_∞	0.218	0.443	0.84	0.88	$C_{L_{\delta_R}}$ (/rad)	0.0219	0.0192	0.0211	0.0217
C_{S_β} (/rad)	−0.8727	−0.6532	−0.7277	−0.7449	C_{N_β} (/rad)	0.1633	0.1232	0.1547	0.1604
$C_{S_{\delta_A}}$ (/rad)	0	0	0	0	C_{N_p} (sec)	−0.0255	−0.0047	−0.0009	−0.0005
$C_{S_{\delta_R}}$ (/rad)	0.1865	0.1865	0.1865	0.1865	C_{N_r} (sec)	−0.0573	−0.0245	−0.0164	−0.0164
C_{L_β} (/rad)	−0.1582	−0.1375	−0.1673	−0.1736	$C_{N_{\delta_A}}$ (/rad)	−0.0106	−0.0035	−0.0037	−0.0040
C_{L_p} (sec)	−0.1125	−0.0632	−0.0445	−0.0443	$C_{N_{\delta_R}}$ (/rad)	−0.0834	−0.0834	−0.0834	−0.0834
C_{L_r} (sec)	0.0725	0.0201	0.0127	0.0120					

Note: All data given for stability axes.

Table B.15 DC-8 Longitudinal Dimensional Derivatives

	Flight Condition			
	1	2	3	4
h, ft	0	15,000	33,000	33,000
M_∞	0.218	0.443	0.84	0.88
$X_u + X_{Pu}$	−0.0291	−0.00714	−0.014	−0.0463
X_α	15.316	15.029	3.544	22.364
X_{δ_E}	0	0	0	0
$Z_u + Z_{Pu}$	−0.2506	−0.1329	−0.0735	0.0622
Z_α	−152.845	−353.959	−664.305	−746.893
$Z_{\dot{\alpha}}$	0	0	0	0
Z_{δ_E}	−10.19	−23.7	−34.6	−38.6
$M_u + M_{Pu}$	−0.0000077	−0.000063	−0.000786	−0.00254
M_α	−2.1185	−5.0097	−9.1486	−12.0021
$M_{\dot{\alpha}}$	−0.2601	−0.3371	−0.4203	−0.4490
M_q	−0.7924	−0.991	−0.924	−1.008
M_{δ_E}	−1.35	−3.24	−4.59	−5.12

Note: All data given for stability axes.

Table B.16 DC-8 Lateral-Directional Dimensional Derivatives

	Flight Condition					Flight Condition			
	1	2	3	4		1	2	3	4
h, ft	0	15,000	33,000	33,000	L'_{δ_A} [1]	0.726	1.62	2.11	2.3
M_∞	0.218	0.443	0.84	0.88	L'_{δ_R}	0.1813	0.392	0.549	0.612
Y_β	−27.102	−47.195	−71.541	−80.388	N'_β	0.757	1.301	2.14	2.43
Y_{δ_A}	0	0	0	0	N'_p	−0.124	−0.0346	−0.0204	−0.0172
Y_{δ_R}	5.795	13.484	18.297	20.119	N'_r	−0.265	−0.257	−0.228	−0.25
L'_β	−1.328	−2.71	−4.41	−5.02	N'_{δ_A}	−0.0532	−0.0188[2]	−0.0652	−0.0788
L'_p	−0.951	−1.232	−1.181	−1.29	N'_{δ_R}	−0.389	−0.864	−1.164[3]	−1.277
L'_r	0.609	0.397	0.334	0.346					

Note: All data given for stability axes.

[1] All values given as negative in Ref. 1.

[2] Believed to be too small. Should be ~ −0.06.

[3] Listed as −0.01164 in Ref. 1.

A LARGE, FLEXIBLE HIGH-SPEED AIRCRAFT (See Section 7.9, Example 12.7, and Refs. 2–3)

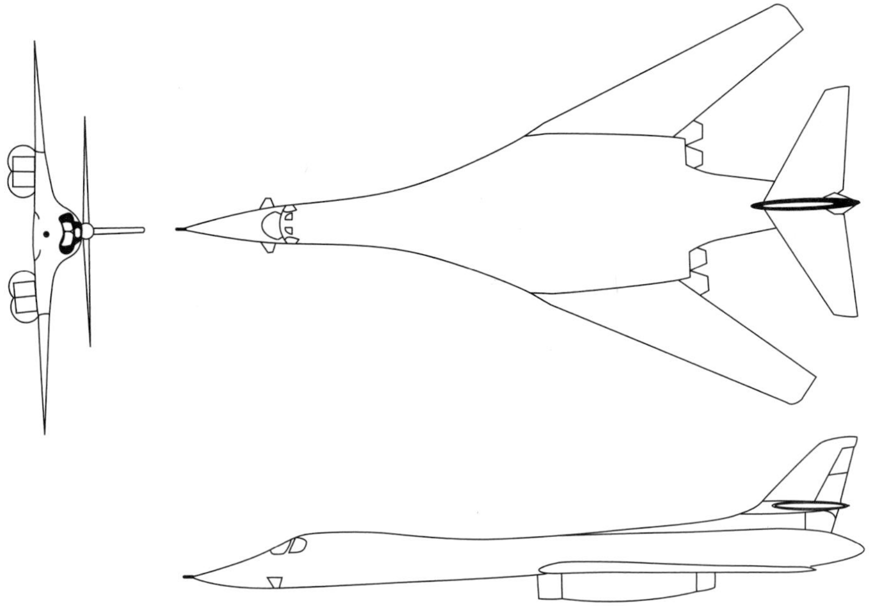

Figure B.9 Vehicle geometry.

Table B.17 Vehicle Configuration and Data

Wing geometry	$S_W = 1{,}950$ ft^2 $\bar{c}_W = 15.3$ ft $b_W = 70$ ft $\Lambda_{LE} = 65$ deg	Inertias	$I_{xx} = 9.5 \times 10^5$ sl-ft^2 $I_{yy} = 6.4 \times 10^6$ sl-ft^2 $I_{zz} = 7.1 \times 10^6$ sl-ft^2 $I_{xz} = -52{,}700$ sl-ft^2
Weight	$W = 288{,}000$ lb ($m = 8{,}944$ sl)	Vehicle length and $c.g.$	$L = 143$ ft $c.g. = 0.25\ \bar{c}_W$ (Fuse. Station 1061)
Modal generalized masses	$\mathcal{M}_1 = 184$ sl-ft^2 $\mathcal{M}_2 = 9{,}587$ sl-ft^2 $\mathcal{M}_3 = 1{,}334$ sl-ft^2 $\mathcal{M}_4 = 436{,}000$ sl-ft^2	Modal frequencies	$\omega_1 = 12.6$ rad/sec $\omega_2 = 14.1$ rad/sec $\omega_3 = 21.2$ rad/sec $\omega_4 = 22.1$ rad/sec
Mode-shape displacement at cockpit	$\nu_{Z_1}(x_{cp}) = 0.32$ ft $\nu_{Z_2}(x_{cp}) = 0.40$ ft $\nu_{Z_3}(x_{cp}) = 0.18$ ft $\nu_{Z_4}(x_{cp}) = 0.14$ ft	Mode-shape slope at cockpit	$\nu_1'(x_{cp}) = 0.027$ rad $\nu_2'(x_{cp}) = 0.027$ rad $\nu_3'(x_{cp}) = 0.032$ rad $\nu_4'(x_{cp}) = 0.032$ rad

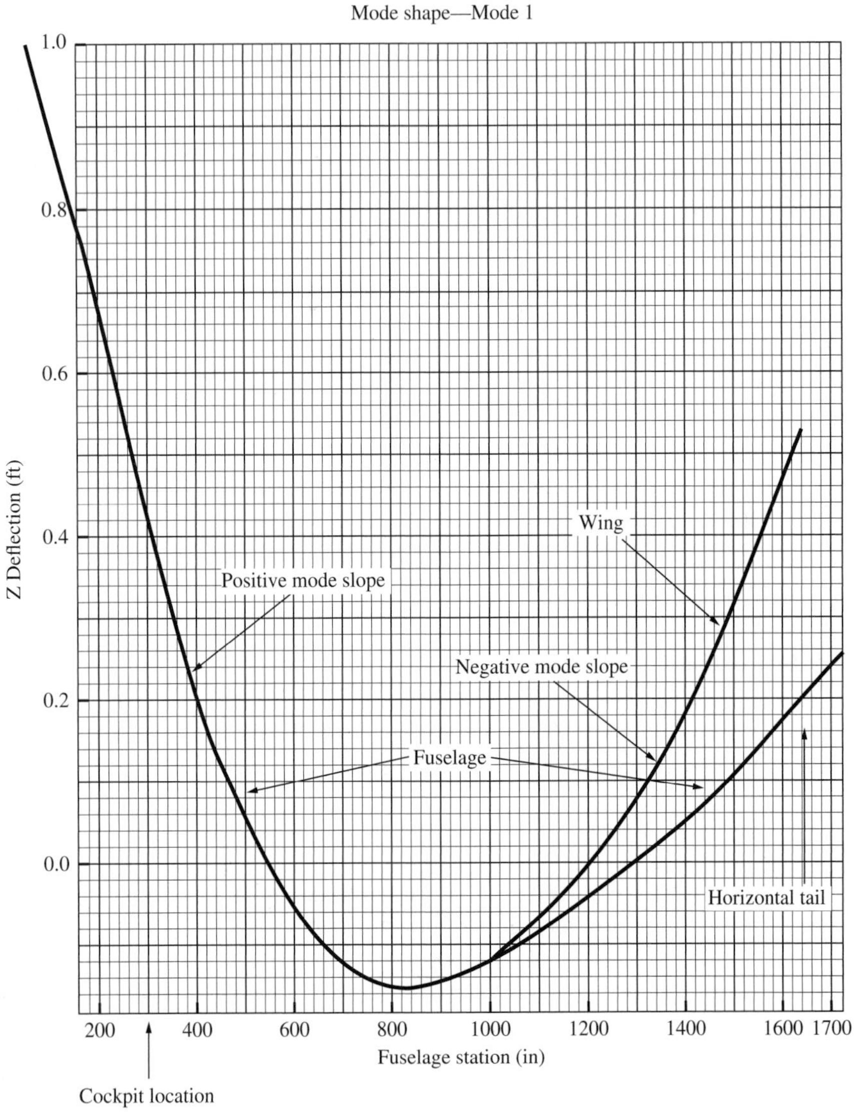

Figure B.10*a* Vibration mode shape—Mode 1.

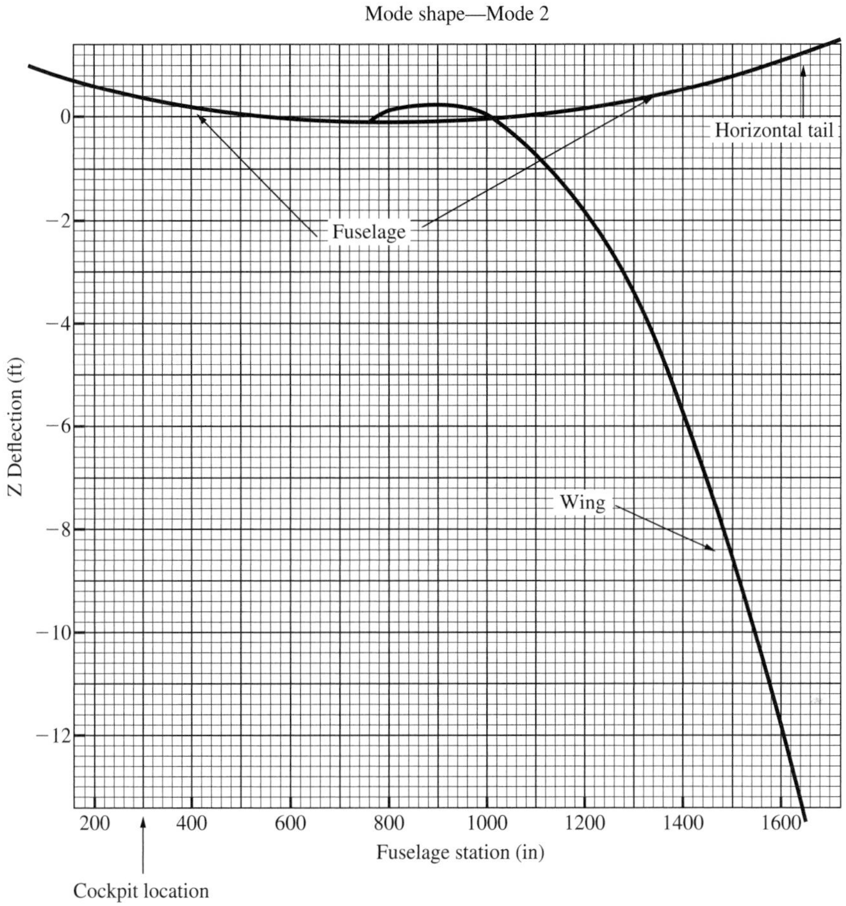

Figure B.10*b* Vibration mode shape—Mode 2.

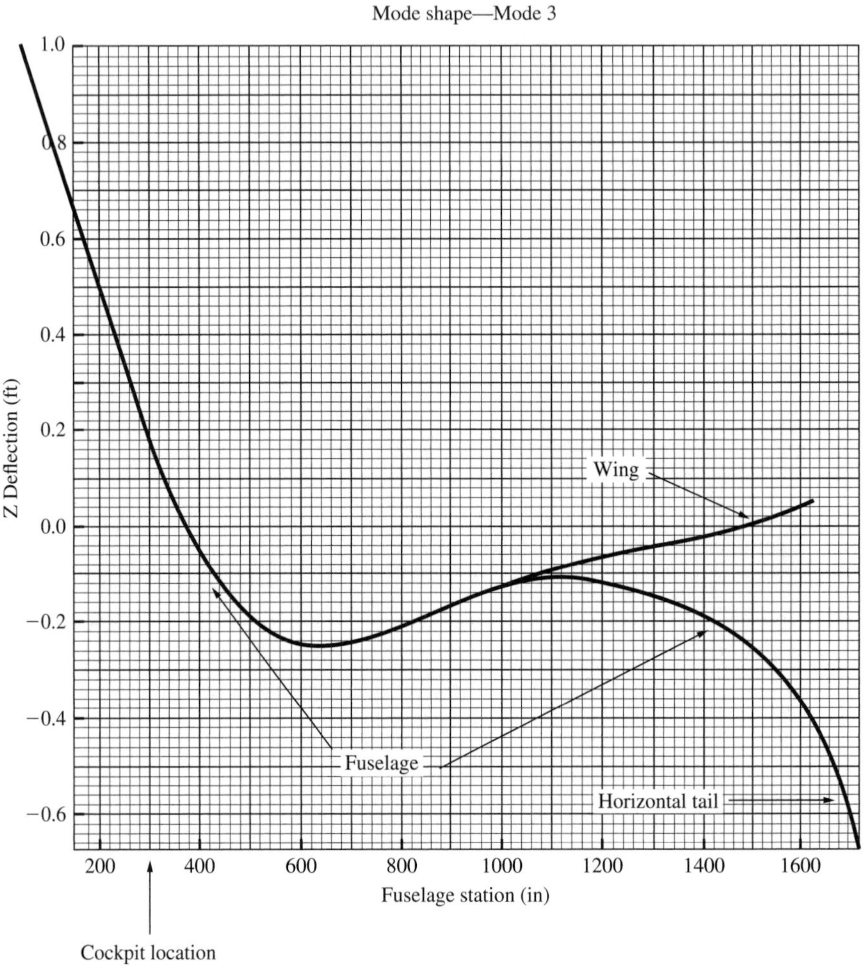

Figure B.10*c* Vibration mode shape—Mode 3.

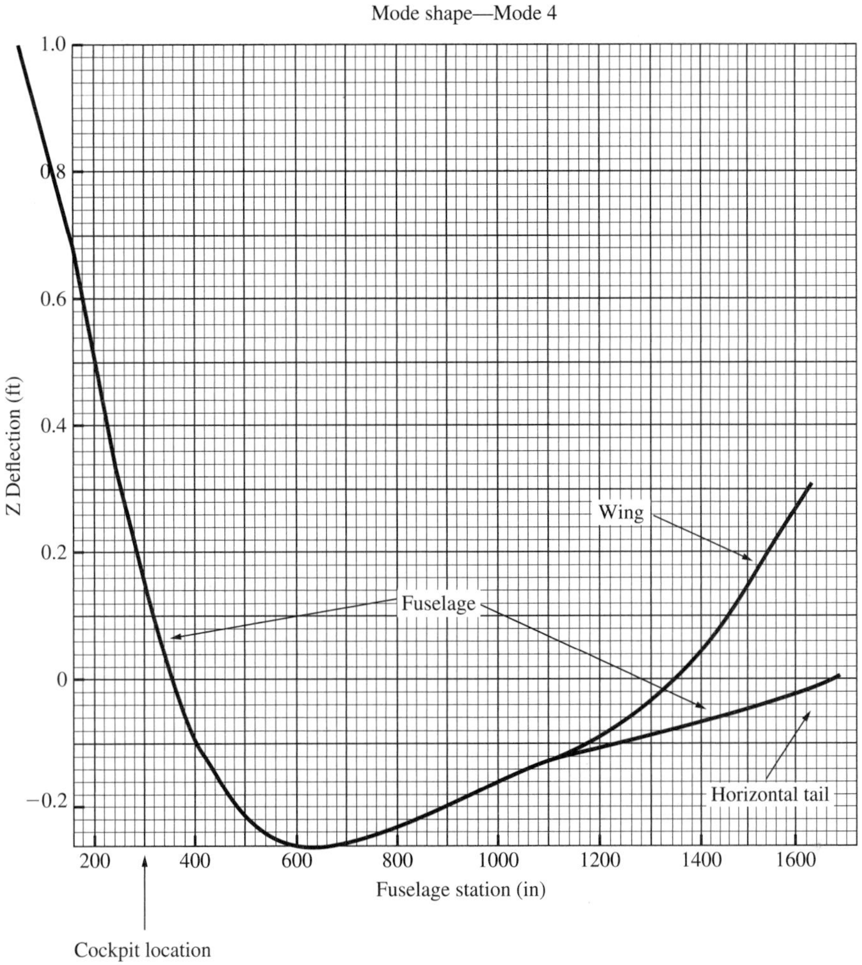

Figure B.10*d* Vibration mode shape—Mode 4.

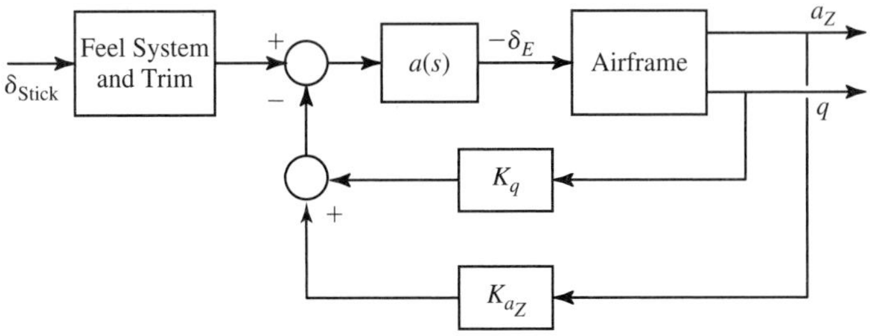

Figure B.11 Stability-augmentation system ($a(s)$ = actuator w. pole at 10 /sec).

Table B.18 Flight Conditions and Rigid-Body Effectiveness Coefficients

	Flight Condition			Flight Condition	
	1	2		1	2
h, ft	5,000	35,000	C_{Z_q} (sec)	0.171	−0.280
M_∞	0.6	0.8	$C_{Z_{\delta_E}}$ (/rad)	−0.435	−0.277
V_V, fps	659	778	C_{X_0}	−0.028	0.027
q_∞, psf	434	223	C_{X_α} (/rad)	0.200	0.229
α_0, deg	0	6.3	C_{X_q} (sec)	−0.020	0.026
U_0, fps	659	773	$C_{X_{\delta_E}}$ (/rad)	0.153	0.115
W_0, fps	0	85	C_{M_0}	−0.252	−0.374
γ_0	0	0	C_{M_α} (/rad)	−1.662	0
$C_{L\,\text{Trim}}$	0.340	0.662	$C_{M_{\dot{\alpha}}}$ (sec)	−0.050	−0.054
$C_{D\,\text{Trim}}$	0.028	0.043	C_{M_q} (sec)	−0.404	−0.342
C_{Z_0}	−0.34	−0.636	$C_{M_{\delta_E}}$ (/rad)	−2.579	−2.636
C_{Z_α} (/rad)	−2.92	−1.38			

Note: All data given for fuselage-referenced axes.

Table B.19 Rigid-Body Dimensional Derivatives

	Flight Condition			Flight Condition	
	1	2		1	2
$X_u + X_{P_u}$	0	0	Z_{δ_E}	−42.22	−13.46
X_α	19.45	11.13	$Z_{\delta_{cv}}$	−2.11	−0.673
X_q	−1.913	1.264	$M_u + M_{P_u}$	0	0
X_{δ_E}	14.83	5.589	$M_\alpha + M_{P_\alpha}$	−3.445	0
$X_{\delta_{cv}}$	0.742	0.279	$M_{\dot{\alpha}}$	−0.1035	−0.0561
$Z_u + Z_{P_u}$	−0.1001	−0.0425	M_q	−0.8363	−0.3554
Z_α	−283.3	67.07	M_{δ_E}	−5.346	−2.739
Z_q	16.55	−13.61	$M_{\delta_{cv}}$	0.376	0.193

Note: All data given for fuselage-referenced axes.

Table B.20 Aeroelastic Effectiveness Coefficients

Coefficient	Mode 1	Mode 2	Mode 3	Mode 4
$C_{Z_{\eta_i}}$	−0.029	0.306	0.015	−0.014
$C_{Z_{\dot{\eta}_i}}$ (sec)	$-0.658/V_\infty$	$7.896/V_\infty$	$0.461/V_\infty$	$-0.132/V_\infty$
$C_{M_{\eta_i}}$	−0.032	−0.025	0.041	−0.018
$C_{M_{\dot{\eta}_i}}$ (sec)	$-1.184/V_\infty$	$9.409/V_\infty$	$1.316/V_\infty$	$-0.395/V_\infty$
$C_{Q_{i_0}}$	0	0	0	0
$C_{Q_{i_\alpha}}$	-1.49×10^{-2}	2.58×10^{-2}	1.49×10^{-2}	3.35×10^{-5}
$C_{Q_{i_q}}$ (sec)	$-0.726/V_\infty$	$0.089/V_\infty$	$0.304/V_\infty$	~ 0
$C_{Q_{i_{\delta_E}}}$	-1.28×10^{-2}	-6.42×10^{-2}	2.56×10^{-2}	1.50×10^{-4}
$C_{Q_{i_{\eta_1}}}$	5.85×10^{-5}	4.21×10^{-3}	2.91×10^{-4}	2.21×10^{-5}
$C_{Q_{i_{\eta_2}}}$	-9.0×10^{-5}	-9.22×10^{-2}	1.44×10^{-3}	-1.32×10^{-4}
$C_{Q_{i_{\eta_3}}}$	3.55×10^{-4}	1.97×10^{-3}	-3.46×10^{-4}	9.68×10^{-6}
$C_{Q_{i_{\eta_4}}}$	1.20×10^{-4}	3.37×10^{-3}	1.44×10^{-4}	1.77×10^{-3}
$C_{Q_{i_{\dot{\eta}_1}}}$ (sec)	$-0.0032/V_\infty$	$0.0665/V_\infty$	$-0.0048/V_\infty$	$-0.0004/V_\infty$
$C_{Q_{i_{\dot{\eta}_2}}}$ (sec)	$-0.0015/V_\infty$	$-2.277/V_\infty$	$0.1494/V_\infty$	$0.0031/V_\infty$
$C_{Q_{i_{\dot{\eta}_3}}}$ (sec)	$0.0050/V_\infty$	$0.0320/V_\infty$	$-0.0001/V_\infty$	$-0.0004/V_\infty$
$C_{Q_{i_{\dot{\eta}_4}}}$ (sec)	$-0.0011/V_\infty$	$0.0317/V_\infty$	$-0.0100/V_\infty$	$0.6112/V_\infty$

Table B.21 Aeroelastic Dimensional Derivatives

	Flight Condition			Flight Condition			Flight Condition			Flight Condition	
	1	2		1	2		1	2		1	2
Z_{η_1}	−2.812	−1.409	$\Xi_{1\alpha}$	−1,075	−538.5	$\Xi_{2\eta_3}$	1.993	0.9989	$\Xi_{3\dot{\eta}_3}$	−1.51e-3	−6.45e-4
Z_{η_2}	29.67	14.87	Ξ_{1q}	−79.44	−33.94	$\Xi_{2\eta_4}$	−0.1826	−0.0915	$\Xi_{3\dot{\eta}_4}$	−6.04e-3	−2.58e-3
Z_{η_3}	1.454	0.7280	$\Xi_{1\eta_1}$	4.219	2.114	$\Xi_{2\dot{\eta}_1}$	−3.15e-3	−1.35e-3	$\Xi_{3_{\delta_E}}$	254.6	127.6
Z_{η_4}	−1.357	−0.6804	$\Xi_{1\eta_2}$	303.6	152.2	$\Xi_{2\dot{\eta}_2}$	−4.782	−2.043	$\Xi_{3_{\delta_{CV}}}$	−114.6	−57.43
$Z_{\dot{\eta}_1}$	−0.0968	−0.0414	$\Xi_{1\eta_3}$	20.98	10.52	$\Xi_{2\dot{\eta}_3}$	0.3137	0.1340	$\Xi_{4\alpha}$	1.02e-3	5.11e-4
$Z_{\dot{\eta}_2}$	1.162	0.4963	$\Xi_{1\eta_4}$	1.594	0.7988	$\Xi_{2\dot{\eta}_4}$	6.51e-3	2.78e-3	Ξ_{4q}	0	0
$Z_{\dot{\eta}_3}$	0.0678	0.0290	$\Xi_{1\dot{\eta}_1}$	−0.3502	−0.1496	$\Xi_{2_{\delta_E}}$	−88.85	−44.53	$\Xi_{4\eta_1}$	3.65e-3	1.83e-3
$Z_{\dot{\eta}_4}$	−0.0194	−8.30e-3	$\Xi_{1\dot{\eta}_2}$	7.277	3.108	$\Xi_{2_{\delta_{CV}}}$	−1.777	−0.8907	$\Xi_{4\eta_2}$	0.1026	0.0514
M_{η_1}	−0.0663	−0.0333	$\Xi_{1\dot{\eta}_3}$	−0.5252	−0.2244	$\Xi_{3\alpha}$	148.2	74.28	$\Xi_{4\eta_3}$	4.38e-3	2.20e-3
M_{η_2}	−0.0518	−0.0260	$\Xi_{1\dot{\eta}_4}$	−0.0438	−0.0187	Ξ_{3q}	4.588	1.960	$\Xi_{4\eta_4}$	0.0539	0.0270
M_{η_3}	0.0850	0.0426	$\Xi_{1_{\delta_E}}$	−923.0	−462.6	$\Xi_{3\eta_1}$	3.531	1.770	$\Xi_{4\dot{\eta}_1}$	−5.08e-5	−2.17e-05
M_{η_4}	−0.0373	−0.0187	$\Xi_{1_{\delta_{CV}}}$	−89.53	−44.87	$\Xi_{3\eta_2}$	19.59	9.821	$\Xi_{4\dot{\eta}_2}$	1.46e-3	6.25e-4
$M_{\dot{\eta}_1}$	−3.72e-3	−1.59e-3	$\Xi_{2\alpha}$	35.71	17.90	$\Xi_{3\eta_3}$	−3.441	−1.725	$\Xi_{4\dot{\eta}_3}$	−4.62e-4	−1.97e-4
$M_{\dot{\eta}_2}$	0.0296	1.26e-3	Ξ_{2q}	0.1869	0.0798	$\Xi_{3\eta_4}$	0.0963	0.0483	$\Xi_{4\dot{\eta}_4}$	0.0282	0.0121
$M_{\dot{\eta}_3}$	4.14e-3	1.77e-3	$\Xi_{2\eta_1}$	−0.1246	−0.0624	$\Xi_{3\dot{\eta}_1}$	0.0755	0.0322	$\Xi_{4_{\delta_E}}$	4.56e-3	2.29e-3
$M_{\dot{\eta}_4}$	−1.24e-3	−5.31e-4	$\Xi_{2\eta_2}$	−127.6	−63.96	$\Xi_{3\dot{\eta}_2}$	0.4830	0.2063	$\Xi_{4_{\delta_{CV}}}$	−1.14e-3	−5.72e-4

Linear Force and Moment Model

$$(f_{A_X} + f_{P_X})/m = \overline{X}_u u + X_\alpha \alpha + X_q q + X_{\delta_E}\delta_E + X_{\delta_{cv}}\delta_{cv}$$

$$(f_{A_Z} + f_{P_Z})/m = \overline{Z}_u u + Z_\alpha \alpha + Z_q q + Z_{\delta_E}\delta_E + Z_{\delta_{cv}}\delta_{cv} + \sum_{i=1}^{4}\left(Z_{\eta_i}\eta_i + Z_{\dot{\eta}_i}\dot{\eta}_i\right)$$

$$(m_A + m_P)/I_{yy} = \overline{M}_u u + \overline{M}_\alpha \alpha + M_q q + M_{\dot{\alpha}}\dot{\alpha} + M_{\delta_E}\delta_E + M_{\delta_{cv}}\delta_{cv} + \sum_{i=1}^{4}\left(M_{\eta_i}\eta_i + M_{\dot{\eta}_i}\dot{\eta}_i\right)$$

$$Q_i/\mathcal{M}_i = \Xi_{i_\alpha}\alpha + \Xi_{i_q} q + \Xi_{i_{\delta_E}}\delta_E + \Xi_{i_{\delta_{cv}}}\delta_{cv} + \sum_{j=1}^{4}\left(\Xi_{i\eta_j}\eta_j + \Xi_{i\dot{\eta}_j}\dot{\eta}_j\right)$$

$$\ddot{\eta}_i + 2\zeta_i\omega_i\dot{\eta}_i + \omega_i^2\eta_i = Q_i/\mathcal{M}_i$$

$$\overline{X}_u \triangleq X_u + X_{P_u} \quad \overline{M}_u \triangleq M_u + M_{P_u}$$

$$\overline{Z}_u \triangleq Z_u + Z_{P_u} \quad \overline{M}_\alpha \triangleq M_\alpha + M_{P_\alpha}$$

$$a_{Z_{cp}} = \dot{w} - V_V q - l_{cp}\ddot{\theta} + \sum_{i=1}^{4}\nu_{Z_i}(x_{cp})\ddot{\eta}_i \quad \text{where } l_{cp} = (1061-300)/12 = 63.4 \text{ ft}$$

AN AIRBREATHING, HYPERSONIC VEHICLE (See Examples 10.3 and 10.5, Section 11.3.3, and Ref. 4)

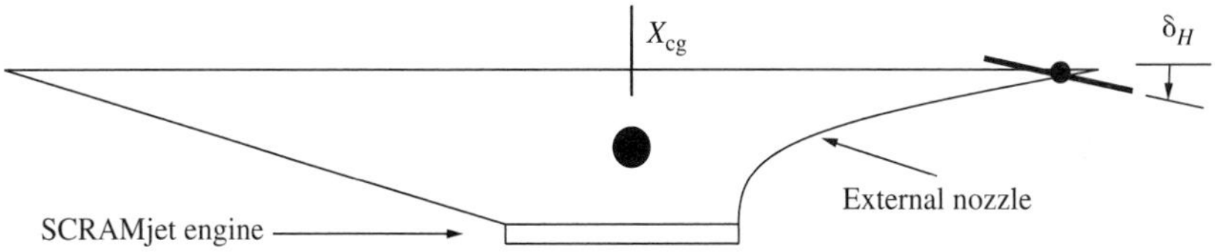

Figure B.12 Hypersonic vehicle.

Table B.22 Vehicle Configuration and Flight Condition

Length, L	150 ft	Generalized mass, $\mathcal{M}$	200 sl
Weight	80,000 lb	Vibration frequency, ω and damping, ζ	18 rad/sec, 0.02
I_{yy}	5.0×10^6 sl-ft^2	Elastic-mode slope at nose, $\nu'(0)$	1 deg
X_{cg}	90 ft aft	Altitude, h	80,000 ft
V_V (U_0)	7,770 fps	Mach number, M_∞	8
W_0	0	Sonic velocity, a	970.8 fps
γ_0	0	Density, ρ_∞	8.45×10^{-5} sl/ft^3

State-Variable Model $\left(X_{\dot{\alpha}} = Z_{\dot{\alpha}} = Z_q = M_{\dot{\alpha}} = \Xi_{\dot{\alpha}} = 0\right)$

$$\mathbf{x}^T = \mathbf{y}^T = \left[u(\text{fps}) \quad \alpha(\text{rad}) \quad \theta(\text{rad}) \quad \dot{\theta}(\text{rad/sec}) \quad \eta(-) \quad \dot{\eta}(/\text{sec})\right], \quad u = \delta_H(\text{rad})$$

$$\mathbf{A} = \begin{bmatrix} X_u + X_{P_u} & X_\alpha + X_{P_\alpha} & -g & X_q + X_{P_q} & X_\eta + X_{P_\eta} & X_{\dot{\eta}} + X_{P_{\dot{\eta}}} \\ \dfrac{Z_u + Z_{P_u}}{U_0} & \dfrac{Z_\alpha + Z_{P_\alpha}}{U_0} & 0 & 1 & \dfrac{Z_\eta + Z_{P_\eta}}{U_0} & \dfrac{Z_{\dot{\eta}} + Z_{P_{\dot{\eta}}}}{U_0} \\ 0 & 0 & 0 & 1 & 0 & 0 \\ M_u + M_{P_u} & M_\alpha + M_{P_\alpha} & 0 & M_q + M_{P_q} & M_\eta + M_{P_\eta} & M_{\dot{\eta}} + M_{P_{\dot{\eta}}} \\ 0 & 0 & 0 & 0 & 0 & 1 \\ \Xi_u + \Xi_{P_u} & \Xi_\alpha + \Xi_{P_\alpha} & 0 & \Xi_q + \Xi_{P_q} & \left(\Xi_\eta + \Xi_{P_\eta}\right) - \omega^2 & \left(\Xi_{\dot{\eta}} + \Xi_{P_{\dot{\eta}}}\right) - 2\zeta\omega \end{bmatrix}$$

$$\mathbf{B} = \begin{bmatrix} X_{\delta_H} \\ \dfrac{Z_{\delta_H}}{U_0} \\ 0 \\ M_{\delta_H} \\ 0 \\ \Xi_{\delta_H} \end{bmatrix}, \mathbf{C} = \mathbf{I}, \mathbf{D} = \mathbf{0}_6$$

Table B.23 Rigid-Body and Aeroelastic Dimensional Derivatives

$X_u + X_{P_u}$	−1.936e-3	$X_\eta + X_{P_\eta}$	−0.2525
$X_\alpha + X_{P_\alpha}$	24.284	$X_{\dot{\eta}} + X_{P_{\dot{\eta}}}$	0.0118
$X_q + X_{P_q}$	0.6168	$Z_\eta + Z_{P_\eta}$	−8.006
X_δ	−0.5562	$Z_{\dot{\eta}} + Z_{P_{\dot{\eta}}}$	0.0457
$Z_u + Z_{P_u}$	−0.0162	$M_\eta + M_{P_\eta}$	0.1901
$Z_\alpha + Z_{P_\alpha}$	−490.3	$M_{\dot{\eta}} + M_{P_{\dot{\eta}}}$	−1.40e-3
$Z_q + Z_{P_q}$	2.331	$\Xi_u + \Xi_{P_u}$	0.1523
Z_δ	−21.46	$\Xi_\alpha + \Xi_{P_\alpha}$	4,731
$M_u + M_{P_u}$	3.385e-4	$\Xi_q + \Xi_{P_q}$	−37.06
$M_\alpha + M_{P_\alpha}$	11.023	$\Xi_\eta + \Xi_{P_\eta}$	82.57
$M_q + M_{P_q}$	−0.0816	$\Xi_{\dot{\eta}} + \Xi_{P_{\dot{\eta}}}$	−0.2682
M_δ	−0.4794	Ξ_δ	245.6

Note: All dimensional derivatives have significant aerodynamic <u>and</u> propulsive contributions.
All data given in fuselage-referenced axes.

TRANSFORMING DIMENSIONAL DERIVATIVES FROM STABILITY TO FUSELAGE-REFERENCED AXES

The transformation of dimensional stability derivatives from stability to fuselage-referenced axes involves the resolution not only of the forces and moments, but also of the perturbations in motion and the changed inertias. After making these conversions, the following expressions provide the relationships between the inertias and the dimensional derivatives derived in the stability axes (listed with no subscript) and those derived in the fuselage-referenced axes (see Ref. 5).

Inertias:

$$I_{xx}|_F = I_{xx}\cos^2\alpha_0 + 2I_{xz}\cos\alpha_0\sin\alpha_0 + I_{zz}\sin^2\alpha_0$$

$$I_{yy}|_F = I_{yy}$$

$$I_{zz}|_F = I_{zz}\cos^2\alpha_0 - 2I_{xz}\cos\alpha_0\sin\alpha_0 + I_{xx}\sin^2\alpha_0$$

$$I_{xz}|_F = \left(I_{zz} - I_{xx}\right)\cos\alpha_0\sin\alpha_0 + I_{xz}\left(\cos^2\alpha_0 - \sin^2\alpha_0\right)$$

Longitudinal Dimensional Derivatives:

$$X_u|_F = X_u\cos^2\alpha_0 - \left(X_w + Z_u\right)\cos\alpha_0\sin\alpha_0 + Z_w\sin^2\alpha_0$$

$$X_w|_F = X_w\cos^2\alpha_0 + \left(X_u - Z_w\right)\cos\alpha_0\sin\alpha_0 + Z_u\sin^2\alpha_0$$

$$X_{\dot{w}}|_F = X_{\dot{w}}\cos^2\alpha_0 - Z_{\dot{w}}\cos\alpha_0\sin\alpha_0$$

$$X_{q\text{ or }\delta}|_F = X_{q\text{ or }\delta}\cos\alpha_0 - Z_{q\text{ or }\delta}\sin\alpha_0$$

$$Z_u|_F = Z_u\cos^2\alpha_0 - \left(Z_w - X_u\right)\cos\alpha_0\sin\alpha_0 - X_w\sin^2\alpha_0$$

$$Z_w|_F = Z_w\cos^2\alpha_0 + \left(Z_u + X_w\right)\cos\alpha_0\sin\alpha_0 + X_u\sin^2\alpha_0$$

$$Z_{\dot{w}}|_F = Z_{\dot{w}}\cos^2\alpha_0 + X_{\dot{w}}\cos\alpha_0\sin\alpha_0$$

$$Z_{q\text{ or }\delta}|_F = Z_{q\text{ or }\delta}\cos\alpha_0 + X_{q\text{ or }\delta}\sin\alpha_0$$

$$M_u|_F = M_u\cos\alpha_0 - M_w\sin\alpha_0$$

$$M_w|_F = M_w\cos\alpha_0 + M_u\sin\alpha_0$$

$$M_{\dot{w}}|_F = M_{\dot{w}}\cos\alpha_0$$

$$M_{q\text{ or }\delta}|_F = M_{q\text{ or }\delta}$$

Lateral-Directional Dimensional Derivatives:

$$Y_{v \text{ or } \delta \text{ or } \dot{v}}|_F = Y_{v \text{ or } \delta \text{ or } \dot{v}}$$

$$Y_p|_F = Y_p \cos\alpha_0 - Y_r \sin\alpha_0$$

$$Y_r|_F = Y_r \cos\alpha_0 + Y_p \sin\alpha_0$$

$$L'_{v \text{ or } \delta \text{ or } \dot{v}}|_F = L'_{v \text{ or } \delta \text{ or } \dot{v}} \cos\alpha_0 - N'_{v \text{ or } \delta \text{ or } \dot{v}} \sin\alpha_0$$

$$L'_p|_F = L'_p \cos^2\alpha_0 - (L'_r + N'_p)\sin\alpha_0 \cos\alpha_0 + N'_r \sin^2\alpha_0$$

$$L'_r|_F = L'_r \cos^2\alpha_0 - (N'_r - L'_p)\sin\alpha_0 \cos\alpha_0 - N'_r \sin^2\alpha_0$$

$$N'_{v \text{ or } \delta \text{ or } \dot{v}}|_F = N'_{v \text{ or } \delta \text{ or } \dot{v}} \cos\alpha_0 + L'_{v \text{ or } \delta \text{ or } \dot{v}} \sin\alpha_0$$

$$N'_p|_F = N'_p \cos^2\alpha_0 - (N'_r - L'_p)\sin\alpha_0 \cos\alpha_0 - L'_r \sin^2\alpha_0$$

$$N'_r|_F = N'_r \cos^2\alpha_0 + (L'_r + N'_p)\sin\alpha_0 \cos\alpha_0 + L'_p \sin^2\alpha_0$$

Finally, note that the transformations for the unprimed lateral derivatives are exactly the same as those for the primed derivatives.

REFERENCES

1. Teper, G. L.: "Aircraft Stability and Control Data," STI Technical Report 176-1, Systems Technology, Inc., Hawthorne, CA, prepared for NASA Ames Research Center, April 1969.
2. Waszak, M. R. and D. K. Schmidt: "Flight Dynamics of Aeroelastic Vehicles," *Journal of Aircraft,* vol. 25, no. 6, June 1988, pp. 563–571.
3. Waszak, M. R., J. D. Davidson, and D. K. Schmidt: "A Simulation Study of the Flight Dynamics of Elastic Aircraft," vols. I and II, NASA Contractor Report 4102, December 1987.
4. Chavez, F. R. and D. K. Schmidt: "An Analytical Model and Dynamic Analysis of an Aeropropulsive/Aeroelastic Hypersonic Vehicle," *Journal of Guidance, Control, and Dynamics,* vol. 17, no. 6, Nov.–Dec. 1994.
5. McRuer, D., I. Ashkenas, and D. Graham: "Aircraft Dynamics and Automatic Control," Princeton University Press, Princeton, N.J., 1973.

APPENDIX C

Models of Atmospheric Turbulence

A summary of the Dryden gust model will be presented first. The theoretical development regarding power spectra and statistics of stochastic processes can be found in Section C.2 after the summary of results.

C.1 THE DRYDEN GUST MODEL (REFS. 1–3)

The gust model includes translational gust velocities u_g, v_g, and w_g. The *gust power spectra* for each of the gust velocities under the Dryden model, or the frequency-domain representation of the gusts, are given below. Note that these spectra are functions of the reference flight velocity U_0, and characteristics lengths $L_\bullet$.

$$\Phi_{u_g}(\omega) = \sigma_u^2 \frac{2L_u}{\pi U_0} \frac{1}{\left(1 + \left(L_u\omega/U_0\right)^2\right)}$$

$$\Phi_{v_g}(\omega) = \sigma_v^2 \frac{L_v}{\pi U_0} \frac{1 + 3\left(L_v\omega/U_0\right)^2}{\left(1 + \left(L_v\omega/U_0\right)^2\right)^2} \tag{C.1}$$

$$\Phi_{w_g}(\omega) = \sigma_w^2 \frac{L_w}{\pi U_0} \frac{1 + 3\left(L_w\omega/U_0\right)^2}{\left(1 + \left(L_w\omega/U_0\right)^2\right)^2}$$

At altitudes $h < 1750$ ft, $L_w = h$, $L_u = L_v = 145h^{1/3}$ ft

At altitudes $h > 1750$ ft, $L_w = L_u = L_v = 1750$ ft

The standard deviations of the gusts σ, or rms gust velocities, are obtained from data like that shown in Figure C.1. This data can be used to define σ_u, σ_v, or σ_w in Equations (C.1). The level of turbulence is categorized in terms of "light," "moderate," and "severe," and regions corresponding to these

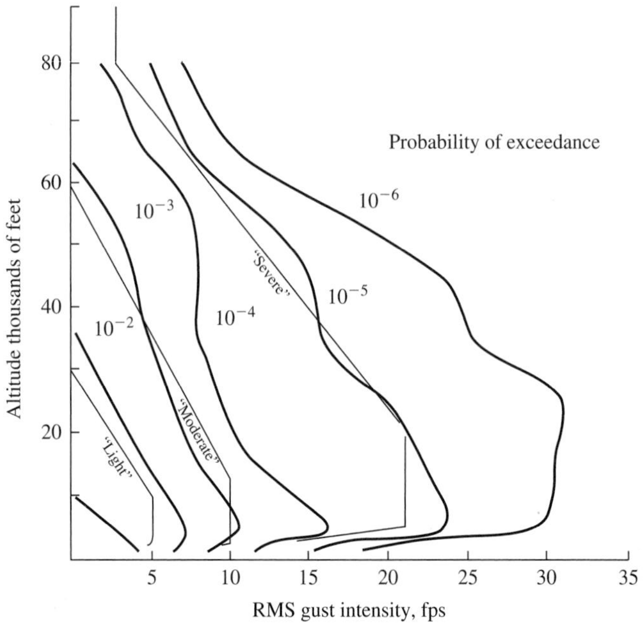

Figure C.1 RMS gust velocities, Ref. 4.

categorizations are indicated. The *probability of exceedance* corresponds to the likelihood the gusts will exceed the given rms velocity, or gust intensity, at a given altitude. For example, the rms gust velocity corresponding to an altitude of 30,000 ft in "severe" turbulence ($P_{exceed} = 10^{-5}$) is about 18–19 fps. So the probability that the rms gust velocity would exceed this value is 10^{-5}.

Although the above gust power spectra are useful in the frequency domain, it is frequently necessary to have time-domain representations for the atmospheric turbulence. Such representations are used in simulations, for example, and in other time-domain analyses. To obtain the time-domain representations, note that as shown in Section C.2, the random processes may be modeled in terms of a linear system excited or driven by white noise. (A white stochastic process has a power spectrum that is constant across all frequencies.) Letting the transfer function of such a linear system be denoted as $g(s)$, from the definition of a transfer function we know that

$$y(s) = g(s)n(s)$$

where y is the response of the linear system (i.e., gust velocity), and n is the white disturbance. For such a system, the power spectrum of the response may be written as

$$\Phi_y(\omega) = |g(j\omega)|^2 \Phi_n(\omega) = g(j\omega)g(-j\omega)\Phi_n \qquad \text{(C.2)}$$

where Φ_n is the (constant) power spectrum of the white disturbance.

Turning now to the gust power spectra in Equations (C.1), and letting $\Phi_n = 1$, we may write

$$\Phi_{u_g}(\omega) = \sigma_u^2 \frac{2L_u}{\pi \mathrm{U}_0} \frac{1}{(1 + j\omega L_u/\mathrm{U}_0)(1 - j\omega L_u/\mathrm{U}_0)} \triangleq g_{u_g}(j\omega) g_{u_g}(-j\omega)(1)$$

$$\Phi_{v_g}(\omega) = \sigma_v^2 \frac{L_v}{\pi \mathrm{U}_0} \frac{(1 + j\omega\sqrt{3}L_v/\mathrm{U}_0)(1 - j\omega\sqrt{3}L_v/\mathrm{U}_0)}{\left((1 + j\omega L_v/\mathrm{U}_0)(1 - j\omega L_v/\mathrm{U}_0)\right)^2} \triangleq g_{v_g}(j\omega) g_{v_g}(-j\omega)(1) \qquad \text{(C.3)}$$

$$\Phi_{w_g}(\omega) = \sigma_w^2 \frac{L_w}{\pi \mathrm{U}_0} \frac{(1 + j\omega\sqrt{3}L_w/\mathrm{U}_0)(1 - j\omega\sqrt{3}L_w/\mathrm{U}_0)}{\left((1 + j\omega L_w/\mathrm{U}_0)(1 - j\omega L_w/\mathrm{U}_0)\right)^2} \triangleq g_{w_g}(j\omega) g_{w_g}(-j\omega)(1)$$

So then we may define

$$g_{u_g}(j\omega) \triangleq \sigma_u \sqrt{\frac{2L_u}{\pi \mathrm{U}_0}} \frac{1}{(1 + j\omega L_u/\mathrm{U}_0)}$$

$$g_{v_g}(j\omega) \triangleq \sigma_v \sqrt{\frac{L_v}{\pi \mathrm{U}_0}} \frac{(1 + j\omega\sqrt{3}L_v/\mathrm{U}_0)}{(1 + j\omega L_v/\mathrm{U}_0)^2} \qquad \text{(C.4)}$$

$$g_{w_g}(j\omega) \triangleq \sigma_w \sqrt{\frac{L_w}{\pi \mathrm{U}_0}} \frac{(1 + j\omega\sqrt{3}L_w/\mathrm{U}_0)}{(1 + j\omega L_w/\mathrm{U}_0)^2}$$

or the three linear systems are

$$g_{u_g}(s) \triangleq \sigma_u \sqrt{\frac{2L_u}{\pi \mathrm{U}_0}} \frac{1}{(1 + sL_u/\mathrm{U}_0)}$$

$$g_{v_g}(s) \triangleq \sigma_v \sqrt{\frac{L_v}{\pi \mathrm{U}_0}} \frac{(1 + s\sqrt{3}L_v/\mathrm{U}_0)}{(1 + sL_v/\mathrm{U}_0)^2} \qquad \text{(C.5)}$$

$$g_{w_g}(s) \triangleq \sigma_w \sqrt{\frac{L_w}{\pi \mathrm{U}_0}} \frac{(1 + s\sqrt{3}L_w/\mathrm{U}_0)}{(1 + sL_w/\mathrm{U}_0)^2}$$

Using the above transfer functions, the time-domain representations for the gust velocities are then

$$
\begin{aligned}
\dot{u}_g(t) &= -\frac{U_0}{L_u}u_g(t) + \sigma_u\left(\frac{2U_0}{\pi L_u}\right)^{1/2} n(t) \\
\dot{v}_g(t) &= -\frac{U_0}{L_v}v_g(t) + \sigma_v(1-\sqrt{3})\left(\frac{U_0}{L_v}\right)^{3/2} v_{g_1}(t) + \sigma_v\left(\frac{3U_0}{L_v}\right)^{1/2} n(t) \\
\dot{v}_{g_1}(t) &= -\frac{U_0}{L_v}v_{g_1}(t) + n(t) \\
\dot{w}_g(t) &= -\frac{U_0}{L_w}w_g(t) + \sigma_w(1-\sqrt{3})\left(\frac{U_0}{L_w}\right)^{3/2} w_{g_1}(t) + \sigma_w\left(\frac{3U_0}{L_w}\right)^{1/2} n(t) \\
\dot{w}_{g_1}(t) &= -\frac{U_0}{L_w}w_{g_1}(t) + n(t)
\end{aligned}
\qquad \text{(C.6)}
$$

Note that these equations are linear with constant coefficients, and that the forcing function $n(t)$ is a zero-mean, Gaussian "white" stochastic process of unit intensity. So the autocovariance of n may be written in terms of the impulse function, or

$$E\{n(t)n(t-\tau)\} = \delta(\tau) \qquad \text{(C.7)}$$

For purposes of simulation, one may numerically integrate Equations (C.6) to simulate the vehicle response to turbulence. In such simulations, the continuous forcing function $n(t)$ may be assumed constant over each integration of step size Δt. That is, the forcing function $n(t)$ is replaced by a sequence of random numbers n_i, which are Gaussian, with zero mean and variance $1/\Delta t$.

Examples of the shapes of the gust power spectra for u_g and w_g (v_g same as w_g), with $U_0 = 824$ fps, $L_u = L_w = 1750$ ft, $b = 142$ ft (wing span), and $\sigma_\bullet = 1$ fps, are plotted in Figure C.2. Referring to the equations for the gust spectra (Equations (C.1)), one can see that the spectra scale with the variance σ^2. Therefore it is straightforward to adjust these plots for other values of rms gust intensity σ.

In Chapter 12, we noted that a feedback-control system is designed to reject external disturbances, and that the designer must therefore be aware of the frequency content of these disturbances. Such disturbances are denoted as D in the generic block diagram shown in Figure C.3 The above plots of the gust spectra, also called *power-spectral densities,* indicate the frequency content of the gusts. The frequency range over which the magnitude of the spectra is large is the frequency range in which the disturbance has the most power, so the gusts consist primarily of harmonic functions in this frequency range. The gust spectra can now be used to estimate the frequency content of the relevant disturbances D, which will be demonstrated next.

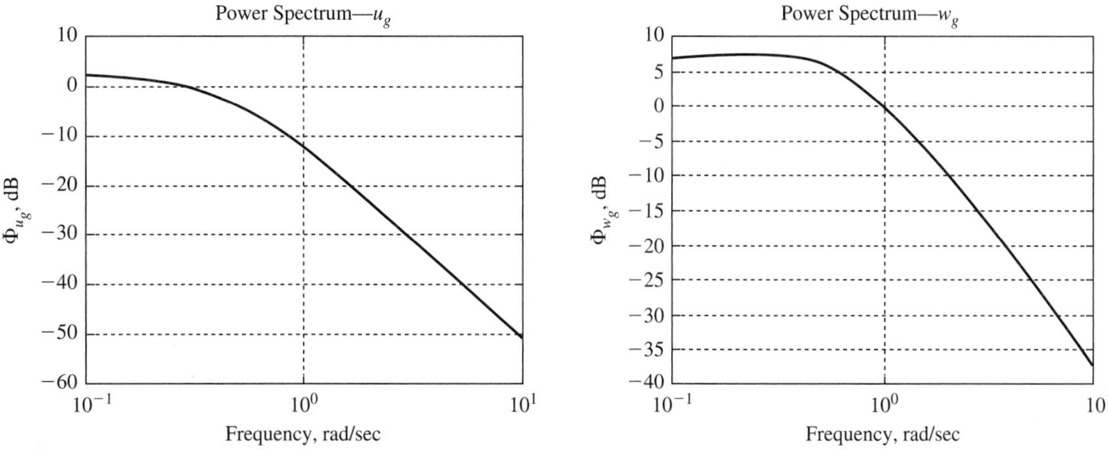

Figure C.2 Power spectra for u_g and w_g.

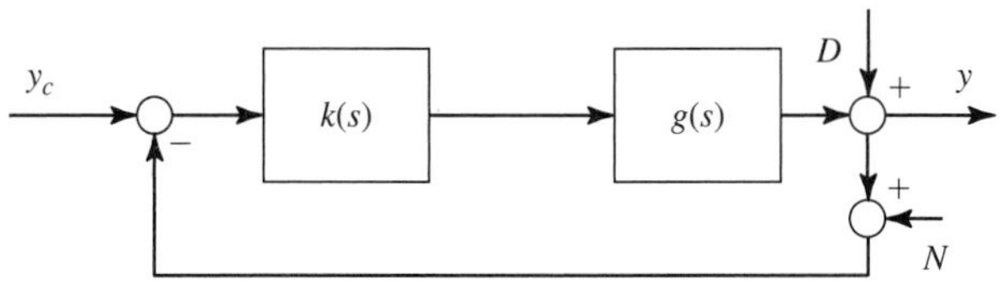

Figure C.3 Generic feedback-control block diagram.

Consider a DC-8 in a cruise flight condition at 33,000 ft altitude, Mach = 0.84 (Flight Condition 3 in Appendix B), and assume the vehicle is operating in moderate turbulence ($\sigma_{w_g} = 6$ fps). The pitch-attitude response θ to atmospheric turbulence (modeled as only w_g for simplicity) is found as follows: Since

$$\theta_g(s)|_{s=j\omega} = \theta_g(j\omega) = \frac{\theta(j\omega)}{n(j\omega)}n(j\omega) = \frac{\theta(j\omega)}{w_g(j\omega)}\frac{w_g(j\omega)}{n(j\omega)}n(j\omega)$$

the spectrum of the pitch-attitude disturbance is

$$|\theta_g(\omega)|^2 \triangleq \theta_g(j\omega)\theta_g(-j\omega) = \left(\frac{\theta(j\omega)}{w_g(j\omega)}\frac{\theta(-j\omega)}{w_g(-j\omega)}\right)\left(\frac{w_g(j\omega)}{n(j\omega)}\frac{w_g(-j\omega)}{n(-j\omega)}n(j\omega)n(-j\omega)\right)$$
$$= \left|\frac{\theta(\omega)}{w_g(\omega)}\right|^2 \Phi_{w_g}(\omega) \qquad \text{(C.8)}$$

And note that $\Phi_n(\omega) = 1$, by definition.

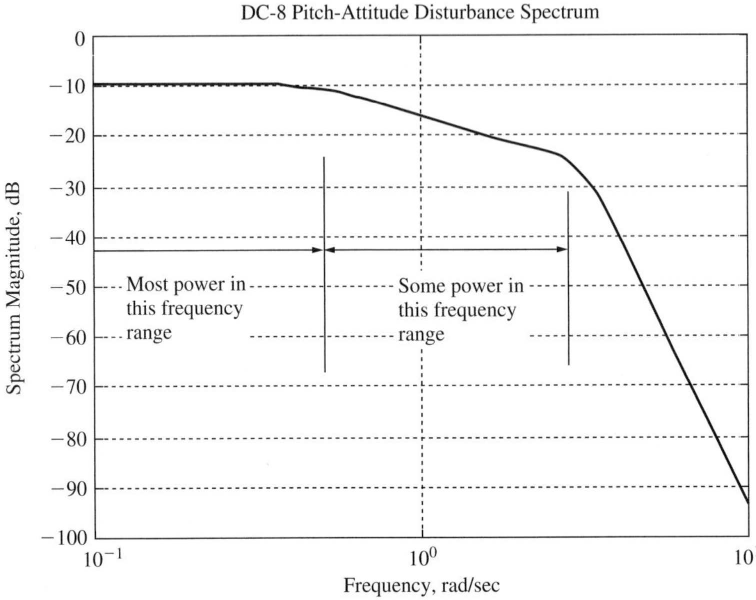

Figure C.4 Example spectrum of pitch-attitude disturbance from w_g.

The frequency content of the pitch-attitude disturbance (deg) is now obtained from the spectrum plotted in Figure C.4. Note that the magnitude of the pitch-disturbance spectrum begins to diminish at frequencies above 0.47 rad/sec. This is due to the shape of the gust spectrum itself, since $U_0/L_w = 0.47$ /sec here. Then the magnitude rolls off more sharply above 3.2 rad/sec, which is the short-period frequency. So clearly the frequency content of the pitch-attitude disturbance from atmospheric turbulence extends out to 0.47 rad/sec, with some additional frequency content out to 3.2 rad/sec. Above that frequency the power in the pitch disturbance diminishes rapidly. Consequently, when attempting to control pitch attitude and reject pitch-attitude disturbances from gusts, the gain-crossover frequency needs to clearly be above 0.47 rad/sec, and near or above the short-period frequency, if possible.

C.2 STOCHASTIC THEORY

We will now present the theoretical background dealing with the modeling of turbulence via the theory of stochastic process. Consider a random process $r(t)$, such as the measured wind speed at some point above the earth. Let a time history of the process, or of these measurements, be plotted as shown in Figure C.5. Note that at each time, or epoch, t there is a probability distribution of r, denoted as $p(r,t)$ associated with the process.

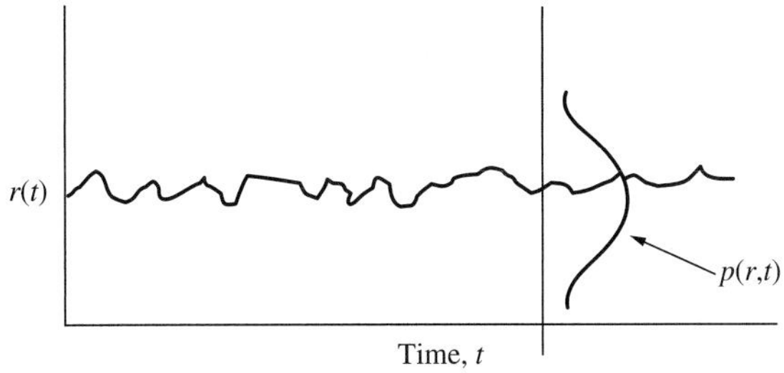

Figure C.5 Random process r and probability distribution p.

Let us now define some important statistics of the process r:

Mean:

$$\bar{r}(t) \triangleq E\{r(t)\} = \int_{-\infty}^{\infty} rp(r,t)dr \tag{C.9}$$

Autocorrelation:

$$R_r(t_1,t_2) \triangleq E\{r(t_1)r(t_2)\} = \iint_{-\infty}^{\infty} r(t_1)r(t_2)p(r(t_1),\, r(t_2),\, t_1,t_2)dr(t_1)dr(t_2) \tag{C.10}$$

Autocovariance:

$$C_r(t_1,t_2) \triangleq E\{(r(t_1) - \bar{r}(t_1))(r(t_2) - \bar{r}(t_2))\} = R_r(t_1,t_2) - \bar{r}(t_1)\bar{r}(t_2) \tag{C.11}$$

Variance:

$$\sigma_r^2(t) \triangleq C_r(t,t) = R_r(t,t) - \bar{r}^2(t), \text{ that is, } t_1 = t_2 = t \tag{C.12}$$

These statistics are defined in terms of the expected-value operation E, which is given in each case in terms of the respective integral expression.

By definition, the process as called *stationary* if the mean and autocorrelation are independent of time t. That is, if

$$\bar{r}(t) = \bar{r} \text{ constant}$$

and

$$R_r(t_1,t_2) = R_r(\tau),\ \tau = t_2 - t_1$$

or the autocorrelation is only a function of the time difference between t_2 and t_1. Furthermore, if the process is a *real* (e.g., physical) process, then the autocorrelation is *real* and *even,* or

$$R_r(\tau) = R_r(-\tau)$$

Finally, the process is *white* if the autocorrelation is an impulse function, or

$$R_r(\tau) = \delta(\tau)$$

This means that the process at time t is uncorrelated with itself at time $t + \Delta t$.

The *power spectrum,* or *power-spectral density,* of the process is the Fourier transform of its autocorrelation function, or

$$\boxed{\Phi_r(\omega) \triangleq \int_{-\infty}^{\infty} R_r(\tau)e^{-j\omega\tau}d\tau} \tag{C.13}$$

And from the Fourier-inversion formula we have

$$R_r(\tau) = \frac{1}{2\pi}\int_{-\infty}^{\infty} \Phi_r(\omega)e^{j\omega\tau}d\omega \tag{C.14}$$

Therefore, when $\tau = 0$ we have

$$R_r(0) \triangleq E\{r^2(t)\} = \frac{1}{2\pi}\int_{-\infty}^{\infty} \Phi_r(\omega)d\omega \tag{C.15}$$

So the area under the power spectrum equals $R_r(0)$, establishing a relationship between the power spectrum and the autocorrelation function. Furthermore when the mean of the process is also zero $(\bar{r} = 0)$, then $R_r(0) = \sigma_r^2$, or the area under the power spectrum equals the variance of the process $r(t)$. Thus the power spectrum is not only an indicator of the frequency content of the process, but of the power as measured by the variance of a zero-mean process.

An important application of interest in flight dynamics is the study of the behavior of a physical system (the flight vehicle) driven by a real random process (atmospheric turbulence). Consider some linear system with an impulse-response function $g(t)$, where $g(t) = 0$ if $t < 0$. The system's transfer function is then $g(s)$, where

$$g(s) = \int_0^{\infty} g(t)e^{-st}dt = \int_{-\infty}^{\infty} g(t)e^{-st}dt \text{ since } g(t) = 0 \text{ if } t < 0 \tag{C.16}$$

Note that the *Fourier transform* of $g(t)$ is also given by

$$g(j\omega) = g(s)|_{s=j\omega} = \int_{-\infty}^{\infty} g(t)e^{-st}dt|_{s=j\omega} \tag{C.17}$$

Now let the system be driven or excited by some random process $r(t)$, and let the system's response be denoted as $y(t)$. As with any time-varying input to a linear system, the system's response $y(t)$ can be found from the convolution of the input $r(t)$ and the system's impulse-response function $g(t)$. That is,

$$y(t) = \int_{-\infty}^{\infty} r(t-\alpha)g(\alpha)d\alpha \triangleq r(t)*g(t) \tag{C.18}$$

where the convolution operation is denoted by the asterisk. Now if the process $r(t)$ is real and $g(t)$ represents a real (e.g., physical) system, then the process $y(t)$ will be real.

Let us now address the statistics of the process $y(t)$. First, its mean is given in general by

$$\bar{y} \triangleq E\{y(t)\} = \int_{-\infty}^{\infty} E\{r(t-\alpha)\}g(\alpha)d\alpha \tag{C.19}$$

But if the input process $r(t)$ is stationary, then the mean of the response may be expressed in terms of the mean of the input and the DC value (or the value at zero frequency) of the frequency response of the system g. Or

$$\bar{y} \triangleq \bar{r}\int_{-\infty}^{\infty} g(\alpha)d\alpha = \bar{r}g(j\omega)|_{\omega=0} \tag{C.20}$$

Now define the *cross-correlation* between the input and response, or $R_{yr}(\tau)$ as

$$R_{yr}(\tau) \triangleq E\{y(t)r(t-\tau)\} = E\{y(t+\tau)r(t)\} \tag{C.21}$$

Using the expected-value operation and the expression for $y(t)$ (Equation (C.18)) we have the cross-correlation given by

$$R_{yr}(\tau) = E\{y(t)r(t-\tau)\} = \int_{-\infty}^{\infty} E\{r(t-\alpha)r(t-\tau)\}g(\alpha)d\alpha \tag{C.22}$$

But since r is stationary,

$$E\{r(t-\alpha)r(t-\tau)\} = R_r(\tau-\alpha)$$

then

$$R_{yr}(\tau) = \int_{-\infty}^{\infty} R_r(\tau-\alpha)g(\alpha)d\alpha = R_r(\tau)*g(\tau) \tag{C.23}$$

Finally, the autocorrelation of the process $y(t)$, or $R_y(\tau)$, is given as

$$R_y(\tau) = E\{y(t+\tau)y(t)\} = \int_{-\infty}^{\infty} E\{y(t+\tau)r(t-\alpha)\}g(\alpha)d\alpha \tag{C.24}$$

But since y is also stationary,

$$E\{y(t+\tau)r(t-\alpha)\} = R_{yr}(\tau+\alpha)$$

and so

$$R_y(\tau) = \int_{-\infty}^{\infty} R_{yr}(\tau+\alpha)g(\alpha)d\alpha = R_{yr}(\tau)^{*}g(-\tau) \tag{C.25}$$

It can also similarly be shown that $R_y(\tau) = R_{ry}(\tau)^{*}g(\tau)$.

Now define the *cross spectrum* $\Phi_{yr}(\omega)$ as

$$\boxed{\Phi_{yr}(\omega) \triangleq \int_{-\infty}^{\infty} R_{yr}(\tau)e^{-j\omega\tau}d\tau} \tag{C.26}$$

or the Fourier transform of $R_{yr}(\tau)$. Note the similarity between this spectrum and the power spectrum given in Equation (C.13). Also recall that the transform (Fourier or Laplace) of two convolved time functions is the product of the transforms of the two time functions. We will show below that since

$$R_{yr}(\tau) = R_r(\tau)^{*}g(\tau)$$

then the input-output cross-spectrum is

$$\boxed{\Phi_{yr}(\omega) = \Phi_r(\omega)g(j\omega)} \tag{C.27}$$

where

$\Phi_{yr}(\omega)$ is the transform of $R_{yr}(\tau)$ (Equation (C.26))

$\Phi_r(\omega)$ is the transform of $R_r(\tau)$ (Equation (C.13))

$g(j\omega)$ is the transform of $g(\tau)$ (Equation (C.17))

Likewise, we will also show below that since

$$R_y(\tau) = R_{yr}(\tau)^{*}g(-\tau)$$

then the output power spectrum is

$$\boxed{\Phi_y(\omega) = \Phi_{yr}(\omega)g^{*}(j\omega)} \tag{C.28}$$

where

$\Phi_y(\omega)$ is the transform of $R_y(\tau)$

$\Phi_{yr}(\omega)$ is the transform of $R_{yr}(\tau)$ (Equation (C.26))

$g^*(j\omega)$ is the conjugate of $g(j\omega) = g(-j\omega) =$ transform of $g(-\tau)$

Directly from Equations (C.27) and (C.28), we note that the output spectrum is related to the input spectrum via the system's frequency response, or

$$\boxed{\Phi_y(\omega) = \Phi_r(\omega)|g(j\omega)|^2} \tag{C.29}$$

We also find that the system's frequency response may be expressed in terms of the input-output cross spectrum and the input spectrum (both of which can be measured experimentally). Or

$$\boxed{g(j\omega) = \Phi_{yr}(\omega)/\Phi_r(\omega)} \tag{C.30}$$

Equation (C.29) was essentially derived in the DC-8 example when developing Equation (C.8), but with less rigor. And Equation (C.30) is critical to the measurement of the describing function of the human operator, for example, as discussed in Chapter 13. Consequently, these two expressions have important practical applications.

To prove Equation (C.27), recall that the cross-correlation is given by

$$R_{yr}(\tau) = \int_{-\infty}^{\infty} R_r(\tau - \alpha)g(\alpha)d\alpha$$

So from the definition of the cross spectrum $\Phi_{yr}(\omega)$ in Equation (C.26) we have

$$\Phi_{yr}(\omega) \triangleq \int_{-\infty}^{\infty} R_{yr}(\tau)e^{-j\omega\tau}d\tau = \int_{-\infty}^{\infty} e^{-j\omega\tau}\left(\int_{-\infty}^{\infty} R_r(\tau - \alpha)g(\alpha)d\alpha\right)d\tau = \int_{-\infty}^{\infty} g(\alpha)\left(\int_{-\infty}^{\infty} R_r(\tau - \alpha)e^{-j\omega\tau}d\tau\right)d\alpha \tag{C.31}$$

by interchanging the order of integration. But the inner integral is

$$\int_{-\infty}^{\infty} R_r(\tau - \alpha)e^{-j\omega\tau}d\tau = e^{-j\omega\alpha}\Phi_r(\omega) \tag{C.32}$$

from the definition of the power spectrum (Equation (C.13)) and the transform property of a time delay of α. So then

$$\Phi_{yr}(\omega) = \left(\int_{-\infty}^{\infty} g(\alpha) e^{-j\omega\alpha} d\alpha \right) \Phi_r(\omega) = g(j\omega)\Phi_r(\omega) \tag{C.33}$$

which is Equation (C.27).

Likewise, from Equation (C.25) we have

$$R_y(\tau) = \int_{-\infty}^{\infty} R_{yr}(\tau + \alpha) g(\alpha) d\alpha$$

Following the approach just taken above, from the definition of the power spectrum (Equation (C.13)) we have

$$\Phi_y(\omega) = \int_{-\infty}^{\infty} e^{-j\omega\tau} \left(\int_{-\infty}^{\infty} R_{yr}(\tau + \alpha) g(\alpha) d\alpha \right) d\tau = \int_{-\infty}^{\infty} g(\alpha) \left(\int_{-\infty}^{\infty} R_{yr}(\tau + \alpha) e^{-j\omega\tau} d\tau \right) d\alpha \tag{C.34}$$

But the inner integral is similarly

$$\int_{-\infty}^{\infty} R_{yr}(\tau + \alpha) e^{-j\omega\tau} d\tau = e^{+j\omega\alpha} \Phi_{yr}(\omega) \tag{C.35}$$

and so the power spectrum of the response may be written as

$$\Phi_y(\omega) = \left(\int_{-\infty}^{\infty} g(\alpha) e^{+j\omega\tau} d\alpha \right) \Phi_{yr}(\omega) \tag{C.36}$$

But

$$\int_{-\infty}^{\infty} g(\alpha) e^{+j\omega\tau} d\alpha = g(-j\omega) = g^*(j\omega), \text{ the conjugate of } g(j\omega) \tag{C.37}$$

Therefore, the output power spectrum may be expressed as

$$\Phi_y(\omega) = g^*(j\omega)\Phi_{yr}(\omega) \tag{C.38}$$

which proves Equation (C.28).

REFERENCES

1. Press, H., M. T. Meadows, and I.Hadlock: "Estimates of Probability Distributions of Root-Mean-Square Gust Velocity of Atmospheric Turbulence from Operational Gust-Load Data by Random-Process Theory," NACA TN 3362, 1955.

2. Press, H. and M. T. Meadows: "A Re-evaluation of Gust-Load Statistics for Applications in Spectral Calculations," NACA TN 3540, 1955.
3. Tolefson, H. B.: "Summary of Derived Gust Velocities Obtained from Measurements Within Thunderstorms," NACA Report 1285, 1956.
4. (U.S.) "Military Specification-Flying Qualities of Piloted Airplanes," MIL-F-8785C, November 1980.
5. Moorhouse, D. and R.,Woodcock: "Background Information and User Guide for MIL-F-8785C, Military Specification—Flying Qualities of Piloted Airplanes," Air Force Wright Aeronautical Labs Report AFWAL-TR-81-3109, Wright Patterson AFB, OH, 45433, July 1982.
6. Papoulis, A.: *Probability, Random Variables, and Stochastic Processes,* McGraw-Hill, New York, 1965.

D

APPENDIX

Cramer's Rule for Solving Simultaneous Equations

Given a set of n linear simultaneous equations governing n unknowns U_i $i = 1 \ldots n$, the equations may always be written in the following matrix format:

$$\begin{bmatrix} C_{1,1} & \cdots & C_{1,n} \\ C_{2,1} & \cdots & C_{2,n} \\ C_{n,1} & \cdots & C_{n,n} \end{bmatrix} \begin{Bmatrix} U_1 \\ \vdots \\ U_n \end{Bmatrix} = \begin{Bmatrix} D_1 \\ \vdots \\ D_n \end{Bmatrix} \text{ or } \mathbf{Cu} = \mathbf{d} \tag{D.1}$$

where $C_{i,j}$ are the coefficients, and D_i are the right-hand sides of the system of equations, independent of the unknowns. Or, in terms of the columns of the $n \times n$ **C** matrix, denoted as $\mathbf{c}_i$, we may also write the equations as

$$\begin{bmatrix} \mathbf{c}_1 & \cdots & \mathbf{c}_n \end{bmatrix} \mathbf{u} = \mathbf{d} \tag{D.2}$$

Cramer's Rule states that the solution for the i'th unknown U_i may be written as the ratio of two matrix determinants, or

$$U_i = \frac{\det \mathbf{M}_i}{\det \mathbf{C}} \tag{D.3}$$

where the matrix $\mathbf{M}_i$ is

$$\mathbf{M}_i \triangleq \begin{bmatrix} \mathbf{c}_1 & \cdots & \mathbf{c}_i \Rightarrow \mathbf{d} & \cdots & \mathbf{c}_n \end{bmatrix} \tag{D.4}$$

That is, the i'th column of **C** is replaced by the vector **d**. Therefore, the solution of simultaneous equations reduces to simply finding a set of determinants.

For example, consider the $n = 2$ system of equations given by

$$\begin{bmatrix} C_{1,1} & C_{1,2} \\ C_{2,1} & C_{2,2} \end{bmatrix} \begin{Bmatrix} U_1 \\ U_2 \end{Bmatrix} = \begin{Bmatrix} D_1 \\ D_2 \end{Bmatrix} \text{ or } \begin{bmatrix} \mathbf{c}_1 & \mathbf{c}_2 \end{bmatrix} \mathbf{u} = \mathbf{d} \tag{D.5}$$

Then

$$\mathbf{M}_1 = \begin{bmatrix}\mathbf{d} & \mathbf{c}_2\end{bmatrix}$$

and

$$\mathbf{M}_2 = \begin{bmatrix}\mathbf{c}_1 & \mathbf{d}\end{bmatrix}$$

Consequently, under Cramer's Rule,

$$U_1 = \frac{D_1 C_{2,2} - C_{1,2} D_2}{C_{1,1} C_{2,2} - C_{1,2} C_{2,1}}$$

and (D.6)

$$U_2 = \frac{D_2 C_{1,1} - C_{2,1} D_1}{C_{1,1} C_{2,2} - C_{1,2} C_{2,1}}$$

Finally, note that the determinant of matrices larger than 2 × 2 may be expressed in terms of the elements of a single row or column and their respective cofactors. Hence, the determinant of a large matrix may be found by repeated application of this fact. For example, consider the 3 × 3 matrix

$$\mathbf{C} = \begin{bmatrix} C_{1,1} & C_{1,2} & C_{1,3} \\ C_{2,1} & C_{2,2} & C_{2,3} \\ C_{3,1} & C_{3,2} & C_{3,3} \end{bmatrix} \tag{D.7}$$

By expanding in terms of the cofactors of the elements of the first row, we have

$$\begin{aligned} \det \mathbf{C} &= C_{1,1} \det\begin{bmatrix} C_{2,2} & C_{2,3} \\ C_{3,2} & C_{3,3} \end{bmatrix} - C_{1,2} \det\begin{bmatrix} C_{2,1} & C_{2,3} \\ C_{3,1} & C_{3,3} \end{bmatrix} + C_{1,3} \det\begin{bmatrix} C_{2,1} & C_{2,2} \\ C_{3,1} & C_{3,2} \end{bmatrix} \\ &= C_{1,1}\left(C_{2,2}C_{3,3} - C_{2,3}C_{3,2}\right) - C_{1,2}\left(C_{2,1}C_{3,3} - C_{2,3}C_{3,1}\right) + C_{1,3}\left(C_{2,1}C_{3,2} - C_{2,2}C_{3,1}\right) \end{aligned} \tag{D.8}$$

If one needed to find the determinant of a 4 × 4 matrix, for example, they could write that determinant in terms of the determinants of 3 × 3 cofactors, which in turn may be written in terms of 2 × 2 cofactors.

INDEX

A

B

C

E

M

O

P

Q

R

S

U

V

W

X

Y

Z